Three-Dimensional Electromagnetics

Three-Dimensional Electromagnetics

Edited by
Michael Oristaglio
Schlumberger-Doll Research

Brian Spies
Cooperative Research Centre for Australian Mineral Exploration Technologies

Geophysical Developments Series Editor
Michael R. Cooper

Published by
the Society of Exploration Geophysicists
P.O. Box 702740, Tulsa, OK 74170-2740

Library of Congress Cataloging-in-Publication Data

Three-dimensional electromagnetics / edited by Michael Oristaglio and Brian Spies.
p. cm. — (Geophysical developments series : v. 7)
Includes bibliographical references (p. –) and index.
ISBN 1-56080-079-8
1. Electromagnetic prospecting. I. Oristaglio, Michael L. II. Spies, Brian R. III. Series: Geophysical development series : v. 7.
TN269.A14 1999
622′.153—dc21 98-30016
CIP

ISBN 0-931830-41-9 (Series)
ISBN 1-56080-079-8 (Volume)

Printed in the United States of America

Contents

Gerald W. Hohmann (1940–1992)

The numerous possibilities for theoretical and programming errors make it necessary to compare results computed by different methods before a numerical solution can be considered valid.

Hohmann (1988)

Preface

In 1975 Jerry Hohmann published a paper[1] that described his numerical implementation of an integral-equation method for three-dimensional electromagnetic (3-D EM[2]) modeling. The matrix equation for the simple model that he studied—a half-space containing a rectangular body discretized into 100 cubic cells—barely fit into the computer (a UNIVAC 1108 at the University of Utah). Coaxing interesting and correct results from the model and method clearly comprised much of the art and fun of the paper. And winding through the paper's 50 or so equations and nearly 20 figures was a clear message: 3-D EM is different!

Three-dimensional electromagnetics is qualitatively different with new phenomena[3] and new challenges to our understanding of how electromagnetic fields interact with Earth and other conductive bodies (including our own). In subsequent years, Jerry with his students and colleagues pursued these challenges across many fields—mining geophysics, geothermal exploration, magnetotelluric crustal studies, environmental geophysics, oil and gas exploration—in both the time and frequency domains. Of his 51 articles[4] in journals and monographs, more than half dealt with three-dimensional electromagnetics.

In 1995, 20 years after Jerry's classic paper (and three years after his death from cancer in May, 1992), nearly 200 scientists from around the world gathered at Schlumberger–Doll Research in Ridgefield, Connecticut, for a symposium in his memory, the (first) International Symposium on Three-Dimensional Electromagnetics. More than 70 papers were presented in oral and poster sessions during three days organized

[1] Hohmann, Gerald W., 1975, Three-dimensional induced polarization and electromagnetic modeling: Geophysics, **40**, 309–324.

[2] The first of many appearances of these two acronyms—3-D, EM—which are used as adjectives (*three-dimensional, electromagnetic*) and nouns (*three dimensions, electromagnetics*) throughout this volume.

[3] Such as *current channeling*, now generally called *galvanic response:* "The surprising result is that decreasing the background resistivity increases the EM response. . . . The same effect has been observed in scale model experiments. However, this enhancement of response with decreasing background conductivity does not occur for a two-dimensional body excited by an infinite line source."

[4] See Appendix at the end of this book.

around the themes: Modeling, Inversion, and Practice. The quality of the work presented, the liveliness of the discussions, and the demand for the symposium proceedings were the impetus for this new volume. We invited the authors to submit longer, more tutorial versions of their articles for a book to be published by the Society of Exploration Geophysicists (SEG) in the series *Geophysical Developments*.

As is evident from the size of this volume, we were overwhelmed by the response. We hope that readers will find the contents equally weighty. The 44 articles collected here are the work of 97 authors, representing 55 different institutions (universities, government or industrial research labs) from 13 countries around the world. All have been reviewed and edited according to the strict standards of SEG's lead journal, *Geophysics*. They represent the state of the art in 3-D EM at the time final revisions were received (from the fall of 1997 through the spring of 1998).

The lead article addresses one of Jerry's favorite subjects—the need for independent checks on any numerical calculation; it shows how far we have come since 1975 and how far we still are from routine, confident use of 3-D EM models. We have grouped the remaining articles into nine sections:

Integral-equation modeling
Finite-difference modeling
Inversion
3-D EM and parallel computers
Magnetotellurics and global induction
Mining and exploration geophysics
Borehole geophysics and logging
Equipment
General

This division into techniques and applications is naturally very rough; many articles could easily appear in two or three different sections. The subjects covered in this volume touch, we believe, on every major technique being used today to compute, analyze, visualize, and understand 3-D EM fields in every major application of electrical geophysics (and in two applications outside geophysics: the interaction of 3-D EM fields with the human body and the non-destructive testing of aircraft). The late 1980's saw the rapid development of 3-D seismics, which has revolutionized exploration for oil and gas in the 1990's. The early years of the new millenium may see another revolution brought about by the rapid advances now occurring in 3-D EM.

Acknowledgments

Many people contributed to the success of 3-D EM Symposium and helped to bring together this volume. At Schlumberger-Doll Research, we thank especially Cathy Corris, the Symposium co-ordinator; Tarek Habashy, Carlos Torres-Verdín, and Vladimir Druskin, who helped with the selection and organization of articles; and Computing and Information Resources (especially Ken Scherwenik, Ray Kocian, Paul Gerardi, and Karyn Muller) for helping us handle the flood of electronic submissions. Schlumberger generously donated use of the facility and the time of all of the above. Eleanor Umali of Techbooks and Judy Hastings of SEG oversaw production of the volume and ensured that it met SEG's high standards. We also thank Tom Oristaglio for collating and compiling the Index. Most of all, we thank the authors for contributing their work and for indulging our insistence on receiving everything electronically (and for patiently helping us get it right).

Frank Morrison first suggested the idea for a symposium honoring Jerry Hohmann; the idea gathered force under the auspices of the G. W. Hohmann Memorial Trust for Teaching and Research in Applied Electrical Geophysics and its trustees: Stanley H. Ward, Phillip Michael Wright, Louise Pellerin, and Charles M. Swift, Jr. The proceeds from this book will be donated to the Trust, which is funding scholarships (through the SEG Foundation) and awards in electrical geophysics.

Michael Oristaglio
Schlumberger-Doll Research
Ridgefield, Connecticut USA

Brian Spies
CSIRO, Australia
23 May 1998

Three-Dimensional Electromagnetics

Three-Dimensional Transient Electromagnetic Modeling—A User's View

Robert Smith[1]
John Paine[2]

Summary. A simple exercise in inverting transient electromagnetic (TEM) data for the layering in a basin ran into difficulties when the inversion failed to converge for about half of the soundings. The failure was caused by the forward modeling program that created artifacts for models with thin, shallow conductive layers. The artifacts were difficult to identify visually on raw plots because of the large dynamic range of most TEM data. They did show up clearly in relative comparisons of the curves from two programs. This experience led to a broader study of the differences between the many layered-earth and 3-D TEM modeling codes available and of the various approximations used. This study has already resulted in improvements in several codes, which generated more confidence in their results.

1 Introduction

This work began with a basin study, the aim of which was to map resistivity variations in a deep layer. The basin had a well-defined layered structure (Table 1), and the environment allowed soundings to be taken to very late times with very low noise. However, when the soundings were inverted with a standard layered-earth inversion program, the mismatch between the model and the field data was above the desired limits for about half of the soundings. The program, which was based on an older circular-loop approximation code, had been modified for arbitrary rectangular transmitter loops and arbitrary transmitter waveforms, and was thus felt to be more accurate than the circular-loop code. But when the circular-loop code was used for the inversion, convergence was obtained for all soundings. The question naturally arose as to why the improved version of the program failed to obtain convergence.

Detailed examination of the soundings did not uncover any noise problems or other variations that could explain the lack of convergence, and so, a study was undertaken of the differences between the forward-model calculations for all layered-earth codes available to CRAE in an attempt to identify the cause of the problem. This study

[1]ET&I Group, CRA Exploration Ltd., Norwood, South Australia, Adelaide 5067, Australia.
[2]Scientific Computing & Applications, Tranmere, South Australia, Adelaide 5073, Australia.

Table 1. Model 1 structure (Basin model)

Layer	Thickness (m)	Resistivity (ohm-m)
1	60	20
2	20	5
3	420	200
4	50	10
5	750	6
6	basement	20

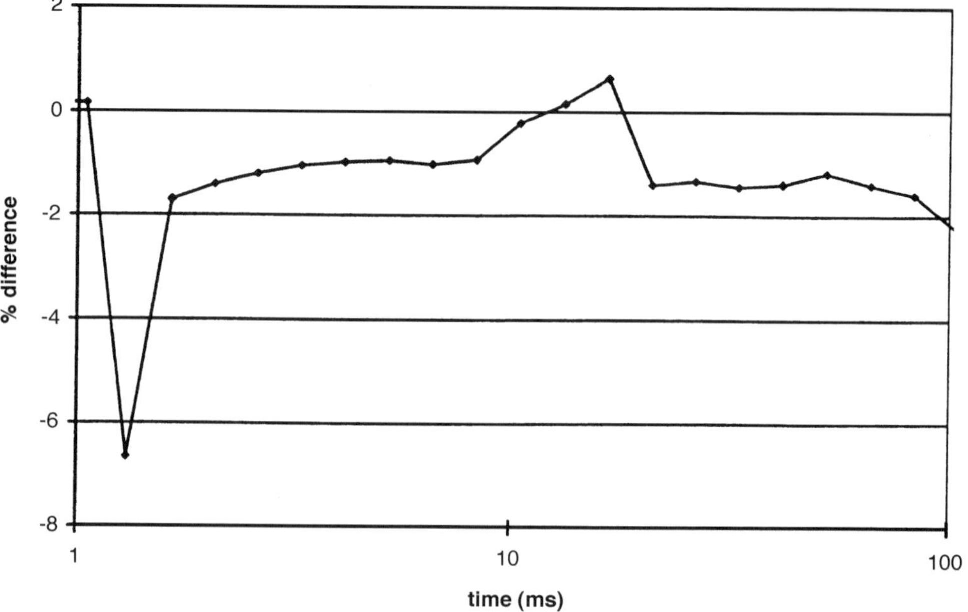

Figure 1. Difference between CRABEO and UBCEM step response for Model 1.

found that, when the square-loop code was used for a model that contained a thin, shallow conductive layer, two artifacts were introduced into the modeled response. It was hypothesized that these artifacts caused the inversion to fail. The presence of the artifacts was first established by plotting the relative difference between curves from the two codes. This is shown in Fig. 1 where the artifacts are clearly visible as spikes in the difference between the modeled responses near 1 ms and near 10 ms. This approach does not, however, identify which method actually contained the artifact because it was based on a single comparison between two different methods. On further investigation it was found that the artifact became stronger as the conductivity of the layer was increased. The layer differential plot in Fig. 2 unequivocally shows the presence of the artifact in the square-loop code (CRABEO) because it is based on a comparison of two forward models by the same program. This figure also explains why the inversions failed because it shows that the artifact near 10 ms became stronger as the conductivity of the layer was increased; this made it impossible for the optimization algorithm to decrease the total error to less than 5%, and so, convergence failed.

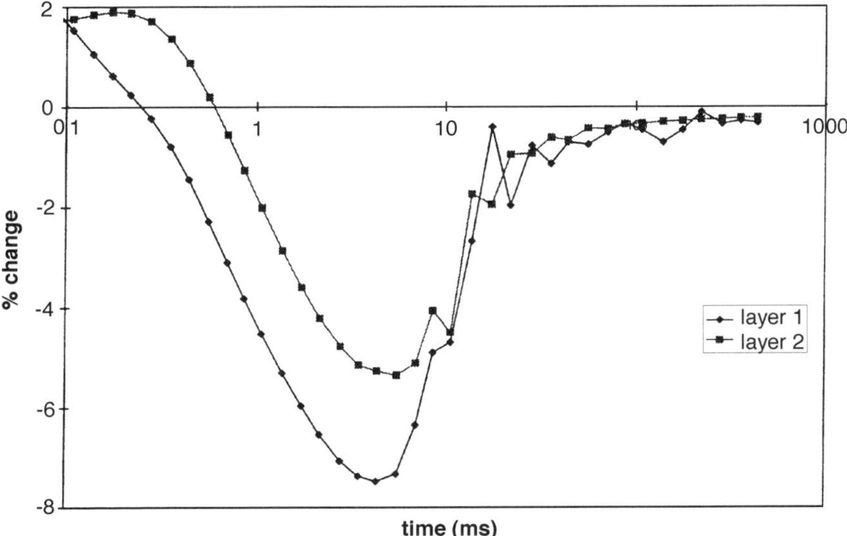

Figure 2. Effect of 3% change in layer resistivity for Model 1 computed using CRABEO.

The artifact is a relatively small component of the total decay, and the primary difficulty with identifying the inaccuracy of the forward-model response was that it was difficult to see by simply plotting the decay (or any transformed quantity such as apparent resistivity or any simple scaling such as dividing the response by the half-space response). The artifact was only identified clearly when the relative difference between two different methods was plotted. Such comparisons between two different methods are difficult because it is not known a priori which method is correct. (A comparison using half-space models would not have identified the problem because such models would not have generated the artifact.)

More detailed comparisons then clearly identified the new code as the source of the problem. The author of the code traced the problem to a particular approximation in computing the layered-earth response and corrected it. Still, many questions had been raised about general consistency of the layered-earth decays computed using the different codes available to CRAE. Some of the differences in implementation are fairly obvious to the user, e.g., the use of circular loops to approximate rectangular loops; others are somewhat less visible, e.g., the method used to convert the underlying frequency-domain solution to the time domain; and still others are almost entirely invisible to the user, e.g., the method used to integrate around the rectangular loop or the Hankel transform used in computing the frequency-domain response. A study was undertaken to identify possible significant sources of error in the computed response for all codes available to CRAE so that the methods could be used with confidence. We also decided to use this framework to compare the results from available 3-D codes to get some sense of their reliability.

Because most of the available programs compute the solution in the frequency domain and then transform to the time domain, we first consider the influence of the frequency-domain to time-domain conversion on the accuracy of the computed decay. Following that, Section 3 briefly examines the differences between circular- and rectangular-loop approximations. Because most 3-D codes model a 3-D body in a layered half-space, it is also useful to compare these codes in the absence of a 3-D body, as is done in Section 4. Finally, Sections 5 and 6 present comparisons of the responses computed using a variety of 3-D methods.

The purely 1-D (layered-earth) codes to be considered in the following sections are

CRABEO (CSIRO, ver. 4.1, 1993) featuring frequency-domain formulation, integral-equation solution, rectangular-loop approximation, and time-domain output.

GRENDL (CSIRO, 1989) featuring frequency-domain formulation, integral-equation solution, circular-loop approximation, and time-domain output.

UBCEM (University of British Columbia, 1994) featuring frequency-domain formulation, integral-equation solution, rectangular-loop approximation, and time- and frequency-domain output.

Not all codes were run for each of the models considered.

2 Conversion of frequency response to transient decays

Underlying almost all transient electromagnetic (TEM) modeling is a conversion of the frequency response to the time domain. This is usually done by an inverse Fourier transform, computed as a lagged convolution that requires the response at a wide range of frequencies. The number and the location of these frequencies varies according to the particular implementation, and the strategy for determining the best set of frequencies is not obvious. Some methods use a predetermined set parameterized by the minimum and maximum frequency, which the user selects. Others fix the frequencies internally in the code, and some force the user to explicitly specify all frequencies. After the response is evaluated at the selected frequencies, it is interpolated (usually with a cubic spline) onto a larger set of discrete frequencies required by the convolution filter.

2.1 Interpolation error

To study the impact of the interpolation process we now restrict attention to the electromagnetic (EM) code from the University of British Columbia (UBC). This program has been modified by CRAE to incorporate arbitrary transmitter waveforms, and this modified version (referred to as UBCEM) was used as the basis for the layered-earth comparisons reported in the preceding section. The version of the code used here computes the response at user-specified frequencies and then passes these values to the double precision code DLAGF0 from Walt Anderson of the US Geological Survey (USGS). Figure 3 displays the relative error in using five to nine frequencies per decade (equispaced logarithmically) relative to the decay computed using 10 frequencies per decade for the UBCEM method for model 1. Figure 3 shows that in the 1- to 10-ms range the differences show the expected reduction in error as the number of frequencies used is increased, but for times greater than 10 ms, there is far less structure and the error actually can increase as the number of frequencies is increased. This means that for these late times the interpolation process can introduce significant errors in the transient curve, which cannot be reduced simply by increasing the number of frequency points used.

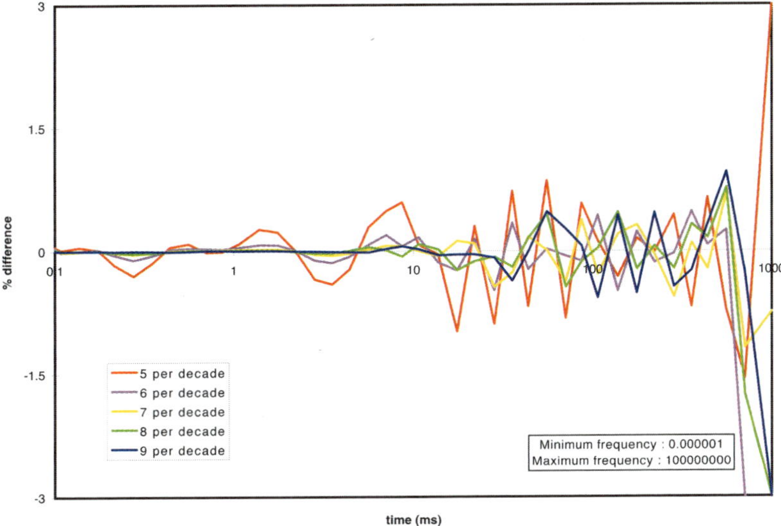

Figure 3. Effect of increasing number of frequencies on decay computed using UBCEM relative to decay computed using 10 frequencies per decade.

It was realized, however, that the interpolation errors could be trivially eliminated by computing the response at the actual frequencies required by the convolution filter. One drawback is that this requires more computation because the convolution filter requires a large number of frequencies. To balance the need for accurate results against speed, there needs to be a further consideration of whether reasonable results can be obtained using fewer frequencies. As a starting point the 251 frequencies given in Appendix A (chosen to cover the extreme range of values required by the DLAGF0 code) are used as the reference from which to study the impact of the various options for using fewer frequencies. Figure 4 displays the effect of using every second, every third, and every fourth frequency. It shows more structure than the results for Fig. 3, especially since the interpolation errors are now almost entirely negligible in the 0.1- to 10-ms interval. In the 10- to 1000-ms range, however, the errors introduced by interpolating the frequency response become noticeable when using every third frequency and exceed the 1% error threshold when using every fourth.

The next stage in studying the influence of the frequencies at which the frequency response is evaluated is to examine the influence of the minimum and maximum frequencies on the accuracy. To do this, Fig. 5 displays the effect of dropping successive decades from the bottom of the frequency range used as the reference. The effect of dropping one decade (which is 20 frequencies) is quite substantial near 1000 ms but is certainly negligible for times <100 ms. For two decades the times for which the difference is negligible shifts down by one decade in time to about 10 ms, whereas for three decades the shift down is somewhat less, with the error dropping to 0.1% at 7 ms. Thus, the acceptability of dropping low frequencies as a means for improving the computation speed depends on the time range to be computed.

The effect of dropping high frequencies (as seen in Fig. 6) is the opposite of that for low frequencies. The effect of dropping three decades from the reference set appears negligible across the entire time range, whereas for four decades the effect is significant near 0.1 ms, and for five decades the effect is unacceptable for times <2 ms.

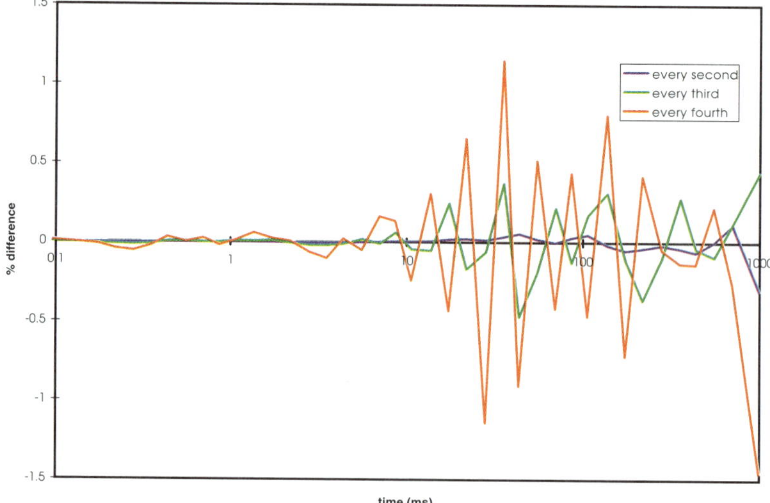

Figure 4. Effect of reducing density of frequencies on the accuracy of decays computed using UBCEM.

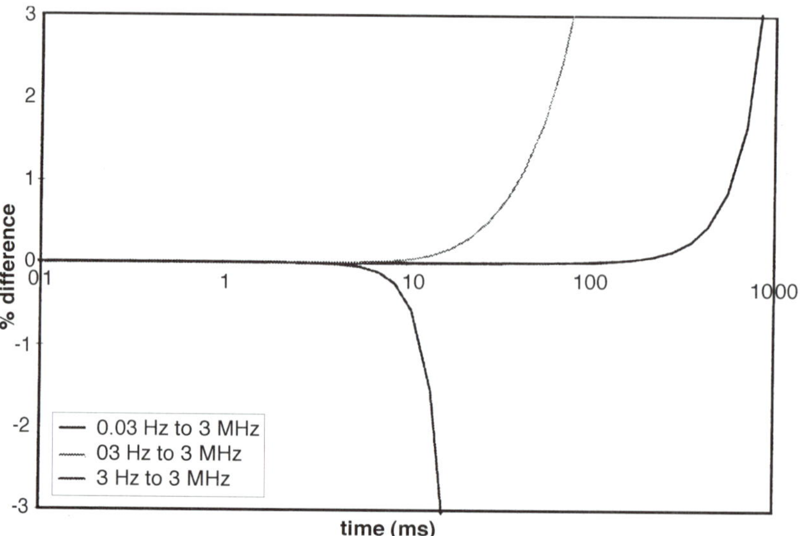

Figure 5. Effect of dropping lower frequencies on the accuracy of decays computed using all frequencies.

3 Circular-loop approximations

The next most important approximation to consider for the inversion method was the difference between the rectangular-loop and equivalent area circular-loop forward modeling methods. Two such methods available to CRAE are the programs GRENDL (circular loop) and CRABEO (rectangular loop). To highlight the differences between these methods, the rectangular-loop–based code UBCEM was selected as the basis for the layered-earth comparisons. These methods were applied to the model described in Table 1 for a 1000 × 1000 m square loop with a dipole receiver having an effective area of 200 000 m^2, located at the center of the loop. The transmitter waveform and receiver times are given in Appendix B.

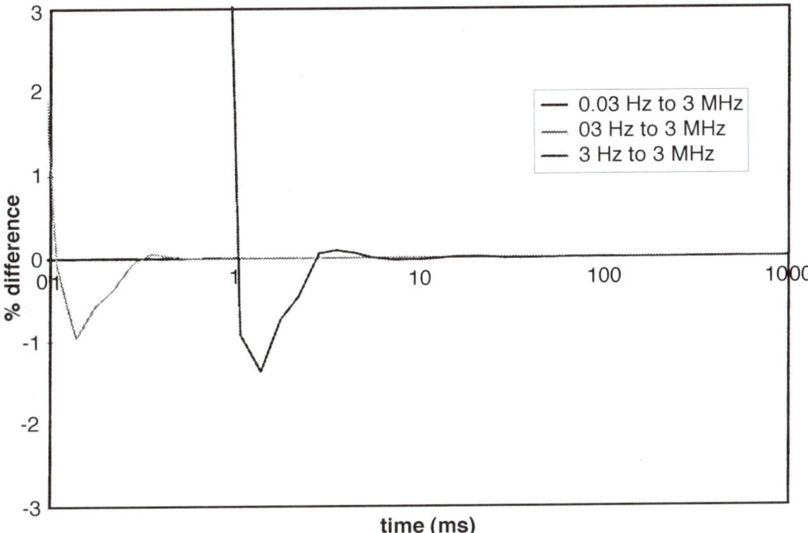

Figure 6. Effect of dropping high frequencies on the accuracy of decays computed using UBCEM.

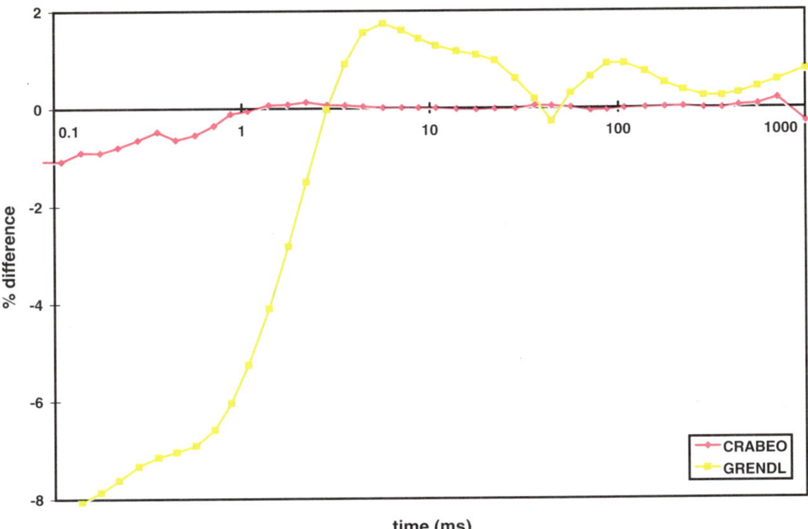

Figure 7. Comparison of decay for Model 1 computed using CRABEO and GRENDL relative to that computed using UBCEM.

The results for Model 1 are displayed in Fig. 7 which clearly shows that the circular-loop code GRENDL deviates substantially from the rectangular-loop codes CRABEO and UBCEM at early times. This is to be expected because the main differences between a circular loop and a square loop will occur at early times before the differences in the field resulting from the different transmitters has had sufficient time to decay. Clearly, the use of circular-loop approximations as the basis for inverting data generated using square transmitter loops is not to be recommended if early times or shallow data are important.

The differences between UBCEM and CRABEO at early times are probably due to differences in the numerical solutions of the analytical frequency-domain solution, but are less than 1% and are not considered to be significant.

4 Comparison of layered-earth models

The primary difficulty in objectively comparing the responses computed using the wide range of available forward modeling codes is the extreme dynamic range of the response itself. The simplest method for overcoming this large dynamic range is to display the decays computed using the available methods relative to a single chosen method. This process has the advantage that it clearly delineates the differences between methods, but has the disadvantage that it appears that the chosen method is somehow being identified as the correct solution. This latter disadvantage is more psychological than actual, especially once more than two methods have been presented in this form. In examining the results of this comparison, remember that the results presented in this way are not absolute and that it is simply being used as a means of identifying consistencies and differences between various methods.

We decided to select the reference method as the one that gave the smoothest results when examining the effects due to changes in layer parameters (i.e., compute a base model, change one layer parameter, recompute the model, and display the difference between the two results as a percentage of the base model). Because this is essentially the process on which inversion methods are based, it is necessary (but not sufficient) that a method should produce smooth variations when model parameters are changed. A full examination of these variations was undertaken; the UBCEM code was (marginally) the best behaved and was selected as the reference model for this section.

4.1 Codes

The 3-D codes for which the layered-earth responses were considered are

EM3D (University of Utah, 1989) featuring frequency-domain formulation, rectangular-prism 3-D model, integral-equation solution, rectangular-loop approximation, and frequency-domain output.

EMIE3D (University of Utah, Ver. 2.1, 1995) featuring frequency-domain formulation, rectangular-prism 3-D model, integral-equation solution, rectangular-loop approximation, and frequency-domain output.

EMIGMA (PetRos Eikon, Ver. 2.0, 1995) featuring frequency-domain formulation, rectangular-prism 3-D model, localized nonlinear-approximation integral-equation solution, rectangular-loop approximation, and frequency-domain output.

LEROI (CSIRO, Ver. 2.0, 1994) featuring frequency-domain formulation, thin-plate 3-D model, integral-equation solution, mixed rectangular- and circular-loop approximation, and frequency- and time-domain output.

SAMAYA (CSIRO, Ver. 10.0, 1995) featuring frequency-domain formulation, rectangular-prism 3-D model, hybrid integral-equation/finite-element solution, mixed rectangular- and circular-loop approximation, and frequency- and time-domain output.

SYSEM (University of Utah, Ver. 3.0, 1994) featuring frequency-domain formulation, rectangular-prism 3-D model, integral-equation solution, rectangular- or circular-loop approximation, and frequency- and time-domain output.

Table 2. Model 2 structure

Layer	Thickness (m)	Resistivity (ohm-m)
1	20	5
2	basement	50

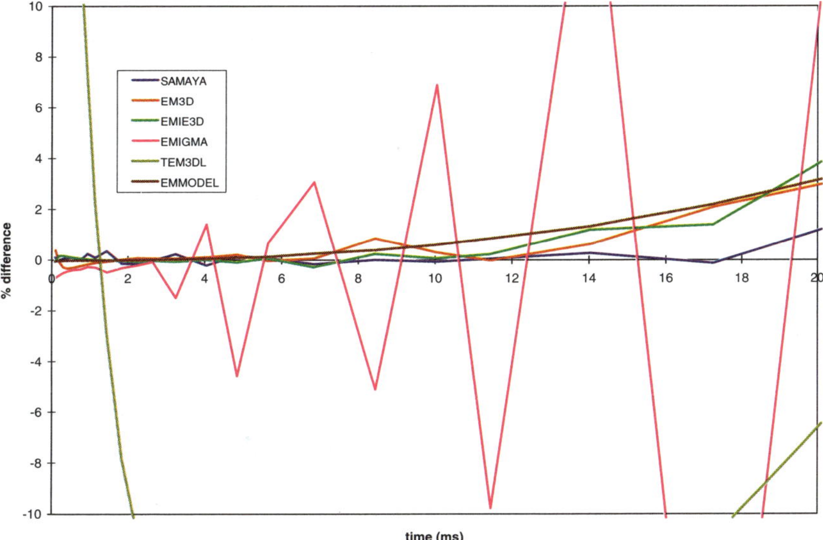

Figure 8. Relative difference between the layered-earth response for Model 2 computed using 3-D programs and that computed using UBCEM.

TELMA (Sniiggims, 1994) featuring time-domain formulation, rectangular-prism full-domain 3-D model, finite-difference solution, circular-loop approximation, and time-domain output.

TEM3DL (University of Utah, Ver. 1.0, 1995) featuring time-domain formulation, rectangular-prism full-domain 3-D model, finite-difference solution, rectangular-loop approximation, and time-domain output.

VHPLATE (PetRos Eikon, Ver. 2.0, 1995) featuring frequency-domain formulation, thin-plate 3-D model, integral-equation solution, rectangular-loop approximation, and frequency-domain output.

Note that not all codes have been run for each of the models considered.

4.2 Layered-earth model

The model chosen to examine the differences between these methods is given in Table 2 and was computed for a 100-m^2 loop with a receiver having an effective area of 10 000 m^2 at the center. The transmitter waveform and receiver times for this model are given in Appendix C. Figure 8 displays the time-domain results computed using these methods expressed as a percentage difference from that generated by the UBCEM code. Note that MARCO, SAMAYA, and LEROI use identical code for the layered-earth component, and so, only the results for SAMAYA are displayed; similarly, EMIGMA and VHPLATE use identical layered-earth code, and so, only the EMIGMA results are displayed. The 3-D component of MARCO was distributed formerly by the University of Utah in the code SYSEM.

The three CSIRO codes (LEROI, SAMAYA, and MARCO) have no facility for the user to specify frequencies and TEM3DL only computes in the time domain, and so, these codes were run with their standard time-domain output. All other codes were run in frequency-domain mode using the frequencies specified in Appendix A. To minimize potential sources of difference, the frequency response then was converted to the time-domain response for the transmitter waveform and receiver times defined in Appendix C by passing the imaginary component of the response to the CRAE program DFTRAN (which is based on Walt Anderson's double-precision lagged convolution code DLAGF0). It can be seen from Fig. 8 that the Utah codes EM3D and EMIE3D agree quite well with UBCEM as do EMMODEL, LEROI, MARCO, and SAMAYA. The main discrepancies between the methods are the significant deviations between the results for UBCEM and those for TEM3DL and EMIGMA.

The differences between TEM3DL and UBCEM arise from TEM3DL's use of a Gaussian pulse to approximate the step turn-on. The width of this pulse is set by the code as a function of the conductivity range in the model and the minimum cell size, and the difference between the pulse and the actual transmitter waveform is enough to explain the early time differences seen in Fig. 8. The late time differences are as yet unexplained, but may be a result of the limited extent of the finite-difference mesh. It would be interesting to transform the TEM3DL response to produce a response that is more comparable with that for the other methods, but this was not considered to be of high enough priority to warrant further work at this stage.

4.3 Frequency response

To get another perspective on the differences between the methods, we now consider the frequency-domain responses generated by the various methods. First, as a point of reference, Fig. 9 displays the frequency response for Model 1 computed using UBCEM. Figure 10 displays the difference between the imaginary component of the frequency

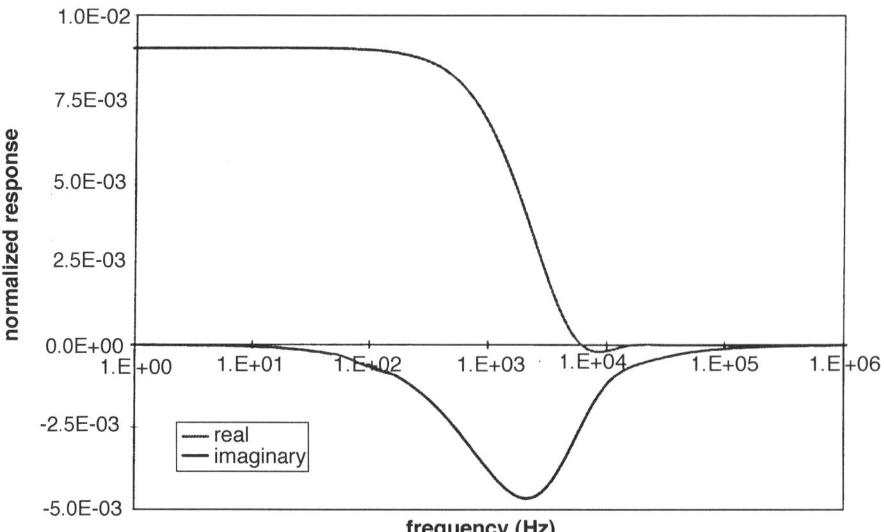

Figure 9. Frequency response of Model 2 computed using UBCEM.

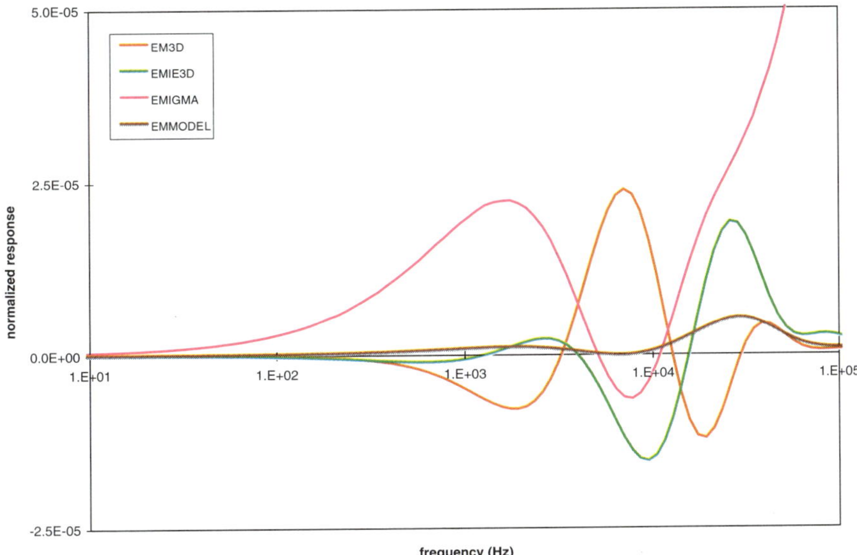

Figure 10. Difference between imaginary component of frequency response for UBCEM and that computed using EM3D, EMIE3D, EMIGMA, and EMMODEL for Model 2.

response computed using UBCEM and the imaginary components for EM3D, EMIE3D, EMIGMA, and EMMODEL. It can be seen in Fig. 10 that there are significant differences between the responses for UBCEM and EMIGMA in the 200- to 2000-Hz range and in the >10 000-Hz range. The latter difference would produce only minor early time differences, whereas the former is consistent with the oscillatory variations seen in Fig. 8 (with the amplitude increasing as a function of time because the decay becomes smaller with time, and so, the relative difference becomes larger). Note that although the differences between the imaginary component of the frequency response for UBCEM and those of EM3D and EMIE3D are just as large in amplitude as the differences between UBCEM and EMIGMA, the actual time-domain differences for EM3D and EMIE3D are much smaller than those for EMIGMA. The reason for this distinction is that the frequency-domain differences for EM3D and EMIE3D occur principally in a higher frequency range than those for EMIGMA, and so, the time-domain effects are restricted to very early times and tend to be damped out by the frequency to time transformation. In general, large differences in the frequency response do not necessarily transform to large differences in the transients and vice versa.

5 Three-dimensional modeling comparisons: Dike model

With this understanding of the simple layered-earth response, the final step is to compare the responses for 3-D bodies within a layered-earth structure. Fortunately, the comparison of the 3-D responses is somewhat easier than the layered-earth case because the anomalous response does not have such an extreme dynamic range. Consequently, it can be displayed either as the actual secondary response or as a percentage of the layered-earth response. In either case, this is an absolute quantity that can be compared across all available methods. To compare the above 3-D codes, a single 3-D body was included in the layered-earth model used in the preceding section. The body is a dike model defined as follows: width, 30 m; strike length, 300 m; depth extent, 150 m; depth to top, 30 m; resistivity, 0.5 ohm-m. In the plan and section view in Fig. 11, the center

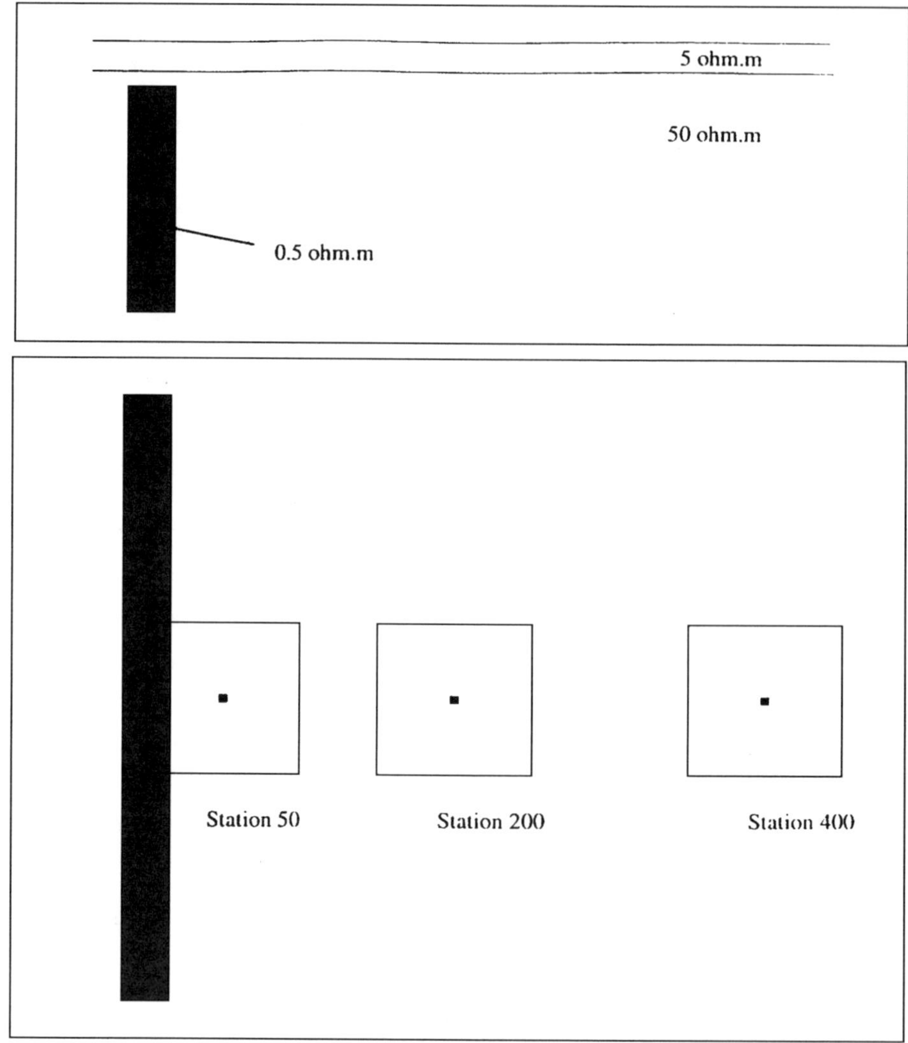

Figure 11. Plan and section schematic for dike model.

of the body was located at the origin, and three transmitter loop locations were used: one with the center placed 50 m east of the origin (or right of the origin in the plan view), the second placed 200 m east of the origin, and a third at 400 m east. These three locations are referred to as stations 50, 200, and 400, respectively. For each station, there is a single dipole receiver located at the center of the loop measuring the vertical component of dB/dt. In computing the response of the 3-D body, all methods require the body to be approximated using a specified number of cells or evaluation points. To simplify the comparison, the body was discretized using 400 points (or cells) in total for all of the prism-based methods. For the thin-plate methods EMMODEL and LEROI, 50-m^2 cells were used; for VHPLATE, no cell size is required because the code uses a fixed discretization.

The imaginary component of the scattered 3-D response computed using each of the methods is displayed in Fig. 12 for station 400. It shows good qualitative agreement among all methods with EM3D giving the largest-amplitude response and VHPLATE the smallest. The response for EMIGMA appears to be marginally shifted toward lower

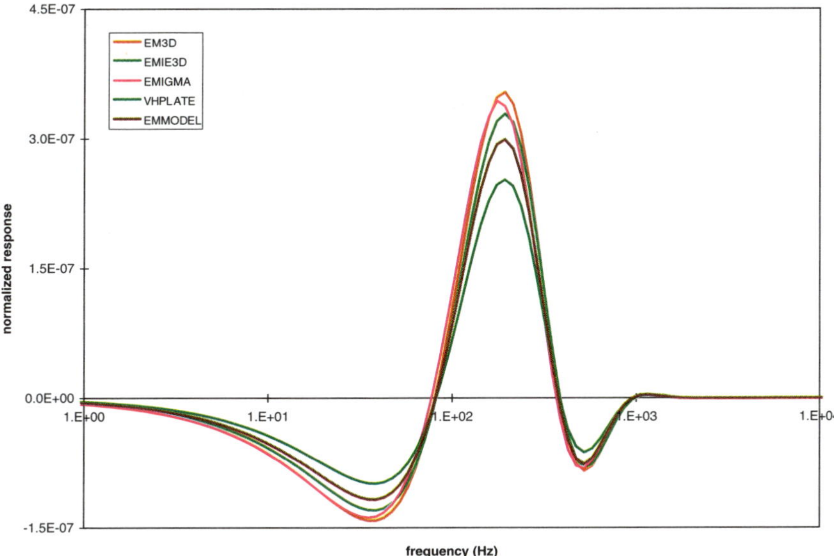

Figure 12. Imaginary component of frequency response at station 400 for 100:1 conductivity contrast dike.

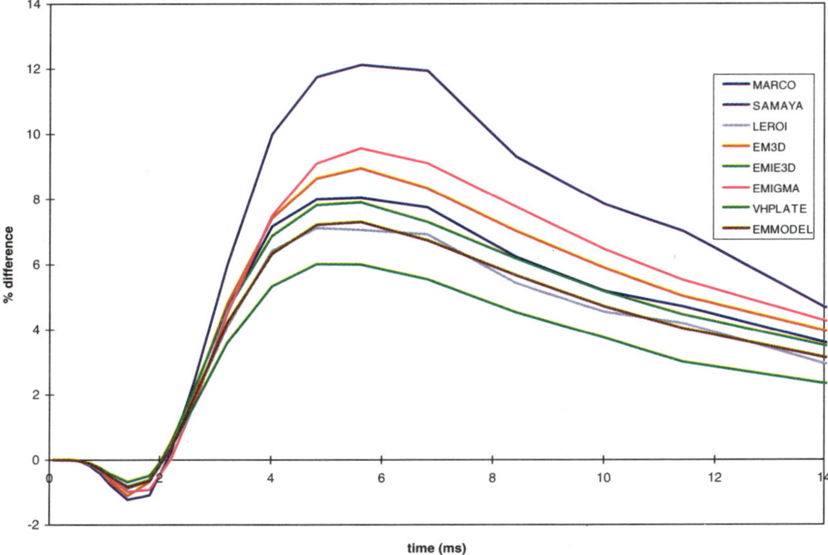

Figure 13. Response at station 400 for dike body with 100:1 conductivity contrast as a percentage of layered-earth response for UBCEM.

frequencies relative to the other methods, and the response for EMIE3D appears to be shifted toward slightly higher frequencies.

The time-domain response for station 400 (which was computed using HFTRAN for the frequency-domain methods EM3D, EMIE3D, EMIGMA, EMMODEL, and VHPLATE) is displayed in Fig. 13, where the scattered response is plotted as a percentage of the layered-earth response for UBCEM. Figure 13 shows that there is very good agreement among MARCO, LEROI, EM3D, EMIE3D, and EMMODEL across the entire time range. The response for VHPLATE is qualitatively similar to that for these five methods, but is somewhat smaller in amplitude, whereas that for SAMAYA

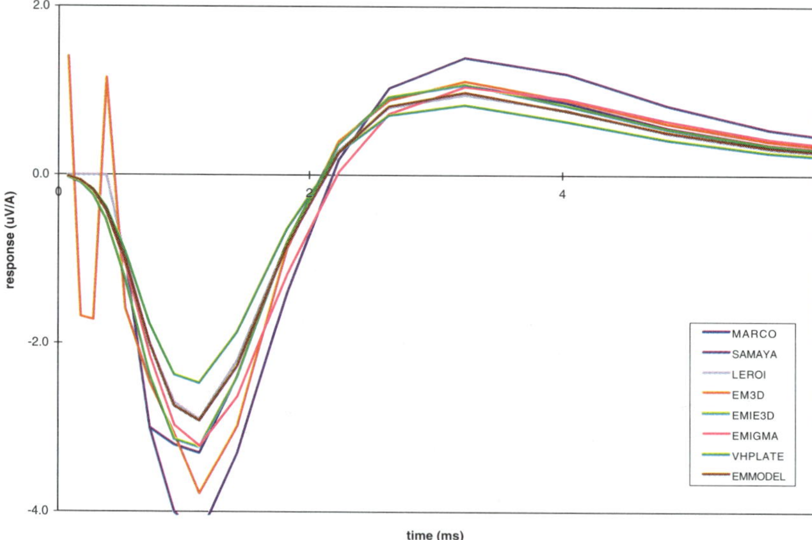

Figure 14. Response at station 400 for dike body with 100:1 conductivity contrast.

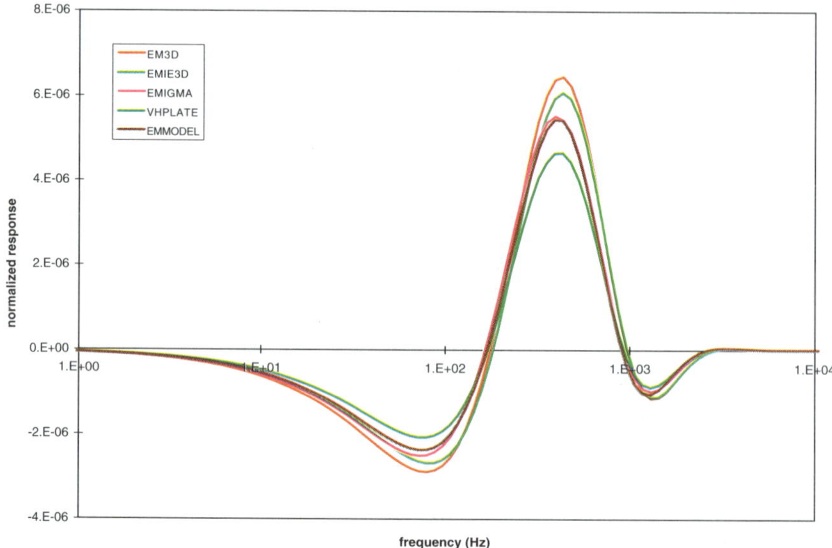

Figure 15. Imaginary component of frequency response at station 200 for 100:1 conductivity contrast dike.

is significantly larger in peak amplitude. Note that station 400 is located on the flank of the dike anomaly and the absolute 3-D response is very small.

Figure 14 displays the actual time-domain scattered response of the 3-D body at station 400, and gives a different perspective to the percentage plot in Fig. 13 because the differences at early times are now much more evident, whereas the late time differences are much less noticeable. Note that the plate-based models give the three smallest-amplitude responses in both presentations of the secondary response.

The results for the imaginary component of the frequency response at station 200 in Fig. 15 shows a structure similar to that for station 400, but with a larger-amplitude signal that peaks at a higher frequency. The percent secondary results displayed in Fig. 16 show that MARCO, LEROI, EM3D, EMIE3D, EMMODEL, and TEM3DL

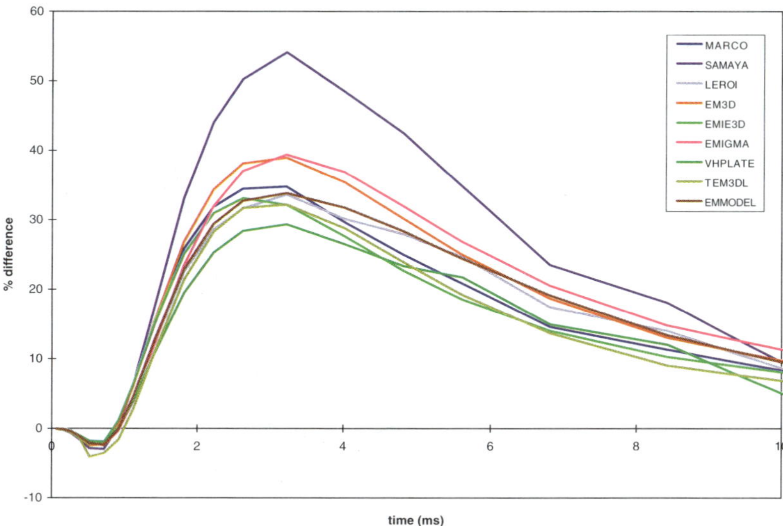

Figure 16. Response at station 200 for dike body with 100:1 conductivity contrast as a percentage of layered-earth response for UBCEM.

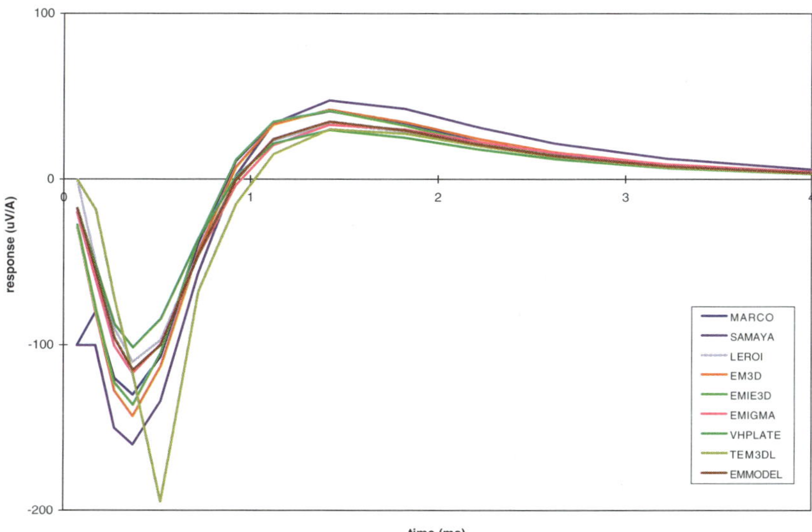

Figure 17. Response at station 200 for dike body with 100:1 conductivity contrast.

produce qualitatively similar responses, but with a larger spread than was evident for station 400. The plate-based methods produce responses that are not as smooth as those for station 400, and the response generated by VHPLATE is still smaller in peak amplitude relative to the other methods, but is now much more consistent with the other methods for times >5 ms. The responses of EM3D and EMIGMA agree quite well, but are marginally larger in amplitude relative to the other methods, whereas the response for SAMAYA is significantly larger and decays faster in the 4- to 8-ms interval with significant nonsmoothness in the 7- to 20-ms interval.

The actual transient scattered response for station 200 shown in Fig. 17 highlights the early time difference between TEM3DL and the other methods, which is probably because of the Gaussian pulse transmitter waveform used in the code. Also now evident

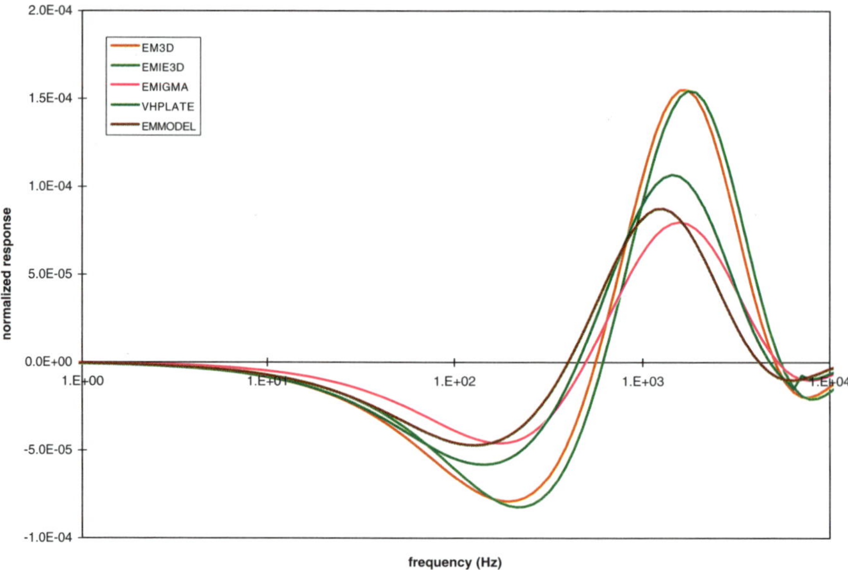

Figure 18. Imaginary component of frequency response at station 50 for 100:1 conductivity contrast dike.

is the close agreement between EMIGMA and VHPLATE; this is not as obvious in the percent secondary plot because the EMIGMA response is shifted to slightly later times in comparison to the other methods, which consequently produces a much larger percentage effect.

The results for the imaginary component of the frequency response for station 50, displayed in Fig. 18, now show a much larger spread than at stations 200 and 400. The prism-based methods EM3D and EMIE3D show good agreement. The responses for EMIGMA, VHPLATE, and EMMODEL are significantly smaller than the EM3D and EMIE3D, with the positions of the peak positive and negative responses being shifted to lower frequencies for EMMODEL and to a lesser extent for VHPLATE, and the position where the response changes sign also shifted lower for EMIGMA, VHPLATE, and EMMODEL.

The percent secondary transient response in Fig. 19 shows that the transmitter loop is now well coupled with the 3-D body and a large anomaly is evident. Figure 19 also shows that there are large differences between the results computed using the different codes. The responses for VHPLATE and EMMODEL are in much better agreement than their frequency responses would indicate, and both appear to be shifted to later time relative to the other codes. The response for EM3D also appears to be shifted slightly toward later time, whereas that for EMIE3D is slightly shifted toward earlier times. LEROI now produces the smallest response, and the results for EMIGMA appear to decay more slowly than do the other codes. The time-domain responses for SAMAYA, VHPLATE, TEM3DL, and LEROI all become negative for delay times in the 10- to 20-ms time range, with those for SAMAYA being the largest and earliest to change sign.

The actual secondary transient responses displayed in Fig. 20 show that the time shifts commented on above are present, but the shifts are quite small. Also, these latter figures show that SAMAYA actually generates the largest peak response for the methods considered in the 0–2-ms range. The results for EMIE3D, EM3D, and MARCO show very good agreement, and the results for TEM3DL again display the

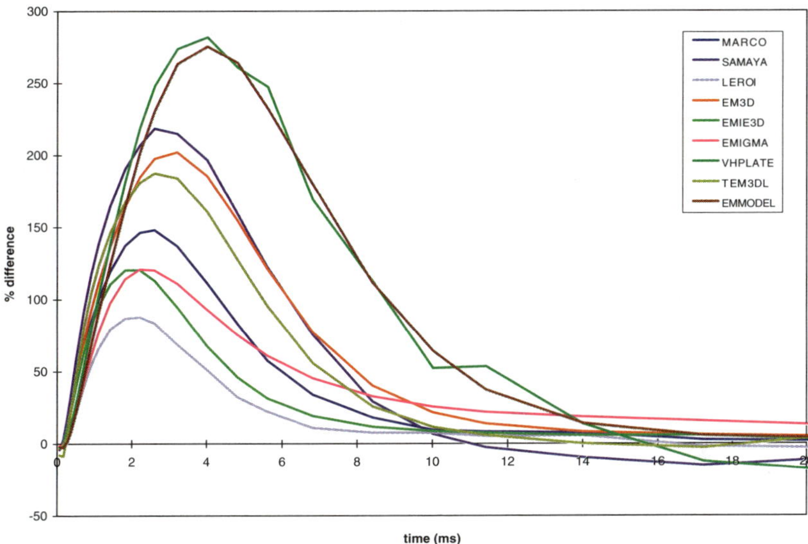

Figure 19. Response at station 50 for dike body with 100:1 conductivity contrast as a percentage of layered-earth response for UBCEM.

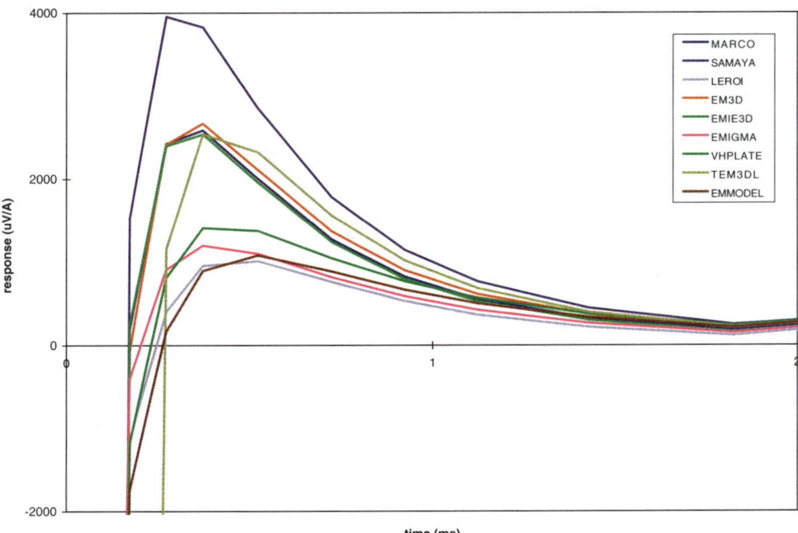

Figure 20. Response at station 50 for dike body with 100:1 conductivity contrast.

early time influence of the Gaussian pulse transmitter waveform. Note that EMIGMA and the plate-based codes produce the smallest-amplitude responses for station 50; for the plate-based codes, VHPLATE produces the smallest response at station 400, but produces the largest response at station 50.

Of the three integral-equation-based codes (EM3D, EMIE3D, and MARCO), EM3D gives the largest absolute and percentage responses and the latest peak percentage response, whereas EMIE3D gives the smallest absolute and percentage responses and the earliest peak percentage response, but all three methods give responses that are qualitatively similar across the entire time range. The differences are larger for station 50 than for stations 200 and 400 because the 100:1 conductivity contrast between the 3-D

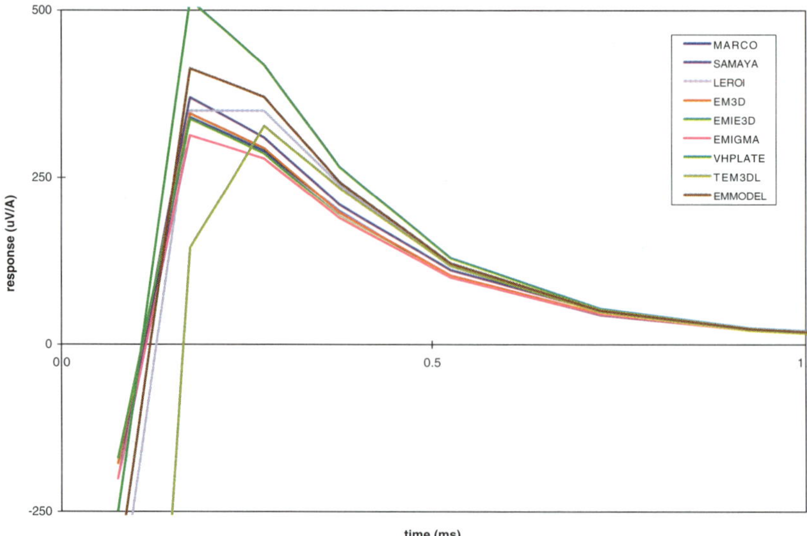

Figure 21. Response at station 50 for dike body with 2:1 conductivity.

body and the host is large enough for the numerical methods used to solve the integral equations to begin experiencing stability problems. This increase in instability results in the amplification of any differences between the methods, such as basis functions, methods used to compute the Green's functions, and matrix solvers.

Because LEROI, EMMODEL, and VHPLATE are based on thin-plate models, it is reasonable to expect that the results for station 50 will be less consistent with methods based on bodies with nonzero thickness than will be the case for stations 200 or 400. This is confirmed by an examination of the results for these codes for stations 50, 200, and 400. For stations 200 and 400, the plate-based codes generate secondary responses that (although marginally smaller) agree quite well with the prism-based codes. For station 50 the responses generated by the plate-based codes are about half of those generated by all prism-based codes except EMIGMA. It also appears that the response generated by VHPLATE decays faster as a function of distance from the 3-D body than do the other plate-based codes.

To provide another perspective on the results for station 50, Fig. 21 displays the scattered response of a dike with a resistivity of 25 ohm-m (i.e., with a conductivity contrast of 2:1). Clearly the methods now show generally good qualitative agreement, although the response for VHPLATE is significantly larger than that of the other methods. The improved agreement between SAMAYA and the other methods is known to be a function of the narrow gap between the top of the 3-D body and the overburden interface. SAMAYA uses a hybrid scheme in which a boundary layer is placed around the 3-D body. The thickness of this boundary layer is a function of the conductivity contrast between the body and the host medium; the greater the conductivity contrast, the greater the thickness required. For the configuration used here, the 100:1 conductivity contrast model cannot use a thick enough boundary layer because of the proximity of the overburden interface, and so, the accuracy deteriorates. Clearly, the plate-based methods also cannot accurately model the response for high-conductivity contrasts and produce significantly smaller responses than the prism-based methods for bodies with nonzero thickness (see Figs. 20 and 21).

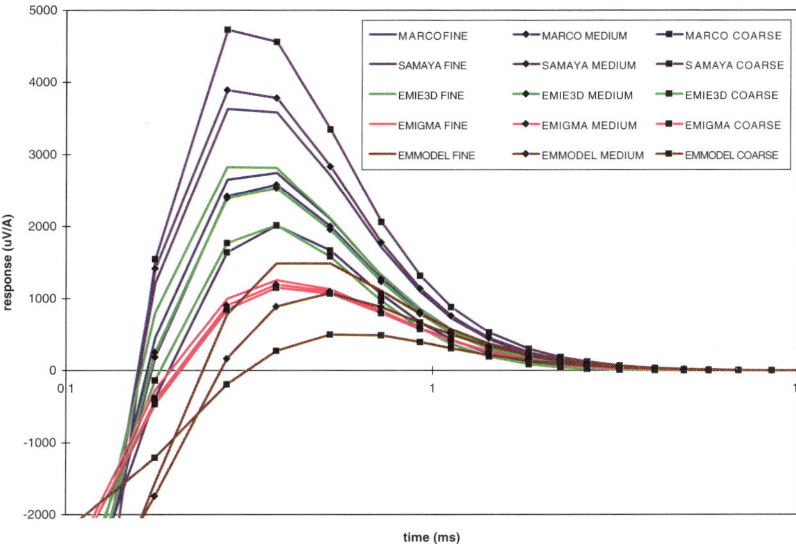

Figure 22. Coarse-, medium-, and fine-mesh 3-D responses for 100:1 conductivity contrast dike model at station 50.

Figure 22 displays the effect of changing the discretization on the secondary response. For each method the coarse mesh is obtained by halving the number of points/cells used, whereas the fine mesh represents a doubling of the number of points. This figure shows that all methods except EMIGMA show significant changes as the mesh changes and that, even for the fine-mesh results, there are still large differences among the responses computed by the various methods. For all of the methods except SAMAYA, the response increases with the number of points used. SAMAYA goes in the opposite direction, decreasing as the discretization becomes finer.

6 Three-dimensional modeling comparisons: Slab model

The second 3-D model is a horizontal resistive slab model for which the layered-earth model is defined as overburden, 400-m thick with resistivity 200 ohm-m; basement, 6 ohm-m. The 3-D body is defined as: width, 1000 m; strike length, 1000 m; depth extent, 50 m; depth to top, 800 m; resistivity, 10 ohm-m. In the plan and section view in Fig. 23, the center of the body is located at the origin and a circular transmitter loop of radius 282 m is centered 1000 m east of the origin (or right of the origin in plan view) with a single receiver located at the center of the loop. The transmitter waveform and receiver times are given in Appendix D.

The response for this model for TELMA was computed by the authors of the program and supplied in tabular form. The programs MARCO, SAMAYA, EMIE3D, EMIGMA, and TEM3DL were used to compute the response for the same model. The program MARCO includes an option for circular loops which was used for this model; however, SAMAYA, EMIE3D, EMIGMA, and TEM3DL do not allow circular loops, and so, an equivalent-area 500-m^2 loop was used instead. Because the 3-D body is more resistive than the layer in which it resides, the thin-plate methods LEROI, EMMODEL, and VHPLATE could not be used.

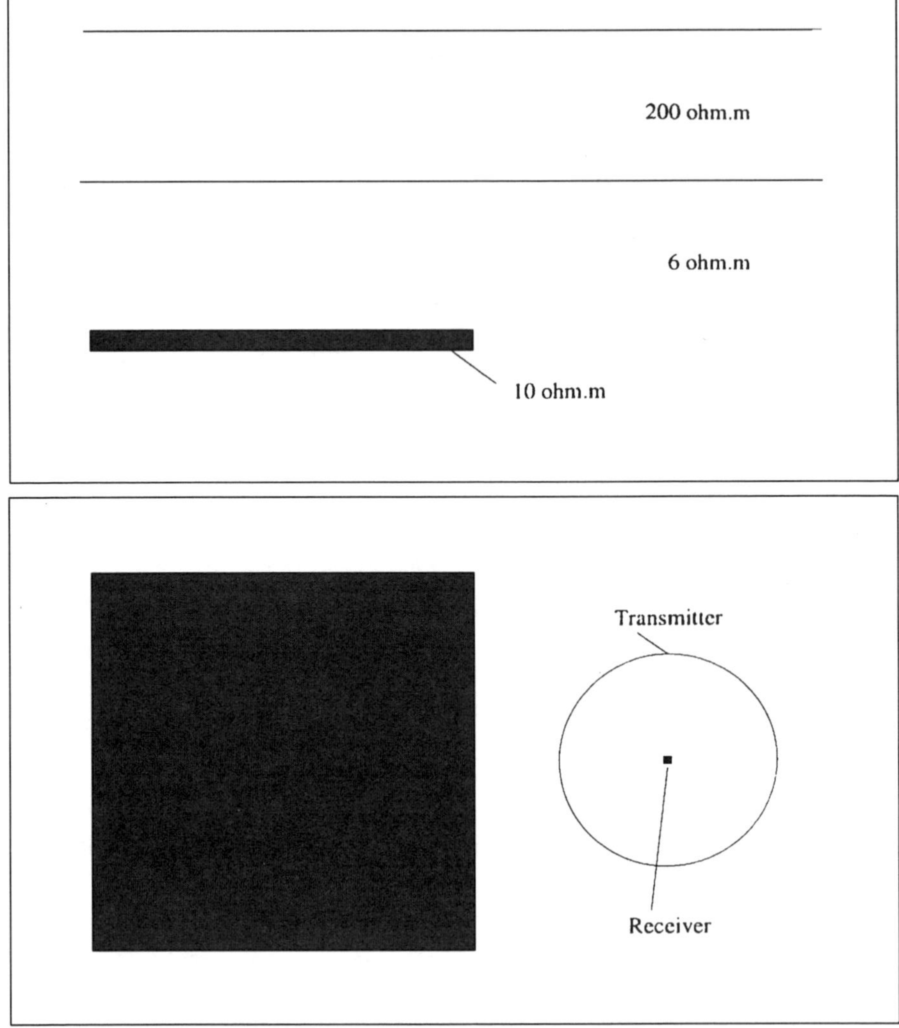

Figure 23. Plan and section schematic for slab model.

Once again the first step is to compare the layered-earth response from each of the methods. Figure 24 displays the differences between the methods (relative to MARCO) as a percentage of the MARCO layered-earth response (UBCEM was not taken as the basis of comparison because it did not have a circular-loop option). Figure 24 shows that the differences between the methods is less than 2% for all methods in the 0.3- to 200-ms time range, except for TEM3DL for which the maximum difference is 30% at 0.5 ms. The cause of this large difference is most probably due to TEM3DL's use of a Gaussian pulse to approximate an impulse. The deconvolution of the Gaussian pulse from the response for TEM3DL was considered but was not done because it fell outside the primary aim of this project.

The difference between TELMA and MARCO increases at late times. This may be caused by the time stepping in TELMA for which accuracy is expected to decrease with increasing delay time. The results for EMIGMA display a similar difference at late time, but the cause is unclear. In most practical cases, these late time differences are

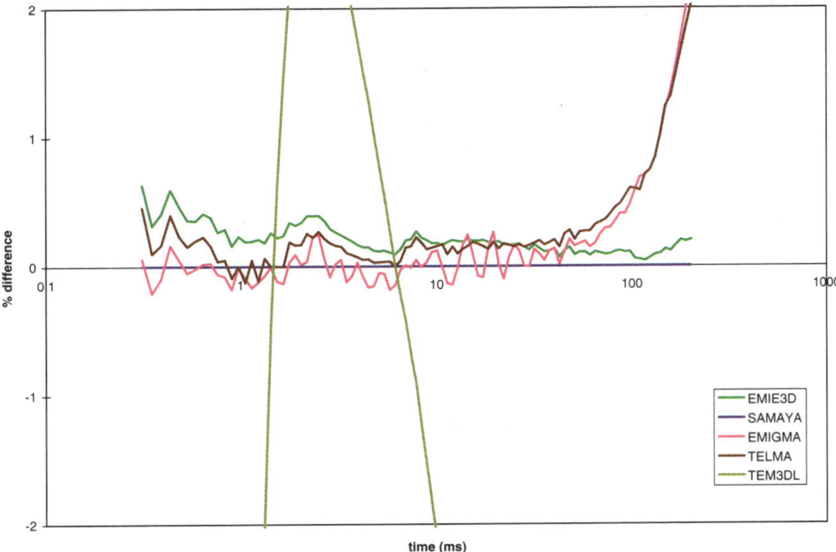

Figure 24. Difference between layered-earth response computed using EMIE3D, SAMAYA, EMIGMA, TELMA, and TEM3DL relative to that computed using MARCO.

of minor interest because the most important time range is between 1 and 100 ms. Note that there appears to be no significant difference between the circular- and square-loop approximation for this model. This insensitivity to loop type probably can be attributed to the thick highly resistive overburden which results in a much more rapid decay in the signal before the first reading is taken.

6.1 *Cell discretization*

To compare the differences between the 3-D responses computed using each of the methods, the 3-D body was discretized using $10 \times 10 \times 2$ cells for MARCO, SAMAYA, and EMIE3D; $12 \times 12 \times 1$ for TEM3DL; and 200 points for EMIGMA. The symmetry options were invoked for each of the programs to reduce the effective number of cells to $5 \times 5 \times 2$ for MARCO and EMIE3D and $5 \times 10 \times 2$ for SAMAYA. It is not known what discretization or symmetry options were used for TELMA. The differing discretizations are partly determined by the different approaches used by each of the methods to define the body and partly determined by practical considerations such as run time and disk space.

Figure 25 displays the absolute response of the 3-D body computed using each of the methods. This figure shows clearly that there is very good overall agreement between the different methods, and also that there is very good agreement between all the methods on the zero crossing time. The TELMA method gives a slightly larger amplitude response near the peak in response at 20 ms and the TEM3DL response is (marginally) the smallest and also displays some instability at late time. Once again, the EMIGMA response is time shifted relative to the other methods, but in this case it is shifted to marginally earlier times. The consistency of the results for the different methods indicates that the different discretizations used for the 3-D body have had little impact on the accuracy of the computed 3-D response.

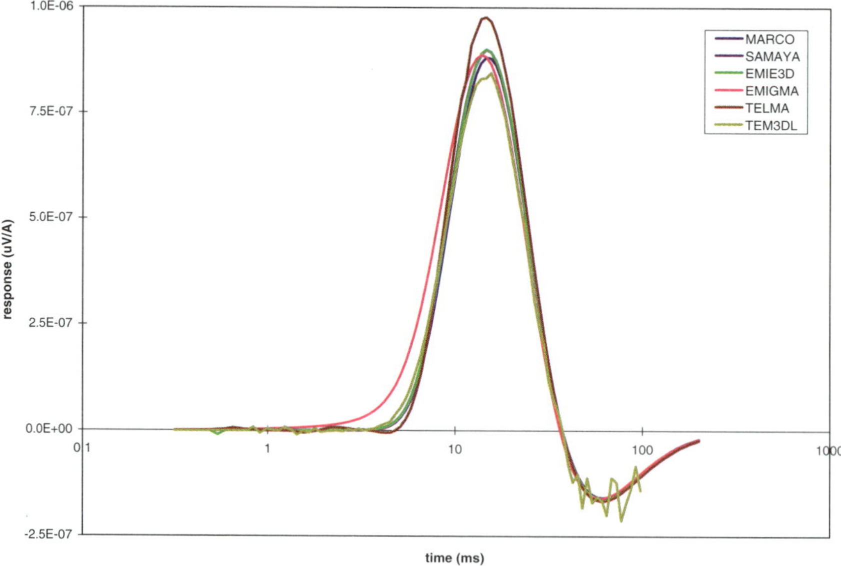

Figure 25. Response of 3-D body computed using MARCO, SAMAYA, EMIE3D, EMIGMA, and TEM3DL.

7 Conclusions

Nearly all of the programs reported in this paper give similar anomaly shapes and they could be used for parametric studies in many situations. They also could contribute to the development of useful qualitative interpretation insights. Detailed modeling of real field situations or 3-D inversion still appears to be unrealistic unless we can have more confidence in our 3-D forward models.

The main specific conclusions are:

1. In areas with conductive surface layers or in which early time readings are required, it is advisable to use programs that implement a rectangular-loop approximation to model data collected with a rectangular loop.
2. For frequency-domain codes, it is better to use the actual frequencies required by the time-domain transformation than to interpolate the frequency response.
3. Substantial computational efficiencies can be achieved for frequency-domain codes by choosing the number of frequencies and the upper and lower frequency limits to achieve the desired accuracy in the time interval of interest.
4. There is generally good agreement on the layered-earth response for all methods except EMIGMA, for which the results appear to be inaccurate.
5. There is good agreement between the 3-D response for the thin-plate methods LEROI, EMMODEL, and VHPLATE and the results generated by the prism-based codes except where the transmitter is close enough to the body for the nonzero thickness to become significant.
6. SAMAYA produces significantly larger 3-D responses than other codes for the dike model.
7. There are large differences among the secondary responses generated for stations close to the dike model when the conductivity contrast is large, and in this case, plate codes generate significantly smaller responses than do the prism codes.

8. The results for TEM3DL differ significantly from the other methods because of the use of a Gaussian pulse to approximate the pulse response.

Clearly, we need more rigorous testing and comparative studies of the different modeling methods currently available. Too often we see a new method or a new approximation tested in one or two simple cases (usually with results plotted on a logarithmic scale) and then unleashed on unsophisticated users. Developers and users must cooperate to support parametric studies and rigorous comparative tests as part of the development cycle. Such studies will provide a better understanding of the TEM method and of the limitations of the various modeling methods, and will inevitably lead to improvements in the programs and (hopefully) to more useful interpretation tools based on these programs.

Acknowledgments

The authors would like to express their sincere thanks to the many code developers and organizations who have assisted in providing information and test runs for the various programs reported here. Thanks especially to Walt Anderson (USGS), Mike Hoversten (University of California, Berkeley), Art Raiche (CSIRO), Fredijanto Sugeng (CSIRO), Peter Walker (PetRos EiKon), Tsili Wang (Schlumberger-Doll Research), Phil Wannamaker (University of Utah), and Zonghou Xiong (CSIRO). Thanks are also due to the CEMI group at the University of Utah, the JACI consortium at the University of British Columbia, and the Sniiggims research group.

Appendix A: Reference list of frequencies

The 251 frequencies (in hertz) used for all calculations requiring explicit choice of frequencies are listed in Table A1.

Table A1. Frequencies for calculations

0.29150245E−02	0.32216003E−02	0.35604190E−02	0.39348714E−02	0.43487055E−02	
0.48060627E−02	0.53115208E−02	0.58701383E−02	0.64875064E−02	0.71698031E−02	
0.79238582E−02	0.87572178E−02	0.96782222E−02	0.10696090E−01	0.11821007E−01	
0.13064234E−01	0.14438211E−01	0.15956691E−01	0.17634870E−01	0.19489545E−01	
0.21539280E−01	0.23804585E−01	0.26308136E−01	0.29074986E−01	0.32132830E−01	
0.35512269E−01	0.39247125E−01	0.43374781E−01	0.47936548E−01	0.52978080E−01	
0.58549833E−01	0.64707570E−01	0.71512923E−01	0.79034008E−01	0.87346084E−01	
0.9653235E−01	0.10668475	0.11790488	0.13030505	0.14400935	
0.15915494	0.17589341	0.19439229	0.21483670	0.23743127	
0.26240215	0.28999922	0.32049870	0.35420585	0.39145800	
0.43262798	0.47812787	0.52841300	0.58398670	0.64540511	
0.71328294	0.78829962	0.87120581	0.96283132	1.0640931	
1.1760048	1.2996863	1.4363755	1.5874405	1.7543930	1.9389042
2.1428204	2.3681829	2.6172469	2.8925052	3.1967125	3.5329137
3.9044735	4.3151107	4.7689347	5.2704878	5.8247900	6.4373884
7.1144147	7.8626442	8.6895657	9.6034546	10.613460	11.729687
12.963308	14.326672	15.833421	17.498636	19.338984	21.372883
23.620687	26.104897	28.850374	31.884594	35.237926	38.943932
43.039700	47.566223	52.568810	58.097519	64.207687	70.960464
78.423447	86.671310	95.786613			

(*Continues*)

Table A1. *(Continued)*

105.86058	116.99403	129.29840	142.89684	157.92543	174.53459
192.89055	213.17703	235.59705	260.37500	287.75888	318.02277
351.46951	388.43387	429.28583	474.43420	524.33087	579.47522
640.41919	707.77264	782.20978	864.47546	955.39313	1055.8727
1166.9198	1289.6459	1425.2791	1575.1770	1740.8398	1923.9255
2126.2666	2349.8879	2597.0278	2870.1597	3172.0171	3505.6208
3874.3103	4281.7749	4732.0933	5229.7720	5779.7920	6387.6577
7059.4536	7801.9028	8622.4365	9529.2656	10531.467	11639.071
12863.163	14215.994	15711.103	17363.453	19189.584	21207.771
23438.211	25903.230	28627.496	31638.275	34965.703	38643.078
42707.207	47198.762	52162.699	57648.699	63711.664	70412.281
77817.602	86001.750	95046.633	105042.77	116090.22	128299.54
141792.92	156705.41	173186.27	191400.42	211530.17	233777.00
258363.53	285535.88	315565.94	348754.31	385433.09	425969.47
470769.06	520280.28	574998.62	635471.75	702304.94	776166.94
857797.12	948012.44	1047715.8	1157905.0	1279683.0	1414268.4
1563008.4	1727391.4	1909062.6	2109840.5	2331734.5	2576965.0
2847986.8	3147512.2	3478539.0	3844380.2	4248697.0	4695536.5
5189370.5	5735141.0	6338311.5	7004917.5	7741631.0	8555825.0
9455649.0	10450109.	11549156.	12763792.	14106171.	15589730.
17229316.	19041340.	21043934.	23257144.	25703120.	28406340.
31393862.	34695584.	38344548.	42377280.	46834136.	51759728.
57203344.	63219472.	69868320.	77216440.	85337360.	94312368.
0.10423129E+09	0.11519339E+09	0.12730838E+09	0.14069752E+09		
0.15549482E+09	0.17184835E+09	0.18992179E+09	0.20989605E+09		

Appendix B: Transmitter and receiver for Model 1

The transmitter waveform used in all calculations is

time (ms)	<0	0	1500	>1500
current (A)	0.0	1.0	1.0	0.0

Three bipolar repetitions of the waveform were used in the calculation of the time-domain response, and the offtime was set at 1020 ms.

The 41 receiver times used for all time-domain calculations and their corresponding window widths are listed in Table B1.

Table B1. Receiver times and window widths for Model 1

Time (ms)	Window width (ms)	Time (ms)	Window width (ms)	Time (ms)	Window width (ms)
0.087	0.008	2.19	0.22	55.4	5.62
0.108	0.011	2.82	0.28	70.4	7.08
0.138	0.014	3.53	0.35	86.5	8.91
0.174	0.017	4.41	0.44	107	11.2
0.216	0.021	5.61	0.56	138	14.1
0.277	0.027	7.06	0.7	175	17.8
0.353	0.035	8.65	0.86	219	20
0.441	0.044	10.7	1.1	282	20
0.561	0.056	13.8	1.41	356	20
0.706	0.071	17.5	1.78	437	20
0.865	0.086	21.9	2.24	554	20
1.07	0.11	28.2	2.82	704	20
1.38	0.14	35.6	3.55	1000	20
1.75	0.18	43.7	4.47		

Appendix C: Transmitter waveform and receiver times for dike model

The transmitter waveform used in all calculations is for the dike model:

time (ms)	0.0	0.1	20.73	20.93
current (A)	0.0	1.0	1.0	0.0

The offtime for the transmitter waveform was 22.3 ms and four bipolar repetitions of the waveform were used in the calculation of the time domain response.

The 23 receiver times used for all time-domain calculations and corresponding window widths are listed in Table C1.

Table C1. Receiver times and window widths for Model 1

Time (ms)	Window width (ms)	Time (ms)	Window width (ms)	Time (ms)	Window width (ms)
0.075	0.0113	1.425	0.2138	6.825	1.024
0.175	0.0263	1.825	0.2738	8.425	1.264
0.275	0.0413	2.225	0.3338	10.025	1.504
0.375	0.0563	2.625	0.3938	11.425	1.714
0.525	0.0788	3.225	0.4838	14.025	2.104
0.725	0.1088	4.025	0.6038	17.225	2.584
0.925	0.1388	4.825	0.7237	20.425	3.064
1.125	0.1688	5.635	0.8438		

Appendix D: Transmitter waveform and receiver times for slab model

The transmitter waveform used in all calculations is for the slab model

time (ms)	0.0	0.0	1000.0	1000.0
current (A)	0.0	1.0	1.0	0.0

The offtime for the transmitter waveform was 1000 ms and four bipolar repetitions of the waveform were used in the calculation of the time-domain response.

The 98 receiver times used for all time-domain calculations and corresponding window widths are listed in Table D1.

Table D1. Receiver times for slab model

0.31303	1.4325	4.4907	12.845	35.670	98.023
0.35240	1.5400	4.7845	13.648	37.862	104.01
0.39416	1.6541	5.0961	14.499	40.188	110.37
0.43846	1.7751	5.4268	15.403	42.656	117.11
0.48546	1.9035	5.7776	16.361	45.274	124.26
0.53533	2.0398	6.1497	17.378	48.052	131.85
0.58823	2.1843	6.5446	18.456	50.998	139.90
0.64435	2.3376	6.9634	19.601	54.124	148.44
0.70389	2.5003	7.4078	20.815	57.441	157.50
0.76706	2.6728	7.8792	22.103	60.959	167.11
0.83407	2.8559	8.3794	23.469	64.692	177.31
0.90517	3.0501	8.9100	24.918	68.652	188.13
0.98059	3.2562	9.4729	26.456	72.853	199.61
1.0606	3.4748	10.070	28.088	77.311	
1.1455	3.7067	10.704	29.819	82.039	
1.2356	3.9527	11.376	31.655	87.055	
1.3322	4.2138	12.089	33.603	92.377	

Window widths (ms) are all zero for this model.

PART I

INTEGRAL-EQUATION MODELING

For too long, electrical prospecting methods have been limited by a lack of adequate interpretation aids. In order to interpret data, geophysicists have had to rely on scale model experiments, analytic solutions for bodies of simple shape, and, more recently, numerical solutions for two-dimensional models. All of these techniques are useful, but they all have limitations.

The purpose of this paper is to describe a numerical solution for calculating the induced polarization (IP) and electromagnetic (EM) responses of three-dimensional bodies buried in the earth. A theoretical solution in the form of an integral equation is derived. . . .

Hohmann (1975)

Transient Diffusive Electromagnetic Field Computation—A Structured Approach Based on Reciprocity

Adrianus T. de Hoop

Summary. The reciprocity theorem for transient diffusive electromagnetic fields is taken as the point of departure for developing computational methods to model such fields. Mathematically, the theorem is representative of any weak formulation of the field problem. Physically, the theorem describes the interaction between (a discretized version of) the actual field and a suitably chosen computational state. The choice of the computational state determines which type of computational method results from the analysis. It is shown that the finite-element method, the integral-equation method, and the domain-integration method can be viewed as particular cases of discretization of the reciprocity relation. The local field representations of the electric- and the magnetic-field strengths in terms of edge-element expansion functions are worked out in some detail.

The emphasis is on time-domain methods. The relationship with complex frequency-domain methods is indicated and used to symmetrize the basic field equations. This symmetrization expresses the correspondence that exists between transient electromagnetic wavefields in lossless media and transient diffusive electromagnetic fields in conductive media where the electric displacement-current contribution to the field can be neglected in the time window of observation. This aspect is also of importance in numerical modeling.

1 Introduction

The local, pointwise behavior in space-time of transient diffusive electromagnetic (EM) fields is governed by a parabolic system of first-order partial differential equations (Maxwell's equations in the diffusive approximation) that represent the EM phenomena on a local scale. When supplemented with boundary conditions that join the field values on either side of the interfaces where the constitutive parameters jump by finite amounts, and with the requirement of causality in the relationship between the field and its generating sources, the problem has a unique solution. A number of properties of this solution, in particular its analyticity and reciprocity properties, follow from this

Laboratory of Electromagnetic Research, Faculty of Electrical Engineering, Delft University of Technology, P.O. Box 5031, 2600 GA Delft, The Netherlands.

description. The computational handling of the field problem, however, often starts from a weak formulation, where the pointwise, or strong, satisfaction of the equality signs in the equations is replaced with requirements on the equality of certain integrated, or weighted, versions of the differential equations. Such weighted versions can be considered as special cases of the global reciprocity theorem that applies to two different admissible field states that are defined in one and the same domain in configuration space. Conceptually, a computational scheme to evaluate the field then is taken to describe the interaction between (a discretized version of) the actual field state and a suitably chosen computational state. The latter is representative of the method at hand (e.g., finite-element method and its related method of weighted residuals, integral-equation method, domain-integration method). Thus, choosing the reciprocity theorem as the point of departure offers the road to a structured approach to constructing computational schemes for evaluating the field. Besides, the standard source/receiver reciprocity properties (which are also consequences of the reciprocity theorem) can serve as a check on the consistency of the numerical results.

The emphasis is on time-domain methods. The relationship with complex frequency-domain methods is indicated, in particular to symmetrize the diffusive EM field equations in such a manner that the correspondence between transient diffusive EM fields in conductive media and EM wavefields in lossless media becomes manifest.

2 Diffusive EM field

The diffusive EM field under consideration is present in 3-D Euclidean space $\mathcal{R}^3$. The distribution of matter in it is assumed to be time invariant and the materials are assumed to be linear in their EM behavior. Position in the configuration is specified by the coordinates $\{x_1, x_2, x_3\}$ with respect to an orthogonal, Cartesian reference frame with the origin $\mathcal{O}$ and the three, mutually perpendicular base vectors $\{\mathbf{i}_1, \mathbf{i}_2, \mathbf{i}_3\}$ of unit length each. In the indicated order, the base vectors form a right-handed system. The corresponding position vector is $\mathbf{x} = x_1\mathbf{i}_1 + x_2\mathbf{i}_2 + x_3\mathbf{i}_3$. The time coordinate is t. The subscript notation for vectors and tensors is used and the summation convention applies. Differentiation with respect to x_m is denoted by ∂_m; ∂_t is a reserved symbol for differentiation with respect to t.

The EM constitutive properties of the media in the configuration are characterized by their (electrical) conductivity $\sigma_{k,r} = \sigma_{k,r}(\mathbf{x})$ and their (magnetic) permeability $\mu_{j,p} = \mu_{j,p}(\mathbf{x})$. The constitutive parameters are taken to be positive definite, symmetric tensors of rank two, thus allowing for anisotropy in the medium. The action of the sources that generate the field is characterized by the volume density of (external) electric current $J_k = J_k(\mathbf{x}, t)$ and the volume density of (external) magnetic current $K_j = K_j(\mathbf{x}, t)$. In each subdomain of the configuration where the constitutive coefficients vary continuously with position, the field quantities electric-field strength $E_r = E_r(\mathbf{x}, t)$ and magnetic-field strength $H_p = H_p(\mathbf{x}, t)$ then satisfy the parabolic system of partial differential equations (Ward and Hohmann, 1989)

$$-\epsilon_{k,m,p}\partial_m H_p + \sigma_{k,r} E_r = -J_k, \tag{1}$$

$$\epsilon_{j,n,r}\partial_n E_r + \mu_{j,p}\partial_t H_p = -K_j, \tag{2}$$

where $\epsilon_{k,m,p}$ is the completely antisymmetric unit tensor of rank three (Levi-Civita tensor): $\epsilon_{k,m,p} = +1$ if $\{k, m, p\}$ is an even permutation of $\{1, 2, 3\}$, $\epsilon_{k,m,p} = -1$ if $\{k, m, p\}$ is an odd permutation of $\{1, 2, 3\}$, $\epsilon_{k,m,p} = 0$ in all other cases. The existence of solutions of these field equations requires satisfaction of the compatibility relations

$$\partial_k(\sigma_{k,r}E_r) = -\partial_k J_k, \tag{3}$$

$$\partial_j(\mu_{j,p}\partial_t H_p) = -\partial_j K_j. \tag{4}$$

Across interfaces where $\sigma_{k,r}$ and/or $\mu_{j,p}$ jump by finite amounts, the field quantities are no longer continuously differentiable and the boundary conditions

$$\epsilon_{k,m,p}\nu_m H_p = \text{continuous}, \tag{5}$$

$$\epsilon_{j,n,r}\nu_n E_r = \text{continuous}, \tag{6}$$

should be satisfied. Here, ν_m is the unit vector along the normal to the interface. If the configuration extends to infinity, it is assumed that outside some bounded closed surface $\partial\mathcal{D}_0$ the medium is homogeneous and isotropic. In this domain, denoted by $\mathcal{D}_0$, the (scalar) conductivity has the value σ_0 and the (scalar) permeability the value μ_0. Because the tensor Green's functions for such a medium are analytically known, analytic source-type integral representations for the field quantities in $\mathcal{D}_0$ exist. The latter play a role in the *contrast source* or *scattering* formulation of the field problem.

In the analysis, the *time convolution* operator is needed. For any two space-time functions $F(\mathbf{x}, t)$ and $Q(\mathbf{x}, t)$, this is defined as

$$\mathrm{C}_t(F, Q; \mathbf{x}, t) = \int_{t' \in \mathcal{R}} F(\mathbf{x}, t')Q(\mathbf{x}, t - t')\,dt' \quad \text{for } t \in \mathcal{R}. \tag{7}$$

It has the properties

$$\mathrm{C}_t(F, Q; \mathbf{x}, t) = \mathrm{C}_t(Q, F; \mathbf{x}, t), \tag{8}$$

$$\partial_t \mathrm{C}_t(F, Q; \mathbf{x}, t) = \mathrm{C}_t(\partial_t F, Q; \mathbf{x}, t) = \mathrm{C}_t(F, \partial_t Q; \mathbf{x}, t). \tag{9}$$

For *causal* space-time functions $F(\mathbf{x}, t)$ and $Q(\mathbf{x}, t)$ having the semiinfinite interval $\{t \in \mathcal{R}; t > 0\}$ as their support, $\mathrm{C}_t(F, Q; \mathbf{x}, t)$ is causal as well, with the same support.

The relation between the time-domain quantities and their complex frequency-domain counterparts is given by the time Laplace transformation, which for any space-time function $F(\mathbf{x}, t)$ is

$$\hat{F}(\mathbf{x}, s) = \int_{t \in \mathcal{R}} \exp(-st)F(\mathbf{x}, t)\,dt \quad \text{for } \mathrm{Re}(s) = s_0, \tag{10}$$

where s_0 is some real value of $s \in \mathcal{C}$ for which the integral on the right-hand side is convergent. For *causal*, bounded, space-time functions $F(\mathbf{x}, t)$ having the semi-infinite interval $\{t \in \mathcal{R}; t > 0\}$ as their support, $\hat{F}(\mathbf{x}, s)$ is analytic in the right half $\{s \in \mathcal{C}; \mathrm{Re}(s) > 0\}$ of the complex s-plane.

From Eqs. (7) and (10), the Laplace transform $\hat{\mathrm{C}}_t(F, Q; \mathbf{x}, s)$ of $\mathrm{C}_t(F, Q; \mathbf{x}, t)$ is found as

$$\hat{\mathrm{C}}_t(F, Q; \mathbf{x}, s) = \hat{F}(\mathbf{x}, s)\hat{Q}(\mathbf{x}, s). \tag{11}$$

Further, from Eq. (10) and a subsequent integration by parts, the Laplace transform $\hat{\partial}_t F(\mathbf{x}, s)$ of $\partial_t F(\mathbf{x}, t)$ is

$$\hat{\partial}_t F(\mathbf{x}, s) = s\hat{F}(\mathbf{x}, s). \tag{12}$$

With the aid of this latter rule, the complex frequency-domain field equations are obtained from Eqs. (1) and (2) and (10) and (12):

$$-\epsilon_{k,m,p}\partial_m \hat{H}_p + \sigma_{k,r}\hat{E}_r = -\hat{J}_k, \tag{13}$$

$$\epsilon_{j,n,r}\partial_n \hat{E}_r + s\mu_{j,p}\hat{H}_p = -\hat{K}_j. \tag{14}$$

The complex frequency-domain compatiblity relations are obtained from Eqs. (3) and (4) and (10) and (12):

$$\partial_k(\sigma_{k,r}\hat{E}_r) = -\partial_k \hat{J}_k, \tag{15}$$

$$s\partial_j(\mu_{j,p}\hat{H}_p) = -\partial_j \hat{K}_j. \tag{16}$$

The boundary conditions across interfaces in jumps of the constitutive coefficients are obtained from Eqs. (5) and (6) and (10) and (12):

$$\epsilon_{k,m,p}\nu_m \hat{H}_p = \text{continuous}, \tag{17}$$

$$\epsilon_{j,n,r}\nu_n \hat{E}_r = \text{continuous}. \tag{18}$$

3 Reciprocity theorem

In the reciprocity theorem that is named after H. A. Lorentz, a certain *interaction quantity* is considered that is representative for the interaction between two admissible solutions (states) of the field equations, where the latter are defined in one and the same (proper or improper) subdomain $\mathcal{D}$ of $\mathcal{R}^3$. The domain $\mathcal{D}$ is assumed to be the union of a finite number of subdomains in each of which the field quantities of the two states are continuously differentiable. Furthermore, each of the two states applies to its own medium and has its own volume source distributions. The two states are indicated by the superscripts A and Z, respectively (Fig. 1). The relevant local interaction quantity is $\epsilon_{m,r,p}\partial_m[\mathrm{C}_t(E_r^A, H_p^Z) - \mathrm{C}_t(E_r^Z, H_p^A)]$ (de Hoop, 1987, 1995). Using the standard rules for the spatial differentiation and employing the field equations of the type (1) and (2) for the two states gives

$$\begin{aligned}\epsilon_{m,r,p}\partial_m&\big[\mathrm{C}_t\big(E_r^A, H_p^Z\big) - \mathrm{C}_t\big(E_r^Z, H_p^A\big)\big]\\ &= -\big(\sigma_{r,k}^Z - \sigma_{k,r}^A\big)\mathrm{C}_t\big(E_r^A, E_k^Z\big) + \big(\mu_{p,j}^Z - \mu_{j,p}^A\big)\partial_t\mathrm{C}_t\big(H_p^A, H_j^Z\big)\\ &\quad + \mathrm{C}_t\big(J_k^A, E_k^Z\big) - \mathrm{C}_t\big(K_j^A, H_j^Z\big) - \mathrm{C}_t\big(J_r^Z, E_r^A\big) + \mathrm{C}_t\big(K_p^Z, H_p^A\big).\end{aligned} \tag{19}$$

Equation (19) is the *local form of the EM reciprocity theorem of the time-convolution type*. The first two terms on the right-hand side are representative of the differences (contrasts) in the EM properties of the media in the two states; these terms vanish at

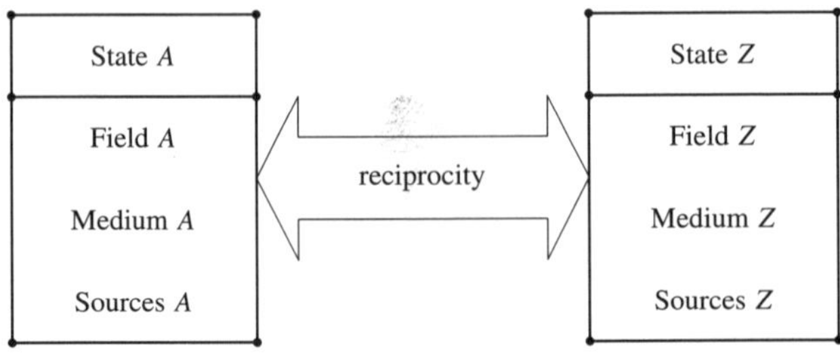

Figure 1. Two admissible states in reciprocity theorem.

those positions where $\sigma^Z_{r,k}(\mathbf{x}) = \sigma^A_{k,r}(\mathbf{x})$ and $\mu^Z_{p,j}(\mathbf{x}) = \mu^A_{j,p}(\mathbf{x})$. At points where these latter conditions hold, the media are denoted as each other's *adjoints*. The last four terms on the right-hand side are representative of the action of the volume sources in the two states; these terms vanish at those positions where the field is source-free.

To arrive at the global form of the reciprocity theorem for some bounded domain $\mathcal{D}$, it is assumed that $\mathcal{D}$ is the union of a finite number of subdomains in each of which the terms in Eq. (19) are continuous. Upon integrating Eq. (19) over each of these subdomains, applying Gauss's integral theorem to the resulting left-hand sides, and adding the results, it follows that

$$\begin{aligned}\epsilon_{m,r,p}\int_{\partial\mathcal{D}} \nu_m\big[\mathrm{C}_t\big(E^A_r, H^Z_p\big) - \mathrm{C}_t\big(E^Z_r, H^A_p\big)\big]\,\mathrm{d}A(\mathbf{x})\\ = \int_{\mathcal{D}} \big[-(\sigma^Z_{r,k} - \sigma^A_{k,r})\mathrm{C}_t\big(E^A_r, E^Z_k\big) + \big(\mu^Z_{p,j} - \mu^A_{j,p}\big)\partial_t \mathrm{C}_t\big(H^A_p, H^Z_j\big)\big]\,\mathrm{d}V(\mathbf{x})\\ + \int_{\mathcal{D}} \big[\mathrm{C}_t\big(J^A_k, E^Z_k\big) - \mathrm{C}_t\big(K^A_j, H^Z_j\big) - \mathrm{C}_t\big(J^Z_r, E^A_r\big) + \mathrm{C}_t\big(K^Z_p, H^A_p\big)\big]\,\mathrm{d}V(\mathbf{x}).\end{aligned} \tag{20}$$

Equation (20) is the *global form, for the domain $\mathcal{D}$, of the reciprocity theorem of the time-convolution type*. Note that in the process of adding the contributions from the subdomains of $\mathcal{D}$, the contributions from common interfaces have canceled in view of the boundary conditions (5) and (6). In view of this, in the left-hand side only a contribution from the outer boundary $\partial\mathcal{D}$ of $\mathcal{D}$ remains.

The complex frequency-domain versions of the local and the global reciprocity theorems follow from their time-domain counterparts by taking the time Laplace transform. Applying the standard rules given in Section 2, the complex frequency-domain version of the local reciprocity theorem follows from Eq. (19) as

$$\begin{aligned}\epsilon_{m,r,p}\partial_m\big(\hat{E}^A_r \hat{H}^Z_p - \hat{E}^Z_r \hat{H}^A_p\big) = -\big(\sigma^Z_{r,k} - \sigma^A_{k,r}\big)\hat{E}^A_r \hat{E}^Z_k + s\big(\mu^Z_{p,j} - \mu^A_{j,p}\big)\hat{H}^A_p \hat{H}^Z_j\\ + \hat{J}^A_k \hat{E}^Z_k - \hat{K}^A_j \hat{H}^Z_j - \hat{J}^Z_r \hat{E}^A_r + \hat{K}^Z_p \hat{H}^A_p,\end{aligned} \tag{21}$$

and the complex frequency-domain version of the global reciprocity theorem from Eq. (20) as

$$\begin{aligned}\epsilon_{m,r,p}\int_{\partial\mathcal{D}} \nu_m\big(\hat{E}^A_r \hat{H}^Z_p - \hat{E}^Z_r \hat{H}^A_p\big)\,\mathrm{d}A(\mathbf{x})\\ = \int_{\mathcal{D}} \big[-\big(\sigma^Z_{r,k} - \sigma^A_{k,r}\big)\hat{E}^A_r \hat{E}^Z_k + s\big(\mu^Z_{p,j} - \mu^A_{j,p}\big)\hat{H}^A_p \hat{H}^Z_j\big]\,\mathrm{d}V(\mathbf{x})\\ + \int_{\mathcal{D}} \big(\hat{J}^A_k \hat{E}^Z_k - \hat{K}^A_j \hat{H}^Z_j - \hat{J}^Z_r \hat{E}^A_r + \hat{K}^Z_p \hat{H}^A_p\big)\,\mathrm{d}V(\mathbf{x}).\end{aligned} \tag{22}$$

3.1 Limiting case of an unbounded domain

In quite a number of cases, the global reciprocity theorems are applied to an unbounded domain. To handle such cases, the embedding provisions of Section 2 are made and the theorem is applied first to the sphere $\mathcal{S}(\mathcal{O}, \Delta)$ with center at the origin $\mathcal{O}$ of the chosen reference frame and radius Δ, after which the limit $\Delta \to \infty$ is taken. From the source-type field integral representations pertaining to the homogeneous, isotropic embedding, it then follows that the contribution from $\mathcal{S}(\mathcal{O}, \Delta)$ vanishes in the limit $\Delta \to \infty$.

In the above procedure, the EM field equations pertaining to the two states have been taken as the point of departure, and the reciprocity theorems have been derived by operating on the equations in the manner indicated. In the realm of the use of the reciprocity theorems as the basis of a structured approach to the computation of the fields, note that, reversely, a necessary and sufficient condition for the global reciprocity theorem for arbitrary EM states Z satisfying equations of types (1) and (2) and boundary conditions of types (3) and (4) to hold is that the field in state A satisfies equations of types (1) and (2) and boundary conditions of types (3) and (4) as well.

4 Embedding procedure and contrast-source formulations

On many occasions the EM field computation in an entire geophysical configuration is beyond the capabilities because of the storage capacity and the computation times involved. In that case, it is standard practice to select a *target region* of bounded support in which a detailed computation is to be carried out, while the medium in the remaining part of the configuration (the *embedding*) is taken to be so simple that the field in it can be determined with the aid of analytical methods. Examples of such embeddings in $\mathcal{R}^3$ as the configuration space are the homogeneous isotropic embedding, and the embedding consisting of a finite number of parallel homogeneous layers. In these cases, combined time Laplace and spatial Fourier transform techniques provide the analytical tools to determine the field or, in fact, construct the relevant Green's tensors. Once the embedding has been chosen, the problem of computing the field in the target region can be formulated advantageously as a *contrast-source* or *scattering* problem (Hohmann, 1989).

To this end, first the *incident field* $\{E_r^i, H_p^i\}$ is introduced as the field that would be generated by the sources as if they were present in the embedding. Let the constitutive parameters of the embedding be $\sigma_{k,r}^{\mathrm{b}} = \sigma_{k,r}^{\mathrm{b}}(\mathbf{x})$ and $\mu_{j,p}^{\mathrm{b}} = \mu_{j,p}^{\mathrm{b}}(\mathbf{x})$; then, the incident field satisfies the basic field equations

$$-\epsilon_{k,m,p}\partial_m H_p^i + \sigma_{k,r}^{\mathrm{b}} E_r^i = -J_k, \tag{23}$$

$$\epsilon_{j,n,r}\partial_n E_r^i + \mu_{j,p}^{\mathrm{b}}\partial_t H_p^i = -K_j. \tag{24}$$

Next, the *scattered field* $\{E_r^{\mathrm{s}}, H_p^{\mathrm{s}}\}$ is defined as the difference between the total field $\{E_r, H_p\}$ and the incident field $\{E_r^i, H_p^i\}$. Hence, $\{E_r, H_p\} = \{E_r^i + E_r^{\mathrm{s}}, H_p^i + H_p^{\mathrm{s}}\}$. The field equations for the scattered field can be written alternatively as

$$-\epsilon_{k,m,p}\partial_m H_p^{\mathrm{s}} + \sigma_{k,r} E_r^{\mathrm{s}} = -\left(\sigma_{k,r} - \sigma_{k,r}^{\mathrm{b}}\right) E_r^i, \tag{25}$$

$$\epsilon_{j,n,r}\partial_n E_r^{\mathrm{s}} + \mu_{j,p}\partial_t H_p^{\mathrm{s}} = -\left(\mu_{j,p} - \mu_{j,p}^{\mathrm{b}}\right) \partial_t H_p^i, \tag{26}$$

or as

$$-\epsilon_{k,m,p}\partial_m H_p^{\mathrm{s}} + \sigma_{k,r}^{\mathrm{b}} E_k^{\mathrm{s}} = -\left(\sigma_{k,r} - \sigma_{k,r}^{\mathrm{b}}\right) E_r, \tag{27}$$

$$\epsilon_{j,n,r}\partial_n E_r^{\mathrm{s}} + \mu_{j,p}^{\mathrm{b}}\partial_t H_p^{\mathrm{s}} = -\left(\mu_{j,p} - \mu_{j,p}^{\mathrm{b}}\right) \partial_t H_p. \tag{28}$$

In both systems, the right-hand sides only differ from zero in the domain where the constitutive properties of the medium differ from those of the embedding. Further, in none of them do the activating source distributions occur. This has the advantage of a smoother behavior of the right-hand sides of the differential equations, a behavior that is due to the fact that the (incident) field variation is smoother in space than its generating source distributions (Hohmann, 1989). Equations (25) and (26) are typically the point of departure for finite-difference or finite-element computations; Eqs. (27) and (28) are typically the point of departure for integral-equation computations and for the construction of absorbing boundary conditions or Dirichlet-to-Neumann maps.

The source-type integral representations for the incident and the scattered fields are of the type

$$E_r^{\mathrm{i,s}}(\mathbf{x},t) = \int_{\mathcal{D}^{\mathrm{i,s}}} \{\mathrm{C}_t[G_{r,k}^{E,J}(\mathbf{x},\mathbf{x}',\cdot), J_k^{\mathrm{i,s}}(\mathbf{x}',\cdot)] + \mathrm{C}_t[G_{r,j}^{E,K}(\mathbf{x},\mathbf{x}',\cdot), K_j^{\mathrm{i,s}}(\mathbf{x}',\cdot)]\}\,\mathrm{d}V(\mathbf{x}'), \tag{29}$$

$$H_p^{\mathrm{i,s}}(\mathbf{x},t) = \int_{\mathcal{D}^{\mathrm{i,s}}} \{\mathrm{C}_t[G_{p,k}^{H,J}(\mathbf{x},\mathbf{x}',\cdot), J_k^{\mathrm{i,s}}(\mathbf{x}',\cdot)] + \mathrm{C}_t[G_{p,j}^{H,K}(\mathbf{x},\mathbf{x}',\cdot), K_j^{\mathrm{i,s}}(\mathbf{x}',\cdot)]\}\,\mathrm{d}V(\mathbf{x}'), \tag{30}$$

where $\mathcal{D}^{\mathrm{i}}$ is the support of the volume-source densities

$$J_k^{\mathrm{i}} = J_k, \tag{31}$$

$$K_j^{\mathrm{i}} = K_j, \tag{32}$$

generating the incident field; $\mathcal{D}^{\mathrm{s}}$ is the support of the contrast volume-source densities

$$J_k^{\mathrm{s}} = (\sigma_{k,r} - \sigma_{k,r}^{\mathrm{b}})\,E_r, \tag{33}$$

$$K_j^{\mathrm{s}} = (\mu_{j,p} - \mu_{j,p}^{\mathrm{b}})\,\partial_t H_p, \tag{34}$$

generating the scattered field; and $G_{r,k}^{E,J}$, $G_{r,j}^{E,K}$, $G_{p,k}^{H,J}$, $G_{p,j}^{H,K}$ are the electric-field/ electric-current, electric-field/magnetic-current, magnetic-field/electric-current, magnetic-field/ magnetic-current Green's tensors of the homogeneous isotropic embedding.

The complex frequency-domain versions of Eqs. (23) and (34) are found from their time-domain counterparts by replacing the operator ∂_t with the multiplying factor s and replacing the time convolutions with the product of their operands.

5 Computational procedures based on reciprocity

In the structured approach to the development of computational procedures based on reciprocity, the first step consists of selecting, in the global reciprocity theorems derived in Section 3, a finite number of linearly independent computational states for the state Z. The relevant states are indicated by the superscript C and their number is taken to be N^C. Next, state A is taken to be an approximation to the scattered field as introduced in Section 4, in the form of an expansion into a sequence of appropriate, linearly independent, known expansion functions and provided with unknown expansion coefficients. The relevant state is indicated by the superscript s and its field representation contains N^{s} terms. Based on the knowledge (see the end of Section 3) that for any number of arbitrary computational states and with an appropriate expansion containing an infinite number of terms for the scattered state, the application of the reciprocity theorem would lead to the unique, exact solution of the field problem, it is now assumed that the procedure with a finite number of computational states and a finite number of terms in the expansion of the scattered state leads to an approximate solution to the field problem. A quantification of the resulting error can be decided only after having introduced an appropriate error criterion. The latter is beyond the scope of the present analysis, which is focused mainly on the construction of both the computational states and the appropriate

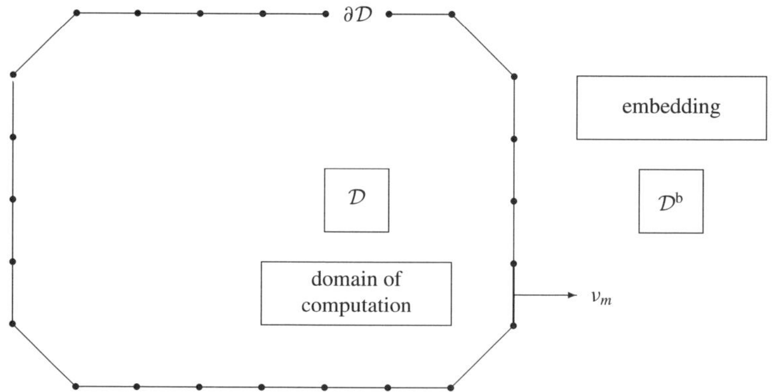

Figure 2. Discretized domain of computation $\mathcal{D}$ with boundary surface $\partial\mathcal{D}$ and embedding $\mathcal{D}^{\mathrm{b}}$.

expansion functions. From the beginning, it is clear that, for $N^C < N^s$, the system of linear algebraic equations in the expansion coefficients is underdetermined and hence cannot be solved, whereas, for $N^C = N^s$, the system of linear algebraic equations in the expansion coefficients has, in principle, a unique solution, whereas for $N^C > N^s$, the system of linear algebraic equations in the expansion coefficients is overdetermined and, hence, is amenable to a minimum norm solution in its residual.

The computations generally are carried out on a geometrically discretized version of the configuration. To this end, first the target region or domain of computation $\mathcal{D}$ is selected and discretized (Fig. 2). The boundary surface $\partial\mathcal{D}$ of this domain is taken to be located in the embedding $\mathcal{D}^{\mathrm{b}}$. Its geometric shape is taken such that it can be handled by a *mesh generator*. Typical cases are the discretization into a union of 3-rectangles or 3-simplices (tetrahedra), all of which have vertices, edges, and faces in common (Naber, 1980). The maximum diameter of the elements of the discretized geometry is denoted as its *mesh size*. The mesh size to be chosen depends on the shape of $\partial\mathcal{D}$, as well as on the spatial variations of the constitutive coefficients and the temporal and spatial variations of the volume-source densities and the field values in $\mathcal{D}$.

The mesh size is first adapted to the spatial variations of the known quantities (constitutive coefficients and volume-source densities in forward problems, volume-source densities and measured field values in inverse problems) and later iteratively adapted to the quantities to be computed (field values in forward problems, constitutive coefficients in inverse problems). Coupled to the mesh are, next, the spatial and temporal representations of the discretized known quantities. Finally, the discretized versions of the computational states and the unknown quantities are selected.

To illustrate the procedure, the forward field problem is discussed in more detail below. Discussion of EM inverse-source and inverse-wave-scattering problems can be found in de Hoop (1991).

It is assumed that the incident field has been determined already, for example, by evaluation of the relevant source-type integral representations containing the known Green's tensors of the embedding (see Section 4). In the forward-field computation problem, the constitutive coefficients and the volume-source distributions are given, and the field values are to be computed. As far as the medium properties are concerned, the analysis is concentrated on the case of strongly heterogeneous media where the constitutive coefficients may, in principle, jump from each subdomain of the discretized

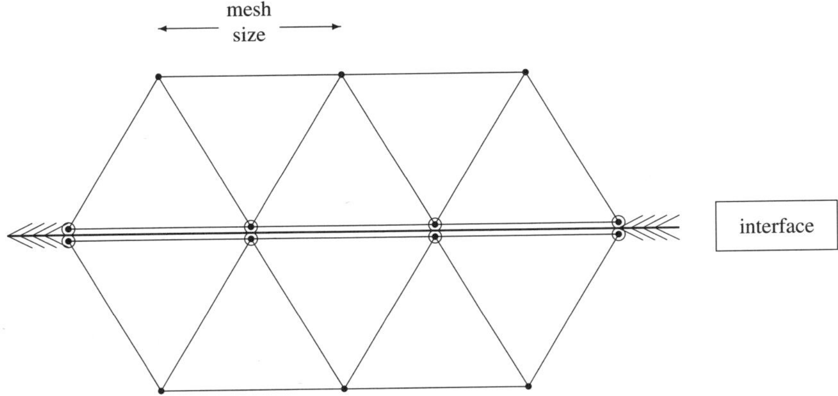

Figure 3. Interface (⋘) and simplicial mesh with multiple nodes (◉) and simple nodes (•).

geometry to any adjacent subdomain. The mesh size is assumed to be chosen so small that *piecewise linear expansions* are accurate enough to locally represent the field values, the constitutive coefficients, and the volume-source densities. A consistent theory then can be developed for a *simplicial mesh* consisting of 3-simplices (tetrahedra) all of which have vertices, edges, and faces in common (Fig. 3).

Consider one of the tetrahedra, Σ say, of the mesh and let $\{x_m(0), x_m(1), x_m(2), x_m(3)\}$ be the position vectors of its vertices. The ordering in the sequence defines the *orientation* of the tetrahedron. Further, let $\{A_m(0), A_m(1), A_m(2), A_m(3)\}$ denote the outwardly oriented vectorial areas of the faces of Σ, where the ordinal number of a face is taken to be the ordinal number of the vertex opposite to it. The position vector $\mathbf{x}$ in Σ then can be expressed in a symmetrical fashion in terms of the *barycentric coordinates* $\{\lambda(0, \mathbf{x}), \lambda(1, \mathbf{x}), \lambda(2, \mathbf{x}), \lambda(3, \mathbf{x})\}$ through

$$x_m = \sum_{I=0}^{3} \lambda(I, \mathbf{x}) x_m(I). \tag{35}$$

Inversely, the barycentric coordinates can be expressed in terms of the position vector via the relation

$$\lambda(I, \mathbf{x}) = 1/4 - (1/3V)(x_m - b_m)A_m(I) \quad \text{for } I = 0, 1, 2, 3, \tag{36}$$

where V is the volume of Σ and

$$b_m = \frac{1}{4} \sum_{I=0}^{3} x_m(I) \tag{37}$$

is the position vector of its *barycenter*. The barycentric coordinates have the property

$$\lambda[I, \mathbf{x}(J)] = \delta(I, J) \quad \text{for } I = 0, 1, 2, 3;\ J = 0, 1, 2, 3, \tag{38}$$

where $\delta(I, J)$ is the Kronecker symbol: $\delta(I, J) = 1$ for $I = J$ and $\delta(I, J) = 0$ for $I \neq J$.

As Eqs. (35) and (38) show, the barycentric coordinates perform a linear interpolation, in the interior of Σ, between the function value 1 at one of the vertices and the function value 0 at the remaining vertices. Consequently, they can be used as the (linear) interpolation functions for any of the quantities occurring in the field computation. As an example, the electric-field strength is considered. This quantity admits the local representation

$$E_r(\mathbf{x}, t) = \sum_{I=0}^{3} A_r^E(I, t)\,\lambda(I, \mathbf{x}) \quad \text{for } \mathbf{x} \in \Sigma, \tag{39}$$

where

$$A_r^E(I, t) = E_r[\mathbf{x}(I), t] \quad \text{for } I = 0, 1, 2, 3. \tag{40}$$

From the *local* representations of type (39), the *global* representations for the domain of computation are constructed. In this process, the values of the constitutive coefficients and the volume-source densities in the interior of the tetrahedron Σ, and hence their limiting values upon approaching (via the interior) the vertices of Σ, have no relation to the values of these quantities in any of the neighbors of Σ. As a consequence, each nodal point of the mesh is, for these quantities, initially considered as a *multiple node*, with multiplicity equal to the number of vertices that meet at that point. Subsequently, the multiple nodes are combined to *simple nodes* in all of those subdomains of the domain of computation where the quantities are known to be continuous. However, for the electric- and the magnetic-field strengths, the situation shows additional features. Here, all components vary continously in space as long as the constitutive coefficients do so (even if the volume-source densities vary only piecewise continuously in space), but across a jump discontinuity in constitutive properties of the medium, the tangential components of the field strengths are to be continuous, whereas their normal components should remain free to jump. A representation that meets these requirements is furnished by the *edge-element representation* (Mur and de Hoop, 1985). In this representation, $A_r^E(I, t) = E_r[\mathbf{x}(I), t]$ is expressed in terms of its projections along the edges that leave the vertex $\mathbf{x}(I)$. Rather than with these projections, we work with the numbers

$$\alpha^E(I, J, t) = E_r[\mathbf{x}(I), t][x_r(J) - x_r(I)] \quad \text{for } I = 0, 1, 2, 3; \quad J = 0, 1, 2, 3, \tag{41}$$

with $\alpha^E(I, I, t) = 0$. In view of the fact that, at the vertex $\mathbf{x}(I)$, the three vectorial edges $\{x_r(J) - x_r(I);\, J \neq I\}$ and the three vectorial faces $\{A_r(K);\, K \neq I\}$ form an (oblique) system of reciprocal base vectors in $\mathcal{R}^3$, the property

$$[x_m(J) - x_m(I)]A_m(K) = -3V[\delta(J, K) - \delta(I, K)] \quad \text{for } I = 0, 1, 2, 3;\ J = 0, 1, 2, 3;\ K = 0, 1, 2, 3 \tag{42}$$

holds. From Eqs. (40)–(42) it follows that

$$E_r[\mathbf{x}(I), t] = -\frac{1}{3V}\sum_{J=0}^{3} \alpha^E(I, J, t) A_r(J) \quad \text{for } I = 0, 1, 2, 3. \tag{43}$$

Because $\alpha(I, I, t) = 0$, we indeed have, through Eq. (41), at each vertex three numbers that, through Eq. (43), represent the expanded electric-field strength. By enforcing the numbers along a particular edge to be the same for all tetrahedra that have this edge in common, the continuity of the tangential components of E_r across edges and faces is guaranteed, and the normal components of E_r across faces are left free to jump. A similar piecewise spatial linear expansion is used for the magnetic-field strength H_p.

The piecewise linear expansions discussed above are used in the context of the different computational methods in existence. These are indicated briefly below.

5.1 Finite-element method

The finite-element method is characterized by taking $\sigma^C_{r,k} = 0$ and $\mu^C_{p,j} = 0$ and choosing either

$$E^C_k \in \{\text{electric-field-strength expansion functions}\}, \tag{44}$$
$$H^C_j = 0, \tag{45}$$
$$J^C_r = 0, \tag{46}$$
$$K^C_p = -\epsilon_{p,n,k}\partial_n E^C_k, \tag{47}$$

or

$$E^C_k = 0, \tag{48}$$
$$H^C_j \in \{\text{magnetic-field-strength expansion functions}\}, \tag{49}$$
$$J^C_r = \epsilon_{r,m,j}\partial_m H^C_j, \tag{50}$$
$$K^C_p = 0. \tag{51}$$

For this method, the choice of the field strengths typifies the computational state.

5.2 Integral-equation method

The integral-equation method is characterized by taking for the constitutive coefficients the values of the embedding, i.e., $\sigma^C_{r,k} = \sigma_0\delta_{r,k}$ and $\mu^C_{p,j} = \mu_0\delta_{p,j}$ and choosing either

$$J^C_r \in \{\text{electric-current volume-source expansion functions}\}, \tag{52}$$
$$K^C_p = 0, \tag{53}$$
$$E^C_k(\mathbf{x}, t) = \int_{\mathcal{D}^J} \{\mathrm{C}_t[G^{E,J}_{k,r}(\mathbf{x}, \mathbf{x}', \cdot), J^C_r(\mathbf{x}', \cdot)]\,\mathrm{d}V(\mathbf{x}'), \tag{54}$$
$$H^C_j(\mathbf{x}, t) = \int_{\mathcal{D}^J} \{\mathrm{C}_t[G^{H,J}_{j,r}(\mathbf{x}, \mathbf{x}', \cdot), J^C_r(\mathbf{x}', \cdot)]\,\mathrm{d}V(\mathbf{x}'), \tag{55}$$

where $\mathcal{D}^J$ is the support of J^C_r, or

$$J^C_r = 0, \tag{56}$$
$$K^C_p \in \{\text{magnetic-current volume-source expansion functions}\}, \tag{57}$$
$$E^C_k(\mathbf{x}, t) = \int_{\mathcal{D}^K} \{\mathrm{C}_t[G^{E,K}_{k,p}(\mathbf{x}, \mathbf{x}', \cdot), K^C_p(\mathbf{x}', \cdot)]\,\mathrm{d}V(\mathbf{x}'), \tag{58}$$
$$H^C_j(\mathbf{x}, t) = \int_{\mathcal{D}^K} \{\mathrm{C}_t[G^{H,K}_{j,p}(\mathbf{x}, \mathbf{x}', \cdot), K^C_p(\mathbf{x}', \cdot)]\,\mathrm{d}V(\mathbf{x}'), \tag{59}$$

where $\mathcal{D}^K$ is the support of K^C_p. For this method, the choice of the volume-source distributions, located in the embedding, typifies the computational state.

5.3 Domain-integration method

The domain-integration method is characterized by taking $\sigma^C_{r,k} = 0$ and $\mu^C_{p,j} = 0$ and choosing either

$$E^C_k = \text{global constant with support } \mathcal{D}, \tag{60}$$
$$H^C_j = 0, \tag{61}$$

$$J_r^C = 0, \tag{62}$$
$$K_p^C = 0, \tag{63}$$

or

$$E_k^C = 0, \tag{64}$$
$$H_j^C = \text{global constant with support } \mathcal{D}, \tag{65}$$
$$J_r^C = 0, \tag{66}$$
$$K_p^C = 0. \tag{67}$$

The value of the constant drops out from the final equations and the latter are equivalent to replacing the field equations with their integrated counterparts over the elementary subdomains of the domain of computation, applying Gauss's integral theorem, and adding the relevant results.

6 Symmetrization of transient diffusive EM field equations

The basic field equations governing the transient diffusive EM field are not symmetric in E_r and H_p, as opposed to their counterparts for transient EM wave propagation in lossless media. Recently, a symmetrization procedure has been developed that shows the interrelation between the transient diffusive EM-field constituents and their suitably defined lossless-medium wavefield counterparts (de Hoop, 1995). The basic idea is to rewrite the time-domain Laplace-transform Eqs. (13) and (14) as

$$-\epsilon_{k,m,p}\partial_m[(s/\alpha)^{1/2}\hat{H}_p] + (\alpha s)^{1/2}[\alpha^{-1}\sigma_{k,r}]\,\hat{E}_r = -(s/\alpha)^{1/2}\hat{J}_k, \tag{68}$$
$$\epsilon_{j,n,r}\partial_n\hat{E}_r + (\alpha s)^{1/2}\mu_{j,p}[(s/\alpha)^{1/2}\hat{H}_p] = -\hat{K}_j, \tag{69}$$

where α is an arbitrary constant. Equations (70) and (71) resemble the time Laplace-transform EM field equations for wavefields in a lossless medium with permittivity $\alpha^{-1}\sigma_{k,r}$, permeability $\mu_{j,p}$, electric-field strength $\hat{E}_r$, magnetic-field strength $(s/\alpha)^{1/2}$ $\hat{H}_p$, volume-source density of electric current $(s/\alpha)^{1/2}\hat{J}_k$, and volume-source density of magnetic current $\hat{K}_j$, but with s replaced with $(\alpha s)^{1/2}$. The Schouten-Van der Pol theorem for the time Laplace transform [Schouten (1934), (1961); Van der Pol (1934), (1960); see also Van der Pol and Bremmer (1950)] provides the tool to establish the relevant interrelation, which for computational purposes can be used to construct, by a simple time-like integration routine, transient diffusive EM field values from their computed wavefield counterparts in a lossless medium once the latter have been determined with the aid of standard software for computing wavefields. Details are given by de Hoop (1996).

7 Conclusions

A structured approach, with reciprocity as the basic principle, has been developed to construct schemes for the computation of transient diffusive EM-fields. It is shown that the known algorithms concerning the finite-element, integral-equation, and domain-integration techniques all can be viewed as particular choices for the computational state with which the interaction of (the approximating expansion of) the actual field to be computed is set equal to zero. It is believed that the approach also can lead to additional types of algorithms.

Acknowledgment

The research presented in this contribution has been supported financially through a Research Grant from the Stichting Fund for Science, Technology and Research (a companion organization to the Schlumberger Foundation in the USA). This support is gratefully acknowledged.

References

de Hoop, A. T., 1987, Time-domain reciprocity theorems for electromagnetic fields in dispersive media: Radio Sci., **22**, 1171–1178.

———1991, Reciprocity, discretization, and the numerical solution of direct and inverse electromagnetic radiation and scattering problems: Proc. IEEE, **95**, 1421–1430.

———1995, Handbook of Radiation and Scattering of Waves: Academic Press London, 814–817.

———1996, A general correspondence principle for time-domain electromagnetic wave and diffusion fields. Geophys. J. Internat., **127**, 757–761.

Hohmann, G. W., 1989, Numerical modeling for electromagnetic methods of geophysics, *in* Nabighiam, M. N., Ed., Electromagnetic methods in applied geophysics, Vol. I. Theory: Soc. Expl. Geophys., 313–363.

Mur, G., and de Hoop, A. T., 1985, A finite-element method for computing three-dimensional electromagnetic fields in inhomogeneous media: IEEE Trans. Magn., **MAG-21**, 2188–2191.

Naber, L., 1980, Topological methods in Euclidean space: Cambridge Univ. Press.

Schouten, J. P., 1934, A new theorem in operational calculus together with an application of it: Physica, **1**, 75–80.

Schouten, J. P., 1961, Operatorenrechnung: Springer-Verlag Berlin, 124–126.

Van der Pol, B., 1934, A theorem on electrical networks with an application to filters: Physica, **1**, 521–530.

———1960, A theorem on electrical networks with an application to filters, *in* Selected scientific papers: North Holland Publ. Co.

Van der Pol, B., and Bremmer, H., 1950, Operational calculus based on the two-sided Laplace transform: Cambridge Univ. Press, 232–236.

Ward, S. H., and Hohmann, G. W., 1989, Electromagnetic theory for geophysical applications, *in* Nabighiam, M. N., Ed., Electromagnetic methods in applied geophysics, Vol. I. Theory: Soc. Expl. Geophys., 131–311.

Integral-Equation Method for Modeling Transient Diffusive Electromagnetic Scattering

Evert C. Slob[1]
Peter M. van den Berg[2]

Summary. Standard 3-D integral-equation methods using pulse expansion functions and point matching have difficulties with large contrasts in conductivity between the body and the host. We formulate a new expansion with rectangular building blocks and trilinear functions that more accurately models general current distributions inside the scattering body without increasing the number of unknowns. The formulation is in the frequency domain; to transform results to the time domain, we solve for the coefficients of an expansion of the decaying field in transients that are determined analytically by the behavior of an iterative solution of the integral equation. The method requires a modest number of frequencies and is stable and accurate even at late times.

1 Introduction

Standard 3-D integral-equation methods in geophysics approximate the unknown fields in a (usually homogeneous) scatterer with pulse functions and use collocation, or point matching, to construct a matrix equation at a fixed frequency (Raiche, 1974; Hohmann, 1975; Wannamaker et al., 1984). These schemes give accurate results up to contrasts in conductivity of about 100 between the host and the scatterer. SanFilipo and Hohmann (1985) extended the range of accuracy of integral-equation methods with a time-domain formulation that included in the expansion a set of concentric current tubes of constant amplitude that could model better (divergence-free) vortex currents in the scatterer. They used piecewise linear functions in time to approximate the evolution of the electric field and solved recursively for the field at successive time steps. Their discretization required time steps smaller than a tenth of the (diffusion) time constant of the body. To speed up the computation, Newman and Hohmann (1988) reformulated this method in the frequency domain and transformed to the time domain by a sine (or cosine) integral that was approximated by a digital filter. This method now is used widely because it gives the most accurate time-domain results. It can, however, give poor results for

[1]Applied Geophysics Section, Department of Applied Earth Sciences, Delft University of Technology, P.O. Box 5028, 2600 GA Delft, The Netherlands.
[2]Laboratory of Electromagnetic Research, Department of Electrical Engineering, Delft University of Technology, P.O. Box 5031, 2600 GA Delft, The Netherlands.

the voltage induced in a loop receiver at late times (Hohmann 1988; Newman and Hohmann, 1988).

We have developed a new expansion that yields a local description of the electric field inside each block as the weighted sum of its values at the vertices of rectangular blocks. Piecing together all of the elementary blocks and applying the boundary conditions for the tangential electric field and the normal electric current reduces the number of global expansion coefficients to three for the electric field connected with each grid point, which is the smallest number possible in a linear expansion. The discretized (matrix) equation is solved with an iterative technique; the scattered electromagnetic (EM) field outside the scatterer is computed with Green's function. To obtain results in the time domain, we use the behavior of the scattered field as a function of frequency (as described by an iterative solution to the scattering problem) to construct suitable temporal expansion functions. The expansion functions have known transients; all that remains is to determine their expansion coefficients.

2 Weak formulation of integral equation

We investigate the diffusive EM field in a configuration that consists of a bounded, arbitrarily shaped, inhomogeneous, isotropic object buried in an unbounded homogeneous and isotropic background medium. To specify position in the configuration, we use the coordinates $\{x_1, x_2, x_3\}$ with respect to a Cartesian reference frame with origin $\mathcal{O}$ and three mutually perpendicular base vectors $\{\boldsymbol{i}_1, \boldsymbol{i}_2, \boldsymbol{i}_3\}$ of unit length each. In the indicated order, the base vectors form a right-handed system. The subscript notation for Cartesian vectors and tensors is used, except when explicitly specified otherwise. The subscript notation applies to repeated lowercase Latin subscripts that range over the values 1, 2, and 3. Whenever appropriate, the position also is specified by the vector $\boldsymbol{x} = x_m \boldsymbol{i}_m$. Differentiation with respect to x_m is denoted ∂_m. The analysis is in the Laplace-transform domain, with real, positive transform parameter s.

Let $\mathcal{D}^s$ be the bounded domain occupied by the scatterer and let $\sigma^s(\boldsymbol{x})$ be its conductivity and μ its permeability. The region (embedding) exterior to $\mathcal{D}^s$ is denoted $\tilde{\mathcal{D}}^s$ and has a conductivity σ and permeability μ. The electric-field integral equation for this configuration is (e.g., Hohmann, 1975),

$$\hat{E}_k^i(\boldsymbol{x}, s) = \hat{E}_k(\boldsymbol{x}, s) - [\partial_k \partial_r - s\sigma\mu\delta_{k,r}] \int_{\mathbf{x}' \in \mathcal{D}^s} \frac{\exp(-\sqrt{s\sigma\mu}\,|\boldsymbol{x} - \boldsymbol{x}'|)}{4\pi\,|\boldsymbol{x} - \boldsymbol{x}'|} \chi^\sigma(\boldsymbol{x}') \hat{E}_r(\boldsymbol{x}', s)\, dV, \tag{1}$$

in which the normalized electric contrast function is $\chi^\sigma = \sigma^s/\sigma - 1$.

Solving this integral equation pointwise (by collocation or point matching) is not easy because of the strong singularity of the kernel. A weak formulation eliminates the singularity by solving the equation in an average sense. The weak formulation multiplies the integral equation by a sequence of weighting functions of finite spatial support, and integrates over their support. Reciprocity theorems (see de Hoop, 1991) indicate that these weighting functions can be interpreted as localized source distributions. Let us therefore denote the weighting functions as $\hat{J}_k^G$ and the domain they occupy as $\mathcal{D}^G$, also denoted as the domain of computation. The desired weak form of the electric-field integral equation is then

$$\int_{\mathbf{x} \in \mathcal{D}^G} \hat{J}_k^G \hat{E}_k^i \, dV = \int_{\mathbf{x} \in \mathcal{D}^G} \hat{J}_k^G \hat{E}_k \, dV - \int_{\mathbf{x} \in \mathcal{D}^s} \hat{E}_r^G \chi^\sigma \hat{E}_r \, dV \quad \text{for all } \hat{J}_k^G, \tag{2}$$

where $\hat{E}_r^G$ denotes the field generated by the localized source distribution, which also is denoted as the weak form of the electric Green's state because it would be a true Green's state if the localized source were a point source. It is

$$\hat{E}_r^G(\boldsymbol{x}, s) = [\partial_r \partial_k - s\sigma\mu\delta_{r,k}] \int_{\mathbf{x}' \in \mathcal{D}^G} \hat{J}_k^G(\boldsymbol{x}', s) \frac{\exp[-\sqrt{s\sigma\mu}|\boldsymbol{x} - \boldsymbol{x}'|]}{4\pi|\boldsymbol{x} - \boldsymbol{x}'|} dV. \tag{3}$$

When $\hat{J}_k^G$ is a continuous function, the Green's state, $\hat{E}^G$, is also continuous in all space.

3 Discretization

The numerical handling of the integral equation requires the discretization of the domain of computation $\mathcal{D}^G$. This implies that the quantities occurring in the integral equation are discretized as well. As a basic building block we use the *rectangle* in $\mathbb{R}^3$ (see Naber, 1980) in the approximation of the computational domain. To describe the rectangle, denoted $\mathcal{R}$, we use a local numbering that applies to any rectangle. The vertices of $\mathcal{R}$ are labeled P with arguments reflecting their relative positions in the Cartesian reference frame: $\{P(0, 0, 0), \ldots, P(1, 1, 1)\}$. Their position vectors are given accordingly as $\{x_m(0, 0, 0), \ldots, x_m(1, 1, 1)\}$. The rectangle then is represented by

$$\mathcal{R} = [x_1(0, 0, 0), x_1(1, 1, 1)][x_2(0, 0, 0), x_2(1, 1, 1)][x_3(0, 0, 0), x_3(1, 1, 1)], \tag{4}$$

and its volume, denoted $|\mathcal{R}|$, is the product of the lengths of its sides, that is,

$$|\mathcal{R}| = \Delta x^{(1)} \Delta x^{(2)} \Delta x^{(3)}. \tag{5}$$

Now that we have a general setup for one rectangle we can use a union of these rectangles to discretize the computational domain and the scattering domain. They are represented by

$$[\mathcal{D}^G] = \sum_{P=0}^{L^{(3)}} \sum_{N=0}^{L^{(2)}} \sum_{M=0}^{L^{(1)}} \mathcal{R}(M, N, P), \tag{6}$$

$$[\mathcal{D}^s] = \sum_{P=1}^{L^{(3)}-1} \sum_{N=1}^{L^{(2)}-1} \sum_{M=1}^{L^{(1)}-1} \mathcal{R}(M, N, P), \tag{7}$$

where the rectangles $\mathcal{R}(M, N, P)$ have vertices, sides, and faces in common. The vertices of the rectangles also are denoted as the nodes of the (geometrical) mesh and the supremum h of the maximum diameters of the rectangles is denoted as the mesh size. In describing the configuration, we employ a global numbering for the nodes according to their Cartesian coordinates, where each node is in the interior of the computational domain $[\mathcal{D}^G]$ or on its boundary. They are given by

$$x(M, N, P) = \{M\Delta x^{(1)}, N\Delta x^{(2)}, P\Delta x^{(3)}\}, \tag{8}$$

where $\{M = 0, \ldots, L^{(1)} + 1; N = 0, \ldots, L^{(2)} + 1; P = 0, \ldots, L^{(3)} + 1\}$ and $\Delta x^{(k)}$ denotes the dimension of the rectangle in the x_k-direction. For nodes in the scattering domain, we use the same indices but with the following, different, ranges: $\{M = 1, \ldots, L^{(1)}; N = 1, \ldots, L^{(2)}; P = 1, \ldots, L^{(3)}\}$. The domain that is occupied by a rectangle $\mathcal{R}(M, N, P)$ is (Eq. 4)

$$\mathcal{R}(M, N, P) \rightarrow \begin{cases} x_1(M, N, P) \leq x_1 \leq x_1(M+1, N+1, P+1), \\ x_2(M, N, P) \leq x_2 \leq x_2(M+1, N+1, P+1), \\ x_3(M, N, P) \leq x_3 \leq x_3(M+1, N+1, P+1). \end{cases} \tag{9}$$

The simplicial star of each node is the collection of the eight rectangles that meet at that node; it is denoted by Star(M, N, P) and is given by

$$\text{Star}(M, N, P) = \sum_{p,n,m=0}^{1} \mathcal{R}(M-m, N-n, P-p). \tag{10}$$

All quantities occurring in the integral equation are expanded in a sequence of local tri-linear functions. In the domain $[\mathcal{D}^G]$, the global representations of the quantities $[\hat{Q}]$ follow, by combining all local expansions and using global indices, as

$$[\hat{Q}](\boldsymbol{x}) = \sum_{P,N,M} \sum_{p,n,m=0}^{1} \hat{B}^Q(m, n, p, M, N, P)\Psi(m, n, p, M, N, P; \boldsymbol{x}) \quad \text{for} \quad \boldsymbol{x} \in [\mathcal{D}^G], \tag{11}$$

in which $\hat{B}^Q(m, n, p, M, N, P)$ denotes the global expansion coefficient and has the value of $\hat{Q}$ at the vertex with global number $\{M, N, P\}$ which has the local number $\{m, n, p\}$ in the rectangle $\mathcal{R}(M-m, N-n, P-p)$ of the simplicial star. The global expansion functions Ψ are defined by

$$\Psi(m, n, p, M, N, P; \boldsymbol{x}) = \begin{cases} \psi(m, n, p; \boldsymbol{x}) & \text{for } \boldsymbol{x} \in \mathcal{R}(M-m, N-n, P-p), \\ 0 & \text{for } \boldsymbol{x} \notin \mathcal{R}(M-m, N-n, P-p), \end{cases} \tag{12}$$

where the local trilinear interpolation function is given by

$$\psi(m, n, p; \boldsymbol{x}) = \left[1 - \frac{|x_1 - x_1(m, n, p)|}{\Delta x^{(1)}}\right]\left[1 - \frac{|x_2 - x_2(m, n, p)|}{\Delta x^{(2)}}\right] \times \left[1 - \frac{|x_3 - x_3(m, n, p)|}{\Delta x^{(3)}}\right]. \tag{13}$$

In case the quantity can be represented at a particular node $\boldsymbol{x}(M, N, P)$ by one global expansion coefficient, i.e., in a simple node expansion, the global expansion is written as

$$[\hat{Q}](\boldsymbol{x}) = \sum_{P,N,M} \hat{B}^Q(M, N, P)\phi(M, N, P; \boldsymbol{x}) \quad \text{for} \quad \boldsymbol{x} \in \text{Star}(M, N, P), \tag{14}$$

where ϕ is defined over a simplicial star as

$$\phi(M, N, P; \boldsymbol{x}) = \left[1 - \frac{|x_1 - x_1(M, N, P)|}{\Delta x^{(1)}}\right]\left[1 - \frac{|x_2 - x_2(M, N, P)|}{\Delta x^{(2)}}\right] \times \left[1 - \frac{|x_3 - x_3(M, N, P)|}{\Delta x^{(3)}}\right]. \tag{15}$$

These representations describe the discretization of the field quantities.

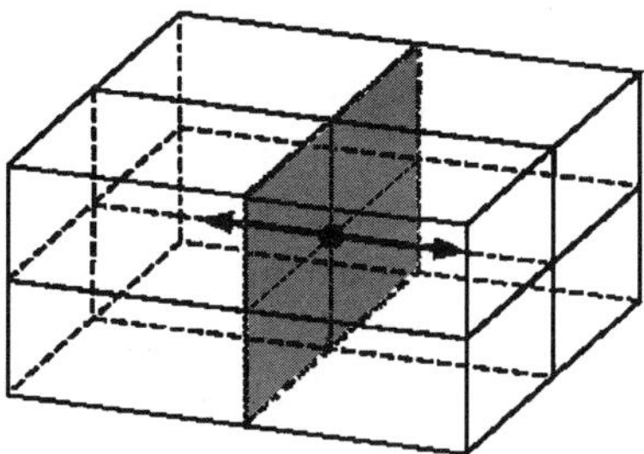

Figure 1. The simplicial star of a node where the two vectors normal to the four shaded faces represent the two edge-expansion coefficients for $\hat{E}_1$ pertaining to the rectangles at the left and right of the shaded faces, respectively.

The global representation of the electrical conductivity and contrast functions are given by (Eq. 11)

$$[\sigma](\boldsymbol{x}) = \sum_{P,N,M} \sum_{p,n,m=0}^{1} \hat{B}^{\sigma}(m, n, p, M, N, P)\Psi(m, n, p, M, N, P; \boldsymbol{x}), \tag{16}$$

$$[\chi^{\sigma}](\boldsymbol{x}) = \sum_{P,N,M} \sum_{p,n,m=0}^{1} \hat{B}^{\chi^{\sigma}}(m, n, p, M, N, P)\Psi(m, n, p, M, N, P; \boldsymbol{x}). \tag{17}$$

The values for the expansion coefficients follow from user-supplied input.

The global representation for the electric field is obtained with the application of boundary conditions. Because each node is connected with eight rectangles, there are 24 unknown local expansion coefficients for the electric field at that node. This is reduced to 6 by enforcing the continuity of the tangential components using the edge expansion (see Mur and de Hoop, 1985). Each vector component now is represented by two edge-expansion coefficients that pertain to four adjacent rectangles at each side of the four faces to which that vector component is normal. In Figure 1, the situation is sketched for the edge expansion in the x_1-direction; the four shaded faces of the simplicial star are the faces to which the x_1-components of the vectorial edge-expansion functions are normal. The two vectors in the figure represent the two unknown expansion coefficients for E_1 at the node. These two are now reduced to one unknown expansion coefficient by requiring the continuity of J_1 across the four shaded faces. The four coefficients $B^{\sigma}(0, n, p)$ correspond to the values of the conductivity at the node used in the four rectangles to the right of the shaded faces, whereas the four coefficients $B^{\sigma}(1, n, p)$ correspond to the values of the conductivity at that node used in the four rectangles to the left of the shaded faces. Because there are four faces and only two edge-expansion coefficients, the continuity of the normal current density cannot be enforced exactly and the problem is solved in the least-squares sense. A similar procedure is used for the x_2 and x_3-components of the electric field. Now we have only three unknown global expansion coefficients for the electric field at each node—one for each vector component. The final global representation for the electric field is given by

$$[\hat{E}_k](\boldsymbol{x}) = \sum_{P,N,M} \hat{B}^{e}_{k}(M, N, P) \sum_{p,n,m=0}^{1} \mathrm{BC}^{(k)}(m, n, p, M, N, P)\Psi(m, n, p, M, N, P; \boldsymbol{x}), \tag{18}$$

where the global coefficients arising from application of the boundary conditions are obtained as

$$\mathrm{BC}^{(1)}(0,n,p)=1,\qquad \mathrm{BC}^{(1)}(1,n,p)=\frac{\sum_{p,n=0}^{1} B^{\sigma}(0,n,p)B^{\sigma}(1,n,p)}{\sum_{p,n=0}^{1} B^{\sigma}(1,n,p)B^{\sigma}(1,n,p)}, \tag{19}$$

$$\mathrm{BC}^{(2)}(m,0,p)=1,\qquad \mathrm{BC}^{(1)}(m,1,p)=\frac{\sum_{p,m=0}^{1} B^{\sigma}(m,0,p)B^{\sigma}(m,1,p)}{\sum_{p,m=0}^{1} B^{\sigma}(m,1,p)B^{\sigma}(m,1,p)}, \tag{20}$$

$$\mathrm{BC}^{(3)}(m,n,0)=1,\qquad \mathrm{BC}^{(1)}(m,n,1)=\frac{\sum_{n,m=0}^{1} B^{\sigma}(m,n,0)B^{\sigma}(m,n,1)}{\sum_{n,m=0}^{1} B^{\sigma}(m,n,1)B^{\sigma}(m,n,1)}. \tag{21}$$

Note that in the case in which the simplicial star at node $\boldsymbol{x}(M,N,P)$ contains a face with a unique unit normal, the continuity of the current density is satisfied exactly across that face.

The global representation of the incident electric field is obtained in terms of a simple node expansion as

$$\left[\hat{E}_k^i\right](\boldsymbol{x}) = \sum_{P,N,M} \hat{B}_k^{E^i}(M,N,P)\phi(M,N,P;\boldsymbol{x}). \tag{22}$$

Because there are three unknown expansion coefficients at each node of the mesh, we can replace the integral equation by a square set of linear equations, by taking three linearly independent choices for the localized source distributions at each node. The support of each global expansion coefficient is the simplicial star which we take as the support of the localized source distributions also. Hence we have

$$\begin{aligned} &\left[\hat{J}_k^G\right](\boldsymbol{x}) = j_k^{(\ell)}(I',J',K')\phi(I',J',K';\boldsymbol{x}), \qquad \ell = 1,2,3,\\ &I' = 1,\cdots,L^{(1)};\qquad J' = 1,\cdots,L^{(2)};\qquad K' = 1,\ldots,L^{(3)}, \end{aligned} \tag{23}$$

where $j_k^{(\ell)}(I',J',K')$ is a vectorial coefficient that is nonzero in the x_ℓ-direction only. By now, assigning the values 1, 2, and 3 successively to the superscript (ℓ), we construct three linear independent weighting moments. We take $j_k^{(\ell)}(I',J',K') = \delta_{k,\ell}$ for all I',J',K'. Because the electric Green's state is a continuous function, we can expand it in a simple node expansion. A localized source defined in the simplicial star of node $\boldsymbol{x}(I',J',K')$ results in

$$\left[\hat{E}_r^{G_\ell}\right](I',J',K';\boldsymbol{x}) = \sum_{K,J,I} \hat{B}_r^{EG_\ell}(I',J',K',I,J,K)\phi(I,J,K;\boldsymbol{x}), \tag{24}$$

where the expansion coefficients are obtained by a simple integration of the scalar Green's function and localized source function over the simplicial star of node $\boldsymbol{x}(I',J',K')$, plus application of a finite-difference rule to approximate the derivatives. We have used a rule of $O(h^2)$, h being the mesh size. The coefficients are

$$\begin{aligned} \hat{B}_1^{EG_1}(I',J',K',I,J,K) = {}& [2\Delta x^{(1)}]^{-2}[\mathrm{IG}(I',J',K',I-1,J,K)\\ &-2\mathrm{IG}(I',J',K',I,J,K)+\mathrm{IG}(I',J',K',I+1,J,K)]\\ &-s\sigma\mu\mathrm{IG}(I',J',K',I,J,K), \end{aligned} \tag{25}$$

$$\begin{aligned} \hat{B}_2^{EG_1}(I',J',K',I,J,K) = {}& [4\Delta x^{(1)}\Delta x^{(2)}]^{-1}[\mathrm{IG}(I',J',K',I-1,J-1,K)\\ &-\mathrm{IG}(I',J',K',I+1,J-1,K)\\ &-\mathrm{IG}(I',J',K',I+1,J-1,K)\\ &+\mathrm{IG}(I',J',K',I+1,J+1,K)]. \end{aligned} \tag{26}$$

The other coefficients are similar in an obvious way and so, there is no need to write them all down. The scalar integral

$$\mathrm{IG}(I', J', K', I, J, K) = \int_{\boldsymbol{x}' \in \mathrm{Star}(I',J',K')} \phi(I', J', K'; \boldsymbol{x}) \frac{\exp[-\sqrt{s\sigma\mu}\,|\boldsymbol{x}(I, J, K) - \boldsymbol{x}'|]}{4\pi\,|\boldsymbol{x}(I, J, K) - \boldsymbol{x}'|}\, dV \tag{27}$$

is computed easily but the expression becomes too lengthy to be presented here. Note the convolutional structure, $\mathrm{IG}(I', J', K', I, J, K) = \mathrm{IG}(|I'-I|, |J'-J|, |K'-K|)$, which is exploited in the numerical solution with the aid of a fast Fourier transform (FFT).

4 Method of solution

The solution to the scattering problem consists of two steps. First, we must compute the total electric field inside the scattering domain for which we use an iterative technique. This is done for a finite number of discrete values of s. With these results the scattered EM field outside the scattering object can be obtained easily using the source-type integral relations, which are not discussed here. The second step is to transform the numerically obtained s-domain scattered EM field back to the time domain. A new method is derived to perform this transformation by inspection. In this section we do not use the summation convention.

4.1 Iterative solution to the integral equation

The integral Eq. (3) is put in a Hilbert-space framework and written in the abstract form

$$\hat{L}\hat{u} = \hat{f}, \tag{28}$$

where $\hat{u}$ denotes the unknown electric field and $\hat{f}$ denotes the incident field weighted over the localized source domains, whereas $\hat{L}$ denotes the non-self-adjoint bounded linear operator acting on $\hat{u}$, with bounded inverse, $\hat{L}^{-1}$, which maps a Hilbert space $\mathbf{H}$ onto itself. The space is equipped with an inner product $\langle\cdot,\cdot\rangle$ and a norm $\|\cdot\|$ as

$$\langle\hat{u}, \hat{v}\rangle = \int_{\mathbf{x}\in\mathcal{D}} \hat{u}(\mathbf{x})\hat{v}^*(\mathbf{x})\, dV, \quad \|\hat{u}\| = \langle\hat{u}, \hat{u}\rangle^{\frac{1}{2}}, \tag{29}$$

where the asterisk denotes complex conjugation. The adjoint operator, $\hat{L}^{\star}$, associated with $\hat{L}$ is defined as the operator satisfying

$$\langle\hat{L}\hat{u}, \hat{v}\rangle = \langle\hat{u}, \hat{L}^{\star}\hat{v}\rangle. \tag{30}$$

With these definitions, we develop the iterative solution of equation (28) of the form

$$\hat{u}_0 \text{ arbitrary}, \quad \hat{u}_n = \hat{u}_{n-1} + \alpha_n\hat{v}_n, \quad n \geq 1, \tag{31}$$

with an associated residual

$$\hat{r}_n = \hat{f} - \hat{L}\hat{u}_n, \quad n \geq 0, \tag{32}$$

in terms of which the functions $\hat{v}_n$ will be defined subsequently. The residual provides us with a quantitative error, for which we take the global rms error

$$\mathrm{ERR}_n = \|\hat{r}_n\|, \tag{33}$$

as the quantity that is minimized. Substitution of equation (31) into equation (32) yields the following iterative relation for the residual

$$\hat{r}_0 = \hat{f} - \hat{L}\hat{u}_0, \qquad \hat{r}_n = \hat{r}_{n-1} - \alpha_n \hat{L}\hat{v}_n, \qquad n \geq 1. \tag{34}$$

Minimization of $\|\hat{r}_n\|$ leads in the conjugate gradient method to

$$\alpha_n = \frac{\|\hat{L}^\star \hat{r}_{n-1}\|^2}{\|\hat{L}\hat{v}_n\|^2}, \quad \text{and} \quad \hat{v}_n = \hat{L}^\star \hat{r}_{n-1} + \frac{\|\hat{L}^\star \hat{r}_{n-1}\|^2}{\|\hat{L}^\star \hat{r}_{n-2}\|^2}\hat{v}_{n-1}. \tag{35}$$

This scheme is known as the conjugate gradient scheme for the nonself-adjoint operator $\hat{L}$ (van den Berg, 1984). This algorithm is simplified further by introducing the substitution

$$\hat{v}_n = \|\hat{L}^\star \hat{r}_{n-1}\|^2 \hat{w}_n \tag{36}$$

into equation (35), yielding

$$\alpha_n = \|\hat{L}^\star \hat{r}_{n-1}\|^{-2}\|\hat{L}\hat{w}_n\|^{-2}, \quad \text{and} \quad \hat{w}_n = \hat{w}_{n-1} + \frac{\hat{L}^\star \hat{r}_{n-1}}{\|\hat{L}^\star \hat{r}_{n-1}\|^2}. \tag{37}$$

Using the relations of equations (36) and (37) in the iteration formulas of equations (31) and (34) gives the desired iterative scheme

$$\begin{aligned}
&\hat{u}_0 \text{ arbitrary}, \quad \hat{w}_0 = 0, \quad \hat{r}_0 = \hat{f} - \hat{L}\hat{u}_0, \\
&\hat{w}_n = \hat{w}_{n-1} + \frac{\hat{L}^\star \hat{r}_{n-1}}{\|\hat{L}^\star \hat{r}_{n-1}\|^2}, \qquad n \geq 1, \\
&\hat{r}_n = \hat{r}_{n-1} - \frac{\hat{L}\hat{w}_n}{\|\hat{L}\hat{w}_n\|^2}, \qquad n \geq 1, \\
&\hat{u}_n = \hat{u}_{n-1} + \frac{\hat{w}_n}{\|\hat{L}\hat{w}_n\|^2}, \qquad n \geq 1.
\end{aligned} \tag{38}$$

A full derivation of this, and related, iterative schemes can be found in Kleinman and van den Berg [1991a; the scheme of Eq. (38) corresponds to their Table 3.7 with $i = 2$ and $T = \hat{L}^\star$]. This method is equivalent to the method given by Le Foll (1971).

4.2 *Diffusion expansion method*

In this section we develop a new method to transform (complex) frequency-domain data back to the time domain. The method exploits the fact that the kernel of the integral equation has a known behavior as a function of frequency and that the solution to the integral equation can be written as a sum of repeated applications of the kernel to the incident field. A set of expansion functions is found that have analytically known time-domain counterparts and that need only a few frequencies for the transformation back to time. To derive the method, we start by writing the solution to the weak form of the integral equation as it would be obtained when using the *stationary overrelaxation method* (Kleinman and van den Berg, 1991b):

$$\hat{u}_N = \sum_{k=0}^{N-1} (I - \alpha \hat{L}^\star \hat{L})^k \alpha \hat{L}^\star \hat{f}, \tag{39}$$

with an associated residual

$$\hat{r}_N = (I - \alpha \hat{L}\hat{L}^\star)^N \hat{f}. \tag{40}$$

For values of α in the range $0 < \alpha < 2/\|\hat{L}^\star\hat{L}\|$, where $\|\cdot\|$ denotes the norm of the operator, Kleinman and van den Berg (1991b) have proven, for an equivalent operator equation, that

$$\lim_{N\to\infty} u_N = u, \qquad \lim_{N\to\infty} r_N = 0.$$

In this proof the coefficient α is a function of s, but because $\|\hat{L}^\star\hat{L}\|$ is assumed to be bounded, there exists some value for α, with $\alpha \neq 0$ and α independent of s, for which the scheme converges for all bounded values of s.

If we would solve Eq. (3) using the discretization procedure described in Section 3, with the iterative procedure of Eq. (39), we would be able to write the solution as an infinite sum of polynomials in $\sqrt{s}$, each multiplied by functions that decay proportionally to $\exp(-\sqrt{s})$. Each power of $\sqrt{s}$ has an unknown coefficient that is independent of s. We can evaluate the integral of Eq. (27) approximately by expanding the exponential term using the trilinear expansion function of Eq. (13) on all of the vertices of the rectangles in the simplicial star that is the support of the integral. Then, the remaining integral is independent of s and can be evaluated exactly. The result is an approximation of the scalar Green potential. Because the Green source is a continuous function, we know the resulting Green state (Green's electric field) is continuous also and the "gradient divergence" can be computed by taking a finite-difference rule on the scalar Green potential. In this way we have a discrete approximation of the operator L as defined in Eq. (28). This procedure is used in the scheme of Eq. (38) to actually solve the integral equation, but here we use it in Eq. (39) to find that the discretized operation $\hat{L}\hat{f}$ for a weighting function defined on the simplicial star of node $\boldsymbol{x}(I, J, K)$ is

$$\begin{aligned}\hat{L}\hat{f}(I, J, K) = \sum_{P,N,M}\sum_{q=0}^{3} h_q(I, J, K; M, N, P)s^{q/2} \\ \times \exp[-\sqrt{s\sigma\mu}\,|\boldsymbol{x}(I, J, K) - \boldsymbol{x}(M, N, P)|]\hat{f}(M, N, P), \qquad (41)\end{aligned}$$

where the coefficients h_q only depend on the spatial coordinates and the EM parameters. The summation over $\{M, N, P\}$ denotes the summation over all of the nodes in the grid. To solve the integral equation, we need weighting functions for all nodes $\{I, J, K\}$ to obtain a squared system of equations. As Eq. (39) shows, the final solution thus can be written, taking into account that the adjoint operator of $\hat{L}$ is similar to $\hat{L}$ itself, as

$$\hat{u}_N^{(m)} \approx \sum_{k=1}^{K}\sum_{j=0}^{4N} \beta_{k,j} s_m^{j/2} \exp(-2\sqrt{s_m \tau_k}), \qquad (42)$$

where we have, for notational convenience, taken the spatial sum over one index and we have explicitly written the solution to be valid for an arbitrary discrete value s_m. All of the coefficients h_q that finally correspond to $s^{j/2}$ in the final solution are added in one coefficient $\beta_{k,j}$. All of the values of $|\boldsymbol{x}(I, J, K) - \boldsymbol{x}(M, N, P)|$ that add up after N iterations in the final solution to the same distance can be put together in parameter τ_k. Now, the highest occurring power of $\sqrt{s}$ in Eq. (42) comes from $(N-1)$ applications of $(\hat{L}^\star\hat{L})$, yielding $4(N-1)$, from a last application of $\hat{L}^\star$, yielding two, and two more from the incident field, a point-source excitation assumed, results in a total of $4N$. Unfortunately, the parameters τ_k are, in principle, unknown but because they depend on the medium parameters of the background medium and the distance from the source to the scatterer and all possible distances between the grid points inside the scatterer, we can make a good guess of the range of reasonable values. The total number, K, of τ_k-values depends on the size of the discretized operator and on the total number of iterations. The simplest way to implement this method is to determine a minimum and maximum value for τ and divide the time range up in $(K - 1)$ equidistant time steps, $\tau_k = \sigma\mu d_{\min}^2/4 + (k - 1)\Delta\tau$, $d_{\min}$ being the smallest distance from source to scatterer and $\Delta\tau$ denoting the time step. Computation of the scattered EM field outside the scatterer does not alter the s-dependency; it simply alters all of the coefficients and adds more terms. In view of the Laplace-transform pair,

$$\exp(-2\sqrt{s\tau}) \quad \stackrel{\mathcal{L}^{-1}}{\longrightarrow} \left(\frac{\tau}{\pi t^3}\right)^{1/2} \exp(-\tau/t), \tag{43}$$

the maximum value of τ depends on the time window of interest, because the expansion functions decay exponentially for $\tau > t$. This means that K can be taken smaller than the number one would find from the algorithm described above. Another approximation is to reduce the highest power of $\sqrt{s}$ involved. This reduction is justified by the observation that higher powers in $\sqrt{s}$ correspond to time-domain functions that decay for $t > \tau$ proportional to higher, odd-integer, inverse powers of $\sqrt{t}$, whereas for $t < \tau$, the exponential damping is stronger. We truncate the sequence by retaining only those powers of $\sqrt{s}$ that result from scattering by a point scatterer. In doing this, we find for a component of the scattered electric and magnetic fields the approximation in the form

$$\hat{E}^s(\mathbf{x}^R, s_m) \approx \sum_{k=1}^{K}\sum_{j=0}^{4} \lambda_{k,j}^e \hat{F}^{(j)}[\tau_k(R), s_m], \tag{44}$$

$$\hat{H}^s(\mathbf{x}^R, s_m) \approx \sum_{k=1}^{K}\sum_{j=0}^{3} \lambda_{k,j}^h \hat{F}^{(j)}[\tau_k(R), s_m], \tag{45}$$

where $\hat{F}^{(j)}(\tau_k, s_m) = s_m^{j/2}\exp[-2\sqrt{s\tau_k(R)}]$ whereas the values of $\tau_k(R)$ are receiver-position dependent. The expansion coefficients are independent of s_m and hence if we solve for these coefficients by matching the expansion functions with the numerically obtained scattered field for a number of discrete s-values, we assume the coefficients to be valid for all values of s, and we have the transient scattered field if we know the time-domain equivalents of $\hat{F}^{(j)}(\tau, s)$. These functions are known analytically in terms of the recurrence relation

$$F^{(j)}(\tau, t) = \frac{\sqrt{\tau}}{t} F^{(j-1)}(\tau, t) - \frac{j}{2t} F^{(j-2)}(\tau, t). \tag{46}$$

With the known functions

$$F^{(-2)}(\tau, t) = \operatorname{erfc}(\sqrt{\tau/t})H(t), \qquad F^{(-1)}(\tau, t) = \frac{\exp(-\tau/t)}{\sqrt{\pi t}} H(t), \tag{47}$$

where erfc denotes the complementary error function, the other functions also are known. Note that if the time source is a step function, the time-domain result is obtained by taking $F^{(j-2)}(\tau, t)$, hence without the need for numerical integration of the delta response result. For the linear ramp source function $F^{(j-4)}(\tau, t)$ should be used. In the implementation of this method, we have used the expansion given in Eq. (45) for the transformation back to time both for the scattered electric and scattered magnetic fields, hence with four unknown expansion coefficients for each value of τ.

5 Numerical results

To test the diffusion expansion method (DEM), we computed the frequency-domain response of a spherical scatterer caused by a magnetic dipole in a conductive unbounded embedding and compared the results obtained with DEM and with a standard FFT. The conductivity of the embedding is $\sigma = 10^{-3}$ S/m. The sphere, with its center at the origin, has a radius of 60 m and a conductivity of $\sigma^s = 1$ S/m. The source is located at $[x_1^S, x_2^S, x_3^S] = [-250\,\text{m}, 0\ \text{m}, -100\,\text{m}]$, while the receiver position is $[x_1^R, x_2^R, x_3^R] = [-200\,\text{m}, 0\ \text{m}, -100\,\text{m}]$. For the DEM we have computed the scattered electric and the vertical component of the scattered magnetic field at the receiver location for 48 frequencies, logarithmically spaced using eight points per decade starting at 1 Hz. The results are obtained with 4 coefficients per τ-value and 12 τ-values, hence with 48 unknown expansion coefficients. To obtain a dynamic range of about four decades in the transient results obtained with the FFT, we needed 2^{19} frequencies with a frequency step of 1 Hz. The results obtained with the DEM and with the FFT for the scattered electric field (Fig. 2), and the vertical component of the scattered magnetic field (Fig. 3), are identical over the time range where the FFT gives reliable results. The late-time response as obtained with the DEM shows the correct asymptotic behavior [see Kaufman and Keller (1985)].

The weak form of the integral equation was tested in the same configuration used for the DEM, but now a sphere radius of 30 m and conductivity contrasts up to 100. The

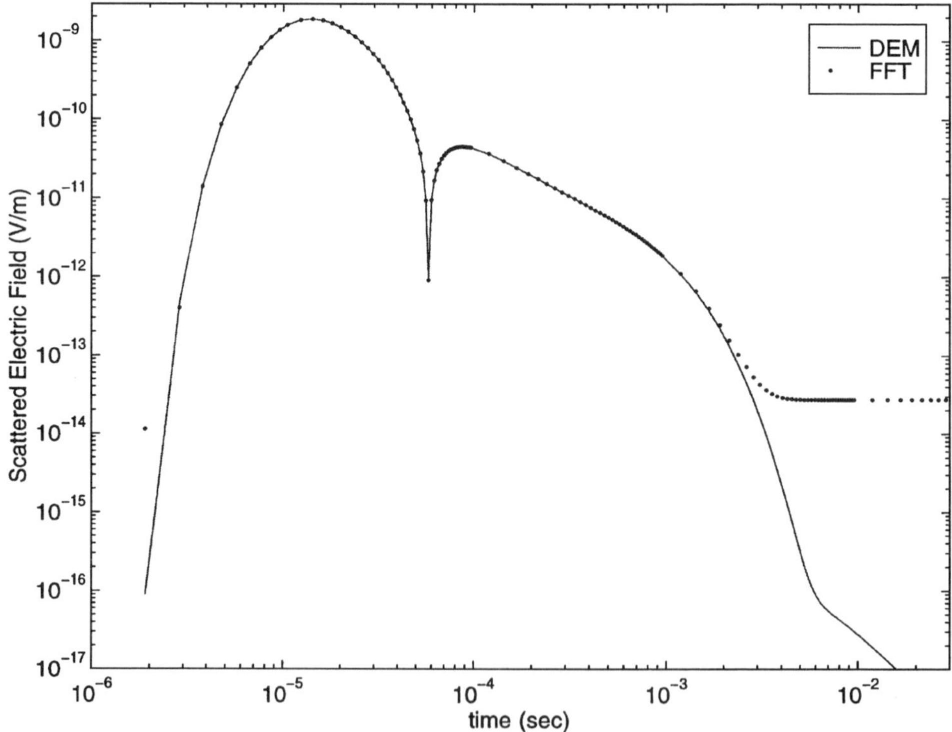

Figure 2. Transient scattered electric field obtained at [−200 m, 0 m, −100 m] with DEM and FFT, for a 60-m-radius sphere, centered around the origin, with a conductivity of $\sigma^s = 1$ S/m in a full space with conductivity $\sigma = 10^{-3}$ S/m. The source is a vertical magnetic dipole at [−250 m, 0 m, −200 m].

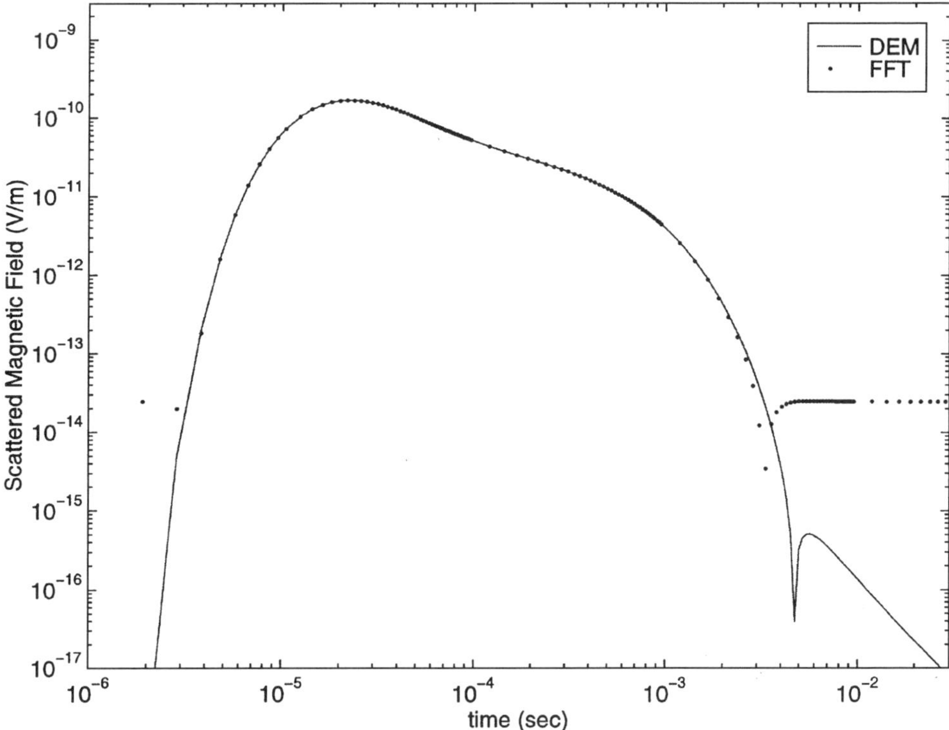

Figure 3. Vertical component of transient scattered magnetic field obtained at [−200 m, 0 m , −100 m] with DEM and FFT, for a 60-m-radius sphere, centered around the origin, with a conductivity of $\sigma^s = 1$ S/m in a full space with conductivity $\sigma = 10^{-3}$ S/m. The source is a vertical magnetic dipole at [−250 m, 0 m, −200 m].

sphere is discretized into 14 × 14 × 14 subcubes to get a reasonable discretized sphere surface. The sphere is one of the most difficult objects to discretize with cubes because the number of cubes associated with the surface of the sphere is $\mathcal{O}(N^2)$, when N denotes the number of cubes in each Cartesian direction, whereas many objects have parts of surfaces that are almost flat. Hence, this is a good test configuration for numerical methods that use discrete volumes. All results shown here are obtained with a background conductivity of 10^{-3} S/m and a sphere conductivity of 10^{-1} S/m. Results obtained for lower-conductivity (not shown here) contrasts are better. We have solved the electric field integral equation (EFIE) for 56 frequencies, using eight values per decade over seven decades starting at 0.1 Hz. Figure 4 shows the normalized difference in amplitude and phase, as a function of frequency, of the scattered electric field obtained with our EFIE method and the analytic solution. The amplitude error is below 3% up to 100 kHz but rises to about 60% at 1 MHz. The phase error is below 0.1% up to 100 kHz and rises to about 5% at 1 MHz. The corresponding transient results for both the analytic solution and our numerical solution is shown in Fig. 5. The early time error is below 1% around the maximum amplitude, whereas the late time error is below 5%. The most dramatic error perhaps is the time shift of the sign change in the scattered field. The time-domain result is obtained using all available frequencies. From Fig. 5, we conclude that the large amplitude errors at high frequencies have no effect on the transformation back to time. The amplitude and phase errors as a function of frequency in the vertical component of the scattered magnetic field are shown in Fig. 6. The amplitude error stays below 1% up to a frequency of about 30 kHz, but rises rapidly to about 80% at 1 MHz, whereas

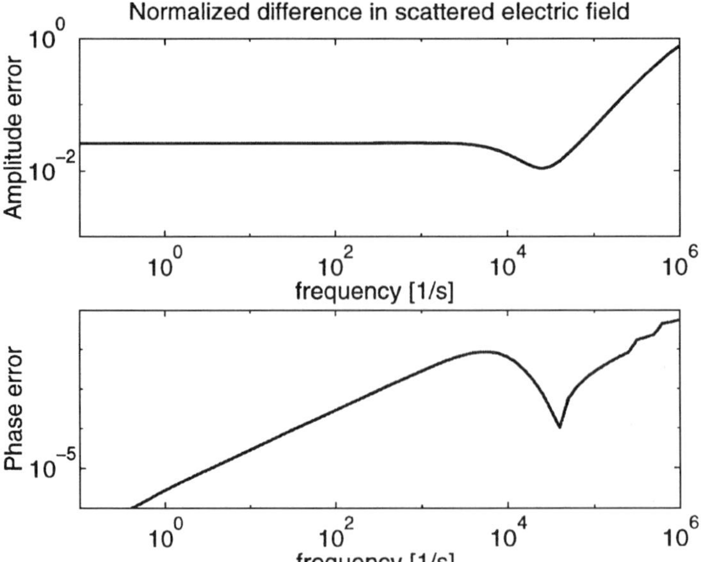

Figure 4. Amplitude and phase error in scattered electric field as a function of frequency, obtained with the weak form of the electric-field integral equation. The source is a vertical magnetic dipole at [−250 m, 0 m, −100 m], the point of observation is at [−200 m, 0 m, −100 m], the conductivity contrast is 100, and the sphere has a 30-m radius.

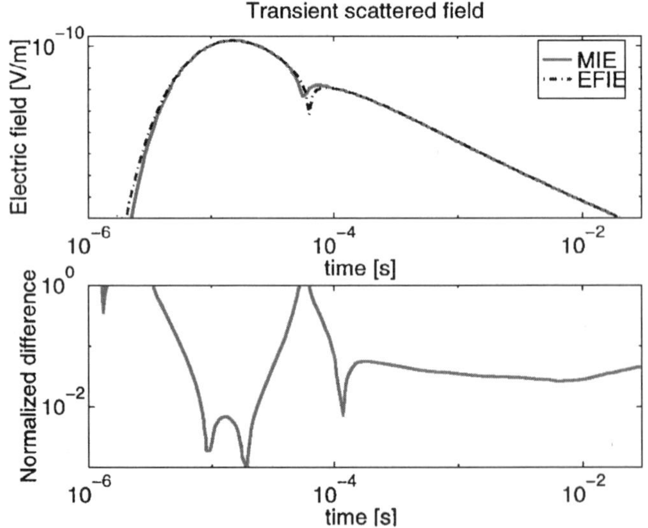

Figure 5. Transient response of scattered electric field and normalized difference with the transient result from the analytic frequency-domain solution shown in Fig. 4.

the phase error is very small for low frequencies, rises to about 3% around 7 kHz but drops to below 1% from 10 to 100 kHz. Notice that the scattered magnetic field at low frequencies is slightly more accurate than the scattered electric field. The corresponding transient result is shown in Fig. 7. Comparing the frequency- and time-domain results obtained with the weak form of the integral equation in combination with DEM, we conclude that the accuracy obtained with the integral-equation method in the frequency domain is almost maintained under the transformation back to time domain with DEM.

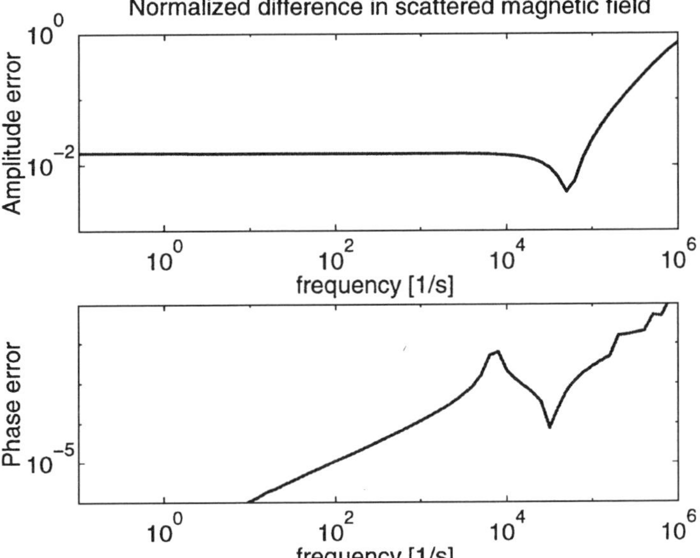

Figure 6. Amplitude and phase error in vertical component of scattered magnetic field as a function of frequency, obtained with the weak form of the electric-field integral equation. The source is a vertical magnetic dipole at [−250 m, 0 m, −100 m], the point of observation is at [−200 m, 0 m, −100 m], the conductivity contrast is 100, and the sphere has a 30-m radius.

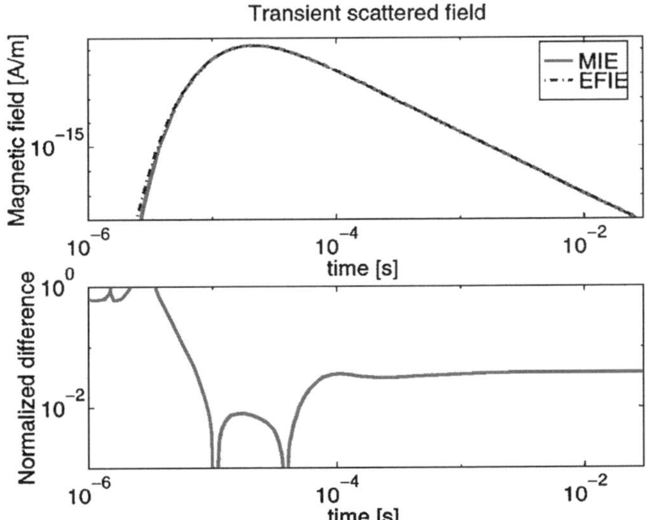

Figure 7. Transient response of vertical component of scattered magnetic field and the normalized difference with transient result from the analytic frequency-domain solution shown in Fig. 6.

These results show that we can obtain transient results with seven decades in amplitude with a reasonable accuracy using few frequencies.

Second, we consider a cube that is 60 m on each side, with center position $\mathbf{x}^C =$ [0 m,0 m,90 m] and conductivity $\sigma^s = 2$ S/m. The embedding has a conductivity of $\sigma = 10^{-2}$ S/m. The source is a large square loop, 160 m on each side and with its center at $[x_1^B, x_2^B, x_3^B] = [-330\text{ m}, 0\text{ m}, 0\text{ m}]$. The vertical component of the scattered voltage is computed at 21 observation points, symmetrically spaced over the scattering cube, in the

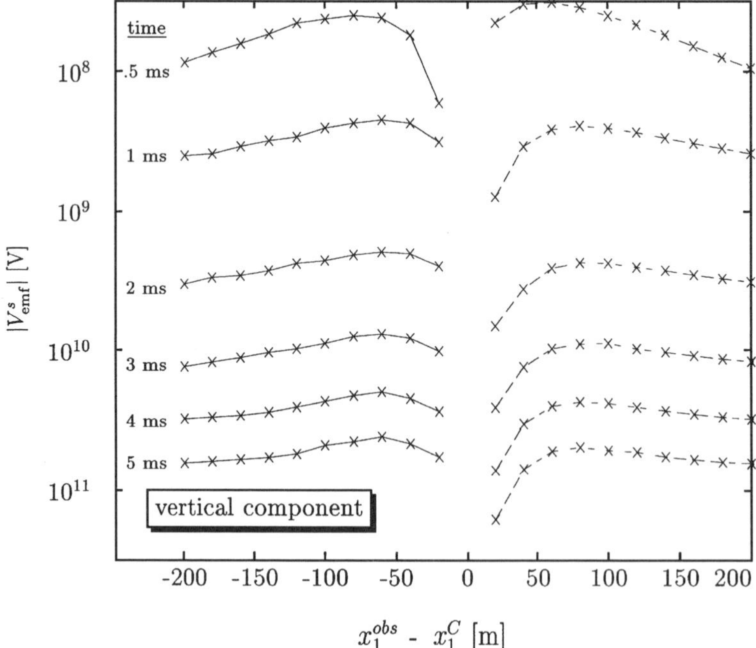

Figure 8. The vertical component of the scattered voltage along the receiver line for a cube that is 60 m on a side, with conductivity $\sigma^s = 2$ S/m in a full space with conductivity $\sigma = 10^{-2}$ S/m.

plane of the source loop, at a horizontal distance of 20 m between the observation points: $\mathbf{x}^{\rm obs} = [-200+(I^R-1)\times 20, 0, 0]$ m, $I^R = 1, \ldots, 21$. To compute the transient result, we have solved the integral equation for 50 real values of s, using 10 points per decade starting at $s = 1$. For the numerical solution the cube was discretized into $6 \times 6 \times 6$ subcubes. The numerical results along the receiver line are plotted at six different instants in time (see Fig. 8). These kinds of plots are shown by SanFilipo and Hohmann (1985) and Newman and Hohmann (1988), where the scattered voltage is symmetric around the horizontal center position of the cube. We surmise that this comes from their fixed-amplitude, concentric current tubes as expansion functions, which force the scattered field to be symmetric around the center position of the scatterer. Because the incident field does not show this symmetry, it is unlikely the scattered field will. Based on the care that we have taken to allow for arbitrary currents and to enforce continuity of both the tangential electric field and normal current components, we expect our scheme to model the response of the scatterer more accurately for this configuration.

6 Conclusion

We have developed a new discretization procedure to solve the electric-field integral equation for diffusive EM scattering problems. This procedure arises naturally using the reciprocity theorem to formulate the integral equation. In this formulation we have introduced weakened Green's states that are generated by localized source distributions (weighting functions). These localized source distributions are taken such that the resulting Green's states can be discretized in a straightforward way, which is achieved using trilinear functions defined on rectangular blocks. The electric field inside the scatterer is approximated by a sequence of these trilinear functions in such a way that the continuity conditions for tangential electric field and normal electric current, across interfaces of discontinuity, are satisfied, which is a least-squares solution because of the use of rectangular blocks and linear expansion functions for the electric field. The DEM that is developed here utilizes the possibility of writing the scattered EM field as a sequence of functions with a known frequency dependency and unknown spatial dependency. These are used as expansion functions, where the unknown spatial dependency occurs in the unknown expansion coefficients, to be solved for, and in the argument of occurring exponentials, where they can be approximated. The expansion functions as functions of frequency have analytically known time-domain equivalents. Once the expansion coefficients in the frequency domain are solved, the transient result is known by inspection. Comparing the results, obtained for the electric and magnetic field scattered by a sphere, using a simple FFT and the DEM, shows that DEM performs excellently on a sparse set of frequency-domain data over a wide time window. Numerical tests for the sphere configuration shows that the combination of our weak formulation of the electric-field integral equation with DEM is a reliable method yielding accurate results for both frequency- and time-domain scattered electric and magnetic fields at least for conductivity contrasts up to 100. Results for the voltage in a receiver loop are obtained analytically from the numerically computed transient magnetic field. The transient voltage scattered by a cube, numerically obtained with the discretized weak form of the integral equation, shows that the response is not symmetric around the horizontal center position of the cube when the source is at a large horizontal distance from the cube.

Acknowledgment

The authors would like to thank Tarek Habashy for providing the computer code for the sphere.

References

de Hoop, A. T., 1991, Reciprocity, discretization and the numerical solution of direct and inverse electromagnetic radiation and scattering problems: Proc. IEEE, **95**, 1421–1430.

Hohmann, G. W., 1975, Three-dimensional induced polarization and electromagnetic modeling: Geophysics, **40**, 309–324.

———1988, Numerical modeling for electromagnetic methods of geophysics, *in* Nabighiam, M. N., Ed., Electromagnetic methods in applied geophysics, Series: Investigations in Geophysics, **3**, Soc. Expl. Geophys.

Kaufman, A. A., and Keller, G. V., 1985, Inductive mining prospecting, Part I. Theory: Elsevier Science Publ. Co.

Kleinman, R. E., and van den Berg, P. M., 1991a, Iterative methods for solving integral equations, *in* Application of conjugent gradient method to electromagnetics and signal analysis, PIER 5: Elsevier Science Publ. Co., Chap. 3.

———1991b, Iterative methods for solving integral equations: Radio Sci., **26**, 175–181.

Le Foll, J., 1971, An iterative procedure for the solution of linear and non-linear equations, *in* Conference on applications of numerical analysis, Series: Lecture notes in mathematics: New York Springer-Verlag.

Mur, G., and de Hoop, A. T., 1985, A finite-element method for computing three-dimensional electromagnetic fields in inhomogeneous media: IEEE Trans. Magn., **MTT-21**, 2188–2191.

Naber, G. L., 1980, Topological methods in Euclidean space: Cambridge Univ. Press.

Newman, G. A., Hohmann, G. W., and Anderson, W. L., 1986, Transient electromagnetic response of a three-dimensional body in a layered earth: Geophysics, **51**, 1608–1627.

Newman, G. A., and Hohmann, G. W., 1988, Transient electromagnetic response of a high-contrast prism in a layered earth: Geophysics, **53**, 691–706.

Raiche, A. P., 1974, An integral equation approach to 3D modeling: Geophys. J. Roy. Astr. Soc., **36**, 363–376.

SanFilipo W. A., and Hohmann, G. W., 1985, Integral equation solution for the transient electromagnetic response of a three-dimensional body in a conductive half-space: Geophysics, **50**, 798–809.

Van den Berg, P. M., 1984, Iterative computational techniques in scattering based upon the integrated square error criterion, IEEE Transactions on Antennas and Propagation, AP-32, 1063–1071.

Wannamaker, W. E., Hohmann, G. W., and SanFilipo, W. A., 1984, Electromagnetic modeling of three-dimensional bodies in layered earths using integral equations: Geophysics, **49**, 60–74.

Physical Expansion Functions for Electromagnetic Integral-Equation Modeling

D.E. Boerner[1]
W. Qian[2]

Summary. Numerical methods of electromagnetic (EM) modeling usually represent spatial and temporal variations of the field with piecewise polynomials of low order. We study the accuracy of several piecewise representations—constant, linear, quadratic, and cubic—using an integral equation to compute EM scattering from a thin wire. The actual scattered field of a uniform conductor should be holomorphic (i.e., have derivatives of all orders). Although holomorphism is impossible to achieve with piecewise polynomial expansions, we do ensure that the scattered electric field is continuous along the uniform conductor and has derivatives continuous to the order allowed by the polynomial basis. The higher-order bases dramatically improve the accuracy to which boundary conditions are satisfied. Higher-order expansions also reduce the degree to which the accuracy depends on the method used to solve the numerical equations. Moreover, convergence can be achieved with fewer discretization cells. Larger cell sizes, in turn, allow more accurate numerical evaluations of the integrals over the (singular) Green's functions. We believe that these higher-order bases offer substantial improvements in 3-D numerical modeling of EM fields.

1 Introduction

Although computationally demanding, the integral-equation (IE) method is still the most popular and powerful technique for modeling the response of confined conductors in Earth (Raiche, 1974; Hohmann, 1975; Weidelt, 1975; Wannamaker et al., 1984; Newman et al., 1986; Hohmann, 1987; Wannamaker, 1991; Xiong, 1992). Much work in IE modeling has been directed toward making the method numerically tractable, e.g., by optimizing computation of Green's function (Wannamaker, 1991) or using specialized methods to compute the matrix inverse (Tripp and Hohmann, 1984; Xiong, 1992; Mackie and Madden, 1993; Xiong and Tripp, 1995). But the accuracy of the IE method remains difficult to appraise, mostly because of the lack of reliable solutions for comparison. Limiting the scatterer to be electrically thin [i.e., such that the

[1]Geological Survey of Canada, Continental Geoscience Division, 1 Observatory Crescent, Ottawa K1A 0Y3, Canada.
[2]Aerodat Inc., 3883 Nashua Drive, Mississauga, Ontario L4V 1R3, Canada.

electromagnetic (EM) fields are not spatially attenuated in one dimension] leads to the thin-sheet solutions (e.g., Vasseur and Weidelt, 1977; Hanneson and West, 1984a,b, Walker and West, 1991) and reduces the computational effort required to solve an intrinsically 3-D problem. Carried further, requiring radially and azimuthally invariant current distributions in a cylindrical body allows EM solutions for the response of thin wires (Qian and Boerner, 1994, 1995). Although the geometry is very limited, thin-wire solutions are easy to compute and provide insight into computational issues. Moreover, analytical solutions are known for some models (Wait, 1972; Tsubota and Wait, 1980), allowing absolute comparisons.

Most IE algorithms represent the scattering currents with piecewise constant (pulse) expansion functions. Pulse functions are easy to program, but require a fine discretization to represent rapid spatial variations. Hanneson and West (1984a,b) employed global expansion functions in computing thin-sheet responses. The form of the global representation was chosen to obey naturally the boundary conditions on the edge of the scattering body (i.e., no current enters the plate through the edges). These physical basis functions proved remarkably successful in representing the EM response of plate conductors with few unknowns. Global basis functions also were adopted by Walker and West (1991) who orthogonalized their polynomial basis. San Filipo and Hohmann (1985) showed that adding induction vortices to the pulse functions improved the numerical behavior of the solution.

We investigate the behavior of low-order polynomial expansion functions for the EM integral equation for thin bodies (Qian and Boerner, 1995). Our philosophy is that the solution should adhere, as far as possible, to the physical behavior of the fields. In particular, because solutions of Maxwell's equations are holomorphic (analytic) at any point not on a property boundary or at a source (Müller, 1969), the scattered electric fields must be continuous and continuously differentiable (to all orders) in these regions. Although this is impossible to achieve with piecewise polynomials, we do ensure that expansion functions of order N have continuous derivatives up to order $N-1$.

The benefits of higher-order bases have been demonstrated many times in antenna modeling at radio or microwave frequencies (Sarkar, 1983; Miller and Burke, 1992). We examine the fields of a thin wire embedded at the surface of a conducting earth and consider the most singular case, that of zero frequency (dc). Our results are directly relevant to the galvanic representation of conductive bodies in a conductive host when induction is negligible. We believe, however, that the conclusions also will apply to more general models.

2 Integral equation for a cylinder

Consider a straight cylindrical conductor in an EM field. Let the conductor axis be oriented along the x-direction and have radius R. Following the procedure outlined by Weidelt (1975), Qian and Boerner (1995) derived the IE for the total horizontal electric field as

$$\mathbf{E}(\mathbf{r}_0) = \mathbf{E}^n(\mathbf{r}_0) + \int_v \underset{\sim}{\mathbf{G}}(\mathbf{r} \mid \mathbf{r}_0) \cdot \mathbf{j}(\mathbf{r})\, d^3\mathbf{r}, \tag{1}$$

where $\mathbf{E}$ is the total observed horizontal electric field. The total electric field is the sum of the horizontal electric field that would exist without the cylindrical conductor ($\mathbf{E}^n$)

plus the electric field scattered from the cylinder. $\underset{\sim}{\mathbf{G}}(\mathbf{r} \mid \mathbf{r}_0)$ is the Green's function, i.e., the horizontal electric field at $\mathbf{r}$ generated by horizontal electric dipoles at $\mathbf{r}_0$, and $\mathbf{j}(\mathbf{r})$ is the scattering electric-current density in the cylindrical conductor. The integration is performed over the conductor volume.

If the cylindrical conductor has a radius much smaller than its length, we can assume that the scattering electric current $\mathbf{j}(\mathbf{r})$ is aligned along the conductor axis and distributed uniformly over the conductor cross-section. Placing $\mathbf{r}_0$ on the axis of the cylindrical conductor gives an IE describing the EM interaction between the scattered current flowing along the conductor and the incident electric field,

$$E_x(x_0) = E_x^n(x_0) + \frac{1}{\pi R^2} \int_0^{2\pi} d\Phi \int_0^R r dr \int_0^a G_{xx}(\mathbf{r} \mid x_0) I(x)\, dx, \qquad (2)$$

where a is the conductor length and R is the conductor radius. On the cross-section of the cylindrical conductor, the electric field can be expressed in terms of the current, $E_x(x_0) = I(x_0)/(S\sigma_a)$ where the anomalous conductivity $\sigma_a = \sigma_{\text{conductor}} - \sigma_{\text{earth}}$ and where S is the cross-sectional area of the conductor. Equation (2) then can be rewritten in terms of just the scattered current and the incident electric field,

$$\frac{I(x_0)}{S\sigma_a} - \frac{1}{\pi R^2} \int_0^{2\pi} d\Phi \int_0^R r dr \int_0^a G_{xx}(\mathbf{r} \mid x_0) I(x)\, dx = E_x^n(x_0). \qquad (3)$$

By allowing the radius to approach zero and making the cylinder a perfect conductor ($\sigma_a = \infty$), IE (3) reduces to Pocklington's equation for the current distribution in a wire antenna (e.g., Miller and Burke, 1992). Pocklington's equation has been solved using many different numerical methods (e.g., Miller and Deadrick, 1975). Of course, the Green's function appropriate for antenna theory describes wave propagation, the antennas are made highly conducting to reduce dissipative losses, and it is often assumed that the scatterer is in a nonconductive whole space. Consequently, the physical behavior required of the expansion functions is different than for the quasi-static case (cf. Balanis, 1992).

At this point we restrict the general form of Eq. (3) and consider that the cylindrical model is embedded at the surface of a homogeneous half-space. Further, we study Green's function at zero frequency where it is most singular. These limitations represent no loss in generality, nor do they reflect real restrictions on our ability to solve the IE. Our purpose is to obtain closed-form expressions for the discrete form of Eq. (3) and thus avoid numerical integrations required by non-dc cases. In particular, it becomes possible to evaluate the Cauchy principal value of the singular integrals, thereby removing an important potential source of error in establishing the discrete form of Eq. (3). For a homogeneous half-space of conductivity σ, the dc Green's function is

$$G_{xx}(\mathbf{r} \mid x_0) = \frac{1}{2\pi\sigma} \frac{\partial^2}{\partial x^2} \frac{1}{\sqrt{[(x - x_0)^2 + r^2]}}, \qquad (4)$$

where $\mathbf{r}$ and x_0 are close to the half-space surface.

3 Discretization

We employ the method of moments (Harrington, 1968) to solve Eq. (3) for the current $I(x)$, given knowledge of the host conductivity structure and the incident electric field

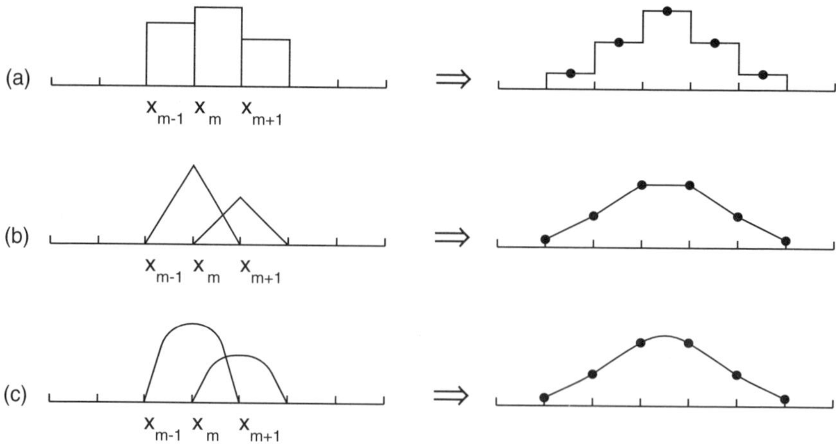

Figure 1. Sketch of various expansion functions: (a) constant basis function, (b) linear expansion function, (c) quadratic basis function. Consider the form of the spatial derivative imposed by the selection of a particular expansion function.

$E_x^n(x)$. This involves selecting expansion functions to represent the scattered current and defining test functions to weight the residual error in the solution to zero (Harrington, 1968). To develop the expansion functions, we divide the length of the cylindrical conductor into N cells with nodes at $x_1, x_2, \ldots, x_N, x_{N+1}$.

Polynomial expansion functions serve to approximate the current in a given cell (shown schematically in Fig. 1),

$$I(x_m \leq x \leq x_{m+1}) \approx \sum_{k=0}^{N} a_k \left(\frac{x - x_m}{x_{m+1} - x_m} \right)^k, \tag{5}$$

where N is the order of the polynomial. This approximation is crucial in representing the physics of the model. For example, constant basis functions ($N = 0$) are commonly used in EM modeling algorithms but implicitly introduce artificial charge elements at cell boundaries to account for the discontinuities required in the representation of spatially variable scattered currents (consider Fig 1a).

Substituting Eq. (5) into Eq. (3) gives the discrete integral equation,

$$(\underset{\sim}{\mathbf{Z}} - \underset{\sim}{\mathbf{\Gamma}})\mathbf{I} + \mathbf{E}^n = \mathbf{R}, \tag{6}$$

where $\underset{\sim}{\mathbf{Z}}$ is a diagonal matrix with Z_{ii} representing the unit length internal resistance of the conductor $(1/S\sigma_a)$ at x_0^i. $\underset{\sim}{\mathbf{\Gamma}}$ is the $M \times M$ Green's function matrix and $\mathbf{I}$ and $\mathbf{E}^n$ are $M \times 1$ column vectors representing the current and normal electric field at points x_0^i. M is the number of expansion functions chosen to represent the current along the conductor length. $\mathbf{R}$ is a vector of residual errors generated by the numerical approximations.

The impedance matrix $\underset{\sim}{\mathbf{\Gamma}}$ is calculated using the fundamental indefinite integral,

$$S_k(x) = \frac{1}{\pi R^2} \int_0^{2\pi} d\Phi \int_0^R r dr \int x^k G_{xx}(\mathbf{r} \,|\, 0)\, dx. \tag{7}$$

For polynomials of order $k = 0, 1, 2, 3$, the indefinite integrals can be evaluated easily from Eqs. (4) and (7) to be

$$S_0(x) = \frac{1}{\pi\sigma R^2}\left[\frac{x}{\sqrt{x^2+R^2}} - \operatorname{sign}(x)\right], \tag{7a}$$

$$S_1(x) = -\frac{1}{\pi\sigma}\frac{1}{\sqrt{x^2+R^2}}, \tag{7b}$$

$$S_2(x) = -\frac{1}{\pi\sigma}\left(\frac{x}{\sqrt{x^2+R^2}} - \operatorname{sign}(x)\ln\left[\frac{|x|+\sqrt{x^2+R^2}}{R}\right]\right), \tag{7c}$$

and

$$S_3(x) = \frac{1}{\pi\sigma R^2}\left[\frac{x^4}{\sqrt{x^2+R^2}} - 3x^2\sqrt{x^2+R^2} + 2\sqrt{(x^2+R^2)^3}\right]. \tag{7d}$$

The integrals (7a)–(7d) are the components from which we assemble a discrete form of IE (3) using polynomials up to order three as a basis.

The exact solution of Eq. (6) has zero residual at all points along the conductor. In practice, the method of moments requires only that an integral of the residual, weighted by a set of test functions, be equal to zero. The general form is written as a Hilbert space inner product,

$$\int_L R(x_0)W(x_0)\,dx_0 = 0.$$

A solution can be obtained by point matching or point collocation (delta function weighting), or by averaging the residual over each discretization interval using pulse weighting functions. A least-squares solution also can be defined within the general framework of the method of moments. Sarkar (1985) and Sarkar et al. (1985) show that the choice of a weighting function strongly affects the accuracy of the solution, particularly for low-order expansion functions. We illustrate this below for our model problem.

3.1 Constant basis: Discontinuous

Zeroth-order expansion functions require that the current $I(x)$ be locally constant in each cell. Using integral (7a) we have the trivial result

$${}_0\Gamma_{nm} = S_0(x_{m+1}-x_0) - S_0(x_m - x_0), \quad \text{where} \quad x_n \le x_0 \le x_{n+1}. \tag{8}$$

The prefix subscript for Γ refers to the order of the basis-function approximation; the postfix subscripts indicate the cells for which the mutual impedance is to be evaluated. The point x_0 is the evaluation point for the inner-product definition used to force the residual to zero. Constant expansion functions result in a scattering-current distribution that is discontinuous at the discretization nodes. This means that the derivative of the scattering current consists of delta functions at the nodes. Except in the trivial case of constant current, continuity of the current between cells is impossible to achieve using constant expansion functions.

3.2 Linear basis: Continuous

A first-order basis allows the current $I(x)$ to vary linearly in each cell,

$$I(x_m \le x \le x_{m+1}) \approx I(x_m) + \frac{(x-x_m)[I(x_{m+1}) - I(x_m)]}{(x_{m+1}-x_m)}. \tag{9}$$

In selecting this expansion function we ensure that the current is continuous at each node, although the first derivative of the current can be discontinuous at the nodes. For linear expansion functions, the impedance matrix elements can be written as

$$
\begin{aligned}
{}_1\Gamma_{nm} = {} & \frac{1}{(x_{m+1} - x_m)} \{(x_{m+1} - x_0)[S_0(x_{m+1} - x_0) - S_0(x_m - x_0)] \\
& - (x_{m-1} - x_0)[S_0(x_m - x_0) - S_0(x_{m-1} - x_0)] \\
& + 2S_1(x_m - x_0) - S_1(x_{m+1} - x_0) - S_1(x_{m-1} - x_0)\}. \qquad (10)
\end{aligned}
$$

In Eq. (10), $x_n \leq x_0 \leq x_{n+1}$. Unless the current is known to be zero at the ends of the conductor, the linear basis functions in the end interval do not form a complete set (see Fig. 1b) and thus must be written

$$
{}_1\Gamma_{1n} = \frac{1}{(x_2 - x_1)} \{(x_2 - x_0)[S_0(x_2 - x_0) - S_0(x_1 - x_0)] + S_1(x_1 - x_0) - S_1(x_2 - x_0)\},
$$

$$
{}_1\Gamma_{N+1,n} = \frac{(x_0 - x_N)[S_0(x_{N+1} - x_0) - S_0(x_N - x_0)] + S_1(x_{N+1} - x_0) - S_1(x_N - x_0)}{(x_{N+1} - x_N)}.
$$

For the thin-wire scatterer, the current should be nearly zero at the ends of the wire. In fact, this condition is enforced sometimes for antenna solutions by selecting appropriate bases (e.g., Sarkar, 1985). However, for more general conductor geometries, Dirichlet boundary conditions may not be appropriate. Thus, we choose to test the IE solution by not enforcing a priori boundary conditions on the expansion functions.

3.3 *Quadratic basis: Continuous first derivative*

The second-order (quadratic) expansion function allows another degree of freedom in selecting the polynomial coefficients in each interval. This extra degree of freedom could be treated as an additional unknown, or can be tailored to be physically appealing by providing continuity to first order. The current in any cell is written as

$$
I(x_m \leq x \leq x_{m+1}) \approx I(x_m) + \frac{(x - x_m)[I(x_{m+1}) - I(x_m)]}{(x_{m+1} - x_m)} + c_m(x - x_m)(x - x_{m+1}), \qquad (11a)
$$

where the unknown coefficient c_m is found by requiring the first-order derivative of the current to be continuous across each discretization node. That is,

$$
\frac{dI}{dx}(x_m \leq x \leq x_{m+1}) = \frac{I(x_{m+1}) - I(x_m)}{x_{m+1} - x_m} + c_m(2x - x_m - x_{m+1}) \qquad (11b)
$$

and

$$
\frac{dI}{dx}(x_{m+1} \leq x \leq x_{m+2}) = \frac{I(x_{m+2}) - I(x_{m+1})}{x_{m+2} - x_{m+1}} + c_{m+1}(2x - x_{m+1} - x_{m+2}), \qquad (11c)
$$

must be equal at $x = x_{m+1}$. This requirement yields an equation that can be evaluated recursively for c_m,

$$
c_{m+1} = \frac{d_m - d_{m+1} + c_m(x_{m+1} - x_m)}{x_{m+1} - x_{m+2}}, \qquad (12)
$$

where

$$
d_m = \frac{I(x_{m+1}) - I(x_m)}{x_{m+1} - x_m}.
$$

The continuity condition provides $N-1$ equations for N unknowns (c_m) to express the current on the whole interval. To find c_m uniquely, an additional condition on the current distribution in the interval is required. For example, we can insist that the derivative of the current be zero at $x=0$,

$$\frac{dI}{dx}=0,$$

which implies that

$$c_1=\frac{I(x=0)-I(x=x_1)}{x_1^2}.$$

This condition essentially means that the current must be a constant in the first cell. Alternatively, enforcing conditions on the second derivative of the current at $x=0$ is possible,

$$\frac{d^2I}{dx^2}=0,$$

which requires that $c_1=0$ and implies that only linear current variations are permitted in the first cell. At present, we have not examined the consequences of choosing between these, or other, boundary conditions. For this study, we arbitrarily adopt the seemingly less restrictive condition that the second derivative be zero at one end of the wire ($x=0$).

From Eq. (12) and the extra conditions on the form of the polynomial, the impedance matrix can be shown to be

$$\begin{aligned}{}_2\Gamma_{nm}={}_1\Gamma_{nm}+\sum_{k=1}^{N}\frac{u_{km}}{\Delta_k^2}\{&[S_2(x_{k+1}-x_0)-S_2(x_k-x_0)]\\&-(x_{k+1}+x_k-2x_0)[S_1(x_{k+1}-x_0)-S_1(x_k-x_0)]\\&+(x_{k+1}-x_0)(x_k-x_0)[S_0(x_{k+1}-x_0)-S_0(x_k-x_0)]\},\end{aligned}\tag{13}$$

where $x_n\le x_0\le x_{n+1}$. To develop this equation, we have used

$$c_m=\frac{1}{\Delta_m^2}\sum_{k=1}^{N+1}u_{km}I_k,$$

where $\Delta_m=x_{m+1}-x_m$, and ${}_1\Gamma_{mn}$ is the Green's function matrix assembled in Eq. (10); u_{km} can be determined from the initial conditions and Eq. (12).

3.4 Cubic basis: Continuous second derivative

The third-order (cubic) expansion function allows another degree of freedom in selecting the polynomial coefficients for each interval. We choose to ensure that the first- and second-order derivatives of the current be continuous at the conductor cell boundaries. The current in each segment then can be expressed as

$$\begin{aligned}I(x_m<x<x_{m+1})\approx I(x_m)+\frac{(x-x_m)[I(x_{m+1})-I(x_m)]}{(x_{m+1}-x_m)}+c_m(x-x_m)(x-x_{m+1})\\+f_m(x-x_m)(x-x_{m+1})\big[x-\tfrac{1}{2}(x_m+x_{m+1})\big],\end{aligned}\tag{14a}$$

where the unknown coefficient c_m and f_m are found by requiring that the first- and second-order derivatives of the current be continuous across each discretization node.

That is,

$$\frac{dI}{dx}(x_m \le x \le x_{m+1}) = \frac{I(x_{m+1}) - I(x_m)}{x_{m+1} - x_m} + c_m(2x - x_m - x_{m+1})$$
$$+ f_m\{(x - x_m)(x - x_{m+1}) + (x - x_m)[x - \tfrac{1}{2}(x_m + x_{m+1})]$$
$$+ (x - x_{m+1})[x - \tfrac{1}{2}(x_m + x_{m+1})]\}, \tag{14b}$$

$$\frac{dI}{dx}(x_{m+1} \le x \le x_{m+2}) = \frac{I(x_{m+2}) - I(x_{m+1})}{x_{m+2} - x_{m+1}} + c_{m+1}(2x - x_{m+1} - x_{m+2})$$
$$+ f_{m+1}\{(x - x_{m+1})(x - x_{m+2})$$
$$+ (x - x_{m+1})[x - \tfrac{1}{2}(x_{m+1} + x_{m+2})]$$
$$+ (x - x_{m+2})[x - \tfrac{1}{2}(x_{m+1} + x_{m+2})]\}, \tag{14c}$$

and

$$\frac{d^2I}{dx^2}(x_m \le x \le x_{m+1}) = 2c_m + 3f_m(2x - x_m - x_{m+1}), \tag{14d}$$

$$\frac{d^2I}{dx^2}(x_{m+1} \le x \le x_{m+2}) = 2c_{m+1} + 3f_{m+1}(x - x_{m+1} - x_{m+2}). \tag{14e}$$

Continuity of the first- and second-order derivatives at $x = x_{m+1}$ yields

$$d_m + c_m\Delta_m + f_m\Delta_m^2/2 = d_{m+1} - c_{m+1}\Delta_{m+1} + f_{m+1}\Delta_{m+1}^2/2, \tag{14f}$$

and

$$2c_m + 3f_m\Delta_m = 2c_{m+1} - 3f_{m+1}\Delta_{m+1}. \tag{14g}$$

The continuity condition provides $2(N-1)$ equations for $2N$ unknowns (c_m and f_m) to express the current on the whole interval. We require two extra conditions to find all of the c_m and f_m uniquely and again could consider that the first-order derivatives of the current are zero at $x = 0$ and $x = x_{N+1}$, which implies that

$$d_1 - c_1\Delta_1 + f_1\Delta_1^2/2 = 0,$$

and

$$d_N + c_N\Delta_N + f_N\Delta_N^2/2 = 0.$$

Also, we could choose to enforce conditions on the second-order derivative of the current,

$$2c_1 - 3f_1\Delta_1 = 0,$$

and

$$2c_N + 3f_N\Delta_N = 0.$$

Another alternative would be to simply set $f_1 = 0$ and $f_N = 0$. From the above conditions, we can determine uniquely the c_m and f_m as

$$c_m = \frac{1}{\Delta_m^2}\sum_{k=1}^{N+1} u_{mk}I_k,$$

and

$$f_m = \frac{1}{\Delta_m^3} \sum_{k=1}^{N+1} v_{mk} I_k.$$

Thus the impedance matrix can be assembled from

$$\begin{aligned} {}_3\Gamma_{nm} = {}_2\Gamma_{nm} + \sum_{k=1}^{N} \frac{v_{km}}{\Delta_k^3} \big\{ & [S_3(x_{k+1} - x_0) - S_3(x_k - x_0)] \\ & \times \left[-\tfrac{3}{2}(x_{k+1} + x_k - 2x_0)[S_2(x_{k+1} - x_0) - S_2(x_k - x_0)] \right] \\ & + \left[(x_{k+1} - x_0)(x_k - x_0) + \tfrac{1}{2}(x_{k+1} + x_k - 2x_0)^2 \right] [S_1(x_{k+1} - x_0) S_1(x_k - x_0)] \\ & - \tfrac{1}{2}(x_{k+1} - x_0)(x_k - x_0)(x_{k+1} + x_k - 2x_0)[S_0(x_{k+1} - x_0) - S_0(x_k - x_0)] \big\} \end{aligned} \tag{15}$$

where $x_n \leq x_0 \leq x_{n+1}$.

Although the form of the polynomial approximations given above seem complicated, they are just combinations of the fundamental integrals given in Eqs. (7a)–(7d). This makes it a simple matter to construct the impedance matrix, particularly when the body is partitioned into equidimensional cells.

4 An example

Testing the properties of various expansion functions for the numerical solution of Eq. (7) is difficult without analytic solutions, because of the potential for numerical errors. To simplify comparisons we derive a synthetic test case in which the current and incident electric field are known in closed form. This permits rigorous comparisons of the exact and numerically estimated current distribution and forms the most stringent test possible of the IE formulation. For a thin cylindrical body, any current channeled into the line conductor is mainly through its lateral area. In other words, we expect the scattering current to be approximately zero at the two ends of the conductor. Representing the current in the thin wire by

$$I(x) = x^2(a - x)^2 \tag{16}$$

mimics the expected current variation and roughly conforms with prior numerical experiments (Qian and Boerner, 1995). Evaluation of Eq. (3) determines the normal electric field required to drive the current described by Eq. (16) along an elongated, small-radius conductor. The corresponding electric-field distribution is given by

$$\begin{aligned} E_x^n(x_0) = \frac{x^2(a - x_0)^2}{\pi \sigma_a R^2} - \frac{1}{\pi \sigma R^2} \Bigg[& -2x_0^2(a - x_0)^2 \\ & - \frac{\sqrt{R^2 + x_0^2}}{2} \left(-8aR^2 - 2a^2 x_0 + 13R^2 x_0 + 4ax_0^2 - 2x_0^3 \right) \\ & + \frac{\sqrt{R^2 + (a - x_0)^2}}{2} \left(-5aR^2 + 13R^2 x_0 + 2ax_0^2 - 2x_0^3 \right) \\ & + \frac{R^2 \left(2a^2 - 3R^2 - 12ax_0 + 12x_0^2 \right)}{2} \ln \left(\frac{\sqrt{R^2 + (a - x_0)^2} + (a - x)}{\sqrt{R^2 + x_0^2} - x_0} \right) \Bigg]. \end{aligned} \tag{17}$$

The current distribution represented by Eq. (16) is artificial, in the sense that the source distribution required to create the field in Eq. (17) has not been specified. Also, the current is a fourth-order polynomial and low-order expansion functions undoubtedly should show improved solution convergence properties as the order of the basis functions increases. The current distribution in a homogeneous conductor should be expected to be a slowly varying spatial function (Qian and Boerner, 1995). Thus, convergence of the IE for more realistic models also should improve with increasing order, except in pathological cases. Although the test example is sufficient to show the effects of the expansion-function selection, we caution that it may offer only qualitative information regarding the modeling of true source fields and current distributions.

Analytic results now can be compared with the numerical solution of Eqs. (3) and (16) using various expansion functions. For the numerical examples, the internal resistance of the conductor is taken to be zero ($S\sigma_a = \infty$), the host conductivity is $\sigma = 0.01$ S/m, the wire is $a = 100$ m long and has a radius $R = 1$ m. Figure 2 shows the distribution of the true current [smooth curve, bottom panel; Eq. (16)] and normal electric field along the conductor [top; Eq. (17)]. Also shown in Fig. 2 (bottom) are the constant expansion-function IE solutions calculated using equally spaced discretizations of 10, 50, and 200 cells. For these calculations, the collocation point $x[W(x_0) = \delta(x)]$ was chosen to be in the center of each cell. Adding more cells (degrees of freedom) to the IE constant-expansion-function formulation permits a more accurate determination of the scattered current. Note, however, that the spatially integrated current is not well constrained until

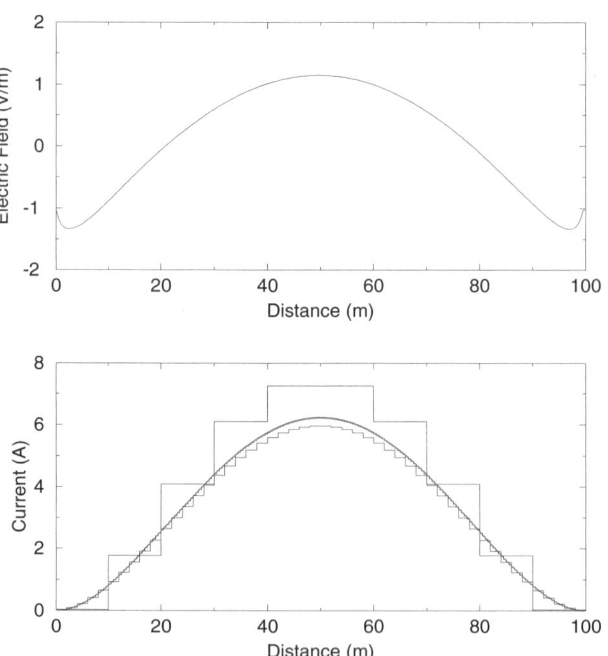

Figure 2. (top) Normal electric field along the conductor [Eq. (16)]; (bottom) convergence of constant expansion-function solution (stepped curves) to the exact solution (smooth curve) for zero internal resistance ($S\sigma_a = \infty$), host conductivity of $\sigma = 0.01$ S/m, wire length $a = 100$ m, and a radius $R = 1$ m. A normalization factor of 10^{-6} is used for both the electric field and the scattering current.

there are 200 cells in the IE approximation. The current is overestimated by using 10 cells, and underestimated by using 50 cells. Although there is a convergence toward the true solution, the L_2 norm of the distance between the approximate response and the exact solution is quite large for coarse discretizations.

The estimated currents shown in Fig. 2 are deliberately shown as represented by constant expansion functions to illustrate the deleterious effects of this basis. Observing the scattered current distribution at a remote observation point could obscure the differences between the solutions having 10 cells or 50 cells and certainly mitigates the nonphysical effects of discontinuous current distributions. The assumed benefits of smoothing (introduced by integrating the current distribution against the Green's function) may be illusory, however. Remote point-convergence tests are a necessary, but not sufficient, condition for guaranteeing solution accuracy. Accuracy of the solution is ensured only when the scattering current obeys the EM boundary conditions. Convergence tests based on the scattering current will always require finer discretization than those based on the remote EM fields.

Figure 3 shows a comparison between discrete IE solution and the true solution versus the total number of discretization cells using the different polynomial bases. The level of fit is defined by the sum of the squares of the difference between the calculated solution and the true solution, normalized by the number of test points. This fit statistic is determined at the center of the cells for the constant expansion functions, but at the node points for the other bases. In using the quadratic and cubic expansion functions, recall that we have arbitrarily set the first- or second-order derivatives in the

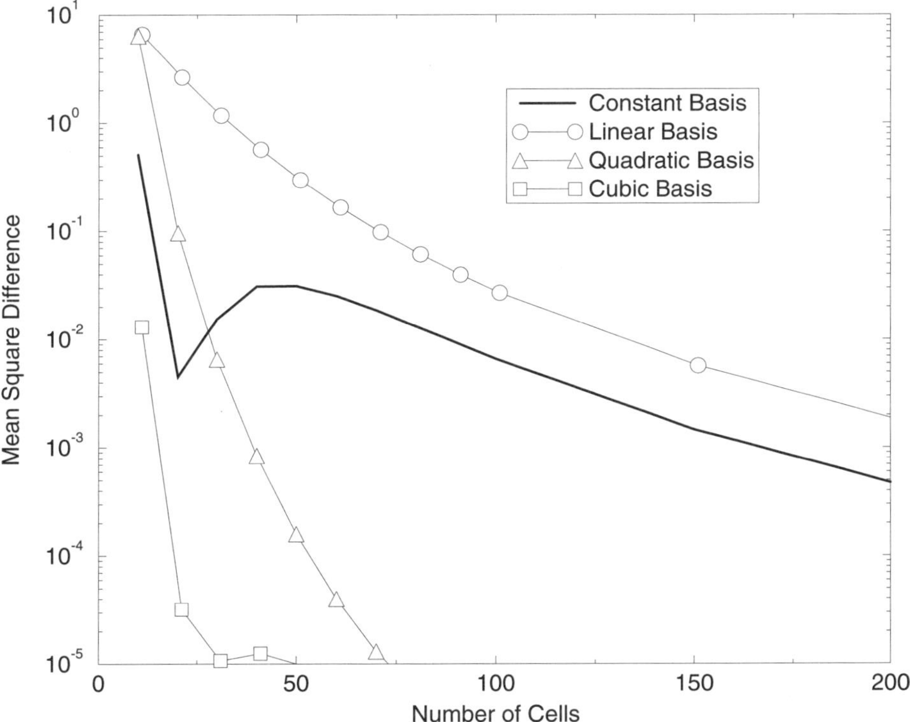

Figure 3. Misfit between true and approximate solutions for total number of cells for various expansion functions. Fit is defined as mean square difference between calculated and true solutions at the node points (linear, quadratic, cubic) and the cell centers (constant).

first cell to be zero. This assumption only restricts the expansion function to be order one less than for the rest of the wire and is much less restrictive than imposing Dirichlet boundary conditions at the wire ends.

Figure 3 illustrates that the constant and linear expansion functions have approximately the same convergence rate as the discretization interval becomes smaller. The quadratic and cubic bases converge to the true solution much more rapidly and may be limited mostly by numerical precision. One disturbing characteristic of Fig. 3 is that the constant expansion functions seem to represent the scattered current more accurately than either the linear or the quadratic bases for discretizations of less than 30 cells. This discrepancy between intuition and experiment can be reconciled by noticing that Fig. 2 suggests that point comparisons between the true and the approximate curves can be dramatically dependent on the location at which the fit statistic is calculated. A proper misfit metric should be based on an L^2 norm in Hilbert space, not on point comparisons. However, the unexpected differences between the lower-order basis functions implied by Fig. 3 depends both upon the definition of the misfit metric and the moment method used to solve the equations.

Consider first the problem of fixing the collocation point needed for the method of moments. Equation (3) is a mathematical representation of the EM boundary conditions on the fields scattered by the cylindrical volume heterogeneity (boundary conditions for the host media are contained within Green's function). An IE implicitly expresses the duality between sources and boundary conditions by adopting equivalent sources to reproduce the effects of the boundaries. Thereby, the physical presence of the scattering body can be eliminated from the modeling, leaving only the equivalent sources. Errors in satisfying Eq. (3) will determine the solution accuracy. Moment methods ensure that the residual has been minimized numerically at the collocation points, but Eq. (3) is not guaranteed to be satisfied at other points on the conductor.

For the particular model used in our test examples, $\sigma_a = \infty$, allowing Eq. (3) to be rewritten as

$$\frac{1}{\pi R^2}\int_0^{2\pi} d\Phi \int_0^R r\,dr \int_0^a G_{xx}(\mathbf{r}\,|\,x')I(x)\,dx + E_x^n(x') = \mathcal{R}(x'). \qquad (18)$$

The residual $\mathcal{R}(x')$ should be zero everywhere $x' \in V$ if the IE is satisfied. In physical terms, the body is a perfect conductor and thus should have no internal field, requiring that the scattering currents exactly cancel the incident field. To calculate $\mathcal{R}(x')$ we represent the current [$I(x)$, Eq. (16)] with the known expansion functions for a given discretization, and then evaluate the integrals and electric field in Eq. (18).

Plots of $\mathcal{R}(x')$ for a 20-cell division of the thin-wire conductor in the above model are shown in Fig. 4 for each of the four expansion functions. Comparing ordinate scales shows that increasing the order in the expansion function strongly reduces the residual. For the constant and linear expansion functions, $\mathcal{R}(x')$ is extremely large near the node points (i.e., at 0, 5, 10, 15, ... m). Clearly, placing the collocation point at cell centers for the constant basis functions is beneficial (nearly optimal), whereas using the node positions as collocation points for the linear expansion function is a poor choice. In effect, Fig. 3 shows a comparison between the solutions using the worst collocation-point locations for the linear expansion with the best case for the constant functions.

The optimal collocation points would be at the zero crossings of the curves shown in Fig. 4. At these points the incident electric field in Eq. (18) can be matched exactly, even using an incomplete basis. Without knowledge of the true current distribution, finding

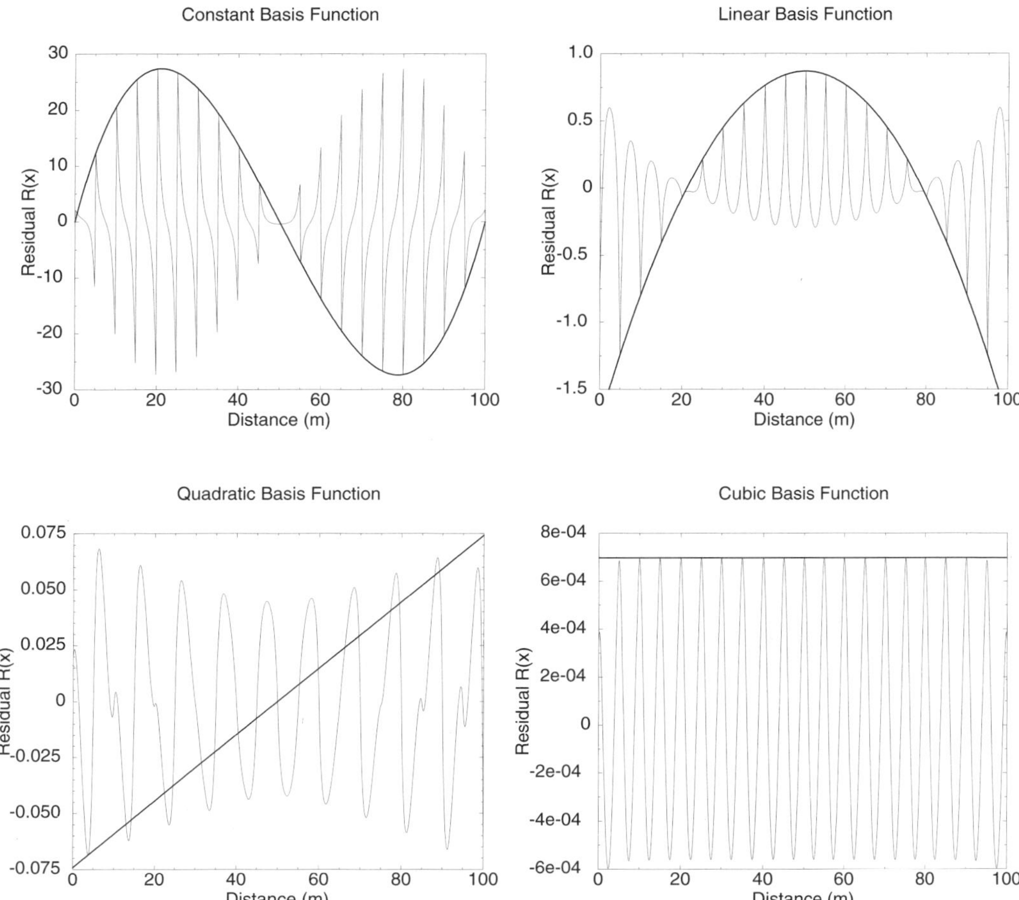

Figure 4. Measure of the residual expressed in Eq. (6) for the various bases as a function of position along the wire conductor (divided into 20 segments). Heavy solid lines generally defining the envelope of the curves are the spatial derivatives of the true current distribution [Eq. (16)]. The derivative order is one less than the expansion-function order and the value is scaled arbitrarily to match one point on the residual functional. The scale factor is roughly the same for all four diagrams.

zero crossings in the residual $\mathcal{R}(x')$ is impossible. Pulse weighting functions [$W(x_0) = 1$] remove the concern of point collocation selection and certainly result in more stable estimations of the current. The disadvantage is that the residual shown in Fig. 4 would be minimized only on average across each interval. Determining collocation point locations that minimize the residual shown in Fig. 4 is important, perhaps even crucial for numerical stability, but it does not compensate for the inadequacy of the numerical approximations. Large residuals occur where the expansion function is inadequate for matching the true spatial variations of the scattered current. In other words, Fig. 4 shows where the numerical approximation to the integral equation is inappropriate. Constant expansion functions are unable to represent any variations in the first-order derivative of the current. Therefore the spatial distribution of the first-order derivative of the true current defines regions of large residuals. Similarly, linear expansion functions have a zero second derivative and so cannot account for x^2 variations in the current. The spatial derivatives of the current then form an envelope defining the residual $\mathcal{R}(x')$. To show this correspondence between the expansion-function order and the residual, the derivatives of $I(x)$ are shown by the heavy lines in Fig. 4.

The importance of the collocation points is apparent in the preceding examples, particularly for low-order expansion basis. To examine further the role of the collocation points, we consider the test model discretized into 20 cells. The integral equation then is solved repeatedly using collocation points offset a distance d from the node point. Denoting the nodes by x_n, x_{n+1}, etc., we choose $x_0 = x_n + d$ where d is expressed as a fraction of the cell size. For the linear and higher-order bases, the additional x_0 point is chosen to be the last node point ($x_0 = a$). For this example, however, we examine the Hilbert L_2 norm as a measure of the true distance between the exact and numerical solutions.

Figure 5 shows the L_2 norm as a function of the offset parameter d. For the constant expansion function, not selecting the cell centers as the test points increases the norm by two orders of magnitude, whereas for the linear expansion function the collocation point is best placed approximately 0.17 cell widths from the node points. For the quadratic and cubic expansion functions, the node point selection is not crucial, although placing the match point near 0.4 is beneficial. Figure 5 shows clearly that increasing the order of the expansion function generally improves the accuracy of the numerical solution. It is also apparent that the minima in the residual curves is broader for higher-order expansion functions. Because the optimal collocation-point distribution is undoubtedly dependent on the conductivity model, source field, and expansion-function order, Fig. 5 suggests that numerical stability can be improved using higher-order expansion functions.

The primary difference between the various expansion-function solutions lies in the impedance matrix ($\underset{\sim}{\Gamma}$). Spatial attenuation by the Green's function in Eq. (3) means that ${}_k\Gamma_{mn}$ becomes independent of the expansion-function order as the distance implied

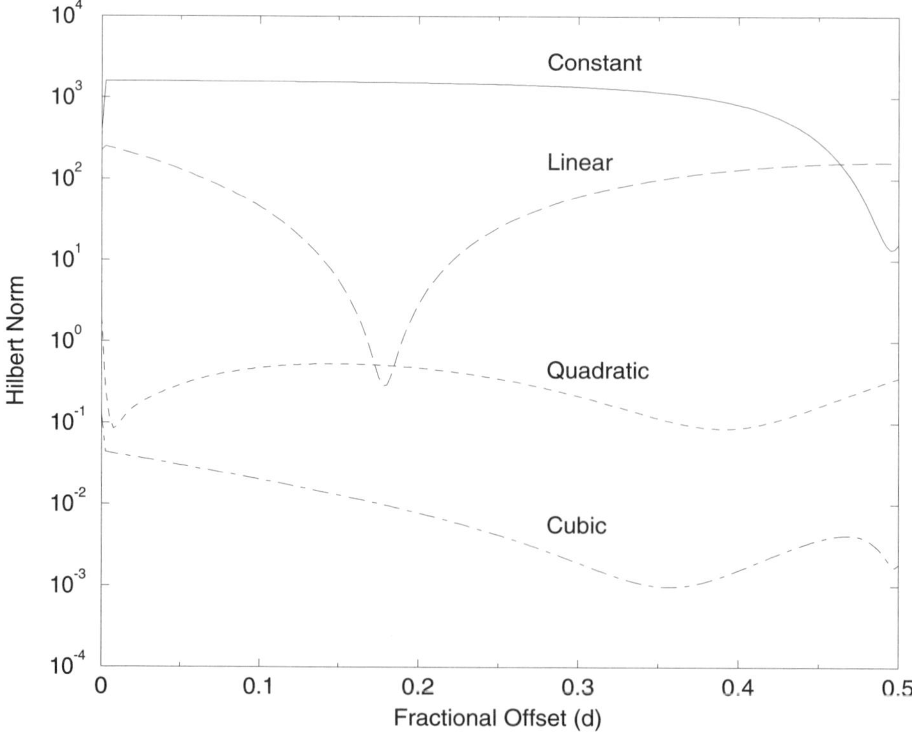

Figure 5. Hilbert norm of true and numerical IE solutions for a discretization of 20 cells. The abscissa represents the offset of the point x_0 from the node points expressed as a fractional cell length.

by mn (that is, $|x - x_0|$) increases. However, the magnitudes of the diagonal and near-diagonal elements of $\underset{\sim}{\Gamma}$ are strongly modified by the choice of the expansion function. Indeed, Γ_{nn}, and possibly the $\Gamma_{n,n+1}$ and $\Gamma_{n,n-1}$ terms contain the Green's function singularity depending upon the collocation-point location and the order of the expansion function. Consider the comparison between the constant and linear expansion-function examples shown in Fig. 5. By virtue of locating the test point x_0 at the node locations for the linear expansion function, Γ_{nn}, $\Gamma_{n,n+1}$, and $\Gamma_{n,n-1}$ all contain a singularity. Contrast this with the constant expansion function where the test point is in the center of the cell and only Γ_{nn} contains the singularity. In essence, the impedance is distributed broadly across the rows and columns of $\underset{\sim}{\Gamma}$ instead of being more sharply focused around the diagonal as in the constant expansion-function case. As a result, when solving Eq. (6), the scattered current is systematically overestimated by the linear expansion-function solution.

5 Discussion and conclusions

Our examples show that higher-order representations of the scattering current can improve modeling efficiency. In homogeneous regions, a suitable set of basis functions are functions with the highest continuity allowed by the order of the representation. Use of such a basis has the benefit of reducing the number of unknowns in the discrete formulation of the IE. It also ensures that the solution conforms to the physical behavior of the fields, which are holomorphic in homogeneous regions.

Accuracy in IE modeling by the method of moments depends strongly on the collocation points. The location of these points is particularly important for low-order expansion functions. Careful testing using different distributions of collocation points for low-order expansion functions may allow for accurate solutions of the IE. Alternatively, using pulse weighting functions can reduce numerical instabilities resulting from a poor expansion basis. The penalty of pulse weighting is that the IE is satisfied only in an average sense across the discretization cell. This may be inappropriate for large cells. The long-wire IE is ideal for this type of study because it requires only one-dimensional integration of the Green's functions, and has an analytic solution for a wire at the surface of a homogeneous half-space (at dc).

Acknowledgments

The authors acknowledge J. A. Craven and D. W. S. Eaton for providing useful comments and criticisms on the manuscript. M. Oristaglio and two anonymous reviewers offered valuable suggestions that helped to clarify ideas and improve the presentation of this material, Geological Survey of Canada contribution number 39795.

References

Balanis, C. A., 1992, Antenna theory: A review, Proc. IEEE, **80**, 7–23.

Hanneson, J. E., and West, G. F., 1984a, The horizontal loop electromagnetic response of a thin plate in a conductive earth, Part I. Computational method: Geophysics, **49**, 411–420.

———1984b, The horizontal loop electromagnetic response of a thin plate in a conductive earth, Part II. Computational results and examples: Geophysics, **49**, 421–432.

Harrington, R. F., 1968, Field computation by moment methods: Macmillan Publ. Co.

Hohmann, G. W., 1975, Three-dimensional induced polarization and electromagnetic modeling: Geophysics, **40**, 309–324.

———1987, Numerical modeling for electromagnetic methods of geophysics, *in* Nabighian M. N., Ed., Electromagnetic methods in applied geophysics series: Investigations in Geophysics, **3**, Soc. Expl. Geophys. 313–363.

Mackie, R. L., and Madden, T. R., 1993, Three-dimensional magnetotelluric inversion using conjugate gradients: Geophys. J. Internat., **115**, 215–229.

Miller, E. K., and Burke, G. J., 1992, Low-frequency computational electromagnetics for antenna analysis, Proc. IEEE, **80**, 24–43.

Miller, E. K., and Deadrick, F. J., 1975, Some computational aspects of thin-wire modeling, *in* Mittra, R., Ed., Numerical and asymptotic techniques in electromagnetics, Springer-Verlag New York, Inc.

Müller, C., 1969, Foundations of the mathematical theory of electromagnetic waves: Springer-Verlag, Berlin.

Newman, G. A., Hohmann, G. W., and Anderson, W. L., 1986, Transient electromagnetic response of a three-dimensional body in a layered earth: Geophysics, **51**, 1608–1627.

Qian, W., and Boerner, D. E., 1994, Electromagnetic response of a discretely grounded circuit—An integral equation solution: Geophysics, **59**, 1680–1694.

Qian, W., and Boerner, D. E., 1995, Electromagnetic modeling of buried line conductors using an integral equation: Geophys. J. Internat., **121**, 203–214.

Raiche, A., 1974, An integral equation approach to 3-D modeling: Geophys. J. Roy. Astr. Soc., **36**, 363–376.

Sarkar, T. K., 1983, A note on the variational method (Rayleigh-Ritz), Galerkin's method and the method of least squares: Radio Sci., **18**, 1207–1224.

———1985, A note on the choice of weighting functions in the method of moments: IEEE Trans. Antennas Propag., **AP-33**, 436–441.

Sarkar, T. K., Djordjević, A. R., and Arvas, E., 1985, On the choice of expansion and weighting functions in the numerical solution of operator equations, IEEE Trans. Antennas Propag., **AP-33**, 988–996.

San Filipo, W. A., and Hohmann, G. W., 1985, Integral equation solution for the transient electromagnetic response at a three-dimensional body in a conductive half-space: Geophysics, **50**, 798–809.

Tripp, A. C., and Hohmann, G. W., 1984, Block diagonalization of the electromagnetic impedance matrix of a symmetric buried body using group theory: Trans. Geosci. Remote Sensing, **22**, 62–69.

Tsubota, K., and Wait, J. R., 1980, The frequency and the time-domain responses of a buried axial conductor: Geophysics, **45**, 941–951.

Vasseur, G., and Weidelt, P., 1977, Bimodal electromagnetic induction in nonuniform thin sheets with an application to the northern Pyrenean induction anomaly: Geophys. J. Roy. Astr. Soc., **51**, 669–690.

Wait, J. R., 1972, Electromagnetic wave propagation along a buried insulated wire, Can. J. Phys., **50**, 2402–2409.

Walker, P. W., and West, G. F., 1991, A robust integral equation solution for electromagnetic scattering by a thin plate in conductive media: Geophysics, **56**, 1140–1152.

Wannamaker, P. E., 1991, Advances in three-dimensional magnetotelluric modeling using integral equations: Geophysics, **49**, 60–74.

Wannamaker, P. E., Hohmann, G. W., and San Filipo, W. A., 1984, Electromagnetic modeling of three-dimensional bodies in layered earths using integral equations: Geophysics, **49**, 60–74.

Weidelt, P., 1975, Electromagnetic induction in three-dimensional structures: J. Geophys., **41**, 85–109.

Xiong, Z., 1992, Electromagnetic modeling of 3-D structures by the method of system iteration using integral equations: Geophysics, **57**, 1556–1561.

Xiong, Z., and Tripp, A. C., 1995, A block iterative algorithm for 3-D electromagnetic modeling using integral equations with symmetrized substructures: Geophysics, **60**, 291–295.

Electromagnetic Modeling with Surface Integral Equations

Eric H. Liu
Yves Lamontagne

Summary. We describe a new method for solving Maxwell's equations that couples scalar surface integral equations (SIEs) with an iterative implementation of the (vector) boundary conditions. The formulation can handle multiple homogeneous regions and, in computational efficiency, compares favorably with the more conventional volume integral method. The system equations are solved iteratively in three stages: (1) iterations to solve the coupled scalar SIE; (2) iterations to implement vector boundary conditions; and (3) iterations to calculate surface fields in three dimensions. Numerical results for a sphere excited by a plane wave and for an oblate spheroid in a half-space excited by a rectangular surface loop show that the method is stable for a wide range of conductivity contrasts and gives reasonable results for the electromagnetic response from the resistive up to the inductive limit.

1 Introduction

Surface integral equations (SIEs) are common in the field of antenna design and microwave (radar) scattering (Müller, 1969; Poggio and Miller, 1973; Colton and Kress, 1983), but have not been used much for electromagnetic (EM) modeling in geophysical applications. We describe a new formulation of the SIE method and present some results that show that this method can be very competitive with volume integral (Raiche, 1974; Hohmann, 1975; Weidelt, 1975) and differential equation (Druskin and Knizhnerman, 1988; Mackie et al., 1993; Wang and Hohmann, 1993; Newman, 1995) methods for geophysical modeling. Our new formulation, which we call the Multiscalar Boundary Element Method or MBEM (Liu and Lamontagne, 1996), solves relatively complex boundary conditions to handle multiple regions even with finite conductivity contrast.

2 Three-region scattering problem

The motivation behind the SIE method is to reduce a 3-D volume integral into a surface integral by representing the fields in the anomalous region by their values on a closed surface, which is usually the boundary surface of an anomalous region. Although SIEs

Lamontagne Geophysics Ltd., Kingston, Ontario K7L 4V4, Canada.

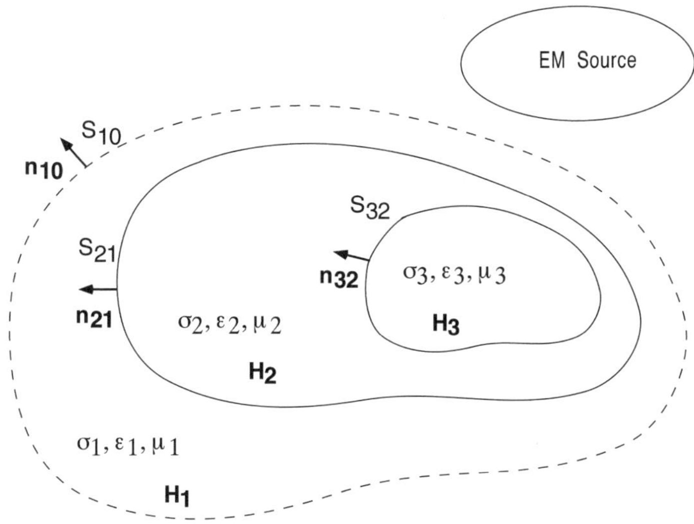

Figure 1. Three-region scattering problem.

are trickier to formulate than volume integral equations, the reduction of nodes from the whole volume of the body to only the surface often more than compensates for the increased complexity.

Consider the three-region scattering problem in Fig. 1. There are three closed surfaces dividing an unbounded space into four regions. Each region is assumed to be enclosed by one or two disconnected closed surfaces. Region 3 ($\mathcal{V}_3$), bounded by surface S_{32}, with dielectric constant ϵ_3, conductivity σ_3, and magnetic susceptibility μ_3, is embedded in region 2 ($\mathcal{V}_2$) with ϵ_2, σ_2, and μ_2, which is itself bounded by a closed surface S_{21} from the outside. Outside region 2 is region 1 which is bounded by surface S_{10}. Sources are located outside S_{10}. We further assume that the region exterior to S_{10} is identical to $\mathcal{V}_1$ in constitutive parameters. The surface S_{10} is drawn with a dashed line indicating that it is different from the other surfaces in that, if there are no sources outside of S_{10}, it recedes to infinity.

In the derivations that follow, subscripts m and ℓ denote different regions; subscript i represents different Cartesian components; and subscripts t and n denote tangential and normal components, respectively. Let $\mathbf{H}_m(\mathbf{x})$ be the magnetic field in region m; also let $\sigma' = \sigma + j\epsilon\omega$ be the complex conductivity ($j = \sqrt{-1}$) and k be the complex propagation constant ($k^2 = -j\mu\omega\sigma'$), where ω is the angular frequency. The following equations hold for $\mathbf{H}$:

$$\nabla^2 \mathbf{H}_m + k_m^2 \mathbf{H}_m = 0, \quad \text{and} \quad \nabla \cdot \mathbf{H}_m = 0, \quad \mathbf{x} \in \mathcal{V}_m \tag{1}$$

with $m = 1, 2, 3$. The boundary conditions on the surfaces separating region m and region ℓ are

$$\mathbf{H}_{m,t} = \mathbf{H}_{\ell,t}, \quad \mathbf{x} \in S_{m\ell}; \tag{2}$$

$$\mu_m H_{m,n} = \mu_\ell H_{\ell,n}, \quad \mathbf{x} \in S_{m\ell}; \tag{3}$$

$$\frac{(\nabla \times \mathbf{H}_m)_t}{\sigma'_m} = \frac{(\nabla \times \mathbf{H}_\ell)_t}{\sigma'_\ell}, \quad \mathbf{x} \in S_{m\ell}; \tag{4}$$

$$\mathbf{n} \cdot \nabla H_{m,n} - \mathbf{n} \cdot \nabla H_{\ell,n} = (H_{\ell,n} - H_{m,n}) \nabla \cdot \mathbf{n}, \quad \mathbf{x} \in S_{m\ell}, \tag{5}$$

where $\mathbf{n} = \mathbf{n}(\mathbf{x})$ is the unit normal vector of surface $S_{m\ell}$. Equation (1) is the vector Helmholtz equation derived from Maxwell's equations satisfying the condition that the magnetic field be divergence free in regions with no variation in magnetic susceptibility. Equations (2) and (3) are continuity conditions of the magnetic flux density $\mathbf{B} = \mu\mathbf{H}$ at the boundary. The continuity condition for tangential $\mathbf{E}$ is stated in Eq. (4), and Eq. (5) can be derived from the general Maxwell equations using Heaviside operators (Bouix, 1966; Liu and Sidaway, 1996). We assume that the solution satisfies the Sommerfeld radiation condition at infinity. [It can be proved that the divergence-free condition of an exterior solution to the Helmholtz equation follows from the Sommerfeld radiation condition at infinity; see Müller (1969).]

If there are no magnetic contrasts ($\mu_m = \mu_\ell$), Eqs. (2) and (3) then can be combined into one equation, and Eq. 5 can be simplified as

$$\mathbf{H}_m = \mathbf{H}_\ell, \quad \mathbf{x} \in S_{m\ell}; \tag{6}$$

$$\mathbf{n} \cdot \nabla H_{m,n} - \mathbf{n} \cdot \nabla H_{\ell,n} = 0, \quad \mathbf{x} \in S_{m\ell}. \tag{7}$$

3 Multiscalar approach

The vector system described by the set of differential equations and the boundary conditions for the field $\mathbf{H}$—Eqs. (1), (6), and (7)—are replaced with a system of three scalar Helmholtz equations and boundary conditions for each of the three (Cartesian) components of the magnetic field. Each Cartesian component of $\mathbf{H}$ can be taken as a scalar field satisfying the scalar Helmholtz equation, but the boundary conditions couple different Cartesian components of $\mathbf{H}$. We assume that the two boundary conditions needed for a scalar second-order differential scattering problem are

$$H_{m,i} = H_{\ell,i}; \tag{8}$$

$$\frac{\partial H_{m,i}}{\partial \mathbf{n}} = \alpha_{m\ell} \frac{\partial H_{\ell,i}}{\partial \mathbf{n}} + \beta_{m\ell,i}, \tag{9}$$

where $i = x, y, z$ and $m\ell = 12, 23$. The contrast parameter $\alpha_{m\ell} = \sigma'_m/\sigma'_\ell$ is a constant and $\beta_{m\ell,i} = \beta_{m\ell,i}(\mathbf{x}, \mathbf{H}_\ell, \partial\mathbf{H}_\ell/\partial\mathbf{n})$ is a surface field depending on all three components of $\mathbf{H}_\ell$ and $\partial\mathbf{H}_\ell/\partial\mathbf{n}$. The dependence of $\beta_{m\ell,i}$ on $\mathbf{x}$ is through its dependence on a local Cartesian coordinate defined on the boundary surface $\{\hat{p}, \hat{q}, \hat{n}\}$. Now, we can write the equivalent set of three scalar Helmholtz scattering problems as

$$\left(\nabla^2 + k_m^2\right) H_{m,i} = 0, \quad \mathbf{x} \in \nu_{\mathrm{m}}, \tag{10}$$

where $m = 1, 2, 3; i = x, y, z$; and the corresponding boundary conditions on the surface separating regions m and ℓ are

$$H_{m,i} = H_{\ell,i}, \quad \mathbf{x} \in S_{m\ell}; \tag{11}$$

$$\frac{\partial H_{m,i}}{\partial \mathbf{n}} = \alpha_{m\ell} \frac{\partial H_{\ell,i}}{\partial \mathbf{n}} + \beta_{m\ell,i}(\mathbf{x}, \mathbf{H}_\ell, \partial\mathbf{H}_\ell/\partial\mathbf{n}), \quad \mathbf{x} \in S_{m\ell}, \tag{12}$$

where $i = x, y, z$. The functional form of β is derived by Liu and Lamontagne (1996, Appendix A); for surface $S_{m\ell}$, it is

$$\begin{pmatrix} \beta_x \\ \beta_y \\ \beta_z \end{pmatrix} = \left(1 - \frac{\sigma'_m}{\sigma'_\ell}\right) \begin{pmatrix} \eta_x^p & \eta_y^p & \eta_z^p \\ \eta_x^q & \eta_y^q & \eta_z^q \\ \eta_x^n & \eta_y^n & \eta_z^n \end{pmatrix}^{-1} \begin{pmatrix} \partial H_n/\partial p \\ \partial H_n/\partial q \\ \partial H_n/\partial n \end{pmatrix}, \tag{13}$$

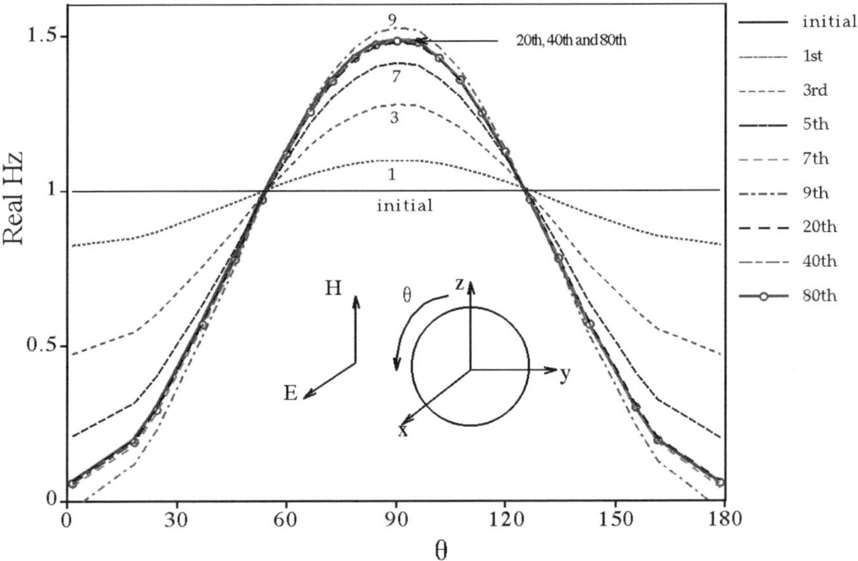

Figure 2. Convergence behavior of multiscalar iterations. The real part of H_z, normalized by the incident field, is plotted along a circumferential arc in the xz plane for a number of multiscalar iterations. The initial **H** is set to the incident field.

where we have defined a rotation $\{\eta_{\hat{x}\hat{y}\hat{z}}^{\hat{p}\hat{q}\hat{n}}\}$ between the global and the local Cartesian coordinate systems such that

$$H_\xi = \sum_{i=x,y,z} \eta_i^\xi H_i, \tag{14}$$

$$\frac{\partial H_\xi}{\partial n} = \sum_{i=x,y,z} \eta_i^\xi \frac{\partial H_i}{\partial n}, \tag{15}$$

and $\xi = p, q, n$.

The three scalar systems are not independent of each other. The discontinuity in the normal derivative of any one Cartesian component depends on all other components of **H** and $\partial\mathbf{H}/\partial\mathbf{n}$. In the multiscalar iteration approach, we start with some initial values of $\mathbf{H}^{[0]}$ and $\partial\mathbf{H}^{[0]}/\partial\mathbf{n}$ and solve the three scalar Helmholtz equations to get $\mathbf{H}^{[1]}$ and $\partial\mathbf{H}^{[1]}/\partial\mathbf{n}$ and then $\mathbf{H}^{[2]}$ and $\partial\mathbf{H}^{[2]}/\partial\mathbf{n}$, and so on until the desired convergence is reached.

Figure 2 illustrates the convergence of the multiscalar iteration for a sphere scattering an incident plane wave. The figure shows the magnetic field along a line on the surface of the target for a number of multiscalar iterations. The target has a radius $r = 100$ m and conductivity $\sigma = 100$ S/m. The operating frequency is $f = 500$ Hz. The host conductivity is 10^{-4} S/m. The orientation of the incident field is specified in Fig. 2. The profiles of the real part of H_z, along a circumferential arc in the xz plane, for the first 80 multiscalar iterations, are given in Fig. 2. Other components display similar behavior. To start the iteration, **H** is set to the incident field. In this example, the field moves toward the correct solution for the first 7 iterations, and then overshoots a bit at iteration 9. After iteration 20, the field remains essentially unchanged.

4 SIEs

In the last section, a general m-region EM scattering problem has been reduced into a set of $3m$ scalar Helmholtz scattering problems with specially defined boundary conditions. In this section, the Kirchhoff integration method will be used to solve the scalar scattering problems. We will first expand the Kirchhoff formulation described in the Appendix B of Liu and Lamontagne (1996) to include three-region scattering. Then an iteration scheme for inter-surface scattering is described.

4.1 System of SIEs

To expedite the derivation, we introduce the operator K,

$$K^r_{S_m}(a,b) = \frac{1}{4\pi}\int_S \left(G_r\frac{\partial a}{\partial \mathbf{n}'} - b\frac{\partial G_r}{\partial \mathbf{n}'}\right) ds', \tag{16}$$

where $a(\mathbf{x}')$ and $b(\mathbf{x}')$ are surface scalar fields defined on S_m, $\mathbf{x}' \in S_m$; $G_r = G_r(\mathbf{x}, \mathbf{x}')$ is the Green's function with the medium parameter k_r. The prime on the normal vector $\mathbf{n}$ indicates that it is taken at the integration point $\mathbf{x}'$, that is, $\mathbf{n} = \mathbf{n}(\mathbf{x}')$. Similarly, $s' = s(\mathbf{x}')$. For a given surface m, r can take one of two values corresponding to two regions on both sides of the surface. The operator K itself is a function of $\mathbf{x}$, $K = K(\mathbf{x})$.

Let us first consider surface 21 and perform the Kirchhoff integration procedure [Liu and Lamontagne, 1996, Eqs. $(b9)$ and $(b10)$] for both region 1 and region 2. One SIE is obtained for each of the two regions. If φ_m is the scalar field in region m, $m = 1, 2$, the two SIEs on surface 21 are

$$\tfrac{1}{2}\varphi_2(\mathbf{x}) = K^2_{S_{21}}(\varphi_2', \varphi_2) - K^2_{S_{32}}(\varphi_2', \varphi_2), \quad \mathbf{x} \in S_{21}; \tag{17}$$

$$\tfrac{1}{2}\varphi_1(\mathbf{x}) = K^1_{S_{10}}(\varphi_1', \varphi_1) - K^2_{S_{21}}(\varphi_1', \varphi_1), \quad \mathbf{x} \in S_{21}, \tag{18}$$

where φ' represents the normal derivative of φ. Because the scalar field is continuous across surface 21, we introduce a surface field $\tau_{21}(\mathbf{x})$ such that $\varphi_1(\mathbf{x}) = \varphi_2(\mathbf{x}) = \tau_{21}(\mathbf{x})$, $\mathbf{x} \in S_{21}$. In general, the normal derivatives of the scalar fields are not continuous and we have to introduce two surface fields on each surface to represent the normal derivatives of the scalar field φ. We use $\tau'(\mathbf{x})$ to represent the normal derivative of φ. The first subscript of $\tau'(\mathbf{x})$ is the surface on which it is defined and the second subscript indicates the region from within which the surface is approached. For example, $\tau'_{21,1}$ is defined as $\partial\varphi_1(\mathbf{x})/\partial\mathbf{n}_{21}$, with $x \to S_{21}$ from within region 1. In terms of τ', the second boundary condition on surface 21 is expressed as $\tau'_{21,2} = \alpha_{21}\tau'_{21,1} + \beta_{21}$, where $\alpha_{21} = \sigma_2'/\sigma_1'$ is the medium parameter contrast and β_{12} is the auxiliary surface field defined in the preceding section.

Similar expressions for surface 32 are readily obtained. In the new notations, the four equations for the two surfaces are:

$$\tfrac{1}{2}\tau_{21}(\mathbf{x}) = K^2_{S_{21}}(\tau'_{21,2}, \tau_{21}) - K^2_{S_{32}}(\tau'_{32,2}, \tau_{32}), \quad \mathbf{x} \in S_{21}; \tag{19}$$

$$\tfrac{1}{2}\tau_{21}(\mathbf{x}) = P_{21} - K^2_{S_{21}}(\tau'_{21,1}, \tau_{21}), \quad \mathbf{x} \in S_{21}; \tag{20}$$

$$\tfrac{1}{2}\tau_{32}(\mathbf{x}) = K^3_{S_{32}}(\tau'_{32,3}, \tau_{32}), \quad \mathbf{x} \in S_{32}; \tag{21}$$

$$\tfrac{1}{2}\tau_{32}(\mathbf{x}) = K^1_{S_{10}}(\tau'_{21,2}, \tau_{21}) - K^2_{S_{32}}(\tau'_{32,2}, \tau_{32}), \quad \mathbf{x} \in S_{32}; \tag{22}$$

where $P_{21}(\mathbf{x}) = K^1_{S_{10}}(\tau'_{10,1}, \tau_{10})$, $\mathbf{x} \in S_{21}$. P_{21} represents a contribution from sources outside of S_{10} (the incident field).

The normal derivatives are related by the boundary conditions

$$\tau'_{21,2} = \alpha_{21}\tau'_{21,1} + \beta_{21}, \quad \mathbf{x} \in S_{21}; \tag{23}$$

$$\tau'_{32,3} = \alpha_{32}\tau'_{32,2} + \beta_{32}, \quad \mathbf{x} \in S_{32}. \tag{24}$$

Equations (19)–(24) are the complete set of system equations describing the three-region scattering problem in Fig. 1. Taking into account Eqs. (23) and (24) leaves four unknown fields, two for each surface: $\tau_{m\ell}$ and $\tau'_{m\ell}$, with $m\ell = 12, 23$. The number of unknowns matches the number of equations. Once the set of system equations is solved, the surface fields τ and τ' are known and the field anywhere in space can be obtained by a surface integral over the proper enclosing surface or surfaces.

4.2 *Interboundary iteration*

An iterative approach can be taken to solve the set of system equations. First, the surface fields τ and τ' for each surface are expanded into series:

$$\tau = \tau^{(0)} + \tau^{(1)} + \tau^{(2)} \cdots \tag{25}$$

$$\tau' = \tau'^{(0)} + \tau'^{(1)} + \tau'^{(2)} \cdots \tag{26}$$

The superscript represents the order of the scattering, which is identified later in this section. The first four interboundary iterations are given below in detail.

- Step 1: Assume $\tau_{32}^{(0)} = 0$ and $\tau_{32,2}'^{(0)} = 0$ on surface 32, and solve for $\{\tau_{21}^{(0)}, \tau_{21,1}'^{(0)}\}$ through equations

$$\tfrac{1}{2}\tau_{21}^{(0)} = K_{S_{21}}^{2}\left(\tau_{21,2}'^{(0)}, \tau_{21}^{(0)}\right), \quad \mathbf{x} \in S_{21}; \tag{27}$$

$$\tfrac{1}{2}\tau_{21}^{(0)} = P_{21} - K_{S_{21}}^{2}\left(\tau_{21,1}'^{(0)}, \tau_{21}^{(0)}\right), \quad \mathbf{x} \in S_{21}; \tag{28}$$

$$\tau_{21,2}'^{(0)} = \alpha_{21}\tau_{21,1}'^{(0)} + \beta_{21}^{(0)}, \quad \mathbf{x} \in S_{21}. \tag{29}$$

- Step 2: Solve for $\{\tau_{32}^{(0)}, \tau_{32,2}'^{(0)}\}$ on surface 32 through equations

$$\tfrac{1}{2}\tau_{32}^{(0)} = K_{S_{32}}^{3}\left(\tau_{32,3}'^{(0)}, \tau_{32}^{(0)}\right), \quad \mathbf{x} \in S_{32}; \tag{30}$$

$$\tfrac{1}{2}\tau_{32}^{(0)} = K_{S_{10}}^{1}\left(\tau_{21,2}'^{(0)}, \tau_{21}^{(0)}\right) - K_{S_{32}}^{2}\left(\tau_{32,2}'^{(0)}, \tau_{32}^{(0)}\right), \quad \mathbf{x} \in S_{32}; \tag{31}$$

$$\tau_{32,3}'^{(0)} = \alpha_{32}\tau_{32,2}'^{(0)} + \beta_{32}^{(0)}, \quad \mathbf{x} \in S_{32}. \tag{32}$$

- Step 3: Solve for $\{\tau_{21}^{(1)}, \tau_{21,1}'^{(1)}\}$ on surface 21 through equations

$$\tfrac{1}{2}\tau_{21}^{(1)} = K_{S_{21}}^{2}\left(\tau_{21,2}'^{(1)}, \tau_{21}^{(1)}\right) - K_{S_{32}}^{2}\left(\tau_{32,2}'^{(0)}, \tau_{32}^{(0)}\right), \quad \mathbf{x} \in S_{21}; \tag{33}$$

$$\tfrac{1}{2}\tau_{21}^{(1)} = -K_{S_{21}}^{2}\left(\tau_{21,1}'^{(1)}, \tau_{21}^{(1)}\right) \quad \mathbf{x} \in S_{21} \tag{34}$$

$$\tau_{21,2}'^{(1)} = \alpha_{21}\tau_{21,1}'^{(1)} + \beta_{21}^{(1)}, \quad \mathbf{x} \in S_{21}. \tag{35}$$

- Step 4: Solve for $\{\tau_{32}^{(1)}, \tau_{32,2}'^{(1)}\}$ on surface 32 through equations

$$\tfrac{1}{2}\tau_{32}^{(1)} = K_{S_{32}}^{3}\left(\tau_{32,3}'^{(1)}, \tau_{32}^{(1)}\right), \quad \mathbf{x} \in S_{32}; \tag{36}$$

$$\tfrac{1}{2}\tau_{32}^{(1)} = K_{S_{10}}^{1}\left(\tau_{21,2}'^{(1)}, \tau_{21}^{(1)}\right) - K_{S_{32}}^{2}\left(\tau_{32,2}'^{(1)}, \tau_{32}^{(1)}\right), \quad \mathbf{x} \in S_{32}; \tag{37}$$

$$\tau_{32}'^{(1)} = \alpha_{32}\tau_{23}'^{(1)} + \beta_{32}^{(1)}, \quad \mathbf{x} \in S_{32}. \tag{38}$$

The interboundary iteration can be terminated at order ℓ if certain criteria are satisfied, for example, if the ratio $\|\tau^{(\ell)}\| / \| \sum_{\ell=1,\cdots,\ell} \tau^{(l)}\|$ is smaller than the desired tolerance.

5 Discussion

Some considerations that are important in the practical implementation of these equations are the following:

- **Primary sources.** The assumption that the sources are located outside S_{10} is not essential for the derivation. If, for example, there is a source located in $\mathcal{V}_2$, the modifications to the set of system equations consist of an additional term in Eq. (19) and another in Eq. (22). The two new terms are known source functions and are independent of the unknown surface fields τ and τ'. In practice, however, there are limitations imposed by the numerics and the hardware. For example, if a magnetic dipole source is located very close to the target, one would have to make sure the discretization of the target surface is fine enough to give a reasonable representation of the variations of the source field yet still remain within the computer hardware capacity.
- **Interboundary iterations.** Depending on the nature of the problem, some of the steps of the interboundary iteration listed in the preceding section often can be eliminated. For example, let us consider a target buried in a conductive earth. The target surface is surface 32 and the earth surface is surface 21. In this case, a half-space analytical solution can be used to bypass step 1. The incident field on the target, including the effect of the conductive earth, can be obtained directly. Otherwise, it will be numerically very difficult to deal with near sources of large extent, such as a large loop transmitter laid on the surface of the earth. One would require an extremely large number of discretization points of the half-space surface to correctly describe the half-space solution.
- **Green's functions.** One also could use, instead of the Green's function for an unbounded homogeneous space, the Green's function for a homogeneous half-space. In that way, there is no need for any backscattering between the surface of the earth and the surface of the target. The choice of Green's function clearly depends on the nature of the problem. It obviously is unwise to use the interboundary iteration method outlined in the preceding section if the coupling between the surface of the half-space and the surface of the target is very strong or if the air–earth boundary is very reflective.
- **System matrices.** One major advantage of solving the vector Helmholtz equation by the multiscalar approach is that each of the three Cartesian components of **H**, along with its normal derivative, satisfies the same set of two coupled linear equations. The two $N_s \times N_s$ matrices are decomposed using the LU method and then the matrix equation is solved. Any subsequent iteration for any of the three components of **H** consists only of a backsubstitution, where N_s is the number of discretization points on the surface of the scatterer.
- **Propagations.** Once the surface fields are obtained by solving the system equations, to calculate the field value for any spatial point other than on the boundaries, a surface integral over one or two boundaries, along with the necessary evaluations of the appropriate Green's function terms, is needed. The procedure is similar to that of volume integral equation method if the observation point is outside of the target, except the difference between an integration over the volume and an integration over the surface. If the observation point is inside the target, however, the two methods differ. Although there is no need for any additional calculation, except perhaps interpolation to obtain field values inside the target, the SIE

will still use the same procedure to obtain field values by performing a surface integral.

- **Open boundary.** In the MBEM, the surface of the earth is considered to be an open boundary. There exist many ways of solving open-boundary problems using finite-element methods. The simplest method is that of truncation. Other methods include ballooning, infinite element, analytical method, etc. For a summary please refer to the review paper by Bettess (1988). The truncation method was used for results presented in the present paper.

6 Numerical results

6.1 Resistive limit and inductive limit

To study the stability of the method for the entire range from the resistive limit to the inductive limit by varying the conductivity of the target, we consider a simple model in which an analytical solution is available (Nabighian, 1970): a conductive sphere with $r = 100$ m in free air under a plane-wave field illumination with the **H** field aligned along the z-axis. The frequency is 500 Hz and the conductivity of the target varies from 0.001 to 15 000 S/m.

An Argand diagram for the z-component of the induced magnetic field is given in Fig. 3, in which the imaginary part of H_z is plotted as a function of its real part. The two curves correspond to two points on the sphere where the field values were taken: one point is on the pole and the other point is on the equator. For a geometry and a primary field alignment like this, the worst discretization error occurs at the pole. For a 300-patch discretization, the error on the pole is about 2%.

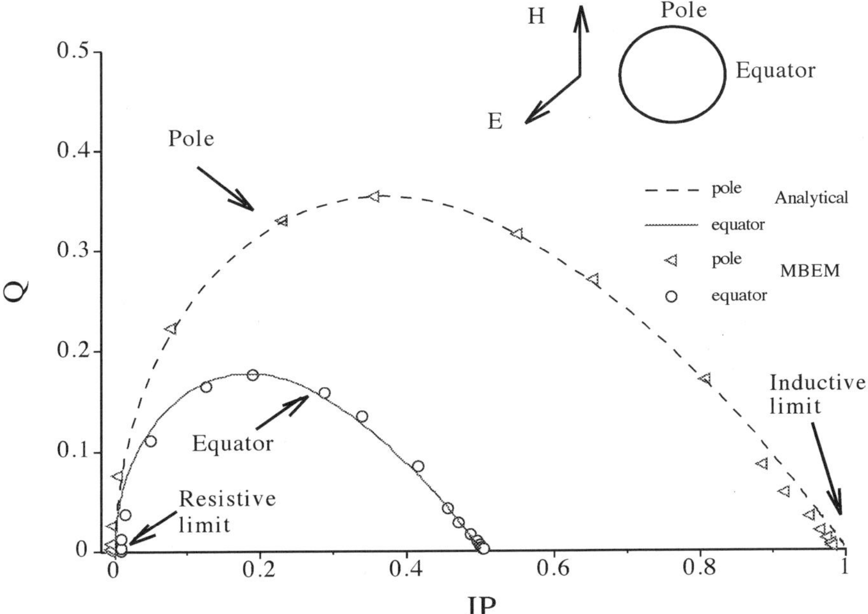

Figure 3. Argand diagram for z-component of induced magnetic field as a percentage of the incident field. The target is a sphere with $r = 100$ m in free air ($\sigma = 10^{-14}$ S/m). The incident field is a plane-wave field with H field aligned along the z-axis. The frequency is 500 Hz and the conductivity of the target varies from 0.001 to 15 000 S/m.

As can be seen from Fig. 3, correct results were obtained even when the skin depth in the conductor was smaller than an average patch size. This is because, when the skin depth is short, the most rapid field variation is along the normal direction of the surface. In MBEM formulation, the $\partial \mathbf{H}/\partial \mathbf{n}$ is solved as fundamental variables rather than derived from the values of **H**, and so, the need for numerical differentiation along the normal direction is eliminated and a better numerical accuracy is achieved in describing the normal variation of the fields. This feature is very useful for solving the kind of diffusive EM problems often encountered in geophysical modeling.

6.2 Geophysical model

In this section we present numerical results for one model of geophysical interest. A 3:1 oblate spheroid with $a = 40$ m and $b = c = 120$ m is buried in a conductive half-space with conductivity 0.005 S/m. The target dips away from the EM source at a 45° angle and its center depth is 155 m. The conductivity of the target takes two values: 10 S/m and 1000 S/m. The frequency range is from 0.1 to 12 800 Hz.

The target was first gridded into 403 nodes, and then, 802 triangular elements were identified. The number of nodes and the number of elements for the earth's surface are, respectively, 365 and 676. Three moment functions are defined on each triangular element which then are assembled into node moment functions. [Details of the surface gridding as well as methods of defining surface elements, moment functions, weight functions are not discussed in this paper; see Harrington (1968), and Wilton et al. (1984).]

The EM source is a large rectangular loop. The geometry of the model is illustrated in Fig. 4. Unless otherwise specified, all results presented in this section have been normalized by the value of the primary field at the normalization point specified in Fig. 4. Figure 5 gives two suites of response curves for the vertical component of the secondary **H**, including the responses of both the buried conductor and the half-space for the lower target conductivity. The field has been computed on the surface of the earth along the x-axis. The frequency ranges from 12 to 12 800 Hz. The response is continuously normalized, that is, normalized on each spatial point by the primary field at that point.

Similar results for a target that is 100 times more conductive are given in Fig. 6. The frequency range has been shifted downward to cover more active features in the

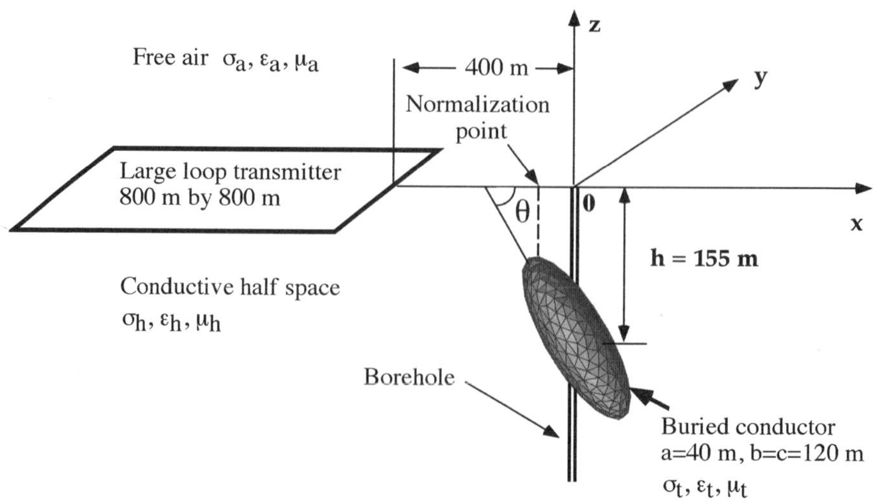

Figure 4. Geometry of geophysical model.

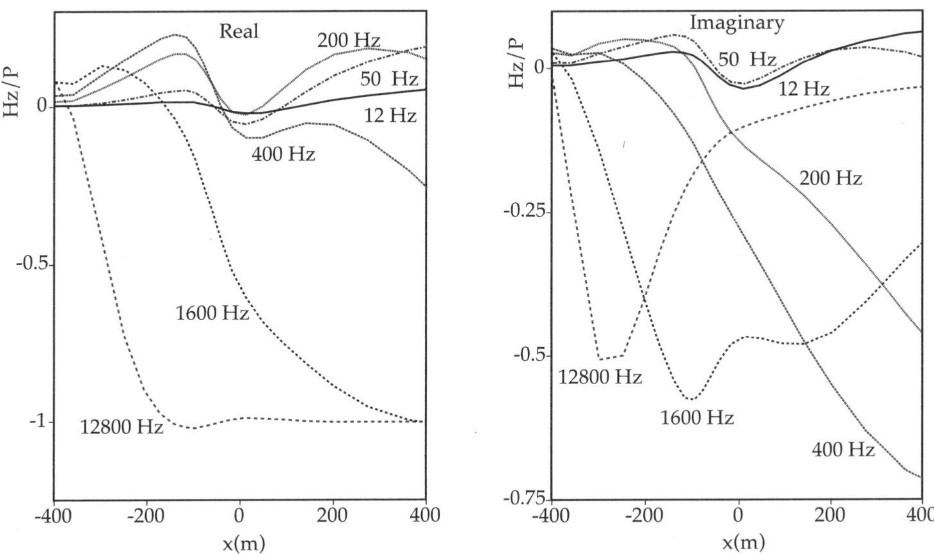

Figure 5. Secondary EM response, including the effects of both the half-space and the buried target, calculated on the surface of the earth along the x-axis for a range of frequencies. The depth $h = 155$ m and the dipping angle $\theta = 45°$. The host conductivity is 0.005 S/m and the target conductivity is 10 S/m.

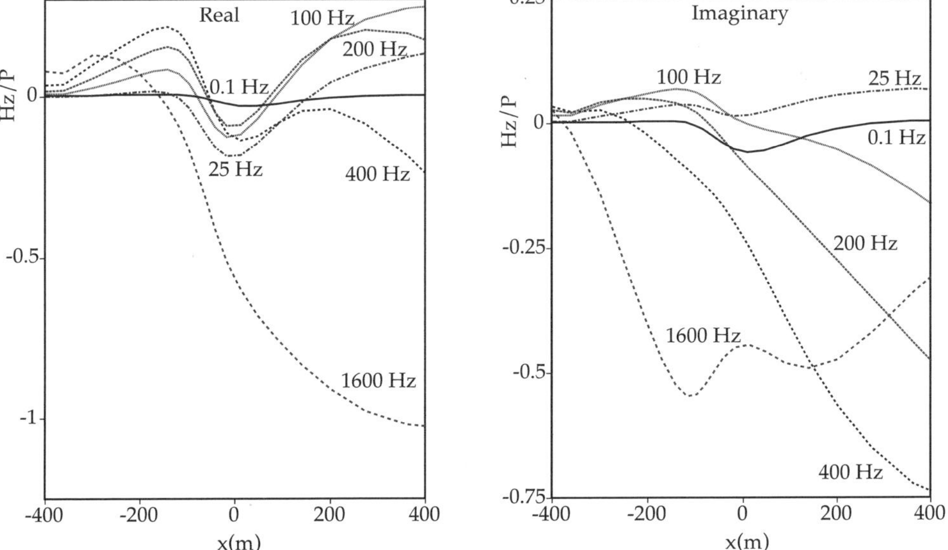

Figure 6. Secondary EM response for 1000-S/m target. All other conditions are the same as in Fig. 5.

response. It now starts at 0.1 Hz. As expected, the onset of the blanking (screening) effect of the host for both models occurs at the same frequency, at around 400 Hz. A complete screening occurs when the frequency reaches 1 600 Hz, and the target becomes undetectable. Within the 200–400-Hz frequency range, there is an appreciable contribution to the response, which we attribute to the gathering of the host currents by the target.

To clearly demonstrate the target contribution to the response measured on the surface of the earth, we plot the target response alone in Fig. 7 and 8 corresponding to a 10-S/m target and a 1000-S/m target. These are the responses one would observe if the

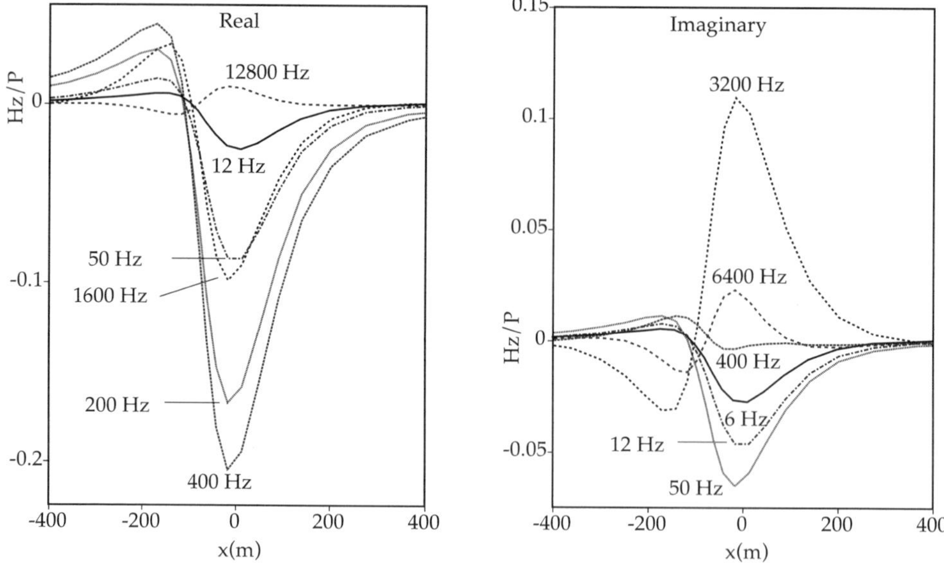

Figure 7. Secondary EM response, including only the effect of the buried target, calculated on the surface of the earth along the x-axis for a range of frequencies. The response is continuously normalized. The depth $h = 155$ m and the dipping angle $\theta = 45°$. The host conductivity is 0.005 S/m and the target conductivity is 10 S/m.

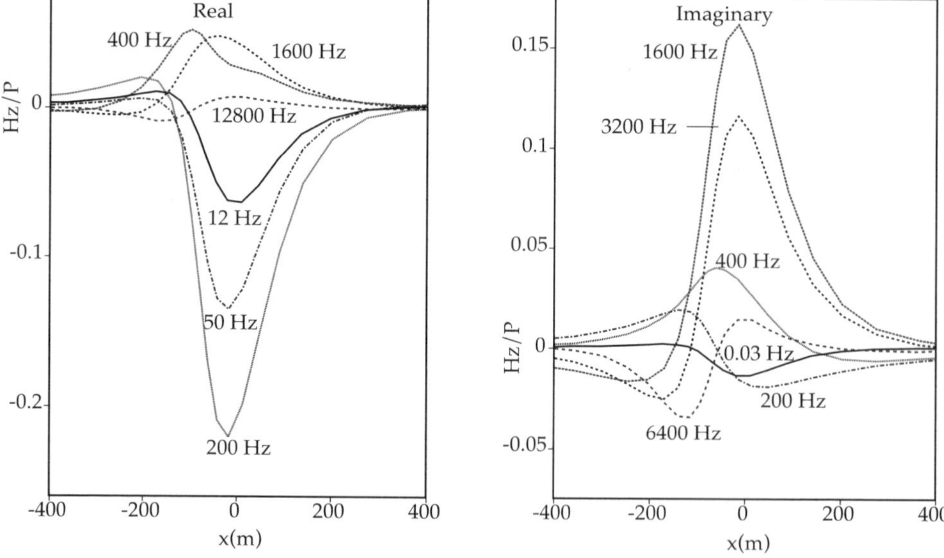

Figure 8. Secondary EM response for 1000-S/m target. All other conditions are the same as in Fig. 7.

host response was subtracted. As the frequency increases, both the in-phase part and the quadrature part of the response go up to a maximum and then start to drop. But they do not peak at the same frequency. This kind of behavior is best illustrated by a peak-to-peak Argand diagram, such as in Liu and Lamontagne (1995, Fig. 5), which is for a similar model but with slightly different depth.

Figures 9 and 10 show the responses calculated along a vertical borehole passing through the center of the target for the two conductivity values. The axial component

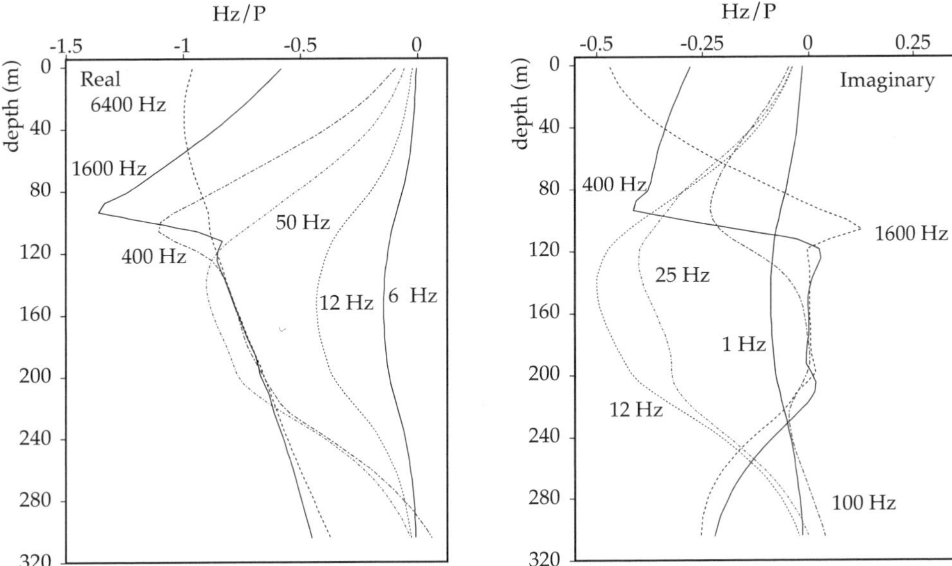

Figure 9. Secondary EM response, including the effects of both the half-space and the buried target, calculated along a vertical borehole passing through the center of the target for a range of frequencies. The response is continuously normalized. The depth $h = 155$ m and the dipping angle $\theta = 45°$. The host conductivity is 0.005 S/m and the target conductivity is 10 S/m.

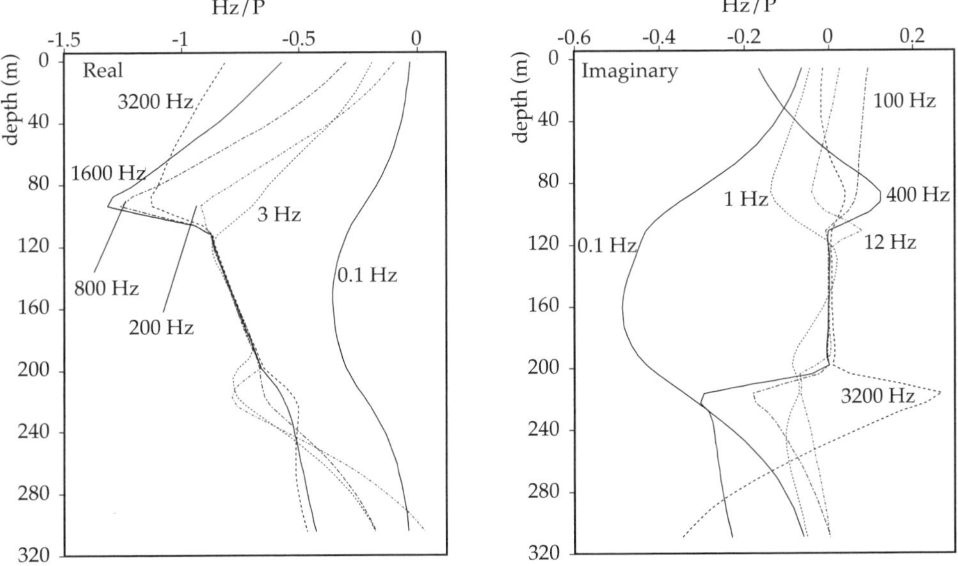

Figure 10. Secondary EM response for 1000-S/m target. All other conditions are the same as in Fig. 9.

of the secondary field, including the contributions from both the target and the half-space, is plotted along a vertical borehole that goes through the center of the tilted target. The borehole position is marked in Fig. 4. Note that at point (0,0,0) in Fig. 4, where the borehole meets the surface of the earth, the responses given in Figs. 5 and 6 for the surface profile are the same as those for the borehole profile in Fig. 9 and 10.

7 Conclusions

We have described a new method for 3-D EM field modeling based on a scalar formulation of the SIE method. The formulation can handle general target geometry, source location, conductivity contrast, and frequency, but the best efficiency is achieved when the target has a small surface area-to-volume ratio.

The equations are solved with three sets of iterations: (1) solve the coupled scalar SIEs, (2) implement the vector boundary conditions, and (3) calculate surface fields on different boundaries. More work is needed on the domains of convergence of these iterations. Though we have not been able to prove the existence of solutions in general cases, we have obtained convergent solutions for a range of geophysical models of practical interest. The method produces stable results even for very high-conductivity contrast. Caution must be exercised at very low frequencies, however. Our method can be expected to behave like a finite-element formulation, where the error in the magnetic field is inversely proportional to the square root of the frequency at low frequencies (Yosida, 1974).

Numerical efficiency is achieved mainly through the use of scalar fields, each of which satisfies that same differential equation. Because solving a matrix is an N^3 process, any reduction in the maximum size of the matrix N that one has to solve will significantly increase the numerical efficiency. Roughly speaking, by reducing a fully coupled vector equation into three scalar equations, the maximum size of the matrix that one has to solve is reduced by a factor of 3, which amounts to a factor-of-27 reduction in matrix solving time. This reduction is usually more than enough to accommodate the extra time needed for the iterations to implement the boundary conditions.

Acknowledgments

The Natural Sciences and Engineering Research Council of Canada supported the research in part by providing a grant in the form of an Postdoctoral Industrial Research Fellowship to E.H.L. during the early stage of the research. Ben Polzer of Lamontagne Geophysics Ltd. provided the code for 3-D display and editing. Elspeth Sidaway of Lamontagne Geophysics Ltd. prepared the figures.

References

Bettess, P., 1988, Finite element modeling of exterior electromagnetic problems: IEEE Trans. Magn., **24**, 238–243.

Bouix, M., 1966, Les discontinuités du rayonnement electromagnétique: Dunod.

Colton, D., and Kress, R., 1983, Integral equation methods in scattering theory: John Wiley & Sons, Inc.

Druskin, V. L., and Knizhnerman, L. A., 1988, Spectral differential-difference method for numerical solution of three-dimensional nonstationary problems of electric prospecting: Izvestiya: Earth Phys., **24**, 641–648.

Harrington, R. F., 1968, Field computation by moment methods: Macmillan, Publ. Co.

Hohmann, G. W., 1975, Three-dimensional induced polarization and electromagnetic modeling: Geophysics, **40**, 309–324.

Liu, E. H., and Lamontagne, Y., 1996, Three-dimensional electromagnetic field computation by the multiscalar boundary element method: Radio Sci., **31**, 423–435.

Liu, E. H., and Sidaway, E., 1996, Maxwell's equations and boundary conditions in curvilinear coordinate systems: Lamontagne Geophys. Res. Rep. 96-01.

Liu, E. H., and Lamontagne, Y., 1998, Geophysical application of a new surface integral equation method for EM modeling: Geophysics, **63**, 1–13, in press.

Mackie, R. L., Madden, T. R., and Wannamaker, P. E., 1993, Three-dimensional magnetotelluric modeling using difference equations—Theory and comparisons to integral equation solutions: Geophysics, **58**, 215–226.

Nabighian, M. N., 1970, Quasi-static transient response of a conductive sphere in a dipolar field: Geophysics, **35**, 303–309.

Newman, G. A., 1995, Crosswell electromagnetic inversion using integral and differential equations: Geophysics, **60**, 899–911.

Müller, C., 1969, Foundations of the mathematical theory of electromagnetic waves: Springer-Verlag, Berlin.

Poggio, A. J., and Miller, E. K., 1973, Integral equation solutions of 3-D scattering problems, *in* R. Mittra, Ed., Computer techniques for electromagnetics: Pergamon Press, Inc.

Raiche, A. P., 1974, An integral equation approach to 3-D modeling: Geophys. J. Roy. Astr. Soc., **36**, 363–376.

Wang, T., and Hohmann, G. W., 1993, A finite-difference, time-domain solution for three-dimensional electromagnetic modeling: Geophysics, **58**, 797–809.

Weidelt, P., 1975, Electromagnetic induction in three-dimensional structures: J. Geophys., **41**, 85–109.

Wilton, D. R., Rao, S. M., Glisson, A. W., Schaubert, D. H., Al-Bundak, O. M., and Butler, C. M., 1984, Potential integrals for uniform and linear source distributions on polygonal and polyhedral domains: IEEE Trans. Antennas Propag., **32**, 276–281.

Yosida, K., 1974, Functional Analysis: Springer-Verlag, New York, Inc.

A Volume-Surface Integral Equation for Electromagnetic Modeling

Zonghou Xiong
Art Raiche
Fred Sugeng

Summary. Conventional methods for solving 3-D volume integral equations in electromagnetic (EM) modeling produce matrices that are ill-conditioned when the conductivity contrast between the target and the host is large or when the host resistivity itself is small. This problem is especially acute when pulse basis functions represent the scattering currents because the artificial discontinuities in the current create spurious charges in homogeneous conductive regions (where there are no physical charges). Basis functions with higher-order continuity do not entirely eliminate this problem. The field of these charges is amplified at high host resistivities by Green's tensor and artificially attenuates the physical vortex currents. We present a new formulation that eliminates these spurious charges analytically by replacing the volume integral for the field caused by charges (on the boundary of homogeneous regions) with a surface integral, while retaining the volume integral for the field of the induced currents. This formulation is mathematically closer to the physics of EM induction and effectively removes a major source of error. Also, condition numbers of matrix systems without artificial charges are also much smaller than those of the conventional method.

1 Introduction

The volume-integral equation method has dominated 3-D electromagnetic (EM) modeling in applied geophysics since its appearance in the early 1970s (Hohmann, 1971, 1975; Raiche, 1974; Weidelt, 1975). Most of the improvements since then have come in the computation of the tensor, Green's function, or in the solution of the matrix system. For example, Tripp and Hohmann (1984) and Tripp (1990) developed group reductions for symmetric models under arbitrary excitations; Wannamaker (1991) improved the numerical integration of the charge terms in Green's functions by reducing the volume integrations to surface integrations; and Xiong (1992) introduced the method of system iteration to overcome the problems of computer storage requirements and solution stability. Xiong and Tripp (1995a,b) extended this to symmetric structures and then to very large structures with tens of thousands of cells through spatial symmetry reductions

Cooperative Research Centre for Australian Mineral Exploration Technologies (CRC AMET), CSIRO, P.O. Box 136, North Ryde, New South Wales 2113, Australia.

(Xiong and Tripp, 1993). There also have been dramatic improvements in the integral equation method for other applications, especially in electrical engineering (Rokhlin, 1990; Sarkar, 1991; Zwamborn and van den Berg, 1991; Chew, 1992; Canning, 1993). However, these developments mostly rely on properties of the whole-space Green's tensor, and do not readily apply to the layered-earth Green's tensor needed for modeling structures in the Earth.

Modeling homogeneous conductive structures with the integral-equation method becomes difficult when the contrast between the structure and the background becomes large. In homogeneous or piecewise homogeneous regions, free charge exists only on the boundaries of regions of different conductivities, not in the interiors. Because numerical solutions are not exact, however, it is difficult to get the terms representing charge effects to sum to zero in the interior. The Green's tensor has two parts: one that propagates induced current and one that propagates charge effects. As host resistivity increases, the error associated with spurious charges is amplified, thus erroneously attenuating the vortex currents which should in fact be dominant. As target conductivity increases, this effect is magnified further. This problem was first elucidated during an Australian, Canadian, and US workshop (G.F. West, personal communication, 1977).

To represent the vortex currents that should dominate the solutions at high-conductivity contrasts or in very resistive hosts, SanFilipo and Hohmann (1985) added to the usual pulse basis functions a second set of divergence-free functions composed of "tubes" of current. They used this representation in a time-stepping solution of the integral equation for transient EM fields and obtained good results for models with large contrasts. Newman and Hohmann (1988) and Hohmann (1988) later adapted the new basis functions to the standard frequency-domain code. Although they improve the solution, current tubes impose symmetry requirements on the scattering currents, which severely limits their application. For example, if the body is very close to the Earth's surface, or has an arbitrary shape, the solutions imposed by the current tube basis functions cannot represent the true solutions.

The problem of spurious charges is less severe for platelike models, where the electric fields in the target are essentially two-dimensional. Hanneson and West (1984) used global basis functions and avoided the creation of spurious charges by electric-field discontinuity. In earlier work, Lajoie and West (1976) approximated the scattering currents in the plate by the curl and gradient of two scalar potentials, with the potential for the curl being multiplied by a unit vector for the curl operation. They used spline basis functions to ensure continuity. Weidelt (1981) modified the Lajoie-West approach by transforming the potentials such that the contribution of the curl potential to the charge term of Green's tensors was cancelled, thus greatly improving the stability of the modeling process. It was this work that inspired the volume-surface integral equation formulation that we describe below.

We reformulate the conventional volume integral equation into a volume-surface integral equation that forces the divergence of the electric field to zero in the target interior. This formulation removes all artificial charges within homogeneous regions. A similar method has been proposed by Wust et al. (1993) for biomedical applications, though their method was introduced for the purpose of handling surfaces. The method of Wust et al. involves only a whole-space Green's tensor. It is known that the volume integral involving the galvanic part (related to the scalar potential) of Green's tensor can be transformed readily into a surface integral (G. W. Hohmann and P. Weidelt, personal communication, late 1980s). We reduce the volume integral involving the galvanic part of Green's tensor of a stratified host to a surface integral and demonstrate its applications to our modeling problems.

2 Conventional integral-equation method

We begin with a brief description of the conventional volume-integral-equation method. For simplicity we assume that the conductivity of the host medium is isotropic. The electric field in a space $\mathbf{E}$ can be divided into a normal part $\mathbf{E}_n$ related to a 1-D structure and a secondary part $\mathbf{E}_s$ due to a 3-D target V:

$$\mathbf{E} = \mathbf{E}_n + \mathbf{E}_s. \tag{1}$$

The normal field $\mathbf{E}_n$ in a layered media can be computed analytically in terms of Hankel integrals [for example, see Weidelt (1975), Stoyer (1977), and Xiong (1989)]. The scattered field $\mathbf{E}_s$ is computed via the scattering currents, denoted by $\mathbf{J}_s$, as (Tai, 1971; Weidelt, 1975)

$$\mathbf{E}_s(\mathbf{r}) = \int_V \underset{\sim}{\mathbf{G}}(\mathbf{r} \mid \mathbf{r}') \cdot \mathbf{J}_s(\mathbf{r}')\, dv, \tag{2}$$

where $\underset{\sim}{\mathbf{G}}$ is the electric Green's tensor. $\mathbf{J}_s$ within the target V is defined as

$$\mathbf{J}_s = \Delta\sigma \mathbf{E}, \tag{3}$$

with the differential conductivity $\Delta\sigma = \sigma - \sigma_n$, where σ is the conductivity of the structure, and σ_n is the conductivity of the host media.

Combining Eqs. (1), (2), and (3), we see that the scattering current $\mathbf{J}_s$ within V satisfies the following Fredholm integral equation of the second kind:

$$\mathbf{J}_s(\mathbf{r})/\Delta\sigma - \int_V \underset{\sim}{\mathbf{G}}(\mathbf{r} \mid \mathbf{r}') \cdot \mathbf{J}_s(\mathbf{r}')\, dv' = \mathbf{E}_n(\mathbf{r}). \tag{4}$$

An integral equation also can be written for the total electric field, but Eq. (4) has the advantage that it can be discretized into a symmetric matrix for variable $\Delta\sigma$. Once $\mathbf{J}_s$ is found from Eq. (4), the scattered EM fields everywhere can be determined by the integration of the scattering current with corresponding Green's functions.

Equation (4) can be discretized in many ways, depending on the choice of basis functions for the representation of $\mathbf{J}_s$. The simplest is to discretize it by dividing the structure V into M cells, using pulse-basis functions (Harrington, 1968), which yields the matrix equation

$$[\Gamma][J_s] = [E_n], \tag{5}$$

with

$$\underset{\sim}{\mathbf{\Gamma}}_{mn}(\mathbf{r}_m|\mathbf{r}_n) = \delta_{mn}/\Delta\sigma_m - \int_{v_n} \underset{\sim}{\mathbf{G}}^E(\mathbf{r}_m|\mathbf{r}_n)\, dv, \tag{6}$$

where

$$\delta_{mn} = \begin{cases} \underset{\sim}{\mathbf{\Theta}}, & m = n, \\ 0, & m \neq n, \end{cases}$$

and $\underset{\sim}{\mathbf{\Theta}}$ is the unit tensor. The coefficient matrix Γ in Eq. (5) is usually referred to as scattering impedance matrix, or for simplicity, scattering matrix. The size of Γ is $N \times N$ where $N = 3 \times M$, because we have three components for each vector quantity.

3 Volume-surface integral-equation formulation

Elements of the electric Green's tensor usually are expressed as the sum of two terms, the current term and the charge term (Wait, 1981, 1982):

$$\mathbf{G}_j = -i\omega\mu\mathbf{A}_j - \nabla\psi_j, \tag{7}$$

with ψ_j being given by

$$\psi_j = -\frac{1}{\sigma_n}\nabla\cdot\mathbf{A}_j, \tag{8}$$

where $j = 1, 2$, and 3 stands for dipole excitations in the three coordinate axis x, y, and z. The components of the Green's tensor obey the reciprocity relation

$$G_{ij}(\mathbf{r}\,|\,\mathbf{r}') = G_{ji}(\mathbf{r}'\,|\,\mathbf{r}). \tag{9}$$

Equation (2) breaks into two parts according to Eq. (7):

$$\mathbf{E}_s = \mathbf{E}_s^a + \mathbf{E}_s^\psi, \tag{10}$$

with

$$\mathbf{E}_s^a(\mathbf{r}) = -i\omega\mu\int_V \underset{\sim}{\mathbf{A}}(\mathbf{r}\,|\,\mathbf{r}')\cdot\mathbf{J}_s(\mathbf{r}')\,dv, \tag{11}$$

$$\mathbf{E}_s^\psi(\mathbf{r}) = -\int_V \nabla\boldsymbol{\psi}(\mathbf{r}\,|\,\mathbf{r}')\cdot\mathbf{J}_s(\mathbf{r}')\,dv, \tag{12}$$

where $\underset{\sim}{\mathbf{A}}$ and $\boldsymbol{\psi}$ are the tensor and vector forms of the potentials $\mathbf{A}_j$ and ψ_j due to the three excitations $j = 1, 2$, and 3. $\mathbf{E}_s^a$ and $\mathbf{E}_s^\psi$ are the vortex (or the current term) and the charge term of the scattered field, respectively.

The dot product in Eqs. (2), (11), and (12) is carried out according to the indices for the directions of the excitations in the tensors. The gradient operator ∇ in Eq. (12) is applied to $\mathbf{r}$. The direction of the observed electric field is given by the direction of the gradient. We begin by using the reciprocity theorem (9) to rewrite Eq. (12) as

$$\mathbf{E}_s^\psi(\mathbf{r}) = -\int_V \nabla_{\mathbf{r}'}\boldsymbol{\psi}(\mathbf{r}'\,|\,\mathbf{r})\cdot\mathbf{J}_s(\mathbf{r}')\,dv, \tag{13}$$

or

$$E_{sj}^\psi(\mathbf{r}) = -\int_V \nabla_{\mathbf{r}'}\psi_j(\mathbf{r}'\,|\,\mathbf{r})\cdot\mathbf{J}_s(\mathbf{r}')\,dv, \qquad j = 1, 2, 3 \tag{14}$$

per component. Equation (14) indicates that the direction of the electric field is given now by the direction of the source direction of Green's tensor. In the above equations, $\nabla_{\mathbf{r}'}$ means that the gradient operates on $\mathbf{r}'$. The dot product is carried out per vector direction of the gradient $\nabla_{\mathbf{r}'}$. Thus $\nabla_{\mathbf{r}'}\psi_j\cdot\mathbf{J}_s(\mathbf{r}')$ obeys vector operation rules. Note that the expression for $\mathbf{E}_s^a$ also must be transformed according to Eq. (9) because reciprocity holds only for Green's tensor as whole.

If the 3-D region V is homogeneous, $\mathbf{E}$ is divergence free, and so is $\mathbf{J}_s$ according to Eq. (3), except on the boundary of the region. This means that there is no charge in the interior of a homogeneous region. Assuming V is homogeneous, we divide it into two parts, a shell-shaped boundary part V_1 and the interior part V_2, as shown in Fig. 1, and further break Eq. (13) into two terms,

$$\mathbf{E}_s^\psi = \mathbf{E}_{s1}^\psi + \mathbf{E}_{s2}^\psi \tag{15}$$

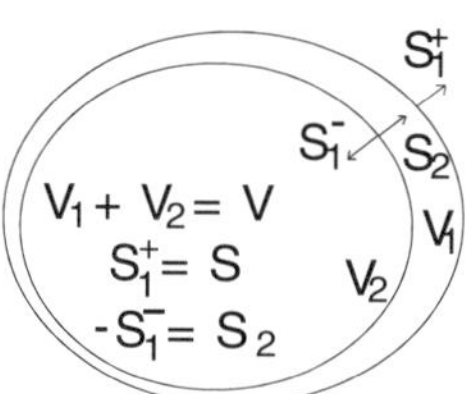

Figure 1. Division of homogeneous anomalous region into a center region and a shell-like surface region.

with

$$\mathbf{E}_{s1}^{\psi}(\mathbf{r}) = -\int_{V_1} \nabla_{\mathbf{r}'}\boldsymbol{\psi}(\mathbf{r}'|\,\mathbf{r}) \cdot \mathbf{J}_s(\mathbf{r}')\,dv, \tag{16}$$

$$\mathbf{E}_{s2}^{\psi}(\mathbf{r}) = -\int_{V_2} \nabla_{\mathbf{r}'}\boldsymbol{\psi}(\mathbf{r}'|\,\mathbf{r}) \cdot \mathbf{J}_s(\mathbf{r}')\,dv. \tag{17}$$

In component form, we can reduce Eq. (17) to a surface integral

$$\begin{aligned} E_{s2j}^{\psi}(\mathbf{r}) &= -\int_{V_2} \nabla_{\mathbf{r}'}\psi_j(\mathbf{r}'\,|\,\mathbf{r}) \cdot \mathbf{J}_s(\mathbf{r}')\,dv, \\ &= -\int_{V_2} \{\nabla_{\mathbf{r}'} \cdot [\psi_j(\mathbf{r}'\,|\,\mathbf{r})\mathbf{J}_s(\mathbf{r}')] - \psi_j(\mathbf{r}'\,|\,\mathbf{r})\nabla_{\mathbf{r}'} \cdot \mathbf{J}_s\}\,dv \\ &= -\oint_{S_2} \psi_j(\mathbf{r}'\,|\,\mathbf{r})\mathbf{J}_s(\mathbf{r}') \cdot d\mathbf{S}. \end{aligned} \tag{18}$$

The last step follows from the fact that $\nabla_{\mathbf{r}'} \cdot \mathbf{J}_s = 0$ in V_2. Thus, for all components of $\mathbf{E}_{s2}^{\psi}$, we have

$$\mathbf{E}_{s2}^{\psi} = -\oint_{S_2} \boldsymbol{\psi}(\mathbf{r}'\,|\,\mathbf{r}) \cdot \mathbf{J}_s(\mathbf{r}') \cdot d\mathbf{S}. \tag{19}$$

To derive the expression for $\mathbf{E}_s^{\psi}$ as $V_2 \to V$ and $V_1 \to 0$, we again consider a component of $\mathbf{E}_{s1}^{\psi}$ as in Eq. (16). If the shell-shaped region V_1 is thin enough, we can assume that $\mathbf{J}_s$ is constant across the shell at any particular point, and that the two surfaces of V_1 are parallel to each other. Using a curvilinear coordinate system with one of the coordinate axes, say axis $\hat{\mathbf{u}}_1$, coinciding with the normal vector of the outer surface of V_1, Eq. (16) reduces to

$$\begin{aligned} E_{s1j}^{\psi}(\mathbf{r}) &= -\int_{V_1} \nabla_{\mathbf{r}'}\psi_j(\mathbf{r}'\,|\,\mathbf{r}) \cdot \mathbf{J}_s(\mathbf{r}')\,dv \\ &= -\int_{V_1} \left[\frac{1}{h_1}\frac{\partial \psi_j}{\partial u_1} J_{su_1} + \nabla_{u_2u_3}\psi_j \cdot \mathbf{J}_{su_2u_3}\right] dv \\ &= -\oint_{S_1^+} \psi_j(\mathbf{r}'\,|\,\mathbf{r})\mathbf{J}_s(\mathbf{r}') \cdot d\mathbf{S}_1^+ - \oint_{S_1^-} \psi_j(\mathbf{r}'\,|\,\mathbf{r})\mathbf{J}_s(\mathbf{r}') \cdot d\mathbf{S}_1^- \\ &\quad -\int_{V_1} \nabla_{u_2u_3}\psi_j \cdot \mathbf{J}_{su_2u_3}dv, \end{aligned} \tag{20}$$

where $\nabla_{u_2u_3}$ and $\mathbf{J}_{su_2u_3}$ are the projections of $\nabla_{\mathbf{r}'}$ and $\mathbf{J}_s$ on the surface formed by the other two coordinate axes u_2 and u_3, respectively, and h_1 is the related transformation coefficient. As V_2 approaches V and $V_1 \to 0$, the volume integral in the above equation vanishes, because $\nabla_{u_2u_3}\psi_j$ is bounded for $\mathbf{r} \notin V_1$, and because the Cauchy principal value of the integral is zero for $\mathbf{r} \in V_1$. Because $\mathbf{S}_1^+ = \mathbf{S}$ and $\mathbf{S}_1^- = -\mathbf{S}_2$, we have from Eqs. (15), (19), and (20)

$$\mathbf{E}_s^{\psi} = -\oint_S \psi(\mathbf{r}' \mid \mathbf{r}) \, \mathbf{J}_s(\mathbf{r}') \cdot d\mathbf{S}. \tag{21}$$

Combining Eqs. (10), (11) transformed by the reciprocity theorem, and (21) gives

$$\mathbf{E}_s(\mathbf{r}) = -i\omega\mu \int_V \underset{\sim}{\mathbf{A}}^{\dagger}(\mathbf{r}' \mid \mathbf{r}) \cdot \mathbf{J}_s(\mathbf{r}') \, dv - \oint_S \psi(\mathbf{r}' \mid \mathbf{r}) \, \mathbf{J}_s(\mathbf{r}') \cdot d\mathbf{S}, \tag{22}$$

where $\underset{\sim}{\mathbf{A}}^{\dagger}$ means that the source-field indices are exchanged. Thus we arrive at a new volume-surface integral equation for the scattering current:

$$\mathbf{J}_s(\mathbf{r}) = \mathbf{E}_n(\mathbf{r})/\Delta\sigma - i\omega\mu \int_V \underset{\sim}{\mathbf{A}}^{\dagger}(\mathbf{r}' \mid \mathbf{r}) \cdot \mathbf{J}_s(\mathbf{r}') \, dv - \oint_S \psi(\mathbf{r}' \mid \mathbf{r}) \, \mathbf{J}_s(\mathbf{r}') \cdot d\mathbf{S}, \tag{23}$$

where $\mathbf{E}_n$ is the incident field. If V consists of piecewise homogeneous regions, the surface integral in Eqs. (22) and (23) contains all of the boundaries of the inhomogeneous regions. Equation (23) represents the physics of EM induction more transparently than strict volume integral formulation because it explicitly separates the induction and the charge accumulation processes. It correctly reduces to the dc scalar surface integral equation as the frequency approaches zero.

4 Numerical tests

4.1 A 1-D no-charge case

The validity and effectiveness of the volume-surface integral-equation formulation can be tested by comparing our 3-D solutions to those of a 1-D earth. For simplicity, we choose a thin conductive layer embedded in a half-space. If the layer is thin enough, the vertical currents through the layer will be negligible compared to the currents flowing in the layer so that we can assume that there is no charge on the surfaces of a very thin conductive layer. Because the 1-D layer is not bounded in the lateral directions, we also can neglect the charges on the layer edges. To further reduce the effects of charges, we use a central-loop excitation. Thus we can drop the surface integral in Eq. (23). We refer to this reduced integral equation with the volume integral only as a no-charge formulation for short.

For comparisons with this special 1-D case, we adapted our 3-D modeling programs both for the conventional method (4) and for the new formulation (23) with the volume integral only. In both codes we neglected the vertical components of the scattering currents and discretized the integral Eqs. (4) and (23) with a pulse basis function. Although in the no-charge formulation all charge effects are removed, the conventional method still retains the charge terms of the Green's functions. Thus there are artificial charges in the lateral meshes for the conventional method.

Figure 2 shows the comparisons of 1-D versus 3-D solutions of a 2-m-thick conductive layer buried 100 m deep in a half-space of 1000 ohm-m. The conductivity of the embedded layer was varied, so as to show the comparisons at various conductivity contrasts. The excitation source is a 200 × 200 m^2 loop. We consider the dB_z/dt response at the loop center. For 3-D computation, we took a plate of 3 × 3 km with the plate center directly below the receiver site. A smaller plate will produce about the same responses for times earlier than 1 ms, but its late time responses drop from those

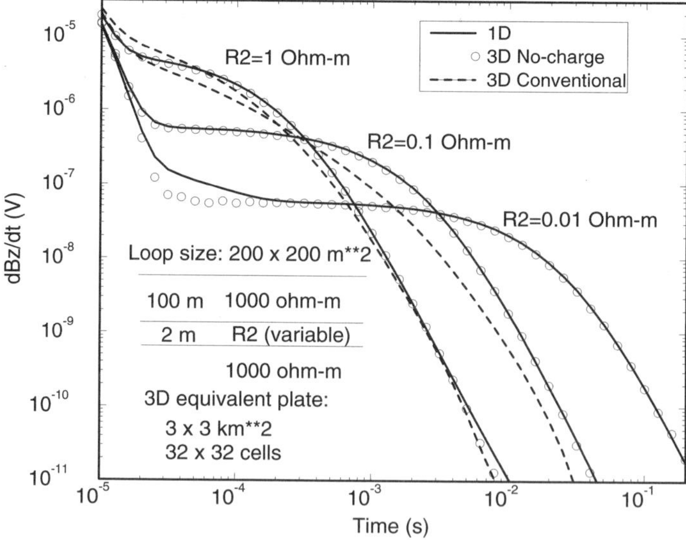

Figure 2. Comparisons of 3-D and 1-D solutions.

of the 1-D solutions. As shown in Fig. 2, 3-D results are computed with 32 × 32 cells. With only half the number of cells in both directions we obtained results that are indistinguishable in the scale of Fig. 2.

Figure 2 shows excellent agreement between the 1-D solutions and the solutions with the no-charge formulation except for the sharp turnabout from 2×10^{-5} to 2×10^{-4} s for plate resistivity of 0.01 ohm-m. This part of the 1-D response could not be reproduced by increasing the number of cells. This may indicate that the overall accuracy of our code stops there because the conductivity contrast of 10 000 to 1 for that model is quite high. For comparison, we also present in the figure the solutions of the conventional method for layer resistivity of 1 and 0.1 ohm-m (dashed line). For the layer resistivity of 1 ohm-m, there are obvious discrepancies between the conventional method and the 1-D solution which can hardly be improved by increasing the number of cells. For the layer resistivity of 0.1 ohm-m, the conventional method fails almost completely.

To study the numerical effects of spurious charges, we examined the condition numbers of the matrix systems rendered by the conventional method and the no-charge formulation. Because we eliminated the vertical components of the scattering currents, the matrices are of the order $2M \times 2M$ for M cells. Condition numbers are computed by

$$\kappa = \|\underset{\sim}{\mathbf{A}}\|\,\|\underset{\sim}{\mathbf{A}}^{-1}\| \tag{24}$$

with the matrix norms being determined by the maximal column norm

$$\|\underset{\sim}{\mathbf{A}}\|_1 = \max_{1\le j\le 2M} \sum_{i=1}^{2M} |a_{ij}|. \tag{25}$$

Using the maximal row norm $\|\underset{\sim}{\mathbf{A}}\|_\infty$ gives the same results because the matrices are symmetric.

Table 1 presents the condition numbers of the matrix systems obtained by the conventional method and the no-charge formulation for the plate with different resistivities and different discretizations at three frequencies. We see that the condition numbers for the no-charge formulation are much smaller, even for a contrast of 100:1, or for a plate resistivity of 10 ohm-m. In general, the condition numbers increase with the numbers

Table 1. Condition numbers of conventional method and new formulation for a 3×3 km horizontal plate.

		Discretizations					
	Resistivity	16 × 16			32 × 32		
Method	(ohm-m)	10 000 Hz	1000 Hz	100 Hz	10 000 Hz	1000 Hz	100 Hz
Conventional	10	3.4	5.4	5.6	7.3	10.5	10.8
	1	6.8	38.1	56.0	23.7	92.2	123
	0.1	7.5	75.3	301	27.9	258	761
	0.01	7.6	80.4	468	28.3	292	1437
No-charge	10	2.4	1.9	1.2	3.3	2.0	1.2
	1	4.1	8.4	3.0	8.5	10.7	3.2
	0.1	4.3	14.6	16.7	9.0	29.4	20.8
	0.01	4.3	14.8	28.5	9.0	30.5	57.0

of cells and the conductivity contrasts, and at lower frequencies. Thus at high contrasts it is impossible for the conventional method to compute accurately simply by increasing the number of cells. Whereas the condition numbers for the conventional method always increase as the frequency decreases, the condition numbers for the no-charge formulation actually drop at lower frequencies for plate resistivities of 10 and 1 ohm-m. This is a numerical verification of the accuracy degradation caused by spurious charges that dominate the low-frequency responses.

4.2 *A 3-D model*

The 1-D model shown above demonstrates the accuracy and the efficiency of the volume-surface formulation where charges can be neglected. Now we present a 3-D case to verify further the correctness of formulation (23). The 3-D target is a 1-ohm-m vertical dike of $20 \times 300 \times 200$ m^3 buried 100 m in a 100-ohm-m half-space. The dike is excited by a 100-m^2 loop with the near wire being 100 m from the center of the dike to the left. The receiver is 100 m away to the right. This model and the comparisons of three different solutions for the secondary fields are shown in Fig. 3. The responses shown in Fig. 3 are simulated as if they were measured by an EM37 system, which is the way our modeling software is designed. Program Leroi is a thin-sheet code based on the Weidelt (1981) formulation of the thin-sheet approximation. Program Marco is a 3-D integral-equation code using the conventional volume integral approach. Results for the volume-surface integral-equation formulation were computed using $2 \times 20 \times 15$ cells, with the cells on the dike edges being much thinner than the cells elsewhere. This effectively creates a shell over the dike so that we can easily modify our existing modeling code effectively to represent the surface integral in Eq. (23) without explicitly implementing it. The excellent agreement among the three codes is an indication of the correctness of the new volume-surface integral formulation.

5 Concluding remarks

We have developed a volume-surface integral-equation method for EM scattering problems. This formulation is potentially much more accurate than the conventional volume integral-equation method because it removes all spurious charges analytically. The

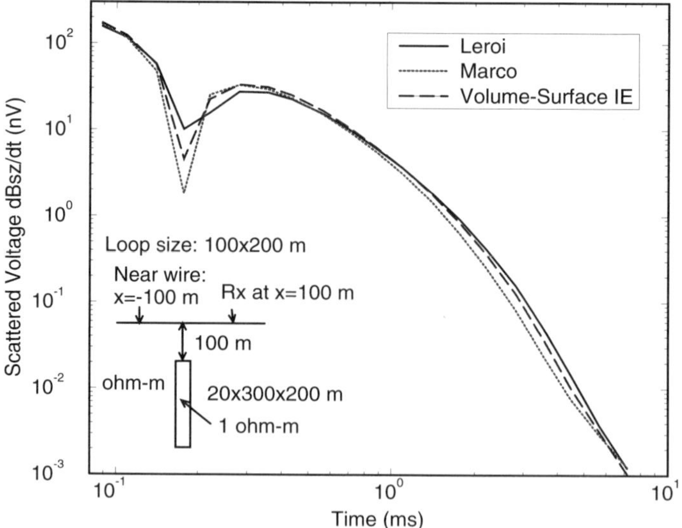

Figure 3. Verification of volume-surface integral formulation against two other integral-equation methods for a 3-D target.

volume and surface integrals in Eq. (23), however, may cause numerical instability at low frequencies unless the two integrals are well coupled—e.g., by basis functions—in the numerical implementation. This is obvious as the frequency approaches zero: the volume integral vanishes and the system can become increasingly ill-conditioned as the conductivity contrast, which determines the remaining diagonal terms of the matrix, increases. Effective implemention of this volume-surface integral-equation formulation requires more research.

Acknowledgments

This work was partially funded by sponsors of AMIRA project P223C. We wish to thank Aberfoyle Resources, BHP Minerals, CRA Exploration, M.I.M. Exploration, North Ltd., Pasminco Exploration, PNC Exploration (Australia), Sumitomo Metal Mining Oceania, Cominco Exploration, Anglovaal, and Gencor for their support. CRC AMET is established and supported under the Australian Government's Cooperative Research Centres Program.

References

Canning, F. X., 1993, Improved impedance matrix localization method: IEEE Trans. Antennas Propagat., **41**, 659–667.

Chew, W. C., 1992, Fast algorithms for wave scattering developed at the University of Illinois' Electromagnetics laboratory, IEEE Antennas and Progagation Magazine, **35**, 22–32.

Hanneson, J. E., and West, G. F., 1984, The horizontal loop electromagnetic responses of a thin plate in a conductive earth; Part I—Computational method: Geophysics, **49**, 411–420.

Harrington, R. F., 1968, Field computation by moment methods: MacMillan Publ. Co.

Hohmann, G. W., 1971, Electromagnetic scattering by conductors in the Earth near a source of current: Geophysics, **36**, 101–131.

———1975, Three-dimensional induced polarization and electromagnetic modeling: Geophysics, **40**, 309–324.

———1988, Numerical modeling for electromagnetic methods in geophysics, *in* Nabighian, M. N., Ed., Electromagnetic methods in applied geophysics, theory: Soc. Expl. Geophys., **1**, 313–363.

Lajoie, J. J., and West, G. F., 1976, The electromagnetic response of a conductive inhomogeneity in a layered earth: Geophysics, **41**, 1133–1156.

Newman, G. A., and Hohmann, G. W., 1988, Transient electromagnetic responses of high-contrast prisms in a layered earth: Geophysics, **53**, 691–706.

Raiche, A. P., 1974, An integral equation approach to three-dimensional modeling: Geophys. J. Roy. Astr. Soc., **36**, 363–376.

Rokhlin, V., 1990, Rapid solution of integral equations of scattering theory in two dimensions: J. Comput. Phys., **86**, 414–439.

SanFilipo, W. A., and Hohmann, G. W., 1985, Integral equation solution for the transient electromagnetic response of a three-dimensional body in a conductive half-space: Geophysics, **50**, 798–809.

Sarkar, T. K., 1991, Application of conjugate gradient method to electromagnetics and signal analysis: Elsevier Science Pub. Co., Inc.

Stoyer, C. H., 1977, Electromagnetic fields of dipoles in stratified media: IEEE Trans. Antennas Propagat., **AP-25**, 547–552.

Tai, C. T., 1971, Dyadic Green's functions in electromagnetic theory: International Textbook Co.

Tripp, A. C., 1990, Group theoretic reduction of the electromagnetic impedance matrix for large-contrast symmetric prisms in a layered earth: Pageoph, **133**, 127–147.

Tripp, A. C., and Hohmann, G. W., 1984, Block diagonalization of the electromagnetic impedance matrix of a symmetric buried body using group theory: IEEE Trans. Geosci. Remote Sensing, **GE-22**, 62–68.

Wait, J. R., 1981, Wave propagation theory: Pergmon Press, Inc.

———1982, Geo-electromagnetism: Academic Press Inc.

Wannamaker, P. E., 1991, Advances in three-dimensional magnetotelluric modeling using integral equations, Geophysics, **56**, 1716–1728.

Weidelt, P., 1975, Electromagnetic induction in three-dimensional structures: J. Geophys., **41**, 85–109.

———1981, Electromagnetic dipole induction of a thin plate in a conductive host with an overburden, BGR report 89727.

Wust, P., Nadobny, J., Seebass, M., Dohlus, J. M., John, W., and Felix, R., 1993, 3-D computation of E fields by the volume-surface integral equation (VSIE) method in comparison with the finite-integration theory (FIT) method: IEEE Trans. Biomed. Eng., **40**, 745–759.

Xiong, Z., 1989, Electromagnetic fields of electrical dipoles embedded in a stratified anisotropic earth: Geophysics, **54**, 1643–1646; errata, **56**, 414 (the $k_i^{\pm}$ in Eqs. (28) and (39) should be $u_i^{\pm}$ and $v_i^{\pm}$, respectively).

———1992, Electromagnetic modeling of three-dimensional structures by the method of system iteration using integral equations: Geophysics, **57**, 1556–1561.

Xiong, Z., and Tripp, A. C., 1993, Scattering matrix evaluation using spatial symmetry in electromagnetic modeling: Geophys. J. Internat., **114**, 459–464,

———1995a, A block iterative algorithm for 3D electromagnetic modeling using integral equations with symmetrized substructures: Geophysics, **60**, 291–295.

———1995b, Electromagnetic scattering of large structures in layered earth using integral equations: Radio Sci., **30**, 921–929.

Zwamborn, A. P. M., and van den Berg, P. M., 1991, The three-dimensional weak form of the conjugate gradient FFT method for solving scattering problems: IEEE Trans. Microwave Theory Techn., **MTT-40**, 1757–1766.

PART II

FINITE-DIFFERENCE MODELING

Solving three-dimensional (3-D) transient electromagnetic (TEM) problems is important in understanding the physics of observed responses, and in providing insight for data interpretation. This paper describes a finite-difference solution to a general 3-D TEM problem. The solution, which is based on time-stepping Maxwell's equations, computes both electric and magnetic responses of arbitrarily complicated earth structures.

. . .The development of a satisfactory finite-difference solution to 3-D time-domain problems has been slow, because of numerical difficulties and computer limitations. . . .

These problems can be overcome by solving the coupled, first-order Maxwell's equations using a staggered-grid scheme. . . .

Wang and Hohmann (1993)

Consistent Discretization of Electromagnetic Fields and Transient Modeling

Knútur Árnason

Summary. Differential forms provide an elegant formulation of electromagnetic field theory. Their geometric structure also leads to a self-consistent scheme for discretizing Maxwell's equations in conducting media. This discretization, which is essentially a staggered-grid scheme, preserves differential operator identities, conservation laws, and physical boundary conditions.

The self-consistent scheme has been implemented in a code to model transient electromagnetic fields in a half-space with a step-function current excitation. For small conductivity contrasts results from the self-consistent scheme agree very well with results from a 3-D integral-equation code and from a staggered-grid finite-difference code. For high contrasts, though, the results from the integral-equation code differ. The difference may be a result of improper discretization of the integral equation.

1 Introduction

Numerical methods have been used for modeling 3-D geoelectrical structures for more than a decade, as a result of work largely pioneered by Jerry Hohmann. Both integral-equation and finite-difference methods are available, but integral-equation methods have been more popular because they require less computer memory and time. Discretization of the electromagnetic (EM) field in either method can be subtle because its dynamics follow coupled equations with fewer degrees of freedom than the field components. Different discretizations often give quite different results (Smith and Paine, 1995). A consistent discretization arises naturally when EM field theory is written with differential forms. The differential-geometric structure of Maxwell's equations in this formulation suggests a discretization on dual grids—a primary and a staggered grid—and thus sheds light on the staggered-grid method invented by Yee (1966).

2 Electromagnetism and differential forms

This section sketches the formulation of EM theory with differential forms; the Appendix contains some fundamental definitions and properties of forms. For more detailed descriptions, see Spivak (1970), Felsager (1981), or Meisner et al. (1973).

National Energy Authority, Reykjavík, Iceland

In 4-D Minkowski space-time, the EM field is a skew-symmetric cotensor (covariant tensor) of rank two. It also can be represented as a differential form of rank two (two-form). The two-form is written

$$\begin{aligned} F &= \tfrac{1}{2} F_{\mu\nu}\, dx^{\mu} \wedge dx^{\nu} \\ &= E_x\, dt \wedge dx + E_y\, dt \wedge dy + E_z\, dt \wedge dz - B_z\, dx \wedge dy \\ &\quad + B_y\, dx \wedge dz - B_x\, dy \wedge dz, \end{aligned} \tag{1}$$

where μ and ν run through {0, 1, 2, 3} representing $\{t, x, y, z\}$ (repeated indices imply summation). By expressing the wedge product of the basis one-forms as a skew-symmetric tensor product (see Appendix), $F_{\mu\nu}$ becomes the usual EM field tensor. The exterior derivative of F is

$$\begin{aligned} dF &= \tfrac{1}{2} \partial_{\gamma} F_{\mu\nu}\, dx^{\gamma} \wedge dx^{\mu} \wedge dx^{\nu} \\ &= (\partial_y E_x - \partial_x E_y - \partial_t B_z)\, dt \wedge dx \wedge dy + (\partial_z E_x - \partial_x E_z + \partial_t B_y)\, dt \wedge dx \wedge dz \\ &\quad + (\partial_z E_y - \partial_y E_z - \partial_t B_x)\, dt \wedge dy \wedge dz - (\partial_x B_x + \partial_y B_y + \partial_z B_z)\, dx \wedge dy \wedge dz. \end{aligned} \tag{2}$$

Setting the exterior derivative to zero,

$$dF = 0, \tag{3}$$

gives two of Maxwell's equations:

$$\nabla \times \mathbf{E} + \partial_t \mathbf{B} = 0; \quad \nabla \cdot \mathbf{B} = 0. \tag{4}$$

The fundamental property of the exterior derivative, $d^2 = 0$, and $dF = 0$ imply that F can be written as the exterior derivative of a one-form $A = A_{\nu}\, dx^{\nu}$; i.e.,

$$F = dA = \partial_{\mu} A_{\nu}\, dx^{\mu} \wedge dx^{\nu}. \tag{5}$$

A is, of course, the four-potential, and the field form F is invariant under the gauge transformation $A \to A' = A + d\chi$, where χ is a scalar function (a zero-form).

The two-form dual to the EM-field form F is

$$\begin{aligned} {}^*F &= c[-B_x\, dt \wedge dx - B_y\, dt \wedge dy - B_z\, dt \wedge dz] \\ &\quad - (1/c)[E_z\, dx \wedge dy - E_y\, dx \wedge dz + E_x\, dy \wedge dz], \end{aligned} \tag{6}$$

where c is the speed of light ($c^{-2} = \mu\epsilon$). The four-current density one-form is

$$J = J_{\nu}\, dx^{\nu} = c^2 \rho\, dt + j_x\, dx + j_y\, dy + j_z\, dz, \tag{7}$$

and its dual ${}^*J$ is the three-form

$${}^*J = c(-\rho\, dx \wedge dy \wedge dz + j_x\, dt \wedge dy \wedge dz - j_y\, dt \wedge dx \wedge dz + j_z\, dt \wedge dx \wedge dy). \tag{8}$$

Taking the exterior derivative of ${}^*F$ and setting

$$d\,{}^*F = \mu\, {}^*J \tag{9}$$

gives the second set of Maxwell's equations,

$$\nabla \cdot \mathbf{E} = (1/\epsilon)\, \rho; \quad \nabla \times \mathbf{B} = (1/c^2)\, \partial_t\, \mathbf{E} + \mu\, \mathbf{j}. \tag{10}$$

Like its vector counterparts, Eq. (9) implies charge conservation; because $d^2 = 0$,

$$d\,{}^*J = -c(\partial_t \rho + \partial_x j_x + \partial_y j_y + \partial_z j_z)\, dt \wedge dx \wedge dy \wedge dz = 0. \tag{11}$$

The factor c^{-2} multiplying the time derivative of the electric field in Maxwell's equations [the second Eq. (10)] can be considered as being of differential-geometric origin, namely from the dual mapping [Eq. (6)].

2.1 Formulation in 3-D space and time

EM-field theory also can be expressed with differential forms in three-space and time. Differential forms are easier to work with in three-space because the metric is simpler (see Appendix). The general EM-field form F can be split into an electric part (time part) and a magnetic part (space part), which are one-forms and two-forms, respectively:

$$E = E_x\,dx + E_y\,dy + E_z\,dz; \quad B = B_x\,dy \wedge dz - B_y\,dx \wedge dz + B_z\,dx \wedge dy. \tag{12}$$

Introducing the current density one-form, $j = j_x\,dx + j_y\,dy + j_z\,dz$, and the charge density zero-form, ρ, gives

$$dB = 0, \tag{13}$$

$$dE = -\partial_t B, \tag{14}$$

$$d\,{}^*E = (1/\epsilon)\,{}^*\rho, \tag{15}$$

$$d\,{}^*B = (1/c^2)\,\partial_t{}^*E + \mu\,{}^*j. \tag{16}$$

Equations (13) and (14) show that the fields B and E can be written in terms of a one-form, A, and a zero-form, ϕ:

$$B = -dA; \quad E = -d\phi + \partial_t A. \tag{17}$$

A and ϕ are vector and scalar potentials; the fields are invariant under the gauge transformation $A \to A' = A + d\chi$ and $\phi \to \phi' = \phi + \partial_t \chi$, where χ is an arbitrary zero-form.

3 Discretization in 3-D space

Equations (1) through (11) are a good starting point for discretization in 4-D space-time. Here three-space will be discretized, but not time and, hence, Eqs. (12) through (17) are used. Let three-space be tiled with a rectangular grid with nodal points P_{ijk} and grid spacings Δx_i, Δy_j, and Δz_k. To discretize the differential forms and the exterior derivative, consider the integral of a differential form T over a domain Ω in three-space

$$\int_\Omega T\,. \tag{18}$$

If T is a one-form,

$$T = T_x\,dx + T_y\,dy + T_z\,dz, \tag{19}$$

the integral is a line integral along a curve Ω. In discretized space the curve is approximated by links between grid nodes, and the integral is approximated by a sum of the components of the one-form along the links. The (continuous) integral is, of course, the limit of such a discrete sum when the grid spacings go to zero. It is therefore natural, in the discretized space, to look at one-forms as discrete values on the links. The integral along a link is equal to the field value on the link, multiplied by the length of the link.

If the one-form is the exterior derivative of a zero-form, $T = d\phi$, then, by the generalized Stokes theorem [Eq. (A22) in the Appendix], the integral is given as the difference between values of ϕ at the end points of the curve. In the discretized space, the integral along a link is therefore the difference between values of ϕ at the end points of the link, i.e., the grid nodes. Discretized zero-forms therefore must live on the grid nodes. The discretized exterior derivative of ϕ is a one-form living on the links extending from the node P_{ijk} toward increasing $\{i, j, k\}$:

$$d\phi_{ijk} = d\phi^x_{ijk}\,dx + d\phi^y_{ijk}\,dy + d\phi^z_{ijk}\,dz$$
$$= \frac{\phi_{i+1jk} - \phi_{ijk}}{\Delta x_i}\,dx + \frac{\phi_{ij+1k} - \phi_{ijk}}{\Delta y_j}\,dy + \frac{\phi_{ijk+1} - \phi_{ijk}}{\Delta z_k}\,dz. \tag{20}$$

If the form T is a two-form

$$T = T_x\,dy \wedge dz - T_y\,dx \wedge dz + T_z\,dx \wedge dy, \tag{21}$$

the integral in Eq. (18) is a surface integral over a 2-D surface. In the discretized approximation, the surface is approximated by rectangular surface patches between grid nodes, and the integral is approximated by a sum of integrals over the patches. The discretized two-form therefore is taken naturally to represent discrete values living on surfaces between grid nodes. As for the line integral, a mathematically rigorous definition of the surface integral is obtained by the limit of vanishing grid spacings. The integral of the two-form over a patch is given as the value of the form on the patch, multiplied by the area of the patch.

If the two-form is the exterior derivative of a one-form, $T = dA$, then the generalized Stokes theorem equates the integral of T over a surface patch to the line integral of A along the oriented boundary of the patch, i.e., the links. The integral of dA over the grid surface patch in the (x, y) plane extending from P_{ijk} toward increasing $\{i, j\}$ is

$$\int_{\Delta x \times \Delta y} dA = dA^z_{ijk}\Delta x_i \Delta y_j = \left(A^y_{i+1jk} - A^y_{ijk}\right)\Delta y_j - \left(A^x_{ij+1k} - A^x_{ijk}\right)\Delta x_i. \tag{22}$$

The value of the two-form dA on the surface is obtained by dividing by the area $\Delta x_i \Delta y_j$. Proceeding in a similar way with surface patches in the (x, z) and (y, z) planes give the discretized version of the exterior derivative of a one-form A:

$$\begin{aligned} dA_{ijk} &= dA^x_{ijk}\,dy \wedge dz - dA^y_{ijk}\,dx \wedge dz + dA^z_{ijk}\,dx \wedge dy \\ &= \left[\frac{A^z_{ij+1k} - A^z_{ijk}}{\Delta y_j} - \frac{A^y_{ijk+1} - A^y_{ijk}}{\Delta z_k}\right] dy \wedge dz \\ &\quad - \left[\frac{A^x_{ijk+1} - A^x_{ijk}}{\Delta z_k} - \frac{A^z_{i+1jk} - A^z_{ijk}}{\Delta x_i}\right] dx \wedge dz \\ &\quad + \left[\frac{A^y_{i+1jk} - A^y_{ijk}}{\Delta x_i} - \frac{A^x_{ij+1k} - A^x_{ijk}}{\Delta y_j}\right] dx \wedge dy. \end{aligned} \tag{23}$$

If A itself is an exterior derivative (of a zero-form), then inserting Eq. (20) into Eq. (23) gives

$$d^2\phi_{ijk} = 0. \tag{24}$$

In vector calculus, this identify reads

$$\nabla \times \nabla \phi_{ijk} = 0. \tag{25}$$

Similar arguments show that discretized three-forms live in the cubic cells between the grid nodes and that the discretized version of the exterior derivative of a two-form B is

$$dB = \left[\frac{B^x_{i+1jk} - B^x_{ijk}}{\Delta x_i} + \frac{B^y_{ij+1k} - B^y_{ijk}}{\Delta y_j} + \frac{B^z_{ijk+1} - B^z_{ijk}}{\Delta z_k} \right] dx \wedge dy \wedge dz. \tag{26}$$

Inserting the discretized exterior derivative of a one-form [Eq. (23)] into Eq. (26) gives

$$d^2 A_{ijk} = 0, \tag{27}$$

which, in vector calculus, is the identity

$$\nabla \cdot \nabla \times \mathbf{A}_{ijk} = 0. \tag{28}$$

Differential forms in discretized three-space thus induce a natural discretization of scalar, vector, pseudovector, and pseudoscalar fields on a grid. Moreover, the discretized exterior derivative d respects the fundamental property $d^2 = 0$. In the discretized space, the exterior derivative of a differential form gives a new form that lives on grid elements (links, surfaces, or volumes) bounded by the elements on which the original form lives. The values of the form involve values of the original form on the boundary elements (the exterior elements, hence the name exterior derivative).

3.1 Dual forms and staggered grids

Maxwell's equations (15) and (16) contain dual forms. Because the dual operation does not commute with the exterior derivative, these equations require discretization of dual forms. The Appendix shows that the dual mapping in three-space transforms a three-form into a zero-form, a two-form into a one-form, a one-form into a two-form, and a zero-form into a three-form. In discretized space, however, three-forms live in cubic cells, two-forms on surfaces, one-forms on links, and zero-forms on grid nodes. The dual mapping therefore maps forms living in cubes into forms on nodes, forms on surfaces into forms on links, forms on links into forms on surfaces, and forms on nodes into forms in cubes. There is only one consistent way this can happen. The primary grid must carry along with it a second grid, which is offset (staggered) from the original grid in space. The nodes of the staggered grid lie at the centers of the cubes between grid nodes in the primary grid. If the values of the discretized forms live at the centers of the links, surfaces, and cubes, then the dual mapping takes forms on the primary grid into corresponding forms on the staggered grid (see Fig. 1). Grid spacings in the staggered grid are the average of the adjacent grid spacings in the primary grid.

The exterior derivative on the staggered grid behaves the same way as it does on the primary grid and satisfies the fundamental property $d^2 = 0$. The formulas in the Appendix for the dual forms in three-space show that taking the dual twice restores the original form. In discretized space, taking the dual twice takes forms from the primary grid to the staggered grid and back to the primary grid.

A differential form, considered as a form on one of the grids, can be differentiated with respect to the other grid. Let T be a form on the primary grid. The operation ${}^*d^*T$ maps the form to the staggered grid, takes the exterior derivative on the staggered grid,

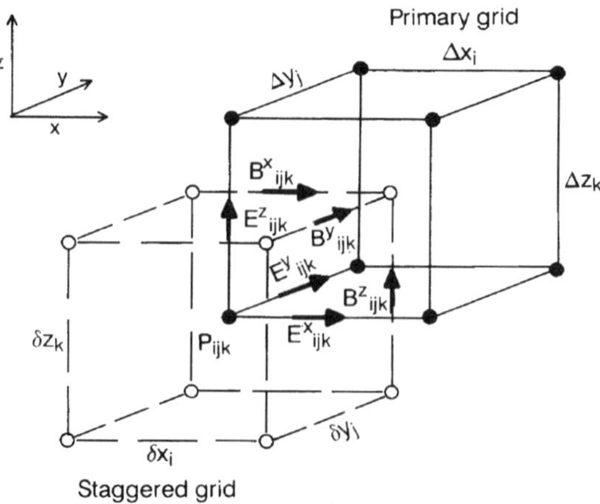

Figure 1. Primary and staggered grids.

and maps back to the primary grid. The same operation differentiates forms on the staggered grid with respect to the primary grid. This is, up to a sign, the discretization of the codifferential, which acts on a k-form in three-space as $\delta = (-1)^k\, {}^*d^*$ (see the Appendix). The codifferential inherits the fundamental property $\delta^2 = 0$ from the property $d^2 = 0$ of the exterior derivative. This prohibits taking second derivatives directly with respect to either of the grids. Nontrivial second derivatives, however, can be obtained by taking one derivative with respect to each grid, i.e., by the operations $\delta dT = (-1)^k\, {}^*d^*dT$ and $d\delta T = (-1)^k d^*d^*T$.

With these concepts in place, discretization of the EM field can proceed as follows. On a primary grid, let charge and the scalar potential (zero-forms) live on grid nodes; electric field, vector potential, and current (one-forms) on links between the nodes; and the magnetic field (a two-form) on surfaces. The dual forms of Maxwell's equations (15) and (16) can live on the primary grid, but the staggered grid is still needed, because the derivatives in Eqs. (15) and (16) refer to the staggered grid.

3.2 Conductivity and discretization

An important question remains. How should the conductivity be discretized? This question concerns the differential-geometric properties of the conductivity. In a general medium, the conductivity is a symmetric tensor $\sigma_{ij}(x)$ that connects the current and electric-field vectors, both of which live on links in the primary grid,

$$j_i = \sigma_{ij} E^j \tag{29}$$

(here i and j take the values $\{1, 2, 3\}$, representing the x-, y-, and z-components, and summation is implied over repeated indices). In the language of differential forms, conductivity is a symmetric cotensor of rank two:

$$\sigma = \sigma_{ij}\, dx^i \otimes dx^j. \tag{30}$$

The conductivity is not, a differential form; differential forms are totally antisymmetric cotensors. However, by a canonical identification of covectors (one-forms) and vectors (see Appendix), the electric-field form E becomes a vector on which the conductivity tensor operates, resulting in a current-density one-form, with the components given by Eq. (29).

Further progress requires consideration of discretization itself. Discretization of a field theory in three-space consists of discretizing the space and approximating the field values in continuous space by values attached to grid elements in the discretized space (for numerical calculations, the number of discretized field values is made finite by confining the solution within a bounded region of space and imposing appropriate boundary values). The discretized field values can be taken as average values of the field in volume elements around the grid elements. A discretized field theory can therefore be thought of as a continuum field theory with constant field values in domains of a partitioned space that supports a modified calculus. In the discretization of the EM field described here, the charge density and the scalar potential (zero-forms) have constant values around the grid nodes (in the primary grid), that is, in the volumes between nodes of the staggered grid. The vector potential, the electric field, and the current density (one-forms) are constant around the links of the primary grid, in the volume sliced out by the surface element of the staggered grid through which the link passes. The magnetic field (a two-form) is constant around surfaces in the primary grid, in volumes sliced out by the surfaces along the links of the staggered grid.

Discretized space itself is a piecewise constant continuum in the following sense. On a general curved differentiable manifold with a metric, the metric

$$g = g_{\alpha\beta}\, dx^{\alpha} \otimes dx^{\beta} \tag{31}$$

is a symmetric cotensor field (a function of position) on the manifold. Such a manifold can be discretized by choosing a discrete set of points and approximating the manifold in between by domains of the (flat) tangent spaces to the manifold. The metric tensor is constant in each tangent space (Cartesian coordinates are assumed) and equal to the average of the metric tensor on the portion of the manifold approximated by the tangent space. In a flat manifold, the metric tensor is constant and the tangent spaces are identical to the manifold. In discretized flat three-space, the volume cubes between nodes in the primary grid can be taken as domains of tangent spaces to the manifold (at the nodes in the staggered grid) with a constant metric in each cube.

The conductivity tensor has the same differential-geometric properties as the metric tensor. In the discretized EM field theory, the conductivity tensor operates on vectors (canonically identical to one-forms) on links in the primary grid. This leads us to the conclusion that the discretized conductivity lives in the volume cells of the primary grid. The conductivity along a link then becomes the average conductivity in the volume along the link with the cross-sectional area defined by the surface element of the staggered grid. The surface element of the staggered grid, taken along the link, slices one-quarter of each of the four volume elements in the primary grid that have the link in common. The discretized conductivity along the link is therefore the average conductivity around the link (see Fig. 2).

To understand these results physically, imagine a conductive three-space made of many small cubes of dimension Δ and with different conductivities. How should a uniform grid, with spacing Δ, cut such a space? The requirement that charge is to live on nodes in the primary grid (volume cells of the staggered grid), and not on links, prohibits links of the primary grid from cutting through boundaries between cubes of different conductivities: The constant electric field along the link would require a charge at the conductivity discontinuity on the link, which is inconsistent with charge only living on nodes and would lead to nonconserved charge. This leaves only one alternative: The cubes of constant conductivity must coincide with the volume elements in the primary

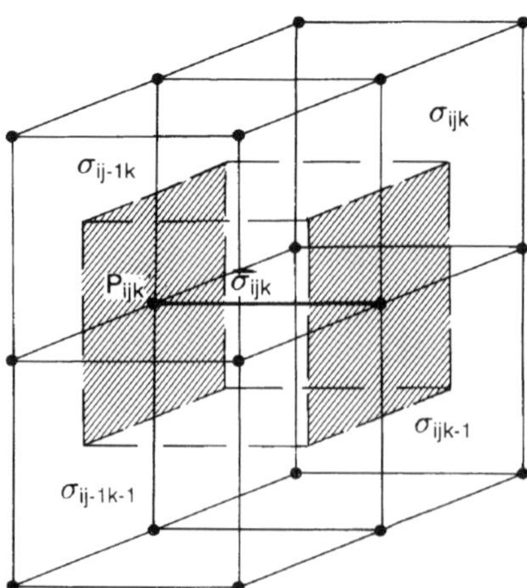

Figure 2. Discretized conductivities; the conductivity along a link is the average of neighboring conductivities.

grid. The discretized conductivity along a link, as the average conductivity through the four volume elements having the link as common boundary, is now obvious. This choice automatically provides the discretized versions of the boundary conditions that tangential components of the electric field and normal components of the magnetic field be continuous across discontinuities in the conductivity. The boundary conditions for the normal components of the electric field and tangential components of the magnetic field also can be shown to be implied by the discretized version of Maxwell's equations (15) and (16).

This completes the discretization of the EM field in conducting media. The scheme is self-consistent. It makes the discretized versions of the curl of a gradient and the divergence of a curl identically zero. If the field strengths are defined in terms of discretized vector and scalar potentials, the discretized Maxwell's equations (13) and (14) are satisfied automatically. The property $d^2 = 0$ of the discretized exterior derivative makes the discretized version of Maxwell's equations (15) and (16) guarantee charge conservation (current conservation in the quasi-stationary approximation). By placing the grid in such a way that cubes of constant conductivity are volume elements in the primary grid, the right boundary conditions for field strengths at conductivity discontinuities also are fulfilled.

4 Application of discretization scheme

As an example of this discretization, consider the computation of the transient EM field at the surface of a conducting half-space, after a step-current excitation at $t = 0$. At later times, the time variation of the fields is moderate and the quasi-stationary approximation applies. Taking the dual of Maxwell's equation (16) gives

$$^{*}d\,^{*}B = \delta B = \mu j, \tag{32}$$

because B is a two-form and $\delta B = (-1)^2\,^{*}d\,^{*}B$. The current is the sum of the external current and the current induced in the isotropic conductive half-space,

$$j = \sigma E + j_{\text{ext}}. \tag{33}$$

Taking the time derivative on both sides of Eq. (32) and using Eq. (14) gives

$$ {}^*d{}^*dE = \delta dE = -\mu\sigma\partial_t E - \mu\partial_t j_{\text{ext}}. \tag{34} $$

Now, take the inner product with the unit one-form $I = dx + dy + dz$ on each side, i.e., take the wedge product with its dual form and integrate over the conducting half-space (refer to the Appendix). Then, we get

$$ \int ({}^*I) \wedge \delta dE = -\mu\partial_t \int ({}^*I) \wedge (\sigma E) - \mu\partial_t \int ({}^*I) \wedge j_{\text{ext}}. \tag{35} $$

Next, discretize the half-space and the forms according to the self-consistent scheme described above (the discretized form ${}^*I$ has the value 1 on the surfaces of the staggered grid). The integrals become sums of integrals over volume elements in the discrete space. The integrands are wedge products of a two-form on the staggered grid (the dual of a one-form on the primary grid) and a one-form on the primary grid so the volume elements are volumes around the links in the primary grid with a cross-sectional area equal to the surface elements of the staggered grid. Demanding that the equation above be satisfied in each volume element generates a vector differential equation for the components of the electric field. Write the discretized electric field as a column vector $E_n = E^{\alpha}_{ijk}$, where α runs over x-, y-, and z-components, $\{i, j, k\}$ run through node numbers in the x-, y-, and z-directions, respectively, and n is a numbering $n(\alpha, i, j, k)$ of the field components. The vector differential equation then is (dropping the subscript ext on the current)

$$ A^m_n E_m = -\mu\bar{\sigma}_n\delta^m_n \frac{d}{dt} E_m - \mu\frac{d}{dt} j_n. \tag{36} $$

The matrix $\bar{\sigma}_n\delta^m_n$ is a diagonal matrix where $\bar{\sigma}_n$ is the average conductivity along the link n, multiplied by the volume element around the link, and j_n is the source current along the link n (with the same numbering as for the electric field), multiplied by the volume element around the link. The discretized operator $\delta d = {}^*d{}^*d$ is now represented by the matrix A^m_n, which can be shown to be symmetric and nonnegative definite.

It is convenient to express the differential operator solely with field values in the discretized half-space, on and below the surface. But the discretized differential operator at the surface requires one grid plane above the surface. To eliminate these nodes, consider Eq. (34) above the surface. There are no currents in the empty space above the half-space, and so, the equation becomes

$$ \delta dE = 0. \tag{37} $$

Also, Maxwell's equation (15) in the empty space above the half-space becomes

$$ d{}^*E = 0. \tag{38} $$

Taking the dual of this equation, followed by an exterior derivative, gives

$$ \delta dE = 0. \tag{39} $$

Equations (37) and (39) now imply that

$$ \delta dE + \delta dE = 0. \tag{40} $$

This may look like a complication, but by the formulas in the Appendix, the one-form δdE is seen to have components given by the vector $\nabla \times (\nabla \times \mathbf{E}) = \nabla(\nabla \cdot \mathbf{E}) - \nabla^2\mathbf{E}$, and $d\delta E$ has components given by $-\nabla(\nabla \cdot \mathbf{E})$, and so, the one-form on the left-hand side of Eq. (40) has the components of $-\nabla^2\mathbf{E}$.

Now take the inner product of Eq. (40) with the unit one-form I and integrate over the x and y variables and discretize the x-y plane with nodes having the same (x, y) values as the grid in the half-space (z remains continuous, and the fields are functions of z). This results in vector differential equations for the x- and y-components:

$$D_n^m E_m^{x,y} = \frac{d^2}{dz^2} E_n^{x,y}, \tag{41}$$

where n is a numbering of the x or y links, respectively. E_z is fixed by the boundary condition that it vanish at $z = 0$. The matrix $\underset{\sim}{\mathbf{D}}$ can be shown to be symmetric and nonnegative definite, and a formal solution to Eq. (41) gives the fields at the plane Δz above the conducting half-space:

$$\mathbf{E}^{x,y}(\Delta z) = e^{-\sqrt{\underset{\sim}{\mathbf{D}}}\Delta z} \cdot \mathbf{E}^{x,y}(0). \tag{42}$$

Boundary conditions demand that the tangential components of the electric field be continuous at the surface of the half-space, and hence Eq. (42) gives the needed values at the grid plane above the surface in terms of those at the surface. The matrix-valued exponential function in Eq. (42) can be evaluated by a singular value decomposition of the matrix $\underset{\sim}{\mathbf{D}}$. For a general grid, we have different matrices $\underset{\sim}{\mathbf{D}}$ for the x- and y-components, but if the grid spacings are the same in the x- and y-directions, the matrices are identical and a lot of computation time is saved.

Consider now the vector differential equation (36) for the electric field in the half-space. It can be symmetrized by defining

$$\mathbf{E}' = \underset{\sim}{\mathbf{S}} \cdot \mathbf{E}; \quad \underset{\sim}{\mathbf{B}} = \frac{1}{\mu} \underset{\sim}{\mathbf{S}}^{-1} \cdot \underset{\sim}{\mathbf{A}} \cdot \underset{\sim}{\mathbf{S}}^{-1}; \quad \mathbf{j}' = \underset{\sim}{\mathbf{S}}^{-1}\mathbf{j}; \tag{43}$$

(subscripts are dropped) where $\underset{\sim}{\mathbf{S}}$ is the diagonal matrix $S_n^m = \sqrt{\bar{\sigma}_n}\delta_n^m$. The equation for the electric field now becomes

$$\underset{\sim}{\mathbf{B}} \cdot \mathbf{E}' = -\frac{d}{dt}\mathbf{E}' - \frac{d}{dt}\mathbf{j}'. \tag{44}$$

For a step-current j_0 turned on at $t = 0$, this equation has the formal solution

$$\mathbf{E}'(t) = -e^{-\underset{\sim}{\mathbf{B}}t} \cdot \mathbf{j}_0'. \tag{45}$$

The electric-field components now are given as $\mathbf{E} = \underset{\sim}{\mathbf{S}}^{-1} \cdot \mathbf{E}'$ and the time derivative of the magnetic field (induction in a receiver loop) can be obtained by the discretized version of Maxwell's equation $dE = -\partial_t B$.

5 Implementation and numerical results

The discretization method described above has been implemented in a computer program called TEMDDD. The matrix $\underset{\sim}{\mathbf{B}}$ containing the discretized differential operator is huge for large 3-D grids, but is also sparse and symmetric. It can be stored efficiently, and special fast routines can be used to multiply vectors onto the matrix. The matrix exponential on the right-hand side of Eq. (45) can be evaluated numerically by the spectral Lanczos decomposition method (Druskin and Knizhnerman, 1988).

An acceptable accuracy is obtained in most cases on grids of from $30 \times 30 \times 20$ to $40 \times 40 \times 25$ nodes ($x \times y \times z$ grids). Grid spacings are usually constant in a central

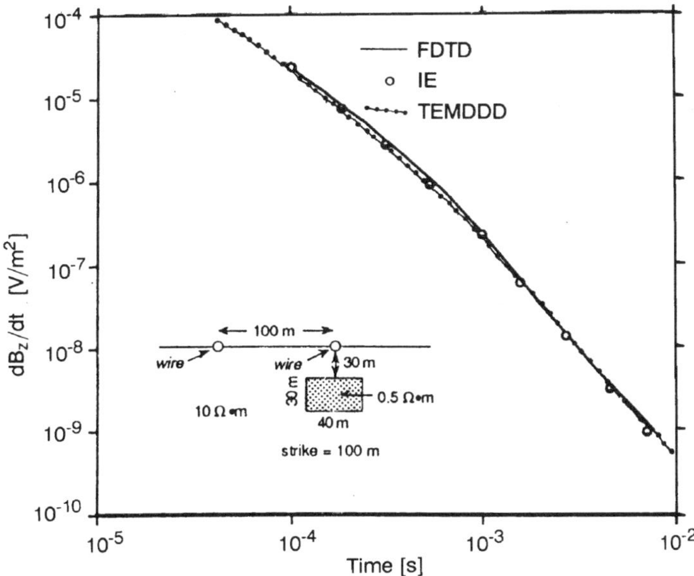

Figure 3. Comparison of results from TEMDDD, the integral-equation (IE) code of Newman et al. (1986), and FDTD (Wang and Hohmann, 1993), for a moderate contrast (20:1). Modified from Wang and Hohmann (1993).

region below the surface, and then increase according to a power law toward the outer boundaries where the field is set to zero. About 1000 basis vectors in the Krylov space of the spectral Lanczos decomposition [1000 applications of the matrix $\underset{\sim}{\mathbf{B}}$; see Druskin and Knizhnerman (1988)] give stable solutions up to 0.1–1 s, depending on the resistivity. For models with the typical range of dimensions, the computing time varies from about 0.5 to 3 hours on a DEC Alpha 3000/600 workstation; the algorithm needs about 10 to 30 Mbytes of memory.

Results from the program compare favorably with 1-D algorithms and published results of existing 3-D codes for moderate resistivity contrasts. Figure 3 shows the calculated central-loop transient electromagnetic (TEM) response of three different programs: The IE-program of Newman et al. (1986), the finite-difference (FD) time-stepping program (FDTD) of Wang and Hohmann (1993), and TEMDDD. They all compare well for this moderate resistivity contrast (20:1).

For higher resistivity contrasts, discrepancies show up. Figure 4 shows a comparison of central-loop TEM response for a relatively high resistivity contrast (1000:1) calculated by three programs: TEMDDD, FDTD (Tsili Wang, personal communication, 1995), and a recent IE program (Zonghou Xiong, personal communication, 1995). The FDTD and TEMDDD compare reasonably well. The results from the IE program are different, but converge toward the results of FDTD and TEMDDD as the inhomogeneity is discretized with increasing number of cells despite some evidence of instability (the actual number of cells is four times the numbers shown in the figure). The TEMDDD response is calculated on a $34 \times 34 \times 20$ grid, where the low-resistivity body is divided into 72 volume cells in the primary grid. A run with a finer grid, with the body divided into 196 cells, gave practically the same result. A similar discrepancy was observed when comparing results, for relatively high contrasts, from TEMDDD and the IE code applied by Newman et al. (1987).

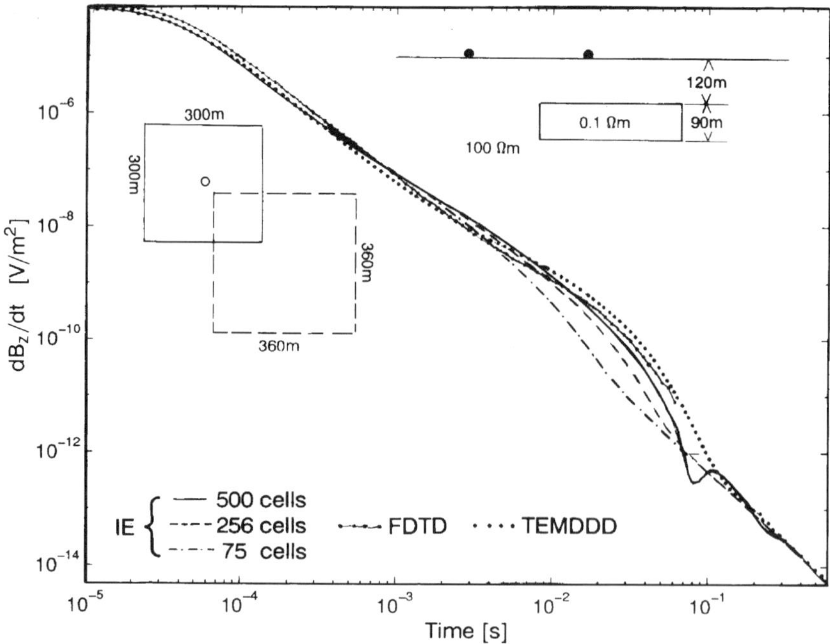

Figure 4. Comparison of results from TEMDDD, the IE code of Xiong (personal communication, 1995) and FDTD (Wang, personal communication, 1995) for a high contrast (1000:1).

The discrepancy between the FD and IE methods may be due to different discretizations. Both TEMDDD and FDTD use staggered grids with constant conductivity cells in volume elements of the primary grid [in FDTD, the first-order differential equations for the electric and the magnetic fields are solved by time stepping; see Wang and Hohmann (1993)]. In the IE approach (Wannamaker et al., 1984; Hohmann, 1988), the anomalous body is discretized into volume cells with the discretized electric field defined inside the cells. This discretization makes it difficult to enforce the boundary condition of continuity of tangential components of the electric field at conductivity discontinuities and thus suppresses induction currents in a conducting body in a resistive host. Newman and Hohmann (1988) tried to correct this behavior by introducing a second divergence-free set of current tubes into the discretization. This is somewhat artificial, because the original discretization has enough degrees of freedom to describe induction currents. The suppression of induction currents and the discrepancy in Fig. 4 are probably the result of defining the electric field inside grid cells with constant conductivity, instead of on their edges.

Acknowledgments

The author wants to thank Dr. Zonghou Xiong and Dr. Tsili Wang for permission to present their results. Dr. Sven Sigurdsson, Dr. Hjálmar Eysteinsson, and Dr. Jón Orn Bjarnason are thanked for reviewing the manuscript.

References

Druskin, V. L., and Knizhnerman, L. A., 1988, Spectral differential-difference method for numerical solution of three-dimensional nonstationary problems of electric prospecting: Izvestiya, Earth Phys., **24**, 641–648.

Felsager, B., 1981, Geometry, particles and fields: Odense Univ. Press.

Hohmann, G. W., 1975, Three-dimensional induced polarization and electromagnetic modeling: Geophysics, **40**, 309–324.

———1988, Numerical modeling for electromagnetic methods of geophysics, *in* Nabighian, M. N., Ed. Electromagnetic methods in applied geophysics: Soc. Expl. Geophys., 314–361.

Meisner, C. W., Thorne, K. S., and Wheeler, J. A., 1973, Gravitation: W. H. Freeman & Co.

Newman, G. A., and Hohmann, G. W., 1988, Transient electromagnetic responses of high-contrast prisms in a layered earth: Geophysics, **53**, 691–706.

Newman, G. A., Hohmann, G. W., and Anderson, W. L., 1986, Transient electromagnetic response of a three-dimensional body in a layered earth: Geophysics, **51**, 2117–2130.

Newman, G. A., Anderson, W. L., and Hohmann, G. W., 1987, Interpretation of transient electromagnetic soundings over three-dimensional structures for the central-loop configuration: Geophys. J. Roy. Astr. Soc., **89**, 889–914.

Smith, R., and Paine, J., 1995, 3D TEM Modeling—A Users' View, *in* Proc. Internat. Symp. on Three-dimensional Electromagnetics: Schlumberger-Doll Research, 13–40.

Spivak, M., 1970, A comprehensive introduction to differential geometry: Publish or Perish.

Wannamaker, P. E., Hohmann, G. W., and SanFilipo, W. A., 1984, Electromagnetic modeling of three-dimensional bodies in layered earth using integral equations: Geophysics, **49**, 60–74.

Wang, T., and Hohmann, G. W., 1993, A finite-difference, time-domain solution for three-dimensional electromagnetic modeling: Geophysics, **58**, 797–809.

Yee, K. S., 1966, Numerical solutions of initial boundary problems involving Maxwell's equations in isotropic media: IEEE Trans. Antennas Propag., **AP-14**, 302–309.

Appendix: Differential Forms

This Appendix reviews the basic geometric objects and operations used in this paper. For detailed treatments, see Spivak (1970) or Felsager (1981). Differential geometry deals with differential and integral calculus on differentiable manifolds. These manifolds have tangent vectors and tangent spaces attached to every point. These are generalized to tensors and tensor spaces. Each tangent space also has a dual space of covectors that map tangent vectors into real numbers. The covector spaces are generalized to cotensor spaces. Vectors and tensors can be related to covectors and cotensors by the canonical identification through the metric tensor, which can lower and raise indices. A tensor (cotensor) field on the manifold associates with each point on the manifold a tensor (cotensor) in the tensor (cotensor) space at that point.

A differential form T of rank p, or a p-form, on a manifold of dimension n, is a totally antisymmetric cotensor field of rank p that maps p vector fields into real-valued functions on the manifold. In a given coordinate system for the manifold, a p-form can be written in terms of the wedge product of the n-basis one-forms dx^i:

$$T = T_{i_1,\ldots,i_p} dx^{i_1} \wedge \ldots \wedge dx^{i_p}. \tag{A-1}$$

Repeated indices imply summation from 1 to n.

The wedge product is the totally antisymmetric tensor product of the basis forms. In three-space, there are only three basis one-forms and hence two types of wedge products:

$$dx^i \wedge dx^j = dx^i \otimes dx^j - dx^j \otimes dx^i; \tag{A-2}$$

$$\begin{aligned} dx \wedge dy \wedge dz = {} & dx \otimes dy \otimes dz - dx \otimes dz \otimes dy + dy \otimes dz \otimes dx \\ & - dy \otimes dx \otimes dz + dz \otimes dx \otimes dy - dz \otimes dy \otimes dx. \end{aligned} \tag{A-3}$$

To each p-form T, on a differentiable manifold of dimension n, there is a corresponding $(n-p)$-form, ${}^*T$, called the dual form. The dual form is obtained by the dual mapping (not to be confused with the dual spaces to the tangent vector and tensor spaces discussed above). This mapping consists of transforming the p-form into a tensor field of rank p by the canonical identification through the metric (raising indices by the metric tensor) and letting the so-called Levi-Civita n-form, ϵ, operate on the tensor field. In Minkowski space, the metric tensor is given as $g_{\mu\nu} = \text{diag}[c^2, -1, -1, -1]$ (c is the speed of light) and the Levi-Civita form is $\epsilon = c\,dt \wedge dx \wedge dy \wedge dz$. This makes the dual mapping quite complicated in Minkowski space. In a flat three-space, the metric tensor is simply $g_{ij} = \text{diag}[1, 1, 1]$, and the Levi-Civita form is $\epsilon = dx \wedge dy \wedge dz$, which generates a simple dual mapping. The dual mapping is, up to a sign, its own inverse, because, for a p-form on an n-dimensional manifold, we have

$${}^{**}T = \frac{g}{|g|}(-1)^{p(n-p)}T, \tag{A-4}$$

where g is the determinant of the metric tensor and $g/|g|$ is $+1$ for a Euclidean metric and -1 for a Minkowski metric.

The exterior derivative of the p-form T on an n-dimensional manifold is a $(p+1)$-form S, obtained by taking the partial derivatives of the cotensor field coefficients and antisymmetrizing in the indices (the antisymmetrization is necessary to maintain covariance under coordinate transformations). In a basis of wedge products of one-forms, the exterior derivative of T is

$$S = dT = \partial_{x_k} T_{i_1,\ldots,i_p} dx^k \wedge dx^{i_1} \wedge \ldots \wedge dx^{i_p}. \tag{A-5}$$

Antisymmetry in the indices and commutativity of partial derivatives forces two successive applications of the exterior derivative to give zero. This is a fundamental property of the exterior derivative

$$d^2 = 0. \tag{A-6}$$

Antisymmetry also makes the exterior derivative of an n-form on an n-dimensional manifold vanish.

The codifferential δ is defined by combining the dual mapping and the exterior derivative. The codifferential of a p-form T on an n-dimensional manifold is the $(p-1)$-form

$$\delta T = \frac{g}{|g|}(-1)^{p(n-p+1)}\,{}^*d^*T. \tag{A-7}$$

As in Eq. (A4), $g/|g|$ is $+1$ for a Euclidean metric and -1 for a Minkowski metric. Equations (A4), (A6), and (A7) show that the codifferential inherits the property $\delta^2 = 0$ from the exterior derivative. Nontrivial second derivatives, however, are given by the operations δd and $d\delta$.

Following are the basic properties of differential forms in three-space, including their dual forms and exterior derivatives:

- A zero-form ϕ is a scalar field:

$$\phi; \quad d\phi = \partial_x\phi\, dx + \partial_y\phi\, dy + \partial_z\phi\, dz; \quad {}^*\phi = \phi\, dx \wedge dy \wedge dz; \tag{A-8}$$

$$d\,{}^*\phi = 0; \quad {}^*\phi \wedge \phi = \phi^2 dx \wedge dy \wedge dz. \tag{A-9}$$

- A one-form A is a vector field ($\mathbf{A}$):

$$A = A_x\, dx + A_y\, dy + A_z\, dz; \tag{A-10}$$

$$dA = (\nabla\times\mathbf{A})_x\, dy \wedge dz - (\nabla\times\mathbf{A})_y\, dx \wedge dz + (\nabla\times\mathbf{A})_z\, dx \wedge dy; \tag{A-11}$$

$$\begin{aligned} {}^*A &= A_x\, dy \wedge dz - A_y\, dx \wedge dz + A_z\, dx \wedge dy; \\ d\,{}^*A &= (\nabla \cdot \mathbf{A})\, dx \wedge dy \wedge dz; \end{aligned} \tag{A-12}$$

$${}^*A \wedge A = \left(A_x^2 + A_y^2 + A_z^2\right) dx \wedge dy \wedge dz. \tag{A-13}$$

- A two-form B is an antisymmetric cotensor of rank 2 or a pseudovector field ($\mathbf{B}$):

$$B = B_x\, dy \wedge dz - B_y\, dx \wedge dz + B_z\, dx \wedge dy;$$

$$dB = (\nabla \cdot \mathbf{B})\, dx \wedge dy \wedge dz; \tag{A-14}$$

$${}^*B = B_x\, dx + B_y\, dy + B_z\, dz; \tag{A-15}$$

$$d\,{}^*B = (\nabla \times \mathbf{B})_x\, dy \wedge dz - (\nabla \times \mathbf{B})_y\, dx \wedge dz + (\nabla \times \mathbf{B})_z\, dx \wedge dy; \tag{A-16}$$

$${}^*B \wedge B = \left(B_x^2 + B_y^2 + B_z^2\right) dx \wedge dy \wedge dz. \tag{A-17}$$

- A three-form C is an antisymmetric cotensor of rank 3 or a pseudoscalar field:

$$C = c\, dx \wedge dy \wedge dz; \quad dC = 0; \quad {}^*C = c; \tag{A-18}$$

$$d\,{}^*C = \partial_x c\, dx + \partial_y c\, dy + \partial_z c\, dz; \quad {}^*C \wedge C = c^2\, dx \wedge dy \wedge dz. \tag{A-19}$$

- The basic one-, two-, and three-forms

$$dx^i, \quad dx^i \wedge dx^j, \quad dx \wedge dy \wedge dz, \tag{A-20}$$

can be interpreted as line, surface, and volume elements, respectively. The integral of a p-form T over a p-dimensional domain Ω,

$$\int_\Omega T, \tag{A-21}$$

is a line integral for $p=1$, a surface integral for $p=2$, and a volume integral for $p=3$. When $T=dS$, this becomes the fundamental theorem of integration (generalized Stokes theorem)

$$\int_\Omega dS = \int_{\partial\Omega} S \tag{A-22}$$

where $\partial\Omega$ is the oriented boundary of the integration domain Ω. This generalizes the integral of the derivative of a function for $p=1$, Stokes theorem for $p=2$, and Gauss's theorem for $p=3$.

- There is a natural inner product for differential forms defined through the so-called Hilbert product. If S and T are p-forms, then their inner (Hilbert) product is defined as

$$\langle S \mid T \rangle = \int ({}^*S) \wedge T \ . \tag{A-23}$$

Since S and T are of the same rank, $({}^*S) \wedge T$ is always a three-form and the integral is a volume integral. The integrals above have an obvious generalization to an n-dimensional differentiable manifold.

3-D Conductivity Models: Implications of Electrical Anisotropy

Peter Weidelt

Summary. Electrical anisotropy in the Earth is usually a scale effect, created by averaging over structures with a preferred orientation. Homogenization theory applied to typical geological structures can help to determine the range of values to be expected in electrically anisotropic formations. The staggered-grid finite-difference method for Maxwell's equations can easily accommodate anisotropic regions, without a significant increase in computational load. A simple magnetotelluric model shows that, in elongated anisotropic structures, the tipper is no longer perpendicular to the strike direction; thus, the electric and magnetic polarizations are mixed.

1 Introduction

Electrical anisotropy has been ignored for a long time by the geoelectromagnetics community because problems with variable *isotropic* electrical conductivity were already abundant. In many examples, however, only *anisotropic* conductors allow a consistent interpretation of electromagnetic (EM) induction data, e.g., when there is considerable deviation between magnetic and electrical preferred directions (Schmucker, 1994). Electrical anisotropy offers new degrees of freedom, which (used with care) should allow a better interpretation of data.

With harmonic time dependence $e^{i\omega t}$ and without displacement currents, Maxwell's equations in anisotropic media are

$$\nabla \times \mathbf{E} = -i\omega\mathbf{B}, \qquad \nabla \times \mathbf{B} = \mu_0\mathbf{J}, \qquad \mathbf{J} = \underset{\sim}{\sigma}\mathbf{E}. \tag{1}$$

Here **E**, **B**, and **J** denote, respectively, the vectors of the electric field, magnetic field, and current density, whereas $\underset{\sim}{\sigma}$ is the 3×3 conductivity tensor. Eliminating **B** and **J**, we obtain

$$\nabla \times \nabla \times \mathbf{E} + i\omega\mu_0\underset{\sim}{\sigma}\mathbf{E} = 0. \tag{2}$$

Institute of Geophysics and Meteorology, Technical University of Braunschweig, D-38106 Braunschweig, Germany.

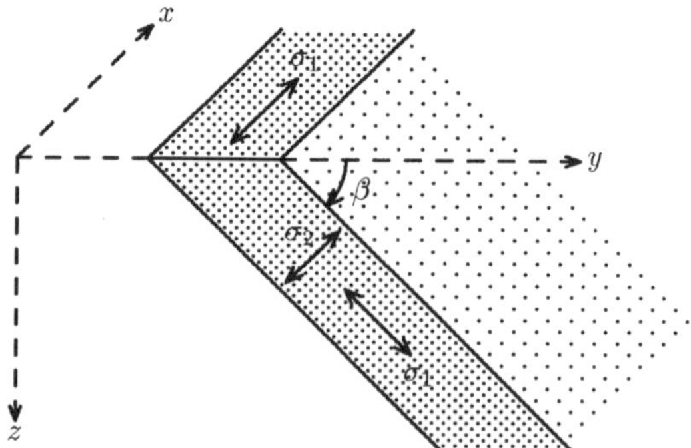

Figure 1. Example of a conductivity distribution with dipping anisotropy.

I focus on magnetotelluric (MT) problems, where the quasi-uniform incident field source is incorporated by inhomogeneous boundary conditions.

The conductivity tensor $\underset{\sim}{\sigma}$ has the following properties:

1. **Symmetry.** The conductivity tensor is symmetric whenever the magnetic field does not play a role in the conduction process (Onsager, 1931). In the presence of Hall currents, as in a plasma, $\underset{\sim}{\sigma}$ is nonsymmetric. Therefore symmetry is granted for the purely ohmic conduction considered here.
2. **Nonnegativity.** The conductivity tensor $\underset{\sim}{\sigma}$ has to be positive semidefinite because the time-averaged specific energy dissipation, $(\frac{1}{2})\mathbf{E}^* \cdot \mathbf{J} = (\frac{1}{2})\mathbf{E}^* \cdot \underset{\sim}{\sigma} \cdot \mathbf{E}$, is nonnegative. Here, the asterisk denotes complex conjugation.

In Cartesian coordinates (x, y, z), z positive downward, $\underset{\sim}{\sigma}$ has the representation

$$\underset{\sim}{\sigma} = \begin{pmatrix} \sigma_{xx} & \sigma_{xy} & \sigma_{xz} \\ \sigma_{xy} & \sigma_{yy} & \sigma_{yz} \\ \sigma_{xz} & \sigma_{yz} & \sigma_{zz} \end{pmatrix}. \tag{3}$$

In the air (half-space $z < 0$) all tensor elements vanish; in the conductor $z \geq 0$ the tensor is assumed to be positive definite. Necessary and sufficient conditions for this property are the positivity of the three major determinants. Interpretation of the off-diagonal elements of $\underset{\sim}{\sigma}$ is obvious: For instance, σ_{xy} can denote a deviation of the direction of regional strike from the horizontal axes of anisotropy; dipping anisotropy is described by σ_{xz} and σ_{yz}. As an example, we consider a dipping slab with σ_1 parallel and σ_2 perpendicular to the foliation. Then, for points inside the slab (cf. Fig. 1),

$$\underset{\sim}{\sigma} = \begin{pmatrix} \sigma_1 & 0 & 0 \\ 0 & \sigma_1 \cos^2\beta + \sigma_2 \sin^2\beta & (\sigma_1 - \sigma_2)\sin\beta\cos\beta \\ 0 & (\sigma_1 - \sigma_2)\sin\beta\cos\beta & \sigma_1 \sin^2\beta + \sigma_2 \cos^2\beta \end{pmatrix}.$$

Outside the slab, the conductivity may be isotropic.

1.1 *Origin and ranges of electrical anisotropy*

The anisotropy of electrical conductivity is essentially a scale effect: Even if the conductivity is isotropic on the microscale, it will become anisotropic on a larger scale if, in the averaging volume, preferred orientations (e.g., layering or lamination) exist. An example is given in Fig. 2.

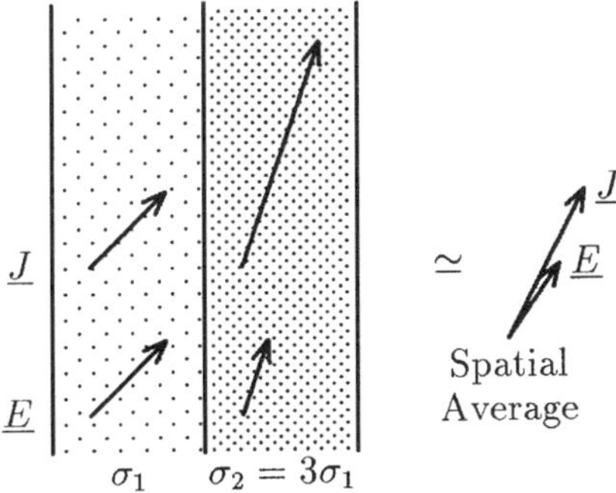

Figure 2. A simple example for the origin of structural anisotropy from a spatial average of **E** and **J** over isotropic conductors with a preferred orientation. The averaged current density is deflected toward the preferential direction.

In both isotropic lamina, electric field **E** and current density **J** are parallel, and the tangential electric field and the current density normal to the interface are continuous at the interface. After spatial averaging over both lamina, however, **E** and **J** are no longer aligned.

On this larger scale, therefore, exists some *structural* anisotropy. It is distinguished from the *intrinsic* anisotropy, caused by ordered inhomogeneities, which may exist already on a still smaller scale. Therefore, the assumption of (structural) anisotropy reflects our inability or reluctance to model adequately the structure on the microscale.

The degree of anisotropy depends on the electrical connections inside the rocks. However, some general bounds are provided by homogenization theory (e.g., Jikov et al., 1994). Although strictly applicable only to direct currents, they give guidelines also for fields varying slowly with time. Let $\sigma_-(>0)$ and σ_+ be lower and upper bounds, respectively, of the isotropic microscale conductivity $\sigma(\mathbf{r})$, $0 < \sigma_- \leq \sigma(\mathbf{r}) \leq \sigma_+$. It is assumed for simplicity that $\sigma(\mathbf{r})$ is periodically repeated, the period cell being an n-dimensional parallelepiped (here $n = 3$) with edges along the Cartesian unit vectors $\hat{\mathbf{e}}_i$, $i = 1, \ldots, n$. Let $\langle\cdot\rangle$ denote the spatial average over this cell. To the cell, we apply external unit electric fields $\hat{\mathbf{e}}_i$ and define the *homogenized* conductivity tensor $\underset{\sim}{\boldsymbol{\sigma}}^0$ as

$$\underset{\sim}{\boldsymbol{\sigma}}^0\hat{\mathbf{e}}_i := \langle\sigma(\mathbf{r})[\hat{\mathbf{e}}_i + \nabla u_i(\mathbf{r})]\rangle\,, \qquad i = 1, \ldots, n, \tag{4}$$

where $\nabla u_i(\mathbf{r})$ is the perturbation of the external field due to charge accumulations, satisfying

$$\nabla \cdot \{\sigma(\mathbf{r})[\hat{\mathbf{e}}_i + \nabla u_i(\mathbf{r})]\} = 0$$

with periodic boundary conditions. Homogenization theory then shows that $\underset{\sim}{\boldsymbol{\sigma}}^0$ is symmetric and positive definite. The latter means geometrically that the averaged current density always subtends an acute angle with the impressed electric field. Let σ_i, $i = 1, \ldots, n$, be the n positive eigenvalues (=principal conductivities) of $\underset{\sim}{\boldsymbol{\sigma}}^0$ and let $\mathcal{A}(\sigma_i)$ and $\mathcal{H}(\sigma_i)$ be their arithmetic and harmonic means, respectively. If $\sigma(\mathbf{r})$ is not a constant, the function

$$G(\alpha) := -\alpha + \left\langle \frac{1}{\sigma + \alpha} \right\rangle^{-1} \tag{5}$$

is monotonically increasing, with $G(0) = \langle 1/\sigma \rangle^{-1}$ and $G(\infty) = \langle \sigma \rangle$ as limits. (The latter is obtained by first expanding the second term of Eq. (5) in terms of σ/α.) If the conducting material is layered, $\langle 1/\sigma \rangle^{-1}$ is the longitudinal conductivity (current flow parallel to $\nabla\sigma$) and $\langle \sigma \rangle$ is the conductivity in the $(n-1)$ transverse directions (current flow perpendicular to $\nabla\sigma$). Only the longitudinal conductivity exists for $n = 1$, where the direction of current flow has to coincide with the direction of layering. (In the other directions, **E** and $\nabla\sigma$ vanish.)

Let $\alpha_{\pm} := (n-1)\,\sigma_{\pm}$. Admitting for competition all possible electrical connections within the period cell, the best possible bounds for the principal conductivities of $\underset{\sim}{\boldsymbol{\sigma}}^0$ are derived by the elegant methods of homogenization theory (Jikov et al., 1994, pp. 187–198). Correcting a few obvious misprints, these bounds can be expressed in compact form as

$$G(0) \le \sigma_i \le G(\infty), \qquad G(\alpha_-) \le \mathcal{H}(\sigma_i) \le \mathcal{A}(\sigma_i) \le G(\alpha_+) \tag{6}$$

and

$$\frac{1}{n}\sum_{i=1}^{n} \frac{1}{\sigma_i - \sigma_-} \le \frac{1}{G(\alpha_-) - \sigma_-}, \qquad \frac{1}{n}\sum_{i=1}^{n} \frac{1}{\sigma_+ - \sigma_i} \le \frac{1}{\sigma_+ - G(\alpha_+)}. \tag{7}$$

As an illustration of Eqs. (6) and (7), consider a two-component conductor with conductivities σ_- and σ_+ and respective volume fractions p_- and p_+, $p_- + p_+ = 1$. With

$$\left\langle \frac{1}{\sigma + \alpha} \right\rangle = \frac{p_-}{\sigma_- + \alpha} + \frac{p_+}{\sigma_+ + \alpha}$$

and

$$s_- := \frac{p_+ \sigma_-}{\sigma_-/(\sigma_+ - \sigma_-) + p_-/n}, \qquad s_+ := \frac{p_- \sigma_+}{\sigma_+/(\sigma_+ - \sigma_-) - p_+/n}$$

then

$$\sigma_- + s_- \le \mathcal{H}(\sigma_i) \le \mathcal{A}(\sigma_i) \le \sigma_+ - s_+ \tag{8}$$

and

$$\frac{1}{n}\sum_{i=1}^{n} \frac{1}{\sigma_i - \sigma_-} \le \frac{1}{s_-}, \qquad \frac{1}{n}\sum_{i=1}^{n} \frac{1}{\sigma_+ - \sigma_i} \le \frac{1}{s_+}. \tag{9}$$

For increasing n, the bounds for the average principal conductivities shift to higher values because of the increased chance of current flow in the better-conducting $(n-1)$ transverse directions and thus the bounds become tighter [and in the formal limit $n \to \infty$, both bounds approach $G(\infty) = \langle \sigma \rangle$].

The geometric content of the 2-D version ($n = 2$) of Eqs. (8) and (9) is displayed in Fig. 3 for a special choice of parameters. The possible values of σ_1 and σ_2 lie in the region bounded by the two segments $\overline{ABC}$ and $\overline{CDA}$, which correspond, respectively, to the first and second inequality of Eq. (9). Individual principal conductivities range from $G(0) = \langle 1/\sigma \rangle^{-1}$ to $G(\infty) = \langle \sigma \rangle$. The isolines of constant arithmetic mean $\mathcal{A}(\sigma_i) = (\sigma_1 + \sigma_2)/2$ are perpendicular to the line $\overline{BD}$. The smallest arithmetic (and harmonic) mean, $\sigma_- + s_-$, is attained at B; the largest, $\sigma_+ - s_+$, at D. Both points lie on the line $\sigma_1 = \sigma_2$. Therefore the extremal average principal conductivities are realized for conductors, which are *isotropic* on the macroscale. On the microscale, these conductors consist of self-similar two-component circles that densely fill the 2-D period cell (Hashin and

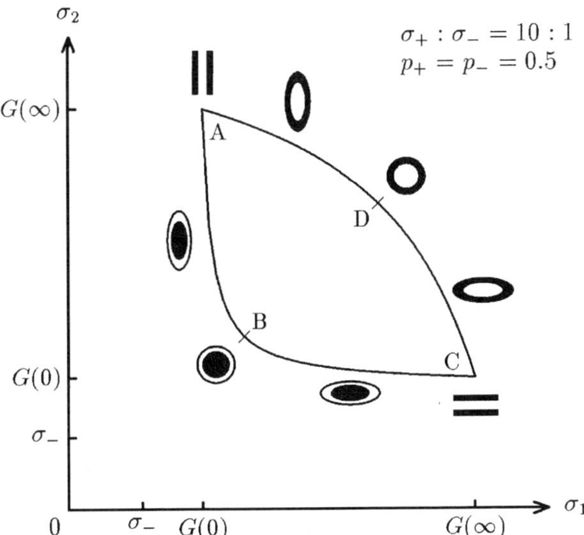

Figure 3. Ranges of principal conductivities of a 2-D two-component conductor. The boundary and the interior of the domain $\overline{ABCDA}$ can be reached. The macroscopic models are isotropic along $\overline{BD}$, from which point anisotropy increases toward A and C. The boundary models (except A and C) are attained by filling the period cell densely with the sketched two-component ellipses with fixed axes ratio and orientation. Black parts are filled with σ_+, white parts with σ_-.

Shtrikman, 1962). The lower (upper) bound is attained for the cell filled with coated circles of core conductivity σ_+ (σ_-) and coat conductivity σ_- (σ_+). The lines parallel to $\overline{BD}$ are lines of constant macroscopic anisotropy $|\sigma_2 - \sigma_1|$. The greatest anisotropy is reached at A and C, where the corresponding microstructure is a one-dimensional layering. Each point of the four segments $\overline{AB}$, $\overline{BC}$, etc. can be reached by filling the cell densely with self-similar ellipses with the orientation shown in Fig. 3. Each point of the segments is associated with a fixed axes ratio. The two-component circles and ellipses have the property that the disturbing electric fields resulting from charge accumulations at the two interfaces are cancelled outside the heterogeneous isotropic conductors when immersed into the macroscopic (anisotropic) background conductivity. In the more interesting 3-D geometry ($n = 3$), circles and ellipses are replaced by spheres and ellipsoids.

2 Implementation of a finite-difference staggered-grid algorithm

Use of staggered grids has revived finite-difference (FD) methods in EM modeling (Druskin and Knizhnerman, 1988; Madden and Mackie, 1989; Smith, 1992, 1996a,b; Wang and Hohmann, 1993; Newman and Alumbaugh, 1995). Staggered grids give a satisfactory discretization of the underlying second-order vector equation, because they automatically yield source-free electric current and magnetic-flux densities; are physically intuitive in view of the integral formulation of Maxwell's equations; lead to a discretization that does not depend on the formulation of the differential equation in terms of **E** or **B**.

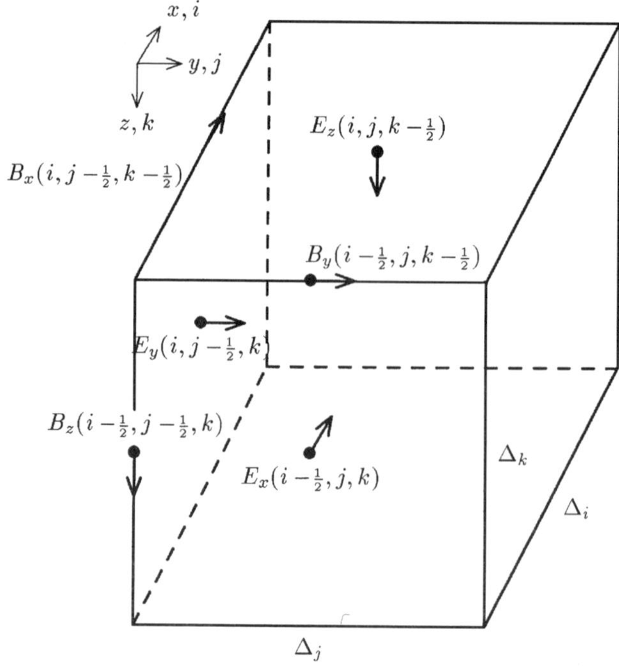

Figure 4. Locations of the EM field components in the unit cell of a staggered grid.

A rectangular product grid, with cell sizes Δ_{xi}, Δ_{yj}, and Δ_{zk} in the x-, y-, and z-directions respectively, is assumed. For convenience, these cell sizes are *abbreviated* as Δ_i, Δ_j, and Δ_k. Moreover, for simplicity the sampling point (x_i, y_j, z_k) associated with a cell is identified with its geometrical center (although, in an irregular grid, offset positions of the sampling point will lead to higher accuracy), and we let

$$\Delta_{i-\frac{1}{2}} := (x_{i-1} - x_i)/2 = (\Delta_{i-1} + \Delta_i)/2, \quad \text{etc.}$$

The center of each cell inside the Earth is assigned the full six-component conductivity tensor (3), which has to be symmetric and positive definite. Figure 4 shows a possible assignment of the EM field components associated with a cell: Magnetic-field components are assigned to the centers of the edges and electric-field components to the centers of the faces (parallel to the normal), where they are, in general, discontinuous. The complementary assignment with electric- and magnetic-field components interchanged also is possible. This assignment has the apparent advantage that the components of $\mathbf{E}$ are localized at positions where they are continuous as tangential components. However, the discontinuities of the first assignment can be handled easily; see Eq. (11) as an example. Because the EM field components are localized at different faces and edges of the cell, the resulting EM field is clearly the average over the cell.

An example for an FD staggered-grid approximation (denoted by a tilde) is

$$\hat{\mathbf{z}} \cdot \tilde{\nabla} \times \mathbf{B} := \frac{B_y\left(i+\frac{1}{2}, j, k-\frac{1}{2}\right) - B_y\left(i-\frac{1}{2}, j, k-\frac{1}{2}\right)}{\Delta_j} - \frac{B_x\left(i, j+\frac{1}{2}, k-\frac{1}{2}\right) - B_x\left(i, j-\frac{1}{2}, k-\frac{1}{2}\right)}{\Delta_j} = \mu_0 J_z\left(i, j, k-\frac{1}{2}\right).$$

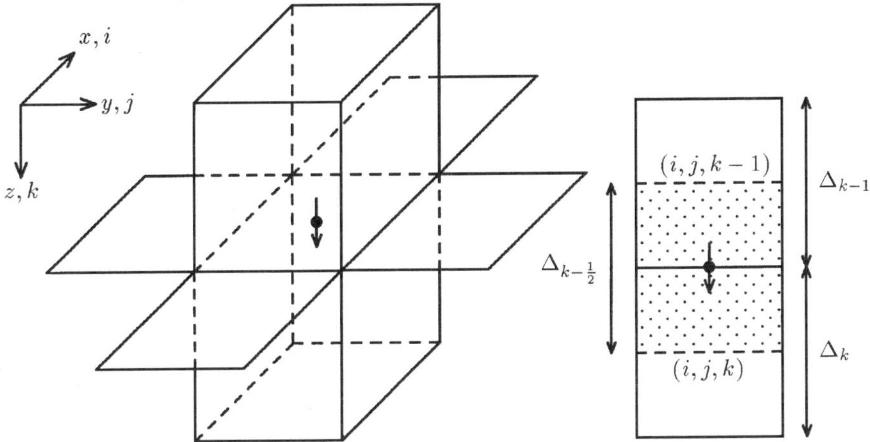

Figure 5. Topology of the $\nabla \times \nabla\times$-operator: The vertical component E_z (denoted by an arrow) is connected with the four vertical components localized at the centers of the adjacent faces in the plane of that component and with the eight horizontal components localized at the centers of vertical faces of the upper and lower prisms. The right figure shows the averaging volume of E_z.

We generally have

$$\mu_0 \tilde{\nabla} \cdot \mathbf{J} = \tilde{\nabla} \cdot \tilde{\nabla} \times \mathbf{B} = 0, \qquad -i\omega \tilde{\nabla} \cdot \mathbf{B} = \tilde{\nabla} \cdot \tilde{\nabla} \times \mathbf{E} = 0.$$

The modeling problem is formulated in terms of the electric field. After defining the grid, the system of linear equations for the electric-field components normal to the faces of the cells is established as follows: Let us consider $E_z(i, j, k - \frac{1}{2})$. The staggered-grid version of the z-component of $\nabla \times \nabla\times$ [left-hand side of Eq. (2)] combines 13 electric-field components. In addition to $E_z(i, j, k - \frac{1}{2})$, these are (cf. Fig. 5, left panel) the four horizontal components of the upper cell $(i, j, k - 1)$, the four horizontal components of the lower cell; (i, j, k), and the four vertical components in the plane $k - \frac{1}{2}$. For an explicit FD approximation, the z-component of Eq. (2) is integrated over the volume between the centers of the cells $(i, j, k - 1)$ and (i, j, k) (cf. Fig. 5, right panel). Whereas J_z is continuous at the interface between $(i, j, k - 1)$ and (i, j, k), the electric-field component E_z is, in general, discontinuous there. By integrating

$$\hat{\mathbf{z}} \cdot \nabla \times \nabla \times \mathbf{E} = -\left(\partial^2_{xx} + \partial^2_{yy}\right) E_z + \partial_z(\partial_x E_x + \partial_y E_y) \tag{10}$$

first over z, it is seen that $E_z(i, j, k - \frac{1}{2})$ has to be interpreted as the volume average in the averaging volume between the centers of cell $(i, j, k - 1)$ and cell (i, j, k). Correspondingly, e.g., $E_x(i - \frac{1}{2}, j, k)$ has to be considered as the volume average of E_x in the volume between the centers of cell $(i - 1, j, k)$ and (i, j, k). Also, E_x as a normal electric-field component is, in general, discontinuous across the interface. The integration of Eq. (10) in x-direction requires the (possibly discontinuous) value of E_x at the x-interface, which is approximated by the volume average of E_x between the centers of the adjacent cells, i.e., by $E_x(i - \frac{1}{2}, j, k)$. Corresponding results hold for E_y.

Therefore, the staggered-grid approximation of Eq. (2) is in a natural way expressed in terms of volume-averaged electric-field components. The upper part of Fig. 6 gives the explicit representation of the z-component of $\tilde{\nabla} \times \tilde{\nabla}\times$, centered at $(i, j, k - \frac{1}{2})$. The dashed frames correspond to the upper and the lower cells; the solid frame corresponds to the plane $k - \frac{1}{2}$, which results from the first right-hand-side term of Eq. (10).

$$i\omega\mu_0 J_z\left(i, j, k-\tfrac{1}{2}\right)\Delta_{k-\frac{1}{2}}\Delta_i\Delta_j =$$

$$+\left[E_x\left(i+\tfrac{1}{2}, j, k-1\right) - E_x\left(i-\tfrac{1}{2}, j, k-1\right)\right]\Delta_j$$
$$+\left[E_y\left(i, j+\tfrac{1}{2}, k-1\right) - E_y\left(i, j-\tfrac{1}{2}, k-1\right)\right]\Delta_i$$

$$-\left[E_x\left(i+\tfrac{1}{2}, j, k\right) - E_x\left(i-\tfrac{1}{2}, j, k\right)\right]\Delta_j$$
$$-\left[E_y\left(i, j+\tfrac{1}{2}, k\right) - E_y\left(i, j-\tfrac{1}{2}, k\right)\right]\Delta_i$$

$$-\left[E_z\left(i, j, k-\tfrac{1}{2}\right) - E_z\left(i-1, j, k-\tfrac{1}{2}\right)\right]\Delta_j\Delta_{k-\frac{1}{2}}/\Delta_{i-\frac{1}{2}}$$
$$-\left[E_z\left(i, j, k-\tfrac{1}{2}\right) - E_z\left(i, j-1, k-\tfrac{1}{2}\right)\right]\Delta_i\Delta_{k-\frac{1}{2}}/\Delta_{j-\frac{1}{2}}$$
$$+\left[E_z\left(i+1, j, k-\tfrac{1}{2}\right) - E_z\left(i, j, k-\tfrac{1}{2}\right)\right]\Delta_j\Delta_{k-\frac{1}{2}}/\Delta_{i+\frac{1}{2}}$$
$$+\left[E_z\left(i, j+1, k-\tfrac{1}{2}\right) - E_z\left(i, j, k-\tfrac{1}{2}\right)\right]\Delta_i\Delta_{k-\frac{1}{2}}/\Delta_{j+\frac{1}{2}}$$

- 13-point formula for $k-\frac{1}{2}$ in the conducting half-space
- 10-point formula for $k-\frac{1}{2}$ at the air–earth interface:

$$[\;\;] = -\left[E_z\left(i, j, k-\tfrac{1}{2}\right) - E_z\left(i, j, k-\tfrac{3}{2}\right)\right]\Delta_i\Delta_j/\Delta_{k-1}$$

- 7-point formula for $k-\frac{1}{2}$ in the air half-space ($=\tilde{\nabla}^2 E_z = 0$):

$$[\;\;] = +\left[E_z\left(i, j, k+\tfrac{1}{2}\right) - E_z\left(i, j, k-\tfrac{1}{2}\right)\right]\Delta_i\Delta_j/\Delta_k$$

Figure 6. Discrete version of z-component of Eq. (2). The framed sections of the upper part contain, respectively, the contributions from the upper and the lower prism and from the central plane of Fig. 5. For the position of E_z in Fig. 5 on the air–earth interface or in the air half-space, this 13-point formula reduces, with the replacements given at the bottom, to a formula connecting only 10 or 7 points.

It remains to express $J_z(i, j, k-\frac{1}{2})$ in terms of the averaged electric-field components. To distinguish in what follows between the averaged field quantity and the true one, the subscript t is added to the latter. The averaging process is denoted by overlining, e.g., $E_z = \overline{E_{zt}}$. From

$$J_{zt} = \sigma_{xz}E_{xt} + \sigma_{yz}E_{yt} + \sigma_{zz}E_{zt}$$

after division by σ_{zz}, averaging, and division by $\overline{1/\sigma_{zz}}$, follows

$$\begin{aligned} J_z &\simeq \{\overline{(\sigma_{xz}/\sigma_{zz})E_{xt}} + \overline{(\sigma_{yz}/\sigma_{zz})E_{yt}} + E_z\}/\overline{(1/\sigma_{zz})} \\ &\simeq \{\overline{(\sigma_{xz}/\sigma_{zz})}\;\overline{E_{xt}} + \overline{(\sigma_{yz}/\sigma_{zz})}\;\overline{E_{yt}} + E_z\}/\overline{(1/\sigma_{zz})}. \end{aligned}$$

It is assumed that E_{xt}, E_{yt}, and J_{zt} as tangential electric fields and normal current density are only slowly variable in the averaging volume (in comparison to E_{zt}). Therefore, in particular, $J_{zt} \simeq J_z$. Finally, $\overline{E_{xt}}$ and $\overline{E_{yt}}$ have to be expressed by averaged field components. First, $\overline{E_{xt}}$ is approximated by the four E_x components defined on the faces of the averaging volume,

$$\begin{aligned} \overline{E_{xt}} = \frac{1}{2(\Delta_{k-1}+\Delta_k)}&\left\{\Delta_{k-1}E_{xt}\left(i-\tfrac{1}{2}+0, j, k-1\right) + \Delta_k E_{xt}\left(i-\tfrac{1}{2}+0, j, k\right)\right. \\ &\left.+\Delta_{k-1}E_{xt}\left(i+\tfrac{1}{2}-0, j, k-1\right) + \Delta_k E_{xt}\left(i+\tfrac{1}{2}-0, j, k\right)\right\}, \end{aligned}$$

where the possible discontinuity of the true values has been taken into account. In the next step, each of the four true values is expressed in terms of the averaged values. Assuming first for simplicity that the x-axis coincides with a principal axis of $\underset{\sim}{\sigma}$, $J_{xt} = \sigma_{xx} E_{xt}$, we obtain

$$J_{xt}/\sigma_{xx} = E_{xt}, \qquad J_x \overline{(1/\sigma_{xx})} \simeq E_x,$$
$$J_x \simeq E_x/\overline{(1/\sigma_{xx})} \simeq \sigma_{xx}(i, j, k) E_{xt}\left(i - \tfrac{1}{2} + 0, j, k\right)$$

and therefore

$$\begin{aligned} E_{xt}\left(i - \tfrac{1}{2} + 0, j, k\right) &\simeq \frac{1}{\sigma_{xx}(i, j, k)\overline{(1/\sigma_{xx})}} E_x\left(i - \tfrac{1}{2}, j, k\right) \\ &= \frac{(\Delta_{i-1} + \Delta_i)/\sigma_{xx}(i, j, k)}{\Delta_{i-1}/\sigma_{xx}(i-1, j, k) + \Delta_i/\sigma_{xx}(i, j, k)} E_x\left(i - \tfrac{1}{2}, j, k\right), \end{aligned} \tag{11}$$

where overlining now refers to averaging between cell $(i - 1, j, k)$ and cell (i, j, k). If arbitrary orientation of principal axes is allowed, $E_x(i - \frac{1}{2}, j, k)$ is augmented by the term

$$\{\overline{(\sigma_{xy}/\sigma_{xx})} - \sigma_{xy}\overline{(1/\sigma_{xx})}\} E_{yt} + \{\overline{(\sigma_{xz}/\sigma_{xx})} - \sigma_{xz}\overline{(1/\sigma_{xx})}\} E_{zt},$$

which contributes only where $\nabla\sigma_{xy}$ or $\nabla\sigma_{xz}$ has a component in the x-direction and is, at present, ignored in the implementation.

The system of linear equations is assembled by applying the $\tilde{\nabla} \times \tilde{\nabla} \times$-operator to all interior faces of the grid cells, at which electric-field components are localized such that one linear equation with, at most, 13 nonzero entries is obtained for each unknown electric-field component. At the outer faces, Dirichlet conditions are imposed. In the MT case, these are the electric fields induced by quasi-uniform external magnetic fields in a layered anisotropic conductor. They were computed with the algorithm of Dekker and Hastie (1980).

The system matrix $\underset{\sim}{\mathbf{A}}$ has the following properties:

- The incorporation of anisotropy described above preserves the symmetry of $\underset{\sim}{\mathbf{A}}$ known from the isotropic case.
- The topological complexity of $\underset{\sim}{\mathbf{A}}$ does not increase when general anisotropic conductors rather than isotropic conductors are considered, i.e., the same 13 electric-field components, which the FD approximation connects in the isotropic case, are connected also in the general anisotropic case. Off-diagonal elements, however, now are, in general, complex. (This also occurs in the isotropic case after preconditioning.)
- $\underset{\sim}{\mathbf{A}}$ is singular if the field in the (insulating) air half-space also is treated by FDs rather than by an integral boundary condition at the air–earth interface.

2.1 Stabilization of linear system

So far, the singularity of the system matrix $\underset{\sim}{\mathbf{A}}$ does not appear to have found explicit attention in the pertinent literature. The probable reason is that it can be circumvented either by allowing in the air half-space the flow of small currents (ohmic currents or displacement currents) or by excluding the air half-space from modeling by applying an integral boundary condition at the air–earth interface.

Now we discuss how the singularity evolves from the assumption of a current-free air half-space and show a way to remove it. Let N be the dimension of the linear system (=number of unknown complex field components) and let the electric-field components be assembled in the complex vector $\mathbf{e} \in \mathcal{C}^N$. Moreover, let $\mathbf{b} \in \mathcal{C}^N$ be the source term (introduced by Dirichlet boundary conditions) and let $\underset{\sim}{\mathbf{A}} \in \mathcal{C}^N \times \mathcal{C}^N$ be the system matrix. Then,

$$\underset{\sim}{\mathbf{A}}\,\mathbf{e} = \mathbf{b}. \tag{12}$$

Assuming $\underset{\sim}{\sigma} = 0$ in the air half-space and neglecting displacement currents, Eq. (2) reduces to

$$\nabla \times \nabla \times \mathbf{E} = 0.$$

Because along with $\nabla \cdot \nabla \times \nabla \times \mathbf{v} = 0$ the staggered-grid approximation $\tilde{\nabla} \cdot \tilde{\nabla} \times \tilde{\nabla} \times \mathbf{v}$ also vanishes for all vectors $\mathbf{v}$, there exists for each cell k in the air half-space, $k = 1, \ldots, M$, a real grid-dependent vector $\mathbf{d}_k \in \mathcal{R}^N$ that forms the staggered-grid divergence for this cell and satisfies $\mathbf{d}_k^T \underset{\sim}{\mathbf{A}} = 0$, where the superscript T denotes transposition. Vector $\mathbf{d}_k$ has only six nonzero entries at the electric-field components corresponding to the faces of cell k. Therefore, the six equations referring to the six faces of cell k are *linearly dependent*, such that the rank of $\underset{\sim}{\mathbf{A}}$ is only $N - M$. Because of this lack of information, $\mathbf{e}$ cannot be obtained uniquely. Invoking the symmetry of the system matrix, $\mathbf{d}_k^T \underset{\sim}{\mathbf{A}} = 0$ implies $\underset{\sim}{\mathbf{A}}\,\mathbf{d}_k = 0$, i.e., $\mathbf{d}_k$ belongs to the null space of $\underset{\sim}{\mathbf{A}}$ such that, along with any solution $\mathbf{e}$ of Eq. (12),

$$\mathbf{e} + \sum_{k=1}^{M} \alpha_k \, \mathbf{d}_k$$

is also a solution, where α_k are arbitrary complex numbers. However, these numbers can be determined by the physical condition that no charges accumulate in the air, and therefore $\nabla \cdot \mathbf{E} = 0$. The enforcement of this condition for each cell ℓ in the air half-space leads to the linear system

$$\sum_{k=1}^{M} \alpha_k \left(\mathbf{d}_\ell^T \mathbf{d}_k\right) = -\mathbf{d}_\ell^T \mathbf{e}, \qquad \ell = 1, \ldots, M.$$

This nonsingular system allows a unique determination of α_k, and thus a unique determination of the electric field.

In practice, a source-free electric field in the air half-space is enforced already when assembling the system matrix. This is outlined at the bottom of Fig. 6: If the level $(i, j, k - \frac{1}{2})$ coincides with the air–earth interface, the contribution from the upper cell (short dashes) is replaced (on account of $\nabla \cdot \mathbf{E} = 0$ for this cell) by the field components $E_z(i, j, k - \frac{3}{2})$ and $E_z(i, j, k - \frac{1}{2})$, thus reducing the 13-point formula to a 10-point formula. If $(i, j, k - \frac{1}{2})$ lies in the air half-space, the contribution of the lower cell (long dashes) also is replaced by $E_z(i, j, k - \frac{1}{2})$ and $E_z(i, j, k + \frac{1}{2})$, implying a further reduction of the 10-point formula to a 7-point formula, which is just the FD approximation of $\nabla^2 E_z = 0$ integrated over the averaging volume. A 10-point hybrid formula also is used at all other boundary cells of the air half-space. All of these replacements preserve the symmetry of $\underset{\sim}{\mathbf{A}}$.

The stabilized linear system of equations with the sparse symmetric complex matrix $\underset{\sim}{\mathbf{A}}$ is solved by conjugate gradient methods. So far, only the simple but efficient equilibration (Jacobi scaling) is used as preconditioner. Because $\underset{\sim}{A}$ is indefinite, the complex version of the ordinary conjugate gradient method (Jacobs, 1981) behaves poorly and leads to an erratic variation of the residual power. Bi-CGSTAB, a stabilized Biconjugate gradients-squared method, did not turn out to be very sucessful, but the quasi-minimal residual (QMR) method of Freund (1992) has shown a very satisfactory behavior. The QMR method minimizes at each iteration the L_2-norm of the coeffcients in an expansion of the residual vector in terms of the orthonormal Lanczos vectors and leads to a smooth, but not necessarily monotonous decrease of the residual power. Very similar experiences were reported by Newman and Alumbaugh (1995).

3 Implications of electrical anisotropy for magnetic and MT transfer functions

As a basic example for the impact of electrical anisotropy on MT transfer functions, we consider the conductivity structure presented in Fig. 7, which is believed to describe to a first approximation the situation around the German continental-depth drilling site. The anomalous conductor consists of two parts:

1. **Regional 2-D conductor.** This thin east-west–striking isotropic conductor lies at a depth of $\simeq$10 km with conductivity increasing northward. In the model of Fig. 7, the conductor is represented by its conductance (=integrated conductivity) τ, ranging approximately from 12 S to 600 S (bottom of Fig. 7).
2. **Anisotropic conductor in upper crust.** This 10-km-thick highly anisotropic conductor overlies the regional conductor and has a north-south extension of $\simeq$30 km. The horizontal axes of anisotropy differ appreciably from the axes defined for the regional conductor: The well-conducting vertical planes point

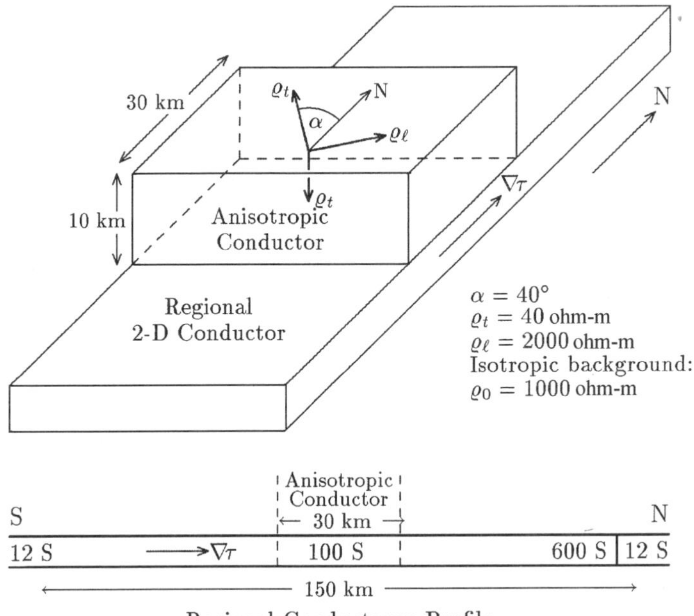

Figure 7. First model of electrical conductivity distribution near German continental-depth drilling site (after M. Eisel, private communication).

approximately in the northwest-southeast direction with a transverse resistivity of $\varrho_t = 40$ ohm-m. The longitudinal resistivity is $\varrho_\ell = 50\varrho_t$.

The conductors are embedded into a uniform background with resistivity $\varrho_0 = 1000$ ohm-m.

Before discussing the EM coupling between these conductors, we briefly introduce the conventional MT and magnetic transfer functions. Source currents in ionosphere and magnetosphere are assumed, providing a quasi-uniform inducing magnetic field. Let $\mathbf{E}_s = \mathbf{E}_s(\mathbf{r}, \omega)$ and $\mathbf{B}_s = \mathbf{B}_s(\mathbf{r}, \omega)$ be the horizontal projections of $\mathbf{E}$ and $\mathbf{B}$ at site $\mathbf{r}$. Then, the (2×2) impedance tensor $\underset{\sim}{\mathbf{Z}} = \underset{\sim}{\mathbf{Z}}(\mathbf{r}, \omega)$ is defined as

$$\mathbf{E}_s = \underset{\sim}{\mathbf{Z}}\mathbf{B}_s.$$

If $Z_{k\ell}$ is an off-diagonal element of $\underset{\sim}{\mathbf{Z}}$ for a particular choice of horizontal axes, then

$$\varrho_{ak\ell} := \frac{\mu_0}{\omega}|Z_{k\ell}|^2$$

is the corresponding *apparent* resistivity. The orientations of the horizontal coordinate system, in which $\varrho_{ak\ell}$ attains its largest (major) and smallest (minor) value, are distinguished.

The magnetic transfer function is the induction vector (tipper), $\mathbf{T}$, defined by

$$B_z = \mathbf{T} \cdot \mathbf{B}_s,$$

where $B_z = B_z(\mathbf{r}, \omega)$ is the downward component of the magnetic field. If we ignore the fact that all fields are complex-valued, vector $\mathbf{T}$ shows that direction into which $\mathbf{B}_s$ generates the greatest (positive) B_z.

$\underset{\sim}{\mathbf{Z}}$ and $\mathbf{T}$ are determined numerically by computing $\mathbf{E}_s$, $\mathbf{B}_s$, and B_z for two linearly independent polarizations of the external magnetic field.

For our first discussion we simplify the model of Fig. 7 by assuming that the anisotropic conductor is unbounded in the horizontal direction. Therefore, it can be represented by a uniform anisotropic layer. The model then degenerates into a 2-D model, where the anisotropy, however, no longer allows a separation of modes in E- and B-polarization. This simple but important class of models has already been studied carefully by Pek (1994), Schmucker (1994), and Pek and Verner (1997).

Figure 8 presents for the simplified version of Fig. 7 the real part of $\mathbf{T}$ (left panel) and the major and minor apparent resistivities (right panel). The latter are aligned along the direction of the corresponding electric field. At the beginning, we consider the first three columns of the left panel, which show (for three periods) the magnetic responses of vertically dipping well-conducting lamina as depicted in Fig. 7. The induction vectors for the long period of 1000 s point southward, i.e., a southward-directed magnetic field produces the greatest B_z. This direction also is expected from the fact that (in isotropic conductors) the induction vectors point away from the well-conducting structure. The physical explanation is given in Fig. 9 (upper panel): An inducing magnetic field pointing southward is produced by ionospheric currents flowing westward. Consequently, the induced currents in the Earth—counteracting their source—will flow eastward. Because of the conductivity increase, the current density in the northern part of the conductor will increase. This gradient then will produce a positive vertical magnetic field. Because of the small induction effect of this long period, the anisotropic overburden remains almost invisible in the induction vector (apart from a tiny deflection to the east).

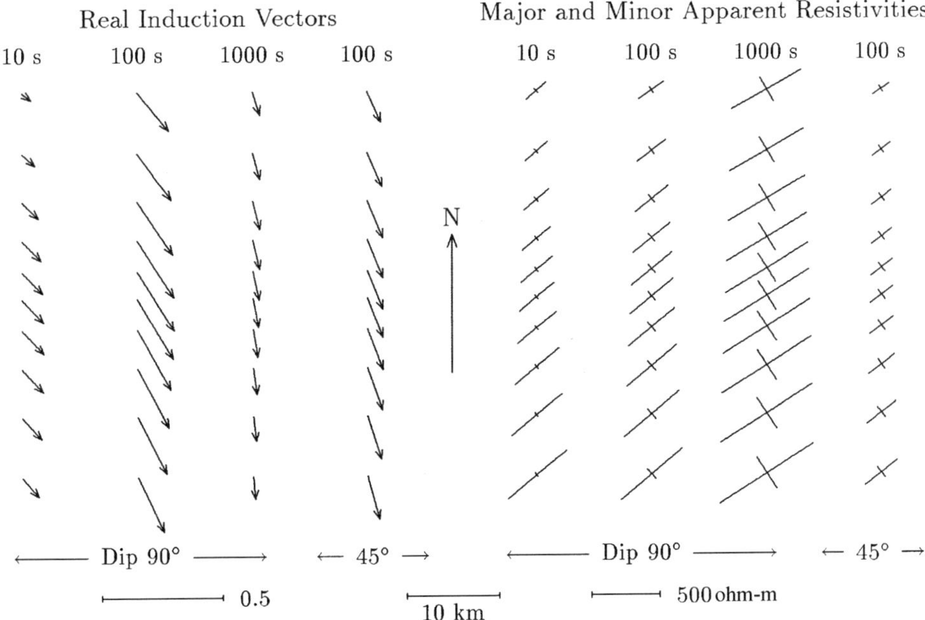

Figure 8. Magnetic and MT transfer functions for a simplified 2-D version of the model of Fig. 7. The data are given along north-south profiles (covering the central part of Fig. 7). Three periods are considered in the case in which the well-conducting lamina of the anisotropic conductor are dipping vertically (as shown in Fig. 7) and for one period in the case in which these lamina have a dip of 45° northeast.

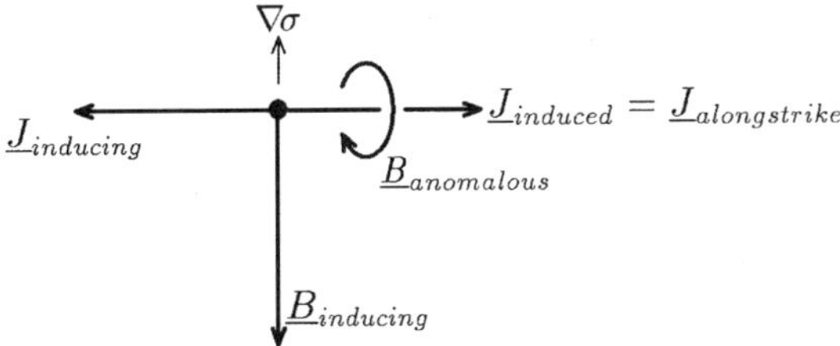

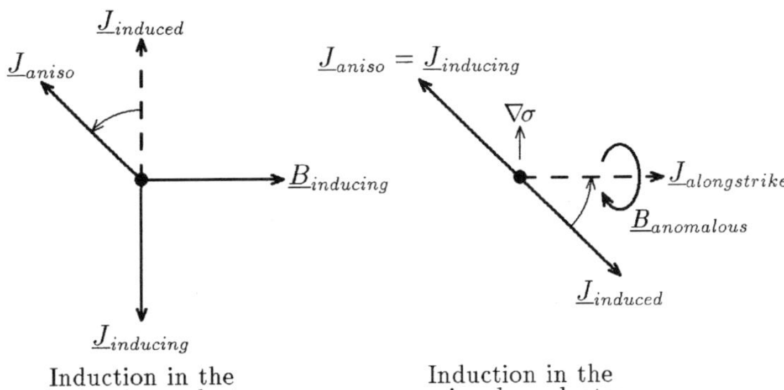

Figure 9. Generation of a vertical magnetic field by an inducing magnetic field (a) *normal* and (b) *parallel* to the strike direction of a regional conductor (plane view). In the latter case the vertical magnetic field is caused by electric-current flow in the anisotropic conductor.

This situation changes for the shorter period of 100 s. The length of the induction vector increases because of the enhanced induction in the regional anomalous conductor. Moreover, the anisotropic conductor, which alone cannot generate a vertical magnetic field, now is coupled to the heterogeneous (isotropic) regional conductor and causes an appreciable eastward deflection of the induction vector, i.e. an inducing magnetic field in the eastward direction (along the strike) now generates a positive vertical magnetic field. Figure 9 (lower panel) gives the physical explanation: The eastward inducing field has been created by southward ionospheric currents; the currents induced in the Earth therefore try to flow northward, but because of the well-conducting lamina striking northwest-southeast, the induced currents in the anisotropic conductor are deflected northwest. These currents now couple with the regional conductor, where they act as inducing currents. The induced currents then flow in a southeast direction and thus they also have an easterly component along the strike. Exactly as in Fig. 9a, the northward increase of electrical conductivity increases the density of induced easterly currents toward the north and thus causes a positive vertical magnetic field. If the period decreases to 10 s the strong coupling between anisotropic and regional conductor still exists, but the skin effect in the overburden now starts to shield the vertical magnetic field, such that the induction vectors become smaller again.

The fourth column of the left panel of Fig. 8 refers to the case in which the well-conducting lamina are dipping 45° northeast, rather than being vertical. The reduction of the dip angle diminishes the impact of anisotropy until the anisotropic conductor for a vanishing dip angle approaches a layer with the (small) horizontally isotropic resistivity ϱ_t and the (high) vertical resistivity ϱ_ℓ. The reduced deflection of induced currents results in a smaller eastern component of the induction vector, and the increased skin effect due to ϱ_t yields a stronger shielding of the regional conductor, i.e., the southward component of the induction vector also is reduced.

Now, we draw our attention to the magnetotelluric (MT) transfer functions, represented by the length and direction of major and minor apparent resistivity (right panel of Fig. 8). As a general feature, these transfer functions reflect the northwest-southeast and southwest-northeast striking directions of low and high resistivity shown in the model of Fig. 7. The resistivity anisotropy decreases with period: The longer periods sense more and more the isotropic resistivity $\varrho_0 = 1000$ ohm-m, in which both conductors are embedded. The decrease of resistivities to the north reflects the increasing conductivity of the regional conductor. The decrease of resistivities with decreasing period is due to the increasing inductive efficiency of the regional conductor, which damps the electric field in the overburden. For still smaller periods, however, the apparent resistivities again will increase until they reach the true values of ϱ_t and ϱ_ℓ, and the regional conductor will be shielded. As explained in the preceding paragraph, the apparent resistivities for a reduced dip angle of 45° (fourth column) already display the trend to horizontal *isotropy* with resistivity ϱ_t.

The most obvious implication of (horizontal) anisotropy in the 2-D example studied above is the occurrence of a component of the induction vector *parallel* to the strike direction. This is, in fact, a quite general phenomenon that occurs whenever an elongated (isotropic) conductor is embedded in an anisotropic host whose horizontal axes do not coincide with those of the elongated conductor. This means that E-polarization (electric field in strike direction) and B-polarization (magnetic field in strike direction) are no longer decoupled, as in the isotropic 2-D case. As a consequence the rotationally invariant trace of the impedance tensor will have a nonzero trace in the anisotropic 2-D

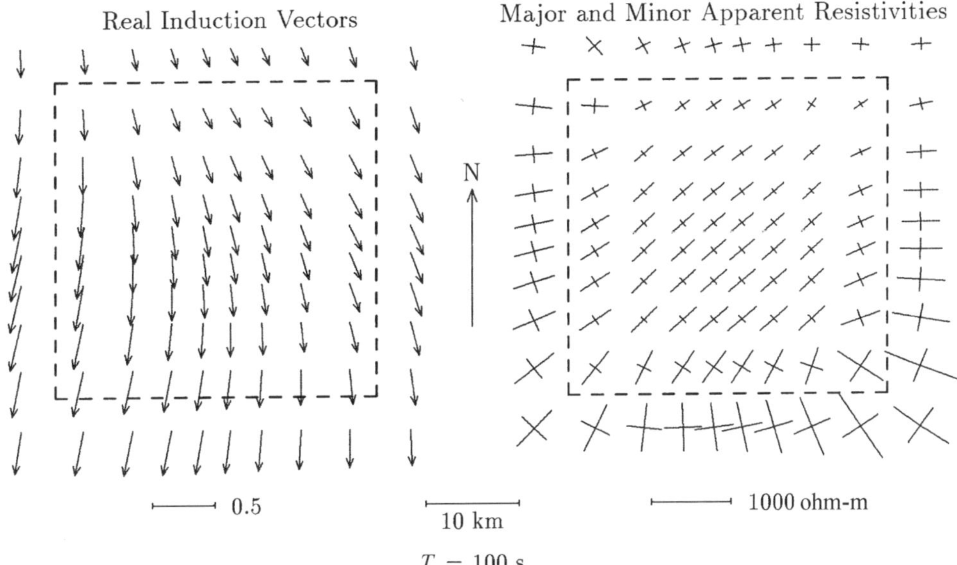

Figure 10. Magnetic and MT transfer functions for the model of Fig. 7 for a period of 100 s. The conductors are embedded into an isotropic background resistivity $\varrho_0 = 1000$ ohm-m. The additional interfaces create new perturbations of the induction vector. In the north, the damping influence of the regional conductor is still visible in the apparent resistivities; in the south the resistivities approach the isotropic background resistivity.

case, such that no coordinate system exists, in which the diagonal elements of $\underset{\sim}{\mathbf{Z}}$ vanish. The latter is characteristic of the isotropic 2-D case. If, in 2-D conductivity models with horizontal anisotropy, the anisotropic conductor is replaced by a sequence of lamina, it becomes obvious that this class of models contains, in fact, hidden 3-D models.

So far, we have treated only a simplified version of the model in Fig. 7, where the bounded anisotropic conductor was replaced by a uniform anisotropic layer. Now we return to the original model and, in Fig. 10, show the transfer functions for the period $T = 100$ s. The isotropic background resistivity of $\varrho_0 = 1000$ ohm-m is approached by the apparent resistivities in the south of the anomalous square. The influence of this isotropic host weakens the strong anomalous eastward component of the induction vector encountered in Fig. 8.

Finally it is worthwhile to observe that anisotropic conductivity models cannot explain data that cannot be interpreted by *isotropic* 3-D models of arbitrary complexity. This is possible because in EM induction studies in which the microstructure of the conduction mechanism is inaccessible, all anisotropy can be explained by *structural* anisotropy resulting from spatial averages over *isotropic* structures with a preferred orientation. The *ab initio* introduction of anisotropy might be simpler than to approach anisotropy, for instance, by isotropic sheeted dykes. Analytical properties of transfer functions that hold within a class of isotropic models of arbitrary complexity should be immediately applicable to those anisotropic models that can be assembled from elements of that class. An example is given in the Appendix.

Conclusions

An FD program is developed for modeling the EM response of 3-D conductors with arbitrary anisotropy of conductivity. The first numerical results are plausible. Compared with the isotropic case, the anisotropy can be taken into account without increasing the computational load. The vanishing conductivity of the air half-space leads to a singular system of equations, which is stabilized by enforcing the condition $\nabla \cdot \mathbf{E} = 0$ in the air half-space. As a particular facet of electrical anisotropy, in elongated structures the tipper is no longer perpendicular to the strike direction and thus E- and B-polarization are mixed.

Acknowledgment

The author is grateful to V. Druskin and M. Schoenberg for their very critical reviews, which finally have improved the manuscript.

References

Dekker, D. L., and Hastie, M. L., 1980, Magneto-telluric impedances of an anisotropic layered Earth model: Geophys. J. Roy Astr. Soc., **61**, 12–20.

Druskin, V. L., and Knizhnerman, L. A., 1988, A spectral semi-discrete method for the numerical solution of 3-D nonstationary problems in electrical prospecting: Phys. Solid Earth, **24**, 641–648.

Freund, R. W., 1992, Conjugate gradient-type methods for linear systems with complex symmetric coefficient matrices, SIAM J. Sci. Stat. Comput., **13**, 425–448.

Hashin, Z., and Shtrikman, S., 1962, A variational approach to the theory of effective magnetic permeability of multiphase materials, J. Appl. Phys., **33**, 3125–3131.

Jacobs, D. A. H., 1981, The exploitation of sparsity by iterative methods, *in* Duff, I. S. Ed., Sparse matrices and their uses: Springer-Verlag Berlin, 191–222.

Jikov, V. V., Kozlov, S. M., and Oleinik, O. A., 1994, Homogenization of differential operators and integral functionals: Springer-Verlag Berlin.

Madden, T. R., and Mackie, R. L., 1989, Three-dimensional magnetotelluric modeling and inversion: Proc. IEEE, **77**, 318–333.

Newman, G. A., and Alumbaugh, D. L., 1995, Frequency-domain modelling of airborne electromagnetic responses using staggered finite differences: Geophys. Prosp., **43**, 1021–1042.

Onsager, L., 1931, Reciprocal relations in irreversible processes: Phys. Rev., **37**, 405–426.

Pek, J., 1994, 2-D numerical modelling of magnetotelluric fields in anisotropic structures—an FD algorithm, *in* Bahr, K., and Junge, A., Eds., Protokoll Kolloq. "Elektromagnetische Tiefenforschung," Höchst im Odenwald, 28.3.-31.3.1994, Deutsche Geophys. Gesellschaft Hannover, 27–37.

Pek, J., and Verner, T., 1997, Finite difference modelling of magnetotelluric fields in 2-D anisotropic media: Geophys. J. Internat., **128**, 505–521.

Schmucker, U., 1994, 2-D Modellrechnungen zur Induktion in inhomogenen dünnen Schichten über anisotropen geschichteten Halbräumen, *in* Bahr, K., and Junge, A., Eds., Protokoll Kolloq. "Elektromagnetische Tiefenforschung," Höchst im Odenwald, 28.3.-31.3.1994, Deutsche Geophys. Gesellschaft Hannover, 3–26.

Smith, J. T., 1992, Conservative modeling of 3-D electromagnetic fields: Presented at 11th Workshop on Electromagnetic Induction in the Earth, Internat. Assn. Geomagn. Aeronomy.

———1996a, Conservative modeling of 3-D electromagnetic fields, Part I: Properties and error analysis: Geophysics, **61**, 1308–1318.
———1996b, Conservative modeling of 3-D electromagnetic fields, Part II: Biconjugate gradient solution and an accelerator: Geophysics, **61**, 1309–1324.
Wang, T., and Hohmann, G. W., 1993, A finite-difference, time-domain solution for three-dimensional electromagnetic modelling: Geophysics, **58**, 797–809.
Weidelt, P., and Kaikkonen, P., 1994, Local 1-D interpretation of magnetotelluric B-polarization impedances: Geophys. J. Internat., **117**, 733–748.

Appendix: B-polarization phase constraints for arbitrary anisotropy

In Section 1, it was shown that conductivity anisotropy can be conceived as a result of a spatial average over an isotropic conductivity that is heterogeneous on the microscale. This holds at least in the dc limit. Therefore, one may suggest that properties of transfer functions that hold for heterogeneous *isotropic* conductors on all scales also apply to anisotropic conductors. This appendix shows that phase constraints on the B-polarization impedance, derived for isotropic conductors, remain valid if an anisotropic conductivity is assumed *ab initio*.

The following results hold for the B-polarization in the case of *arbitrary* electrical 2-D anisotropy:

1. The phase φ of the MT impedance lies strictly in the range $0 < \varphi < 90^\circ$.
2. Dispersion relations exist connecting impedance phase and apparent resistivity.

Sketch of the Proof

This is a generalization of the proof given by Weidelt and Kaikkonen (1994) for the isotropic case. Let $\hat{\mathbf{x}}$ be the strike direction and let the conductor occupy the half-space $z \geq 0$. Then, $\mathbf{B}(\mathbf{r}) = B(\mathbf{r})\hat{\mathbf{x}}$. In the air half-space $z < 0$, the magnetic field $B(\mathbf{r})$ assumes the constant value B_0. In $z \geq 0$, the currents are flowing in the (y, z)-plane. In B-polarization the induction problem is formally simpler when formulated in terms of the resistivity tensor $\underset{\sim}{\varrho} = \underset{\sim}{\sigma}^{-1}$ rather than the conductivity tensor $\underset{\sim}{\sigma}$. Therefore, we assume the spatially variable anisotropic resistivity

$$\underset{\sim}{\varrho} = \begin{pmatrix} \varrho_{yy} & \varrho_{yz} \\ \varrho_{yz} & \varrho_{zz} \end{pmatrix},$$

where positive definiteness requires

$$\varrho_{yy} > 0 \quad \text{and} \quad \varrho_{yy}\varrho_{zz} - (\varrho_{yz})^2 > 0.$$

Then, $\varrho_{yy} > 0$ implies also that $\varrho_{zz} > 0$. Let the rotated resistivity tensor be defined by

$$\tilde{\underset{\sim}{\varrho}} := \begin{pmatrix} 0 & -1 \\ 1 & 0 \end{pmatrix} \underset{\sim}{\varrho} \begin{pmatrix} 0 & 1 \\ -1 & 0 \end{pmatrix} = \begin{pmatrix} \varrho_{zz} & -\varrho_{yz} \\ -\varrho_{yz} & \varrho_{yy} \end{pmatrix}.$$

Assuming again a time factor $\exp(+i\omega t)$, $\omega > 0$, Eq. (1), in connection with $B(\mathbf{r}) = B(\mathbf{r})\hat{\mathbf{x}}$ and $\mathbf{E} = \underset{\sim}{\varrho}\mathbf{J}$ yields as differential equation for B,

$$\nabla \cdot (\tilde{\underset{\sim}{\varrho}} \nabla B) = i\omega\mu_0 B,$$

which, for

$$B/B_0 =: b\exp(i\psi),$$

disintegrates into two coupled nonlinear equations for modulus b and phase ψ:

$$\nabla \cdot (\tilde{\varrho} \nabla b) = b(\nabla \psi)^T \tilde{\varrho} \nabla \psi \geq 0, \tag{A-1}$$

$$\nabla \cdot (b^2 \tilde{\varrho} \nabla \psi) = \omega \mu_0 b^2 > 0. \tag{A-2}$$

The fact that $\tilde{\varrho}$ is positive definite ensures that the right-hand side of Eq. (A-1) is nonnegative. The surface impedance $Z(y,0) := -E_y(y,0)/B_x(y,0)$ is given by

$$Z(y,0) = -[\varrho_{yy}(y,0)/\mu_0][\partial_z B(\mathbf{r})/B_0]|_{z=0} = -[\varrho_{yy}(y, 0/\mu_0)][\partial_z b(\mathbf{r}) + i\partial_z \psi(\mathbf{r})]|_{z=0}. \tag{A-3}$$

First it is seen that $f := b$ or $f := \psi$ cannot have a maximum at an *interior* point $\mathbf{r}_0$ of the half-space $z > 0$. This requires as necessary conditions that, at this point, $\nabla f = 0$ and that

$$\begin{pmatrix} \partial^2_{yy} f & \partial^2_{yz} f \\ \partial^2_{yz} f & \partial^2_{zz} f \end{pmatrix} \tag{A-4}$$

be *negative* definite. This condition is necessary because f has to *decrease* at all points in an infinitesimal neighborhood of $\mathbf{r}_0$. The necessary and sufficient conditions that Eq. (A-4) be negative definite are

$$\partial^2_{yy} f < 0 \quad \text{and} \quad \partial^2_{yy} f \partial^2_{zz} f - \left(\partial^2_{yz} f\right)^2 > 0.$$

Then, $\partial^2_{yy} f < 0$ implies also that $\partial^2_{zz} f < 0$. At $\mathbf{r}_0$ ($\nabla f = 0$), the left-hand sides of Eqs. (A-1) and (A-2) read explicitly

$$L := \varrho_{zz} \partial^2_{yy} f + \varrho_{yy} \partial^2_{zz} f - 2\varrho_{yz} \partial^2_{yz} f.$$

Because of

$$|\varrho_{yz}| < \sqrt{\varrho_{yy}\varrho_{zz}} \quad \text{and} \quad \left|\partial^2_{yz} f\right| < \sqrt{\partial^2_{yy} f \partial^2_{zz} f},$$

we have

$$L < \varrho_{zz} \partial^2_{yy} f + \varrho_{yy} \partial^2_{zz} f + 2\sqrt{\varrho_{yy}\varrho_{zz} \partial^2_{yy} f \partial^2_{zz} f} = -\left\{\sqrt{\varrho_{zz}\left|\partial^2_{yy} f\right|} - \sqrt{\varrho_{yy}\left|\partial^2_{zz} f\right|}\right\}^2 \leq 0,$$

such that, at the position of a maximum, the left-hand sides of Eqs. (A-1) and (A-2) would be negative, whereas the right-hand sides of Eqs. (A-1) and (A-2) are in fact nonnegative. Consequently, no interior maxima exist. Assuming a layered resistivity $\tilde{\varrho}(z)$ for $y \to \pm\infty$ (which may be different at both sides), we obtain from Eqs. (A-1) and (A-2)

$$b'(z) = -\frac{1}{\varrho_{yy}(z)} \int_z^\infty \varrho_{yy}(\zeta) b(\zeta) [\psi'(\zeta)]^2 \, d\zeta < 0,$$

$$\psi'(z) = -\frac{\omega\mu_0}{\varrho_{yy}(z) b^2(z)} \int_z^\infty b^2(\zeta) \, d\zeta < 0.$$

Therefore, b and ψ can only decrease with depth at the left and the right sides of the half-space $z > 0$, and at the lower boundary ($z \to \infty$) we have $b \to 0$ and $\psi \to -\infty$. Hence, the maxima of b and ψ are attained at $z = 0$, where $b = 1$ and $\psi = 0$, such that the derivatives in Eq. (A-3) are nonpositive. They are in fact strictly negative: The converse assumption $\partial_z \psi|_{z=0} = 0$ along with $\partial_y \psi \equiv 0$ at $z = 0$ would lead via Eq. (A-2) to

$$\partial^2_{zz} \psi\Big|_{z=0} = \omega\mu_0/\varrho_{yy}(y, 0) > 0,$$

implying an *increase* of ψ when moving into the conductor. In turn, the assumption $\partial_z b|_{z=0} = 0$ along with $\partial_y b \equiv 0$ at $z = 0$ implies via Eq. (A-1) that

$$\partial^2_{zz} b\big|_{z=0} = \left(\partial_z \psi\big|_{z=0}\right)^2 > 0,$$

which would again mean that b *increases* when penetrating into the conductor.

The existence of dispersion relations between apparent resistivity and impedance phase requires that Z have no zeroes in the lower frequency plane. Assuming the complex frequency $\omega - ip$, ω real, $p > 0$, the differential Eq. (A-1) changes to

$$\nabla \cdot (\tilde{\underset{\sim}{\varrho}} \nabla b) = b[p\mu_0 + (\nabla\psi)^T \tilde{\underset{\sim}{\varrho}} \nabla\psi] > 0.$$

By the same reasoning as for real frequencies, it then is shown that the real part of Z is positive in the lower frequency plane. Therefore, Z does not vanish there, $\log Z$ is analytical, and dispersion relations exist. Let

$$Z(\omega, y) =: \sqrt{\omega\mu_0\varrho_a(\omega)}\, \exp[i\varphi(\omega)]$$

Then,

$$\log \frac{\varrho_a(\omega)}{\varrho_a(\infty)} = \frac{4}{\pi} PV \int_0^\infty \left[\frac{\pi}{4} - \varphi(x)\right] \frac{x\,dx}{x^2 - \omega^2}$$

$$\varphi(\omega) = \frac{\pi}{4} + \frac{\omega}{\pi} PV \int_0^\infty \log \frac{\varrho_a(x)}{\varrho_a(\infty)} \frac{dx}{x^2 - \omega^2},$$

where $\varrho_a(\infty) = \varrho(y, 0)$ and PV denotes the Cauchy principal value of the integral.

All results also hold for an arbitrary (smooth) topography of the air–earth interface.

Staggered Grid for Maxwell's Equations in 3-D Anisotropic Media

Sofia Davydycheva[1]
Vladimir Druskin[2]

Summary. The standard staggered grid for Maxwell's equations is awkward for anisotropic media because the different components of the electric and magnetic fields are located at different nodes. There is, however, a natural alternative that places all components of the electric field at each node of one grid and all components of the magnetic field at each node of a staggered grid. This staggering allows a conservative finite-difference approximation for Maxwell's equations with arbitrary 3-D tensor electrical conductivity, magnetic permeability, and dielectric permittivity. An example of the time-domain solution using spectral Lanczos decomposition is considered.

1 Introduction

The use of Yee's (1966) staggered grid has revived finite-difference modeling of Maxwell's equations in many different fields, including applied geophysics (Wang and Hohmann, 1993). The conventional staggered grid, however, is awkward for modeling anisotropic media because different components of the electric and magnetic fields are located at different points in space, but the constitutive tensors— conductivity, magnetic permeability, and dielectric permittivity—relate these components at the same point. An alternative grid, based on a general approach of Lebedev (1964), avoids this difficulty. The new grid places all components of the electric field at all points of one spatial grid and all components of the magnetic field at all points of a second (staggered) grid. As on Yee's grid, centred-difference approximations of the operators $\nabla \cdot \nabla\times$ and $\nabla \times \nabla$ cancel exactly on the grid (in exact arithmetic). Therefore, the discretized versions of Maxwell's equations retain the properties of their differential counterparts. Especially important is that conservation of current and Ohm's law are valid locally on the grid. In isotropic media, the formulation splits into four uncoupled grids, each of which is a standard Yee grid.

[1]Mathematical Department, Central Geophysical Expedition, Narodnogo Opolchenia 40-3, Moscow 123298, Russia.
[2]Electromagnetics Department, Schlumberger-Doll Research, Ridgefield, CT 06877-4108, USA.

2 Lebedev's staggered grid

In 3-D anisotropic media, with arbitrary symmetric nonnegative definite 3×3 tensors of conductivity $\underset{\sim}{\mathbf{S}}(x, y, z)$, magnetic permeability $\underset{\sim}{\mathbf{M}}(x, y, z)$, and dielectric permittivity $\underset{\sim}{\mathbf{W}}(x, y, z)$, Maxwell's equations are

$$\nabla \times \mathbf{E} = -\underset{\sim}{\mathbf{M}} \frac{\partial \mathbf{H}}{\partial t}, \tag{1}$$

$$\nabla \times \mathbf{H} = \underset{\sim}{\mathbf{S}} \mathbf{E} + \underset{\sim}{\mathbf{W}} \frac{\partial \mathbf{E}}{\partial t} + \mathbf{J}, \tag{2}$$

with the boundary conditions

$$\mathbf{E}, \mathbf{H} \to 0 \quad \text{as} \quad (x, y, z) \to \infty, \tag{3}$$

where $\mathbf{E} \equiv (E_x, E_y, E_z)$ and $\mathbf{H} \equiv (H_x, H_y, H_z)$ are the electric- and magnetic-field vectors, and $\mathbf{J} \equiv (J_x, J_y, J_z)$ is the source current density.

Lebedev's (1964) staggering begins with a 3-D grid defined by the set of points

$$\begin{aligned} Q = \{r_n : r_n &= (x_i, y_j, z_k), \quad n = (i, j, k), \quad i = 0, \ldots, M_x, \\ j &= 0, \ldots, M_y, \quad k = 0, \ldots, M_z\}, \\ M &= M_x M_y M_z, \quad \|n\| = i + j + k. \end{aligned}$$

Next, define two subgrids as the sets of points

$$\begin{aligned} P &= \left\{r_n : \|n\| = 0, 2, 4, \ldots, 2\left[\tfrac{1}{2}M\right]\right\}, \\ R &= \left\{r_n : \|n\| = 1, 3, 5, \ldots, 2\left[\tfrac{1}{2}(M-1)\right] + 1\right\}. \end{aligned}$$

Thus, the subgrid P contains nodes with three even numbers or one even number and two odd ones, whereas the subgrid R contains the rest of the grid Q, i.e., nodes with three odd numbers and nodes with one odd number and two even ones. In addition, define the subgrid functions

$$f^P = f(r_n)\big|_{r_n \in P}, \qquad f^R = f(r_n)\big|_{r_n \in R}.$$

All functions with superscripts P or R are considered as approximations on corresponding subgrids.

Define finite differences along the x-axis,

$$\begin{aligned} f_x^P &= \left(f_x^P{}_{i,j,k}\right), \quad f_x^P{}_{i,j,k} = \frac{f_{i+1,j,k}^R - f_{i-1,j,k}^R}{x_{i+1} - x_{i-1}}, \\ f_x^R &= \left(f_x^R{}_{i,j,k}\right), \quad f_x^R{}_{i,j,k} = \frac{f_{i+1,j,k}^P - f_{i-1,j,k}^P}{x_{i+1} - x_{i-1}}, \end{aligned} \tag{4}$$

and similarly along the y- and z-axes. To make these formulas meaningful, additional fictitious nodes x_{-1}, x_{M_x+1}, y_{-1}, y_{M_y+1}, z_{-1}, z_{M_z+1} are assumed to be present at the boundaries of the grid, where (arbitrarily) $f \equiv 0$. The finite differences defined in this way map functions on the grid P to those on R and back.

3 Gridded Maxwell system and its stationary limits

On Lebedev's grid, let the (discrete) electric-field vector be defined on the subgrid R and the magnetic-field vector on the subgrid P. Then the tensor constitutive relations are naturally defined by

$$\mathbf{J}^R = \underset{\sim}{\mathbf{S}}^R\mathbf{E}^R, \quad \mathbf{B}^P = \underset{\sim}{\mathbf{M}}^P\mathbf{H}^P, \quad \mathbf{D}^R = \underset{\sim}{\mathbf{W}}^R\mathbf{E}^R,$$

where (discrete) **J**, **D**, and $\underset{\sim}{\mathbf{S}}$ occupy the R grid with **E**, whereas $\underset{\sim}{\mathbf{M}}$ and **H** occupy the P grid with **B**. The corresponding discrete approximation of Maxwell's equations is Eqs. (1) and (2),

$$\tilde{\nabla} \times \mathbf{E}^R = -\underset{\sim}{\mathbf{M}}^P \frac{\partial \mathbf{H}^P}{\partial t},$$

$$\tilde{\nabla} \times \mathbf{H}^P = \underset{\sim}{\mathbf{S}}^R\mathbf{E}^R + \underset{\sim}{\mathbf{W}}^R \frac{\partial \mathbf{E}^R}{\partial t} + \mathbf{J}^R, \tag{5}$$

where $\tilde{\nabla}\times$ is the grid *curl* naturally defined using Eq. (4). From Eq. (5), one can easily obtain the stationary-limit anisotropic equations for the dc electric and magnetic potentials:

$$\begin{aligned} \mathbf{E}^R &= -\tilde{\nabla} u^P, \\ \tilde{\nabla} \cdot (\underset{\sim}{\mathbf{S}}^R \tilde{\nabla} u^P) &= -\tilde{\nabla} \cdot \mathbf{J}^R \end{aligned} \tag{6}$$

$$\begin{aligned} \mathbf{H}^P &= \tilde{\nabla} v^R, \\ \tilde{\nabla} \cdot (\underset{\sim}{\mathbf{M}}^P \tilde{\nabla} v^R) &= 0 \end{aligned} \tag{7}$$

The following system for **E** can be derived from Eq. (5):

$$-\tilde{\nabla} \times [(\underset{\sim}{\mathbf{M}}^P)^{-1} \tilde{\nabla} \times \mathbf{E}^R] = \underset{\sim}{\mathbf{S}}^R \frac{\partial \mathbf{E}^R}{\partial t} + \underset{\sim}{\mathbf{W}}^R \frac{\partial^2 \mathbf{E}^R}{\partial t^2} + \frac{\partial \mathbf{J}^R}{\partial t}. \tag{8}$$

When $\mathbf{J} = -\varphi_0(x, y, z)\chi(t)$, where φ_0 is a finite function and χ is Heaviside's function (i.e., the current is switched on instantaneously), the derivative $\partial \mathbf{J}^R/\partial t$ vanishes in Eq. (8) for $t > 0$. Furthermore, the initial condition for **E** becomes:

$$\mathbf{E}^R\Big|_{t=0} = \varphi_0^R. \tag{9}$$

The full solution $\mathbf{E}^R(t)$ approaches the solution of Eq. (6) as $t \to +\infty$ (Druskin and Knizhnerman, 1988). The condition (3) can be replaced by the condition

$$\mathbf{E}^R \times \boldsymbol{\nu} = 0, \quad i = 0, M_x; \quad j = 0, M_y; \quad k = 0, M_z,$$

where $\boldsymbol{\nu}$ is the normal vector to the boundary of the region Q. It is equivalent to surrounding the model with a region of infinite electrical conductivity. This embedding does not spoil the accuracy, if the region of the grid Q is large enough.

If the function $\mathbf{E}(x, y, z)$ is smooth and can be expanded in a Taylor series—and if the grid steps are constant—the difference operator on the left-hand side of Eq. (8) is accurate to second order.

4 Connection with the isotropic case

The gridding of Maxwell's equations with Lebedev's scheme actually consists of four separate Yee grids, which are uncoupled in an isotropic medium. Consider the subset

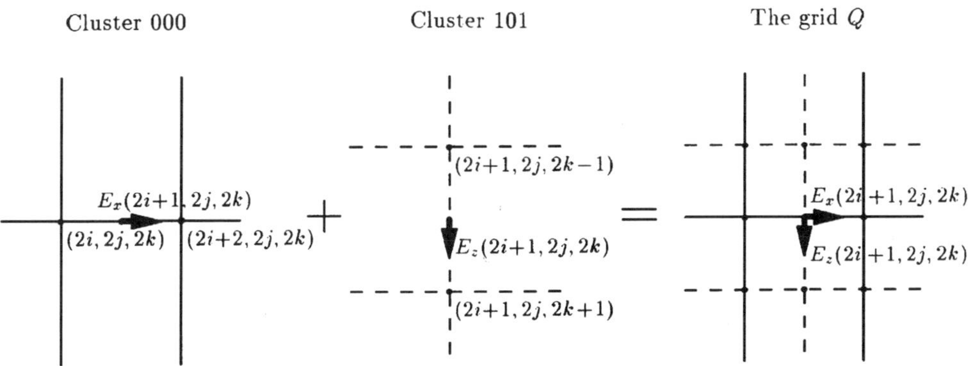

Figure 1. Two-dimensional cross-section (in the plane Oxz, for even nodes y) of Lebedev's staggered grid consisting of two clusters. The crossings of the lines of the same type form the subgrid R, whereas the crossings of the lines of different types form the subgrid P.

of elements of the subgrid functions $\mathbf{E}^R$ and $\mathbf{H}^P$ with the following indices:

$$E_x^R(1, 0, 0), \quad E_y^R(0, 1, 0), \quad E_z^R(0, 0, 1),$$
$$H_x^P(0, 1, 1), \quad H_y^P(1, 0, 1), \quad H_z^P(1, 1, 0),$$

where, for brevity, $(1, 0, 0)$ denotes $(2i+1, 2j, 2k)$, $(0, 1, 0)$ means $(2i, 2j+1, 2k)$ and so on. It is easy to see that these are just the elements of the standard Yee scheme for the isotropic problem (Yee, 1966; Druskin and Knizhnerman, 1988, 1994). This group is called cluster 000. One can show easily that the finite-difference operator $\tilde{\nabla}$ defined by formulas (4) performs mapping only within this cluster.

Now consider the elements of the same components $\mathbf{E}^R$ and $\mathbf{H}^P$ with changed indices, so that the first and the third indices become even instead of odd, and vice versa:

$$E_x^R(0, 0, 1), \quad E_y^R(1, 1, 1), \quad E_z^R(1, 0, 0),$$
$$H_x^R(1, 1, 0), \quad H_y^R(0, 0, 0), \quad H_z^R(0, 1, 1).$$

We denote this group as the cluster 101. Figure 1 shows a 2-D cross-section of the grid Q containing the clusters 000 and 101. It is easy to see that the components E_x^R and E_z^R belonging to the different clusters are defined at the same nodes $(2i+1, 2j, 2k)$.

Then, let cluster 110 be the group of the elements with the first and the second indices changed analogously, i.e.,

$$E_x^R(0, 1, 0), \quad E_y^R(1, 0, 0), \quad E_z^R(1, 1, 1),$$
$$H_x^P(1, 0, 1), \quad H_y^P(0, 1, 1), \quad H_z^P(0, 0, 0),$$

whereas cluster 011 is the group of the elements with the second and the third indices changed, i.e.,

$$E_x^R(1, 1, 1), \quad E_y^R(0, 0, 1), \quad E_z^R(0, 1, 0),$$
$$H_x^P(0, 0, 0), \quad H_y^P(1, 1, 0), \quad H_z^P(1, 0, 1).$$

Therefore, all components of the electrical field are defined at every node of subgrid R, whereas all components of the magnetic field are defined at every node of subgrid P. Note that this approach differs from the one proposed by Lebedev (1964) in which both $\mathbf{E}$ and $\mathbf{H}$ are defined at every node of the grid.

The finite-difference operators defined above map only within each cluster independently, i.e., a grid function in one cluster is mapped to a grid function in the same cluster. Coupling between clusters in Eq. (8) comes only from nondiagonal elements of the matrices $\underset{\sim}{\mathbf{S}}^R$, $\underset{\sim}{\mathbf{M}}^P$, $\underset{\sim}{\mathbf{W}}^R$. They connect elements defined at the same node of the grid but belonging to different clusters. So, if these matrices are diagonal (isotropic case), Eq. (8) can be divided into four independent sets. Each vector Eq. (8) at each cluster coincides with the standard Yee system, whereas potential Eqs. (6) and (7) become the standard seven-point 3-D difference schemes.

5 Approximation of grid coefficients

Consider a homogeneous isotropic medium containing rectangular anisotropic blocks $B_\ell, \ell = 1, \ldots, L$, without intersections, inclined in an arbitrary direction, with constant tensors $\underset{\sim}{\mathbf{S}}_\ell$, $\underset{\sim}{\mathbf{M}}_\ell$, $\underset{\sim}{\mathbf{W}}_\ell$. For simplicity, let us consider the case in which the tensors are diagonal in the coordinate system $O'x'y'z'$ defined by the axes of the block B_l. Let this block have nonempty intersection with an elementary cell of the grid with the center at the point i, j, k. Then the problem of calculating the conductivity, the magnetic permeability, and dielectric permittivity of this elementary cell, denoted by $\bar{\underset{\sim}{\mathbf{S}}}'_{ijk}$, $\bar{\underset{\sim}{\mathbf{M}}}'_{ijk}$ and $\bar{\underset{\sim}{\mathbf{W}}}'_{ijk}$, respectively (here and below we omit the superscripts P, R), can be reduced to a conventional isotropic averaging. Namely, each diagonal component of these tensors can be calculated as the average value of the corresponding component over an elementary cell. For example, when computing $(\bar{\underset{\sim}{\mathbf{S}}}'_{ijk})_{x'x'}$, we take the arithmetic mean over y' and z', and the harmonic mean over x', in accordance to the rules of addition of resistances in electric circuits. Other components of this tensor can be computed analogously. In this way, all these tensors—$\bar{\underset{\sim}{\mathbf{S}}}'_{ijk}$, $\bar{\underset{\sim}{\mathbf{M}}}'_{ijk}$, $\bar{\underset{\sim}{\mathbf{W}}}'_{ijk}$—remain diagonal. The final algorithm for the solution of the discrete equations requires calculation of 3×3 matrix functions $\underset{\sim}{\mathbf{M}}^{-1}$, $\underset{\sim}{\mathbf{S}}^{-1/2}$, and $\underset{\sim}{\mathbf{W}}^{-1/2}$ that can be easily done in the coordinate system where these matrices are diagonal. Returning to the initial coordinate system $Oxyz$ by means of an orthogonal transformation G (which is a combination of planar rotations), we find

$$\bar{\underset{\sim}{\mathbf{S}}}_{ijk}^{-1/2} = \underset{\sim}{\mathbf{G}}\left(\bar{\underset{\sim}{\mathbf{S}}}'_{ijk}\right)^{-1/2}\underset{\sim}{\mathbf{G}}^*,$$

where the asterisk means transposition, and analogous formulas for $\underset{\sim}{\mathbf{M}}^{-1}$, $\underset{\sim}{\mathbf{W}}^{-1/2}$.

This way of calculating the tensors of conductivity, magnetic permeability, and dielectric permittivity does not apply (or should be modified) in the case in which geometric axes of anisotropic blocks do not coincide with the principal axes of their anisotropy tensors.

6 A way to solve the problem

First, consider the quasi-static approximation, $\underset{\sim}{\mathbf{W}} \equiv 0$. Following Druskin and Knizhnerman (1988), we rewrite Eq. (8) in a form convenient for calculation by first reducing it to one involving a symmetric operator. Let

$$\mathbf{E}_{ijk}^R = \frac{1}{\sqrt{\rho_{ijk}^R}}[(\underset{\sim}{\mathbf{S}}^R)^{-1/2}\tilde{\mathbf{E}}^R]_{ijk},$$

where the weight ρ_{ijk}^R is defined as

$$\rho_{ijk}^R = (x_{i+1} - x_{i-1})(y_{j+1} - y_{j-1})(z_{k+1} - z_{k-1}),$$
$$i = 0, \ldots, M_x, \quad j = 0, \ldots, M_y, \quad k = 0, \ldots, M_z.$$

Then, Eqs. (8) and (9) can be rewritten as

$$\begin{aligned} \underset{\sim}{\mathbf{A}}\tilde{\mathbf{E}}^R + \frac{d\tilde{\mathbf{E}}^R}{dt} &= 0, \quad t > 0, \\ \tilde{\mathbf{E}}^R\Big|_{t=0} &= \tilde{\varphi}_0^R, \end{aligned} \tag{10}$$

where the operator $\underset{\sim}{\mathbf{A}}$ is defined by

$$\underset{\sim}{\mathbf{A}} = \sqrt{\rho^R}(\underset{\sim}{\mathbf{S}}^R)^{-1/2}\tilde{\nabla} \times \left\{(\underset{\sim}{\mathbf{M}}^P)^{-1}\tilde{\nabla} \times \left[\frac{1}{\sqrt{\rho^R}}(\underset{\sim}{\mathbf{S}}^R)^{-1/2}\right]\right\}.$$

The symmetry and nonnegative definiteness of the tensors $\underset{\sim}{\mathbf{M}}^P$, $\underset{\sim}{\mathbf{S}}^R$ ensure that the operator $\underset{\sim}{\mathbf{A}}$ is symmetric and nonnegative definite in R^N (where N is the number of nodes of the subgrid R that are responsible for the electrical field) with the usual scalar product. From here, it follows that the difference-differential scheme is stable.

The solution of Eqs. (10) is (symbolically)

$$\tilde{\mathbf{E}}^R(t) = \exp(-t\underset{\sim}{\mathbf{A}})\,\tilde{\varphi}_0^R. \tag{11}$$

This vector can be calculated efficiently by the spectral Lanczos decomposition method (SLDM) (Druskin and Knizhnerman 1995), which is a modification of Lanczos's method (Parlett, 1980, ch. 13) for computation of matrix functions. To calculate Eq. (11), one can generate an orthonormal basis $q_1, q_2, \ldots, q_m$ by the Gram–Schmidt orthogonalization of elements of the Krylov subspace $K^m = \text{span}(\tilde{\varphi}_0^R, \underset{\sim}{\mathbf{A}}\tilde{\varphi}_0^R, \ldots, \underset{\sim}{\mathbf{A}}^{m-1}\tilde{\varphi}_0^R)$. The orthogonalization can be carried out using the Ritz method (Parlett, 1980, ch. 11). This process gives a tridiagonal symmetric matrix $\underset{\sim}{\mathbf{H}}$ of dimension $m \times m$. Even if $m \ll N$ (the dimension of the matrix $\underset{\sim}{\mathbf{A}}$), the right-hand side of Eq. (11) can be approximated by the expression $\|\tilde{\varphi}_0^R\|\underset{\sim}{\mathbf{Q}}\exp(-t\underset{\sim}{\mathbf{H}})\,e_1$. Here, $\underset{\sim}{\mathbf{Q}}$ is the $m \times N$ matrix of the basis vectors $\mathbf{q}_i$. The matrix function $\exp(-t\underset{\sim}{\mathbf{H}})$ can be calculated by a rational approximation.

When $\underset{\sim}{\mathbf{W}}(x, y, z) \neq 0$, but $\underset{\sim}{\mathbf{S}}(x, y, z) \equiv 0$, one also can obtain, by a simple substitution, an equation analogous to Eq. (10):

$$\underset{\sim}{\mathbf{A}}\mathbf{u} + \frac{d^2\mathbf{u}}{dt^2} = 0; \quad \mathbf{u}\Big|_{t=0} = \varphi_0; \quad \frac{d\mathbf{u}}{dt}\bigg|_{t=0} = \varphi_1.$$

The solution of this equation can be written

$$\mathbf{u}(t) = \cos(t\sqrt{\underset{\sim}{\mathbf{A}}})\,\varphi_0 + \underset{\sim}{\mathbf{A}}^{-1/2}\sin(t\sqrt{\underset{\sim}{\mathbf{A}}})\,\varphi_1,$$

where SLDM can be used to calculate the matrix functions.

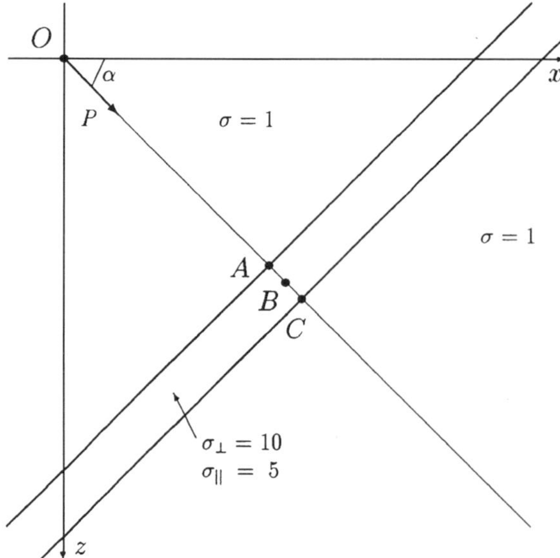

Figure 2. Model containing an inclined anisotropic block.

It also is not difficult to find the solution of the corresponding equation for the electric field in the frequency domain (Druskin and Knizhnerman, 1995). The matrix function to be calculated in this case is $i\omega(\underset{\sim}{\mathbf{A}} + i\omega)^{-1}$, where ω is the frequency of the excited field.

We have written a Fortran77-program, called MAXANIS, to implement the algorithm described here for computing transient EM fields in 3-D anisotropic media. MAXANIS computes fields for a quasi-stationary approximation $\underset{\sim}{\mathbf{W}} \equiv 0$. The number of spatial nodes for $\mathbf{E}^R$ and, therefore, the program's CPU and memory requirements (without taking into account the CPU and memory requirements for calculating and storing the matrices $\underset{\sim}{\mathbf{S}}^R$, $\underset{\sim}{\mathbf{M}}^P$.) are four times higher than the ones for the isotropic case examined by Druskin and Knizhnerman (1988). The program requires 45 Mbytes RAM for a grid of $80 \times 80 \times 80$ nodes; half of the nodes contain elements $\mathbf{E}^R$ whereas the other half contains $\mathbf{H}^P$ (it corresponds to a grid for the isotropic case aiming at the same accuracy with $40 \times 40 \times 40$ nodes).

With MAXANIS, we have calculated the electric field in an inclined anisotropic stratum (shown in Fig. 2) after the current in a electric dipole P is switched off. The dipole axis is perpendicular to the stratum, and the angle between this axis and Ox is equal to α. At this dipole axis, we calculate the colinear component $E_{\|}$ of the electric field. The distance between the dipole and the center of the stratum is $OB = 7.07$ km, whereas the width of the stratum is $AC = 1$ km. The results of these calculations are shown in Fig. 3. Both curves show $E_{\|}$ versus the distance to the dipole, r. The upper curve corresponds to the time $t = 0$ (DC case), whereas, for the lower one, $t = 100$ s (after switching off the current). The dots correspond to the case $\alpha = 0$ (both the dipole and the stratum are not inclined with respect to the grid), but the crosses show the results at $\alpha = \pi/4$. The results show that the required indepedence of the inclination angle of the model with respect to the grid is achieved with good accuracy. Thus, the MAXANIS program makes it possible to deal, with a sufficient accuracy, with anisotropic models containing blocks inclined in arbitrary directions.

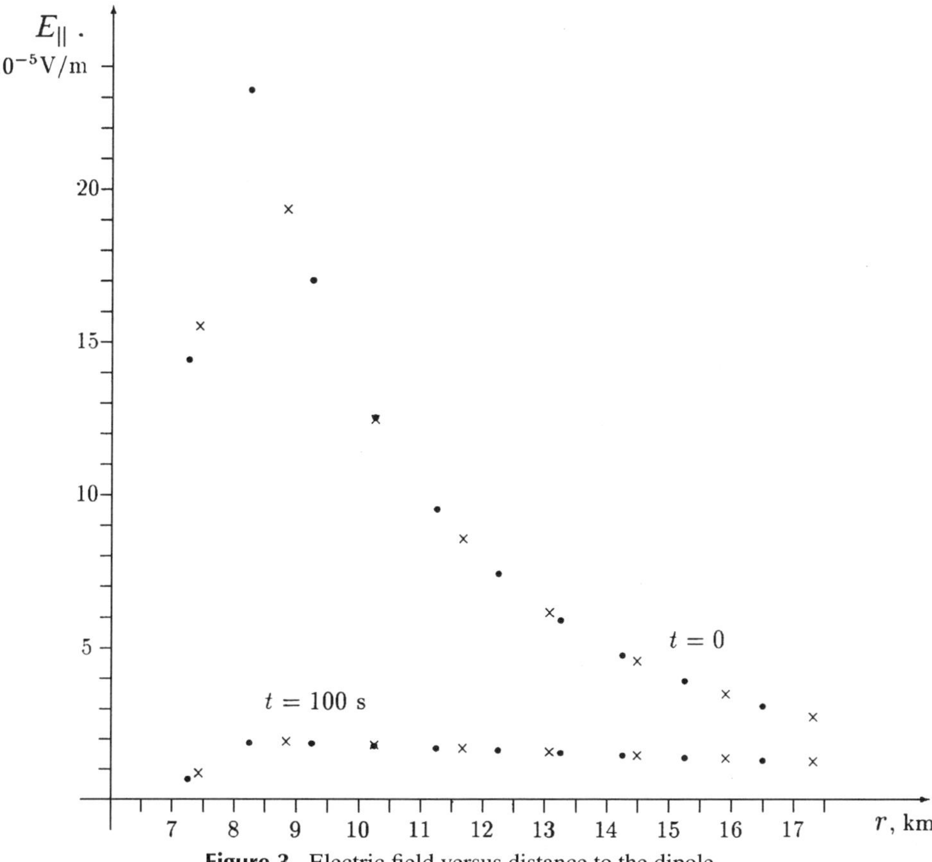

Figure 3. Electric field versus distance to the dipole.

Acknowledgments

The authors are indebted to L. A. Knizhnerman for useful discussions and to the Department of Physics, University of Bergen, for the offer to use the computer facilities.

References

Druskin, V. L., and Knizhnerman, L. A., 1988, A spectral semi-discrete method for the numerical solution of 3-D nonstationary problems in electrical prospecting: Phys. Solid Earth, **24**, 641–648.

———1994, Spectral approach to solving three-dimensional Maxwell's equations in the time and frequency domains: Radio Science, **29**, 937–953.

———1995, Krylov subspace approximation of eigenpairs and matrix functions in exact and computer arithmetic: Num. Lin. Alg. Appl., **2**, 205–217.

Lebedev, V. I., 1964, Difference analogies of orthogonal decompositions of basic differential operators and some boundary value problems, I: USSR. Comput. Maths. Math. Phys., **4**, 449–465 (in Russian).

Parlett, B. N., 1980, The symmetric eigenvalue problem: Prentice-Hall, Inc.

Wang, T., and Hohmann, G., 1993, A finite-difference, time-domain solution for three-dimensional electromagnetic modeling: Geophysics, **6**, 797–810.

Yee, K. S., 1966, Numerical solution of initial boundary value problems involving Maxwell's equations in isotropic media: IEEE, **AP-14**, 302–307.

Finite-Difference Modeling of 3-D EM Fields with Scalar and Vector Potentials

Douglas J. LaBrecque

Summary. Scalar and vector potentials provide a convenient formulation for finite-difference modeling of harmonic electromagnetic (EM) fields in 3-D media. Because the EM potentials are continuous everywhere, the finite-difference method can handle large contrasts in conductivity or dielectric constant. In the Coulomb gauge, the method is stable at low frequencies because the potentials approach those for a static (dc) formulation. The linear system of finite-difference equations can be solved iteratively with a biconjugate-gradient method that is diagonally preconditioned. An impedance boundary condition terminates the grid.

Modeling of a crosswell EM survey with this method shows that 2-m electric dipoles radiating at 5 MHz can detect a small zone of dense nonaqueous-phase liquid (DNAPL) with high resistivity and low dielectric constant between wells 10 m apart. Because the responses to contrasts in dielectric constant and resistivity are distinctly different, both quantities could be estimated. The response changes significantly with the position of the DNAPL zone, the weakest response occurring when the zone is just below the water table.

1 Introduction

The goal of the research described here is to develop a method that will combine the strengths of electrical resistivity tomography (ERT; see LaBrecque et al., this volume) and those of electromagnetic (EM) tomography. At frequencies of tens to hundreds of megahertz, EM tomography must be modeled with the (vector) wave equation (Daily, 1984; Shope and Greenfield, 1988). The response depends on both resistivity and dielectric constant (permittivity). The method under development combines resistivity and EM measurements in the 0.05- to 5-MHz range. The hope is that the combination will have much better resolution than just ERT and also will map the dielectric distribution. EM fields in the low-megahertz range cannot be modeled with either the potential equation used for ERT or with ray-tracing techniques used for higher frequency EM tomography (Daily, 1984). An efficient and robust modeling algorithm to

SteamTech Environmental Services, Bakersfield, CA 93309, USA.

model both high and low frequencies, however, can be developed with scalar and vector potentials.

2 Theory

Maxwell's equations for electric sources in an isotropic, heterogeneous region are

$$\nabla \times \mathbf{E} = \hat{z}\mathbf{H}, \tag{1}$$

$$\nabla \times \mathbf{H} = \hat{y}\mathbf{E} + \mathbf{J}_e, \tag{2}$$

where $\mathbf{E}$ is the electric field, $\mathbf{H}$ is the magnetic field, and $\mathbf{J}_e$ is the source current. The quantities $\hat{y}$ and $\hat{z}$ are

$$\hat{y} = \sigma + i\omega\epsilon, \tag{3}$$

and

$$\hat{z} = -i\omega\mu, \tag{4}$$

where σ is electrical conductivity, ω is radian frequency, ϵ is dielectric permittivity, and μ is magnetic permeability.

A magnetic vector potential, $\mathbf{A}$, can be defined by its curl

$$\nabla \times \mathbf{A} = \mu\mathbf{H} \tag{5}$$

and divergence

$$\nabla \cdot \mathbf{A} = 0. \tag{6}$$

Equation (6) is called the Coulomb gauge, which is one of many. Morisue (1993) compares the Coulomb and Lorentz gauges for boundary integral methods and concludes that both provide the same accuracy. Bardi et al. (1991) apply a modified version of the Coulomb gauge. Boyse et al. (1992, 1993) describe formulations with Lorentz-like gauges. The latter formulation (Boyse et al., 1993) requires simultaneous solution for four coupled potentials, as does the solution discussed here. Some advantages of the present solution are discussed in Section 5.

Substituting Eq. (5) into Eq. (1) gives

$$\nabla \times \mathbf{E} = -i\omega\nabla \times \mathbf{A}. \tag{7}$$

Because their curls are proportional, $\mathbf{E}$ and $\mathbf{A}$ can differ only by the gradient of a scalar potential ($\mathbf{E}$ is not necessarily divergence free); therefore,

$$\mathbf{E} = -i\omega\mathbf{A} - \nabla V. \tag{8}$$

The electric field thus is decomposed into a curl-free part, $-\nabla V$, and a divergence-free part, $-i\omega\mathbf{A}$.

Substituting Eq. (5) into Eq. (2) gives

$$\nabla \times (1/\mu)\nabla \times \mathbf{A} = \mathbf{J}_e + \hat{y}\mathbf{E}. \tag{9}$$

When the magnetic permeability is a constant, μ_0, the vector identity

$$\nabla \times \nabla \times \mathbf{F} = \nabla(\nabla \cdot \mathbf{F}) - \nabla^2\mathbf{F} \tag{10}$$

gives

$$\nabla^2 \mathbf{A} = -\mu_0 \mathbf{J}_e - \mu_0 \hat{y} \mathbf{E} \tag{11}$$

(the divergence of **A** is zero). Substituting Eq. (8) into Eq. (11) and rearranging gives

$$\nabla^2 \mathbf{A} - \hat{z}\hat{y}\mathbf{A} = \mu_0 \hat{y} \nabla V - \mu_0 \mathbf{J}_e. \tag{12}$$

This vector equation gives three equations in the four unknown potentials. The fourth equation needed to give a unique solution follows from taking the divergence of Eq. (12):

$$\nabla \cdot \hat{y} \nabla V = I - i\omega \mathbf{A} \cdot \nabla \hat{y}, \tag{13}$$

where I represents the divergence (grounding) terms of the electric-source currents.

Equation (13) has the same left-hand side as the (scalar) potential equation used to model steady-state electric-current flow. Here, it is coupled to the magnetic vector potential by the source-like terms on the right-hand side. These terms vanish at low frequencies. Equation (12) can be viewed as three Helmholtz equations that are not coupled directly to each other but that are coupled to the scalar potential by terms on the right-hand side.

3 Numerical implementation

The standard finite-difference solution to Eqs. (12) and (13) divides the region into a rectangular grid. I chose not to use a staggered mesh but to locate both scalar and vector potentials at the same nodes. This simplifies the creation and storage of meshes. Finite-difference equations were derived by taking the volume integral about the region surrounding each node. The derivatives are approximated by central, finite differences. The method resembles one used by Dey and Morrison (1979) for steady-state electrical resistivity modeling.

The finite-difference equations are four coupled linear systems. The equations for each individual potential are symmetric, but crosscoupling terms between the scalar and the vector potentials are not. Therefore, the overall system is nonsymmetric. This system of equations is solved iteratively with a biconjugate-gradient method that is diagonally preconditioned (Press et al., 1992, Chapter 2).

4 Boundary conditions

Boundary conditions are needed for the scalar potential and for all three components of the vector potential on all six boundaries of the finite-difference region. Three different types have been implemented: E-parallel symmetric, E-perpendicular symmetric, and absorbing boundary conditions. The first two assume a plane of symmetry in both the media and the source. Applying a plane of symmetry is possible for many problems and reduces the storage requirements and execution time by a factor of two or more. Absorbing boundaries are used to approximate an unbounded region using a finite mesh.

4.1 E-parallel symmetry

E-parallel symmetry applies to a geometry in which the electric field parallel to a plane is symmetric about that plane. There is E-parallel symmetry if an electric dipole source or a small loop lies in a plane of symmetry. If the potentials go to zero at infinity, then the electric field is symmetric only if

$$\frac{\partial}{\partial n} A_{\parallel} = 0, \tag{14}$$

$$\frac{\partial}{\partial n} V = 0, \tag{15}$$

and

$$A_{\perp} = 0, \tag{16}$$

where $\partial/\partial n$ is the derivative taken in the outward direction normal to the boundary.

4.2 E-perpendicular symmetry

E-perpendicular symmetry applies to a geometry in which an electric dipole or a small loop is perpendicular to, and bisected by, the plane of symmetry. Again, if the potentials go to zero at infinity, then

$$V = 0, \tag{17}$$

$$A_{\parallel} = 0, \tag{18}$$

and

$$\frac{\partial}{\partial n} A_{\perp} = 0. \tag{19}$$

4.3 Absorbing boundary conditions

Geophysical problems require that fields in an unbounded region be modeled on a finite mesh. To model an unbounded region, the boundary conditions discussed above can be used sensibly on, at most, three of the six sides of the finite-difference mesh. At low frequencies and in lossy media, boundary conditions were treated by moving them as far from the sources as possible. For modeling lossless media at high frequencies, this cannot be done. Figure 1 shows an example of the problems in modeling boundaries at high frequencies in a resistive medium. The coupling between a pair of square, coaxial loops is modeled. The transmitting loop is in the corner of a $3 \times 3 \times 3$ m region with a dielectric constant of 1.0 and a resistivity of 10^6 $\Omega \cdot$m. Both loops are 0.6 m on a side and 1.2 m apart. The transmitting loop lies in the uppermost boundary of the finite-difference region; this boundary is taken as a plane of E-parallel symmetry. Both loops are bisected by boundary planes with E-perpendicular symmetry. The remaining planes assume E-parallel = 0 (this is the same as E-perpendicular symmetry and represents the boundary of a perfectly conducting region). The size of the finite-difference mesh is $10 \times 10 \times 10$ cells.

In Fig. 2, the finite-difference results calculated for this closed region are compared with the analytical results for a whole space. The whole-space results show a smooth, continuous increase in amplitude and a gradual rotation of phase with increased frequency. Below 50 MHz, the finite-difference results (squares) have roughly the same amplitude as the whole-space case, but a uniform phase of 90°. Above 50 MHz, the phase and the amplitude oscillate wildly. This occurs because the finite-difference solution models an enclosed cavity. When the length of the sides of the cavity are more than

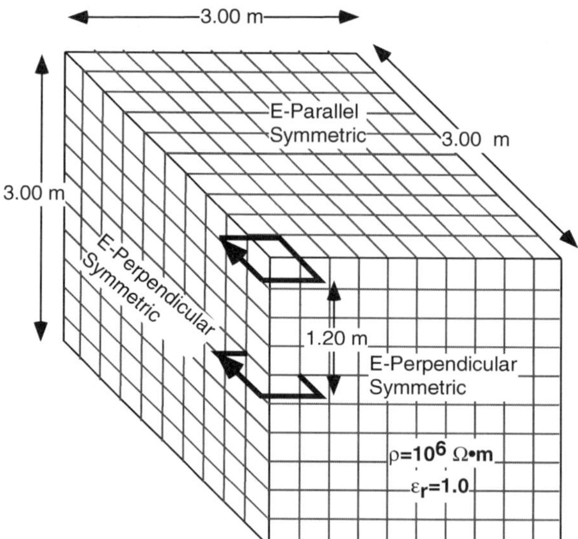

Figure 1. Finite-difference mesh used to model a 3 × 3 × 3-m region.

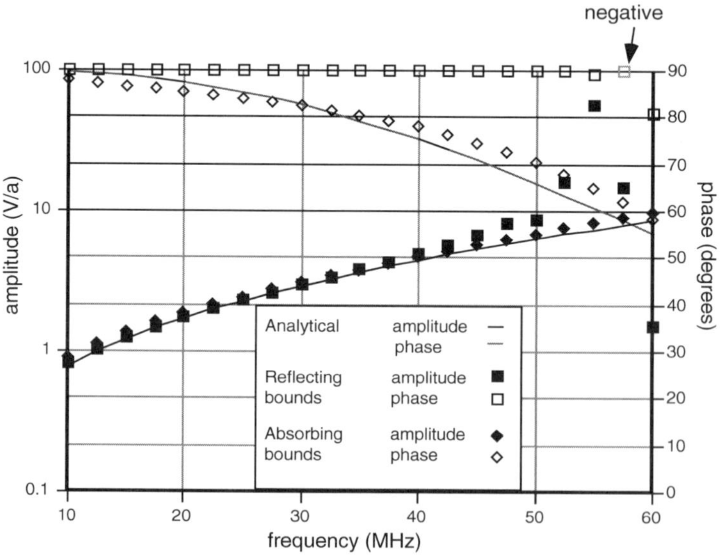

Figure 2. Comparison of mutual coupling of a pair of coaxial loops (Fig. 1) in a 3 × 3 × 3-m volume with the analytical solution for the coupling of coaxial loops in a whole space.

a quarter wavelength, we start to see cavity resonances. If we increase the size of the finite-difference region, the resonances shift to lower frequencies. Therefore, moving the boundaries farther away gives a poorer approximation to a whole space.

Approximating unbounded regions requires absorbing boundary conditions. There has been a great deal of research on absorbing-boundary conditions. Popular methods include the Mur (1981), the perfectly matched layer (PML; Fang and Wu, 1995) and the Liao (Wang and Tripp, 1996). I have chosen to implement the terminal-impedance absorbing-boundary method. This method is much simpler to implement than either the Liao or PML methods.

Implementation of the Mur boundary conditions also is simple but can create numerical instability. In the frequency domain, the first-order Mur boundary condition reduces to an inhomogeneous Neumann boundary condition, and the second-order Mur boundary condition involves the outward normal second derivative. To maintain numerical stability for the steady-state case, Dirichlet boundary conditions must be enforced over at least part of the boundary. The terminal-impedance method used here is a Dirichlet-type boundary condition and thus has an advantage over the Mur method.

The terminal-impedance method can be derived by considering the analogy between the finite-difference mesh and a jointed transmission line. The boundary conditions are equivalent to connecting each of the outer nodes to an electrical ground through a circuit element. The admittivity, $\hat{Y}_n$, of this circuit element must be chosen such that the potential and outward current flow at the boundary nodes match those of some analytical model. The admittance normal to the boundary is of the form

$$\hat{Y}_n = -a\frac{\frac{\partial}{\partial n}F(r)}{F(r)}, \tag{20}$$

where $F(r)$ is an analytical solution for the scalar or vector potential.

Although it is possible to match the boundary conditions to any analytical model, the difficulty is choosing a model that is robust and provides good results for a wide range of conductivity, dielectric constant, and source distributions. The present algorithm uses

$$F(r) = 1/4\pi r \tag{21}$$

for scalar potential V, and

$$F(r) = \frac{e^{-ikr}}{4\pi r} \tag{22}$$

for the vector potentials, where r is the distance from the center of the mesh to the boundary and $k^2 = -\hat{z}\hat{y}$. These equations were easy to implement and tests showed that they gave good results for a variety of models. The constant a depends on the cell size and admittivity. For example, on the $+Z$ boundary,

$$a = \int_{x_{i-1/2}}^{x_{i+1/2}}\int_{y_{j-1/2}}^{y_{j+1/2}} \frac{\hat{y}(x, y)}{z_k - z_{k-1}}\,dy\,dx \tag{23}$$

for scalar potentials, where $x_{i+1/2} = (x_{i+1} - x_i)/2, x_{i-1/2} = (x_i - x_{i-1})/2$ (and similarly for y); for vector potentials,

$$a = \frac{(x_{i+1} - x_{i-1})(y_{j+1} - y_{j-1})}{4(z_k - z_{k-1})}, \tag{24}$$

where x_i, y_j, and z_k are the Cartesian coordinates of node (i, j, k).

Calculating the response for the configuration shown in Fig. 1 using absorbing boundaries gives a good approximation to the whole space (Fig. 2). The amplitude is close to the whole-space results for the entire frequency range. The phase response also is close but not perfect. The maximum error in the phase is 4 degree at 50 MHz.

5 Spurious modes

Incorrectly (or incompletely) posed discretizations of Maxwell's equations can have eigenvectors (with nonzero eigenvalues) that are not close to eigenvectors of the continuous solution (Schroeder and Wolff, 1994); these are called "spurious modes."

There is confusion about the origin of spurious modes. For example, the belief is widespread that the sole source of spurious modes is poor enforcement of a divergence condition on the electric field or potentials. According to Sun et al. (1995), this misconception may have originated with Konrad, who speculated in his Ph.D. thesis that spurious modes are caused by solving

$$\nabla \times \nabla \times \mathbf{E} + k^2\mathbf{E} = 0, \tag{25}$$

alone, without explicitly enforcing the solenoidal (divergence-free) nature of the field (in charge-free regions). Many researchers have attempted to find methods that give fields with zero divergence in the finite-element method (Sun et al., 1995). Some confusion also may have come from early work that just ignored the divergence of the field in numerical simulations (Schroeder and Wolff, 1994; Sun et al., 1995; Webb, 1995). What is surprising about these incomplete formulations is that they can give decent approximations in special circumstances. For example, incomplete vector potential formulations have successfully modeled magnetostatic fields (Preis et al., 1991, 1992; Mesquita and Bastos, 1992). Other incomplete formulations worked for homogeneous waveguides (Webb, 1995) but failed when discontinuities were present. Sun et al. (1995) state that incomplete formulations often give adequate results in the far-field, but are "drastically wrong" in the near-field.

Spurious modes are not confined to finite-difference and finite-element methods. Schroeder and Wolff (1994) discuss their occurrence in the method of moments (for integral equations) and in spectral methods. Methods based on the vector potential are not particularly prone to spurious modes, even though the numerical formulation should include an explicit divergence condition. In fact, an implicit Coulomb gauge has been shown to eliminate spurious solutions (Morisue, 1993). Webb (1995) advises, however, that "it is better to impose the divergence condition earlier, at the level of the differential equations." Simply imposing a divergence condition on the fields, however, does not remove spurious modes (Schroeder and Wolff, 1994); the penalty method, for example, has problems even though it tries to kill the divergence of the electric field by adding penalty terms to the functional minimized by the finite-element formulation (Boyse et al. 1992; Schroeder, 1994; Sun et al., 1995).

Wong and Cendes (1988) show that the cause of spurious modes is improper approximation of the null-space of the curl operator. Sun et al. (1995) point out that Eq. (25) enforces the zero divergence of the flux for all but the static, $\omega = 0$, case. However, the system of equations derived solely from Eq. (25) will have eigenvectors with eigenvalues equal to 0 at $\omega = 0$. In other words, the system of equations is singular at zero frequency. For exact theory, this is not a problem with inductive sources; these modes would be present, but not excited (Druskin and Knizhnerman, 1994). However, when Eq. (25) is approximated numerically, the eigenvalues of these modes are shifted slightly, such that they appear as spurious modes at low frequencies.

Bardi et al. (1991) present a symmetric formulation based on the vector potentials with Coulomb gauge and a modified scalar potential. They make the substitution

$$V = i\omega V', \tag{26}$$

which allows them to write the divergence condition as

$$\nabla \cdot [k^2(\mathbf{A} + \nabla V')] = 0. \tag{27}$$

Their method requires solution of a sparse, symmetric system of equations of the form

$$\underset{\sim}{\mathbf{M}} + k^2\underset{\sim}{\mathbf{N}} = \underset{\sim}{\mathbf{J}}, \tag{28}$$

where $\underset{\sim}{\mathbf{N}}$ is positive definite and $\underset{\sim}{\mathbf{M}}$ is singular. As ω approaches zero, the system of equations becomes increasingly ill-conditioned and, finally, singular at zero frequency. Bardi et al. (1991) state that this solution does not create any spurious solutions. This is true for an exact solution; however, the system of equations is singular at zero frequency. Therefore, there may be zero-frequency modes present that can be shifted in frequency on discretization and appear as low-frequency spurious modes.

Although there are a number of methods of eliminating these low-frequency spurious solutions (Wong and Cendes, 1988; Schroeder and Wolff, 1994; Sun et al., 1995), the approach here was to choose a formulation that correctly incorporates the zero-frequency response as well as the higher frequency response in the solution. At $\omega = 0$, Eq. (12) becomes the classic scalar potential equation and is no longer dependent on Eq. (13). Equation (13) separates into individual Poisson's equations for the three vector potentials. The system of equations is well conditioned as long as the boundary conditions are chosen properly. To eliminate the homogeneous solutions to Laplace's equation, for each potential, there must be Dirichlet boundary conditions on at least one point on the boundaries and either Dirichlet or Neumann conditions on the remaining boundaries. As long as the boundaries meet these criteria, at steady state these equations are all positive definite with real positive eigenvalues. Because there are no modes with zero eigenvalues at zero frequency, this eliminates the low-frequency spurious modes inherent in formulations that are equivalent to Eq. (25). Many test cases were run with the method described here for scalar and vector potentials. All gave accurate results.

6 Checks on the solution

One advantage of this formulation is its ability to calculate results over a very broad frequency range. In Fig. 3, the coupling of a pair of 1.0-m parallel electric dipoles

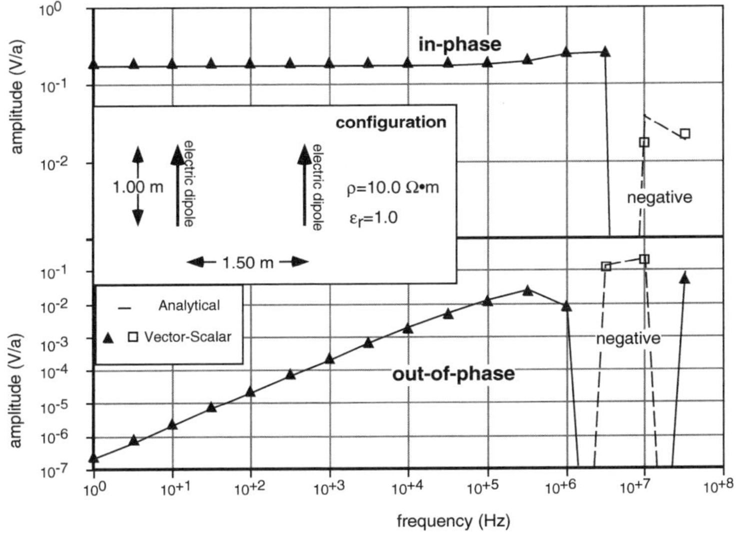

Figure 3. Coupling of parallel electric dipoles in a homogeneous whole space calculated analytically (lines) and using the vector–scalar finite-difference method (symbols).

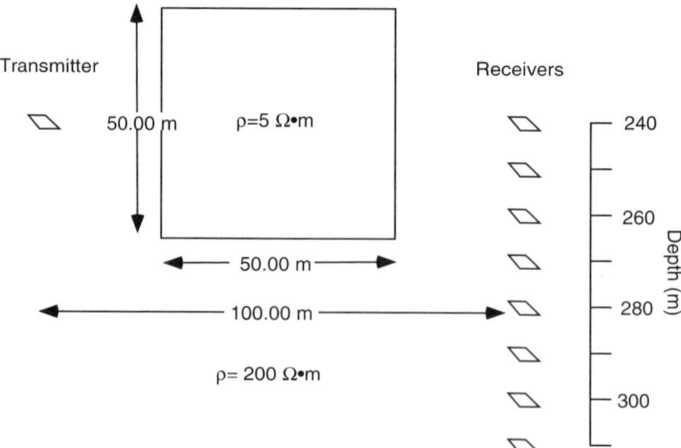

Figure 4. Configuration of model used for comparison with the results of Newman (1995). A 50-m cube is centered between the receiving and the transmitting holes. Both the transmitters and the receivers are 5 × 5-m horizontal loops.

separated by 1.5 m was calculated for a whole space with resistivity of $10\,\Omega\cdot\text{m}$ and dielectric constant of 10. The same mesh, with 0.25-m cells, was used to calculate the response from 1 Hz to 32 MHz. Figure 3 compares the calculated responses with those for the whole space. Responses are very close throughout the entire frequency range.

The response of the body shown in Fig. 4 was calculated as a check of the ability of the code to calculate 3-D responses of bodies with relatively high contrasts. The model consists of a 50-m cube with a resistivity of $5\,\Omega\cdot\text{m}$ in a 200-$\Omega\cdot\text{m}$ whole space. The source is a vertical magnetic dipole at the depth of the center of the body and has a frequency of 20 kHz. The vertical magnetic field was calculated at several depths in a borehole on the opposite side of the body. The two boreholes were in the same plane as the center of the body and were 100 m apart (Fig. 4). The finite-difference model used 5 × 5-m horizontal, square loops to approximate the source and receivers. This model was chosen because it could be compared with the results of Newman (1995), calculated using both an integral-equation solution and a staggered-grid finite-difference solution of Maxwell's equations.

Amplitude and phase of the vertical magnetic fields calculated using the three solutions are shown in Fig. 5. The amplitudes for the scalar–vector solution were nearly identical to those calculated by Newman (1995) using staggered-grid finite differences. The amplitudes for both solutions were a few percent greater than those for the integral-equation solution. The phase values for the scalar–vector solution are about 4° greater than those for the integral-equation solution and about 1.5° greater than those for the staggered-grid finite-difference solution. It is not clear which of the solutions is correct. However, the scalar–vector solution is closer to the staggered-grid solution than the agreement between the two solutions published by Newman (1995).

One disadvantage of the staggered-grid finite-difference solution reported by Newman (1995) is the difficulty in approximating the air–earth interface. This is not a problem with the scalar–vector formulation. Figure 6 compares the analytical and scalar–vector finite-difference solutions for the colinear dipole–dipole array with 100-m dipoles at the surface of a half-space. The half-space has a resistivity of 100 and the air has a resistivity of 10^{14}. Results are calculated for a frequency of 100 Hz. The

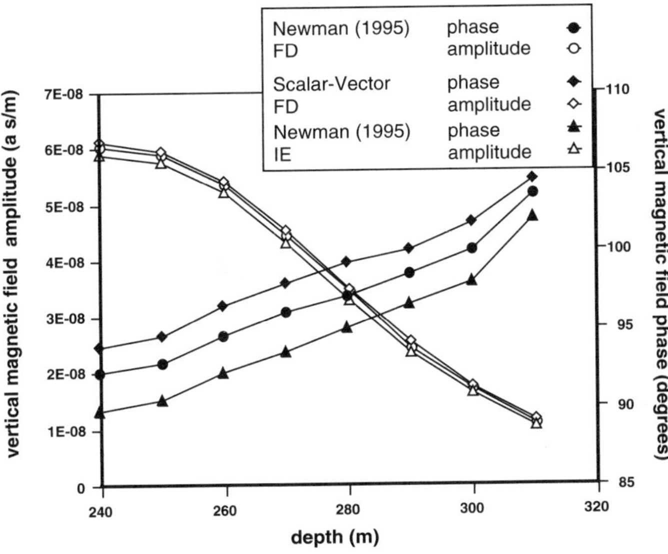

Figure 5. Comparison of cross-borehole magnetic-dipole response for the body shown in Fig. 4, calculated using scalar–vector finite-differences and two solutions from Newman (1995).

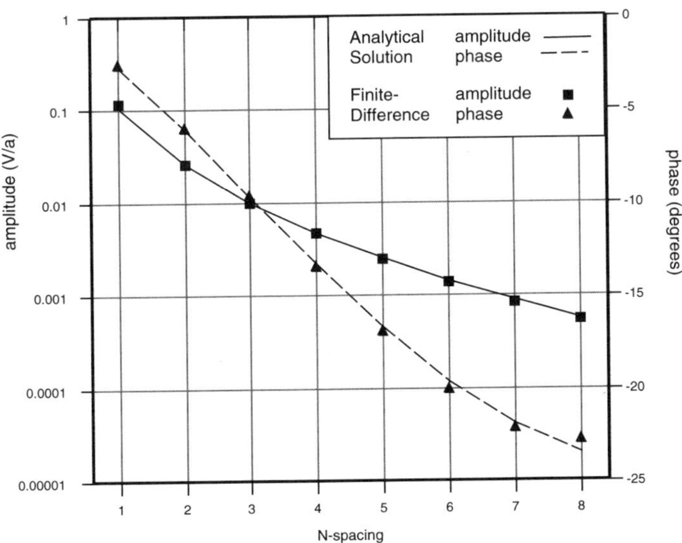

Figure 6. Comparison of calculated and analytical amplitudes and phases for the colinear dipole–dipole array at $n=1$ to $n=8$ with 100-m dipoles, on the surface of a 100-ohm-m half-space at 100 Hz.

finite-difference mesh used three cells per dipole. Dipole separations from $n=1$ to $n=8$ are calculated. The greatest error is at $n=1$; the finite-difference amplitude is 9.6% greater than the analytical value. This is about the same error seen in 3-D, steady-state, finite-difference calculations with the same mesh discretization. Interestingly, the phase values are more accurate than the amplitude values. In Fig. 6, the phase values are plotted on a much larger scale than the amplitude values and thus appear somewhat noisier. Actually, the greatest error in the phase is 4%, also at $n=1$.

7 Feasibility of detecting a DNAPL zone

The proposed EM technique must be able to make measurements at borehole separations of several meters for background resistivities typical of shallow sedimentary environments. Figure 7b shows the test case used in this study. The values were chosen to give conservative estimates for the resistivities and dielectric constants of clay-free sediments. The model considers three zones within the earth: the saturated zone with a resistivity of 100 $\Omega \cdot$ m and a dielectric constant of 7.3; the vadose zone with a resistivity of 1000 $\Omega \cdot$ m and a dielectric constant of 24.5; and a DNAPL zone with resistivity of 100 $\Omega \cdot$ m and a dielectric constant of 8. The air layer above the earth was modeled with a resistivity of 10^{10} $\Omega \cdot$ m and a dielectric constant of 1. The transmitters and receivers are vertical, co-planar electric dipoles 2 m long. The current flow is assumed constant in the transmitters. The boreholes are 20 m long and 10 m apart. All three layers have a magnetic permeability equal to that of free space. A frequency of 5 MHz is used, giving a wavelength of 10.3 m and a skin depth of 3.1 m in the lower layer.

The DNAPL zone is rectangular, 4 m wide, 4 m across, and 1 m thick and is centered between the boreholes. Three cases are considered: (1) The DNAPL is just below the water table, (2) the DNAPL is in the center of the section, and (3) the DNAPL is near the bottom of the boreholes.

The purpose of this model study was to answer a number of questions that determine the feasibility of such a system:

1. Will the received signals be large enough to be measured reliably?
2. Will the DNAPL zone create significant changes in the response?
3. Are responses distinctly different for different DNAPL depths?
4. Are the responses dependent on the dielectric constant?

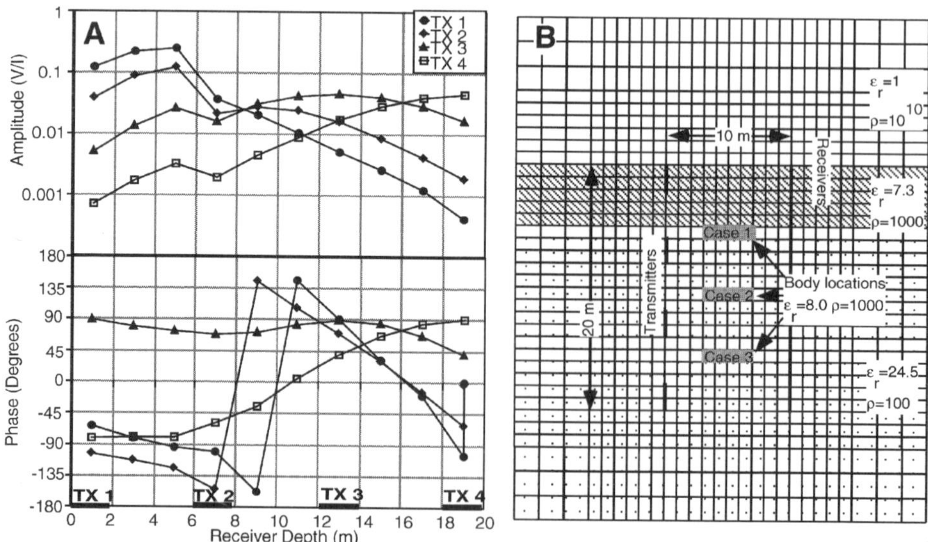

Figure 7. Phase and amplitude responses for the background case are shown for all for transmitting dipoles (a) and a cross-sectional view through the finite-difference mesh (b). Results were calculated for background (no DNAPL zone) and three different depths (cases 1–3) of the zone using four different transmitter depths.

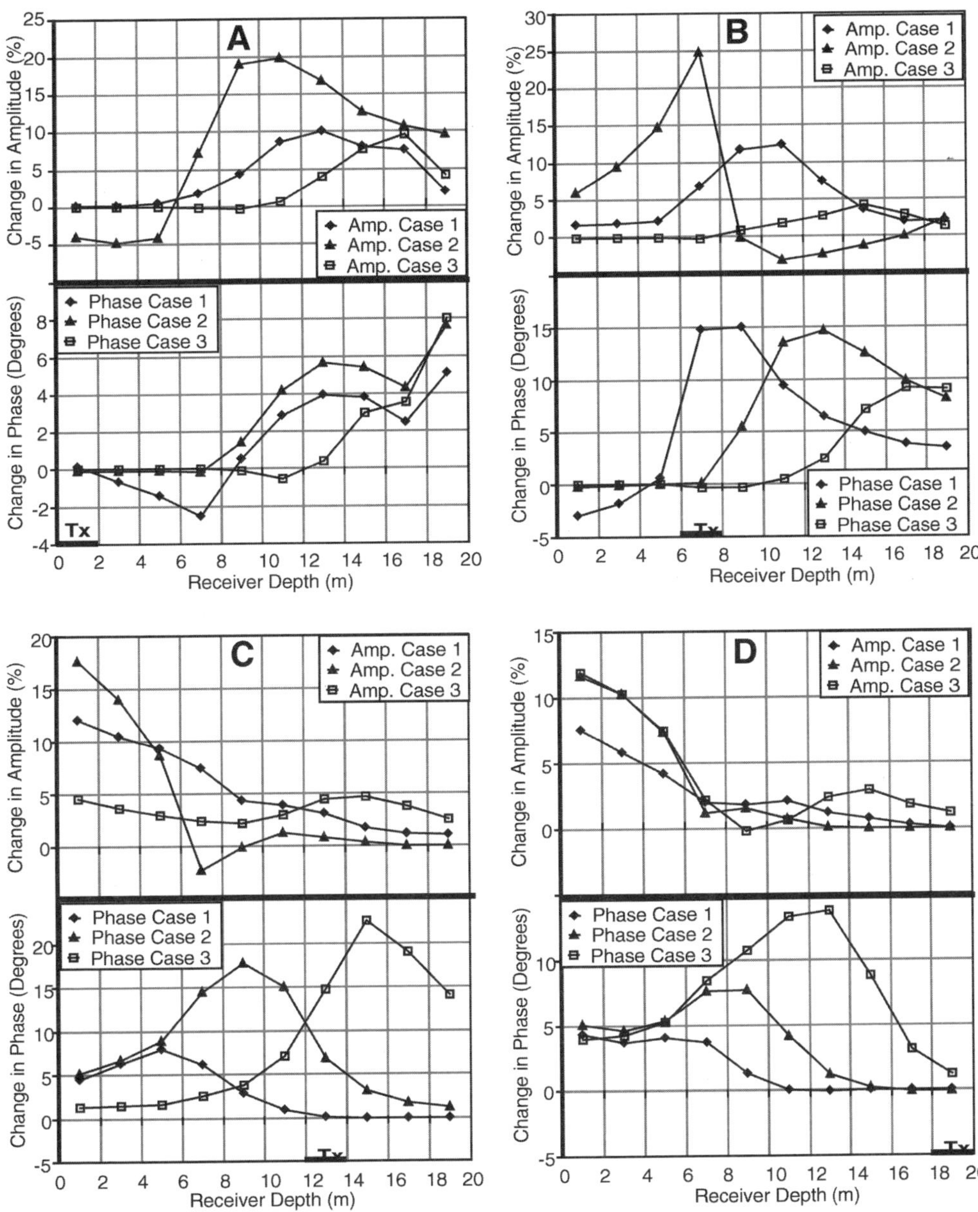

Figure 8. Change in response from the background case caused by the DNAPL zone is shown for (a) the transmitter just below the surface, (b) a transmitter centered 7 m below the surface, (c) a transmitter centered 13 m below the surface, and (d) the deepest transmitter 19 m below the surface.

For the background case with no DNAPL zone present, Fig. 7a shows the crosshole responses for transmitters centered at depths of 1, 7, 13, and 19 m. The magnitudes of the responses vary from around 0.9 to 200 mV/A of source current. The signal levels appear to be large enough to be measured accurately. For the shallow transmitters, the amplitudes at the deep receiving points are almost three orders of magnitude smaller than the amplitudes of receivers near the surface. It may be difficult to make accurate phase measurements at this point.

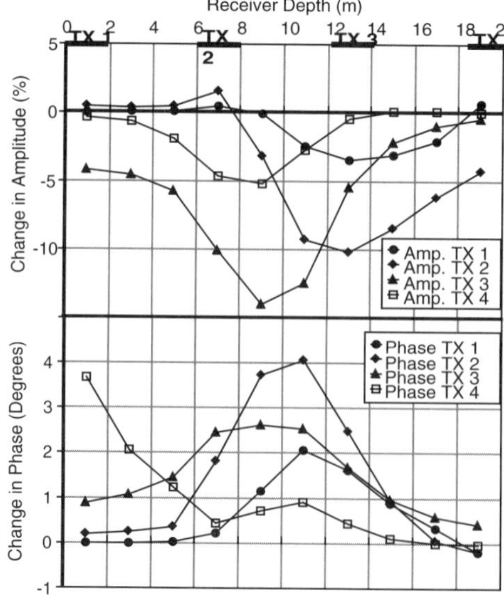

Figure 9. Change in response at the four transmitter locations when the dielectric constant of the DNAPL zone is decreased from 8 to 1 for Case 2. Amplitudes are given as percent change from the values in Fig. 8 and phases as differences from the values of Fig. 8.

To show the anomalous response caused by the DNAPL, the amplitudes in Fig. 8 are shown as a percent change from the background values given in Fig. 7a, and the phase values shown are the differences between the measured phase with the DNAPL present and the background (Fig. 7a) value. For each of the three cases, the DNAPL zone caused significant changes in amplitude, greater than 10%, at one or more transmitter locations. Case 1, where the DNAPL is just below the water table, shows the weakest change in amplitude response. When the DNAPL is deeper (Cases 2 and 3), we see anomalous responses of 25%.

In all three cases, the phase values change by 15° or more. We see the strongest phase responses when the transmitter, receiver, and DNAPL lie roughly along a straight line. These models also show the advantage of cross-borehole measurements over surface or surface-to-borehole measurements. The phase response is largest when both the transmitter and the receiver are at depth (Fig. 8). For the near-surface transmitter, the anomalous phase due to the DNAPL zone is small except for the largest receiver depths (Fig. 8). Although the change in phase is large, these data points have small amplitudes and large phase shifts (Fig. 7) and are the most difficult data points to measure accurately. These results show that the anomalies are distinct enough that it is possible to discriminate between these three cases, and all of the cases produce significant changes in response.

Figure 9 shows the change in response for Case 2 (Fig. 8), with the DNAPL in the middle of the section, when the dielectric constant is changed from 8 to 1.0. Although this is not a realistic value for dielectric constant in the earth, it shows that the response is a strong function of the dielectric constant. The decrease in the dielectric constant causes a large (about 15%) change in the amplitude and a rather modest change, 4°, in the phase. Comparing these anomalies with those in Fig. 8, we see that the shapes are different and that changing the dielectric constant creates a distinctly different anomaly pattern from changing both the resistivity and the dielectric constant together. Therefore, the response is a strong function of dielectric constant and there is hope of measuring the dielectric constant quantitatively.

8 Conclusions

The finite-difference method with scalar and vector potentials provides a flexible and robust formulation for EM modeling. The formulation can handle static (dc) fields, diffusive fields at induction frequencies, and vector wavefields. It also can handle large conductivity contrasts.

Modeling unbounded regions, such as the air layer above a half-space, requires absorbing boundary conditions. A terminal-impedance absorbing boundary method provides acceptable results, even if the grid covers a fraction of a wavelength. This is important because, in the frequency range of interest, boreholes usually will be less than a wavelength in the earth. Because the dielectric constant is much larger in the earth than in air, a large number of elements would be needed to move the edge of the mesh several wavelengths away in the air layer.

A crosswell measurement with 2-m electric dipoles at 5 MHz should detect a small zone of high resistivity and low dielectric constant, with properties similar to those of a DNAPL zone. The effects of dielectric and resistivity constrasts are large and different; both contrasts could be detected. The response also changes significantly with the position of the DNAPL zone; the weakest response appeared when the zone was just below the water table.

Acknowledgments

Much of this work was funded by the US Environmental Protection Agency Robert S. Kerr Laboratory under Cooperative Agreement CR-821517. Support also was provided by the University of Arizona Laboratory for Advanced Subsurface Imaging.

References

Bardi, I., Biro, O., and Preis, K., 1991, Finite element scheme for 3-D cavities without spurious modes: IEEE Trans. Magn., **27**, 4036–4039.

Boyse, W. E., Lynch, D. R., Paulsen, K. D., and Minerbo, G. N., 1992, Nodal-based finite-element modeling of Maxwell's equations: IEEE Trans. Antennas Propag., **40**, 642–651.

Boyse, W. E., Minerbo, G. N., Paulsen, K. D., and Lynch, D. R., 1993, Applications of potentials to finite element modeling of Maxwell's equations: IEEE Trans. Magn., **29**, 1333–1336.

Daily, W., 1984, Underground oil shale retort monitoring using geotomography: Geophysics, **49**, 1701–1707.

Daily, W., Ramirez, A., LaBrecque, D., and Nitao, J., 1992, Electrical resistivity tomography of vadose water movement: Water Resour. Res., **28**, 1429–1442.

Dey, A., and Morrison, H. F., 1979, Resistivity modeling for arbitrarily shaped three-dimensional structures: Geophysics, **44**, 753–780.

Druskin, V., and Knizhnerman, L., 1994, Spectral approach to solving three-dimensional Maxwell's diffusion equations in the time and frequency domains: Radio Sci., **29**, 937–953.

Fang, J., and Wu, Z., 1995, Generalized perfectly matched layer—an extension of Berenger's perfectly matched layer boundary condition: IEEE Microwave Guided Wave Lett., **5**, 451–453.

Mesquita, R. C., and Bastos, J. P. A., 1992, An incomplete gauge formulation for 3-D nodal finite-element magnetostatics: IEEE Trans. Magn., **28**, 1044–1047.

Morisue, T., 1993, A comparison of the Coulomb gauge and Lorentz gauge magnetic vector potential formulations for 3-D eddy current calculations: IEEE Trans. Magn., **29**, 1372–1375.

Mur, G., 1981, Absorbing boundary conditions for the finite-difference approximation of the time-domain electromagnetic-field equations: IEEE Trans. Electromagn. Compat., **23**, 377–382.

Newman, G., 1995, Crosswell electromagnetic inversion using integral equation and differential equations: Geophysics, **60**, 899–911.

Preis, K., Bardi, I., Biro, O., Magele, C., Renhart, W., Richter, K. R., and Vrisk, G., 1991, Numerical analysis of 3-D magnetostatic fields: IEEE Trans. Magn., **27**, 3798–3803.

Preis, K., Bardi, I., Biro, O., Magele, C., Vrisk, G., and Richter, K. R., 1992, Different finite element formulations of 3-D magnetostatic fields: IEEE Trans. Magn., **28**, 1056–1059.

Press, W. H., Teukolsky, S. A., Vetterling, W. T., and Flannery, B. P., 1992, Numerical recipes, 2nd ed.: Cambridge Univ. Press.

Schima, S. A., LaBrecque, D. J., and Lundegard, P., 1994, Using resistivity tomography to track air sparging, *in* Proceedings of the 1994 Symposium on the Application of Geophysics to Engineering and Environmental Problems: 757–774.

Schima, S. A., LaBrecque, D. J., and Miletto, M., 1993, Tracking fluid flow in the unsaturated zone using cross-borehole resistivity and IP, *in* Proceedings of the Symposium on the Application of Geophysics to Engineering and Environmental Problems: 527–544.

Schroeder, W., and Wolff, I., 1994, The origin of spurious modes in numerical solutions of electromagnetic field eigenvalue problems: IEEE Trans. Microwave Theory Tech., **42**, 644–653.

Shope, S., and Greenfield, R. J., 1988, Electromagnetic cross-hole tomography for tunnel detection, Proceedings of the Third Technical Symposium on Tunnel Detection: 123–135.

Sun, D., Manges, J., Xingchao, Y., and Cendes, Z., 1995, Spurious modes in finite-element methods: IEEE Antennas Propag. Magazine, **37**, nos. 12–24.

Wang, T., and Tripp, A. C., 1996, FDTD simulation of EM wave propagation in 3-D media: Geophysics, **61**, 110–120.

Webb, J. P., 1995, Application of the finite-element method to electromagnetic and electrical topics: Rep. Prog. Phys., **58**, 1673–1712.

Wong, S. H., and Cendes, Z. J., 1988, Combined finite-modal solution of three-dimensional eddy current problems: IEEE Trans. Magn., **24**, 2685–2687.

Speed and Accuracy in 3-D Resistivity Modeling

K. Spitzer[1]
B. Wurmstich[2]

Summary. Accuracy in finite-difference (FD) modeling is closely related to the discretization scheme, whereas speed depends mainly on the equation solvers. We compare the accuracy of five FD discretization schemes for 3-D resistivity modeling. Three schemes yield good results: a method using volume-weighted averages from conductivities assigned to neighboring grid cells, a method that integrates over elemental volumes, and a resistivity network approach. Discretization by elemental volume leads to coupling coefficients that are similar to those derived from the volume-weight method. The coefficients only differ by a real factor. In the second section, the cumulative amount of numerical work as a measure of speed is compared for five different equation solvers with and without preconditioning. The most efficient equation solver for symmetric matrices is the preconditioned conjugate gradient method. General matrix solution methods for both symmetric and nonsymmetric matrices—such as ORTHOMIN and the methods of stabilized biconjugate gradients and squared conjugate gradients—also achieve satisfactory convergence rates.

1 Introduction

Direct-current (dc) resistivity modeling is governed by Poisson's (or Laplace's) equation, which is a second-order elliptic partial differential equation. Different approaches may be used to derive finite-difference (FD) discretizations of this equation (Dey and Morrisson, 1979b; Scriba, 1981). We describe and analyze five common discretizations, which are based on a seven-point FD operator in three dimensions. These schemes are popular because of their simplicity and relatively small storage requirements. We discuss possibilities of deriving a linear set of FD equations that approximate the analytical formulation as accurately as possible.

[1]Niedersächsisches Landesamt für Bodenforschung—Geowissenschaftliche Gemeinschaftsaufgaben, Stilleweg 2, 30655 Hannover, Germany.
[2]Deceased. Formerly Western Atlas Logging Services, 10201 Westheimer Rd., Houston, TX 77042, USA.

The linear FD equations have a coefficient matrix that is sparse, banded, and diagonally dominant. The matrix also may be symmetric, depending on the method of incorporating boundary conditions. Conjugate gradient and conjugate residual methods are among the most efficient solvers for FD equations because they take advantage of the sparse banded structure of the coefficient matrix. These methods also are guaranteed to converge in the absence of roundoff errors. We use the total amount of numerical work (operation counts) as a hardware-independent measure of the performance of different equation solvers. [A similar study of the performance of equation solvers in electromagnetic (EM) modeling has been done by Agarwal and Weaver (1994).]

2 Accuracy: Comparison of five discretization schemes

Most 3-D FD approximations are based on seven-point schemes. The discretization involves a grid center point and its six direct neighbors along the main coordinate axes. Electrical conductivity values may be distributed arbitrarily within space. Variable grid spacing is incorporated easily in each case. More elaborate schemes are possible. For example, when all diagonal neighbors of an FD grid point are taken into account, a 27-point star is obtained with 27 coupling coefficients (Rodemann, 1994). The accuracy should increase with the number of coefficients, because grid-orientation effects can be reduced (Rossen and Dalton, 1990). These schemes, however, require more memory per grid point, which inevitably limits the number of grid points in the model and reduces the accuracy. This trade-off is subject to further investigations.

Section 2.1 describes the five seven-point operators. In each case, Neumann boundary conditions ($\partial V/\partial z = 0$) are applied at the surface of the Earth (half-space) and Dirichlet boundary conditions ($V = 0$) at all other boundaries, which are taken to be far away from the central region of interest.

2.1 *Methods*

The governing partial differential equation is the equation of continuity,

$$\nabla \cdot (\sigma \nabla V) = Q, \tag{1}$$

with σ as the electric conductivity, V as the electric potential, and Q as the source term. Q is defined by the electric current I and Dirac's delta function δ:

$$Q = -I\,\delta(x - x_q)\,\delta(y - y_q)\,\delta(z - z_q). \tag{2}$$

2.1.1 Brewitt-Taylor and Weaver (1976). This discretization scheme was originally designed for magnetotelluric model calculations by Brewitt-Taylor and Weaver (1976) and was adopted for two-dimensional (2-D) dc resistivity modeling by Mundry (1984). Spitzer (1995) adapted this approach to 3-D models and combined it with efficient preconditioned conjugate gradient (CG) solvers.

Because σ is a scalar function of space, Eq. (1) can be written

$$\sigma\left(\frac{\partial^2 V}{\partial x^2} + \frac{\partial^2 V}{\partial y^2} + \frac{\partial^2 V}{\partial z^2}\right) + \frac{\partial \sigma}{\partial x}\frac{\partial V}{\partial x} + \frac{\partial \sigma}{\partial y}\frac{\partial V}{\partial y} + \frac{\partial \sigma}{\partial z}\frac{\partial V}{\partial z}$$
$$= -I\,\delta(x - x_q)\,\delta(y - y_q)\,\delta(z - z_q). \tag{3}$$

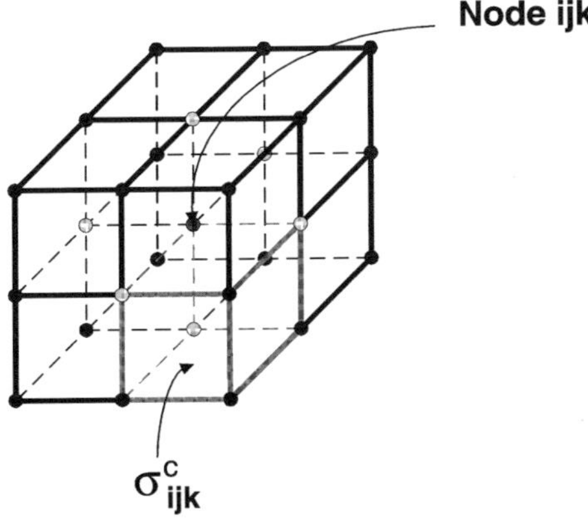

Figure 1. Discretization according to Brewitt–Taylor and Weaver (1976).

The central FD discretization for the potential and its derivatives at the nodes is constructed by a second-order Taylor-series expansion. Conductivities $\sigma_{i,j,k}$ at the grid points are calculated by a volume-weighted arithmetic average from conductivities assigned to grid cells $\sigma^c_{i,j,k}$ (Fig. 1). FD expressions for the conductivity gradients $\partial\sigma/\partial x$, $\partial\sigma/\partial y$, $\partial\sigma/\partial z$ are derived analogously. The source term is discretized through a finite source volume

$$\tau = \frac{(\Delta x_{i_q-1} + \Delta x_{i_q})(\Delta y_{j_q-1} + \Delta y_{j_q})\Delta z_1}{8} \tag{4}$$

for the special case of a source at the surface, yielding

$$Q \approx -\frac{I}{\tau}, \tag{5}$$

where i_q, j_q, and 1 are the source indices and Δx_i, Δy_j, and Δz_k are the grid spacings in x-, y-, and z-directions, respectively. The discretized FD formulation of Eq. (3) is

$$\begin{aligned} &C1_{i,j,k}V_{i-1,j,k} + C2_{i,j,k}V_{i+1,j,k} + C3_{i,j,k}V_{i,j-1,k} + C4_{i,j,k}V_{i,j+1,k} \\ &\quad + C5_{i,j,k}V_{i,j,k-1} + C6_{i,j,k}V_{i,j,k+1} = C0_{i,j,k}V_{i,j,k} + Q \end{aligned} \tag{6}$$

and the resulting coefficients read as follows [only for x–direction and for the nodes $(i-1, j, k)$ and (i, j, k)]:

$$C1^{\text{BT\&W}}_{i,j,k} = \frac{2\sigma_{i,j,k} - \frac{\partial\sigma_{i,j,k}}{\partial x}\Delta x_i}{\Delta x_{i-1}(\Delta x_{i-1} + \Delta x_i)}. \tag{7}$$

2.1.2 Discretization by points according to Dey and Morrison (1979a). This method was introduced for 2-D models and was called discretization by points. For 2-D conductivity structures in a 3-D space, the problem was solved after a Fourier transformation along the invariant direction. Here, we consider the fully 3-D case. The equation of continuity is discretized after modification to the following form using elementary vector identities:

$$\nabla^2(\sigma V) + \sigma\nabla^2 V - V\nabla^2\sigma = -2Q. \tag{8}$$

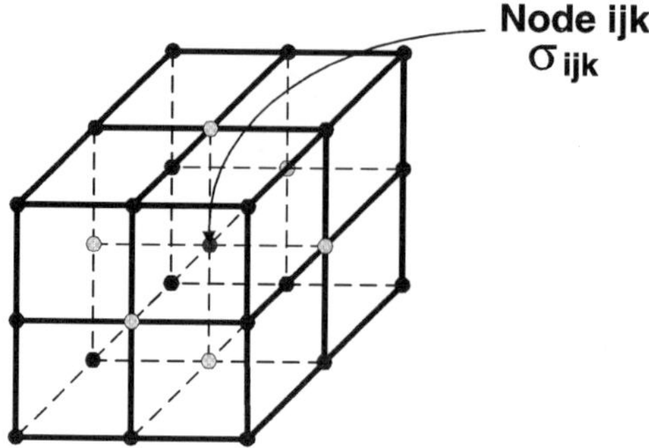

Figure 2. Discretization by points according to Dey and Morrison (1979a).

Conductivities $\sigma_{i,j,k}$ are assigned to the nodes (i, j, k) right from the start (Fig. 2). At any node, the ∇^2 operator on any distribution $P_{i,j,k}$ is approximated by the following FD equation:

$$\nabla^2 P_{i,j,k} = \frac{2}{\Delta x_i + \Delta x_{i-1}} \left(\frac{P_{i-1,j,k} - P_{i,j,k}}{\Delta x_{i-1}} + \frac{P_{i+1,j,k} - P_{i,j,k}}{\Delta x_i} \right) + \frac{2}{\Delta y_j + \Delta y_{j-1}} \left(\frac{P_{i,j-1,k} - P_{i,j,k}}{\Delta y_{j-1}} + \frac{P_{i,j+1,k} - P_{i,j,k}}{\Delta y_j} \right) + \frac{2}{\Delta z_k + \Delta z_{k-1}} \left(\frac{P_{i,j,k-1} - P_{i,j,k}}{\Delta z_{k-1}} + \frac{P_{i,j,k+1} - P_{i,j,k}}{\Delta z_k} \right). \tag{9}$$

$P_{i,j,k}$ stands for either $V_{i,j,k}$, $\sigma_{i,j,k}$, or $(\sigma_{i,j,k} V_{i,j,k})$. The coupling coefficients are derived as the mean value of the grid-point conductivities weighted by the grid spacing (again only for x):

$$C1^{\text{D\&M1}}_{i,j,k} = \frac{\sigma_{i-1,j,k} + \sigma_{i,j,k}}{\Delta x_{i-1}(\Delta x_{i-1} + \Delta x_i)} \tag{10}$$

2.1.3 Discretization by elemental volume according to Dey and Morrison (1979b). Dey and Morrison (1979a) described another FD scheme called discretization by area, which was taken to be the basis for their 3-D approach (Dey and Morrison, 1979b) labeled discretization by elemental volume. Conductivities are assigned to grid cells (Fig. 3). The governing differential equation is the volume-integrated form of the equation of continuity [Eq. (1)]:

$$\iiint\limits_{\Delta v_{i,j,k}} \nabla \cdot (\sigma \nabla V)\, dv_{i,j,k} = -\iiint\limits_{\Delta v_{i,j,k}} I\delta(x - x_q)\delta(y - y_q)\delta(z - z_q)\, dv_{i,j,k}. \tag{11}$$

The volume integral of the left side is converted into a surface integral along the six sides of the elemental volume $\Delta v_{i,j,k}$ using Green's theorem. The integral of the right side is approximated by I/τ [according to Eqs. (4) and (5)], yielding

$$\iint\limits_{s_{i,j,k}} \sigma \frac{\partial V}{\partial \eta}\, ds_{i,j,k} = -\iiint\limits_{\Delta v_{i,j,k}} I/\tau\, dv_{i,j,k}, \tag{12}$$

where η is the outward normal. The right-hand side of Eq. (12) becomes $-2I$ at the location of the sources. The factor 2 appears because of the halved source volume at the surface. Using central finite differences for $\partial V/\partial \eta$, we obtain the following coupling

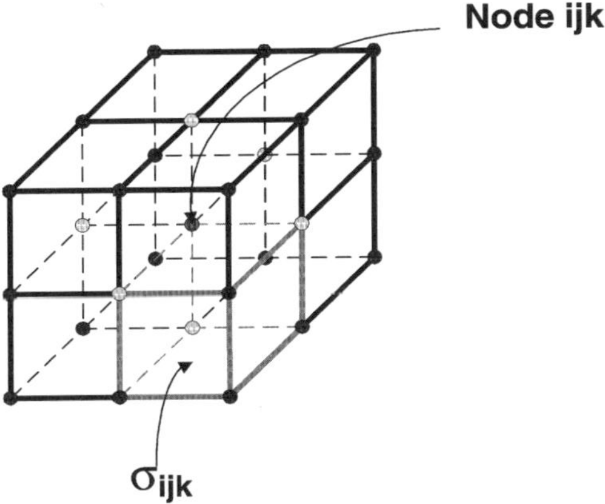

Figure 3. Discretization by elemental volume according to Dey and Morrison (1979b).

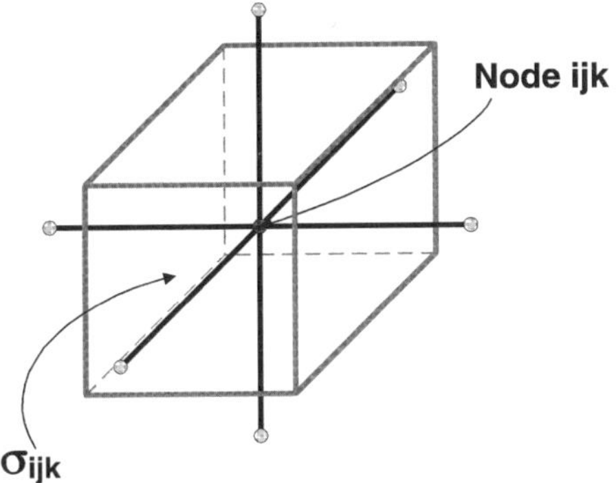

Figure 4. Discretization according to Wurmstich and Morgan (1994).

coefficients, again as an example only for the nodes $(i-1, j, k)$ and (i, j, k):

$$C1_{i,j,k}^{\text{D\&M2}} = \frac{1}{\Delta x_{i-1}} \left(\sigma_{i-1,j-1,k-1} \frac{\Delta y_{j-1} \Delta z_{k-1}}{4} + \sigma_{i-1,j,k-1} \frac{\Delta y_j \Delta z_{k-1}}{4} + \sigma_{i-1,j-1,k} \frac{\Delta y_{j-1} \Delta z_k}{4} + \sigma_{i-1,j,k} \frac{\Delta y_j \Delta z_k}{4} \right). \tag{13}$$

2.1.4 Discretization according to Wurmstich and Morgan (1994). A further FD approach is described by Wurmstich and Morgan (1994). This approach was adopted from reservoir simulation and is based on a conductivity network, using the principle of conservation of charge. Conductivities are assigned to the grid cells; nodal points are located in the center of each cell (Fig. 4). According to Kirchhoff's law, the partial currents $I_\ell = A_\ell j_\ell$, $\ell = 1, \ldots, 6$, are summed at each node to balance the source current I_S:

$$\sum_{\ell=1}^{6} A_\ell j_\ell = I_S. \tag{14}$$

A_ℓ is the ℓth face of the grid cell through which the current of the density j_ℓ flows. The conductivity $\bar{\sigma}_{i,j,k}$ between two nodes is calculated by a harmonic mean value of conductivities of neighboring grid cells, here, for example, in x-direction for node

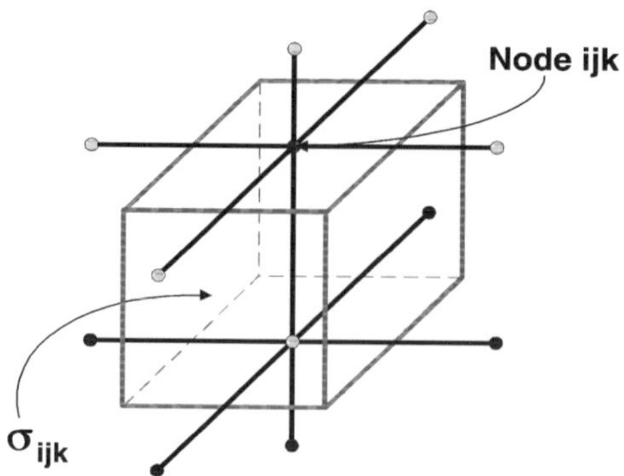

Figure 5. Discretization according to Zhang et al. (1995).

(i, j, k) and $(i+1, j, k)$:

$$\bar{\sigma}_{i,j,k} = 2\,\frac{\sigma_{i,j,k}\sigma_{i+1,j,k}}{\sigma_{i,j,k} + \sigma_{i+1,j,k}}. \tag{15}$$

I_S is only nonzero in volumes containing sources or sinks. The corresponding coupling coefficient is

$$C1^{\text{W\&M}}_{i,j,k} = 4\,\frac{\bar{\sigma}_{i,j,k}}{\Delta x_{i-1}(\Delta x_{i-1} + \Delta x_i)} \tag{16}$$

2.1.5 Discretization according to Zhang et al. (1995). This approach is based on a resistivity network scheme by Madden (1972) and discretizes Kirchhoff's law,

$$\sum_{\ell=1}^{6} I_\ell = I_S, \tag{17}$$

by defining impedances R_x, R_y, and R_z. For example, R_x—in fact, its reciprocal value—is given by

$$\frac{1}{R_x} = \frac{1}{\frac{\Delta x_i \rho_{i,j,k}}{\Delta y_j \Delta z_k} + \frac{\Delta x_{i-1}\rho_{i-1,j,k}}{\Delta y_j \Delta z_k}} + \frac{1}{\frac{\Delta x_i \rho_{i,j,k-1}}{\Delta y_j \Delta z_{k-1}} + \frac{\Delta x_{i-1}\rho_{i-1,j,k-1}}{\Delta y_j \Delta z_{k-1}}}, \tag{18}$$

defining the coupling coefficient $C1^{\text{ZM\&M}}_{i,j,k}$. I_S is the resulting source current summing the six partial currents of each branch along the main axes. Conductivities are assigned to each grid cell; voltage nodes are located at the top and bottom center of each block (Fig. 5).

2.2 Results

The accuracy of the five FD discretization schemes is compared for three models: a homogeneous half-space (Fig. 6), a dike model (Fig. 7), and a three-layered earth (Fig. 8). The grid has 63 × 63 × 30 nodes. The modeling-domain boundaries are located at ±2520 m in x- and y-directions and 2520 m in the z-direction. Grid spacings increase by a factor of approximately 1.3 toward the outer boundaries. The two sources are located at (±1 m, 0 m, 0 m). The comparison is of calculated apparent resistivities $\rho_a = k\Delta V/I$, where k is the geometric factor derived from Schlumberger arrangements,

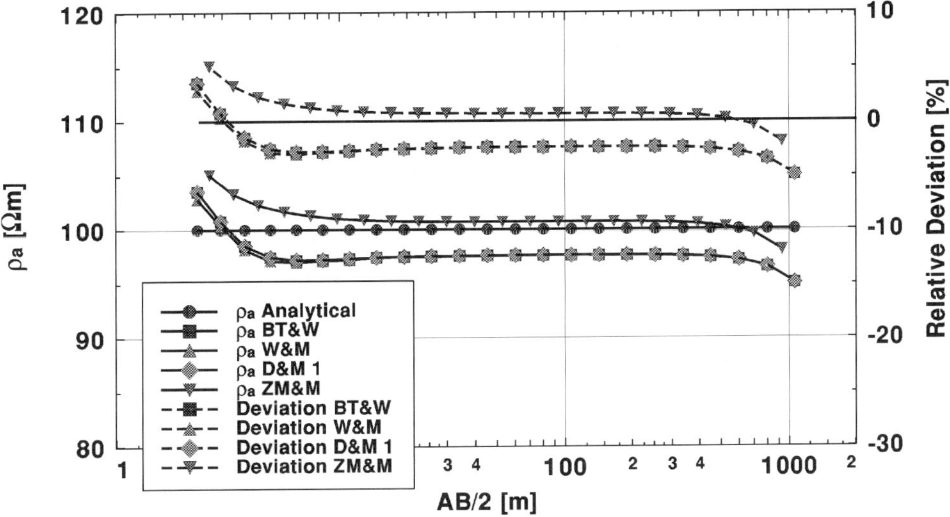

Figure 6. Comparison for a homogeneous half-space.

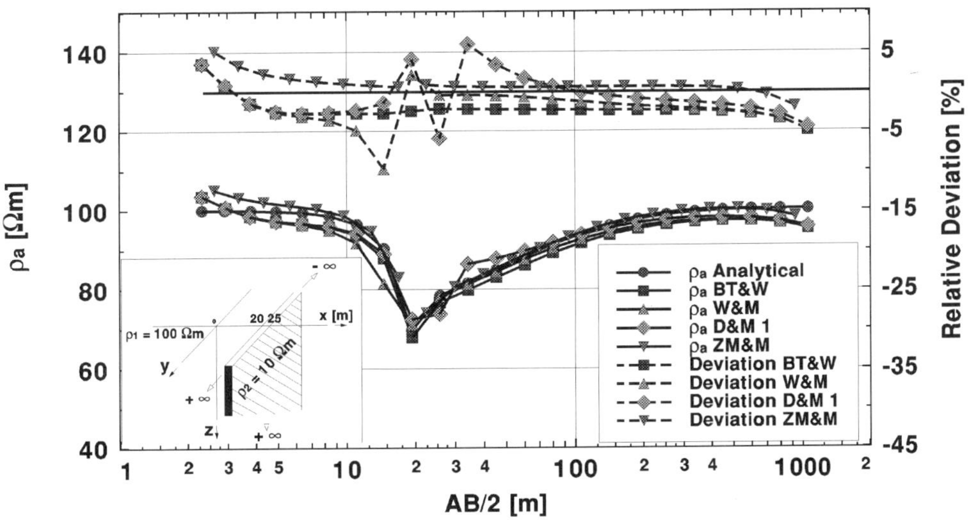

Figure 7. Comparison for a dike model.

I is the source current, and ΔV is the corresponding potential difference. By exploiting the principle of reciprocity, ρ_a, which also could be regarded as a normalized potential, is calculated and plotted as a function of the electrode half-spacing AB/2. In each plot there is a right ordinate, indicating the relative deviation to the analytically determined values or, in the case of the three-layered earth, to the results obtained by applying a 1-D linear filter method (e.g., Koefoed, 1979).

Because the results for BT&W are identical to those for D&M2, we omit all D&M2 curves in the figures. In fact, it can be shown that the two approaches are equal, when the source terms and the coefficients for BT&W are divided by the elemental volume $8/[(\Delta x_{i-1}+\Delta x_i)(\Delta y_{j-1}+\Delta y_j)(\Delta z_{k-1}+\Delta z_k)]$. Mundry (1984) stated a higher accuracy for BT&W at conductivity contrasts of more than 2:1 compared to the approach of Dey and Morrison (1979a), not specifying D&M1 or D&M2. This is at least imprecise. As shown here, only D&M1 is less accurate than BT&W.

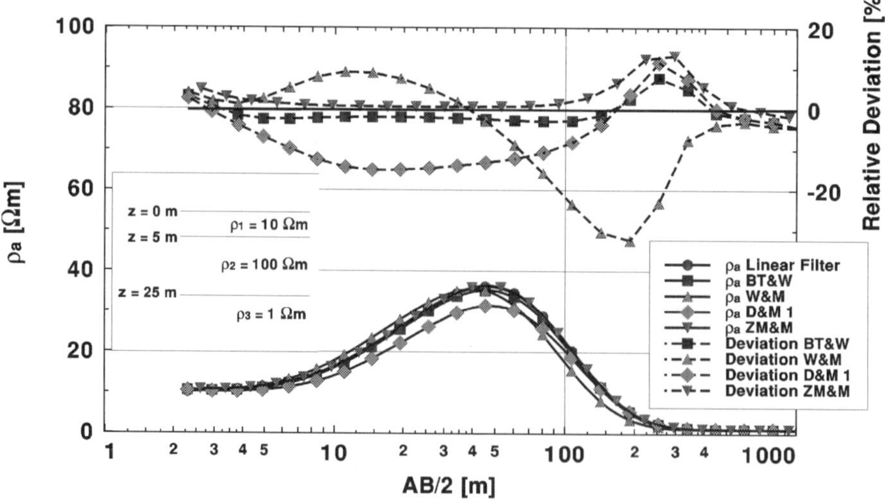

Figure 8. Comparison for a three-layered earth.

For the homogeneous half-space of $\rho_1 = 100$ ohm-m, BT&W, D&M1, D&M2, and W&M yield the same results (Fig. 6), but ZM&M yields better results. Errors at large electrode spacings (AB/2 > 800 m) are caused by the boundaries at finite distance; error at short spacings (AB/2 < 3 m) is caused by the close singularity at the source locations. The three outmost nodal points have been omitted because of the boundary effects.

The accuracy of the methods is affected by conductivity contrasts. As an example, a model with a conductive dike of $\rho_2 = 10$ ohm-m is discussed. The dike is 5 m wide and is 20 m offset from the origin of the coordinate system. It extends to infinity in $\pm y$- and $+z$-directions. In this case, an analytical solution exists (Telford et al., 1990). D&M1 and W&M show significant deviations when approaching the lateral conductivity contrasts, whereas BT&W (D&M2) and ZM&M stay nearly unaffected (Fig. 7).

The third model, a stratified medium, corroborates the preceding result. The model consists of three layers. Layer boundaries are at 5 m and 25 m, respectively. The layer resistivities are $\rho_1 = 10$ ohm-m, $\rho_2 = 100$ ohm-m, and $\rho_3 = 1$ ohm-m. Again, more accurate results are obtained for BT&W (D&M2) and ZM&M when layer boundaries are encountered.

The discretization error can be decomposed into a spatial and a conductivity part. The spatial part is the same for BT&W, D&M1, D&M2, and W&M, whereas the method of discretizing and averaging conductivities differs among the methods. The assignment of conductivities to the grid also differs for all methods. BT&W and D&M2 use conductivity values assigned to grid cells, resulting in well-defined conductivity boundaries along the grid lines. Because of the assignment of conductivities to grid points and the selected averaging schemes, conductivity boundaries for D&M1 and W&M are not as well defined as for BT&W and D&M2, resulting in decreased accuracy for D&M1 and W&M. It is not clear why ZM&M is more accurate for the homogeneous half-space in the absence of conductivity contrasts (Fig. 6). One reason could be that the ZM&M approach results in a different spatial and conductivity discretization when compared to the other four methods. The voltage nodes for ZM&M are located between the standard grid nodes. It is difficult to adapt a standardized grid in a comparable manner for each method. Generally, the accuracy of all methods may be improved when a finer discretization at conductivity contrasts is used.

3 Speed: Comparison of five equation solvers

The efficiency of iterative matrix solution methods, such as the CG-type (CGM) or the conjugate residual type methods (CRM) is best expressed in terms of cumulative numerical work. The cumulative work can be defined as floating point operations (FPOs) per iteration and grid point. Thus, the cumulative numerical work is a hardware-independent measure of performance that allows us to compare algorithms with a different number of FPOs, per iteration. After outlining some general principles of the methods, and preconditioning, some specifics of the five compared methods are explained.

3.1 CG or CR equation solvers

The set of FD equations can be expressed in matrix form,

$$\underset{\sim}{\mathbf{A}}\mathbf{x} = \mathbf{b}, \tag{19}$$

where the matrix $\underset{\sim}{\mathbf{A}}$ contains the coefficients $C0, \ldots, C6$ [Eq. (6)], the vector $\mathbf{x}$ is made up of the unknown potentials V, and the vector $\mathbf{b}$ contains the source terms [Eq. (5)]. Generally, the matrix $\underset{\sim}{\mathbf{A}}$ is sparse, banded, and diagonally dominant. Depending on the method of incorporating Neumann boundary conditions, the matrix also may be symmetric.

Methods for solving Eq. (19) are classified as either direct or iterative. Direct methods consist of some form of Gaussian elimination or closely related procedures such as LU decomposition (Press et al., 1989). For sparse, banded matrices, the efficiency of direct methods is inversely proportional to the bandwidth of the matrix. The bandwidth is given by the smaller dimension of a 2-D model and the product of the two smallest dimensions of a 3-D model, respectively. Direct methods always solve the matrix equations, but they become inefficient for large problems because of the work and storage requirements. Iterative methods are usually more efficient for large problems than direct methods because they take advantage of the sparseness of matrix $\underset{\sim}{\mathbf{A}}$. The amount of numerical work of iterative methods depends on the number of equations to be solved and the number of iterations. However, iterative methods may fail to converge. Conjugate-gradient methods and Conjugate-residual methods are an iterative equivalent of Gaussian elimination. In other words, it may be shown that CGMs and CRMs theoretically converge within N iterations, where N denotes the number of linear equations. In addition, CGMs and CRMs are also efficient, because they take advantage of the sparseness of the matrix $\underset{\sim}{\mathbf{A}}$ by performing multiplications of vectors only with the nonzero elements of the matrix $\underset{\sim}{\mathbf{A}}$ or its transpose (Press et al., 1989).

Iterative methods update an initial guess of the solution vector until the solution is found. The iterative process may be written as

$$\underset{\sim}{\mathbf{A}}(\mathbf{x}^{\ell} + \Delta\mathbf{x}^{\ell+1}) - \mathbf{b} = \mathbf{r}^{\ell} - \Delta\mathbf{r}^{\ell+1}, \tag{20}$$

where ℓ counts the iterations and $\Delta\mathbf{x}$ and $\Delta\mathbf{r}$ denote update vectors for $\mathbf{x}$ and $\mathbf{r}$, respectively. The initial guess for the solution of Eq. (19) is $\mathbf{x}_0 = 0$ in each case.

CGMs and CRMs converge to a minimum of a function $f(\mathbf{x})$. Choices of $f(\mathbf{x})$ differ for both methods. The functions to be minimized are (Press et al., 1989)

$$f(\mathbf{x}) = 0.5\ \mathbf{x}^T \underset{\sim}{\mathbf{A}} x - \mathbf{b}^T \mathbf{x} = (\mathbf{r}, \mathbf{e}), \tag{21}$$

and

$$f(\mathbf{x}) = 0.5\ (\underset{\sim}{\mathbf{A}}\mathbf{x} - \mathbf{b},\ \underset{\sim}{\mathbf{A}}\mathbf{x} - \mathbf{b}) = 0.5\ (\mathbf{r}, \mathbf{r}) \tag{22}$$

for the CGM and the CRM, respectively. The superscript T indicates the transposed property. The error vector is defined as $\mathbf{e} = \mathbf{x} - \mathbf{x}^{\ell}$. These functions are minimized when

$$\nabla f(\mathbf{x}) = \underset{\sim}{\mathbf{A}}\mathbf{x} - \mathbf{b} = 0\,, \tag{23}$$

and

$$\nabla f(\mathbf{x}) = \underset{\sim}{\mathbf{A}}^T(\underset{\sim}{\mathbf{A}}\mathbf{x} - \mathbf{b}) = 0\ , \tag{24}$$

respectively. In case of the CGM, the minimum of the function $f(x)$ corresponds to the solution of the matrix system, which can be obtained only for positive-definite and symmetric (PDS) matrices. In case of the CRM, the minimum of the function $f(x)$ corresponds to a minimized inner product of the residual vector.

For both, CGM and CRM, the update vector $\Delta\mathbf{x}^{\ell+1}$ now can be found as the product of the factor λ^{ℓ} and a vector $\mathbf{n}^{\ell}$, giving

$$\mathbf{x}^{\ell+1} = \mathbf{x}^{\ell} + \lambda^{\ell}\mathbf{n}^{\ell} \tag{25}$$

with different factors λ^{ℓ} for CGM and CRM

$$\lambda^{\ell} = \frac{(\mathbf{r}^{\ell}, \mathbf{r}^{\ell})}{(\mathbf{n}^{\ell}, \underset{\sim}{\mathbf{A}}\mathbf{n}^{\ell})}\quad \text{(CGM)}, \qquad \lambda^{\ell} = \frac{(\mathbf{r}^{\ell}, \underset{\sim}{\mathbf{A}}\mathbf{n}^{\ell})}{(\underset{\sim}{\mathbf{A}}\mathbf{n}^{\ell}, \underset{\sim}{\mathbf{A}}\mathbf{n}^{\ell})}\quad \text{(CRM)}. \tag{26}$$

For both methods, the updated residuals are determined using $\mathbf{n}^{\ell}$ and the corresponding λ^{ℓ}:

$$\mathbf{r}^{\ell+1} = \mathbf{r}^{\ell} - \lambda^{\ell}\underset{\sim}{\mathbf{A}}\mathbf{n}^{\ell}. \tag{27}$$

The vectors $\mathbf{n}^{\ell}$ are conjugate search directions in error space [directions are called conjugate when the minimization along one direction does not interfere with a minimization along another direction; i.e., if $(\mathbf{n}^{k}, \underset{\sim}{\mathbf{A}}\mathbf{n}^{\ell}) = 0$, $k \neq \ell$]. The search direction for the following iteration can be found as that part of the residual that is normal to all previous search directions, leading to two different expressions for CGM and CRM:

$$\mathbf{n}^{\ell+1} = \mathbf{r}^{\ell+1} + \kappa^{\ell}\mathbf{n}^{\ell} \quad \text{(CGM)}, \qquad \mathbf{n}^{\ell+1} = \mathbf{r}^{\ell+1} + \sum_{j=0}^{\ell} \kappa_j^{\ell}\mathbf{n}^{j} \quad \text{(CRM)}, \tag{28}$$

where the orthogonality coefficients are given as

$$\kappa^{\ell} = \frac{(\mathbf{r}^{\ell+1}, \mathbf{r}^{\ell+1})}{(\mathbf{r}^{\ell}, \mathbf{r}^{\ell})}\quad \text{(CGM)} \quad \text{and} \quad \kappa_j^{\ell} = \frac{(\underset{\sim}{\mathbf{A}}\mathbf{n}^{j}, \underset{\sim}{\mathbf{A}}\mathbf{r}^{\ell+1})}{(\underset{\sim}{\mathbf{A}}\mathbf{n}^{j}, \underset{\sim}{\mathbf{A}}\mathbf{n}^{j})}\quad \text{(CRM)}. \tag{29}$$

The above outlined sequence of vectors satisfies the conjugacy condition

$$(\mathbf{n}^{k}, \underset{\sim}{\mathbf{A}}\mathbf{n}^{\ell}) = 0, \quad k \neq \ell, \tag{30}$$

as well as the orthogonality condition

$$(\mathbf{r}^{\ell}, \mathbf{r}^{k}) = 0, \quad k \neq \ell. \tag{31}$$

3.2 Preconditioning

The rate of convergence is related to the condition number and the eigenspectrum of the matrix $\underset{\sim}{\mathbf{A}}$. The rate of convergence of these methods can be accelerated by preconditioning of the matrix $\underset{\sim}{\mathbf{A}}$, i.e., solving the equivalent system

$$(\underset{\sim}{\mathbf{H}}^{-1}\underset{\sim}{\mathbf{A}})\mathbf{x} = \underset{\sim}{\mathbf{H}}^{-1}\mathbf{b},$$

where $\underset{\sim}{\mathbf{H}}^{-1}$ is an approximate inverse of $\underset{\sim}{\mathbf{A}}$, and $(\underset{\sim}{\mathbf{H}}^{-1}\underset{\sim}{\mathbf{A}}) \approx \underset{\sim}{\mathbf{I}}$. Note that the identity matrix $\underset{\sim}{\mathbf{I}}$ has a condition number of 1.

As an example, we discuss the symmetric successive overrelaxation (SSOR)-preconditioning method according to Schwarz (1991), which goes back to the work of Evans (1968). The preconditioner has been applied to all of our tested equation solvers after the matrix was scaled and was chosen because no additional storage for the preconditioning matrix is required. The symmetrized and scaled coefficient matrix $\underset{\sim}{\mathbf{A}}$ can be written as the sum of a lower triangular matrix $\underset{\sim}{\mathbf{E}}$, an upper triangular matrix $\underset{\sim}{\mathbf{F}}$, and the identity matrix $\underset{\sim}{\mathbf{I}}$, respectively,

$$\underset{\sim}{\mathbf{A}} = \underset{\sim}{\mathbf{E}} + \underset{\sim}{\mathbf{I}} + \underset{\sim}{\mathbf{F}}, \tag{32}$$

with $\underset{\sim}{\mathbf{F}} = \underset{\sim}{\mathbf{E}}^T$. The preconditioning matrix $\underset{\sim}{\mathbf{H}}$ is defined by

$$\underset{\sim}{\mathbf{H}} = \underset{\sim}{\mathbf{C}}\underset{\sim}{\mathbf{C}}^T. \tag{33}$$

For $\underset{\sim}{\mathbf{C}}$ we choose a matrix similar to the above-defined matrix $\underset{\sim}{\mathbf{E}}$, giving $\underset{\sim}{\mathbf{C}}$ the nonzero pattern of the strictly lower part of $\underset{\sim}{\mathbf{A}}$. Then, $\underset{\sim}{\mathbf{H}}$ reads as follows:

$$\underset{\sim}{\mathbf{H}} = (\underset{\sim}{\mathbf{I}} + \omega\underset{\sim}{\mathbf{E}})(\underset{\sim}{\mathbf{I}} + \omega\underset{\sim}{\mathbf{F}}), \tag{34}$$

where $\underset{\sim}{\mathbf{C}} = \underset{\sim}{\mathbf{I}} + \omega\underset{\sim}{\mathbf{E}}$, and $\omega \in I\!R$ denotes a relaxation factor. It is chosen to be $\omega = 1.4$ (Eq. 34). When ω is plotted as a function of the convergence rate, it describes a broad minimum around the chosen value of 1.4. Numerous tests have confirmed that this value ensures rapid convergence. Slight alterations of ω do not affect the relaxation process significantly. The unpreconditioned algorithm is obtained for $\omega = 0.0$.

SSOR preconditioning has the advantage of not requiring any further memory because the preconditioning matrix $\underset{\sim}{\mathbf{H}}$ is not explicitly constructed and stored. Incomplete factorization methods are usually efficient alternatives, although the preconditioning matrix requires additional storage. However, they may break down because of attempted division by zero pivot. Substituting an arbitrary positive value for the zero pivot avoids breakdown [e.g., incomplete Cholesky preconditioning; see Kershaw (1978)]. Although factorization requires additional FPOs per equation and iteration, the cumulative amount of numerical work is reduced by accelerating the convergence. A more detailed comparison of different preconditioners is to be the subject of future investigations.

3.3 Methods

3.3.1 CG method. The CG method (Hestenes and Stiefel, 1952) uses 25 (39 with preconditioning) $\times N$ FPOs per iteration, where N denotes the number of equations and the value in brackets is the respective number of iterations for the SSOR-preconditioned version. Storage requirements are $10 \times N$ elements for the SSOR-preconditioned version. As stated above, the CG method is limited to symmetric matrices because of the functional chosen to be minimized.

3.3.2 Biconjugate gradient (BiCG) method. For the solution of nonsymmetric matrix equation systems ($\underset{\sim}{\mathbf{A}}^{\mathrm{T}} \neq \underset{\sim}{\mathbf{A}}$) the BiCG method (Fletcher, 1976) can be used. BiCG uses two similar conjugate search-direction vectors $\mathbf{n}$ and $\tilde{\mathbf{n}}$, resulting in two residual vectors $\mathbf{r}$ and $\tilde{\mathbf{r}}$. These quantities correspond to $\underset{\sim}{\mathbf{A}}$ and $\underset{\sim}{\mathbf{A}}^{\mathrm{T}}$. The update vector $\Delta\mathbf{x}$ is a function of both search-direction vectors $\mathbf{n}$ and $\tilde{\mathbf{n}}$. BiCG is more general than CG and includes CG for $\mathbf{r} = \tilde{\mathbf{r}}$ and $\mathbf{n} = \tilde{\mathbf{n}}$.

Compared to CG, the number of FPOs increases from 25 (39 with preconditioning) $\times N$ to 42 (70 with preconditioning) $\times N$ per iteration. Storage requirements are $14 \times N$ elements.

3.3.3 Conjugate gradients squared (CGS). The CGS method (Sonneveld, 1989) is applicable to nonsymmetric matrices and constitutes a variant of the BiCG method avoiding the use of the transpose $\underset{\sim}{\mathbf{A}}^{\mathrm{T}}$. It is derived from a polynomial description of the CG and the BiCG method regarding the residual vector $\mathbf{r}_i$ as the product of $\mathbf{r}_0$ and an ith-degree polynomial in $\underset{\sim}{\mathbf{A}}$ termed $P_i(\underset{\sim}{\mathbf{A}})$: $\mathbf{r}_i = P_i(\underset{\sim}{\mathbf{A}})\, \mathbf{r}_0$. Satisfying also $\tilde{\mathbf{r}}_i = P_i(\underset{\sim}{\mathbf{A}}^{\mathrm{T}})\, \tilde{\mathbf{r}}_0$, we obtain

$$(\tilde{\mathbf{r}}_i, \mathbf{r}_i) = [P_i(\underset{\sim}{\mathbf{A}}^{\mathrm{T}})\, \tilde{\mathbf{r}}_0,\, P_i(\underset{\sim}{\mathbf{A}})\, \mathbf{r}_0] = \left[\tilde{\mathbf{r}}_0,\, P_i^2(\mathbf{A})\, \mathbf{r}_0\right]. \tag{35}$$

The contraction operator $P_i(\underset{\sim}{\mathbf{A}})$ is applied twice on $\mathbf{r}_n$ instead of exploiting the quasi-residuals $\tilde{\mathbf{r}}_n$ ($\rightarrow$ CG Squared); 45 (73 with preconditioning) FPOs per iteration and equation and $15 \times N$ elements for storage are required.

3.3.4 Biconjugate gradients stabilized (BiCGSTAB). BiCGSTAB (Van der Vorst, 1992) is a variant of CGS avoiding its unstable convergence behavior. It is applicable on nonsymmetric matrices without using the transpose $\underset{\sim}{\mathbf{A}}^{\mathrm{T}}$. Instead of using the contraction operator twice $P_i^2(\underset{\sim}{\mathbf{A}})\mathbf{r}_0$, it computes a sequence $Q_i(\underset{\sim}{\mathbf{A}})\, P_i(\underset{\sim}{\mathbf{A}})\, \mathbf{r}_0$ using the polynomial Q_i as a description of a steepest descent update. BiCGSTAB needs 48 (76 with preconditioning) FPOs per iteration and equation and $14 \times N$ elements for storage.

3.3.5 ORTHOMIN. The ORTHOMIN method (Behie and Vinsome, 1982; Vinsome, 1976) is a truncated CRM that is restarted every four or five iterations using a shift vector $\mathbf{v}$ to obtain a new search direction. The shift vector is defined as the solution of a residual equation

$$\mathbf{v}^{\ell+1} = \underset{\sim}{\mathbf{H}}^{-1}\mathbf{r}^{\ell} \tag{36}$$

that incorporates the preconditioning step. The approximate inverse $\underset{\sim}{\mathbf{H}}^{-1}$ of $\underset{\sim}{\mathbf{A}}$ can be obtained using the SSOR preconditioner; 51 (65 with preconditioning) FPOs per iteration and equation and $15 \times N$ elements for storage are needed.

3.4 Summary of storage and FPOs

Table 1 summarizes the storage requirements and FPOs per equation and iteration. Generalization to nonsymmetric matrices and preconditioning generally add to the storage requirements and the number of FPOs per iteration and equation, respectively.

3.5 Results

Figures 9, 10, and 11 show the convergence behavior of the preconditioned methods described above in terms of FPOs per equation for the homogeneous half-space, the

Table 1. Comparison of FPOs per iteration and storage requirements.

Method	FPOs[a,b]	Storage[a]	Remarks
CG	25 (39) $\times N$	10 $\times N$	PDS matrices only
BiCG	42 (70) $\times N$	14 $\times N$	
CGS	45 (73) $\times N$	15 $\times N$	
BiCGSTAB	48 (76) $\times N$	14 $\times N$	
ORTHOMIN	51 (65) $\times N$	15 $\times N$	

[a] N number of equations.
[b] Value in parentheses stands for SSOR-preconditioned method.

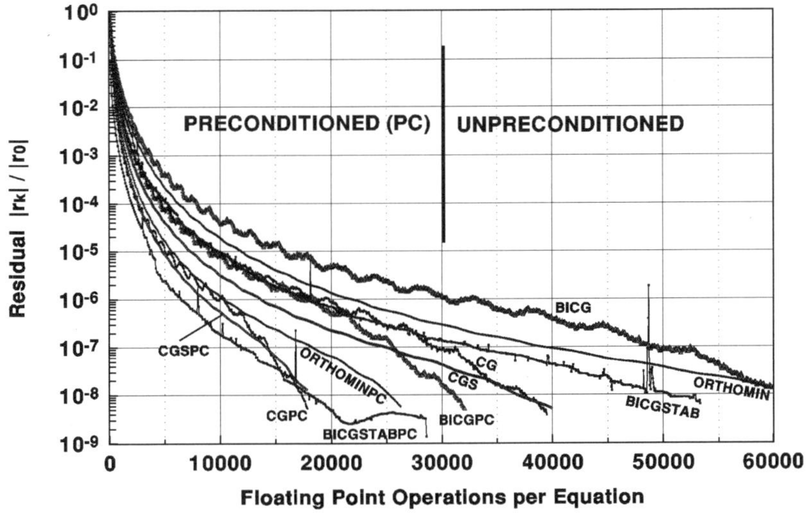

Figure 9. Comparison for a homogeneous half-space.

dike model, and the three-layered earth from Section 2.2. The ordinate denotes the normalized residual $|\mathbf{r}_k|/|\mathbf{r}_0|$ with $|\mathbf{r}_0|$ as the residual for the starting value $\mathbf{x}_0$, and the abscissa indicates the number of FPOs. As an example, Fig. 9 additionally displays the results of the unpreconditioned versions with the relaxation factor $\omega = 0$ [Eq. (34)]. As is expected, they converge considerably slower. Therefore, and for reasons of clarity, the other two diagrams are restricted to the preconditioned methods (indicated by the extension PC). In future work, the efficiency of other preconditioners needs to be included in the comparison.

In all three cases, CG-PC and CGS-PC are among the fastest equation solvers, although CGS-PC may show unstable behavior (Fig. 10). Also, BiCGSTAB-PC sometimes shows spikes (especially in Fig. 9), but generally is relatively steady. The convergence behavior of ORTHOMIN-PC is very smooth in each case. Finally BiCG-PC yields the slowest convergence rates for each of the investigated models.

4 Conclusions

The examples show that accuracy of modeling depends on the method of discretization. It is demonstrated that three (BT&W, D&M2, and ZM&M) of the five discretization methods yield accurate results, especially when conductivity contrasts (e.g., dikes, layers) are introduced into the half-space. For four of the five methods (BT&W, D&M1,

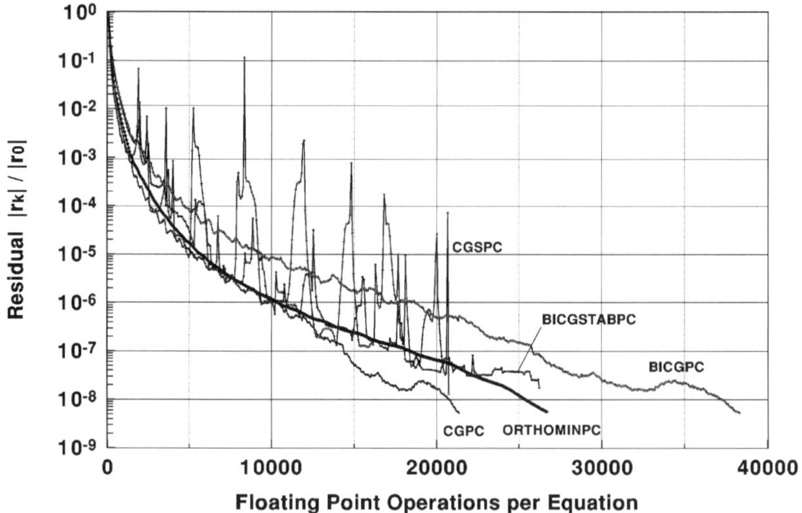

Figure 10. Comparison for a dike model.

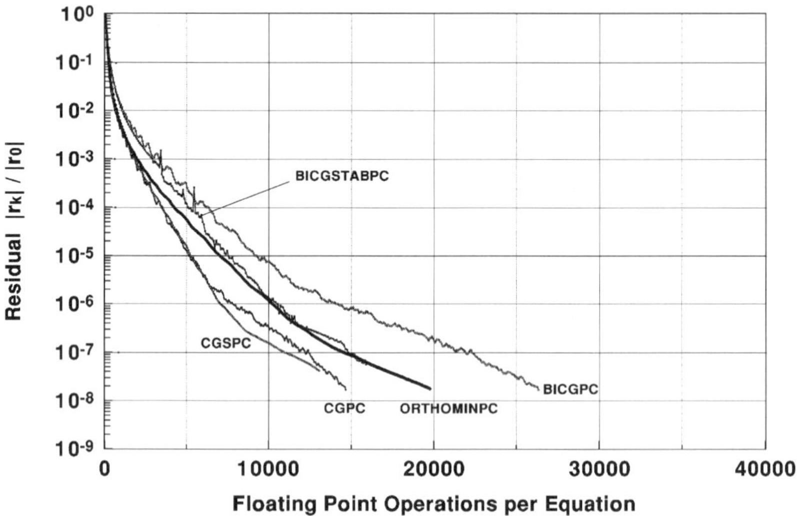

Figure 11. Comparison for a three-layered earth.

D&M2, and W&M), the spatial discretization is the same but the discretization and method of averaging conductivities differ. The accuracy of these methods clearly depends on the method of discretizing conductivities. The ZM&M approach cannot be decomposed into spatial and conductivity discretization.

The speed of convergence of the equation solvers depends on the selected CGM or CRM as well as on preconditioning. Therefore, preconditioning is highly recommended to improve convergence. The preconditioned CG method (CG-PC) and the preconditioned CGS (CGS-PC) generally show the fastest convergence. BiCGSTAB and ORTHOMIN achieve good convergence but they are somewhat slower than CG-PC and CGS-PC. The smoothest convergence behavior is shown by ORTHOMIN, indicating robustness and stability. However, CG-PC is only applicable for symmetric matrices. The spikes in the convergence behavior of CGS-PC and BiCGSTAB may indicate that the algorithms can fail to converge. In conclusion, we recommend CG-PC for symmetric matrices and ORTHOMIN as a more general equation solver.

Acknowledgments

We would like to acknowledge valuable discussions with A. Mezzatesta, M. Rabinovich, T. Tamarchenko (all at Western Atlas Logging Services, Houston); R. A. Wattenbarger (Texas A&M University); R. Schulz and H. Rodemann (both at Niedersächsisches Landesamt für Bodenforschung–Geowissenschaftliche Gemeinschaftsaufgaben, Hannover). Thanks to the anonymous reviewers for helpful comments and suggestions.

References

Agarwal, A. K., and Weaver, J. T., 1994, A parallel implementation of conjugate gradient methods over a network of UNIX-based workstations: Abstract proc. 12th EM Induction Workshop, iAGA, 82.

Behie, A., and Vinsome, P. K. W., 1982, Block iterative methods for fully implicit reservoir simulation: Soc. Ptr. Eng. J., **22**, 658–668.

Brewitt–Taylor, C. R., and Weaver, J. T., 1976, On the finite difference solution of two-dimensional induction problems: Geophys. J. Roy. Astr. Soc., **47**, 375–396.

Dey, A., and Morrison, H. F., 1979a, Resistivity modelling for arbitrarily shaped two-dimensional structures: Geophys. Prosp., **27**, 106–136.

———1979b, Resistivity modelling for arbitrarily shaped three-dimensional structures: Geophysics, **44**, 753–780.

Evans, D. J., 1968, The use of preconditioning in iterative methods for solving linear equations with symmetric positive definite matrices: J. Inst. Math. and Its Appl., **4**, 295–314.

Fletcher, R., 1976, Conjugate gradient methods for indefinite systems: Lect. Notes Math., **506**, 73–89.

Hestenes, M. R., and Stiefel, E., 1952, Method of conjugate gradients for solving linear systems: J. Res. Nat. Bur. Stand., **49**, 409–436.

Kershaw, D. S., 1978, The incomplete Cholesky-conjugate gradient method for the iterative solution of systems of linear equations: J. Comput. Phys., **26**, 43–65.

Koefoed, O., 1979, Geosounding principles 1: Elsevier Science Publ. Co., Inc.

Madden, T. R., 1972, Transmission systems and network analogies to geophysical forward and inverse problems: Dept. of Earth and Planetary Sciences, Report No. **72-3** Massachusetts Inst. of Technology.

Mundry, E., 1984, Geoelectrical model calculations for two–dimensional resistivity distributions: Geophys. Prosp., **32**, 124–131.

Press, W. H., Flannery, B. P., Teukolsky, S. A., and Vetterling, W. T., 1989, Numerical recipes: Cambridge Univ. Press.

Rodemann, H., 1994, Einige Diskretisierungs–Effekte bei Modellrechnungen zur Geoelektrik nach der Methode der finiten Differenzen, *in* Bahr, K. and Junge, A., Eds., Protokoll 15. Kolloquium "Elektromagnetische Tiefenforschung" Deutsche Geophysikalische Gesellschaft, iSSN 0946-7467, 200–203.

Rossen, R. H., and Dalton, R. L., 1990, Selecting grid and timestep sizes, *in* Reservoir Simulation: Mattox, C. C., and Dalton, R. L., Eds. Soc. Petr. Eng. Mono. **13**, 44–56.

Schwarz, H. R., 1991, Methode der finiten Elemente: Teubner.

Scriba, H., 1981, Computation of the electric potential in three-dimensional structures: Geophys. Prosp., **29**, 790–802.

Sonneveld, P., 1989, CGS: A fast Lanczos–type solver for nonsymmetric linear systems: SIAM J. Sci. Stat. Comput., **10**, 36–52.

Spitzer, K., 1995, A 3-D finite difference algorithm for DC resistivity modeling using conjugate gradient methods: Geophys. J. Internat., **123**, 903–914.

Telford, W. M., Geldart, L. P., and Sheriff, R. E., 1990, Applied Geophysics, 2nd ed., Cambridge Univ. Press.

Vinsome, P. K. W., 1976, Orthomin, an iterative method for solving sparse sets of simultaneous linear equations: Proceedings 4th Symposium on Numerical Simulation of Reservoir Performance, Soc. Petr. Eng., Am. Inst. Min., 149–159.

Van der Vorst, H., 1992, Bi–CGSTAB: A fast and smoothly converging variant of Bi–CG for the solution of nonsymmetric linear systems: SIAM J. Sci. Stat. Comput., **13**, 631–644.

Wurmstich, B., and Morgan, F. D., 1994, Modeling of streaming potential responses caused by oil well pumping: Geophysics, **59**, 46–56.

Zhang, J., Mackie, R. L., and Madden, T. R., 1995, 3-D resistivity forward modeling and inversion using conjugate gradients: Geophysics, **60**, 1313–1325.

PART III

INVERSION

We describe here an inverse solution for interpreting the electromagnetic (EM) response of a three-dimensional (3-D) inhomogeneity residing in a conductive layered earth due to a grounded-wire or loop source. In this solution an integral-equation formulation posed in the frequency domain is used to efficiently compute forward model responses and to accurately calculate elements of the sensitivity matrix of partial derivatives. A nonlinear least-squares algorithm is then utilized to refine estimates of the inhomogeneity's geometry and anomalous conductivity.

Eaton and Hohmann (1988)[1]

[1]Eaton, P. A., and Hohmann, G. W., 1988, Three-dimensional electromagnetic inversion: 58th Ann. Internat. Mtg. Soc. Expl. Geophys., Expanded Abstracts, 229–233.

Joint 3-D Electromagnetic Inversion

Robert G. Ellis

Summary. The electromagnetic (EM) inverse problem in geophysics consists of deducing the Earth's conductivity from measurements of electric or magnetic fields near the surface. It requires fast and accurate forward modeling, a method of solving nonlinear equations, and a method of controlling nonuniqueness. Four modeling methods are considered for simulation of airborne EM data over a 3-D conductivity model: a one-dimensional (1-D) approximation, the 3-D Born approximation, a finite-element (FE) method with approximate boundary conditions, and an FE method with exact boundary conditions. The 1-D and Born approximations are fast, but not very accurate. The two FE methods give similar responses, and the method with approximate boundary conditions requires much less computer resources. Three methods are considered for solving the EM inverse problem by unconstrained nonlinear optimization: conjugate gradient (CG), quasi-Newton, and Gauss-Newton (GN) methods. For the simple models considered, the GN method required the least CPU time; the CG method, the most. An alternative to regularization for controlling nonuniqueness is joint inversion of several different surveys to produce a single model that fits all of the data. An example with airborne EM data and pole-pole dc resistivity data shows that such joint inversion can effectively reduce nonuniqueness in EM inversion.

1 Introduction

Electromagnetic (EM) inversion, like most practical inverse problems, is ill-posed. In simple terms, this means that there are usually many different models that can fit the data. One way to control the nonuniqueness is regularization, which selects the model with an extremal property, such as the model with maximum smoothness. Regularization, however has little practical value unless the property chosen is characteristic of the physical model—in geophysical inversion, the Earth. Another way to control nonuniqueness is to invert more than one kind of data. Simultaneous inversion of two or more different data sets is sometimes called joint inversion.

Department of Geophysics and Astronomy, University of British Columbia, Vancouver, British Columbia V6T 1Z4, Canada.

To study the joint inversion of 3-D EM data, I begin with the three major elements of any inversion: first, forward modeling or calculating the response from a conductivity model; second, calculating sensitivities or partial derivatives of the response with respect to model parameters; and third, using the sensitivities to iteratively adjust the model to fit the data. I then consider the inversion of synthetic data from an airborne EM survey and from a dc resistivity survey. Inversion of just the EM data with minimum-structure regularization produces a model with little resemblance to the true model; inversion of just the dc data produces a model that resembles neither the true model nor the result of the EM inversion. In contrast, joint inversion of the EM and dc data produces a model that closely resembles the true model.

2 Forward modeling

Maxwell's equations in an inhomogeneous medium are, with harmonic time dependence,

$$\begin{aligned} \nabla \times \mathbf{E} &= -\hat{z}\mathbf{H} - \mathbf{M}_s \\ \nabla \times \mathbf{H} &= \hat{y}\mathbf{E} + \mathbf{J}_s, \end{aligned} \tag{1}$$

where the impedivity is given by $\hat{z} = i\omega\mu$, the admittivity by $\hat{y}(\mathbf{x}) = \sigma(\mathbf{x}) + i\omega\epsilon$, and the electric and magnetic current densities by $\mathbf{J}_s$ and $\mathbf{M}_s$. In a typical geophysical model the Earth is divided into a number of cells of constant conductivity, and sources and receivers are specified at or near the Earth's surface. Integral-equation methods traditionally have been used for 3-D EM modeling in geophysics, but advances in the staggered-grid finite-difference (FD) method and the edge finite-element (FE) method have made them competitive. These methods produce sparse banded matrices, have zero divergence within each element (or cell), and avoid the specification of discontinuous fields at cell corners. They do require, however, boundary conditions on a grid that usually must be extended beyond the domain of interest. For example, the FE method, with the 3-D region of interest divided into $n_x \times n_y \times n_z = 20 \times 20 \times 10$ conductivity cells and bounded by three padding cells, requires solution of a large, sparse, complex linear system with 37 179 unknowns: 14 553 unknowns in the region of interest and 22 626 extra unknowns in the region of padding.

Below, I concentrate on the FE method with two types of boundary conditions:

1. *Approximate*, which include simple Dirichlet, Newman, or mixed boundary conditions, and absorbing boundary conditions.
2. *Exact*, which use surface or volume integral representations to specify the boundary field exactly.

I compare the FE method with two approximate methods of forward modeling that often are used in geophysical inversion:

1. *The 1-D approximation*, in which the response is computed for a layered model whose conductivity versus depth matches that of the region directly below the midpoint of the source and receiver. The analytic expression for the layered response (Ward and Hohmann, 1988) can be computed easily with a digital Hankel transform (Anderson, 1979).
2. *The Born approximation*, in which the response is calculated from the volume integral representation by assuming that the incident field drives the scattering

currents. The Born approximation can be written

$$\mathbf{H}(\mathbf{r}) = \mathbf{H}_p(\mathbf{r}) + \int_V \Gamma^H(\mathbf{r};\mathbf{r}') \cdot \mathbf{E}_p(\mathbf{r}')\Delta\hat{y}\, dv, \tag{2}$$

where $\mathbf{E}_p(\mathbf{r}')$ is the primary electric field in the reference earth, $\mathbf{H}_p(\mathbf{r})$ is the primary magnetic field at the observation location $\mathbf{r}$, $\Gamma^H(\mathbf{r};\mathbf{r}')$ is the magnetic Green's dyadic, and $\Delta\hat{y}$ is the admittivity perturbation. Although the Born approximation works best with low conductivity contrasts, there has been progress recently on extending it to handle large contrasts (Habashy et al., 1993). I use the standard Born approximation as the prototype of such methods.

2.1 FE Methods: FE-1 and FE-2

The standard FE method for Maxwell's equations begins with the second-order equation for the scattered field, $\mathbf{E}_s$,

$$\nabla \times \frac{\nabla \times \mathbf{E}_s}{\hat{z}} + \hat{y}\mathbf{E}_s = -\Delta\hat{y}\mathbf{E}_p, \tag{3}$$

where $\hat{y}$ and $\hat{z}$ are the admittivity and impedivity of a background medium, and $\Delta\hat{y}$ is the difference in admittivity between the actual and the background medium. Next, Eq. (3) is written in weak form by integration against a test function, and Green's first vector identity is used to give

$$\int_V \left[\frac{1}{\hat{z}}(\nabla \times \mathbf{E}') \cdot (\nabla \times \mathbf{E}_s) + \hat{y}\mathbf{E}_s \cdot \mathbf{E}' + \Delta\hat{y}\mathbf{E}_p\right] dv + \int_{\partial V} \frac{1}{\hat{z}}[\hat{\mathbf{n}} \cdot (\nabla + \mathbf{E}_s) \times \mathbf{E}']\, ds = 0, \tag{4}$$

where $\mathbf{E}'$ is a square integrable vector test function, usually with localized support.

Boundary conditions are needed to evaluate these integrals. Simple Dirichlet or Neumann conditions give good results if the boundaries are far enough away from the scatterer. More sophisticated absorbing boundary conditions rewrite the surface integral in Eq. (4) in terms of $\mathbf{H}_s$ and apply an mth-order operator that annihilates the first $(2m+1)$ powers of r. The zeroth-order absorbing boundary condition is the Sommerfeld radiation condition. Exact values of $\mathbf{E}_s$ on ∂V can be specified by using the volume integral representation

$$\mathbf{E}_s(\mathbf{r}) = \int_V \Gamma^E(\mathbf{r};\mathbf{r}') \cdot [\mathbf{E}_p(\mathbf{r}') + \mathbf{E}_s(\mathbf{r}')]\Delta\hat{y}\, dv, \tag{5}$$

where $\Gamma^E(\mathbf{r};\mathbf{r}')$ is the electric Green's dyadic.

I investigate both exact and approximate boundary conditions using the edge FEs shown in Fig. 1. The E_x field in the cell given by

$$E_x = \frac{(b-y)(c-z)}{bc}E_{x1} + \frac{y(c-z)}{bc}E_{x2} + \frac{(b-y)z}{bc}E_{x3} + \frac{yz}{bc}E_{x4}, \tag{6}$$

where E_{xi} is the field coefficient and b, c are the y, z dimensions of the cell. In the case denoted FE-1, we take the simplest boundary condition by setting the scattered field to zero on the boundary. In the case denoted FE-2, we use Eq. (5). Note that for FE-1 several layers of padding cells *must* be included in the FE mesh to extend the boundary away from the scattering region. For FE-2, the mesh boundary may be considerably closer to the scatterer; however, the resultant matrix system in less sparse

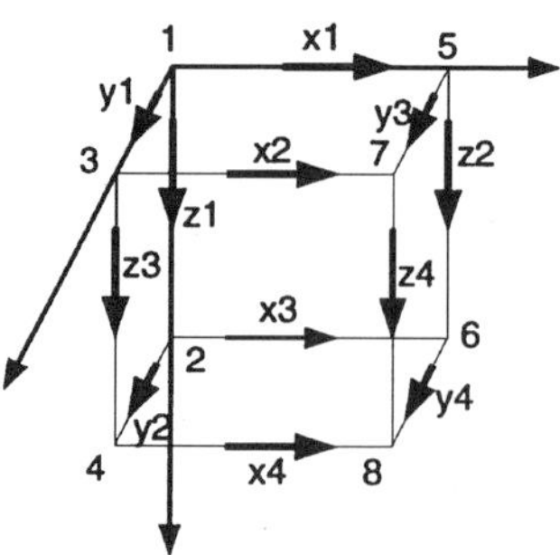

Figure 1. Representation of FE edge elements. The dimensions of the cell are $\Delta x = a$, $\Delta y = b$, $\Delta z = c$. The origin of the local coordinate system is at corner 1.

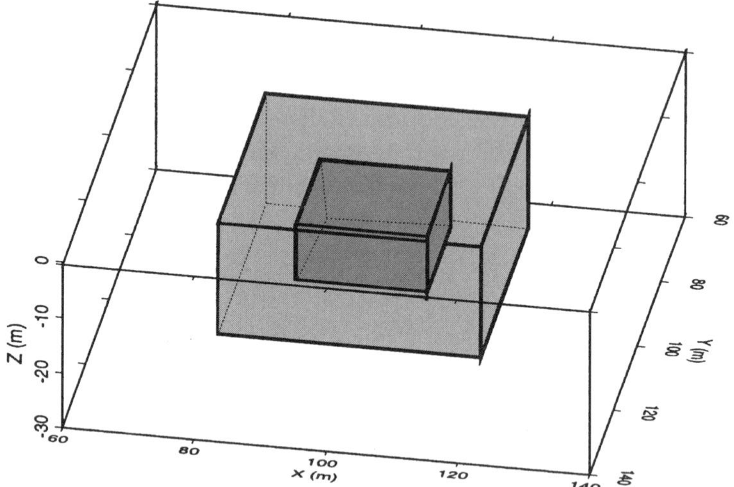

Figure 2. Prism model shown in 3-D perspective. The two prisms are buried in a host conductivity of 0.01 S/m; the inner prism has conductivity 1 S/m and the outer right prism has conductivity 0.001 S/m.

than for FE-1. For both FE-1 and FE-2, the secondary electric field $\mathbf{E}_s$ is used with the magnetic Green's dyadic $\Gamma^H(\mathbf{r};\mathbf{r}')$ to compute the magnetic field at the measurement location:

$$\mathbf{H}(\mathbf{r}) = \mathbf{H}_p(\mathbf{r}) + \int_V \Gamma^H(\mathbf{r};\mathbf{r}') \cdot [\mathbf{E}_p(\mathbf{r}') + \mathbf{E}_s(\mathbf{r}')]\Delta\hat{y}\, dv. \tag{7}$$

2.2 *Test model and comparison*

Figure 2 shows a 3-D test model—the prism model—used to compare the different methods. It consists of a conducting prism inside a resistive prism and is of interest for in situ vitrification experiments (e.g., Spies and Ellis, 1995). The larger resistive prism has a conductivity of 0.001 S/m and extends from $(x, y, z) = (80, 80, 5)$ to $(x, y, z) = (120, 120, 25)$; the inner conducting prism has a conductivity of 1 S/m and extends from $(x, y, z) = (90, 90, 10)$ to $(x, y, z) = (110, 110, 20)$ all in a host of conductivity 0.01 S/m.

Table 1. CPU time and memory requirements for prism model using four different forward-modeling methods.

Method	Disk (Mbytes)	CPU time[a] (s)
1-D	0	2
Born	6.3	120
FE-1	6.3	841
FE-2	65.3	31 200 (3,300)

[a]Number in parentheses gives CPU time required for a forward modeling if intermediate results have been generated previously. The CPU time was measured on a Sun 4 workstation.

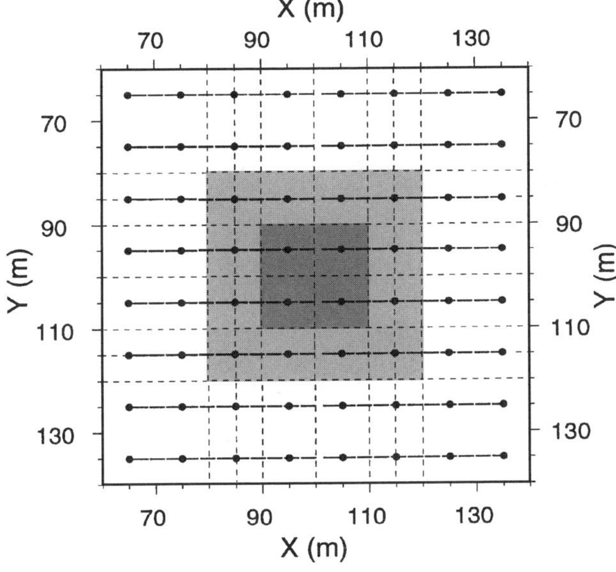

Figure 3. Prism model flight lines shown in plan view. Solid circles represent observation points and heavy dashed lines joining them represent the flight path. The projection of the two buried prisms is shown in gray.

An airborne (helicopter) EM survey was simulated over the prism model with a bird altitude of 5 m, a vertical magnetic dipole source with frequency 28 kHz, and a 1-m coil separation. The vertical component of the magnetic field was collected at the locations shown in Fig. 3. Eight flight lines were simulated with line separation of 10 m and with the magnetic field estimated at 10-m intervals along the flight lines.

Table 1 shows CPU times for the 1-D, Born, FE-1, and FE-2 methods of computing the response of a model with $8 \times 8 \times 8$ cells. The vertical magnetic field was computed at 64 flight locations and 1 frequency. All times are in CPU seconds on a Sun 4/670 workstation. For this test, the 1-D approximation is almost instantaneous whereas the Born approximation takes only 120 s. The FE-1 method takes 841 s whereas the FE-2 method takes 31 200 s. For FE-2, the controlling factor is the calculation of the electric Green's dyad. Because the Green's dyad depends only on the geometry and background

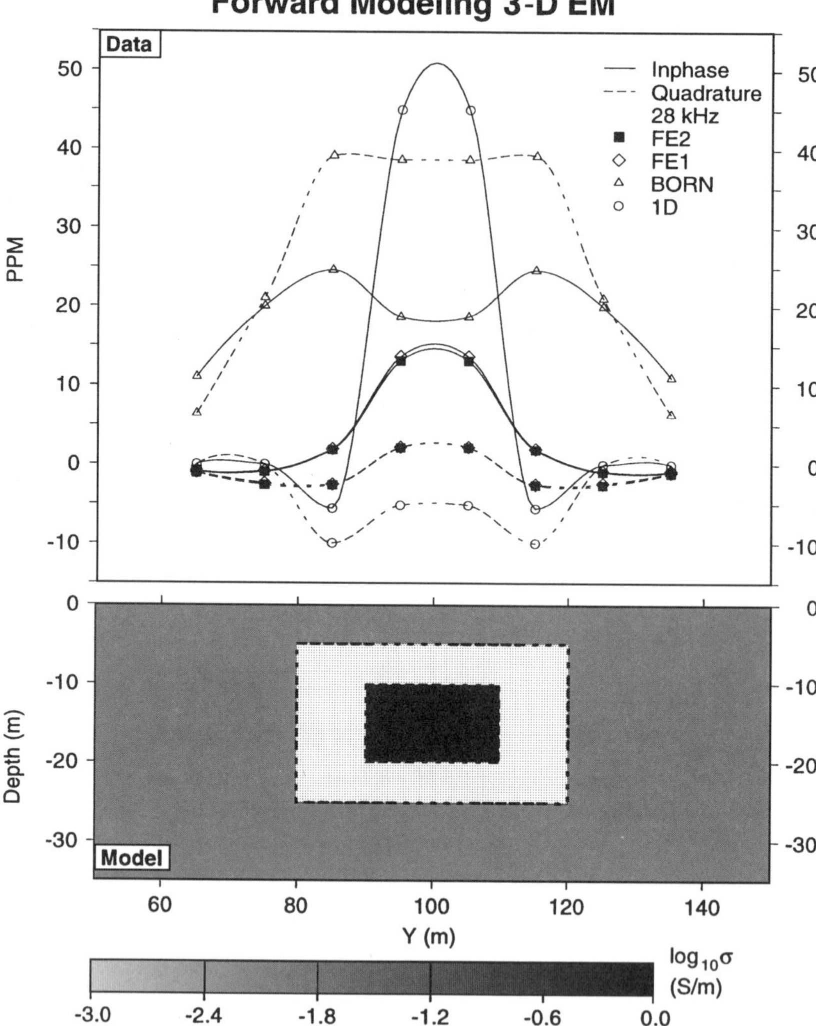

Figure 4. Anomalous in-phase and quadrature response relative to a 0.01-S/m background for the prism model using the four different modeling methods for a flight line at $x = 95$ m. The upper panel consists of the 28 000-Hz response for the 1-D approximation (1-D), the Born approximation (Born), and the FE methods (FE-1, FE-2). The bottom panel shows a slice through the prism conductivity model directly below the flight line. The conductivities are 0.01, 0.001, and 1.0 S/m for the host, the outer prism, and the inner prism, respectively. For reference, the background in-phase and quadrature responses are 9 ppm and 39 ppm, respectively.

conductivity, it is often convenient to store it for subsequent forward modelings. In this example, storage of $\Gamma^E(\mathbf{r}; \mathbf{r}')$ requires 59 Mbytes and reduces the forward modeling time by an order of magnitude to 3300 s. Similar storage of $\Gamma^H(\mathbf{r}; \mathbf{r}')$ requires 6.3 Mbytes and yields only a mimimal reduction in CPU time.

Figure 4 compares the simulated responses from the four methods for the anomalous in-phase and quadrature component for a line at $x = 95$ m. It can be assumed that the FE-2 method gives the most accurate response and is considered to be indistinguishable from the true response. This line chosen is typical of the eight flight lines modeled. The Born and 1-D approximations give rather poor approximation to the true response; however, their simplicity and speed of computation are attractive. Finally, the FE methods with approximate and exact boundary conditions show very similar responses, suggesting that the substantial extra CPU time and disk storage requirements associated with the FE-2 method are not needed in this example.

3 Sensitivity calculation

Inversion algorithms usually require the calculation of sensitivities (partial or Fréchet derivatives), which show how the data change after a small change in the model. Sensitivities can always be computed by additional forward modeling:

1. For inversions using the 1-D forward modeling approximation, the sensitivities can be computed by FDs. Every FD sensitivity calculation requires a forward modeling and must be performed for every cell and datum pair. This would be an arduous task in three dimensions, but is favorable for the 1-D approximation.
2. The Born approximation is itself the expression for the sensitivity about the background medium (usually taken to be a homogeneous whole space or half-space),

$$\delta \mathbf{H}(\mathbf{r}) = \int_V \Gamma^H(\mathbf{r};\mathbf{r}') \cdot \mathbf{E}_p(\mathbf{r}')\, \delta\sigma\, dv. \tag{8}$$

   The sensitivity thus is identified easily as the product of the magnetic Green's dyad and the primary electric field. Of course, the inverse problem under the Born approximation is linear (if it is formulated in terms of the conductivity).
3. For inversions using the FE method, manipulation of Eq. (7) gives gives

$$\delta \mathbf{H}(\mathbf{r}) = \int_V \Gamma^H(\mathbf{r};\mathbf{r}';\Delta\hat{y}) \cdot [\mathbf{E}_p(\mathbf{r}') + \mathbf{E}_s(\mathbf{r}')]\, \delta\sigma\, dv, \tag{9}$$

   where $\Gamma^H(\mathbf{r};\mathbf{r}';\Delta\hat{y})$ is the magnetic Green's dyadic in the medium with admittivity $\hat{y}+\Delta\hat{y}$ and the second-order $\delta\mathbf{E}_s\, \delta\sigma$ term has been dropped. From reciprocity, the magnetic Green's dyadic, which gives the magnetic field at the receiver location $\mathbf{r}$ caused by an electric current at $\mathbf{r}'$, can be replaced by the electric field $\mathbf{E}_R(\mathbf{r}')$ in the model resulting from a magnetic dipole source of strength $-i\omega\mu$ at the receiver location. Integrating $\mathbf{E}_R(\mathbf{r}')$ with the total electric field produces the sensitivity

$$\delta \mathbf{H}(\mathbf{r}) = \frac{-1}{i\omega\mu} \int_V \mathbf{E}_R(\mathbf{r}') \cdot \mathbf{E}(\mathbf{r}')\, \delta\sigma\, dv. \tag{10}$$

   This usually is referred to as the adjoint method for calculating the sensitivity.

The CPU cost differs for each of these methods: For the FD method, an extra run of forward modeling gives the sensitivity; for the Born approximation, no extra computation is required because the necessary elements can be extracted from the forward modeling; for the adjoint method, an extra forward modeling is required to compute the adjoint field. For the geometry of the airborne experiment, however, the adjoint field $\mathbf{E}_R(\mathbf{r}')$ is essentially the same as the primary field $\mathbf{E}(\mathbf{r}')$ and does not require extra forward modeling.

4 Inversion

The airborne EM inverse problem can be cast as a problem of regularized, nonlinear optimization: Given N magnetic-field measurements, H^o_{zi}, with errors, δH^o_{zi}, at locations $\mathbf{x}_i$, find a logarithmic conductivity distribution $m = \ln \sigma$ for the earth that minimizes an objective function $S[m]$ measuring the data misfit and some (overall) feature of the model:

$$S[m] = \sum_{i=1}^{N} \left\{ \left(\frac{\Re H_{zi}^o - \Re H_{zi}[m]}{|\delta \Re H_{zi}^o|} \right)^2 + \left(\frac{\Im H_{zi}^o - \Im H_{zi}[m]}{|\delta \Im H_{zi}^o|} \right)^2 \right\} + \mu \int_V \{W[m(\mathbf{x})]\}^2 \, d^3x. \tag{11}$$

Here, H_{zi} is the magnetic field calculated for the model, and μ is a parameter that controls the relative weight of the two terms. Usually, some scheme is implemented to find a suitable value of μ such that the data misfit is satisfactory, e.g., $\chi^2 \sim N$. The second term regularizes the inversion (Tikhonov and Arsenin, 1977) by controlling the model structure. Typical choices are $W = m$, measuring the size of the model, or $W = \|\nabla m\|$, measuring its smoothness. Models selected in this way are generically called minimum-structure models. In practical inversion, the continuous model, m, is expanded in (usually pulse) basis functions with the coefficients denoted by the vector $\mathbf{m}$.

Common iterative techniques to solve nonlinear (unconstrained) optimization problems are the conjugate gradient (CG) method, the quasi-Newton (QN) method, and the Gauss-Newton (GN) method. [A detailed description of these methods can be found in any standard text on optimization, e.g., Gill et al. (1981).] Each computes a correction or perturbation $\mathbf{p}$ of the current model $\mathbf{m}$,

$$\mathbf{m}^{(k+1)} = \mathbf{m}^{(k)} + \lambda^{(k)} \mathbf{p}^{(k)} \quad k = 0, 1, \ldots, \tag{12}$$

where $\langle \nabla S[\mathbf{m}^{(k)}], \mathbf{p}^{(k)} \rangle < 0$ and $\lambda^{(k)}$ is chosen such that $S[\mathbf{m}^{(k+1)}] < S[\mathbf{m}^{(k)}]$. Newton's method itself uses $\lambda^{(k)} = 1$, $\mathbf{p}^{(k)} = -\{\nabla^2 S[\mathbf{m}^{(k)}]\}^{-1} \nabla S[\mathbf{m}^{(k)}]$. Newton's method converges quadratically near a local minimium, but the difficulty of computing the Hessian matrix makes the direct implementation of Newton's method impractical. (Also, the Hessian usually contains misleading information far from a minimum.)

4.1 CG method

The CG method chooses $\lambda^{(k)}$ and $\mathbf{p}^{(k)}$ as follows:

$$\text{Choose } \mathbf{m}^{(1)} \text{ and set } \mathbf{r}^{(1)} = -\nabla S\big[\mathbf{m}^{(1)}\big], \quad \mathbf{p}^{(1)} = \mathbf{r}^{(1)} \tag{13}$$

Perform the following iteration over k:

$$\text{Choose } \alpha = \lambda^{(k)} \text{ to minimize } S(\alpha) = S\big[\mathbf{m}^{(k)} + \alpha \mathbf{p}^{(k)}\big] \tag{14}$$

$$\text{Set } \mathbf{m}^{(k+1)} = \mathbf{m}^{(k)} + \lambda^{(k)} \mathbf{p}^{(k)}, \quad \mathbf{r}^{(k+1)} = -\nabla S\big[\mathbf{m}^{(k+1)}\big] \tag{15}$$

$$\mathbf{p}^{(k+1)} = \mathbf{r}^{(k+1)} + b^{(k)} \mathbf{p}^{(k)}, \quad b^{(k)} = \frac{\left|\mathbf{r}^{(k+1)}\right|^2 - \mathbf{r}^{(k)} \cdot \mathbf{r}^{(k+1)}}{\left|\mathbf{r}^{(k)}\right|^2} \tag{16}$$

Terminate iterations when $|\mathbf{r}^{k+1}|$ is sufficently small.

The advantage of this method is its simplicity: It requires only the gradient of the objective function (not all of the sensitivities) and a line search. It is the natural choice for problems with extremely large numbers of cells. The gradient of the objective function can be calculated by one additional forward modeling run using the adjoint sensitivities discussed in Section 3. The line search, if done accurately, can require a substantial number of forward runs; often it is done approximately.

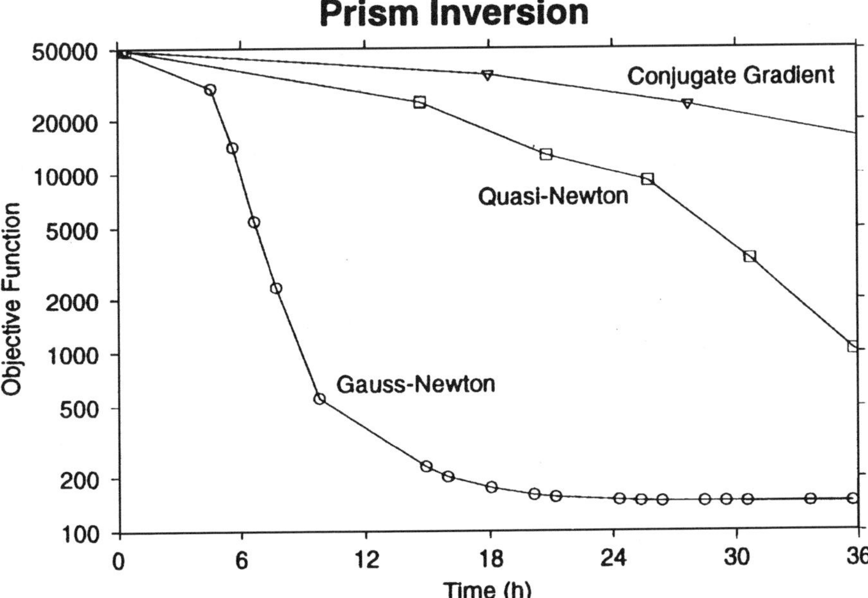

Figure 5. Convergence of GN, QN, and the CG methods with FE-2 forward-modeling method for inversion of prism-model data. The objective function is shown as a function of CPU hours.

4.2 QN methods

In the QN method, Eq. (12) becomes

$$\mathbf{m}^{(k+1)} = \mathbf{m}^{(k)} + \lambda^{(k)}\left[\underset{\sim}{\mathbf{B}}^{(k)}\right]^{-1}\nabla S\left[\mathbf{m}^{(k)}\right]; \quad k = 0, 1, \ldots, \tag{17}$$

where $\underset{\sim}{\mathbf{B}}^{(k)}$ is a symmetric positive-definite matrix which, in the neighborhood of a local minimizer, resembles $\nabla^2 S[\mathbf{m}^{(k)}]$ and at early iterations resembles a steepest-descent direction. Although this method requires more storage than the CG method and is more difficult to implement, it has the advantage of behaving like Newton's method in the neighborhood of a minimizer. The essence of the method is in the construction of $\underset{\sim}{\mathbf{B}}^{-1}$ for the detail of which we refer the reader to the literature and here note two points. First, like the CG method, only the gradient of the objective function is required, not the sensitivities; second, $\underset{\sim}{\mathbf{B}}^{-1}$ usually is constructed by FDs.

4.3 GN method

The GN method is another step closer to Newton's method. It uses essentially the same iterative equation (17) as the QN method, but takes as $\underset{\sim}{\mathbf{B}}^{(k)}$ the Gauss approximation to the Hessian, i.e., $\nabla^2 S \sim \underset{\sim}{\mathbf{J}}^T \underset{\sim}{\mathbf{J}}$, where $\underset{\sim}{\mathbf{J}}$ is the sensitivity matrix whose elements are the partial derivatives (sensitivities) of data points with respect to model parameters.

4.4 Test inversion and comparison

Synthetic airborne EM data over the prism model were computed (using FE-2) at the bird (receiver) locations shown in Fig. 3, and Gaussian noise was added. The CG, QN, and GN methods then were used to solve the same inverse problem (with forward modeling done by FE-2).

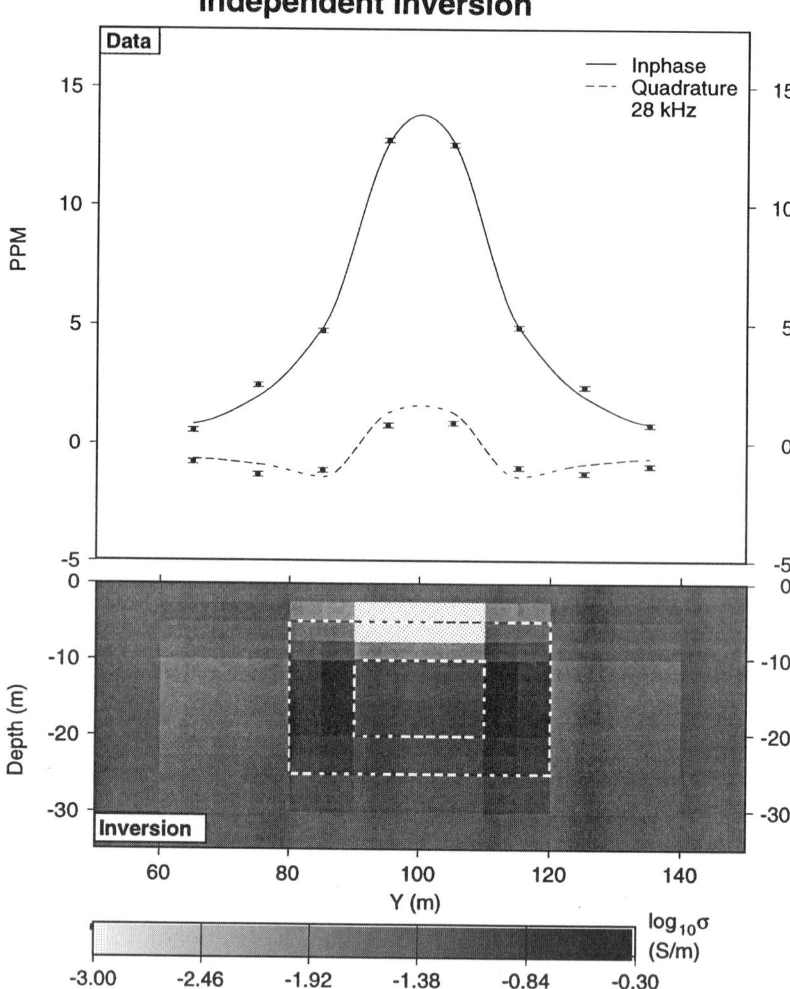

Figure 6. True and predicted data (upper panel) and recovered model (lower panel) for inversion of airborne H_z data. The data (in ppm) shown consists of that subset of the total data with $x = 95$ m. The lower panel shows a conductivity section through the recovered model with $x = 95$ m with the boundary of the true model shown in dashed lines. The gray scale beneath the lower panel gives the conductivity of the recovered model in $\log_{10}$ S/m.

Figure 5 shows the rates of convergence (the objective function versus CPU hours). The GN method requires the least number of forward modelings and the least CPU time of the three inversion methods. The reason for this is that the GN method gives the best approximation to the Hessian and, therefore, gives the most rapid convergence. This result can be generalized to any EM inversion in three dimensions. Given a model with $n_x \times n_y \times n_z$ cells, there will be approximately $6 \times (n_x + 1) \times (n_y + 1) \times (n_z + 1)$ real unknowns (field components) to be solved for in each FE forward modeling. The number of field components will be considerably greater if padding cells are used to satisfy the boundary conditions. In contrast, the GN optimization problem involves a Hessian matrix of size $n_x \times n_y \times n_z$. The cost of solving the linear system for forward modeling generally will exceed the cost of solving the linear system for the perturbation. If, as with airborne EM (because of its nearly coincident source and receivers), the sensitivities

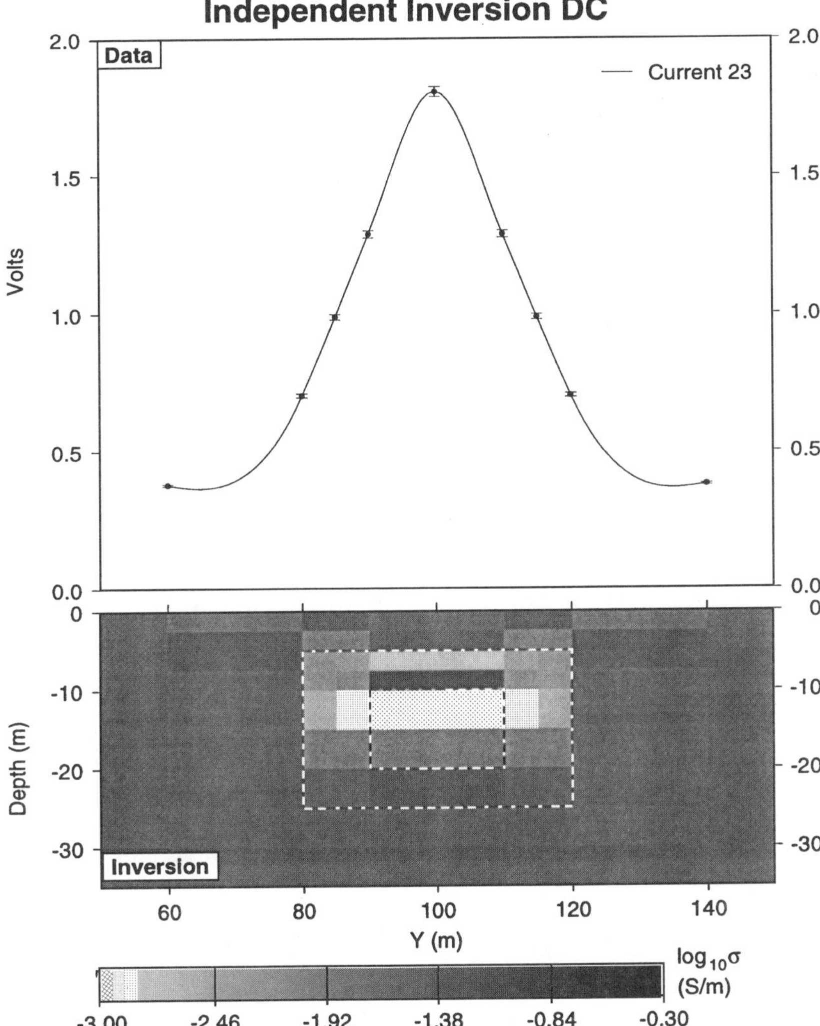

Figure 7. True and predicted data (upper panel) and recovered model (lower panel) for inversion of dc resistivity data. The dc resistivity data shown consist of that subset of the total data with $x = 95$ m. The current source was located at surface coordinate (85, 100). The lower panel shows a conductivity section through the recovered model with $x = 95$ m, with the boundary of the true model shown in dashed lines. The gray scale beneath the lower panel gives the conductivity of the recovered model in $\log_{10}$ S/m.

(elements of $\underset{\sim}{\mathbf{J}}$) can be obtained without additional forward modeling, then the GN method generally will be the most efficient way of solving the optimization problem.

Figure 6 shows the model recovered by this inversion. The upper panel shows a subset of the predicted response (solid and dashed lines) and the observed data (points with error bars). Even though the predicted data and the observed data agree quite well, the recovered model bears little resemblance to the model that generated the data (see Fig. 4). This reflects the nonuniqueness of the inverse problem. Special weighting operators—tinkering with W in Eq. (11)—can force the recovered model to look more like the true model. Another approach is to perform a joint inversion of two or more geophysical data sets.

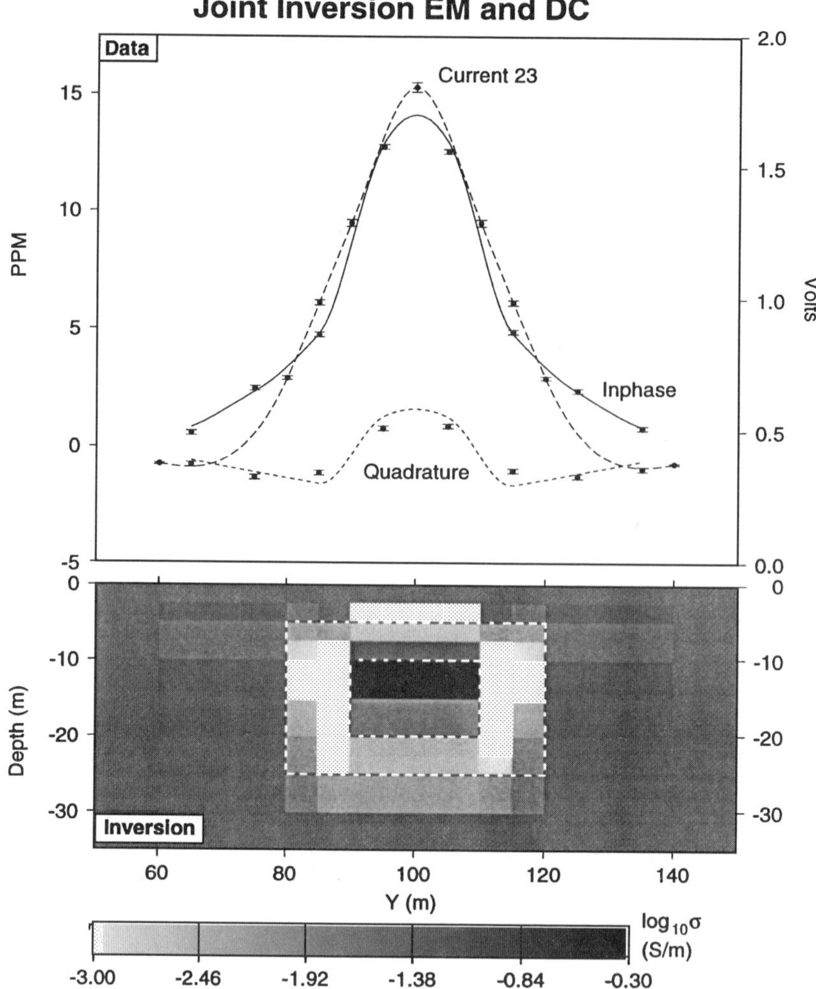

Figure 8. True and predicted data (upper panel) and recovered model (lower panel) for joint inversion of EM and dc resistivity data. The EM data (in ppm) and dc resistivity data (in volts) shown consists of that subset of the total data with $x = 95$ m. The current source for the dc data was located at surface coordinate (95, 100). The lower panel shows a conductivity section through the recovered model with $x = 95$ m with the boundary of the true model shown in dashed lines. The gray scale beneath the lower panel gives the conductivity of the recovered model in $\log_{10}$ S/m.

5 Joint inversion

Consider a pole-pole dc resistivity survey over the prism model of Fig. 2 with a 9×9 grid of surface electrodes. In turn, each electrode is used as a 1-A current source and the potential is measured at the remaining electrodes at a distance greater than 15 m from the current source, yielding 6000 potentials. Figure 7 shows the potentials (synthetic data plus noise) measured along a profile at $x = 95$ m with the current source at $(x, y) = (85, 100)$. The entire dc data set can be inverted using methods similar to those discussed above (Ellis and Oldenburg, 1994). Figure 7 shows the recovered model and the predicted response (solid). Again, because of nonuniqueness, the predicted response fits the data, but the recovered model differs from the model that generated the original (synthetic) data.

The recovered models for the independent inversions of EM and dc data over the prism model shown in Figs. 6 and 7 are very different. It would be quite a challenge for the geophysicist to mentally integrate them to recover the true model, even for this simple synthetic case. The methods developed above, however, can be extended easily to joint inversion of these data. First, the objective function, Eq. (11), is modified to include the misfit of the dc data:

$$S[m] = \sum_{i=1}^{N_{em}} \left\{ \left(\frac{\Re H_{zi}^{o} - \Re H_{zi}^{p}[m]}{\delta \Re H_{zi}^{o}} \right)^2 + \left(\frac{\Im H_{zi}^{o} - \Im H_{zi}^{p}[m]}{\delta \Im H_{zi}^{o}} \right)^2 \right\} + \sum_{i=1}^{N_{dc}} \left(\frac{V_i^{o} - V_i^{p}[m]}{\delta V_i^{o}} \right)^2 + \mu \int_V \{W[m(\mathbf{x})]\}^2 \, dx^3, \tag{18}$$

where N_{dc} is the number of dc data and V represents the measured voltage that constitutes the ith dc resistivity datum.

Minimization of the joint objective function (18) may be performed by any of the methods discussed previously. For this example, the QN method was chosen to avoid calculation of the large Jacobian associated with the dc resistivity measurements and yields the recovered model and predicted responses shown in Fig. 8. The predicted responses show satisfactory agreement with the true responses; more important is that the recovered model shows a structure similar to that of the true model—a conducting core surrounded by a resistive shell.

6 Conclusions

Most geophysical inverse problems are nonunique. In practice, this difficulty is overcome by performing several different surveys over the region of interest. It is unlikely that independent inversion of data from different surveys will produce consistent models; however, a single model whose responses agree with all data can be derived by joint inversion. To demonstrate this, airborne EM data and pole-pole dc resistivity data were independently and jointly inverted. Independent inversion produced recovered models that were neither consistent with each other nor showed significant resemblance to the true model. In contrast, the model recovered from a joint inversion had responses in satisfactory agreement with the dc resistivity and EM data and was close to the true model.

For the examples in this paper, the FE method with Dirichlet boundary conditions (FE-1) on a padded grid gave a very good approximation to the FE method using exact boundary conditions. This result will not necessarily hold for general cases, but does show the usefulness of approximate boundary conditions and the associated CPU savings. When forward modeling is expensive—and the source-receiver geometry allows sensitivities to be computed without additional forward modeling—the preferred optimization method is GN. Its approximation to the Hessian matrix generally gives faster convergence than the CG and QN methods.

Acknowledgments

The author gratefully acknowledges funding for this research from the Joint and Cooperative Inversion consortium and an National Science and Engineering Research Council Industry Oriented Research grant. The JACI consortium consists of the following companies: BHP Minerals, CRA Exploration, Cominco Exploration, Falconbridge, Hudson Bay Exploration and Development, INCO Exploration & Technical Services, Kennecott Exploration Company, Newmont Gold Company, Noranda Exploration, Placer Dome, and Western Mining Corporation. The author also acknowledges K. Lee for providing a copy of the computer program HYBRID (Lee et al., 1981) from which several elements were extracted to produce the programs used in this work, A. Raiche for making available a copy of the computer program SAMAIR (Gupta et al., 1989) used in the early stages of this research, G. Newman for supplying benchmark EM responses, and D. Oldenburg for coordinating the JACI consortium.

References

Anderson, W. L., 1979, Numerical integration of related Hankel transforms of orders 0 and 1 by adaptive digital filtering: Geophysics, **44**, 1287–1305.

Ellis, R. G., and Oldenburg, D. W., 1994, The pole-pole 3D dc resistivity inverse problem: A conjugate gradient approach: Geophys. J. Internat., **119**, 187–194.

Gill, P. E., Murray, W., and Wright, M. H., 1981, Practical optimization: Academic Press, Inc.

Gupta, P. K., Raiche, A. P., and Sugeng, F., 1989, Three-dimensional time-domain electromagnetic modeling using a compact finite-element frequency stepping method: Geophys. J., **96**, 457–468.

Habashy, T. M., Groom, R. W., and Spies, B. R., 1993, Beyond the Born and Rytov approximations: A non-linear approach to electromagnetic scattering: J. Geophys. Res., **98**, 1759–1775.

Lee, K. H., Pridmore, D. F., and Morrison, H. F., 1981, A hybrid three-dimensional electromagnetic modeling scheme: Geophysics, **46**, 796–805.

Spies, B. R., and Ellis, R. G., 1995, Cross-borehole resistivity tomography of a pilot-scale, in-situ vitrification test: Geophysics, **60**, 886–898.

Tikhonov, A. N., and Arsenin, V. Y., 1977, Solution of ill-posed problems: John Wiley & Sons, Inc.

Volakis, J. L., Chatterjee, A., and Gong, J., 1994, A class of hybrid finite element methods for electromagnetics: A review: J. Elect. Waves Appl., **8**, 1095–1124.

Ward, S. H., and Hohmann, G. W., 1988, Electromagnetic methods in applied geophysics, Series: Investigations in Geophysics, **3**, Soc. Expl. Geophys.

A New Algorithm for 3-D Nonlinear Electromagnetic Inversion

Ganquan Xie
Jianhua Li

Summary. We propose a new algorithm for 3-D electromagnetic inversion that works with the magnetic-field integral equation (instead of the traditional electric-field integral equation). The forward and inverse integral equations are discretized by the finite element method; solution of the matrix system involves alternating conjugate-gradient and biconjugate-gradient iterations. The solution is regularized by a novel external annealing technique. A coupled-domain decomposition allows a very efficient (96%) implementation of the algorithm on massively parallel machines. Tests on both synthetic and field data from environmental sites yield good subsurface images in a reasonable amount of computational time.

1 Introduction

The distribution of electrical conductivity in the Earth is important in geophysical exploration, oil reservoir management, and environmental site characterization, because the conductivity often is determined mainly by the pore fluids, porosity, and saturation of the rocks. Three-dimensional inversion of electromagnetic (EM) data is, however, ill-posed, strongly nonlinear, and computationally demanding. In our work on this problem (Lee et al., 1995; Xie and Lee, 1995; Xie et al., 1995a, 1995b), we have found that use of the EM integral equation for the magnetic field has certain advantages over standard approaches that use the integral equation for the electric field. We outline the reasons here, beginning with a short derivation of the magnetic-field integral equation for forward modeling and recasting the (nonlinear) EM inverse problem in terms of this equation. We then describe a finite-element method for its discretization and an efficient parallel algorithm for its solution. We conclude with sample inversions of synthetic and field data.

2 Magnetic-field integral equation

Maxwell's equations

$$\nabla \times \mathbf{E} = -i\omega\mu(\mathbf{H} + \mathbf{M}) \tag{1}$$

Earth Sciences Division, Bldg. 90, Lawrence Berkeley National Laboratory, 1 Cyclotron Road, Berkeley, CA 94720, USA.

and

$$\nabla \times \mathbf{H} = (\sigma + i\omega\epsilon)\mathbf{E} + \mathbf{J} \tag{2}$$

allow derivation of integral equations for both the electric and the magnetic fields (Habashy et al., 1993; Torres-Verdin and Habashy, 1994). The more familiar electric-field integral equation is

$$\mathbf{E}(\mathbf{r}) = \mathbf{E}_b(\mathbf{r}) - i\omega\mu \int_{V_s} \underset{\sim}{\mathbf{G}}_b^E(\mathbf{r}, \mathbf{r}')[(\sigma - \sigma_b) + i\omega(\epsilon - \epsilon_b)]\mathbf{E}(\mathbf{r}')\,d\mathbf{r}'. \tag{3}$$

The different quantities are defined in the standard way: $\mathbf{E}$ is the total electric field; $\mathbf{E}_b$ is the background electric field; $\mathbf{H}$ is the total magnetic field; μ is the magnetic permeability; σ is the electric conductivity; ϵ is the electric permittivity; ω is the (angular) frequency; σ_b is the background electric conductivity; ϵ_b is the background electric permittivity; $\mathbf{J}$ is an electric current source; $\mathbf{M}$ is a magnetic current source; points $\mathbf{r}$ and $\mathbf{r}'$ are in the domain V_s, where $\sigma - \sigma_b + i\omega(\epsilon - \epsilon_b) \neq 0$; finally, $\underset{\sim}{\mathbf{G}}_b^E(\mathbf{r}, \mathbf{r}')$ is the Green dyadic for the electric field in the background medium. In a layered medium, the Green dyadic can be split into a whole-space part and a contribution from the layering:

$$\underset{\sim}{\mathbf{G}}_b^E(\mathbf{r}, \mathbf{r}') = \left(I + \frac{1}{k_b^2}\nabla\nabla\right) g_b(\mathbf{r}, \mathbf{r}') + \underset{\sim}{\mathbf{G}}_s^E(\mathbf{r}, \mathbf{r}'), \tag{4}$$

where

$$g_b(\mathbf{r}, \mathbf{r}') = \frac{e^{-ik_b|\mathbf{r}-\mathbf{r}'|}}{4\pi|\mathbf{r}-\mathbf{r}'|}; \tag{5}$$

$$k_b^2 = -i\omega\mu(\sigma_b + i\omega\epsilon_b), \tag{6}$$

$$|\mathbf{r}-\mathbf{r}'| = \sqrt{(x-x')^2 + (y-y')^2 + (z-z')^2}. \tag{7}$$

Taking $\nabla\times$ on both sides of Eq. (3) and using the second of Maxwell's equations gives the following integral equation for the magnetic field:

$$\begin{aligned}\mathbf{H}(\mathbf{r}) = \mathbf{H}_b(\mathbf{r}) &- \int_{V_s} \nabla_{r'} g_b(\mathbf{r}, \mathbf{r}') \times \frac{(\sigma - \sigma_b) + i\omega(\epsilon - \epsilon_b)}{\sigma + i\omega\epsilon}(\nabla_{r'} \times \mathbf{H} - \mathbf{J})\,d\mathbf{r}' \\ &+ \int_{V_s} \nabla_r \times \underset{\sim}{\mathbf{G}}_s^E(\mathbf{r}, \mathbf{r}')\frac{(\sigma - \sigma_b) + i\omega(\epsilon - \epsilon_b)}{\sigma + i\omega\epsilon}(\nabla_{r'} \times \mathbf{H} - \mathbf{J})\,d\mathbf{r}',\end{aligned} \tag{8}$$

where $\mathbf{H}_b$ is the background magnetic field, ∇_r is the gradient in variable $\mathbf{r}$, and $\nabla_{r'}$ is the gradient in variable $\mathbf{r}'$. The magnetic and electric integral equations are equivalent theoretically, but not numerically. The advantages of the integral equation (8) is described in Section 8.

3 New nonlinear EM inversion using integral equations

EM inversion amounts to solving the following nonlinear Fredholm integral equation of the first kind:

$$\begin{aligned}\mathbf{H}_d(\mathbf{r}) = \mathbf{H}_b(\mathbf{r}) &- \int_{V_s} \nabla_{r'} g_b(\mathbf{r}, \mathbf{r}') \times \frac{(\sigma - \sigma_b) + i\omega(\epsilon - \epsilon_b)}{\sigma + i\omega\epsilon}(\nabla_{r'} \times \mathbf{H} - \mathbf{J})\,d\mathbf{r}' \\ &+ \int_{V_s} \nabla_r \times \underset{\sim}{\mathbf{G}}_s^E(\mathbf{r}, \mathbf{r}')\frac{(\sigma - \sigma_b) + i\omega(\epsilon - \epsilon_b)}{\sigma + i\omega\epsilon}(\nabla_{r'} \times \mathbf{H} - \mathbf{J})\,d\mathbf{r}',\end{aligned} \tag{9}$$

where $\mathbf{H}_d$ is the measured magnetic-field data; $\mathbf{H}_b$ is the background magnetic field; $\mathbf{H}$ is a magnetic field inside the domain V_s that satisfies the magnetic integral equation (8). Equation (9) is similar to Eq. (8) in form, but is a nonlinear integral equation for the electric conductivity and permittivity in V_s when data are given at points $\mathbf{r}$ outside the domain. The first variation of the nonlinear operator is

$$\begin{aligned}\delta\mathbf{H}(\mathbf{r}) = & \int_{V_s} \nabla_{r'} g_b(\mathbf{r}, \mathbf{r}') \times \frac{\delta(\sigma + i\omega\epsilon)}{\sigma + i\omega\epsilon} \frac{(\sigma_b + i\omega\epsilon_b)}{\sigma + i\omega\epsilon} (\nabla_{r'} \times \mathbf{H} - \mathbf{J})\, d\mathbf{r}' \\ & - \int_{V_s} \nabla_r \times \underset{\sim}{\mathbf{G}}{}_s^E(\mathbf{r}, \mathbf{r}') \frac{\delta(\sigma + i\omega\epsilon)}{\sigma + i\omega\epsilon} \frac{(\sigma_b + i\omega\epsilon_b)}{\sigma + i\omega\epsilon} (\nabla_{r'} \times \mathbf{H} - \mathbf{J})\, d\mathbf{r}' \\ & - \int_{V_s} \nabla_{r'} g_b(\mathbf{r}, \mathbf{r}') \times \frac{(\sigma - \sigma_b) + i\omega(\epsilon - \epsilon_b)}{\sigma + i\omega\epsilon} \nabla_{r'} \times \delta\mathbf{H}\, d\mathbf{r}' \\ & + \int_{V_s} \nabla_r \times \underset{\sim}{\mathbf{G}}{}_s^E(\mathbf{r}, \mathbf{r}') \frac{(\sigma - \sigma_b) + i\omega(\epsilon - \epsilon_b)}{\sigma + i\omega\epsilon} \nabla_{r'} \times \delta\mathbf{H}\, d\mathbf{r}'. \end{aligned} \tag{10}$$

Because the nonlinear integral equation (9) is ill-posed, we transform it into the minimization of the following regularized functional:

$$\begin{aligned} \sum_{\ell=1}^{n_{fsr}} \Bigg\| & \mathbf{H}_\ell(\mathbf{r}) - \mathbf{H}_{b,\ell}(\mathbf{r}) + \int_{V_s} \nabla_{r'} g_b(\mathbf{r}, \mathbf{r}') \times \frac{(\sigma - \sigma_b) + i\omega(\epsilon - \epsilon_b)}{\sigma + i\omega\epsilon} (\nabla_{r'} \times \mathbf{H}_\ell - \mathbf{J})\, d\mathbf{r}' \\ & - \int_{V_s} \nabla_r \times \underset{\sim}{\mathbf{G}}{}_s^E(\mathbf{r}, \mathbf{r}') \frac{(\sigma - \sigma_b) + i\omega(\epsilon - \epsilon_b)}{\sigma + i\omega\epsilon} (\nabla_{r'} \times \mathbf{H}_\ell - \mathbf{J})\, d\mathbf{r}' \Bigg\|^2 \\ & + \alpha(\mathbf{R}(\sigma + i\omega\epsilon), \sigma - i\omega\epsilon) = \min! \end{aligned} \tag{11}$$

where $n_{fsr} = n_f \times n_s \times n_r$; n_f is the number of frequencies; n_s is the number of sources; n_r is the number of receivers; $\mathbf{R}$ is a positive-definite regularizing operator; and α is the regularizing parameter.

We use the modified Gauss-Newton iterative method to solve the minimization problem defined by Eq. (11). Let $\mathbf{J}_\ell$ be the Jacobian operator for Eq. (9) ($\mathbf{J}_\ell$ maps a perturbation in $\sigma + i\omega\epsilon$ to a change in magnetic field $\delta\mathbf{H}$; in discrete form, it is a matrix of partial derivatives),

$$\mathbf{J}_\ell \delta(\sigma + i\omega\epsilon) = \delta\mathbf{H}. \tag{12}$$

We have

$$\begin{aligned} [\mathbf{J}_\ell^T \mathbf{J}_\ell + \alpha\mathbf{R}]\delta(\sigma + i\omega\epsilon) = & -\mathbf{J}_\ell^T \Bigg[\mathbf{H}_\ell(\mathbf{r}) - \mathbf{H}_{b,\ell}(\mathbf{r}) + \int_{V_s} \nabla_{r'} g(\mathbf{r}, \mathbf{r}') \\ & \times \frac{(\sigma - \sigma_b) + i\omega(\epsilon - \epsilon_b)}{\sigma + i\omega\epsilon} (\nabla_{r'} \times \mathbf{H}_\ell - \mathbf{J})\, d\mathbf{r}' \\ & - \int_{V_s} \nabla_r \times \underset{\sim}{\mathbf{G}}{}_s^E(\mathbf{r}, \mathbf{r}') \frac{(\sigma - \sigma_b) + i\omega(\epsilon - \epsilon_b)}{\sigma + i\omega\epsilon} \\ & \times (\nabla_{r'} \times \mathbf{H}_\ell - \mathbf{J})\, d\mathbf{r}' \Bigg] - \alpha\mathbf{R}(\sigma + i\omega\epsilon). \end{aligned} \tag{13}$$

4 Regularizing method

4.1 External regularizing method

The regularizing operator $\mathbf{R}$ (Tikhonov and Arsenin, 1977) is a positive-definite operator; for example, $\mathbf{R} = -\Delta$, where Δ is the Laplacian operator. This external regularizing approach is easy to construct, but the optimum regularizing parameter α is difficult to select. Let

$$\begin{aligned}\|\mathbf{H}_{\text{data}} - \mathbf{H}_{\text{model}}\|^2 = \sum_{\ell=1}^{n_{fsr}} \Bigg\| \mathbf{H}_\ell(\mathbf{r}) - \mathbf{H}_{b,\ell}(\mathbf{r}) \\ + \int_{V_s} \nabla_{r'} g_b(\mathbf{r}, \mathbf{r}') \times \frac{(\sigma - \sigma_b) + i\omega(\epsilon - \epsilon_b)}{\sigma + i\omega\epsilon} (\nabla_{r'} \times \mathbf{H}_\ell - \mathbf{J})\, d\mathbf{r}' \\ - \int_{V_s} \nabla_r \times \underset{\sim}{\mathbf{G}}{}_s^E(\mathbf{r}, \mathbf{r}') \frac{(\sigma - \sigma_b) + i\omega(\epsilon - \epsilon_b)}{\sigma + i\omega\epsilon} (\nabla_{r'} \times \mathbf{H}_\ell - \mathbf{J})\, d\mathbf{r}' \Bigg\|^2 .\end{aligned} \tag{14}$$

The optimization (11) will be

$$\|\mathbf{H}_{\text{data}} - \mathbf{H}_{\text{model}}\|^2 + \alpha(\mathbf{R}(\sigma + i\omega\epsilon), \sigma - i\omega\epsilon) = \min! \tag{15}$$

Let $\mathbf{H}_e$ be the exact data and

$$\|\mathbf{H}_e - \mathbf{H}_{\text{data}}\| \le \delta, \tag{16}$$

$$f(\alpha) = \|\mathbf{H}_{\text{data}} - \mathbf{H}_{\text{model}}\|^2, \tag{17}$$

$$g(\alpha) = (\mathbf{R}(\sigma + i\omega\epsilon), \sigma - i\omega\epsilon), \tag{18}$$

$$h(\alpha) = \|\mathbf{H}_{\text{data}} - \mathbf{H}_{\text{model}}\|^2 - \delta^2. \tag{19}$$

It can be proved that $f(\alpha)$ is a continuous and almost monotonic nondecreasing function, $g(\alpha)$ is a continuous and almost monotonic nonincreasing function of α, and $h(\alpha)$ is a continuous and almost monotonic function. When $\mathbf{R} = \mathbf{I}$, Yagola (1980) proved a similar result. The minimum root of Eq. (19) will be an optimum regularizing parameter. Xie et al. (1987) proved that the regularized solution is convergent when α goes to zero.

In nonlinear 3-D EM inversion for practical data, δ can only be estimated crudely because it includes physical system data noise and numerical operator error, etc. Therefore, Eq. (19) is solved approximately. We used a local annealing regularizing process to modify the global discrepancy approach. The data-noise preestimation is very important for the inverse problem. For given noise bound δ, because the discrepancy function $h(\alpha)$ is continuous and almost monotonic, we use the quasi-Newton and bisection mixed method to find the optimum regularizing parameter. We also use the bisection method to estimate the error bound δ.

4.2 Internal regularizing method

We used two internal regularizing approaches to solve the nonlinear magnetic integral equation.

4.2.1 High-order variation operator. From the variational operator formula (10), we can obtain first-, second-, and higher-order approximate variational operators:

$$\delta^{(1)}\mathbf{H}(\mathbf{r}) = -\int_{V_s} \nabla_{r'} g_b(\mathbf{r},\mathbf{r}') \times \frac{\delta(\sigma+i\omega\epsilon)}{\sigma+i\omega\epsilon}\frac{(\sigma_b+i\omega\epsilon_b)}{\sigma+i\omega\epsilon}(\nabla_{r'}\times\mathbf{H}-\mathbf{J})\,d\mathbf{r}'$$
$$+\int_{V_s} \nabla_r \times \underset{\sim}{\mathbf{G}}{}_s^E(\mathbf{r},\mathbf{r}')\frac{\delta(\sigma+i\omega\epsilon)}{\sigma+i\omega\epsilon}\frac{(\sigma_b+i\omega\epsilon_b)}{\sigma+i\omega\epsilon}(\nabla_{r'}\times\mathbf{H}-\mathbf{J})\,d\mathbf{r}', \tag{20}$$

and

$$\delta^{(2)}\mathbf{H}(r) = -\int_{V_s} \nabla_{r'} g_b(\mathbf{r},\mathbf{r}') \times \frac{\delta(\sigma+i\omega\epsilon)}{\sigma+i\omega\epsilon}\frac{(\sigma_b+i\omega\epsilon_b)}{\sigma+i\omega\epsilon}(\nabla_{r'}\times\mathbf{H}-\mathbf{J})\,d\mathbf{r}'$$
$$+\int_{V_s} \nabla_r \times \underset{\sim}{\mathbf{G}}{}_s^E(\mathbf{r},\mathbf{r}')\frac{\delta(\sigma+i\omega\epsilon)}{\sigma+i\omega\epsilon}\frac{(\sigma_b+i\omega\epsilon_b)}{\sigma+i\omega\epsilon}(\nabla_{r'}\times\mathbf{H}-\mathbf{J})\,d\mathbf{r}'$$
$$-\int_{V_s} \nabla_{r'} g_b(\mathbf{r},\mathbf{r}') \times \frac{(\sigma-\sigma_b)+i\omega(\epsilon-\epsilon_b)}{\sigma+i\omega\epsilon}\nabla_{r'}\times\delta^{(1)}\mathbf{H}\,d\mathbf{r}'$$
$$+\int_{V_s} \nabla_r \times \underset{\sim}{\mathbf{G}}{}_s^E(\mathbf{r},\mathbf{r}')\frac{(\sigma-\sigma_b)+i\omega(\epsilon-\epsilon_b)}{\sigma+i\omega\epsilon}\nabla_{r'}\times\delta^{(1)}\mathbf{H}\,d\mathbf{r}'. \tag{21}$$

In this nonlinear inversion, $\delta^{(1)}\mathbf{H}$ is used in the first few iterations and then $\delta^{(2)}\mathbf{H}$ is used.

4.2.2 Log-scale approximation

$$\frac{\delta\sigma+i\omega\delta\epsilon}{\sigma+i\omega\epsilon} \cong \delta\log(\sigma+i\omega\epsilon). \tag{22}$$

This formula is a good approximation for low frequency but rough for high frequency. It can be a natural internal regularizing term for high-frequency noise, which is an advantage of using the nonlinear magnetic integral equation (9).

4.3 Annealing regularizing process

There exists a large number of local minima in the nonlinear regularizing magnetic integral optimization (11). In particular, there are accumulative points of the local minimum set because Eq. (11) is ill-posed. The regularizing term can isolate local minima. We use an annealing process to find the global minimum of the regularizing magnetic integral optimization that will provide a high-resolution EM imaging (Xie et al., 1995a, 1995b). Let

$$\Delta_1(\sigma+i\omega\epsilon) = \sum_{\ell=1}^{n_{fsr}} \Bigg\| \mathbf{H}_{d,\ell}(\mathbf{r}) - \mathbf{H}_{b,\ell}(\mathbf{r})$$
$$+\int_{V_s} \nabla_r g_b(\mathbf{r},\mathbf{r}) \times \frac{(\sigma-\sigma_b)+i\omega(\epsilon-\epsilon_b)}{\sigma+i\omega\epsilon}(\nabla_r\times\mathbf{H}-\mathbf{J})\,d\mathbf{r}$$
$$-\int_{V_s} \nabla_r \times \underset{\sim}{\mathbf{G}}{}_s^E(\mathbf{r},\mathbf{r})\frac{(\sigma-\sigma_b)+i\omega(\epsilon-\epsilon_b)}{\sigma+i\omega\epsilon}(\nabla_r\times\mathbf{H}-\mathbf{J})\,d\mathbf{r}\Bigg\|^2$$
$$+\alpha(\mathbf{R}(\sigma+i\omega\epsilon),\sigma-i\omega\epsilon), \tag{23}$$

$$\Delta_2(\sigma+i\omega\epsilon) = A(\sigma+i\omega\epsilon), \tag{24}$$

where A is a positive functional (related to the internal regularizing and constraint conditions); the annealing function is

$$f(\lambda) = e^{-\lambda T}, \tag{25}$$

with decay coefficient λ and pseudotemperature T.

Given an initial $\sigma_0 + i\omega\epsilon_0$ and pseudotemperature T_0, the annealing process is as follows:

1. Perform a quasi-Newton iteration for solving Eqs. (9)–(13) and run a random-process *RAN* simultaneously.
2. Suppose $\sigma_n + i\omega\epsilon_n$ is known and $\delta(\sigma_n + i\omega\epsilon_n)$ is obtained; calculate

$$\begin{aligned}\lambda_n = {} & \Delta_1[\sigma_n + i\omega\epsilon_n + \delta(\sigma_n + i\omega\epsilon_n)] - \Delta_1(\sigma_n + i\omega\epsilon_n) \\ & - \Delta_2[\sigma_n + i\omega\epsilon_n + \delta(\sigma_n + i\omega\epsilon_n)]\end{aligned} \tag{26}$$

3. Update

$$\sigma_{n+1} + i\omega\epsilon_{n+1} = \sigma_n + i\omega\epsilon_n + \delta(\sigma_n + i\omega\epsilon_n) \tag{27}$$

   if

$$f(\lambda_n) > RAN. \tag{28}$$

4. Change $T = \mu T_0$, and go back to step 1.

The annealing-regularizing method presented here is robust and is useful for interpreting practical data. The optimum-regularizing parameter cannot be chosen accurately because of noise in practical data. Using the annealing-regularizing method we can, usually obtain reasonable results.

5 Finite-element method

We use the finite-element method to discretize the forward magnetic integral equation (8) and the nonlinear magnetic integral equation (11) (Xie et al., 1995c). The cubic domain is divided into a set of finite cubic block elements. There are eight vertex nodes in each element. Let (x_i, y_i, z_i) be coordinate of the vertex i, and let ℓ, h, and v be the length of the side of the element in the x-, y-, and z-directions, respectively. The trilinear finite-element space $\mathbf{H}^h$ can be constructed with basis functions ϕ_i.

$$\mathbf{H}^h(x, y, z) = \sum_{j=1}^{8} \mathbf{H}_j \phi_j(x, y, z), \tag{29}$$

where

$$\mathbf{H}^h(x, y, z) = \begin{bmatrix} H_x^h(x, y, z) \\ H_y^h(x, y, z) \\ H_z^h(x, y, z) \end{bmatrix}, \tag{30}$$

and

$$\mathbf{H}_j = \begin{pmatrix} H_{xj} \\ H_{yj} \\ H_{zj} \end{pmatrix}. \tag{31}$$

In each cubic element, the basis function is a trilinear function

$$\varphi_i(x, y, z) = \frac{(x - x_i^*)(y - y_i^*)(z - z_i^*)}{(\ell - 2x_i^*)(h - 2y_i^*)(v - 2z_i^*)}, \tag{32}$$

where

$$x_i + x_i^* = \ell, \quad y_i + y_i^* = h, \quad z_i + z_i^* = v, \tag{33}$$

and

$$(x_1, y_1, z_1) = (0, 0, 0), \ldots, (x_8, y_8, z_8) = (0, h, v). \tag{34}$$

Upon substituting Eqs. (29)–(31) and their derivatives into Eq. (8), we obtain the finite-element equation for the discrete magnetic field:

$$\underset{\sim}{\mathbf{K}}\mathbf{H} = \mathbf{S}. \tag{35}$$

This matrix $\underset{\sim}{\mathbf{K}}$ is a full matrix, composed of element matrices

$$\underset{\sim}{\mathbf{K}} = \sum_{e=1}^{M} (\underset{\sim}{\mathbf{C}}^e)^T \underset{\sim}{\mathbf{K}}^e \underset{\sim}{\mathbf{C}}^e \tag{36}$$

where

$$\underset{\sim}{\mathbf{K}}^e = \left(K_{i,j}^e\right), \quad i, j = 1, 2, \ldots, 8, \tag{37}$$

and $\underset{\sim}{\mathbf{C}}^e$ is a connection matrix between local and global nodes. Similarly, we can make a finite-element approximation for the nonlinear inverse magnetic integral equation (9) and its derivative operator (20) or (21).

6 Parallel algorithm

We solve the integral equation by a domain-decomposition coupling the global integral equation and a local Galerkin finite-element method. The total numerical-model domain is divided into 2^n subdomains. Two adjacent subdomains should have overlapping strips as in Fig. 1. An algorithm based on finite-element solution or the differential equation for the magnetic field is performed in each subdomain (Xie and Zuo, 1991;

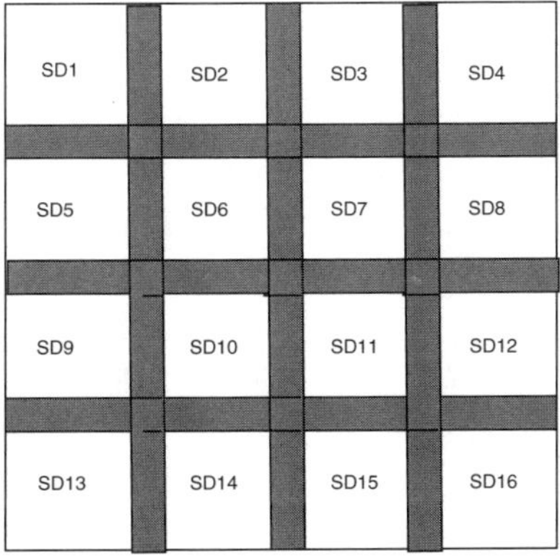

Figure 1. The domain is decomposed into 2^4 subdomains.

Xie et al., 1995c; and Xie et al., 1997) and the integral magnetic forward and inversion algorithm is performed in the overlapping strips. Details of the decomposition method are given in the references. Here we give a brief description. The total discrete matrix is decomposed into 2^n submatrices, each one coupled to its surrounding subdomains. A second-order preconditioned biconjugate (SOR-BCG) iteration is used in parallel to solve the submatrix equations. The 2^n jobs are distributed uniformly into 2^n nodes in a massively parallel computer. A modified first-order global preconditioned biconjugate iteration (PBCG) is used in parallel to solve the reduced global matrix equation. A preconditioned SOR-BCG iteration (Varga, 1962; Wilkinson, 1965; Golub and Van Loan, 1989) is used to solve the inverse integral matrix equation. Note that second-order elements are used in the local subdomain and first-order elements are used in the reduced global iteration. In the 3DEMITINV parallel program, the shared data, shared do loop, and message passing are used to communicate and distribute subdomain field data and matrix data. In this algorithm and parallel program, distribution of the jobs in the parallel processing is uniform and the parallel arrangement is done appropriately. The new domain decomposition approach also has been used for nonlinear integral inversion using Special Parallel Processing (SPP) on the CRAY-A.NERSC.GOV and massively parallel processing (MPP) on the computer T3D.

7 Numerical modeling and inversion

We tested our nonlinear inversion algorithm on two synthetic models. Model 1 is a $90 \times 90 \times 80$ m cubic frame conductor of 0.1 S/m conductivity that includes a $30 \times 30 \times 50$ m cubic conductor of 0.25 S/m conductivity inside (Fig. 2A). The geometry of model 2 is the same as that of model 1; the conductivity in the cubic frame is 0.25 S/m, and the conductivity in the cubic conductor is 0.1 S/m (Fig. 2B). The 18 frequencies (10, 18, 31, 55, 96, 180, 300, 530, 938, 1658, 3000, 5000, 10 000, 16 000, 20 000, 28 000, 38 000, and 50 000 Hz), 64 vertical magnetic dipole sources on the surface, and 768 receivers on the surface were used to make synthetic surface data by solving the forward magnetic integral equation. The geometry of the one source and 12 receivers is shown in Fig. 3. In each receiver point, the three magnetic components, H_x, H_y, and H_z are measured. The amplitude of the vertical magnetic surface data excited by the vertical magnetic source in the center of the surface is shown in Fig. 4 by a solid line; the total field, incident field, and scattered field at 50 000 Hz are shown in plots 1.1, 1.2, and 1.3; the total field, incident field, and scattered field at 10 000 Hz are plotted in 2.1, 2.2, and 2.3; the total field, incident field, and scattered field at 10 Hz are plotted in the 3.1, 3.2, and 3.3. The phase of the vertical magnetic surface data excited by the vertical magnetic source in the center of the surface is shown in Fig. 5 by a solid line. In testing of the inversion, the above model was imbedded in the large cubic domain [−90 m, 90 m; −90 m, 90 m; 0,120 m] and the initial conductivity is 0.05 S/m. After 18 iterations, the conductivity image was obtained. The conductivity imaging of model 1 is shown in (C) of Fig. 2 and the conductivity imaging of model 2 is shown in (D) of Fig. 2. The amplitude and phase of the vertical magnetic field of inversion of the model 1 are shown in Fig. 4 and Fig. 5 by using a dashed line.

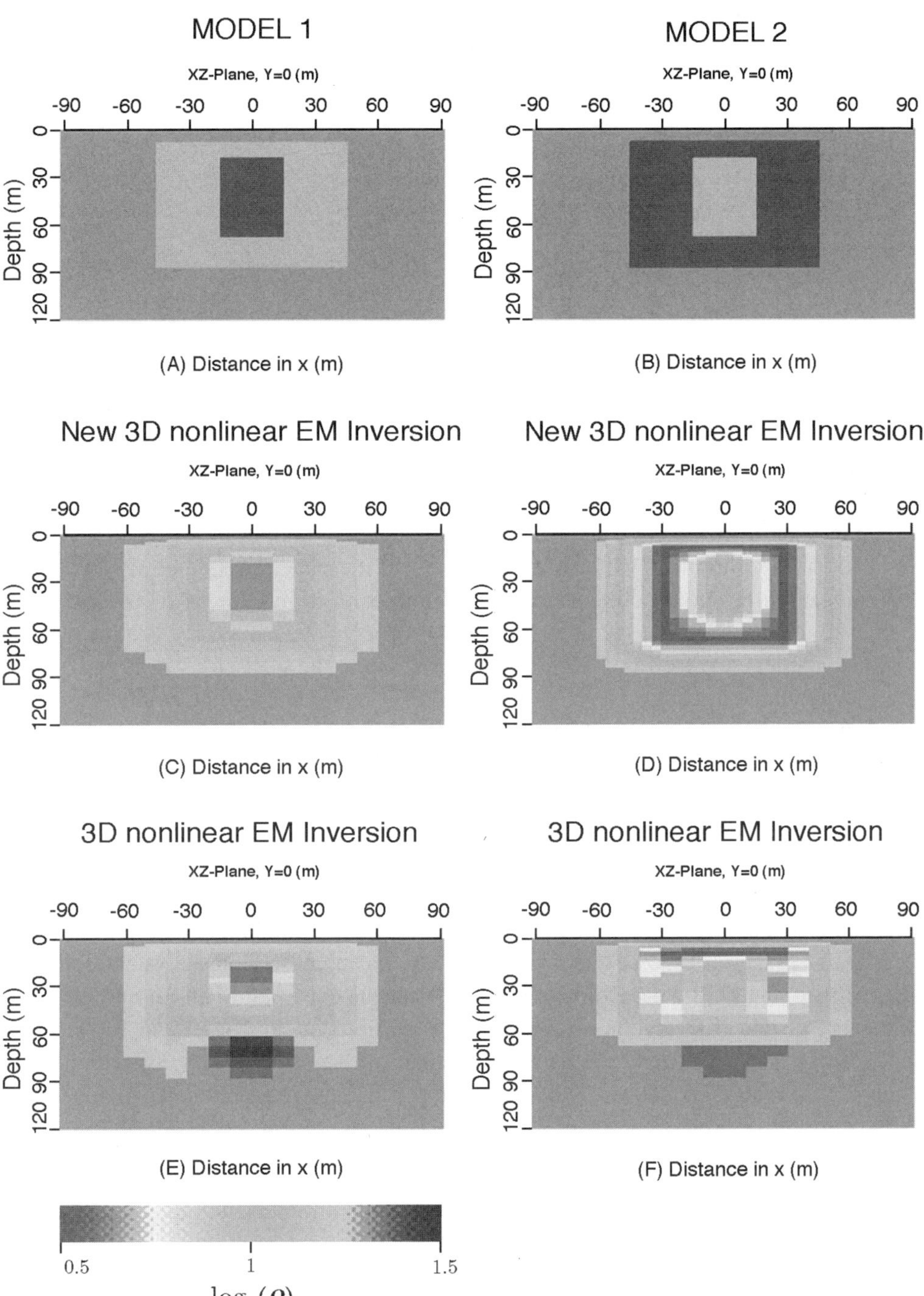

Figure 2. Resistivity imaging of the 3-D EM inversion for synthetic data.

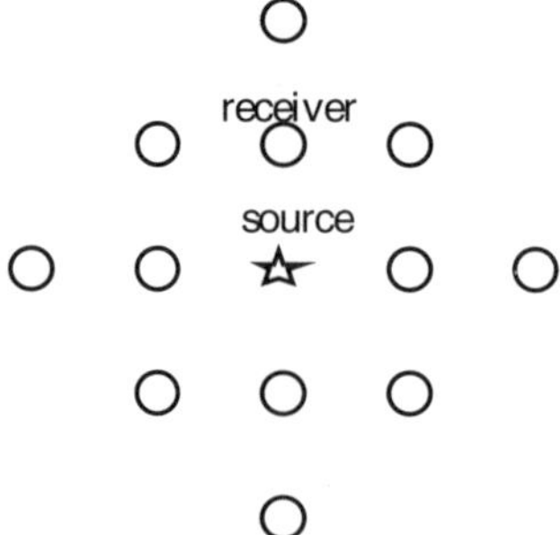

Figure 3. Pattern of the one source and 12 receivers.

For comparison, we used our electric integral inversion program to run the above models and the results are shown in (E) and (F) of Fig. 2. The regularizing parameter of the new inversion is 1.07639X10-5 for model 1, and 78561X10-5 for model 2. We used our new magnetic integral inversion code and electrical integral inversion code to invert VETEM data; the conductivity images are shown in Fig. 6. The data configuration of VETEM is presented in the paper by Lee et al. (1996). A new integral-differential coupled domain decomposition, by Xie et al. (1996c), has been used to parallelize the magnetic integral inversion. We used Cray-C90 to run the synthetic models. On a Cray-C90 using 16 processors, the wall clock time is 30 min, and the parallel efficiency is 96%. The new 3-D nonlinear EM inversion is stable and convergent; its normalized residual reduced to 1.0e-3 from 1.0 after 18 iterations.

8 Conclusion

A new 3-D nonlinear inversion, which works with the magnetic-field integral equation, has been tested on synthetic data and field data from environmental sites. We obtained very good images (Fig. 2) from synthetic data and a reasonable subsurface image (Fig. 6) from the field data. Comparison of the results (Figs. 2C with 2E and 2D with 2F) suggests that 3-D inversion of the magnetic-field integral equation for electric conductivity and permittivity behaves better than (the more familiar) inversion of the electric-field integral equation. The advantages of the new magnetic integral inversion are

1. The magnetic field in Eq. (8) is continuous when electric conductivity is discontinuous, which is convenient for the finite-element method.
2. The kernel function in Eq. (8) is weakly singular.
3. There is a natural internal regularizing term.
4. The integral equation (8) can be used easy to construct a new integral differential coupled parallel domain decomposition.
5. The annealing regularizing for Eq. (8) can be used to find a global minimum of Eq. (11).
6. The magnetic integral equations (8) and (9) are consistent for nonlinear inversion.

Our new algorithm is not suitable for $\omega = 0$. A nonlinear resistivity inversion for direct curent data has been developed by Li et al. (1995).

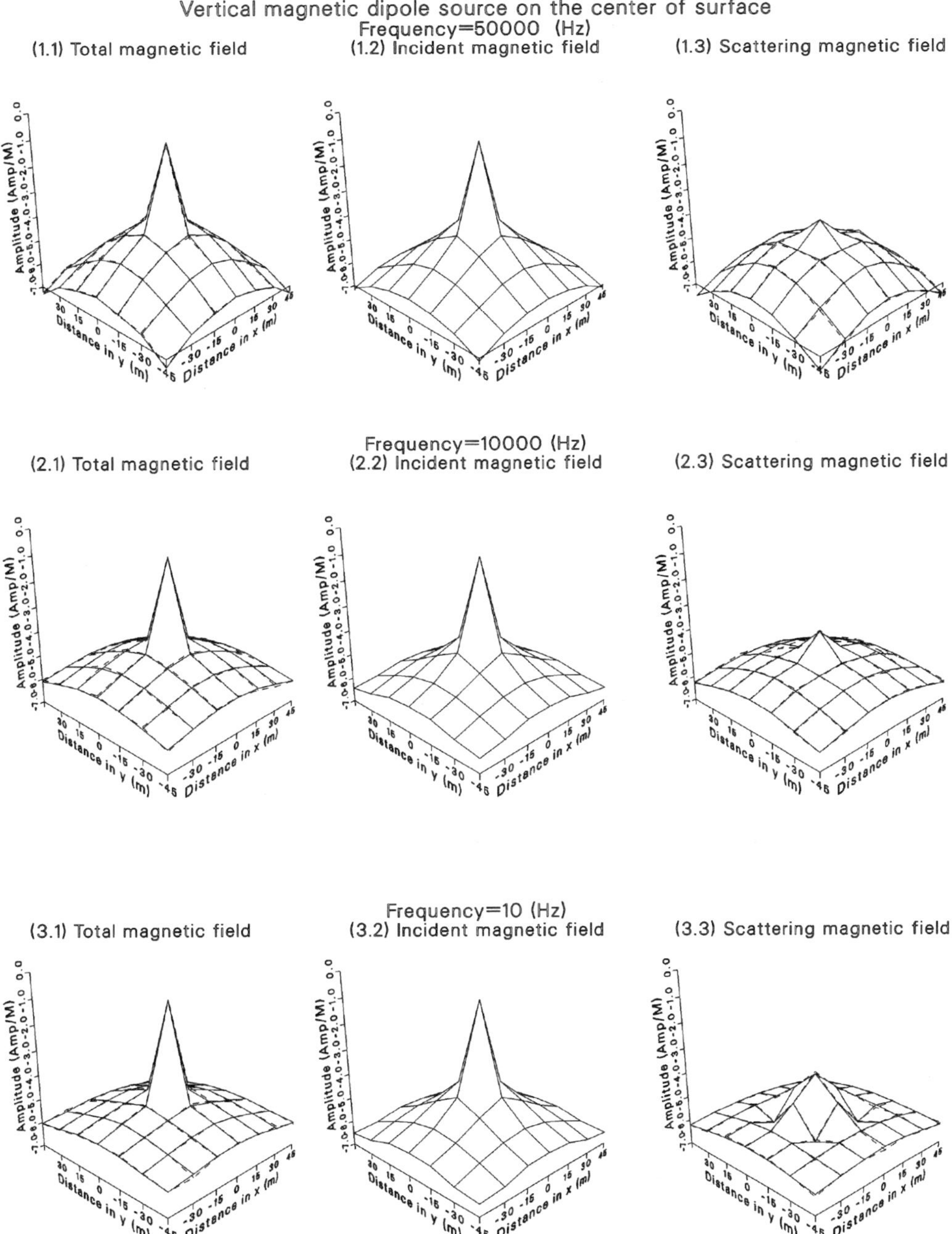

Figure 4. Amplitude of vertical magnetic field on the surface: (solid line) magnetic field of modeling, (dashed line) magnetic field of inversion.

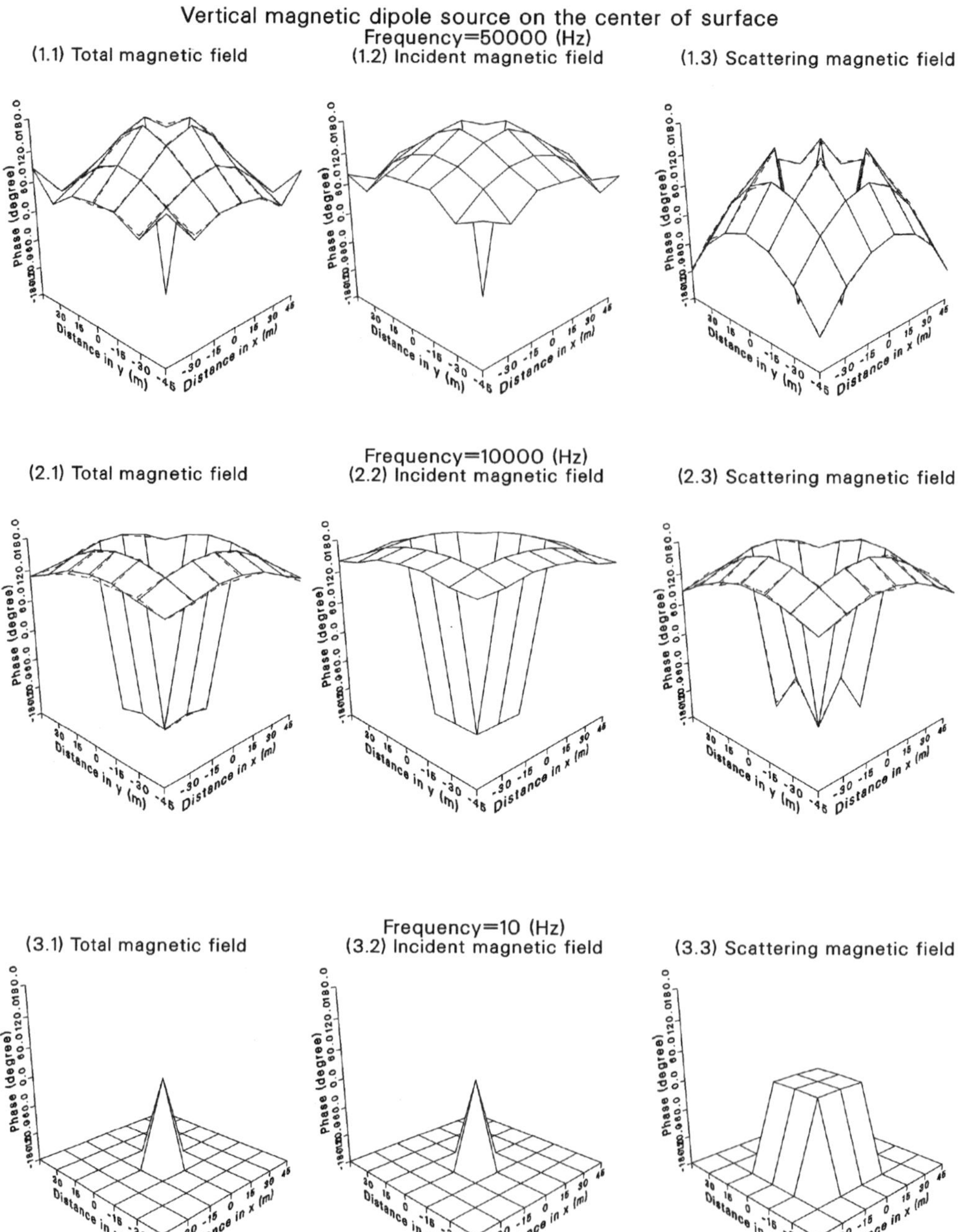

Figure 5. Phase of vertical magnetic field on the surface: (solid line) magnetic field of modeling, (dashed line) magnetic field of inversion.

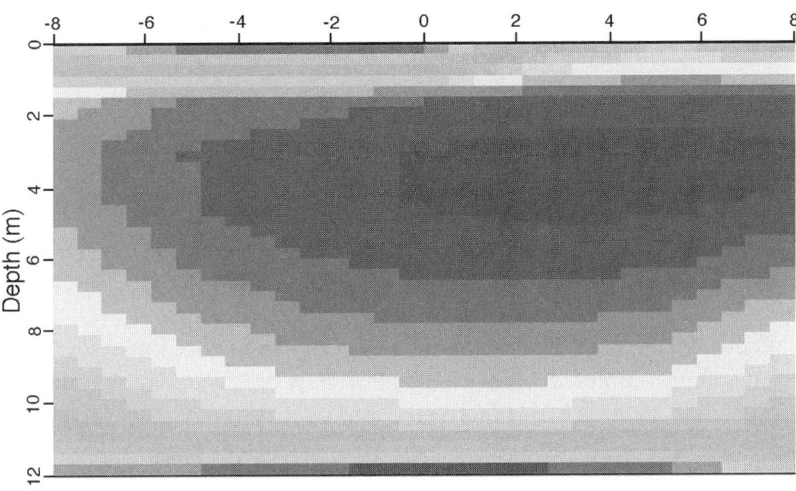

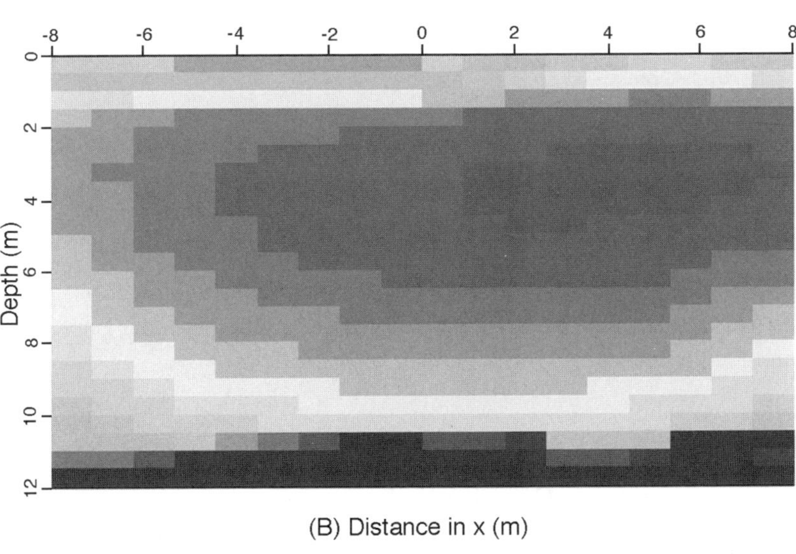

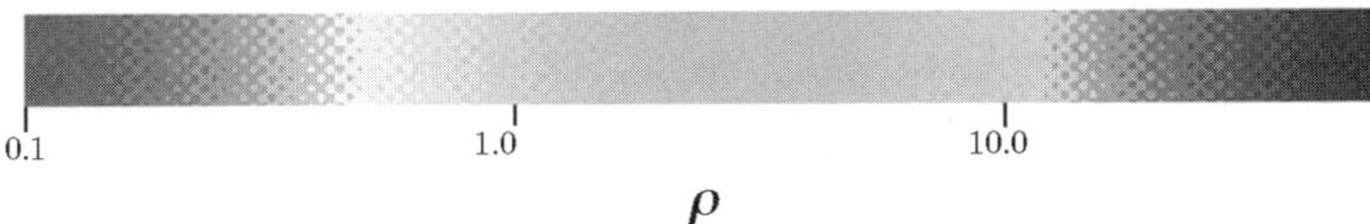

ρ

Figure 6. Resistivity imaging of 3-D EM inversion for real data. (A) magnetic integral equation inversion. (B) electric integral equation inversion.

For magnetic permeability inversion, we have developed a new electric integral equation (Li and Xie, 1997),

$$\mathbf{E}(\mathbf{r}) = \mathbf{E}_b(\mathbf{r}) - \int_{V_s} \nabla_{r'} g_b(\mathbf{r}, \mathbf{r}') \times \frac{\mu - \mu_b}{\mu} (\nabla_{r'} \times \mathbf{E} - \mathbf{M})\, d\mathbf{r}' + \int_{V_s} \nabla_r \times \underset{\sim}{\mathbf{G}}{}_s^H(\mathbf{r}, \mathbf{r}') \frac{\mu - \mu_b}{\mu} (\nabla_{r'} \times \mathbf{E} - \mathbf{M})\, d\mathbf{r}', \tag{38}$$

where μ is the magnetic permeability to be defined, μ_b is the background magnetic permeability, $\underset{\sim}{\mathbf{G}}{}_s^H(\mathbf{r}, \mathbf{r}')$ is the background secondary magnetic Green's function, $\mathbf{M}$ is a magnetic source term, $\mathbf{H}_b^E$ is the background magnetic field exited by the electric dipole source, V_s is the scattering integral domain in which $\mu - \mu_b \neq 0$, and σ, ϵ is constant.

The magnetic-field integral equation (8) is dual of the electric integral equation (38) respectively. A new complete EM inversion for σ, ϵ, μ is developed by joining Eqs. (8) and (38) (Xie et al., 1996c). Historically, the approaches for the forward problem and the inverse problem were developed independently of each other. In general, the matrix of the forward problem is sparse and well-posed when the differential equation is discretized by the finite-element or the finite-difference method, but require artificial radiating or absorbing boundary conditions. A merit of the integral-equation method for inversion is that the artificial boundary condition is not needed, but the matrix is full. Coupling integral and differential methods through a domain decomposition gives a local sparse matrix in which boundary conditions are handled by the integral equation.

Acknowledgments

The new 3-D nonlinear magnetic integral inversion algorithm is supported by the Office of Basic Energy Sciences, VETEM program, Engineering and Geosciences Division, Office of Technology Development, and the Office of Oil, Gas and Shale Technologies, Fossil Energy Division, of the U.S. Department of Energy under Contract No. DE-AC03-76SF00098 and DOE Massively Parallel computer allocation. The authors would like to thank Dr. Bruce Curtis and consultants of the National Energy Research Supercomputer Center, and Carol Taliaferro for their help.

References

Golub, G. H., and Van Loan, C. F., 1989, Matrix computations: Johns Hopkins Univ. Press.

Habashy, T. M., Groom, R. W., and Spies, B. R., 1993, Beyond the Born and Rytov approximation: J. Geophys. Res., **98**, no. B2, 1759–1775.

Lee, K. H., and Xie, G., 1995, Electrical and EM methods for high-resolution subsurface imaging: 3rd Soc. Expl. Geophys. Jap./Soc. Expl. Geophys. Internat. Symposium on Geotomography.

Lee, K. H., Xie, G., Hoversten, M., and Pellerin, L., 1995, EM imaging for environmental site characterization: Internat. Symposium on Three-Dimensional Electromagnetics, Schlumberger-Doll Research.

Li, J., Lee, K. H., Javandel, I., and Xie, G., 1995, Nonlinear three-dimensional inverse imaging for direct current data: 65th Annual Internat. Mtg. Soc. Expl. Geophys., Expanded Abstracts, 250–253.

Li, J., and Xie, G., 1997, A new 3D magnetic permeability inversion: Lawrence Berkeley National Laboratory Report.

Tikhonov, A. N., and Arsenin, V. Y., 1977, Solutions to ill-posed problems: John Wiley & Sons, Inc.

Torres-Verdin, C., and Habashy, T. M., 1994, Rapid 2.5-D forward modeling and inversion via a new nonlinear scattering approximation: Radio Sci., **29**, 1051–1079.

Varga, R. S., 1962, Matrix iterative analysis: Prentice–Hall, Inc.

Wilkinson, J. H., 1965, The algebraic eigenvalue problem: Clarendon Press.

Xie, G., Li, J., and Chen, Y. M., 1987, Gauss-Newton-regularizing method for solving coefficient inverse problem of PDE and its convergence: J. Comput. Math., **5**, 38–49.

Xie, G., and Zou, Q., 1991, A parallel algorithm for solving the 3-D inverse scattering problem: Comput. Phys. Commun., **65**, 320–326.

Xie, G., and Lee, K. H., 1995, Nonlinear inversion of 3-D electromagnetic data, *in* Progress in Electromagnetics Research Symposium, Proc., Univ. of Washington, 323.

Xie, G., Lee, K. H., and Li, J., 1995c, A new parallel 3-D numerical modeling of the electromagnetic field: 65th Ann. Mtg, Soc. Expl. Geophys., Expanded Abstracts, 821–824.

Xie, G., Li, J., and Lee, K. H., 1995a, New 3-D nonlinear electromagnetic inversion: Internat. Symposium on Three-Dimensional Electromagnetics, Schlumberger-Doll Research, 405–414.

———1995b, Annealing regularization for high resolution geophysical tomography: Proc. of 3rd Internat. Symposium on Geotomography, Soc. Expl. Geophys. Jap./Soc. Expl. Geophys., 102–109.

Xie, G., Lee, K. H., Li, J., Pellerin, L., and Zuo, D., 1996, 3-D fast finite element Born accelerating electromagnetic imaging using integral equation: 66th Ann. Mtg., Soc. Expl. Geophys., Expanded Abstracts, 261–264.

Xie, G., and Li, J., 1997, A new 3-D parallel high resolution electromagnetic nonlinear inversion based on a global integral and local differential decomposition: Lawrence Berkeley National Laboratory Report, LBNL-40265.

Yagola, A. G., 1980, On the choice of regularization parameter when solving ill-posed problems in reflexive spaces: USSR Comput. Math. Math. Phys., **20**, 40–52.

Iterative Algorithm for 3-D Microwave Imaging

Hong Gan[1]
Weng Cho Chew[2]

Summary. We develop an inverse scattering algorithm for imaging 3-D dielectric scatterers with microwaves. The algorithm uses the biconjugate gradient and fast Fourier transform method (BCG-FFT) for forward modeling and a conjugate gradient algorithm to solve the inverse problem, which is formulated as a nonlinear optimization problem. The BCG-FFT method reduces the computational complexity of forward modeling to order $N \log_2 N$, where N is the number of unknowns used to represent the vector fields in 3-D space. The conjugate gradient method requires the computation of the Fréchet derivative (matrix of partial derivatives) operating on a vector of residuals, which is equivalent to solving another forward problem. Simulations show the efficiency of this algorithm, especially when the number of illumination angles is limited.

1 Introduction

Microwave inverse scattering, which has applications in medical imaging (Albanese et al., 1994; Borup et al., 1992; Broquetas et al., 1991; Carvicchi et al., 1988; Johnson and Tracy, 1983), target identification (Borden, 1994), geophysical exploration (Collins and Kuperman, 1994; Tarits, 1994; Khruslov and Shepelsky, 1994), and remote sensing (Jordan and Veysoglu, 1994), is a tremendous computational challenge. In most of these applications, the size of the electromagnetic (EM) wavelength at microwave frequencies is comparable to the size of scatterer, and so, forward modeling requires a full-wave (vector) formulation. Also, contrasts in material properties (permittivity and conductivity) can be large, which makes the inverse problem very nonlinear (Colton and Kress, 1992; Gan and Chew, 1995; Gan et al., 1994; Gutman and Klibanov, 1993; Moghaddam and Chew, 1993; Otto and Chew, 1994; Wang and Chew, 1989). There are some general techniques that can be applied to such problems. For example, Shima (1992) and Sasaki (1994) describe an approach (for low-frequency scattering) in which a finite-element method is used for the forward problem and the inverse problem is

[1]Electromagnetics Laboratory, Department of Electrical and Computer Engineering, University of Illinois, Urbana, IL 61801, USA.
[2]Advanced Product Division, Cellular Infrastructure Group, Motorola, Arlington Heights, IL 60004, USA.

solved iteratively by calculating the Jacobian matrix of partial derivatives of data with respect to the model parameters [see also Oristaglio and Worthington (1980)]. Sasaki (1994) compares this approach, which is equivalent to Newton's method for nonlinear equations, with methods (quasi-Newton) that approximate the Jacobian. His simulations show that Newton's method gives better resolution, but its computational complexity is a large barrier to its use in 3-D problems.

Chew and Liu (1994) addressed the problem of computational efficiency for two-dimensional (2-D) problems by combining fast-forward modeling with Newton's method. This paper extends their ideas to 3-D microwave inverse scattering. For forward modeling, we use a new 3-D algorithm (Gan and Chew, 1994) that combines the biconjugate gradient method with the fast Fourier transform (BCG-FFT). Its computational complexity is of order $N_f \log_2 N_f$ per iteration, where N_f is the number of unknowns used to represent the vector fields in the 3-D space. It is very efficient for computing the field scattered from large, weak scatterers. For instance, it can solve a forward scattering problem with more that 80 000 unknowns in about 35 minutes (10 iterations) on a Sun Sparc-10 workstation (Gan and Chew, 1995). To reduce the computational complexity of the inverse problem, we define an error functional that is optimized by a conjugate gradient algorithm.

2 Forward modeling

A dielectric scatterer can be represented by a function $O(\mathbf{r})$ defined as the (normalized) difference between the permittivity $\epsilon(\mathbf{r})$ of the scatterer and the permittivity ϵ_0 of a background medium:

$$O(\mathbf{r}) = \frac{\epsilon(\mathbf{r}) - \epsilon_0}{\epsilon_0} = \epsilon_r(\mathbf{r}) - 1. \tag{1}$$

The total electric field satisfies the integral equation (Chew, 1990)

$$\mathbf{E}(\mathbf{r}) = \mathbf{E}_{\text{inc}}(\mathbf{r}) + k_0^2 \int_{V_s} \underset{\sim}{\mathbf{G}}(\mathbf{r}, \mathbf{r}') \cdot O(\mathbf{r}')\mathbf{E}(\mathbf{r}')\, d^3\mathbf{r}', \tag{2}$$

where $\underset{\sim}{\mathbf{G}}$ is the dyadic Green's function,

$$\underset{\sim}{\mathbf{G}}(\mathbf{r}, \mathbf{r}') = \left(\underset{\sim}{\mathbf{I}} + \frac{\nabla\nabla}{k_0^2}\right) g(\mathbf{r}, \mathbf{r}');$$

$$g(\mathbf{r}, \mathbf{r}') = \frac{e^{ik_0|\mathbf{r}-\mathbf{r}'|}}{4\pi|\mathbf{r} - \mathbf{r}'|}.$$

Here, $\mathbf{r}$ is the observation location, $\mathbf{r}'$ is the source location, k_0 is the wavenumber in the background medium, and V_s is the support of the function $O(\mathbf{r})$. The inverse problem is to reconstruct $O(\mathbf{r})$ from measurements of the field in a region outside the scatterer, which we call the domain V_m. In practice, these measurements will be a finite number N_m.

Information about the scatter is contained in the scattered field, which is defined as the difference between the total field and the incident field,

$$\mathbf{E}_s(\mathbf{r}) = \mathbf{E}(\mathbf{r}) - \mathbf{E}_{\text{inc}}(\mathbf{r}).$$

The scattered field has the integral representation,

$$\mathbf{E}_s(\mathbf{r}) = k_0^2 \int_{V_s} \underset{\sim}{\mathbf{G}}(\mathbf{r}, \mathbf{r}') \cdot O(\mathbf{r}')\mathbf{E}(\mathbf{r}')\, d^3\mathbf{r}'. \tag{3}$$

2.1 BCG-FFT algorithm

We first describe briefly the BCG-FFT algorithm for forward modeling; more details can be found in Gan and Chew (1995). The algorithm starts with Galerkin's method for the numerical solution of the integral Eq. (2). The field in the support region of the object function O is discretized with 3-D rooftop basis functions, which also serve as testing functions. The integral equation then can be represented as discrete convolutions of kernel functions with the expansion coefficients of the total field, modulated by the the object function. The final equation has the form

$$\begin{aligned} b^q(k) = & \sum_m a_{k,m} d_m^q - k_0^2 h_{1q}(k) * t_1^q(k) - k_0^2 z_q^{-1} h_{2q}(k) * t_2^q(k) \\ & + \sum_p \{(1 - z_q)[I(k) - J_p(k)]\} * t_1^p(k) \\ & + \sum_p \left\{(z_q - 1)\left[z_p^{-1} I(k) - J_p(k)\right]\right\} * t_2^p(k), \end{aligned} \tag{4}$$

for $q = x, y, z$. Here z_p $(p = x, y, z)$ are one-step-forward shift operators with respect to the direction $\hat{p}$; z_p^{-1} is the corresponding backward shift operator;

$$t_1^p(k) = \chi(k) d_k^p,$$

$$t_2^p(k) = [z_p \chi(k)] d_k^p,$$

where the $\chi(k)$ are related to samples of the object function; and $I(k)$ and $J_p(k)$ are kernel functions. $b^q(k)$ is related to the incident field and d_k^p is the unknown to be solved for. Explicit formulas for these quantities are given by Gan and Chew (1995).

The discrete formulation (4) can be written in a matrix form

$$\begin{aligned} \mathbf{B}_q &= \underset{\sim}{\mathbf{A}}_q \cdot \mathbf{D}_q + \sum_p \underset{\sim}{\mathbf{G}}_{1qp} \cdot \mathbf{T}_{1p} + \sum_p \underset{\sim}{\mathbf{G}}_{2qp} \cdot \mathbf{T}_{2p}, \\ \mathbf{T}_{1p} &= \underset{\sim}{\mathbf{\Lambda}}_{1p} \cdot \mathbf{D}_p \\ \mathbf{T}_{2p} &= \underset{\sim}{\mathbf{\Lambda}}_{2p} \cdot \mathbf{D}_p, \end{aligned} \tag{5}$$

where $\underset{\sim}{\mathbf{G}}_{1qp}$ and $\underset{\sim}{\mathbf{G}}_{2qp}$, for $p, q = x, y, z$, are Toeplitz matrices. They can be evaluated from the discrete kernels described above. And $\underset{\sim}{\mathbf{\Lambda}}_{1p}$ and $\underset{\sim}{\mathbf{\Lambda}}_{2p}$ are diagonal matrices with the parameters of the scatterers, $\chi(k)$ and $z_p\chi(k)$ as their diagonal elements, respectively. $\mathbf{D}_p$ is a vector that includes the unknown coefficients of the face-based elements with normal pointing in the $\hat{p}$ direction. $\mathbf{B}_q$ is the vector corresponding to the incident field in the $\hat{q}$ direction.

More simply, the matrix form can written

$$\mathbf{B} = \underset{\sim}{\mathbf{A}} \cdot \mathbf{D} + \underset{\sim}{\mathbf{G}}_1 \cdot \mathbf{T}_1 + \underset{\sim}{\mathbf{G}}_2 \cdot \mathbf{T}_2, \tag{6}$$

where $\underset{\sim}{\mathbf{A}}$ is a block diagonal sparse matrix, $\underset{\sim}{\mathbf{G}}_i$, $i = 1, 2$ are 3×3 block Toeplitz matrices. The corresponding elements can be found easily from the above component Eqs. (5). The BCG-FFT algorithm combines the generalized biconjugate gradient algorithm (Sarkar, 1987) with FFT to solve Eq. (5) for field $\mathbf{D}$. The total number of unknowns of field $\mathbf{D}$ is N_f. The computational complexity is of order $N_f \log_2 N_f$ per iteration. One

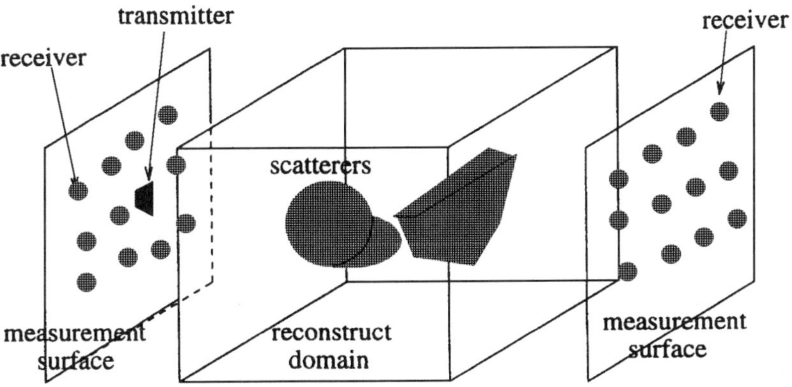

Figure 1. Configuration of 3-D scatterers and measurement surfaces for inversion.

also observes that this algorithm can be implemented in a parallel form. It allows for the possibility of combining hardware with software to calculate the 3-D scattered fields in real time.

3 Inversion from limited illumination angles

The inverse problem is to reconstruct the function O, or ϵ_r, from the measurements of the total field outside the scatterer: $\{\mathbf{Z}(\mathbf{r}), \mathbf{r} \in V_m\}$. We assume that the support of function O is a cuboid. The measurement domain is limited to six surfaces which are parallel to the corresponding six faces of the cuboid as shown in Fig. 1. The incident fields in different experiments are plane waves directed at the scatterer from a few different angles.

To obtain the function O with limited measurements, we use a maximum *a posteriori estimation* method. When the measurement noise and the *a priori* distribution of the function O obey Gaussian distribution, it is well known (Ochi, 1990) that the maximum *a posteriori estimation* is equivalent to minimizing the functional

$$J(O) = \int_{V_m} d^3\mathbf{r} \|\mathbf{Z} - \mathbf{E}(O)\|^2_{\Sigma^{-1}} + \gamma \|O\|^2, \tag{7}$$

which is the weighted squared residual between the measurements and the predicted field based on solving Eq. (2) plus a Tikhonov regularization term.

The necessary condition of minimizing the functional $J(O)$ is $\partial J/\partial O = 0$ for all $\mathbf{r} \in V_m$ which can be written as

$$\left(\frac{\partial \mathbf{E}}{\partial O}\right)^H \Sigma^{-1}[\mathbf{Z} - \mathbf{E}(O)] - \gamma O = 0. \tag{8}$$

3.1 *Fréchet derivative operator*

The nonlinear optimization problem given by Eq. (7) or Eq. (8) can be solved iteratively by updating the object function along a sequence of search directions. The search directions involve partial or Fréchet derivatives of the modeled response with respect to the model parameters (Tarantola, 1987).

An efficient method for computing the Fréchet derivative is important to both linearized inversion and imaging. For distributed-model parameters, however, computation of the Fréchet derivatives dominates the computational complexity. Boerner and Holladay (1990) illustrated some aspects of the Fréchet derivative for 1-D conductivity inversion in EM response function. Potthast (1994) theoretically analyzed the existence of a Fréchet derivative operator for boundary integral operators with respect to 3-D Helmholtz equation. Chew and Liu (1994) addressed computational efficiency issues for the 2-D problem. We describe some of the issues related to the computation of the Fréchet derivative operator for 3-D inverse scattering problems to reduce the computational complexity.

The Fréchet derivative, in fact, is related closely to the integral in equation (3). The field $\mathbf{E}(\mathbf{r})$ can be viewed as a functional of $\epsilon_r(\mathbf{r})$ for a given incident field $\mathbf{E}_{\text{inc}}(\mathbf{r})$:

$$\mathbf{E}(\mathbf{r}) = \mathcal{E}[\epsilon_r(\mathbf{r})]. \tag{9}$$

For a small perturbation of $\epsilon_r(\mathbf{r})$, denoted by $\delta\epsilon$, around the background media $\epsilon_b(\mathbf{r})$, the above formulation can be expanded as

$$\mathbf{E}(\mathbf{r}) = \mathcal{E}[\epsilon_b(\mathbf{r})] + \mathcal{F}[\epsilon_b(\mathbf{r})]\delta\epsilon(\mathbf{r}) + \mathbf{O}(\|\delta\epsilon\|^2). \tag{10}$$

In the above equation, $\mathcal{F}$ is a linearization of the nonlinear equation (9) at point $\{\epsilon_b, \mathcal{E}[\epsilon_b(\mathbf{r})]\}$. $\mathcal{F}$ is also the Fréchet derivative operator (Tarantola, 1987). Denoting the functional spaces Θ and Υ, the Fréchet derivative operator maps an element in Θ onto the functional space Υ.

To describe and efficiently evaluate operations related to $\mathcal{F}$ in the 3-D inverse scattering problem, we consider the perturbation of the total field due to $\delta\epsilon(\mathbf{r})$, denoted by

$$\delta\mathbf{E}(\mathbf{r}) = \mathcal{E}[\epsilon_r(\mathbf{r})] - \mathcal{E}[\epsilon_b(\mathbf{r})].$$

The difference field $\delta\mathbf{E}(\mathbf{r})$ satisfies the vector wave equation

$$\nabla \times \nabla \times \{\mathcal{E}[\epsilon_b(\mathbf{r})] + \delta\mathbf{E}(\mathbf{r})\} - k_0^2[\epsilon_b(\mathbf{r}) + \delta\epsilon(\mathbf{r})]\{\mathcal{E}[\epsilon_b(\mathbf{r})] + \delta\mathbf{E}(\mathbf{r})\} = 0. \tag{11}$$

Because the field $\mathcal{E}[\epsilon_b(\mathbf{r})]$ itself satisfies the wave equation with background media $\epsilon_b(\mathbf{r})$, the equation for the difference field becomes

$$\nabla \times \nabla \times \delta\mathbf{E}(\mathbf{r}) - k_0^2\epsilon_b\delta\mathbf{E}(\mathbf{r}) - k_0^2\delta\epsilon(\mathbf{r})\mathcal{E}[\epsilon_b(\mathbf{r})] - k_0^2\delta\epsilon(\mathbf{r})\delta\mathbf{E}(\mathbf{r}) = 0. \tag{12}$$

When the second-order bilinear term in the above equation is small enough compared to the other terms, we have the first-order approximation of the perturbed field

$$\nabla \times \nabla \times \delta\mathbf{E}(\mathbf{r}) - k_0^2\epsilon_b\delta\mathbf{E}(\mathbf{r}) \approx k_0^2\delta\epsilon(\mathbf{r})\mathcal{E}[\epsilon_b(\mathbf{r})]. \tag{13}$$

This differential equation can be cast into an integral equation (Chew, 1990)

$$\delta\mathbf{E}(\mathbf{r}) = k_0^2 \int_{V_s} d^3r' \underset{\sim}{\mathbf{G}}_b(\mathbf{r}, \mathbf{r}') \cdot \mathcal{E}[\epsilon_b(\mathbf{r}')]\delta\epsilon(\mathbf{r}') + \mathbf{O}(\|\delta\epsilon\|^2), \tag{14}$$

where $\underset{\sim}{\mathbf{G}}_b(\mathbf{r}, \mathbf{r}')$ is the inhomogeneous dyadic Green's function defined by the wave equation

$$\nabla \times \nabla \times \underset{\sim}{\mathbf{G}}_b(\mathbf{r}, \mathbf{r}') - k_0^2\epsilon_b(\mathbf{r})\underset{\sim}{\mathbf{G}}_b(\mathbf{r}, \mathbf{r}') = \mathbf{I}\delta(\mathbf{r} - \mathbf{r}'). \tag{15}$$

Equation (14) is the general formulation of the distorted-wave Born approximation (or simply "distorted Born approximation"). It has the same form as the Born approximation—in that the incident field is used in place of the total field in the integral over the scatterer—but uses Green's function and incident field for a general background medium, not simply a homogeneous medium.

Comparing Eq. (14) with the definition of the Fréchet derivative operator in Eq. (10) shows that

$$\mathcal{F}[\epsilon_b(\mathbf{r})] = k_0^2 \int_{V_s} d^3\mathbf{r}' \underset{\sim}{\mathbf{G}}_b(\mathbf{r}, \mathbf{r}') \cdot \mathcal{E}[\epsilon_b(\mathbf{r}')](\cdot), \tag{16}$$

which maps $\delta\epsilon \in \Theta$ to $\delta\mathbf{E} \in \Upsilon$.

If the inhomogeneity of the background media is also considered as an equivalent source, Eq. (13) can be expressed as

$$\nabla \times \nabla \times \delta\mathbf{E}(\mathbf{r}) - k_0^2\delta\mathbf{E}(\mathbf{r}) = k_0^2\delta\epsilon(\mathbf{r})\mathcal{E}[\epsilon_b(\mathbf{r})] + k_0^2(\epsilon_b - 1)\delta\mathbf{E}(\mathbf{r}) + \mathbf{O}(\|\delta\epsilon\|^2). \tag{17}$$

Then, the perturbed field can be described using the homogeneous background dyadic Green's function:

$$\delta\mathbf{E}(\mathbf{r}) = k_0^2 \int_{V_s} d^3\mathbf{r}' \underset{\sim}{\mathbf{G}}_0(\mathbf{r} - \mathbf{r}') \cdot \mathcal{E}[\epsilon_b(\mathbf{r}')]\delta\epsilon(\mathbf{r}') + k_0^2 \int_{V_s} d^3\mathbf{r}' \underset{\sim}{\mathbf{G}}_0(\mathbf{r} - \mathbf{r}') \cdot (\epsilon_b - 1)\delta\mathbf{E}(\mathbf{r}) + \mathbf{O}(\|\delta\epsilon\|^2). \tag{18}$$

When both the second term and the third term on the right-hand side in the above equation are ignored, one gets the approximation

$$\delta\mathbf{E}(\mathbf{r}) \approx k_0^2 \int_{V_s} d^3\mathbf{r}' \underset{\sim}{\mathbf{G}}_0(\mathbf{r}, \mathbf{r}') \cdot \mathcal{E}[\epsilon_b(\mathbf{r}')]\delta\epsilon(\mathbf{r}'), \tag{19}$$

or

$$\mathcal{F}[\epsilon_b(\mathbf{r})] \approx k_0^2 \int_{V_s} d^3\mathbf{r}' \underset{\sim}{\mathbf{G}}_0(\mathbf{r}, \mathbf{r}') \cdot \mathcal{E}[\epsilon_b(\mathbf{r}')](\cdot).$$

Obviously, when the difference between the actual medium and a homogeneous background is large, the distorted Born approximation (about an intermediate medium) is much more accurate than the ordinary Born approximation. Consequently, it can handle a large range of perturbation, $\delta\epsilon$. In general, the distorted Born approximation is the Fréchet derivative operator about any background medium.

Use of approximation (19) for the Fréchet derivative leads to an iterative scheme called the Born Iterative Method (BIM), whereas use of Eq. (18) is the full distorted-Born Iterative Method (DBIM).

3.2 Evaluation of the Fréchet derivative operator

Direct computation of the Fréchet derivative operator requires inversion of a large dense matrix. For 3-D problems, the computation is rather expensive. When the inverse problem is formulated as an optimization problem, however, a gradient (or conjugate gradient) algorithm will not require computation of the full Fréchet derivative operator. One only needs the operation of the Fréchet derivative operator on a function (and the operation of its conjugate transpose on a function). The computation of the Fréchet derivative operator acting on a function is equivalent to solving a forward scattering problem.

3.2.1 Fréchet derivative operating on a function. Given function $\zeta \in \Theta$, the Fréchet derivative operator maps ζ to a vector field $\mathbf{b} \in \Upsilon$

$$\mathbf{b} = \mathcal{F}[\epsilon_b(\mathbf{r})]\zeta.$$

From Eq. (17), we can write $\mathbf{b}$ as

$$\nabla \times \nabla \times \mathbf{b}(\mathbf{r}) - k_0^2\mathbf{b}(\mathbf{r}) = k_0^2\zeta(\mathbf{r})\mathcal{E}[\epsilon_b(\mathbf{r})] + k_0^2(\epsilon_b - 1)\mathbf{b}(\mathbf{r}), \tag{20}$$

or in an integral formulation

$$\mathbf{b}(\mathbf{r}) = k_0^2 \int_{V_s} d^3\mathbf{r}' \underset{\sim}{\mathbf{G}}_0(\mathbf{r}-\mathbf{r}') \cdot \mathcal{E}[\epsilon_b(\mathbf{r}')]\zeta(\mathbf{r}') + k_0^2 \int_{V_s} d^3\mathbf{r}' \underset{\sim}{\mathbf{G}}_0(\mathbf{r}-\mathbf{r}') \cdot (\epsilon_b - 1)\mathbf{b}(\mathbf{r}). \tag{21}$$

One observes from the above equation that solving for $\mathbf{b}(\mathbf{r})$ is equivalent to solving a forward scattering problem with equivalent incident field

$$\mathbf{b}_{\text{inc}}(\mathbf{r}) = k_0^2 \int_{V_s} d^3\mathbf{r}' \underset{\sim}{\mathbf{G}}_0(\mathbf{r}-\mathbf{r}') \cdot \mathcal{E}[\epsilon_b(\mathbf{r}')]\zeta(\mathbf{r}').$$

Applying the BCG-FFT algorithm, we first solve the above integral equation for $\mathbf{b}(\mathbf{r})$, $\mathbf{r} \in V_s$, then map $\mathbf{b}(\mathbf{r})$, $\mathbf{r} \in V_s$ to $\mathbf{b}(\mathbf{r})$, $\mathbf{r} \in V_m$. This procedure can be accomplished in the following three steps:

1. Evaluating the vector $\mathbf{B}$ which corresponds to the equivalent incident field
$$\mathbf{B} = \underset{\sim}{\mathbf{G}}_1 \cdot \mathbf{T}_1 + \underset{\sim}{\mathbf{G}}_2 \cdot \mathbf{T}_2, \tag{22}$$
where $\underset{\sim}{\mathbf{G}}_i$ and $\underset{\sim}{\mathbf{T}}_i$ for $i = 1, 2$ are the same as defined in the foward problem with background parameter ζ.
2. Using BCG-FFT to solve the following equation for $\mathbf{D} = \epsilon_b \mathbf{b}$,
$$\mathbf{B} = \underset{\sim}{\mathbf{A}} \cdot \mathbf{D} + \underset{\sim}{\mathbf{G}}_1 \cdot \mathbf{T}_1 + \underset{\sim}{\mathbf{G}}_2 \cdot \mathbf{T}_2 \tag{23}$$
3. Mapping $\mathbf{b}(\mathbf{r}')$, $\mathbf{r}' \in V_s$, to $\mathbf{b}(\mathbf{r})$, $\mathbf{r} \in V_m$, via the following:
$$\mathbf{b}(\mathbf{r}) = k_0^2 \int_{V_s} d^3\mathbf{r}' \underset{\sim}{\mathbf{G}}_0(\mathbf{r}-\mathbf{r}') \cdot \{\mathcal{E}[\epsilon_b(\mathbf{r}')]\zeta(\mathbf{r}') + (\epsilon_b - 1)\mathbf{b}(\mathbf{r}')\}. \tag{24}$$

3.2.2 Action of the conjugate transpose Fréchet derivative. Given a vector field $\mathbf{b} \in \Upsilon$, the conjugate transpose of the Fréchet derivative operator maps $\mathbf{b}$ to a function $\zeta \in \Theta^H$

$$\zeta = \mathcal{F}^H[\epsilon_b(\mathbf{r})]\mathbf{b}.$$

From Eq. (16), the conjugate transpose of the Fréchet derivative operating on the vector field $\mathbf{b} \in \Upsilon$ in the measurement domain can be written as

$$\zeta(\mathbf{r}') = k_0^2 \mathcal{E}^H[\epsilon_b(\mathbf{r}')] \cdot \int_{V_m} d^3\mathbf{r}\, \underset{\sim}{\mathbf{G}}_b^H(\mathbf{r}, \mathbf{r}') \cdot \mathbf{b}(\mathbf{r}). \tag{25}$$

To evaluate the above operation with the inhomogeneous Green's function, we define

$$\mathbf{q}(\mathbf{r}') = k_0^2 \int_{V_m} d^3\mathbf{r}\, \underset{\sim}{\mathbf{G}}_b^H(\mathbf{r}, \mathbf{r}') \cdot \mathbf{b}(\mathbf{r}),$$

where $\mathbf{q}(\mathbf{r}')$ can be considered as the result from the back-propagation of the field $\mathbf{b}$ in the measurement domain. Using the differential equation for the inhomogeneous dyadic Green's function, Eq. (15), and reciprocity gives the following equation for $\mathbf{q}(\mathbf{r}')$:

$$\nabla \times \nabla \times \mathbf{q}(\mathbf{r}) - k_0^2 \epsilon_b^* \mathbf{q}(\mathbf{r}) = k_0^2 \mathbf{b}(\mathbf{r}). \tag{26}$$

This can be converted to an integral equation with the homogeneous dyadic Green's function:

$$\mathbf{q}(\mathbf{r}') = k_0^2 \int_{V_m} d^3\mathbf{r}\, \underset{\sim}{\mathbf{G}}_0^*(\mathbf{r}'-\mathbf{r}) \cdot \mathbf{b}(\mathbf{r}) + k_0^2 \int_{V_s} d^3\mathbf{r}\, \underset{\sim}{\mathbf{G}}_0^*(\mathbf{r}'-\mathbf{r}) \cdot (\epsilon_b - 1)^* \mathbf{q}(\mathbf{r}), \tag{27}$$

and the problem is transformed into solving a conjugate forward problem with equivalent incident field:

$$\mathbf{q}_{\text{inc}}(\mathbf{r}') = k_0^2 \int_{V_m} d^3\mathbf{r}\, \underset{\sim}{\mathbf{G}}_0^*(\mathbf{r}' - \mathbf{r}) \cdot \mathbf{b}(\mathbf{r}).$$

The following process can be employed to solve for $\mathbf{q}(\mathbf{r}')$, $\mathbf{r}' \in V_s$ using the BCG-FFT algorithm:

1. Evaluating the vector $\mathbf{B}$ which corresponds to the complex conjugate of the back-propagated field:

$$\mathbf{B}^* = \left\langle \mathbf{f} \cdot k_0^2 \int_{V_m} d^3\mathbf{r}\, \underset{\sim}{\mathbf{G}}_0^*(\mathbf{r}' - \mathbf{r}) \cdot \mathbf{b}(\mathbf{r}) \right\rangle. \tag{28}$$

2. Using BCG-FFT to solve the following equation for $\mathbf{D} = \epsilon_b \mathbf{q}^*$:

$$\mathbf{B} = \underset{\sim}{\mathbf{A}} \cdot \mathbf{D} + \underset{\sim}{\mathbf{G}}_1 \cdot \mathbf{T}_1 + \underset{\sim}{\mathbf{G}}_2 \cdot \mathbf{T}_2. \tag{29}$$

3. Evaluating the function $\zeta(\mathbf{r}') = \mathcal{F}^H \mathbf{b}(\mathbf{r})$ for $\mathbf{r}' \in V_s$ and $\mathbf{r} \in V_m$

$$\begin{aligned} \zeta(\mathbf{r}') &= \mathcal{F}^H \mathbf{b}(\mathbf{r}) \\ &= \mathcal{E}^H[\epsilon_b(\mathbf{r}')] \cdot \mathbf{q}(\mathbf{r}'). \end{aligned}$$

3.3 *Inverse algorithm using distorted Born iterations and BCG-FFT*

The scheme for evaluating the action of the Fréchet derivative operator leads to an iterative solution of the optimization problem. The field $\mathbf{E}$ as a functional of the function O is expanded up to the first-order of a small perturbation of O near a previous estimation O_b. The Fréchet derivative operator from the preceding section approximates the field in the measurement domain by a linear expansion in the small perturbation δO:

$$\mathbf{E}(O) \approx \mathbf{E}(O_b) + \mathcal{F}[\epsilon_b(\mathbf{r})]\delta O.$$

Here we used the fact that $\delta O = \delta\epsilon$. The optimal estimation of δO in each iteration is that δO satisfies the linear equation:

$$\{\mathcal{F}[\epsilon_b(\mathbf{r})]\}^H \Sigma^{-1}[\mathbf{Z} - \mathbf{E}(O_b)] = \{\mathcal{F}[\epsilon_b(\mathbf{r})]\}^H \Sigma^{-1}\{\mathcal{F}[\epsilon_b(\mathbf{r})]\}\delta O + \gamma\delta O, \tag{30}$$

and

$$O = O_b + \delta O.$$

In summary, the inversion algorithm can be implemented in the following steps:

1. Design the experiment and collect measurement data for all illumination angles;
2. Estimate an initial guess of the function O;
3. Predict the scattered field using BCG-FFT with respect to each incident field based on the last estimation of O;
4. Solve Eq. (30) for δO using BCG-FFT to implement the operations of the Fréchet derivative operator with a function as discussed in the preceding section.
5. Update O by: $O_{n+1} = O_n + \delta O$, where n denotes the number of iterations.
6. Check the error criteria; if satisfied, stop; if not, continue to the third step.

Table 1. Comparison of computational complexity of DBIM and BIM

Operation	Computational complexity	
	DBIM	BIM
$\mathbf{E}(O_b)$	$N_t N_{\text{BCG}} N_f \log_2 N_f$	$N_t N_{\text{BCG}} N_f \log_2 N_f$
$\mathcal{F}\zeta$	$N_t N_f \log_2 N_m + N_t(N_{\text{BCG}} + 1) N_f \log_2 N_f$	$N_t N_f \log_2 N_m$
$\mathcal{F}^H \mathbf{b}$	$(N_t + 1) N_f \log_2 N_m + N_t N_{\text{BCG}} N_f \log_2 N_f$	$N_t N_f \log_2 N_m$
CGNR	$N_{\text{CGNR}}(\mathcal{F}\zeta + \mathcal{F}^H + N_o)$	$N_{\text{CGNR}}(\mathcal{F}\zeta + \mathcal{F}^H + N_o)$

3.4 Computational complexity analysis of DBIM and BIM

For convenience, in the discussion of the computational complexity of the above iterative inverse algorithm, we denote by N_t the number of incident waves (transmitters), N_{BCG} the average number of iterations in BCG-FFT algorithm, and N_o the number of the voxels of the function O over V_s.

We have shown (Gan and Chew, 1995) that the memory requirement of the BCG-FFT algorithm is of order N_f and the computational complexity is of order $N_{\text{BCG}} N_f \log_2 N_f$. By choosing the parallel measurement surface, the mapping of the field from the domain of the scatterer to the domain of measurement or vice versa also can be represented by Toeplitz matrices with respect to the two corresponding coordinates on the surface. Therefore, the memory requirements for these is of order N_f. The computational complexity of evaluating Eq. (24) and Eq. (28) is of order $N_f \log_2 N_m$. The computation of Eq. (22) is of order $N_f \log_2 N_f$.

The comparison of the computational complexity of DBIM and BIM for the major steps in the inverse algorithm for each iteration is summarized in Table 1.

4 Numerical simulations

We have tested the DBIM combined with BCG-FFT by reconstructing a canonical 3-D scatterer using computer-simulated synthetic data. The results are compared with those from BIM. The convergence and accuracy are measured using two criteria. One is the

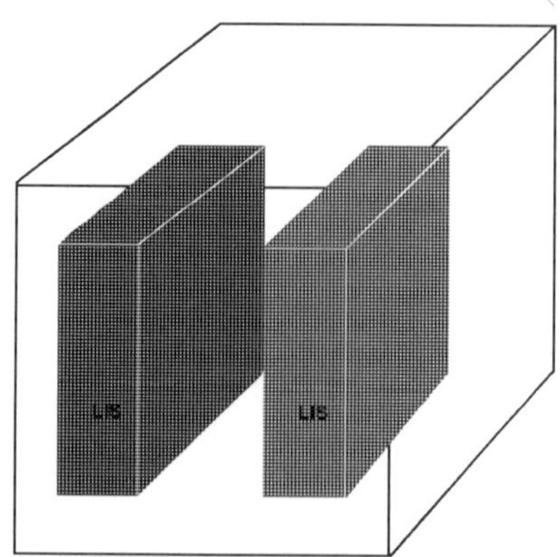

Figure 2. Configuration of 3-D scatterers to be reconstructed.

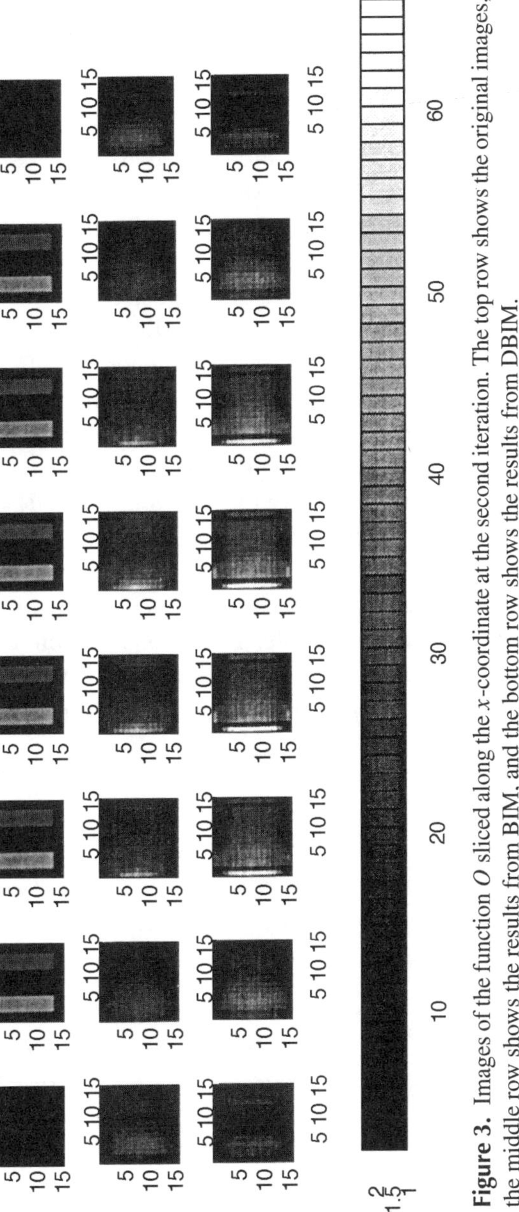

Figure 3. Images of the function O sliced along the x-coordinate at the second iteration. The top row shows the original images, the middle row shows the results from BIM, and the bottom row shows the results from DBIM.

ratio of the power of the residual to the power of the measured synthetic data, which measures the misfit of the model and is defined as

$$\varepsilon_m = \frac{\|\mathbf{Z} - \mathcal{E}[\hat{\epsilon}_r(\mathbf{r})]\|^2}{\|\mathbf{Z}\|^2}.$$

Another is the relative change rate of the reconstructed function O:

$$\varepsilon_o = \frac{\| O^{n+1} \|^2}{\| O^n \|^2},$$

where n is the number of iterations. The initial value of $\| O^n \|^2$ is identity.

The reconstruction domain is a cuboid. At present we assume that the measurements can be collected over all six surfaces outside the cuboid and parallel to the faces of the cuboid. The incident frequency is 1 MHz. The reconstruction domain is a $30 \times 30 \times 30$ m cube. The measurement surfaces are 5 m from the corresponding surface of the cube. The reconstruction domain is discretized using a $16 \times 16 \times 16$ grid. The number of unknowns for the field over the cube are 10 800, the number of voxels that represents the function O is 3 375. There are six illuminations perpendicular to the six surfaces of the cube, respectively. Over each measurement surface, there are 16×16 receivers. The configuration of the scatterer is shown in Fig. 2.

The original and reconstructed images of the function O sliced along the x-coordinate are illustrated in Fig. 3. The convergence and accuracy parameters versus the number of iterations are shown in Fig. 4.

Next, the reconstruction from a surface radar measurements is simulated, where only one measurement surface is available. The reconstruction domain is the same as that in the Fig. 2. The original scatterer is a layered spherical object. The original and reconstructed images of the function O sliced along the x-coordinate are shown in Fig. 5.

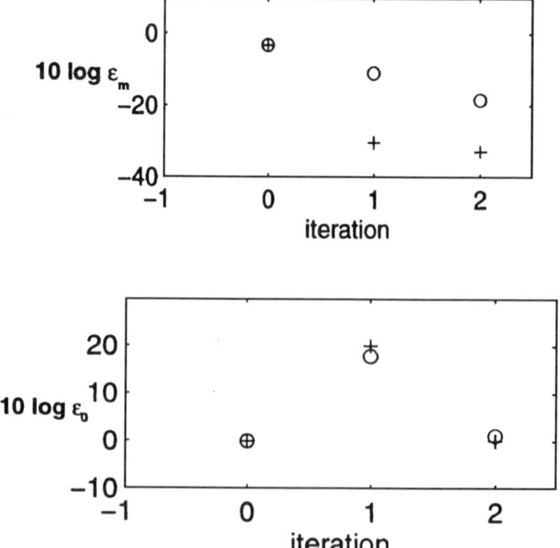

Figure 4. Convergence measure of the inverse algorithms: (+) DBIM (○) BIM.

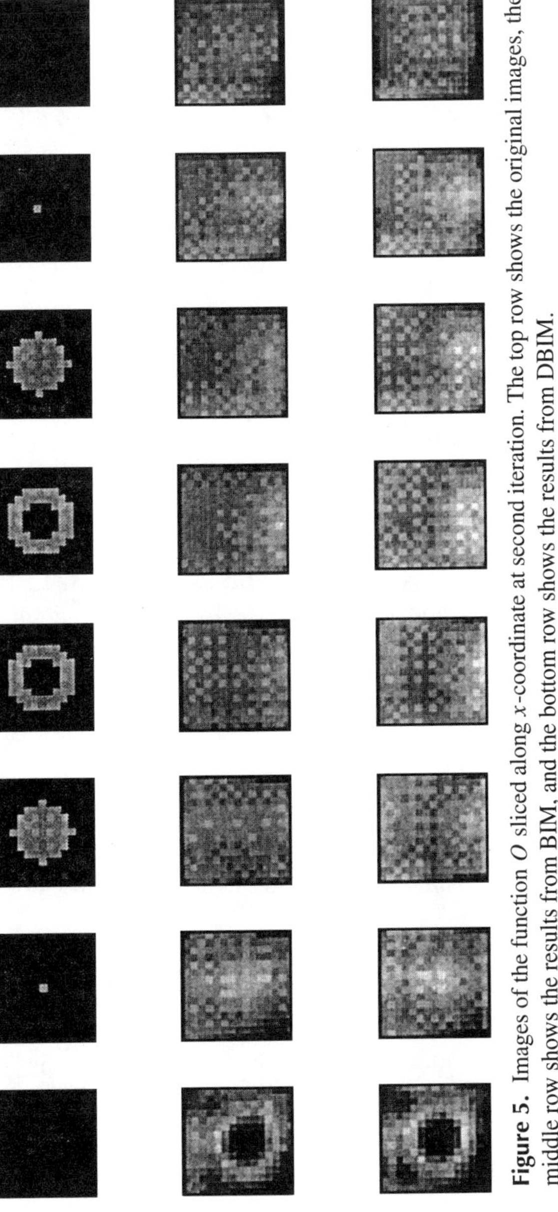

Figure 5. Images of the function O sliced along x-coordinate at second iteration. The top row shows the original images, the middle row shows the results from BIM, and the bottom row shows the results from DBIM.

5 Conclusion

We have developed a fast algorithm for 3-D microwave inversion using the distorted Born approximation and the BCG-FFT iterative method. Using BCG-FFT algorithm to implement the operation of the Fréchet derivative (and its complex conjugate) acting on a vector reduces the computational complexity of each inversion step to that of solving the forward problem. For practical applications, further work is need on the choice of the incident wave and the design of experiment. The choice of preconditioners in the BCG and CG process and more efficient iterative solver also should be studied.

Acknowledgments

This work is supported by the National Science Foundation under contract NSF-ECS-92-24466, Army Research Office under contract DAAL03-91-G-0339, and the Office of Naval Research under contract N00014-89-J-1286.

References

Albanese, R. A., Medina, R. L., and Penn, J. W., 1994, Mathematics, medicine and microwaves: Inverse Problems, **10**, 995–1007.

Borden, B., 1994, Problems in airborne radar target recognition: Inverse Problems, **10**, 1009–1022.

Boerner, D. E., and Holladay, J. S., 1990, Approximate Fréchet derivative in inductive electromagnetic soundings: Geophysics, **55**, 1589–1595.

Borup, D. T., Johnson, S. A., Kim, W. W., and Berggren, M. J., 1992, Nonperturbation diffraction tomography via Gauss-Newton iteration applied to the scattering integral equation: Ultrasonic Imaging, **14**, 69–85.

Broquetas, A., Romeu, J., Rius, J. M., Elias-Fuste, R., Cardama, and Jofre, L., 1991, Cylindral geometry: A further step in active microwave tomography: IEEE Trans. Microwave Theory Tech., **39**, 836–844.

Carvicchi, T. J., Johnson, S. A., and O'Brien, W. D., Jr., 1988, Application of the sinc basis moment method to the reconstruction of infinite circular cylinders: IEEE Trans. Ultrasonics, Ferroelectrics, Frequency Control, **35**, 22–33.

Chew, W. C., 1990, Waves and field in inhomogeneous media: Van Nostrand Reinhold, New York.

Chew, W. C., and Liu, Q. H., 1994, Inversion of induction tool measurements using the distorted-Born iterative method and CG-FFHT: IEEE Trans. Geosci. Remote Sensing, **32**, 878–884.

Collins, M. D., and Kuiperman, W. A., 1994, Inverse problems in ocean acoustics: Inverse Problems, **10**, 1023–1040.

Colton, D., and Kress, R., 1992, Inverse acoustic and electromagnetic scattering theory: Springer Pub. Co., Inc.

Gan, H., and Chew, W. C., 1995, A discrete BCG-FFT algorithm for 3D inhomogeneous scatterer problems: J. Elec. Waves Appl., Vol. 9, No. 10, 1339–1357, 1995.

Gan, H., Chew, W., Leven, P., and Ludwig, R., 1994, A model based recursive inverse scattering algorithm for microwave imaging; *in* IEEE AP-S Internat. Symp., Inst. Electr. Electron. **3**, 1660–1663.

Gutman, S., and Klibanov, M., 1993, Regularized quasi-Newton method for inverse scattering problems: Math. Comput. Modeling, **1**, 5–31.

Johnson, S. A., and Tracy, M. L., 1983, Inverse scattering solution by a sinc basis, multiple source, moment method, Part I, Theory: Ultrasonic Imaging, **5**, 361–375.

Jordan, A. K., and Veysoglu, M. E., 1994, Electromagnetic remote sensing of sea ice: Inverse Problems, **10**, 1041–1058.

Khruslov, E. T., and Shepelsky, D. G., 1994, Inverse scattering method in electromagnetic sounding theory: Inverse Problems, **10**, 1–37.

Moghaddam, M., and Chew, W. C., 1993, Study of some practical issues in inversion with the Born iterative method using time-domain data: IEEE Trans. Antennas Propag., **41**, 177–184.

Ochi, M. K., 1990, Applied probability and stochastic processes: John Wiley & Sons, Inc.

Oristaglio, M. L., and Worthington, M., 1980, Inversion of surface and base hole electromagnetic data for two-dimensional electrical conductivity structures: Geophys. Prosp., **28**, 633–657.

Otto, G. P., and Chew, W. C., 1994, Microwave inverse scattering—local shape function for improved resolution of strong scatterers: IEEE Trans. Microwave Theory Tech., **42**, 137–141.

Potthast, R., 1994, Fréchet differentiability of boundary integral operators in inverse acoustic scattering: Inverse Problems, **10**, 431–447.

Sarkar, T. K., 1987, On the application of the generalized biconjugate gradient method: J. Elec. Waves Appl., **1**, 223–242.

Sasaki, Y., 1994, 3-D resistivity inversion using the finite-element method: Geophysics, **59**, 1939–1948.

Shima, H., 1992, 2-D and 3-D resistivity image reconstruction using crosshole data: Geophysics, **57**, 1270–1281.

Tarantola, A., 1987, Inverse problem theory, Oxford University Press.

Tarits, P., 1994, Electromagnetic studies of global geodynamic processes: Surv. Geophys., **15**, 209–238.

Wang, Y. M., and Chew, W. C., 1989, An iterative solution of the two-dimensional electromagnetic inverse scattering problem: Internat J. Imaging Sys. Technol., **1**, 100–108.

Parameter Estimation for 3-D Geoelectromagnetic Inverse Problems

Oleg Portniaguine
Michael S. Zhdanov

Summary. Parameter estimation in geoelectromagnetics aims to obtain the most important parameters of a well-defined conductivity model of the Earth. These parameters are features of typical geological structures, such as depth and size of conductive or resistive targets, angle of dike inclination and its length, and conductivity of anomalous bodies. We develop this approach through regularized nonlinear optimization. We use finite differences of forward computations and Broyden's updating formula to compute sensitivities (Frechet or partial derivatives) for each parameter. To estimate the optimal step length, we apply line search, with a simple and fast parabolic correction. Our inversion also includes Tikhonov's regularization procedure. We use our method to study measurements of the magnetic fields from a conductive body excited by a loop source at the surface. Keeping the depth of the body constant, we estimate the horizontal coordinates of the body from three components of the magnetic field measured in a borehole. These measurements accurately determine the direction to the conductive target.

1 Introduction

In the past decade, many advances have occurred in multidimensional inversion of dc resistivity data (Shima, 1992; Oldenburg and Li, 1993; Sasaki, 1994; Zhang et al., 1994), and both transient and harmonic electromagnetic (EM) data (Eaton, 1989; Madden and Mackie, 1989; Smith and Booker, 1991; Xiong and Kirsch, 1992; Lee and Xie, 1993; Pellerin et al., 1993; Tripp and Hohmann, 1993; Nekut, 1994; Torres-Verdin and Habashy, 1994; and Zhdanov and Fang, 1995). Most of the advances came in inversion for models with many cells of constant conductivity, in which an optimization algorithm finds a distribution of conductivity whose response matches the original data. These methods all face the difficulties of large-scale inversion: Computer power and memory capacity grow exponentially with the number of cells, and the stability of the inverse problem gets worse (Tikhonov and Arsenin, 1977).

When interpreting EM data, however, one often can construct several possible geoelectrical models on the basis of prior geological and geophysical information. All of

University of Utah, Department of Geology and Geophysics, Salt Lake City, UT 84112.

these models could contain the same geological structure, but with different specific parameters—say, depth and size of conductive or resistive targets, angle of dike inclination and its length, and conductivity of the anomalous bodies. The goal of inversion then becomes the estimation of a few important parameters of the model. Inversion for only a few parameters is, of course, more efficient than a general inversion. The first EM inversions (in the 1970s) were parametric; however, they were limited to one-dimensional (1D) layer thicknesses and conductivities. We take up this approach, but with all of the advantages of modern 3-D forward modeling.

2 Inversion scheme

2.1 Minimization problem

A general approach to ill-posed inverse problems is based on minimization of the Tikhonov parametric functional (Tikhonov and Arsenin, 1977),

$$P^{\alpha}(\mathbf{m}) = \phi(\mathbf{m}) + \alpha s(\mathbf{m}) = \min, \tag{1}$$

where ϕ is a misfit functional,

$$\phi(\mathbf{m}) = \|\mathbf{r}(\mathbf{m})\|^2, \qquad \mathbf{r}(\mathbf{m}) = \mathbf{A}(\mathbf{m}) - \mathbf{d}^o; \tag{2}$$

$\mathbf{d}^o$ is the vector of N observed EM data; $\mathbf{m}$ is the vector of M model parameters; $\mathbf{A}(\mathbf{m})$ is the vector of theoretical (predicted) EM data; $\mathbf{r}(\mathbf{m})$ is the residual vector; and $s(\mathbf{m})$ is the stabilizing functional

$$s(\mathbf{m}) = \|\mathbf{m} - \mathbf{m}_{\text{apr}}\|^2. \tag{3}$$

Minimizing Eq. (1) replaces the original ill-posed inverse problem with the family of well-posed problems, which tend to the original problem as the regularization parameter α goes to zero (Tikhonov and Arsenin, 1977). Eventually, we want to find the model that best fits the observed data. The stabilizing functional (3) is designed to keep the inverse model relatively close to some prior reference model $\mathbf{m}_{\text{apr}}$. The minimization problem (1) is solved for different values of the regularization parameter α. We can select the quasi-optimal value of α by using prior information about the accuracy of the original data.

2.2 Optimization method

Our inversion code has options for using conjugate gradient, steepest descent, and Newtonian methods. We usually use only a few free parameters, so that the Hessian matrix has a small size. This allows us to use Newton's method which has a superior conversion rate.

The method iteratively updates the model at the ith iteration according to formulas

$$\mathbf{m}_{i+1} = \mathbf{m}_i + \delta\mathbf{m}_i, \tag{4}$$

$$\delta\mathbf{m}_i = k\delta\mathbf{m}_i', \tag{5}$$

$$\delta\mathbf{m}_i' = -[\underset{\sim}{\mathbf{H}}(\mathbf{m}_i) + \alpha\underset{\sim}{\mathbf{I}}]^{-1}\boldsymbol{\ell}^{\alpha}(\mathbf{m}_i), \tag{6}$$

$$\boldsymbol{\ell}^{\alpha}(\mathbf{m}_i) = \underset{\sim}{\mathbf{F}}^*(\mathbf{m}_i)\,\mathbf{r}(\mathbf{m}_i) + \alpha(\mathbf{m}_i - \mathbf{m}_{\text{apr}}), \tag{7}$$

$$\underset{\sim}{\mathbf{H}}(\mathbf{m}_i) = \underset{\sim}{\mathbf{F}}^*(\mathbf{m}_i)\,\underset{\sim}{\mathbf{F}}(\mathbf{m}_i), \tag{8}$$

where $\delta\mathbf{m}_i'$ is the Newtonian step, $\delta\mathbf{m}_i$ is the corrected Newtonian step, k is the correction factor, $\ell^\alpha(\mathbf{m}_i)$ is the regularized direction of the steepest ascent, $\underset{\sim}{\mathbf{F}}(\mathbf{m}_i)$ is the Frechet derivative matrix of size $N \times M$, $\underset{\sim}{\mathbf{H}}(\mathbf{m}_i)$ is the Hessian matrix of size $N \times M$, and $\underset{\sim}{\mathbf{I}}$ is the unit matrix. An asterisk denotes the conjugate transposed matrix.

The length of the Newtonian step $\delta\mathbf{m}_i'$ is determined by assuming that the parameteric functional is a perfect quadratic which is only true for a linear inverse problem. To improve convergence for nonlinear functionals, the step length should be chosen by a search for a minimum along the direction of the Newtonian step (Fletcher, 1981):

$$P^\alpha\left(\mathbf{m}_i + k\delta\mathbf{m}_i'\right) = \min!. \tag{9}$$

We apply the simplest one-step search that assumes parabolic behavior of the residuals $\mathbf{r}(\mathbf{m}_i)$ at point $\mathbf{m}_i$:

$$\mathbf{r}\left(\mathbf{m}_i + k\delta\mathbf{m}_i'\right) = \mathbf{c}k^2 + \mathbf{g}(\mathbf{m}_i)k + \mathbf{r}(\mathbf{m}_i).$$

The case $k = 1$ corresponds to the classical Newtonian step without correction. We compute the residual $\mathbf{r}(\mathbf{m}_i + \delta\mathbf{m}_i')$ at the destination point of the Newtonian step; then, knowing the gradient along the step direction $\mathbf{g}(\mathbf{m}_i) = \underset{\sim}{\mathbf{F}}(\mathbf{m}_i)\delta\mathbf{m}_i'$ and the residual $\mathbf{r}(\mathbf{m}_i)$ at the current point, we can estimate the vector $\mathbf{c}$ which consists of the second derivative of the residuals:

$$\mathbf{c} = \mathbf{r}\left(\mathbf{m}_i + \delta\mathbf{m}_i'\right) - \mathbf{g}(\mathbf{m}_i) - \mathbf{r}(\mathbf{m}_i). \tag{10}$$

Equation (9) thus can be replaced by the fourth-order polynomial with respect to k, if we know the residual $\mathbf{r}(\mathbf{m}_i + \delta\mathbf{m}_i')$ at the destination point of the Newtonian step:

$$\|ck^2 + \mathbf{g}(\mathbf{m}_i)k + \mathbf{r}(\mathbf{m}_i)\|^2 + \alpha\left\|\mathbf{m}_i - \mathbf{m}_{\text{apr}} + k\delta\mathbf{m}_i'\right\|^2 = \min!. \tag{11}$$

The norm of any vector $\mathbf{B}$ is $\|B\|^2 = \mathbf{B}^*\mathbf{B}$. We can rewrite Eq. (11) in the form of the scalar fourth-order polynomial minimization problem with respect to parameter k as

$$p_0 + p_1k + p_2k^2 + p_3k^3 + p_4k^4 = \min!, \tag{12}$$

where polynomial coefficients are defined as

$$p_0 = \|\mathbf{r}(\mathbf{m}_i)\|^2 + \alpha\|\mathbf{m}_i - \mathbf{m}_{\text{apr}}\|^2,$$

$$p_1 = 2\,\text{Re}\left[\mathbf{g}(\mathbf{m}_i)^*\mathbf{r}(\mathbf{m}_i) + \alpha(\mathbf{m}_i - \mathbf{m}_{\text{apr}})^*\delta\mathbf{m}_i'\right],$$

$$p_2 = \|\mathbf{g}(\mathbf{m}_i)\|^2 + \alpha\left\|\delta\mathbf{m}_i'\right\|^2 + 2\,\text{Re}[\mathbf{c}^*\mathbf{r}(\mathbf{m}_i)],$$

$$p_3 = 2\,\text{Re}[\mathbf{c}^*\mathbf{g}(\mathbf{m}_i)], \qquad p_4 = \mathbf{c}^*\mathbf{c}.$$

We solve Eq. (12) numerically using the secant root-finding method and select the smallest positive root as an optimal step length, because we have to be conservative and stay close to the previous iteration.

2.3 Frechet derivatives

The elements $F^{(k\ell)}$ of the Frechet (partial) derivative matrix, which are required in formulas (7) and (8) to compute the Newtonian step, can be estimated with finite differences:

$$F^{(k\ell)} = \frac{\partial d^{(k)}}{\partial m^{(\ell)}} \approx \frac{A^{(k)}\left[m + \delta m^{(\ell)}\right] - A^{(k)}(m)}{\delta m^{(\ell)}}, \tag{13}$$

where $d^{(k)}$ is the kth element of the vector of data and $\delta m^{(\ell)}$ is a small perturbation of the lth element of the vector of parameters. In numerical calculations we select a perturbation equal to 1% of corresponding parameter value. To fill out the whole matrix, we have to apply formula (13) for each parameter.

To save computational time, the Frechet matrix on the next step, $\underset{\sim}{\mathbf{F}}_{i+1}$, can be estimated from the Frechet matrix on the previous step, $\underset{\sim}{\mathbf{F}}_i$, using the approximate Broyden updating formula (Fletcher, 1981; Gill et al., 1981). To derive the Broyden formula, we express the Frechet derivative $\underset{\sim}{\mathbf{F}}_{i+1}$ at the point $\mathbf{m}_{i+1}$ as a difference between the forward solution $\mathbf{A}(\mathbf{m}_{i+1})$ at the subsequent iteration $\mathbf{m}_{i+1} = \mathbf{m}_i + \delta\mathbf{m}_i$ and the forward solution for the current iteration $\mathbf{A}(\mathbf{m}_i)$:

$$\underset{\sim}{\mathbf{F}}_{i+1}\delta\mathbf{m}_i \approx \mathbf{A}(\mathbf{m}_{i+1}) - \mathbf{A}(\mathbf{m}_i). \tag{14}$$

However, knowing the current Frechet derivative $\underset{\sim}{\mathbf{F}}_i$, we also can express its variation $\Delta\underset{\sim}{\mathbf{F}}_i$ as

$$\Delta\underset{\sim}{\mathbf{F}}_i \cong \underset{\sim}{\mathbf{F}}_{i+1} - \underset{\sim}{\mathbf{F}}_i. \tag{15}$$

Let $\mathbf{F}^{(k\cdot)}$ stand for the kth row of the Frechet derivative matrix. Then, combining Eqs. (14) and (15) gives the underdetermined system of N equations with respect to $N \times M$ elements of the matrix $\Delta\underset{\sim}{\mathbf{F}}_i$:

$$\Delta\underset{\sim}{\mathbf{F}}_i^{(k\cdot)}\delta\mathbf{m}_i = B_i^{(k)}, \qquad k = 1, 2, \ldots N, \tag{16}$$

where

$$B_i^{(k)} = A^{(k)}(m_{i+1}) - A^{(k)}(m_i) - F_i^{(k\cdot)}\delta m_i. \tag{17}$$

This system of equations has a unique solution under the additional condition that the vectors $\Delta\mathbf{F}_i^{(k\cdot)}$ have the minimum norm,

$$\left\|\Delta\mathbf{F}_i^{(k\cdot)}\right\| = \min. \tag{18}$$

According to the Riesz representation theorem (Parker, 1994), the solution of Eqs. (16) under condition (18) can be written as

$$\Delta\mathbf{F}_i^{(k\cdot)} = f_i^{(k)}\delta\mathbf{m}_i^T, \qquad k = 1, 2, 3, \ldots, N, \tag{19}$$

where $f_i^{(k)}$ are unknown constants determined from the equation

$$f_i^{(k)}\delta\mathbf{m}_i^T\delta\mathbf{m}_i = B_i^{(k)}, \tag{20}$$

and $\delta\mathbf{m}_i^T$ is a row vector of the parameter perturbation (transpose-of-column vector $\delta\mathbf{m}_i$). Solving Eq. (20) and substituting the result into Eq. (19) gives

$$\Delta\mathbf{F}_i^{(k\cdot)} = \frac{B_i^{(k)}\delta\mathbf{m}_i^T}{\delta\mathbf{m}_i^T\delta\mathbf{m}_i}. \tag{21}$$

Using formula (15) for the Frechet derivative $\underset{\sim}{\mathbf{F}}_{i+1}$ and expression (17) gives the first-order Broyden updating formula

$$\underset{\sim}{\mathbf{F}}_{i+1} = \underset{\sim}{\mathbf{F}}_i + [\mathbf{A}(\mathbf{m}_{i+1}) - \mathbf{A}(\mathbf{m}_i) - \underset{\sim}{\mathbf{F}}_i\delta\mathbf{m}_i]\frac{\delta\mathbf{m}_i^T}{\delta\mathbf{m}_i^T\delta\mathbf{m}_i}. \tag{22}$$

At the starting point of the iteration process, we apply formula (13) to estimate the Frechet matrix, take a Newtonian step using formula (6), solve the forward problem at this point, and estimate a correction factor k, solving Eq. (11). Then, we take the corrected step, using formula (5). At the arrival point, we estimate a new Frechet derivative, using Eq. (22), and take a new Newtonian step. If the correction fails to make progress (the parametric functional increases), the Frechet derivative is reevaluated using expression (13).

When the correction factor k is close to zero, we assume that we have reached the minimum of the problem, and we adjust the regularization parameter using the expression $\alpha_{\text{new}} = \alpha_{\text{old}}/2$, and continue with the new value of α. Global iterations stop after the misfit functional drops below the given accuracy level. An application of this method for the simple nonlinear inverse problem is shown in Fig. 1. The nonlinear problem to be solved is described by the following system of equations:

$$x^3 + y^2 = 5, \qquad x^2 - y = -1, \qquad -2x + 2y^2 = 6.$$

We define the misfit funtional $\phi(x, y)$ as

$$\phi(x, y) = (x^3 + y^2 - 5)^2 + (x^2 - y + 1)^2 + (-2x + 2y^2 - 6)^2.$$

The inversion path is shown by the dashed line in Fig. 1. The solid line shows isolines of the misfit functional. It has a minimum at the solution point ($x = 1$, $y = 2$). Iteration starts from the point $x = 0.4$, $y = 1$, which is marked by the asterisk in Fig. 1. At this point the Frechet matrix is estimated using a finite-difference method. The iteration step

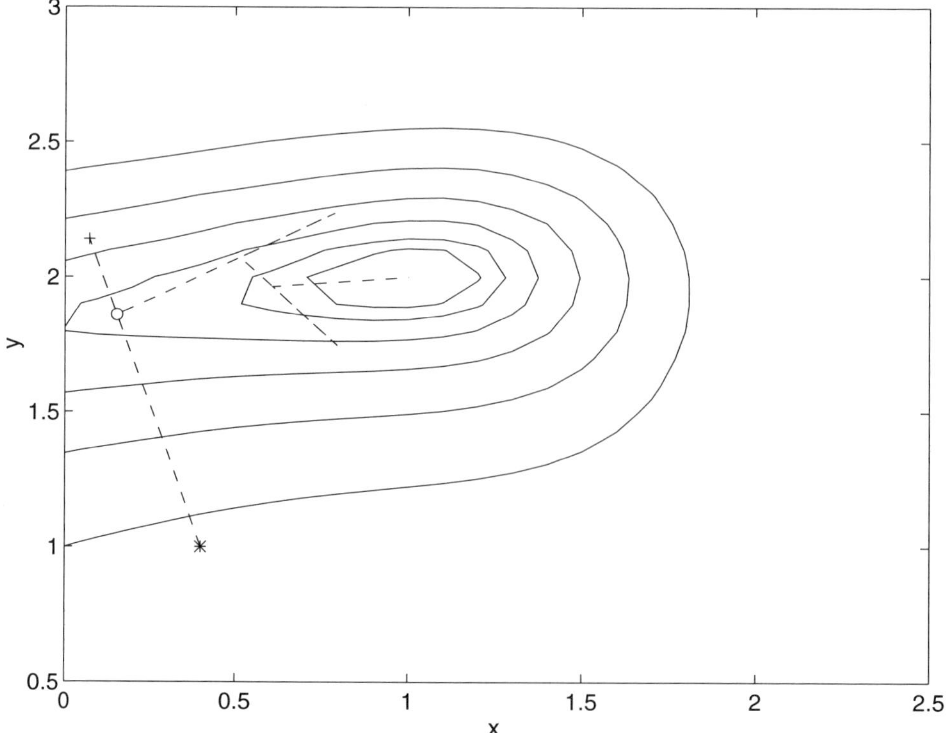

Figure 1. Example of optimization for nonlinear problem with two parameters.

brings us to the point shown by the cross. Note that the step length is overestimated. A parabolic correction reduces the step to the local minimum, shown by the circle. At this point the Frechet derivative is estimated using the Broyden formula, and the next step is performed in a new direction. Iterations converge rapidly to the global minimum.

The main advantages and disadvantages of the numerical computation of the sensitivities are well known. The disadvantage is that, for a problem with N_m parameters, we have to solve the forward problem $N_m + 1$ times, whereas algorithms based on the quasi-analytic solution for Frechet derivatives require computing efforts equivalent to two forward modeling runs for each estimation.

One advantage of our approach is the possibility of choosing nontrivial inversion parameters, e.g., depth and coordinates of the anomalous body and its resistivity, size of the conductive or resistive target, and angle of inclination. In the next section, we demonstrate the effectiveness of our inversion scheme on a synthetic model.

3 Directional sensitivities of three-component magnetic data

EM observations in a single borehole that can provide direction to the target are potentially interesting both for mining and oil and gas applications. In mining exploration, it is important to give accurate direction to off-hole conductors. In oil and gas applications, a system with directional sensitivity can be used for navigation of the bit during horizontal drilling. Today, there are numerous borehole tools built for the downhole measurement of three components of a magnetic field (Crone Geophysics & Exploration Ltd., 1995. Three component borehole survey: Flying Doctor Prospect, Broken Hill, Australia). Studying a model of a 3-D conductor, we demonstrate that three-component measurements have good directional sensitivity.

Consider the model of a conductive body located at a depth of 150 m, 80 m away from a borehole in the x-direction (Fig. 2). The transmitter is a circular loop 200 m in diameter with the center at the coordinate origin. Eleven receivers are located in the borehole and are spaced equally within the depth range from 100 to 200 m. The body is a cube with a side of 60 m. Conductivity of the body is 1 ohm-m, whereas background conductivity is 1000 ohm-m. The theoretical time-domain magnetic field in this model was simulated within the time range from 1 μs to 1000 μs using TEM3-DL finite-difference code (Wang and Hohmann, 1993).

The data are three components of the magnetic field measured along the single observation line (borehole). It is obvious that the depth of the body can be determined by the location of the maximum of the secondary field in the vertical profile. However, our goal is more complicated. We would like to determine the distance and the direction from the borehole to the conducting body.

Thus, we can fix the depth of the body and introduce the polar coordinates of the body center: the distance R from borehole to the body and the angle θ between the x-axis and the direction to the body center (Fig. 2). The actual polar coordinates of the conductive body are $R = 80$ m and $\theta = 0$. The synthetic data for this model ($\partial H_x^0/\partial t$ for all receivers, with 5% random noise added) are shown in Fig. 3.

The inverse problem is reduced in this case to determining (R, θ) for given EM data. We introduce the misfit functionals ϕ_x, ϕ_y, ϕ_z, defined as the norm of the difference between the corresponding x, y, or z components of the predicted $\partial \mathbf{H}/\partial t$ and actual

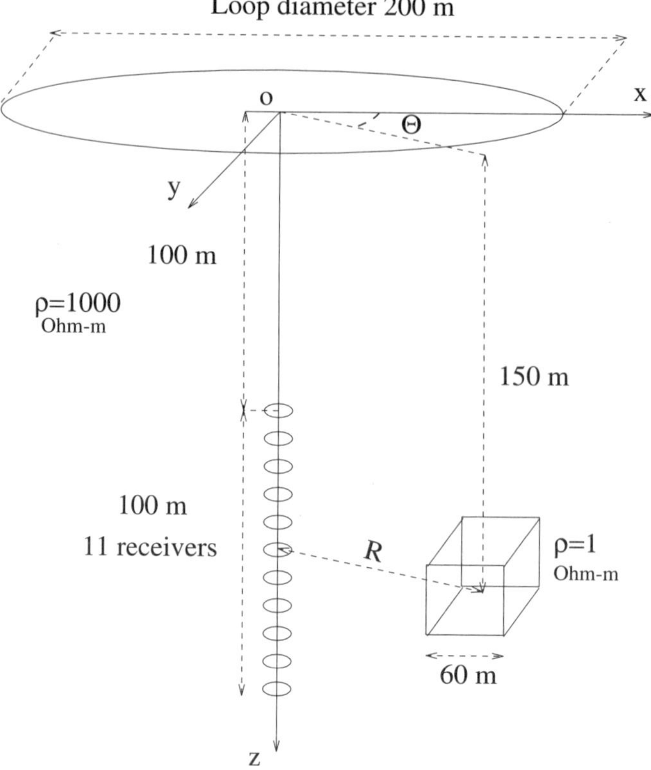

Figure 2. Survey design and model used for directional sensitivity investigation.

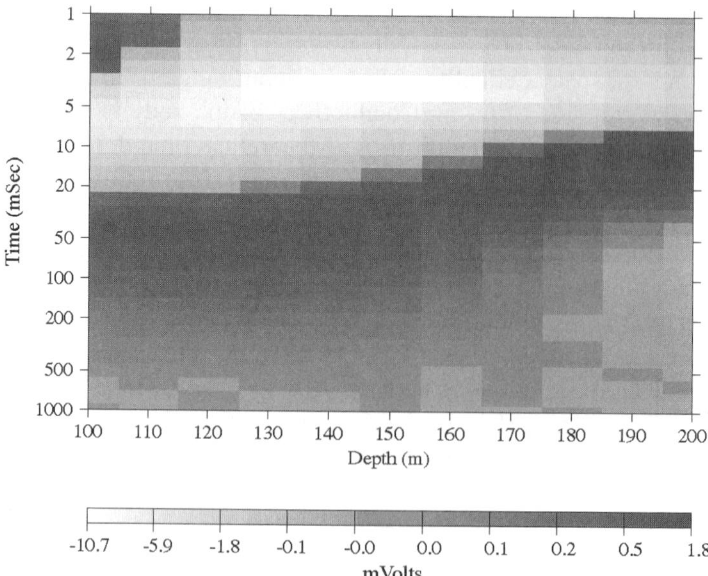

Figure 3. Time derivative of the magnetic field (x-component) from the actual model (5% random noise added).

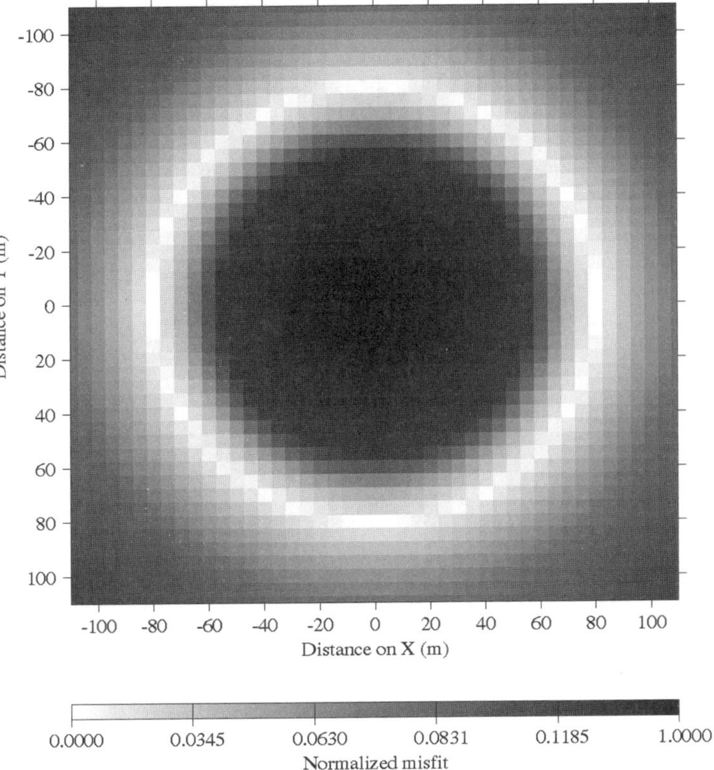

Figure 4. Misfit functional for z-component versus horizontal coordinates of the body.

$\partial \mathbf{H}^0/\partial t$ magnetic field:

$$\phi_x = \left\| \partial H_x/\partial t - \partial H_x^0/\partial t \right\|^2, \qquad \phi_y = \left\| \partial H_y/\partial t - \partial H_y^0/\partial t \right\|^2,$$

$$\phi_z = \left\| \partial H_z/\partial t - \partial H_z^0/\partial t \right\|^2,$$

and the misfit functional ϕ_Σ is defined for all three components,

$$\phi_\Sigma = \phi_x + \phi_y + \phi_z,$$

where we use the L_2 norm over the time interval of the magnetic-field observation.

The plots of misfit functionals ϕ_z, ϕ_y, ϕ_x, and ϕ_Σ as functions of the horizontal coordinates of the body are presented in Figs. 4, 5, 6, and 7, respectively. We expect that the misfit functionals have minima at the location of the body. However, the modeling results show that the z-component is sensitive only to the distance to the body R, but is not sensitive to the direction θ. The map of the misfit functional for this component has a circular structure with the circular minimum corresponding to an 80-m radius (Fig. 4). At the same time, the ϕ_y misfit functional corresponding to the y-component of the magnetic field has a minimum everywhere along the x-axis, but it gives no information about the distance to the body (Fig. 5). The map of the ϕ_x misfit functional is rather complicated; however, it has a weak and flat minimum in the vicinity of the body location (Fig. 6). Only the combination of three components produces a clear minimum on the map of ϕ_Σ at the true location of the body (Fig. 7).

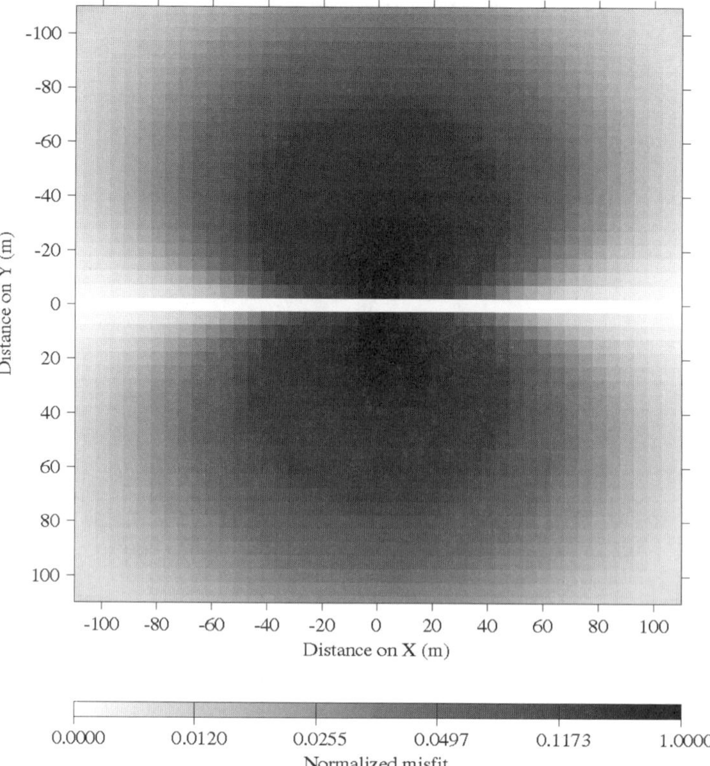

Figure 5. Misfit functional for y-component versus horizontal coordinates of the body.

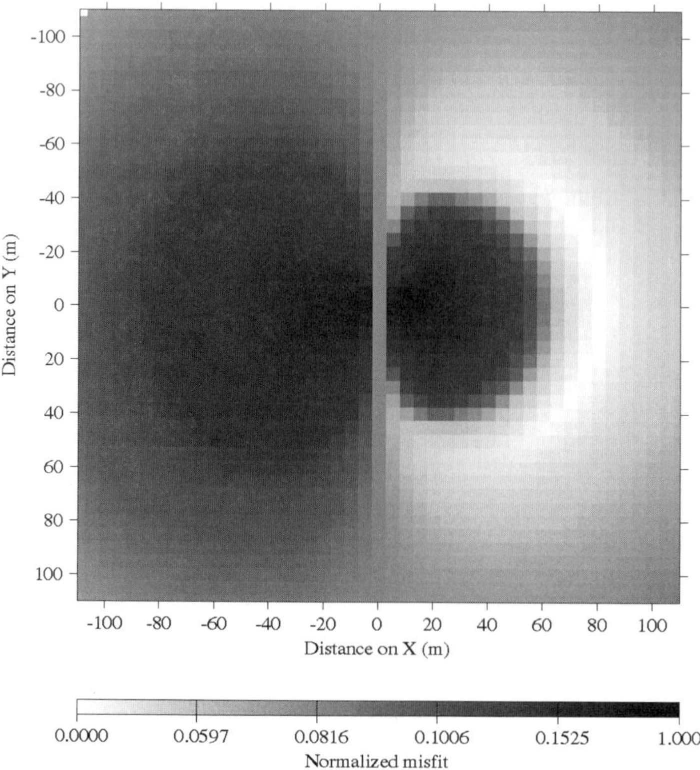

Figure 6. Misfit functional for x-component versus horizontal coordinates of the body.

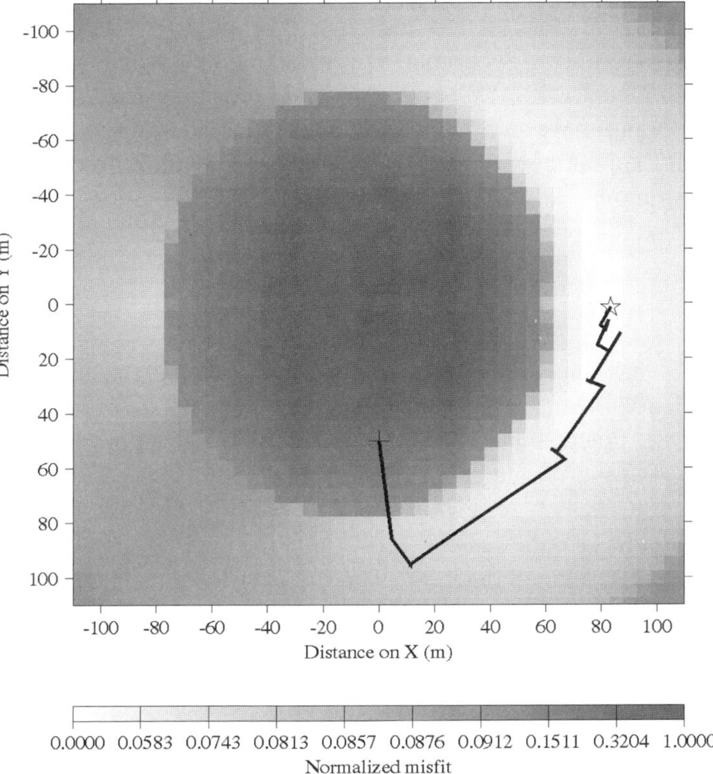

Figure 7. Misfit functional for three components versus horizontal coordinates of the body. Solid line shows inversion path.

Now we can apply the minimization technique developed in the preceding sections to locate the position of the conductive body by the EM field observed in the vertical borehole. In this model test, we have chosen the starting body location at $R = 50$ m, $\theta = 90^\circ$. It is shown in Fig. 7 by a cross. We started the optimization process with the regularized Newtonian method as described above and, after a few iterations, finally arrived at the minimum at the actual location of the body (marked by the star). Solid lines show the inversion path. On some iterations the method produces overshooting which was corrected by the line search. This can be seen on the plot in places where the next iteration starts at the middle of the line, describing the preceding step, rather than from the head of the line. This example shows that directional information can be extracted from noisy three-component observations.

4 Conclusions

Parametric inversion permits easy utilization of existing forward modeling codes. Newton's method, combined with Broyden's updating formula and a parabolic line search, leads to an efficient algorithm with a fast convergence rate. Tikhonov regularization helps to stabilize the inversion.

We demonstrate our parameteric inversion scheme with simulations of three-component transient EM data collected in a borehole. The study shows that three-component observations in a vertical well have good directional sensitivity, even with only one source position.

Acknowledgments

We thank the Consortium on Electromagnetic Modeling and Inversion at the Department of Geology and Geophysics, University of Utah, including CRA Exploration Ltd., Newmont Exploration, Western Mining, Kennecott Exploration, Schlumberger-Doll Research, Shell Exploratie en Produktie Laboratorium, Western Atlas, US Geological Survey, Zonge Engineering, MIM Exploration, BHP Exploration, and Mindeco for providing additional support for this work.

References

Eaton, P., 1989, 3-D electromagnetic inversion using integral equations: Geophys. Prosp., **37**, 407–426.

Fletcher, R., 1981, Practical methods of optimization: John Wiley & Sons, Inc.

Gill, P., Murray, W., and Wright, M., 1981, Practical optimization: Academic Press Inc.

Lee, K., and Xie, G., 1993, A new approach to imaging with low frequency electromagnetic fields: Geophysics, **58**, 780–796.

Madden, T. R., and Mackie, R. L., 1989, Three-dimensional magnetotelluric modeling and inversion: Proc. IEEE, **77**, No. 2, 318–332.

Nekut, A., 1994, Electromagnetic ray-trace tomography: Geophysics, **59**, 371–377.

Oldenburg, D., and Li, Y., 1993, Inversion of induced polarization data: Soc. Expl. Geophys., Expanded Abstracts, 396–399.

Parker, R., 1994, Geophysical inverse theory: Princeton Univ. Press.

Pellerin, L., Johnston, J., and Hohmann, G., 1993, Three-dimensional inversion of electromagnetic data: Soc. Expl. Geophys., Expanded Abstracts, 360–363.

Sasaki, Y., 1994, 3-D resistivity inversion using the finite-element method: Geophysics, **59**, 1839–1848.

Shima, H., 1992, 2-D and 3-D resistivity image reconstruction using crosshole data: Geophysics, **57**, 1270–1281.

Smith, J. T., and Booker, J. R., 1991, Rapid inversion of two- and three-dimensional magnetotelluric data: J. Geophys. Res., **96**, 3905–3922.

Tikhonov, A. N., and Arsenin, V. Y., 1977, Solution of ill-poised problems: W. H. Winston and Sons.

Torres-Verdin, C., and Habashy, T., 1994, Rapid 2.5-dimensional forward modeling and inversion via a new nonlinear scattering approximation: Radio Sci., **29**, 1051–1079.

Tripp, A. C., and Hohmann, G. W., 1993, Three-dimensional electromagnetic crosswell inversion: IEEE Trans. Geosci. Remote Sensing, **31**, 121–126.

Wang, T., and Hohmann, G. W., 1993, A finite-difference time domain solution for three-dimensional electromagnetic modeling: Geophysics, **58**, 797–809.

Xiong, Z., and Kirsch, A., 1992, Three-dimensional Earth conductivity inversion: J. Comput. Appl. Math., **42**, 109–121.

Zhang, J., Mackie, R., and Madden, T., 1994, 3-D resistivity forward modeling and inversion using conjugate gradients: Soc. Expl. Geophys., Expanded Abstracts, 377–380.

Zhdanov, M. S., and Fang, S., 1995, Quasi linear approximation in 3-D electromagnetic modeling: Geophysics, **61**, 646–665.

Three-Dimensional Quasi-linear Electromagnetic Modeling and Inversion

Michael S. Zhdanov
Sheng Fang

Summary. The quasi-linear (QL) approximation replaces the (unknown) total field in the integral equation of electromagnetic (EM) scattering with a linear transformation of the primary field. This transformation involves the product of the primary field with a reflectivity tensor, which is assumed to vary slowly inside inhomogeneous regions and therefore can be determined numerically on a coarse grid by a simple optimization. The QL approximation predicts EM responses accurately over a wide range of frequencies for conductivity contrasts of more than 100 to 1 between the scatterer and the background medium. It also provides a fast-forward model for 3-D EM inversion. The inversion equation is linear with respect to a modified material property tensor, which is the product of the reflectivity tensor and the anomalous conductivity. We call the (regularized) solution of this equation a *quasi-Born inversion.* The material property tensor (obtained by inversion of the data) then is used to estimate the reflectivity tensor inside the inhomogeneous region and, in turn, the anomalous conductivity. Solution of the nonlinear inverse problem thus proceeds through a set of linear equations. In practice, we accomplish this inversion through gradient minimization of a cost function that measures the error in the equations and includes a regularization term. We use synthetic experiments with plane-wave and controlled sources to demonstrate the accuracy and speed of the method.

1 Introduction

There has been great progress recently in 3-D electromagnetic (EM) modeling and inversion with both integral-equation (Eaton, 1989; Xiong, 1992; Xiong and Kirsch, 1992; Tripp and Hohmann, 1993; Xiong and Tripp, 1993; Xie and Lee, 1995) and finite-difference methods (Madden and Mackie, 1989; Newman and Alumbaugh, 1995). These "exact" methods, however, usually require too large a computational effort to allow their routine use. We have been developing a practical 3-D inversion based on a fast new method of forward modeling called the quasi-linear (QL) approximation (Zhdanov and Fang, 1996a). In the QL approximation, the anomalous field inside the

Department of Geology and Geophysics, University of Utah, Salt Lake City, UT 84112, USA.

inhomogeneous region is written as the product of the incident field and a reflectivity tensor. This tensor is assumed to be slowly varying, and therefore can be computed on a much coarser grid than the field itself. Our experience is that the QL approximation is easy to compute and very accurate. For the inverse problem, we recast the QL approximation as a linear integral equation for a modified material property tensor, which then is estimated from the data. The results are used to determine the reflectivity tensor and the anomalous conductivity.

Our method resembles inversion based on the extended Born approximation (Habashy et al., 1993; Torres-Verdín and Habashy, 1994, 1995a,b), but there are some important differences. For example, the extended Born approximation also replaces the (unknown) total field inside the scatterer with a product of the incident field and a tensor, but this scattering tensor is defined explicitly through a weighted integral of the anomalous conductivity. In the QL approximation, in contrast, the reflectivity tensor itself is determined by the solution of an optimization problem.

We develop QL inversion for 3-D EM fields using a gradient algorithm to solve a set of coupled linear inverse problems. The inversion is stabilized by Tikhonov regularization (Tikhonov and Arsenin, 1977; Zhdanov, 1993). Synthetic examples, with and without random noise, indicate that the algorithm for inverting 3-D EM data is fast and stable.

2 Approximations to EM scattering

Consider a 3-D geoelectric model with the normal (or background) complex conductivity $\tilde{\sigma}_n$ and local inhomogeneity D with conductivity $\tilde{\sigma} = \tilde{\sigma}_n + \Delta\tilde{\sigma}$. Complex conductivity includes the effect of displacement currents: $\tilde{\sigma} = \sigma - i\omega\varepsilon$, where σ and ε are electrical conductivity and dielectric permittivity. We assume that $\mu = \mu_0 = 4\pi \times 10^{-7}$ H/m, the free-space magnetic permeability. The model is excited by an EM field generated by an arbitrary source. This field is time harmonic as $e^{-i\omega t}$. The EM fields in this model can be split into normal and anomalous fields:

$$\mathbf{E} = \mathbf{E}^n + \mathbf{E}^a, \quad \mathbf{H} = \mathbf{H}^n + \mathbf{H}^a, \tag{1}$$

where the normal field is the field generated by the given sources in the model with the background distribution of conductivity $\tilde{\sigma}_n$, (i.e., in the model without the inhomogeneity) and the anomalous field is the difference between the total field and the normal field.

The anomalous field can be expressed as an integral over the excess (scattering) currents in the inhomogeneous domain D (Hohmann, 1975; Weidelt, 1975):

$$\mathbf{E}^a(\mathbf{r}_j) = \iiint_D \underset{\sim}{\mathbf{G}}^n(\mathbf{r}_j \mid \mathbf{r})\Delta\tilde{\sigma}(\mathbf{r})[\mathbf{E}^n(\mathbf{r}) + \mathbf{E}^a(\mathbf{r})]\, dv, \tag{2}$$

where $\underset{\sim}{\mathbf{G}}^n(\mathbf{r}_j \mid \mathbf{r})$ is the EM Green's tensor for the medium with the normal conductivity $\tilde{\sigma}_n$.

2.1 *Born and extended Born approximations*

If the anomalous field is small inside D (in comparison with the normal or incident field), then the anomalous field can be neglected inside the integral in Eq. (2), giving the Born approximation for the scattering (Born, 1933):

$$\mathbf{E}^B(\mathbf{r}_j) = \iiint_D \underset{\sim}{\mathbf{G}}^n(\mathbf{r}_j \mid \mathbf{r})\Delta\tilde{\sigma}(\mathbf{r})\mathbf{E}^n(\mathbf{r})\, dv. \tag{3}$$

This approximation, however, is not very accurate for EM scattering by the large conductivity contrasts (or large bodies) that are typical of geophysical problems. Habashy et al. (1993) and Torres-Verdın and Habashy (1994) developed the extended Born approximation, which replaces the internal field in the integral (2) not by the normal field, but by its projection onto a scattering tensor $\underset{\sim}{\mathbf{\Gamma}}(r)$:

$$\mathbf{E}(\mathbf{r}) = \underset{\sim}{\mathbf{\Gamma}}(\mathbf{r})\mathbf{E}^n(\mathbf{r}). \tag{4}$$

An expression for the scattering tensor is derived by rewriting Eq. (2) as an integral equation for the total field,

$$\mathbf{E}(\mathbf{r}_j) = \mathbf{E}^n(\mathbf{r}_j) + \iiint_D \underset{\sim}{\mathbf{G}}^n(\mathbf{r}_j \mid \mathbf{r})\Delta\tilde{\sigma}(\mathbf{r})\mathbf{E}(\mathbf{r})\, dv, \tag{5}$$

and then approximating $\mathbf{E}(\mathbf{r})$ in the integral by its value at the point $\mathbf{r}_j$.

$$\mathbf{E}(\mathbf{r}_j) \approx \mathbf{E}^n(\mathbf{r}_j) + \mathbf{E}(\mathbf{r}_j) \iiint_D \underset{\sim}{\mathbf{G}}^n(\mathbf{r}_j \mid \mathbf{r})\Delta\tilde{\sigma}(\mathbf{r})\, dv, \tag{6}$$

or

$$\mathbf{E}(\mathbf{r}_j) \approx \left[\mathbf{I} - \iiint_D \underset{\sim}{\mathbf{G}}^n(\mathbf{r}_j \mid \mathbf{r})\Delta\tilde{\sigma}(\mathbf{r})\, dv\right]^{-1} \mathbf{E}^n(\mathbf{r}_j). \tag{7}$$

The expression in brackets is the scattering tensor; it does not depend on the illuminating sources and is an explicit nonlinear functional of the anomalous conductivity. In forward modeling with the extended Born approximation, the scattering tensor can be calculated directly; in inversion, the scattering tensor is calculated for an (assumed) initial model, and then updated iteratively after solving an inverse problem for the anomalous conductivity. Torres-Verdin and Habashy (1994) also showed that, for some models, the iterative procedure could be collapsed into a simple two-step inversion.

2.2 QL approximation

In Zhdanov and Fang (1996a), we developed ideas that can be considered an extension of Torres-Verdin and Habashy's (1994) method. Expression (2) can be rewritten in operator form:

$$\mathbf{E}^a = \mathbf{C}[\mathbf{E}^a], \tag{8}$$

where $\mathbf{C}[\mathbf{E}^a]$ is an integral operator on the anomalous field $\mathbf{E}^a$:

$$\mathbf{C}[\mathbf{E}^a] = \mathbf{A}[\mathbf{E}^n] + \mathbf{A}[\mathbf{E}^a], \tag{9}$$

and $\mathbf{A}$ is a linear scattering operator:

$$\mathbf{A}[\mathbf{E}] = \iiint_D \underset{\sim}{\mathbf{G}}^n(\mathbf{r}_j \mid \mathbf{r})\Delta\tilde{\sigma}(\mathbf{r})\mathbf{E}(\mathbf{r})\, dv. \tag{10}$$

The solution of the integral equation (8) for the anomalous field $\mathbf{E}^a$ is a fixed point of the operator $\mathbf{C}$. This solution therefore can be obtained by the method of successive iterations,

$$\mathbf{E}^{a(N)} = \mathbf{C}\big[\mathbf{E}^{a(N-1)}\big], \quad N = 1, 2, 3 \ldots, \tag{11}$$

which converges if $\mathbf{C}$ is a contraction operator; that is, if $\|\mathbf{C}\| < 1$.

The Born approximation is simply the first iteration of this scheme when the initial approximation $\mathbf{E}^{a(0)}$ is set to zero,

$$\mathbf{E}^B = \mathbf{E}^{a(1)} = \mathbf{C}[0] = \mathbf{A}[\mathbf{E}^n]. \tag{12}$$

We try to obtain a more accurate approximation by assuming that the anomalous field inside the inhomogeneous domain is linearly related to the normal field by a tensor $\underset{\sim}{\boldsymbol{\lambda}}$, which we call an *electrical reflectivity tensor:*

$$\mathbf{E}^a(\mathbf{r}) \approx \underset{\sim}{\boldsymbol{\lambda}}(\mathbf{r})\mathbf{E}^n(\mathbf{r}). \tag{13}$$

If expression (13) is taken as the zeroth-order approximation for the scattered field inside the inhomogeneity $[\mathbf{E}^{a(0)} = \underset{\sim}{\boldsymbol{\lambda}}\mathbf{E}^n]$, then the first-order approximation is

$$\mathbf{E}^{a(1)} = \mathbf{C}[\underset{\sim}{\boldsymbol{\lambda}}\mathbf{E}^n] = \mathbf{A}[\mathbf{E}^n + \underset{\sim}{\boldsymbol{\lambda}}\mathbf{E}^n] = \mathbf{A}[(\underset{\sim}{\mathbf{I}} + \underset{\sim}{\boldsymbol{\lambda}})\mathbf{E}^n] = \mathbf{E}^a_{q\ell}. \tag{14}$$

We call this a *QL approximation* $\mathbf{E}^a_{q\ell}$ for the anomalous field. Written out explicitly, the approximation is

$$\mathbf{E}^a_{q\ell} \approx \mathbf{A}[(\mathbf{I} + \underset{\sim}{\boldsymbol{\lambda}})\mathbf{E}^n] = \iiint_D \underset{\sim}{\mathbf{G}}^n(\mathbf{r}_j \mid \mathbf{r})\Delta\tilde{\sigma}(\mathbf{r})[\underset{\sim}{\mathbf{I}} + \underset{\sim}{\boldsymbol{\lambda}}(\mathbf{r})]\mathbf{E}^n(\mathbf{r})\, dv. \tag{15}$$

The accuracy of the QL approximation obviously depends on the accuracy of the representation (13). The actual anomalous field $\mathbf{E}^a$ is equal to

$$\mathbf{E}^a = \mathbf{A}[\mathbf{E}^n] + \mathbf{A}[\mathbf{E}^a] \tag{16}$$

The error is therefore

$$\left\|\mathbf{E}^a - \mathbf{E}^a_{q\ell}\right\| = \|\mathbf{A}(\mathbf{E}^a - \underset{\sim}{\boldsymbol{\lambda}}\mathbf{E}^n)\| \le \|\mathbf{A}\|\|\mathbf{E}^a - \underset{\sim}{\boldsymbol{\lambda}}\mathbf{E}^n\| \tag{17}$$

or,

$$\left\|\mathbf{E}^a - \mathbf{E}^a_{q\ell}\right\| \le \|\mathbf{A}\|\varepsilon, \tag{18}$$

where $\varepsilon = \|\mathbf{E}^a - \underset{\sim}{\boldsymbol{\lambda}}\mathbf{E}^n\|$, and $\|....\|$ is an L_2 norm. If the electrical reflectivity tensor $\underset{\sim}{\boldsymbol{\lambda}}(\mathbf{r})$ is allowed to be a general function of $\mathbf{r}$, it is clear that ε can be made arbitrarily small. In fact, the error is zero if the reflectivity tensor is taken as the normalized dyadic product

$$\underset{\sim}{\boldsymbol{\lambda}}(\mathbf{r}) = [\mathbf{E}^{n*}(\mathbf{r}) \cdot \mathbf{E}^n(\mathbf{r})]^{-1}\mathbf{E}^a(\mathbf{r})\mathbf{E}^{n*}(\mathbf{r}),$$

where the center dot is the (real) inner product $\mathbf{a} \cdot \mathbf{b} = \sum_i a_i b_i$ and the asterisk indicates complex conjugate. Of course, this expression is not very useful in practice because it involves the unknown total field.

Zhdanov and Fang (1996a) analyze different methods of determining an optimal $\underset{\sim}{\lambda}$. They show that one can use the following condition to determine $\underset{\sim}{\lambda}$:

$$\left\| \underset{\sim}{\lambda}(\mathbf{r}_j)\mathbf{E}^n(\mathbf{r}_j) - \iiint_D \underset{\sim}{\mathbf{G}}^n(\mathbf{r}_j \mid \mathbf{r})\Delta\tilde{\sigma}(\mathbf{r})[(\underset{\sim}{\mathbf{I}} + \underset{\sim}{\lambda})\mathbf{E}^n(\mathbf{r})]\, dv \right\| = \varphi(\underset{\sim}{\lambda}) = \min! \quad (19)$$

In numerical calculations we usually asume that $\underset{\sim}{\lambda}(\mathbf{r})$ is a slowly varying (tensor) function inside the anomalous domain D (the simplest form is a constant). Equation (19) then can be treated as an overdetermined problem and solved numerically by a least-squares method (Zhdanov and Fang, 1996a). After the $\underset{\sim}{\lambda}$ is found, the QL approximation to the field is calculated using

$$\mathbf{F}^a \approx \iiint_D \underset{\sim}{\mathbf{G}}^F(\mathbf{r}_j \mid \mathbf{r})\Delta\tilde{\sigma}(\mathbf{r})[\underset{\sim}{\mathbf{I}} + \underset{\sim}{\lambda}(\mathbf{r})]\mathbf{E}^n(\mathbf{r})\, dv. \quad (20)$$

where $\mathbf{F}^a$ stands for the anomalous electric ($\mathbf{E}^a$) or magnetic ($\mathbf{H}^a$) field observed outside the scatterer (e.g., at surface of the Earth), and $\underset{\sim}{\mathbf{G}}^F$ is the appropriate (electric or magnetic) Green's function.

2.3 Comparison

In their roles relating unknown anomalous or total fields to the incident field, the electrical reflectivity tensor $\underset{\sim}{\lambda}$ of the QL approximation and the scattering tensor $\underset{\sim}{\Gamma}$ of the extended Born approximation are themselves related by the simple formula:

$$\underset{\sim}{\lambda} = \underset{\sim}{\Gamma} - \underset{\sim}{\mathbf{I}}. \quad (21)$$

The two approximations differ significantly, however, in computing these tensors. The scattering tensor $\underset{\sim}{\Gamma}$ is defined explicitly by expression (7). The accuracy of the extended Born approximation depends on how well the integral in Eq. (5) is approximated by taking the constant value for the field $\mathbf{E}(\mathbf{r}_j)$. Because the Green's dyadic is strongly peaked for values $\mathbf{r} \approx \mathbf{r}_j$, the approximation should be good if the field itself is not varying rapidly at $\mathbf{r}_j$. Habashy et al. (1993) called this the "localized approximation."

The QL approximation determines the electrical reflectivity tensor by solving a minimization problem (Eq. 19) on a coarse grid. The accuracy of QL approximation depends only on the accuracy of this discretization of $\underset{\sim}{\lambda}$ and, in principle, can be made arbitrarily good, though care may be needed with a fine discretization, because Eq. (19) can become underdetermined.

3 Numerical examples of the QL approximation

This section compares the fields obtained by solving the integral equation (2) numerically, by computing the Born approximation (3), and by computing the QL approximation (15). Figure 1 shows the 3-D geoelectrical model, which consists of a homogeneous half-space of resistivity 100 ohm-m and a conductive rectangular inclusion with resistivity 1 ohm-m. The EM field in the model is excited by a horizontal rectangular loop, which is 10 × 10 m, carries a current of 1 A, and is 50 m to the left of the model, We have used the full integral-equation (IE) code, SYSEM (Xiong, 1992), and QL code, SYSEMQL (Zhdanov and Fang, 1996a) for computing the frequency-domain response of the complex conductivity structure along profiles parallel to the x-axis.

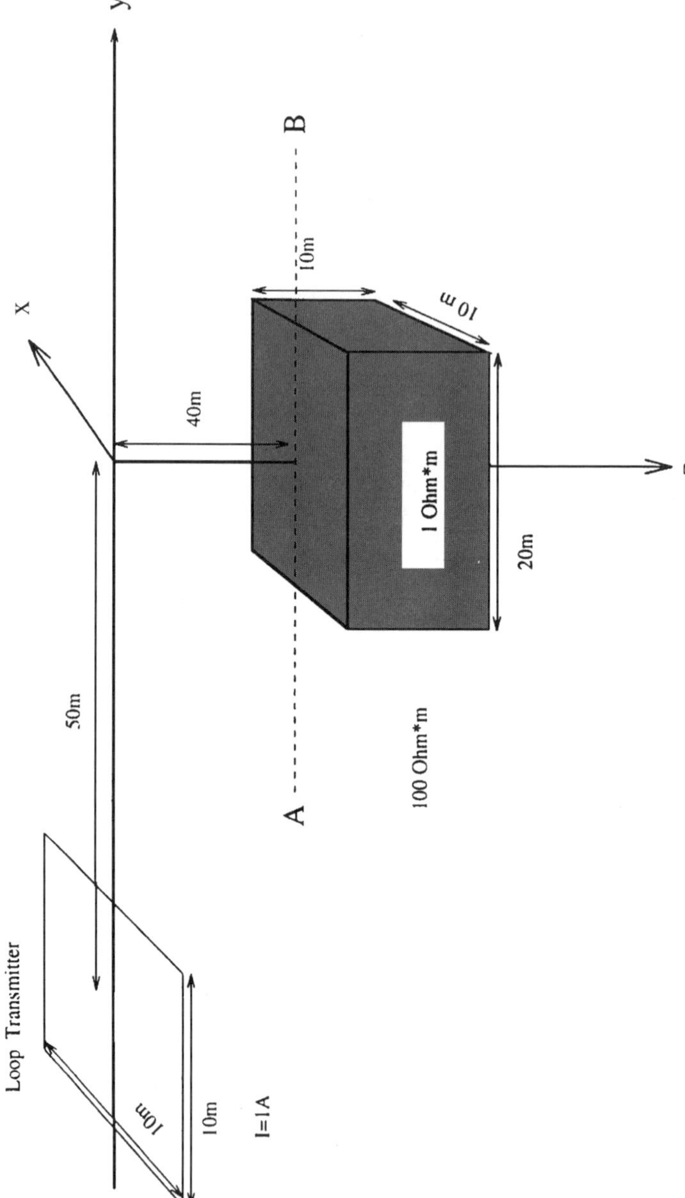

Figure 1. Three-dimensional geoelectric model, containing one conductive body in a homogeneous half-space, with rectangular loop excitation (Model 1).

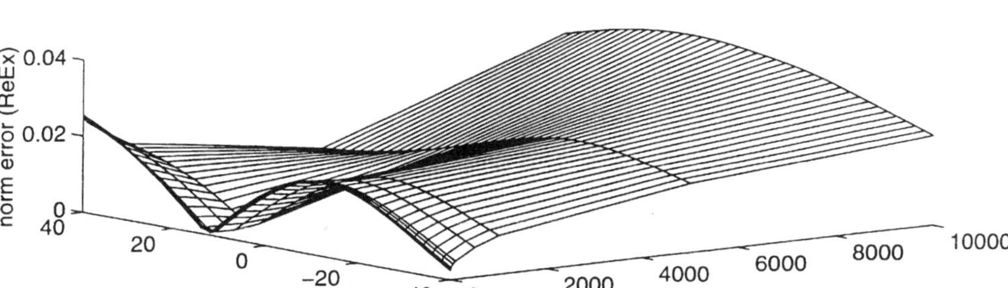

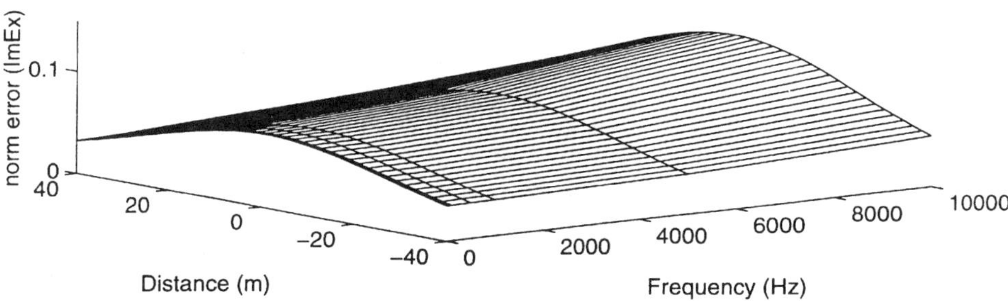

Figure 2. Numerical comparison of full IE solution and QL approximation computed for Model 1 (Fig. 1) at the frequency range from 0.1 Hz to 10 kHz. Calculations were performed for the receivers located along profiles parallel to the y-axis on the surface. Plots show the differences between IE solution and QL approximation for x-component of the secondary electric field normalized by the value of corresponding component of the field at the point $y = 40$ (normalized error).

Figures 2 and 3 compare the different solutions for real and imaginary parts of the anomalous electrical field $\mathbf{E}_x^a$ for different frequencies. The point $x = 0$ along each profile corresponds to the location of the conductive rectangular inclusion center. Figure 2 shows the differences between IE solution and QL approximation, normalized by the value of the corresponding component of the field at the point $y = 40$. The accuracy of the QL approximation for the electric-field components is within 5% for frequencies from 0.1 Hz to 10 kHz. Figure 3 presents the differences between the IE solution and Born approximation, normalized by the value of the corresponding component of the field at the point $y = 40$. The QL approximation produces a reasonable result, whereas the conventional Born approximation is far off the mark.

The next set of comparisons uses the same geometric model, but varies the body's conductivity. We selected four different resistivities of the inclusion: 1 ohm-m, 0.1 ohm-m, 0.01 ohm-m, and 0.001 ohm-m. Figure 4 shows the differences between the IE solution and the QL approximation, at a frequency 0.1 Hz, normalized by the value of the corresponding component of the field at the point $y = 40$. One of the horizontal axes on Fig. 4 is the resistivity contrast $C = \rho_i/\rho_b$, where $\rho_b = 100$ ohm-m is the resistivity of the background, and ρ_i is the resistivity of the conductive inclusion. The errors of QL approximation are generally small and grow only for very-high-conductivity contrasts, equal to $1/C = 10^5$, reaching about 10% in extremum point for the electric field. For lower-conductivity contrasts, the relative errors are below 5%.

Table 1. Comparison of CPU time (s) for the frequency-domain EM modeling, using different methods

Method	Cells in anomalous domain		
	250 cells	400 cells	800 cells
Full IE solution	1029.1	2995.0	13127.0
QL approximation	382.4	530.4	1170.1

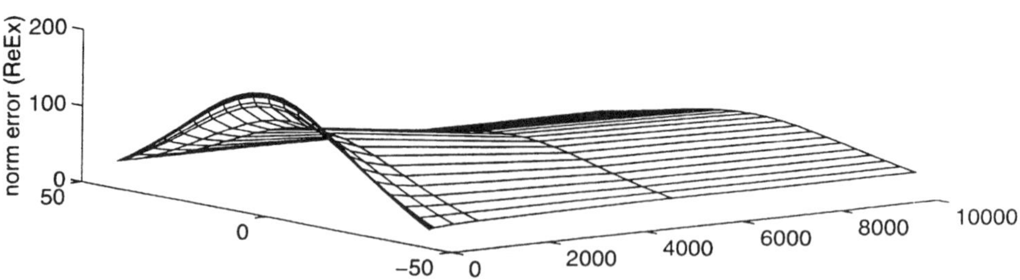

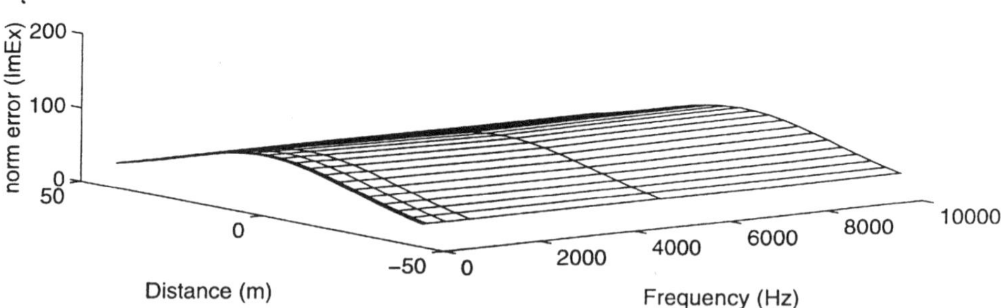

Figure 3. Numerical comparison of full IE solution and Born approximation computed for Model 1 (Fig. 1) at the frequency range from 0.1 Hz to 10 kHz. Calculations were performed for the receivers located along profiles parallel to the y-axis on the surface. Plots show the differences between IE solution and Born approximation for x-component of secondary electric field normalized by the value of corresponding component of the field at the point $y = 40$ (normalized error).

Table 1 shows the computation times needed for full IE solution and for QL approximation in the frequency domain. One can see from this table that the CPU time required for QL approximation grows much more slowly with the number of cells, than the CPU time required for the full IE solution.

4 QL inversion

For the inverse problem, we introduce a new tensor function,

$$\underset{\sim}{\mathbf{m}}(\mathbf{r}) = \Delta\tilde{\sigma}(\mathbf{r})[\underset{\sim}{\mathbf{I}} + \underset{\sim}{\boldsymbol{\lambda}}(\mathbf{r})], \tag{22}$$

which we call a *modified material property tensor*. Equation (20) then takes the form

$$\mathbf{F}^a(\mathbf{r}_j) \approx \iiint_D \underset{\sim}{\mathbf{G}}^F(\mathbf{r}_j \mid \mathbf{r})\underset{\sim}{\mathbf{m}}(\mathbf{r})\mathbf{E}^n(\mathbf{r})\,dv, \tag{23}$$

Quasi–linear approximation (full tensor)

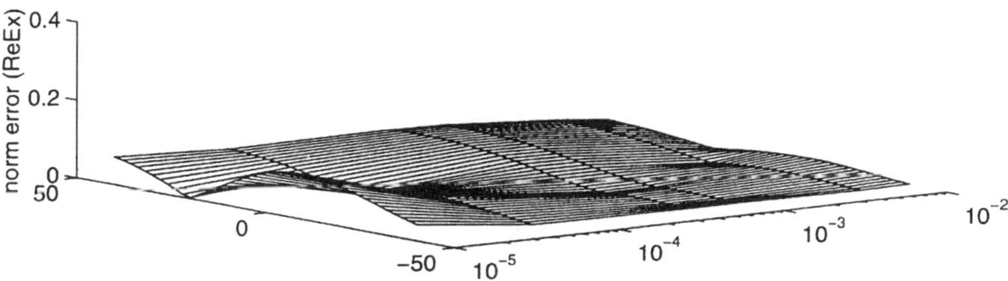

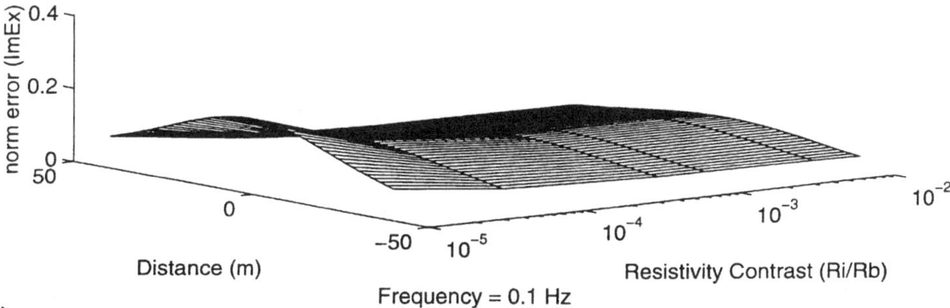

Figure 4. Numerical comparison of full IE solution and QL approximation computed for Model 1 (Fig. 1) at the resistivity ratio of inclusive body to the background range from 0.00001 to 0.01 (or -5 to -2 in log scale). Calculations were performed for the receivers located along profiles parallel to the y-axis on the surface. Plots show the differences between IE solution and QL approximation for x-component of the secondary electric field at the frequency 0.1 Hz normalized by the value of corresponding component of the field at the point $y = 40$ (normalized error).

which is linear with respect to $\underset{\sim}{\mathbf{m}}(\mathbf{r})$ (the original Eq. 20 is nonlinear with respect to $\Delta\tilde{\sigma}$ because the reflectivity tensor depends implicity on $\Delta\tilde{\sigma}$). It has the same structure as the Born approximation for the anomalous field, with the modified material property tensor $\underset{\sim}{\mathbf{m}}(\mathbf{r})$ replacing the anomalous conductivity $\Delta\tilde{\sigma}(\mathbf{r})$. We call Eq. (23) a *quasi-Born approximation*, and its solution (for $\underset{\sim}{\mathbf{m}}$), a *quasi-Born inversion.*

The reflectivity tensor $\underset{\sim}{\boldsymbol{\lambda}}$ can be computed from $\underset{\sim}{\mathbf{m}}$, because

$$\mathbf{E}^a(\mathbf{r}_j) \approx \iiint_D \underset{\sim}{\mathbf{G}}^E(\mathbf{r}_j \mid \mathbf{r})\underset{\sim}{\mathbf{m}}(\mathbf{r})\mathbf{E}^n(\mathbf{r})\, dv \approx \underset{\sim}{\boldsymbol{\lambda}}(\mathbf{r}_j)\mathbf{E}^n(\mathbf{r}_j). \tag{24}$$

Once $\underset{\sim}{\mathbf{m}}$ and $\underset{\sim}{\boldsymbol{\lambda}}$ are known, the anomalous conductivity $\Delta\tilde{\sigma}$ follows from Eq. (22). This inversion scheme reduces the original nonlinear inverse problem to three linear steps:

- inversion of the quasi-Born equation (23) for $\underset{\sim}{\mathbf{m}}$;
- computation of the integral (24) to obtain $\underset{\sim}{\boldsymbol{\lambda}}$; and
- (local) inversion of Eq. (22) to obtain the conductivity $\Delta\tilde{\sigma}$.

We call this procedure a *QL inversion.* As we explain further below, these three steps *do not* solve the full nonlinear inverse problem for $\Delta\tilde{\sigma}$ (mainly because the inversion in the first step is intrinsically nonunique), but they do provide the basis for an effective iterative solution. This iterative scheme resembles the source-type IE method of Habashy et al. (1994) and the modified gradient method of Kleinman and van den Berg (1993).

4.1 Discrete QL equations

The first step of the three-step QL inversion—the solution of Eq. (23)—is nonunique. The equation has a null space, because there exist $\underset{\sim}{\mathbf{m}}$ distributions that produce a zero external field, similar to the nonradiating current distributions that generate zero field outside their domain of support (Habashy et al., 1994; Svetov and Gubatenko, 1985). Although the linear inverse (source) problem for $\underset{\sim}{\mathbf{m}}$ is nonunique, the nonlinear inverse problem for the conductivity can be unique in the class of piecewise-analytic functions (for certain experimental configurations), according to a theorem of Gusarov (1981).

To make our QL inversion unique, we therefore have to develop a unified (iterative) approach for simultaneously finding $\underset{\sim}{\mathbf{m}}$, $\underset{\sim}{\boldsymbol{\lambda}}$, and $\Delta\tilde{\sigma}(\mathbf{r})$. We do this with a constrained inversion that generalizes a method developed by Zhdanov and Chernyak (1987) for two-dimensional (2-D) models. A similar approach to the 2-D inverse scattering problem was discussed by Kleinman and van den Berg (1993). We first discretize the equations by dividing the domain D into substructures (subdomains) $D = \bigcup_{k=1,K} D_k$ and assume that the material property tensor is constant in each substructure, so that Eq. (22) becomes

$$\mathbf{F}^a(\mathbf{r}_j) = \sum_{k=1,N} \iiint_{D_k} \underset{\sim}{\mathbf{G}}^F(\mathbf{r}_j \mid \mathbf{r}) \underset{\sim}{\mathbf{m}}_k \mathbf{E}^n(\mathbf{r})\, dv. \tag{25}$$

where $\underset{\sim}{\mathbf{m}}_k$ depends only on k. We also assume that the reflectivity tensor is constant in each substructure so that the equation for determining $\underset{\sim}{\boldsymbol{\lambda}}$ is

$$\underset{\sim}{\boldsymbol{\lambda}}_k \mathbf{E}^n(\mathbf{r}_j) \approx \sum_{\ell=1,N} \iiint_{D_\ell} \underset{\sim}{\mathbf{G}}^E(\mathbf{r}_j \mid \mathbf{r}) \underset{\sim}{\mathbf{m}}_\ell \mathbf{E}^n(\mathbf{r})\, dv, \qquad \mathbf{r}_j \in D_k. \tag{26}$$

Finally,

$$\underset{\sim}{\mathbf{m}}_k = \Delta\tilde{\sigma}_k[\underset{\sim}{\mathbf{I}} + \underset{\sim}{\boldsymbol{\lambda}}_k] \tag{27}$$

gives the relationship between the conductivity and the material property and reflectivity tensors in each substructure.

To proceed further, we write these equations in matrix form, for the simplest case when the modified material property and reflectivity tensors are scalars (i.e., proportional to the unit tensor). Let $\mathbf{m} = [m_1, m_2, \ldots, m_N]^T$ and $\Delta\boldsymbol{\sigma} = [\Delta\tilde{\sigma}_1, \Delta\tilde{\sigma}_2, \ldots, \Delta\tilde{\sigma}_N]^T$ be column vectors whose elements are the modified material properties and conductivities in the substructures; and let $\underset{\sim}{\boldsymbol{\Lambda}} = \mathrm{diag}[\lambda_1, \lambda_2, \ldots, \lambda_N]$ be a diagonal matrix of the reflectivities. Equation (25) becomes

$$\mathbf{F} = \underset{\sim}{\mathbf{G}}^F \mathbf{m}, \tag{28}$$

where $\mathbf{F}$ is a column vector of data, and $\underset{\sim}{\mathbf{G}}^F$ is a matrix representation of the linear operator defined by formula (25). Equation (27) becomes

$$\mathbf{m} = (\underset{\sim}{\mathbf{I}} + \underset{\sim}{\boldsymbol{\Lambda}})\Delta\boldsymbol{\sigma}. \tag{29}$$

This equation is essentially Ohm's law,

$$\mathbf{j}^D = \Delta\tilde{\sigma}\mathbf{E} = \Delta\tilde{\sigma}(\mathbf{E}^n + \mathbf{E}^a) = \Delta\tilde{\sigma}[\underset{\sim}{\mathbf{I}} + \underset{\sim}{\boldsymbol{\lambda}}^k(\mathbf{r})]\mathbf{E}^n. \tag{30}$$

Equation (26) is overdetermined and can be inverted directly (in least-squares sense) by

$$\lambda_k = [\mathbf{E}^{n*}(\mathbf{r}_j) \cdot \mathbf{E}^{n}(\mathbf{r}_j)]^{-1}\mathbf{E}^{n*}(\mathbf{r}_j) \cdot \sum_{\ell=1,N} \iiint_{D_\ell} \underset{\sim}{\mathbf{G}}^E(\mathbf{r}_j \mid \mathbf{r}) m_\ell \mathbf{E}^n(\mathbf{r})\, dv, \quad \mathbf{r}_j \in D_k, \tag{31}$$

or, in matrix form,

$$\boldsymbol{\lambda} = (\underset{\sim}{\mathbf{E}}^{n*}\underset{\sim}{\mathbf{E}}^{n})^{-1}\underset{\sim}{\mathbf{E}}^{n*}\underset{\sim}{\mathbf{G}}^{E}\mathbf{m}, \tag{32}$$

where $\boldsymbol{\lambda}$ is now a column vector of the reflectivities; $\underset{\sim}{\mathbf{E}}^n$ is a block diagonal matrix whose diagonal blocks are the (3×1 complex) vectors $\mathbf{E}^n(\mathbf{r}_j)$; and the asterisk indicates conjugate transpose.

With multifrequency data, both $\mathbf{m}$ and $\boldsymbol{\lambda}$ will depend on frequency. We assume, however, that $\Delta\tilde{\sigma} = \Delta\sigma - i\omega\Delta\varepsilon$, where $\Delta\sigma$ and $\Delta\varepsilon$ do not depend on frequency. In the absence of any constraints, the least-squares solution of Eq. (29) for the real and imaginary parts of $\Delta\boldsymbol{\sigma}$ is

$$\mathrm{Re}(\Delta\boldsymbol{\sigma}) = \mathrm{Re}\left\{\left[\sum_{\omega}(\underset{\sim}{\mathbf{I}} + \underset{\sim}{\boldsymbol{\Lambda}})^*(\underset{\sim}{\mathbf{I}} + \underset{\sim}{\boldsymbol{\Lambda}})\right]^{-1} \sum_{\omega}(\underset{\sim}{\mathbf{I}} + \underset{\sim}{\boldsymbol{\Lambda}})^*\mathbf{m}\right\}, \tag{33}$$

and

$$\mathrm{Im}(\Delta\boldsymbol{\sigma}) = \omega\,\mathrm{Im}\left\{\left[\sum_{\omega}\omega(\underset{\sim}{\mathbf{I}} + \underset{\sim}{\boldsymbol{\Lambda}})^*(\underset{\sim}{\mathbf{I}} + \underset{\sim}{\boldsymbol{\Lambda}})\right]^{-1} \sum_{\omega}(\underset{\sim}{\mathbf{I}} + \underset{\sim}{\boldsymbol{\Lambda}})^*\mathbf{m}\right\}. \tag{34}$$

4.2 Regularized QL inversion

QL inversion requires the solution of Eq. (28) for $\mathbf{m}$, computation of λ_k by Eq. (31), and solution of Eq. (29) for $\Delta\tilde{\sigma}_j$. To obtain a stable, regularized solution, we introduce the functional

$$P^\alpha(\mathbf{m}) = \phi(\mathbf{m}) + \alpha S(\mathbf{m}), \tag{35}$$

where the misfit functional is specified as

$$\begin{aligned}\phi(\mathbf{m}) &= \|\underset{\sim}{\mathbf{G}}^F\mathbf{m} - \mathbf{F}\|^2 + \|\mathbf{m} - (\underset{\sim}{\mathbf{I}} + \underset{\sim}{\boldsymbol{\Lambda}})\Delta\boldsymbol{\sigma}\|^2 \\ &= (\underset{\sim}{\mathbf{G}}^F\mathbf{m} - \mathbf{F})^*(\underset{\sim}{\mathbf{G}}^F\mathbf{m} - \mathbf{F}) + [\mathbf{m} - (\underset{\sim}{\mathbf{I}} + \underset{\sim}{\boldsymbol{\Lambda}})\Delta\boldsymbol{\sigma}]^*[\mathbf{m} - (\underset{\sim}{\mathbf{I}} + \underset{\sim}{\boldsymbol{\Lambda}})\Delta\boldsymbol{\sigma}].\end{aligned} \tag{36}$$

The misfit functional tracks the solution of both equations (28) and (29). The stabilizer is

$$S(\mathbf{m}) = \|\mathbf{m} - \mathbf{m}_p\|^2 = (\mathbf{m} - \mathbf{m}_p)^*(\mathbf{m} - \mathbf{m}_p). \tag{37}$$

The prior model $\mathbf{m}_p$ is some reference model, selected on the basis of all available geological and geophysical information about the area under investigation. The scalar multiplier α is a regularization parameter.

The misfit functional provides the solution that best fits the observed data $\mathbf{F}$, whereas the stabilizing functional ties the solution to the prior model $\mathbf{m}_p$. The regularization parameter α controls the trade-off between these two goals. Principles for determining the regularization parameter α are discussed by Tikhonov and Arsenin (1977) and Zhdanov and Keller (1994). We use a simple numerical method to determine the parameter α. Consider the progression of numbers

$$\alpha_k = \alpha_0 q^k; \quad k = 0, 1, 2, \ldots, n; \quad q > 0. \tag{38}$$

For any number α_k, we can find an element m_{α_k}, minimizing $P^{\alpha_k}(\mathbf{m})$, and calculate the misfit $\|\underset{\sim}{\mathbf{G}}^F\,\mathbf{m}^{\alpha_k} - \mathbf{F}\|^2$. The optimal value of the parameter α is the number α_{k0}, for which

$$\|\underset{\sim}{\mathbf{G}}^F\mathbf{m}^{\alpha_{k0}} - \mathbf{F}\|^2 = \delta, \tag{39}$$

where δ is the level of noise in observed data. The equality (39) is called the misfit condition. To avoid divergence, we begin an iteration from a big value of α (e.g. $\alpha_0 = 100$), then reduce α ($\alpha = \alpha_0/10$) on each subsequent iteration and continuously iterate until the misfit condition is reached.

The inversion thus is reduced to the solution of the minimization problem for the parametric functional,

$$P^{\alpha}(\mathbf{m}) = \min! \tag{40}$$

which we do by a regularized steepest-descent method (Appendix). The solution m^{α} of the regularized problem (40) is a continuous function of the data (and so, it is stable) and uniformly tends to the actual solution of the original inverse problem when $\alpha \to 0$.

5 Examples of QL inversion

5.1 Plane-wave excitation model

To test the algorithm, we have computed an EM field for two conductive rectangular structures in a homogeneous half-space, excited by a plane wave (Fig. 5a). The observed data on the surface were simulated by forward modeling using a full IE code (Xiong, 1992). Figures 6a and 6b show the comparison of the full IE solutions (solid line) and QL approximation (dashed line) for apparent resistivities computed for TM on (Transverse Magnetic) mode (ρ_{yx}) at the frequencies (in Hz): 10, 1, 0.5, and 0.2. Calculations are performed for the receivers at the surface located along profiles parallel to the y-axis. Figures 7a and 7b present the amplitude and the phase of the apparent resistivity distribution, calculated from the observed EM field on the surface of the Earth for the frequency equal to 1 Hz. The inversion used EM data collected along 15 profiles on the surface of the Earth at the four frequencies listed above. Displacement currents at these frequencies are negligibly smaller than the conductive currents, and so the inversion was applied only for the conductivity distribution. In the numerical test, we selected 144 substructures for inversion, shown in Fig. 5b, and we used the additional simplification that the reflectivity tensor $\underset{\sim}{\boldsymbol{\lambda}}$ is scalar and constant within every substructure. The results of inversion for the data with 5% random noise added are shown on the following figures.

Figure 8 presents a vertical slice along the line $x = 0$ of the results of inversion of the noisy data. One can see clearly the cross-sections of conductive body on this picture; however, the upper part of the body is resolved slightly better than the deeper parts, which can be explained by the fact that EM field is less sensitive to the lower parts of anomalous structures. Figure 9 shows the vertical slice along the line $x = 300$ m that passes outside the body with anomalous conductivity. We can see now only background conductivity on this cross-section with a very weak variation, which corresponds very well to the original model. Figure 10 presents the volume image of the inverted model. The result clearly shows the anomalous body. The relationship between the misfit functional and the number of iterations is shown in Fig. 11. One can see that the noise practically didn't affect the result. This can be explained simply by using a regularized solution.

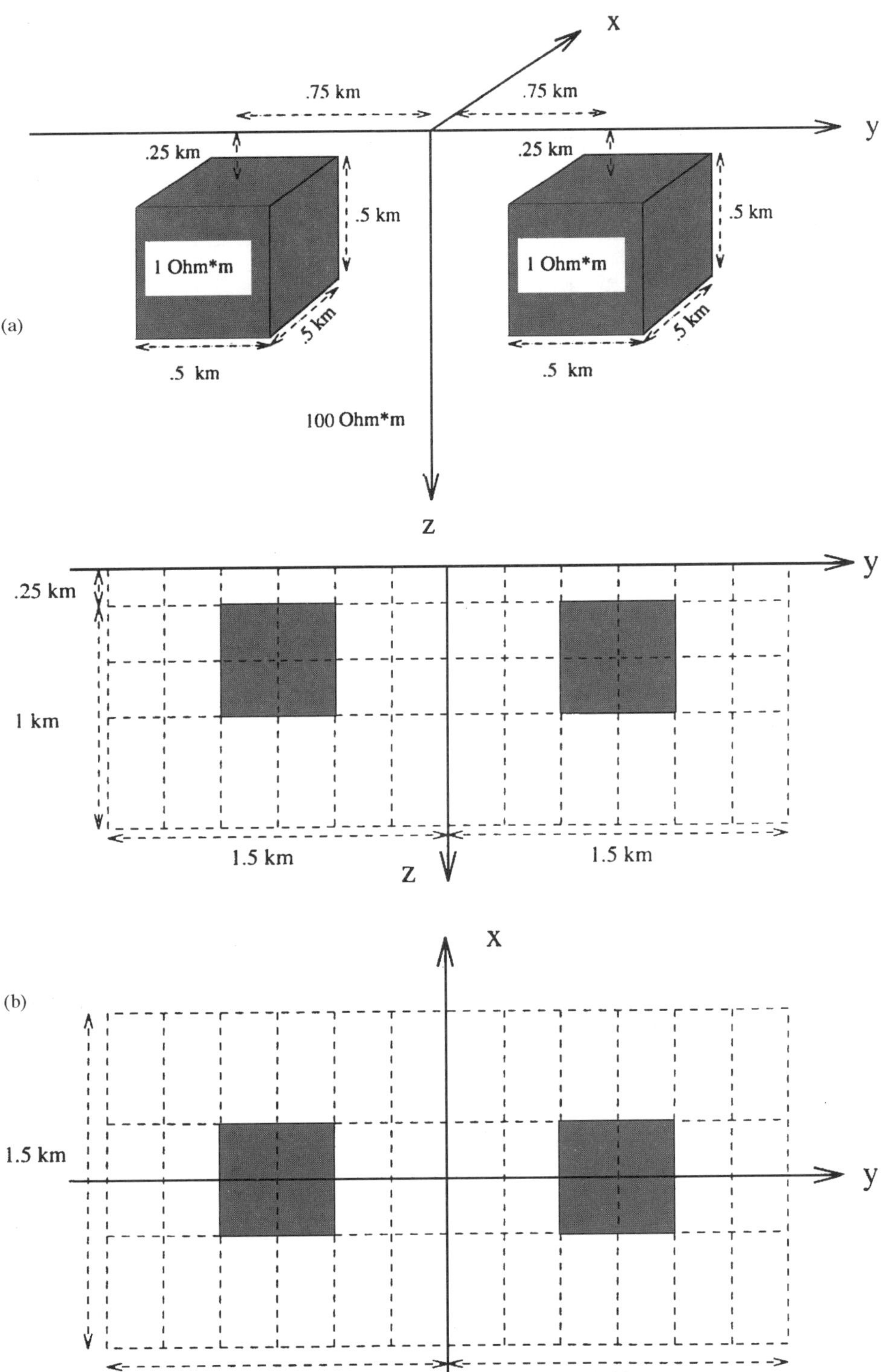

Figure 5. (a) Three-dimensional model of two rectangular conductive structures in a homogeneous half-space, excited by a plane wave (Model 2); (b) division of model into substructures used for inversion.

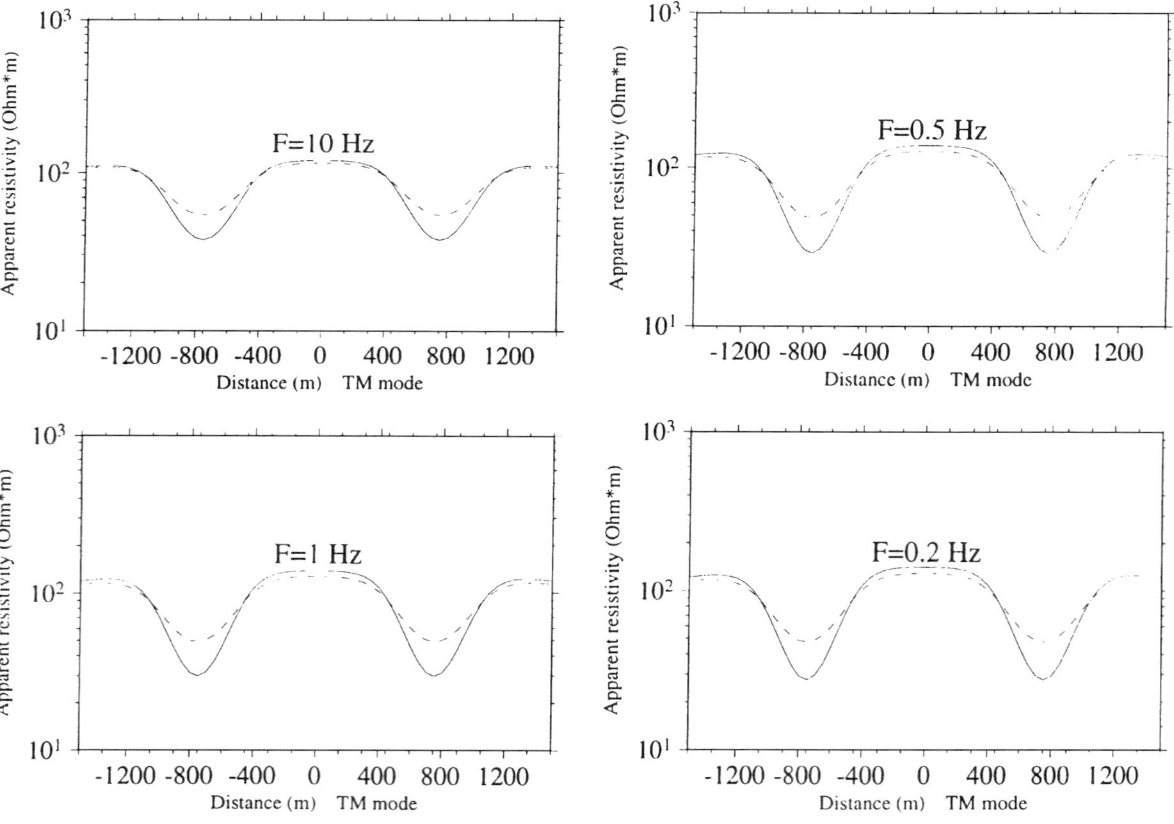

Figure 6. Numerical comparison of full IE solution (a), and QL approximation (b) computed for Model 2 (Fig. 5) at the frequencies 10, 1, 0.5, and 0.2 Hz. Calculations are performed for receivers located along profiles parallel to the y-axis on the surface. Plots present apparent resistivities computed for TM mode (ρ_{yx}).

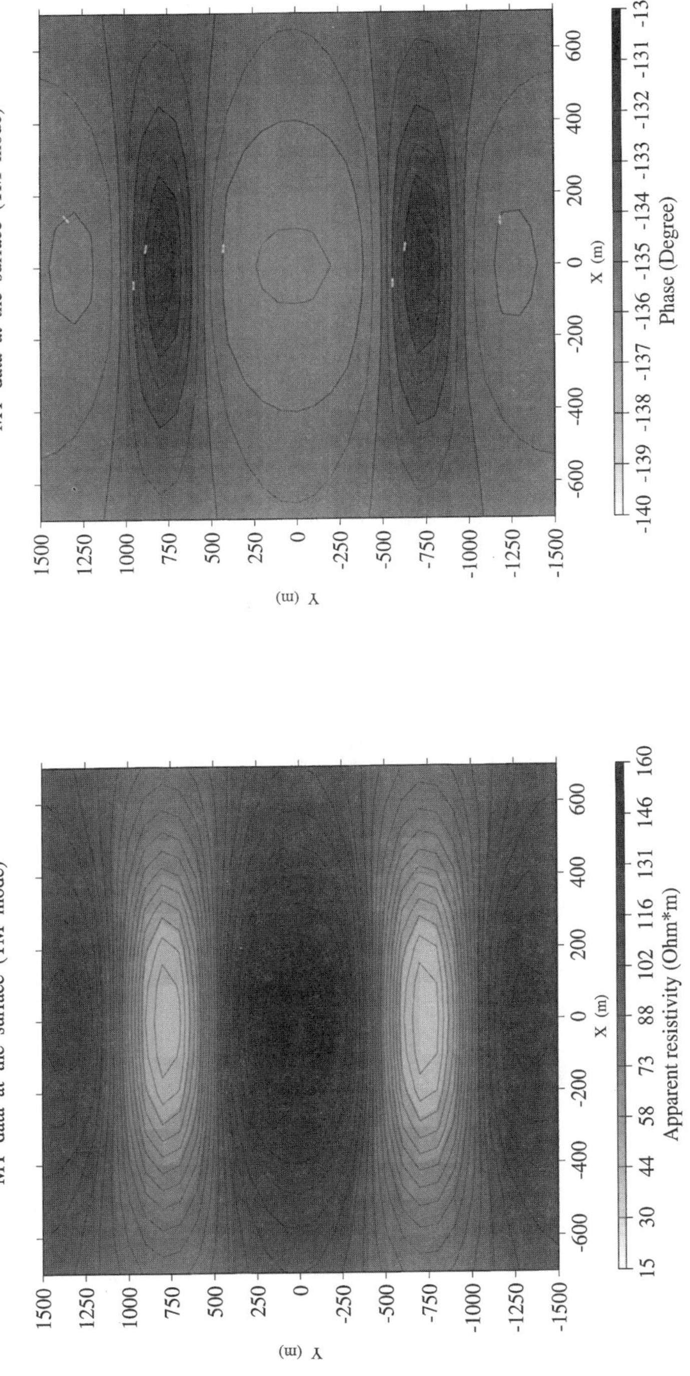

Figure 1. (a) Apparent resistivity amplitude distribution for the model shown in Fig. 5, calculated on the surface of the Earth for a frequency of 1 Hz. (b) Impedance phase distribution for the model shown in Fig. 5, calculated for a frequency of 1 Hz.

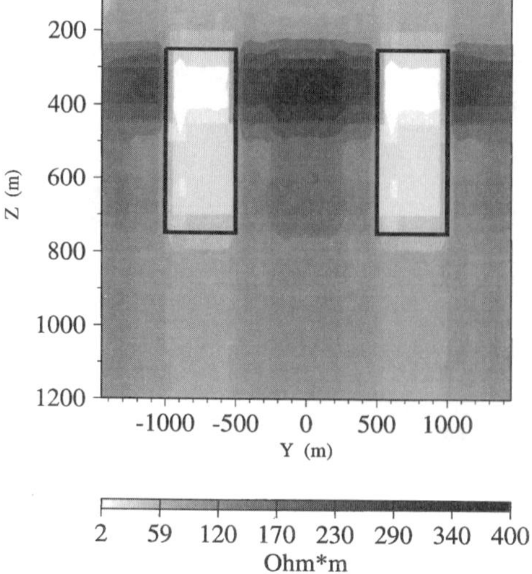

Figure 8. Vertical slice along the line $x = 0$ of the results of inversion of the data with 5% noise added for the model shown in Fig. 5.

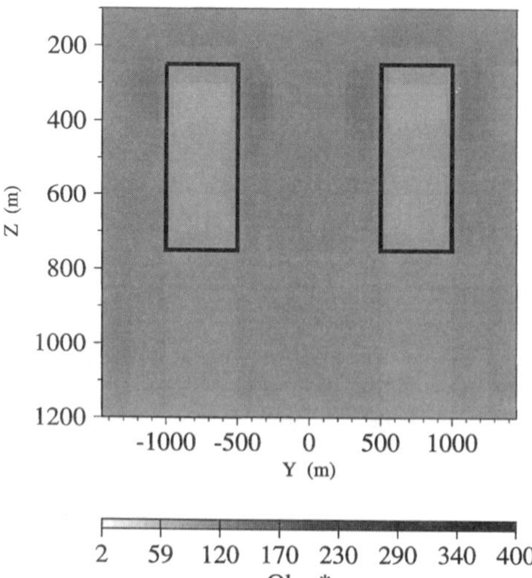

Figure 9. Vertical slice along the line $x = 300$ m of the results of inversion of the data with 5% noise added for the model shown in Fig. 5.

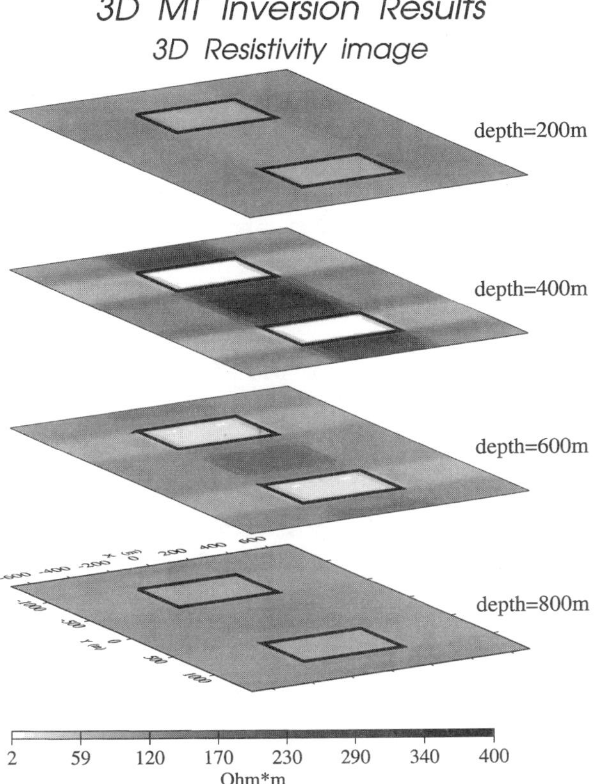

Figure 10. The volume image of the inverted model computed from EM data (with 5% noise added) collected along 15 profiles on the surface of the Earth for four frequencies (10, 1, 0.5, and 0.2 Hz) for the model shown in Fig. 5.

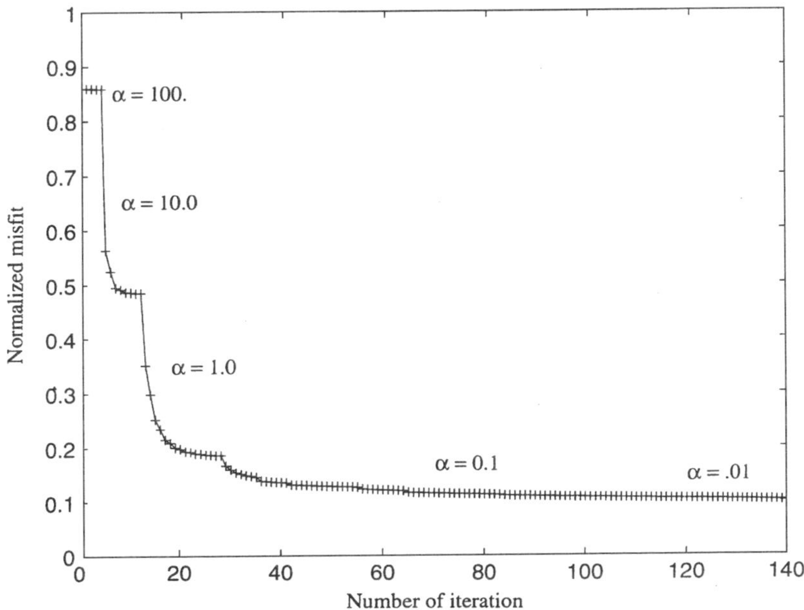

Figure 11. Plot of misfit functional as a function of the number of iterations, calculated during inversion.

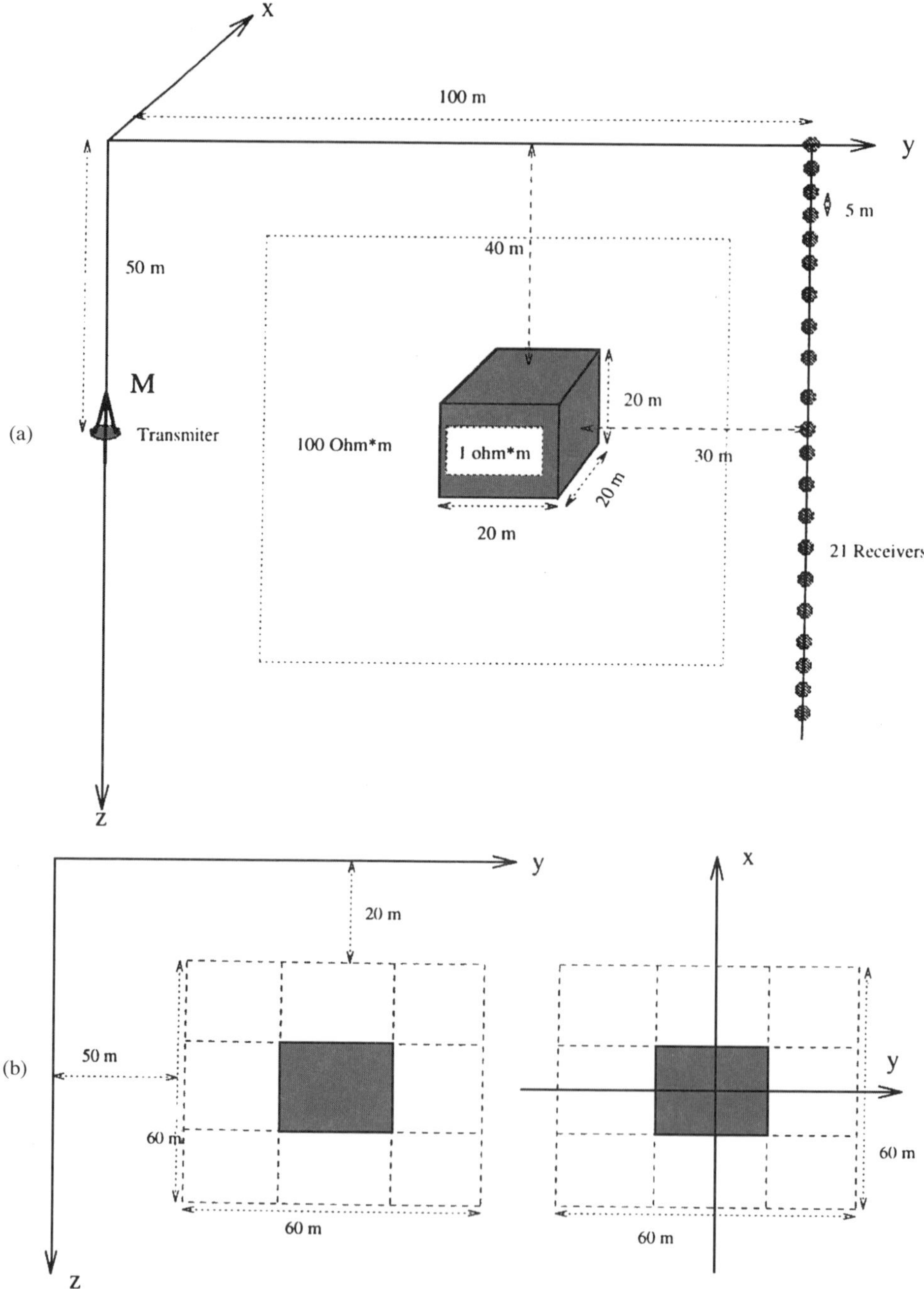

Figure 12. (a) Model of conducting body excited by the vertical dipole in the borehole (Model 3). (b) Inverse area is subdivided into 27 substructures; the size of the substructures is selected to be equal to the size of the actual conducting body.

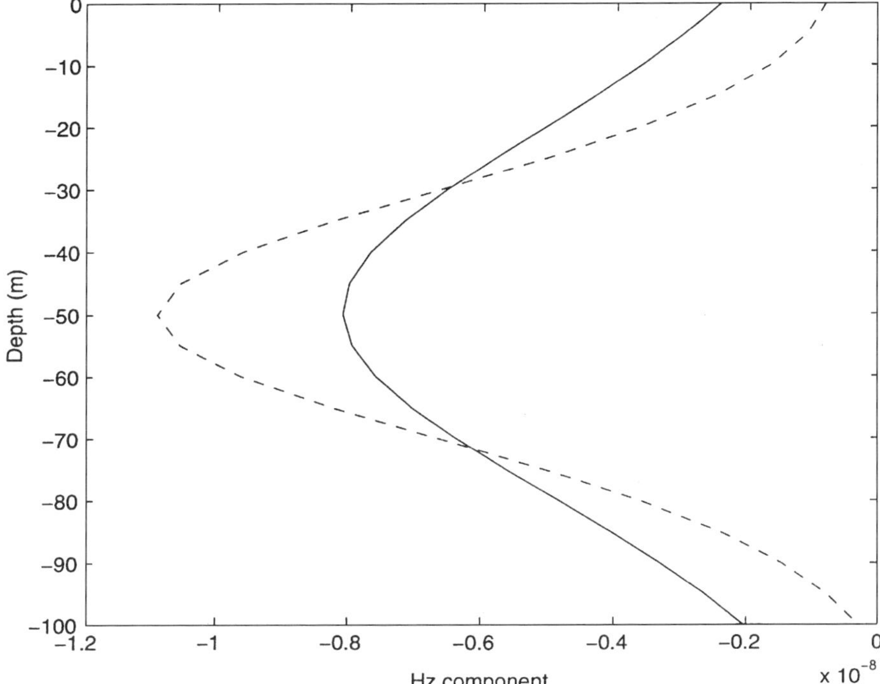

Figure 13. Plots of vertical anomalous magnetic field H_z^a,(real and imaginary parts) calculated at the frequency 50 kHz along the borehole.

5.2 *Cross-borehole vertical magnetic dipole excitation model*

Let us consider a model simulating an orebody (Bertrand and McGaughey, 1994). We present the orebody as a cube with side 20 m and resistivity 1 ohm-m embedded in a homogeneous media with resistivity 100 ohm-m. This model simulates typical massive-sulfide deposits. The orebody is located exactly in the middle of the two boreholes at a depth of 40 m. The distance between the boreholes is 100 m (Fig. 12a). Cross-borehole EM surveys can be conducted by the frequency-domain vertical magnetic dipole system. The transmitter (vertical magnetic dipole) is located at a depth of 50 m in the first borehole, and 21 receivers, observing the vertical magnetic field, are in the second borehole, from a depth of 0 to 100 m. The plots of the vertical anomalous magnetic field H_z^a (real and imaginary parts) calculated for the frequency 50 kHz along the borehole are presented in Fig. 13.

The unknown region is subdivided into 27 substructures: The size of the substructures is selected to be equal to the size of the actual conducting body (Fig. 12b). The vertical slices of the geoelectrical model obtained as the result of the inversion for borehole data are presented at Fig. 14. Comparison of these results with the original model (Fig. 12a) shows that QL inversion produces a reasonable model of the target.

6 Conclusion

We have developed a fast algorithm for 3-D EM inversion based on the QL approximation of forward modeling. The method works for models with various sources of excitation, including plane waves for magnetotellurics, horizontal bipoles, vertical bipoles, horizontal rectangular loops, vertical magnetic dipoles, and the loop-loop system for surface (and airborne) electromagnetics.

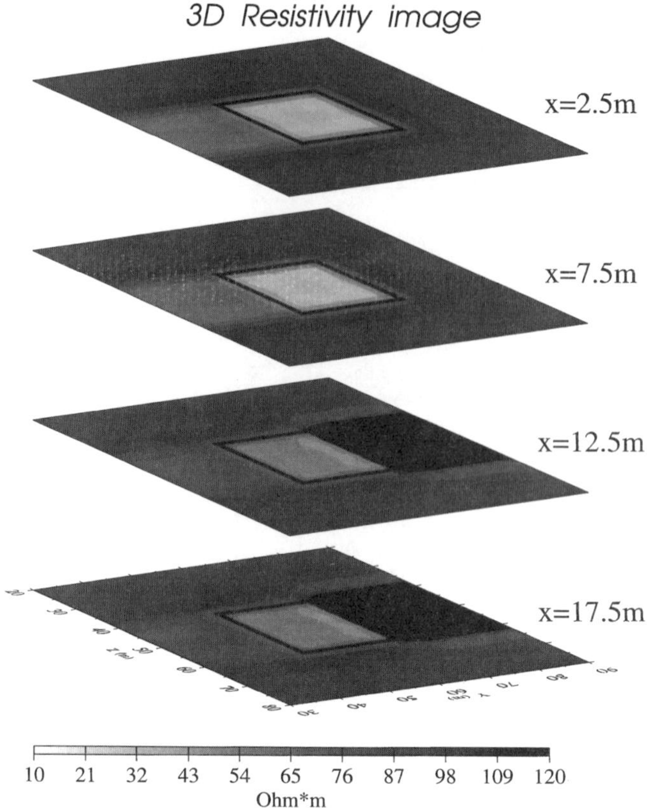

Figure 14. The vertical slices of the geoelectrical model (at the x positions 2.5 m, 7.5 m, 12.5 m, 17.5 m) obtained as the result of the inversion for one borehole profile vertical magnetic data, presented at Fig. 13. The size of the substructures is selected to be equal to the size of the actual conducting body (Fig. 12(b)).

The main advantage of the method is that it reduces the original nonlinear inverse problem to a set of linear inverse problems to obtain a rapid 3-D conductivity inversion. The QL inverse problem is solved by a regularized gradient-type method that ensures stability and rapid convergence.

Acknowledgments

The authors acknowledge the support of the University of Utah Consortium of Electromagnetic Modeling and Inversion (CEMI), which includes CRA Exploration Ltd., Newmont Exploration, Western Mining, Kennecott Exploration, Schlumberger-Doll Research, Shell Exploratie en Produktie Laboratorium, Western Atlas, the United States Geological Survey, Zonge Engineering, Mindeco, MIM Exploration, and BHP Exploration.

We thank Dr. Z. Xiong for providing the forward-modeling IE code and Dr. Michael Oristaglio and Dr. Carlos Torres-Verdin for their useful discussions which helped to improve the manuscript.

We also thank Weidong Li for his help with numerical calculations.

References

Born, M., 1933, *Optics*: Springer Pub. Co., Inc.

Berdichevsky, M. N., and Zhdanov, M. S., 1984, Advanced theory of deep geomagnetic sounding: Elsevier Science Publ. Co., Inc.

Bertrand, M. J., and McGaughey, J., 1994, Crosshole EM detectability of high-conductivity contrast anomalies: 64th Ann. Mt., Soc. Expl. Geophys. Expanded Abstracts, 407–410.

Druskin, V., and Knizhnerman, L., 1994, Spectral approach to solving three-dimensional Maxwell's diffusion equations in the time and frequency domains: Radio Sci., **29**, 937–953.

Eaton, P. A., 1989, 3-D electromagnetic inversion using integral equations: Geophys. Prosp., **37**, 407–426.

Gusarov, A. L., 1981, About the uniqueness of the solution of the magnetotelluric inverse problem, *in* Mathematical models in geophysical problems: Moscow Univ. Press, 31–80.

Habashy, T. M., Groom, R. W., and Spies, B. R., 1993, Beyond the Born and Rytov approximations: A nonlinear approach to EM scattering: J. Geophys. Res., **98**, 1759–1775.

Habashy, T. M., Oristaglio, M. L., and de Hoop, A. T., 1994, Simultaneous nonlinear reconstruction of two-dimensional permittivity and conductivity: Radio Sci., **29**, 1101–1118.

Hohmann., G. W., 1975, Three-dimensional induced polarization and EM modeling: Geophysics, **40**, 309–324.

Kleinman, R. E., and van den Berg, P. M., 1993, An extended range-modified gradient technique for profile inversion: Radio Sci., **28**, 877–884.

Madden, T. R., and Mackie, R. L., 1989, Three-dimensional magnetotelluric modeling and inversion: Proc. IEEE, **77**, 318–332.

Newman, G. A., and Alumbaugh, D. L., 1995, 3-D massively parallel electromagnetic inversion: PIERS Proc., Univ. of Washington, 324.

Smith, J .T., and Booker, J. R., 1991, Rapid inversion of two- and three-dimensional magnetotelluric data: J. Geophys. Res., **96**, 3905–3922.

Svetov, B. S., and Gubatenko, V. P., 1985, About the equivalence of the system of the extraneous electric and magnetic currents: Radiotechniques and electronics, No. 4, 31–40.

Tikhonov, A. N., and Arsenin, V. Y., 1977, Solution of ill-posed problems: W. H. Winston and Sons.

Torres-Verdín, C., and Habashy, T. M., 1994, Rapid 2.5-dimensional forward modeling and inversion via a new nonlinear scattering approximation: Radio Sci., **29**, 1051–1079.

———1995a, Two-steps linear inversion: Radio Sci., **29**, 1051–1079.

———1995b, An overview of the extended Born approximation as a nonlinear scattering approach, and its application to cross-well resistivity imaging: PIERS Proc., Univ. of Washington, 321.

Tripp, A. C., 1990, Group theoretic reduction of the EM impedance matrix for large-contrast symmetric prisms in a layered earth: Pageoph, **133**, 127–147.

Tripp, A. C., and Hohmann, G. W., 1993, Three-dimensional EM cross-well inversion: IEEE Trans. Geosci. Remote Sensing, **31**, 121–126.

Wannamaker, P. E., 1991, Advances in 3-D magnetotelluric modeling using integral equations: Geophysics, **56**, 1716–1728.

Weidelt, P., 1975, EM induction in three-dimensional structures: J. Geophys., **41**, 85–109.

Xie, G., and Lee, K. H., 1995, Nonlinear inversion of 3-D electromagnetic data: PIERS Proc., Univ. of Washington, 323.

Xiong, Z., 1992, EM modeling of three-dimensional structures by the method of system iteration using integral equations: Geophysics, **57**, 1556–1561.

Xiong, Z., and Kirsch, A., 1992, Three-dimensional earth conductivity inversion: J. Comput. Appl. Math., **42** , 109–121.

Xiong, Z., and Tripp, A. C., 1993, Scattering matrix evaluation using spatial symmetry in EM modeling: Geophys. J. Internat., **114**, 459–464.

Zhdanov, M. S., 1993, Tutorial: Regularization in inversion theory: Colorado School of Mines.

Zhdanov, M. S., and Chernyak, V. V., 1987, An automated method of solving the two-dimensional inverse problem of electromagnetic induction within the earth: Trans. (Doklady) USSR Acad. Sci., Earth Sci., **296**, 59–63.

Zhdanov, M. S., and Keller, G., 1994, The geoelectrical methods in geophysical exploration: Elsevier Science Pub. Co., Inc.

Zhdanov, M. S., and Fang, S., 1996a, Quasi-linear approximation in 3-D EM modeling: Geophysics, **61**, 646–665.

———1996b, Three-Dimensional quasi-linear electromagnetic inversion: Radio Sci., **31**(4), 741–754.

Appendix: Regularized steepest-descent method for minimizing the parametric functional

To get a stable solution of Eqs. (28) and (31), we introduced the parametric functional:

$$P^{\alpha}(\mathbf{m}) = \phi(\mathbf{m}) + \alpha\, S(\mathbf{m}),$$

where functionals $\phi(\mathbf{m})$ and $S(\mathbf{m})$ were determined by Eqs. (36) and (37). To solve the minimization problem (40) we calculate the first variation of the parametric functional under the assumption that $\Delta\tilde{\sigma}$ and λ are, temporarily, constants:

$$\delta P^{\alpha}(\mathbf{m}) = 2\,\mathrm{Re}\,\{\delta\mathbf{m}^{*}[\underset{\sim}{\mathbf{G}}^{F*}(\underset{\sim}{\mathbf{G}}^{F}\mathbf{m} - \mathbf{F}) + [\mathbf{m} - (\underset{\sim}{\mathbf{I}} + \underset{\sim}{\mathbf{\Lambda}})\Delta\sigma] + \alpha(\mathbf{m} - \mathbf{m}_p)]\}.$$

Let us select $\delta\mathbf{m}$ as

$$\delta\mathbf{m} = -k^{\alpha}\,\ell^{\alpha}(\mathbf{m}), \quad 0 < k^{\alpha} < \infty, \tag{A1}$$

where

$$\ell^{\alpha}(\mathbf{m}) = \underset{\sim}{\mathbf{G}}^{F*}(\underset{\sim}{\mathbf{G}}^{F}\mathbf{m} - \mathbf{F}) + [\mathbf{m} - (\underset{\sim}{\mathbf{I}} + \underset{\sim}{\mathbf{\Lambda}})\Delta\sigma] + \alpha\,(\mathbf{m} - \mathbf{m}_p). \tag{A2}$$

This selection makes $\delta P^{\alpha}(\mathbf{m}) = -2\,k^{\alpha}\,\mathrm{Re}\{\ell^{\alpha*}(\mathbf{m})\,\ell^{\alpha}(\mathbf{m})\} < 0$. That means that the parametric functional is reduced if we apply perturbation (A1) to the model parameters. We construct an iteration process as follows:

$$\mathbf{m}^{\alpha}_{N+1} = \mathbf{m}^{\alpha}_{N} + \delta\mathbf{m}_N = \mathbf{m}^{\alpha}_{N} - k^{\alpha}_{N}\,\ell^{\alpha}(\mathbf{m}_N), \tag{A3}$$

where

$$\ell^{\alpha}(\mathbf{m}_N) = \underset{\sim}{\mathbf{G}}^{F*}(\underset{\sim}{\mathbf{G}}^{F}\mathbf{m}_N - \mathbf{F}) + [\mathbf{m}_N - (\underset{\sim}{\mathbf{I}} + \underset{\sim}{\mathbf{\Lambda}}_N)\Delta\sigma_{N-1}] + \alpha\,(\mathbf{m}_N - \mathbf{m}_p). \tag{A4}$$

The reflectivity $\boldsymbol{\lambda}_N$ is determined from m_N using Eq. (31):

$$\underset{\sim}{\mathbf{E}}^n \boldsymbol{\lambda}_N = \underset{\sim}{\mathbf{G}}^E \mathbf{m}_N, \tag{A5}$$

Note that the anomalous conductivity has to satisfy the condition (componentwise)

$$\mathrm{Re}(\Delta\tilde{\boldsymbol{\sigma}}_{N-1}) \geq -\sigma_n; \quad \mathrm{Im}(\Delta\boldsymbol{\sigma}_{N-1}) \leq \omega\varepsilon_n \tag{A6}$$

because the electrical conductivity and dielectric permittivity have to be positive. Therefore, the conductivity $\Delta\tilde{\boldsymbol{\sigma}}_{N-1}$ can be found by using Eqs. (33) and (A6) with the following conditions (componentwise):

$$\mathrm{Re}(\Delta\tilde{\sigma}_{N-1}) = \mathrm{Re}\left[\sum_{\omega}[(1+\lambda_{N-1})^*(1+\lambda_{N-1})]\right]^{-1}\left[\sum_{\omega}(1+\lambda_{N-1})^* m_{N-1}\right]$$
$$= a_{N-1} \quad \text{for } a_{N-1} \geq -\sigma_n \tag{A7}$$

and

$$\mathrm{Re}(\Delta\tilde{\sigma}_{N-1}) = -\tfrac{1}{2}\sigma_n, \quad \text{for } \mathbf{a}_{N-1} \leq -\sigma_n. \tag{A8}$$

Similarly,

$$\mathrm{Im}(\Delta\tilde{\sigma}_{N-1}) = -\omega\,\mathrm{Im}\left\{\left[\sum_{\omega}\omega(1+\lambda_{N-1})^*(1+\lambda_{N-1})\right]^{-1}\right.$$
$$\left.\times\left[\sum_{\omega}(1+\lambda_{N-1})^* m_{N-1}\right]\right\} = b_{N-1},$$

$$\text{for } b_{N-1} < \omega\varepsilon_n$$
$$\mathrm{Im}(\Delta\tilde{\sigma}_{N-1}) = \tfrac{1}{2}\omega\varepsilon_n, \quad \text{for } b_{N-1} \geq \omega\varepsilon_n.$$

The initial iteration should be done using the formula

$$\mathbf{m}_1^{\alpha} = \mathbf{m}_0^{\alpha} + \delta\mathbf{m}_0 = \mathbf{m}_0^{\alpha} - k_0^{\alpha}\,[\underset{\sim}{\mathbf{G}}^{F*}(\underset{\sim}{\mathbf{G}}^F\mathbf{m}_p - \mathbf{F})], \tag{A9}$$

where

$$\mathbf{m}_p = (\underset{\sim}{\mathbf{I}} + \underset{\sim}{\boldsymbol{\Lambda}}_p)\Delta\boldsymbol{\sigma}_p$$

The second iteration is

$$\mathbf{m}_2^{\alpha} = \mathbf{m}_1^{\alpha} + \boldsymbol{\delta}\mathbf{m}_1 = \mathbf{m}_1^{\alpha} - k_1^{\alpha}\,\ell^{\alpha}(\mathbf{m}_1), \tag{A10}$$

where

$$\ell^{\alpha}(\mathbf{m}_1) = \underset{\sim}{\mathbf{G}}^{F*}(\underset{\sim}{\mathbf{G}}^F\mathbf{m}_1 - \mathbf{F}) + [\mathbf{m}_1 - (\underset{\sim}{\mathbf{I}} + \underset{\sim}{\boldsymbol{\Lambda}}_1)\Delta\boldsymbol{\sigma}_p] + \alpha\,(\mathbf{m}_1 - \mathbf{m}_p). \tag{A11}$$

The coefficient k_n^{α} can be determined from the condition

$$P^{\alpha}\left(\mathbf{m}_{N+1}^{\alpha}\right) = P^{\alpha}\left[\mathbf{m}_N^{\alpha} - k_N^{\alpha}\,\ell^{\alpha}(\mathbf{m}_N)\right] = f\left(k_N^{\alpha}\right) = \min!$$

Solution of this minimization problem gives the following best estimation for the length of the step:

$$k_N^{\alpha} = \frac{\ell^{\alpha*}\left(\mathbf{m}_N^{\alpha}\right)\ell^{\alpha}\left(\mathbf{m}_N^{\alpha}\right)}{\ell^{\alpha*}\left(\mathbf{m}_N^{\alpha}\right)(\underset{\sim}{\mathbf{G}}^{F*}\underset{\sim}{\mathbf{G}}^F + \alpha\underset{\sim}{\mathbf{I}})\ell^{\alpha}\left(\mathbf{m}_N^{\alpha}\right)}. \tag{A12}$$

Using Eqs. (A2), (A3), and (A12), we can obtain $\mathbf{m}$ iteratively.

Approximate Sensitivities for Multidimensional Electromagnetic Inversion

Colin G. Farquharson
Douglas W. Oldenburg

Summary. Linearized, iterative inversion of electromagnetic data requires computation of partial derivatives (sensitivities) with respect to parameters of the model, e.g., the conductivities of cells. Methods based on the adjoint equation compute these sensitivities by integrating, over each cell, the scalar product of an adjoint electric field with the electric field produced by the forward modeling in the preceeding iteration. We approximate the partial derivative by computing the adjoint field in either a homogeneous or a layered half-space. Computation of the approximate adjoint field is significantly faster than that of the true adjoint field, with the relative efficiency increasing with the size of the problem. Approximate sensitivities compare well with exact values for sample controlled-source surveys in 2-D and 3-D models. We demonstrate that approximate sensitivities can be sufficiently accurate to drive an iterative algorithm by inverting synthetic magnetotelluric data. Approximate sensitivities should enable the solution of inverse problems larger than those now practical.

1 Introduction

Most inversions of nonlinear problems require computation of partial derivatives with respect to the parameters of the model. These often appear in the form of the Jacobian or sensitivity matrix,

$$J_{ij} = \frac{\partial d_i}{\partial m_j}, \quad i = 1, \ldots, M, \quad j = 1, \ldots, N, \tag{1}$$

where M is the number of measurements and N is the number of model parameters. Partial derivatives always can be computed by additional forward modeling: by brute force (perturbing each parameter and forming a finite difference), by solving a new equation for the sensitivity (see Section 2.1), or by solving an adjoint equation (McGillivray and Oldenburg, 1990). If a forward modeling run gives the field everywhere in space, then each extra run with the brute-force and sensitivity methods gives partial derivatives of all data with respect to one (more) parameter. In contrast, each run with the adjoint

UBC—Geophysical Inversion Facility, Department of Earth & Ocean Sciences, University of British Columbia, Vancouver, BC V6T 1Z4, Canada.

method gives the partial derivative of one data point with respect to all model parameters. If there are M_o measurement locations and M_f frequencies or measurement times ($M_f \times M_o = M$), then the brute-force and sensitivity methods require $N \times M_f$ runs to give all derivatives, whereas the adjoint method requires $M_o \times M_f$ runs. When the model is overparameterized ($N > M$), the adjoint method is the most efficient one.

Features of a particular algorithm can reduce the time needed to compute sensitivities. For example, one can store a factored finite-element or finite-difference matrix to speed up further forward solutions for sensitivities (Oristaglio and Worthington, 1980), or one can use a gradient or conjugate-gradient search, which requires only the product of the Jacobian matrix with a vector (Mackie and Madden, 1993). With reciprocity, this product can be computed in a single forward modeling run. Iterative methods that use the full Jacobian matrix generally converge at a faster rate than gradient methods, but computation of exact sensitivities quickly becomes prohibitive for large problems. We propose an approximate method that is much faster than any exact method, and accurate enough to drive an iterative inversion algorithm.

2 Approximate sensitivities

Boerner and Holladay (1990) demonstrated that sensitivities with respect to (logarithms of) layer conductivities in a 1-D model can be closely approximated by sensitivities computed in a homogeneous half-space and scaled by the conductivities in the layered model. Smith and Booker (1991) and Ellis et al. (1993) also used approximate sensitivities for inversion of 2-D and 3-D magnetotelluric (MT) data. Their approximate sensitivities were based on the 1-D inverse problem, were easy to compute, and were sufficiently accurate in many cases. These sensitivities, however, were only defined for cells directly below the location of a particular measurement, and so, the Jacobian matrix was incomplete. The approximation that we present here gives a complete Jacobian matrix for any source–receiver configuration. It uses the adjoint equation, but replaces the exact adjoint field with one that is easy to calculate.

2.1 *Sensitivities with the adjoint equation*

Adjoint equations (or reciprocity) provide a convenient method for calculating sensitivities (Weidelt, 1975; Park, 1988; Madden and Mackie, 1989; and Madden, 1990). When the number of parameters is greater than the number of data, this method is the most efficient. We review the relevant equations as given by McGillivray et al. (1994).

Let domain D have conductivity σ that varies as a function of position, but constant electrical permittivity ϵ and magnetic permeability μ. Let $\mathbf{J}_e$ and $\mathbf{J}_m$ represent the physical electric or magnetic current source used in the geophysical experiment. The electric and magnetic fields, $\mathbf{E}$ and $\mathbf{H}$, generated by these sources satisfy Maxwell's equations,

$$\nabla \times \mathbf{E} = -i\omega\mu\mathbf{H} + \mathbf{J}_m, \tag{2}$$

$$\nabla \times \mathbf{H} = (\sigma + i\omega\epsilon)\mathbf{E} + \mathbf{J}_e \tag{3}$$

and appropriate boundary conditions. In practice, the conductivity is represented as a finite linear combination of basis functions ψ_j,

$$\sigma(\mathbf{r}) = \sum_{j=1}^{N} \sigma_j \psi_j(\mathbf{r}), \tag{4}$$

where σ_j are the coefficients. Substituting this into Eqs. (2) and (3), and differentiating with respect to σ_k yields Maxwell's equations for the sensitivities,

$$\nabla \times \frac{\partial \mathbf{E}}{\partial \sigma_k} = -i\omega\mu \frac{\partial \mathbf{H}}{\partial \sigma_k}, \tag{5}$$

$$\nabla \times \frac{\partial \mathbf{H}}{\partial \sigma_k} = (\sigma + i\omega\epsilon)\frac{\partial \mathbf{E}}{\partial \sigma_k} + \psi_k \mathbf{E}. \tag{6}$$

The partial derivatives $\partial \mathbf{E}/\partial \sigma_k$ and $\partial \mathbf{H}/\partial \sigma_k$ are the sensitivities of the electric and magnetic fields with respect to σ_k. These sensitivities satisfy the homogeneous form of the boundary conditions that $\mathbf{E}$ and $\mathbf{H}$ satisfy (i.e., the boundary values of the fields are fixed and do not vary).

Consider now an auxiliary problem in which electric and magnetic fields, $\mathbf{E}^\dagger$ and $\mathbf{H}^\dagger$, are generated in domain D by new electric and magnetic sources $\mathbf{J}_e^\dagger$ and $\mathbf{J}_m^\dagger$. These auxiliary fields satisfy

$$\nabla \times \mathbf{E}^\dagger = -i\omega\mu \mathbf{H}^\dagger + \mathbf{J}_m^\dagger, \tag{7}$$

$$\nabla \times \mathbf{H}^\dagger = (\sigma + i\omega\epsilon)\mathbf{E}^\dagger + \mathbf{J}_e^\dagger. \tag{8}$$

Take the fields to satisfy homogeneous boundary conditions which, if D is finite, are the same as those satisfied by the sensitivities, but need not be the same if D extends to infinity.

The fields defined by Eqs. (5), (6), (7), and (8) satisfy the (Lorentz) reciprocity relation,

$$\int_D \left(\mathbf{J}_m^\dagger \cdot \frac{\partial \mathbf{H}}{\partial \sigma_k} + \mathbf{J}_e^\dagger \cdot \frac{\partial \mathbf{E}}{\partial \sigma_k} \right) dv = \int_D \mathbf{E}^\dagger \cdot \mathbf{E} \psi_k \, dv. \tag{9}$$

This fundamental equation relates the sensitivities for the electric and magnetic fields (on the left-hand side) to the inner product of original and adjoint fields (right-hand side). Individual sensitivities can be obtained by appropriate choice of auxiliary sources. For example, to obtain the sensitivity for the x-component of the electric field at an observation location $\mathbf{r}_o$, choose $\mathbf{J}_e^\dagger = \delta(\mathbf{r} - \mathbf{r}_o)\hat{\mathbf{e}}_x$ and $\mathbf{J}_m^\dagger = 0$. Substituting into Eq. (9) gives

$$\left. \frac{\partial E_x}{\partial \sigma_k} \right|_{\mathbf{r}_o} = \int_D \mathbf{E}^\dagger \cdot \mathbf{E} \psi_k \, dv. \tag{10}$$

The auxiliary electric field $\mathbf{E}^\dagger$ in this example is the electric field in D caused by an x-directed unit electric dipole at the location $\mathbf{r}_o$ where the original field E_x is measured. The field $\mathbf{E}$ is the electric field in the domain caused by the physical source ($\mathbf{J}_e$ and $\mathbf{J}_m$) used in the actual experiment.

2.2 Approximate adjoint field

Equation (9) requires (1) calculation of the electric field $\mathbf{E}$ to model the geophysical experiment, (2) calculation of the adjoint field $\mathbf{E}^\dagger$ of the appropriate adjoint source, and (3) evaluation of the volume integral of the scalar product of these two fields. The numerical integration is negligible compared to any forward modeling. Also, in an iterative inversion algorithm, the electric field $\mathbf{E}$ will already have been computed to form the data misfit for the previous iteration. The extra computation needed for this method is just the calculation of the adjoint field.

To speed up this computation, we use an approximation to $\mathbf{E}^{\dagger}$. For example, we use the electric field in a homogeneous half-space or in a horizontally layered half-space. Computation of these fields is much faster than that of the true adjoint field. The approximate sensitivities obtained in this way are similar to those in the Born iterative method (e.g., Sena and Toksöz, 1990; Alumbaugh and Morrison, 1995). However, unlike the Born iterative method—in which Green's function [or the adjoint field $\mathbf{E}^{\dagger}$ in the notation of Eq. (10)] remains unchanged throughout the iterations and only the electric field $\mathbf{E}$ is updated—our approximate adjoint field is modified at each iteration in response to the current conductivity model. The homogeneous or layered half-space in which the approximate adjoint field is computed is obtained from a weighted average of the conductivities (or the logarithms of the conductivities) of all cells in the current conductivity model, or from the weighted averages of the cells in each horizontal row or plane of the current model.

Approximate adjoint fields obviously will not contain all features of the true adjoint field. The true adjoint field, however, is dominated by its decay away from the dipole source. Suitable choices of conductivities for the homogeneous or layered half-space result in an approximate adjoint field that has the same dominant behavior. The approximate sensitivities seem sufficiently accurate to allow an iterative inversion procedure to converge. In the following sections, we compare our approximate sensitivities with the true sensitivities for 2.5-D and 3-D models.

2.3 A 2.5-D example

For the 2-D conductivity model shown in Fig. 1, consider a controlled-source electromagnetic (EM) experiment in which an electric dipole source perpendicular to the section is located at $x_s = -11\,000$ m and $y_s = z_s = 0$. Suppose measurements are made

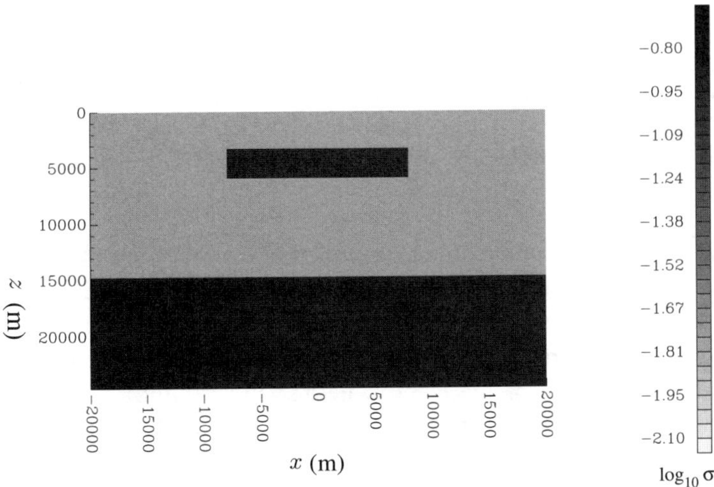

Figure 1. Conductivity model to illustrate approximate sensitivities for the 2.5-D problem.

Table 1. Computation times for exact and approximate sensitivities for 2.5-D example

Sensitivity	Calculation method	Computation time, min
Exact	Brute-force	5000
	Adjoint equation	25
Approximate	Layered half-space	5

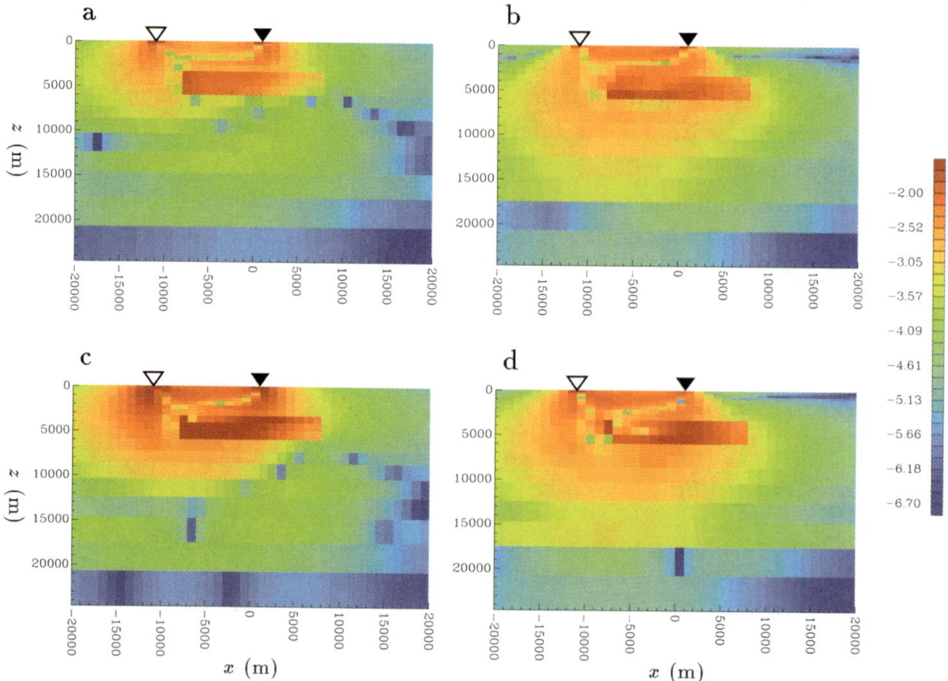

Figure 2. Sensitivities for the 2.5-D example. The conductivity model is that shown in Fig. 1. The source and observation locations are indicated by open and solid triangles, respectively. Panels (a) and (c) show $\log_{10}|\partial \ln |E|/\partial \ln \sigma|$, and panels (b) and (d) show $\log_{10}|\partial \phi/\partial \ln \sigma|$. The color bar refers to all four panels. Panels (a) and (b) were produced by the adjoint-equation method, and panels (c) and (d) are the approximate sensitivities calculated using an adjoint field in a layered half-space.

of the electric field in the source direction at the location $x_o = 1100$ m and $y_o = z_o = 0$, at frequency 0.2 Hz. "Exact" sensitivities for this example were calculated by brute force using the modeling program of Unsworth, et al. (1993) and by an adjoint solution (Unsworth and Oldenburg, 1995). Figures 2a and 2b show the exact sensitivities (from the adjoint method) of the electric field's (logarithmic) amplitude and phase with respect to the logarithms of the conductivities of each of the 800 cells in the model.

Approximate sensitivities were calculated using the adjoint field computed in a layered half-space obtained by averaging the logarithm of the conductivities of the cells, weighted by their area, in each horizontal layer of the model in Fig. 1. The approximate sensitivities shown in Figs. 2c and 2d agree well with the true sensitivities. Table 1

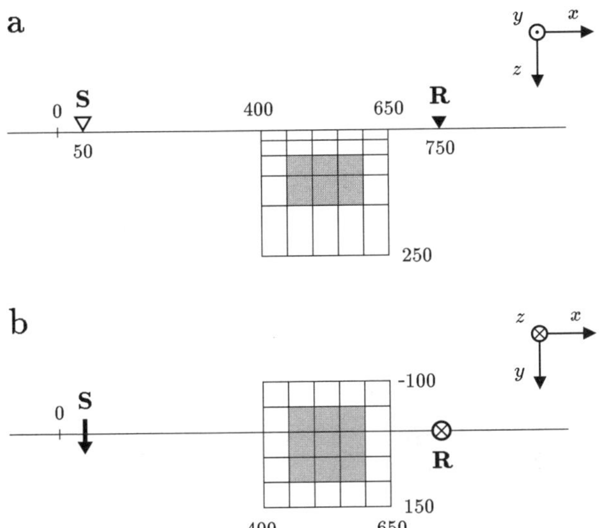

Figure 3. Geometry for 3-D example: (a) side view; (b) plan view. S indicates the y-directed grounded-wire source and R indicates the observation location. Shaded cells show the position of the conductive block. Numbers correspond to distances in meters.

gives the computation times for this example. These times are for derivatives of the amplitude and phase of the along-strike electric field at six observation locations at a frequency of 0.2 Hz and for the 800 cells in the model.

2.4 A 3-D example

Consider the experimental configuration shown in Fig. 3 consisting of a grounded-wire source extending from $(50, -5, 0)$ m to $(50, 5, 0)$ m, and measurements of the vertical component of the H-field at $(0, 750, 0)$ m. The frequency was 100 Hz. The region extending from $x = 400$ to 650 m, from $y = -100$ to 150 m, and from $z = 0$ to 250 m was discretized into $5 \times 5 \times 5$ cuboidal cells as shown in Fig. 3. The conductivity model comprised a conductive block of 0.1 S/m in a background of 0.01 S/m. The location of the conductive block is indicated by the shaded region in Fig. 3. Sensitivities were calculated by brute force using the modeling program SAMAYA (Gupta et al., 1989). The resulting sensitivities for the third vertical plane of cells from the source are shown in Figs. 4a and 4b.

Approximate sensitivities, calculated for this model using an adjoint field in a layered half-space, are shown in Figs. 4c and 4d. The approximate sensitivities agree well with the exact values, although there are differences in the sensitivities of the phase for cells outside the conductive block, especially in the second row. The resolution of this comparison is poor, however, because of the limited number of cells that could be included in the model. Table 2 gives the computation times. Six observation locations were considered at the single frequency of 100 Hz. An estimated time for the adjoint-equation method also is given. (This estimate is equal to the time required to carry out six forward modelings for this test case using SAMAYA.) Even for this small example, computation of the approximate sensitivities is almost two orders of magnitude quicker

Table 2. Computation times for exact and approximate sensitivities for 3-D inverse problem.

Sensitivity	Calculation method	Computation time
Exact	Brute force	260 min
	Adjoint equation	12 min
Approximate	Layered half-space	10 s

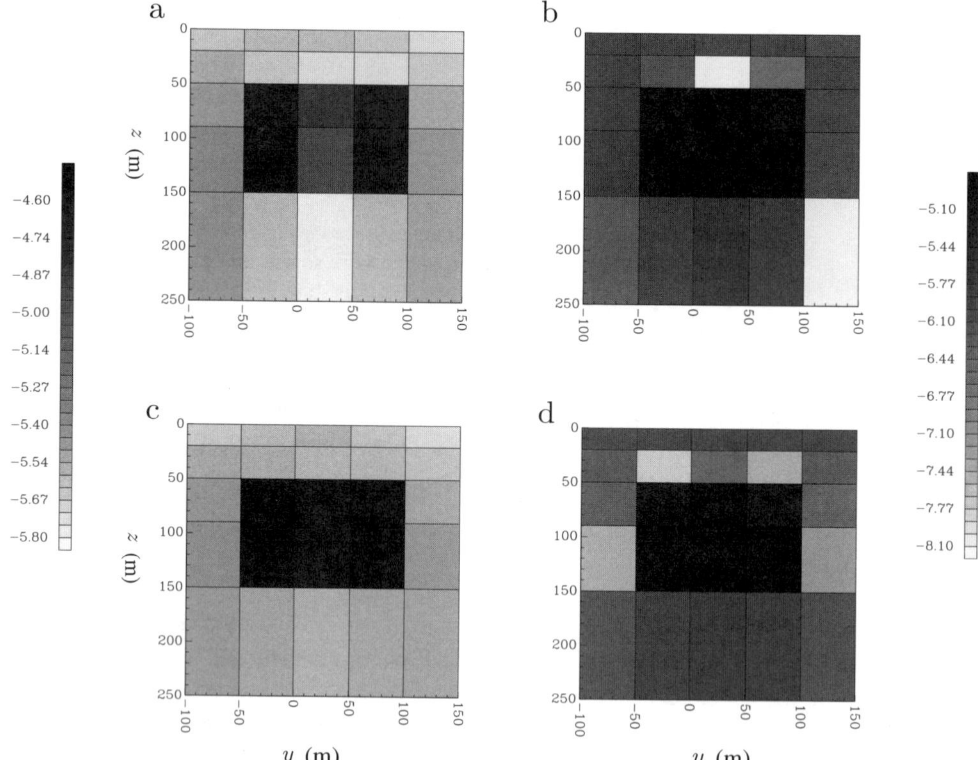

Figure 4. Sensitivities for the 3-D problem. The conductivity model comprises a homogeneous background of 0.01 S/m with a conductive block of 0.1 S/m coinciding with the shaded cells in Fig. 3. Panels (a) and (c) and the gray scale on the left side of the figure show $\log_{10}|\partial \ln |H|/\partial \ln \sigma|$, and panels (b) and (d) and the gray scale on the right side of the figure show $\log_{10}|\partial \phi/\partial \ln \sigma|$. The sensitivities shown are those for the cells in the vertical plane central to the discretized region and perpendicular to the x-direction. Panels (a) and (b) were produced by the brute-force method, and panels (c) and (d) are the approximate sensitivities calculated using an adjoint field in a layered half-space.

than that of the exact sensitivities. This time difference will increase further as the number of observation locations increases.

3 Inversion with approximate sensitivities

We tested our approximate sensitivities by inversion of synthetic magnetotelluric data. Values of apparent resistivity and phase for E-polarization and H-polarization were calculated for the model in Fig. 1 at four observation locations ($x_o = -11\,000$, -3000,

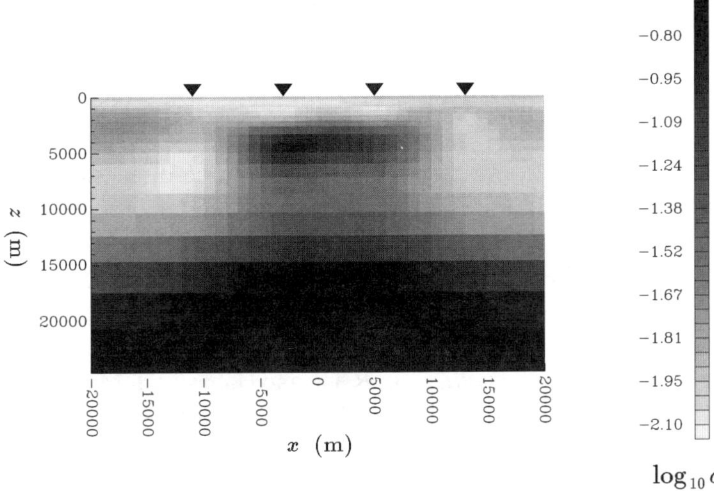

Figure 5. Model produced by inversion with approximate sensitivities. Triangles indicate locations where synthetic data were computed.

5000, and 13 000 m) for five frequencies (1, 0.5, 0.2, 0.1, and 0.05 Hz). Gaussian random noise was added to these data. The standard deviation of the noise added to the apparent resistivity was 5%, and that of the noise added to the phase was 2 degrees. These 80 data were inverted by iterative minimum-structure inversion, in which the system of equations at each iteration was solved using a subspace technique (Oldenburg and Ellis, 1993). The final model is shown in Fig. 5. The predicted data from this model reproduce the synthetic data to the desired misfit. The conductive block has approximately the correct amplitude and is situated at the correct depth and horizontal location. The conductive basement also has been recovered. Other differences between this model and the true model in Fig. 1 are attributable to finding a smooth conductivity structure that reproduces the rather small data set with which we worked here.

4 Conclusions

We propose a method of approximating the sensitivities used in linearized, iterative inversion of EM data. The approximation is general for any source–receiver geometry. We show that the approximation is good for sample 2.5-D and 3-D problems; it appears to be sufficiently accurate to drive standard inversion procedures. Computation of these approximate sensitivities is significantly quicker than that of exact sensitivities, by at least two orders of magnitude for 3-D problems. Additionally, the relative difference in computation times increases as the size of the problem increases. The use of these approximate sensitivities hopefully will expand the size of inverse problems that can be tackled on today's computers.

References

Alumbaugh, D. L., and Morrison, H. F., 1995, Theoretical and practical considerations for crosswell electromagnetic tomography assuming a cylindrical geometry, Geophysics, **60**, 846–870.

Boerner, D. E., and Holladay, J. S., 1990, Approximate Fréchet derivatives in inductive electromagnetic soundings, Geophysics, **55**, 1589–1595.

Ellis, R. G., Farquharson, C. G., and Oldenburg, D. W., 1993, Approximate inverse mapping inversion of the COPROD2 data, J. Geomagn. Geoelectr., **45**, 1001–1012.

Gupta, P. K., Raiche, A. P., and Sugeng, F., 1989, Three-dimensional time-domain electromagnetic modelling using a compact finite-element frequency-stepping method, Geophys. J. Internat., **96**, 457–468.

McGillivray, P. R., and Oldenburg, D. W., 1990, Methods for calculating Fréchet derivatives and sensitivities for the non-linear inverse problem: a comparative study, Geophys. Prosp., **38**, 499–524.

McGillivray, P. R., Oldenburg, D. W., Ellis, R. G., and Habashy, T. M., 1994, Calculation of sensitivities for the frequency-domain electromagnetic problem, Geophys. J. Internat., **116**, 1–4.

Mackie, R. L., and Madden, T. R., 1993, Three-dimensional magnetotelluric inversion using conjugate gradients, Geophys. J. Internat., **115**, 215–229.

Madden, T. R., 1990, Inversion of low frequency electromagnetic data, *in* Oceanographic and Geophysical Tomography: Desaubies, Y., Tarantola, A., and Vinn-Justin, J., Eds., Elsevier, Science Publ. Co., Inc., 379–408.

Madden, T. R., and Mackie, R. L., 1989, Three-dimensional magnetotelluric modeling and inversion, Proc. IEEE, **77**, 318–333.

Oldenburg, D. W., and Ellis, R. G., 1993, Efficient inversion of magnetotelluric data in two dimensions, Phys. Earth Planet. Internat., **81**, 177–200.

Oristaglio, M. L., and Worthington, M. H., 1980, Inversion of surface and borehole electromagnetic data for two-dimensional electrical conductivity models, Geophys. Prosp., **28**, 633–657.

Park, S. K., 1988, Inversion of magnetotelluric data for multidimensional structure, Institute of Geophysics and Planetary Physics Report 87/6, University of California, San Diego.

Sena, A. G., and Toksöz, M. N., 1990, Simultaneous reconstruction of permittivity and conductivity for crosshole geometries, Geophysics, **55**, 1302–1311.

Smith, J. T., and Booker, J. R., 1991, Rapid inversion of two- and three-dimensional magnetotelluric data, J. Geophys. Res., **96**, 3905–3922.

Unsworth, M. J., and Oldenburg, D. W., 1995, Subspace inversion of electromagnetic data—application to mid-ocean ridge exploration, Geophys. J. Internat., **123**, 161–168.

Unsworth, M. J., Travis, B. J., and Chave, A. D., 1993, Electromagnetic induction by a finite electric dipole source over a 2-D Earth, Geophysics, **58**, 198–214.

Weidelt, P., 1975, Inversion of two-dimensional conductivity structures, Phys. Earth Planet. Internat., **10**, 282–291.

Linearized 3-D Electromagnetic Vector Wave Inversion

Karl J. Langenberg
Michael Brandfaß
Andreas Fritsch
Bernd Potzkai

Summary. Generalized (scalar) diffraction tomography is a linear inverse scattering method that can be extended to electromagnetic vector fields with complete polarization information. Its essential equation is a vector form of the Porter–Bojarski integral equation linearized in the material parameters through the Kirchhoff approximation to scattering. This vector equation can be inverted with dyadic algebra and the standard techniques of diffraction tomography using data from multiple frequencies (frequency diversity) or multiple angles of incidence (angle diversity). An algorithm using scattering at multiple frequencies to reconstruct perfectly conducting objects is discussed in detail and checked against synthetic data generated with the MAFIA code for an airplane model. The results are overwhelmingly superior to those obtained by scalar inversion.

1 Introduction

Multidimensional inverse scattering with acoustic, elastic, and electromagnetic (EM) waves has important applications in radar, medical diagnostics, geophysical exploration, and nondestructive testing. Inverse scattering theory has concentrated primarily on a scalar approach (with acoustic waves; for reviews, see, e.g., Herman et al., 1987; Langenberg, 1989). Scalar algorithms also have been applied to microwaves (Langenberg et al., 1993a), treating one component of the scattered field and ignoring information in the complete vector field. The scalar theory can be modified into a vector theory with polarized EM waves (Langenberg et al., 1994a) or elastic waves (Langenberg et al., 1993b).

2 Formulation of the problem

We consider the following problem: A source volume Q generates a known EM field $\mathbf{E}_i(\mathbf{R}, t)$, $\mathbf{H}_i(\mathbf{R}, t)$, which is incident onto a scatterer with volume V_c and surface S_c. The space–time coordinates $(\mathbf{R}, t)$ are denoted as t for time and $\mathbf{R}$ for position vector. The scattered field $\mathbf{E}_s(\mathbf{R}, t)$, $\mathbf{H}_s(\mathbf{R}, t)$ is measured on a closed measurement surface S_M surrounding the scatterer completely; its outward normal is $\mathbf{n}$. The total field anywhere

Department of Electrical Engineering, University of Kassel, 34109 Kassel, Germany.

in space is composed of the incident and the scattered field according to

$$\mathbf{E}(\mathbf{R}, t) = \mathbf{E}_i(\mathbf{R}, t) + \mathbf{E}_s(\mathbf{R}, t). \tag{1}$$

Here, the scatterer is considered to be perfectly conducting; the case of a lossless dielectric material is dealt with by Langenberg et al. (1994a). The medium surrounding the scatterer is assumed as homogeneous, isotropic, lossless, and time independent; i.e., it is described by a real constant scalar permittivity ϵ, and a similar permeability μ, which together define the wave speed as $c = 1/(\epsilon\mu)^{1/2}$ and the wavenumber as $k = \omega/c$.

With time-harmonic fields, we obtain the following integral representation for points $\mathbf{R}$ outside Q and outside $V_c \cup S_c$

$$\mathbf{E}(\mathbf{R}, \omega) = j\omega\mu \int_{-\infty}^{+\infty}\int_{-\infty}^{+\infty}\int_{-\infty}^{+\infty} [\mathbf{J}_q(\mathbf{R}', \omega) + \mathbf{J}_c(\mathbf{R}', \omega)] \cdot \underset{\sim}{\mathbf{G}}(\mathbf{R} - \mathbf{R}', \omega)\, d^3\mathbf{R}', \tag{2}$$

which is a solution of the inhomogeneous vector wave equation

$$\nabla \times \nabla \times \mathbf{E}(\mathbf{R}, \omega) - k^2\mathbf{E}(\mathbf{R}, \omega) = j\omega\mu[\mathbf{J}_q(\mathbf{R}, \omega) + \mathbf{J}_c(\mathbf{R}, \omega)]. \tag{3}$$

$\underset{\sim}{\mathbf{G}}(\mathbf{R} - \mathbf{R}', \omega)$ is the electric dyadic Green's function given by

$$\underset{\sim}{\mathbf{G}}(\mathbf{R} - \mathbf{R}', \omega) = \left(\underset{\sim}{\mathbf{I}} + \frac{1}{k^2}\nabla'\nabla'\right) G(\mathbf{R} - \mathbf{R}', \omega) \tag{4}$$

and $G(\mathbf{R} - \mathbf{R}', \omega)$ is the free-space scalar Green's function

$$G(\mathbf{R} - \mathbf{R}', \omega) = \frac{e^{jk|\mathbf{R}-\mathbf{R}'|}}{4\pi|\mathbf{R} - \mathbf{R}'|}. \tag{5}$$

The quantities $\mathbf{J}_q(\mathbf{R}, \omega)$, and $\mathbf{J}_c(\mathbf{R}, \omega)$, respectively, denote volume current densities of either the primary sources in Q or the secondary equivalent sources on S_c. An appropriate representation for $\mathbf{J}_c(\mathbf{R}, \omega)$ is

$$\mathbf{J}_c(\mathbf{R}, \omega) = \gamma(\mathbf{R})\mathbf{n} \times \mathbf{H}(\mathbf{R}, \omega), \tag{6}$$

where the singular function $\gamma(\mathbf{R})$ of the surface S_c has the property of reducing a volume integral of some function $\Phi(\mathbf{R})$ to a surface integral

$$\int_{-\infty}^{+\infty}\int_{-\infty}^{+\infty}\int_{-\infty}^{+\infty} \Phi(\mathbf{R})\gamma(\mathbf{R})\, dV = \iint_{S_c} \Phi(\mathbf{R})\, dS. \tag{7}$$

Equation (6) gives rise to the well-known physical-optics or Kirchhoff linearization,

$$\mathbf{J}_c^{\mathrm{PO}}(\mathbf{R}, \omega) = \begin{cases} 2\boldsymbol{\gamma}(\mathbf{R}) \times \mathbf{H}_i(\mathbf{R}, \omega) & \text{on the illuminated side} \\ 0 & \text{on the shadow side,} \end{cases} \tag{8}$$

where we have introduced the vector singular function

$$\boldsymbol{\gamma}(\mathbf{R}) = \gamma(\mathbf{R})\mathbf{n}. \tag{9}$$

The validity of the physical-optics approximation is restricted primarily to smoothly curved objects that are large compared to the wavelength; it works remarkably better than the Born approximation for penetrable objects (Herman et al., 1987). Even objects with sharp edges—e.g., a perfectly conducting strip—do not really present problems.

The incident field now is supposed to be a linearly polarized plane wave with frequency spectrum $F(\omega)$, unit polarization vector $\hat{\mathbf{E}}_0$, and unit propagation vector $\hat{\mathbf{k}}_i$:

$$\mathbf{E}_i(\mathbf{R}, \omega) = \hat{\mathbf{E}}_0 F(\omega)\, e^{jk\hat{\mathbf{k}}_i \cdot \mathbf{R}}, \tag{10}$$

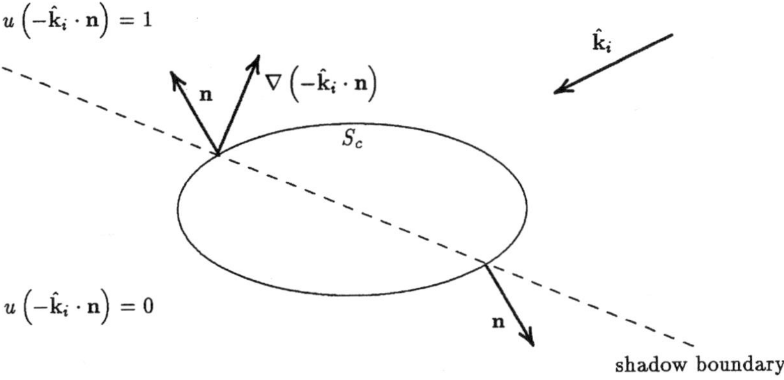

Figure 1. Illustration of the shadow boundary appearing in the physical-optics approximation.

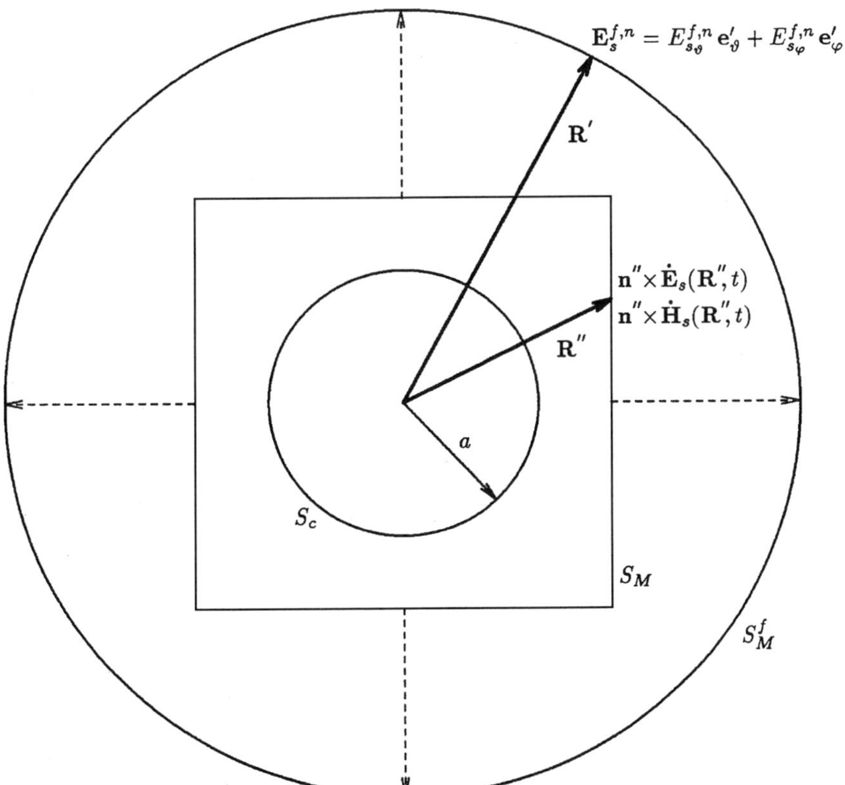

Figure 2. Near-field far-field transformation of scattering data.

yielding

$$\mathbf{J}_c^{\mathrm{PO}}(\mathbf{R}, \omega) = \frac{2F(\omega)}{Z} \mathbf{J}_c^{\mathrm{PO}}(\mathbf{R})\, e^{jk\hat{\mathbf{k}}_i \cdot \mathbf{R}}, \tag{11}$$

with the wave impedance $Z = (\mu/\epsilon)^{1/2}$ and

$$\begin{aligned} \mathbf{J}_c^{\mathrm{PO}}(\mathbf{R}) &= \gamma_u(\mathbf{R}) \times (\hat{\mathbf{k}}_i \times \hat{\mathbf{E}}_0) \\ &= \hat{\mathbf{k}}_i(\gamma_u \cdot \hat{\mathbf{E}}_0) - \hat{\mathbf{E}}_0(\gamma_u \cdot \hat{\mathbf{k}}_i). \end{aligned} \tag{12}$$

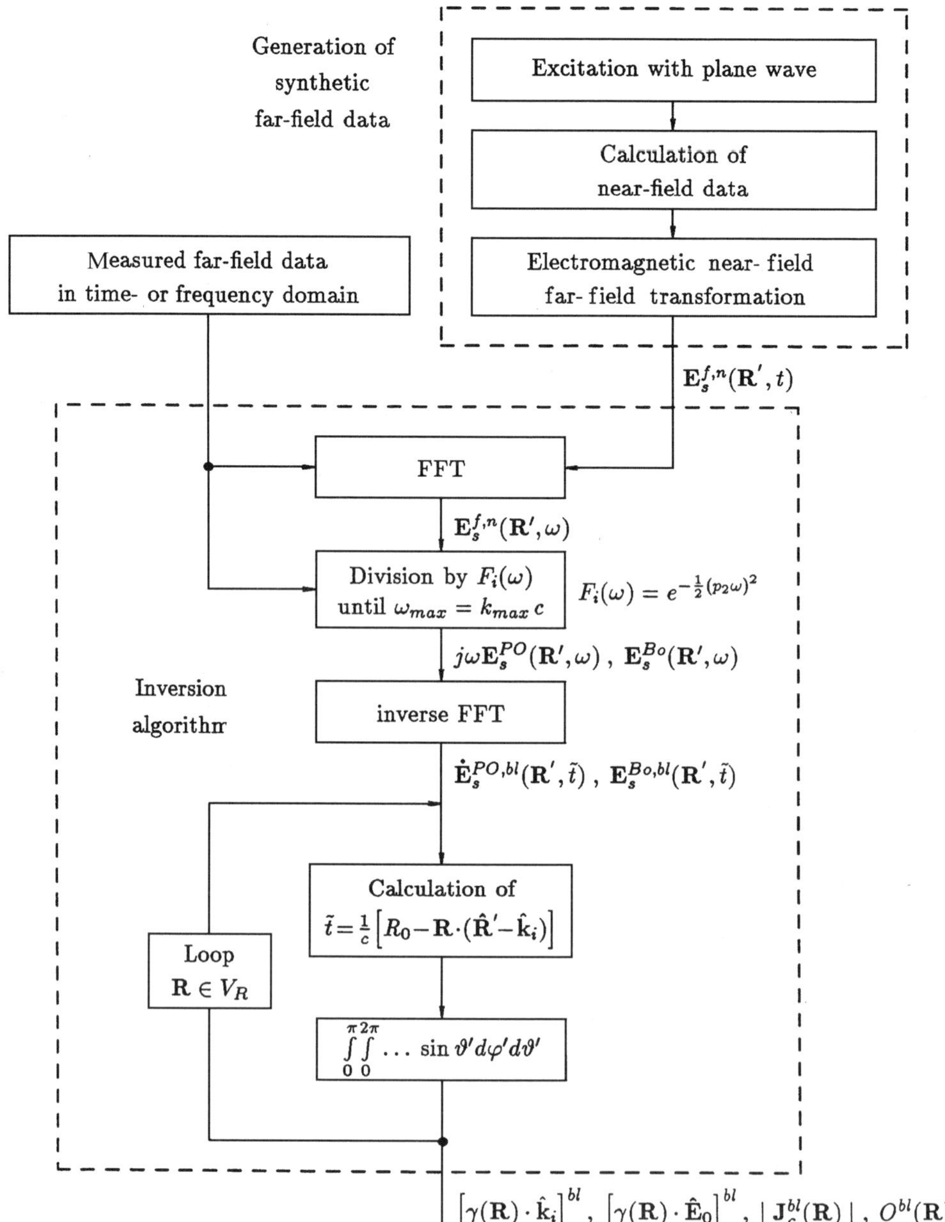

Figure 3. Flowchart of polarimetric inversion.

The physical-optics vector singular function $\gamma_u(\mathbf{R}) = \gamma(\mathbf{R})\mathbf{n}\,u(-\hat{\mathbf{k}}_i \cdot n)$ includes the unit-step function u, accounting for the physical-optics shadow region (see Fig. 1).

It is our goal to recover $\gamma_u \cdot \hat{\mathbf{E}}_0$ and $\gamma_u \cdot \hat{\mathbf{k}}_i$, or even γ_u itself, from polarimetric measurements. $\mathbf{J}_c(\mathbf{R})$ can be rewritten in terms of the orthogonal polarization $\hat{\mathbf{E}}_{\text{orth}}$(i.e., $\hat{\mathbf{E}}_{\text{orth}} \cdot \hat{\mathbf{E}}_0 = 0$)

$$\mathbf{J}_c(\mathbf{R}) = \hat{\mathbf{E}}_{\text{orth}} \times \gamma_u(\mathbf{R}); \tag{13}$$

then, taking the divergence of $\mathbf{J}_c(\mathbf{R})$ gives

$$\nabla \cdot \mathbf{J}_c(\mathbf{R}) = -[\nabla \times \gamma_u(\mathbf{R})] \cdot \hat{\mathbf{E}}_{\text{orth}}. \tag{14}$$

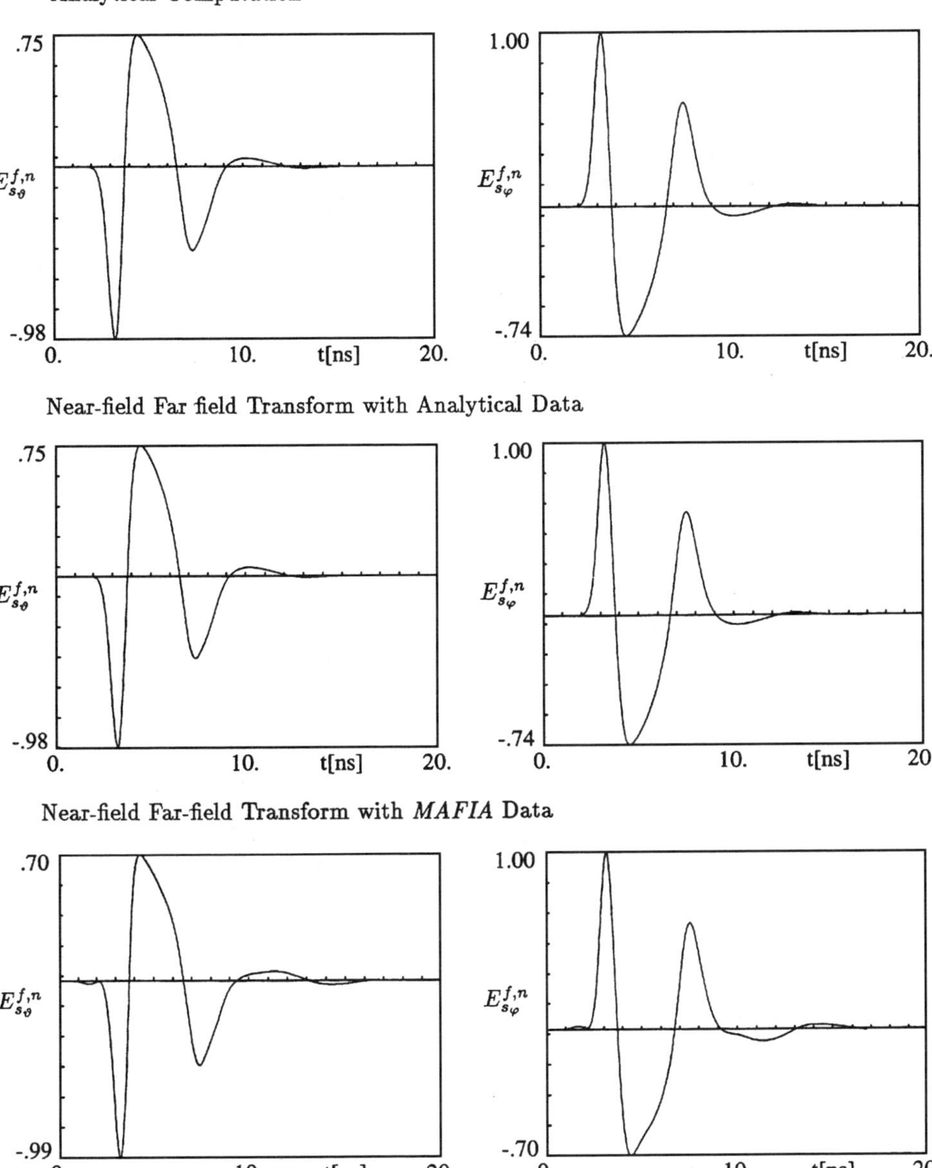

Figure 4. Components of scattered electric far-field in the time domain for the observation angles $\vartheta' = 30^o$, $\varphi' = 45^o$.

Because $\gamma(\mathbf{R})$ is—in terms of $\mathbf{n}$—proportional to the gradient of a potential function describing the surface S_c, we have

$$\nabla \times \gamma(\mathbf{R}) = 0, \tag{15}$$

and, hence,

$$\nabla \times \gamma_u(\mathbf{R}) = \delta(-\hat{\mathbf{k}}_i \cdot \mathbf{n})\gamma(\mathbf{R}) \times \nabla(\hat{\mathbf{k}}_i \cdot \mathbf{n}) \tag{16}$$

with the δ-distribution $\delta(-\hat{\mathbf{k}}_i \cdot \mathbf{n})$ being singular only on the shadow boundary *line*. Because this line is somehow artificially introduced via the physical optics argument,

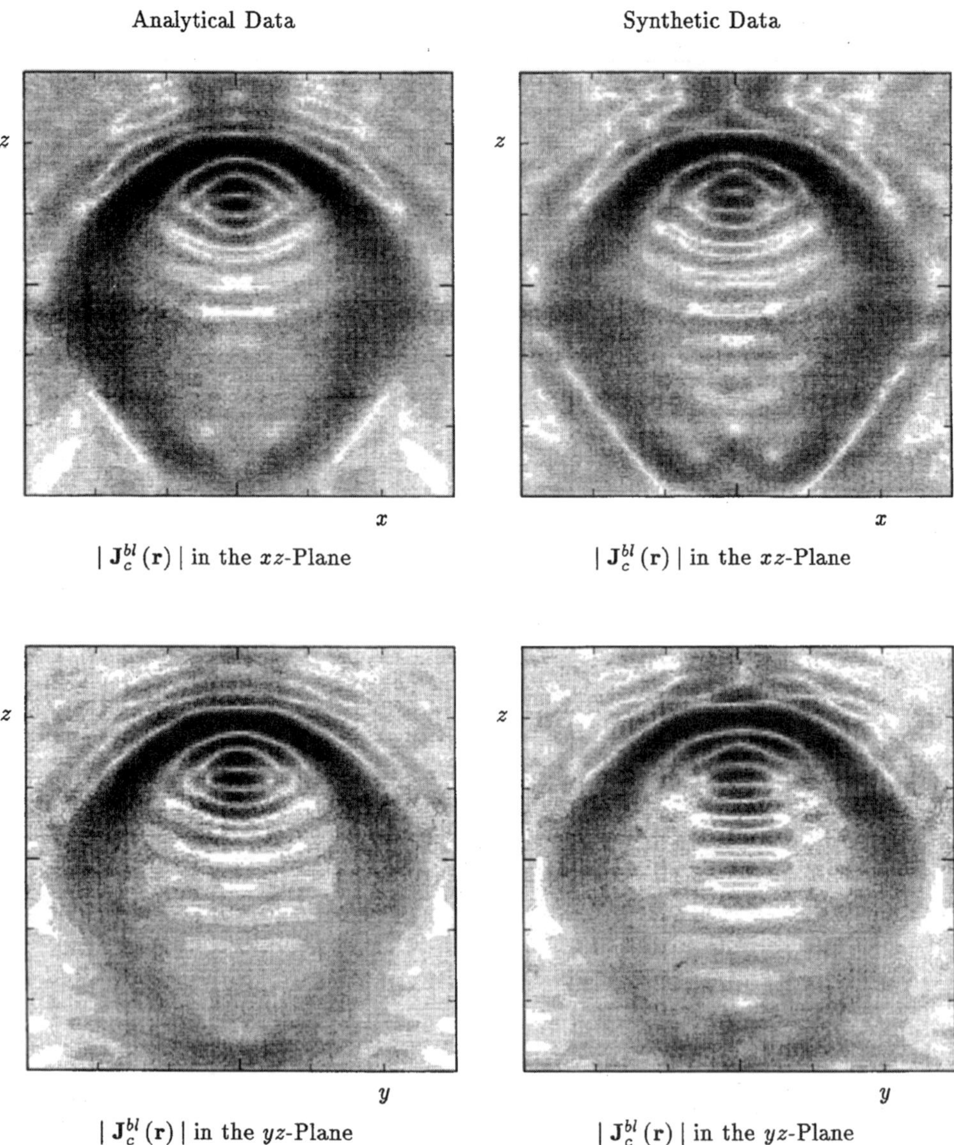

Figure 5. Magnitude of equivalent current density: Two orthogonal slices through the image volume for (left) analytical data and (right) MAFIA data.

we set the curl of $\gamma_u(\mathbf{R})$ to zero:

$$\nabla \times \gamma_u(\mathbf{R}) = 0. \tag{17}$$

According to Eq. (13), this has the consequence of making $\mathbf{J}_c(\mathbf{R})$ divergence free; i.e., we have

$$\nabla \cdot \mathbf{J}_c(\mathbf{R}) = 0, \tag{18}$$

and, as a further consequence, Eq. (18) allows for polarimetric inversion. Looking at the results, for example, for the perfectly conducting sphere, it is obvious that Eq. (17) should hold.

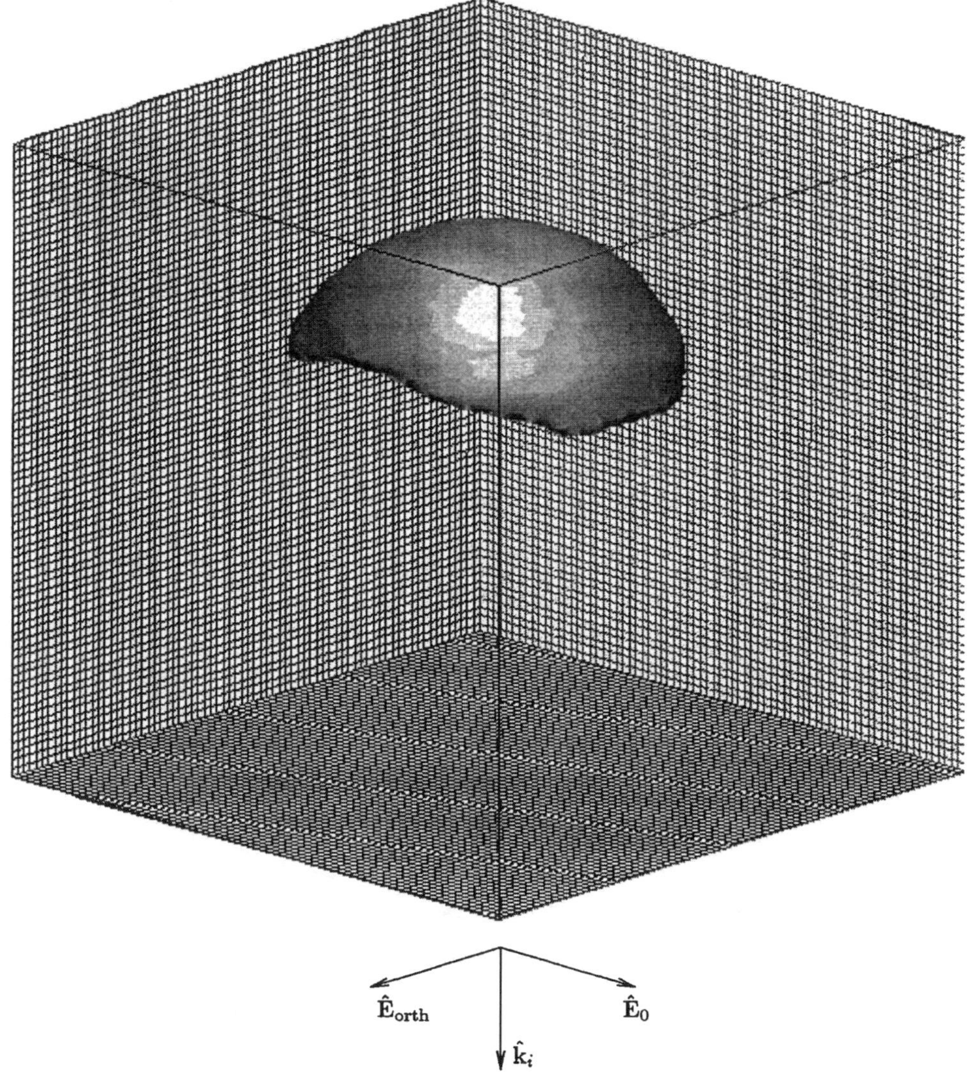

Figure 6. A 3-D representation of magnitude of equivalent current density for $\hat{\mathbf{E}}_0$-polarization.

3 Dyadic backpropagation: Frequency diversity

Following the scalar theory of Herman et al. (1987), we define a generalized vector holographic field $\boldsymbol{\Theta}_H^{\mathbf{E}}(\mathbf{R}, \omega)$ for all points $\mathbf{R}$ according to

$$\boldsymbol{\Theta}_H^{\mathbf{E}}(\mathbf{R}, \omega) = -2\omega\mu \int_{-\infty}^{+\infty}\int_{-\infty}^{+\infty}\int_{-\infty}^{+\infty} \mathbf{J}_c(\mathbf{R}', \omega) \cdot \underset{\sim}{\mathbf{G}}^I(\mathbf{R} - \mathbf{R}', \omega)\, d^3\mathbf{R}', \tag{19}$$

with the imaginary part $\underset{\sim}{\mathbf{G}}^I$ of the dyadic Green's function. Equation (19) can be transformed with the help of the vector Green's theorem into

$$\boldsymbol{\Theta}_H^{\mathbf{E}}(\mathbf{R}, \omega) = -\iint_{S_M} \{j\omega\mu[\mathbf{n}' \times \mathbf{H}_s(\mathbf{R}', \omega)] \cdot \underset{\sim}{\mathbf{G}}^*(\mathbf{R} - \mathbf{R}', \omega)$$
$$+ [\mathbf{n}' \times \mathbf{E}_s(\mathbf{R}', \omega)] \cdot \nabla' \times \underset{\sim}{\mathbf{G}}^*(\mathbf{R} - \mathbf{R}', \omega)\}\, dS', \tag{20}$$

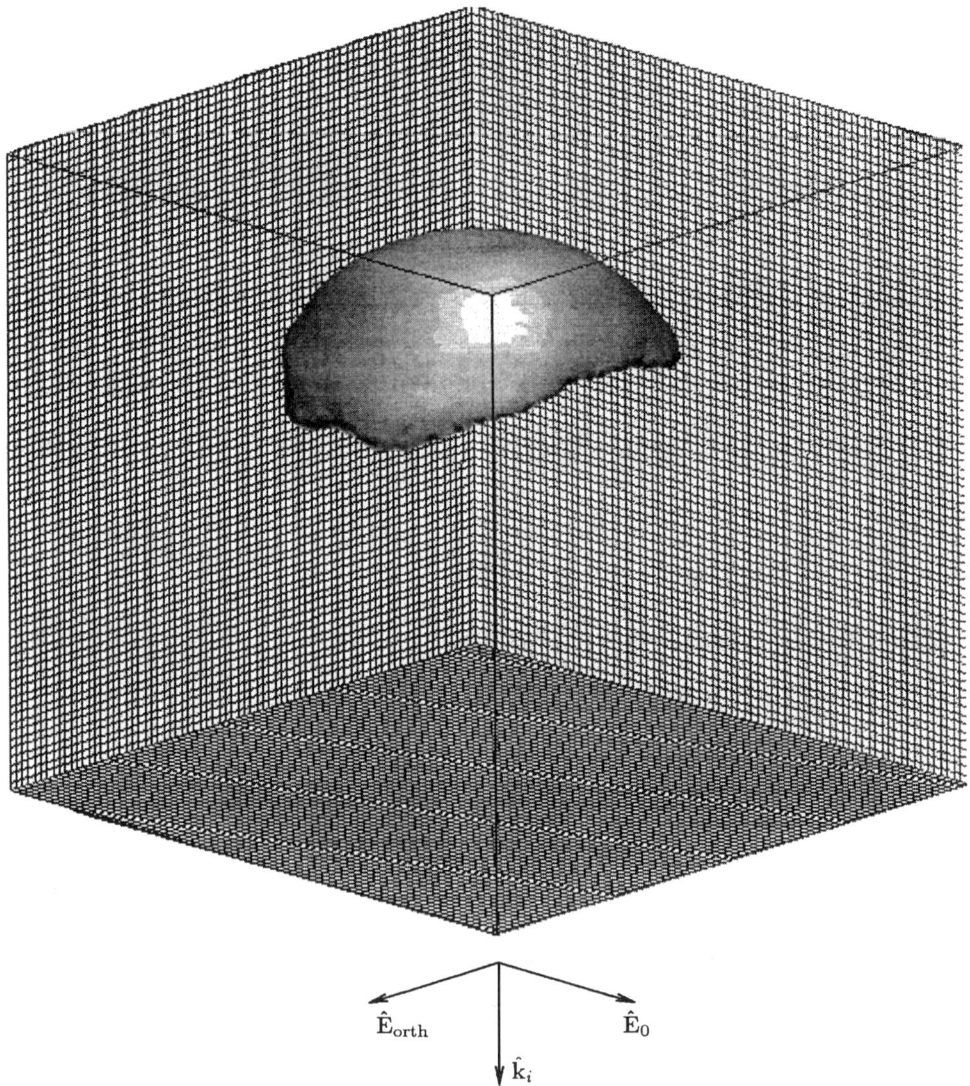

Figure 7. A 3-D representation of magnitude of equivalent current density for $\hat{\mathbf{E}}_{\text{orth}}$-polarization.

where S_M denotes a closed measurement surface completely surrounding the scatterer, and $\underset{\sim}{\mathbf{G}}^*$ is the complex conjugate of $\underset{\sim}{\mathbf{G}}$. Therefore, $\boldsymbol{\Theta}_H^{\mathbf{E}}$ is identified as a polarimetrically measurable quantity, and Eq. (19) turns out to be an integral equation—a vector form of the Porter–Bojarski integral equation—for the equivalent sources. In Langenberg et al. (1994a), the vector holographic field is *defined* according to Eq. (20), but with $\underset{\sim}{\mathbf{G}}^I$ instead of $\underset{\sim}{\mathbf{G}}^*$; this does not significantly change the Porter–Bojarski equation, but it has the slight advantage that the spatial Fourier transform can be applied directly to the respective Eq. (20), whereas in the present formulation, which closely follows the scalar case, only Eq. (19) is subject to a spatial Fourier transform.

Define the 3-D Fourier transform according to

$$\tilde{\boldsymbol{\Phi}}(\mathbf{K}, \omega) = \int_{-\infty}^{+\infty}\int_{-\infty}^{+\infty}\int_{-\infty}^{+\infty} \boldsymbol{\Phi}(\mathbf{R}, \omega)e^{-j\mathbf{K}\cdot\mathbf{R}}\, d^3\mathbf{R}, \tag{21}$$

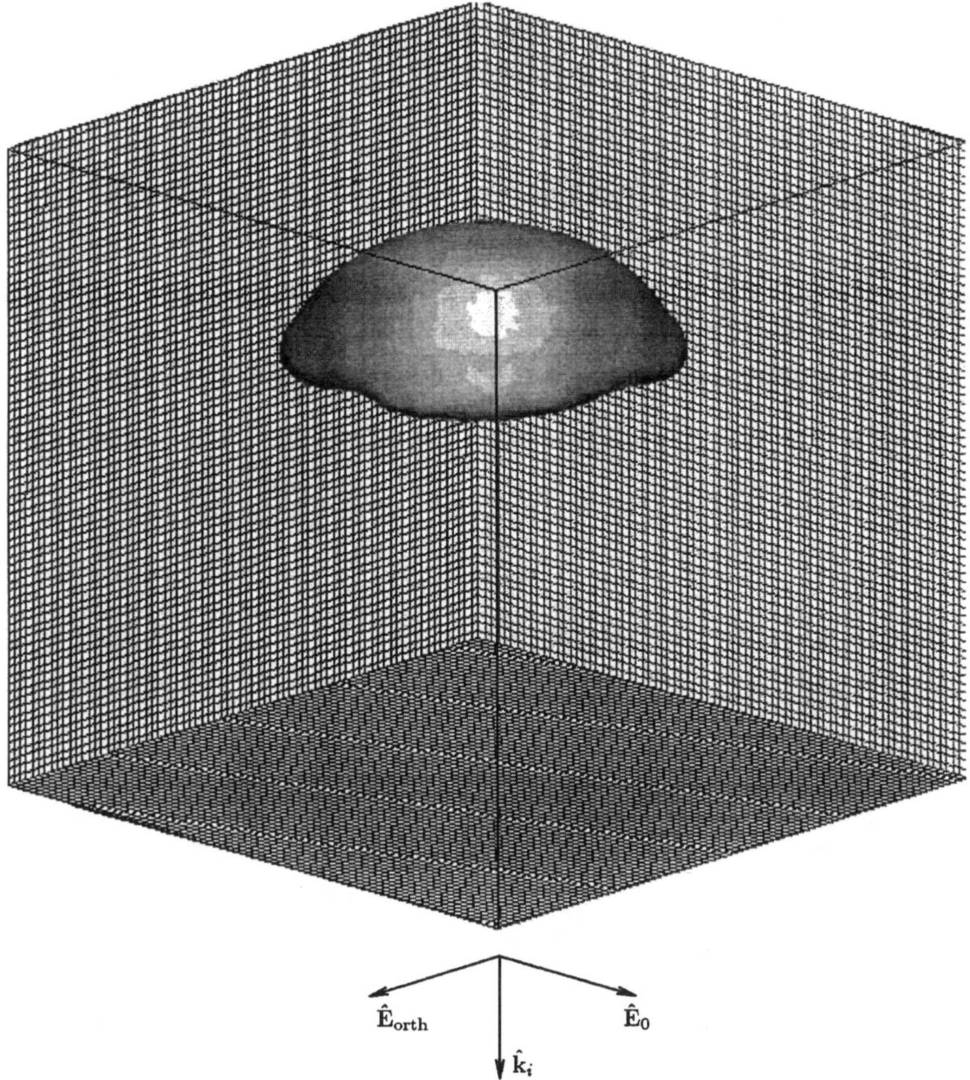

Figure 8. A 3-D representation of magnitude of singular function.

where $\mathbf{K}$ denotes the Fourier vector. Multiplication of Eq. (19) by $e^{-jk\hat{\mathbf{k}}_i\cdot\mathbf{R}}$ and integration with regard to frequency (i.e., a broadband experiment at a single illumination angle is performed) yields

$$\int_0^\infty \frac{1}{F(\omega)}\tilde{\boldsymbol{\Theta}}_H^{\mathbf{E}}(\mathbf{K}+k\hat{\mathbf{k}}_i,\omega)\,dk$$
$$= -2\pi\frac{K^2}{2(\mathbf{K}\cdot\hat{\mathbf{k}}_i)^2}u(-\mathbf{K}\cdot\hat{\mathbf{k}}_i)[\hat{\mathbf{k}}_i(\tilde{\gamma}_u\cdot\hat{\mathbf{E}}_0)-\hat{\mathbf{E}}_0(\tilde{\gamma}_u\cdot\hat{\mathbf{k}}_i)]\cdot\underset{\sim}{\tilde{\mathbf{M}}}(\mathbf{K},\hat{\mathbf{k}}_i,k_j). \tag{22}$$

Here, we have used (Herman et al., 1987; Langenberg et al., 1994a)

$$\underset{\sim}{\mathbf{G}}^I(\mathbf{K},\omega)=\left(\underset{\sim}{\mathbf{I}}-\frac{1}{k^2}\mathbf{KK}\right)\frac{\pi}{2k}\delta(K-k)\quad \text{for } k\geq 0 \tag{23}$$

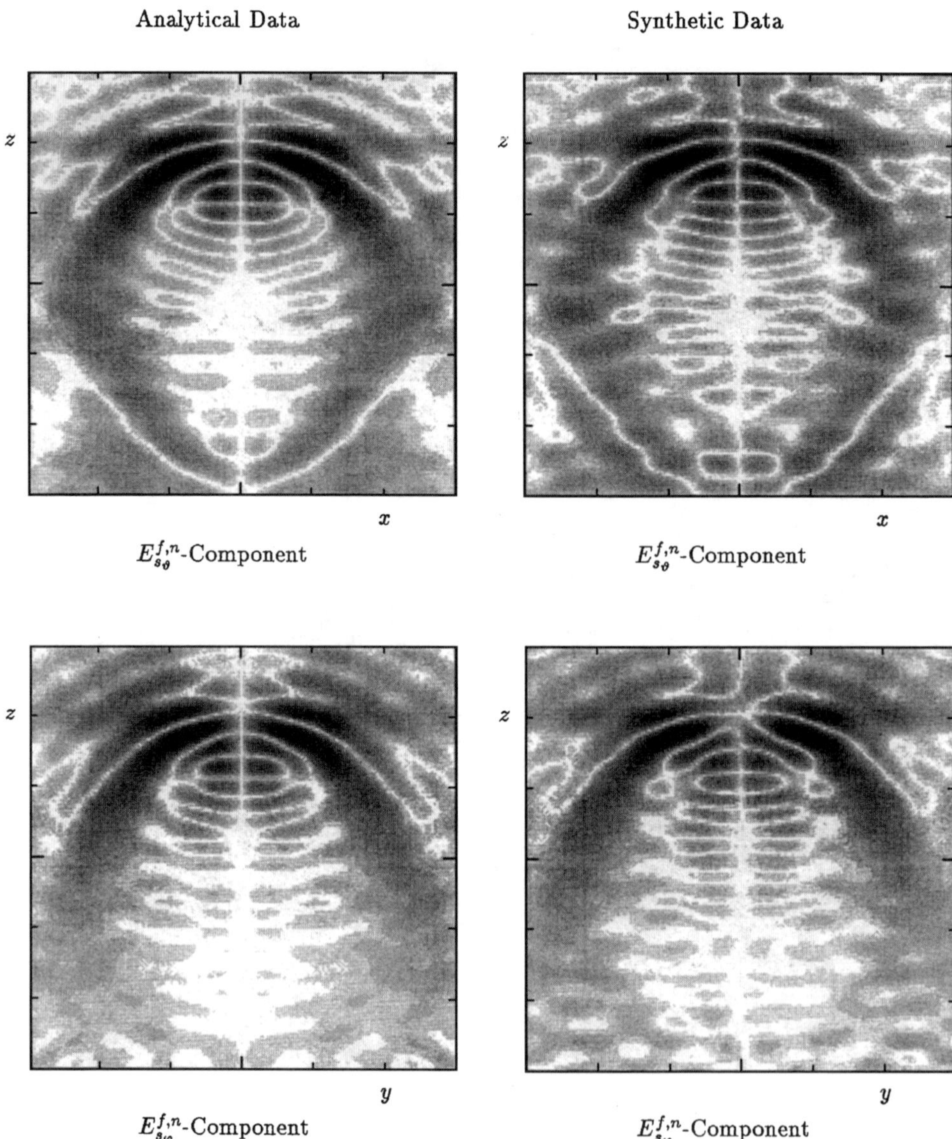

Figure 9. Magnitude of scalar inversion data: Two orthogonal slices through image volume for (left) analytical data and (right) MAFIA data, for electric field components (top) ϑ and (bottom) φ.

and

$$\int_0^\infty \delta(|\mathbf{K} + k\hat{\mathbf{k}}_i| - k)\, dk = \frac{K^2}{2(\mathbf{K} \cdot \hat{\mathbf{k}}_i)^2} u(-\mathbf{K} \cdot \hat{\mathbf{k}}_i), \tag{24}$$

which results in the dyadic

$$\underset{\sim}{\mathbf{M}}(\mathbf{K}, \hat{\mathbf{k}}_i, k) = \left[\underset{\sim}{\mathbf{I}} - \frac{1}{k^2}(\mathbf{K} + k\hat{\mathbf{k}}_i)(\mathbf{K} + k\hat{\mathbf{k}}_i) \right] \tag{25}$$

and

$$k_j = -\frac{K^2}{2\mathbf{K} \cdot \hat{\mathbf{k}}_i}. \tag{26}$$

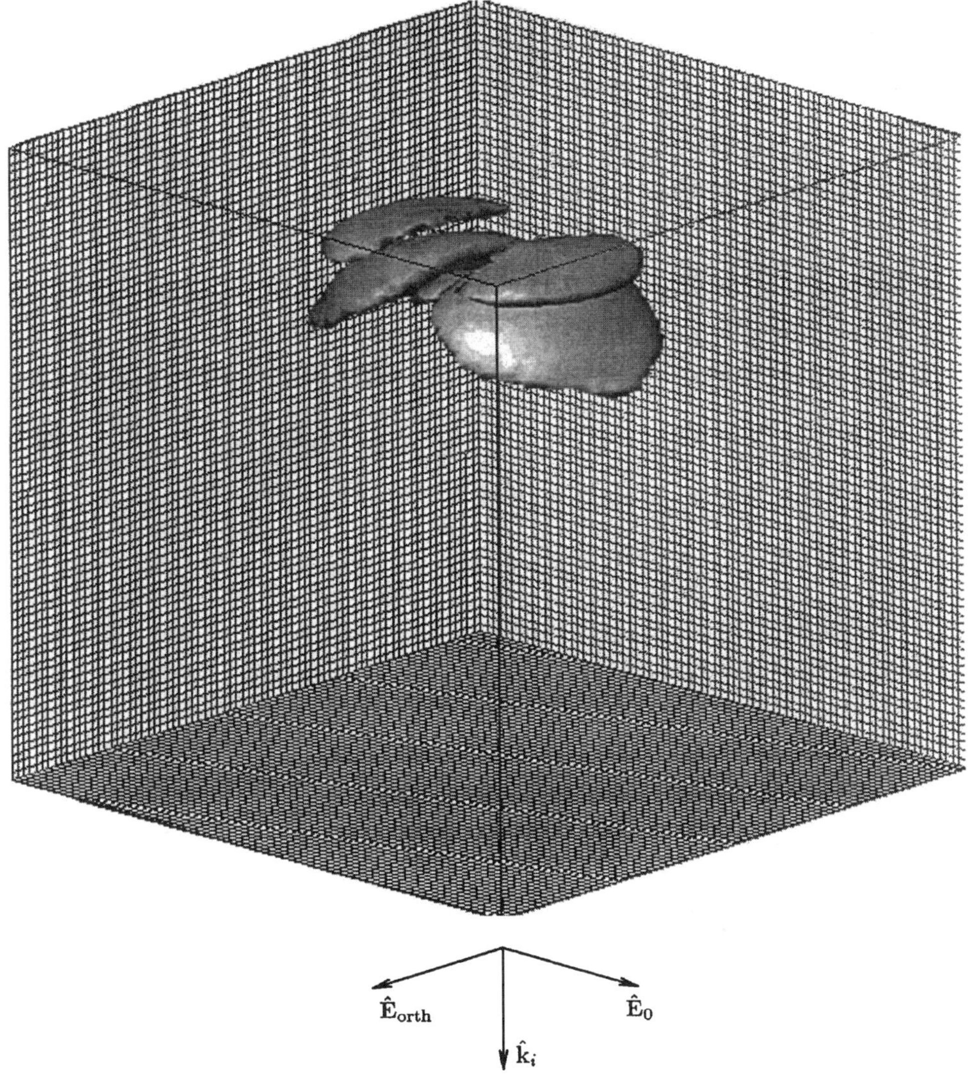

Figure 10. A 3-D representation of scalar inversion data for ϑ-component of the electric field.

Unfortunately, since det $\underset{\sim}{\mathbf{M}} \equiv 0$, the dyadic $\underset{\sim}{\mathbf{M}}$ is not invertible; but, from Eq. (18), we have

$$\mathbf{K} \cdot \tilde{\mathbf{J}}_c(\mathbf{K}) = 0 \tag{27}$$

and, hence,

$$\begin{aligned} \underset{\sim}{\mathbf{M}}(\mathbf{K}, \hat{\mathbf{k}}_i, k_j) \cdot \tilde{\mathbf{J}}(\mathbf{K}) &= \left[\underset{\sim}{\mathbf{I}} - \frac{1}{k_j^2}(\mathbf{K} + k_j\hat{\mathbf{k}}_i)(\mathbf{K} + k_j\hat{\mathbf{k}}_i) \right] \cdot \tilde{\mathbf{J}}_c(\mathbf{K}) \\ &= \left[\underset{\sim}{\mathbf{I}} - \frac{1}{k_j}(\mathbf{K} + k_j\hat{\mathbf{k}}_i)\hat{\mathbf{k}}_i \right] \cdot \tilde{\mathbf{J}}_c(\mathbf{K}), \end{aligned} \tag{28}$$

where the dyadic on the right-hand side becomes invertible. Standard—though a bit tedious—dyadic algebra (Chen, 1983) then gives an explicit expression for $\tilde{\mathbf{J}}_c(\mathbf{K})$ in terms of $\tilde{\Theta}_H^{\mathbf{E}}(\mathbf{K}, \omega)$, which can be formally Fourier inverted into the spatial domain

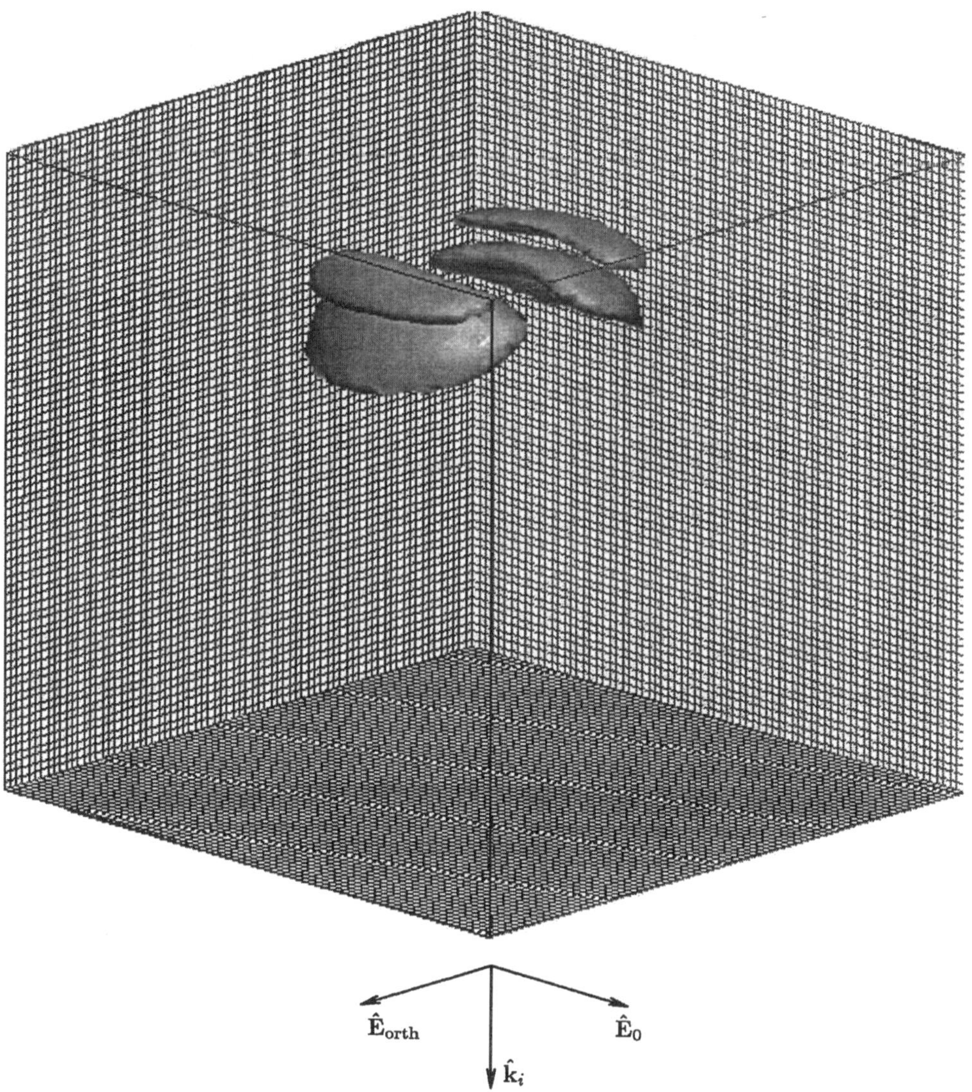

Figure 11. A 3-D representation of scalar inversion data for φ-component of the electric field.

to yield $\mathbf{J}_c(\mathbf{R})$. The expressions are given by Langenberg et al. (1994a), Langenberg et al. (1994b), and Brandfaß (1989) and are not repeated here because they are rather voluminous and not really suited for numerical evaluation. We point out, however, that the derivation given here, based upon Eq. (27), is much simpler than the one available in Langenberg et al. (1994a), Langenberg et al. (1994b), and Brandfaß (1989).

4 Far-field Fourier inversion or time-domain back projection: Results for a perfectly conducting sphere

4.1 Far-field Fourier inversion

It is easy to show (Langenberg et al., 1994a) that the generalized vector holographic field can be represented by the far-field scattering amplitude $\mathbf{C}(\hat{\mathbf{R}}, \omega)$ of the

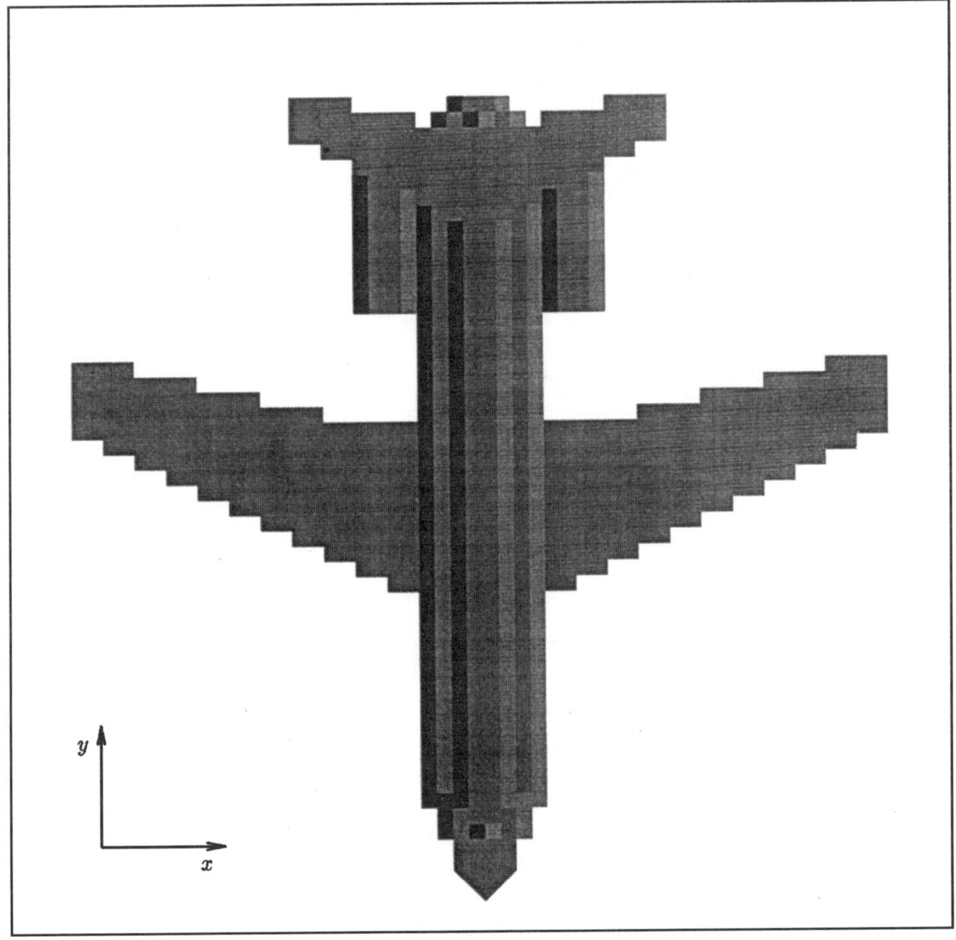

Figure 12. Airplane model as designed with MAFIA.

equivalent currents

$$\boldsymbol{\Theta}_H^{\mathbf{E}}(\mathbf{R}, \omega) = \frac{jk}{2\pi} \iint_{S^2} (\underset{\sim}{\mathbf{I}} - \hat{\mathbf{R}}'\hat{\mathbf{R}}') \cdot \mathbf{C}(\hat{\mathbf{R}}', \omega) e^{jk\mathbf{R}\cdot\hat{\mathbf{R}}'} d^2\hat{\mathbf{R}}', \tag{29}$$

where

$$\mathbf{C}(\hat{\mathbf{R}}', \omega) = \frac{j\omega\mu}{4\pi} \int_{-\infty}^{+\infty} \int_{-\infty}^{+\infty} \int_{-\infty}^{+\infty} \mathbf{J}_c(\mathbf{R}', \omega) e^{-jk\hat{\mathbf{R}}\cdot\mathbf{R}'} d^3\mathbf{R}' \tag{30}$$

resulting in

$$\mathbf{E}_s^{\text{far}}(\mathbf{R}, \omega) = \frac{e^{jkR}}{R} (\underset{\sim}{\mathbf{I}} - \hat{\mathbf{R}}\hat{\mathbf{R}}) \cdot \mathbf{C}(\hat{\mathbf{R}}, \omega). \tag{31}$$

In (29), $d^2\hat{\mathbf{R}}'$ denotes the differential solid-angle element of the unit-sphere S^2 with the unit-vector $\hat{\mathbf{R}}'$ directed toward the measurement surface. Inserting Eq. (29) into the reconstruction equation for $\mathbf{J}_c(\mathbf{R})$, we obtain (Brandfaß, 1989; Langenberg et al., 1994a; Langenberg et al., 1994b)

$$\begin{aligned}\mathbf{J}_c(\mathbf{R}) = -\Re\Bigg\{ &\frac{1}{2\pi^2 Z} \int_0^{\infty} \frac{jk}{F(\omega)} \iint_{S^2} R' e^{-jkR'} \mathbf{E}_s^{\text{far}}(\mathbf{R}', \omega) \\ &\times [\hat{\mathbf{k}}_i\hat{\mathbf{k}}_i + \hat{\mathbf{E}}_0\hat{\mathbf{E}}_0 + (\hat{\mathbf{k}}_i\hat{\mathbf{E}}_0\hat{\mathbf{E}}_0 - \hat{\mathbf{E}}_0\hat{\mathbf{E}}_0\hat{\mathbf{k}}_i) \cdot \hat{\mathbf{R}}'] e^{jk\hat{\mathbf{R}}'\cdot\mathbf{R}} d^2\hat{\mathbf{R}}' e^{-jk\hat{\mathbf{k}}_i\cdot\mathbf{R}} dk \Bigg\}.\end{aligned} \tag{32}$$

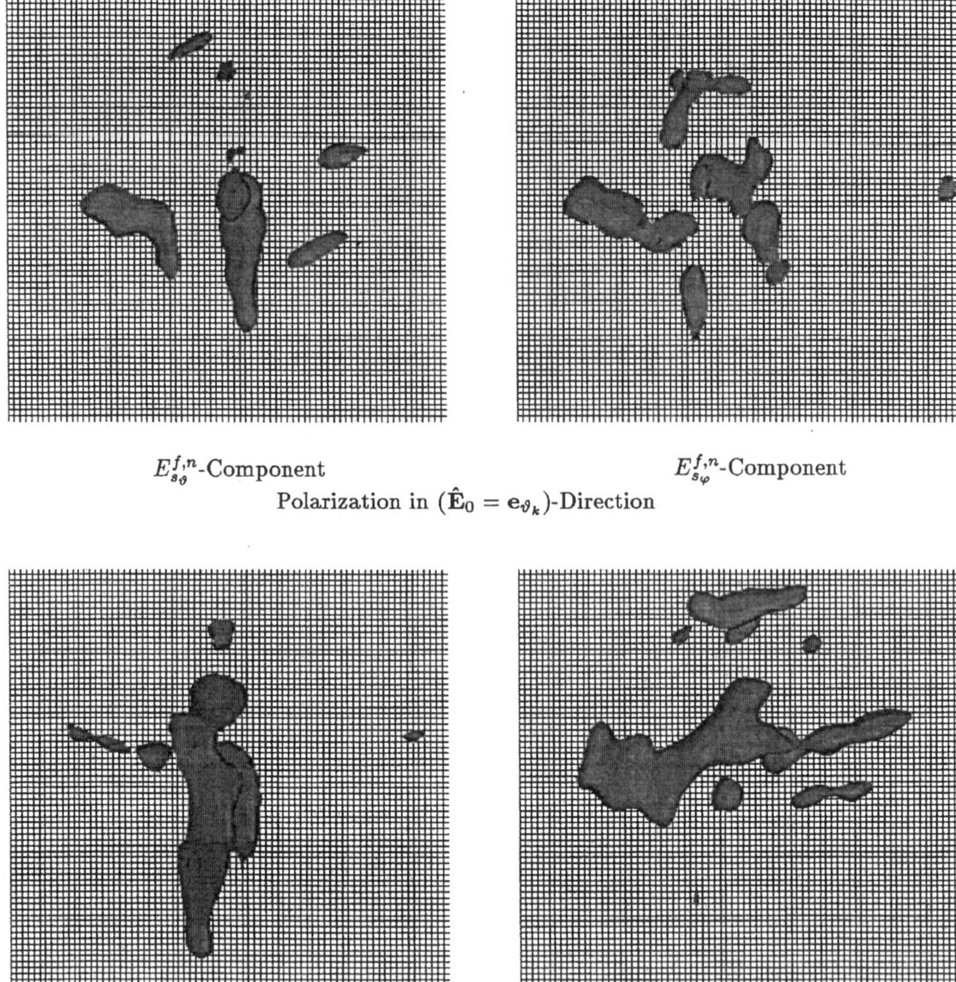

Figure 13. Result of scalar inversion for the airplane model: electric-field component (left) ϑ- and (right) φ-component for (top) ($\hat{\mathbf{E}}_0 = e_x$)-polarization and (bottom) ($\hat{\mathbf{E}}_{\text{orth}} = e_y$)-polarization.

Inversion equations of this kind are very well known for the scalar case and are summarized as "far-field Fourier inversion" (Herman et al., 1987). Setting

$$\mathbf{K} = k(\hat{\mathbf{R}}' - \hat{\mathbf{k}}_i) \tag{33}$$

and computing a Jacobian according to

$$d^3\mathbf{K} = k^2(1 - \hat{\mathbf{k}}_i \cdot \hat{\mathbf{R}}')\, d^2\hat{\mathbf{R}}'\, dk \tag{34}$$

reveals that Eq. (32) is an inverse Fourier integral: For an arbitrary $\hat{\mathbf{k}}_i$, points in $\mathbf{K}$-space are identified by the "experimental" variables k and $\hat{\mathbf{R}}'$, where the data $\mathbf{E}_s^{\text{far}}(\mathbf{R}', \omega)$ have to be allocated.

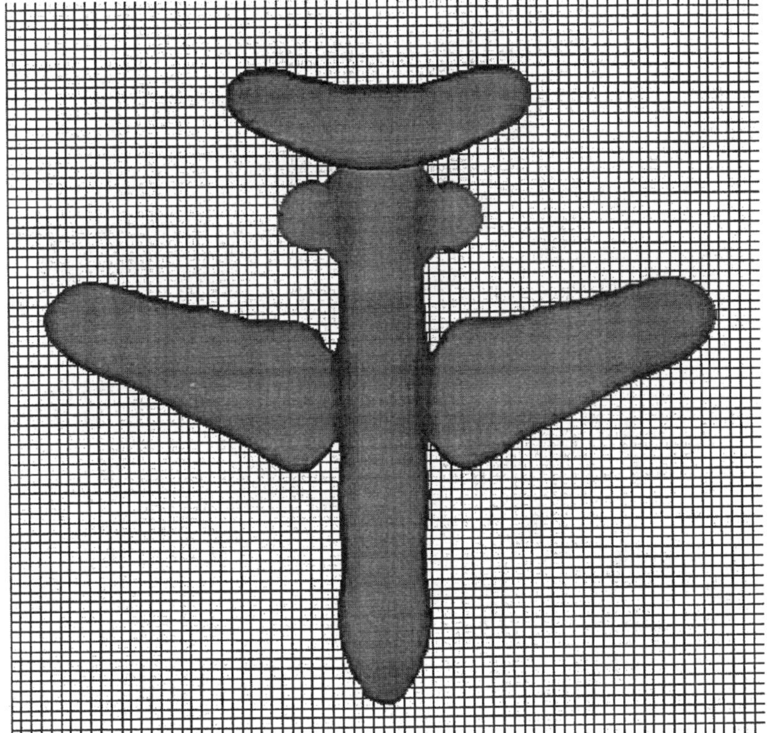

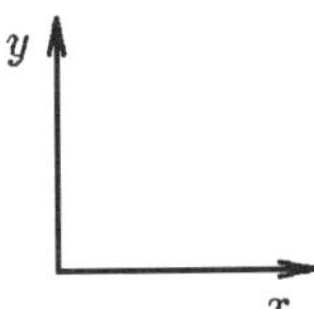

Figure 14. Magnitude of singular function.

4.2 Time-domain back projection

For the case of a perfectly conducting sphere, $\mathbf{E}_s^{\text{far}}(\mathbf{R}, \omega)$ can be computed via vector eigenfunction expansion, and it turns out that the azimuth-ϕ'-angle integration on the unit-sphere can be performed analytically. Computing the remaining θ', k-integrations numerically, in Langenberg et al. (1994a) and Langenberg et al. (1994b), we presented results for $\mathbf{J}_c(\mathbf{R})$ for $0 \leq ka \leq 200$, a being the radius of the sphere. As compared to similar scalar inversion treating either the ϕ- or the θ-component of $\mathbf{E}_s^{\text{far}}$ as scalar input data, the polarimetric inversion yielded something like a coherent superposition of scalar images. Therefore, the question arises as to how the polarimetric scheme would perform for more complex targets. Of course, synthetic data then would have to be computed with an appropriate numerical code: We have chosen MAFIA (MAFIA, 1994) which contains a module for 3-D time-domain scattering simulations. The procedure is as follows (Fig. 2).

Tangential components of the electric- and magnetic-field strengths for broadband plane-wave scattering by a perfectly conducting target are computed on the six surfaces of a box surrounding the target; the time dependence of the incident wave is base-band

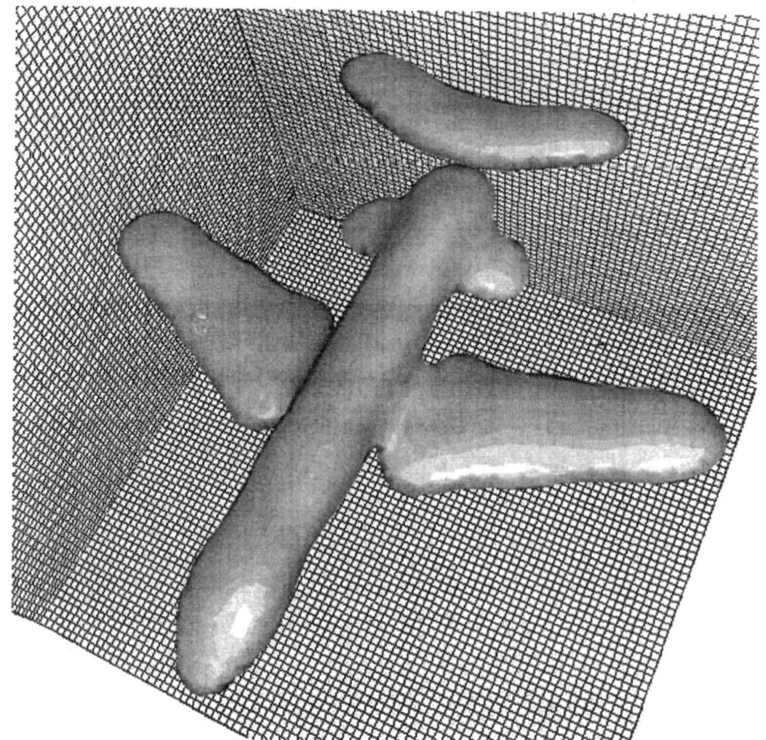

Figure 15. A 3-D representation of singular-function magnitude.

Gaussian. To be used in Eq. (32), these near-field data have to be transformed into the far-field according to a time-domain near-field far-field transform based on the EM Huygens principle,

$$\mathbf{E}_s^{f,n}(\mathbf{R}', t) = -\frac{1}{4\pi c} \iint_{S_M} \{Z(\underline{\mathbf{I}} - \hat{\mathbf{R}}'\hat{\mathbf{R}}') \cdot [\mathbf{n}'' \times \dot{\mathbf{H}}_s(\mathbf{R}'', t')] - [\mathbf{n}'' \times \dot{\mathbf{E}}_s(\mathbf{R}'', t')] \times \hat{\mathbf{R}}'\} \, dS'', \tag{35}$$

for the R'-multiplied (normalized) far-field on a surface S_M^f, where

$$t' = t + \frac{\hat{\mathbf{R}}' \cdot \mathbf{R}'' - R'}{c}. \tag{36}$$

The $j\omega$-factor in the time-harmonic Huygens principle accounts for the time (dot) derivative of the scattered field.

Up to now we are working fully in the time domain and it seems natural to interpret Eq. (32) in the time domain also, which is done easily by recognizing the k-integral as an inverse Fourier-integral with $\omega = 0$. The result is

$$\mathbf{J}_c(\mathbf{R}) = -\frac{1}{2\pi c^2 Z} \iint_{S^2} \dot{\mathbf{E}}_s^{f,n,F}\left(\mathbf{R}', t = R' - \frac{\mathbf{R} \cdot (\hat{\mathbf{R}}' - \hat{\mathbf{k}}_i)}{c}\right) \cdot [\hat{\mathbf{k}}_i\hat{\mathbf{k}}_i + \hat{\mathbf{E}}_0\hat{\mathbf{E}}_0 + (\hat{\mathbf{k}}_i\hat{\mathbf{E}}_0\hat{\mathbf{E}}_0 - \hat{\mathbf{E}}_0\hat{\mathbf{E}}_0\hat{\mathbf{k}}_i) \cdot \hat{\mathbf{R}}'] \, d^2\hat{\mathbf{R}}'. \tag{37}$$

This equation is a time-domain dyadic backprojection. The quantity to be backprojected, $\dot{\mathbf{E}}_s^{f,n,F}(\mathbf{R}', t)$, is the inverse Fourier transform of $jk\mathbf{E}_s^{f,n}(\mathbf{R}', \omega)/F(\omega)$; i.e., a deconvolution with the spectrum of the incident field also is involved.

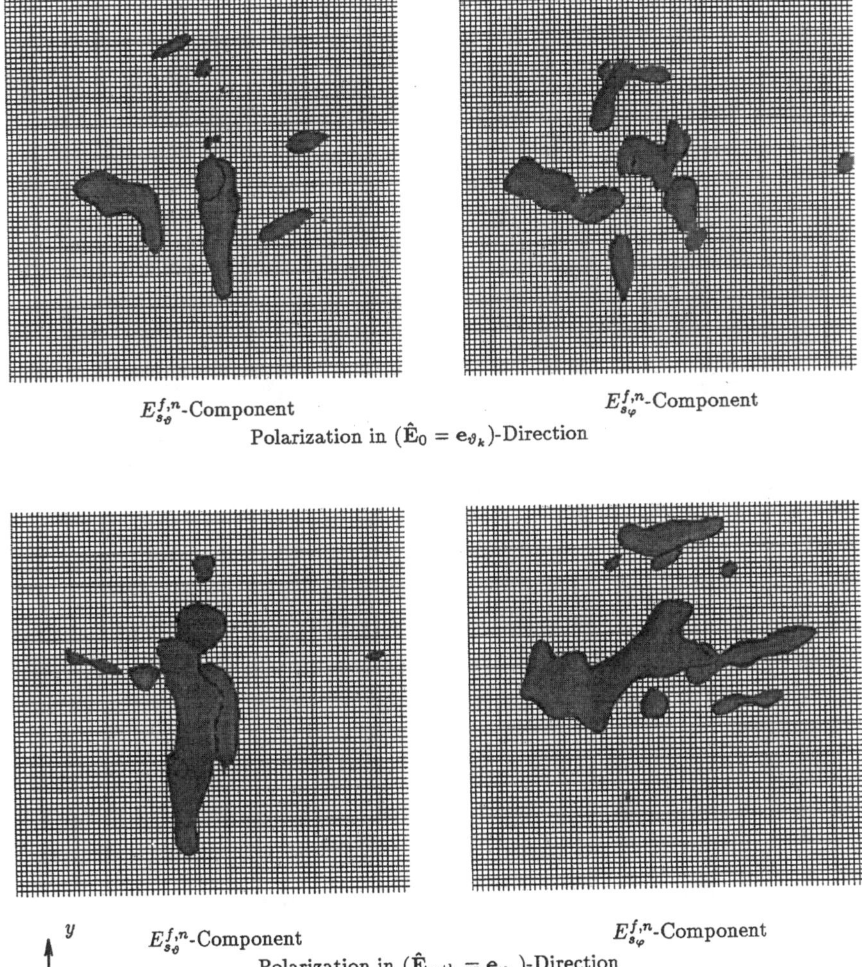

Figure 16. Result of scalar inversion for airplane model with incidence from $\vartheta_i = 37°$, $\varphi_i = 232°$: electric-field component (left) ϑ and (right) φ for (top) ($\hat{\mathbf{E}}_0 = \vartheta_i$)-polarization and (bottom) ($\hat{\mathbf{E}}_{\text{orth}} = \varphi_i$)-polarization.

4.3 Implementation of the algorithm

The above procedure has been implemented according to the flowchart of Fig. 3. Presently, we are only using synthetic data, but the algorithm is ready to work on experimental data also. The R'-multiplied transformed data $\mathbf{E}_s^{f,n}$ are Fourier transformed with respect to t and deconvolved by simple division with the Gaussian spectrum of the incident plane wave; the upper index *PO* indicates data for the physical-optics algorithm under the Kirchhoff approximation for perfectly conducting scatterers, whereas *Bo* stands for its counterpart if the Born approximation is applied to dielectric scatterers. Multiplication with $j\omega$, followed by an inverse fast Fourier transform (FFT), results in band-limited (bl) time functions, which then are subject to the dyadic backpropagation (37) for every point $\mathbf{R}$ out of the reconstruction or image volume V_R. The angular ϕ'- and θ'-integrations over the far-field sphere of radius R_0 are performed numerically. The result are the vector components of the equivalent current $\mathbf{J}_c(\mathbf{R})$ or its magnitude $|\mathbf{J}_c(\mathbf{R})|$, where an upper index bl indicates that we have used band-limited time-domain

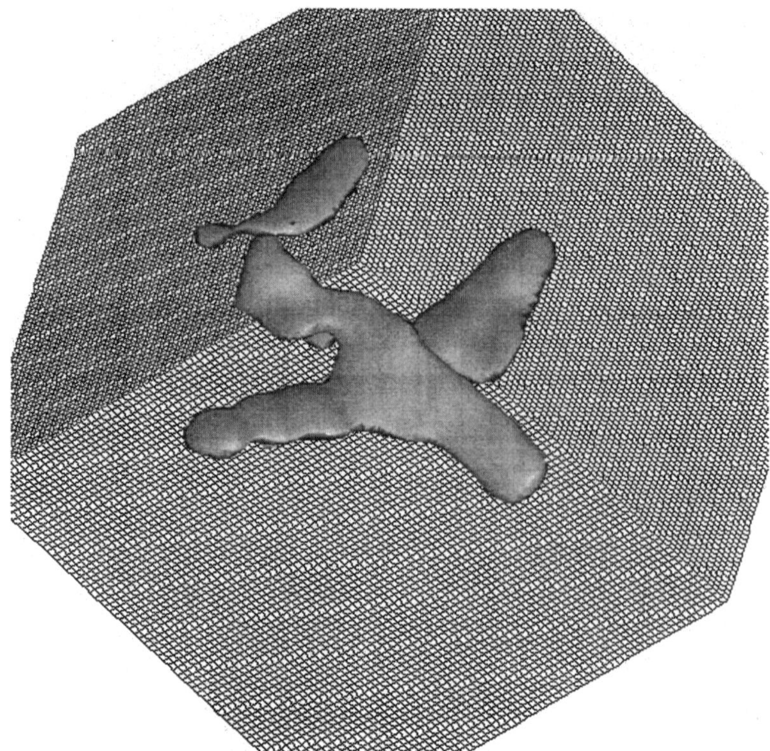

Figure 17. A 3-D representation of singular function magnitude with incidence from $\vartheta_i = 37^\circ$, $\varphi_i = 232^\circ$.

data, whose spectrum does only extend to a maximum frequency ω_{max}. If a Born-type scatterer has been under investigation, the output is the so-called object function $O(\mathbf{R})$ (Langenberg et al., 1994a; Fritsch, 1994).

If we take the projections of $\mathbf{J}_c(\mathbf{R})$ on the known orthogonal directions $\hat{\mathbf{k}}_i$ and $\hat{\mathbf{E}}_0$ we can find the projections $\hat{\mathbf{E}}_0 \cdot \boldsymbol{\gamma}_u(\mathbf{R})$ and $-\hat{\mathbf{k}}_i \cdot \boldsymbol{\gamma}_u(\mathbf{R})$ of the vector singular function; switching from the incident polarization $\hat{\mathbf{E}}_0$ to $\hat{\mathbf{E}}_{\text{orth}}$, we can find the third component $\hat{\mathbf{E}}_{\text{orth}} \cdot \boldsymbol{\gamma}_u(\mathbf{R})$ of the vector singular function in the orthonormal set $\hat{\mathbf{k}}_i$, $\hat{\mathbf{E}}_0$, $\hat{\mathbf{E}}_{\text{orth}}$, and, hence, $\boldsymbol{\gamma}_u(\mathbf{R})$ itself.

4.4 *Perfectly conducting sphere*

Because we had the analytical data for the perfectly conducting sphere already available, this provided a means to check the relatively complex combination of the MAFIA simulation code with a nontrivial inversion algorithm. Therefore, we carefully compared the output if either MAFIA or analytical data were used as input. Fig. 4 shows electric far-field components in the time domain for a selected observation point if

- computed from the eigenfunction expansion directly in the far-field, or
- computed via the near-field far-field transform from analytical data, or, finally,
- computed from the near-field far-field transform applied to MAFIA data.

(In absolute units we have the following parameters: The diameter of the sphere is 0.5 m sitting in the middle of a MAFIA grid box of side lengths 1 m; the incident Gaussian impulse has a duration of 0.4 ns extending to a maximum frequency of 2.4

GHz, which therefore relates to a wavelength of 0.13 m.) The results are convincing. Next, Fig. 5 exhibits two orthogonal slices through the 3-D image volume for the magnitude $|\mathbf{J}_c(\mathbf{R})|$ of the equivalent current density, once obtained through Eq. (37) from analytical data, and once from MAFIA data; the incident wave is coming from the top. Again, there is hardly any difference. Of course, only the top of the sphere is "visible" for a single illumination angle, because there is practically no current in the shadow region for such a short time signal. The oscillatory structure of the images is a result of the band-limitation of the excitation, and the artifacts originate from the Kirchhoff approximation, which does *not* account for shadow-region effects such as creeping waves. These facts have already been thoroughly discussed for the scalar case (Herman et al., 1987); here, we want to concentrate on polarization effects. Therefore, Figs. 6 and 7 display the sphere results—this time, MAFIA only—in a more intuitive, 3-D way for two orthogonal linear polarizations of the incident field—Fig. 6 for $\hat{\mathbf{E}}_0$ as indicated, and Fig. 7 for $\hat{\mathbf{E}}_{\mathrm{orth}}$. Both results can be superimposed to the magnitude of the singular function, which is given in Fig. 8. This is obvioulsy the most complete top of the sphere.

As in Langenberg et al. (1994b), we ask whether the polarimetric image is superior to a purely scalar image. Again, we take either the θ- or the ϕ-component of the electric far-field as a scalar quantity and apply the scalar time-domain back-projection scheme (Herman et al., 1987); the results are given in terms of two orthogonal slices through the 3-D volume in Fig. 9 and as 3-D representations for either the θ- (Fig. 10) or the ϕ-component (Fig. 11) keeping the polarization fixed to $\hat{\mathbf{E}}_0$. The polarimetric result of Fig. 6 is not only a coherent superposition of Figs. 10 and 11, it additionally avoids the artifact caps, and hence, polarimetric inversion surely can be considered as better-than-scalar inversion, even for the highly symmetrical case of a sphere. [In Langenberg et al. (1994b) these artifacts were not observed, because we displayed the image data *on* the spherical surface, having the a priori information available that the target *was* a sphere.]

5 Results for an airplane model

Figure 12 displays an airplane model as designed by MAFIA; it has the same relative length with regard to the duration of the incident Gaussian pulse as the sphere, i.e. the spatial duration of the pulse is approximately one-fifth of the airplane extension from nose to tail. For shorter pulses the MAFIA memory requirements for 3-D modeling are beyond the capacity of present-day workstations.

Let us first present the scalar results in Fig. 13, when the plane wave comes vertically from the top: Obviously, the scalar images (top views into the 3-D image volume) have complimentary information about the target—as was the case for the sphere—and switching to the orthogonal polarization is approximately equivalent to a switch to the orthogonal field component. Figure 14 presents the magnitude of the singular function as output of the polarimetric inversion algorithm in the same way as in Fig. 13, and, as a matter of fact, the airplane is completely recovered (at least its top surface). This is even more impressive if the view into the 3-D image volume is displayed under a different angle (Fig. 15).

We did not hesitate to investigate the polarimetric imaging scheme critically. First, we chose a nonvertical illumination angle, and Fig. 16 reveals that scalar imaging is nearly useless, whereas the polarimetric scheme (Fig. 17) still delivers a flying aircraft,

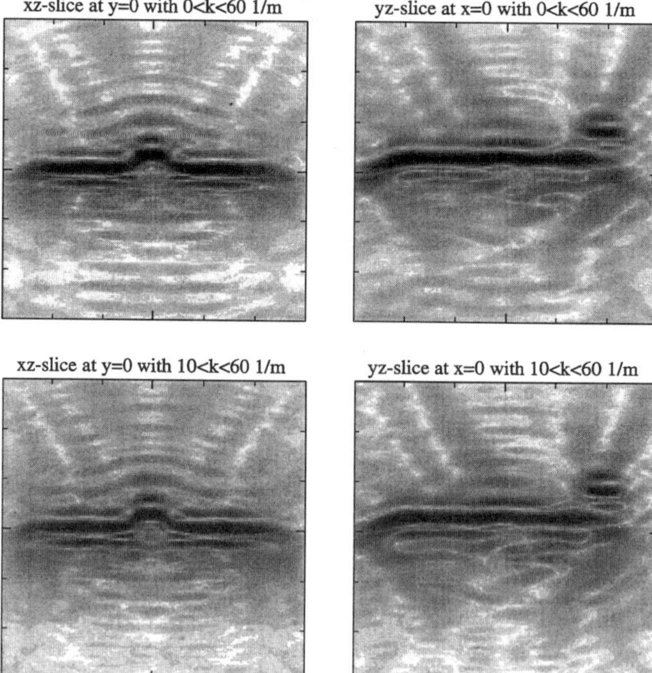

Figure 18. Two orthogonal slices through singular function image volume with (top) full bandwidth $0 \leq k \leq 60$ m^{-1} and (bottom) restricted bandwidth $10 \leq k \leq 60$ m^{-1}.

maybe not in the sense of an aerodynamical point of view but from an imaging point of view.

Figures 18 and 19 report numerical simulations with restricted frequency bandwidths retaining the complete "measurement" aperture of 4π radians. For a proper display, we have chosen two orthogonal slices out of the 3-D volume of Fig. 15 for the full bandwidth of $0 \leq ka \leq 60$ (Fig. 18), which then is to be compared to the restricted bandwidths of $10 \leq ka \leq 60$ (Fig. 18), $20 \leq ka \leq 40$, and $40 \leq ka \leq 60$ (Fig. 19). As already well known for the scalar case (Herman et al., 1987), bandwidth reduction results in an increasing oscillatory structure of the image, and it becomes difficult to display these images three-dimensionally because a predefined isocontour is not appropriate. Potzkai (1994) also carried out simulations for measurement aperture limitations, and as always for algorithms based on physical optics, specular reflections of airplane components must be recorded to provide their images properly. Polarization does not replace this missing information.

6 Conclusions

We have demonstrated with the help of simulations for a rather complex target that EM polarimetric imaging is far superior to scalar imaging. This was done for the frequency diversity mode of operation using a single illumination angle. In Rohrmoser (1995), the counterpart of single frequency and angular diversity has been formulated and investigated. In addition, to complete the series of algorithms, Brandfaß (1996) contains the derivation of the polarimetric Fourier Diffraction Slice Theorem [compare

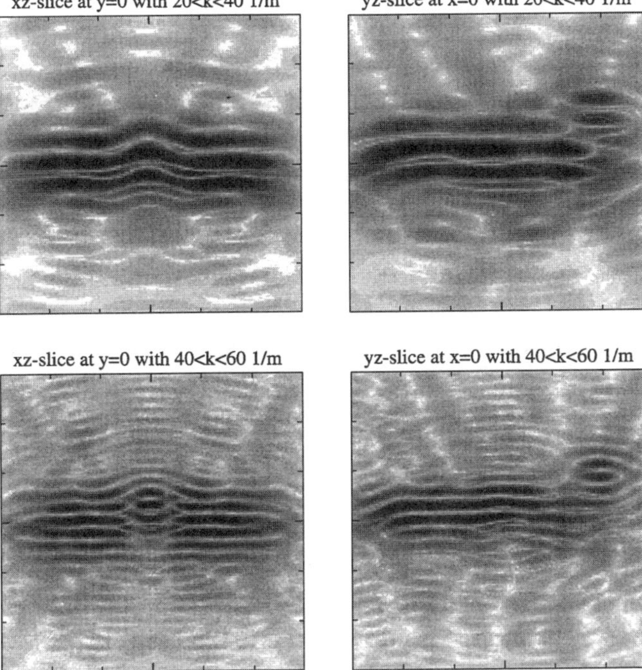

Figure 19. Two orthogonal slices through singular-function image volume for (top) bandwidths of $20 \leq k \leq 40$ m^{-1} and (bottom) restricted bandwidth $40 \leq k \leq 60$ m^{-1}.

Herman et al. (1987) for the scalar version], which allows for a Fourier inversion of *near-field* data.

References

Brandfaß, M., 1989, Breitbandige polarimetrische Holographie mit elektromagnetischen Wellen am Beispiel einer idealleitenden Kugel: M.S. thesis, Univ. of Kassel.

——— 1996, Inverse Beugungstheorie elektromagnetischer Wellen: Algorithmen und numerische Realisierung: Ph.D. thesis, Univ. of Kassel.

Chen, H. C., 1983, Theory of electromagnetic waves: McGraw-Hill Book Co.

Fritsch, A., 1994, Elektromagnetische Fernfeldinversion zur dreidimensionalen Rekonstruktion beliebiger Streukörpergeometrien: M.S. thesis, Univ. of Kassel.

Herman, G. T., Tuy, H. K., Langenberg, K. J., and Sabatier, P., 1987, Basic methods of tomography and inverse problems: Adam Hilger, Bristol 1987.

Langenberg, K. J., 1989, Introduction to the special issue on inverse problems, Wave Motion, 11, 99–112.

Langenberg, K. J., Brandfaß, M., Mayer, K., Kreutter, T., Brüll, A., Fellinger, P., and Huo, D., 1993a, Principles of microwave imaging and inverse scattering, EARSeL Advances in Remote Sensing, **2**, 163–186.

Langenberg, K. J., Brandfaß, M., Fellinger, P., Gurke, T., Kreutter, T., 1994a, A unified theory of multidimensional electromagnetic vector inverse scattering within the Kirchhoff or Born approximation, *in* Boerner W.-M., and Überall: H. Eds., Radar target imaging, Springer-Verlag, 113–151.

Langenberg, K. J., Brandfaß, M., and Fritsch, A., 1994b, Vector diffraction tomography: Algorithmic check against FDTD data; presented at Prog. in Electromagn. Res. Symp., European Space Agency, Noordwijk, The Netherlands.

Langenberg, K. J., Fellinger, P., Marklein, R., Zanger, P., Mayer, K., and Kreutter, T., 1993b, Inverse methods and imaging, *in* J. D. Achenbach, Ed., Evaluation of materials and structures by quantitative ultrasonics, Springer–Verlag, 317–398.

MAFIA User Guide, 1991, Release 3.1: CST GmbH.

Potzkai, B., 1994, Elektromagnetische Inversion zur dreidimensionalen Rekonstruktion beliebiger Streukörpergeometrien im Zweischichtmodell: M.S. thesis, Univ. of Kassel.

Rohrmoser, A., 1995, Dreidimensionale Rekonstruktionen mit Hilfe von Winkeldiversitätsformulierungen in Born- und Kirchhoff-Näherung mit elektromagnetischen Fernfeldstreudaten für beliebige Streukörper am Beispiel des Kugelstreuers: M.S. thesis, Univ. of Kassel.

Theoretical Inverse Problems for 3-D Electromagnetic Fields

P. S. Martyshko

Summary. Theoretical inverse problem is the terminology (in the Russian literature) for a geophysical inverse problem in which the field is given by an explicit expression. This can arise, for example, when data are approximated by singular sources (monopoles, dipoles, etc.) in a half-space. This chapter derives explicit integrodifferential equations of theoretical inverse problems for 3-D electromagnetic fields satisfying the Helmholtz, telegraphic, and diffusion equations. Functional equations for a perfect conductor are presented, and some numerical examples are given.

1 Introduction

Inversion of electromagnetic (EM) data in geophysical prospecting involves solution of a nonlinear operator equation of the first kind (with an implicit, often ill-conditioned, operator). Numerical solution of such equations can require considerable computer time (Martyshko, 1983a). For the theoretical inverse problem (TIP) in electrical prospecting with a dc current, however, the author was able to obtain explicit integrodifferential equations for the electric and magnetic fields (Martyshko, 1983b, 1986a,b), develop effective algorithms for solving these equations, and construct examples of equivalent regions. A TIP is one in which the governing fields are specified explicitly, usually as the field of singular sources lying in a half-space. Solution of a TIP can be the last step of interpretation methods that first approximate observed data with the fields of singular sources (Strakhov, 1974; Nikonova and Tsirul'skiy, 1978). It also makes possible the construction of geologically meaningful equivalents for different classes of singular sources. The TIP equations and their derivation both merit attention. TIP equations are derived for EM fields satisfying the Helmholtz and telegrapher's equations. The derivation uses representations of fields through their values and derivatives at the boundary of the anomalous object (Martyshko, 1986a), the main formula of the theory of harmonic functions (Martyshko, 1983b, 1986b), and the Stratton-Chu formulas (Stratton, 1948).

Geophysics Institute, Ural Department, Russian Academy of Sciences, Amundsen Str 100. 620016 Ekaterinburg, Russia.

2 Formulation of the inverse problem

Assume that in a linear isotropic medium with conductivity σ_1, permeability μ_1, and permittivity ϵ_1, there is an inclusion, a body T with parameters σ_2, μ_2, and ϵ_2. Also assume that in the medium there are sources generating electromagnetic fields, $(\mathbf{H}_1, \mathbf{E}_1)$ and $(\mathbf{H}_2, \mathbf{E}_2)$, outside and inside the conducting inclusion, respectively. We assume that T is a 3-D region, S is its boundary, and $\mathbf{r}$ is the radius-vector of a point in R^3.

In a linear isotropic medium (σ, ϵ, μ are constant) without impressed sources, the magnetic and electric fields $(\mathbf{H}, \mathbf{E})$ satisfy Maxwell's equations:

$$\begin{aligned} \nabla \times \mathbf{H} &= \mathbf{j} + \frac{\partial \mathbf{D}}{\partial t}, \\ \nabla \times \mathbf{E} &= \frac{\partial \mathbf{B}}{\partial t}, \\ \nabla \cdot \mathbf{B} = 0, &\qquad \nabla \cdot \mathbf{D} = 0, \end{aligned} \tag{1}$$

where the consititutive relations are $\mathbf{D} = \epsilon \mathbf{E}, \mathbf{B} = \mu \mathbf{H}, \mathbf{j} = \sigma \mathbf{E}$. On boundary S between the body and the surrounding region, the following jump conditions hold:

$$\begin{aligned} [\mathbf{H}_2 - \mathbf{H}_1, \mathbf{n}] = 0, &\qquad (\mu_2 \mathbf{H}_2 - \mu_1 \mathbf{H}_1, \mathbf{n}) = 0, \\ [\mathbf{E}_2 - \mathbf{E}_1, \mathbf{n}] = 0, &\qquad (\epsilon_2 \mathbf{E}_2 - \epsilon_1 \mathbf{E}_1, \mathbf{n}) = \eta, \end{aligned} \tag{2}$$

where η is surface density of the electrical charge. In this notation, $[\mathbf{a}, \mathbf{b}] = \mathbf{a} \times \mathbf{b}$ is the cross product of two vectors, and $(\mathbf{a}, \mathbf{b}) = \mathbf{a} \cdot \mathbf{b}$ is the scalar product.

Assume that $\mathbf{E}_1^\alpha$, $\mathbf{H}_1^\alpha$ is the scattered field outside of T induced by impressed sources, whose fields (in the absence of T) are $\mathbf{H}^S$ and $\mathbf{E}^S$. Then,

$$\mathbf{H}_1 = \mathbf{H}_1^\alpha + \mathbf{H}^S, \qquad \mathbf{E}_1 = \mathbf{E}_1^\alpha + \mathbf{E}^S,$$

$$\left[\mathbf{H}^S + \mathbf{H}_1^\alpha, \mathbf{n}\right] = [\mathbf{H}_2, \mathbf{n}], \qquad \left[\mathbf{E}^S + \mathbf{E}_1^\alpha, \mathbf{n}\right] = [\mathbf{E}_2, \mathbf{n}], \tag{3}$$

$$(\mathbf{H}_2, \mathbf{n}) = \frac{\mu_1}{\mu_2}\left(\mathbf{H}_1^\alpha + \mathbf{H}^S, \mathbf{n}\right), \qquad (\mathbf{E}_2, \mathbf{n}) = [\epsilon_1 (\mathbf{E}_1, \mathbf{n}) + \eta]/\epsilon_2. \tag{4}$$

Equations (1) and the constitutive relations combine to give the telegrapher's equation in homogeneous regions without impressed sources,

$$\Delta \mathbf{F} - \mu\epsilon \frac{\partial^2 \mathbf{F}}{\partial t^2} - \mu\sigma \frac{\partial \mathbf{F}}{\partial t} = 0. \tag{5}$$

Thus $\mathbf{H}_1^\alpha$ and $\mathbf{E}_1^\alpha$ satisfy Eq. (5), with $\sigma = \sigma_1, \mu = \mu_1, \epsilon = \epsilon_1$, outside T; and $\mathbf{H}_2$ and $\mathbf{E}_2$ satisfy Eq. (5), with $\sigma = \sigma_2, \mu = \mu_2, \epsilon = \epsilon_2$, inside T.

Monochromatic $[\exp(i\omega t)]$ fields satisfy

$$\nabla \times \mathbf{H} = \sigma^* \mathbf{E}, \qquad \nabla \times \mathbf{E} = i\omega\mu \mathbf{H}, \qquad \nabla \cdot \mathbf{H} = 0, \qquad \nabla \cdot \mathbf{E} = 0, \tag{6}$$

where $\sigma^* = \sigma - i\omega\epsilon$ is complex conductivity; in this case Eq. (5) becomes the Helmholtz equation

$$\Delta \mathbf{F} + k^{*2} \mathbf{F} = 0, \tag{7}$$

where the (squared) wave number of the medium is $k^{*2} = i\omega\mu\sigma + \omega^2 \mu\epsilon$.

Quasi-stationary fields (i.e., without displacement currents) satisfy equations

$$\nabla \times \mathbf{H} = \sigma \mathbf{E}, \qquad \nabla \times \mathbf{E} = -\mu \frac{\partial \mathbf{H}}{\partial t}, \tag{8}$$

which combine to give diffusion equations for **E** and **H**:

$$\Delta \mathbf{F} - \mu\sigma \frac{\partial \mathbf{F}}{\partial t} = 0. \tag{9}$$

For the time-dependent Eqs. (5) and (6), there are, of course, appropriate initial conditions on $\mathbf{E}_1^\alpha$ and $\mathbf{H}_1^\alpha$.

The inverse problem can be formulated in the following way: Given the vector-function $\mathbf{E}_1^\alpha$ (or $\mathbf{H}_1^\alpha$), outside T satisfying Eqs. (5) [(6) or (7)], and the boundary conditions (4), find the region T or boundary S.

3 Integrodifferential equations for inverse problem

Tsirul'skiy (1974) proposed that the inverse problem be called theoretical if $\mathbf{H}_1^\alpha$ (or $\mathbf{E}_1^\alpha$) is given in explicit form. This situation can arise in practice after approximating measured data with functions having singularities belonging to the lower half-space. (As a rule they are the fields of singular sources: monopole, dipole, etc.)

Assume that **F** and P are arbitrary functions twice continuously differentiable in T (up to the boundary S). Then, the following integral relation holds (Zhdanov, 1984):

$$\begin{aligned}&\int_r (\Delta P \cdot \mathbf{F} + \nabla P \nabla \cdot \mathbf{F} + P \nabla \times \nabla \times \mathbf{F})\, dv \\ &\quad = \int_S \{(\mathbf{n}, \mathbf{F})\nabla P + [\mathbf{n}, \mathbf{F}] \times \nabla P + [\mathbf{n}, \nabla \times \mathbf{F}]P\}\, ds.\end{aligned} \tag{10}$$

Formula (10) plays an important role in the theory of geophysical fields because it provides a representation of a field through its boundary values. [To show this, it is sufficient to take a fundamental solution of the equation for **F** in Eq. (10).] Such representations in turn can be used for deriving TIP equations.

The Stratton-Chu formulas derived for a monochromatic field from Eq. (10) are The Stratton-Chu formulas derived for a monochromatic field from Eq. (10) are

$$\int_S \{(\mathbf{n}, \mathbf{E}_2)\nabla G_2 + [\mathbf{n}, \mathbf{E}_2] \times \nabla G_2 + i\omega\mu_2[\mathbf{n}, \mathbf{H}_2]G_2\}\, ds = \begin{cases} \mathbf{E}_2(\mathbf{r}'), & \mathbf{r}' \in T \\ 0, & \mathbf{r}' \in C\bar{T} \end{cases}, \tag{11}$$

$$\int_S \left\{(\mathbf{n}, \mathbf{E}_1^\alpha)\nabla G_1 + \left[\mathbf{n}, \mathbf{E}_1^\alpha\right] \times \nabla G_1 + i\omega\mu_1\left[\mathbf{n}, \mathbf{H}_1^\alpha\right]G_1\right\} ds = \begin{cases} 0, & \mathbf{r}' \in T \\ -\mathbf{E}_1^\alpha(\mathbf{r}'), & \mathbf{r}' \in C\bar{T} \end{cases}, \tag{12}$$

$$\int_S \{(\mathbf{n}, \mathbf{H}_2)\nabla G_2 + [\mathbf{n}, \mathbf{H}_2] \times \nabla G_2 + \sigma_2^*[\mathbf{n}, \mathbf{E}_2]G_2\}\, ds = \begin{cases} \mathbf{H}_2(\mathbf{r}'), & \mathbf{r}' \in T \\ 0, & \mathbf{r}' \in C\bar{T} \end{cases}, \tag{13}$$

$$\int_S \left\{(\mathbf{n}, \mathbf{H}_1^\alpha)\nabla G_1 + \left[\mathbf{n}, \mathbf{H}_1^\alpha\right] \times \nabla G_1 + \sigma_1^*\left[\mathbf{n}, \mathbf{E}_1^\alpha\right]G_1\right\} ds = \begin{cases} -\mathbf{H}_1^\alpha(\mathbf{r}'), & \mathbf{r}' \in T \\ 0, & \mathbf{r}' \in C\bar{T} \end{cases}, \tag{14}$$

where

$$G_{1,2} = -\frac{e^{ik_{1,2}^*|\mathbf{r}-\mathbf{r}'|}}{4\pi|\mathbf{r}-\mathbf{r}'|}. \tag{15}$$

The notation $\mathbf{r} \in C\bar{T}$ means that **r** lies outside T (and its boundary S).

Subtracting Eq. (12) from Eq. (11) when $\mathbf{r}' \in C\bar{T}$ and using Eqs. (4) and (6) gives

$$\mathbf{E}_1^\alpha(\mathbf{r}') = \int_S \Big\{ (\mathbf{n}, \mathbf{E}_1^\alpha)\nabla(G_2 - G_1) + [\mathbf{n}, \mathbf{E}^S] \times \nabla G_2 + i\omega[\mathbf{n}, \mathbf{H}_1^\alpha](\mu_2 G_2 - \mu_1 G_1) + \mu_2[\mathbf{n}, \mathbf{H}^S]G_2 + (\mathbf{n}, \mathbf{E}^\alpha)\Big[\nabla\Big(\frac{\epsilon_1}{\epsilon_2}G_2 - G_1\Big)\Big] + \frac{\nabla G_2}{\epsilon_2}[\eta + (\mathbf{n}, \mathbf{E}^S)\epsilon_1] \Big\} ds, \tag{16}$$

where $\eta = (\epsilon_1/\sigma_1^* - \epsilon_2/\sigma_2^*)\nabla_S \cdot [\mathbf{n}, \mathbf{H}_1]$, since $(\mathbf{n}, \mathbf{E}|_S) = -(1/\sigma^*)\nabla_S \cdot [\mathbf{n}, \mathbf{H}]$. Similarly, Eqs. (13) and (14) give

$$\mathbf{H}_1^\alpha(\mathbf{r}') = \int_S \Big\{ (\mathbf{n}, \mathbf{H}_1^\alpha)\nabla\Big(\frac{\mu_1}{\mu_2}G_2 - G_1\Big) + \frac{\mu_1}{\mu_2}(\mathbf{n}, \mathbf{H}^S)\nabla G_2 + [\mathbf{n}, \mathbf{H}_1^\alpha] \times \nabla(G_2 - G_1) + [\mathbf{n}, \mathbf{H}^S] \times \nabla G_2 + [\mathbf{n}, \mathbf{E}_1^\alpha](\sigma_2^* G_2 - \sigma_1^* G_1) + [\mathbf{n}, \mathbf{E}^S]\sigma_2^* G_2 \Big\} ds. \tag{17}$$

Relations (16) and (17) are the equations of the TIP for a monochromatic field (relative to the boundary S). The material properties of the anomalous region are assumed to be parameters; i.e., the solution of the TIP holds for various values $\sigma_2, \epsilon_2, \mu_2$. The result is an equivalent family of bodies that generate the same electric or magnetic field. For a stationary field ($\omega = 0$, $k^* = 0$), equations of the inverse problem for dc electrical prospecting already have been derived from Eqs. (16) and (17) (Martyshko, 1983a, 1986a):

$$\mathbf{E}_1^\alpha(\mathbf{r}') = \frac{1}{4\pi}\frac{\sigma_1 - \sigma_2}{\sigma_2}\int_S (\mathbf{n}, \mathbf{E}_1)\nabla\frac{1}{|\mathbf{r} - \mathbf{r}'|}\, ds, \tag{18}$$

$$\mathbf{H}_1^\alpha(\mathbf{r}') = \frac{1}{4\pi}\int_S \Big\{ \frac{\mu_1 - \mu_2}{\mu_2}(\mathbf{n}, \mathbf{H}_1)\nabla\frac{1}{|\mathbf{r} - \mathbf{r}'|} + \frac{\sigma_2 - \sigma_1}{\sigma_1}\frac{[\mathbf{n}, \nabla \times \mathbf{H}_1^\alpha]}{|\mathbf{r} - \mathbf{r}'|} \Big\} ds. \tag{19}$$

If modified Stratton-Chu integrals (Khenl et al., 1964) are used for a monochromatic field, it is possible to derive simpler equations not containing the normal components of $\mathbf{E}$ and $\mathbf{H}$ under the integral:

$$\mathbf{E}_1^\alpha(\mathbf{r}') = \nabla' \times \nabla' \times \int_S \Big\{ [\mathbf{n}, \mathbf{H}_1^\alpha]\Big(\frac{G_2}{\sigma_2^*} - \frac{G_1}{\sigma_1^*}\Big) + \frac{G_2}{\sigma_1^*}[\mathbf{n}, \mathbf{H}^S] \Big\} ds + \nabla' \times \int_S \{[\mathbf{n}, \mathbf{E}_1^\alpha](G_2 - G_1) + G_2[\mathbf{n}, \mathbf{E}^S]\}\, ds, \tag{20}$$

$$\mathbf{H}_1^\alpha(\mathbf{r}') = \frac{1}{i\omega}\nabla' \times \nabla' \times \int_S \Big\{ [\mathbf{n}, \mathbf{E}_1^\alpha]\Big(\frac{G_2}{\mu_2} - \frac{G_1}{\mu_1}\Big) + \frac{G_2}{\mu_2}[\mathbf{n}, \mathbf{E}^S] \Big\} ds + \nabla' \times \int_S \{[\mathbf{n}, \mathbf{H}_1^\alpha](G_2 - G_1) + G_2[\mathbf{n}, \mathbf{H}^S]\}\, ds. \tag{21}$$

Equations (16), (17), (20), (21) are all TIP equations (relative to the boundary S) for a monochromatic field. In their numerical solution, it is possible to use the algorithm formulated in Martyshko (1986a).

The case $\sigma_2 = \infty$ (T is a perfect conductor) is of special interest. Then, at the boundary, we have

$$(\mathbf{H}_1, \mathbf{n}) = 0, \qquad [\mathbf{E}_1, \mathbf{n}] = 0. \tag{22}$$

These are TIP functional equations for determining the surface of a perfect conductor (Martyshko, 1986a) and can be a good starting approximation in solving Eqs. (16), (17),

(20), (21). In Zhdanov (1984), the theory of Stratton-Chu integrals for a monochromatic field is generalized to inhomogeneous media. [Specifically, a magnetically homogeneous medium with a piecewise-continuous distribution of conductivity $\sigma^*(\mathbf{r})$ is considered, and the region T is bounded by a smooth surface S.] An expression was derived for the EM field on the basis of its values at the surface using Green's electromagnetic tensors:

$$\mathbf{E}(\mathbf{r}') = \int_S \{i\omega\mu \underset{\sim}{\mathbf{G}}^e(\mathbf{r}' \mid \mathbf{r})[\mathbf{n}, \mathbf{H}] + \tilde{\Delta} \times \underset{\sim}{\mathbf{G}}^e(\mathbf{r}' \mid \mathbf{r})[\mathbf{n}, \mathbf{E}]\}\, ds, \tag{23}$$

$$\mathbf{H}(\mathbf{r}') = \int_S \{i\omega\mu \underset{\sim}{\mathbf{G}}^m(\mathbf{r}' \mid \mathbf{r})[\mathbf{n}, \mathbf{H}] + \tilde{\Delta} \times \underset{\sim}{\mathbf{G}}^m(\mathbf{r}' \mid \mathbf{r})[\mathbf{n}, \mathbf{E}]\}\, ds, \tag{24}$$

where $\underset{\sim}{\mathbf{G}}^e, \underset{\sim}{\mathbf{G}}^m$ are tensor functions of electrical and magnetic types, and $\mathbf{r}' \in T$. If $\mathbf{r}' \in C\bar{T}$, the integrals in Eqs. (23) and (24) are equal to zero.

In an arbitrary medium, determination of $\underset{\sim}{\mathbf{G}}^e$ and $\underset{\sim}{\mathbf{G}}^m$ is difficult. However, for some models, such as a layered medium, this problem can be solved, and accordingly it is possible to write the TIP equation.

Comment. If T is bounded by a cylindrical surface S whose generatrix is parallel to the y axis and the EM field $(\mathbf{E}, \mathbf{H})$ is uniform along this axis, the problem becomes two-dimensional (2-D). The 2-D equations retain the structure of the 3-D equations: Integration is carried out along a closed contour and $G(\mathbf{r}' \mid \mathbf{r}) = -(i/4)H_0^{(1)}(k^*|\mathbf{r} - \mathbf{r}'|)$, where $H_0^{(1)}$ is a zero-order Hankel function of the first kind.

For a quasi-stationary field, the Stratton-Chu integrals have the form

$$\int_{-\infty}^{t'} \int_S \left\{ (\mathbf{n}, \mathbf{E})\nabla G^d + [\mathbf{n}, \mathbf{E}] \times \nabla G^d + \mu[\mathbf{n}, \mathbf{H}] \frac{\partial G^d}{\partial t} \right\} ds\, dt = \mathbf{E}, \tag{25}$$

$$\int_{-\infty}^{t'} \int_S \{(\mathbf{n}, \mathbf{H})\nabla G^d + [\mathbf{n}, \mathbf{H}] \times \nabla G^d + \sigma[\mathbf{n}, \mathbf{E}]G^d\}\, ds\, dt = \mathbf{H}, \tag{26}$$

where G^d is the Green's function for the diffusion equation. With Eqs. (25) and (26), it is possible to derive a TIP equation for a quasi-stationary field:

$$\begin{aligned}
\mathbf{E}_1^a(\mathbf{r}', t') = \int_{-\infty}^{t'} \int_S \Bigg\{ & \left[\mathbf{n}, \mathbf{E}_1^a\right] \times \nabla\left(G_2^d - G_1^d\right) + \left[\mathbf{n}, \mathbf{H}_1^a\right]\left(\mu_2 \frac{\partial G_2^d}{\partial t} - \mu_1 \frac{\partial G_1}{\partial t}\right) \\
& + \left(\mathbf{n}, \mathbf{E}_1^a\right)\left[\nabla\left(\frac{\epsilon_1}{\epsilon_2} G_2^d - G_1^d\right)\right] + \frac{\nabla G_2}{\epsilon_2}[\eta + \epsilon_1(\mathbf{n}, \mathbf{E}^S)] \Bigg\}\, ds\, dt,
\end{aligned} \tag{27}$$

$$\begin{aligned}
\mathbf{H}_1^a(\mathbf{r}', t') = \int_{-\infty}^{t'} \int_S \Bigg\{ & \left(\mathbf{n}, \mathbf{H}_1^a\right)\left[\nabla\left(\frac{\mu_1}{\mu_2} G_2^d - G_1^d\right)\right] + \frac{\mu_1}{\mu_2}(\mathbf{H}^S, n)\nabla G_2^d + \left[\mathbf{n}, \mathbf{H}_1^a\right] \\
& \times \nabla\left(G_2^d - G_1^d\right) + \left[\mathbf{n}, \mathbf{E}_1^a\right]\left(\sigma_2^* G_2^d - \sigma_1^* G_1^d\right) \Bigg\}\, ds\, dt, \quad \mathbf{r}' \in C\bar{T}.
\end{aligned} \tag{28}$$

Similar representations for $\mathbf{E}_1$ and $\mathbf{H}_1$, when $\mathbf{r}' \in C\bar{T}$, give a TIP equation for nonstationary EM fields.

4 Algorithm for stellate bodies

Bodies stellate in relation to a certain internal point are a broad class important in practical terms. To solve for such bodies, construct a spherical system of coordinates with the center at that point. Performing in Eq. (16), a spherical substitution of variables gives

$$\mathbf{E}_1^\alpha(\mathbf{r}') = \int_0^{2\pi}\int_0^{\pi} \mathbf{U}_{\sigma_2,\epsilon_2}\left[\rho(\Theta,\phi),\Theta,\phi,\mathbf{r}'\right] d\Theta\, d\phi, \tag{29}$$

where **U** is the function under the integral in Eq. (16), and $\rho(\Theta,\phi)$ is the right-hand side of the equation for the body's surface in spherical coordinates. Equation (29) can be regarded as an implict equation for the function $\rho(\Theta,\phi)$.

The right-hand side of the equation for the surface of the body T can be represented by a double Fourier series,

$$\rho(\Theta,\phi) = \sum_{k=-\infty}^{\infty}\sum_{j=-\infty}^{\infty} a_{kj} e^{i(k\Theta+j\phi)}. \tag{30}$$

Solution of Eq. (29) thus can be attempted in the form of a (partial) double Fourier series,

$$\rho_{nm}(\Theta,\phi) = \sum_{k=-n}^{n}\sum_{j=-m}^{m} a_{kj} e^{i(k\Theta+j\phi)}. \tag{31}$$

Coefficients are determined by minimizing the functional,

$$f(\gamma) = \sum_{i=1}^{M}\left[\mathbf{E}_1^\alpha(\mathbf{r}_i') - \mathbf{E}_1^{\alpha,nm}(\mathbf{r}_i';\gamma)\right]^2, \tag{32}$$

where the points $\mathbf{r}_i'$, which supply information about the field, can be chosen; γ is the vector of coefficients of the function $\rho(\Theta,\phi)$, for which minimization is performed; and $\mathbf{E}_1^{\alpha,nm}$ is the right-hand side of Eq. (29) substituting by ρ_{nm}.

As was noted by Martyshko (1986a), the function $\rho(\Theta,\phi)$, which defines the boundary of the simply connected 3-D region, must have the following property:

$$\rho(0,\phi) = \text{const}, \qquad \rho(\pi,\phi) = \text{const}. \tag{33}$$

These conditions impose relations on the coefficients of ρ_{nm}:

$$\sum_k a_{kj} = 0, \qquad \sum_{k_1} a_{k_1 j} = 0,\ j \in [-m,m]; \tag{34}$$

$$k, k_1 \in [-n,n], \qquad k = \pm 1, \pm 3, \ldots,$$

$$k_1 = \pm 2, \pm 4, \ldots. \tag{35}$$

The solution is the minimum of the functional f and is subject to these equality constraints. In practice, the functional f is combined with the functional

$$f_1 = \sum_j P_j\left(\left|\sum_k a_{kj}\right|^2 + \left|\sum_{k_1} a_{k_1 j}\right|^2\right),$$

where P_j are the penalty coefficients. Regularization of Eq. (29) is done by Martyshko (1986a) with a first-order smoothing Tikhonov normalizer (Tikhonov and Arsenin, 1977):

$$\Omega(\rho) = \int_0^{2\pi}\int_0^{\pi}\left(\left|\rho_\Theta'\right|^2 + \left|\rho_\phi'\right|^2\right) d\Theta\, d\phi = 2\pi^2 \sum_{k=-n}^{n}\sum_{j=-m}^{m}(k^2+j^2)|a_{kj}|^2. \tag{36}$$

In solving Eq. (29) to test the TIP, the functional

$$f + f_1 + \alpha\Omega, \tag{37}$$

was minimized, where the regularization parameter α was determined from the residual error. Powell's method was used for minimization (Himmelblau, 1975); the integral in Eq. (29) was computed by using formulas of a high trigonometric accuracy, which in this case are formulas with equal weights and equidistant nodes. In solving Eq. (22), a sphere of a minimum radius, including all of the singularities $\mathbf{E}_1^\alpha$, was taken as an initial approximation. The initial approximation then was defined for Eq. (29)—which includes the conductivity σ_2 as parameter, making it possible to construct an entire family of equivalent bodies generating with different conductivities the same field—for a certain value of $\sigma_2 = \sigma_0$. At each subsequent step, the body T_{n-1} was taken as the initial approximation for the conductivity $\sigma^n < \sigma^{n-1}$. A sphere known to contain the solution was chosen for the location of the points $\mathbf{r}_i$.

5 Numerical examples

The algorithm of the preceding section for solving Eqs. (16–22) in the class of stellate bodies was developed by Martyshko (1986a). The TIP solution was represented by a double Fourier series with 25 coefficients. The following theoretical examples were computed ($\mu_1 = \mu_2$):

1. Figure 1 shows the cross-sections on the coordinate planes of the solutions of Eq. (18), where

$$V_1 = \frac{Q_1}{R_1} + \frac{Q_2}{R_2}, \qquad \mathbf{E}_1^\alpha = \nabla V_1, \qquad \mathbf{E}_1 = \nabla\left(V_1 + \frac{Q_3}{R_3}\right)$$

and $Q_1 = 1$, $Q_2 = -1$, $Q_3 = 1$. Solutions of TIP are shown for $\sigma_1/\sigma_2 = 1/5$, $\sigma_1/\sigma_2 = 1/10$ (constructed by dc current charge method).

2. Figure 2 shows the cross-sections on the coordinate planes of the TIP solutions of the Helmholtz equation for $\mathbf{E}_1^\alpha = (E_x, E_y, E_z)$, where

$$E_x = Q_{1x}\frac{e^{ik_2r_1}}{r_1} + Q_{2x}\frac{e^{ik_2r_2}}{r_2} + Q_{3x}\frac{e^{ik_2r_3}}{r_3},$$

$$E_y = Q_{1y}\frac{e^{ik_2r_1}}{r_1} + Q_{2y}\frac{e^{ik_2r_2}}{r_2} + Q_{3y}\frac{e^{ik_2r_3}}{r_3},$$

$$E_z = Q_{1z}\frac{e^{ik_2r_1}}{r_1} + Q_{2z}\frac{e^{ik_2r_2}}{r_2} + Q_{3z}\frac{e^{ik_2r_3}}{r_3},$$

and

$$\mathbf{E}^H(P) = \left(Q_{4x}\frac{e^{ik_1r_4}}{r_4}, Q_{4y}\frac{e^{ik_1r_4}}{r_4}, Q_{4z}\frac{e^{ik_1r_4}}{r_4}\right),$$

where

$$P_1, P_2, P_3 \in T^+; \quad P_4, P \in T^-, r_i = |PP_i|, i = \overline{1,4}; \quad P_4 = (0,0,9);$$

and

$$Q_{1x} = Q_{1y} = Q_{1z} = 1, \qquad Q_{2x} = Q_{2y} = Q_{2z} = 2,$$
$$Q_{3x} = Q_{3y} = Q_{3z} = -3, \qquad Q_{4x} = Q_{4y} = Q_{4z} = -5.$$

Solutions of TIP for $\sigma_1/\sigma_2 = 1/5$, $\sigma_1/\sigma_2 = 1/10$ are shown.

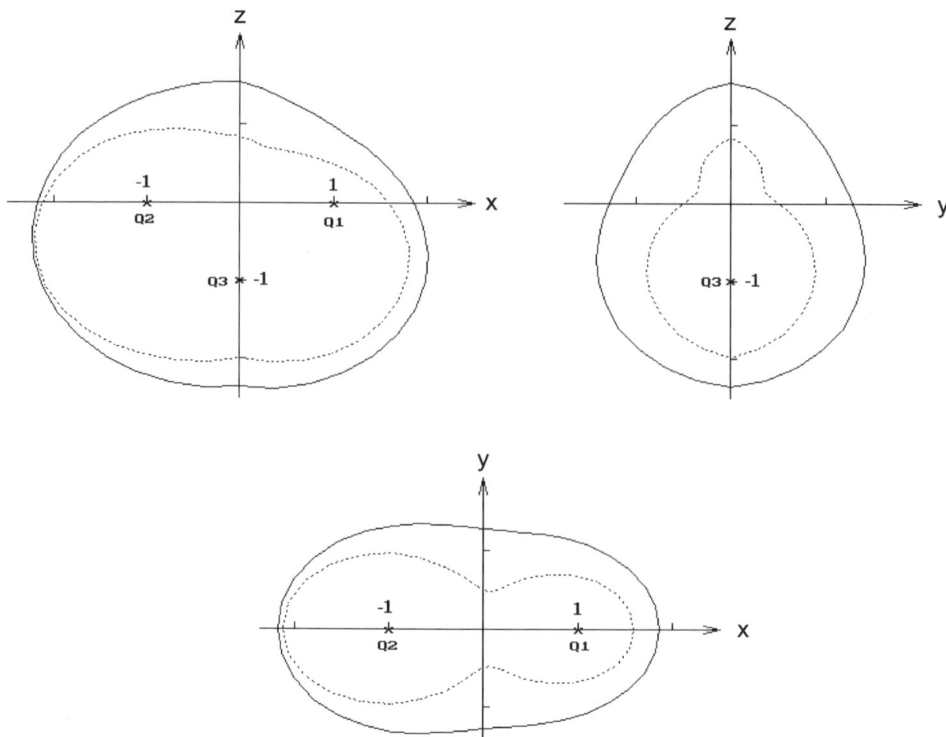

Figure 1. Cross-sections on the coordinate planes for TIP solutions of Eq. (18): Solid line is contour of the body for solution with $\sigma_2/\sigma_1 = 5$; dashed line is for $\sigma_2/\sigma_1 = 10$.

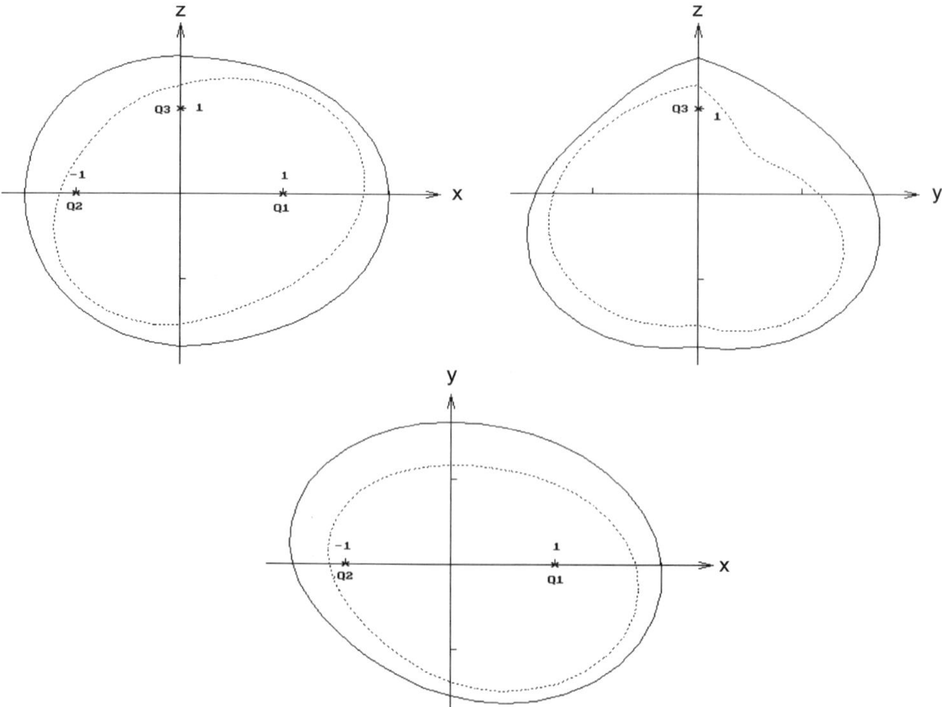

Figure 2. Cross-sections on the coordinate planes for solutions of the Helmholtz equation: Solid line is contour of the body for solution with $\sigma_2/\sigma_1 = 5$; dashed line is for $\sigma_2/\sigma_1 = 10$.

6 Conclusions

Explicit integrodifferential equations have been derived for TIPs for EM fields which satisfy the telegrapher's, diffusion, and Helmholtz equations. The algorithms developed in Martyshko (1986a) can be effective in solving these equations (good results were obtained for monochromatic fields). The algorithms can be used on experimental data together with a method for approximating data with the fields of singular sources. TIP solutions make it possible to develop methods for interpreting EM fields that explicitly take into account the possibility of equivalent solutions.

References

Himmelblau, D., 1975, Applied nonlinear programming: Mir (Russian translation).

Khenl, Kh., Maue, P., and Vestpfal', K., 1964, Teorya difraktsii (Diffraction theory): Mir.

Martyshko, P. S., 1983a, Solution of direct and inverse three-dimensional problems in artificial magnetization method in parametric classes: Izv. Akad. Nauk. SSSR. Fiz. Zemli, **3**, 52.

———1983b, Integrodifferential equations of inverse problem for magnetic field of spreading currents, *in* Metody interpretatsii i matematichescoye modelirovaniye geofizicheskikh poley (Interpretation methods and mathematical simulation of geophysical fields), Sverdlovsk: Uralski Nauchny Tsentr Akad. Nauk. SSSR, 49.

———1986a, Solution of inverse problem in electrical prospecting with dc current for arbitrary classes of potentials: Izv. Akad. Nauk. SSSR. Fiz. Zemli, **1**, 87.

———1986b, Solution of inverse problem for magnetic field of spreading currents, *in* Geologiya i poleznyye iskopayemyye Urala (Geology and Minerals in Urals), Sverdlovsk, Uralski Nauchny Tsentr Akad. Nauk. SSSR, 55.

Nikonova, F. I., and Tsirul'skiy, A. V. , 1978, Interpretation of gravimagnetic anomalies on basis of classes of potentials for which the inverse problem is solvable in finite form: Izv. Akad. Nauk. SSSR Fiz. Zemli, **2**, 49.

Strakhov, V. N., 1974, Functional equations of plane inverse potential problem and numerical approximate solutions of this problem. Dokl. Akad. Nauk. SSSR, 213, **1**, 863.

Stratton, J., 1949, Dzh. Teoriya elektromagnetizma (Electromagnetic theory): Gostoptekhizdat.

Tikhonov, A. N., and Arsenin, V. Y., 1977, Solution of ill-posed problems: W. H. Winston and Sons.

Tsirul'skiy, A. V., 1974, Solution of direct and inverse problems in gravimetric prospecting: Izv. Akad. Nauk. SSSR Fiz. Zemli, **7**, 84–90.

Zhdanov, M. S., 1984, Analogi integrala tipa Koshi v teorii geofizicheskikh poley (Analogs of integral of Cauchy type in theory of geophysical fields): Nauka.

PART IV

3-D EM AND PARALLEL COMPUTERS

This work suggests that using ample computing resources it is possible. . . .

Eaton and Hohmann (1988)

Electromagnetic Modeling and Inversion on Massively Parallel Computers

Gregory A. Newman
David L. Alumbaugh

Summary. A numerical method has been developed to simulate the electromagnetic (EM) response of a 3-D earth to a dipole source at frequencies ranging from 100 Hz to 100 MHz. The problem is formulated in the frequency domain with a modified vector Helmholtz equation for the scattered electric fields. The differential equation is approximated on a staggered finite-difference grid, giving a sparse complex symmetric matrix equation. The system is solved by a preconditioned quasi-minimum–residual method.

Dirichlet boundary conditions are imposed at the edges of the mesh by setting the tangential electric fields equal to zero. At frequencies less than 1 MHz, grid stretching reduces reflections off the grid boundaries. At higher frequencies, absorbing boundary conditions are imposed by making the stretching parameters of the modified vector Helmholtz equation complex to introduce loss at the boundaries.

Iterative solution to the nonlinear 3-D EM inverse problem proceeds by linearized model updates using the method of conjugate gradients. Full wave equation modeling is employed to compute model sensitivities and predicted data in the frequency domain with the 3-D finite-difference algorithm. Both the forward and inverse solutions are implemented on a massively parallel computing platform which allows forward and inverse models with millions and ten of thousands of parameters, respectively.

1 Introduction

Great strides have been made over the past decade in electromagnetic (EM) forward modeling using staggered 3-D finite differences. Druskin and Knizhnerman (1988, 1994), Smith (1992), Wang and Hohmann (1993), and Newman (1995) all employ some type of staggered finite-difference grid (Yee, 1966) to solve for the EM fields in the time or frequency domain. Still, the complexity of the models that can be simulated on traditional serial computers is limited by memory and flop rate of the processor. Moreover, 3-D inversion on serial computers is still a dream. Rapid advancements in massively parallel (MP) computers, however, are removing the limitations posed by

Sandia National Laboratories, P.O. Box 5800 MS 0750, Albuquerque, NM 87185-0750, USA.

serial computers. The rate at which MP simulations can proceed is dramatic when thousands of processors operate on the problem simultaneously. This computational efficiency makes possible a realistic attack even on the 3-D inverse problem.

We describe an approach to solving 3-D forward and inverse problems on an MP platform. For the forward problem we examine the implementation of a frequency-domain finite-difference (FDFD) scheme on a staggered grid. We then describe an inversion algorithm built around the forward modeling and the method of conjugate gradients. We briefly describe how to implement these schemes on an MP computer. Finally, we demonstrate the forward code's usefulness over a wide frequency range for different types of geophysical scenarios and provide some tests of the inversion scheme.

2 Theoretical development of the forward problem

2.1 *Finite-difference formulation*

We numerically solve the frequency-domain version of the vector Helmholtz equation for the scattered electric fields using a finite-difference approximation on a staggered grid (Yee, 1966). The FDFD solution has been designed to compute the 3-D EM response for a wide variety of earth properties at frequencies ranging from approximately 100 Hz up to 100 MHz. This scheme is similar to those outlined by Alumbaugh and Newman (1994) and Newman and Alumbaugh (1995), but has been extended as described by Alumbaugh et al. (1996) to include both variable magnetic permeability and absorbing boundary conditions (ABCs). The ABCs are required to reduce reflections from grid boundaries at frequencies greater that 10 MHz. We have chosen to employ the perfectly matched layer (PML) ABCs originally developed by Berenger (1993) for 2-D time-domain calculations and later modified for 3-D calculations by Katz et al. (1994) and Chew and Weedon (1994). This method uses a modified form of the Helmholtz equation in which the absorption is incorporated through complex grid stretching.

We have chosen to work with the scattered-field versions of the governing equations. Often this allows a coarser discretization about the source location than would be feasible with a total-field solution and helps to limit storage overhead. In addition, because the scattered-field versions of Maxwell's equations numerically decouple in the air at low frequencies (<10 MHz), we work with the modified Helmholtz equation for the scattered electric field instead of the coupled Maxwell system. The vector Helmholtz equation for $\mathbf{E_s}$ is

$$\nabla_h \times \frac{\mu_p}{\mu}\nabla_e \times \mathbf{E_s} = -i\omega\mu_p(\sigma + i\omega\epsilon)\mathbf{E_s} - i\omega\mu_p[(\sigma - \sigma_p) + i\omega(\epsilon - \epsilon_p)]\mathbf{E_p} - i\omega\mu_p\nabla_h \times \left[\left(\frac{\mu - \mu_p}{\mu}\right)\mathbf{H_p}\right] \tag{1}$$

where

$$\nabla_e = \mathbf{i}\frac{1}{e_x}\frac{\partial}{\partial x} + \mathbf{j}\frac{1}{e_y}\frac{\partial}{\partial y} + \mathbf{k}\frac{1}{e_z}\frac{\partial}{\partial z} \tag{2}$$

and

$$\nabla_h = \mathbf{i}\frac{1}{h_x}\frac{\partial}{\partial x} + \mathbf{j}\frac{1}{h_y}\frac{\partial}{\partial y} + \mathbf{k}\frac{1}{h_z}\frac{\partial}{\partial z}. \tag{3}$$

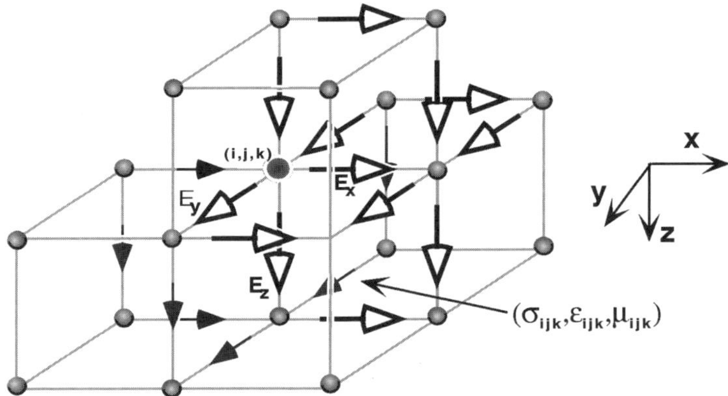

Figure 1. Finite-difference stencil for solving the scattered-electric-field Helmholtz equation. The arrows represent the position of the electric fields, with the open arrows representing the unknowns needed to form the equation for E_x.

In these equations, the electrical conductivity, magnetic permeability, and dielectric permittivity are denoted by σ, μ, and ϵ, respectively, with **p** designating a whole-space background value; $\mathbf{E_s}$ and $\mathbf{E_p}$ are the scattered and primary electric field whose sum is equal to the total electric field; $\mathbf{H_p}$ is the primary magnetic field; and e_i and h_i for $i = x, y, z$ are coordinate stretching variables that stretch the x, y, and z coordinates. As shown by Chew and Weedon (1994), when e_i and h_i are complex, the medium can be perfectly absorbing. Note that the terms at the end of Eq. (1) are equivalent-source terms wherever the properties of the medium are different from that of the assumed background. The boundary conditions employed are Dirichlet conditions, i.e., the tangential component of $\mathbf{E_s}$ is set to zero on the grid boundary.

The scattered electric fields are assigned to each cell following a staggered grid scheme (Fig. 1). For node (i, j, k), the x, y, and z components of the electric field are sampled at $(i + \frac{1}{2}, j, k)$, $(i, j + \frac{1}{2}, k)$ and $(i, j, k + \frac{1}{2})$, respectively. This corresponds to assigning the electric fields to the edges of the cell and the magnetic field to its faces, and requires that the conductivity and dielectric permittivity be computed halfway along a given cell edge and the magnetic permeability be computed in the center of the cell face. This is accomplished with averaging schemes described by Alumbaugh et al. (1996).

Discretizing Eq. (1) gives a linear system

$$\underset{\sim}{\mathbf{K}}\mathbf{E_s} = \mathbf{s} \tag{4}$$

where $\underset{\sim}{\mathbf{K}}$ is the stiffness matrix containing the numerical approximations to the derivatives as well as the electrical properties of the medium, $\mathbf{E_s}$ is the unknown vector for the scattered electric field, and **s** is the equivalent-source vector. Alumbaugh et al. (1996) show that $\underset{\sim}{\mathbf{K}}$ is complex symmetric, even when complex grid stretching is employed. The solution vector can be obtained using the quasi-minimum residual (QMR) technique (Freund, 1992) with preconditioning to iteratively determine the solution within a predetermined error level, which is defined here as

$$\text{error} = \frac{\|\underset{\sim}{\mathbf{K}}\mathbf{E_s} - \mathbf{s}\|^2}{\|\mathbf{s}\|^2}. \tag{5}$$

Tests with different types of incomplete decomposition and polynomial preconditioners have found that Jacobi scaling is an efficient preconditioner.

After the scattered fields at the grid points have been determined, the fields at the receivers must be calculated. The electric field is simply calculated using bilinear interpolation; the magnetic field is calculated by first taking a numerical approximation of Faraday's law for the scattered electric fields on the grid surrounding the receiver,

$$\nabla_e \times \mathbf{E_s} = -i\omega\mu\mathbf{H}_s + (\mu - \mu_p)\mathbf{H_p} \tag{6}$$

and then interpolating the result to the point of interest. In this expression $\mathbf{H}_s$ is the scattered magnetic field. For each new source position or frequency, a new system must be solved, although some time savings can be implemented by using the previous solution vector as an initial guess in the QMR solver.

2.2 *Properties of the PML ABC*

Chew and Weedon (1994) develop lossy, nonreflecting conditions along mesh boundaries using Berenger's PML material. The complex stretching parameters are assigned a value of the form $1 + a - ib$. On the internal portion of the mesh, $a = b = 0$ such that the modified Helmholtz equation reduces to the normal form. Near the edges of the mesh, e_i and h_i are allowed to vary over several cells, but only in the direction that is perpendicular to the boundary. For example, along the $+z$ boundary $e_x = e_y = h_x = h_y = 1$ and only e_z and h_z are allowed values of a and b that are not equal to zero.

We have found that, to incorporate a given amount of attenuation across a number of cells serving as the absorbing boundary, it is better to set a and b constant rather than gradually increasing their value toward the mesh boundaries as suggested by Berenger (1993); gradually increasing their value results in a greater number of iterations needed to achieve convergence. Simple MATLAB experiments have shown that the condition number of $\underset{\sim}{\mathbf{K}}$ increases as the ratio between the complex amplitudes of the largest and smallest cell dimensions in the mesh increases. Thus, gradually increasing the stretching parameters outward will produce a cell along the edge of the mesh that is effectively much larger than any of the cells employing constant stretching. Because the smallest cell size is the same in either case, the solution of the model that employs the gradual stretching will take longer to converge.

We are studying how to choose stretching parameters that provide adequate attenuation but do not slow convergence. First define the background wave number for a given frequency as

$$k_p = \sqrt{-i\omega\mu_p(\sigma_p + i\omega\epsilon_p)} = \alpha - i\beta, \tag{7}$$

where α and β are both real. Using the wavenumber domain analysis of Chew and Weedon (1994), which was modified for the FDFD Helmholtz equation by Alumbaugh et al. (1996), allows us to incorporate the stretching parameters into the background wavenumber to yield a pseudowavenumber defined as

$$k'_p = \sqrt{-i\omega\mu_p(\sigma_p + i\omega\epsilon_p)e_i h_i}. \tag{8}$$

With this assumption, we can develop a pseudo-skin-depth in each cell, defined by

$$\delta^{ps} = \frac{1}{[\alpha b + \beta(1 + a)]} \tag{9}$$

and a pseudowavelength defined as

$$\lambda^{ps} = \frac{2\pi}{[\alpha(1+a) - \beta b]}. \tag{10}$$

We have found that, for frequencies greater than 1 MHz, accurate results and quick solution convergence are achieved when a and b are chosen such that five pseudo-skin-depths of attenuation are present across the stretching region and the pseudowavelength is adjusted to be 10 times the cell dimension. At frequencies below 100 kHz, the analysis seems to become more difficult because the manner in which the grid is stretched can significantly alter the convergence of the system. In general, at these frequencies, we have obtained good results using only real grid stretching, i.e., setting b equal to zero and varying only a.

3 Theoretical development of the inverse problem

3.1 Regularized least squares

The parameterization used in the 3-D inverse solution is kept fine because we are interested in reconstructions that do not underparameterize the Earth. The 3-D inverse problem then is underdetermined. The inversion for model parameters $\mathbf{m}$ is stabilized with regularization (Tikhonov and Arsenin, 1977) that removes solutions that are too rough by imposing constraints on the model.

Given an earth model, $\mathbf{m}^{(i)}$, the following functional can provide smooth reconstructions if it is minimized with respect to the model parameters, $\mathbf{m}$, which can include electrical conductivity, dielectric permittivity, and magnetic permeability:

$$\begin{aligned} S = {} & [\underset{\sim}{\mathbf{D}}\{\mathbf{d} - \mathbf{d}^{\mathbf{p(i)}} - \underset{\sim}{\mathbf{A}}^{\mathbf{p(i)}}[\mathbf{m} - \mathbf{m}^{(i)}]\}]^{\mathbf{H}}[\underset{\sim}{\mathbf{D}}\{\mathbf{d} - \mathbf{d}^{\mathbf{p(i)}} - \underset{\sim}{\mathbf{A}}^{\mathbf{p(i)}}[\mathbf{m} - \mathbf{m}^{(i)}]\} - \chi^2] \\ & + \lambda(\underset{\sim}{\mathbf{W}}\mathbf{m})^{\mathbf{t}}(\underset{\sim}{\mathbf{W}}\mathbf{m}). \end{aligned} \tag{11}$$

Here, we are going to minimize model roughness $(\underset{\sim}{\mathbf{W}}\mathbf{m})^{\mathbf{t}}(\underset{\sim}{\mathbf{W}}\mathbf{m})$ subject to a specified square error, χ^2. The superscripts $\mathbf{H}$ and $\mathbf{t}$ denote Hermitian transpose and transpose. The matrix $\underset{\sim}{\mathbf{W}}$ is the roughness matrix consisting of a finite-difference approximation to the Laplacian ∇^2 operator and is sparse. In Eq. (11), the observed data are represented by the vector $\mathbf{d}$ and the predicted data arising from the model $\mathbf{m}^{(i)}$ are denoted by $\mathbf{d}^{\mathbf{p(i)}}$. The data weighting matrix $\underset{\sim}{\mathbf{D}}$ is diagonal, consisting of the reciprocal of the data variances, the reciprocal of the data amplitude, or, in some instances, an identity matrix if data weighting is unwarranted. The Jacobian or model sensitivities matrix, $\underset{\sim}{\mathbf{A}}^{\mathbf{p(i)}}$, is determined from the frequency-domain forward-solver algorithm described above. Note that we assume that the estimated model parameters are real. Determining and manipulating the elements of the Jacobian matrix in the most efficient manner is critical for a robust 3-D inverse solution, because calculation and use of these elements can be a bottleneck in the inversion. Derivation and efficient use of $\underset{\sim}{\mathbf{A}}^{\mathbf{p(i)}}$ in the inversion is given below.

The parameter λ is the trade-off parameter between model smoothness and data fit. Its selection requires special care if the inverse solution is to provide acceptable results. There is no universal strategy for selecting λ in Eq. (11). Trade-off parameters that are too small can produce models that are physically unreasonable. Trade-off parameters that are too large produce highly smoothed models, but these models show poor dependence on the data. We defer further discussion of this parameter until we discuss the iterative nature of Eq. (11).

Minimization of Eq. (11) yields the model step, where

$$\mathbf{m} = [(\underset{\sim}{\mathbf{D}}\underset{\sim}{\mathbf{A}}^{\mathbf{p(i)}})^{\mathbf{H}}(\underset{\sim}{\mathbf{D}}\underset{\sim}{\mathbf{A}}^{\mathbf{p(i)}}) + \lambda \underset{\sim}{\mathbf{W}}^{\mathbf{t}}\underset{\sim}{\mathbf{W}}]^{-1}(\underset{\sim}{\mathbf{D}}\underset{\sim}{\mathbf{A}}^{\mathbf{p(i)}})^{\mathbf{H}}(\underset{\sim}{\mathbf{D}}\delta\mathbf{d}^{(\mathbf{i})}) \tag{12}$$

with

$$\delta\mathbf{d}^{(\mathbf{i})} = (\mathbf{d} - \mathbf{d}^{\mathbf{p(i)}} + \underset{\sim}{\mathbf{A}}^{\mathbf{p(i)}}\mathbf{m}^{(\mathbf{i})}). \tag{13}$$

Because negative parameter estimates can arise from Eq. (12), it is advisable that Eq. (11) be reformulated to invert for the logarithm of the parameters, instead of the parameters themselves. This causes the imaged properties to be always positive. Using the log parameterization, it is also possible to incorporate a lower-bound positivity constraint in the inverse solution, where

$$\delta[m^{(i)}] = [m^{(i)} - \epsilon]\delta\ln[m^{(i)} - \epsilon], \tag{14}$$

with $\delta m^{(i)} = [m - m^{(i)}]$ along with $\delta\ln[m^{(i)} - \epsilon] = \ln\{(m - \epsilon)/[m^{(i)} - \epsilon]\}$, and $m > \epsilon$ and $\epsilon > 0$.

3.2 Model update via conjugate gradients

Using Eq. (12) to compute the updated model directly is not feasible for the full 3-D problem because direct matrix inversion is prohibitive (even on an MP platform) when the number of unknowns exceed several thousand. Instead, we opt for an iterative solution. Because Eq. (11) satisfies the normal equations, the linear system is symmetric semidefinite even when regularized. Thus, the conjugate gradient (CG) method of Hestenes and Stiefel (1952) can be used to get the solution. More important, following Mackie and Madden (1993) and Zhang et al. (1995), it is possible to avoid explicitly forming the Jacobian matrix, $\underset{\sim}{\mathbf{A}}^{\mathbf{p(i)}}$ altogether with this approach, thus saving considerable computer storage. In the CG methods, all one needs is one matrix-vector multiplication per relaxation step. Because the matrix in question is by the product of $[\underset{\sim}{\mathbf{D}}\underset{\sim}{\mathbf{A}}^{\mathbf{p(i)}}]^{\mathbf{H}}$ with $[\underset{\sim}{\mathbf{D}}\underset{\sim}{\mathbf{A}}^{\mathbf{p(i)}}]$, we really require two matrix-vector multiplications; in addition, two matrix-vector multiplications arise in the CG routine from the regularization matrix $\underset{\sim}{\mathbf{W}}$ and its transpose. However, these multiplications are easy to compute and need no further elaboration until the MP implementation. Explicitly, we have

$$\mathbf{y} = (\underset{\sim}{\mathbf{D}}\underset{\sim}{\mathbf{A}}^{\mathbf{p(i)}})\,\mathbf{u} \tag{15}$$

and

$$\mathbf{z} = (\underset{\sim}{\mathbf{D}}\underset{\sim}{\mathbf{A}}^{\mathbf{p(i)H}})\,\mathbf{y}, \tag{16}$$

where $\mathbf{u}$ is an arbitrary real vector, $\mathbf{y}$ is a complex vector, and $\mathbf{z}$ is a real vector.

The CG algorithm requires efficient computations of the matrix-vector products, which in turn requires efficient manipulation of the Jacobian matrix. Consider a single data measurement defined for a given transmitter, ℓ, where

$$d_{j(\ell)} = d^b_{j(\ell)} + \mathbf{g}^t_{\mathbf{j}(\ell)}\mathbf{E}_{\mathbf{s}(\ell)}. \tag{17}$$

In Eq. (17), $d^b_{j(\ell)}$ is the background field at location $j(\ell)$ and is specified by either $\mathbf{E}_{\mathbf{p}(\ell)}$ or $\mathbf{H}_{\mathbf{p}(\ell)}$; $\mathbf{E}_{\mathbf{s}(\ell)}$ is the scattered electric-field vector with dimension of $NT \times 1$ and is determined from our forward solver at the node points from Eq. (4); NT represents the number of field unknowns; and the vector $\mathbf{g}^t_{j(\ell)}$ is an interpolation vector for the

$j(\ell)$th measurement point and is of dimension $1 \times NT$. This vector interpolates the field values on the staggered grid to the measurement point of interest and also can numerically approximate the curl of the electric field so that magnetic-field measurements are allowed for in Eq. (17). With this definition, an element of the Jacobian matrix is

$$\frac{\partial d_{j(\ell)}}{\partial m_k} = \mathbf{g}^t_{j(\ell)} \frac{\partial \mathbf{E}_{\mathbf{s}(\ell)}}{\partial m_k}. \tag{18}$$

From the forward problem, we know that the scattered electric fields are related to the source vector of a given transmitter, $\mathbf{s}_{(\ell)}$, by the linear system

$$\underset{\sim}{\mathbf{K}}\mathbf{E}_{\mathbf{s}(\ell)} = \mathbf{s}_{(\ell)}, \tag{19}$$

where $\underset{\sim}{\mathbf{K}}$ is the sparse finite-difference stiffness matrix. Thus,

$$\frac{\partial \mathbf{E}_{\mathbf{s}(\ell)}}{\partial m_k} = \underset{\sim}{\mathbf{K}}^{-1}\left[\frac{\partial \mathbf{s}_{(\ell)}}{\partial m_k} - \frac{\partial \underset{\sim}{\mathbf{K}}}{\partial m_k}\mathbf{E}_{\mathbf{s}(\ell)}\right] \tag{20}$$

and an element of the Jacobian matrix can be written as

$$\frac{\partial d_{j(\ell)}}{\partial m_k} = \mathbf{g}^t_{j(\ell)}\underset{\sim}{\mathbf{K}}^{-1}\left[\frac{\partial \mathbf{s}_{(\ell)}}{\partial m_k} - \frac{\partial \underset{\sim}{\mathbf{K}}}{\partial m_k}\mathbf{E}_{\mathbf{s}(\ell)}\right]. \tag{21}$$

We now determine the $j(\ell)$th element of the first matrix-vector multiplication in Eq. (15) to be

$$\begin{aligned} y_{j(\ell)} = \text{Complex}\Bigg(&\text{Re}\left\{\mathbf{g}^t_{j(\ell)}\underset{\sim}{\mathbf{K}}^{-1}\sum_{k=1}^{M} u_k\left[\frac{\partial \mathbf{s}_{(\ell)}}{\partial m_k} - \frac{\partial \underset{\sim}{\mathbf{K}}}{\partial m_k}\mathbf{E}_{\mathbf{s}(\ell)}\right]\right\}\text{Re}[D_{j(\ell)j(\ell)}], \\ &\text{Im}\left\{\mathbf{g}^t_{j(\ell)}\underset{\sim}{\mathbf{K}}^{-1}\sum_{k=1}^{M} u_k\left[\frac{\partial \mathbf{s}_{(\ell)}}{\partial m_k} - \frac{\partial \underset{\sim}{\mathbf{K}}}{\partial m_k}\mathbf{E}_{\mathbf{s}(\ell)}\right]\right\}\text{Im}[D_{j(\ell)j(\ell)}]\Bigg) \end{aligned} \tag{22}$$

where M is the total number of parameters to be estimated and $D_{j(\ell)j(\ell)}$ is the $j(\ell)$th diagonal entry of the matrix $\underset{\sim}{\mathbf{D}}$. Using the same approach, one also can show that, for the second matrix-vector multiplication,

$$\begin{aligned} z_k = \text{Re}\Bigg\{ &\sum_{\ell=1}^{N_{tx}}\sum_{j=1}^{N_{d(\ell)}} \text{Complex}\{\text{Re}[D_{j(\ell)j(\ell)}]\text{Re}[y_{j(\ell)}], \text{Im}[D_{j(\ell)j(\ell)}]\text{Im}[y_{j(\ell)}]\}^* \\ &\mathbf{g}^t_{j(\ell)}\underset{\sim}{\mathbf{K}}^{-1}\left(\frac{\partial \mathbf{s}_{(\ell)}}{\partial m_k} - \frac{\partial \underset{\sim}{\mathbf{K}}}{\partial m_k}\mathbf{E}_{\mathbf{s}(\ell)}\right)\Bigg\}, \end{aligned} \tag{23}$$

where N_{tx} is the total number of transmitter positions and $N_{d(\ell)}$ is the amount of data used in the inversion for a given transmitter. In addition, the asterisk stands for complex conjugation. Note that, in Eqs. (22) and (23), the term $(\partial \mathbf{s}_{(\ell)}/\partial m_k - \partial \underset{\sim}{\mathbf{K}}/\partial m_k \mathbf{E}_{\mathbf{s}(\ell)})$ is easy to compute; the vector $\partial \mathbf{s}_{(\ell)}/\partial m_k$ has 12 nonzero entries and the matrix $\partial \underset{\sim}{\mathbf{K}}/\partial m_k$ can have 12 nonzero entries if m_k represents the conductivity or permittivity or 84 nonzero entries if m_k represents the magnetic permeability. When magnetic permeability properties are being estimated, we restrict the measurement point to being in uniform background medium, such as air. In that way the interpolator vector $\mathbf{g}^t_{j(\ell)}$ will not depend on changes in the magnetic permeability and Eq. (22) remains valid. Recall that this vector operator approximates Faraday's law and thus requires that the magnetic permeability be determined at the measurement point in order to compute the magnetic field.

In addition to the ℓ forward solutions needed for the different transmitters where $\ell = 1, \ldots N_{tx}$, we can efficiently carry out the matrix-vector multiplications in Eqs. (22) and (23) by solving a series of forward problems equal to the total number of data measurement locations (Rodi, 1976),

$$\mathbf{v}^t_{j(\ell)} = \mathbf{g}^t_{j(\ell)} \underset{\sim}{\mathbf{K}}^{-1}, \tag{24}$$

or

$$\underset{\sim}{\mathbf{K}} \mathbf{v}_{j(\ell)} = \mathbf{g}_{j(\ell)}, \tag{25}$$

since $\underset{\sim}{\mathbf{K}}^t = \underset{\sim}{\mathbf{K}}$ (matrix $\underset{\sim}{\mathbf{K}}$ is symmetric). The number of solutions needed is equal to the number of different measurements, where each measurement is a specific field component at a site for each transmitter positions and frequency. Thus to get the total number of forward solutions needed for each model update, we have $N_{rx} + N_{tx}$, where N_{rx} is the total number of unique receiver positions used in the inversion; multiple frequency data will require additional forward solves for the sources and receiver positions. Handling the Jacobian matrix elements in this manner is much more efficient than computing them directly using Eq. (21) and then using the results to form the matrix-vector multiplications. For example, if we are estimating over 30 000 parameters, this would require 30 000 separate forward solves, which is impractical. On the other hand, because the amount of data used in the inversion is limited, we anticipate no more than several hundred to several thousand forward solutions per model update with our approach. This approach also has been recommended by McGillivray and Oldenburg (1990) and Oldenburg (1990) because of its efficiency. It has been used by Park (1983) and Zhang et al. (1995) in their constructions of the inverse solution.

There is another approach to the scheme just outlined that can limit the number of forward solutions in the inversion. Following Mackie and Madden (1993), we regroup the sums in Eqs. (22) and (23),

$$\mathbf{p}_{(\ell)} = \sum_{k=1}^{M} u_k \left[\frac{\partial \mathbf{s}_{(\ell)}}{\partial m_k} - \frac{\partial \underset{\sim}{\mathbf{K}}}{\partial m_k} \mathbf{E}_{s(\ell)} \right] \tag{26}$$

and

$$\mathbf{r}_{(\ell)} = \sum_{j=1}^{N_{d(\ell)}} \text{Cmplx}\{\text{Re}[D_{j(\ell)j(\ell)}]\text{Re}[y_{j(\ell)}], \text{Im}[D_{j(\ell)j(\ell)}]\text{Im}[y_{j(\ell)}]\}^* \mathbf{g}^t_{j(\ell)}. \tag{27}$$

To evaluate the multiplication of these sums with $\underset{\sim}{\mathbf{K}}^{-1}$ in expressions (22) and (23), we solve

$$\underset{\sim}{\mathbf{K}} \mathbf{v}_{(\ell)} = \mathbf{p}_{(\ell)}, \tag{28}$$

and

$$\underset{\sim}{\mathbf{K}} \mathbf{w}_{(\ell)} = \mathbf{r}_{(\ell)}, \tag{29}$$

at each CG relation step and substitute in to produce

$$y_{j(\ell)} = \text{Cmplx}\left\{\text{Re}\left[\mathbf{g}^t_{j(\ell)} \mathbf{v}_{(\ell)}\right]\text{Re}[D_{j(\ell)j(\ell)}], \text{Im}\left[\mathbf{g}^t_{j(\ell)} \mathbf{v}_{(\ell)}\right]\text{Im}[D_{j(\ell)j(\ell)}]\right\} \tag{30}$$

and

$$z_k = \text{Re}\left\{ \sum_{\ell=1}^{N_{tx}} \mathbf{w}_{(\ell)} \left[\frac{\partial \mathbf{s}_{(\ell)}}{\partial m_k} - \frac{\partial \underset{\sim}{\mathbf{K}}}{\partial m_k} \mathbf{E}_{s(\ell)} \right] \right\}. \tag{31}$$

One can show that the total number of forward solutions with this approach for each frequency is $2 * \text{nrel}(N_{tx}) + N_{tx}$, where nrel is the number of CG relaxation steps. It is easy to see that as the number of transmitters or relaxation steps increases, this method becomes less efficient than the former. Thus, for the inverse problems we plan to consider, the fewest forward solutions at each frequency correspond to $N_{tx} + N_{rx}$.

3.3 *Iterative solution*

Because of the computation of an exact forward solution in the inversion, we cannot afford to slowly reduce the tradeoff parameter or determine an optimal λ at a given iteration to exclude rough models. However, experience indicates that smooth models can be produced with the following strategy.

We initiate an inversion, assuming an initial background model and compute the predicted data for all transmitter and receiver locations. At the first iteration, we determine the matrix-vector multiplications in the CG algorithm and compute the model update via Eq. (12). This model is determined once the tradeoff parameter, λ, is selected. To ensure a smooth model at the first iteration, we select the tradeoff parameter as

$$\lambda = \text{Max Row Sum } (\underset{\sim}{\mathbf{D}}\underset{\sim}{\mathbf{A}}^{\mathbf{p(i)}})^{\mathbf{H}}(\underset{\sim}{\mathbf{D}}\underset{\sim}{\mathbf{A}}^{\mathbf{p(i)}})/2^{i-1}, \tag{32}$$

where $i = 1$ for the first iteration. We have selected this method of choosing λ because it is an estimate of the largest eigenvalue of the nonregularized least-squares system matrix. Thus, weighting $\underset{\sim}{\mathbf{W}}^{\mathbf{t}}\underset{\sim}{\mathbf{W}}$ by this amount allows only the largest eigenvalues to influence the solution. The maximum row sum is easy to compute and follows from Eqs. (22) and (23) with $\mathbf{u}$ selected to be the unit vector.

We proceed to the next iteration if the data error (sum-of-square errors) is above χ^2, or if the error has not reached a minimum. If this is true, the model is linearized again about the new model $\mathbf{m}$; the predicted data and electric fields are computed from the updated background model; and the new model update is determined once the tradeoff parameter is specified with Eq. (32). We have found that, for the first few iterations, this method of selecting the tradeoff parameter reduces the error by about a factor of 2. The iterative procedure just outlined is continued until the data error is below χ^2, convergence occurs, or a prespecified number of iterations has taken place.

Even this procedure can drive the tradeoff parameter down too quickly, especially when one attempts to fit the data to an unrealistic noise level or uses an excessive number of iterations. However, if the tradeoff parameter is not relaxed sufficiently, the inversion can stall far above the estimated noise level in the data. Our solution to this difficulty is to have a good estimate of the data noise and monitor the tradeoff parameter and squared error in the inversion. If excessive model structure is appearing in the image, we stop the inversion and relaunch it using an acceptable reconstruction and tradeoff parameter at some previous iteration. After this restart, the tradeoff parameter is kept fixed for the rest of the inversion. Although this strategy is somewhat subjective, it has yielded good results.

At each iteration, we restrict the number of relaxation steps in the CG routine, because only a modest number of steps is sufficient to produce an accurate model update, especially during the early stages (Zhang et al., 1995). For the first and second iterations, 20 and 40 relaxation steps are used, respectively. Subsequent iterations use 60 steps.

4 Implementation on MP computers

To image more complex structures than previously has been possible, the original serial versions of the forward and inverse codes have been modified to run on MP MIMD machines. Such machines can have thousands of processors and are programmed by assigning a given number of processors in each direction of the forward and inverse modeling domain (nx in x, ny in y, and nz in z) and then breaking up the model across the processor bank such that each individual processor is in charge of a 3-D subset, with all processors sharing the same data. Because each processor needs only to make the necessary calculations for this subset, and because all of the processors are making their calculations simultaneously, the solution time is reduced by a factor that is approximately equal to the total number of processors employed ($nx * ny * nz$).

An issue that needs to be addressed is the manner in which the model is input and output; input could constitute a starting model to launch the inversion or a restart model if the inversion needs to be restarted midway through the process. We have decomposed the input data into two different sets: global and local. Global data are those variables that each processor needs to know such as the source and receiver positions, the frequencies, what type of solver is being employed, and the location of the mesh nodes. These form a fairly small data set that can easily be read in by a lead processor and then broadcast to all other processors. The second type of input is the local data, or local model parameters (conductivity, dielectric permittivity, and magnetic permeability) that are assigned to each cell within the model. Because each processor needs only a small subset of this data and contains only a small amount of local memory, the local data is broken up into multiple files, one for each processor, which are then individually read in from or written out to a parallel disk system which allows several files to be accessed simultaneously.

Even with the increased performance of an MP platform, memory considerations will dictate the largest model that can be simulated. A 3-D EM inversion can easily require the solution of at least several hundred forward problems per iteration. We also anticipate that each solution could constitute over a million field unknowns. Nevertheless, it is still possible on large-scale platforms, such as the 1840-node Intel Paragon, to execute all solutions without writing to disk. A significant portion of the storage required to perform the inversion is taken up by the electric-field solution vectors that are obtained from the forward solver and are needed to complete matrix-vector multiplications in the CG routine.

4.1 Message passing required in the forward problem

To complete the calculations required in the forward problem, information will need to be exchanged between processors by message passing. Consider the forward problem after the data have been accessed and each processor has constructed its own portion of the stiffness matrix $\underset{\sim}{\mathbf{K}}$ and the source vector $\mathbf{s}$. Each processor proceeds to solve for its portion of the solution vector in Eq. (4). However, each iteration within the QMR solver requires one matrix-vector multiplication and several vector dot products. These operations require information to be exchanged between all of the processors as well as between small subsets of processors. The dot products are fairly easy to implement because they involve (1) a local calculation in which each processor computes the dot product of its portion of the vector and (2) a global calculation in which all of the

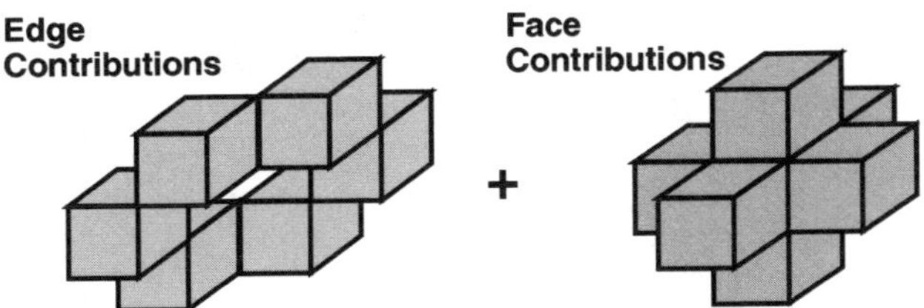

Figure 2. Processor stencil employed for message passing in order to correctly complete the matrix-vector multiplication for the forward problem.

local calculations are gathered by the lead processor, summed, and the result broadcast across the machine.

The matrix-vector multiplication is more difficult to implement because it requires that each processor communicate with those neighboring processors that are solving for the scattered electric fields in adjacent portions of the model. Determining these neighboring processors and the actual unknowns that need to be communicated is accomplished in the following manner: Assume that each processor contains only a single node, and imagine it as a cubic shape enclosing node (i, j, k) in Fig. 1. Careful examination then indicates that there are two types of communication that each processor needs to execute with its neighbors. The first type of communication will occur across the faces of the cube. For node (i, j, k), this implies communication with those nodes directly connected to it by the gray lines of the finite-difference stencil, i.e., nodes $(i-1, j, k)$, $(i+1, j, k)$, $(i, j-1, k)$, $(i, j+1, k)$, $(i, j, k-1)$, and $(i, j, k+1)$. For these communications, either two or three unknowns are exchanged per nodal position. The second type of communication occurs across certain edges of the cube, and involves those nodes that are not directly connected to (i, j, k) by the stencil lines, for example, node $(i+1, j, k-1)$. This type of communication requires only one unknown per node being communicated each way. If we now expand the idea such that each processor cube contains a 3-D distribution of nodal points, then we can develop the processor communication stencil shown in Fig. 2.

The last point to be addressed is the message passing needed for data output. Because for any given source we only need to know the results at a limited number of receiver positions, all of which may lie on the same processor, the data output is inherently nonparallel and is accomplished in the following steps:

1. Each processor determines which processor holds the portion of the model that contains the receiver position.
2. This receiver processor then determines if it needs any values from adjacent processors, completes the necessary point-to-point communication with those processors, and then does the necessary bilinear interpolation.
3. The results then are sent to the lead processor that outputs them to disk.

4.2 *Message passing required in the inverse problem*

There is additional message passing needed for inversion. Of primary importance for efficiency is to limit the amount of interprocessor communication within the CG routine; that incorporates the matrix-vector multiplication for both the Jacobian and regularization matrices. We only show one processor stencil for local communications needed

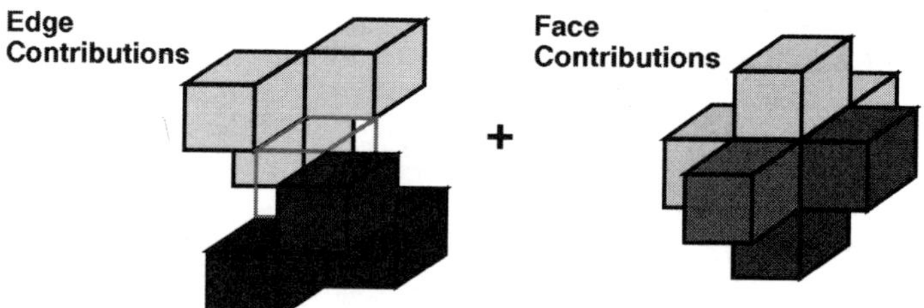

Figure 3. Processor stencil employed for message passing in order to correctly complete the matrix-vector multiplications in the inverse problem.

within the inverse. The local communication pattern for a given processor is illustrated in Fig. 3.

The communication needed to complete the matrix-vector multiplications involving the Jacobian matrix is along the faces of processors as well as along edges. Thus, before the CG routine is called, electric-field values arising from the forward solver are exchanged which provide each processor with the necessary values to complete the calculations in Eqs. (22) and (23). Specifically, information is passed from the central processor to those neighbors designated by the lighter colors in Fig. 3. Likewise, neighboring processors that are darker pass information to the central processor.

Local communication for multiplications with the regularization matrix and its transpose involve only communication along processor faces in Fig. 3. Here, all such processors send elements of the CG vectors to the central processor as well as receive from it. Local communication occurs every time the matrix-vector multiplication is encountered in the CG routine because we have explicitly formed the regularization matrix and the CG vectors are constantly updated for each relaxation step. In addition, there are five global dot products within a generic CG routine and one in Eq. (23) that incur additional global communication overhead at each iteration.

After exiting the CG routine, additional message passing is needed. Electrical properties for cells along processor boundaries need to be communicated with neighboring processors to accurately calculate the correct average conductivity, dielectric permittivity, and magnetic permeability for subsequent forward modeling in the inversion. Those face and edge processors designated with lighter colors send information to the central processor, whereas those that are darker receive information from it.

4.3 Message-passing software

We have chosen to employ the message passing interface (MPI; Skjellum et al., 1993) rather than using machine-specific commands. This code will be able to run on any parallel machine or distributed network of machines on which public-domain MPI library is available. Both forward and inverse codes have been implemented on the 1840 processor Intel Paragon at Sandia National Laboratories.

5 Demonstration of the finite-difference forward solution

We have simulated two different models that represent measurement configurations that might be employed in the field. The first simulation involves frequencies in the low ground-penetrating radar (GPR) range; the second simulates a portion of a helicopter

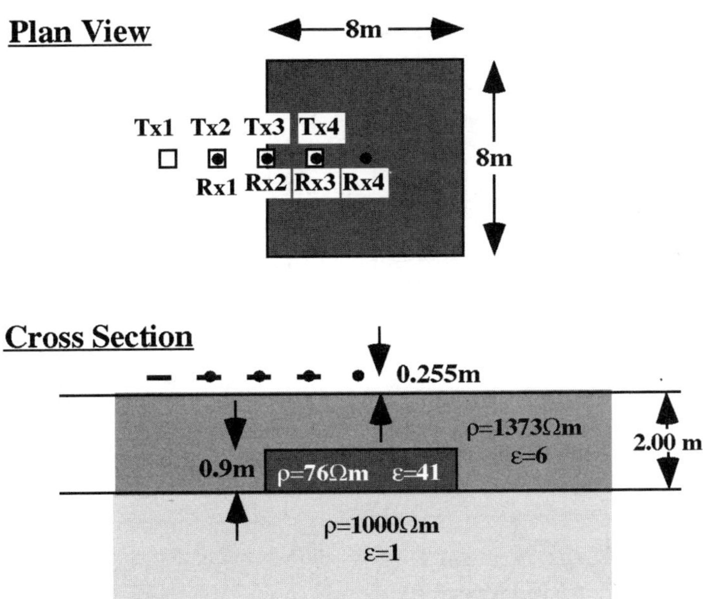

Figure 4. Colorado School of Mines 3-D (CSM3D) model.

EM survey. In the two cases, the solution is assumed to have converged to an adequate error level when Eq. (5) is found to be less than or equal to 10^{-7} and 10^{-8}, respectively. These error levels are empirical and are based on extensive comparisons of the solution with other numerical solutions and scale-model experiments (Alumbaugh and Newman, 1994).

5.1 High-frequency simulation for the VETEM project

The Very Early Time ElectroMagnetic (VETEM) project is an attempt to build an EM prospecting system that operates above traditional geophysical induction frequencies (100 kHz) yet below GPR frequencies (100 MHz) (Pellerin et al., 1995). To illustrate the ability of the code to simulate the EM response at these frequencies, variations of the model shown in Fig. 4 have been employed. This example was designed to simulate a site at the Colorado School of Mines, where a prototype of the VETEM system known as the High-Frequency Sounder (HFS) (Stewart et al., 1994) was first tested. The model is particularly difficult to simulate because of two conflicting conditions that are imposed by the material properties: (1) the wavelength in the block at 28.5 MHz is approximately 1.6 m, which requires a maximum cell dimension of 0.16 m to avoid grid dispersion (Chew, 1990, p. 244); and (2) the skin depth in the first layer at that same frequency is 17.8 m, which requires that the boundaries be placed very far away to avoid reflections off the grid. The small cell size coupled with the large distance to the boundaries produce a very large mesh if no ABCs are present.

To simulate this example, a $120 \times 120 \times 120$-cell mesh was employed with a constant cell size of 0.15 m in the x and y directions. The total distance across the mesh was 18 m. In z, the maximum cell size was also 0.15 m, with a minimum cell size of 0.13 m to accommodate the layer thicknesses. This mesh produces a total of 5×10^6 unknowns for which to solve, which is too large a problem for most computers. The vertical magnetic dipole (VMD) source was placed at the center of the mesh in x and

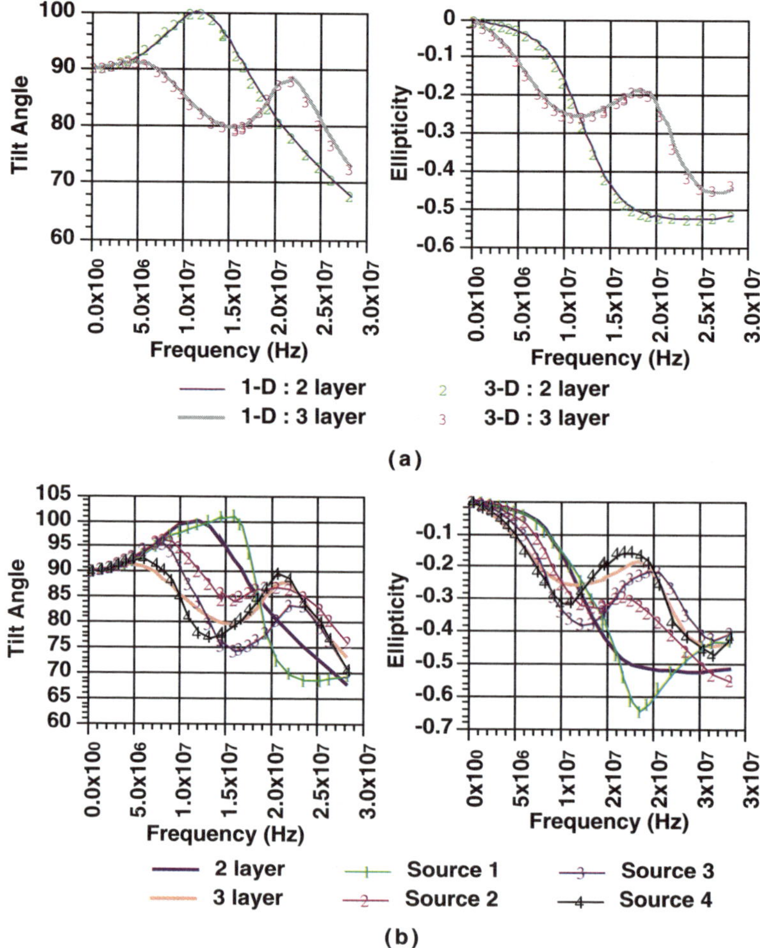

Figure 5. Results for the CSM3D model: (a) 1-D comparisons for two- and three-layer models; (b) 3-D comparison for different source positions.

y, i.e., 9 m from each boundary, and a background conductivity of $\sigma = 10^{-16}$ S/m was assumed. To incorporate loss and thus avoid reflections off the edges of the mesh, b was set equal to 0.6 over 20 cells along each edge of the mesh.

In the first case, we simulate two 1-D models: a two-layer model that assumes that the block is absent and a three-layer model that assumes that the block extends to infinity in the x and y directions. This allows us to make comparisons to a 1-D code developed by Ki Ha Lee at Lawrence Berkeley Laboratory. Because HFS directly measures tilt angle and ellipticity of the magnetic field (Smith and Ward, 1974), the results are plotted in terms of these parameters rather than amplitude and phase of the different components. As can be seen in Fig. 5a for the layered models, the 3-D code reproduces the 1-D calculations extremely well.

In Fig. 5b, 3-D results for the block for four different source positions are plotted with the results for the two 1-D models. Notice that the 3-D responses never reproduce the 1-D results, even when the source-receivers are completely outside or within the block. This indicates that 3-D effects are measurable at greater distances than is immediately evident and that 1-D inversions probably would not accurately reproduce the structure of the subsurface.

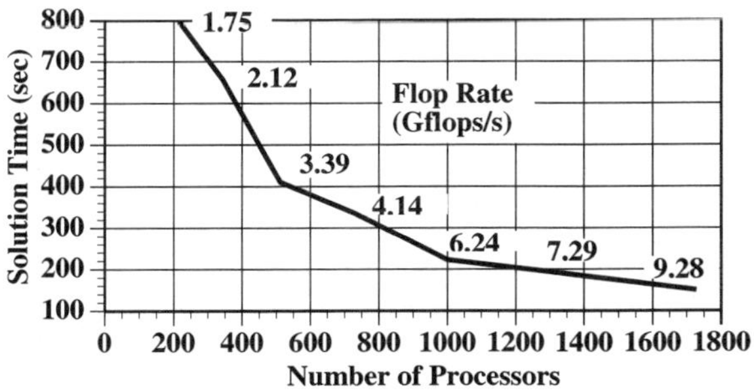

Figure 6. Run time versus number of processors employed for the three-layer model at 10.1 MHz.

To demonstrate some of the questions that must be answered when using the parallel machines, the solution time as well as the flop rate have been plotted against the number of processors employed for the three-layer model at 10.1 MHz on the Intel Paragon. Figure 6 shows that a large decrease in run time occurs with an increasing number of processors from 200 up to 1000. This corresponds to solving for 24 000 to 3000 unknowns per processor and indicates that the processors are spending the majority of their time performing calculations rather than communicating. However, the relatively small decrease in run time with increasing number of processors over 1000 indicates that the solution time is beginning to be dominated by message passing if fewer than 1000 unknowns are being solved for on each processor. Thus we are left with a decision to make. If we wish to use the machine most efficiently, we would employ fewer than 1000 processors such that the internal computations are dominating the solution time. We then could run several jobs simultaneously such that the efficiency increases proportionally to the number of jobs. On the other hand, if we desire as quick a turnaround time as possible for a single computation, then we would want to operate near the right end of the curve.

5.2 *Airborne simulation*

The second example simulates a helicopter EM survey flown to define the location of a buried paleochannel through which conductive saltwater is migrating, and is designed after a survey flown in Australia in the early 1990s. Figure 7 shows a plan view of the model at 5-m depth below the earth surface as well as a cross-section through the model. The flight lines are 30 m above the surface, spaced at 200-m intervals from top to bottom in Fig. 7. Along each line, the sampling interval is 100 m. This yields a total of 187 source positions. A VMD source is operating at 0.9 kHz, 7.2 kHz, and 56 kHz, with the receiver located 8 m to the right of the source. The three frequencies coupled with the 187 positions yields a total of 561 forward solves.

To calculate this with the 3-D finite-difference code, the earth and air were divided into a $208 \times 184 \times 49$-cell grid that yields a total of 5.6×10^6 unknowns. To reduce reflections off the mesh boundaries, normal grid stretching (i.e., $b = 0$) moved them out to 400 m from the nearest sampling point. The smallest cell size was 5 m $\times$ 5 m $\times$ 2.5 m at the air–earth interface underneath each source array. The largest cell size was in the corners of the mesh and was 20 m $\times$ 20 m $\times$ 20 m. A background conductivity of $s = 10^{-16}$ S/m simulated the electrical properties of the air.

In Fig. 8, for all three frequencies, the channel is clearly defined, although its resistivity is not defined as accurately at higher frequencies. This is because the increased sensitivity to the surface resistive layer at higher frequencies. To run this model on 1360

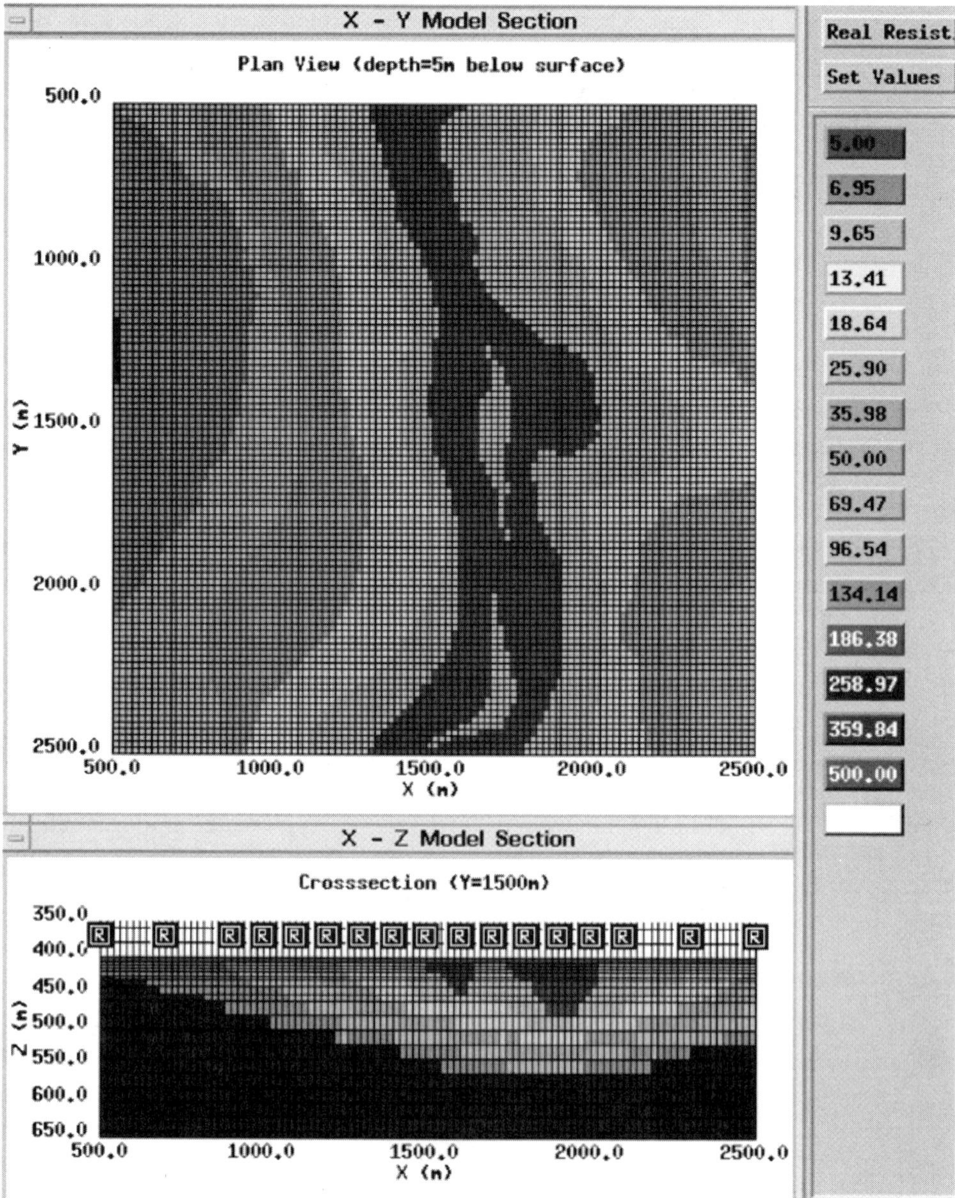

Figure 7. Subsurface channel model employed for the airborne simulation: (top) plan view at the top of the channel; (bottom) vertically exaggerated cross-section at $y = 1500$ m. Although it is difficult to see, a 5-m-thick 500-ohm-m layer exists from Earth's surface down to the top of the channel. The gray scale varies logarithmically from 5 ohm-m (black) to 697 ohm-m (white).

processors of the Intel Paragon took approximately two days. This illustrates the utility of these machines for solving realistic geologic problems.

6 The 3-D data inversion

6.1 *Synthetic example*

Figure 9 shows a model used to test the 3-D inversion. The data from this model were generated from the integral-equation solution of Newman et al. (1986). The test

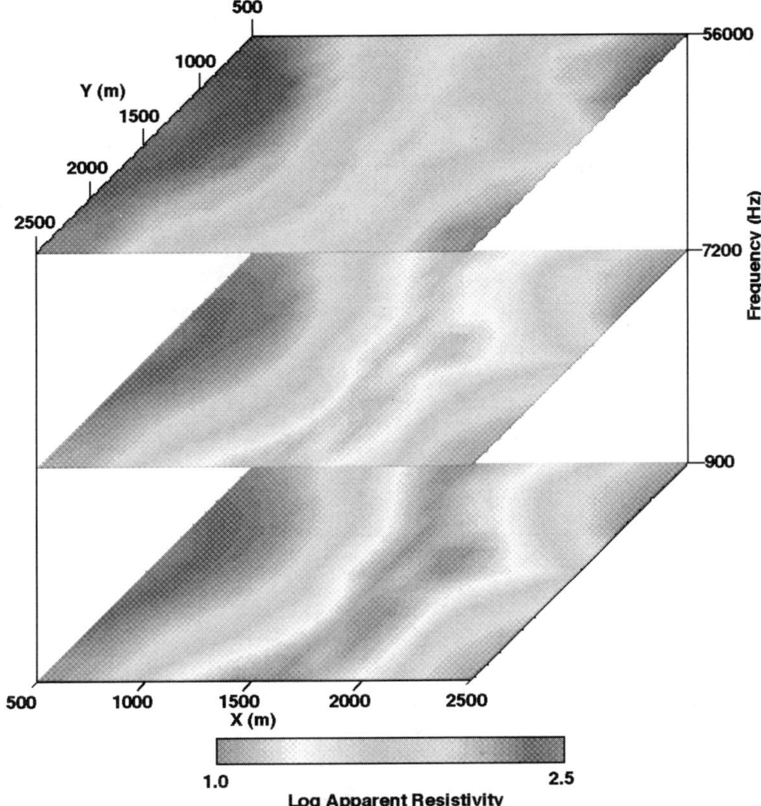

Figure 8. Calculated apparent resistivities for subsurface channel model.

model consists of a 0.2-S/m cube, 50 m on a side, residing in a 0.005-S/m background. Eight wells surround the target, each containing 15 VMD transmitters at 10-m intervals straddling the target. The vertical magnetic fields were calculated in all other wells, excluding the transmitter well, at 10-m intervals. At the frequency of excitation used in this test (20 kHz), the dielectric properties of the target and host are not important in the simulation, and only the conductivity properties need to be estimated; the magnetic permeability is assumed to be constant and is set to free space throughout the model.

Gaussian noise equal to 2% of the data amplitude was added to each data point. The data then were weighted by the noise, before inversion. In total, they comprise 12 600 transmitter-receiver pairs. The inversion domain consists of 29 791 cells, but only 13 824 cells are shown in the interwell region in Fig. 10; cells outside this region are used to keep the boundary of the inversion domain at distance so as not to affect the conductivity estimates in the interwell region. The inversion has recovered the location and geometry of the cube fairly well, but has only approximately reconstructed the conductivity; the estimates vary from 0.1 to 0.75 S/m. In addition, the conductivity estimates of the background are as low as 0.0015 S/m. Improved estimates on the conductivity can be obtained by tightening the lower-bound positivity constraint. In this example, the conductivity estimates were restricted to be greater than 0.001 S/m.

Fifteen iterations were needed to obtain this reconstruction; the reduction in relative error against iteration count is illustrated in Fig. 11. After the 15th iteration the error begins to increase. With Gaussian noise of zero mean, the relative error should approach a value of one. Because this did not happen, we assume there is bias in the data caused

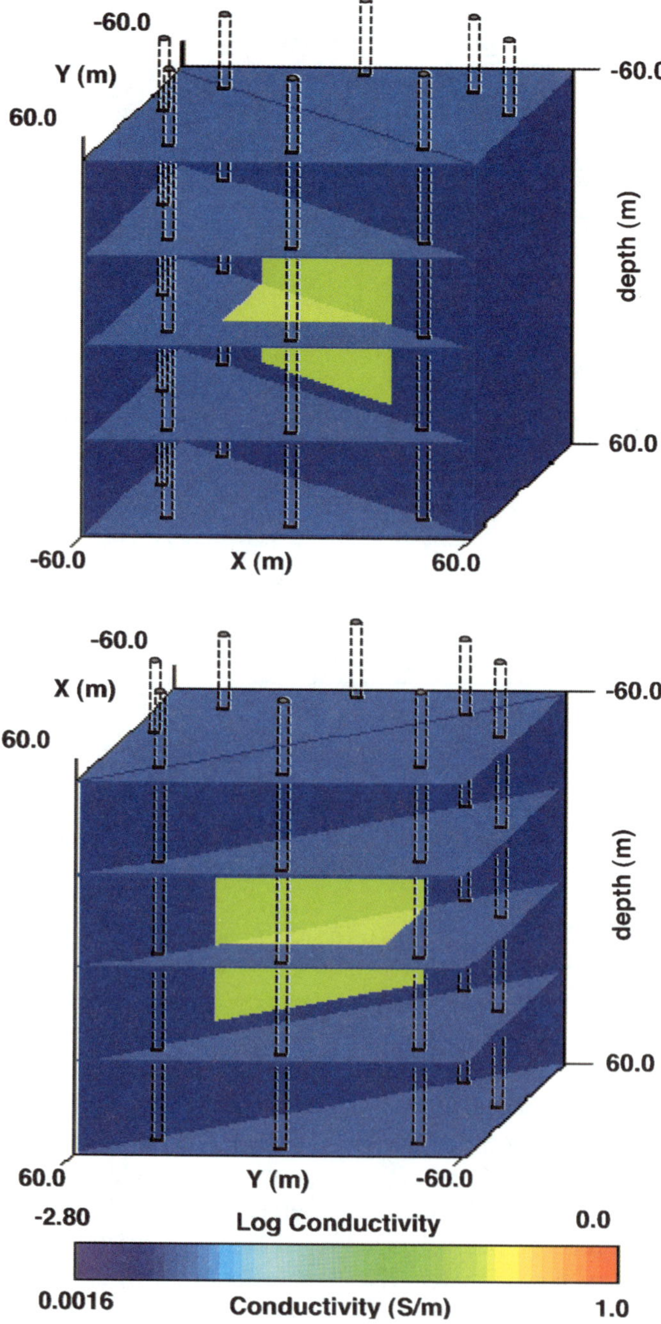

Figure 9. Synthetic example, with wellbores, used to test the inversion algorithm. Data were calculated from this model using an integral-equation solution. Different slices of the model are shown from different perspectives.

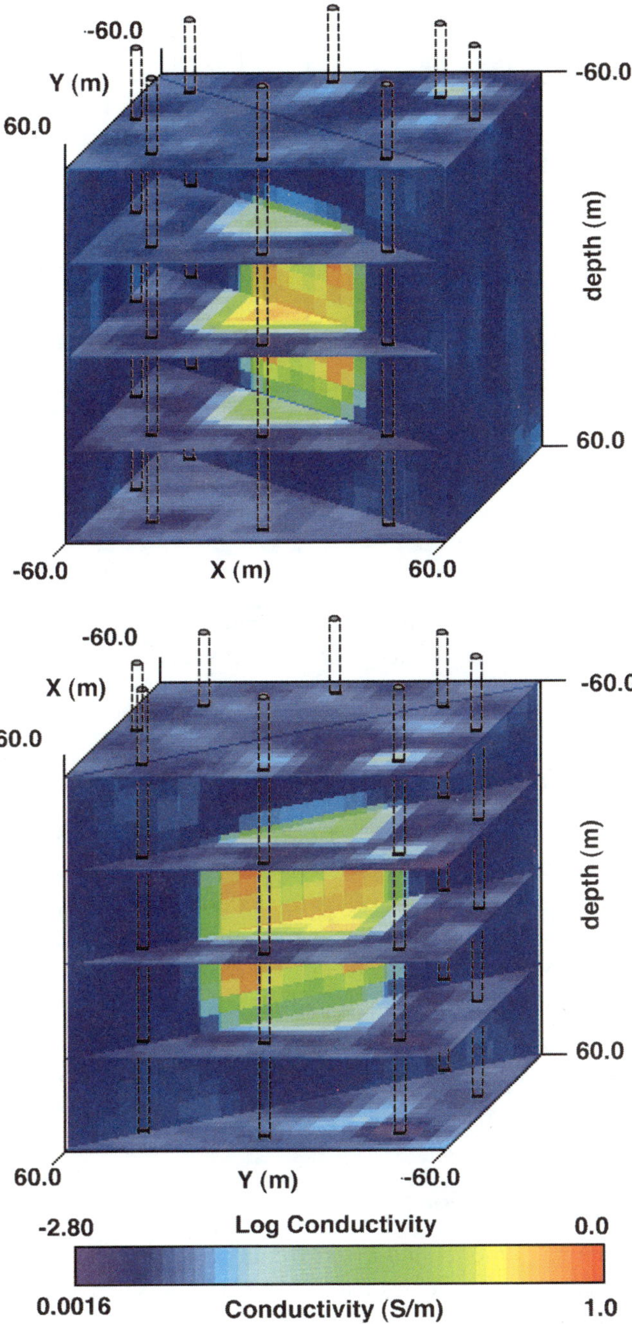

Figure 10. Reconstructed conductivity for the synthetic example illustrated in Fig. 9 for different slices from two different perspectives. The wellbores used in the simulation also are indicated.

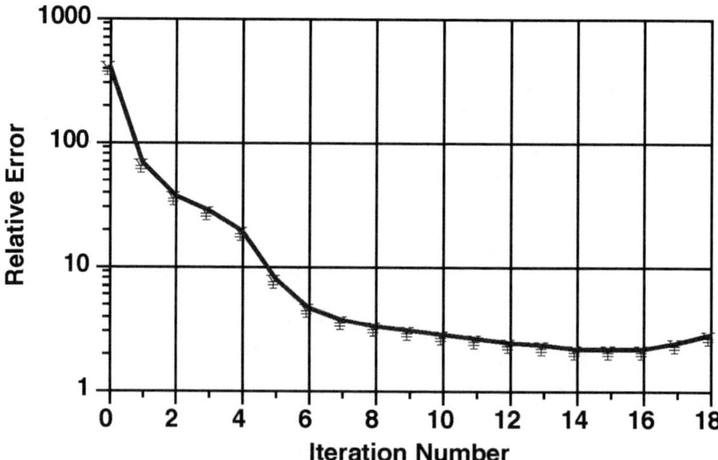

Figure 11. Sum of squared error plotted against iteration number for the 0.2-S/m test body shown in Fig. 9. The squared error has been normalized by the variance of the noise.

by the forward modeling algorithm being different than the one used in the inverse. Finally, the processing time needed to produce the image in Fig. 10 was approximately 21 hours on the Paragon, with 512 processors utilized.

6.2 *Analysis of field data*

The 3-D inversion scheme also has been used to analyze a crosswell EM data set collected at the University of California's Richmond Field Station test site north of Berkeley, California [see Wilt et al. (1995) for a description of the experiment]. The scheme was able to image the data that were collected before and after the injection of saltwater at depth, and allowed an experimental design study as well as a postinversion error analysis (Alumbaugh and Newman, 1997). Movies showing ther results of the 3-D imaging experiment can be found at the SEG World Wide Web site: http://www.seg.org/books/3dem/newman/newman.html. These movies are available in both *.pics and *.avi formats for Macintoshes and PCs with the proper drivers.

7 Discussion and Conclusions

We presented a scheme to model and invert frequency-domain EM response of a 3-D earth over a wide band of frequencies using MP computers. Two simulations demonstrate the versatility of the forward code. The inversion code has been implemented such that reconstructions can be produced with tens of thousands of cells. Because the 3-D MP inversion also includes rigorous 3-D forward modeling for computing model sensitivities and predicted data, we hope that this solution also will serve as an accuracy benchmark on approximate inverse methods now being implemented on workstations. Such solutions are beginning to see widespread use in the EM geophysical community [see Torres-Verdin and Habashy (1994, 1995); Zhdanov and Fang (1996)].

We hope to improve the solution through the use of multigrid preconditioners and methods to separately treat the real and imaginary components of the matrix system. In addition, a scheme to accelerate the convergence for low-frequency simulations in which channeling currents are dominant needs to be developed to simulate natural field measurements as well as to extend the frequency band down below 100 Hz. Smith (1996) has found that a static correction can be incorporated to accommodate this.

Finally, better ways are needed to manage the memory needed to carry out a 3-D inversion on large data sets. Because electric-field vectors need to be computed and stored in memory for all of the different transmitters and receivers, there are limits on the model size that can be inverted. To overcome this limitation, we are investigating methods that skeletonize the inversion domain but still retain the fine parameterization level in the forward solves for accuracy. With the coarser grid, the electric-field vectors needed in the inverse will be interpolated from a fine grid based on the forward modeling to a sparser grid as needed in the inversion, thus reducing the storage overhead of the electric-field vectors and increasing dramatically the amount of data that can be inverted. In a problem where the inversion grid is eight times coarser than the forward modeling grid. If 120^3 nodes are used in the forward calculations, the skeletonized inversion grid which still has 216 000 cells, and the number of transmitter and receivers can be increased from 700 to over 3000 on the Intel Paragon.

8 Acknowledgments

G.A. Newman warmly remembers his late professor, G. W. Hohmann, and the knowledge and inspiration he inspired in his students. We express our thanks to Dr. Ki Ha Lee of Lawrence Berkeley Laboratory for the use of the 1-D layered-earth code used in the model comparisons. Funding for this project was provided by the U.S. Department of Energy (DOE) Office of Basic Energy Sciences, Division of Engineering and Geoscience under contract DE-AC04-94AL85000; the Sandia National Laboratory Director's Research funds; and the VETEM project, which is supported by the DOE Office of Technology and Development.

References

Alumbaugh, D. L., and Newman G. A., 1994, Fast, frequency domain electromagnetic modeling using finite differences: 64th Ann. Internat. Mtg., Soc. Expl. Geophys., Expanded Abstracts, 369–373.

———1997, 3-D Massively parallel electromagnetic inversion, Part b: Analysis of a crosswell EM experiment: Geophys. J. Internat., **128**, 355–363.

Alumbaugh, D. L., and Newman, G. A., Prevost, L., and Shadid, J. N., 1996, Three dimensional wide band electromagnetic modeling on massively parallel computers: Radio Sci., **31**, 1–23.

Berenger, J., 1993, A perfectly matched layer for the absorption of electromagnetic waves: J. Comput. Phys., **114**, 185–200.

Chew, W. C., 1990, Waves and Fields in Inhomogeneous Media: Van Norstrand Reinhold, New York.

Chew, W. C., and Weedon, W. H., 1994, A 3-D perfectly matched medium from modified Maxwell's equations with stretched coordinates: Microwave and Optical Tech. Lett., **7**, 599–604.

Druskin, V., and Knizhnerman, L., 1988, A spectral semi-discrete method for the numerical solution of three-dimensional nonstationary problems of electric prospecting: Izvestiya, Earth Phys., **24**, 641–648.

———1994, A spectral approach to solving three-dimensional diffusion Maxwell's equations in the time and frequency domains: Radio Sci., **29**, 937–953.

Freund, R., 1992, Conjugate gradient type methods for linear systems with complex symmetric coefficient matrices: SIAM J. Sci. Statist. Comput., **13**, 425–448.

Hestenes, M. R., and Stiefel, E., 1952, Methods of conjugate gradients for solving linear systems: J. Res. Nat. Bur. Stand., **49**, 409–435.

Katz, D. S., Thiele, E. T., and Taflove, A.,1994, Validation and extension to three dimensions of the Berenger PML absorbing boundary condition for FD-TD meshes: IEEE Microwave and Guided Wave Lett., **4**, 268–270.

McGillivray, P. R., and Oldenburg, D. W., 1990, Methods for calculating Frechet derivatives and sensitivities for the non-linear inverse problem: Geophys. Prosp., **38**, 499–524.

Mackie, R. L., and Madden, T. R., 1993, Three-dimensional magnetotelluric inversion using conjugate gradients: Geophys. J. Internat., **115**, 215–229.

Newman, G. A., 1995, Crosswell electromagnetic inversion using integral and differential equations: Geophysics, **60**, 899–910.

Newman G. A., and Alumbaugh, D. L., 1995, Frequency-domain modeling of airborne electromagnetic responses using staggered finite differences: Geophys Prosp., **43**, 1021–1042.

Newman, G. A., Hohmann, G. W., and Anderson, W. L., 1986, Transient electromagnetic response of the three-dimensional body in a layered earth: Geophysics, **51**, 1606–1627.

Oldenburg, D. W., 1990, Inversion of electromagnetic data: An overview of new techniques: Surveys in Geophysics, **11**, 231–270.

Park, S. K., 1983, Three-dimensional magnetotelluric modeling and inversion: Ph.D. thesis, Massachusetts Institute of Technology.

Pellerin, L., Labson, V. F., and Pfeifer, M. C., 1995, VETEM—a very early time electromagnetic system: Proceedings of the Symposium on the Application of Geophysics to Environmental and Engineering Problems, 725–732.

Rodi, W. L., 1976, A technique for improving the accuracy of finite element solutions for magnetotelluric data: Geophys. J. Roy. Astr. Soc., **44**, 483–506.

Skjellum, A., Doss, N. E., and Bangalore, P. V., 1993, Writing libraries in MPI, *in* Skjellum, A. and Reese, D., Ed., Proceedings of the scalable parallel libraries conference: Inst. Electr. Electron. Eng. Comput. Sci. Press, 166–173.

Smith, T. J., 1996, Conservative modeling of 3-D electromagnetic fields; Part II: Biconjugate gradient solution and an accelerator: Geophysics, 1319–1324.

Smith, B. D., and Ward, S. H., 1974, On the computation of polarization ellipse parameters: Geophysics, **39**, 867–869.

Stewart, D. C., Anderson, W. L., Grover, T. P., and Labson, V. F., 1994, Shallow subsurface mapping by electromagnetic sounding in the 300 KHz to 30 MHz range: Model studies and prototype system assessment: Geophysics, **59**, 1201–1210.

Tikhonov, A. N., and Arsenin, V. Y., 1977, Solutions to ill-posed problems: John Wiley & Sons, Inc.

Torres-Verdin, C., and Habashy, T. M., 1994, Rapid 2.5-dimensional forward modeling and inversion via a new nonlinear scattering approximation: Radio Sci., **29**, 1051–1079.

———1995. A two step linear inversion of two dimensional electrical conductivity: IEEE Trans. Antenna Propagat., **43**, 405–415.

Wang, T., and Hohmann, G. W., 1993, A finite difference time-domain solution for three-dimensional electromagnetic modeling: Geophysics, **58**, 797–809.

Wilt, M. J., Alumbaugh, D. L., Morrison, H. F., Becker, A., Lee, K. H., and Deszcz-Pan, M., 1995, Crosshole electromagnetic tomography: System design considerations and field results: Geophysics, **60**, 871–885.

Yee, K. S., 1966, Numerical solution of initial boundary problems involving Maxwell's equations in isotropic media: IEEE Trans. Antenna. Propaqat., **AP-14**, 302–309.

Zhdanov, M. S., and Fang, S., 1996, 3-D quasi-linear electromagnetic inversion: Radio Science, **31**, 741–754.

Zhang, J., Mackie, R. L., and Madden, T. R., 1995, Three-dimensional resistivity forward modeling and inversion using conjugate gradients: Geophysics, **60**, 1313–1325.

Electromagnetic Modeling on Parallel Computers

Andrew J. S. Wilson[1]
Kenneth MacDonald[2]
Liming Yu[3]
Bill Day[4]
Hamish Mills[5]

Summary. We have experimented with running three electromagnetic (EM) modeling codes on parallel machines: a 3-D integral-equation code, a finite-difference code for axisymmetric models, and a 3-D finite-difference code. All three codes calculate EM responses at several frequencies and then transform the results into a transient response. Our first method of parallelization uses a task farm where the work is divided into independent subtasks, which are distributed across the processors of a parallel computer. Each subtask calculates the response for one frequency, 30 to 40 of which are required to calculate a transient. This method gives excellent speedups on systems ranging from a departmental workstation cluster to a Cray T3D massively parallel supercomputer.

The 3-D finite-difference program was also ported onto a Thinking Machines CM200 and a DEC mpp 12000/SX and run in data-parallel mode. These single-instruction, multiple-data (SIMD) parallel machines employ several thousand simple processors and offer built-in support for simple operations on data arrays. They are easier to program than clusters of larger, more powerful, processors, but their simplicity limits flexibility in programming. With emerging parallel software standards, however, it soon may be possible to run the same software on clusters of workstations or massively parallel supercomputers with little change.

1 Introduction

Modern multichannel electromagnetic (EM) systems now can easily collect large volumes of data requiring 3-D interpretation (Hördt et al., 1995; Schnegg and Sommaruga,

[1]Edinburgh Parallel Computing Centre, University of Edinburgh, JCMB, KB, Mayfield Road, Edinburgh, EH9 3JZ, UK; E-mail:andrew.wilson@bgtech.co.uk. Formerly at the Department of Geology and Geophysics, University of Edinburgh.
[2]Department of Geology and Geophysics, University of Edinburgh, Edinburgh EH9 3JW, UK.
[3]Departement de Genie Mineral, Ecole Polytechnique, Case postale 6079, succ. Centre-ville, Montreal, Quebec H3C 3A7, Canada. Formerly at the Geophysics Laboratory, Department of Physics, University of Toronto.
[4]University of Minnesota Supercomputer Institute, 1200 Washington Avenue South, Minneapolis, MN 55 415-1227, USA.
[5]Edinburgh Parallel Computing Centre, University of Edinburgh, Edinburgh EH9 3JZ, UK.

1995; Wilt et al., 1995). However, 3-D modeling still requires too much computer power to be used routinely for this purpose. Consequently, the production of pseudosections by merging one-dimensional (1-D) inversions into two-dimensional (2-D) or 3-D images remains common, despite its limitations. Research in new algorithms for EM modeling may one day crack this problem (Walker and Groom, 1994), but in the absence of a real breakthrough it remains necessary to bring more computing power to the task. We have experimented with running three EM modeling programs on parallel machines using two different strategies. The first uses a modest number of processors, but is suitable for facilities ranging from a workstation cluster to a parallel supercomputer. The second adapts an iterative solver to run on a massively parallel computer. This paper describes our experience with these strategies. We first briefly review the modeling codes.

2 Integral-equation formulation

The EM3D package from the University of Utah calculates the EM response of a model at fixed frequency by solving an integral equation (Newman et al., 1986). Different sources can be constructed from finite-length grounded electric dipoles. Receivers can be placed in or over a layered Earth model containing a body of finite extent that is discretized into cuboidal cells.

The integral-equation formulation in EM3D divides the total EM field $\mathbf{E}_t$ into the normal field $\mathbf{E}_n$ of a layered background model, σ_n, and the anomalous field $\mathbf{E}_a$ of (3-D) heterogeneities, σ_a. The total conductivity of the model is simply

$$\sigma(\mathbf{x}) = \sigma_n(\mathbf{x}) + \sigma_a(\mathbf{x}). \tag{1}$$

The anomalous field $\mathbf{E}_a$ may be written in integral form as the sum of contributions from the scattering currents $\mathbf{J}_a = \sigma_a \mathbf{E}_t$ in the body,

$$\mathbf{E}_t(\mathbf{x}) = \mathbf{E}_n(\mathbf{x}) + \mathbf{E}_a(\mathbf{x}) \tag{2}$$

$$= \mathbf{E}_n(\mathbf{x}) + \int_V \underset{\sim}{\mathbf{G}}(\mathbf{x}, \mathbf{x}')\mathbf{J}_a(\mathbf{x}')\, d\mathbf{x}', \tag{3}$$

where $\underset{\sim}{\mathbf{G}}(\mathbf{x}, \mathbf{x}')$ is the tensor Green's function, which gives the electric field at a point $\mathbf{x}$ caused by an infinitesimal electric-current dipole at $\mathbf{x}'$ (see Fig. 1).

The unknown on the right-hand side of Eq. (3) is the scattering current, $\mathbf{J}_a$, which itself has the integral representation

$$\frac{\mathbf{J}_a(\mathbf{x})}{\sigma_a(\mathbf{x})} = \mathbf{E}_t(\mathbf{x}) \tag{4}$$

$$= \mathbf{E}_n(\mathbf{x}) + \int_V \underset{\sim}{\mathbf{G}}(\mathbf{x}, \mathbf{x}')\mathbf{J}_a(\mathbf{x}')\, d\mathbf{x}'. \tag{5}$$

This is actually a Fredholm integral equation of the second kind for $\mathbf{J}_a$, because it appears as an unknown on the left-hand side and inside the integral. To solve it, EM3D uses a series of approximations that reduces the equation to a linear algebraic system.

First, the anomalous body is divided into regions V^n, in which the scattering current is approximated by a constant J_a^n (pulse function):

$$\frac{\mathbf{J}_a(\mathbf{x})}{\sigma_a(\mathbf{x})} = \mathbf{E}_n(\mathbf{x}) + \int_V \underset{\sim}{\mathbf{G}}(\mathbf{x}, \mathbf{x}')\mathbf{J}_a(\mathbf{x}')\, d\mathbf{x}' \tag{6}$$

$$\approx \mathbf{E}_n(\mathbf{x}) + \sum_{n=1}^{N} \int_{V^n} \underset{\sim}{\mathbf{G}}(\mathbf{x}, \mathbf{x}')\mathbf{J}_a^n\, d\mathbf{x}'. \tag{7}$$

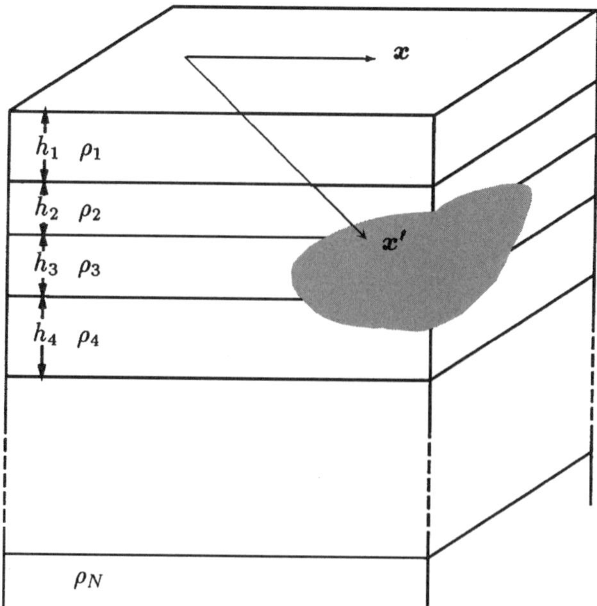

Figure 1. Green's tensor $\underset{\sim}{\mathbf{G}}(\mathbf{x}, \mathbf{x}')$ describes the field at a point $\mathbf{x}$ due to a infinitesimal scattering anomaly at $\mathbf{x}'$. The additional field $\mathbf{E}_a$ due to an anomaly may be written as the integral over the scattering body of the product of Green's tensor $\underset{\sim}{\mathbf{G}}$, the incident (normal) field $\mathbf{E}_n$, and the conductivity anomaly σ_a.

Evaluating this equation at some set of points $\mathbf{x}^m$, which is representative of the location of each cell V^m, gives

$$\frac{\mathbf{J}_a(\mathbf{x}^m)}{\sigma_a(\mathbf{x}^m)} \approx \mathbf{E}_n(\mathbf{x}^m) + \sum_{n=1}^{N} \int_{V^n} \underset{\sim}{\mathbf{G}}(\mathbf{x}^m, \mathbf{x}')\mathbf{J}_a^n \, d\mathbf{x}' \tag{8}$$

$$= \mathbf{E}_n(\mathbf{x}^m) + \sum_{n=1}^{N} \Gamma(\mathbf{x}^m, \mathbf{x}^n)\mathbf{J}_a^n, \tag{9}$$

where

$$\Gamma(\mathbf{x}^m, \mathbf{x}^n) = \int_{V^n} \underset{\sim}{\mathbf{G}}(\mathbf{x}^m, \mathbf{x}') \, d\mathbf{x}'. \tag{10}$$

Let $\underset{\sim}{\Gamma}^{m,n}$ denote the approximation of $\Gamma(\mathbf{x}^m, \mathbf{x}^n)$ resulting from the numerical integration of Green's tensor over the cell surrounding $\mathbf{x}^n$. The integral equation thus is reduced to a set of linear equations, which can be rearranged into the matrix system to give

$$\frac{\mathbf{J}_a^m}{\sigma_a^m} = \mathbf{E}_n^m + \sum_{n=1}^{N} \underset{\sim}{\Gamma}^{m,n}\mathbf{J}_a^n \quad m = 1, \ldots, N \tag{11}$$

$$\sum_{n=1}^{N} \left[\underset{\sim}{\Gamma}^{m,n} - \frac{\delta_{mn}}{\sigma_a^m}\right]\mathbf{J}_a^n = -\mathbf{E}_n^m \quad m = 1, \ldots, N \tag{12}$$

where N is the number of cells.

Calculation of the total field begins with computation of the matrices $\underset{\sim}{\Gamma}^{m,n}$, followed by solution of the matrix system for the (discrete) scattering currents. The total field at the receivers then is computed (by the same integral representation) as the sum of the anomalous field and the normal (layered Earth) field, which can be computed separately.

3 Axisymmetric finite-difference formulation

The electrical conductivity of an axisymmetric structure varies in the radial ρ- and vertical z-direction, but is independent of the azimuthal angle ϕ. If the electric dipole transmitter is placed arbitrarily with respect to the origin of symmetry, then the EM field must be described fully in three dimensions. The combination of an axisymmetric 2-D structure and a finite source—the 2.5-D problem—has been investigated by Goldman and Stoyer (1983), Chang and Anderson (1984), Liu (1993), Pai et al. (1993), and others. Most compute results for a source located on the axis of rotational symmetry.

The EMAFD package was first developed by Liming Yu and Nigel Edwards at the Geophysics Laboratory of the Department of Physics, University of Toronto (Yu, 1994). EMAFD uses a finite-difference approximation to calculate the response of an axisymmetric target in a double half-space background. The formulation of EMAFD also divides the EM field into a normal field (subscript n) and an anomalous field (subscript a). The normal field is the response of a background layered Earth, whereas the anomalous field is the additional field caused by an anomaly.

The EMAFD code works with the Laplace transform of the response. Maxwell's equations for the anomalous electric and magnetic fields can be written, after a Laplace transformation,

$$\nabla \times \mathbf{B}_a = \mu_0 \sigma \mathbf{E}_a + \mu_0 \sigma_a \mathbf{E}_n, \tag{13}$$

$$\nabla \times \mathbf{E}_a = -s \mathbf{B}_a, \tag{14}$$

where σ is the total conductivity and σ_a is anomalous conductivity, the difference between the background and total conductivities.

Because the variable ϕ is periodic, it is convenient to expand the fields in a Fourier series,

$$F(\rho, \phi, z) = \sum_{n=-\infty}^{\infty} f_n(\rho, z) \exp(-in\phi). \tag{15}$$

Consequently, we have

$$\frac{\partial}{\partial \phi} = -in. \tag{16}$$

In the wavenumber domain, the ϕ components of the electric and magnetic fields satisfy two coupled second-order differential equations,

$$\begin{aligned} &\frac{\partial}{\partial \rho}\left[\frac{\rho\mu\sigma}{\gamma}\frac{\partial(i\rho E_{a\phi})}{\partial \rho}\right] + \frac{\partial}{\partial z}\left[\frac{\rho\mu\sigma}{\gamma}\frac{\partial(i\rho E_{a\phi})}{\partial z}\right] - \frac{\mu\sigma}{\rho}(i\rho E_{a\phi}) = \\ &\quad + \frac{n}{\gamma^2}\frac{\partial\gamma}{\partial\rho}\frac{\partial(\rho B_{a\phi})}{\partial z} - \frac{n}{\gamma^2}\frac{\partial\gamma}{\partial z}\frac{\partial(\rho B_{a\phi})}{\partial\rho} - \frac{\partial}{\partial\rho}\left(\frac{n\rho\mu\sigma_a}{\gamma}E_{n\rho}\right) \\ &\quad + \frac{\partial}{\partial z}\left(\frac{n\rho\mu\sigma_a}{\gamma}E_{nz}\right) + \mu\sigma_a(iE_{n\phi}), \end{aligned} \tag{17}$$

and

$$\begin{aligned} &\frac{\partial}{\partial \rho}\left[\frac{s\rho}{\gamma}\frac{\partial(\rho B_{a\phi})}{\partial \rho}\right] + \frac{\partial}{\partial z}\left[\frac{s\rho}{\gamma}\frac{\partial(\rho B_{a\phi})}{\partial z}\right] - \frac{s}{\rho}(\rho B_{a\phi}) = \\ &\quad + \frac{n}{\gamma^2}\frac{\partial\gamma}{\partial\rho}\frac{\partial(i\rho E_{a\phi})}{\partial z} - \frac{n}{\gamma^2}\frac{\partial\gamma}{\partial z}\frac{\partial(i\rho E_{a\phi})}{\partial\rho} + \frac{\partial}{\partial\rho}\left(\frac{\rho^2 s\mu\sigma_a}{\gamma}E_{nz}\right) \\ &\quad - \frac{\partial}{\partial z}\left(\frac{\rho^2 s\mu\sigma_a}{\gamma}E_{n\rho}\right), \end{aligned} \tag{18}$$

where $\gamma = n^2 + \rho^2 s\mu\sigma$.

EMAFD approximates these equations with finite differences and uses an iterative technique to solve for the anomalous field. For a given azimuthal harmonic n and a Laplace variable value s, the electric field first is computed from Eq. (17), neglecting the magnetic-field terms. The magnetic field then is computed from Eq. (18) using the calculated electric field. The next iteration of Eq. (17) includes the calculated magnetic field, and so on until the fields stabilize, usually in three to eight iterations. The MUDPACK finite-difference package from NCAR (Adams, 1989) is used to solve the differential equations at each iteration. The ρ and z components of the electric and magnetic fields then are calculated from the ϕ component and its derivatives.

Finally, the fields are converted back to the time-space domain by applying inverse Fourier and Laplace transforms. The Gaver–Stehfest technique is used to invert the Laplace transform. This technique is neither the most accurate nor the most generally applicable, but it is fast, easy to code, and perhaps most important, only requires values of $F(s)$ for *real* values of s. A detailed description of the algorithm can be found in Stehfest (1970), Knight and Raiche (1982), Villinger (1984), and Edwards and Cheesman (1987).

4 A 3-D finite-difference formulation with thin sheets

The program developed by Xinghua Pu (1994) at the University of Victoria, British Columbia, Canada, allows a general 3-D conductivity structure, constrained by 2-D vertical boundaries, a uniform half-space below, and a thin sheet above, as shown in Fig. 2. A finite-difference scheme is used to solve for the magnetic-field vector, $\mathbf{B} = (X, Y, Z)$. The complete finite-difference equations are too lengthy to give here, but a brief description highlights the key points for the parallelization of the scheme. A

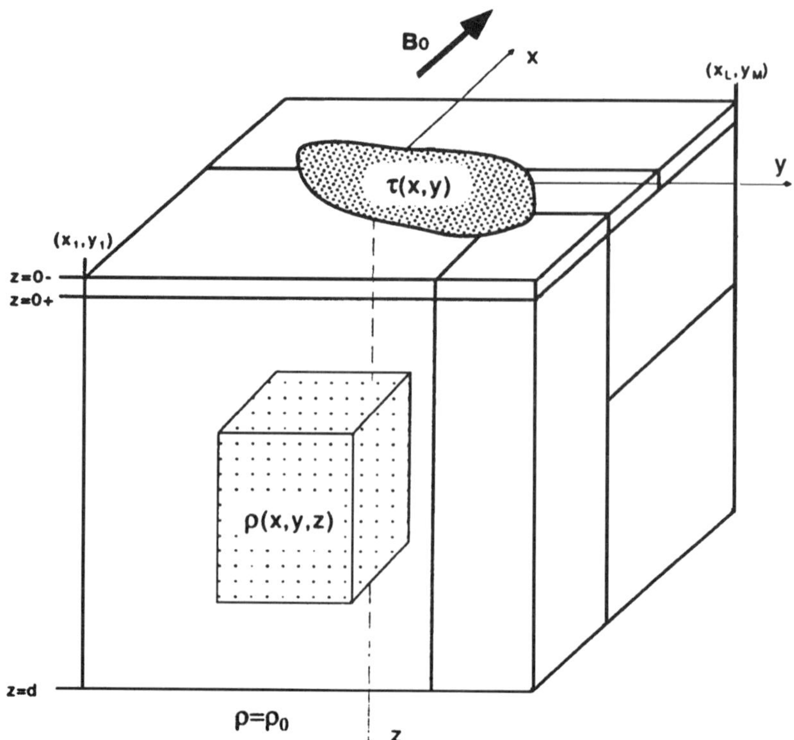

Figure 2. General 3-D model.

nine-point finite-difference stencil is written for both the E-polarized and B-polarized (TE-mode and TM-mode) 2-D magnetic fields around the vertical faces of the model, leading to fixed Dirichlet boundary conditions.

The thin sheet at the upper surface of the model allows shallow features to be incorporated without requiring many grid nodes for their representation. The approximation holds as long as it is much thinner than the skin depth inside it, and that high-conductivity layers are set deep in the model. The Z-component remains unchanged across the thin sheet ($Z- = Z+$), and is used to calculate the horizontal components,

$$X(\mathbf{r}, 0-) = B_0 - M_1 Z(\mathbf{r}, 0-), \quad Y(\mathbf{r}, 0-) = -M_2 Z(\mathbf{r}, 0-). \tag{19}$$

where B_0 is the source field and

$$M_2 Z = \frac{1}{2\pi} \int_{-\infty}^{\infty} \int_{-\infty}^{\infty} Z(u, v) \frac{y - v}{[(x-u)^2 + (y-v)^2]} \, du \, dy. \tag{20}$$

The top and the bottom of the model are constrained by integral boundary conditions, which define the field components at each point as a combination of the nearest-neighbor components and a surface integral over the whole boundary. The internal governing differential equation is obtained by setting $\mathbf{E} = \rho \nabla \times \mathbf{B}$, and integrating Faraday's law over the cuboid with surface S and volume V, surrounding a grid point:

$$\oint_S \rho (\nabla \times \mathbf{B}) \times d\mathbf{S} = i\omega\mu_0 \int_V \mathbf{B} \, dv. \tag{21}$$

The finite-difference equations relate each field component at any non-boundary grid node to all field components at the surrounding 27 nearest neighbors, requiring (at most) 243 coefficients. These equations then are solved by an iterative relaxation method, serial Gauss–Seidel, or data-parallel point Jacobian.

5 Parallel computation

It is possible to increase the power of a computer either by adding a faster processor or by adding more processors (and developing the software necessary to use them). For many problems it is cheaper to take the latter course. Conventional computers employ the single instructions, single data (SISD) stream, or Von Neumann programming model (Flynn, 1972). Symmetric multiprocessing (SMP) computers give several independent processors access to global memory store (Fig. 3). Distributed-memory, multiple-instructions, multiple-data (DM-MIMD) computers are analogous to a network of workstations, but with an unusually fast interconnect (Fig. 4). For applications with

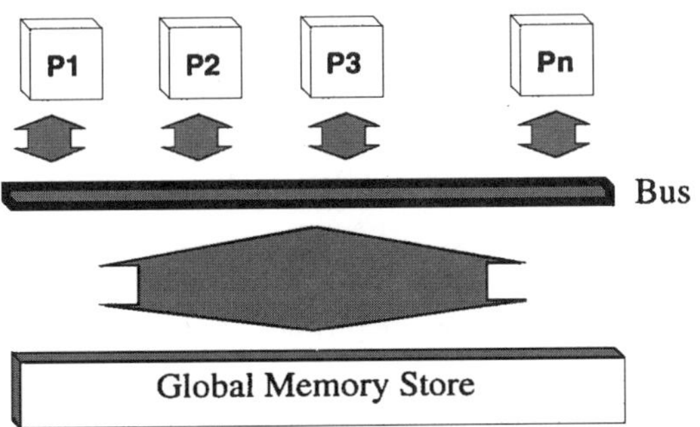

Figure 3. Schematic of SMP architecture.

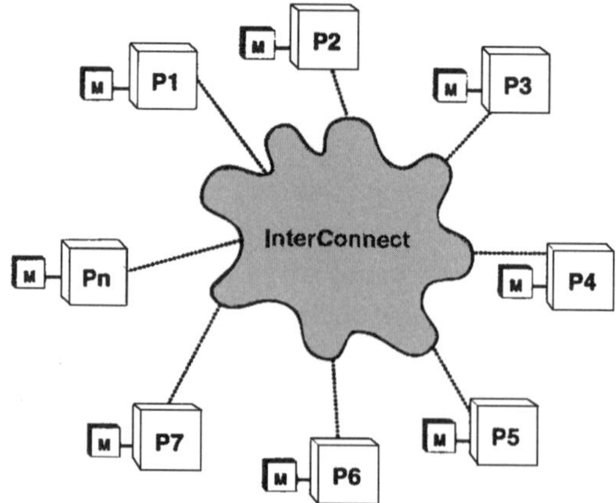

Figure 4. Schematic of DM-MIMD architecture.

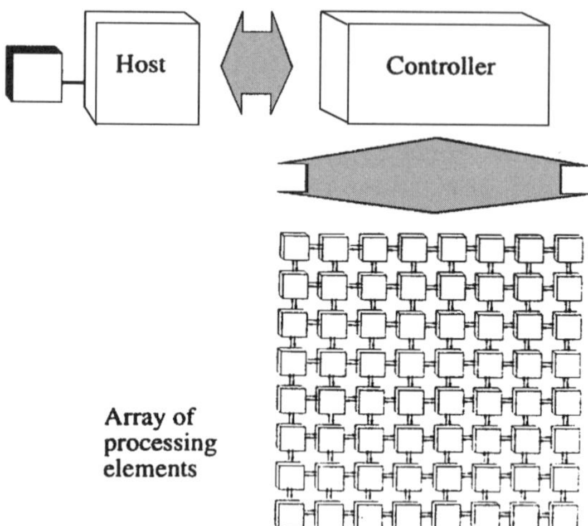

Figure 5. Schematic of SIMD architecture.

low communications requirements, the speed of the interconnecting network is not a restriction, and networks of workstations also may be effectively used as DM-MIMD parallel computers. Single-instruction multiple-data (SIMD) machines retain a single instruction stream, but have multiple data streams, all of which are acted upon at the same time, and in the same way, by a large number of processors (Fig. 5).

At the present time, the use of SIMD parallel computers for general-purpose supercomputing is declining. Non-uniform memory access (NUMA) computers, which blur the distinction between SMP and DM-MIMD machines, are an emerging trend. Through a combination of hardware and software, NUMA architectures support global access to a distributed memory, some of which is local (fast access) and some remote (relatively slower access).

6 Fine- and coarse-grain parallelization

To take advantage of a parallel computer, a problem must be decomposed into *grains* that can be executed simultaneously on different processors. Parallelization techniques can be classified according to whether the problem is divided on the basis of functionality or data, and on the number or "size" of the grains into which the problem is decomposed.

A coarse-grained parallelization technique divides the problem into a few large grains. This limits the number of processors that can be employed profitably. A fine-grained parallelization divides the problem into very many grains, which offers a high degree of parallelism. However, the computation of each grain must be managed and this overhead can grow as the number of grains increases.

It is important to match the parallelization technique with the problem and the computer being targeted. SIMD computers often provide very effective support for fine-grained parallelization of mesh-based problems. However, because this essentially is provided by hardware, it often can support only very simple computations. DM-MIMD computers are more flexible but more difficult to program.

6.1 Parallelization by task farming

If the main stage of a calculation can be decomposed into several independent tasks, then there is no need for communication between processes executing each task and *task farming* may be employed. A classical task farm consists of one *source* process that generates tasks, one or more *worker* processes that service them to produce results, and one *sink* process that handles results (Fig. 6). Each worker requests a task from the source process, executes it and passes the results on to the sink process, and then requests the next task from the source process. This cycle repeats until all tasks have been completed. Task farming generally is considered a coarse-grain approach, well suited to DM-MIMD computers.

6.2 Parallelization by data decomposition

Many simulations apply similar or identical operations to every member of a large data set. A common example is the solution of partial differential equations using a mesh discretization and finite-difference approximations. If the calculation for several mesh sites can proceed at the same time, then the simulation can be parallelized by data

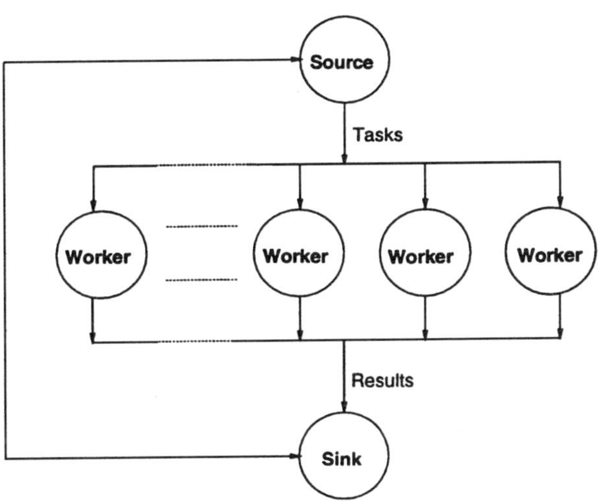

Figure 6. Classical task farm.

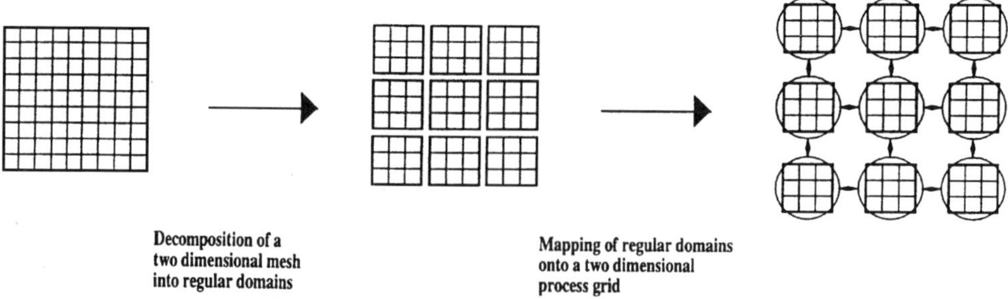

Figure 7. Geometric domain decomposition in two dimensions.

decomposition. In data decomposition, such a mesh is decomposed over processes, so that every process is responsible for storage and computation of one portion of the mesh (Fig. 7).

7 Task-farm parallelization of EM codes

The EM3D, EMAFD, and Pu codes all work one frequency at a time. Many frequencies are required to calculate accurate transient responses. Obviously, one can increase performance by running the calculation for different frequencies simultaneously on different processors. Normally, a series of different Earth models would be investigated, which further extends the number of processors that can work in parallel.

EM3D, EMAFD, and the program of Poll and Weaver have been parallelized using a task farm. PUL-TF, a parallel utility library written at the Edinburgh Parallel Computer Centre (Bruce et al., 1995), was chosen as a ready-built task-farm implementation. Subroutines were written to perform the source, sink, and worker tasks and were linked to the PUL-TF libraries; the main program was altered to call the PUL-TF utility which then controls and coordinates the operation of the task farm (see Figs. 8, 9 and 10).

In this parallel implementation of EM3D, every process in the task farm contains the full functionality of the application. At run time, the first process to register with the message-passing system is selected as the master process. The master process reads in the model input file and counts how many tasks (models and frequencies) there are before passing control to PUL-TF. During operation of the task farm, the master process will act as source and sink but also may act as a worker if this would help spread computation evenly over the processors. This single-program, multiple-data programming model is required by some DM-MIMD computers. Note that PUL-TF is very flexible and does not force the user to use a SPMD programming model.

The source subroutine packs into an array a description of the next task to be executed. The worker subroutine unpacks the task description and calls the main calculation routine. This picks the specified model and frequency from the model description file and calculates the response. Because many processes were run at the same time, all I/O had to be modified so that each process would write to a different file. This was achieved easily by using the number of the task currently being worked upon as an extension to filenames. After the main calculation for each task, the worker subroutine packs into an array a confirmation that the task has been completed correctly. This is passed to the sink subroutine, which in this task farm only undertakes simple bookkeeping of which tasks have been completed.

```
c  ---------------------------------------------------------------------
c@ $Id$
c@ em3d: EM simulation using integral equations
c~
c~ References:
c~ G.A. Newman, G.W. Hohmann and W.L. Anderson),
c~ Transient electromagnetic response of a three-dimensional body in
c~ a layered earth. Geophysics, V51(8) 1608-1627, 1986.
c  ---------------------------------------------------------------------

#include "em3dfarm.inc"

        program em3d

c ---   variable declarations

        dimension freqs(MAX_FREQS)
        integer i, nfreq

c ---   open and read input file
        open(8,file='freqs',status='old')
        read(8,*) nfreq
        print *,'reading ',nfreq,' frequencies'
        do i=1,nfreq
          read(8,*) freqs(i)
        enddo
        close(8)

c ---   for each frequency
        do i=1,nfreq
c ---     call main fortran routines
          call em3d1(freqs(i), 's')
          call em3d2(freqs(i), 's')
          call em3d3(freqs(i), 's')
          call em3d4(freqs(i), 's')
        enddo

        end
c ---   end of em3d main program
c  ---------------------------------------------------------------------
```

Figure 8. Main program section for sequential version of EM3D.

Reusable utilities such as PUL-TF minimize porting effort. Porting one of the simulation programs required 100 lines of new code and around 60 modified lines to port a total of 5500 lines. However, utilities such as PUL-TF cannot mitigate against poor software design. Another of the simulations required dramatic restructuring before it could be parallelized sensibly.

In addition to reducing coding effort, PUL-TF increases portability by hiding some system dependencies. Versions of the PUL-TF library are available for all of the main parallel systems, and our parallel programs have been tested on networks of Sun and SGI workstations, T800- and i860-based Meiko Computing Surfaces and a Cray T3D.

8 Fine-grained parallelization

The computational cost of the Pu 3-D finite-difference code is dominated by an iterative Gauss–Seidel solver. The simulation was adapted to run on SIMD architecture parallel computers by implementing the solver routines in variants of High-Performance Fortran (HPF). The same code, with minor differences in compiler directives and HPF intrinsics, runs on the Connection Machine CM200 at Edinburgh, and the DEC MasPar at the British Columbia Provincial Government computing facility.

Both machines allowed the construction of the coefficient matrices and 2-D boundary-condition solutions to be performed on their workstation-class front ends. These steps take an insignificant time to perform, compared to the solution of the system of equations, and so are not re-implemented in parallel.

Iterative solvers repeatedly improve an estimate of the solution until the desired accuracy is achieved, and are well suited to fine-grained parallelization. A parallel point Jacobian solver was integrated into the original Pu code.

```
c ------------------------------------------------------------------
c@ $Id$
c@ em3dfarm: task farm parallel version of em3d
c@ EM simulation using integral equations
c~
c~ References:
c~ G.A. Newman, G.W. Hohmann and W.L. Anderson),
c~ Transient electromagnetic response of a three-dimensional body in
c~ a layered earth. Geophysics, V51(8) 1608-1627, 1986.
c ------------------------------------------------------------------

#include "em3dfarm.inc"

      program em3dfarm

c --    include header files for MPI, PUL and PUL-TF
#include <mpif.h>
#include <pul-tf.inc>
#include <pul.inc>

c ---   variable declarations

c --    task farm
        integer status, farm, rank, masrank, thefarm, ierror
        parameter (masrank=0)

c --    model description
        common /blkfreqs/ freqs(MAX_FREQS)
        common /blkcount/ icount, nfreq
        integer i

c --    declare the subroutines which implement the task generation,
c --    task execution and result collation
c --    a cpp macro is used to hide cross-calling formatting differences
        EXTERNAL TFFFN(maketask)
        EXTERNAL TFFFN(dotask)
        EXTERNAL TFFFN(processresult)

c --    Initialize the MPI world communicator.
        call MPI_INIT(ierror)
        if (ierror.ne.MPI_SUCCESS) then
          print *, 'em3dfarm: error in MPI initialization'
        endif
        thefarm=MPI_COMM_WORLD

c --    create handle for the task farm using PUL-TF for MPI.
        farm = TFinitMPI(thefarm)

c ---   this task farm operates in SPMD mode and each process
c ---   decides at run-time if it will be a source, worker or sink
c ---   we choose one ''master'' process to operate as a source

        call MPI_COMM_RANK(thefarm,rank,ierror)
        if (ierror.ne.MPI_SUCCESS) then
          print *, 'em3dfarm: error identifying master process'
        endif

c ---   if this is the master process then
        if (rank.eq.masrank) then
c ---     open and read input file
          open(8,file='freqs',status='old')
          read(8,*) nfreq
          PRINT *,'reading ',nfreq,' frequencies'
          do i=1,nfreq
            read(8,*) freqs(i)
          enddo
          close(8)
          icount=1

c ---     the master can operate as source, worker or sink
          status = TFopen(farm, TFSRCWRKSNK,
     &                SIZEOF_PARAM_STRUCT, SIZEOF_PARAM_STRUCT)
        else
c ---     other processes can only operate as workers
          status = TFopen(farm, TFWRK,
     &                SIZEOF_PARAM_STRUCT, SIZEOF_PARAM_STRUCT)
        endif

        if (status.ne.TFOK) then
          print *,'em3dfarm: TFopen failed with error value ',status
        elseif (rank.eq.masrank) then
          print *,'em3dfarm: TFopen succeeded.'
        endif

        status = TFoperate(farm, TFFFN(maketask),
     &          TFFFN(dotask), TFFFN(processresult))

        if (status .NE. TFOK) then
          print *,'em3dfarm: TFoperate failed with error value ',status
        elseif (rank.eq.masrank) then
          print *,'em3dfarm: TFoperate succeeded.'
        endif

        status=TFclose(farm)

        if (status .NE. TFOK) then
          print *,'em3dfarm: TFclose failed with error value ',status
        elseif (rank.eq.masrank) then
          print *,'em3dfarm: TFclose succeeded.'
        endif

c ---   Finalize and shut-down MPI system.
        call MPI_FINALIZE(ierror)
        if (ierror.NE.MPI_SUCCESS) then
          print *,'em3dfarm: error in MPI_FINALIZE'
        endif

        end
c ---   end of em3dfarm main program
c ------------------------------------------------------------------
```

Figure 9. Main program section for task-farm parallel version of EM3D.

```
c ---------------------------------------------------------------------
c@ $Id: tasks.F,v 1.1 1996/07/03 16:21:46 ajsw Exp ajsw $
c@ tasks.F: subroutines which implement the task generation,
c@         task execution and result collation for em3d task farm

c       key for parameter descriptions
c
c       I       must be set in input
c       O       will be set on output
c ---------------------------------------------------------------------

#include "em3dfarm.inc"

c ---------------------------------------------------------------------
        subroutine maketask(taskout, outlength, retval)
c@      subroutine maketask: packs a description of a task into an array

      real      taskout(2)
      integer   outlength, retval

      common /blkfreqs/ freqs(MAX_FREQS)
      common /blkcount/ icount,nfreq

c~      variable        I/O     description
c~
c~      taskout         O       task description packed into an array
c~      length          O       length of array in integers
c~      retval          O       length of array in bytes, 0 if finished
c ---------------------------------------------------------------------

        print *,'maketask: making task ',icount,' of ',nfreq
        if (icount.GT.nfreq) then
          print *,'maketask: last frequency completed'
          retval = 0
        else
          taskout(1)=icount
          taskout(2)=freqs(icount)
          icount=icount +1
         retval = SIZEOF_PARAM_STRUCT
c -     usually I would now set the length of the task description array
c -     however in this task farm this will always be the same length
c -     and so this is not used
c       outlength=2
        endif
        return
        end
c ---   end of subroutine maketask

c ---------------------------------------------------------------------
        subroutine dotask(taskin, inlength, taskout, outlength, retval)
        implicit none

c@      subroutine dotask: process each task as it arrives
c@      by making a call to the main fortran routines

c~      variable        I/O     description
c~
c~      taskin          I       task description packed into an array
c~      inlength        I       length of input task array in integers
c~      taskout         O       task results packed into an array
c~      outlength       O       length of output task array in integers
c~      retval          O       length of output task array in bytes
c ---------------------------------------------------------------------

      real      taskin(2), taskout(2)
      integer   inlength, outlength, retval
      integer   i, count, zero
      character suffix(SUFFIX_LEN)

c --  we simply copy the input task description into the output task
c --  this allows the processresult task to know which task has just
c --  been completed

      taskout(1)=taskin(1)
      taskout(2)=taskin(2)

c --- generate unique suffix for output files
      count=taskin(1)
      zero=ICHAR('0')
      do i=SUFFIX_LEN-1,0,-1
        suffix(SUFFIX_LEN-i)=CHAR(zero+count/(10**i))
        count=count-10**i*INT(count/(10**i))
      enddo
      count=taskin(1)
      PRINT *,'dotask: doing task ',count,' suffix ',suffix
#ifndef FARMTEST
      call em3d1(taskin(2), suffix)
      call em3d2(taskin(2), suffix)
      call em3d3(taskin(2), suffix)
      call em3d4(taskin(2), suffix)
#endif
      retval = SIZEOF_PARAM_STRUCT
      return
      end
c --- end of subroutine dotask

c ---------------------------------------------------------------------
        subroutine processresult(taskin, length, retval)
c@      subroutine processresult: calls out as tasks are finished
c~      variable        I/O     description
c~
c~      taskin          I       task description packed into an array
c~      length          I       length of input task array in integers
c~      retval          O       -ve on error

      real taskin(2)
      integer   length, retval, count
      common /blkcount/ icount,nfreq

c --- print out as each task is finished
      count=taskin(1)
      print *,'em3dfarm: completed frequency ',count,' of ',nfreq
      retval = 1
      return
      end
c --- end of subroutine processresult
```

Figure 10. Subroutines to implement task generation, task execution, and result collation.

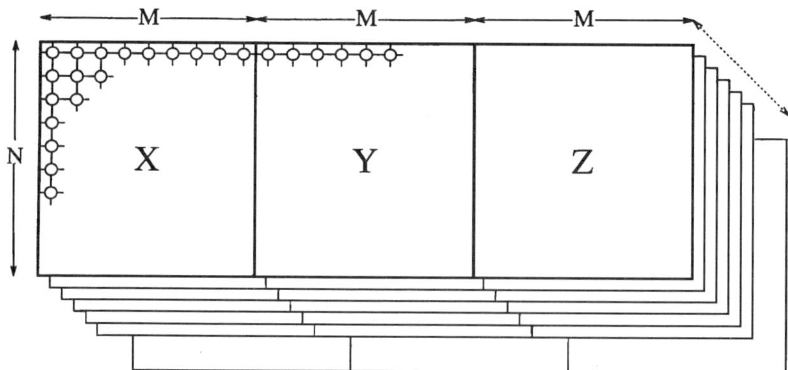

Figure 11. Storage of the three-vector field and coefficient components over a horizontal model plane.

The data-parallel iterative solution assigns each grid location to one virtual processor, the ultimate in data decomposition as described in Fig. 7. Low-level firmware and software map these to however many physical processing nodes are actually available. Controlling this mapping is essential to maintain an efficient parallel execution, because the definition of a SIMD computer states that the processors all must be performing the same tasks, or be idle. In the worst-case scenario of $n+1$ virtual processors being mapped to n physical processors, the machine will be operating at approximately 50% efficiency.

The three components (X, Y, and Z) are stored on offset nodes, as shown in Fig. 11. Each set of $3M \times N$ planes stores the field and the coefficients associated with a single horizontal $M \times N$ plane in the model. This storage scheme guarantees symmetry between the components, allowing the whole relaxation to be performed on the parallel processing unit.

The pth iterative step for internal grid points then can be written in a simple dependency form,

$$X^{p+1} \leftarrow A_x X^p, B_x Y^p, C_x Z^p$$
$$Y^{p+1} \leftarrow A_y Y^p, B_y Z^p, C_y X^p$$
$$Z^{p+1} \leftarrow A_z Z^p, B_z X^p, C_z Y^p.$$

Each term actually represents a series of nearest-neighbor elemental combinations. The HPF `CSHIFT()` function translates the entire field array along the coordinate axes, using high-bandwidth regular communication channels. Thus every element of the array is updated in one time step. The pattern of offset storage allows all three components to be updated simultaneously. The pth field values are translated by M elements after each local set of terms is calculated, using the `CSHIFT()` function with a displacement argument of M. Implimented in this way, the parallel code is a little longer than the serial version (Figs. 13, 14).

The integral boundary conditions and the thin-layer calculations require a rearrangement of field values. Multiple copies of the field values are made to form an array conformant with the coefficient arrays, M, in Eq. (20). The integral is reduced to a simple elemental multiplication and a summation along the rows, using the HPF `SUM()` reduction operator (see Figs. 12 and 15).

The range of active grid nodes for the relaxation are controlled by the HPF construct `WHERE (MASK)...END WHERE`. Processors that are masked out are inactive, and so, they do not update their field value.

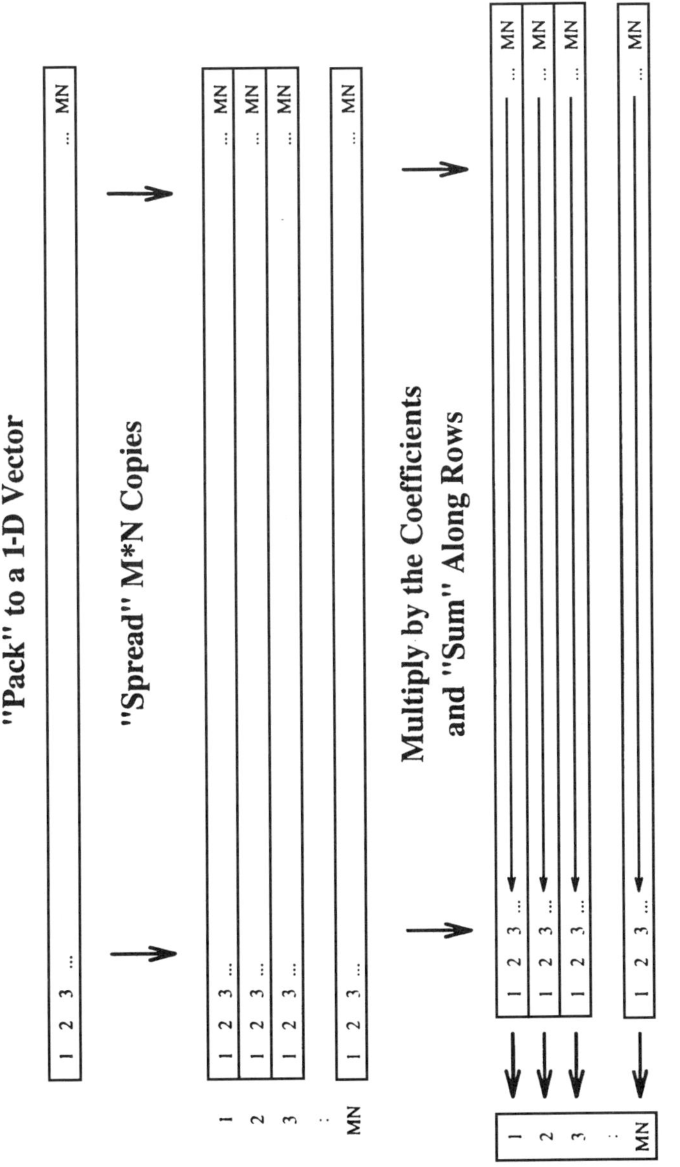

Figure 12. Parallel calculation of surface integrals.

```
      DO 10 K=2,NN-1
         DO 20 J=2,MM-1
            DO 30 I=2,LL-1
               DO 40 II=1,18
                  C(II)=CEL(II,I,J,K)
40             CONTINUE
               X010 = XX(I-1,J  ,K-1)
               X110 = XX(I  ,J  ,K-1)
               X210 = XX(I+1,J  ,K-1)
               X001 = XX(I-1,J-1,K  )
               X101 = XX(I  ,J-1,K  )
               X201 = XX(I+1,J-1,K  )
               X011 = XX(I-1,J  ,K  )
               X111 = XX(I,  J,  K  )
               X211 = XX(I+1,J  ,K  )
               X021 = XX(I-1,J+1,K  )
               X121 = XX(I  ,J+1,K  )
               X221 = XX(I+1,J+1,K  )
               X012 = XX(I-1,J  ,K+1)
               X112 = XX(I  ,J  ,K+1)
               X212 = XX(I+1,J  ,K+1)
               Y100 = YY(I  ,J-1,K-1)
               Y110 = YY(I  ,J  ,K-1)
               Y120 = YY(I  ,J+1,K-1)
               Y001 = YY(I-1,J-1,K  )
               Y101 = YY(I  ,J-1,K  )
               Y201 = YY(I+1,J-1,K  )
               Y011 = YY(I-1,J  ,K  )
               Y111 = YY(I,  J,  K  )
               Y211 = YY(I+1,J  ,K  )
               Y021 = YY(I-1,J+1,K  )
               Y121 = YY(I  ,J+1,K  )
               Y221 = YY(I+1,J+1,K  )
               Y102 = YY(I  ,J-1,K+1)
               Y112 = YY(I  ,J  ,K+1)
               Y122 = YY(I  ,J+1,K+1)
               Z100 = ZZ(I  ,J-1,K-1)
               Z010 = ZZ(I-1,J  ,K-1)
               Z110 = ZZ(I  ,J  ,K-1)
               Z210 = ZZ(I+1,J  ,K-1)
               Z120 = ZZ(I  ,J+1,K-1)
               Z101 = ZZ(I  ,J-1,K  )
               Z011 = ZZ(I-1,J  ,K  )
               Z111 = ZZ(I,  J,  K  )
               Z211 = ZZ(I+1,J  ,K  )
               Z121 = ZZ(I  ,J+1,K  )
               Z102 = ZZ(I  ,J-1,K+1)
               Z012 = ZZ(I-1,J  ,K+1)
               Z112 = ZZ(I  ,J  ,K+1)
               Z212 = ZZ(I+1,J  ,K+1)
               Z122 = ZZ(I  ,J+1,K+1)
               C1C4=C(1)+C(4)
               C2C5=C(2)+C(5)
               C3C6=C(3)+C(6)
               A11= CI+C2C5+C3C6
               A22= CI+C1C4+C3C6
               A33= CI+C1C4+C2C5
               A12= C(15)+C(18)-C(9 )-C(12)
               A13= C(17)+C(14)-C(8 )-C(11)
               A23= C(13)+C(16)-C(7 )-C(10)
               CX = C(2 )*X121 + C(5 )*X101 + C(3 )*X112 + C(6 )*X110
     &              -C(9 )*Y221 + C(15)*Y201 + C(18)*Y021 - C(12)*Y001
     &              -C(8 )*Z212 + C(17)*Z210 + C(14)*Z012 - C(11)*Z010
     &              +(C(15)-C(9 ))*Y211
     &              +(C(18)-C(12))*Y011
     &              +(C(9 )-C(18))*Y121
     &              +(C(12)-C(15))*Y101
     &              +(C(17)-C(8 ))*Z211
     &              +(C(14)-C(11))*Z011
     &              +(C(8 )-C(14))*Z112
     &              +(C(11)-C(17))*Z110
               CY = C(1 )*Y211 + C(4 )*Y011 + C(3 )*Y112 + C(6 )*Y110
     &              -C(9 )*X221 + C(15)*X201 + C(18)*X021 - C(12)*X001
     &              -C(7 )*Z122 + C(13)*Z120 + C(16)*Z102 - C(10)*Z100
     &              -(C(15)-C(9 ))*X211
     &              -(C(18)-C(12))*X011
     &              -(C(9 )-C(18))*X121
     &              -(C(12)-C(15))*X101
     &              +(C(13)-C(7 ))*Z121
     &              +(C(16)-C(10))*Z101
     &              +(C(7 )-C(16))*Z112
     &              +(C(10)-C(13))*Z110
               CZ = C(1 )*Z211 + C(4 )*Z011 + C(2 )*Z121 + C(5 )*Z101
     &              -C(8 )*X212 + C(17)*X210 + C(14)*X012 - C(11)*X010
     &              -C(7 )*Y122 + C(13)*Y120 + C(16)*Y102 - C(10)*Y100
     &              -(C(17)-C(8 ))*X211
     &              -(C(14)-C(11))*X011
     &              -(C(8 )-C(14))*X112
     &              -(C(11)-C(17))*X110
     &              -(C(13)-C(7 ))*Y121
     &              -(C(16)-C(10))*Y101
     &              -(C(7 )-C(16))*Y112
     &              -(C(10)-C(13))*Y110
               CX   = CX-A12*Y111-A13*Z111
               CY   = CY-A12*X111-A23*Z111
               CZ   = CZ-A13*X111-A23*Y111
               ALPHA=ALPIJK(I,J,K)
               G0=G(I-1)*ALPHA
               G2=G(I  )*ALPHA
               R2G0=2D0/G0
               R2G2=2D0/G2
               RGS =1D0/(G0+G2)
               H0=H(J-1)*ALPHA
               H2=H(J  )*ALPHA
               R2H0=2D0/H0
               R2H2=2D0/H2
               RHS =1D0/(H0+H2)
               K0=FK(K-1)*ALPHA
               K2=FK(K  )*ALPHA
               R2K0=2D0/K0
               R2K2=2D0/K2
               RKS =1D0/(K0+K2)
               D1=(  X211*R2G2+X011*R2G0
     &              +(Y221-Y201-Y021+Y001)*RHS
     &              +(Z212-Z210-Z012+Z010)*RKS   )*RGS
               D2=(  Y121*R2H2+Y101*R2H0
     &              +(X221-X201-X021+X001)*RGS
     &              +(Z122-Z120-Z102+Z100)*RKS   )*RHS
               D3=(  Z112*R2K2+Z110*R2K0
     &              +(X212-X210-X012+X010)*RGS
     &              +(Y122-Y120-Y102+Y100)*RHS   )*RKS
               F1=2D0/(G0*G2)
               F2=2D0/(H0*H2)
               F3=2D0/(K0*K2)
               XNEW=(CX+D1)/(A11+F1)
               YNEW=(CY+D2)/(A22+F2)
               ZNEW=(CZ+D3)/(A33+F3)
               IF(MCHECK.EQ.0) THEN
                  ERR=CDABS(XNEW-X111)+CDABS(YNEW-Y111)+CDABS(ZNEW-Z111)
                  IF(ERR.GT.ERROR) THEN
                     ERROR = ERR
                     IP = I
                     JP = J
                     KP = K
                  ENDIF
               ENDIF
               XX(I,J,K) = XNEW
               YY(I,J,K) = YNEW
               ZZ(I,J,K) = ZNEW
30          CONTINUE
20       CONTINUE
10    CONTINUE
```

Figure 13. Serial iteration step for internal cells.

9 Results and performance analysis

9.1 *Task-farm parallelization of EM3D*

A series of experiments was undertaken to test the effectiveness of the task-farm parallelization of the EM3D simulation on a heterogeneous workstation cluster. To keep the computational cost of these simulations to a reasonable level, a very simple test model was chosen from the literature (Fig. 9, Newman et al., 1986). This model consists of a single anomalous body in a half-space illuminated by a loop source and with a line of receivers. Symmetry is exploited to allow discretization of the body using just five cells.

To measure the success of a parallelization, we might plot the total execution times of an n-processor implementation against n. Because the workstation cluster employed is heterogeneous, and each workstation has a different performance, it is more informative to plot against total capacity of the workstations employed. The execution time for the simple test model was measured on each workstation to gauge its relative performance, which was then normalized relative to the performance on a Sun Sparc 2 (see Table 1).

Table 1. Timing Results for dike4 model using one frequency

Machine	Abbreviation	Workstation model	Elapsed time (min)	Normalized performance
Briar	b	Sparc 10	1.8	2.36
Fungi	f	Sparc 2	4.3	0.99
Gorse	g	Sparc 2	4.25	1.00
Rose	r	Sparc 2	5.0	0.85
Sitka	s	Sun ELC	5.333	0.797
Sage	sg	Sun ELC	5.75	0.748
Oak	o	Sun IPC	7.133	0.596

```
      SUBROUTINE ParITECEL (Field, A, B, C, DeltaField, NX, NY, NZ)

C     Do one iteration over the central points.  The coefficients are store
d
C     in three arrays (A, B, and C).  Each consists of several 'planes'
C     conformable with the field array.

C     A links corresponding components
C     B links the component to component + 1
C     C links the component to component + 2

C     Essentially, we have the following three equations for the three
C     components.

C     X = A(1:NX) * X + B(1:NX) * Y + C(1:NX) * Z
C     Y = A(NX+1:2NX) * Y + B(NX+1:2NX) * Z + C(NX+1:2NX) * X
C     Z = A(2NX+1:3NX) * Z + B(2NX+1:3NX) * X + C(2NX+1:3NX) * Y

      IMPLICIT NONE

C     Arguments

      INTEGER NX, NY, NZ          ! The size of the model
C     OMPLEX*16  Field(3*NX, NY, NZ) ! The B field (Bx|By|Bz)
      REAL*8  A(3*NX, NY, NZ, 7) ! Same component coefficients
      REAL*8  B(3*NX, NY, NZ, 19) ! Coefficients for component+1
      REAL*8  C(3*NX, NY, NZ, 19) ! Coefficients for component+2
C     OMPLEX*16  DeltaField(3*NX, NY, 0:NZ)! The field changes

C     Local Variables

C     OMPLEX*16  NewField(3*NX, NY, NZ)  ! The updated field
C     OMPLEX*16  FieldShift(3*NX, NY, NZ)! The shifted field
      LOGICAL Central(3*NX, NY, NZ) ! Central points
C     OMPLEX*16  i   ! The square root of (-1)

C     Compiler directives

#ifdef MasPar_Arch
CMPFONDPU NewField
CMPFONDPU FieldShift
CMPFONDPU Central
#endif
#ifdef CM200_Arch
CMF$LAYOUT Field(:NEWS, :NEWS, :NEWS)
CMF$LAYOUT A(:NEWS, :NEWS, :NEWS, :SERIAL)
CMF$LAYOUT B(:NEWS, :NEWS, :NEWS, :SERIAL)
CMF$LAYOUT C(:NEWS, :NEWS, :NEWS, :SERIAL)
CMF$LAYOUT DeltaField(:NEWS, :NEWS, :NEWS)
CMF$LAYOUT NewField(:NEWS, :NEWS, :NEWS)
CMF$LAYOUT FieldShift(:NEWS, :NEWS, :NEWS)
CMF$LAYOUT Central(:NEWS, :NEWS, :NEWS)
#endif

C     Begin Code

C     First Set 'i'

      i = (0.0, 1.0)

C     Define the central points

C     entral = .FALSE.   ! All off by default
C     entral(2:NX-1, 2:NY-1, 2:NZ-1) = .TRUE.! X components on
C     entral(NX+2:2*NX-1, 2:NY-1, 2:NZ-1) = .TRUE.   ! Y components on
C     entral(2*NX+2:3*NX-1, 2:NY-1, 2:NZ-1) = .TRUE. ! Z components on

C     Calculate the new field values for all the points, and only worry
C     about which ones we are interested in at the end, when we copy back t
o
C     old field.

C     Multiply the same components

      NewField = A(:, :, :, 2) * CSHIFT (Field, DIM=1, SHIFT=-1) + ! 011
     &     A(:, :, :, 3) * CSHIFT (Field, DIM=1, SHIFT=1) + ! 211
     &     A(:, :, :, 4) * CSHIFT (Field, DIM=2, SHIFT=-1) + ! 101
     &     A(:, :, :, 5) * CSHIFT (Field, DIM=2, SHIFT=1) + ! 121
     &     A(:, :, :, 6) * CSHIFT (Field, DIM=3, SHIFT=-1) + ! 110
     &     A(:, :, :, 7) * CSHIFT (Field, DIM=3, SHIFT=1) ! 112

C     Now the component to the right. Eg. X -> Y, Y -> Z, Z -> X
C     Copy the field to the shifted field

      FieldShift = CSHIFT (Field, DIM=1, SHIFT=NX)

      NewField = NewField +     ! Remember the same components
     &     B(:, :, :, 1) * FieldShift + ! 111
     &     B(:, :, :, 2) * CSHIFT (FieldShift, DIM=1, SHIFT=-1) + ! 011
     &     B(:, :, :, 3) * CSHIFT (FieldShift, DIM=1, SHIFT=1) + ! 211
     &     B(:, :, :, 4) * CSHIFT (FieldShift, DIM=2, SHIFT=-1) + ! 101
     &     B(:, :, :, 5) * CSHIFT (FieldShift, DIM=2, SHIFT=1) + ! 121
     &     B(:, :, :, 6) * CSHIFT (FieldShift, DIM=3, SHIFT=-1) + ! 110
     &     B(:, :, :, 7) * CSHIFT (FieldShift, DIM=3, SHIFT=1) + ! 112
     &     B(:, :, :, 8) *
     &     CSHIFT(CSHIFT(FieldShift,DIM=1,SHIFT=-1),DIM=2,SHIFT=-1)+ ! 001
     &     B(:, :, :, 9) *
     &     CSHIFT(CSHIFT(FieldShift,DIM=1,SHIFT=-1),DIM=2,SHIFT=1)+ ! 021
     &     B(:, :, :, 10) *
     &     CSHIFT(CSHIFT(FieldShift,DIM=1,SHIFT=-1),DIM=3,SHIFT=-1)+ ! 010
     &     B(:, :, :, 11) *
     &     CSHIFT(CSHIFT(FieldShift,DIM=1,SHIFT=-1),DIM=3,SHIFT=1)+ ! 012
     &     B(:, :, :, 12) *
     &     CSHIFT(CSHIFT(FieldShift,DIM=2,SHIFT=-1),DIM=3,SHIFT=-1) ! 100
      NewField = NewField +
     &     B(:, :, :, 13) *
     &     CSHIFT(CSHIFT(FieldShift,DIM=2,SHIFT=-1),DIM=3,SHIFT=1)+ ! 102
     &     B(:, :, :, 14) *
     &     CSHIFT(CSHIFT(FieldShift,DIM=2,SHIFT=1),DIM=3,SHIFT=-1)+ ! 120
     &     B(:, :, :, 15) *
     &     CSHIFT(CSHIFT(FieldShift,DIM=2,SHIFT=1),DIM=3,SHIFT=1)+ ! 122
     &     B(:, :, :, 16) *
     &     CSHIFT(CSHIFT(FieldShift,DIM=1,SHIFT=1),DIM=2,SHIFT=-1)+ ! 201
     &     B(:, :, :, 17) *
     &     CSHIFT(CSHIFT(FieldShift,DIM=1,SHIFT=1),DIM=2,SHIFT=1)+ ! 221
     &     B(:, :, :, 18) *
     &     CSHIFT(CSHIFT(FieldShift,DIM=1,SHIFT=1),DIM=3,SHIFT=-1)+ ! 210
     &     B(:, :, :, 19) *
     &     CSHIFT(CSHIFT(FieldShift,DIM=1,SHIFT=1),DIM=3,SHIFT=1) ! 212

C     Now for the next components, shift by another NX in the X direction

      FieldShift = CSHIFT (FieldShift, DIM=1, SHIFT=NX)

C     Now sum up the products of each direction

      NewField = NewField +     ! Remember the same components
     &     C(:, :, :, 1) * FieldShift + ! 111
     &     C(:, :, :, 2) * CSHIFT (FieldShift, DIM=1, SHIFT=-1) + ! 011
     &     C(:, :, :, 3) * CSHIFT (FieldShift, DIM=1, SHIFT=1) + ! 211
     &     C(:, :, :, 4) * CSHIFT (FieldShift, DIM=2, SHIFT=-1) + ! 101
     &     C(:, :, :, 5) * CSHIFT (FieldShift, DIM=2, SHIFT=1) + ! 121
     &     C(:, :, :, 6) * CSHIFT (FieldShift, DIM=3, SHIFT=-1) + ! 110
     &     C(:, :, :, 7) * CSHIFT (FieldShift, DIM=3, SHIFT=1) + ! 112
     &     C(:, :, :, 8) *
     &     CSHIFT(CSHIFT(FieldShift,DIM=1,SHIFT=-1),DIM=2,SHIFT=-1)+ ! 001
     &     C(:, :, :, 9) *
     &     CSHIFT(CSHIFT(FieldShift,DIM=1,SHIFT=-1),DIM=2,SHIFT=1)+ ! 021
     &     C(:, :, :, 10) *
     &     CSHIFT(CSHIFT(FieldShift,DIM=1,SHIFT=-1),DIM=3,SHIFT=-1)+ ! 010
     &     C(:, :, :, 11) *
     &     CSHIFT(CSHIFT(FieldShift,DIM=1,SHIFT=-1),DIM=3,SHIFT=1)+ ! 012
     &     C(:, :, :, 12) *
     &     CSHIFT(CSHIFT(FieldShift,DIM=2,SHIFT=-1),DIM=3,SHIFT=-1) ! 100
      NewField = NewField +
     &     C(:, :, :, 13) *
     &     CSHIFT(CSHIFT(FieldShift,DIM=2,SHIFT=-1),DIM=3,SHIFT=1)+ ! 102
     &     C(:, :, :, 14) *
     &     CSHIFT(CSHIFT(FieldShift,DIM=2,SHIFT=1),DIM=3,SHIFT=-1)+ ! 120
     &     C(:, :, :, 15) *
     &     CSHIFT(CSHIFT(FieldShift,DIM=2,SHIFT=1),DIM=3,SHIFT=1)+ ! 122
     &     C(:, :, :, 16) *
     &     CSHIFT(CSHIFT(FieldShift,DIM=1,SHIFT=1),DIM=2,SHIFT=-1)+ ! 201
     &     C(:, :, :, 17) *
     &     CSHIFT(CSHIFT(FieldShift,DIM=1,SHIFT=1),DIM=2,SHIFT=1)+ ! 221
     &     C(:, :, :, 18) *
     &     CSHIFT(CSHIFT(FieldShift,DIM=1,SHIFT=1),DIM=3,SHIFT=-1)+ ! 210
     &     C(:, :, :, 19) *
     &     CSHIFT(CSHIFT(FieldShift,DIM=1,SHIFT=1),DIM=3,SHIFT=1) ! 212

C     Now divide by the singular point coefficient

      NewField = NewField / (A(:, :, :, 1) + i) ! Add 'i' first

C     Only update DeltaField, and Field for the central grid points.

      WHERE (Central)

C     Update the central elements in the difference array

        DeltaField(:, :, 1:NZ) = NewField - Field

C     Copy New Field to the Old Field

        Field = NewField

      END WHERE

C     Done

      RETURN
```

Figure 14. Parallel iteration step for internal cells.

```
C ParITEBBC.f - Subroutine to perform one iteration over all the grid
C points on the bottom plane of the model.  Calculates the new field at
C all the bottom plane points, and only updates the old field and array of
C changes at the points of interest, ie.  not on the boundaries.

SUBROUTINE ParITEBBC (Field, S, Ao, A, B, C, SI4to9,
 &DeltaField, NX, NY, NZ)

C Do one iteration on the bottom boundary of the model.  This differs
C from the central points in that there is a surface integral to be
C calculated for each component, and an extra term for the X component too
C (SI4to9).

IMPLICIT NONE

C Arguments

INTEGER NX, NY, NZ  ! The size of the model
COMPLEX*16 Field(3*NX, NY, NZ)  ! The 3 components of the field
COMPLEX*16 S(NX*NY, NX*NY)  ! The surface integral coefficients
COMPLEX*16 Ao(3*NX, NY) ! Singular point coefficient
REAL*8  A(3*NX, NY, 6)  ! Same component coefficients
REAL*8  B(3*NX, NY, 6)  ! Component+1 coefficients
REAL*8  C(3*NX, NY, 6)  ! Component+2 coefficients
COMPLEX*16 SI4to9(NX, NY)   ! Extra to add to X comp
COMPLEX*16 DeltaField(3*NX, NY, 0:NZ)  ! The field changes

C Local Variables

COMPLEX*16 FieldShift(3*NX, NY) ! Shifted field
COMPLEX*16 NewField(3*NX, NY)   ! The new field
COMPLEX*16 Integrands(NX*NY, NX*NY) ! Surface integral (B * S)
COMPLEX*16 StretchField(NX*NY)  ! Bottom plane in a vector
LOGICAL OffDiagonal(NX*NY, NX*NY)   ! True for off the diagonal
LOGICAL All2D(NX, NY)   ! True everywhere
INTEGER I, J! Grid point indices
REAL*8  Pi  ! Just Pi

C Compiler directives

#ifdef MasPar_Arch
CMPFONDPU FieldShift
CMPFONDPU NewField
CMPFONDPU Integrands
CMPFONDPU StretchField
CMPFONDPU OffDiagonal
CMPFONDPU All2D
#endif
#ifdef CM200_Arch
CMF$LAYOUT Field(:NEWS, :NEWS, :NEWS)
CMF$LAYOUT S(:NEWS, :NEWS)
CMF$LAYOUT Ao(:NEWS, :NEWS)
CMF$LAYOUT A(:NEWS, :NEWS, :SERIAL)
CMF$LAYOUT B(:NEWS, :NEWS, :SERIAL)
CMF$LAYOUT C(:NEWS, :NEWS, :SERIAL)
CMF$LAYOUT SI4to9(:NEWS, :NEWS)
CMF$LAYOUT DeltaField(:NEWS, :NEWS, :NEWS)
CMF$LAYOUT FieldShift(:NEWS, :NEWS)
CMF$LAYOUT NewField(:NEWS, :NEWS)
CMF$LAYOUT Integrands(:NEWS, :NEWS)
CMF$LAYOUT StretchField(:NEWS)
CMF$LAYOUT OffDiagonal(:NEWS, :NEWS)
CMF$LAYOUT All2D(:NEWS, :NEWS)
#endif


C First of all we calculate each integral for each component in
C sequence.  The 2D bottom surface is reshaped to a 1D vector, which is
C spread over the 2D array of coefficients (S).  These two are multiplied
C elementally, and then summed over one dimension to calculate the
C integral at each point.  These values are then reshaped back to the 2D
C surface grid.  The code for each component is given explicitly, no loop
C over the three components.

C Initialise Pi

Pi = 4.0 * ATAN (1.0)

C Set the pack / unpack masks to true everywhere

All2D = .TRUE.

C Set the mask to true on the off diagonal elements

OffDiagonal = .TRUE.! Default is true
FORALL (I=1:NX*NY, J=1:NX*NY)
 &OffDiagonal(I, J) = (I.NE.J)  ! False on diagonal

C First reshape the X component at the bottom plane of the model

StretchField = PACK (ARRAY=Field(1:NX, :, NZ), MASK=.TRUE.,
 &VECTOR=StretchField)

C Spread this and multiple by the coefficients to get the integrands,
C under control of the off diagonal mask

WHERE (OffDiagonal)
  Integrands = S * SPREAD (StretchField, DIM=2, NCOPIES=NX*NY)
END WHERE

C Now sum up the integrands at each grid point

StretchField = SUM (Integrands, MASK=OffDiagonal, DIM=1)

C Reshape the stretched field back to the 2D grid shape

NewField(1:NX, :) = UNPACK (VECTOR=StretchField,
 &MASK=All2D, FIELD=NewField(1:NX, :))

C This is the X component, so add on the extra terms

NewField(1:NX, :) = NewField(1:NX, :) + SI4to9

C Now divide by (2Pi)

NewField(1:NX, :) = NewField(1:NX, :) / (2D0 * Pi)

C The same needs to be done for Y component now

StretchField = PACK (ARRAY=Field(NX+1:2*NX, :, NZ),
 &MASK=.TRUE., VECTOR=StretchField)

WHERE (OffDiagonal)
  Integrands = S * SPREAD (StretchField, DIM=2, NCOPIES=NX*NY)
END WHERE

C Sum integrands

StretchField = SUM (Integrands, MASK=OffDiagonal, DIM=1)

NewField(NX+1:2*NX, :) = UNPACK (VECTOR=StretchField,
 &MASK=All2D, FIELD=NewField(NX+1:2*NX, :))

NewField(NX+1:2*NX, :) = NewField(NX+1:2*NX, :) / (2D0 * Pi)

C And finally the Z component

StretchField = PACK (ARRAY=Field(2*NX+1:3*NX, :, NZ),
 &MASK=.TRUE., VECTOR=StretchField)

WHERE (OffDiagonal)
  Integrands = S * SPREAD (StretchField, DIM=2, NCOPIES=NX*NY)
END WHERE

C Sum integrands

StretchField = SUM (Integrands, MASK=OffDiagonal, DIM=1)

NewField(2*NX+1:3*NX, :) = UNPACK (VECTOR=StretchField,
 &MASK=All2D, FIELD=NewField(2*NX+1:3*NX, :))

NewField(2*NX+1:3*NX, :) = NewField(2*NX+1:3*NX, :) / (2D0 * Pi)

C Now do the rest of the terms in parallel

C The extra self compnents.  The sixth coefficient is for the NZ-1 field
C value, so the (NZ-1) plane is used.

NewField = NewField +
 &A(:, :, 2) * CSHIFT (Field(:, :, NZ), DIM=1, SHIFT=-1) +  ! 011
 &A(:, :, 3) * CSHIFT (Field(:, :, NZ), DIM=1, SHIFT=1) +   ! 211
 &A(:, :, 4) * CSHIFT (Field(:, :, NZ), DIM=2, SHIFT=-1) +  ! 101
 &A(:, :, 5) * CSHIFT (Field(:, :, NZ), DIM=2, SHIFT=1) +   ! 121
 &A(:, :, 6) * Field(:, :, NZ-1)! 110

C Copy the bottom plane to the shifted array for the component+1 terms

FieldShift = CSHIFT (Field(:, :, NZ), DIM=1, SHIFT=NX)

C Add on the component+1 terms.  The sixth coefficient is multiplied by
C the (NZ-1) plane shifted by (+NX).

NewField = NewField +
 &B(:, :, 1) * FieldShift + ! 111
 &B(:, :, 2) * CSHIFT (FieldShift, DIM=1, SHIFT=-1) +   ! 011
 &B(:, :, 3) * CSHIFT (FieldShift, DIM=1, SHIFT=1) +! 211
 &B(:, :, 4) * CSHIFT (FieldShift, DIM=2, SHIFT=-1) +   ! 101
 &B(:, :, 5) * CSHIFT (FieldShift, DIM=2, SHIFT=1) +! 121
 &B(:, :, 6) * CSHIFT (Field(:, :, NZ-1), DIM=1, SHIFT=NX)  ! 110

C Shift the bottom plane of the field by another component (NX)

FieldShift = CSHIFT (FieldShift, DIM=1, SHIFT=NX)

C Add on the component+2 terms.  This time the sixth coefficient is
C multiplied by the (NZ-1) plane shifted by (-NX), which is the same as
C (+2NX), but faster.

NewField = NewField +
 &C(:, :, 1) * FieldShift + ! 111
 &C(:, :, 2) * CSHIFT (FieldShift, DIM=1, SHIFT=-1) +   ! 011
 &C(:, :, 3) * CSHIFT (FieldShift, DIM=1, SHIFT=1) +! 211
 &C(:, :, 4) * CSHIFT (FieldShift, DIM=2, SHIFT=-1) +   ! 101
 &C(:, :, 5) * CSHIFT (FieldShift, DIM=2, SHIFT=1) +! 121
 &C(:, :, 6) * CSHIFT (Field(:, :, NZ-1), DIM=1, SHIFT=-NX) ! 110

C Divide through by the singular point coefficient, but being careful of
C Ao=0

WHERE (Ao.NE.0)
  NewField = NewField / Ao
END WHERE

C Update the array of changes in the field, at the bottom plane.  Update
C all the points on the plane, even the boundaries.  These should then be
C ignored when the residuals are being calculated.

DeltaField(:, :, NZ) = NewField - Field(:, :, NZ)

C Copy the new values to the old field now, component at a time

Field(2:NX-1, 2:NY-1, NZ) = NewField(2:NX-1, 2:NY-1)! X component
Field(NX+2:2*NX-1, 2:NY-1, NZ) =
 &NewField(NX+2:2*NX-1, 2:NY-1) ! Y component
Field(2*NX+2:3*NX-1, 2:NY-1, NZ) =
 &NewField(2*NX+2:3*NX-1, 2:NY-1)   ! Z component

C Done

RETURN
END
```

Figure 15. Parallel iteration step for bottom cells.

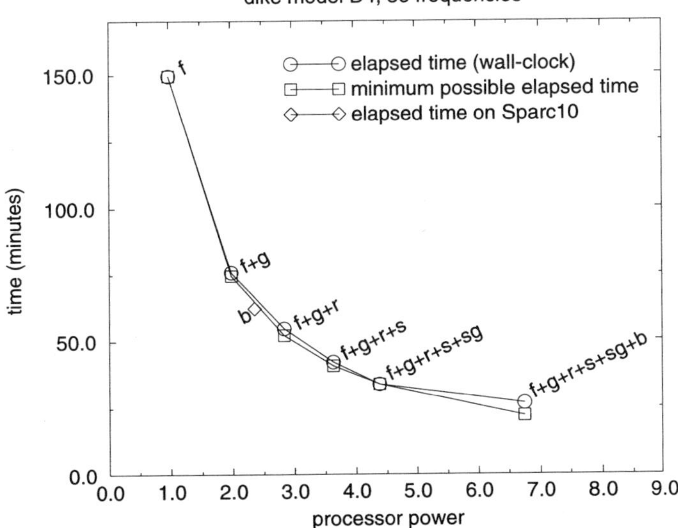

Figure 16. Execution times of task-farm version of EM3D running dike4 model using 36 frequencies.

Figure 16 plots wall-clock execution time against total capacity of the workstations employed. The wall-clock execution time includes all overheads associated with startup, I/O, load imbalance, shutdown, and sorting of output files. The performance of the sequential code on a single Sparc 10 workstation (b) was exceeded by the performance of the sequential program on three Sparc 2 workstations [(f), (g), and (r)], which collectively had a much lower purchase price.

Also plotted in Fig. 16 is the minimum time in which the program could have been expected to be completed. A more revealing measure of success is to plot the *speedup* of the program, defined as the relative speeds of an n-processor and a single-processor implementation. A perfect result would give a straight line, but Fig. 17 reveals that, eventually, using yet more processors does not yield as great an increase in performance as might be expected.

Because the task farm waits for a processor to finish its current task before allocating further work to it, variations in workstation performance are compensated for, provided there are sufficiently many tasks. As the number of processors is increased, it becomes progressively harder to divide the finite number of tasks among them such that each processor is fully loaded for the entire time of the calculation.

9.2 Fine-grained data parallelization of finite-difference model

The point Jacobian algorithm is simple to implement and analyze but well understood to have a slow rate of convergence. Table 2 reveals that although the parallel point Jacobian relaxation method requires more iterations than the original Gauss-Seidel solver, the time taken for each iteration is greatly reduced and the overall performance is increased.

The major problem with this approach is the large memory requirement. An $x \times y \times z$ model grid requires $8(4x^2y^2 + 141xyz + 131xy)$ bytes to store the field and the

Table 2. Performance of 3-D model (29 × 29 × 9)

Computer	Method	Iterations	Time for each iteration (min:s)
Sun 2000	Gauss–Seidel	404	52:12
DEC AXP	Gauss–Seidel	404	26:27
CM200	Jacobian	671	8:7

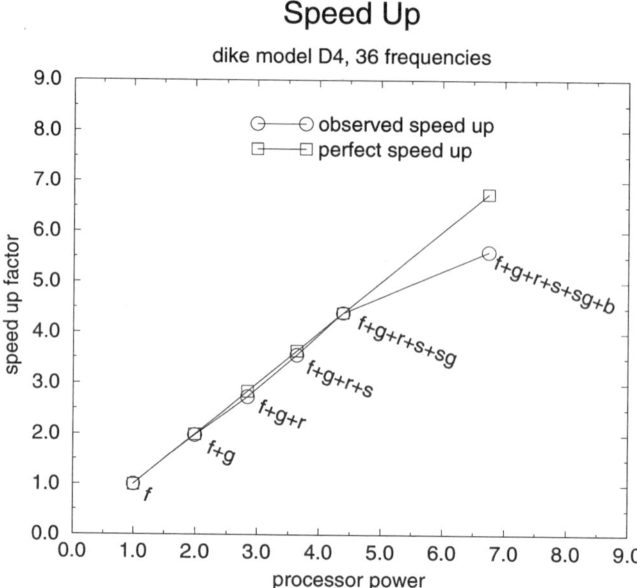

Figure 17. Speedup of task-farm version of EM3D running dike4 model using 36 frequencies.

coefficients, disregarding any working space for temporary arrays. Unfortunately, the MasPar processor topology is 2-D, requiring a single processor's memory to store the field and coefficients for all the grid nodes with the same (x, y) coordinates. As mentioned before, the low-level software emulates virtual processors to give the illusion that the processor array is multidimensional, but this illusion is strictly limited by the local memory available to each processor. A similar, although less severe, restriction is presented by the Connection Machine, which has double the local memory per processor.

10 Large-scale feasibility studies

The EMAFD package calculates the response of a dipole source in an axisymmetric Earth model. The calculation proceeds in a transform domain in which a Laplace transform is taken with respect to time and a Fourier transform with respect to azimuth. Parallelizing over both temporal and azimuthal components results in several hundred independent tasks. Each task requires approximately the same amount of computation. This allows the task-farm version of EMAFD to make effective use of a massively parallel computer.

The task-farm version of EMAFD was ported to run on the Edinburgh Cray T3D, which contains 512 DEC Alpha processors, each running at 150 MHz with 64 Mbytes

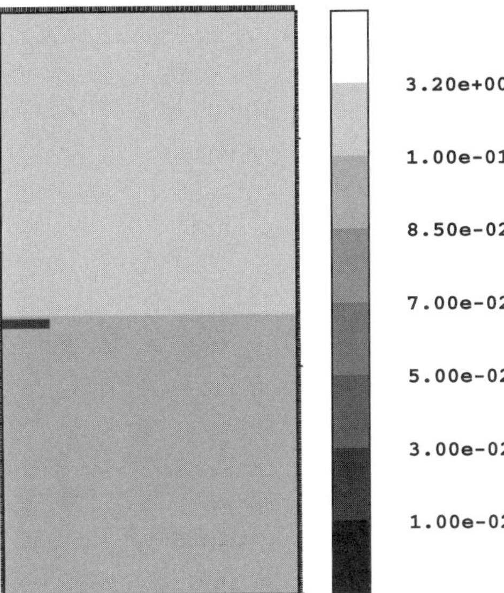

Figure 18. Smoothed conductivity section for submarine model ilm3000b.

of RAM. This allowed feasibility studies to be undertaken with large-scale models in a reasonable amount of time.

10.1 Earth model

As an example, we present a simulation of the detectability of the lateral extent of an offshore oil reservoir using submarine transient EM. The situation is modeled as a background of two uniform, isotropic, half-spaces with the interface between the half-spaces corresponding to the seabed and the reservoir forming a resistive anomaly in the lower half-space. The upper half-space, which models the seawater, has a resistivity of 0.3125 ohm-m (or equivalently, a conductivity of 3.2 S m^{-1}). The lower half-space, which models the seabed, has a resistivity of 10 ohm-m. The reservoir is modeled as a moderately resistive disk with a depth centered at 500 m below the seabed, a vertical thickness of approximately 200 m, and a horizontal radius of approximately 1000 m.

Computation of the anomalous field takes place on a regular finite-difference grid extending 6400 m above and 6400 m below the seabed and with a radius of 6400 m. The grid is discretized using a 50 m $\times$ 50 m mesh size. Large discontinuities in the conductivity model can cause instabilities in finite-difference simulations of diffusive EM propagation. To avoid this, the conductivity distribution is smoothed under user control. This smoothed conductivity section is used only to calculate the secondary field; the primary field is calculated using the analytic formula for a double half-space.

Figure 18 presents the smoothed conductivity section on the finite-difference grid. Figures 19 and 20 present a vertical profile of conductivity at the reservoir.

10.2 Source and receiver configuration

The time-domain response is calculated for a switch-off transmitter current profile and a colinear electric dipole–dipole configuration. A horizontal electric dipole source is

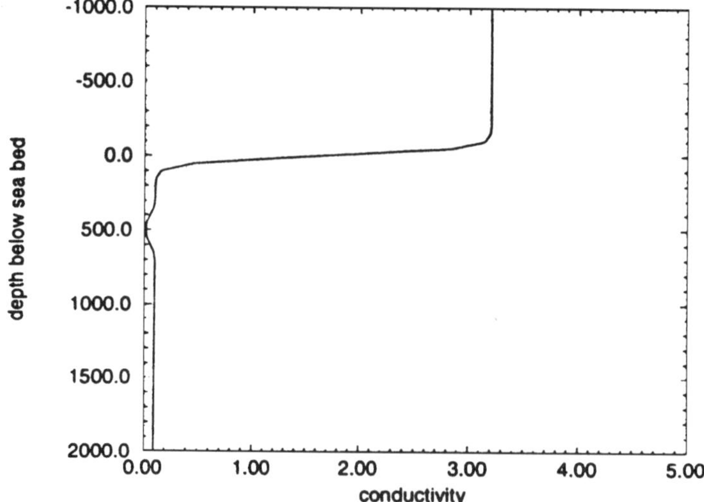

Figure 19. Smoothed vertical conductivity profile for submarine model ilm3000b.

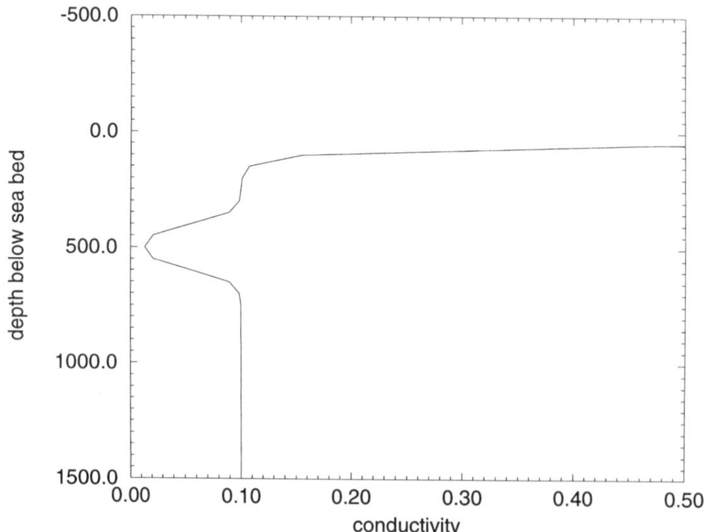

Figure 20. Zoom on vertical profile of resistive anomaly. Maximum resistivity is 75 ohm-m.

modeled lying on the seabed, 1500 m off the axis of rotational symmetry of the Earth model. The direction of the dipole is oriented along the radius from the axis of symmetry. Electric dipole receivers, aligned colinearly with the source dipole, are placed from an offset of 50 m at intervals of 50 m. The results are shown in Fig. 21, normalized by the field in the background model.

Depending upon the transmitter and receiver configuration, a single simulation with this Earth model required 120 to 160 processor hours, the equivalent of 5 to 7 days of constant computation on a single DEC Alpha processor. Use of the Edinburgh Cray T3D enabled this simulation to be run in under an hour, and allowed a suite of models to be simulated in a few days.

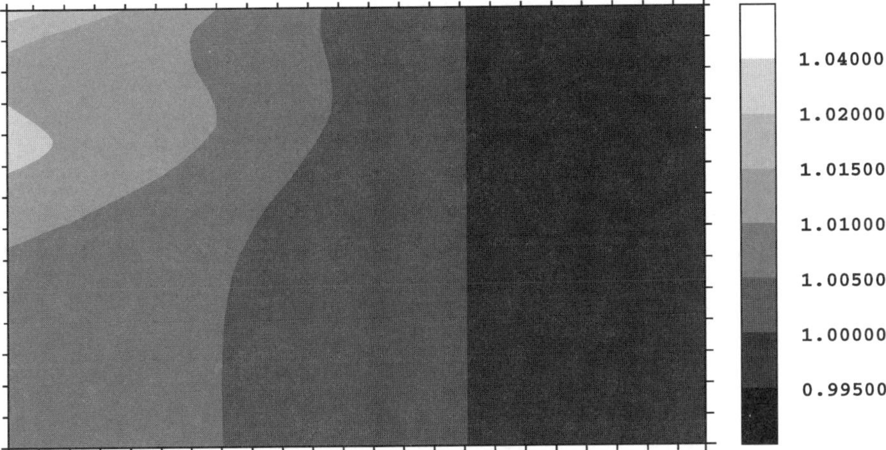

Figure 21. Ratio of total radial electric field to normal radial electric field presented as a contour plot. Receivers are plotted against time with receiver offset from the origin increasing from left to right and time increasing down the page. Receiver offsets from the origin vary from 50 m to 1200 m in intervals of 50 m; the time range is on a logarithmic scale from 29 ms to 100 s. The in-line horizontal electric dipole transmitter is located beyond the right-most receiver, at an offset of 1500 m from the origin.

11 Discussion and conclusions

Three different EM modeling programs have been parallelized. In these programs the interaction of the impinging EM fields with the Earth model is formulated in the frequency domain, parameterized and then discretized to yield a system of linear equations that must be solved. Two very different paths of parallelization have been explored; a coarse-grained task farming of the calculation for each frequency response and a fine-grained parallelization of the iterative solver.

A fine-grained data-parallel implementation, suitable for use on a SIMD parallel machine, required use of a SIMD-parallelizable iterative solver algorithm. This was implemented using standard HPF constructs which gave portability among SIMD platforms from different vendors. The slow convergence of the simple iterative solver that was implemented is well understood, but was offset by the efficiency of the massively parallel SIMD machines, when performing these algorithms. Future work should concentrate on implementing accelerated iterative schemes.

For all programs the task-farm implementation allowed an efficient numerical algorithm to be retained, and was able to run on an existing departmental workstation cluster. These computers often have spare CPU cycles waiting to be used, and this has proved an ideal method of speeding up computationally expensive tasks without purchasing expensive equipment. Despite the use of workstations of widely varying performance, the task farm did a good job of balancing the load among processors without requiring detailed analysis of the relative performance of processors or the amount of work to be done.

The PUL-TF utility (Bruce et al., 1995), which provides a ready-made task-farm framework into which the application can be bolted, reduces porting effort and increases portability. Our task-farm codes are now in use on several platforms ranging from workstation clusters to a massively parallel Cray T3D supercomputer.

Despite these results, the coarse-grained task-farm approach has little inherent parallelism which limits the number of processors that can run in parallel. The parallel EMAFD program can make good use of hundreds of processors but would not sensibly scale to use thousands of processors. A highly scalable implementation might be obtained by parallelizing the work of each task. Such a mixed-mode approach would ease the search for a suitable linear solver because the parallel solver need only scale efficiently to a modest number of processors, rather than to a massive number of processors. This approach would be well suited to the growing number of parallel NUMA computers.

Acknowledgments

Edinburgh Parallel Computing Centre is an interdisciplinary focus for high-performance parallel computing projects involving groups within the University of Edinburgh, industrial partners, and academic users throughout Europe. The center is supported by major grants and contracts from industry, the Commission of the European Communities, the Department of Trade and Industry, the Engineering and Physical Sciences Research Council, the Joint Information Systems Committee of the Higher Education Funding Councils and Scottish Enterprise.

Many thanks go to Shari Trewin, Simon Chapple, and the entire PUL team at EPCC for writing, developing, and supporting PUL-TF.

Andrew J. S. Wilson was supported by European Community THERMIE Research Grant No. OG/0305/92/NL-UK and Elf Enterprise Caledonia Contract No. CA5527. Kenneth MacDonald was supported by NERC Studentship Award GT4/91/GS/41. These authors are indebted to Anton Ziolkowski, Bruce Hobbs, and John Weaver for their support and supervision. Bill Day was supported by the EPCC Summer Scholarship Programme.

This work would not have been possible without the generosity of others who made available EM modeling programs into which years of energy and expertise had been invested. The authors thank Professor Alan Tripp of the University of Utah for permission to use the EM3D program, and would also like to thank Dean Livelybrooks, Louise Pellerin, and Andreas Hördt for instruction and help in its use. We also are indebted to Helena Poll, Ashok Agarwal, and Xinghua Pu at the University of Victoria for making their finite-difference programs available and giving in-depth introduction to their operation.

References

Adams J. C., 1989, MUDPACK: Multigrid portable Fortran software for the efficient solution of linear elliptic partial differential equations. App. Math. Comput., 54–62.

Agarwal, A. K., Pu, X. H., Weaver, J. T., and MacDonald, K. J., 1994, Electromagnetic modeling in three dimensions using massively parallel processing, Annal. Geophys. part 1, (suppl. 12), C9 (abstract).

Bruce, R. A. A., Chapple, S., MacDonald, N. B., Trew, A. S., and Trewin, S., 1995, CHIMP and PUL: Support for portable parallel computing: J. Future Generation Comput. Sys.

Chang, S. K., and Anderson, B., 1984, Simulation of induction logging by the finite-element method: Geophysics, **49**, 1943–1958.

Edwards, R. N., and Cheesman, S. J., 1987. Two-dimensional modeling of a towed transient magnetic dipole–dipole seafloor electromagnetic system: J. Geophys. **61**, 110–121.

Hördt, A., Vozoff, K., and Neubauer, F. M., 1995, Multichannel transient electromagnetics for underground gas storage monitoring, *in* Proceedings of the 57th Meeting, Eur. Assn. Expl. Geophys., Abstract No. D051.

Flynn, M. J., 1972, Some computer organisations and their effectiveness: IEEE Trans. Comput., **C-21**, 948–960.

Goldman, M. M., and Stoyer, C. H., 1983, Finite-difference calculations of the transient field of an axially symmetric earth for vertical magnetic dipole excitation: Geophysics, **48**, 953–963.

Newman, G. A., Hohmann, G. W., and Anderson, W. L., 1986, Transient electromagnetic response of a three-dimensional body in a layered earth: Geophysics, **51**, 1608–1627.

Knight, J. H., and Raiche, A. P., 1982, Transient electromagnetic calculations using the Gaver–Stehfest algorithm: Geophysics, **47**, 47–50.

Liu, Q. H., 1993, Electromagnetic field generated by an off-axis source in a cylindrically layered medium with an arbitrary number of horizontal discontinuities: Geophysics, **58**, 616–625.

Pai, D. M., Ahmad, J., and Kennedy, W. D., 1993, Two-dimensional induction log modeling using a coupled-mode, multiple-reflection series method: Geophysics, **58**, 466–474.

Pu, X., 1994, Three Dimensional Geomagnetic Forward Modeling: University of Victoria.

Schnegg P.-A., and Sommaruga, A., 1995, Constraining seismic parameters with a CSAMT method: Geophys. J. Internat., **122**, 152–160.

Stehfest, H., 1970, Algorithm 368, numerical inversion of Laplace transforms: Commun. ACM, **94**, 13,879–13,893.

Vilinger, H., 1984, Solving cylindrical geothermal problems using the Gaver–Stehfest inverse Laplace transform: Geophysics, **50**, 1581–1587.

Walker, P. W., and Groom, R. W., 1994, A rapid solution for electromagnetic scattering—the first step to 3D imaging, *in* Proceedings of 56th Meeting, Eur. Assn. Expl. Geophys., Abstract No. IO27.

Wilt, M., Morrison, H. F., Becker, A., Teng, H. W., Lee, K. H., Torres-Verdin, C., and Alumbaugh, D., 1995, Crosshole electromagnetic tomography: A new technology for oil field characterization: The Leading Edge, **14**, No. 1, 173–177.

Yu, L., 1994, Computation of the electrical responses of mid-ocean ridge structures: Research in Marine Geophysics, No. 5., University of Toronto.

PART V

MAGNETOTELLURICS AND GLOBAL INDUCTION

The magnetotelluric (MT) method, which makes use of naturally occurring electromagnetic (EM) fields, is one of the most widely used electrical prospecting techniques due to its potential for very deep exploration. However, MT has been hampered severely by a lack of interpretational capability. Inappropriate one-dimensional (1-D) or two-dimensional (2-D) interpretation models are often used mainly because the necessary three-dimensional (3-D) models are not readily available.

Ting and Hohmann (1981)

Affordable Magnetotellurics: Interpretation in Natural Environments

Philip E. Wannamaker

Summary. Areal magnetotelluric (MT) surveys and their 3-D interpretation are expensive. It therefore can be valuable to know when a line profile and a 2-D (or quasi-2-D) interpretation may suffice. Some factors that determine this are the effects of electric charges that arise at resistivity contrasts, the similitude in the electromagnetic (EM) responses of small and large structures, and the immunity of certain MT tensor elements to 3-D effects. The most important task in designing a 2-D survey, of course, is to determine geoelectric trends accurately from existing information. Electric dipoles should be laid out parallel and perpendicular to a profile that is orthogonal to the strike of the terrane; most of the dipoles should be oriented across strike for adequate sampling of the more rapidly varying response. Two-dimensional modeling should concentrate on the impedance in the transverse magnetic (TM) mode because of its low sensitivity to the effects of finite strike. The vertical magnetic field and the phase of transverse electrical impedance should be checked for consistency with a 2-D model. Impedances in the Groom-Bahr decomposition can help fix geoelectrical trends, but care is needed to preserve longer spatial wavelengths of the response amplitudes. Profile interpretation is not applicable to all MT prospects. Successes and limitations are illustrated with MT transects over the Juan de Fuca subduction zone and the Long Valley caldera magmatic system.

1 Introduction

The magnetotelluric (MT) method is the preferred technique for mapping resistivity below a few kilometers in the Earth (Vozoff, 1991). The MT method uses as its incident source field electromagnetic (EM) plane waves that are generated by worldwide lightning activity and solar wind-ionosphere interactions. Better instrumentation and data processing have improved dramatically the precision of tensor MT data (Wannamaker and Hohmann, 1991; Jiracek, 1995). There also has been tremendous progress in numerical simulation and inversion of MT data (Hohmann, 1988; Wannamaker and Hohmann, 1991; Whittal and Oldenburg, 1992; Mackie and Madden, 1993). Tensor

Energy & Geoscience Institute, University of Utah, 423 Chipeta Way, Salt Lake City, Utah 84108, USA.

data are, however, expensive. Commercial MT surveys typically cost about $2000 US per sounding, so that a modest profile of 25 sites costs about $50,000. More than 100 sites might be needed for 3-D coverage, and even then it is difficult to obtain adequate sampling of near-surface variations.

One of my goals in this review is to show that two-dimensional (2-D) modeling or inversion can still lead to good MT interpretations in 3-D geometries (e.g., Fig. 1). This is important because a well-sampled MT profile is often affordable when an areal survey is not, and because 2-D modeling and inversion have become powerful and widely available (e.g., DeGroot-Hedlin and Constable, 1990; Smith and Booker, 1991). Three features in particular are useful for understanding 3-D effects in natural data profiles. The first is the creation of electrical charge that occurs when there is a component of the electric field along a conductivity gradient. The second is the similitude of EM responses, which relates the variation with frequency in the response of a large target

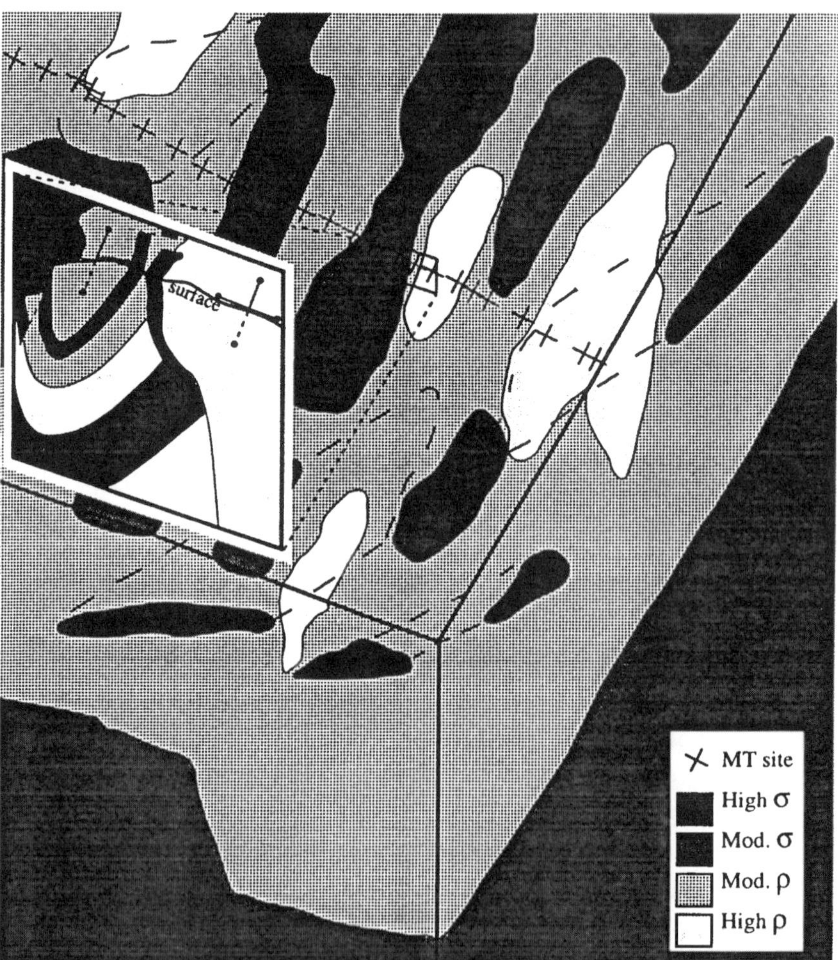

Figure 1. Multilevel resistivity heterogeneity showing an average structural trend. Moderately dense profile of MT sites is deployed to cross strike of upper-crustal structure. Conductive structures may include surficial sedimentary basins, bedrock strata, and hydrothermal alteration. Resistive structures may include igneous intrusives and high-grade metamorphic rocks. Inset signifies that scale dimensions can range from hundreds of meters to hundreds of kilometers.

to that of a small source of geological noise. The third is the relative insensitivity of certain elements of the MT tensor to structural variations along (the interpreted) strike. I illustrate these concepts with data from two large-scale MT transects.

2 3-D MT responses

Signatures characteristic of 3-D structures are present even in the MT response of a single body in a one-dimensional (1-D) host. In analyzing these characteristics, it is useful to separate the contributions of the inhomogeneity from the host by writing the impedance and vertical magnetic-field tensors in terms of scattered and incident fields. EM scaling theory then helps to differentiate the responses of large structures, which may represent an exploration target, from those of small structures, which may constitute geological noise. This leads to an approximation that underlies various tensor decompositions.

2.1 Tensor field relations

Helmholtz equations relating the incident (subscript i) and scattered (subscript s) electric and magnetic fields of a 3-D body in a 1-D earth are (Hohmann, 1988)

$$\nabla^2 \mathbf{E}_s + \nabla\left(\mathbf{E}_s \cdot \frac{\nabla \hat{y}}{\hat{y}}\right) + k^2 \mathbf{E}_s = -k_a^2 \mathbf{E}_i - \nabla\left(\mathbf{E}_i \cdot \frac{\nabla \hat{y}_a}{\hat{y}}\right), \tag{1}$$

$$\nabla^2 \mathbf{H}_s + \hat{y}(\nabla \times \mathbf{H}_s) \times \nabla(1/\hat{y}) + k^2 \mathbf{H}_s = -k_a^2 \mathbf{H}_i - \hat{y}\nabla(\hat{y}_a/\hat{y}) \times \mathbf{E}_i, \tag{2}$$

where $\hat{y} = i\omega\epsilon + \sigma$ is admittivity, $\hat{z} = i\omega\mu_o$ is impedivity, ϵ is dielectric permittivity, σ is electrical conductivity, and μ_o is magnetic permeability of free space; $\omega = 2\pi f$, where f is the frequency (in Hertz). Also, $k^2 = -\hat{y}\hat{z}$ is the local wavenumber (squared), and $k_a^2 = -(\hat{y}_a - \hat{y})\hat{z}$ is the anomalous wavenumber (squared), which is nonzero in the body only. In this representation, the inhomogeneity is an equivalent source of anomalous current.

The source of the secondary electric field in Eq. (1) has a volume current component, the term $k_a^2\mathbf{E}_i$, and a free-charge component, the divergence term (Wannamaker et al., 1984). The free charge preserves continuity of normal total current density across conductivity boundaries (i.e., along gradients in the conductivity), but in doing so makes the normal total electric field discontinuous or divergent (Price, 1973; Jiracek, 1990). This condition holds at all frequencies (even at dc). Because $\mathbf{E}_i$ is continuous, $\nabla \cdot \mathbf{E}_s = \nabla \cdot \mathbf{E}$.

Because the Helmholtz equations are linear with respect to the incident electric field, the horizontal (subscript h) fields obey

$$\mathbf{E}^0(\mathbf{r}) = \mathbf{E}_i^0 + \left[\underset{\sim}{\mathbf{P}}{}_s^0(\mathbf{r})\right] \cdot \mathbf{E}_i^0 \tag{3}$$

and

$$\mathbf{H}^0(\mathbf{r}) = \mathbf{H}_i^0 + \left[\underset{\sim}{\mathbf{Q}}{}_s^0(\mathbf{r})\right] \cdot \mathbf{E}_i^0, \tag{4}$$

where superscript zero indicates that we are considering $\mathbf{r}$ at the surface of the earth over which the incident field $\mathbf{E}_i$ is constant (Wannamaker et al., 1984). $[\underset{\sim}{\mathbf{P}}{}_s^0(\mathbf{r})]$ and $[\underset{\sim}{\mathbf{Q}}{}_s^0(\mathbf{r})]$ are 3×2 tensors that relate the scattered and incident fields, and which depend on the 3-D body, the host, and the frequency.

Combining the secondary field tensors with the layered-earth impedance, the total field MT impedance tensor, which is defined by

$$\mathbf{E}_h^0(\mathbf{r}) = [\underset{\sim}{\mathbf{Z}}] \cdot \mathbf{H}_h^0(\mathbf{r}), \tag{5}$$

can be written

$$[\underset{\sim}{\mathbf{Z}}(\mathbf{r})] = \left\{[\underset{\sim}{\mathbf{I}}] + \left[\underset{\sim}{\mathbf{P}}_{hs}^0(\mathbf{r})\right]\right\} \cdot [\underset{\sim}{\mathbf{Z}}_{1D}] \cdot \left\{[\underset{\sim}{\mathbf{I}}] + \left[\underset{\sim}{\mathbf{Q}}_{hs}^0(\mathbf{r})\right] \cdot [\underset{\sim}{\mathbf{Z}}_{1D}]\right\}^{-1}, \tag{6}$$

(Wannamaker et al., 1984). Quantities $\underset{\sim}{\mathbf{I}}$ and $\underset{\sim}{\mathbf{Z}}_{1D}$ are the identity tensor and the 1-D host impedance tensor, respectively. In a similar manner, the vertical H-field transfer tensor, defined by

$$\mathbf{H}_z^0(\mathbf{r}) = [\underset{\sim}{\mathbf{K}}_z] \cdot \mathbf{H}_h^0(\mathbf{r}), \tag{7}$$

becomes

$$[\underset{\sim}{\mathbf{K}}_z(\mathbf{r})] = \left[\underset{\sim}{\mathbf{Q}}_{vs}^0(\mathbf{r})\right] \cdot [\underset{\sim}{\mathbf{Z}}_{1D}] \cdot \left\{[\underset{\sim}{\mathbf{I}}] + \left[\underset{\sim}{\mathbf{Q}}_{hs}^0(\mathbf{r})\right] \cdot [\underset{\sim}{\mathbf{Z}}_{1D}]\right\}^{-1}. \tag{8}$$

2.1.1 Low frequencies. At very low frequencies, when the distance from the body is short compared to wavelengths in the host layers, the governing Helmholtz equations reduce (to a high degree of approximation) to Laplace's or Poisson's equations. In this limit, the volume current source term in Eq. (1) vanishes, because it is proportional to f. Thus, $\mathbf{E}_s(\mathbf{r})$ and $[\underset{\sim}{\mathbf{P}}_{hs}^0(\mathbf{r})]$ at low frequencies are determined solely by boundary charge. The charge itself is determined by the electric field across a conductivity gradient, and so, it also depends on the electric field inside the inhomogeneity, where the wavelengths are shorter if it is a conductor (Ward and Hohmann, 1988). This is true for $\mathbf{H}_s(\mathbf{r})$, too. Frequencies therefore must be low enough (or periods large enough) that wavelengths within the body are long compared to its dimensions, before $[\underset{\sim}{\mathbf{P}}_{hs}^0(\mathbf{r})]$ and $[\underset{\sim}{\mathbf{Q}}_{hs}^0(\mathbf{r})]$ will be real and independent of period.

Furthermore, it is easy to show that any layered-earth impedance $[\underset{\sim}{\mathbf{Z}}_{1D}]$ decreases monotonically with decreasing frequency (Wannamaker et al., 1984). Hence, even though $[\underset{\sim}{\mathbf{Q}}_{hs}^0(\mathbf{r})]$ possesses a nonzero low-frequency limit,

$$\mathbf{H}_{hs}^0(\mathbf{r}) = \left[\underset{\sim}{\mathbf{Q}}_{hs}^0(\mathbf{r})\right] \cdot [\underset{\sim}{\mathbf{Z}}_{1D}] \cdot \mathbf{H}_{hi}^0$$

will vanish as $f \to 0$. Equation (6) thereby reduces to

$$[\underset{\sim}{\mathbf{Z}}(\mathbf{r})] = \left\{[\underset{\sim}{\mathbf{I}}] + \left[\underset{\sim}{\mathbf{P}}_{hs}^0(\mathbf{r})\right]\right\} \cdot [\underset{\sim}{\mathbf{Z}}_{1D}]. \tag{9}$$

Each of the four elements of $[\underset{\sim}{\mathbf{Z}}(\mathbf{r})]$ is related to $\underset{\sim}{\mathbf{Z}}_{1D}$ by real constants so that all four apparent resistivities are related to layered-earth apparent resistivities by these constants squared (Wannamaker et al., 1984). For the general 3-D case, all four elements of $[\underset{\sim}{\mathbf{P}}_{hs}^0(\mathbf{r})]$ may be nonzero because charge would form regardless of the polarization of the incident electric field. Impedance phases, in contrast, are not affected by such real, multiplicative distortions. Smith (1995) has pointed out from symmetry arguments that any $[\underset{\sim}{\mathbf{P}}_{hs}^0(\mathbf{r})]$ caused by a 3-D body can be reproduced by an obliquely oriented 2-D body that causes exactly the same galvanic distortion at the measurement point.

For a 2-D body along the x-axis, an incident electric field along x induces no boundary or space charge so that $P_{xx} \to 0$ and $\rho_{xy} \to \rho_{1D}$ as $f \to 0$. Thus, the transverse electric (TE) mode includes no boundary charge effects. In the transverse magnetic (TM) mode, however, an incident electric field E_y crosses boundaries, so that charges are induced, and anomalies persist to arbitrarily long period. In this way, the TM mode is fundamentally more akin to 3-D responses.

In the low-frequency limit also, each element of $[\underset{\sim}{\mathbf{K}}_z(\mathbf{r})]$ is proportional to Z_{1D} so that their magnitudes approach zero as $f \to 0$. The rate at which they diminish is greater for bodies over a conductive basement than for those over a resistive basement (Wannamaker et al., 1984).

2.2 EM similitude in MT

Earth has geological structures of various scales (and resistivities). Large, deep bodies can be discriminated from small, shallow bodies by the frequency dependence of their MT responses. Consider two inhomogeneities A and B in a half-space of resistivity ρ_h (conductivity σ_h) (Fig. 2). The bodies are identical in conductivity ($\sigma_A = \sigma_B$) and geometry except for a dimensional scaling by a factor of ξ. They are excited by incident fields ($\mathbf{E}_{iA}$, $\mathbf{H}_{iA}$) and ($\mathbf{E}_{iB}$, $\mathbf{H}_{iB}$) of different frequencies, ω_A and $\omega_B = \omega_A/\xi^2$. When displacement currents are negligible, an induction number

$$\theta = \omega_A \mu_0 \sigma_A r_A^2 |\hat{\mathbf{r}}_A| = \omega_A \mu_0 \sigma_B r_B^2 |\hat{\mathbf{r}}_B|, \tag{10}$$

is preserved, where $\hat{\mathbf{r}}$ is a unit vector in the $\mathbf{r}$ direction (Grant and West, 1965).

Under such scaling, secondary field tensors $[\underset{\sim}{\mathbf{P}}_s^0(\mathbf{r})]$ and $[\underset{\sim}{\mathbf{R}}_s^0(\mathbf{r})] = [\underset{\sim}{\mathbf{Q}}_{hs}^0(\mathbf{r})] \cdot [\underset{\sim}{\mathbf{Z}}_{1D}]$ remain invariant. It follows from Eqs. (3), (4), and (6) that

$$[\underset{\sim}{\mathbf{Z}}_A(\mathbf{r}_A)] = \xi [\underset{\sim}{\mathbf{Z}}_B(\mathbf{r}_B)], \tag{11}$$

leading to

$$\rho_{ijA} = \rho_{ijB} = \rho_{ij}(\theta), \tag{12}$$

$$\phi_{ijA} = \phi_{ijB} = \phi_{ij}(\theta), \tag{13}$$

and

$$[\underset{\sim}{\mathbf{K}}_{zA}(\mathbf{r}_A)] = [\underset{\sim}{\mathbf{K}}_{zA}(\mathbf{r}_B)] = [\underset{\sim}{\mathbf{K}}_z(\theta)]. \tag{14}$$

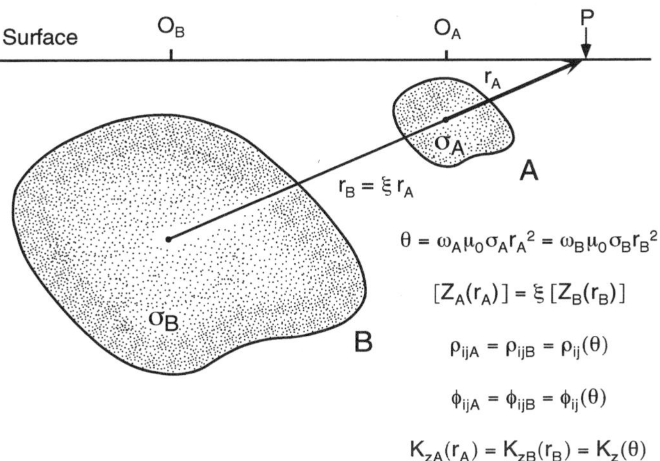

Figure 2. Pair of simple structures identical in geometry but separated in scale to illustrate the concept of EM scaling.

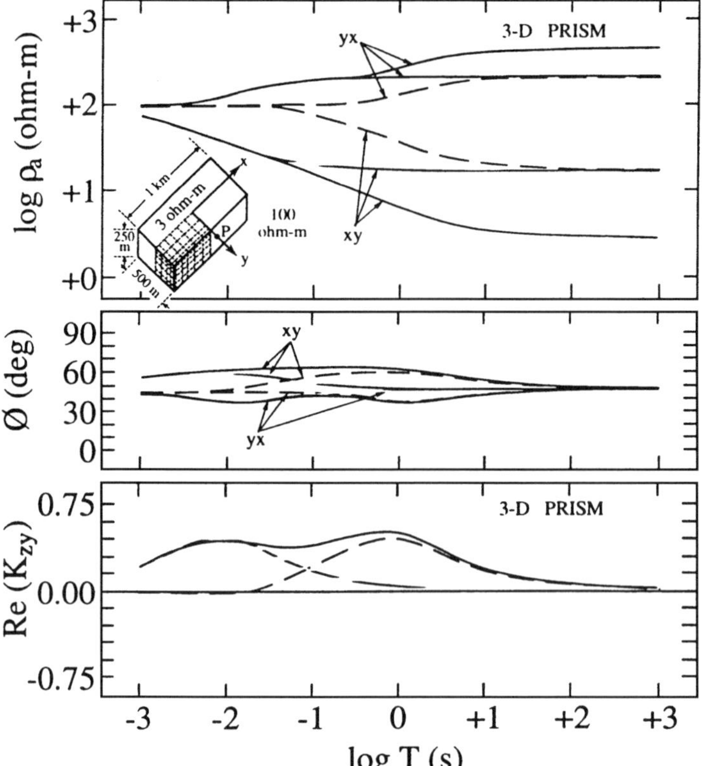

Figure 3. Tensor off-diagonal MT responses at observation point P outside an outcropping conductive prism in a half-space. Responses are shown for prism dimensions given in the inset (dotted curves), for dimensions scaled up by 10 (dashed curves), and for the sum of the small- and large-body responses (solid curves). Body discretization is shown for the integral-equation code of Wannamaker (1991).

Basically, small inhomogeneities can have MT responses as strong as those of large ones, although the small- and large-body responses have dependencies on period (frequency) that are, all else being equal, separated as the square of the geometric scale factor distinguishing the different bodies. Calculations for a conductive, outcropping prism in a half-space using the integral-equation code of Wannamaker (1991) are shown in Fig. 3. The T^2 separation of responses of the 500-m-wide (short dashes) and 5-km-wide (long dashes) bodies is clear. The overshoot in ρ_{yx} and the undershoot in ρ_{xy} is characteristic of what here is essentially an electric dipole scatterer (Wannamaker et al., 1984). With both bodies together in the half-space situated so that the receiver P is at an equivalent location, the joint response is approximately the sum of the two individual responses [solid curves, coupling is negligible; see, e.g., Groom and Bahr (1992)].

Telluric distortion of the apparent resistivity response by boundary charges in near-surface structures can mask the response of a larger, deeper target at arbitrarily long periods. But, because of EM similitude, the effects of small-scale structures on the phases of all impedance elements, plus the elements of $[\underset{\sim}{\mathbf{K}}_z(\mathbf{r})]$, are band limited to short periods. Therefore, one may "see through" such structures with these functions to target responses occurring at longer periods (Wannamaker et al., 1984).

2.3 Quantification of near-surface distortion

The impedance phase—or $\underset{\sim}{\mathbf{K}}_z(\mathbf{r})$—allows only qualitative structural interpretations because of its poor or nonexistent depth and resistivity discrimination (Wannamaker et al., 1984; Gough, 1989). We must include the ρ_a response to get a resistivity-depth model.

2.3.1 Invariants and averages. At an observation point P outside the body (as in Fig. 3), ρ_{xy} is decreased whereas ρ_{yx} is enhanced relative to the host ρ_a of 100 ohm-m. This behavior has lead many workers to propose some average of the two ρ_a curves to yield a corrected response. Such averages are carried out on a site-by-site basis. Common examples are the arithmetic average of Z_{xy} and Z_{yx} (Z_{av}) and the impedance determinant Z_{det}, both of which are invariant under coordinate rotation. Simple 3-D model studies have suggested that such averages are useful for soundings located outside confined conductive inhomogeneities (e.g., Ingham, 1988). Groom and Bahr (1992) advocate Z_{av} because its estimate may be more stable in the presence of noise. Averaging has two important limitations: First, impedance averages are not useful inside the inhomogeneity where both ρ_a curves are biased in the same direction. In the field, one may not know the location of sites with respect to structural transitions. Second, the degree of distortion of ρ_{xy} and ρ_{yx} relative to ρ_{1D} about a resistive 3-D body differs from that about a conductive one, and is a strong function of the body's aspect ratio. As a function of contrast and aspect, Berdichevskiy and Dmitriev (1976) show that the degree of distortion can range from appearing solely in ρ_{xy} to appearing solely in ρ_{yx}.

2.3.2 Spatial averaging. Spatial averaging of the impedance also has been proposed to reduce effects of variability in near-surface structure. The most general approach is the electromagnetic array processing (EMAP) technique (Torres-Verdin and Bostick, 1992a,b). A fundamental concept in EMAP, derived from Faraday's law, is that the secondary electric field along a traverse across a 3-D body has no zero-wavenumber component, i.e.,

$$\int_{-\infty}^{\infty} \mathbf{E}_s \cdot d\ell = 0, \tag{15}$$

where $d\ell$ is an increment along the (possibly crooked) profile (Fig. 4). This is equivalent to a zero-mean property for P_{yy}, and thus the zero-wavenumber component of Z_{yx} approaches Z_{1D} (Torres-Verdin and Bostick, 1992a). Element Z_{xx}, on the other hand, is zero-mean. In principle, the layered-earth impedance could be recovered by a simple spatial average, provided that the lateral extent of any heterogeneity was several times smaller than the profile length. In practice, EMAP combines an adaptive spatial filter the length of which increases with period, with local 1-D inversion to obtain a lateral resolution versus depth that is consistent with the physics of diffusive EM fields. Contiguous electric bipoles along the profile are recommended to ensure proper sampling (Torres-Verdin and Bostick, 1992b; Fig. 4),

EMAP processing can be viewed as an attempt to remove all effects of galvanic (boundary charge) contributions from the measured data and leave a response akin to a TE mode. Synthetic test examples verify the effectiveness of this approach (Torres-Verdin and Bostick, 1992b). However, some loss of resolution is inevitable because galvanic effects, of course, contain independent structural information. A rigorous 2-D

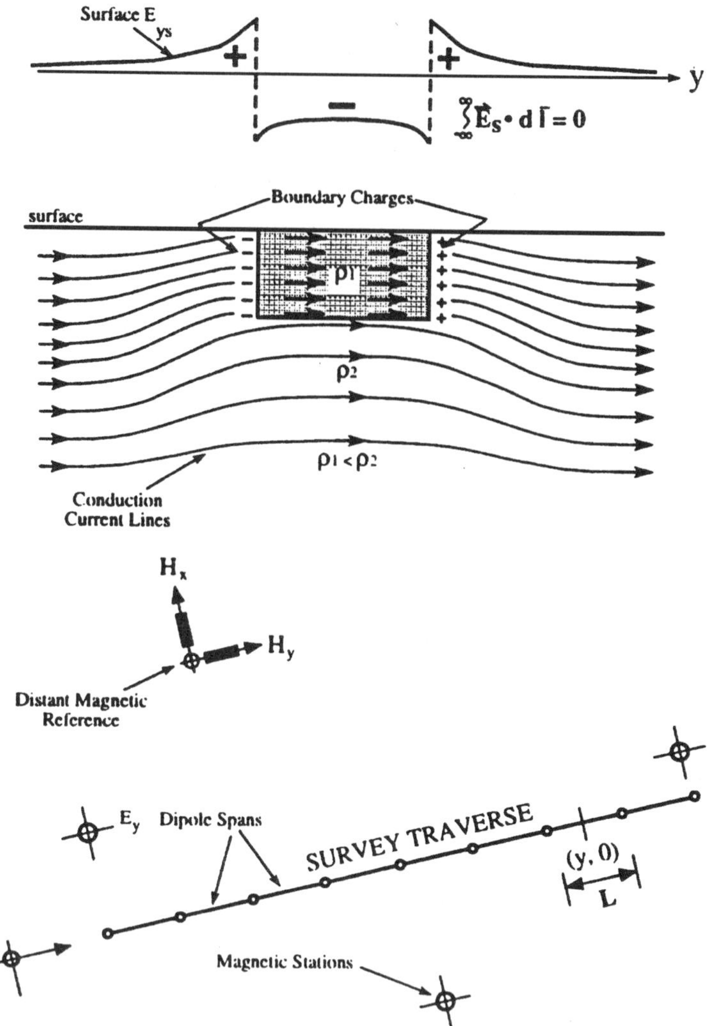

Figure 4. Schematic portrayal of low-frequency electric field along a profile over a near-surface conductor (top). Anomaly has no zero wavenumber component. Close or contiguous bipole deployment allows recovery of appropriate long-wavelength averages of resistivity (e.g., EMAP). Magnetic station deployments are more widely spaced because of the smoother, band-limited nature of the H-field. Modified from Torres-Verdin and Bostick (1992a,b).

inversion of ρ_{yx} and ϕ_{yx} (the TM mode with the x-axis as strike) thus may be better for a straight profile across an effectively 2-D target overlain by small-scale 3-D geologic noise.

Equation (15) applies only to the electric field along the profile (integration path of E_y); no such relation exists for E_x, which is normal to the profile. Thus, no matter how many bipoles are deployed normal to the transect, no meaningful average of E_x can be guaranteed. Furthermore, there can never be perfect lateral sampling of E_x because bipoles normal to the strike cannot be deployed continuously. For a 2-D target beneath a near-surface 3-D structure, joint 2-D inversion of ρ_{yx} and ϕ_{yx} (TM to the 2-D target) together with ϕ_{xy} and K_{zy} may be possible. Only the properties of ρ_{yx}, however, allow it to be relied upon for amplitude information. In practice, enough measurements of E_y

are needed to characterize its variation statistically. Short-wavelength filtering of Z_{yx} to suppress the near-surface galvanic variation (but not that of the target) may be done prior to inversion. Alternatively, one may invert for an equivalent 2-D, near-surface variation over the target structure. This assumes, as discussed later in 3-2 and 3-3, that an equivalent variation exists.

2.3.3 Impedance tensor decomposition. Several authors have extended the model of 1-D impedance with electric galvanic distortion [Eq. (9)] to a distorted, regional-scale 2-D impedance, written as

$$[\underset{\sim}{\mathbf{Z}}(\mathbf{r})] = \left\{[\mathbf{I}] + \left[\underset{\sim}{\mathbf{P}}^0_{hs}(\mathbf{r})\right]\right\} \cdot [\underset{\sim}{\mathbf{Z}}_{2D}], \tag{16}$$

where $[\underset{\sim}{\mathbf{Z}}_{2D}]$ here is in the coordinates of the regional 2-D structure and thus is anti-diagonal (Bahr, 1988; Groom and Bailey, 1989). Parsing the elements of Eq. (16) reveals a similar distortion as represented in Eq. (9) with the 2-D principal impedances shifted by real, multiplicative factors but no change in the phases. However, if the measurement x-axis does not correspond to a 2-D strike, a complication arises in that the elements of $[\underset{\sim}{\mathbf{Z}}(\mathbf{r})]$ contain arbitrary proportions of (2-D) Z_{xy} and Z_{yx} [(impedance mode mixing; Bahr (1988)]. An inappropriate strike definition, in particular, could lead to phase anomalies whose spatial scales of variation reflect the small-scale, near-surface structure, but whose period range reflects that of the deep 2-D structure. It may be impossible to find a 2-D model that duplicates such behavior. Of course, the assumption of a 2-D target should be tested for real field situations (Bahr, 1991).

Smith (1995), following Bahr (1988), decomposes the electric distortion term of Eq. (16) as

$$[\underset{\sim}{\mathbf{I}}] + \left[\underset{\sim}{\mathbf{P}}^0_{hs}(\mathbf{r})\right] = [\underset{\sim}{\mathbf{D}}(\mathbf{r})] = \left[g_x \begin{pmatrix} \cos\beta_x \\ \sin\beta_x \end{pmatrix} \; g_y \begin{pmatrix} -\sin\beta_y \\ \cos\beta_y \end{pmatrix} \right], \tag{17}$$

where g_x and g_y are gains on E_x and E_y, and β_x and β_y are clockwise angles by which the regional E_x and E_y rotate because of a near-surface scatterer. Groom and Bailey (1989) present a more elaborate decomposition:

$$[\underset{\sim}{\mathbf{D}}(\mathbf{r})] = g \begin{bmatrix} 1 & -t \\ t & 1 \end{bmatrix} \begin{bmatrix} 1 & e \\ e & 1 \end{bmatrix} \begin{bmatrix} 1+s & 0 \\ 0 & 1-s \end{bmatrix}, \tag{18}$$

where g is an average site gain, t and e are termed twist and shear, and s is the differential gain. Smith (1995) explores transformations relating parameters of the two decompositions and discusses their least-squares estimation. The model of Eq. (16) is defined in the coordinates of the large-scale 2-D structure. If the measurement coordinates differ significantly, large residuals in estimation of the parameters of Eqs. (17) or (18) can be caused by impedance mode mixing. Smith (1995) recommends incremental coordinate rotation to find an apparent strike direction that minimizes the parameter misfit.

The distortion of regional impedance elements can be estimated only to within two unknown scale factors. These frequency-independent terms are g_x and g_y in Eq. (17) and $g(1+s)$ and $g(1-s)$ in Eq. (18), and also have been termed static shift factors [e.g., Groom and Bahr (1992)]. Independent determination of the unknown gains typically is sought by the usual methods such as invariant averages, shallow time-domain EM soundings, or conductive marker horizons of known depth. All of these have their limitations, even without the added complications of anisotropy. Therefore, impedance decompositions may be most valuable for uncovering deep geoelectrical orientations and for estimating impedance phases more free of mode mixing. The Southern Appalachians MT transect is an example of successful trend confirmation using impedance decompositions (Wannamaker et al., 1996).

None of the scale factors, including g_y and $g(1+s)$, equals P_{yy}. Thus, such factors do not possess the long-wavelength property of P_{yy} (and hence Z_{yx}) implicit in Eq. (15). Curves of ρ_{yx} scaled by the former factors therefore should not be used in 2-D TM mode inversion of a profile across a 2-D target structure overlain by a 3-D, small-scale inhomogeneity. Fixed-axis quantities aligned with the profile trend are more appropriate for inversion. However, when the strike of the underlying 2-D structure is difficult to identify and the 3-D distortion is severe, decomposed ϕ_{xy} (nominal TE mode quantity) may be superior for joint TM-TE inversion because of its immunity to regional impedance mode mixing. Decomposition may be particularly useful if ρ_{xy} is substantially less than ρ_{yx} (which I have often observed).

Chave and Smith (1994) note also that the H-field tensors $[\tilde{\mathbf{Q}}^0_{vs}(\mathbf{r})]$ and $[\tilde{\mathbf{Q}}^0_{hs}(\mathbf{r})]$ around a small-scale, shallow structure tend to a galvanic limit. The precise estimation of the tensors requires knowledge of the 2-D regional impedances, but the specific frequency band which they influence may be downweighted in a strike estimation or inversion using $[\mathbf{K}_z(\mathbf{r})]$.

Inversion of (nominal) TM mode quantities (ρ_{yx} and ϕ_{yx}) is more problematic when the profile is not aligned with regional strike. There is a conflict between accurate representation of medium- to long-wavelength electric fields and minimal mode mixing. It is best of course to avoid this problem from the outset by careful survey design incorporating reconnaissance EM data, other geophysics, and a clear understanding of the geological framework. Otherwise, presuming that the lateral variation in mode mixing is characteristic of the near-surface 3-D structure, it is better perhaps to retain the fixed-axis definition along the profile initially and to filter ρ_{yx} and ϕ_{yx} spatially to remove the short-wavelength components. This may yield a filtered ρ_{yx} and ϕ_{yx} that reflect locally isotropic conditions at short periods. These short-period values of ρ_{yx} may define static shift factors with which to correct the decomposed impedance elements. Such corrections, however, are more complicated in the presence of resistivity anisotropy, at either local or regional scales (e.g., Kellett et al., 1993; Mareschal et al., 1995).

3 Contrasting 2-D and 3-D MT responses

Most of the recommendations above apply to a large 2-D target, buried under smaller, 3-D geological noise. When the target structure itself is 3-D, other considerations are important. Consider first modest 3-D effects.

3.1 Prismatic sedimentary-basin model

MT data in the Basin and Range can be analyzed with a simple, prismatic model of conductive sediments in a regional resistivity layering (Wannamaker et al., 1984). This 3-D model and its response in ρ_{yx} and ϕ_{yx} for profiles at $x = 0$ and $x = 9$ km are shown in Fig. 5 for comparison with corresponding 2-D TM mode calculations (Wannamaker et al., 1986). The x- and y-coordinates of the responses are coincident with those of the 3-D body for all sites and all periods. The 3-D responses were computed with an earlier version of the algorithm described in Wannamaker (1991), and are reasonably accurate.

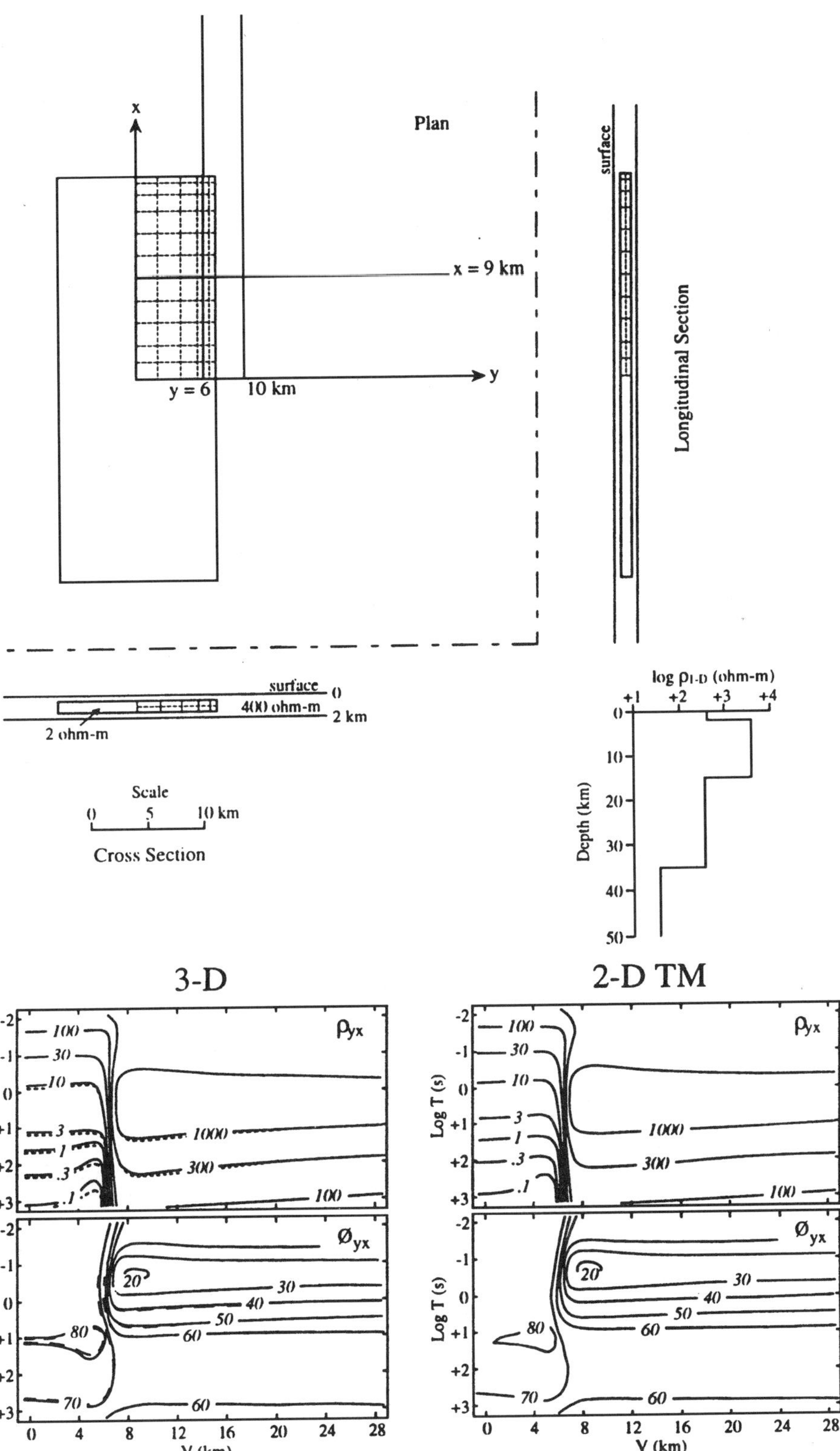

Figure 5. Plate-like conductive prism representing conductive sedimentary basin plus 3-D yx-mode and corresponding 2-D TM response pseudosections for a profile along the y-axis (solid contours) and for $x = 9$ km (dashed contours) (redrawn from Wannamaker et al., 1984). Profile locations are shown at $y = 6$ km and $y = 10$ km for computations here and in Fig. 6. Layered background structure is drawn toward center right.

Anomalies in ρ_{yx} and ϕ_{yx} (nominal TM mode) are essentially the same for both 3-D profiles and 2-D profiles at all periods. This is because both formulations include boundary charges on the sides of the basin. Vertical current gathering appears as a roughly dipolar variation in the electric field over the basin as period increases. The frequency dependence of the response is much lower than that of the small-scale feature in Fig. 3, in keeping with expectations from EM similitude. Figure 5 suggests that accurate cross-sections of Earth resistivity may be interpreted from profiles of MT measurements across elongate, geometrically regular 3-D conductors using a 2-D TM mode algorithm. For such 2-D modeling, a fixed-coordinate system compatible with target trends should be used instead of a variable one, e.g., based on principal axes of $[\underset{\sim}{\mathbf{Z}}(\mathbf{r})]$. The latter approach also suffers from some impedance mode mixing, possibly generating long-period impedance phase anomalies of high spatial wavenumber that no purely 2-D model can fit (Wannamaker et al., 1984).

Comparison of nominal TE mode quantities ρ_{xy} and ϕ_{xy}, and of the magnitude and phase of K_{zy}, also were carried out by Wannamaker et al. (1984). Striking differences were demonstrated between the 3-D and 2-D TE responses because boundary charge terms, prominent on the ends of the 3-D body, are entirely absent from the 2-D TE formulation (Fig. 6). Note that the 3-D ρ_{xy} becomes increasingly depressed relative to the 2-D case toward the longest periods whereas the 3-D ϕ_{xy} generally takes on higher values than its 2-D counterpart. Consequently, in attempting to replicate the 3-D response in ρ_{xy} and ϕ_{xy} using a 2-D TE algorithm, one would need to place additional (false) low resistivities at depth below the true basin. This type of 3-D effect has led to concerns that many published results based on 1-D or 2-D TE modeling of 3-D data have experienced a bias toward shallow, low resistivities.

A discrepancy of similar importance was observed between 3-D and 2-D TE responses in K_{zy} (Wannamaker et al., 1984), although the character differs significantly. For the basin model, the 3-D values became greatly subdued relative to the 2-D for periods about 1 s and longer (Fig. 6). Boundary charges on the ends of the 3-D model depress the interior electric field, and hence the current density, giving smaller magnetic anomalies. The band-limited nature of $[\underset{\sim}{\mathbf{K}}_z(\mathbf{r})]$ anomalies is clear here, in comparison with Fig. 3, showing that survey design to determine the optimum period range for induction arrow or tipper strike definition is of value.

Simulating the 3-D response in K_{zy} with a 2-D TE algorithm requires a 2-D basin model with a conductance substantially less than that of the true structure (because of the reduced 3-D response). This is the reverse of the error made in a 2-D TE interpretation of ρ_{xy} and ϕ_{xy} (above). The model requires a joint interpretation of the TE impedance (particularly ϕ_{xy}) and vertical magnetic-field functions. Finding a single 2-D model consistent with both TE quantities and, of course, the TM data should be evidence that an essentially 2-D target geometry has been surveyed.

A complementary 3-D structure where the ends of the basin model connect to regional, conductive "oceans" has been simulated by D. McKirdy and described by Jones (1983). This model, as expected, showed strongly enhanced $[\underset{\sim}{\mathbf{K}}_z(\mathbf{r})]$ anomalies relative to the 2-D case because of current concentration in the basin from the oceans. An amplification of ρ_{xy} also would be expected, but was not presented. Furthermore, some 2-D/3-D discrepancies in ρ_{yx} and ϕ_{yx} might be expected because of sideswipe from the conductive oceans off-profile.

Emphasizing the (2-D) TM mode is not always appropriate; there are structures that simply do not respond strongly for this mode [e.g., narrow buried conductive dikes; see

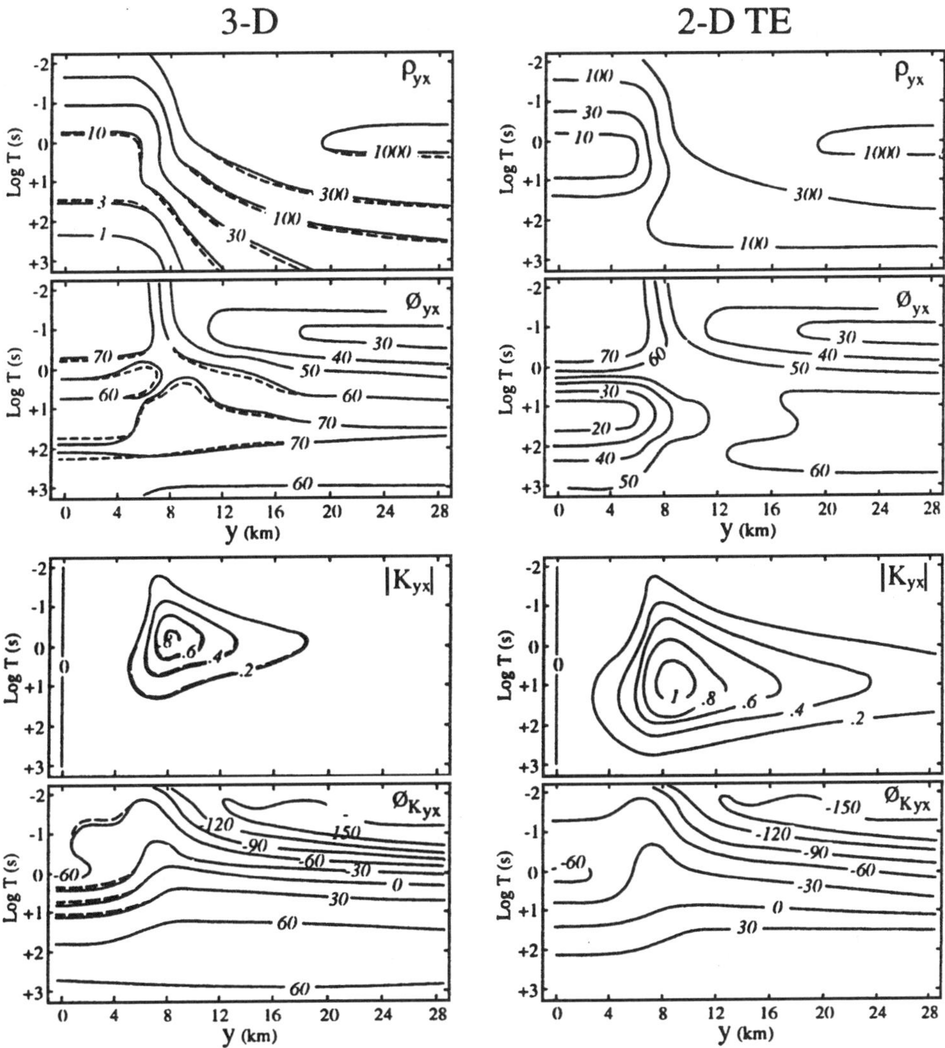

Figure 6. Pseudosections of 3-D xy-mode and corresponding 2-D TE response pseudosections for a profile along the y-axis (solid contours) and along $x = 9$ km (dashed contours) of the model of Fig. 5 (redrawn from Wannamaker et al., 1984).

Wannamaker et al. (1984), Vozoff (1991)]. Consideration of the full tensor response, if possible, is always recommended.

Conductive heterogeneities have been modeled here also because they are more representative of sedimentary bodies, graphitic or mineralized zones, or hydrothermal alteration. Certain resistive structures, such as isolated horsts or salt domes, apparently need to be more elongate than conductive ones for precise agreement between 3-D and 2-D TM modes (Berdichevsky et al., 1998). Moreover, boundary charge effects on the nominal TE mode quantities appear to be less severe for resistive structures than conductive ones. Nevertheless, previous spatial average concepts still apply, in particular the zero-mean property of the TM response.

3.2 *Noncentrally located data profiles*

A regular profile over a target structure, as shown in Fig. 5, may not always be possible because of logistical considerations or unanticipated structures. In a study of noncentral

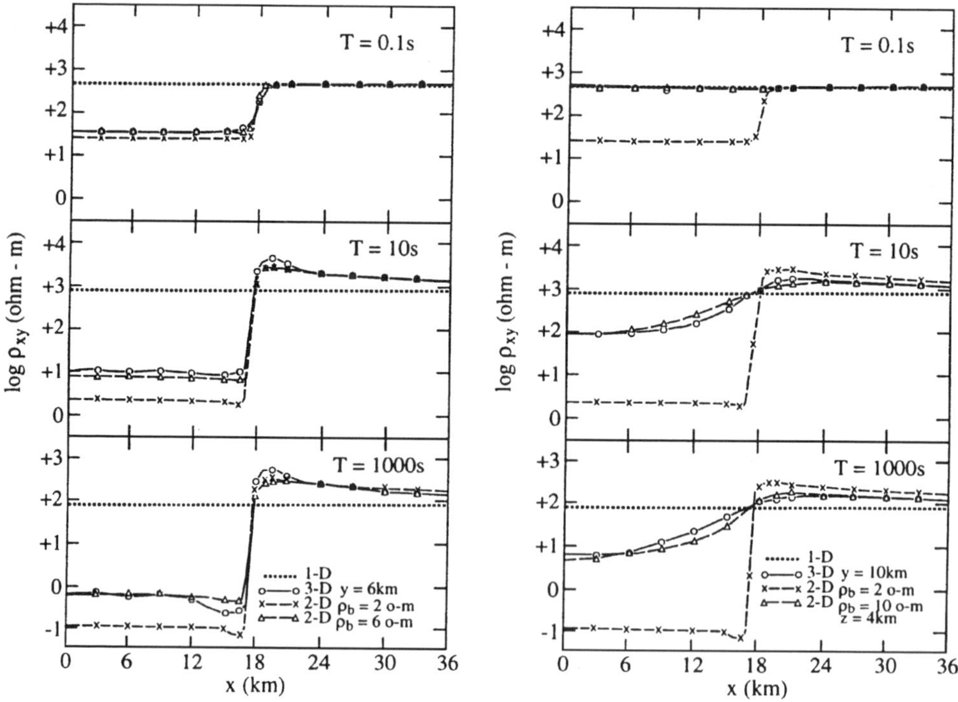

Figure 7. Calculated ρ_a profiles at $T = 0.1$, 10, and 1000 s along $y = 6$ km (left) and $y = 10$ km (right) over the plate-like body of Fig. 5 and over comparable 2-D bodies described in the text. The apparent resistivity of the layered host is ρ_{1D}, whereas ρ_b refers to the resistivity of the 3-D plate, or the equivalent 2-D plate [redrawn from Kariya (1986)].

profiles, Kariya (1986) computed responses in ρ_{xy} and ϕ_{xy} at three frequencies (10, 0.1, and 0.001 Hz) for profiles parallel to the model x-axis at offsets along y of 6 km and 10 km (Fig. 7). Both profiles showed an elongate, electric dipolar response; as expected, the profile at 10 km was smoother and showed less variation.

Both profiles computed by Kariya (1986) could be fitted reasonably well by the TM response of 2-D plate-like models (Fig. 7). For the nearer profile (at 6-km offset), the depth, thickness, and width of the 2-D model were identical to the 3-D, although the 2-D body resistivity was 6 ohm-m instead of the original 2 ohm-m. This overestimate of body resistivity is not surprising because the profile location is near the body's edge. For the far offset (at 10 km), the thickness and width of the 2-D model were identical to the 3-D, but the 2-D model depth was 4 km (instead of the original 500 m) with resistivity of 10 ohm-m.

Hence, 2-D TM modeling of profiles parallel to the long axis of simple conductive structures, but offset from the center, usually will underestimate the conductivity of the body and overestimate the depth of bodies off to the side. Strike estimates from K_{zy} and the yx-mode data can indicate whether the structure lies completely off the profile. However, since the full dipolar response of the 3-D conductor was covered by the profiles, the regional layering of the 2-D models was consistent with the true layering.

3.3 *Profiling across irregular structures*

Natural structures are never perfectly regular, and so, it is helpful to know the limitations of 2-D modeling in such cases. Studies of an offset 3-D basin overlying an infinitely resistive basement have been done with (dc) thin-sheet modeling (e.g., Wannamaker,

1990) and integral-equation methods (Wannamaker, 1991). The dc thin-sheet model demonstrated that profiles of ρ_{yx} across the offset differ significantly from the TM response of a 2-D body of identical cross-section. One can conclude that errors in estimates of the upper crustal structure could result from straightforward 2-D (TM) approach. Wannamaker (1990) also showed analytically, however, that any TM profile across a 3-D thin-sheet structure of this class may be fitted in the long-period (dc) limit by an equivalent 2-D structure embedded in the same regional layering. Demonstrating that shallow 3-D structures cause errors in deep structural interpretations requires multifrequency simulations.

Wannamaker (1991) simulated the offset basin model at periods of 0.32 to 10^4 s. The apparent resistivities ρ_{yx} and ϕ_{yx}, defined with respect to a fixed set of coordinates across the basin, were inverted as TM mode data with the (minimum-structure) inversion program of DeGroot-Hedlin and Constable (1990) called Occam-2. Errors were assigned to the 3-D data on the basis of discrepancy between 2-D (finite-element) and 3-D (integral-equation) modeling of similar 2-D structures. The first 2-D TM inversion attempted to fit the 3-D results by confining lateral variations to lie within the upper 2 km as in the original 3-D model. This attempt failed, leaving a (normalized) rms misfit of greater than 4 (Wannamaker, 1991). Only by allowing structure in the deeper, resistive basement could a fit within an acceptable rms misfit ($\simeq$1) be obtained. Resistivities of about 100 ohm-m, spreading laterally with depth under the offset basin, allowed currents in the upper 2 km to leak deeper; this prevented the severe depression in ρ_{yx} seen in the purely 2-D TM case. It was not possible to limit the perturbation in the model resistivity to the same depth range as the original structure that caused the 3-D effect (unlike in the dc thin-sheet model). This may have been caused in part by the (in effect) infinitely resistive basement in the original model. However, indiscriminate 2-D TM mode modeling cannot be carried out reliably over structures that depart substantially from a preferred structural orientation. Care should be taken in design of real profiles to avoid structural irregularities by use of independent geological and geophysical data. The success of the design may be judged at least in part by checking geoelectric strike and 3-D indicators.

4 Field MT profiles

When the interpretation methods recommended here are applied to actual field profiles, what confidence can be put in the interpreted structures and what cannot be resolved without additional 3-D analysis? In the remaining sections, I discuss field examples analyzed in the TM mode and also consider the value of modeling both TE and TM modes.

4.1 EMSLAB—Juan de Fuca Experiment

The EMSLAB project, funded by the U.S. National Science Foundation, collected 39 broadband and 15 long-period MT soundings at periods from 10^{-2} to 10^4 s along a 200-km profile on land (Lincoln Line) from the northern Oregon coast eastward over the Juan de Fuca subduction system and Cascadia volcanic arc (Fig. 8). The evident north-south grain of this compressional tectonic zone dictated the survey design; the final resistivity model on land was derived almost entirely from 2-D modeling of the TM mode impedance functions. Modeling K_{zy}, however, clarified resistivities in the offshore sedimentary sections and in a conductive zone below the High Cascades volcanic front. The

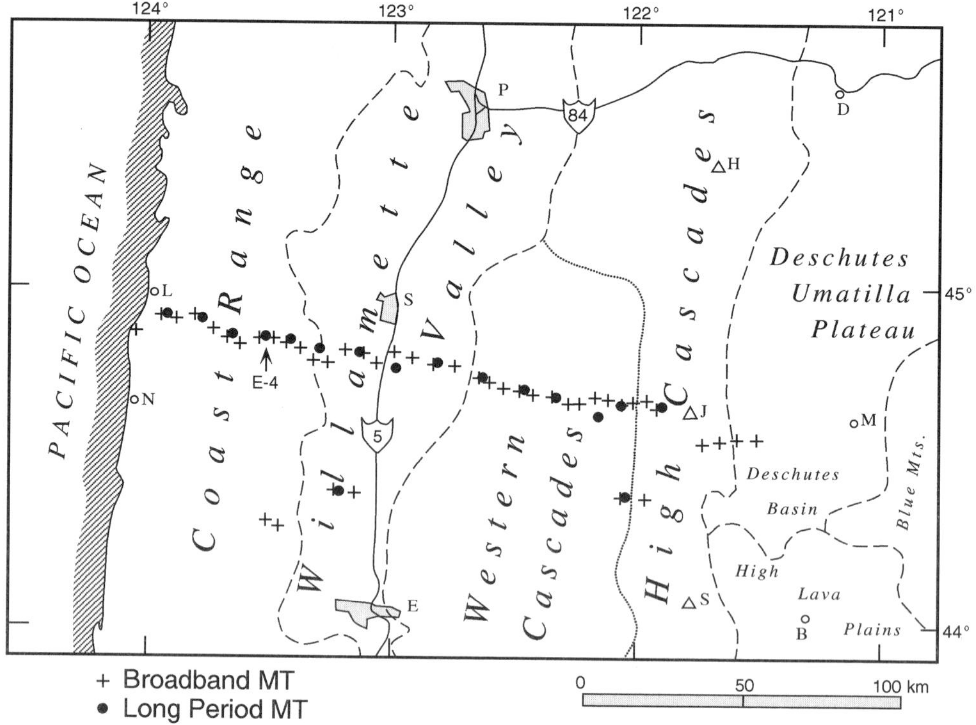

Figure 8. Location map of land MT sites occupied during the EMSLAB project. EMSLAB Site 4 discussed with Fig. 11 is denoted E-4 (from Wannamaker et al., 1989).

existence of an inland-dipping conductive layer nearly coincident with the subduction slip plane, the extension of Willamette Basin sediments under the Western Cascades volcanics, plus a low-resistivity oceanic asthenosphere were corroborated by active and passive source seismic surveying (Rasmussen and Humphries, 1988; Trehu et al., 1994).

Conversely, ρ_{xy} in Fig. 9 often showed pronounced finite strike (boundary charge) effects of short lateral scales; these persist to arbitrarily long periods (Wannamaker et al., 1989). Narrow anomalies of both high and low values of ϕ_{xy} at long periods also could not be fitted by the 2-D TE models considered by the authors (Fig. 10). Phase mixing appears to be the cause. Figure 11 shows a sample sounding curve, which has apparent resistivity and impedance phase data at long-period site 4. At the top are the data in the original fixed coordinates showing an extreme upward excursion in ϕ_{xy} as T exceeds about 2000 s. Groom-Bailey decomposition shows that regional strike is within a few degrees of geographic north down to this period, but swings to nearly 30° west of north at the longest periods. When rotated to this new strike direction, the unusual character in ϕ_{xy} at long periods disappears, but other quantities hardly change. In the original sounding, a portion of the dominant Z_{yx} has been mixed into the smaller Z_{xy}, resulting in dramatic effects. Other Groom-Bailey decompositions farther to the east confirmed a fundamentally north-south trend to the deep crustal-scale structures (Curtis, 1989).

Decomposed TE phase data constrained by ρ_{yx} may contain information about structure near the coast or beneath the ancestral and modern volcanic arcs. Decomposition may allow more accurate estimation of the phase response at larger scales. The appropriateness of 2-D modeling may be assessed by the method discussed earlier. That K_{zy} was so much more consistent with the TM mode data than with the TE data implies a

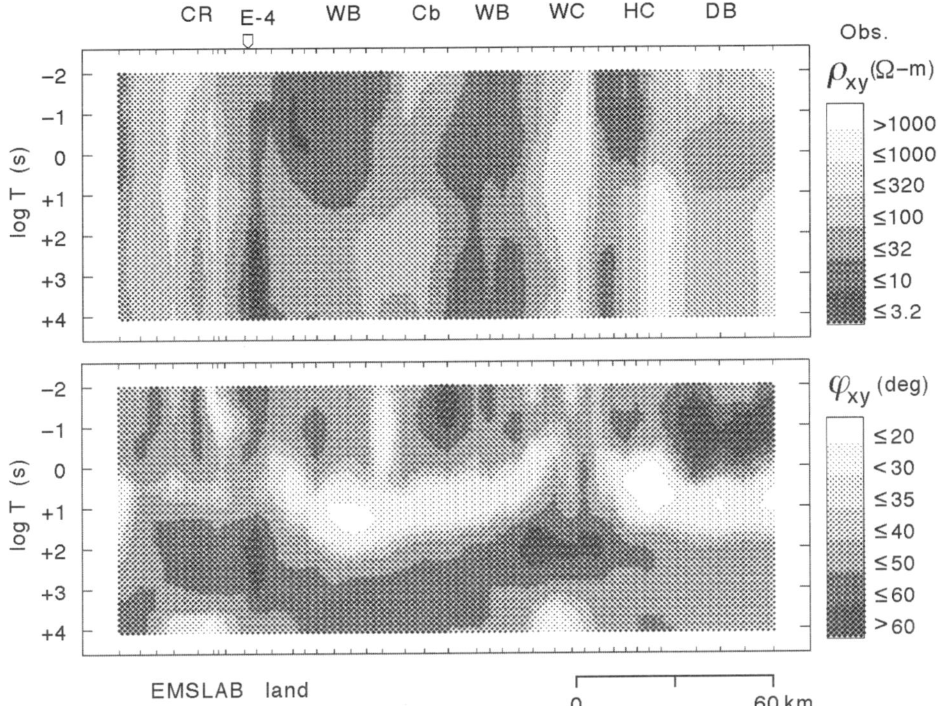

Figure 9. Pseudosections of observed ρ_{xy} and ϕ_{xy} along the land profile in the EMSLAB project. The x-axis is geographic (true) north. Site discussed with Fig. 11 is labeled E-4.

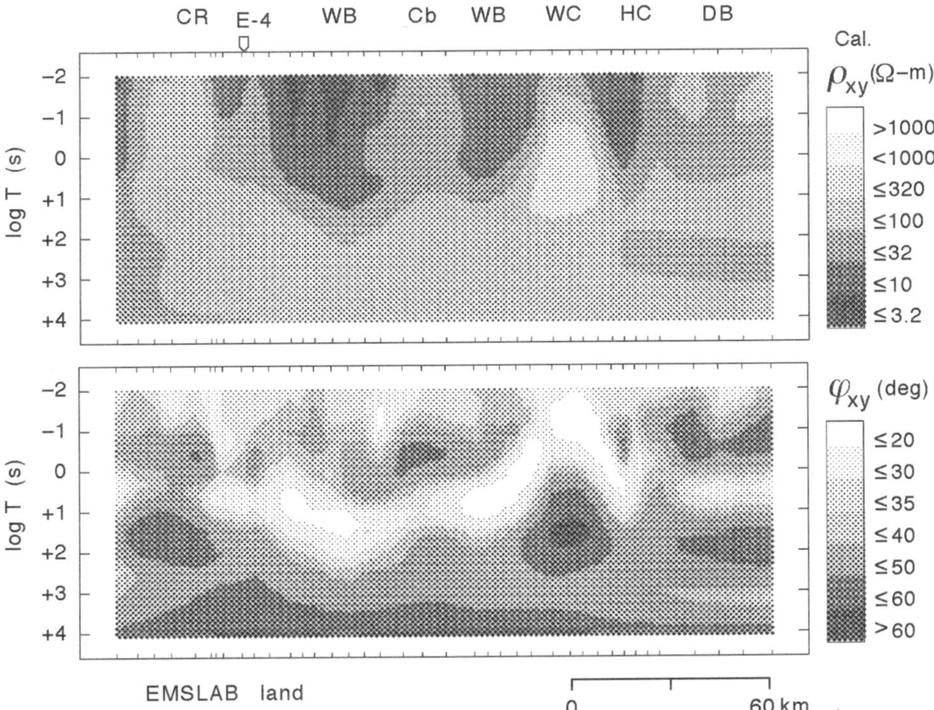

Figure 10. Pseudosections of modeled ρ_{xy} and ϕ_{xy} along the land profile in the EMSLAB project. The x-axis is geographic (true) north.

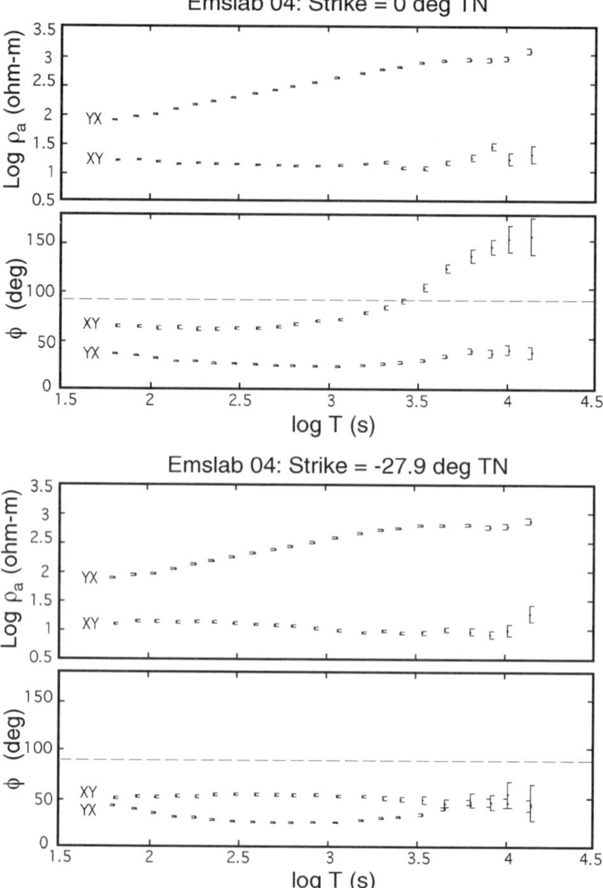

Figure 11. Original and rotated apparent resistivities and impedance phases at site E-4 of Fig. 8. Rotated values are for regional strike estimations by Groom-Bailey decomposition which change from about 2° true down to 2000-s period, to nearly 30° west of north at the longest periods.

greater continuity along the strike of the current density variations relative to those of the E-field.

On the easternmost end of the Lincoln Line (Deschutes Basin), a large-scale east-west boundary is evident off-line. At least one cross-line survey and 3-D modeling are needed to quantify the structure (Wannamaker et al., 1989). Modeling of K_{zx} or the on-diagonal impedance functions (not yet attempted) might give more information on continuity of structures along the strike. These are fully 3-D considerations that are beyond the scope of single-profile interpretation.

4.2 Long Valley caldera magmatic system

A University of Utah project funded by the U.S. Department of Energy collected 24 MT soundings at periods of 10^{-2} to 5×10^2 s in an east-west profile across the center of Long Valley caldera (Wannamaker et al., 1991). The design and interpretation of the profile were based on the observed north-northwest trends in regional-scale geology of the caldera and in geophysical anomalies. Current filament imaging of a profile of contiguous, central-loop time-domain soundings indicated minor to negligible difficulty

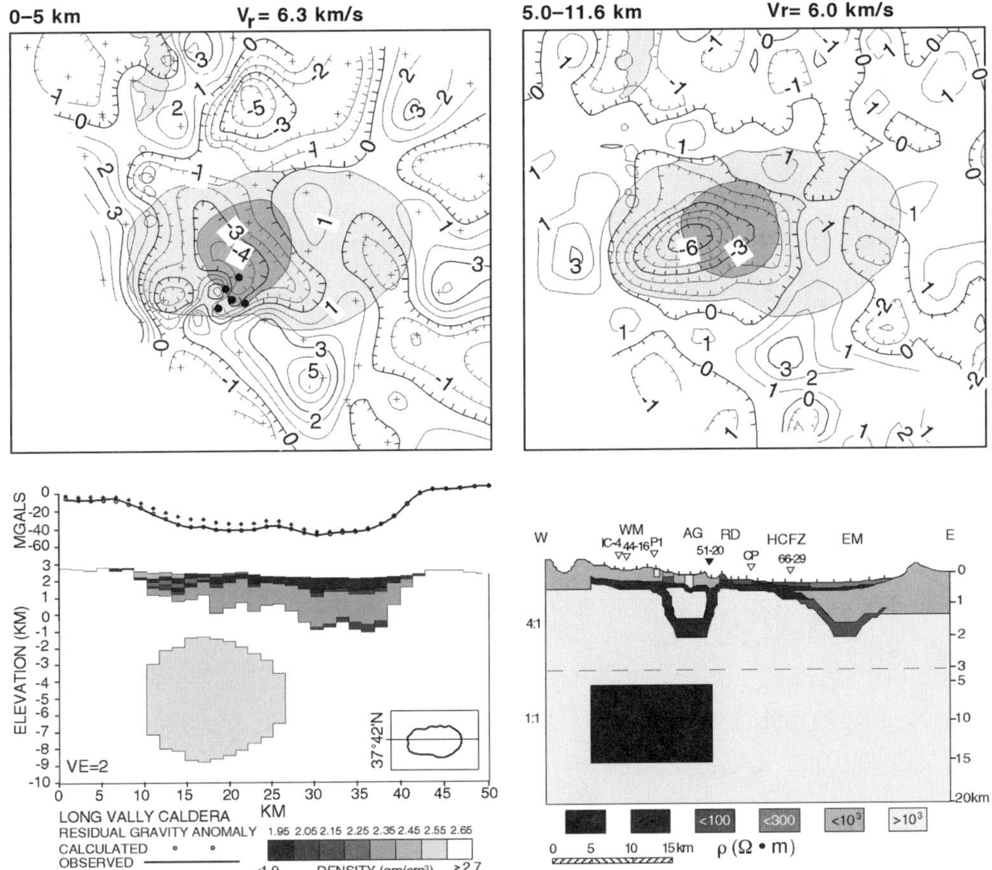

Figure 12. Comparison of east-west MT resistivity model of Long Valley caldera [lower right panel, Wannamaker et al. (1991)] with models of teleseismic *P*-wave velocity [upper panels, Dawson et al. (1990)] and density [lower left panel, Carle (1988)] models. MT profile is very close to caldera-dissecting gravity profile. Velocity models are for labeled depth interval and are expressed as percent deviation from listed reference velocity V_r.

with static distortions in the more active west moat of the caldera. Of relevance to the search for deep magmatic and hydrothermal energy, a conductive body at 5-km depth under the west moat was detected (Fig. 12). It appears in the data as relatively high ϕ_{yx} and steeply falling ρ_{yx} in the 10- to 300-s-period range (Fig. 13). Midcrustal thermal activity in the west moat involving magma or hydrothermal brecciation has been interpreted from teleseismic anomalies and gravity [Fig. 12; for review, see Sanders et al. (1995)]. High-salinity fluids are probably the principal cause of midcrustal low resistivity there (Wannamaker et al., 1991).

Wannamaker et al. (1991) studied the TE mode—ρ_{xy} and ϕ_{xy}, and K_{zy}—by comparing the data to 2-D TE calculations for the model obtained by interpretation of the TM mode. The observed ϕ_{xy} is substantially greater than the computed values, whereas the observed K_{zy} was much less than computed values (Figs. 14 and 15). The comparison suggests that both the upper-crustal conductive caldera and the midcrustal west moat have limited strike extent (as in Figs. 5 and 6).

Figure 16 shows a simple, trial-and-error 3-D model of the caldera, which supports this interpretation. The model has one plane of symmetry and 462 cells on each side of the x-z plane; the integral-equation solution required about 30 minutes per frequency

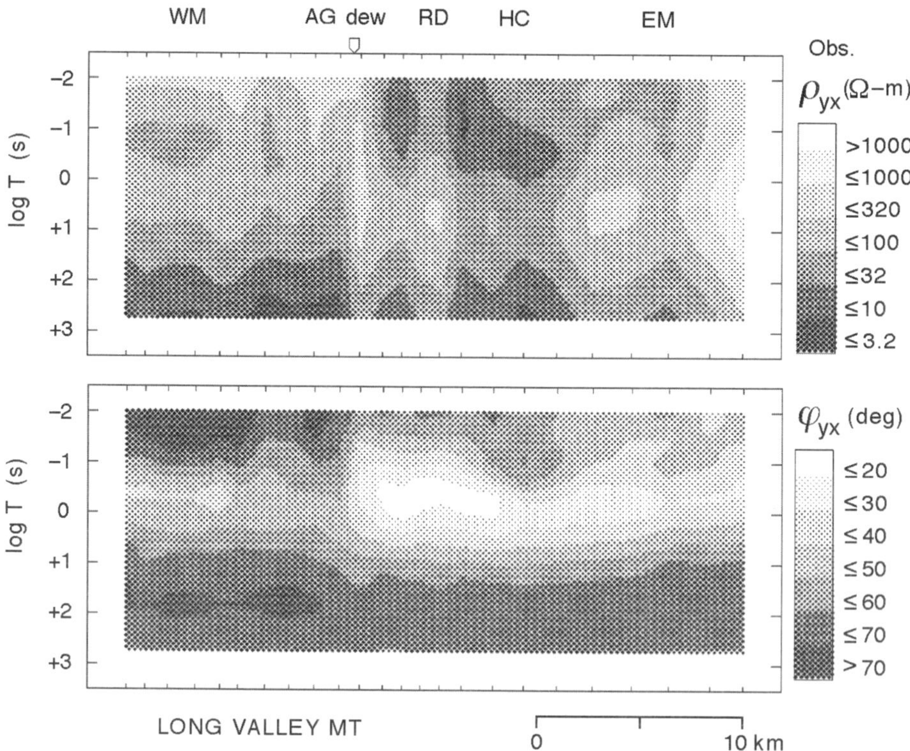

Figure 13. Observed pseudosections of nominal TM mode ρ_{yx} and ϕ_{yx} on east-west profile (top) compared to calculations from 2-D model (bottom) derived by Wannamaker et al. (1991, Fig. 12). The x-axis is N15°E for all sites and all periods.

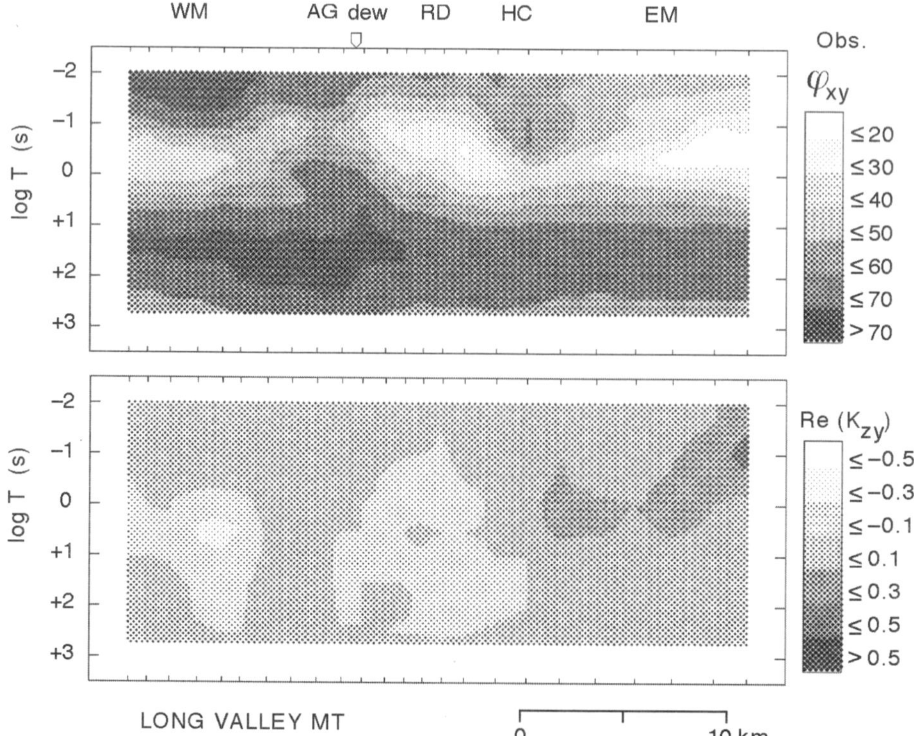

Figure 14. Observed pseudosections of nominal TE mode ϕ_{xy} and $\mathrm{Re}(K_{zy})$ on east-west MT profile in Long Valley. The x-axis is N20°W for all sites and all periods.

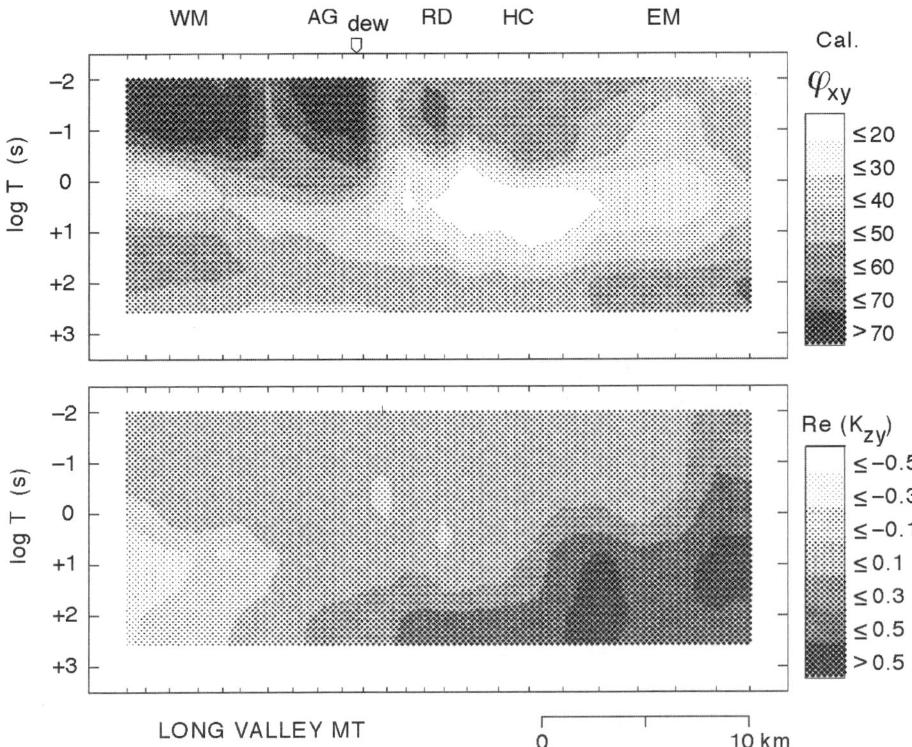

Figure 15. Calculated pseudosections of TE mode ϕ_{xy} and Re(K_{zy}) on E-W MT in Long Valley for 2-D model derived by Wannamaker et al. (1991).

with a Pentium-120 processor. Pseudosections of ρ_{yx} and ϕ_{yx}, and of ϕ_{xy} and Re(K_{zy}), from the 3-D model appear in Figs. 17 and 18. The simulations fit TM mode data reasonably well; more importantly, the 3-D model with a finite strike reproduces much better than 2-D models the observed behavior of the TE phase and the vertical magnetic field.

In particular, a much lower ρ_{xy} and a higher ϕ_{xy} are present in the 1- to 100-s-period range over the central area of the 3-D model because of the finite strike length of the axial graben and of the deeper west moat conductor (Figs. 14 and 18). Values of K_{zy} in the 3-D model at periods longer than a few seconds are moderate and closer to the data than values in the 2-D model. The vertical magnetic-field response caused by the 3-D west moat conductor nowhere exceeds 0.08; all contours visible in Fig. 18 result from the shallow caldera structure. For the 2-D model, values of K_{zy} can exceed 0.5, particularly in the east-central caldera at periods of 10–100 s (Fig. 15).

The deep conductive half-space, which starts at 25-km depth and is 35 ohm-m, is only slightly shallower and less resistive in the 3-D model than in the original 2-D model (30 km and 40 ohm-m). The limited apparent effect of the finite strike length of the caldera seems to support the validity of 2-D modeling of ρ_{yx} and ϕ_{yx}, but other structural features and a better fit might be obtained by more sophisticated 3-D inversion incorporating the on-diagonal impedance elements and K_{zx}.

5 Summary and recommendations

Every MT survey requires careful design. Recent work has shown the importance of near-surface structures in MT interpretation and emphasizes the need for dense lateral

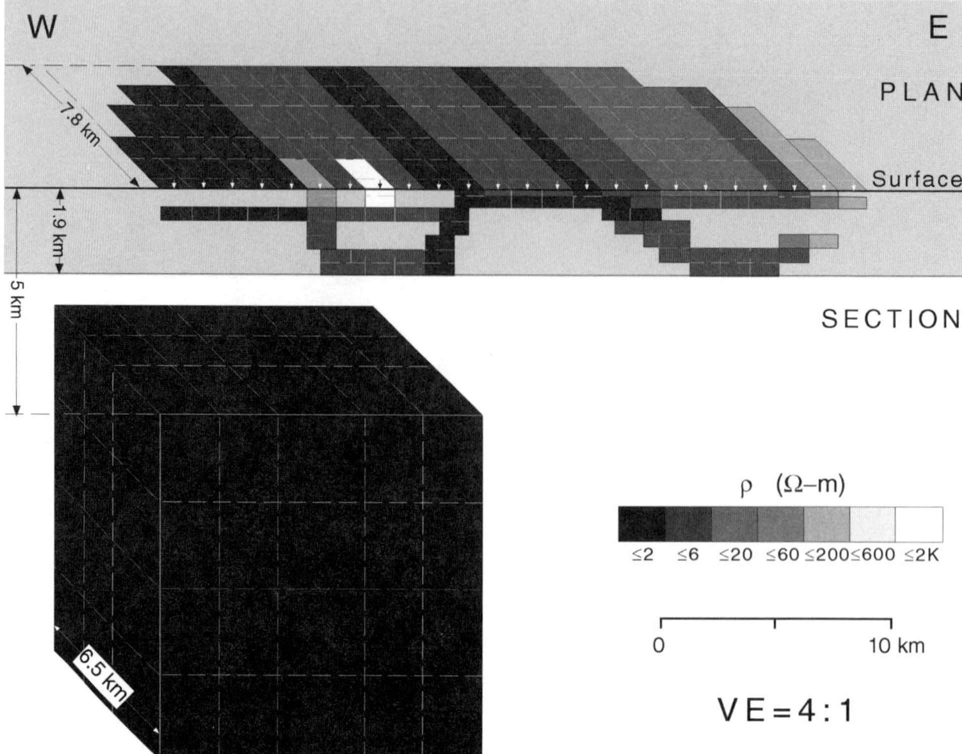

Figure 16. Three-dimensional resistivity model of Long Valley caldera discretized for simulation by integral-equation code of Wannamaker (1991). Layered background also includes 35-ohm-m basement at 25-km depth. For simplicity, the 24 response sites are equispaced over an east-west distance of about 31 km. Also, the strike direction is approximated by true north.

sampling and a wide frequency bandwidth in survey data. I suggest that the planning and interpretion of an MT traverse can proceed along these lines:

1. Begin by ascertaining the geoelectric trends in the survey area from geological or other geophysical data, including reconnaissance EM.
2. Deploy the profiles orthogonal to the chosen trends, with electric bipole orientations parallel and perpendicular to the profile. Expend most of the electric-field bipole measurements on the component along the profile.
3. Ensure that the acquisition coordinate system matches the strike coordinates for the impedance data at all sites and all periods; this will most closely approximate the desired spatial-average attributes.
4. Two-dimensional modeling or inversion should emphasize the presumed TM mode impedance because of its long-wavelength attributes and because it is more immune to finite strike effects.
5. Subsequent to the TM modeling, inclusion of TE mode data should start with K_{zy} because of greater preservation of current flow along strike compared to the E-field. Next, TE mode phase may be included, at least for the limited frequency range representing the target structures. A danger in excluding ρ_{xy} is the possibility of intrinsic anisotropy.
6. Impedances decomposed by the Groom-Bahr method are valuable for geoelectrical trend analysis, but quantitative 2-D modeling or inversion must be careful of

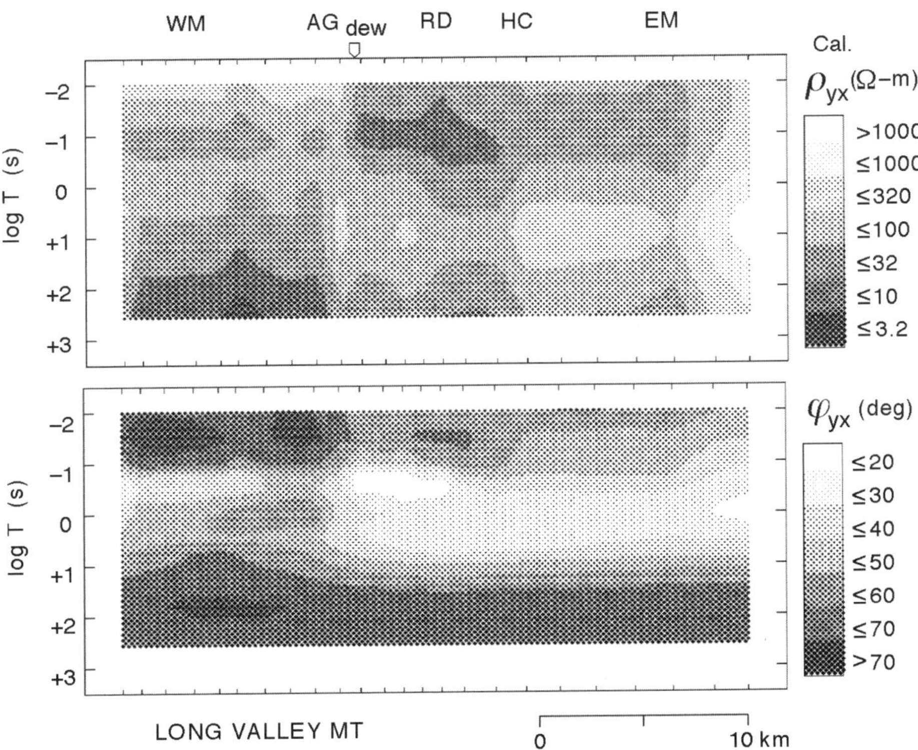

Figure 17. Calculated pseudosections along east-west MT profile in Long Valley of TM mode ρ_{yx} and ϕ_{yx} from 3-D model of Fig. 16.

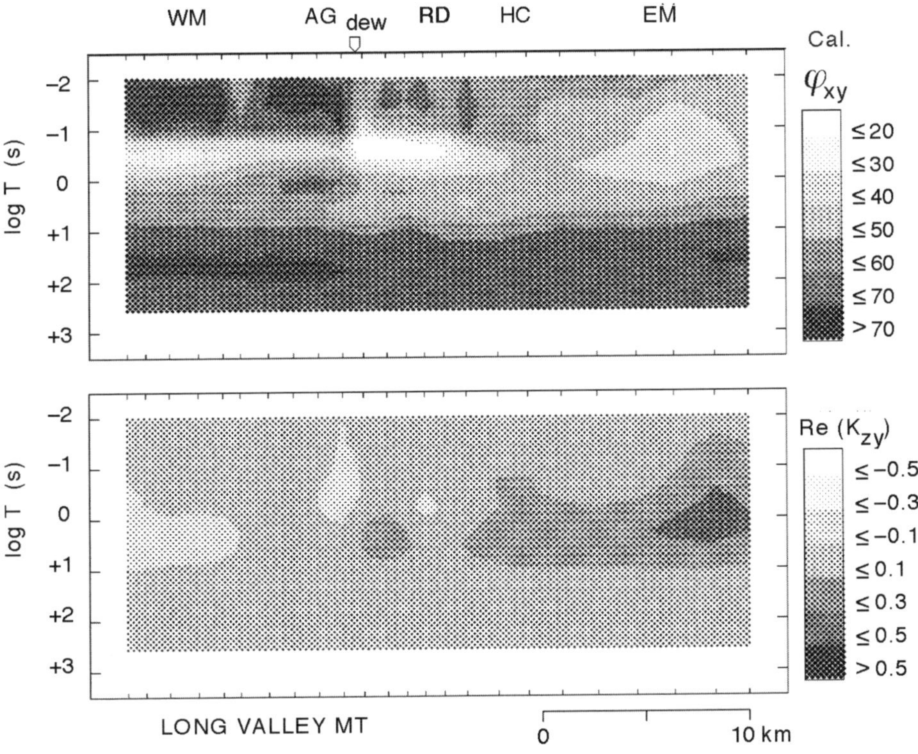

Figure 18. Calculated pseudosections along east-west MT profile in Long Valley of TE mode ϕ_{xy} and Re(K_{zy}) from 3-D model of Fig. 16.

uncertainties in any long-wavelength averages. Such trends should be consistent with those implied by $[\mathbf{K}_z(\mathbf{r})]$.

7. TE mode results decomposed by the Groom-Bahr method, with amplitude levels constrained by ϕ of the TM mode, may be superior to TM impedances in areas of moderate misalignment of the profile, of serious near-surface distortion, or of moderate departures from 2-D by the deep target structure.
8. If strike misalignment or different trends at different depths become apparent after data collection, initial interpretation probably should stick with the original coordinate direction. This will allow quantitative treatment of the shallowest structures whose response is likely to be the strongest.

Acknowledgments

My personal impression of the progress in MT interpretation over the past 20 years has been influenced strongly by Jerry Hohmann, Stan Ward, Francis Bostick, and Ted Madden. The U.S. Department of Energy, Geothermal Technology program, has supported long-term development of MT modeling and interpretation algorithms and the (short-term) synthesis of this paper (currently under contract DE-AC07-95ID13274). John Booker provided the decomposed, long-period EMSLAB data. I am grateful to the editors for their patience in arranging the 3D/EM symposium and in collecting this volume. Robert Turner and Mickey Begent created several of the figures, and Dave Johnson helped considerably with LaTex and postscript production.

References

Bahr, K., 1988, Interpretation of the magnetotelluric impedance tensor: Regional induction and local telluric distortion: J. Geophys., **62**, 119–127.

———1991, Geological noise in magnetotelluric data: A classification of distortion types: Phys. Earth Plan. Int., **66**, 24–38.

Berdichevsky, M. N., and Dmitriev, V. I., 1976, Basic principles of interpretation of magnetotelluric curves, *in* Adam, A., Ed., Geoelectric and geothermal studies, Akademini Kiado, 165–221.

Berdichevskiy, M. N., Dmitriev, V. I., and Pozdnjakova, E. E., 1998, On two-dimensional interpretation of magnetotelluric soundings: Geophys. J. Internat., in press.

Carle, S. F., 1988, Three-dimensional gravity modeling of the geological structure of Long Valley caldera: J. Geophys. Res., **93**, 13 237–13 250.

Chave, A. D., and Smith, J. T., 1994, On electric and magnetic galvanic distortion tensor decompositions: J. Geophys. Res., **99**, 4669–4682.

Curtis, J. H., 1989, Magnetotelluric imaging of the subsurface beneath the EMSLAB Lincoln Line in western Oregon: M.S. thesis, San Diego State Univ.

Dawson, P. B., Evans, R., and Iyer, H. M., 1990, Teleseismic tomography of the compressional wave velocity structure beneath the Long Valley region, California: J. Geophys. Res., **95**, 11 202–11 050.

DeGroot-Hedlin, C., and Constable, S., 1990, Occam's inversion to generate smooth, two-dimensional models from magnetotelluric data: Geophysics, **55**, 1613–1624.

Gough, D. I., 1989, Magnetometer array studies, earth structure, and tectonic processes: Rev. Geophys., **27**, 141–157.

Grant, F. S., and West, G. F., 1965, Interpretation theory in applied geophysics: McGraw-Hill Book Co. (Div. of McGraw-Hill, Inc.)

Groom, R. W., and Bahr, K., 1992, Corrections for near surface effects: Decomposition of the magnetotelluric impedance tensor and scaling corrections for regional resistivities: A tutorial: Surv. Geophys., **13**, 341–379.

Groom, R. W., and Bailey, R. C., 1989, Decomposition of magnetotelluric impedance tensors in the presence of local three-dimensional galvanic distortion: J. Geophys. Res., **93**, 1913–1925.

Hohmann, G. W., 1988, Numerical modeling for electromagnetic methods of geophysics, *in* M. N. Nabighian, Ed., Electromagnetic methods in applied geophysics, **1**, Soc. Expl. Geophys., 313–364.

Ingham, M. R., 1988, The use of invariant impedances in magnetotelluric interpretation: Geophys. J., **92**, 165–169.

Jiracek, G. R., 1990, Near-surface and topographic distortions in magnetotellurics: Surv. Geophys., **11**, 163–203.

———1995, Geoelectromagnetics charges on: U.S. national report to IUGG, Rev. Geophys., supplement, 169–176.

Jones, A. G., 1983, The problem of current-gathering: A critical review: Geophys. Surv., **6**, 79–122.

Kariya, K. A., 1986, Limitations of the magnetotelluric method as applied to the Pioche-Marysvale trend, Utah: M.S. thesis, Univ. of Utah.

Kellett, R., Bishop, J., and Van Reed, E., 1993, The effects of source polarization in CSAMT data over two massive sulfide deposits in Australia: Geophysics, **58**, 1764–1772.

Mackie, R. L., and Madden, T. R., 1993, Three-dimensional magnetotelluric inversion using conjugate gradients: Geophys. J. Internat., **115**, 215–229.

Mareschal, M., Kellett, R. L., Kurtz, R. D., Ludden, J. N., Ji, S., and Bailey, R. C., 1995, Archean cratonic roots, mantle shear zones and deep electrical anisotropy: Nature, **375**, 134–137.

Price, A. T., 1973, The theory of geomagnetic induction: Phys. Earth Plan. Int., **7**, 227–233.

Rasmussen, J., and Humphries, E., 1988, Tomographic image of the Juan de Fuca plate beneath Washington and western Oregon using teleseismic P-wave travel times: Geophys. Res. Lett., **15**, 1417–1420.

Sanders, C. O., Ponko, S. C., Nixon, L. D., and Schwartz, E. A., 1995, Seismological evidence for magmatic and hydrothermal structure in Long Valley caldera from local earthquake attenuation and velocity tomography: J. Geophys. Res., **100**, 8311–8326.

Smith, J. T., 1995, Understanding telluric distortion matrices: Geophys. J. Internat., **122**, 219–226.

Smith, J. T., and Booker, J. R., 1991, Rapid inversion of two- and three-dimensional magnetotelluric data: J. Geophys. Res., **96**, 3905–3922.

Torres-Verdin, C., and Bostick, F. X., Jr., 1992a, Implications of the Born approximation for the magnetotelluric problem in three-dimensional environments: Geophysics, **57**, 587–602.

———1992b, Principles of spatial surface electric field filtering in magnetotellurics: Electromagnetic array profiling (EMAP): Geophysics, **57**, 603–622.

Trehu, A. M., Asudeh, I., Brocher, T. M., Luetgert, J. H., Mooney, W. D., Nabelek, J. L., and Nakamura, Y., 1994, Crustal architecture of the Cascadia Forearc: Science, **266**, 237–243.

Vozoff, K., 1991, The magnetotelluric method, *in* M. N. Nabighian, Ed., Electromagnetic methods in applied geophysics, **2B**, Soc. Expl. Geophys., 641–711.

Wannamaker, P. E., 1990, On thin-layer telluric modeling of magnetotelluric responses: Geophysics, **55**, 372–375.

———1991, Advances in three-dimensional magnetotelluric modeling using integral equations: Geophysics, **56**, 1716–1728.

Wannamaker, P. E., and Hohmann, G. W., 1991, Electromagnetic induction studies: U.S. national report to IUGG: Rev. Geophys., supplement, 405–415.

Wannamaker, P. E., Hohmann, G. W., and Ward, S. H., 1984, Magnetotelluric responses of three-dimensional bodies in layered earths: Geophysics, **49**, 1517–1534.

Wannamaker, P. E., Stodt, J. A., and Rijo, L., 1986, A stable finite element solution for two-dimensional magnetotelluric modeling: Geophys. J. Roy. Astr. Soc., **88**, 277–296.

Wannamaker, P. E., Booker, J. R., Jones, A. G., Chave, A. D., Filloux, J. H., Waff, H. S., and Law, L. K., 1989, Resistivity cross-section through the Juan de Fuca subduction system and its tectonic implications: J. Geophys. Res., **94**, 14, 127–14, 144.

Wannamaker, P. E., Wright, P. M., Zhou, Z.-X., Li, X. B., and Zhao, J.-X., 1991, Magnetotelluric transect of Long Valley Caldera: Resistivity cross section, structural implications, and the limits of a two-dimensional analysis: Geophysics, **56**, 926–940.

Wannamaker, P. E., Chave, A. D., Booker, J. R., Jones, A. G., Filloux, J. H., Ogawa, Y., Unsworth, M., Tarits, P., and Evans, R., 1996, Magnetotelluric experiment probes deep physical state of southeastern U.S.: EOS, Trans. Am. Geophys. Union, **77**, 329, 332–333.

Ward, S. H., and Hohmann, G. W., 1988, Electromagnetic theory for geophysical applications, *in* M. N. Nabighian, Ed., Electromagnetic methods in applied geophysics, **1**, Soc. Expl. Geophys., 131–312.

Whittal, K. P., and Oldenburg, D. W., 1992, Inversion of magnetotelluric data for a one-dimensional conductivity, *in* Fitterman, D. V., Ed., Geophys. Monog. **5**: Soc. Expl. Geophys.

Comparison of 2-D and 3-D Models of a Magnetotelluric Survey in Southern Portugal

F. W. Jones[1]
A. Correia[2]

Summary. We constructed 2-D and 3-D models for an areal magnetotelluric (MT) survey covering about 2500 km^2 in southern Portugal. The geology is complex because the Messejana fault and the Ferreira-Ficalho overthrust intersect within the region, but it was possible to choose some profiles crossing areas that could be considered 2-D. Data from 34 measurement sites were processed and inverted for 1-D models (below each site); 2-D models were constructed from the 1-D models toward the boundary of the study area, where the geology appears to be 2-D. A 3-D electromagnetic model was constructed for the central region of the study area. Comparison of the 2-D and 3-D results shows that some of the geoelectric characteristics found in the 3-D model are not obvious, or even visible, in the 2-D results. The 2-D analysis does not give any indication of the 3-D nature of the nearby region. Furthermore, comparison between field data and the 3-D model results indicates that the main geoelectric characteristics of the region are evident in the response of the 3-D model, despite its simplicity. For this area of southern Portugal, 2-D MT profiles do not seem to present a good picture the region's geoelectric structure.

1 Introduction

A magnetotelluric (MT) survey was carried out to study the geoelectric structure in a complex geological area in southern Portugal (Jones et al., 1992; Correia et al., 1993). During the field work, 34 MT sites were occupied over an area of about 2500 km^2. In this area, located in the southwest part of the Iberian Hercynian belt (Ribeiro et al., 1979), the Ossa-Morena and South Portuguese geotectonic units are separated by the Ferreira-Ficalho overthrust (Fig. 1). The Ossa-Morena Zone is characterized by Precambrian and Lower Paleozoic rocks which exhibit intense deformation and widespread magmatism. The Upper Paleozoic is more fully developed in the South-Portuguese zone where little plutonism occurs and the metamorphism is low grade and where some volcanic and sedimentary deposits that were deformed during the Hercynian orogeny exist.

[1]Department of Physics, University of Alberta, Edmonton, Alberta T6G 2J1, Canada.
[2]Department of Physics, University of Évora, Évora 7000, Portugal.

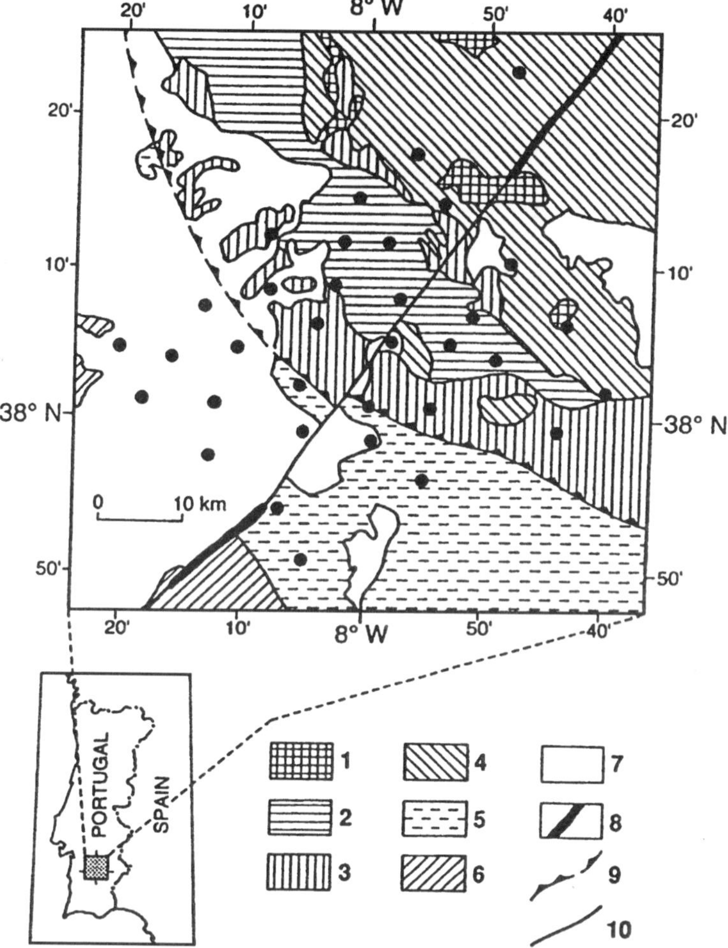

Figure 1. Geological sketch of the study area: Dots represent MT sites. (1) granite, (2) porphyry, (3) gabbro-diorite complex, (4) undifferentiated Precambrian rocks, (5) undifferentiated Devonian rocks, (6) undifferentiated Carboniferous rocks, (7) Cenozoic terranes, (8) doleritic dyke associated with the Messejana fault, (9) Ferreira-Ficalho overthrust, (10) Messejana fault (redrawn from the geological map of Portugal, Geological Survey of Portugal, 1968).

The study area is crossed by another major tectonic feature known as the Messejana fault. This left-lateral strike-slip fault cuts and offsets, up to 4 km, Hercynian structures, and trends approximately northeast-southwest. A doleritic dyke system was emplaced along the Messejana fault by a multiple intrusive process during Early and Middle Jurassic, possibly starting in Late Triassic (Fig. 1).

Data from the MT survey indicate that the study area consists of several high-resistivity blocks that appear to have deep roots in the crust. These blocks are separated by zones of low resistivity that coincide with the general trends of the Ferreira-Ficalho overthrust and the Messejana fault. Furthermore, the Ossa-Morena zone shows a resistivity structure more complex and with higher resistivity than the South-Portuguese zone (Correia et al., 1993).

2 Apparent resistivity and phase pseudosections

Figure 1 suggests that the region has both two-dimensional (2-D) and 3-D structures, depending on the location and the scale. The geology of the central part of the region is certainly 3-D. However, if the MT sites along the edges of the study area are used to construct MT profiles, these apparently cross 2-D geologic structures. With this in mind, four MT profiles passing through those MT stations were considered and their locations are shown in Fig. 2.

Apparent resistivity and phase-data pseudosections for the four profiles were constructed for the study area for the E-polarization (TE) and H-polarization (TM) modes. For these pseudosections, the field data were rotated toward the strike direction. The approximate strike direction was determined by measuring the angle between the general trend of the geological structures and the north-south direction in a clockwise sense (the recording sensors were oriented in the north-south and east-west directions). The data (i.e., the impedance tensors) for each MT station included in the profile were mathematically rotated by an amount equal to the measured angle. The apparent resistivity and phase pseudosections parallel (E-polarization) and perpendicular (H-polarization) to the geological strike were then plotted. The resulting rotated-field apparent resistivity and phase pseudosections are shown for profiles AA′ and BB′ in Fig. 3 and for profiles CC′ and DD′ in Fig. 4.

In the four profiles in Figs. 3 and 4, the apparent resistivity is generally low and varies between 30 and 1000 ohm-m, with a few zones of about 3000 ohm-m. In these figures (and in Figs. 6 and 7), the contour values are the logs of the apparent resistivities. Furthermore, the Messejana fault does not exhibit a strong electrical signature in the profiles considered, despite the fact that the resistivity decreases near it. On the contrary,

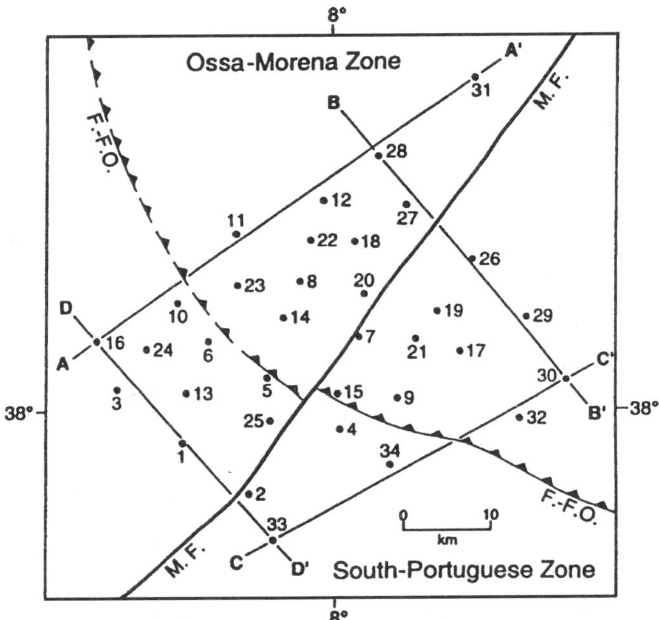

Figure 2. Tectonic units of the study area. M.F. indicates the Messejana fault and F.-F.O. indicates Ferreira-Ficalho overthrust. The MT sites, their numbers, and the four profiles considered in this paper are also shown.

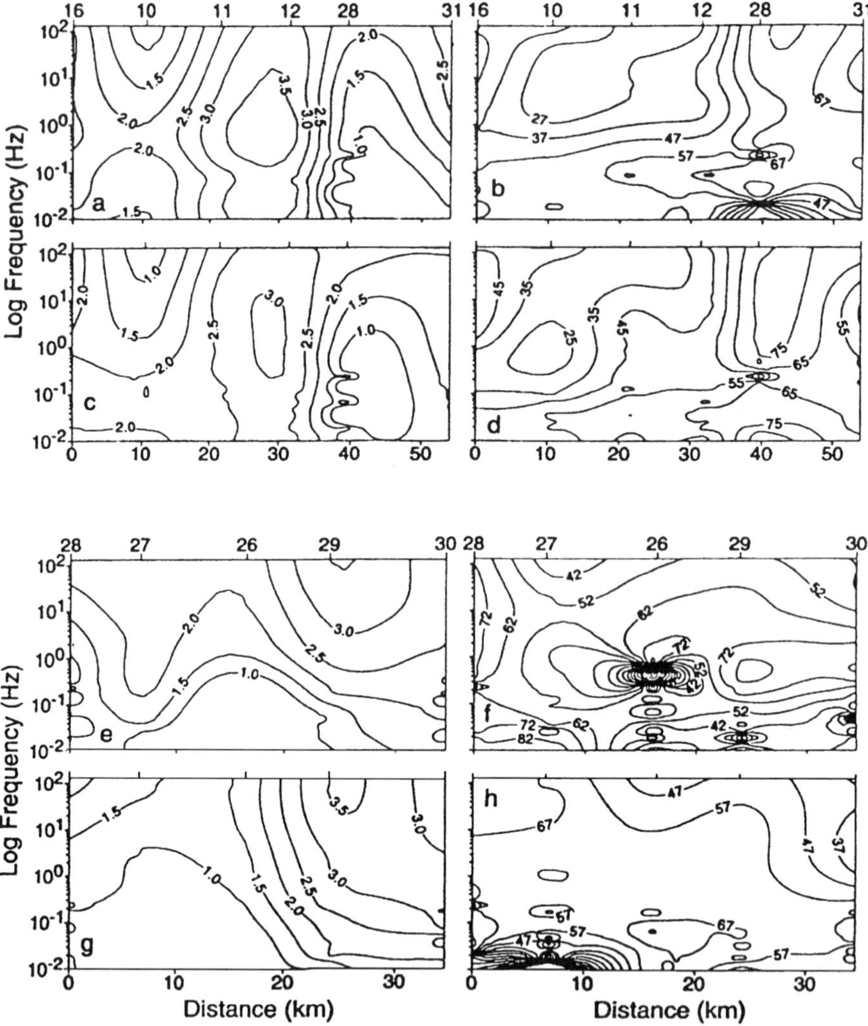

Figure 3. Field data pseudosection for profiles AA′ and BB′ of Fig. 2. The first column of figures corresponds to logarithms of apparent resistivities and the second to phases (contours in degrees). Blocks a, b, c, and d correspond to profile AA′ of Fig. 2 with a and b being the TE mode and c and d being the TM mode; e, f, g, and h correspond to profile BB′ of Fig. 2 with e and f being the TE mode and g and h being the TM mode.

the Ferreira-Ficalho overthrust appears to be well identified in the profiles and there are indications that it separates two different geoelectric domains, which is expected on the basis of knowledge of the regional geology. These results show that different kinds of faults can have different electrical signatures and that the MT method is appropriate for locating them and possibly identifying the different types.

3 2-D electromagnetic modeling

We constructed 2-D resistivity models for each of the profiles described above. The 2-D models were based on 1-D inversions obtained at each of the sites that fall on the profiles. The 1-D models were calculated with a Monte Carlo hedgehog inversion procedure (see Jones et al., 1992) using the invariant impedance concept of Berdichevsky and Dmitriev (1976).

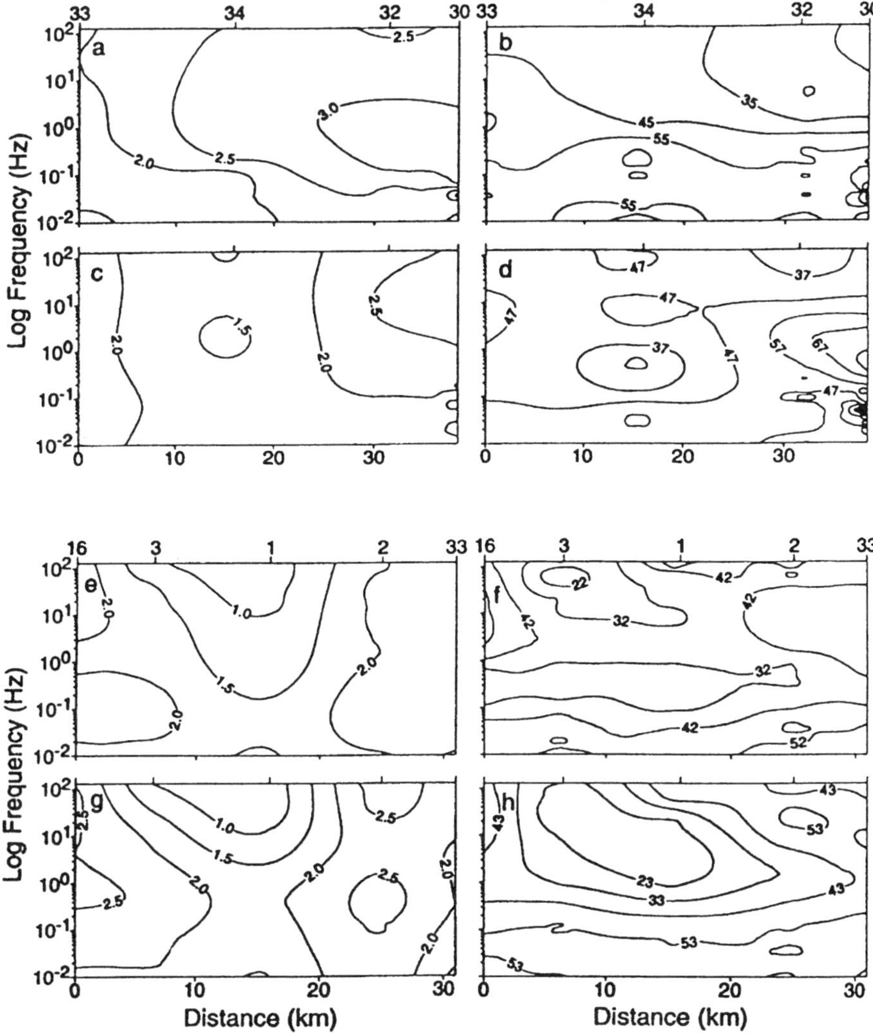

Figure 4. Field data pseudosections for profiles CC′ and DD′ of Fig. 2. See Fig. 3 caption for explanation.

Along each profile, each station was represented by a sequence of layers with resistivities and thicknesses from the 1-D inversions. Vertical planes perpendicular to the profiles, and passing through the center points between adjacent pairs of MT stations, were considered as the boundaries between their resistivity distributions given by the 1-D inversions. As a requirement of the 2-D computer program that calculates the 2-D model responses, the resistivity distributions as functions of depth for the MT stations at the ends of each profile were extended to great distances, so that the model boundary values, which are held fixed during the calculation, do not influence the field perturbations due to the conductivity variations along the profiles. Figure 5 shows the four 2-D models constructed using the MT stations along the edges of the study area (Fig. 2). The number within each block gives the resistivity of that block in ohm-meters. To simplify the numerical models, these values are based on the 1-D inversions, but are rounded-off average values of close groups of resistivities. Model calculations for different resistivity distributions based on the 1-D inversions were made, and it was found that the rounding and averaging processes did not significantly influence the

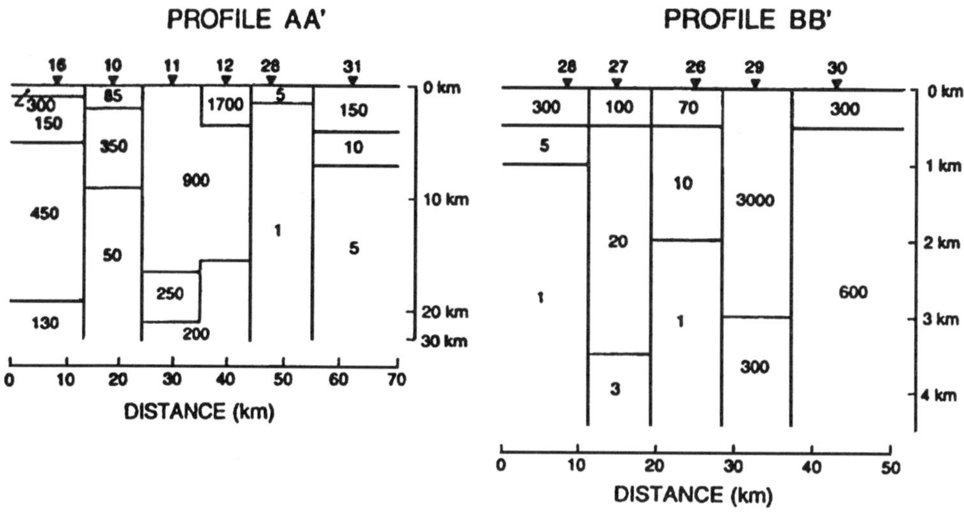

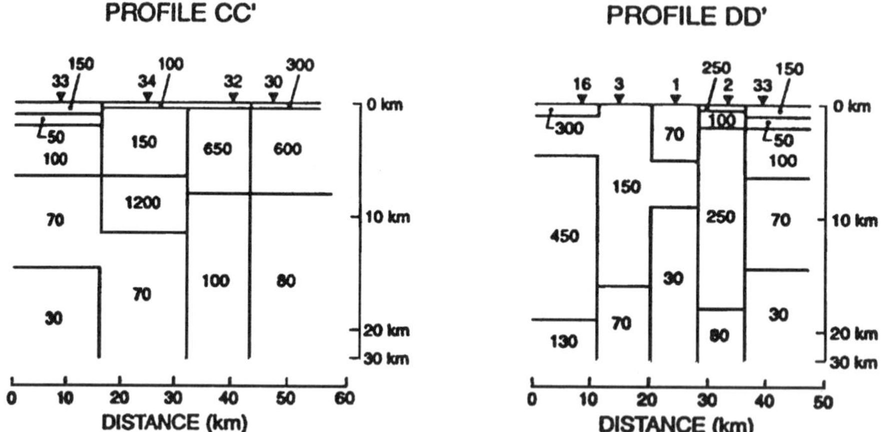

Figure 5. Two-dimensional EM models for profiles AA′, BB′, CC′, and DD′ of Fig. 2.

model results for each profile considered and, therefore, the simplest 2-D models were adopted.

To calculate the electromagnetic (EM) responses of the 2-D resistivity distributions, the finite-difference method developed by Jones and Price (1970) and programmed by Jones and Pascoe (1971) with modifications as indicated by Williamson et al. (1974) and discussed by Jones and Thomson (1974) [see also Brewitt-Taylor and Weaver (1976)] was used over a mesh of 80 × 80 cells.

Figures 6 and 7 show calculated apparent resistivity and phase pseudosections for TE and TM modes for the models constructed for each profile of Fig. 5.

The results of the model for profile AA′ (Fig. 6) show that the main vertical discontinuities, as observed in the field data, are also observed in the model resistivity pseudosections, in both TE and TM modes. This is also true for the model TE phase pseudosection. However, in the TM case, some misfit exists in phase between field data and model results.

The results of the model of profile BB′ (Fig. 6) show the same general behavior as those for profile AA′ and the main electrical discontinuities, as seen in the field data, are

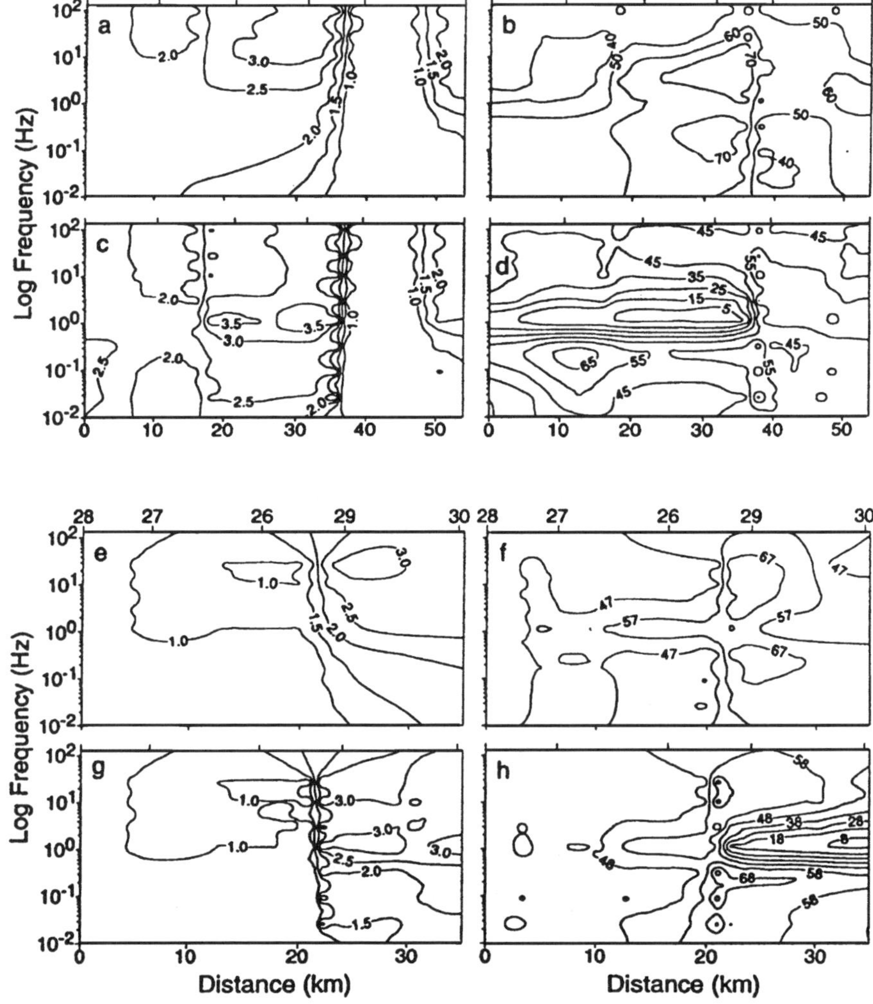

Figure 6. Model results for TE and TM modes for profiles AA′ and BB′. See Fig. 3 caption for explanation. Blocks a, b, c, and d correspond to profile AA′ of Fig. 2 with a and b being the TE mode and c and d being the TM mode; e, f, g, and h correspond to profile BB′ of Fig. 2 with e and f being the TE mode and g and h being the TM mode.

observed in the model apparent resistivity pseudosections. However, large differences between field data and model results are apparent in both pseudosections.

The situation for profile CC′ (Fig. 7) is similar to that for profile BB′. In this case, the greatest misfit is observed in the TM phase pseudosections.

In the results for profile DD′ (Fig. 7), the main trends can be identified in both apparent resistivity pseudosections and phase pseudosections. The field data and the model results for this profile are simpler than those for the other three profiles and this indicates that profile DD′ crosses a region of less electrical complexity than the other profiles. This is consistent with the geological sketch of Fig. 1, which indicates that the MT stations located in the southwest part of the study area lie in a less geologically complicated region.

The set of models shown in Fig. 5 is one of a number of different sets that were tested. The results obtained for other models were not substantially different from the results in Figs. 6 and 7. This indicates that a 2D model probably cannot completely

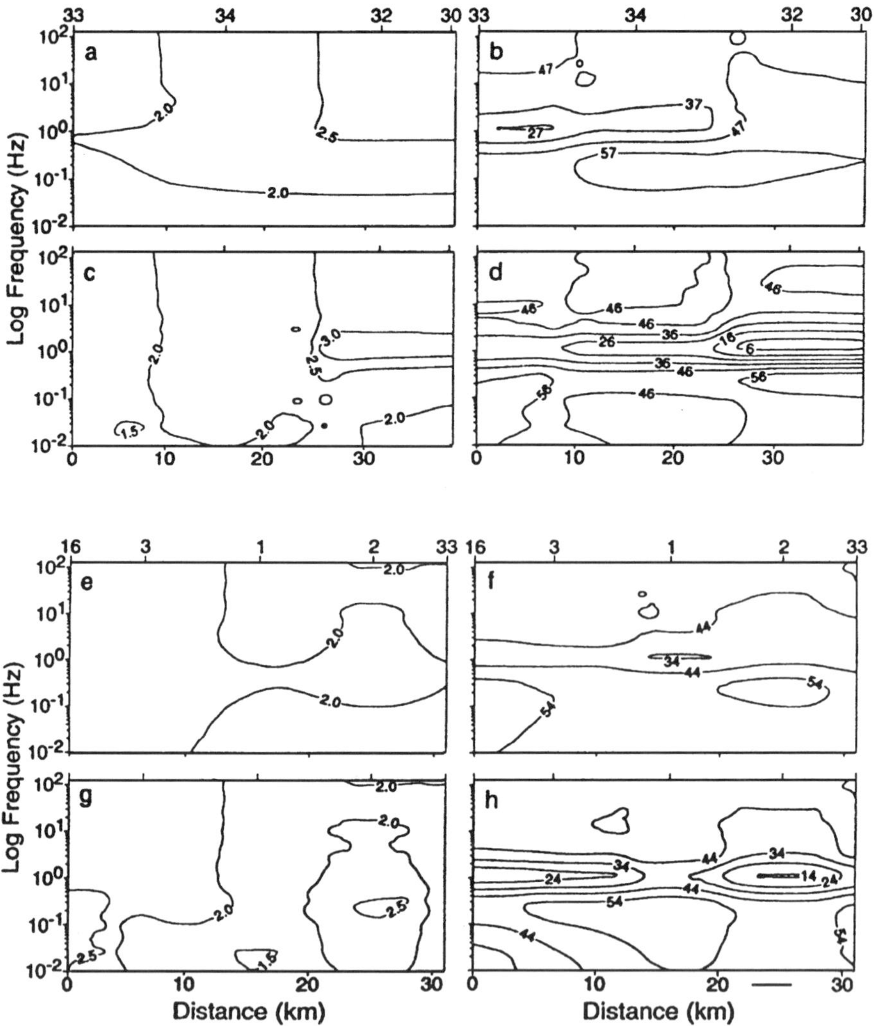

Figure 7. Model results for TE and TM modes for profiles *CC′ and DD′*. See Figs. 3 and 6 captions for explanation.

satisfy the observed data, and that 3-D effects are present. However, there are large resistivity contrasts observed in the area and they can be represented by the models. Furthermore, it is evident that resistivity contacts associated with the Messejana fault and Ferreira-Ficalho overthrust, the two major tectonic features, occur.

The field data and modeling also show that a structure previously ignored appears in both the field data and the model results of profile AA′. Between stations 12 and 28 there is an apparent vertical contact that coincides with the location of a fault that crosses the northern edge of the study area in an approximately east-west trend—the Vidigueira fault.

4 3-D EM modeling

We tried 3-D modeling to quantify the local resistivity structure. The general 3-D EM induction problem was approached solving Maxwell's equations within a 3-D region by a finite-difference method (Jones and Pascoe, 1972; Lines, 1972; Lines and Jones,

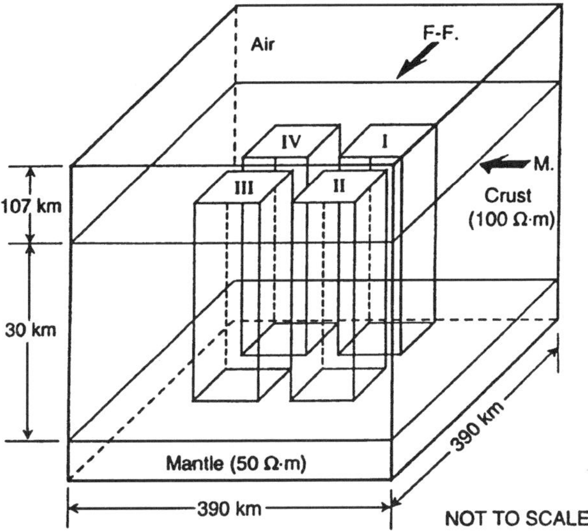

Figure 8. Three-dimensional EM model of study area. F-F indicates the Ferreira-Ficalho overthrust and M indicates the Messejana fault.

1973a,b; and Jones, 1974). The model of the study area, constructed using a 3-D mesh of 40 × 40 × 40 (64 000 cells), was based on knowledge of the geology of the region of the MT survey (Figs. 1 and 2) and the results obtained from the field data. In the simplified model of the area, the curvature of the Ferreira-Ficalho overthrust was not taken into account, nor was its inclination of 60° to northeast. Another simplification was that the deep low-resistivity layers were assumed to extend to the Moho at 30-km depth. Furthermore, it was assumed that the study area could be modeled by considering the high-resistivity blocks, as obtained in the resistivity maps constructed for the study area (Correia et al., 1993), to be embedded in a crust of 100-ohm-m resistivity. The upper mantle was assumed to have a resistivity of 50 ohm-m, which is consistent with values reported by Jones (1992) and Haak and Hutton (1986). Figure 8 shows the model with the four blocks and Fig. 9 shows the details of each block. Nine frequencies were calculated for the 3-D model (0.01, 0.03, 0.1, 0.3, 1, 3, 10, 30, 100 Hz), which covered the frequency range of the MT survey.

Examples of apparent resistivities calculated from the 3-D model are presented in Fig. 10, where they are compared with results from field data. The comparison is good, considering the simplicity of the model and the complexity of the actual geology. Additional comparisons between field results and 3-D model results are given by Correia (1994).

Site 22 is located in the region that corresponds to block I in the 3-D model shown in Fig. 8. Comparison between the field data and the 3-D model results confirm the general high-resistivity character of that area to approximately Moho depths. Furthermore, a high-resistivity character in the northeast part of the study area can be inferred from the comparison between the field data and the 3-D model results for site 19 which lies within block II of Fig. 8. However, it is apparent that the resistivity values chosen for the model are higher than appropriate for this area, and further refinements should be made.

Block III of Fig. 8 corresponds to a relatively low-resistivity area and its character can be identified in the resistivity curve of site 33. The resistivity at this location is

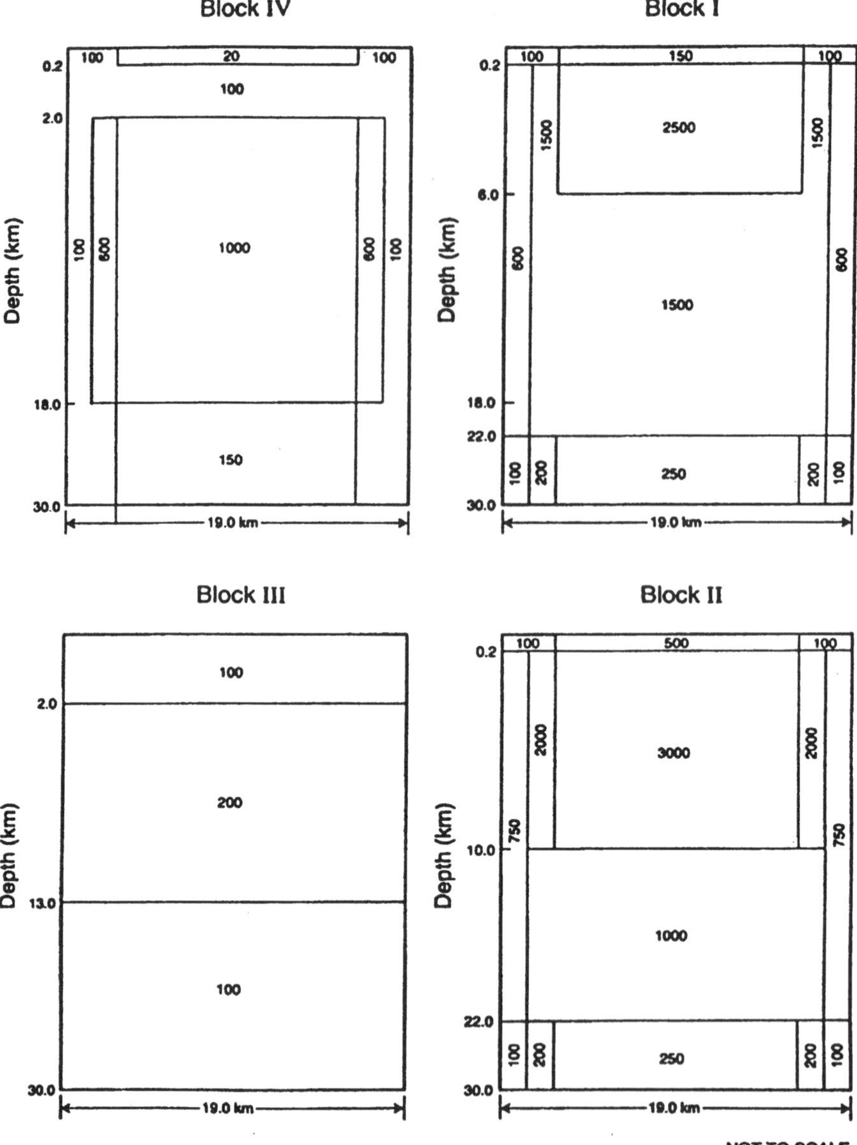

Figure 9. Details of the four blocks considered in Fig. 8. Values inside the blocks are resistivities in ohm-meters. The structure of each block is the same in two perpendicular directions that are parallel to the vertical faces and that pass through the vertical axis of the block.

fairly constant throughout the whole frequency range, which indicates that the electrical structure varies little with depth.

Block IV of the 3-D model corresponds to another high-resistivity region as is seen in the resistivity curve for site 6. The resistivity in this zone decreases significantly with depth from about 1000 ohm-m at intermediate depths to about 150 ohm-m below 18 km.

It should be possible to refine the simple 3-D model so that good fits are obtained for most of the MT sites. A more quantitative analysis and interpretation for the study area might be achieved if a 3-D model with higher resolution was used. This means that models with more than $40 \times 40 \times 40$ cells will be required, with the number of cells increasing with the increase in complexity of the structure to be represented. It

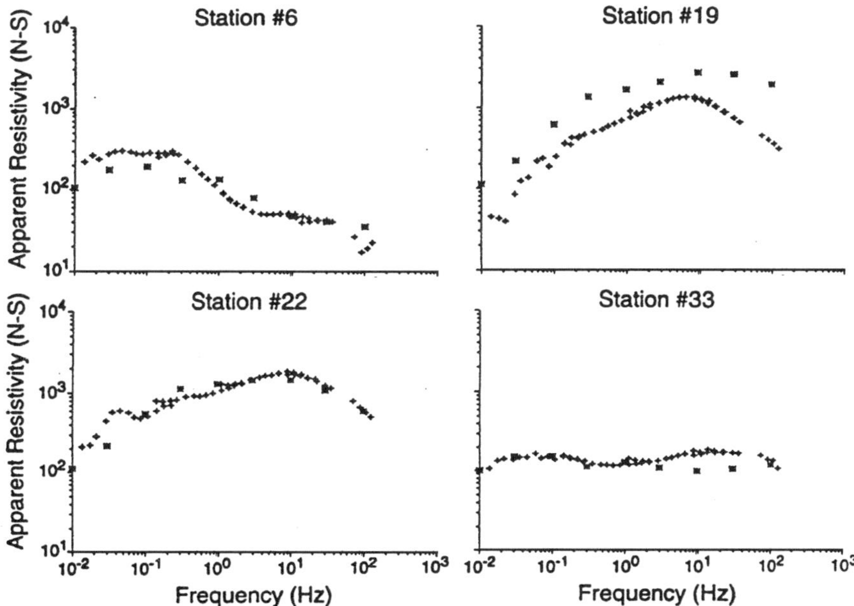

Figure 10. Comparison of some of the apparent resistivity curves of the 3-D EM model (stars) with the corresponding field curves (small crosses).

must be emphasized, however, that even though the 3-D model used here is only an approximation of the regional geology, the results from it support the interpretation that the region of the MT survey is characterized by the existence of high-resistivity blocks embedded in more conductive formations.

5 Conclusions

Three-dimensional EM modeling can be used to establish the validity of 1-D and 2-D models to interpret MT measurements. For the Portuguese data, the results show that the use of 1-D inversion models and the Berdichevsky invariant were good approximations from which to construct EM models. It is also apparent that recording and analyzing MT data along profiles can be inappropriate for inferring resistivity in areas with 3-D character. In fact, the 2-D profiles considered here gave no indication of the high-resistivity blocks that are located in the central region of the study area, only a few kilometers away.

Acknowledgments

The authors would like to thank the University of Alberta, Junta Nacional de Investigação Científica de Tecnológica, the University of Évora, Fundação Calouste Gulbenkian, and the Natural Sciences and Engineering Research Council of Canada for their support during various stages of this work.

References

Berdichevsky, M. N., and Dmitriev, V. I., 1976, Basic principles of interpretation of magnetotelluric sounding curves, *in* Adam, A. Ed., Geoelectrical and geothermal studies: KAPG Geophys. Mono., 164–221.

Brewitt-Taylor, C. R., and Weaver, J. T., 1976, On the finite difference solution of two-dimensional induction problems: Geophys. J. Roy. Astr. Soc., **47**, 375–396.

Correia, A., 1994, A magnetotelluric study in the region of the intersection of the Messejana fault and the Ferreira-Ficalho overthrust in Portugal: Ph.D. thesis, Univ. of Alberta.

Correia, A., Jones, F. W., Dawes, G., and Hutton, V. R. S., 1993, A magnetotelluric deep crustal study in south-central Portugal: Stud. Geophys. Geod., **37**, 331–344.

Haak, V., and Hutton, R., 1986, Electrical resistivity in the continental lower crust, *in* Dawson, J. B., Carswell, D. A., Hall, J., and Wedepohl, K. H., Eds., The nature of the lower continental crust: Geol. Soc. Special Publication, **24**, 35–49.

Jones, A. G., 1992, Electrical conductivity of the continental crust, *in* Fountain, D. M., Arculus, R. J., and Kay, R. W., Eds., Continental lower crust: Elsevier Science Publ. Co., Inc., 81–143.

Jones, F. W., 1974, The perturbation of slowly varying electromagnetic fields by three-dimensional conducting bodies: Can. J. Phys., **52**, 1195–1202.

Jones, F. W., and Pascoe, L. J., 1971, A general computer programme to determine the perturbation of alternating electric currents in a two-dimensional model of a region of uniform conductivity with an embedded inhomogeneity: Geophys. J. Roy. Astr. Soc., **24**, 3–30.

———1972, The perturbation of alternating geomagnetic fields by three-dimensional conductivity inhomogeneities: Geophys. J. Roy. Astr. Soc., **27**, 479–485.

Jones, F. W., and Price, A. T., 1970, The perturbation of alternating geomagnetic fields by conductivity anomalies: Geophys. J. Roy. Astr. Soc., **20**, 317–334.

Jones, F. W., and Thomson, D. J., 1974, A discussion of the finite difference method in computer modelling of electrical conductivity structures: Geophys. J. Roy. Astr. Soc., **3**, 537–543.

Jones, F. W., Correia, A., Dawes, G. K., Hutton, V. R. S., Jones, P., and MacDonald, K., 1992, Preliminary results of a magnetotelluric survey over a geothermal anomaly in Portugal: Phys. Earth Planet. Inter., **73**, 274–281.

Lines, L. R., 1972, A numerical study of the perturbation of alternating geomagnetic fields near island and coastline structures: M.Sc. thesis, University of Alberta.

Lines, L. R., and Jones, F. W., 1973a. The perturbation of alternating geomagnetic fields by three-dimensional island structures: Geophys. J. Roy. Astr. Soc., **32**, 133–154.

———1973b, The perturbation of alternating geomagnetic fields by an island near a coastline: Can. J. Earth Sci., **10**, 510–518.

Ribeiro, A., Antunes, M. T., Ferreira, M. P., Rocha, R. B., Soares, A. F., Zbyszewski, G., Moitinho de Almeida, F., de Carvalho, D., and Monteiro, J. H., 1979, Introduction à la Géologie Générale du Portugal, Serviços Geológicos de Portugal.

Williamson, K., Hewllet, C., and Tammemagi, H. Y., 1974, Computer modeling of electrical conductivity structures: Geophys. J. Roy. Astr. Soc., **37**, 533–536.

Three-Dimensional Modeling of a Magnetotelluric Survey over Chaves Graben in Northeast Portugal

F. A. Monteiro Santos[1]
A. Dupis[2]
A. R. Andrade Afonso
L. A. Mendes-Victor[3]

Summary. Magnetotelluric soundings have been made at 41 locations across the main tectonic structures in the Chaves region in northeast Portugal, near the frontier with Spain. The zone studied is dominated by a graben bounded by Hercynian granite and Silurian metamorphic schist. It is an elemental unit of a broad hydrothermal province, in which the low-enthalpy system of Chaves is the most important in the Portuguese mainland. Calculated impedance tensors at 27 sites were converted into invariant (determinant) apparent resistivity and phase curves and used to construct a regional 3-D model of the electrical resistivity in the crust. A conductive zone (resistivity of 100 ohm-m) was detected at a depth between 7 and 11 km. Based on geological information and on results from a previous 3-D dc model, a more detailed 3-D model of the graben area was constructed, by forward modeling. The detailed model of the graben was imbedded in the regional electrical model obtained from the invariant curves. The main shallow structure—a conductive body (resistivity of 12 ohm-m) at a depth ranging from 250 to 1000 m—appears to be related to the Hercynian and neotectonic activities. Our results indicate that a valuable picture of the regional electrical model can be obtained from 3-D interpretation of the determinant of the impedance tensors.

1 Introduction

During 1992 and 1993, magnetotelluric (MT) data were collected at 41 sites along four profiles across the main tectonic structures of the Chaves area in northeastern Portugal as part of a project, supported by European Community, to assess the geothermal potential of the region. The regional basin is part of the Galicia Trás-os-Montes subzone where several mineral and thermal springs can be found (Fig. 1). Close to Chaves city, a hot spring reaching a temperature of 78°C has been known at least since Roman times. Chaves thermal waters are the most important external manifestation of the

[1]Departamento de Física da Universidade de Lisboa, Centro de Geofísica da Universidade de Lisboa, 58, R. Escola Politécnica, Lisboa 1250, Portugal; e-mail: dfams@fc.ul.pt
[2]Centre de Recherches Géophysiques (CNRS), Garchy 58150, France.
[3]Departamento de Física da Universidade de Lisboa, Centro de Geofísica da Universidade de Lisboa, 58, R. Escola Politécnica, Lisbo 1250, Portugal.

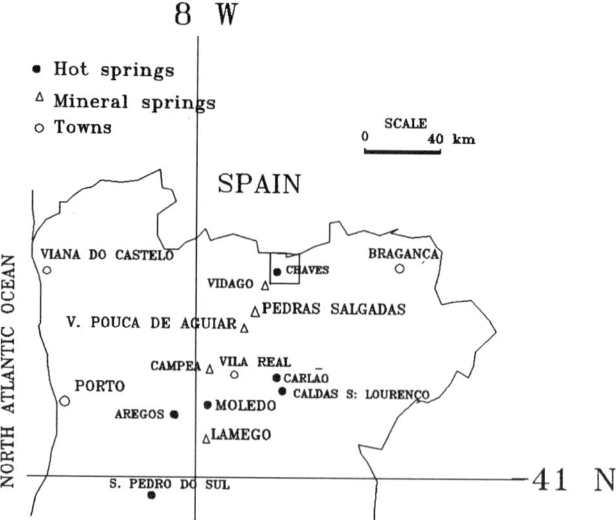

Figure 1. Map showing location of area studied (rectangle) and indicating the main hot and mineral springs, in northeastern Portugal.

low-temperature geothermal fields on the Portuguese mainland which are related to crystalline granitic rocks (Aires-Barros, 1989).

Gravity, resistivity, self-potential (SP), audio-magnetotelluric (AMT), and MT data—as well as temperature in boreholes—were measured in the Chaves region to study the main features of the regional and local structures (Andrade Afonso et al., 1994a,b; Monteiro Santos et al., 1995, 1996, 1997; INMG 1992). The interpretation of the complex geology has been based largely on surface observations with extrapolation for deep structure (UTAD, 1992).

One purpose of our study was to use the MT method to impose geophysical constraints on the geological settings of the region. We aimed to obtain a 3-D image of the conductivity distribution in the upper and deep crust in the region and to understand its connection with the geothermal field. We describe here the procedures used in the 3-D modeling of the Chaves area. We first present a regional 3-D resistivity model and discuss its geological implications; we then present a more detailed 3-D model of the Chaves graben based on the previous data and the new MT study.

2 MT measurements

Using a CNRS-Garchy single-station system, MT data were recorded at 41 locations along four profiles in the Chaves region. Figure 2 shows the MT sites on the geological map. The data were acquired in the frequency range of 180 to 0.008 Hz, in four selected frequency bands. The measurement directions of the horizontal fields were roughly N36E and N126E, in accordance with the graben structural direction. The time series have been processed by cascade decimation and an algorithm developed by Sutarno and Vozoff (1991). For periods greater than 10 s, the data are poor, mainly in the sites within the graben, because of the weakness of the signal level and man-made noise.

The vertical displacement of the apparent resistivity curves shows a static shift, mostly in the data acquired over granitic formations. By comparison with vertical electrical soundings (Schlumberger array) carried out at several sites inside the graben

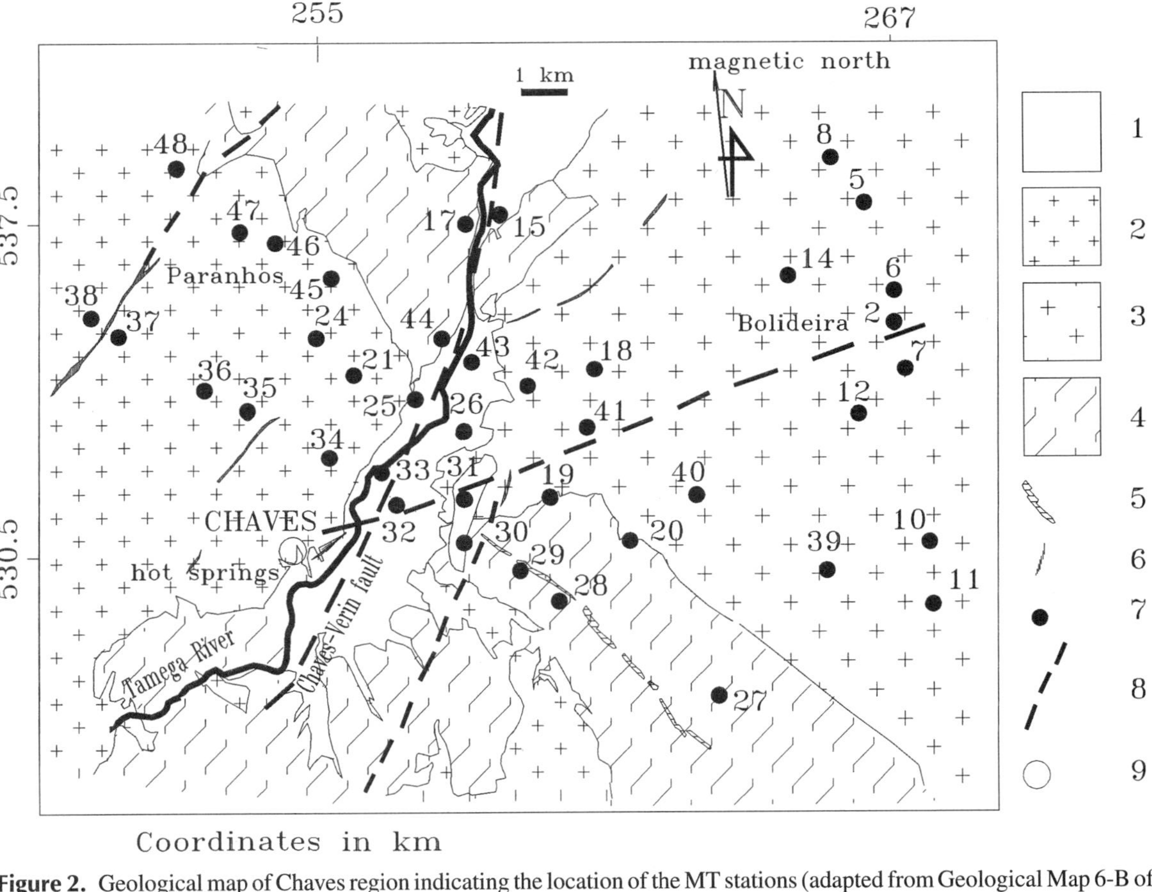

Figure 2. Geological map of Chaves region indicating the location of the MT stations (adapted from Geological Map 6-B of the Servicos Geológicos de Portugal): (1) sediments; (2) Syn-tectonic granite; (3) posttectonic granite; (4) schist-graywacke complex; (5) graphitic slates; (6) dikes; (7) MT station; (8) main normal faults; (9) main towns.

(Monteiro Santos et al., 1994), we conclude that the data acquired in the basin are not affected by static shift in a dramatic way. Static shift removal has been examined by several workers but no absolute method has yet been found. Berdichevsky and Dimitriev (1976) suggested the use of a rotationally invariant impedance reducing the effects of local structures. In this work we have used the invariant impedance tensor defined as $Z_{\text{det}} = (Z_{xx}\, Z_{yy} - Z_{xy}\, Z_{yx})^{1/2}$ (Ranganayaki, 1984). Because Z_{det} is an average of all four impedance tensor elements, it may well be preferable to other invariants that are averages of only the off-diagonal elements.

Figure 3 shows the determinant apparent resistivity and phase curves from 17 sites located on the main geological formations. As can be seen, except for sites close to or within the basin, the curves follow a similar pattern. The determinant apparent resistivity curves have a maximum between 10 and 60 s. The change in the slope of the apparent resistivity curves at around 1 s, correlated to the maximum of the phase curves, between 0.3 and 3 s, indicates the presence of a deep conductor.

Figure 4 shows plots of the rotated impedance tensor components Z_{xx} and Z_{xy} for two frequencies (7.8 and 0.078 Hz). Patterns in the shape and orientation of these elements can be associated with one-dimensional (1-D), two-dimensional (2-D), or 3-D structures. Generally, Z_{xy} is larger than Z_{xx}, but at some sites the magnitude and asymmetric form of Z_{xx} indicate either 3-D effects [see Hermance (1982) for details] or increased noise in the data. Because the scale is not the same for all sites, no conclusions concerning the variations of apparent resistivity can be derived from such maps.

At 7.8 Hz the depth of penetration does not exceed 2 km outside the graben and 1 km inside it (considering a 150- and 30-ohm-m homogeneous earth). Therefore, the directions of the polar diagrams are strongly influenced by the structures of the surface—the visible faults (e.g., at sites 38, 37, and 35) and conductive graben-fill sediments (e.g., at sites 31, 33, and 43). The pattern (Fig. 4) at sites within and near the graben indicates a complex conductive structure oriented approximately north-northeast–south-southwest [see also the synthetic examples by Hermance (1982) and Wannamaker et al. (1984)]. There is clearly much structural complexity in the southeast part of the graben. The polar diagrams at sites 28 and 29 are influenced strongly by the graphitic slates outcropping in the area. The general features revealed by polar diagrams at 7.8 Hz are maintained at the frequency of 0.078 Hz (penetration depths of 22 and 10 km, for homogeneous earths of 150 and 30 ohm-m). At sites 24, 31, and 32, there is a severe degradation of the Z_{xx} diagram, mainly because of the high noise level.

These features suggest that 3-D modeling would be preferable to a 2-D analysis. Indeed, there are two main reasons to believe that the structure is 3-D: (1) The basin presents a strong offset geometry along north-south and east-west directions; (2) the basement of the basin itself is divided into blocks with minor grabens and horsts (Andrade Afonso et al., 1994a).

3 3-D MT modeling

3.1 The 3-D regional model

The modeling presented in this paper is just a first attempt at a global interpretation of the MT data acquired in the region. Only 27 soundings around and inside the graben were used in the regional modeling (Fig. 4). Because the geologic and tectonic structures of the area are quite complex, we chose to interpret the apparent resistivity given by the determinant of the impedance tensor. We modeled the region with the 3-D modeling

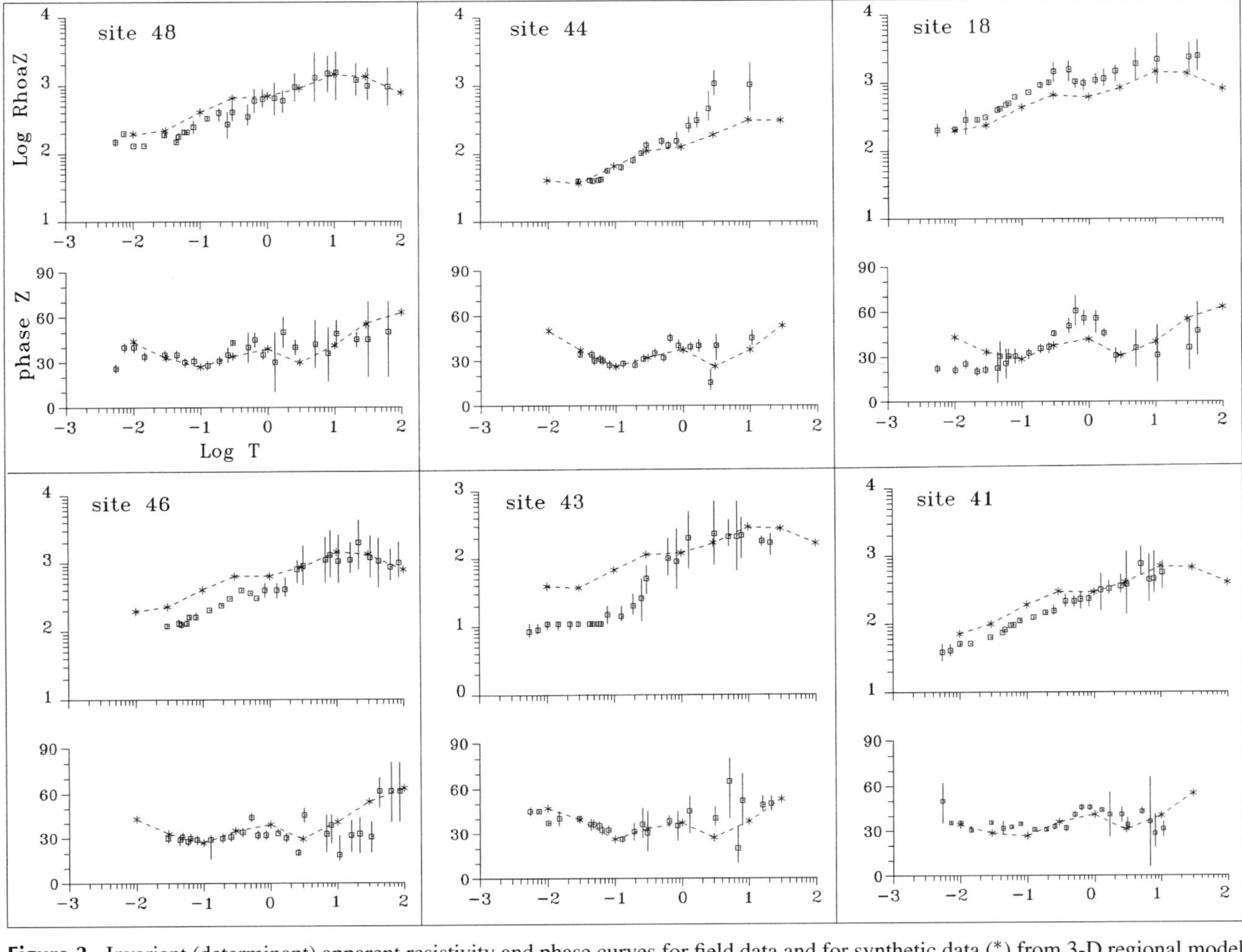

Figure 3. Invariant (determinant) apparent resistivity and phase curves for field data and for synthetic data (*) from 3-D regional model.

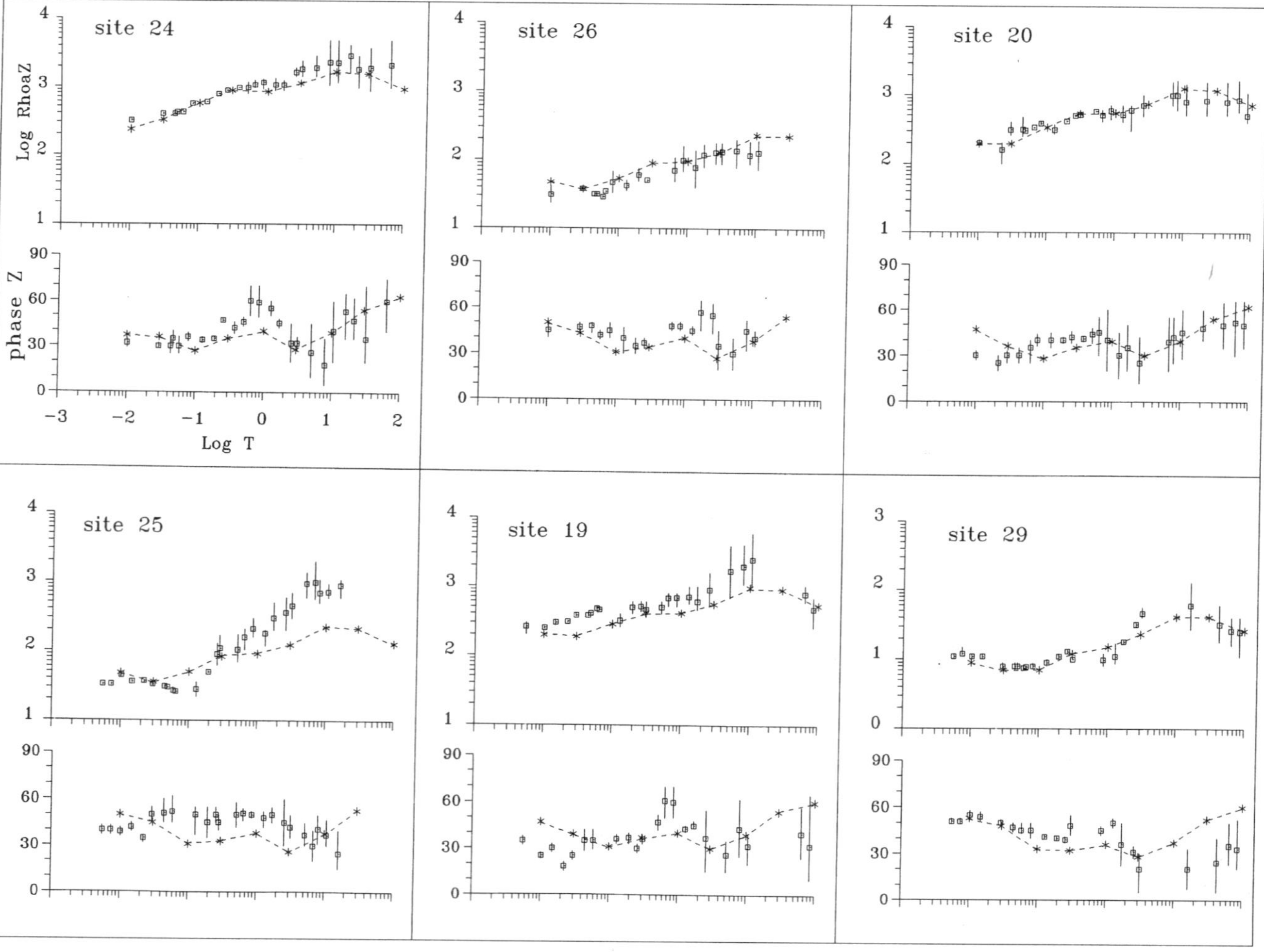

Figure 3. *(Continues)*

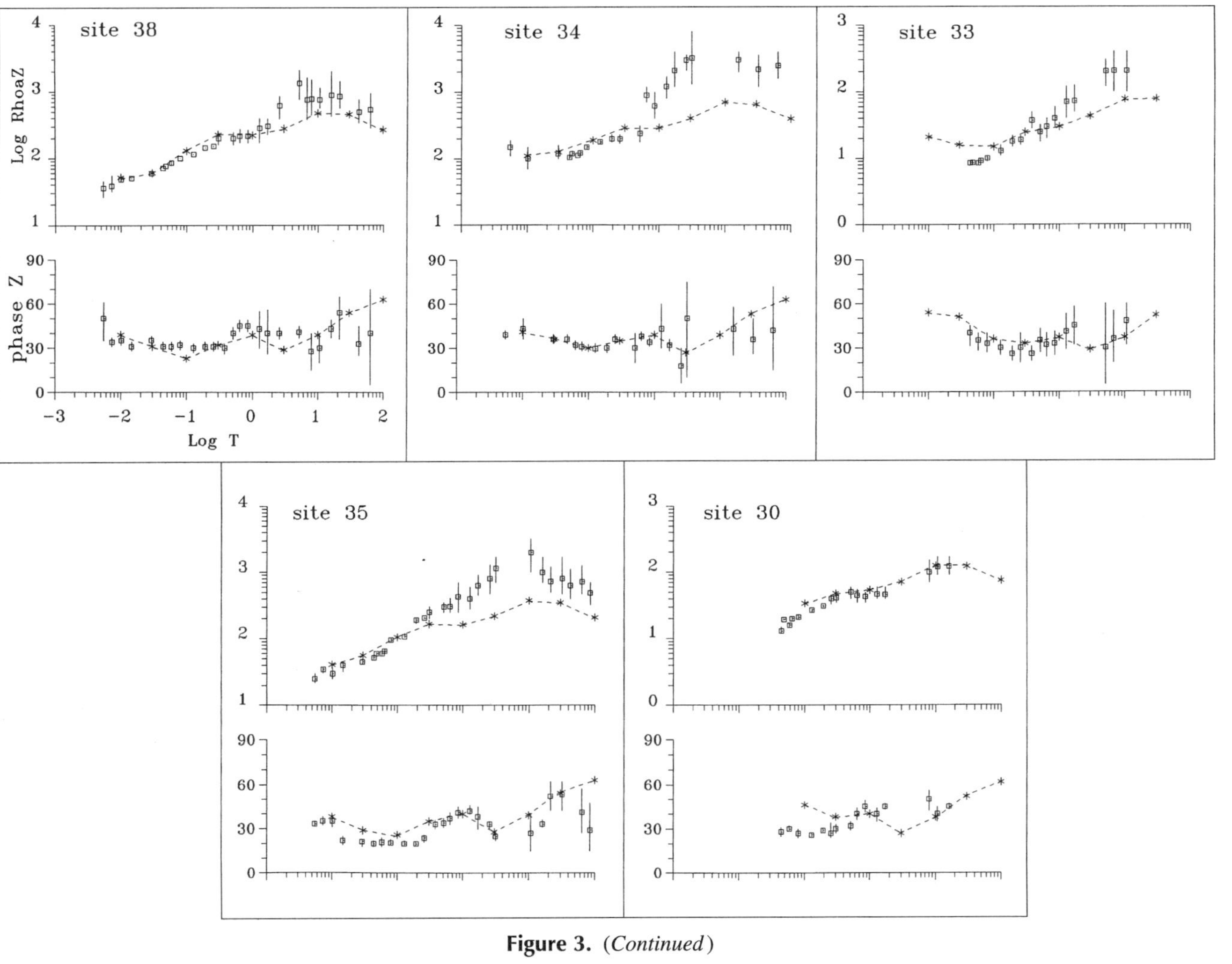

Figure 3. *(Continued)*

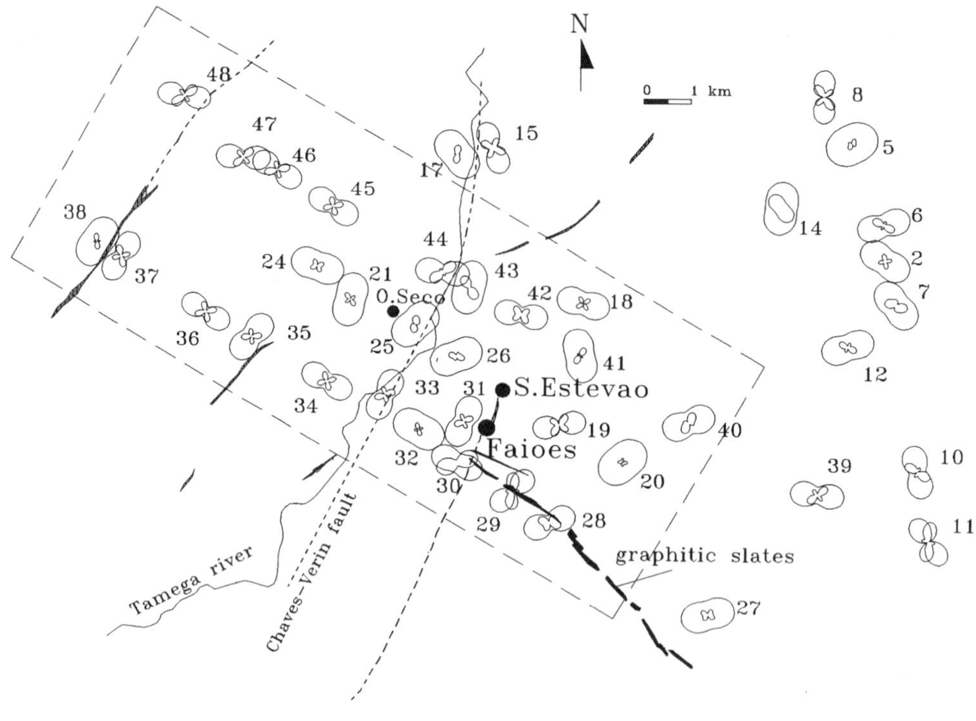

a) Polar diagrams for 7.8 Hz

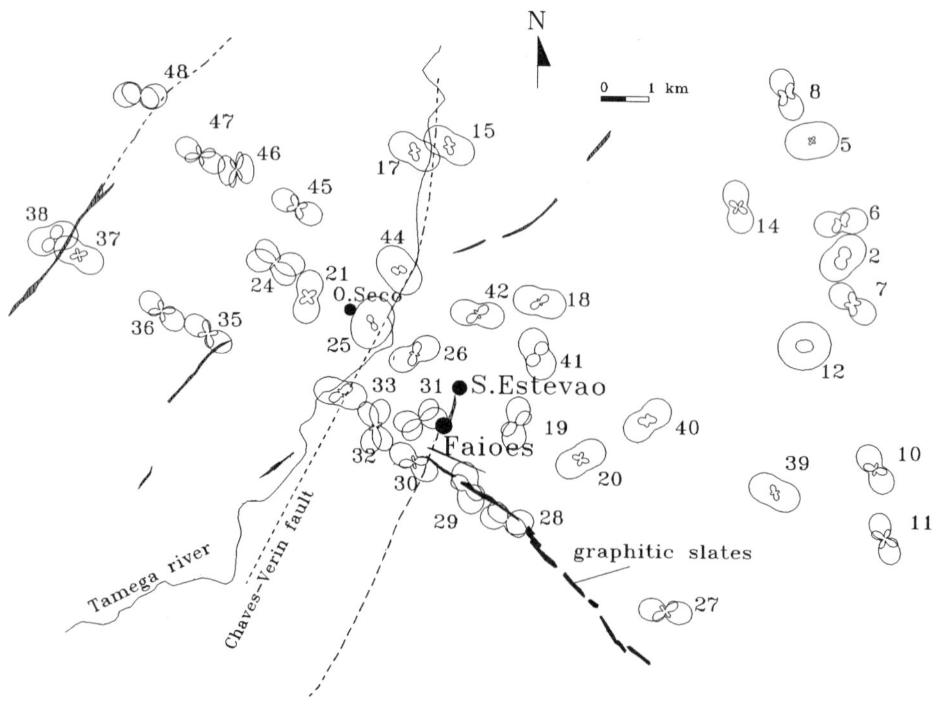

b) Polar diagrams for 0.078 Hz

Figure 4. Polar diagrams of impedance elements Z_{xx} and Z_{xy} at two frequencies. Generally, Z_{xy} appears as a two-lobed figure and Z_{xx} appears as a four-lobed figure. The dashed rectangle outlines the main part of the 3-D regional model shown in Fig. 5, as well as the area shown in Fig. 7.

program developed by Mackie et al. (1993), which is based on the integral forms of Maxwell's equation, and it is able to model arbitrarily complex media.

The region was discretized into 58 blocks in the x-direction (N126E), 38 blocks in the y-direction (N36E), and 24 earth layers in the z-direction. Even on such a grid, the complex geology of the area is represented crudely. The model was graded in the horizontal and vertical directions using the geometric factors proposed by Mackie et al. (1993), i.e., factors of 2 in the vertical direction and factors of 3 in the horizontal direction. The model was more finely discretized near regions of strong resistivity gradient, mainly in the graben zone. The outer boundaries of the model are actually up to 100 km from the origin at the center of the basin. The 3-D forward modeling algorithm assigns a tangential magnetic field in the boundaries of the model. Those boundary values are calculated previously by a 2-D TM algorithm in which each vertical plane of the 3-D mode is taken as the 2-D model of a large-scale (Mackie et al., 1993).

An initial regional model was constructed on the basis of the geology of the region and the 1-D MT modeling (Monteiro Santos et al., 1995). One simplification was that the graben was modeled as a conductive structure with an L-shape. The general parameters of the graben (length, width, thickness, and resistivity) were taken from a detailed 3-D resistivity model (Monteiro Santos et al., 1997). No attempt, however, was made to include detailed shallow structures that are suggested by the MT and the other surveys, at some sites inside and outside of the depression.

Figures 5 and 6 show the 3-D final model obtained by trial and error, with emphasis on the fitting of the invariant apparent resistivity curves. Figure 5 shows four horizontal slices, together with a geological sketch of the area. Figure 6 shows two vertical cross-sections of resistivity versus depth. The following tentative observations can be made from this model:

- The low electrical resistivity structure (15–30 ohm-m) at a depth between 250 and 1000 m corresponds to the Chaves graben. Under the main conductive body, there is a block of 150 ohm-m.
- West of the graben, the resistivity of the upper altered granite decreases from north (220 ohm-m) to south (100 ohm-m). Sites 38, 37, and 35 are located near two north-northeast–south-southwest faults. The uppermost weathered zones associated with those faults were represented by two blocks (resistivity of 30 ohm-m).
- The graphitic slates, southeast of Faiões, are represented by the structure of 15 ohm-m down to a depth of 750 m.
- At a depth between 7 and 11 km, there is a relatively conductive zone (100 ohm-m), which is required to explain the Z_{det} phase data values around 1 Hz. We have studied the effect on the model response by perturbing the resistivity of this layer. These sensitivity tests have shown that the resistivity must be less than 500 ohm-m (otherwise phases do not increase at 0.8 s) and greater than 50 ohm-m.

  The presence of this conductor below the shistose complex is not clear in the MT data. The usual explanations for conductive zones in the upper part of the middle crust are partial melting, water, or presence of graphite or other conductive materials (Mareschal et al., 1992). Several seismic profiles in the Galicia-Trás-os-Montes did not show any evidence of low-velocity layers associated with partial melting (Téllez et al., 1993). Because the zone is crossed by several deep faults, the cause of the relatively high conductivity has been interpreted as water infiltrating through deep faults (Monteiro Santos et al., 1995). This layer also could represent

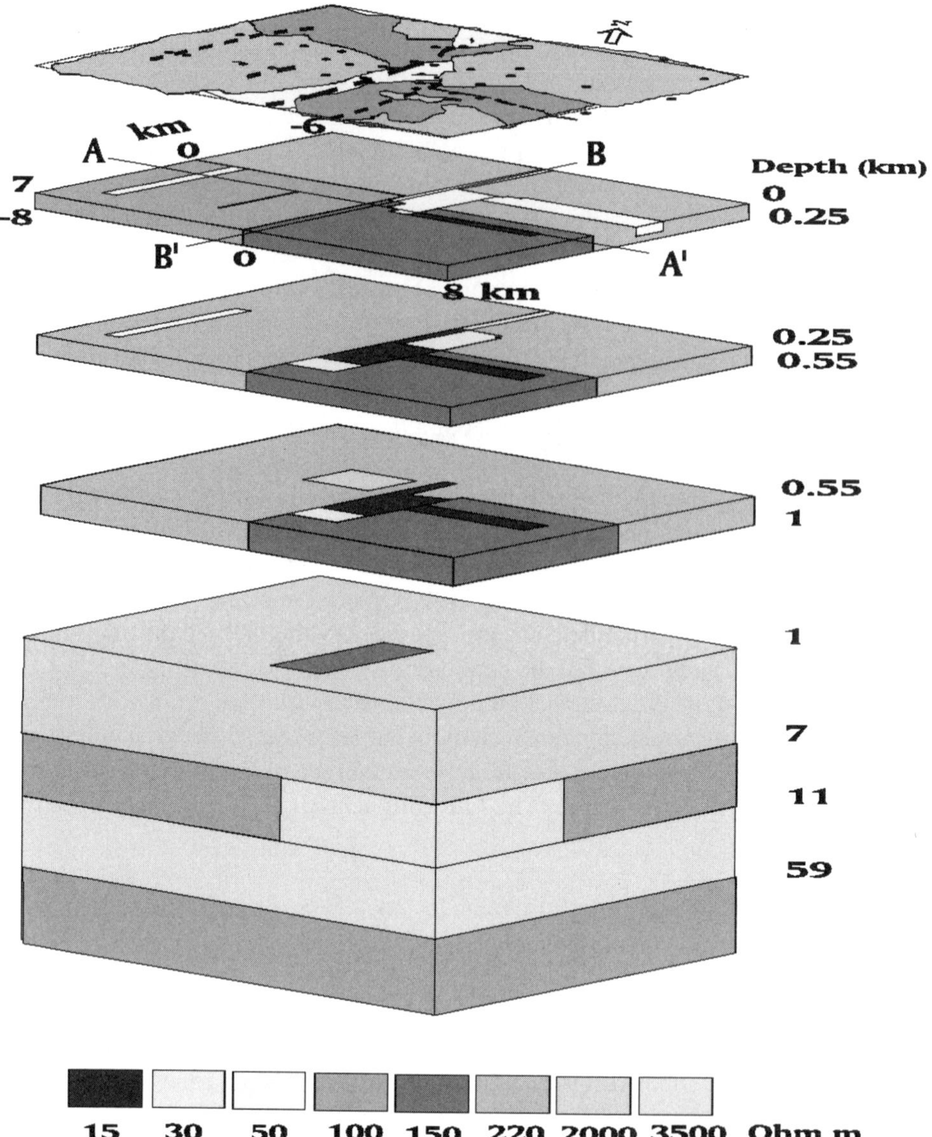

Figure 5. Horizontal slices of electrical model recovered from modeling of the determinant apparent resistivity data. The top panel shows a geological sketch of the studied area.

a change in crust behavior from brittle to ductile, which is important when dealing with mechanical and thermal behavior of the crust in this region.

At depths greater than 11 km, a uniform substratum of 3500 ohm-m was used for the whole region until 59-km depth. The model is floored by an electrical body of 100 ohm-m, which represents the upper mantle. This is a compromise in the fit between the determinant phases and the apparent resistivity data. At increasing depth, the maximum of the apparent resistivity curves shifts to periods exceeding 100 s, which does not agree with the majority of the data. On the other hand, a decrease in this depth leads to a large misfit between the phases. The poor quality of the data at longest periods means that the deeper structure (at depths greater than 30 km) is probably only indicative of true structure.

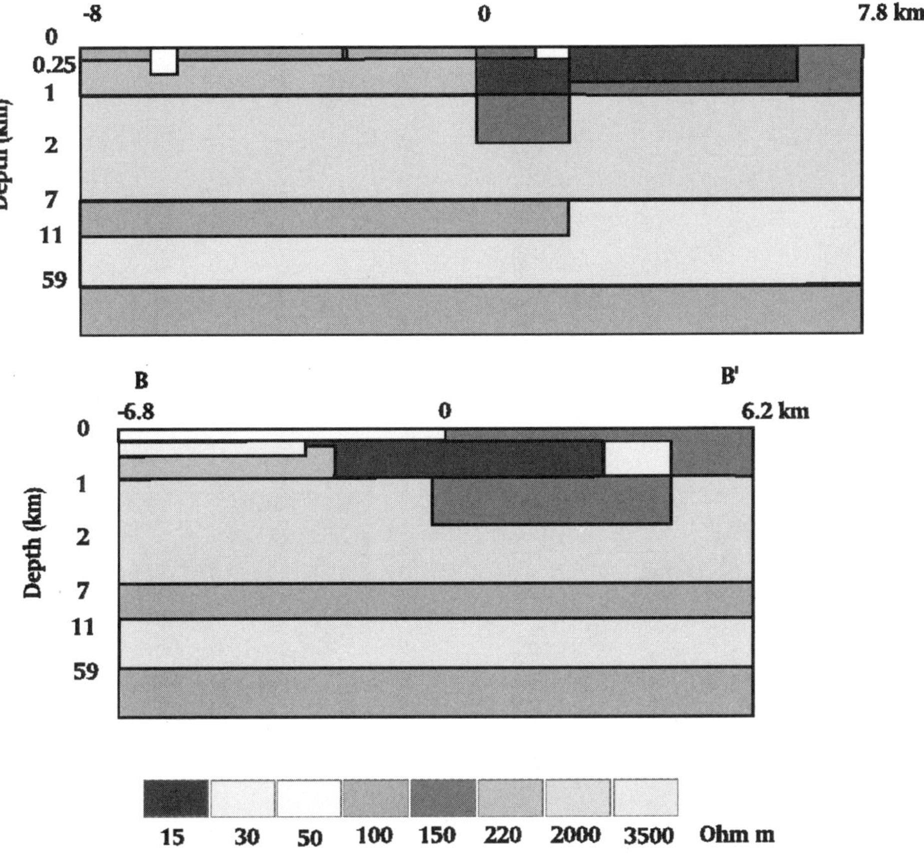

Figure 6. Two vertical cross-sections of the 3-D electrical model.

3.2 *Three-dimensional model responses*

We compared the model's response with the data—the invariant apparent resistivity and the phase—to check its validity. Responses were calculated for nine frequencies covering the range of the MT data. Impedance tensors at each site were calculated from the electric and magnetic fields at the nearest-neighbor nodes using linear interpolation. The invariant apparent resistivity data maps and model response for two frequencies (100 and 1 Hz) are shown in Fig. 7. The fit between both is acceptable but not perfect, especially for the short periods. The figure also shows a map of the ratio of predicted and observed (determinant) apparent resistivities. The misfit is substantial at some sites in the graben, and could not be reduced with such a general model. The model does not attempt to reproduce the superficial features of the conductivity distribution. The degree of the misfit on both the eastern and the western sides of the basin allows only a general conclusion on the geometry and the geoelectrical parameters of the filled graben. Note, however, that the interpreted model is a good regional model, explaining the major features of the MT data.

Figure 8 shows the invariant apparent resistivity and phase curves (model responses and data) for 17 sites. There is a reasonable fit to the observed data. Comparison between the 3-D model responses and the field data at sites located in the region that corresponds to syn- and post-tectonic granitoids confirms the general high resistivity character of

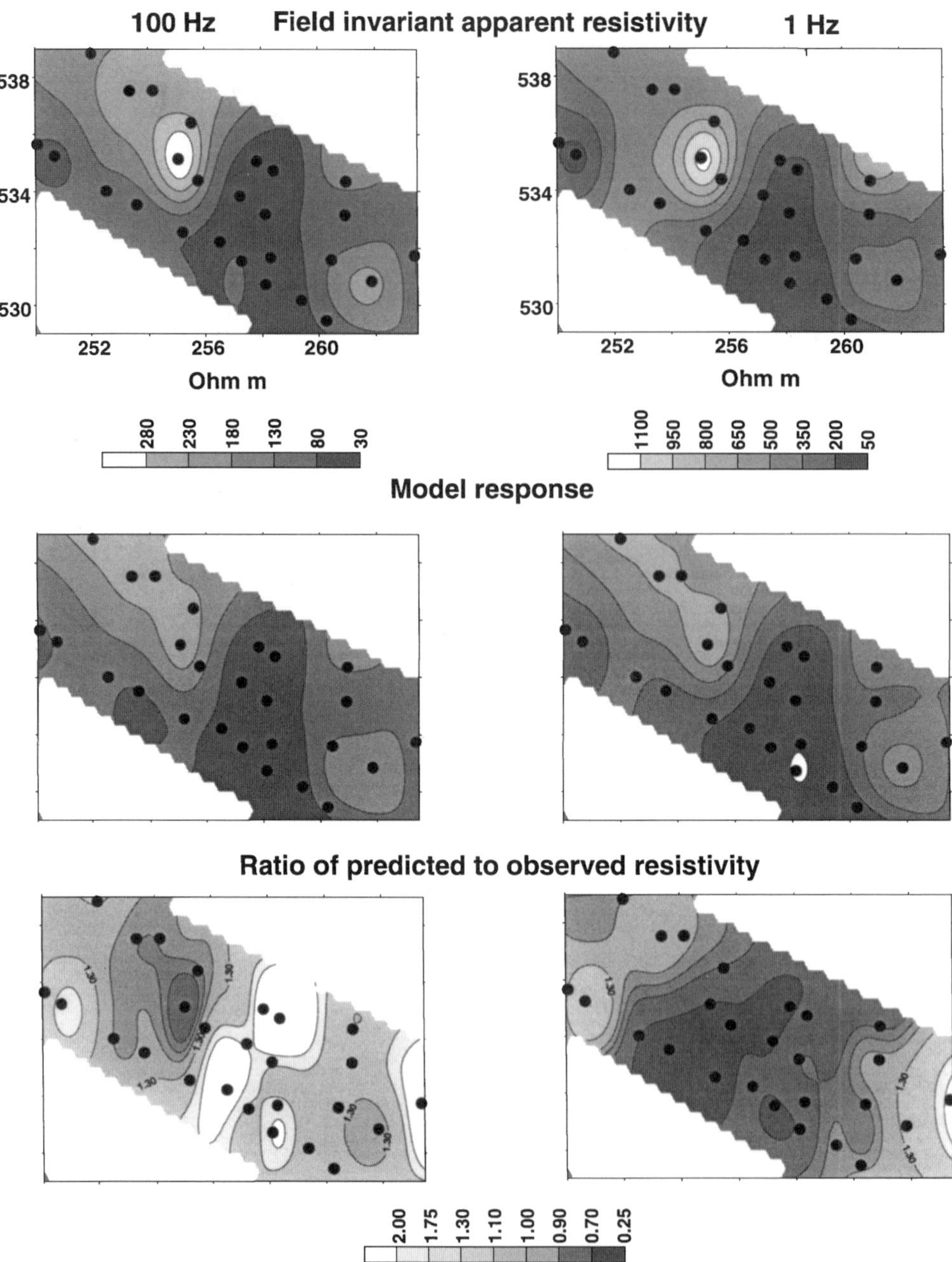

Figure 7. Maps of determinant apparent resistivity for 100 and 1 Hz. The map in bottom panel is generated using the ratio of predicted to observed determinant apparent resistivity values.

those geologic units at a depth between 2 and 7 km. Sites 35, 37, and 38 are located near two faults with distinct regional importance. The fit at these sites is quite good above 0.3 s. The model responses do not fit the data at frequencies greater than 1 s, partly because of a wrong estimation of the limited strike extent of those structures, mostly because of the large spatial sampling of the MT data. The resistivity structure of the graben is characterized by a relatively thick (1 km) conductive (15–30 ohm-m) block. According to exposed geology, this corresponds to Miocene-Pleistocene sedimentary series, overlaying altered metamorphic and granitic rocks. At sites located inside the graben (e.g., sites 26 and 33), the fit between data and model responses is acceptable. However, the model does not fit the data well from sites located close to the edges of the basin (e.g., sites 43, 44, and 25). Site 29 is located on an exposure of graphitic slates. Even though the model used is only an approximation of the regional geology, the fit between both the field invariant apparent resistivity and the model response curves is good.

The results are encouraging and suggest that a more refined model is possible, particularly for the graben area, if a 3-D model with higher resolution is used.

4 Refined 3-D model of the graben

We created a more detailed 3-D model of the basin by trial-and-error fitting of both measured apparent resistivity curves from 11 MT soundings inside and near to the Chaves graben (Fig. 8). For survey lines in a different direction from N36E-N126E, the impedance tensor was rotated into the field measurement direction. Because the quality of the data is poor at periods greater than 10 s, only data up to that period were considered for modeling. The aim of this modeling was to construct a model from the near-surface structure, taking into account information from previous studies on dc resistivity and AMT.

The graben model was considered to be imbedded in the regional electrical model obtained in the preceding section, i.e., structures at a depth greater than 2 km were maintained from the regional model. Interpretation of resistivity data provided an image of the resistivity distribution in the graben, and its relation to main geological structures was derived from geological interpretation (Andrade Afonso et al., 1994a; Monteiro Santos et al., 1997).

Based on that information a 3-D model was constructed, using a mesh of $42\times46\times20$ cells. The boundary values came from a 2-D TM mode calculation and were imposed at 60 km (east-west edges) and at 80 km (north-south edges), from the center of the model. The 1-D boundary condition was imposed at the bottom of the earth model at a depth of 100 km. Considering the lack of data and its quality, it was necessary to simplify the model. One simplification that had to be introduced was that the irregular geometry of the Chaves-Verin Fault was not taken into account. Another simplification was that the electrical structure to the south of the hot springs was not considered. These simplifications must be responsible for some of the misfit between data and model responses, particularly for periods greater than 1 s.

The upper 2 km of the derived 3-D resistivity structure, together with a geological sketch of the zone, is shown in Fig. 9. The main features of this structure can be summarized as follows:

- Beneath the overburden (resistivity of 30 ohm-m, thickness of 250 m, associated with Pleistocene sedimentary formations), a 750-m-thick conductor (resistivity of 12 ohm-m) is underlain by a more resistive substratum (150 ohm-m). That zone has been associated with the shallow geothermal reservoir.
- To the north of the main body, a relatively large zone of 30 ohm-m occurs with its bottom at a depth of 550 m. This region corresponds to the sedimentary fill of the graben, with some clay and freshwater content.
- To the south, the main conductive structure is partially limited by a resistive block (200 ohm-m). This structure appears to be associated with a small horst, occurring within the graben (block A in the geological section shown in Fig. 10).

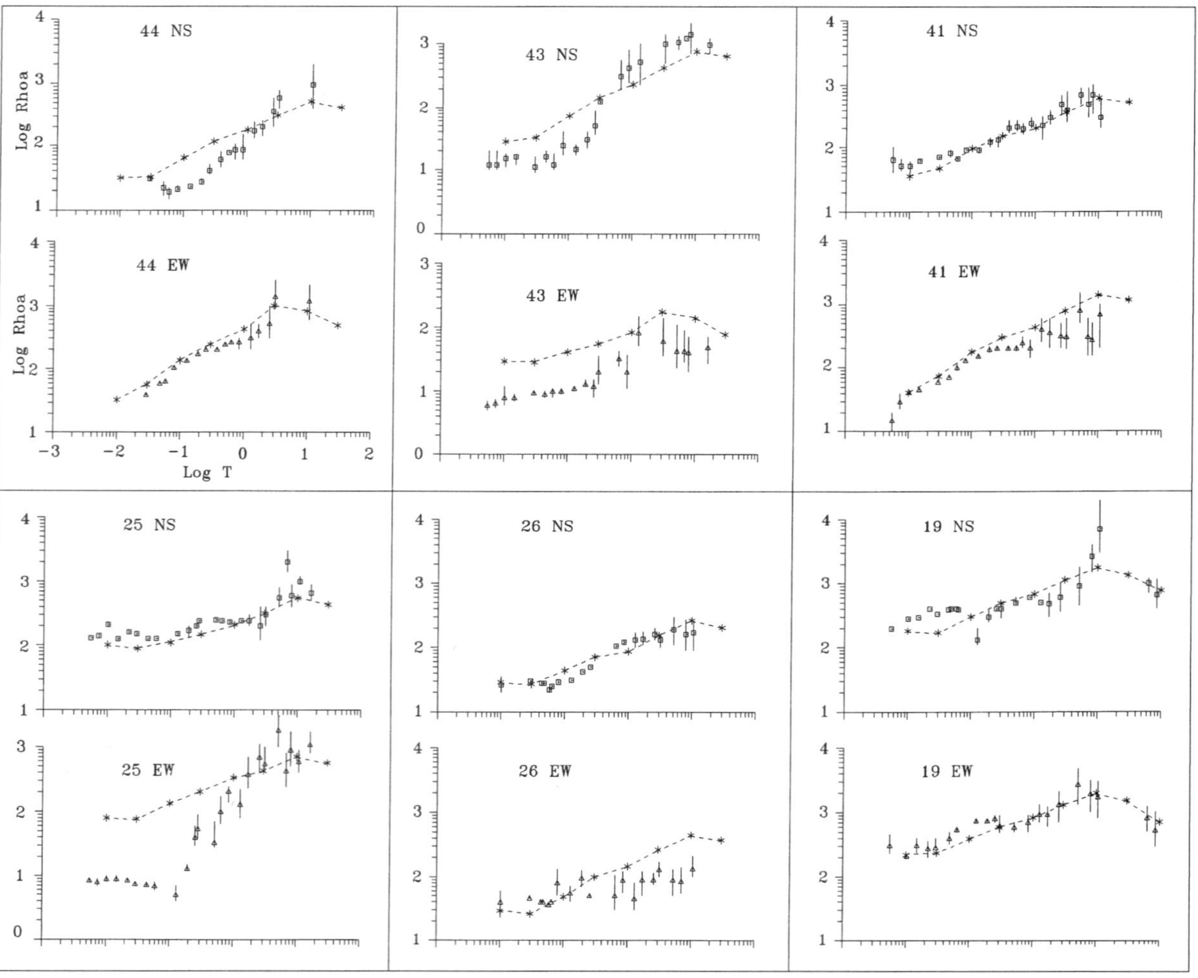

Figure 8. Data and 3-D model responses (*) for 11 sites used in the modeling of graben. *(Continues)*

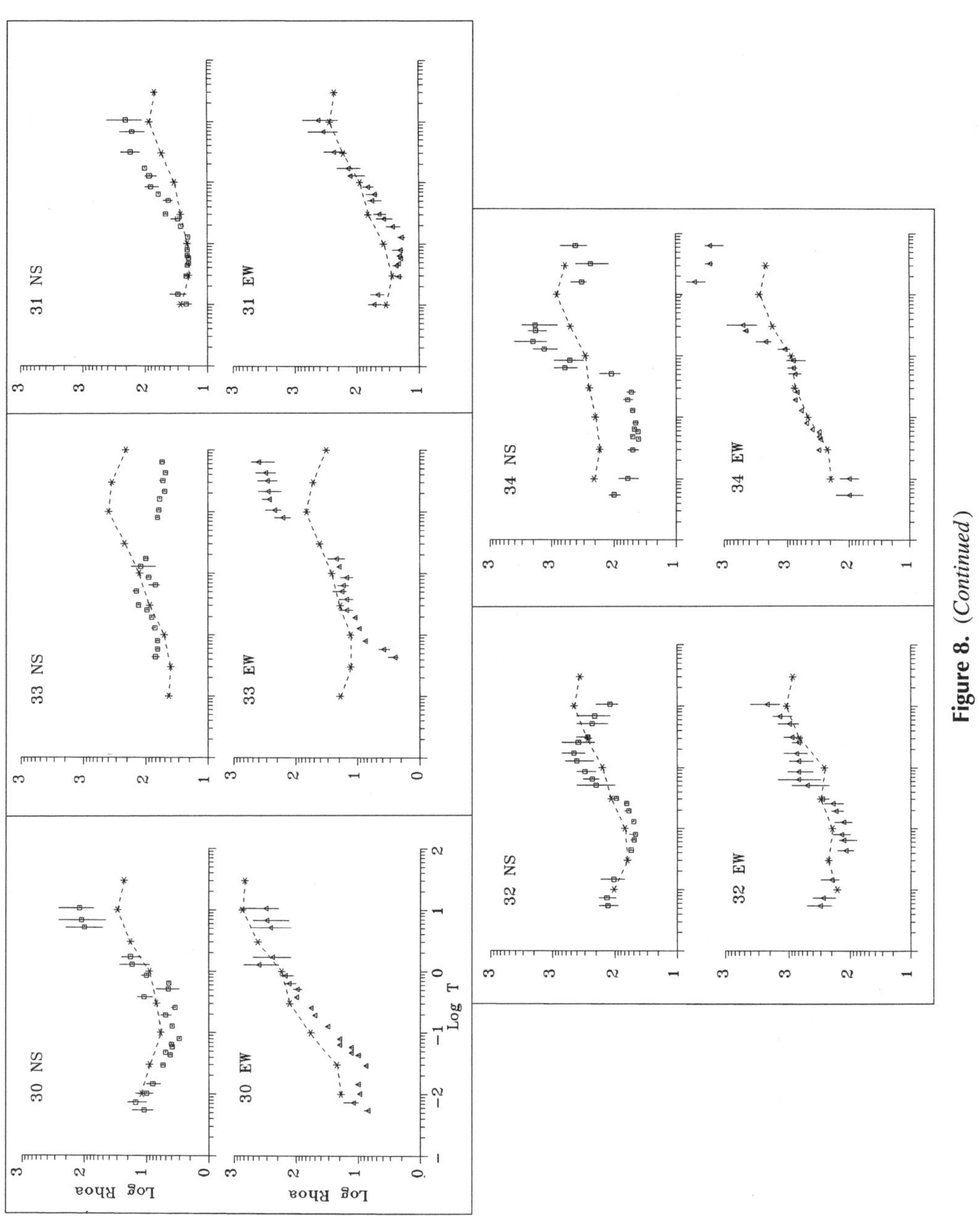

Figure 8. *(Continued)*

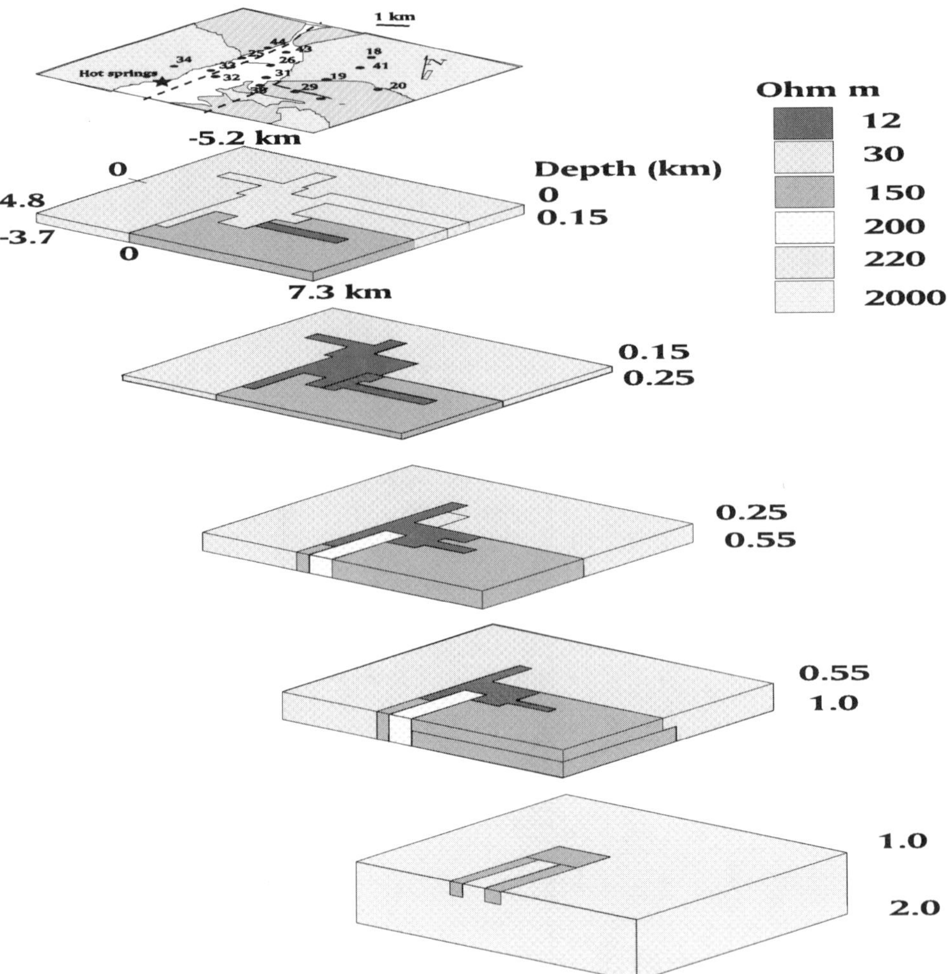

Figure 9. Horizontal slices of 3-D resistivity model to 2-km depth obtained for the graben.

- From north to south the resistivity of the deeper part of the basin decreases from 220 to 150 ohm-m. Such a feature reveals the change in the basement nature—granite and metamorphic rocks, respectively.

As can be seen in Figs. 9 and 10, the complex geometry of the electrical structure is controlled by the fault systems that cross the graben in the north-northeast–south-southwest, east-northeast–west-southwest, and west-northwest–east-southeast directions. This highlights the importance of neotectonic activities in this zone of the Portuguese mainland. Measurement directions were calculated to simulate the observed responses and are shown in Fig. 8, together with the field data. In general, the fit of the modeled structure to the data is quite good. Data acquired in the graben strike direction from sites located in resistive regions, but close to the graben (e.g., sites 34 and 44), may be affected by channeling, and the fit between data and model responses is more problematic. At sites where 3-D effects are severe, such as sites 30, 31, and 32, the fit is reasonable, at least, for the east-west direction. Sites 43 and 44 are located close to the north edge of the graben, and the data at site 43 shows strong anisotropy. If we try to fit

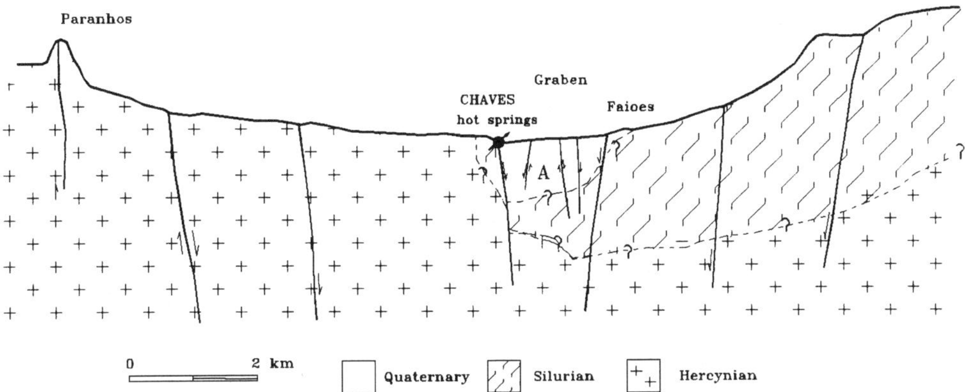

Figure 10. Conceptual geological model for the Chaves graben (adapted from UTAD, 1992). Note the complex pattern of faults in the sedimentary filling and the horst of minor-scale (marked by letter A).

the low apparent resistivity exhibited by the data in the east-west direction, above 1 Hz, then site 43 must be underlain by a more conductive layer. However, this modification causes the fit of site 44 (east-west direction) to deteriorate. Without more closely spaced soundings, a better definition of the electrical structures is not possible in this area. Station 25 is located on or close to the granite-metamorphic rock contact, which is made by a fault trending west-northwest to east-southeast. This relatively narrow structure, impossible to model at the present scale, could be responsible for the feature of the east-west curve at high frequencies and for the poorer fit in the east-west direction. At station 33, the anisotropy at the periods above 3 s may be caused by the presence, on the intersection of the Chaves-Verin Fault with the west-northwest–east-southeast fault, of an upward flow area (Monteiro Santos et al., 1996).

5 Conclusions

Three-dimensional modeling of MT data acquired in the Chaves region in northeastern Portugal has led to a characterization of the electrical conductivity structure of the upper-middle crust in this region. The model indicates that there is a low-resistivity zone at middle depth beneath the granitic rocks and high-conductivity zones at shallow depths in the graben area. For a full investigation, the complexity of the structures related to the graben requires more closely spaced sites and higher quality data as well as extension of the work to the south of the graben.

The results indicate that a preliminary image of the regional structures can be obtained from 3-D modeling of the apparent resistivity and phase data of the invariant impedance tensors. However, for shallow structures, a more detailed model is only possible if both measured curves are considered in the modeling.

Acknowledgments

The authors are grateful to A. Choquier for his assistance with field work and to R. Mackie for the donation of the computer program used in the 3-D MT modeling. The constructive reviews by J. F. Hermance and one anonymous referee are greatly appreciated. This work was supported by the European Community (Programme Joule I, contract JOUG-0009-C) and by JNICT (Junta Nacional de Investigação Científica e Tecnológica).

References

Aires-Barros, L., 1989, Geothermal resources in Portugal: Annales Universidade de Trás-os-Montes e Alto Douro (UTAD), **2**, 11–22.

Andrade Afonso, A. R., Monteiro Santos, F. A., and Mendes Victor, L. A., 1994a, The resistivity method in the evaluation of geothermal resources—Application to Chaves (Portugal) geothermal field. Proc. Symp. Geothermics 94 in Europe, (Edited by BRGM), **203**, 43–50.

Andrade Afonso, A. R., Rio, M. I., Monteiro Santos, F. A., and Mendes Victor, L. A., 1994b, The application of the self-potential method to a low enthalpy geothermal field (Chaves-Portugal): Proc. Symp. Geothermics 94 in Europe, (Edited by BRGM), **203**, 59–65.

Berdichevsky, M. N., and Dimitriev, V. I., 1976, Basic principles of interpretation of magnetotelluric curves, *in* Ádam, Á. Ed., Geoelectric and geothermal studies, Akademini Kiado, Budapest.

Hermance, J. F., 1982, The asymptotic response of three-dimensional basin offsets to magnetotelluric fields at long periods: The effects of current channeling: Geophysics, **47**, 1562–1573.

INMG (Instituto Nacional de Meteorologia e Geofísica), 1992, Evaluation of geothermal resources between Lamego and Vila Verde da Raia. Gravity Final Report of Programme Joule/ECE-JOUG-0009-C.

Mackie, R., Madden, T. R., and Wannamaker, P. E., 1993, Three-dimensional magnetotelluric modelling using difference equations—Theory and comparisons to integral equation solutions: Geophysics, **58**, 215–226.

Mareschal, M., William, S. F., Percival, J., and Chan, T., 1992, Grain-boundary graphite in Kapuskasing gneisses and implications for lower-crustal conductivity: Nature, **357**, 674–676.

Monteiro Santos, F. A., Andrade Afonso, A. R., Dupis, A., and Mendes Victor, L. A., 1994, Joint inversion of AMT and DC data. Application to the Chaves geothermal field, *in* Proc. Symp. Geothermics 94 in Europe (Edited by BRGM), **203**, 51–57.

Monteiro Santos, F. A., Dupis, A., Andrade Afonso, A. R., and Mendes Victor, L. A., 1995, Magnetotelluric observations over the Chaves geothermal field (NE Portugal)—Preliminary results: Phys. Earth Plan. Int. **91**, 203–211.

Monteiro Santos, F. A., Andrade Afonso, A. R., and Mendes Victor, L. A., 1997, Study of the Chaves geothermal field using 3-D resistivity modeling: Journal of Applied Geophysics, **37**, 85–102.

Monteiro Santos, F. A., Dupis, A., Andrade Afonso, A. R., and Mendes Victor, L. A., 1996, An audiomagnetotelluric survey over the Chaves geothermal field (NE Portugal): Geothermics, **25**, 389–406.

Ranganayaki, R. P., 1984, The interpretive analysis of magnetotelluric data: Geophysics, **49**, 1730–1748.

Sutarno, D., and Vozoff, K., 1991, Phase-smoothed robust M-estimation of magnetotelluric impedance functions: Geophysics, **56**, 1999–2007.

Téllez, J., Matias, L. M., Córdoba, D., and Mendes Victor, L. A., 1993, Structure of the crust in the schistose domain of Galicia-Trás-os-Montes (NW Iberian Peninsula). Tectonophysics, **221**, 81–93.

UTAD (Universidade de Trás-os-Montes e Alto Douro), 1992, Evaluation of geothermal resources between Lamego and Vila Verde da Raia: Geological Final Report of Programme Joule/ECE-JOUG-0009-C.

Wannamaker, P. E., Hohmann, G. W., and Ward, S. H., 1984, Magnetotelluric responses of three-dimensional bodies in layered earths: Geophysics, **49**, 1517–1533.

Three-Dimensional Inversion of MT Fields Using Bayesian Statistics

Vjacheslav Spichak[1,2]
Michel Menvielle[2]
Michel Roussignol[3]

Summary. Bayesian statistics provide a formalism for inversion of magnetotelluric (MT) data in 3-D structures composed of elementary homogeneous domains. Available information (including assumptions about the model) is put in a probability density function (PDF) for *prior* values of the conductivities in the region of the search; the parameters to be found are the *posterior* values of the conductivity.

A stochastic algorithm called a Gibbs sampler estimates the posterior PDF. The outer cycle of the iterative inversion consists of scanning the homogeneous domains in the region of the search; the inner cycle involves solution of the forward problem for a set of models. This process represents a Markov chain, whose transition law converges to the marginal PDF of the parameters.

The inner cycle uses the finite-difference program FDM3D-MT, which computes electromagnetic responses in the frequency domain for 1-D, 2-D, or 3-D models. Iterative solution of the finite-difference equations is very fast and reduces greately the total CPU time since the results of the previous cycle are used as starting points for the next forward model. In most of our tests, the outer iteration converges in 15–20 iterations, which allows an attack on 3-D problems even on microcomputers. Examples show how the quantity and quality of data and the prior information affect the results of inversion.

1 Introduction

Inversion of magnetotelluric (MT) data aims to find a distribution of conductivity in the Earth that accounts for insufficient and noisy data, which often is spaced very irregularly on the surface. This ill-posed problem—whether solved stochastically or deterministically—requires regularization or the use of constraints. Questions about the existence and uniqueness of solutions become more critical for 3-D models with a large number of unknowns. Unfortunately, we do not know in advance how the

[1]Geophysical Research Center, Varshavskoe sh.8, Moscow 113105, Russia.
[2]Laboratoire de Physique de la Terre et des Planètes (URA 1369), Université Paris Sud, Bat. 504, 91405 ORSAY CEDEX, France.
[3]Equipe de Mathématiques Appliquées, Université de Marne la Vallée, 2 rue de la Butte Verte, Noisy le Grand Cedex 93166, France.

data—the number of sites and their locations, the components of the electromagnetic (EM) field measured, the frequencies used, the noise, etc.—and the prior information available will affect the results of an inversion. It is thus not difficult to understand why inversions can give poor results. Attempts to improve the results usually involve increasing the volume of data or decreasing the number of parameters. This, in turn, can lead to overestimates of the amount of data needed to resolve the geoelectrical structure.

Despite some success with optimization methods for 3-D inversion of synthetic MT data (Mackie and Madden, 1993), inversion of real data will require careful attention to the issues mentioned above. Bayesian statistics provides an appropriate framework for studying these issues (see, e.g., Tarantola and Valette, 1982; Backus, 1988).

2 Bayesian inversion

In MT inversion, the conductivity model of the Earth can be divided into two kinds of regions: regions with known (or fixed) conductivity values and regions with (unknown) values to be determined by fitting the MT measurements at the Earth's surface, subject to contraints imposed by prior information. Each of the latter regions can be considered, in turn, as composed of homogeneous domains (cells). Let K be the total number of domains $\{P_k; k = 1, \ldots, K\}$, whose conductivities $\sigma = (\sigma_k; k = 1, \ldots, K)$ are to be determined.

Let $\mathbf{E}(M_i, \omega_j, \sigma)$ be the electric field and $\mathbf{H}(M_i, \omega_j, \sigma)$ the magnetic field measured at given points $\{M_i; i = 1, \ldots, I\}$, on the Earth's surface for discrete frequencies $\{\omega_j; j = 1, \ldots, J\}$. Also, let $y_{i,j}$ be values, derived from measurements, of a known function $F(\mathbf{E}, \mathbf{H})$ of these fields. The function F could be, for example, the impedance ratio of specific components of $\mathbf{E}$ and $\mathbf{H}$. We assume that

$$y_{i,j} = F[\mathbf{E}(M_i, \omega_j, \sigma), \mathbf{H}(M_i, \omega_j, \sigma)] + \epsilon_{i,j}, \tag{1}$$

where $\{\epsilon_{i,j}; i = 1, \ldots, I; j = 1, \ldots, J\}$ are noise functions, taken as realizations of independent random variables with probability density functions (PDFs) $p_{i,j}$ and zero-mean values.

An inversion for conductivity in the Earth usually involves simplifying assumptions (made from experience or convenience) about the regions of interest. The assumptions can be that the region is homogeneous, or that the conductivity variations are confined to a thin sheet, or are 1-D or 2-D in character. In the Bayesian approach, these prior assumptions and any prior knowledge are incorporated into the inversion through a probability law q, called the "prior PDF," on the set of possible values of the conductivities. The support of q (i.e, the set of all possible values of conductivity) is the first choice to be made. Numerical computations make it necessary to define the support of q as a finite set. For simplicity, we suppose that the number of possible conductivities consists of L different values $\{c_1, \ldots, c_L\}$ for each homogeneous domain. The support of q is then the set A of the L^K possible elements (referred to below as images of the conductivity) $a = (a_k; k = 1, \ldots, K)$, where a_k belongs to $\{c_1, \ldots, c_L\}$. The size of L determines the attainable precision of the inversion, and also the length of the computation (the search). We use the same set L of possible conductivities (called a "palette") for each domain, but it is possible to narrow the range of prior conductivities in some domain by setting appropriate prior probabilities equal to zero. If no information is available, it is still possible to limit the prior palette to lie between reasonable minimum and maximum values with a uniform PDF.

In the statistical point of view, both observations and model parameters (conductivities) are random variables. Our Bayesian analysis determines the posterior PDF of the conductivities—i.e., the conditional probabilities of the conductivities given the data y, prior information in terms of a conductivity palette $(c_1, \ldots, c_L)$ and PDF q, and the noise level ϵ.

$$P(\sigma = a/Y = y) = \frac{f(y/a)q(a)}{\sum_{b\in A} f(y/b)q(b)}, \tag{2}$$

where $q(a)$ is the prior probability of the image a and $f(y/a)$ is a conditional probability of the variable $y = (y_{i,j},\ i = 1, \ldots, I,\ j = 1, \ldots, J)$ given the values of the conductivities. It is a function of $a = (a_k,\ k = 1, \ldots, K)$ through $\mathbf{E}$ and $\mathbf{H}$ and could be calculated directly as follows:

$$f(y/a) = \prod_{i=1}^{I}\prod_{j=1}^{J} p_{i,j}\{y_{i,j} - F[\mathbf{E}(M_i, \omega_j, a), \mathbf{H}(M_i, \omega_j, a)]\}, \tag{3}$$

where $p_{i,j}$ is the probability density of the noise $\epsilon_{i,j}$.

If the probability densities $p_{i,j}$ are Gaussian with zero mean and covariances $(\varrho_{i,j})^2$, the above formula can be rewritten

$$f(y/a) = Z \exp\left(-\sum_{i,j} \frac{\{y_{i,j} - F[\mathbf{E}(M_i, \omega_j, a), \mathbf{H}(M_i, \omega_j, a)]\}^2}{2(\varrho_{i,j})^2}\right), \tag{4}$$

where Z is a normalizing constant. We assume the latter case, but note that the method presented also could be used without this assumption.

If $A(k, c_j)$ is a set of images that have the conductivity c_j in the domain P_k, the kth marginal posterior probability p_k is

$$p_k(c_j) = P[\sigma \in A(k, c_j)/Y = y] = \frac{\sum_{a\in A(k,c_j)} f(y/a)q(a)}{\sum_{b\in A} f(y/b)q(b)}. \tag{5}$$

We can take as estimator of the conductivity in the domain P_k either the mean value or the value corresponding to the maximum probability of the kth marginal posterior probability. However, in the above formula, there is a hidden difficulty: The denominator requires computation of $f(y/b)q(b)$ for all possible images b of the conductivity, or L^K times, which is unrealistic.

2.1 Markov chains

To overcome this difficulty we use a stochastic algorithm called a Gibbs sampler, which consists of an outer and an inner cycle. The outer cycle scans all K homogeneous domains in the regions of search; the inner cycle solves the forward problem for L prior values of the conductivity. Let $[\sigma_k^{(n)}; k = 1, \ldots, K]$ be the conductivities in the homogeneous domains in the regions of search after n iterations of the outer loop. If the domain $k(n)$ is scanned at iteration $n + 1$, the image of conductivities is updated by changing only the conductivity of this domain to the new value chosen at random with the following probability:

$$P\left(\sigma_{k(n)}^{(n+1)} = c_j\right) = \frac{f(y/a(\sigma^{(n)}, k(n), c_j))q(a(\sigma^{(n)}, k(n), c_j))}{\sum_{j=1}^{L} f(y/a(\sigma^{(n)}, k(n), c_j))q(a(\sigma^{(n)}, k(n), c_j))}, \tag{6}$$

where $a(\sigma, k, c_j)$ denotes the image equal to σ in all domains other than P_k and equal to c_j in the domain P_k. Computation of P in equation (6) requires calculation of

$$f(y/a(\sigma^{(n)}, k(n), c_j))q(a(\sigma^{(n)}, k(n), c_j))$$

L times. Thus, the total number of forward modelings per iteration of the outer cycle is just $L * K$.

The sequence of images $[\sigma^{(n)}, n \geq 0]$ forms a random process, which is a Markov chain over the finite space of all possible images. The conditional probability, calculated in the kth domain at nth iteration is

$$p_k^n(c_j) = \frac{f(y/a(\sigma^{(k+nK)}, k, c_j))q(a(\sigma^{(k+nK)}, k, c_j))}{\sum_{j=1}^{L} f(y/a(\sigma^{(k+nK)}, k, c_j))q(a(\sigma^{(k+nK)}, k, c_j))}. \quad (7)$$

It can be proved that the posterior PDF is an invariant of this Markov chain and that, in each domain of the region of search, the sequence of the mean conditional probabilities converges toward the corresponding marginal probability,

$$p_k(c_j) = \lim_{N \to \infty} \frac{1}{N+1} \sum_{n=0}^{N} p_k^n(c_j). \quad (8)$$

This gives an estimate of the mean posterior conductivities in each homogeneous domain of the region of search:

$$\sigma_k = \sum_{j=1}^{L} c_j p_k(c_j), \qquad k = 1, \ldots, K. \quad (9)$$

The solution of the inverse problem thus is reduced to the search for the a-posteriori conductivity distribution by means of successive solution of the forward problem for the prior values of the conductivities in the homogeneous domains.

2.2 *Forward modeling*

For forward modeling, we use the program FDM3D-MT (Spichak, 1983a), which has been checked against quasi-analytical solutions for a sphere buried in a free space (Spichak, 1983b) and against other numerical solutions (Zhdanov et al., 1990; Zhdanov and Spichak, 1992). The algorithm uses a seven-point finite-difference approximation of the integral "balance" equation for the electric field at each node of the grid. The system of sparse linear equations is solved by the block over-relaxation with adaptive correction of the relaxation factor. The vertical magnetic field is calculated from Maxwell's equation and the horizontal field by Hilbert transforms. (See Spichak, 1998, this volume). The program is fast and allows calculation of the electromagnetic responses in the frequency domain in the earth or in the air for 1-D, 2-D, or 3-D structures with a relief surface.

3 Numerical experiments

We tested the Bayesian algorithm in numerical experiments with synthetic data. The model used for the synthetic data was a prism ($1 \times 1 \times 1$ km, 0.1 S/m), embedded at a depth 0.5 km in a halfspace of 0.01 S/m. With the two planes of symmetry, only one-quarter of the prism, discretized into $3 \times 3 \times 4$ cells (Fig. 1), was needed for numerical modeling. The data were values of the in-phase and quadrature electric-field

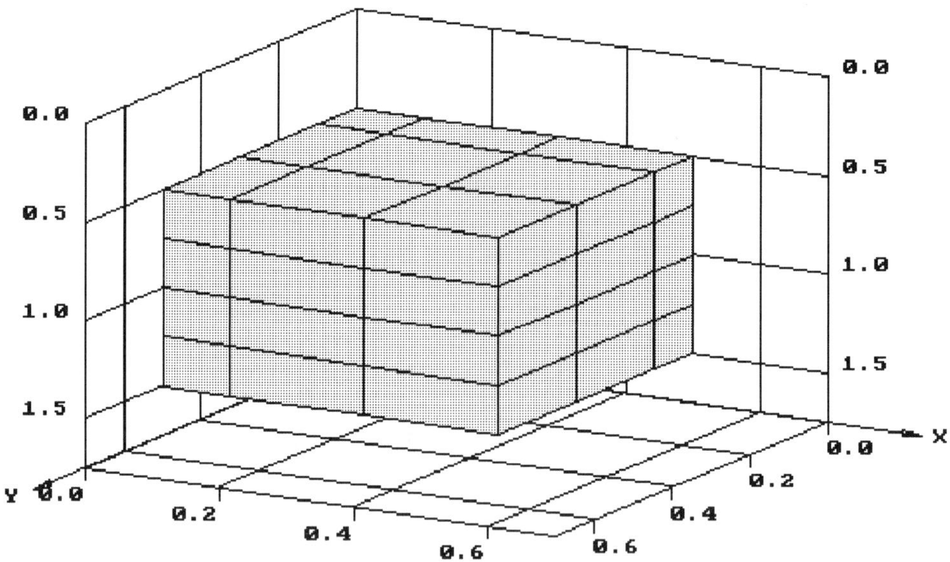

Figure 1. One-prism model used for generation of synthetic data.

components at 36 points at the Earth's surface for various periods and polarizations of the primary field.

To accelerate the inversion process we tried to use an approximate solution of the forward problem in the inner cycle of the inversion. We found, however, that this often led to bad values of conductivities in the domains scanned and could give an erroneous final conductivity distribution. So, to solve the forward problem accurately, we fixed the threshold for the rms error in the inner iteration process at the level 10^{-4}, which usually gave correct solutions of the forward problem. The inner iteration process converged in most cases in 20–50 iterations, depending on the a priori conductivity range preset, and the total CPU time required for one iteration of the outer cycle was quite reasonable.

Another important problem to solve was selection of an appropriate rule for stopping the outer iteration. The standard rule based on the misfit between the forward modeling results and the data is not suitable in our algorithm. In principle (statistically), more iterations of the outer cycle will always bring the result closer to the theoretical posterior conductivity distribution. On the other hand, we always have to settle for a finite number of iterations. The compromise found consists of stopping the iterations once the marginal mean posterior conductivities in the region of search are stabilized:

$$\left(\frac{1}{K}\sum_{k=1}^{K}\left(\frac{\sigma_k^{(n+1)}-\sigma_k^{(n)}}{\sigma_k^{(n)}}\right)^2\right)^{1/2} < \epsilon. \tag{10}$$

Numerical experiments show that the level of 1% for the left-hand side of this inequality is usually achieved in 15–20 iterations (Fig. 2).

The inversion code was tested on data synthesized for a false anomaly (with conductivity equal to those of the surrounding homogeneous half-space), which were inverted to search for the posterior conductivity distribution in the bounded region (Fig. 1) at periods $T = 0.1$, 1, 10, 100, and 1000 s and for various prior conductivities ranging from 0.001 S/m to 0.1 S/m. Regardless of the prior PDF, the posterior mean conductivity in

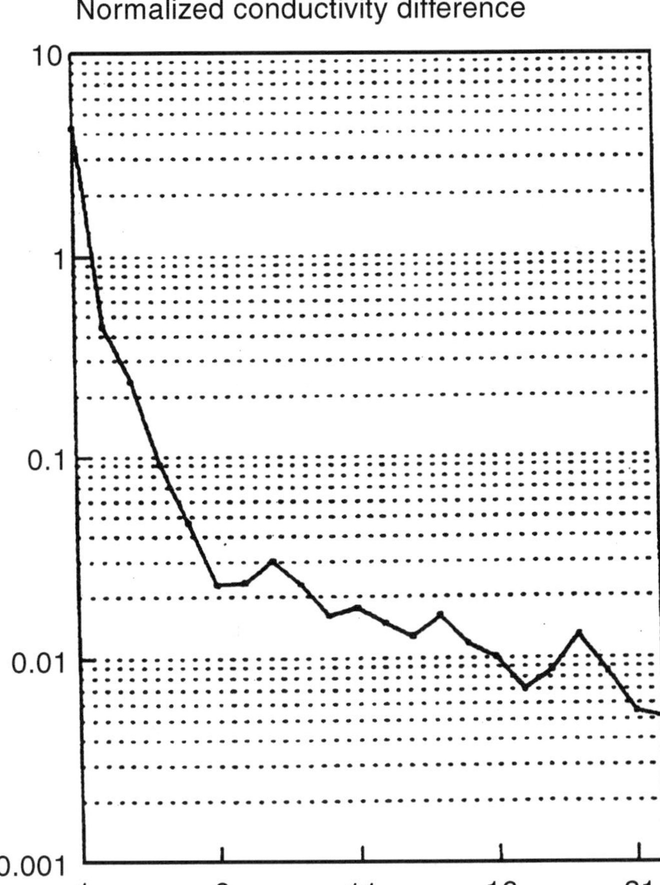

Figure 2. Normalized averaged conductivity difference as a function of iteration number for the model shown in Fig. 1.

the region of search practically coincided with the true value, or with the value from the prior conductivity range closest to the true one in cases when it was outside the prior palette.

3.1 Effect of prior conductivity palette

To study the effect of the prior conductivity palette and its range on the results of the inversion, we simulated two models with conductive ($\sigma = 0.1$ S/m) and resistive ($\sigma = 0.001$ S/m) prisms embedded in the homogeneous half-space (Fig. 1). The region of search coincided with the position of the prism so that the assumption about the homogeneity of the posterior conductivity distribution was valid. In the first experiment, the prior conductivity palette ranged from 0.01 S/m to 0.1 S/m for the conductive target and from 0.001 S/m to 0.01 S/m for the resistive one. In the second experiment, the range (and, correspondingly, the sampling) was increased 10 times—from 0.001 S/m to 0.1 S/m—for both conductive and resistive prisms (with a uniform PDF). The results of the inversion showed that the conductivities of the targets could be reconstructed with accuracy 4% for conductive body and with accuracy 20% for resistive body regardless of the prior conductivity range and the periods used.

3.2 Relative importance of prior information and data

Another experiment was carried out to study the relative importance of the prior PDF and the data. As in the previous experiment, reconstructions of two different target conductivities were made from the synthetic electric-field components for the periods 1 s and 10 s. The prior conductivities were 0.1 S/m and 0.01 S/m in the case of the conductive body and 0.001 S/m and 0.01 S/m in the case of the resisitive one. Standard deviations of the noise added to synthetic data were 3%, 5%, and 10%. For each input data set and each level of noise, the prior PDF was changed so that the probability attributed to the conductivity that was equal to the true target conductivity ranged from 0.1 to 1.0.

The results of the experiments indicate that, up to a 5% level of noise, the prior PDF has no effect on the posterior one in either case. When the level of the noise increases to 10%, however, the reconstruction of the resistive target becomes sensitive to the prior PDF; reconstruction of the conductive target remains stable. Figure 3 shows the graphs of the posterior mean conductivity, normalized by the true value in the case of reconstruction of the resistive body for the periods 1 s and 10 s. (The straight line denotes a normalized mean prior value of the target conductivity, based on the prior palette and PDF.) Figure 3 shows that, in the worst case, when the prior probability that the target conductivity equals the true conductivity is only 0.1, the posterior conductivity can be overestimated by a factor of two or even three, depending on the period of the input data. The differences between the straight and the curved lines can be interpreted as the contribution of the input data to the inversion: The more accurate the prior estimation of the target conductivity, the less important is the contribution of the input data (and vice versa). Figure 3 illustrates this effect quantitatively.

Another useful characteristic of the Bayesian inversion is the statistical uncertainty of the posterior values of the parameters. Nonuniqueness of the inversion can be measured by the standard deviations of the posterior conductivities. Figure 4 shows the graph of the posterior normalized mean conductivity as the prior probability of the target conductivity coinciding with the true value varies, for the case of the resistive prism (period $T = 1$ s, Gaussian noise 10% is added to synthetic data). The uncertainty of the result decreases when the prior estimates of the target conductivity become closer to their true values.

3.3 Effect of data

Some experiments were done to study the influence of the structure and the volume of the input data on the results of the inversion. Here we describe only one result, which is important for understanding the mechanism of the inversion. Synthetic electric-field components were generated for conductive and resistive prisms of the same geometry (Fig. 1) at the period 10 s and inverted using only one polarization of the primary field (along the x-axis). Each cell of the grid was considered as a homogeneous domain in the region of search. Figure 5 shows the results of the inversions after 15 iterations for conductive ((a) and (b)) and resistive ((c) and (d)) targets. With a conductive anomaly, the conductivities are reconstructed with the relative error up to 9% in the

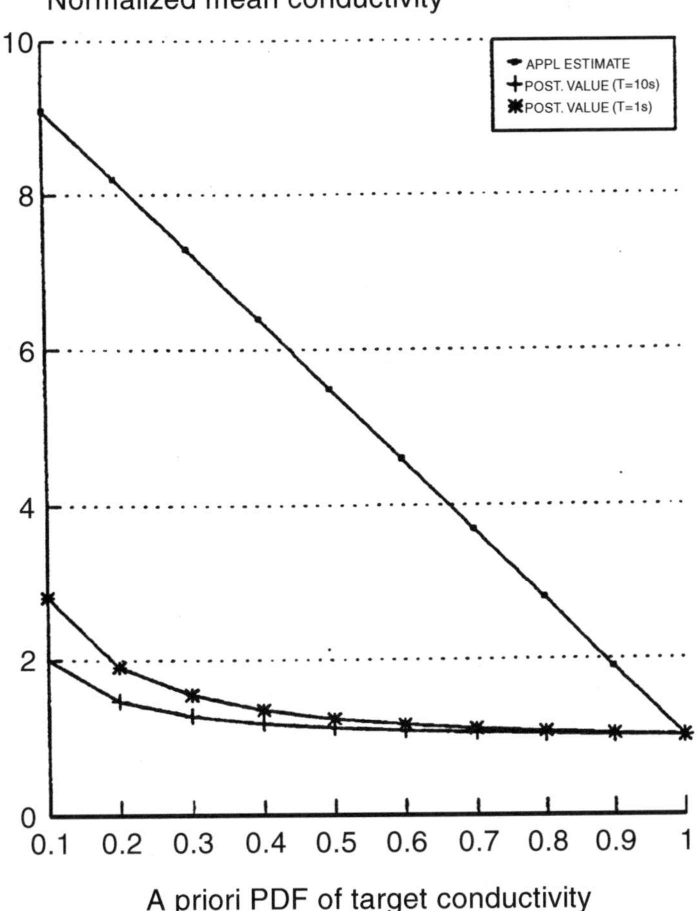

Figure 3. Relative contribution of prior PDF estimation and data to results of inversion for the model shown in Fig. 1; 10% Gaussian noise is added to synthetic data.

region with dimensions $0.3 \times 0.3 \times 0.7$ km (25% of the volume of the region of search), whereas with a resistive anomaly, the values in the same region are reconstructed with the relative error up to 90%.

There is also a different accuracy of the conductivity reconstruction at the boundaries parallel and perpendicular to the direction of the primary field polarization. With a conductive anomaly, the best results are achieved at the parallel boundaries (TE mode) (Fig. 5b), whereas with a resistive anomaly, the results are evidently better at perpendicular boundaries (TM mode) (Fig. 5c). We believe that these results are caused by different mechanisms dominating in the inversion when a conductive or a resistive prior conductivity palette is used. Indeed, because of galvanic effects, the electric field in both cases is sensitive to the perpendicular boundaries; however, because of inductive effects, the electric field becomes sensitive to parallel boundaries only with conductive regions in the palette. Thus, when using the electric fields for the inversion, it is important to match in advance the prior conductivity palette and the range of periods used to get accurate results in a minimum number of iterations.

In another experiment, the volume of data was increased 4 times. The data consisted of synthetic electric fields mixed with 1% Gaussian noise for two polarizations of the

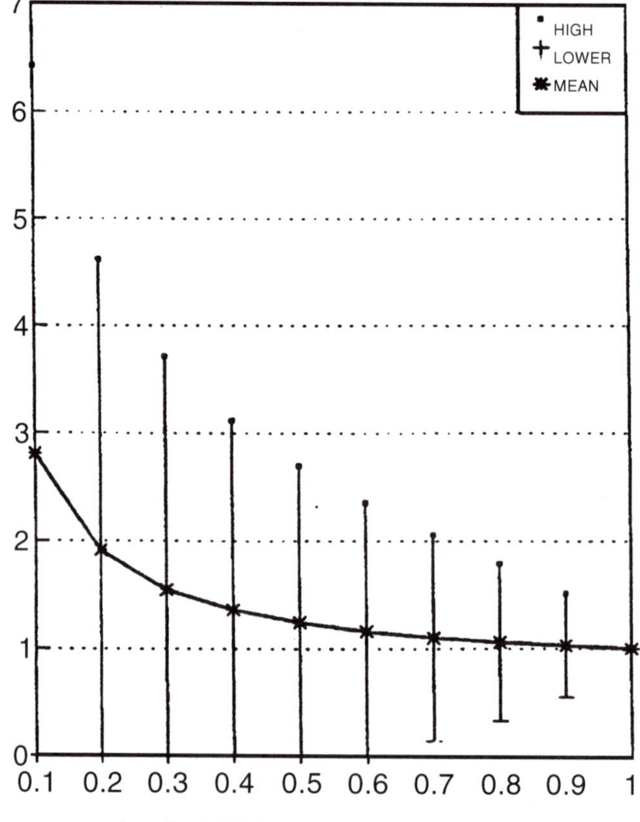

Figure 4. Effect of prior PDF on the uncertainty in posterior conductivity estimation for model shown in Fig. 1. Period $T = 1$ s; 10% Gaussian noise is added to synthetic data.

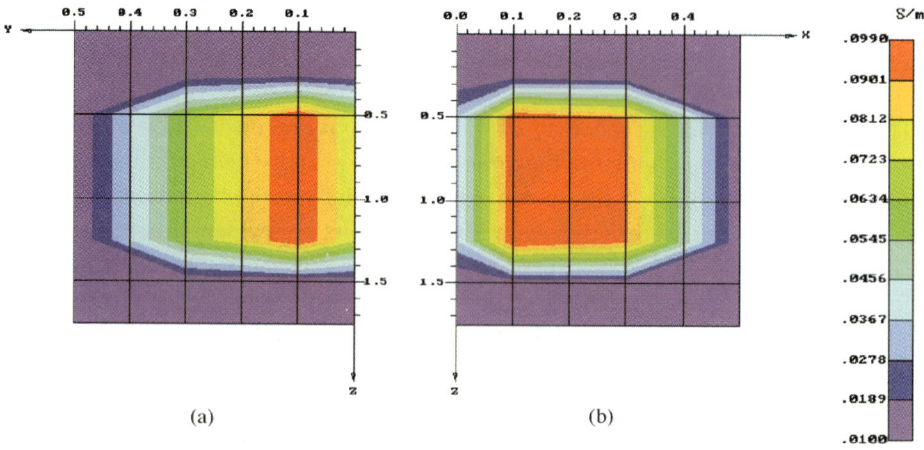

Figure 5. Resolution of conductive ((a) and (b)) and resistive ((c) and (d)) prisms shown in Fig. 1, embedded in the homogeneous half-space ($\sigma = 0.01$ S/m): (a,c) yz cross-sections ($x = 0.3$ km); (b,d) xz cross-sections ($y = 0.3$ km). Data inverted: Synthetic electrical field components, mixed with 1% Gaussian noise; primary field is polarized in x-axis, $T = 10$ s. Prior information used: $\sigma_1^c = 0.01$ S/m, $\sigma_2^c = 0.055$ S/m, $\sigma_3^c = 0.1$ S/m; $\sigma_1^r = 0.0$ S/m, $\sigma_2^r = 0.003$ S/m, $\sigma_3^r = 0.00$ S/m; $p_1 = 0.333$, $p_2 = 0.333$, $p_3 = 0.333$.

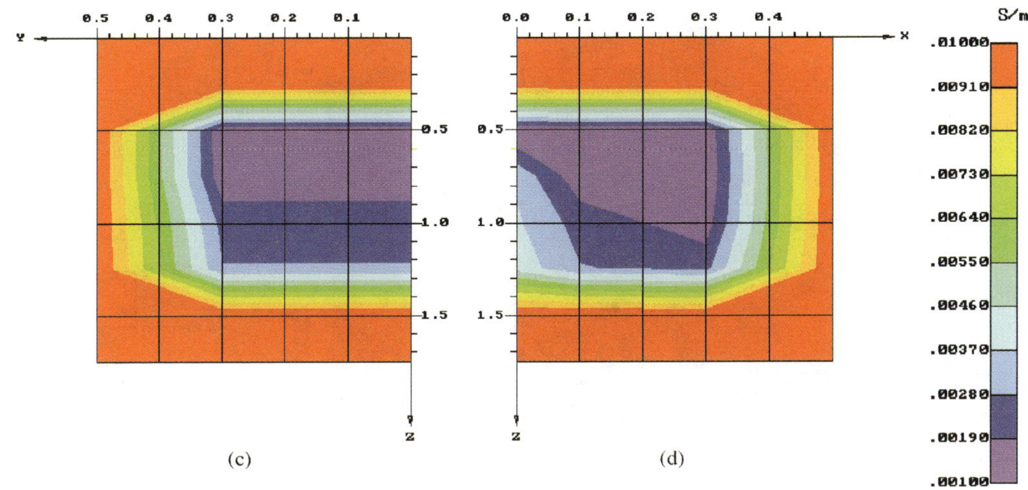

(c) (d)

Figure 5. *(Continued.)*

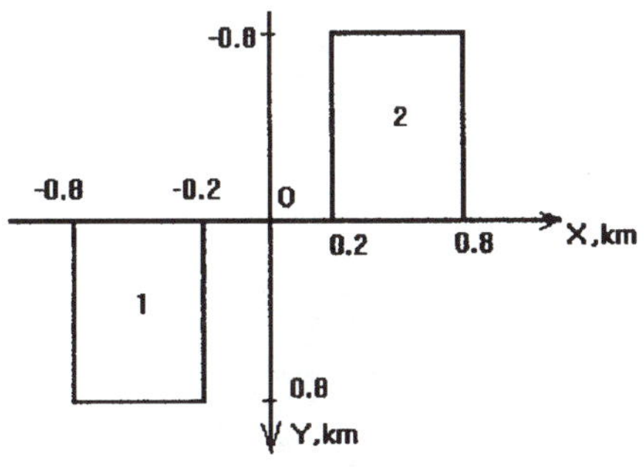

CROSS SECTION

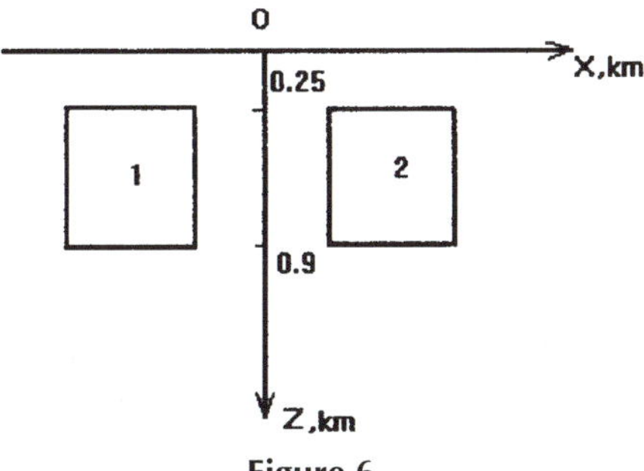

Figure 6.

primary field at two periods ($T = 1$, 10 s). A two-prism model consisting of conductive (1) and resistive (2) prisms, buried in the homogeneous half-space was used for generation of synthetic data (Fig. 6). The conductivities of the targets were 0.1 S/m and 0.001 S/m, correspondingly, and the conductivity of the homogeneous half-space was equal to 0.01 S/m.

The region of search consisted from $5 \times 5 \times 7$ (=175) cells and ranged from -1.0 km to 1.0 km along each horizontal coordinate axis and from the earth surface to the depth2

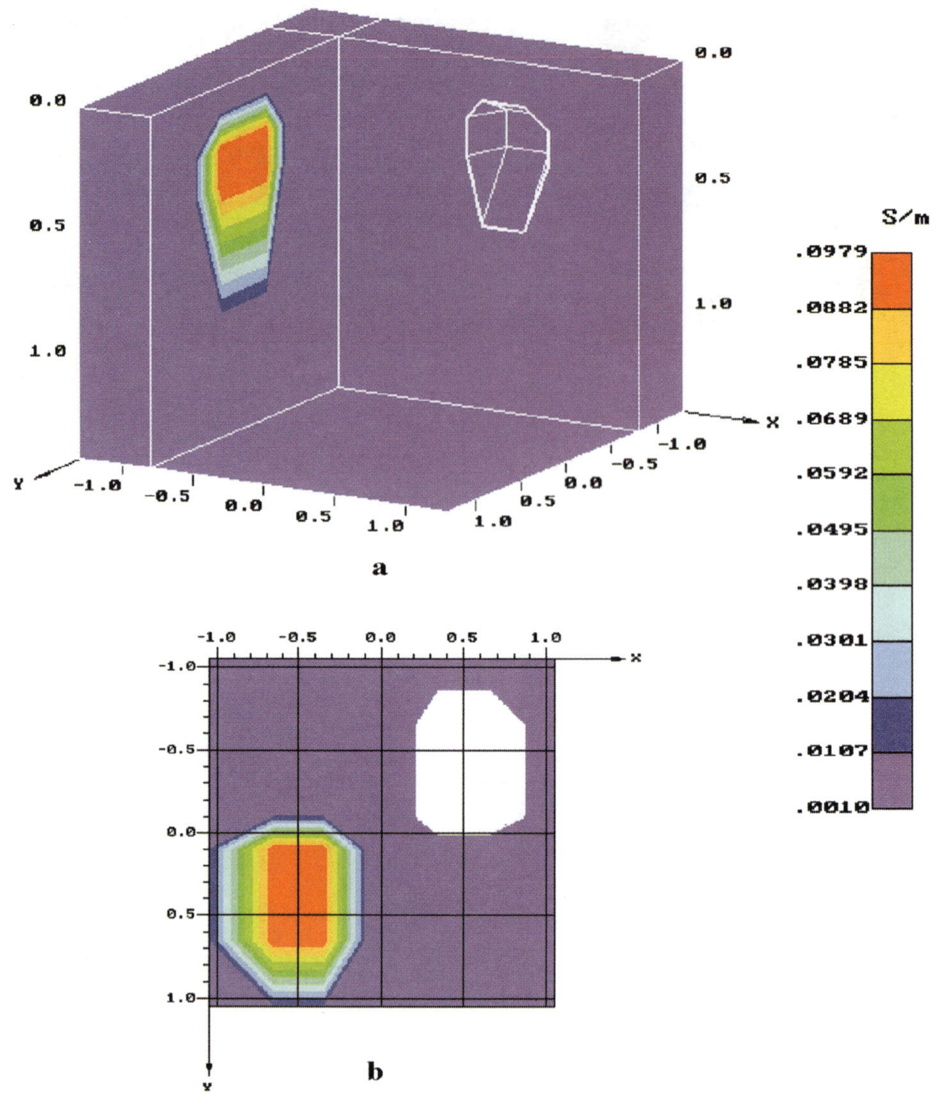

Figure 7.

km. The prior conductivity palette ranged from $\sigma_{\min} = 0.001$ S/m to $\sigma_{\max} = 0.1$ S/m with sampling $0.1\,(\sigma_{\max} - \sigma_{\min})$ and uniform PDF.

Figure 7(a, b) shows the result of the inversion. The posterior conductivity values in the domains occupied by prisms differ from the model ones by not more than 7% for conductive prism and by 45% for resistive one. The comparison of these results with the previous ones clearly indicates the effect of the volume of data used for inversion.

Comparison of these results with those discussed in the Section 3.1 (inversions for conductive and resistive prisms separately) suggests that the additional hypothesis that the region of search consists only of domains occupied by homogeneous prisms would result in improvement of the accuracy of reconstruction, especially in the case of a resistive target (from 45% to 20%).

4 Conclusions

A Bayesian approach has proved to be an efficient tool for 3-D inversion of MT data. It incorporates the prior information into the inversion procedure in a flexible way and converts the problem of nonuniqueness to the practical task of estimating posterior uncertainties. Results of synthetic experiments show that 3-D inversion can be done in a reasonable time on a PC 486. However, to use this method for inversion of real data, we need to accelerate even more the iterations and devise better interpretation strategies for different types of input data.

Acknowledgments

This work was made in the frames of P.A.S.T. position of one of the authors (V. Spichak) at the Université Paris Sud (Laboratoire de Physique de la Terre et des Planetes). The authors wish to thank the anonymous referees and the editors for their valuable suggestions and comments.

References

Backus, G. E., 1988, Bayesian inference in geomagnetism: Geophys. J., **92**, 125–142.

Geman, S., and Geman, D., 1984, Stochastic relaxation, Gibbs distribution, and the Bayesian restoration of images: IEEE Trans. Pattern Analysis and Machine Intelligence, **6**, 721–741.

Grandis, H., 1994, Imagerie électromagnétique Bayesienne par la simulation d'une chaine de Markov: Doctorat d'Université, Université Paris VII.

Jouanne, V., 1991, Application des techniques statistiques Bayesiennes à l'inversion de données électromagnétiques: Doctorat d'Université, Université Paris VII.

Mackie, R. L., and Madden, T. R., 1993, Three-dimensional magnetotelluric inversion using conjugate gradients: Geophys. J. Internat., **115**, 215–229.

Press, S. J., 1989, Bayesian statistics: principle, models and applications: John Wiley & Sons.

Roussignol, M., Jouanne, V., Menvielle, M., and Tarits, P., 1993, Bayesian electromagnetic imaging: *in* Hardle, W., and Siman, L., Eds., Computer intensive methods, Physical Verlag, 85–97.

Spichak, V., 1983a, The FDM3D program package for numerical modeling of the three-dimensional electromagnetic fields: *in* Algorithms and programs for solution of the direct and inverse problems of electromagnetic induction in the Earth, Ed. Zhdanov, M. S., IZMIRAN Publ., 58–68 (in Russian).

Spichak, V., 1983b, Mathematical modeling of EM fields in 3D inhomogeneous media: Ph.D. Thesis, IZMIRAN, 215 (in Russian).

Tarantola, A., and Valette, B., 1982, Inverse problems: Quest for information: J. Geophys., **50**, 159–170.

Zhdanov, M., Varentsov, Iv., Golubev, N., and Krilov, V., 1990, Methods of modeling of the EM fields: Nauka Publ., 200 (in Russian).

Zhdanov, M., and Spichak V., 1992, Mathematical modeling of EM fields in 3-D inhomogeneous media: Nauka Publ., 188 (in Russian).

Imaging Volcanic Interiors with MT Data

Vjacheslav Spichak

Summary. Three-dimensional (3-D) finite-difference modeling shows the sensitivity of the MT method to the electrical structure of volcanic zones. Modeling is done with the program FDM3D-MT, which allows relief topography, mixed conductivity structures (1-D, 2-D, and 3-D), different layered structures at infinity, and calculation of MT fields at arbitrary levels in the earth and atmosphere.

Numerical experiments with different field components and their transforms—including impedances, apparent resistivities, and magnetic transfer functions—indicate that conductivity gradients in volcanic zones can be identified on plots of isosurfaces of the complex electric field and on isosurfaces of the amplitudes and phases of tensor impedances. The actual conductivity values, however, are constrained poorly.

Interpretation is aided by reduction of the data to an artificial reference plane, positioned in the atmosphere above the top level of the relief surface, followed by the construction of 3-D pseudogeoelectrical structure. Examples show the possibility of 3-D imaging of volcanic environments with MT data measured not only at the Earth's surface, but also at different levels in the atmosphere.

1 Introduction

Magnetotelluric (MT) sounding has been used to monitor volcanic activity and to help understand the processes leading to eruptions (Fitterman et al., 1988; Mogi and Nakama, 1990). Modeling of these applications, however, has been relatively crude. For example, Newman et al. (1985) modeled a homogeneous prism in a layered earth with a 3-D integral equation method to study the detectability of a magma chamber, whereas Moroz et al. (1988) built a more elaborate scale model to study the distortion of MT fields by a volcanic cone, but the conductivity of all conductive elements of the structure was equal to 10^6 S/m. I have studied the resolving power of the MT method in volcanic zones with software developed at the Geophysical Research Center in Moscow for use on small computers (Spichak, 1983a, 1985; Zhdanov and Spichak, 1989, 1992). This software allows modeling and analysis of MT fields in a general 3-D model of the Earth, including topography at the Earth's surface. After a brief description of the

Geophysical Research Center, Varshavskoe sh.8, Moscow, 113105, Russia.

modeling algorithm, I present some results aimed at determining those components of the natural electromagnetic (EM) fields (and their transforms) that are most sensitive to the internal structure of volcanoes.

2 The 3-D finite-difference algorithm

Let some domain V in the earth be isotropic, nonmagnetic ($\mu = \mu_0$), and characterized by a 3-D distribution of the electrical conductivity $\sigma(x, y, z)$. At the periods used in MT sounding, the field satisfies the quasi-static Maxwellian equations:

$$\nabla \times \mathbf{H} = \sigma \mathbf{E}, \tag{1}$$

$$\nabla \times \mathbf{E} = i\omega\mu \mathbf{H}, \tag{2}$$

with harmonic time-dependence $\exp(-i\omega t)$. Combining these equations with the identity (take the divergence of the first equation),

$$\sigma \nabla \cdot \mathbf{E} = -\mathbf{E} \cdot \nabla \sigma,$$

gives the following equation for the electric field:

$$\nabla^2 \mathbf{E} + \nabla(\mathbf{E} \cdot \nabla\sigma/\sigma) + k^2 \mathbf{E} = 0, \tag{3}$$

where $k = (i\omega\mu_0\sigma)^{1/2}$; $\operatorname{Re} k > 0$. To determine the EM field in the domain V, first, the partial differential Eq. (3) for the electric field $\mathbf{E}$ is solved; then, the magnetic field $\mathbf{H}$ can be determined from Eq. (2).

Let the continuous vector function $\mathbf{E}$ be replaced by the discrete vector function $\mathbf{U}$, defined only at the nodes of a rectangular grid. Integrating Eq. (3) over the elementary volume in the vicinity of each node (l, m, n) gives

$$\oint_{S_{\ell,m,n}} \nabla \mathbf{U}_{\ell,m,n}\, ds + \oint_{S_{\ell,m,n}} \left(\mathbf{U}_{\ell,m,n} \frac{\nabla \sigma_{\ell,m,n}}{\sigma_{\ell,m,n}} \right) ds + \int_{V_{\ell,m,n}} k^2_{\ell,m,n} \mathbf{U}_{\ell,m,n}\, dv = 0 \tag{4}$$

($\ell = 1, 2, \ldots L; m = 1, 2, \ldots M; n = 1, 2, \ldots N$).

Approximating the derivatives of $\mathbf{U}$ and σ in Eq. (4) by finite differences and the integrals by the trapezoid rule gives the linear algebraic equations,

$$\begin{aligned} \mathbf{U}_{\ell,m,n} = \hat{D}^{(0)^{-1}}_{\ell,m,n} \big(& \hat{D}^{(1)}_{\ell,m,n} \mathbf{U}_{\ell,m,n-1} + \hat{D}^{(2)}_{\ell,m,n} \mathbf{U}_{\ell,m-1,n} + \hat{D}^{(3)}_{\ell,m,n} \mathbf{U}_{\ell-1,m,n} \\ & + \hat{D}^{(4)}_{\ell,m,n} \mathbf{U}_{\ell+1,m,n} + \hat{D}^{(5)}_{\ell,m,n} \mathbf{U}_{\ell,m+1,n} + \hat{D}^{(6)}_{\ell,m,n} \mathbf{U}_{\ell,m,n+1} \big) \end{aligned} \tag{5}$$

($\ell = 1, 2, \ldots L$; $m = 1, 2, \ldots M$; $n = 1, 2, \ldots N$), where $\hat{D}^{(i)} (i = 1, 2, \ldots 6)$ are 3×3 matrices determined by the grid geometry, the conductivity distribution, and the frequency.

For boundary conditions at the edges of the modeling domain, the following differential boundary condition is usually accurate, particularly for models with topography at the earth–air interface (Spichak, 1985):

$$\left(1 - ikr + r\frac{\partial}{\partial r} \right) (\mathbf{E} - \mathbf{E}^n) = 0, \tag{6}$$

where $\mathbf{E}^n$ is a normal electrical field, r is the distance to the points on the boundary of the modeling domain.

The resulting matrix of a system of linear equations has a block-banded shape and is very sparse. To solve it efficiently, a block overrelaxation technique is used:

$$\mathbf{U}^{(t+1)} = (1-\nu)\mathbf{U}^{(t)} + \nu\mathbf{U}^{(t+1/2)}, \quad (0 < \nu < 2), \tag{7}$$

where t is a number of the iteration. To minimize the total number of iterations, the relaxation ratio, ν, is automatically corrected during the iterations.

In principle, the magnetic field $\mathbf{H}$ can be calculated readily from the electric field by differentiating it using the finite-difference approximation of Eq. (2). In regions with large conductivity contrasts, however, this can generate large errors. In particular, it is practically impossible to correctly determine the horizontal components H_x and H_y at the Earth's surface because of numerical instability in derivatives $\partial E_x/\partial z$ and $\partial E_y/\partial z$. The horizontal components can be obtained accurately from the vertical component (whose finite-difference computation is stable) by the Hilbert transforms,

$$H_x(x_0, y_0, 0) = H_x^n - (2\pi)^{-1} \iint\limits_S \frac{H_z(x, y, 0)(x - x_0)}{r^3}\, dx\, dy, \tag{8}$$

$$H_y(x_0, y_0, 0) = H_y^n - (2\pi)^{-1} \iint\limits_S \frac{H_z(x, y, 0)(y - y_0)}{r^3}\, dx\, dy, \tag{9}$$

where $r = [(x - x_0)^2 + (y - y_0)^2]^{1/2}$; S is the Earth's surface (restricted in practice by the boundaries of the modeling domain).

The program package FDM3D-MT (Spichak, 1983b), which realizes the above algorithm, runs very efficiently on personal computers (a model with 32 000 grid nodes runs in 30–40 min on a IBM PC AT 486), requires modest core memory, and gives accurate results. The program allows models with a relief topography, mixed type of conductivity structure (1-D, 2-D, or 3-D regions in the same model), different 1-D layering at the boundaries of the model, and calculation of MT fields at arbitrary levels in the earth and atmosphere.

3 Imaging volcanic interiors

3.1 Model of the volcano

To study the MT response in volcanic environments, I used a 3-D geoelectrical model of a typical volcano of Hawaiian type, based on a model contructed in 1989 (with Prof. George Keller) originally to determine whether the MT method could detect internal processes in the Kilauea volcano that had been found in previous EM measurements in this area (Jackson and Keller, 1972).

The model represents a shield volcano, characterized by a low and flat summit formed by homogeneous basaltic rocks (Fig. 1). Its flanks stretch down into the ocean; the conductivity of the ocean water was taken to be 3.6 S/m. The volcano's summit, 0.5-km thick, is formed by basaltic lavas with conductivity $\sigma = 0.001$ S/m. There is a small layer, 0.8-km thick, with conductivity $\sigma = 0.01$ S/m at the boundary between the air and the ocean. Below are porous volcanic lavas, which are characterized by high content of salty water (this zone is 1.7 km thick and has a conductivity $\sigma = 0.17$ S/m). Then, at a 3-km depth from the volcano summit, there are dense lava formations 5.5-km thick with conductivity $\sigma = 0.01$ S/m, underlined by the crystalline crust with conductivity

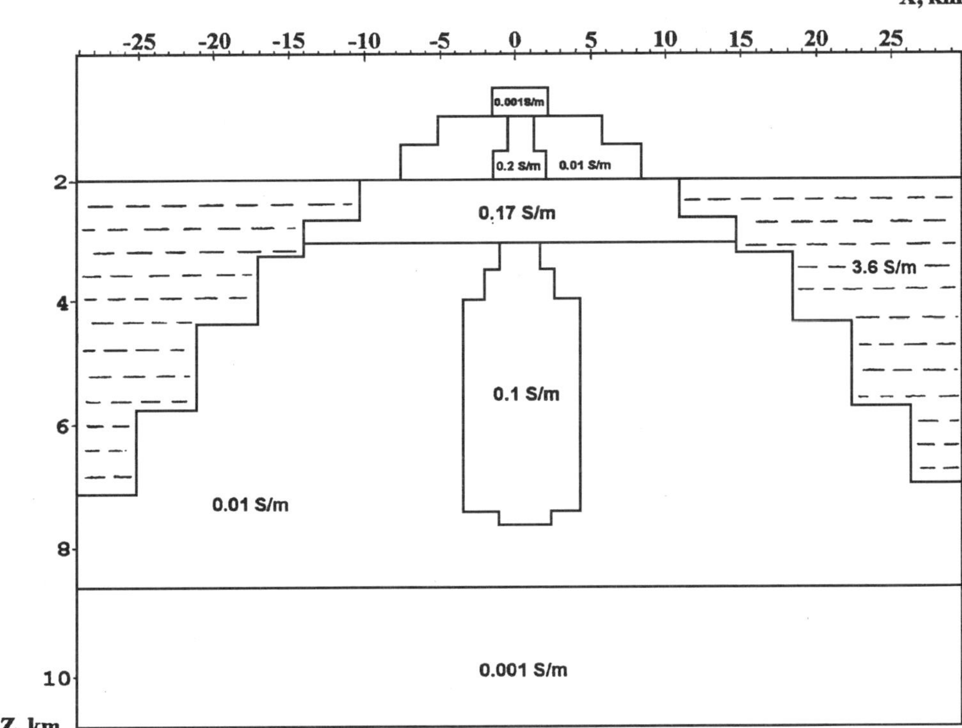

Figure 1. Model of Hawaiian-type volcano.

$\sigma = 0.001$ S/m. Other details of the geoelectrical structure are not important for this study.

The conductivity distribution in the model was considered to be symmetrical with two vertical planes of symmetry. So, only one-quarter of the 3-D grid was used for calculations.

3.2 Synthetic MT pseudosections

MT fields for this model were synthesized for two polarizations of the primary field at periods $T = 0.1, 1, 10$, and 100 s. Then, a number of the MT field transformations were calculated and analyzed at different levels in the atmosphere to find those that are most sensitive to the parameters of the model. The analysis used an algorithm for 3-D imaging of geolectrical structure, which uses transformations of the data recorded at the same level for several periods to create the isosurfaces in the system of coordinates $(X, Y, \log_{10} T)$.

For this model, direct estimation of the conductivity distribution from pseudosections of the apparent resistivity is very difficult. In contrast, 3-D isosurfaces or 2-D maps of the isolines of the transforms based on the impedance phases and on the in-phase and quadrature parts of the horizontal electric fields are the best for imaging the complicated geoelectrical structure of the volcano (Figs. 2–5).

In particular, Figs. 2 and 3 show the vertical cross-sections of the volcano overlapped by the maps of isolines of the transformed impedance phases (ϕ_{xy} and $\phi_{\det}$, correspondingly), constructed for the plane at a height 0.5 km above the summit of the volcano. Although the values attached to the isosurfaces have little to do with the

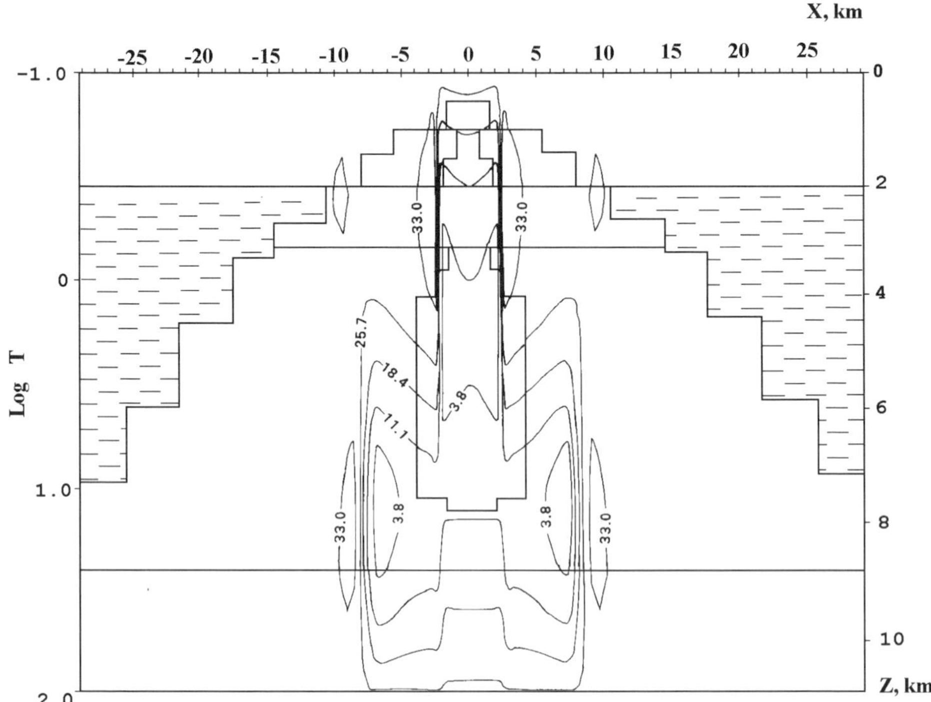

Figure 2. Pseudosection of $\phi_{xy}(\times 10^{-1})$ for volcano model shown in Fig. 1.

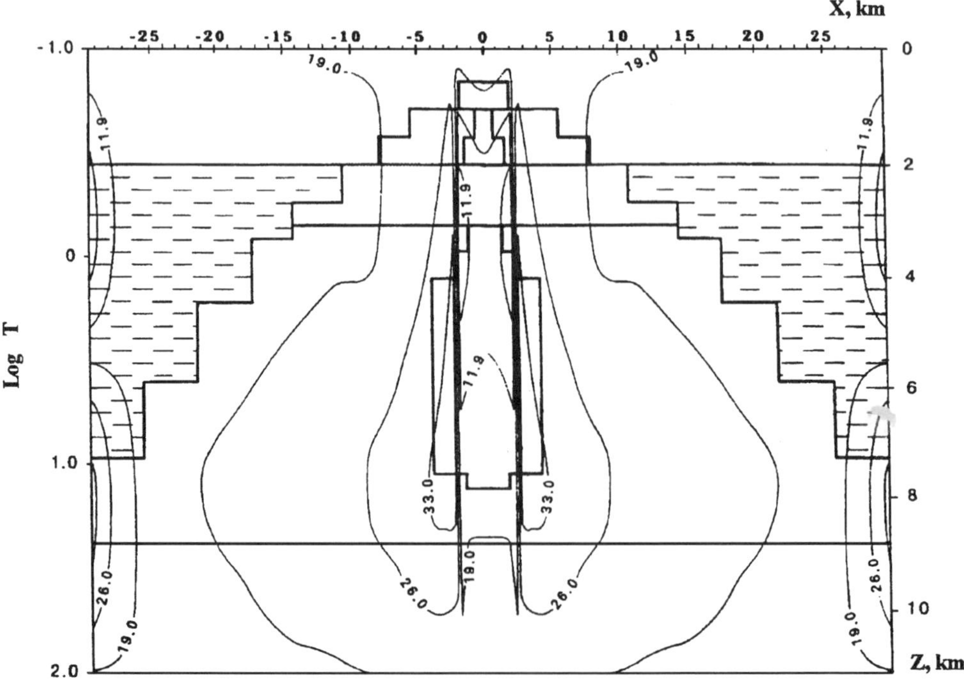

Figure 3. Pseudosection of $\phi_{\text{det}}(\times 10^{-1})$ for volcano model shown in Fig. 1.

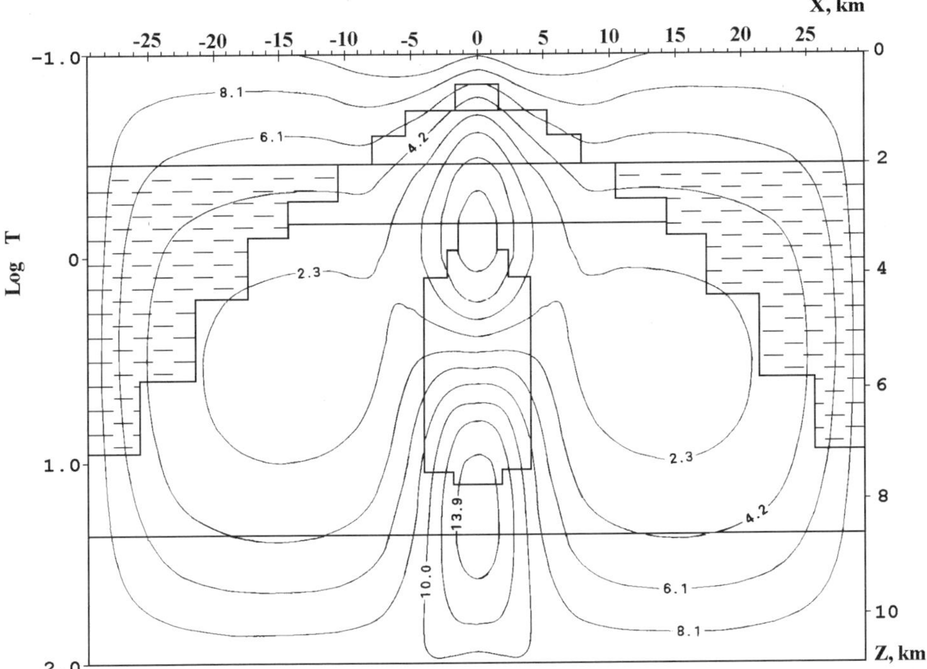

Figure 4. Pseudosection of Re $E_y(\times 10^1)$ for volcano model shown in Fig. 1. Result is based on an incident electrical field, parallel to y-axis.

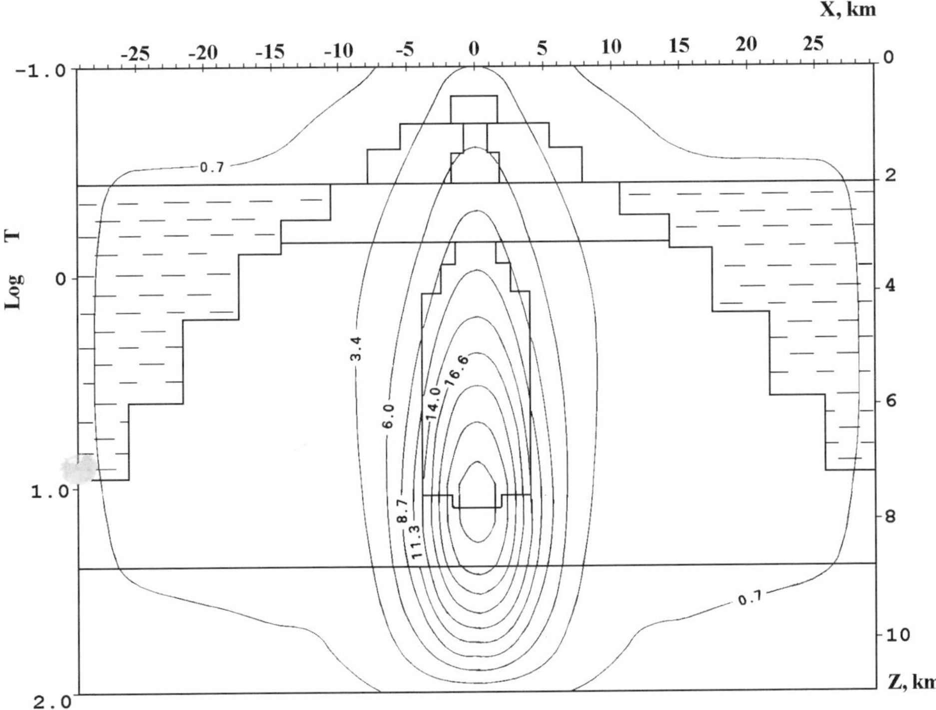

Figure 5. Pseudosection of Im $E_y(\times 10^1)$ for volcano model shown in Fig. 1. Result is based on an incident electrical field, parallel to y-axis.

actual conductivities, the spatial gradients clearly indicate the location of the magma chamber and of the conductive formation above it. The gradient of ϕ_{det} delineates not only the magma chamber but also the flanks of the volcano (Fig. 3). This result matches the findings of Park and Torres-Verdin (1988) and Zhdanov and Spichak (1992) from the interpretation of the phases in array MT data.

Transforms of the in-phase and quadrature parts of the horizontal electrical-field component parallel to the incident electrical field are even more sensitive to gradients of conductivity. Figure 4 shows the vertical cross-section of the model overlapped by the map of the isolines of the Re E_y transformation. The bunching of isolines correlates with gradients of the conductivity; the local extrema mark upper and lower edges of the magma chamber.

Isolines of the transformations of Im E_y are shown in Fig. 5. There is strong extremum located at the lower boundary of the magma chamber. No other features related to the model are visible.

Construction of the 3-D pseudostructures from transforms of the in-phase and quadrature parts of the horizontal electrical field and from the impedance phases appears to be a useful tool for delineating the geometric parameters of the complex volcanic environments. The analytical findings of Szarka and Fischer (1989), explaining the behavior of the MT field transformations at the Earth's surface in terms of the distribution of subsurface currents, support this conclusion.

Calculation of the quantities used above can be difficult in practice. One approach might be the analytical continuation of the EM fields, measured at the relief surface S, by means of the well-known integral transformation [see, e.g., Zhdanov (1988)],

$$\mathbf{F}(\mathbf{r}) = \iint_S \{(\mathbf{n} \cdot \mathbf{F})\nabla' G + [[\mathbf{n} \times \mathbf{F}] \times \nabla' G]\}\, ds' \tag{10}$$

where $\mathbf{F}$ could be $\mathbf{E}$ or $\mathbf{H}$; $G = 1/(4\pi|\mathbf{r} - \mathbf{r}'|)$ is a Green's function of a free space; surface S covers all sites, in which the measurements were made; and $\mathbf{n}$ is a unit normal to the surface pointing to the atmosphere.

Depending on the MT field components measured at the surface of the volcano, different integral transformations (Zhdanov, 1988) could be used to convert the field data into the appropriate horizontal electrical-field transform at the artificial plane. The numerical calculations required are rather straightforward and stable, and so, the use of this approach should not cause any problems.

Another important item to be mentioned concerns the selection of the appropriate height of the reference plane. From many experiments with the models of volcanoes, I found that the best height is 250–500 m above the summit. At lower heights, the pseudo-sections become distorted mostly by the nearest parts of the topography and by noise (natural and artificial), whereas at greater heights, details in geoelectrical structure may be lost. Choice of the ideal height for a given structure requires further investigations.

Conclusions

Synthetic MT fields and their transforms, calculated for a 3-D model of a volcano of Hawaiian type, indicate that the phases of the impedance and the in-phase and quadrature parts of the electrical-field components are the most sensitive to the structure (the latter being the most informative in TE mode).

Good qualitative images of the internal structure of the volcano were obtained by transforming the data on a reference plane located slightly higher than the summit of the volcano. The MT fields measured at the Earth's surface could be mapped to this plane by continuation upward. It also might be possible to measure the field components directly in the air, which would allow MT sounding of regions difficult to access by land.

Acknowledgments

The author is thankful to M-me Genevieve Roche and Mrs. Tilda Fleischhacker for kind assistance during preparation of this manuscript.

References

Fitterman, D. V., Stanley, W. D., and Bisdorf, R. J., 1988, Electrical structure of Newberry Volcano, Oregon: J. Geophys. Res., **93**, 10119–10134.

Jackson, D. B., and Keller, G. V., 1972, An electromagnetic sounding survey of the summit of Kilauea Volcano, Hawaii: J. Geophys. Res., **77**, 4957–4965.

Mogi, T., and Nakama, K., 1990, Three-dimensional geoelectrical structure of geothermal system in Kuju volcano and its interpretation: Geoth. Res. Council Trans., **14**, 1513–1515.

Moroz, Y. F., Kobzova, V. I., Moroz, I. P., and Senchina, A. F., 1988, Analogue modeling of MT-fields of volcano: Vulkanologiya i seismologiya, **3**, 98–104 (in Russian).

Newman, G. A., Wannamaker, P. E., and Hohmann, G. W., 1985, On the detectability of crustal magma chambers using the magnetotelluric method: Geophysics, **50**, 1136–1143.

Park, S. K., and Torres-Verdin, C., 1988, A systematic approach to the interpretation of magnetotelluric data in volcanic environments with applications to the quest for magma in Long Valley, California: J. Geophys. Res., **93**, 13265–13283.

Spichak, V. V., 1983a, Mathematical modeling of EM-fields in the 3D inhomogeneous media: Ph.D. thesis, IZMIRAN, 1983.

———1983b, The program package FDM3D for numerical modeling of the three-dimensional electromagnetic fields, *in* Algorithms and programs for solution of the direct and inverse problems of electromagnetic induction in the Earth, Ed. Zhdanov, M., IZMIRAN, 58–68 (in Russian).

———1985, Differential boundary conditions for the electrical and magnetic fields in the unbounded conducting medium, *in* Electromagnetic sounding of the Earth, Ed. Zhdanov, M., IZMIRAN, 13–22 (in Russian).

———1990, EM-field transformations and their use in interpretation: Surv. Geophys., **11**, 271–301.

Szarka, L., and Fischer, G., 1989, Electromagnetic parameters at the surface of conductive halfspace in terms of the subsurface current distribution: Geophys. Trans., **35**, 157–172.

Zhdanov, M. S., 1988, *Integral transforms in geophysics:* Springer–Verlag.

Zhdanov, M. S., and Spichak, V. V., 1989, Computer simulation of three-dimensional quasi-stationary electromagnetic fields in geoelectrics: Dokl. AN USSR, **309**, 57–60 (in Russian).

———1992, Mathematical modeling of electromagnetic fields in three-dimensional inhomogeneous media: Nauka Publ. (in Russian).

3-D Finite-Difference Modeling of the Magnetic Field in Geoelectromagnetic Induction

J. T. Weaver
A. K. Agarwal
X. H. Pu

Summary. A versatile finite-difference program has been developed for solving 3-D magnetotelluric problems in which the inducing field is uniform and horizontal. Finite-difference equations for the magnetic-field components are obtained by integrations over rectangular volume elements surrounding each node of the grid. This procedure avoids the ambiguity about whether to define weighted average conductivities or resistivities at the nodes of the numerical grid. The algorithm can handle an anomalous thin sheet at the surface of the general 3-D structure, thereby modeling both near-surface and deep-seated conductivity anomalies for certain ranges of period. The anomalous structure is allowed to approach different 2-D configurations at infinity in all four horizontal directions, with the result that both E-polarization and B-polarization solutions arise as boundary conditions on the grid. Integral boundary conditions on the surface of the Earth and at the top of the basement half-space (of uniform conductivity) reduce the overall size of the grid by eliminating the air and basement layers from the numerical solution. Recent developments include an automatic grid generator designed for geoelectromagnetic induction problems, the treatment of difficulties that can occur when the 2-D E-polarization problems on the side boundaries of the grid are solved in terms of the magnetic field, and a new method for calculating the electric-field components.

1 Introduction

Modeling of 3-D structures for the forward and inverse problems of magnetotellurics dates back to the early 1970s when Jones and Pascoe (1972), Lines and Jones (1973), and Jones (1974) first applied finite-difference methods to some simple 3-D models. Because computer storage was very limited in those days, the finite-difference method soon was superseded by the integral equation method, introduced into the field of geoelectromagnetic (EM) induction by Raiche (1974), Weidelt (1975), and Hohmann and coworkers (e.g., Hohmann, 1975; Ting and Hohmann, 1981; Wannamaker et al., 1984a;

Department of Physics & Astronomy and School of Earth & Ocean Sciences, University of Victoria, Victoria, British Columbia V8W 3P6, Canada.

Wannamaker et al., 1984b; San Filipo and Hohmann, 1985; Newman and Hohmann, 1988; Wannamaker, 1991) and by thin-sheet methods (Vasseur and Weidelt, 1977; Dawson and Weaver, 1979; McKirdy et al., 1985), which are applicable when the anomalous regions are confined to a thin (in an electrical sense) layer.

Powerful desktop workstations and new efficient methods for solving large systems of linear equations have made the finite-difference method competitive again. Although the usual approach is to solve for the electric field, there are several advantages in solving for the magnetic field. In many applications—geomagnetic depth sounding and induction vector mapping, for example—the magnetic field is of principal interest. Moreover, it is divergence-free; it satisfies well-known boundary conditions at the surface of the Earth which can eliminate the air layer above the Earth from further consideration; and it is continuous across conductivity boundaries. Problems arise, however, if the 3-D model is allowed to approach 2-D configurations at infinity in all four horizontal directions. The boundary values then are given by E-polarization solutions on one pair of boundaries, and obtaining these solutions from magnetic-field components is difficult, because in the dc limit in two dimensions, the magnetic field is arbitrary to the extent of the gradient of a scalar function (Pu et al., 1993; Weaver et al., 1996).

The use of staggered grids in 3-D modeling alleviates many of these problems (Mackie et al., 1993; Wang and Hohmann, 1993; Mackie et al., 1994; Smith, 1996). Staggered grids, however, are less convenient than traditional fixed grids when computing quantities, such as induction vectors or apparent resistivities, that involve the ratios of field components, because each field component is defined at a different node on the grid. Moreover, only the horizontal components of the magnetic field are given directly on the surface of the Earth when a staggered grid is used. Weaver et al. (1996) compare fixed and staggered grids in 2-D modeling. We continue to use a fixed grid despite the numerical obstacles it presents, but our method nevertheless is adapted readily to a staggered grid, and we intend to make the necessary modifications in due course.

2 Basic equations

Let the Earth of conductivity σ and free-space permeability μ_0 occupy the half-space $z > 0$ in a right-handed Cartesian coordinate system $(x, y, z) \equiv (\mathbf{r}, z)$ where $\mathbf{r}$ is the position vector in a horizontal plane. Unit vectors along the axes are denoted by $\hat{\mathbf{x}}$, $\hat{\mathbf{y}}$, and $\hat{\mathbf{z}}$, respectively. The magnetic- and electric-field vectors are written as

$$\mathbf{B} = (\mathbf{b}, Z) = (X, Y, Z), \quad \mathbf{E} = (\mathbf{e}, W) = (U, V, W)$$

in component form, where $\mathbf{b}$ and $\mathbf{e}$ are the horizontal (vector) components of the respective fields. Each component, as well as the conductivity σ (or resistivity $\rho = 1/\sigma$), is a function of the three space coordinates (x, y, z). With a harmonic time dependence $e^{i\omega t}$ understood, the induction equations governing the field are

$$\operatorname{curl} \mathbf{E} = -i\omega\mathbf{B}, \quad \rho \operatorname{curl} \mathbf{B} = \mu_0\mathbf{E}, \tag{1}$$

or, on elimination of $\mathbf{E}$,

$$\operatorname{curl} (\rho \operatorname{curl} \mathbf{B}) + i\omega\mu_0\mathbf{B} = 0. \tag{2}$$

Let the anomalous region of the Earth be confined to the region $0 < z < d$; the region $z < 0$ is the air layer of vanishing conductivity in which the external inducing

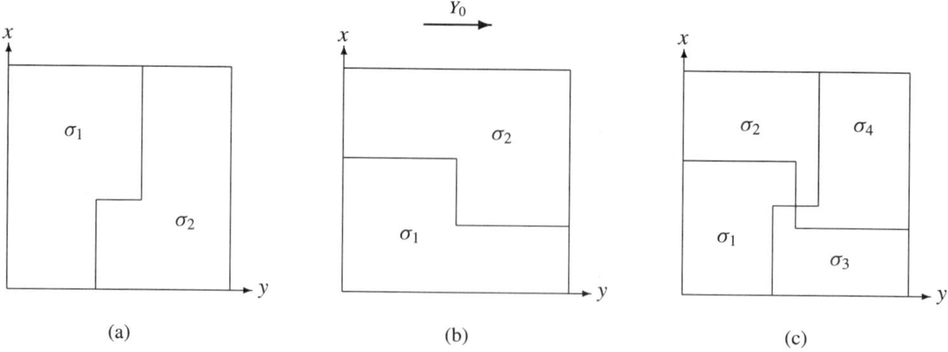

Figure 1. Plan view of 3-D models that approach (a) E-polarization, (b) B-polarization, and (c) both polarizations on the boundaries at infinity.

source is located; and we assume that the region $z > d$ is a uniform half-space of conductivity $\sigma_0 = 1/\rho_0$. Let the inducing magnetic field be aligned along the y-direction so that

$$\mathbf{B} \to Y_0\hat{\mathbf{y}}, \quad \text{as } z \to -\infty,$$

where Y_0 is the uniform regional (or normal) field above a 1-D normal structure at infinity. Figure 1 depicts schematically three possible configurations of the 3-D model at infinity when the anomalous region is unbounded. In Fig. 1a, the model becomes completely 1-D as $|y| \to \infty$, but a 2-D configuration in the E-polarization mode is approached as $|x| \to \infty$; in Fig. 1b, it is the structure at $x = \pm\infty$ that is strictly 1-D, whereas a 2-D B-polarization problem is approached as $|y| \to \infty$ (for which, incidentally, the external field is Y_0 everywhere); and in Fig. 1c, 2-D configurations are approached in both directions, giving rise to respective E- and B-polarization problems at infinity.

It is well known (Weaver, 1964, 1994) that the horizontal components of the magnetic field at the surface of the Earth are related to the vertical component by the integral formulas

$$X(x, y, 0-) = \frac{1}{2\pi}\int_{-\infty}^{\infty}\int_{-\infty}^{\infty} \frac{(u-x)Z(u, v, 0)}{[(x-u)^2 + (y-v)^2]^{3/2}}\,du\,dv$$

$$Y(x, y, 0-) = Y_0 + \frac{1}{2\pi}\int_{-\infty}^{\infty}\int_{-\infty}^{\infty} \frac{(v-y)Z(u, v, 0)}{[(x-u)^2 + (y-v)^2]^{3/2}}\,du\,dv$$

which can be combined into the single vector boundary condition

$$\mathbf{b}(\mathbf{r}, 0-) = Y_0\hat{\mathbf{y}} + \frac{1}{2\pi}\int_{-\infty}^{\infty} \frac{(\mathbf{s}-\mathbf{r})Z(\mathbf{s}, 0)}{|\mathbf{s}-\mathbf{r}|^3}\,d\mathbf{s} \tag{3}$$

where $\mathbf{s} = u\hat{\mathbf{x}} + v\hat{\mathbf{y}}$. At the surface of the basement half-space, the boundary condition

$$\mathbf{b}'(\mathbf{r}, d+) + \alpha_0\sqrt{i}\,\mathbf{b}(\mathbf{r}, d) = \frac{1}{2\pi}\int_{-\infty}^{\infty} [\mathbf{b}(\mathbf{s}, d) - \mathbf{b}(\mathbf{r}, d)]Q(|\mathbf{s}-\mathbf{r}|)\,d\mathbf{s} \tag{4}$$

holds (Weaver, 1994), where we have introduced the notation $F'(\mathbf{r}, z) \equiv \partial F(\mathbf{r}, z)/\partial z$, and have defined

$$Q(r) = r^{-3}(1 + r\alpha_0\sqrt{i})\exp(-r\alpha_0\sqrt{i}), \quad \alpha_0 = \sqrt{\omega\mu_0\sigma_0}. \tag{5}$$

The advantage of using boundary conditions (3) and (4) is that they confine the region of numerical solution to the anomalous region between the planes $z = 0$ and $z = d$ and hence permit the use of a smaller grid than would otherwise be needed.

Finally, in models that include surface inhomogeneities, such as a sedimentary basin or an ocean coastline, it is convenient to include a thin anomalous sheet at the surface of the model without using a dense grid. If the sheet has a conductance $\tau(\mathbf{r})$ and is regarded as infinitesimally thin, then the boundary condition relating the surface field to the field on the underside of the sheet at $z = 0+$ is

$$\mathbf{b}(\mathbf{r}, 0+) - \mathbf{b}(\mathbf{r}, 0-) = -\mu_0 \tau(\mathbf{r})\hat{\mathbf{z}} \times \mathbf{e}(\mathbf{r}, 0). \tag{6}$$

It simply states that the discontinuity in the horizontal magnetic field across the sheet is proportional to the density of the surface current flowing in the sheet.

3 Finite-difference equations

A 3-D grid is assumed to cover the region $x_1 < x < x_L$, $y_1 < y < y_M$, $0 \equiv z_1 < z < z_N$, where z_N is located in the uniform half-space of resistivity ρ_0, and where x_1, x_L, y_1, y_M are chosen to be sufficiently far removed from 3-D variations in conductivity that the field can be assumed to be essentially 2-D on the horizontal boundaries of the grid, $x = x_1$, $x = x_L$, $y = y_1$ and $y = y_M$. Node (ℓ, m, n) in the 3-D grid is located at (x_ℓ, y_m, z_n); the node separations in the x-, y-, and z-directions are, respectively, $g_\ell = x_{\ell+1} - x_\ell$ $(\ell = 1, 2, \ldots L - 1)$, $h_m = y_{m+1} - y_m$ $(m = 1, 2, \ldots M - 1)$ and $k_n = z_{n+1} - z_n$ $(n = 1, 2, \ldots N - 1)$. The resistivity assigned to the cell whose center is at $(x_\ell + g_\ell/2, y_m + h_m/2, z_n + k_n/2)$ is denoted by $\rho_{\ell+1/2, m+1/2, n+1/2}$. In particular, we note that, because the bottom array of cells in the grid lies entirely within the underlying uniform half-space, $\rho_{l+1/2, m+1/2, N-1/2} = \rho_0$.

A less cumbersome subscript notation for the local problem around a particular node is illustrated in Fig. 2. If F is any field component, then its values at the central and neighboring nodes in the local problem are denoted by F_{111}, F_{011}, and F_{211}, respectively, parallel to the x-axis, and similarly in the other two directions. Node separations are g_2, h_2, and k_2 on the positive side of the central node in the x-, y-, and z-directions, respectively, and g_0 etc. on the negative side. Resistivities also are assigned numerical

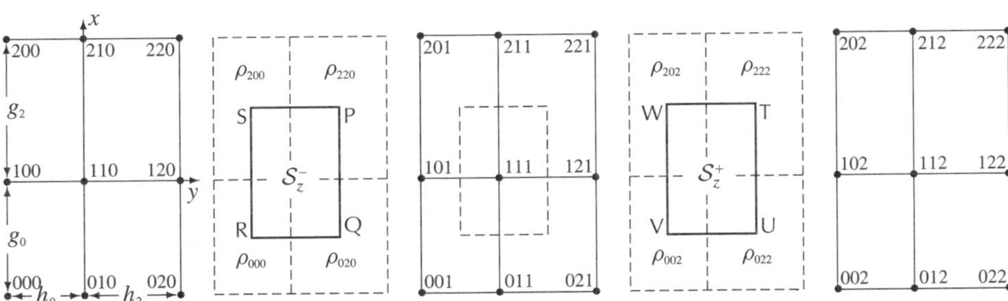

Figure 2. Local region of 3-D grid centered on node 111 at some depth $z = z_n$. The z-axis is directed downward into the plane of the paper. The diagrams depict successively, from left to right, horizontal planes at depths $z_n - k_0$, $z_n - \frac{1}{2}k_0$, z_n, $z_n + \frac{1}{2}k_2$, and $z_n + k_2$. The vertices of the volume element $\mathcal{V}$ surrounding the central node are labeled PQRSTUVW; its bottom and top surfaces in the planes $z = z_n + \frac{1}{2}k_2$ and $z = z_n - \frac{1}{2}k_0$ are marked as $\mathcal{S}_z^+$ and $\mathcal{S}_z^-$, respectively; and its rectangular cross-section in the plane $z = z_n$ is indicated by the broken outline in the middle diagram. (Weaver, 1994, p. 275.)

subscripts of 2 or 0 depending on whether they belong to a cell on the positive or negative side of the central node in the dimension to which the subscript refers. This notation is more compact and leaves subscript 1 available to denote the weighted mean resistivity on the boundary between two cells. For example, we write

$$\rho_{012} = (h_0\rho_{002} + h_2\rho_{022})/h^+, \quad \rho_{011} = (k_0\rho_{010} + k_2\rho_{012})/k^+, \tag{7}$$

where $h^+ = h_0 + h_2$, and similarly for k and g.

Finite-difference equations for the magnetic field are derived by integrating Eq. (2) over the volume element whose vertices are PQRSTUVW in Fig. 2 and applying special cases of the divergence theorem. In the special case $\omega = 0$, however, Eq. (2) involves only the curl of $\mathbf{B}$; the extra condition div $\mathbf{B} = 0$, which is required to define $\mathbf{B}$ uniquely, is not specified in Eq. (2). It follows that a near indeterminacy may arise for small frequencies with a resulting instability in numerical calculations. It is therefore not only possible, but also appropriate, to augment Eq. (2) with the vanishing term grad (ρ div $\mathbf{B}$) as follows:

$$\text{grad}\,(\rho\,\text{div}\,\mathbf{B}) - \text{curl}\,(\rho\,\text{curl}\,\mathbf{B}) = i\omega\mu_0\mathbf{B}. \tag{8}$$

Let $\mathcal{V}$ represent the volume element PQRSTUVW of dimensions $g^+/2$, $h^+/2$, and $k^+/2$ enclosing the central node in Fig. 2 and let $\mathcal{S}$ be its surface area. With $\hat{\mathbf{n}}$ denoting a unit vector along the outward normal from the element of area $d\mathcal{S}$, special manifestations of the divergence theorem can be obtained (Jones, 1964, p. 112; Weaver, 1994, p. 275) in the form

$$\iiint_{\mathcal{V}} \text{grad}\,\psi\, d\mathcal{V} = \iint_{\mathcal{S}} \psi\hat{\mathbf{n}}\, d\mathcal{S}, \quad \iiint_{\mathcal{V}} \text{curl}\,\mathbf{C}\, d\mathcal{V} = \iint_{\mathcal{S}} \hat{\mathbf{n}} \times \mathbf{C}\, d\mathcal{S} \tag{9}$$

which applied to the volume integral of equation (8) yield

$$\iint_{\mathcal{S}} \rho(\hat{\mathbf{n}}\,\text{div}\,\mathbf{B} - \hat{\mathbf{n}} \times \text{curl}\,\mathbf{B})\, d\mathcal{S} = i\omega\mu_0 \iiint_{\mathcal{V}} \mathbf{B}\, d\mathcal{V}$$

whose z-component is

$$\left(\iint_{\mathcal{S}_z^+} - \iint_{\mathcal{S}_z^-}\right)\rho\left[\frac{\partial X}{\partial x} + \frac{\partial Y}{\partial y} + \frac{\partial Z}{\partial z}\right]dx\,dy + \left(\iint_{\mathcal{S}_x^+} - \iint_{\mathcal{S}_x^-}\right)\rho\left[\frac{\partial Z}{\partial x} - \frac{\partial X}{\partial z}\right]dy\,dz$$
$$+ \left(\iint_{\mathcal{S}_y^+} - \iint_{\mathcal{S}_y^-}\right)\rho\left[\frac{\partial Z}{\partial y} - \frac{\partial Y}{\partial z}\right]dz\,dx = i\omega\mu_0 \iiint_{\mathcal{V}} Z\,dx\,dy\,dz \tag{10}$$

where we have written, for example, $\mathcal{S}_z^+$ and $\mathcal{S}_x^-$ to denote the respective faces of $\mathcal{V}$ whose outward normals are in the directions of $\hat{\mathbf{z}}$ and $-\hat{\mathbf{x}}$, i.e., TUVW and UVRQ in Fig. 2. The extension of this notation to the other faces of $\mathcal{V}$ is obvious.

Replacement of the terms involving the field components in the integrands by suitable finite-difference expressions enables the integrals to be evaluated in discrete form. Thus, in the volume integral on the right-hand side of Eq. (10), we represent the field component Z by its nodal value Z_{111} throughout $\mathcal{V}$. Hence,

$$i\omega\mu_0 \iiint_{\mathcal{V}} Z\,dx\,dy\,dz \approx \frac{1}{8} i\omega\mu_0 g^+h^+k^+Z_{111}.$$

There are eight cells depicted in Fig. 2. In the cell whose corners are the nodes 222, 122, 112, 212, 221, 121, 111, and 211, (i.e., the one bounded above and below by the upper right quadrants in the third and fifth diagrams of Fig. 2), the derivatives of any field component F are represented by the expressions

$$\begin{aligned}\partial F/\partial x &\approx \tfrac{1}{4}(F_{211} - F_{111} + F_{221} - F_{121} + F_{212} - F_{112} + F_{222} - F_{122})/g_2 \\ \partial F/\partial y &\approx \tfrac{1}{4}(F_{121} - F_{111} + F_{221} - F_{211} + F_{122} - F_{112} + F_{222} - F_{212})/h_2 \\ \partial F/\partial z &\approx \tfrac{1}{4}(F_{112} - F_{111} + F_{122} - F_{121} + F_{212} - F_{211} + F_{222} - F_{221})/k_2\end{aligned} \tag{11}$$

which are the averages of the central differences along the four sides of the cell lying parallel to the direction in which the derivative is taken. Similar expressions for the derivatives in the other seven cells can be written down immediately by analogy. Each surface integral on the left-hand side of Eq. (10) embraces four of the cells so that the derivatives appearing in their integrands will have different representations over the four subsets of the total surface area that lie in the different cells. Making the appropriate substitutions, collecting all terms, and simplifying by introducing the weighted average resistivities, we obtain a 27-point formula of the form

$$i\omega\mu_0 g^+ h^+ k^+ Z_{111} = \sum_{i=0}^{2}\sum_{j=0}^{2}\sum_{k=0}^{2}(a_{ijk}X_{ijk} + b_{ijk}Y_{ijk} + c_{ijk}Z_{ijk}). \tag{12}$$

The coefficients a, b, and c are too complicated and numerous to list in this short paper. Corresponding equations for Y_{111} and X_{111} are readily obtained from Eq. (12) by making the respective transformations

$$Z \to Y, \quad Y \to X, \quad X \to Z, \quad k \to h, \quad h \to g, \quad g \to k$$

while rotating the subscripts one place backward, and

$$Z \to X, \quad X \to Y, \quad Y \to Z, \quad k \to g, \quad g \to h, \quad h \to k$$

while rotating the subscripts one place forward.

There are $3(L-2)(M-2)(N-2)$ equations of type (12), one for each component at the interior nodes. Note that they differ in several aspects from those given by Weaver (1994). Because the fields are assumed to be strictly 2-D on the horizontal boundaries, we must have $X = 0$ and $Z = 0$ along the vertical corner edges of the grid (x_1, y_1), (x_1, y_M), (x_L, y_1), and (x_L, y_M), and the resulting 1-D solution for Y can be easily generated analytically from its surface value Y_0. The 2-D E-polarization solutions for the magnetic-field components Y and Z then must be calculated on the sides $x = x_1$ and $x = x_L$. This can either be done as a separate exercise to generate fixed boundary values or one can incorporate the degenerate form of Eq. (12) and the equivalent equation for Y (obtained, for example, by putting $F_{0jk} = F_{1jk}$ if one wants the equations on $x = x_1$) directly into the general iteration scheme. Finally, the B-polarization problems arising on $y = y_1$ and $y = y_M$ must be solved to determine the remaining side boundary values. This poses no difficulty at all because the magnetic field is the natural vehicle for obtaining B-polarization solutions.

There remain the boundary conditions on the surface and base of the grid. Eq. (3) evaluated at the surface node $(\ell, m, 1)$ $(2 \le \ell \le L-1;\ 2 \le m \le M-1)$ gives

$$X_{\ell,m,1} = \frac{1}{2\pi}\int_{y_1}^{y_M}\int_{-\infty}^{\infty}\frac{(u - x_\ell)Z(u, v, 0)}{[(x_\ell - u)^2 + (y_m - v)^2]^{3/2}}\,du\,dv \tag{13}$$

$$Y_{\ell,m,1} = Y_0 + \frac{1}{2\pi}\int_{y_1}^{y_M}\int_{-\infty}^{\infty}\frac{(v - y_m)Z(u, v, 0)}{[(x_\ell - u)^2 + (y_m - v)^2]^{3/2}}\,du\,dv\,. \tag{14}$$

The integrals over v vanish beyond the edges of the grid because $Z = 0$ in the regions $y \le y_1$ and $y \ge y_M$ where the model reduces to two dimensions in the B-polarization mode. On the edge $x = x_L$, equations (13) and (14) are replaced by the E-polarization conditions,

$$X_{L,m,1} = 0, \quad Y_{L,m,1} = Y_0 + \frac{1}{\pi} \fint_{y_1}^{y_M} \frac{Z(x_L, v, 0)}{v - y_m} dv$$

and, if we replace L by 1 we obtain the analogous conditions on $x = x_1$. The double integrals (13) and (14) are discretized by expanding Z as a bilinear function over each cell in the surface plane, and by putting $Z(x, y, 0) = Z_{L,m+1,1} + (y_{m+1} - y)(Z_{L,m,1} - Z_{L,m+1,1})/h_m$ for $x > x_L$, $y_m < y < y_{m+1}$ $(m = 1, 2, \ldots, M - 1)$, and similarly for $x < x_1$ with $L = 1$. The resulting coefficients of the values $Z_{\ell,m,1}$ at the surface nodes are given in terms of double integrals, all of which can be reduced to just four basic types, and are evaluated analytically as combinations of elementary algebraic and logarithmic functions which are, nevertheless, too complicated to reproduce here.

One other equation is required at the surface nodes. We have found a simple statement that $\operatorname{div}\mathbf{B} = 0$ to be effective. Thus, using the finite-difference representations (11) for the derivatives, and applying $\operatorname{div}\mathbf{B} = 0$ in the cell whose corners are $(\ell, m, 1)$, $(\ell, m, 2)$, $(\ell+1, m, 1)$, $(\ell+1, m, 2)$, $(\ell, m+1, 1)$, $(\ell, m+1, 2)$, $(\ell+1, m+1, 1)$, and $(\ell+1, m+1, 2)$, we can state that

$$\begin{aligned}(X_{\ell+1,m+1,1} &- X_{\ell,m+1,1} + X_{\ell+1,m,1} - X_{\ell,m,1} + X_{\ell+1,m+1,2} - X_{\ell,m+1,2} \\ &+ X_{\ell+1,m,2} - X_{\ell,m,2})/g_\ell + (Y_{\ell+1,m+1,1} - Y_{\ell+1,m,1} + Y_{\ell,m+1,1} - Y_{\ell,m,1} \\ &+ Y_{\ell+1,m+1,2} - Y_{\ell+1,m,2} + Y_{\ell,m+1,2} - Y_{\ell,m,2})/h_m + (Z_{\ell+1,m+1,2} - Z_{\ell+1,m+1,1} \\ &+ Z_{\ell+1,m,2} - Z_{\ell+1,m,1} + Z_{\ell,m+1,2} - Z_{\ell,m+1,1} + Z_{\ell,m,2} - Z_{\ell,m,1})/k_2 = 0.\end{aligned}$$

This equation can be regarded as defining $Z_{\ell,m,1}$ in terms of the other components, but it may bias the solution in favor of field components in the direction of increasing x and y with respect to the node $(\ell, m, 1)$. In practice, therefore, we defined $Z_{\ell,m,1}$ by the average of the four such equations in the four cells surrounding the node $(\ell, m, 1)$.

Similar equations in which the third subscripts 1 and 2 are replaced by $N - 1$ and N, respectively, and k_2 is replaced by k_{N-1}, can be used to define the boundary values $Z_{\ell,m,N}$. It only remains, therefore, to obtain the discrete form of Eq. (4) (with z_N replacing d) at the nodes (ℓ, m, N) $(\ell = 2, 3, \ldots, L - 1; m = 2, 3, \ldots, M - 1)$ on the base of the grid to complete the system of equations to be solved for the nodal values of the magnetic-field components. The integrals are handled in the same manner as Eqs. (13) and (14) but decay much faster at infinity because of the presence of exponential terms in the integrand. It is no longer possible, however, to calculate all of the coefficients analytically; in most cases, one of the repeated integrals has to be evaluated numerically. Because the grid plane $z = z_N$ lies entirely within the uniform region, $\mathbf{b}'$ is continuous across $z = z_N$ and we can write equation (4) at the node (ℓ, m, N) in the discrete form

$$\mathbf{b}'_{\ell,m,N} + \alpha_0\sqrt{i}\,\mathbf{b}_{\ell,m,N} = \frac{1}{2\pi}\int_{-\infty}^{\infty} [\mathbf{b}(\mathbf{s}, z_N) - \mathbf{b}_{\ell,m,N}]Q(|\mathbf{s} - \mathbf{r}_{\ell,m}|)\,d\mathbf{s}$$

where $\mathbf{r}_{\ell,m} = x_\ell\hat{\mathbf{x}} + y_m\hat{\mathbf{y}}$. The final step is to express the z-derivative in finite-difference form. This is done by expanding $\mathbf{b}$ in a Taylor series upward from the node (ℓ, m, N) to second order and eliminating the second derivative with the aid of Eq. (2) and the fact that $\rho \equiv \rho_0$ is constant just above the plane $z = z_N$. Thus we have

$$\mathbf{b}'_{\ell,m,N} = \frac{\mathbf{b}_{\ell,m,N} - \mathbf{b}_{\ell,m,N-1}}{k_{N-1}} + \frac{k_{N-1}}{2}\left[\frac{i\omega\mu_0}{\rho_{\ell,m,N}} - \frac{\partial^2}{\partial x^2} - \frac{\partial^2}{\partial y^2}\right]\mathbf{b}_{\ell,m,N}. \tag{15}$$

Through this equation, $\mathbf{b}'$ is expressed entirely in terms of horizontal derivatives which can be represented in terms of the nodal values of the field components by standard finite-difference formulas.

The boundary equations applicable at the nodes on the base of the grid are clearly not simple and their inclusion in the system may not be warranted if it is feasible to extend the grid (without making it impossibly large) to sufficient depth in the half-space that the elementary boundary conditions $X_{\ell,m,N} = Y_{\ell,m,N} = Z_{\ell,m,N} = 0$ are applicable. Alternatively, simple one-sided differences could be used to represent the vertical derivatives on the base of the grid.

4 Additional features

4.1 Thin sheet at the surface

If the geoelectric model under investigation includes near-surface anomalies as well as deep-lying structures, it may be convenient to represent the surface anomalies by a thin sheet of variable conductance in the plane $z = 0$. The conditions under which this approximation holds are that the thickness of the surface layer be small compared with both the skin depth in the material in the layer and with the inductive scale length of the underlying structure. In a thin-sheet model, there are two nodes at the point (x_l, y_m), namely, the principal node $(\ell, m, 1)$ on the upper surface $z = 0-$ of the sheet (the actual surface of the Earth), and a secondary node $(\ell, m, 1+)$ on the underside $z = 0+$ of the sheet. The boundary conditions (13) and (14) continue to be valid at $(\ell, m, 1)$, but the 27-point formulas of type (12) at nodes $(\ell, m, 2)$ now include the components $X_{\ell,m,1+}$ and $Y_{\ell,m,1+}$ rather than $X_{\ell,m,1}$ and $Y_{\ell,m,1}$ (note that $Z_{\ell,m,1} = Z_{\ell,m,1+}$ because Z is continuous across the sheet), and similarly with the equation defining $Z_{\ell,m,1}$ itself for which we used the average of div $\mathbf{B} = 0$ in the four cells adjoining the node $(\ell, m, 1+)$ as explained in Section 3. The connecting relation is given by Eq. (6) which, with the aid of Eq. (1), now reads

$$\mathbf{b}_{\ell,m,1+} = \mathbf{b}_{\ell,m.1} - \tau_{\ell,m}\rho_{\ell,m,1}\left[(\partial Z/\partial \mathbf{r})_{\ell,m,1} - \mathbf{b}'_{\ell,m,1+}\right] \tag{16}$$

where $\rho_{\ell,m,1} \equiv \rho_{\ell,m,\frac{3}{2}}$. The vertical derivative $\mathbf{b}'_{\ell,m,1+}$ can be expressed in terms of $\mathbf{b}_{\ell,m,2}$ and horizontal derivatives at $(\ell, m, 1+)$ by expanding $\mathbf{b}$ in a Maclaurin series to second order in z and evaluating it at $z = k_1$. This process leads to appropriate modifications of Eq. (15). Thus, the presence of the thin sheet is equivalent to introducing $2(L-2)(M-2)$ new variables $X_{\ell,m,1+}$ and $Y_{\ell,m,1+}$ ($\ell = 2, 3, \ldots, L-1; m = 2, 3, \ldots, M-1$) and the same number of new equations of type (16).

4.2 Automatic gridding

Manual grid design is time-consuming, error-prone, and tedious. Moreover, when several forward calculations for different periods are performed with the same fixed conductivity model, the grid must be modified for each new period—drastically so, if the periods are orders of magnitude apart. Clearly, this is a major drawback if the forward modeling program is used in an inversion scheme in which many forward calculations over a range of periods have to be made. Some help is provided by the application of the surface and basement integral boundary conditions described in Section 3, because they eliminate the need to decide how high and deep the grid should extend into the air layer and basement half-space, respectively, but design of the main body of the grid remains a major task.

If the grid is carelessly designed, two independent users can obtain quite different outputs from the same modeling program when computing the response of identical conductivity structures. This user dependence of computed responses (as opposed to program dependence) is known to occur in practice and often can be attributed to poor grid design, a common mistake being the failure to change the grid when calculations are performed on the same model but for a wide range of periods. In fact, node spacings and the overall size of the grid are parameters that are very much dependent on the period of the field variations because the dimensions of the grid cells must be kept small compared with the skin depth in the local material and the edges of the grid must be well removed from 3-D anomalies. Otherwise, the various finite-difference approximations made in Section 3 will cease to be valid, and the assumption of two-dimensionality on the boundaries will break down.

The following guidelines have been found helpful in designing a satisfactory grid (Weaver, 1994, p. 193).

1. Within two skin depths of a boundary separating regions of different conductivities, the node separation should be no more than (and, very close to the boundary, preferably much less than) one-quarter of a skin depth. This condition can be relaxed at greater distances where the field gradient is smaller.
2. The vertical sides of the grid should be placed at least two to three skin depths beyond the nearest conductivity gradient perpendicular to the plane of the side in question. Here, skin depth refers to the skin depth in the most resistive region of the 2-D structure on the relevant vertical side. This condition is difficult to satisfy in 3-D modeling if the structure is complicated because available computing power often places a severe restriction on the maximum number of grid points allowed.
3. The separations of adjacent nodes should be kept as nearly equal as possible near conductivity boundaries, and preferably exactly equal across the boundaries themselves. When a rapid change in the node separation from one part of the model to another is required, the transition should be made as smoothly as possible, preferably in regions where field gradients are expected to be small and with no more than a doubling (or halving) of adjacent node separations.

Grid design is clearly an important part of forward modeling. Ideally, therefore, all modeling programs should include an automatic grid generator that eliminates the types of error referred to above, while also relieving the program user of a very time-consuming and tedious task. In designing an automatic grid generator for the 3-D modeling method described in this paper, we have relied heavily on a code written by Poll (1994), which designs a grid for 2-D finite-difference modeling that meets all of the criteria listed in the preceding paragraph and also allows for nodes to be inserted at prescribed station sites on the surface of the Earth. The user is required only to visualize a rectangular parallelepiped or box extending laterally into the 2-D regions on the four sides of the model and vertically into the uniform basement so

Table 1. Data input for 3-D model illustrated in Fig. 3.

x	y	z	Δx	Δy	Δz	ρ
−100	−100	0	100	200	2	10
0	−100	0	100	100	2	1
0	0	0	100	100	2	10
−100	−100	2	200	200	3	100
−100	−100	5	80	200	25	100
−20	−100	5	40	80	25	100
−20	−20	5	40	40	25	1
−20	20	5	40	80	25	100
20	−100	5	80	200	25	100
−100	−100	30	200	200	10	1

that it completely encloses the anomalous structure, and to subdivide the structure inside this box into the least possible number of distinct conductive blocks of various sizes. The input data that uniquely describe the model then are entered successively as: (a) the coordinates (x, y, z) of that vertex of the first block that has the minimum coordinate values among the four vertices; (b) the dimensions Δx, Δy, and Δz of the block; (c) the resistivity of the block; and in (d), steps (a), (b), and (c) are repeated for each separate conductive block comprising the model. For example, suitable input data for the 3-D model shown in Fig. 3 are those given in Table 1. The imaginary box enclosing the structure has been chosen to have dimensions of $200 \times 200 \times 40$ km in this example. The automatic gridder first maps the edges of all of the blocks in the three coordinate directions, and then the numerical grid is constructed in such a way that the node spacing near conductivity boundaries is always less than one-quarter of a skin depth in the local material and that the transitions between different node spacings are kept smooth according to criterion (3), above. The routine also takes care of the placement of the side boundaries of the grid (which are not necessarily coincident with the edges of the imaginary box) and either locates the depth d of the uniform basement ready for an application of integral boundary conditions there, or extends the grid to sufficient depth in the underlying half-space if the condition of vanishing field is used instead. The code is very fast and efficient. The central portion of the grid that it produces for the model in Fig. 3 when the period is 10 s is shown in Fig. 4.

4.3 Surface electric-field calculations

Calculating the electric field on the surface $z = 0$ by means of a simple numerical differentiation of the magnetic field according to the second of equations (1) is ineffective for the reasons stated by Weaver et al. (1996). Computational inaccuracies can arise because one-sided derivatives are required vertically at the surface and also horizontally when the cell boundary is a discontinuity in resistivity; further accuracy can be lost in the subtraction of two marginally accurate derivatives and then multiplication by a possibly large value of ρ as required by the second of equations (1). Following a procedure similar to that proposed by Mackie et al. (1993), we can therefore first compute the electric field at the *center* of a typical surface cell by integrating the second of equations (1) over the cell volume and applying the integral formula of Section 3 to the volume integral of curl **B**. After the integrals have been discretized in the usual

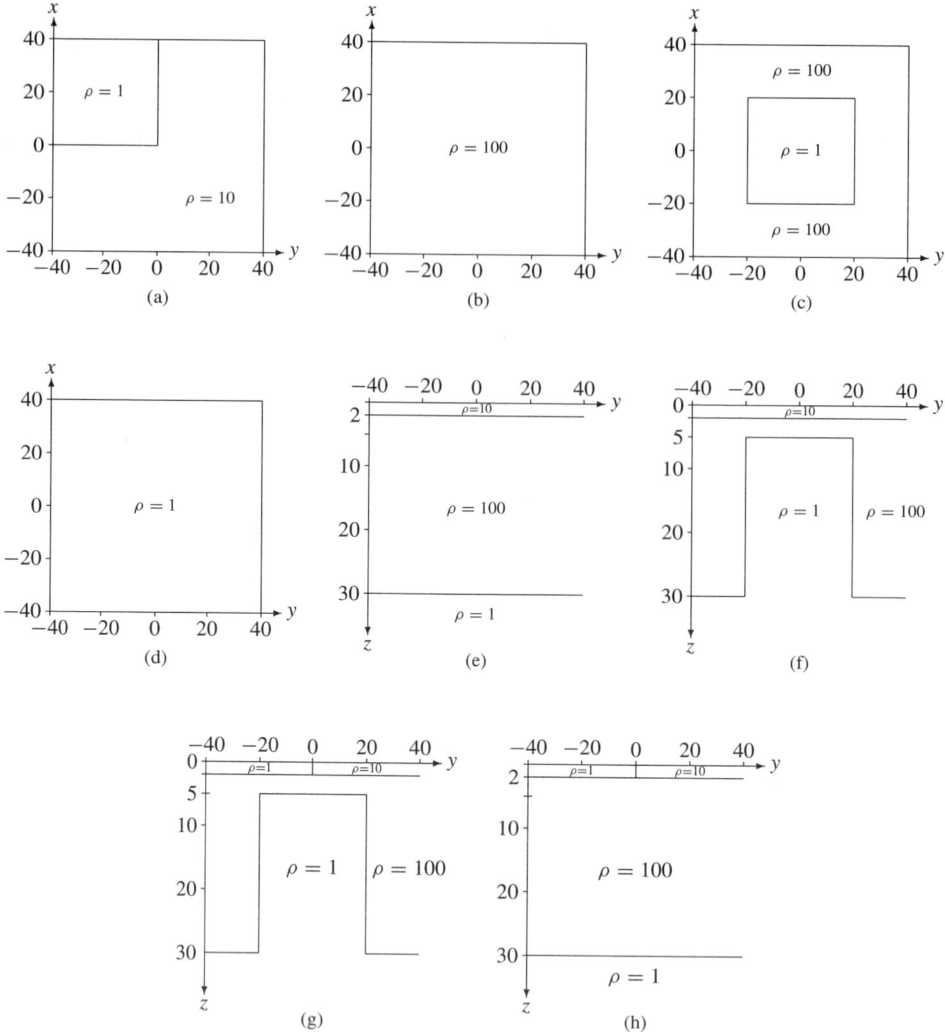

Figure 3. Horizontal cross-sections of 3-D model: (a) between the planes $z=0$ and $z=2$, (b) between $z = 2$ and $z = 5$, (c) between $z = 5$ and $z = 30$, and (d) below $z = 30$. Vertical cross-sections of the model: (e) for $x < -20$, (f) between the planes $x = -20$ and $x = 0$, (g) between the planes $x = 0$ and $x = 20$, and (h) for $x > 20$. All resistivity values are given in ohm-meters and all distances in kilometers. Note that the vertical scale in diagrams (e) to (h) is elongated.

manner, the x-component of the resulting equation is

$$\begin{aligned}(Y_{\ell,m,1} &+ Y_{\ell+1,m,1} + Y_{\ell,m+1,1} + Y_{\ell+1,m+1,1} - Y_{\ell,m,2} - Y_{\ell+1,m,2}\\ &- Y_{\ell,m+1,2} - Y_{\ell+1,m+1,2})/k_n + (Z_{\ell,m+1,1} + Z_{\ell+1,m+1,1} + Z_{\ell,m+1,2}\\ &+ Z_{\ell+1,m+1,2} - Z_{\ell,m,1} - Z_{\ell+1,m,1} - Z_{\ell,m,2} - Z_{\ell+1,m,2})/h_m\\ &= 4\mu_0 U_{\ell+1/2,m+1/2,n+1/2}/\rho_{\ell+1/2,m+1/2,3/2} \end{aligned} \tag{17}$$

with a similar equation for the y-component defining V at the center of the cell. The values of the horizontal components U an V then are raised to the surface by integrating the first of Eqs. (1) over the upper half of the appropriate vertical plane bisecting the cell. The left-hand side is transformed into a line integral by Stokes's theorem, and

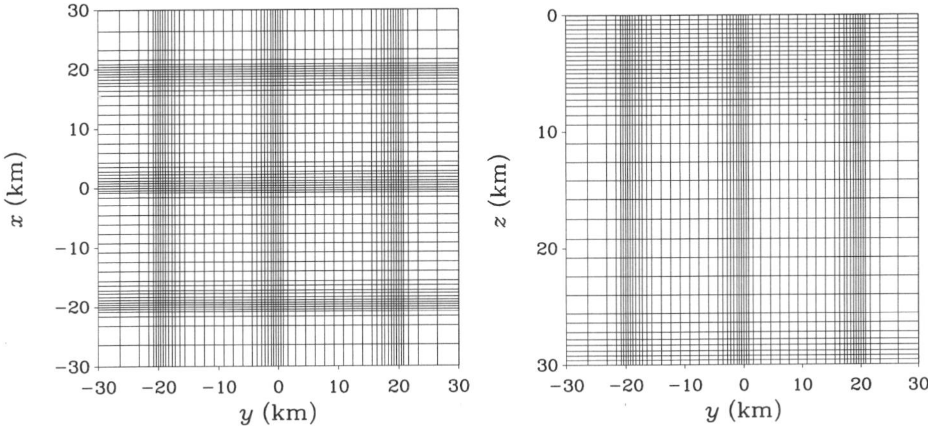

Figure 4. Horizontal (left) and vertical (right) sections of central part of grid generated for model depicted in Fig. 3 when the period of the field is 10 s.

the assumption is made that the contributions on the vertical sections of the contour are negligible because $W = 0$ at $z = 0+$. Thus the surface electric field is simply related to the calculated value at the center of the cell by the flux of magnetic field across the area enclosed by the contour. This flux is approximated to the area multiplied by the magnetic field at the center of the contour, the latter being estimated as a weighted average of the values at the eight corners of the cell. The procedure gives the surface electric field components at the centers of the rectangles defined by adjacent surface nodes. In general, it is not possible to define the electric field uniquely at the surface nodes themselves because they possibly are located at the junction of regions of different conductivities.

Although this approach is quite satisfactory when the conductivity contrasts in the model are not too extreme, previous investigations into methods for solving directly for the magnetic field in 2-D E-polarization problems suggested that the electric-field solutions obtained as above may become erratic in behavior in the presence of very large conductivity variations. In the 2-D case, this difficulty was surmounted by integrating over the full depth of the model rather than just the surface cell (Weaver et al., 1996). The extension of the method to three dimensions entails an integration of the first equation (2) over the volume of a column stretching from the surface to the base of the grid whose rectangular cross-section is defined by the region between four adjacent nodes (ℓ, m), $(\ell+1, m)$, $(\ell, m+1)$, and $(\ell+1, m+1)$. Let $\mathcal{V}$ be the volume of the rectangular column and $\mathcal{S}$ its surface area. Then, an integration over $\mathcal{V}$ and an application of the second result in Eq. (9) gives

$$\iint_{\mathcal{S}} \hat{\mathbf{n}} \times \mathbf{E}\, d\mathcal{S} = -i\omega \iiint_{\mathcal{V}} \mathbf{B}\, d\mathcal{V}. \tag{18}$$

For simplicity, we assume that the base of the grid is sufficiently deep that the electric and magnetic fields vanish on the surface $z = z_N$. Then, the y-component of Eq. (18) becomes

$$\int_{x_\ell}^{x_{\ell+1}} \int_{y_m}^{y_{m+1}} U(x, y, 0)\, dx\, dy + \int_{y_m}^{y_{m+1}} \int_0^{z_N} [W(x_{\ell+1}, y, z) - W(x_\ell, y, z)]\, dy\, dz$$
$$= i\omega \int_{x_\ell}^{x_{\ell+1}} \int_{y_m}^{y_{m+1}} \int_0^{z_N} Y(x, y, z)\, dx\, dy\, dz. \tag{19}$$

The volume integral on the right-hand side is evaluated by assigning the average values $Y_{\ell+1/2,m+1/2,n}$ $(n = 1, 2, \ldots, N)$ at the centers of the horizontal planes defining the top and bottom of each cell in the column, and assuming the usual linear variation over the cell thickness for the integration with respect to z. The resulting expression is

$$\int_{x_\ell}^{x_{\ell+1}} \int_{y_m}^{y_{m+1}} \int_0^{z_N} Y(x, y, z)\, dx\, dy\, dz$$
$$\approx \frac{1}{2} g_\ell h_m \left[k_1 Y_{\ell+1/2,m+1/2,1} + \sum_{n=2}^{N-1} (k_n + k_{n-1}) Y_{\ell+1/2,m+1/2,n} \right]$$

where

$$Y_{\ell+1/2,m+1/2,n} = \tfrac{1}{4}(Y_{\ell,m,n} + Y_{\ell+1,m,n} + Y_{\ell,m+1,n} + Y_{\ell+1,m+1,n}).$$

The first integral in Eq. (19) is approximated to $g_\ell h_m U_{\ell+1/2,m+1/2,1}$ which gives the surface electric field being sought.

There remains only the second integral in Eq. (19) involving the vertical electric field. One way of evaluating this integral is first to compute $W_{\ell+1/2,m+1/2,n+1/2}$ by taking the z-component of the vector equation whose x-component is Eq. (17), and then to interpolate linearly in the x-direction between the centers of adjacent cells to obtain $W_{\ell,m+1/2,n+1/2}(\ell = 2, 3, \ldots, L-1)$. Note that $W_{1,m+1/2,n+1/2}$ and $W_{L,m+1/2,n+1/2}$ vanish in the 2-D E-polarization solutions assumed to hold on the side boundaries of the grid that are normal to the x-axis. It then follows that

$$\int_{y_m}^{y_{m+1}} \int_0^{z_N} [W(x_{\ell+1}, y, z) - W(x_\ell, y, z)]\, dy\, dz$$
$$\approx h_m \sum_{n=1}^{N-1} k_n (W_{\ell+1,m+1/2,n+1/2} - W_{\ell,m+1/2,n+1/2}).$$

The other surface component of electric field $V_{\ell+1/2,m+1/2,1}$ is obtained in like manner from the x-component of Eq. (18) with the additional observation that the boundary values $W_{\ell+1/2,1,n+1/2}$ and $W_{\ell+1/2,M,n+1/2}$ are nonvanishing in this case, but are given by the solutions of the 2-D B-polarization problems that arise on the side boundaries of the grid normal to the y-axis.

5 Model calculations

The algorithm described above was checked by comparing its results for simple test models with those from established programs. One such test was with a model in which an anomalous thin-sheet overlies a two-layer half-space. The results of this calculation could be compared with those provided by the thin-sheet program of McKirdy et al. (1985) and found to be in good qualitative agreement. Another test was to compute the response of a 2-D structure to compare with results obtained by reliable 2-D modeling programs. Because the 2-D results are used as boundary conditions on the sides of the model, the comparison is more a test of the proper functioning of the algorithm than of its accuracy. Nevertheless, it was reassuring to find that the algorithm returned the correct 2-D solution across the whole model with errors of less than 1%—well within the limits of expected numerical noise. Neither of these two types of model tests the true 3-D capability of the program, of course, but serve merely as useful checks. The results of these calculations are not shown here.

A full 3-D model that exercises the special features of our algorithm is shown in Fig. 3. It consists of a shallow conductive quarter-slab at the Earth's surface above a bounded 3-D anomaly in the form of a square upwelling from the conductive basement. It reduces to different 2-D structures, respectively E-polarization and B-polarization, as $x \to \infty$ and $y \to -\infty$, and to a 1-D structure as $x \to -\infty$ and $y \to \infty$. At sufficiently long periods, the anomalous surface layer can be represented by a thin sheet. For automatic gridding the model is defined by the input given in Table 1. The model shows the applicability of the algorithm to quite general 3-D configurations. The calculations were performed for a period of 1000 s at which the skin depths in material of resistivities 1 and 10 ohm-m are about 16 and 50 km, respectively, and the inductive scale length in the underlying structure at the center of the model is roughly 14 km (and greater, of course, away from the upwelling). Thus the conditions required to treat the surface layer of 2-km thickness as a thin sheet of conductance 2000 S in the anomalous quarter-slab and 200 S elsewhere are well satisfied, and we have used this approximation as an alternative method of calculation.

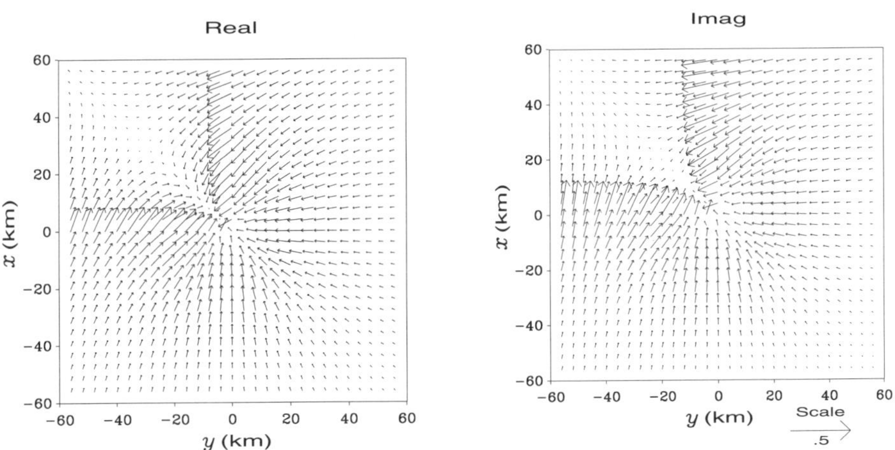

Figure 5. Induction vectors at the surface $z = 0$ for the model shown in Fig. 3 and a period of 1000 s. Real and imaginary vectors are shown in the left and right diagrams, respectively.

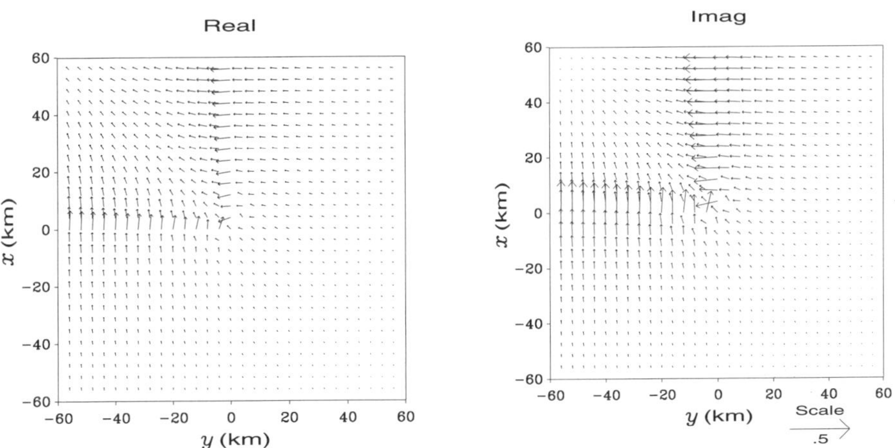

Figure 6. Induction vectors at the surface $z = 0$ for the model shown in Fig. 3 and a period of 1000 s with the conductive upwelling removed, i.e., with a uniform layer $2 < z < 30$ of resistivity 100 ohm-m.

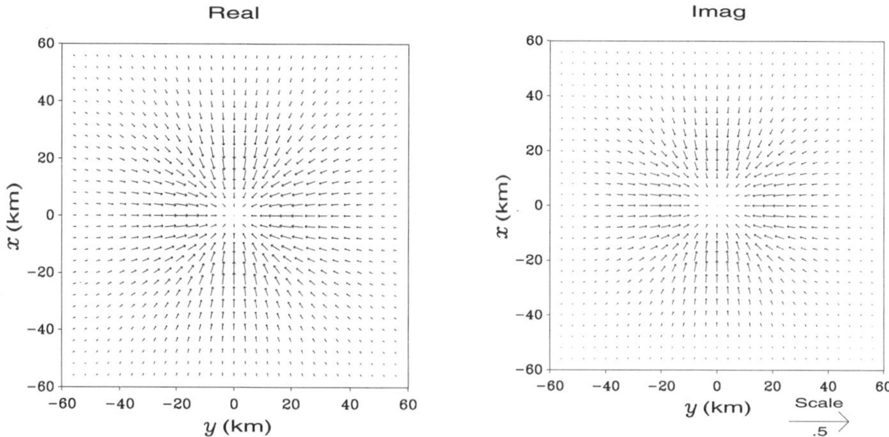

Figure 7. Induction vectors at the surface $z = 0$ for the model shown in Fig. 3 and a period of 1000 s with the conductive quarter-slab at the surface removed, i.e., with a uniform surface layer $0 < z < 2$ of resistivity 10 ohm-m.

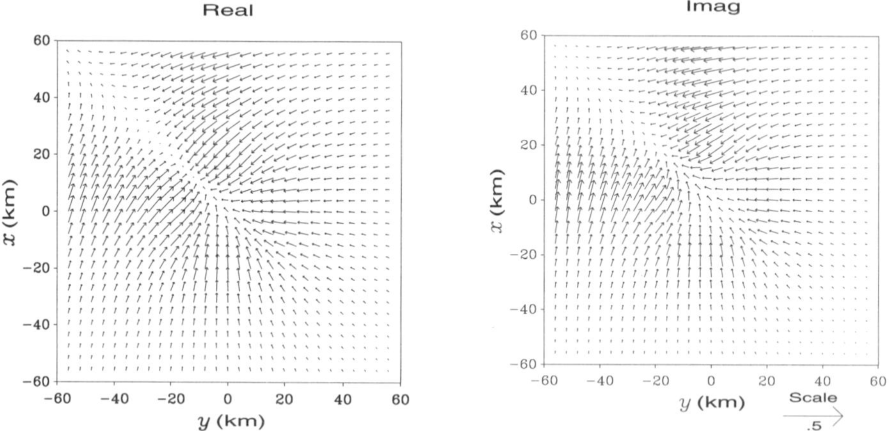

Figure 8. Induction vectors at the surface $z = 0$ for the model shown in Fig. 3 and a period of 1000s with the anomalous surface layer in the model represented by a thin sheet in the calculations.

The results of the magnetic-field calculations are displayed in Fig. 5 in the form of real and imaginary induction vectors $-(\text{Re}\, A)\,\hat{\mathbf{x}} - (\text{Re}\, B)\,\hat{\mathbf{y}}$ and $-(\text{Im}\, A)\,\hat{\mathbf{x}} - (\text{Im}\, B)\,\hat{\mathbf{y}}$, respectively, where $Z = AX + BY$ on $z = 0$. Characteristically, the real vectors point toward the region of high conductivity at the surface, but the distortions caused by the deeper-lying upwelling are also visible. The directions of the imaginary vectors are dependent on the period of the field variations; it is known that, for a given geometry and resistivity structure, there exists a characteristic period at which the imaginary induction vector reverses direction (Agarwal and Dosso, 1990, 1993). For the period of 1000 s, it is seen that for this model the imaginary vectors are directed toward the good conductor.

To better understand the relative contributions to the total response from the separate anomalous bodies, induction vectors are shown in Figs. 6 and 7 for the quarter-slab alone and for the upwelling alone, respectively. Their comparison with Fig. 5 reveals more obviously the types of distortion that have occurred in the combined model. Note that the real induction vectors point almost perpendicularly toward the boundaries of

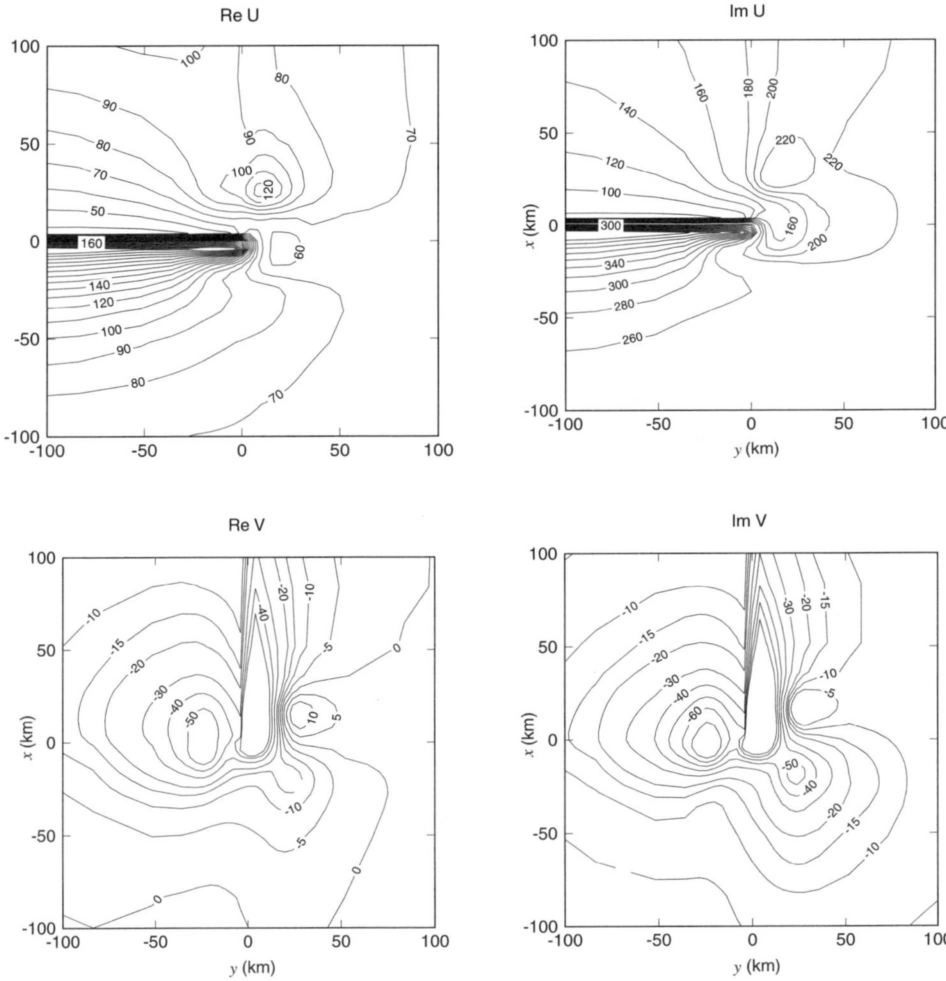

Figure 9. Contour plots of real and imaginary parts of the surface electric-field components U and V in the model of Fig. 3 for a period of 1000 s.

the conductive slab in Fig. 6, and that likewise they point inward toward the upwelling in Fig. 7. The results of a calculation for the general model in which the thin-sheet assumption and the approximate equation (16) have been used are shown in Fig. 8. The general behavior of the induction vectors is similar to that in Fig. 5, but the influence of the buried upwelling appears more prominent when the surface layer is contracted to a thin sheet. Finally, contour plots of the surface electric field as calculated by the formulas given in Section 4.3 are plotted in Fig. 9.

Acknowledgments

Research grants from the Natural Sciences and Engineering Research Council of Canada and the University of Victoria are gratefully acknowledged.

References

Agarwal, A. K., and Dosso, H. W., 1990, On the behaviour of the induction arrows over a buried conductive plate—A numerical model study: Phys. Earth Plan. Int., **60**, 265–277.

———1993, The characteristic periods of the induction arrows for a conductive-resistive vertical interface—A numerical study: Phys. Earth Plan. Int., **76**, 67–74.

Dawson, T. W., and Weaver, J. T., 1979, Three-dimensional induction in a non-uniform thin sheet at the surface of a uniformly conducting earth: Geophys. J. Roy. Astr. Soc., **59**, 445–462.

Hohmann, G. W., 1975, Three-dimensional induced polarization and electromagnetic modeling, Geophysics: **40**, 309–324.

Jones, D. S., 1964, The theory of electromagnetism: Pergamon Press, Inc.

Jones, F. W., 1974, The perturbations of slowly varying electromagnetic fields by three-dimensional conducting bodies: Can. J. Phys., **52**, 1195–1202.

Jones, F. W., and Pascoe, L. J., 1972, The perturbation of alternating geomagnetic fields by three-dimensional conductivity inhomogeneities: Geophys. J. Roy. Astr. Soc., **27**, 479–484.

Lines, L. R., and Jones, F. W., 1973, The perturbation of alternating geomagnetic fields by three dimensional island structures: Geophys. J. Roy. Astr. Soc., **32**, 133–154.

Mackie, R. L., Madden, T. R., and Wannamaker, P. E., 1993, Three-dimensional magnetotelluric modeling using difference equations—Theory and comparisons to integral equation solutions: Geophysics, **58**, 215–226.

Mackie, R. L., Smith J. T., and Madden, T. R., 1994, Three-dimensional electromagnetic modeling using finite difference equations: The magnetotelluric example: Radio Sci., **29**, 923–935.

McKirdy, D. M., Weaver, J. T., and Dawson, T. W., 1985, Induction in a thin sheet of variable conductance at the surface of a stratified earth—II: Three-dimensional theory: Geophys. J. Roy. Astr. Soc., **80**, 177–194.

Newman, G. A., and Hohmann, G. W., 1988, Transient electromagnetic responses of high-contrast prisms in layered earth: Geophysics, **53**, 691–706.

Poll, H. E., 1994, Automatic forward modeling of two-dimensional problems in electromagnetic induction: Ph.D. thesis, Univ. of Victoria.

Pu, X. H., Agarwal, A. K., and Weaver, J. T., 1993, Magnetic field solutions of E-polarization induction problems: J. Geomagn. Geoelectr., **45**, 859–872.

Raiche, A. P., 1974, An integral equation approach to three-dimensional modeling: Geophys. J. Roy. Astr. Soc., **36**, 363–376.

San Filipo, W. A., and Hohmann, G. W., 1985, Integral equation solution for the transient electromagnetic response of a three-dimensional body in a conductive half-space: Geophysics, **50**, 798–809.

Smith, J. T., 1996, Conservative modeling of 3-D electromagnetic fields, Part 1: Properties and error analysis: Geophysics, **61**, 1308–1318.

Ting, S. C., and Hohmann, G. W., 1981, Integral equation modeling of three-dimensional magnetotelluric response: Geophysics, **46**, 182–197.

Vasseur, G., and Weidelt, P., 1977, Bimodal electromagnetic induction in non-uniform thin sheets with an application to the northern Pyrenean induction anomaly: Geophys. J. Roy. Astr. Soc., **51**, 669–690.

Wang, T., and Hohmann, G. W., 1993, A finite-difference time-domain solution for three-dimensional electromagnetic modeling: Geophysics, **58**, 797–809.

Wannamaker, P. E., 1991, Advances in three-dimensional magnetotelluric modeling using integral equations: Geophysics, **56**, 1716–1728.

Wannamaker, P. E., Hohmann, G. W., and San Filipo, W. A., 1984a, Electromagnetic modeling of three-dimensional bodies in layered earths using integral equations: Geophysics, **49**, 60–74.

Wannamaker, P. E., Hohmann, G. W., and Ward, S. H., 1984b, Magnetotelluric responses of three-dimensional bodies in layered earths: Geophysics, **49**, 1517–1533.

Weaver, J. T., 1964, On the separation of local geomagnetic fields into external and internal parts: Z. Geophys., **30**, 29–36.

———1994, Mathematical methods for geo-electromagnetic induction: Research Studies Press.

Weaver, J. T., Pu, X. H., and Agarwal, A. K., 1996, Improved methods for solving for the magnetic field in E-polarization induction problems with fixed and staggered grids: Geophys. J. Internat, **126**, 437–446.

Weidelt, P., 1975, Electromagnetic induction in three-dimensional structures: J. Geophys., **41**, 85–109.

Finite-Element Formulation of Electromagnetic Induction with Coupled Potentials

Mark E. Everett

Summary. Electric and magnetic potentials in the Coulomb gauge provide a convenient formulation for finite-element modeling of electromagnetic (EM) induction in heterogeneous conductivity structure. The formulation uses standard linear finite elements defined on tetrahedra. The finite-element equations are solved by a sparse, ILU decomposition that ignores fill-in to achieve maximum storage efficiency. The formulation can handle models of interest to oil and gas exploration, mining, and environmental studies. It is applied here in spherical geometry to model EM induction in the upper mantle due to a magnetospheric ring current.

1 Introduction

Integral equations (e.g., Newman et al., 1986; Xiong, 1992) and (staggered-grid) finite-difference methods (e.g., Mackie et al., 1994) currently dominate 3-D electromagnetic (EM) modeling for geophysical applications. But despite their limited use, finite-element methods remain attractive because of the ease with which they handle complex geometries and because they are supported by a large body of rigorous mathematical results. An important breakthrough in the finite-element method for 3-D EM modeling has been the development of special elements to accurately represent electric and magnetic fields (Mur and de Hoop, 1985). Use of these elements, however, often leads to a fair amount of programming complexity. An alternative is to use gauged potentials for the fields, which require only standard linear elements. Everett and Schultz (1996) give a complete description of this formulation for 2-D, azimuthally symmetric conductivity structures. Here, I review the formulation and present some results for 3-D models.

2 Formulation

Equations for the coupled-potential formulation of EM induction in an arbitrary 3-D electrical conductivity structure are developed for applications in electrical engineering by Biro and Preis (1990) and for geophysical applications by Everett and Schultz (1996).

Department of Geology and Geophysics, Texas A&M University, College Station, TX 77843, USA.

The formulation uses a magnetic vector potential **A** and an electric scalar Φ potential to represent fields inside the Earth, and a magnetic scalar potential Ψ to represent fields outside the earth (in the air). The Coulomb gauge is enforced explicitly to ensure the uniqueness of the potentials. The source currents are presumed to lie outside the conducting earth. The equations have the following structure:

$$\Psi = \Psi_0 \quad \text{on } \Gamma_2 \qquad \text{(source region)}, \tag{1}$$

$$\nabla^2 \Psi = 0 \quad \text{in } \Omega_2 \qquad \text{(nonconductor)}, \tag{2}$$

$$\left.\begin{aligned} \nabla \times \mathbf{A} &= -\nabla\Psi \\ \partial\Phi/\partial n &= 0 \\ \mathbf{A}\cdot\hat{n} &= 0 \end{aligned}\right\} \quad \text{on } \Gamma_1 \quad \text{(interface between } \Omega_1 \text{ and } \Omega_2\text{)}, \tag{3}$$

$$\left.\begin{aligned} \nabla^2\mathbf{A} + i\omega\mu_0\sigma\mathbf{A} - \mu_0\sigma\nabla\Phi = 0 \\ \nabla\cdot(i\omega\sigma\mathbf{A} - \sigma\nabla\Phi) = 0 \end{aligned}\right\} \quad \text{in } \Omega_1 \quad \text{(conductor)}. \tag{4}$$

The differential Eqs. (2) and (4) are driven by the source term Ψ_0, which is presumed to be known. Coupling of the source to the conducting region (the Earth) is achieved by the application of boundary conditions to potentials and their spatial derivatives, as described by Eq. (3). The Earth's electrical structure is specified by the electrical conductivity function $\sigma(x, y, z)$ which can vary arbitrarily. The equations are solved for the potentials **A**, Φ, and Ψ by the finite-element method (see below). The EM fields $\{\mathbf{E}, \mathbf{B}\}$ then are found by suitable numerical differentiation of the potentials, according to the formulas $\mathbf{B} = \nabla \times \mathbf{A}$ and $\mathbf{E} = i\omega\mathbf{A} - \nabla\Phi$ (inside the Earth), and $\mathbf{B} = -\nabla\Psi$ (outside). This so-called **A**, Φ–Ψ formulation has been used extensively in electrical engineering for 3-D eddy-current calculations (Biro and Preis, 1990).

A standard finite-element analysis is applied to solve the governing equations. The volumes Ω_1 and Ω_2 are discretized into tetrahedra with nodes at the vertices. A Cartesian-coordinate frame is used. The 3-D mesh is defined by a list of node coordinates and a connectivity map. The mesh can have an arbitrary shape. To model global EM induction in the Earth's mantle, a polyhedral approximation to a sphere is used. A small, inner spherical region is excluded from the mesh so that the origin $r = 0$ is not part of the solution domain. This avoids the problem of small tetrahedra near the origin. The radius of the inner region is chosen small enough (normally less than that of the core–mantle boundary) that the potentials can be set to zero there without appreciable effect on the solution in the mantle.

Piecewise linear representations are used for **A**, $\Phi - \Psi$:

$$\mathbf{A}(\mathbf{r}) = \sum_j (a_j\hat{\mathbf{x}} + b_j\hat{\mathbf{y}} + c_j\hat{\mathbf{z}})\,\alpha_j(\mathbf{r}), \tag{5}$$

$$\Phi(\mathbf{r}) = \sum_j \phi_j\alpha_j(\mathbf{r}), \tag{6}$$

$$\Psi(\mathbf{r}) = \sum_k \psi_k\alpha_k(\mathbf{r}) + \sum_l \Psi_0(\mathbf{r})\,\alpha_l(\mathbf{r}). \tag{7}$$

The index j sums over the nodes inside and on the surface of the Earth; k sums over nodes outside and on the Earth; and l sums over nodes on the boundary Γ_2, where the external sources are located.

Inserting Eqs. (5)–(7) into the governing Eqs. (1)–(4), multiplying by appropriate test functions, and integrating by parts over the whole solution domain $\Omega = \Omega_1 + \Omega_2$, gives a sparse, symmetric, block system of complex linear algebraic equations. Everett and Schultz (1996) give details of this procedure and explicit expressions for the matrix entries. The linear system is solved iteratively by ILU factorization. A complex biconjugate gradient technique (Jacobs, 1986) for solving the forward equations was tried but proved very unreliable. Iterative ILU factorization reduces storage requirements by simply ignoring the fill-in that is generated during standard Gaussian elimination. The approximate factorization is used in the back-substitution step of Gaussian elimination to generate a first approximation to the solution. Iterative refinement successively improves the approximation (Everett and Schultz, 1996).

The results described in the next section were obtained on a mesh containing 14 444 nodes (25 938 complex degrees of freedom) and 63 504 tetrahedra. To assemble the matrix and obtain a solution at a single frequency required 129 Mbytes of memory and about 45 CPU minutes on a dedicated Hewlett Packard 9000 series 715/100 workstation. The ILU solver ran for 300 iterations, and exponential convergence of the residual was observed. At termination, the l_2 norm of the residual vector was approximately 10^{-5} times its value after the first iteration.

2.1 Coulomb-gauge condition

The condition $\mathbf{A} \cdot \hat{\mathbf{n}} = 0$, where $\hat{\mathbf{n}}$ is the outwardly directed unit normal to the surface Γ_1 of the conducting region, enforces the Coulomb gauge (defined by $\nabla \cdot \mathbf{A} = 0$) throughout the conductor Ω_1. This can be understood heuristically by recalling the divergence theorem from vector calculus:

$$\int_{\Gamma_1} \mathbf{A} \cdot \hat{\mathbf{n}} \, d\Gamma_1 = \int_{\Omega_2} \nabla \cdot \mathbf{A} \, d\Omega_2, \tag{8}$$

which holds for an arbitrary vector $\mathbf{A}$ field, not just a vector potential. The divergence theorem states that if the outward normal component of $\mathbf{A}$ vanishes everywhere on the closed surface of a conductor, then the corresponding volume integral of the divergence, $\nabla \cdot \mathbf{A}$, also vanishes. Experience with 3-D finite-element algorithms in electrical engineering has shown (Biro and Preis, 1990) that if $\mathbf{A} \cdot \hat{\mathbf{n}} = 0$ everywhere on the surface, then $\nabla \cdot \mathbf{A} = 0$ everywhere in the interior.

Practically, it is difficult to explictly enforce the boundary condition $\mathbf{A} \cdot \hat{\mathbf{n}} = 0$, especially when the normal vector $\hat{\mathbf{n}}$ is not aligned with one of the coordinate axes of the problem. The enforcement of $\mathbf{A} \cdot \hat{\mathbf{n}} = 0$ is equivalent to adding a number of linear constraints to the finite-element system of equations. The same number of degrees of freedom therefore must be removed to keep the system well determined. The mathematics is straightforward, but the programming requires some careful attention and has not been implemented fully in my solution. But even without explicit enforcement of $\mathbf{A} \cdot \hat{\mathbf{n}} = 0$, the Coulomb gauge is approximately satisfied, as shown below. For most of the conductivity models that I have considered, the normal component of $\mathbf{A}$ on Γ_1 is much smaller in magnitude than the tangential components.

The satisfaction of the Coulomb gauge can be monitored by calculating the ratio of l_p norms:

$$\epsilon_p = \frac{\|\nabla \cdot \mathbf{A}\|_p}{\|\mathbf{A}\|_p}. \tag{9}$$

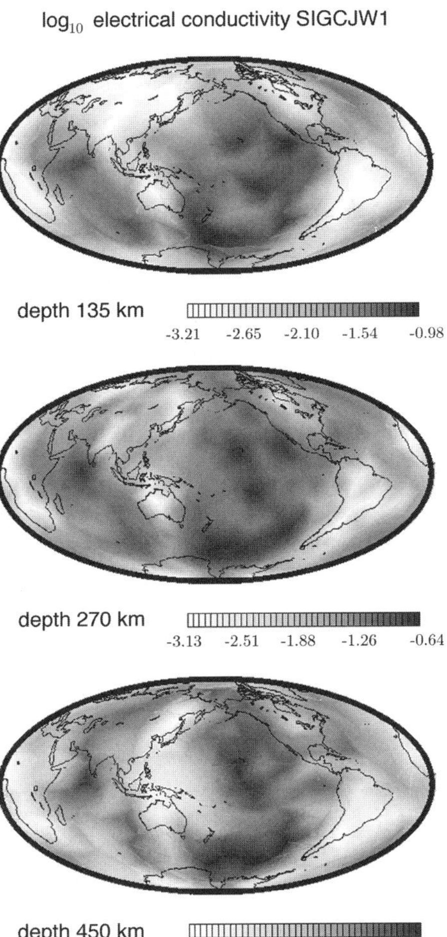

Figure 1. A 3-D map of the electrical conductivity model SIGCJW1 as derived from seismic tomography model S12–WM13 and the Schultz–Akaogi geoelectrotherm $\sigma(T)$. An Aitoff equal-area projection (Snyder and Voxland, 1989) is used.

The norms in Eq. (9) are evaluated by summing the pth power of the magnitude of the vector potential $|\mathbf{A}|^p$ over all nodes and its divergence $|\nabla \cdot \mathbf{A}|^p$ over all tetrahedra in the interior region Ω_1, and then taking the pth root. A small value, i.e., ϵ_p much less than unity, is an indicator that the Coulomb gauge is approximately satisfied. The finite-element computations described in sections yielded values of $\epsilon_1 = 0.00003$, $\epsilon_2 = 0.0001$, and $\epsilon_\infty = 0.002$.

3 Application: Geomagnetic Induction in the mantle

The **A**, Φ–Ψ method is applied here to a classic problem of geomagnetic induction in a heterogeneous sphere. Global studies of the Earth's internal electrical structure began at the end of previous century (Schuster, 1889; Lahiri and Price, 1939) and continue up to the present day (Schultz and Larsen, 1987; Schultz, 1990). The assumption of

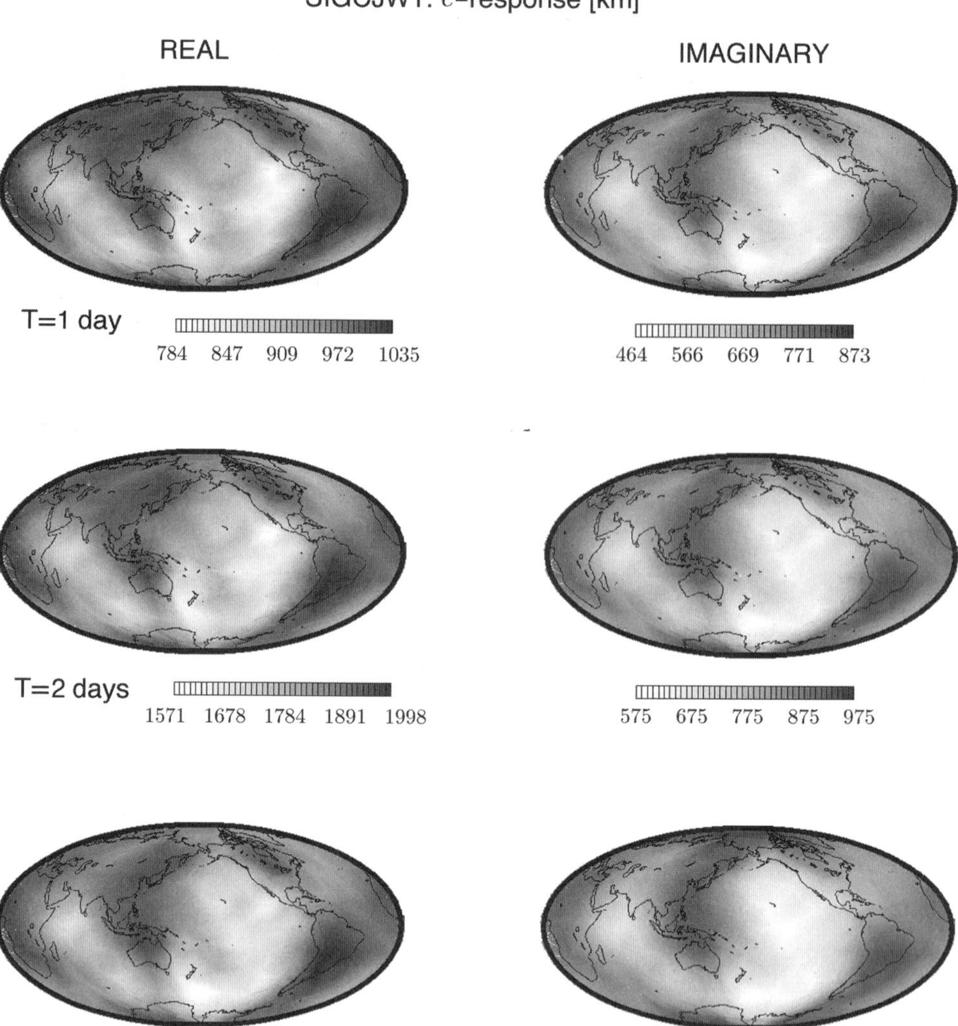

Figure 2. The EM response of model SIGCJW1 expressed as real and imaginary components of Schmucker's c-response, for periods T = 1 day, 2 days, and 1 week.

global radial symmetry is normally invoked to make computations tractable. However, long-period EM induction data from the global observatory network are capable of resolving large-scale lateral heterogeneity (Schultz 1990; Zinger et al., 1993; Tarits, 1994).

In this paper, the 3-D EM response of a mantle electrical conductivity model derived from seismic tomography is computed. The purpose is primarily to demonstrate the finite-element code as a tool for modeling realistic, 3-D structural variations within the Earth. The mantle conductivity model is derived by converting the upper-mantle shear wave velocity model S12–WM13 (Su et al., 1994) into a "seismic temperature" map using laboratory data (Anderson et al., 1992) on the temperature dependence of the elastic properties of olivine. A geoelectrotherm $\sigma(T)$ is constructed by combining the Akaogi et al. (1989) geotherm $T(z)$ with the magnetotelluric/GDS (geomagnetic depth sounding) electrical conductivity profile $\sigma(z)$ of Schultz et al. (1993). Inserting the seismic temperature map into the geoelectrotherm yields a 3-D electrical conductivity map expressed in the basis of S12–WM13, which are products of radial Chebychev polynomials and surface harmonics. This model, denoted SIGCJW1, is shown in Fig. 1.

The electrical model SIGCJW1 was input into the finite-element code and run for a zonal P_1^0 magnetospheric ring-current excitation. The computations produced an inductive scale-length, or Schmucker's, c-response (Schultz and Larsen, 1987),

$$c = -\frac{E_\phi}{10^3 i\omega B_\theta}, \tag{10}$$

which is shown in Fig. 2. The factor of 1000 in the denominator turns the c-response into units of kilometers. Note that the computed EM responses are spatially correlated with the electrical conductivity structure of model SIGCJW1. This demonstrates that the finite-element code produces responses that have a reasonable spatial distribution. Although the code has been carefully checked against two independent solutions for 2-D azimuthally symmetric structures (Everett and Schultz, 1996), such structures excited by zonal sources do not contain galvanic terms. Validation of the finite-element code against an independent, fully 3-D solution has recently been performed (Weiss and Everett, Geophys. J. Int., submitted).

4 Discussion

Potential fields in the Coulomb gauge provide a convenient formulation of the finite-element method for EM induction. The formulation allows use of conventional linear elements. The method has been applied here for a simple problem in solid earth geophysics; its adaptation for solving induction problems in applied geophysics should be investigated.

Acknowledgments

I would like to thank Adam Schultz, Geoff Pritchard, and Chet Weiss for many stimulating discussions.

References

Akaogi, M., Ito, E., and Navrotsky, A., 1989, Olivine-modified spinel–spinel transitions in the system Mg_2SiO_4–Fe_2SiO_4: Calorimetric measurements, thermochemical calculation, and geophysical application: J. Geophys Res. **94**, 15671–15685.

Anderson, O. L., Isaak, D., and Oda, H., 1992, High-temperature elastic constant data on minerals relevant to geophysics: Rev. Geophys. **30**, 57–90.

Biro, O., and Preis, K., 1990, Finite element analysis of 3D eddy currents: IEEE Trans. Magn. **26**, 418–423.

Everett, M. E., and Schultz, A., 1996, Geomagnetic induction in a heterogeneous sphere: Azimuthally symmetric test computations and the response of an undulating 660-km discontinuity: J. Geophys. Res. **101**, 2765–2783.

Jacobs, D. A. H., 1986, A generalization of the conjugate-gradient method to solve complex systems: IMA J. Num. Anal., **6**, 447–452.

Lahiri, B., and Price, A., 1939, Electromagnetic induction in non-uniform conductors, and the determination of the conductivity of the earth from terrestrial magnetic variations: Philos. Trans. Roy. Soc. London, **237**, 509–540.

Mackie, R. L., Smith, J. T., and Madden, T. R., 1994, Three-dimensional electromagnetic modeling using finite difference equations: The magnetotelluric example: Radio Sci., **29**, 923–935.

Mur, G., and de Hoop, A. T., 1985, A finite-element method for computing three-dimensional electromagnetic fields in inhomogeneous media: IEEE Trans. Magn., **21**, 2188–2191.

Newman, G. A., Hohmann, G. W., and Anderson, W. L., 1986, Transient electromagnetic response of a three-dimensional body in a layered earth: Geophysics, **51**, 1608–1627.

Schultz, A., 1990, On the vertical gradient and associated heterogeneity in mantle electrical conductivity: Phys. Earth Planet. Internat., **64**, 68–86.

Schultz, A., and Larsen, J. C., 1987, On the electrical conductivity of the mid-mantle, I. Calculation of equivalent scalar magnetotelluric response functions: Geophys. J. Roy. Astr. Soc., **88**, 733–761.

Schultz, A., Kurtz, R. D., Chave, A. D., and Jones, A. G., 1993, Conductivity discontinuities in the upper mantle beneath a stable craton: Geophys. Res. Lett., **20**, 2941–2944.

Schuster, A., 1889, The diurnal variation of terrestrial magnetism, with an appendix by H. Lamb, On the currents induced in a spherical conductor by variation of an external magnetic potential: Philos. Trans. Roy. Soc. London A, **180**, 467–518.

Snyder, J. P., and Voxland, P. M., 1989, An album of map projections: U.S. Geol. Surv. Prof. Paper 1453.

Su, W.-J., Woodward, R. L., and Dziewonski, A. M., 1994, Degree 12 model of shear velocity heterogeneity in the mantle: J. Geophys. Res., **99**, 6945–6980.

Tarits, P., 1994, Electromagnetic studies of global geodynamic processes: Surv. Geophys., **15**, 209–238.

Xiong, Z., 1992, Electromagnetic modeling of three-dimensional structures by the method of system iterations using integral equations: Geophysics, **57**, 1556–1561.

Zinger, B., Sh., Kuvshinov, A. V., Mishina, L. P., and Faynberg, E. B., 1993, Global geomagnetic sounding: New methodology and results: Izv., Earth Sci., **29**, 35–43.

Three-Dimensional Inversion for Large-Scale Structure in a Spherical Domain

Adam Schultz
Geoffrey Pritchard[1]

Summary. Our inversion method for global conductivity structures uses a forward solver that can represent arbitrary conductivity distributions in a sphere. The inversion uses new methods of grid generation and integration on a convex hull in spherical domains and exploits the relatively high speed and modest memory requirements of the spherical basis expansion. The method reduces the forward problem to the solution of a set of stiff ordinary differential equations, which are solved by finite differences on an adaptive grid. The model is parameterized by a set of regional 1-D models, which then are interpolated into a fully 3-D spherical model using a Delaunay triangulation.

The inversion proceeds in two stages: the first is a nonlinear search in which parameters are varied sequentially to minimize the cost function; the second uses a linearized inversion about the model obtained in the first stage. We have used this method to invert data at 16 frequencies from 20 geomagnetic observatories around the world. The 3-D model is generated from radial profiles (five layers) at 56 locations on the globe.

1 Introduction

No single 1-D electrical conductivity model is consistent with regional inductive response functions around the world (Schultz, 1990; Schultz and Larsen, 1990; Schultz and Semenov 1993). These data sets require global 3-D conductivity structures. We describe a method designed to invert for such structures. It is based on a forward solver originally developed by Zhang and Schultz (1992), which can handle arbitrary conductivity distributions in a sphere, provided that lateral heterogeneity is small. Although other forward solvers, such as those of Tarits et al. (1995) and Everett and Schultz (1996), do not have this constraint, improvements in the original method of Zhang and Schultz have increased its accuracy and reduced the computational effort by approximately one to two orders of magnitude. Such improvements make this forward solver particularly attractive as the kernel of a practical inverse method.

Our methods also may be useful for studies of regional or basin-scale induction. Advances in solving large boundary-value problems on adaptive grids (Braun and

[1]Institute of Theoretical Geophysics, University of Cambridge, Cambridge CB2 3EQ, UK.

Sambridge, 1994, 1995; Sambridge et al., 1995) and meshes (e.g., Golias and Tsiboukas, 1992) may eventually allow the embedding of regional and basin structures in global models. A global (spherical) model domain requires boundary conditions only at the Earth's center (or at the core–mantle boundary where the conductivity jumps by orders of magnitude); the Earth's surface; and at the external source (usually, the magnetospheric ring current). There are no lateral boundaries, and thus no need to impose boundary conditions on the sides of the domain. These boundaries are a problem with local Cartesian models, because the influence of conductivity structures quite distant from the boundaries may be extremely strong (Heinson and Constable, 1992). Although the degree to which distant structure affects local models is controversial—some authors argue for modest effects (Weaver and Dawson, 1992; Tarits et al., 1993)—setting lateral boundary conditions remains a persistent problem for Cartesian models.

1.1 The 3-D spherical Earth

Evidence for the existence of large-scale 3-D Earth structure comes principally from analysis of regional 1-D electromagnetic (EM) response functions. These usually are calculated from magnetic observatory records rather than from magnetotelluric (MT) data.

1.1.1 The c response. If Earth's deep conductivity structure is radially symmetric, inversion for the underlying 1-D structure can be based on the c inductive scale-length response (Schmucker, 1970). This response is defined as follows: With a distant induction source external to the Earth, the magnetic induction $\mathbf{B}$ is taken as the curl of the vector potential $\hat{\mathbf{a}}_r r\psi(r, \theta, \phi)$, where $\hat{\mathbf{a}}_r$, $\hat{\mathbf{a}}_\theta$, and $\hat{\mathbf{a}}_\phi$ are direction vectors in the spherical basis, and ψ is the scalar potential. The induction $\mathbf{B}$ satisfies the vector Helmholtz equation,

$$\nabla \times (\nabla \times \mathbf{B}) = i\omega\mu_0\sigma(r)\mathbf{B} + \frac{\nabla\sigma(r)}{\sigma(r)} \times (\nabla \times \mathbf{B}), \tag{1}$$

where σ is the electrical conductivity and ω is the angular frequency. All harmonic terms have the time dependence $e^{-i\omega t}$.

The potential ψ can be decomposed into a set of eigenfunctions, and the electric and magnetic fields then are decomposed into parts corresponding to each eigenpotential. These can be as written $\mathbf{E}^{mn}$ and $\mathbf{B}^{mn}$, respectively, where mn represents the degree and order of the spherical harmonic coefficients in which the eigenfunctions are expressed, and ψ_{mn} is the eigenpotential associated with each eigenfunction. A geomagnetic response function for each eigenmode can be defined as

$$T^{mn} = B_r^{mn} / B_\theta^{mn}. \tag{2}$$

The conventional MT scalar impedance (units, ms^{-1}) is written

$$I^{mn} = E_\phi^{mn} / B_\theta^{mn} = -E_\theta^{mn} / B_\phi^{mn}. \tag{3}$$

Substituting the eigenpotentials ψ_{mn} into E^{mn} and B^{mn} gives

$$I^{mn} = \alpha_{mn} T^{mn}, \tag{4}$$

where α_{mn} is (Schultz and Larsen, 1987)

$$\alpha_{mn} = \frac{i\omega r_e\left[dP_n^m(\cos\theta)/d\theta\right]}{(n+1)nP_n^m(\cos\theta)}, \tag{5}$$

and where P_n^m are the spherical harmonic functions, r_e is the Earth's radius, and θ is the geomagnetic colatitude. For the first zonal harmonic P_1^0, the c response, $c = I^{10}/i\omega$, is

$$c = \frac{-r\ \tan\theta}{2}\frac{B_r^{10}}{B_\theta^{10}}. \tag{6}$$

1.1.2 Evidence for 3-D structure. The most compelling evidence for global 3-D mantle structure comes from attempts to invert jointly regional 1-D response functions. Parker's (1980) D^+ algorithm gives a proof of the existence of a 1-D inverse solution for a given response. If a set of two or more responses is drawn from the same 1-D Earth, and the standard errors are unbiased, then it should be possible to invert them jointly by D^+ for a 1-D model that satisfies all of the data. Conversely, failure to obtain an acceptable misfit demonstrates that the two or more responses are incompatible with any shared 1-D model.

All responses published by Schultz and Larsen (1987) were consistent simultaneously with excitation by the first-order zonal harmonic (e.g., P_1^0) of the external field and with local 1-D Earth structure. All possible combinations of these regional 1-D response functions were inverted jointly using D^+ (Schultz and Larsen, 1990), and no single radially symmetric model was found to be compatible with the full set of data. Geographical groupings of observatories exist, based roughly on tectonic setting, where joint regional 1-D interpretation is possible (Schultz and Semenov, 1993). More recent analysis of very deep long-period MT data obtained from the central Canadian Shield (Schultz et al., 1993), the Northeast Pacific (Lizarralde et al., 1995), and to a lesser degree the Tucson magnetic observatory (Egbert and Booker, 1992; Egbert et al., 1992) show a consistency with deep regional 1-D interpretation, but when contrasted with each other show significant levels of heterogeneity in upper mantle structure.

Data from midlatitude observatories found to be incompatible with local 1-D analysis also tend to exhibit anomalous structure in the B_r component, but normal (i.e., roughly cosinusoidal variation with geomagnetic latitude) in B_θ. This is consistent with deep laterally heterogeneous structure in the regions beneath these sites. The conclusion from this body of work is that no global 1-D model is consistent simultaneously with all of the regional 1-D response functions; that there appears to be a gross relationship between the form of these response functions and the tectonic setting; and that the source of these heterogeneities appears to lie, in large part, within the mantle.

2 Perturbation-expansion forward solution

We solve the forward problem of EM induction in a 3-D heterogeneous sphere by considering perturbations of the vector Helmholtz equation. These solutions lead to an efficient numerical method for broad-scale lateral variations in mantle electrical conductivity.

The vector Helmholtz equation (1) is written for an Earth with 3-D conductivity structure

$$\nabla\times[\nabla\times\mathbf{B}(r,\theta,\phi)] = i\omega\mu_0\sigma(r,\theta,\phi) + \frac{\nabla\sigma(r,\theta,\phi)}{\sigma(r,\theta,\phi)}\times\nabla\times\mathbf{B}(r,\theta,\phi). \tag{7}$$

As mentioned in the Introduction, in a spherical model there are only three boundaries: the outer boundary of the core, which we assume to be infinitely conducting; the air–Earth interface, in which we assume the air is perfectly resistive; and an interface beneath a magnetospheric source, which we take to be an equatorial ring current in the form of the spherical harmonic P_n^m, located a distance of three Earth radii from the center of the core.

We use a perturbation expansion to construct an efficient numerical solution by reducing the vector Helmholtz equation to a series of coupled 1-D second-order ordinary differential equations (ODEs). Use of a perturbation expansion means, however, that the class of admissible models is limited to those with only mild dependence on tangential coordinates; i.e., the perturbation at any given depth from a radially symmetric reference model cannot exceed a critical threshold or the solution will be invalid. This restriction is probably acceptable for the broadly resolved models expected from the inversion of presently available regional 1-D c responses.

Consider the Mie representation of the magnetic induction written in terms of toroidal magnetic Ψ and poloidal magnetic Π potential functions

$$\mathbf{B} = \nabla \times (\Psi \hat{\mathbf{a}}_r) + \nabla \times \nabla \times (\Pi \hat{\mathbf{a}}_r). \tag{8}$$

By representing the magnetic induction $\mathbf{B}$ in the form above, and by applying a standard perturbation expansion method to the vector Helmholtz equation (7), a simplified set of coupled equations results. It can be shown after considerable algebraic manipulation (Zhang and Schultz, 1992) that solutions for poloidal potential mode may take the form

$$\Pi = \Pi^0 + \Pi^1 + \cdots, \tag{9}$$

and similarly for the toroidal mode Ψ.

For the source function, we use the P_1^0 spherical harmonic. Although the perturbation-expansion solution can accommodate any form, the data presently under consideration [c responses from Schultz and Larsen (1987), Schultz and Semenov (1993), and Schultz et al. (1994)] were calculated by extracting this component, which is the dominant part of the stationary element of the magnetospheric equatorial ring current. We apply the appropriate boundary conditions and write

$$\Pi = \sum_{\ell=1}^{L} \sum_{m=0}^{\ell} R_{\ell m}^0 Y_\ell^m(\theta, \phi) + \sum_{i=1}^{n} \sum_{\ell=1}^{L} \sum_{m=0}^{\ell} R_{\ell m}^i(r) Y_\ell^m(\theta, \phi), \tag{10}$$

where R is a function of radius only and for zeroth order

$$\frac{\partial^2 R_{\ell m}^0(r)}{\partial r^2} + R_{\ell m}^0(r)\left[i\omega\mu_0\sigma_0(r) - \frac{\ell(\ell+1)}{r^2}\right] = 0 \tag{11}$$

and for kth order ($1 \leq k \leq n$),

$$\frac{\partial^2 R_{\ell m}^k(r)}{\partial r^2} + R_{\ell m}^k(r)\left[i\omega\mu_0\sigma_0(r) - \frac{\ell(\ell+1)}{r^2}\right] = \mathcal{F}[R^{k-1}(r), \sigma(r, \theta, \phi)], \tag{12}$$

where

$$\sigma(r, \theta, \phi) = \sigma_0(r) + \sigma_1(r, \theta, \phi). \tag{13}$$

The term $\mathcal{F}$ is a complex surface integral that varies only with depth r. Note that the zeroth-order equation is the solution to the radially symmetric Earth problem.

For an efficient numerical solution of these equations, we use a Thiessen tesselation (also called the Dirichlet or Voronoi tesselation) (Tsai, 1993) of the surface of each model shell in the 3-D model. The tesselation is used to construct a Delaunay triangulation (Hoschek and Lasser, 1993) of the point set, and this in turn is used to define a set of spherical triangles on the surface of each model shell, which we use for surface integration of $\mathcal{F}$. This method is described in Section 3.5. This allows $\mathcal{F}$ to be integrated accurately but also extremely rapidly. Numerical solution of Eq. (12) is performed accurately using finite-difference methods.

The 3-D dependency of the solutions is contained within the higher-order terms of the perturbation expansion. The magnitude of the terms of the expansion decrease rapidly with increasing order. The zeroth-order (radially symmetric) term is much larger than all other terms. For these reasons, extremely accurate solutions are required to prevent the 3-D part of the solution from being overwhelmed by numerical noise. This has proven to be a very difficult part of the problem, and it has only been possible to solve it with a high degree of confidence since the advent of the quasi-analytic azimuthally symmetric D'Yakanov forward solution of Everett and Schultz (1995) which provides a set of test cases against which solutions may be compared.

It has not proved sufficiently accurate to treat Eq. (12) directly as a second-order ODE with complex coefficients on a fixed grid (Zhang and Schultz, 1992). The ODE is stiff because the real and imaginary parts of the radial function can have very different decay rates with respect to r, particularly in the presence of conductive anomalies. To solve this problem accurately, we treat the second-order ODE as a set of four coupled first-order ODEs. Each of these may be solved on its own adaptive grid, thus taking into account the different functional forms of the real and imaginary parts of $R^n_{\ell m}$. Furthermore, we use different grids for each order n and spherical harmonic term ℓm to take into account the dependence of the radial function R on these factors.

Consider a second-order ODE,

$$\frac{\partial^2 x}{\partial r^2} + ikx = (a + ib), \tag{14}$$

where $x = y + iz$ and k, a, b, y and z are real. Then, by substitution,

$$\begin{aligned} \frac{\partial^2 y}{\partial r^2} - kz &= a \\ \frac{\partial^2 z}{\partial r^2} + ky &= b, \end{aligned} \tag{15}$$

which are equivalent to the four coupled equations

$$\tilde{y} = \frac{\partial y}{\partial r}, \quad \tilde{z} = \frac{\partial z}{\partial r} \tag{16}$$

$$\frac{\partial \tilde{y}}{\partial r} - kz = a, \quad \frac{\partial \tilde{z}}{\partial r} + ky = b. \tag{17}$$

By solving the above coupled equations (16) and (17) by an adaptive finite-difference method, we can obtain a solution accurate to machine precision. The adaptive-grid technique is crucial to solving for $\mathcal{F}$, but it is possible to accelerate the calculations substantially. We permit the grid to adapt such that the solution for the lowest frequency that we consider is accurate to a very high tolerance. For frequencies of $\mathcal{O}(10^{-6})$ Hz, we find that a typical realization of the adaptive solver will distribute 3073 nodes unevenly spaced with depth in a domain spanning the center of the Earth out to a distance of three Earth radii (i.e., the ring current source). The solution of these equations, with mesh adaption, takes of $\mathcal{O}(3\text{ s})$ on a high-performance workstation. We find that once this grid is established, it is generally unnecessary to readapt the grid for the remaining set of higher frequencies as long as the tolerance is reduced slightly. In this event, the solution is accelerated dramatically and is still of sufficient accuracy that the 3-D part of the model is not overwhelmed by numerical instabilities.

3 Minimizing complexity of the 3-D parameterization

Three-dimensional inversions benefit from a compact parameterization of the inverse model. We have devised a scheme for reducing the number of free parameters while retaining all of the features of a fully 3-D model. We represent the Earth as a sparse set of independent 1-D regional deep conductivity models. These models are represented either by a layered parameterization or a low-degree Chebychev polynomial. It must be emphasized that although the *parameterization* of the model space is in terms of regional 1-D models, the inversion is 3-D.

A 3-D model is generated by interpolating laterally the set of regional profiles at a set of discrete depths. The misfit between the response of the interpolated 3-D model and a set of observations is used to define an objective function for the inversion, and sensitivity functions appropriate to the 3-D problem are used. The behavior of the interpolator is critical in permitting a stable inversion; a new interpolator developed for this purpose is described in Section 3.3.

3.1 Regional 1-D parameterization

We wish to construct a set of geographically distributed regional 1-D conductivity profiles such that the intersections of the various profiles with the Earth's surface coincide with control points, some of which include locations of magnetic observatories. We make the assumption that the spatial variations of conductivity within the Earth are smooth. This acts as a form of regularization on the inversion and tends to inhibit unjustified levels of structure in the resulting models. Our intent is to obtain broad spatial averages of heterogeneous mantle structure. (In the future, we intend to construct full tensor global response functions more suitable than scalar c responses for inversion of global 3-D structure. This will lead to a commensurate improvement in resolving power.)

The parameterization discussed within this section is applied to the 3-D heterogeneous part of the model $\sigma_1(r, \theta, \phi)$. The perturbation-expansion forward solution involves both this term and a zeroth order term caused by an underlying 1-D conductivity structure $\sigma_0(r)$ upon which the 3-D perturbation is superimposed (see Section 2), $\sigma(r, \theta, \phi) = \sigma_0(r) + \sigma_1(r, \theta, \phi)$.

This underlying 1-D structure is completely independent of, and should not be confused with, the regional 1-D profiles that we use to generate the 3-D model. This is merely a convenient tool for enabling us to define a 3-D structure with a minimum number of parameters. The total conductivity at a given point within the model domain is the sum of the underlying zeroth-order 1-D conductivity structure and the 3-D perturbation interpolated at that point from the set of regional profiles.

Each 1-D regional profile is parameterized as follows: Beneath each point on the Earth's surface where a regional conductivity profile is to be defined [typically we might define 50–100 such profiles], the conductivity $\sigma(r, \theta_j, \phi_j)$, where θ_j, ϕ_j are the tangential coordinates of the jth profile, is represented as N layers or spherical shells

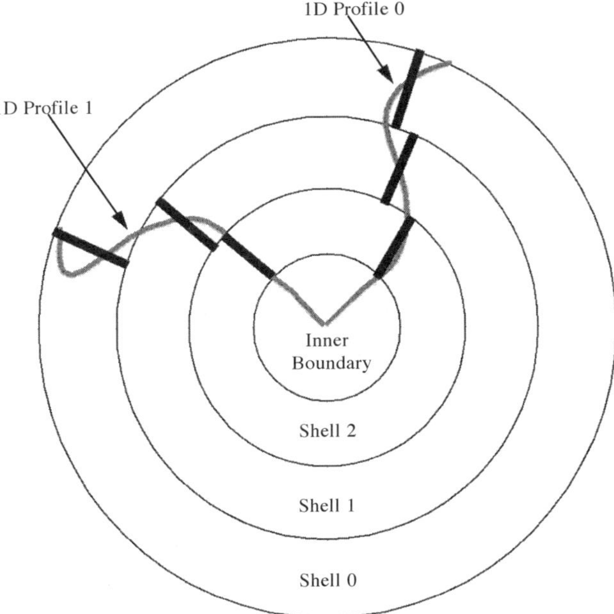

Figure 1. Idealized circular cross-section through Earth parameterized as two regionally 1-D conductivity profiles $\sigma(r, \theta_j, \phi_j)$ and the intersection of that parameterization with a set of heterogeneous nested spherical shells used in the interpolation of the model to three dimensions. The 1-D profiles are represented as thick gray curves. The thick black lines represent the mean value of the intersection of the profiles with the top and the bottom boundaries of each shell. It is these mean values that are interpolated laterally around each shell to form the 3-D model.

of fixed thickness. Typically, we divide the Earth into 5–25 such shells. The value of conductivity assigned to the intersection of each shell with the corresponding regional profile is set to the mean value of the profile at the top and the bottom of each shell. With 50 profiles and 25 shells, there are a total of 625 shell–profile intersections at which the conductivity is defined. Alternatively, there will be fifty 1-D vertical line segments through any given shell where the conductivity is defined. A view of this arrangement is seen in Fig. 1.

A final step is to interpolate between the 1-D profiles such that a fully 3-D conductivity distribution is defined. Within each shell we permit the conductivity to vary only with lateral position and not with radial depth, i.e., $\sigma_1 = \sigma_1(r_k, \theta, \phi)$, where r_k represents the index of the kth layer. The problem therefore is reduced from 3-D interpolation within the volume of a sphere to a more manageable problem of interpolating on the surface of a unit sphere, where each shell can be considered a separate unit sphere.

We have considered and tested a variety of methods of spherical surface interpolation, including surface splines (Dierckx, 1993), and a regularized inversion for spherical harmonic coefficients (Schultz and Zhang, 1994). All published methods that we tried introduced spurious features in the 3-D conductivity structure, particularly at points distant from the original regional profiles. The problem is exacerbated by the extreme sparseness of the observation points on the surface of the Earth. We describe in Section 3.3 a new interpolator more suitable for our purpose.

3.2 Optimal placement of surface nodes

The forward solver uses as basis functions a set of poloidal and toroidal potential functions expressed in terms of spherical harmonic coefficients. By carefully selecting integration nodes on the surface of the unit sphere that is used to represent each heterogeneous shell of the model, we find that the forward equations can be solved to a high degree of accuracy and with far less computational overhead than presented in the original work (Zhang and Schultz, 1992) in which the conductivity also was represented by spherical harmonics. It is essential, however, that the nodes be placed carefully on the surface so that sufficient accuracy is achieved. We have found it acceptable to use identical surface-node distributions for all shells within the model. This simplifies the calculation of the many surface integrals used in the solution of the forward problem and substantially reduces the overall computational overhead.

The inverse solution likewise benefits from the same method of node placement. Consider the case in which there may be 20 locations on the Earth's surface where acceptable response functions exist. A model parameterized by twenty 1-D regional profiles, where the intersection of each with the Earth's surface was one of the 20 observatories, is too restrictive. It is essential that lateral variations in structure, particularly inflections in the lateral gradient of conductivity, be permitted at points between observatories.

Given a set of 20 observatory locations associated with acceptable regional 1-D response functions, we have determined that to position at least one surface node approximately midway between the great-circle arc connecting each of the 20 constrained points with it's nearest neighbors, a total of approximately 56 points must be defined and distributed in a manner described below. By inserting such a set of nodes between nearest-neighboring-constrained points, and by defining for all 56 (20 constrained, and 36 unconstrained) points a regional 1-D parameterization, it is possible for the inversion to change freely between constrained points. By expressing the model in this form, any reasonably smooth and broad-scale 3-D conductivity distribution may be generated.

The problem of optimal placement of nodes on the surface of a sphere is a well-known exercise in computational geometry. The classical problem involves determining the number and placement of nodes, given the desired tesselation, e.g., a set of spherical tetrahedra of given geometry. The problem that we face is the inverse of this situation; we specify the number of surface nodes, some of which are fixed and some of which are free, and we require a placement of the free nodes that an optimal geometry is obtained. We determine the placement of nodes on the surface of the unit sphere through a simple optimization procedure. Initially, the specified number of nodes are distributed stochastically on the surface of the sphere. An iterative process adjusts the position of each node to maximize an objective function. After the positions of all nodes have been adjusted in sequence, the process repeats until either a desired value of the objective function has been obtained or the maximum specified number of iterations has been reached.

We use as the objective function the ∞-norm of a metric in a spherical basis. The great-circle distance γ_{kj} between any two nodes on the surface of the unit sphere is

$$\gamma_{kj} = \cos^{-1}(\cos\theta_k \cos\theta_j + \sin\theta_k \sin\theta_j \cos(\phi_k - \phi_j)), \tag{18}$$

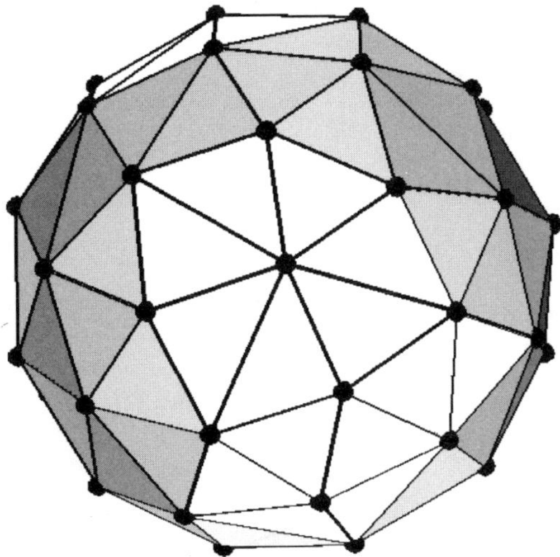

Figure 2. Optimal distribution of 100 points on the surface of the unit sphere. This is an unconstrained optimization, where all points are free to move such that objective function Φ is maximized. The Delaunay triangulation of this optimal distribution of points is seen as a set of facets, and each node has been connected to its nearest neighbors defined on the convex hull as described in Section 2.3.

where θ_k, ϕ_k is the geographic position of node k on the surface of the unit sphere, and θ_j, ϕ_j is the position of node j on the same surface. We define an objective function Φ

$$\Phi = \|[\gamma^{-1}]\|_\infty, \tag{19}$$

where $[\gamma^{-1}]$ is a vector containing the reciprocals of every possible combination of great-circle distance between all nodes on the unit sphere. Thus the optimization problem is to adjust the positions of the nodes such that Φ is maximized. Because Φ is merely the minimum separation between all possible combinations of great circles between nodes on the surface of the sphere, the problem is to maximize the *minimum* great-circle separation between any two points on the sphere.

A particular advantage of maximizing this objective function is that it is well adapted to constrained optimization. We permit the locations of each of the free parameters, i.e., the positions of each of the nodes, to be bounded between minimum and maximum permissible values. For unconstrained points, the bounds are $0 \le \theta_k \le \pi$, $0 \le \phi_k \le 2\pi$. For constrained points, the bounds are set explicitly to the desired geographic coordinates of the point. For instance, the constrained points may be set to the positions of magnetic observatories. An example of unconstrained optimization is seen in Fig. 2. Here, we have allowed 100 points to move freely on the surface of the unit sphere. We describe the triangulation seen in this figure in the next section. Observe, however, that the facets that connect the various nodes form nearly equilateral triangles. For specific numbers of points, the tesselation can lead to perfect equilaterals or other well-defined classes of polyhedra [see Everett and Schultz (1996)], but there is no closed-form solution for completely arbitrary numbers of points.

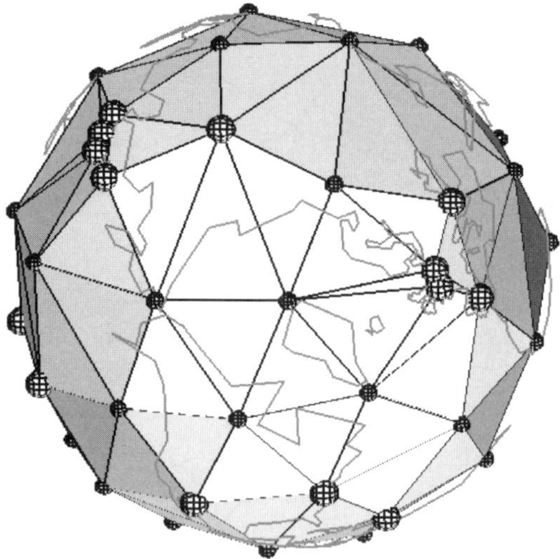

Figure 3. Optimal distribution of 52 points on the surface of the unit sphere. This is a constrained optimization, where 16 points are not free to move. Fixed nodes are denoted by large-diameter spheres, free nodes by small-diameter spheres. This solution produces the maximally equilateral set of all possible triangulations. The correspondence of the fixed nodes with magnetic observatory locations is seen by referring to the continental outlines displayed in the figure.

An example of a constrained optimization is found in Fig. 3. Here, a total of 52 nodes are defined, but the locations of 16 of these are set to the positions of specific magnetic observatories. The optimization method is simple. A table is built containing the indices of the end points of all possible great-circle combinations on the surface of the unit sphere. The algorithm is

```
For Iteration i; i = 1,Maximum number of iterations
   For Great Circle k; k = 1,Number of great circle combinations
      α = ||Great Circle k||₂ × 1/2
      For EndPoint j; j = 1, 2
         For l = 1, 4
            Move EndPoint j away from other EndPoint along Great
               Circle k by α;
            If new location outside bounds for given point, return it
               to previous position;
            If Φ increases then
               continue,
            Else
               α = −α × 3/4;
            Endif
         End;
      End;
   End;
End;
```

3.3 Convex-hull interpolation on a spherical domain

Solution of the forward equations requires many thousands of integration nodes within the spherical domain of the model. It is impractical to allow the conductivity at each of the integration nodes to be a free parameter. Calculation of the functional derivatives for a single iteration of the inversion would take many thousands of forward model evaluations. Many iterations per free parameter can be expected, requiring tens or hundreds of thousands of forward solutions. This would severely tax the current practical limits of any computer. Forward model construction therefore becomes a problem of (sparse) model parameterization and interpolation. This section describes the interpolation.

Much work has gone into using surface spline representations for interpolation on a sphere (Dierckx, 1993; Pevnyi, 1995). Our experience, however, is that spline representations tend to concentrate structure in the vicinity of the spline knots. There is no physical reason why models consisting of localized inflections on a bland background should be preferred for the case of global conductivity modeling. Spherical harmonic expansions are also popular, but suffer from the introduction of potentially unbounded structure at points distant from constraints. Our experience with a regularized inversion for spherical harmonic coefficients (Schultz and Zhang, 1994), designed to minimize spurious structure, showed that even the small residual oscillatory features introduced into the resulting 3-D models had a deleterious effect on the inversion. A new interpolation more resistant to introduction of spurious structure therefore was required.

The figures in the preceding section have illustrated the Delaunay triangulation of a set of points on the surface of the unit sphere. The Delaunay triangulation consists of the set of (spherical) triangles that, out of all possible sets, contains maximally small largest angles (i.e., maximally equilateral triangles). Thiessen tesselation is used to find the perpendicular bisectors to the great circles connecting the points. These, in turn, are used to construct the Delaunay triangulation [see Hoschek and Lasser (1993)].

Our new method of interpolation uses a Delaunay triangulation to construct a closed set of connected spherical triangles about each interpolation point, for which the interpolation point forms the vertex. The nearest neighbors to the interpolation point are defined as those connected to it by that set of adjacent triangles. The union of all spherical triangles is the convex hull of the point set. The interpolation works by taking the sum of its nearest neighbors from the set of points on the uninterpolated grid, weighted by a functional of the great-circle distances. The interpolated models generated by this method tend to be broad-scale, free from spurious structure far from observation points, and resistant to localizing structure immediately around individual observation points.

The value of the natural interpolant on a spherical convex hull for interpolating position θ_k, ϕ_k is

$$f_k = \sum_{j=1}^{NN} \frac{r_j}{\gamma_{kj}^2} \Bigg/ \sum_{j=1}^{NN} \frac{1}{\gamma_{kj}^2}, \tag{20}$$

where NN is the number of nearest neighbors to position θ_k, ϕ_k; r_j is the value of the function taken on at position j, i.e., where j is an index to the position of one of the nearest neighbors; and γ_{kj} is the great-circle distance from the interpolating position to the position of nearest neighbor j.

The interpolation algorithm is as follows:

For Interpolated Position θ_i, ϕ_i; i = 1,Number of interpolation points
  Augment Table Of Observation Points By Interpolated Position i;
  Generate Delaunay Triangulation On Surface Of Sphere For All Points In Table;
  Scan Connection List To Determine Nearest Neighbors of Interpolated Position i;
  For j = 1,Number Of Nearest Neighbors Of Interpolated Position i
    rin $=$ Sum$[r[Position_j]/\gamma[Position_i, Position_j]^2]$;
    weight $=$ Sum$[\gamma[Position_i, Position_j]^{-2}]$
  End;
  Interpolated Value At Position θ_i, ϕ_i $=$ rin/weight;
  Restore Original Table Of Observation Points;
End;

To construct a regular grid of interpolated values (as seen in Fig. 4), it is necessary to carry out completely independent interpolations for each point. Starting from a defined

Figure 4. Natural interpolation on a spherical convex hull. Colored spheres represent the locations where the surface of the model Earth intersects published individual 1-D regional conductivity profiles (Schultz, 1990). The color of a sphere indicates the departure of the profile from a global 1-D (radially symmetric) reference model, at the point at which the particular profile intersects a radial depth of 660 km. Red spheres represent anomalously high conductivity; blue spheres anomalously low conductivity. The colored facets making up the surface of the model Earth represent the value of the natural interpolant on the convex hull (according to the same color scale as the spheres). Observe the smoothness of the interpolated conductivity distribution, and the lack of spurious structure even at substantial distances from fixed points.

grid of interpolation positions, we augment the original array of observation positions by each of these interpolation points in turn. We carry out Delaunay triangulation on the surface of the unit sphere to determine those observation points that are the nearest neighbors, on a convex hull consisting only of the original points and the one interpolation point that we have added. As seen in Eq. (20) the value of the function to be interpolated (i.e., σ) is the weighted sum of the values that the function takes on at each of the nearest neighbors. The weighting is the square of the great-circle distance from the interpolation point to each of the nearest neighbors. We repeat this process, independently, for all other requested interpolation points.

This is a simple distance-based interpolating function, and the value taken on by the interpolant is more strongly influenced by local points than distant points. The degree by which the influence of distant points may be felt is modified easily by changing the weight from the square of the great-circle distances to some other power. For instance, the one-norm tends to accentuate more distant points relative to near points compared with the two-norm. We find that using such a simple and natural interpolator on the surface of the unit sphere leads to a very stable and smooth interpolated function well suited for representing the 3-D conductivity perturbation of our models.

The definition of the interpolant in Eq. (20) allows us to define conveniently a differential operator on the surface of the unit sphere. Consider, for instance, the partial derivative

$$\frac{\partial f_k}{\partial \theta_k} = \left(2 \Big/ \sum_{j=1}^{NN} \gamma_{kj}^{-2}\right)\left[f_k \sum_{j=1}^{NN} \gamma_{kj}' \gamma_{kj}^{-3} - \sum_{j=1}^{NN} r_j \gamma_{kj}' \gamma_{kj}^{-3}\right], \tag{21}$$

where

$$\gamma_{kj}' = \frac{\partial \gamma_{kj}}{\partial \theta_k} = \frac{-a_{kj}}{\sin(\gamma_{kj})}, \tag{22}$$

and $a_{kj} = -\cos\theta_j \sin\theta_k + \cos\theta_k \cos(\phi_j - \phi_k)\sin\theta_j$. The derivative $\partial f_k/\partial \phi_k$ can be similarly derived. These differential operators have the property that the derivative is defined at all points, including the nearest-neighbor nodes.

The solution of the 3-D forward problem of Section 2 requires the derivatives $\partial\sigma/\partial\theta$ and $\partial\sigma/\partial\phi$ to be defined everywhere within the model domain. To make use of the natural differential operator for representing these derivatives, high levels of accuracy are required. This requires us to tesselate the Earth very finely. One could attempt this by repeating the method of Section 3.2 but, for the thousands of surface nodes required, the problem becomes computationally difficult. We have seen that it is possible to carry out direct constrained (or unconstrained) optimization of the positions of (100) nodes on a high-performance workstation, but the number of operations grows geometrically with the number of free parameters. As a means of accelerating the process while still retaining optimal-like properties of the tesselation, we have devised a method of self-similar facet subdivision that we describe in Section 3.4. We find that by subdividing the facets into smaller units, we can rapidly obtain rms misfits of 0.1% or less, as required, for the analytical test cases that we have tried thus far.

3.4 *Self-similar spherical Delaunay facet subdivision*

Each facet of the Delaunay triangularization is identified by connecting each set of three adjacent points in the connection list. To subdivide each facet once, the three great-circle arcs bounding a given facet are subdivided by locating the three midpoints

Table 1. Sequence of subdivisions of triangulation

Subdivision	No. of nodes	No. of facets
0	100	196
1	394	784
2	1570	3136
3	6274	12544

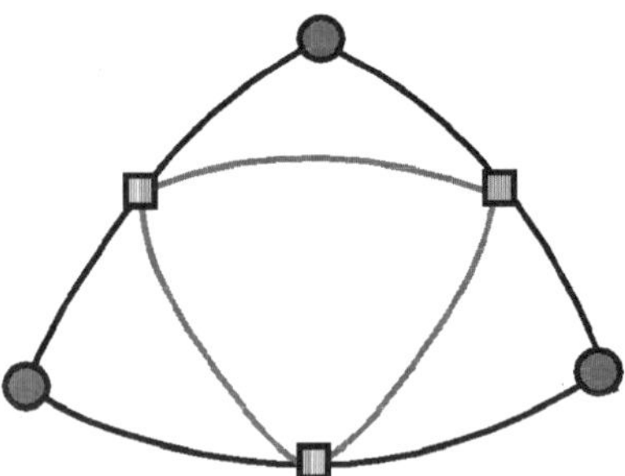

Figure 5. Self-similar subdivision of idealized spherical triangular facet. The outer boundary of a nearly equilateral spherical triangular facet resulting from Delaunay triangularization is shown by the thick lines connecting the three primary nodes denoted by solid circles. The primary facet is stellated by inserting secondary nodes midway along the great-circle arcs connecting the primary nodes. The secondary nodes are connected by great-circle arcs. The resulting figure is a set of four spherical triangles of identical geometry. The process can be repeated iteratively until the desired scale of tesselation is reached. All facets so generated will retain the geometry of the original facet.

of the great-circle arcs, and then by introducing these as new subsidiary nodes. These nodes augment the existing nodes, and the new superset of nodes is retriangularized by updating the connection list of the Delaunay triangularization.

Given a set of nodes on a spherical convex hull, none of which is a boundary node, the number of spherical triangular facets is given by

$$N_t = 2N_n - 4, \tag{23}$$

where N_t is the number of triangular facets and N_n is the number of nodes. If we return to the tesselation of Fig. 2, i.e., 100 optimally located surface nodes, we see that this will result in 196 spherical triangular surface facets. Each time the triangulation is subdivided as described above, the number of nodes will increase by $4N_n - 6$. Consider the sequence of subdivisions of the 100-point triangulation in Table 1. A schematic view of the facet subdivision method appears in Fig. 5. A global triangulation after one subdivision is seen in Fig. 6.

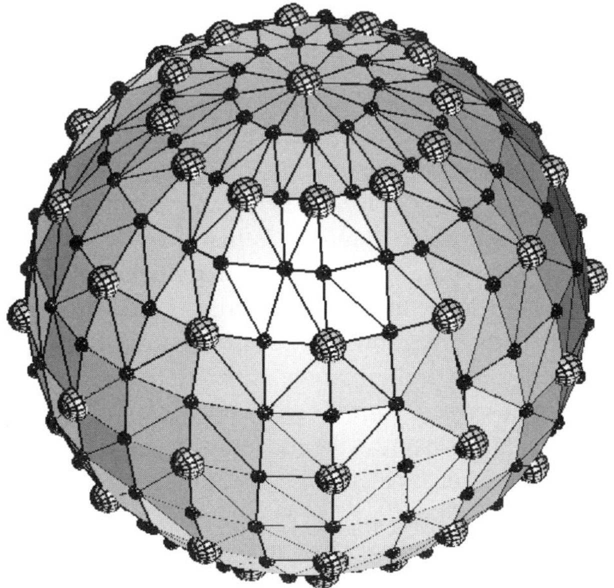

Figure 6. Self-similar subdivision of regularly spaced grid of primary nodes (large spheres). The subsidiary nodes generated by the subdivision process appear as small spheres, and the spherical triangular surface facets of the Delaunay triangulation of the augmented set of nodes is displayed.

For the 100-point set displayed in Fig. 2, we find that to obtain sufficient accuracy for the calculation of the differential operator described above, and for the surface and volume integrator described in Section 3.5, it is usually sufficient to subdivide the grid three times. The method of facet subdivision seen here imposes little computational burden on these calculations because it is inherently fast and needs be carried out only once for all subsequent calculations.

3.5 Surface and volume integration on the convex hull

The Delaunay triangulation of optimally located surface nodes lends itself to a convenient method of surface quadrature. The surface integral of any function defined on the unit sphere is approximated within the area of any individual spherical triangular facet by taking the mean of the values of the function at the three vertex nodes of the triangle normalized by its surface area. To determine the surface integral over the unit sphere, the integrals for all the facets are summed.

This method is extended easily for volume integration. Because all heterogeneous shells describing our 3-D model are tesselated identically, the surface nodes for one shell connect directly along the radial direction to the surface nodes for the underlying shell. This allows one to trivially construct a set of five-sided solid spherical triangular polyhedra. The volume of each polyhedron is known exactly, and the values of the function on each of the six vertices defining the polyhedron also are known. It is therefore trivial to extend the surface integral as defined above into a volume integral, and this method is used extensively in the solution of the forward equations of Section 2. We find that this approach allows one to easily define a sufficient number of nodes within the model domain for an accurate solution to be obtained.

4 Inversion through spectral expansion

Given the linearized nature of the 3-D forward solution, it is reasonable to carry out a linearized inversion. We follow the formalism of Backus and Gilbert (1970) and adopt the spectral expansion method of Parker (1994). This approach defines resolution kernels through which the true conductivity distribution is averaged to yield the given inverse model. It is also possible to estimate error bounds on the model (Parker, 1994). The drawback to such a linearized approach is that the resolution kernels and error bounds are valid only if the starting model is linearly close to the true conductivity distribution. It is difficult to define the maximum excursion from the true distribution to which one may go prior to the breakdown of the linearization. This uncertainty provides the rationale for carrying out a preliminary nonlinear model search, prior to initiating the linearized inversion described below.

Spectral expansion methods express the model in the basis of the data. The model, data, and data kernels are taken as elements of a Hilbert space, where the data are a linear functional on the model, expressed as an inner product with a data kernel. In this case, this data kernel is the functional derivative appropriate to the space of 3-D models under consideration. This inner product is written as

$$\tilde{\mathcal{Z}}_j = \int_V \tilde{\mathcal{G}}_j(\rho) m(\rho)\, d\rho, \tag{24}$$

where $\tilde{\mathcal{Z}}_j$ is the jth datum normalized by the standard error, $\tilde{\mathcal{G}}$ is the functional derivative, m is the model, $d\rho$ is a volume element, and j is the index of the frequency.

The Gram matrix is written

$$\Gamma_{ij} = \int_V \tilde{\mathcal{G}}_i(\rho) \tilde{\mathcal{G}}_j(\rho)\, d\rho. \tag{25}$$

An orthogonal decomposition is used to produce the spectral eigenvalues,

$$\underset{\sim}{\mathbf{O}}^H \underset{\sim}{\mathbf{\Gamma}} \underset{\sim}{\mathbf{O}} = \underset{\sim}{\mathbf{\Lambda}} \tag{26}$$

where the superscript H denotes Hermitian transpose, and $\mathbf{\Lambda}$ is a matrix of eigenvalues λ_j such that $\lambda_j > \lambda_{j+1}$ for all j. The model is taken to be the linear combination of data kernels

$$m(\rho) = \sum_{i=1}^{N} a_i g_i(\rho), \tag{27}$$

where

$$a_i = \lambda_i^{-1/2} \sum_j O_{ji} \tilde{\mathcal{Z}}_j, \tag{28}$$

$$g_i(\rho) = \lambda_i^{-1/2} \sum_j O_{ji} \tilde{\mathcal{G}}_i(\rho). \tag{29}$$

This is the general spectral solution. One may regularize the inversion by making use of the projection theorem, and truncating the expansion as:

$$\tilde{m}(\rho) = \sum_{i=1}^{M} a_i g_i(\rho), \tag{30}$$

where $M \leq N$. The norm of the model is given

$$\|\tilde{m}\|^2 = \sum_{j=1}^{M} \frac{\tilde{\mathcal{Z}}_j^2}{\lambda_j}. \tag{31}$$

The point in the series at which the λ are truncated can be determined by examining the trade-off between model–response misfit and the norm of the model.

4.1 Functional derivatives

The linearized inverse formulation requires the calculation of functional derivatives (although Fréchet differentiability has not been proven formally for the 3-D EM induction problem, it is almost certainly true that the Gateaux-derivable functional derivatives, or sensitivity functions, presented below are Fréchet derivatives as well). The specific form of the functional derivatives depends very critically on the parameterization. The simplest approach is to follow Smith and Booker (1991) or Oldenburg and Ellis (1991) and to calculate pseudo-Fréchet derivatives sensitive to changes in the c response at every surface node resulting from changes in any of the model parameters. This formulation, however, would require us to take the set of response functions at each of the observation points, and to interpolate them laterally on the Earth's surface such that there exists an interpolated response associated with the intersection of each regional 1-D profile with the Earth's surface. This introduces bias into the inversion (the interpolated response will not necessarily satisfy Maxwell's equations at all points). We prefer to calculate fully 3-D derivatives, in which we calculate changes in the c response at the observatories (i.e., only those locations where there are data) resulting from changes in any model parameter. Such a sensitivity function obviates the need for interpolating the response function laterally.

The change in c at each of the observatories, δc with respect to the change in the model, $\delta\sigma$ is

$$\delta c|_{r_e} = \int_0^{r_e} dr\, \delta\sigma\, D[\delta\sigma, c], \tag{32}$$

where D is a functional derivative defined below. In the potential mode formulation of the Helmholtz equation, c is given by Zhang and Schultz (1992),

$$c = \left[\frac{1}{\sin\theta}\frac{\partial}{\partial\phi}\frac{\partial(r\Psi)}{\partial r} + \frac{\partial}{\partial\theta}\frac{\partial^2(r\Pi)}{\partial r^2}\right] \Big/ \left[\frac{1}{\sin\theta}\frac{\partial\Psi}{\partial\phi} + \frac{1}{r}\frac{\partial}{\partial\theta}\frac{\partial(r\Pi)}{\partial r}\right]. \tag{33}$$

A small change in the conductivity, $\delta\sigma$, results in changes to the poloidal and toroidal potential modes such that

$$\Pi \to \Pi + \delta\Pi, \tag{34}$$

and

$$\Psi \to \Psi + \delta\Psi. \tag{35}$$

Substitution of Eqs. (34) and (35) into Eqs. (33) and (8) leads to the functional derivative D:

$$
\begin{aligned}
D[\delta\sigma, c] &= \left\{ \frac{\dfrac{2}{\delta\sigma}\left[\dfrac{1}{\sin\theta}\dfrac{\partial}{\partial\phi}\dfrac{\partial(r\Psi)}{\partial r} + \dfrac{\partial}{\partial\theta}\dfrac{\partial^2(r\Pi)}{\partial r^2} - 1\right]}{\dfrac{r}{\sin\theta}\dfrac{\partial\Psi}{\partial\phi} + \dfrac{\partial}{\partial\theta}\dfrac{\partial(r\Pi)}{\partial r}} \right\} \left\{ \frac{1}{\sin\theta}\frac{\partial}{\partial\phi}\frac{\partial(r\delta\Psi)}{\partial r} + \frac{\partial}{\partial\theta}\frac{\partial^2(r\delta\Pi)}{\partial r^2} \right\} \\
&+ \frac{1}{\delta\sigma}\frac{\partial \ell n\sigma}{\partial r}\left\{ \frac{1}{\sin\theta}\frac{\partial}{\partial\phi}\frac{\partial(r\delta\Psi)}{\partial r} + \frac{\partial}{\partial\theta}\frac{\partial^2(r\delta\Pi)}{\partial r^2} \right\} - i\omega\mu_0 r \\
&- \frac{\left\{ \dfrac{\partial}{\partial\phi}\nabla^2\Psi - \dfrac{\partial \ell n\sigma}{\partial\phi}\nabla^2_{\theta\phi}\Psi + \dfrac{\partial}{\partial\theta}\dfrac{\partial}{\partial r}\left(\dfrac{1}{r}\nabla^2_{\theta\phi}\Pi\right) - \dfrac{\partial \ell n\sigma}{\partial r}\dfrac{1}{r}\dfrac{\partial}{\partial\theta}\nabla^2_{\theta\phi}\Pi \right\}\left[\dfrac{r}{\sin\theta}\dfrac{\partial\delta\Psi}{\partial\phi} + \dfrac{\partial}{\partial\theta}\dfrac{\partial(r\delta\Pi)}{\partial r} \right]}{r\delta\sigma\sin\theta\left[\dfrac{r}{\sin\theta}\dfrac{\partial\Psi}{\partial\phi} + \dfrac{\partial}{\partial\theta}\dfrac{\partial(r\Pi)}{\partial r} \right]\left\{ \dfrac{r}{\sin\theta}\dfrac{\partial\Psi+\delta\Psi}{\partial\phi} + \dfrac{\partial}{\partial\theta}\dfrac{\partial[r(\Pi+\delta\Pi)]}{\partial r} \right\}} \\
&- \frac{\left(\dfrac{\partial}{\partial\phi}\nabla^2_{\theta\phi}\delta\Psi - \dfrac{\partial \ell n\sigma}{\partial r}\nabla^2_{\theta\phi}\delta\Psi \right)}{r\delta\sigma\sin\theta\left[\dfrac{r}{\sin\theta}\dfrac{\partial\Psi}{\partial\phi} + \dfrac{\partial}{\partial\theta}\dfrac{\partial(r\Pi)}{\partial r} \right]\left\{ \dfrac{r}{\sin\theta}\dfrac{\partial(\Psi+\delta\Psi)}{\partial\phi} + \dfrac{\partial}{\partial\theta}\dfrac{\partial[r(\Pi+\delta\Pi)]}{\partial r} \right\}} \\
&+ \frac{\dfrac{\partial}{\partial\theta}\dfrac{\partial}{\partial r}\left(\dfrac{1}{r}\nabla^2_{\theta\phi}\delta\Pi - \dfrac{\partial \ell n\sigma}{\partial r}\dfrac{1}{r}\dfrac{\partial}{\partial\theta}\nabla^2_{\theta\phi}\delta\Pi\right)\left[\dfrac{r}{\sin\theta}\dfrac{\partial\Psi}{\partial\phi} + \dfrac{\partial}{\partial\phi}\dfrac{\partial(r\Pi)}{\partial\theta} \right]}{\left[\dfrac{r}{\sin\theta}\dfrac{\partial\Psi}{\partial\phi} + \dfrac{\partial}{\partial\theta}\dfrac{\partial(r\Pi)}{\partial r} \right]\left\{ \dfrac{r}{\sin\theta}\dfrac{\partial(\Psi+\delta\Psi)}{\partial\phi} + \dfrac{\partial}{\partial\theta}\dfrac{\partial[r(\Pi+\delta\Pi)]}{\partial r} \right\}}, \qquad (36)
\end{aligned}
$$

where

$$\delta\sigma \equiv \delta\sigma(r, \theta, \phi), \qquad (37)$$

$$\delta c \equiv \delta c(\omega, r, \theta, \phi), \qquad (38)$$

$$\Psi \equiv \Psi(\omega, r, \theta, \phi), \qquad (39)$$

$$\Pi \equiv \Pi(\omega, r, \theta, \phi), \qquad (40)$$

and the surface Laplacian is written as

$$\nabla^2_{\theta\phi} = \frac{1}{\sin\theta}\frac{\partial}{\partial\theta}\left(\sin\theta\frac{\partial}{\partial\theta}\right) + \frac{1}{\sin^2\theta}\frac{\partial^2}{\partial\phi^2}. \qquad (41)$$

In Eq. (36), Ψ and Π are the total potential functions of Eqs. (9) and (10), and $\delta\Psi$ and $\delta\Pi$ are the changes in the total potential functions resulting from change $\delta\sigma$ in the model.

The numerical calculation of the functional derivative is tied intrinsically to the model search engine. Our inversion procedure involves two stages: In the first, we do not calculate D, the sensitivity of the reponse with respect to changes in the model. Rather, we carry out a nonlinear model search in which we minimise an objective function (typically the rms misfit to the data) in an iterative process. This strategy follows the same algorithm used in the nonlinear optimization for optimal location of points on the surface of the sphere. We change a single parameter of the model. If the objective function improves, we continue in the same search direction. If the value of the objective function degrades, we reverse the search direction, but reduce the stride of the model perturbation. This repeats four times for each parameter; thus, for a single iteration of a 280-parameter inversion (e.g., 56 regionally 1-D profiles each described by five heterogeneous shells), 1120 forward solutions are calculated. The nonlinear optimization continues until a predetermined misfit level has been set, or until we are satisfied that a valley in the misfit surface has been entered.

Following the nonlinear model search, we initiate the Backus–Gilbert linearized spectral inversion described in Eqs. (24–31). In this case, the functional derivative of Eq. (36) is used to construct the Gram matrix of Eq. (25), i.e., $\tilde{\mathcal{G}}_i(\rho) \equiv D[\delta\sigma(\rho), c]$.

5 Results and discussion

We have carried out a series of inversions using the c response functions. The primary purpose of these inversions has been to demonstrate that our algorithm converges in a tractable timescale. To this end, only small parameterizations have been used thus far, limiting the potential for extracting the maximum information from the data. The inversions were performed for a set of 280 free parameters, consisting of 56 five-layered 1-D profiles, in which the depths of shell boundaries were fixed, but the conductivities were free to vary. Preliminary results are presented for the inversion of data from 20 geomagnetic observatories and 16 discrete frequencies. The data used in the inversion were c responses obtained from three sources: 16 observatories of data from Schultz and Larsen (1987), 3 from Schultz and Semenov (1993), and 1 from Schultz et al. (1993).

With the D'Yakanov solution of Everett and Schultz (1995) as a benchmark, the accuracy of the perturbation forward solution was tested for anomalies in the skin-depth and physical-depth ranges of interest. Setting a limit of 2% as the maximum acceptable error in the model anomalous c response (i.e., in that part of the response generated by the 3-D part of the perturbation expansion), perturbations of up to 200% were modeled. The starting model $\sigma_1(r, \theta, \phi)$ used in the inversion was set to vary with r only. [This should not be confused with the zeroth-order radially symmetric model $\sigma_0(r)$ about which the forward solver calculates the 3-D perturbation; see Section 3.] Although many choices are possible, we initially set the underlying 1-D model $\sigma_0(r)$ to have excessive conductance in the upper mantle. This was a good test for the initial nonlinear model search engine rather than a practical attempt to generate a model with geodynamical significance. For this test of convergence, we also set the starting model to be identical to $\sigma_0(r)$, i.e., the starting perturbation was zero. The depths to the spherical shell boundaries for this model were set to 200, 400, 600, 1000, and 2000 km. The corresponding conductivities of these layers were 0.17, 0.28, 0.44, 0.81, and 1.12 S/m. Below 2000 km, the model was terminated by a high conductor.

The rms misfit between the response of the underlying and inappropriate 1-D model and the observations was 5.05. The nonlinear part of the inversion for the 3-D perturbations to this underlying 1-D model reduced the initial misfit by 43%. The linearized spectral expansion part of the inversion used the result of the initial nonlinear model search as a starting model. This reduced the rms misfit by an additional 5%. The underlying $\sigma_0(r)$ was held constant during this process. The longest inversion runs lasted for 32 iterations. The time of such a run on a workstation (99-MHz Hewlett Packard PA-RISC) is about two weeks. Note that this run time corresponds to an execution time of 2.5 min per 16 frequency-forward-problem solution.

Although a good test for the convergence properties of the algorithm, the model produced by this test is of little practical significance. The starting model is grossly inappropriate. The result of the nonlinear model search is also grossly inappropriate as a starting model for the linearized part of the inversion. This is because $\sigma_0(r)$ was not

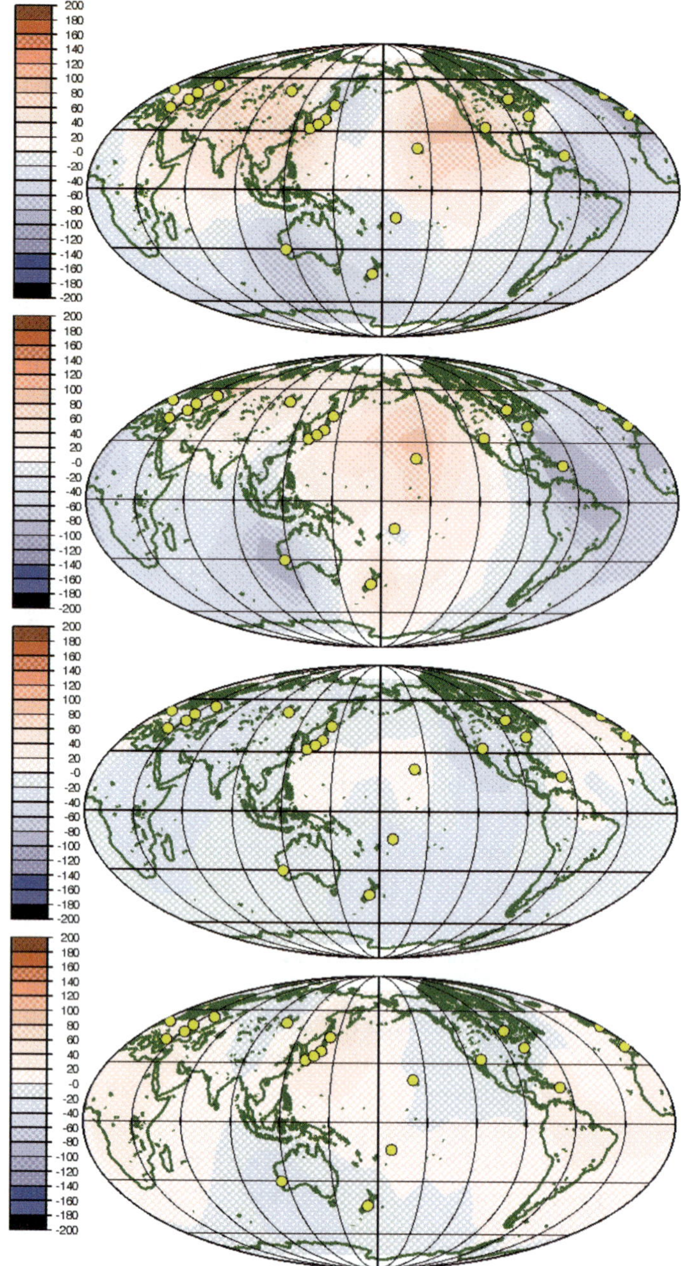

Figure 7. Model for the 3-D distribution of electrical conductivity in a set of four laterally heterogeneous shells extending (from top to bottom) 0–436, 436–774, 774–1630, and 1630–2628 km from the Earth's surface. Each surface (plotted using a Molliweide equal-area projection) represents the percentage deviation away from a globally averaged 1-D electrical conductivity model at the given depth. Yellow dots are geomagnetic observatories; green dots are extra parameter points. At all dots, a local 1-D model is defined and these are interpolated using the method described in Section 3.3 to produce the 3-D surface.

permitted to change during the inversion. One should either permit the value of $\sigma_0(r)$ to be a set of free parameters in the inversion, or alternatively one should ensure that an appropriate starting model is used.

We present below a 3-D conductivity model generated with a more appropriate $\sigma_0(r)$ terminated by a highly conductive core located at the depths of the core–mantle boundary. In this case, the layer thicknesses and conductivities of $\sigma_0(r)$ were determined by permitting them to be free parameters in a genetic algorithm (Schultz et al., 1994). This was used to find a global 1-D model with minimum rms misfit. The conductivities were 0.001, 0.311, 1.397, 2.136 S/m, and the corresponding shell depths were 436, 774, 1630, 2628, and 2891 km. The model was terminated by a conducting core of 11548.73 S/m.

Figure 7 contains a set of color-coded contour plots of the heterogeneities seen in each laterally heterogeneous shell k of the 3-D model, i.e., $\sigma_1(r_k, \theta, \phi)$. The rms misfit of this model was 2.1. Noted that the observed variation of the conductivity is no greater than 200% in any of the shells. Thus the size of the perturbation lies within the range that we have found to be acceptable for the perturbation approach, based on validations against the quasi-analytic D'Yakanov solution (Everett and Schultz, 1995). The sparse parameterization of the model, and the relatively low resolution of the data, has lead to broad-scale smooth structures.

6 Conclusion

We have described an inverse solution for the electrical conductivity structure of a heterogeneous sphere. The need for such a 3-D solver is now apparent from global and regional EM induction studies. Our preliminary inversions suggest that the methodology leads to a tractable inverse problem, but it is also clear that a more exhaustive parameterization is required to satisfy the data fully. To increase the parameterization, it would be desirable to ensure that all great-circle arcs connecting observation points have an intervening parameter point, such that inflections in the lateral conductivity gradients can be accommodated. We believe that the Witteveen (Holland) and Panagyurishte (Bulgaria) observatories are associated with rapid lateral structural changes not presently permitted in the model search. The increase in parameter points (to approximately 100) to accommodate these changes will require us to move to a new class of computer, which we are currently undertaking.

We are also working toward providing practical examples of inversion of synthetic data sets, as well as global sets of experimental data of use in constraining the state of the Earth's mantle.

Acknowledgments

The authors acknowledge the support of (UK) Natural Environment Research Council Grant No. GR3/8705 under which this work was carried out. G.P.'s work was further supported by a Shell (UK) Postgraduate Studentship. Our code for generating a convex hull on the surface of the unit sphere incorporates public-domain source code by Robert Renka of Oak Ridge National Laboratories (US). Our conversations with Mark Everett during the preparation of this manuscript have proved most useful. The perturbation-expansion forward solution used here was based on the original version written by Tianshan Zhang, whom we gratefully acknowledge. Contribution number

4864, Department of Earth Sciences and Institute of Theoretical Geophysics, University of Cambridge.

References

Backus, G. E., Gilbert, J. F., 1970, Uniqueness in the inversion of inaccurate gross Earth data: Philos. Trans. R. Soc. London Ser. A **266**, 187–269.

Braun, J., and Sambridge, M., 1994, Dynamical Lagrangian remeshing (DLR): A new algorithm for solving large strain deformation problems and its application to fault-propagation folding: Earth and Planet. Sci. Lett., **124**, 211–220.

———1995, A numerical method for solving partial differential equations on highly irregular evolving grids: Nature, **376**, 655–660.

Dierckx, P., 1993, Curve and surface fitting with splines. Monographs on numerical analysis: Oxford Univ. Press, Inc.

Egbert, G., and Booker, J. R., 1992, Very long period magnetotellurics at Tucson observatory: Implications for mantle conductivity: J. Geophys. Res., **97**, 15 099–15 112.

Egbert, G., Booker, J. R., and Schultz, A., 1992, Very long period magnetotellurics at Tucson observatory: Estimation of impedances: J. Geophys. Res., **97**, 15 113–15 118.

Everett, M. E., and Schultz, A., 1995, Geomagnetic induction in eccentrically nested spheres: Phys. Earth Planet. Int., **92**, 189–198.

———1996, Geomagnetic induction in a heterogeneous sphere: Azimuthally symmetric test computations and the response of an undulating 660-km discontinuity: J. Geophys. Res., **101**, 2765–2783.

Golias, N. A., Tsiboukis, T. D., 1992, 3-Dimensional automatic adaptive mesh generation: IEEE Trans. Magn., **28**, 1700–1703.

Heinson, G. S., and Constable, S. C., 1992, The electrical conductivity of the oceanic upper mantle: Geophys. J. Internat., **110**, 159–179.

Hoschek, J., and Lasser, D., 1993, Fundamentals of computer-aided geometric design (English transl. by L. Schumaker): A.K. Peters.

Lizarralde, D., Chave, A. D., Hirth, G., and Schultz, A., 1995, Long period magnetotelluric study using Hawaii-to-California submarine cable data: Implications for mantle conductivity: J. Geophys. Res., **100**, 17 837–17 854.

Oldenburg, D. W., and Ellis, R. G., 1991, Inversion of geophysical data using an approximate inverse mapping: Geophys. J. Internat., **105**, 325–353.

Parker, R. L., 1980, The inverse problem of electromagnetic induction: Existence and construction of solutions based on incomplete data: J. Geophys. Res., **85**, 4421–4428.

———1994, Geophys. inverse theory: Princeton Univ. Press.

Pevnyi, A. B., 1995, Spherical splines and interpolation on a sphere: Comput. Math. Math. Phys., **35**, 109–112.

Sambridge, M., Braun, J., and McQueen, H., 1995, Geophysical parametrization and interpolation of irregular data using natural neighbours: Geophys. J. Internat., **122**, 837–857.

Schmucker, U., 1970, Anomalies of geomagnetic variations in the Southwestern United States: Bull. Scripps Inst. Oceanogr., **13**, 1–165.

Schultz, A., 1990, On the vertical gradient and associated heterogeneity in mantle electrical conductivity: Phys. Earth Planet. Int. **64**, 68–86.

Schultz, A., and Larsen, J. C., 1987, On the electrical conductivity of the mid-mantle—I. Calculation of equivalent scalar magnetotelluric response functions: Geophys. J. Roy. Astron. Soc., **88**, 733–761.

———1990, On the electrical conductivity of the mid-mantle—II. Delineation of heterogeneity by application of extremal inverse solutions: Geophy. J. Internat., **101**, 565–580.

Schultz, A., and Semenov, V. Y., 1993, Modeling the electrical conductivity structure of the Earth's mantle (in Russian), Phyz. Zemli, **10**, 39–44.

Schultz, A., and Zhang, T. S., 1994, Regularised spherical harmonic analysis and the three dimensional electromagnetic response of the earth: Geophys. J. Internat., **116**, 141–156.

Schultz, A., Everett, M. E., Pritchard, G., and Smith, J. T., 1994, Nonlinear optimisation strategies in electromagnetic inversion: Proc. 12th Workshop on electromagnetic induction in the Earth, Brest, France.

Schultz, A., Kurtz, R. D., Chave, A. D., and Jones, A. G., 1993, Conductivity discontinuities in the upper mantle beneath a stable craton: Geophys. Res. Lett., **20**, 2941–2944.

Smith, J. T., and Booker, J. R., 1991, Rapid inversion of two- and three-dimensional magnetotelluric data: J. Geophys. Res., **96**, 3905–3922.

Tarits, P., Chave, A. D., and Schultz, A., 1993, Comment On "The electrical conductivity of the oceanic upper mantle" Heinson, G., and Constable, S., Geophys. J. Internat., **114**, 711–716.

Tarits, P., Wahr, J., and Lognonne, P., 1995, A solution to the problem of internal and external electromagnetic induction in a spherical heterogeneous earth, EOS Trans. Amer. Geophys. Union Fall Mtg., **76**, 165–166.

Tsai, V. J. D., 1993, Fast topological constructions of Delaunay triangulations and voronoi diagrams: Comput. and Geosci., **19**, 1463–1474.

Weaver, J. T., 1992, Adjustment distance in TM mode electromagnetic induction: Geophys. J. Internat., **108**, 293–300.

Zhang, T. S., and Schultz, A., 1992, A three-dimensional perturbation solution for the EM induction problem in a spherical earth—the forward problem: Geophys. J. Internat., **111**, 319–334.

PART VI

MINING AND EXPLORATION GEOPHYSICS

Electromagnetic (EM) techniques are widely used in exploration for massive sulphides, other conductive targets, and for depth sounding.

McCracken, Oristaglio, and Hohmann (1986a)

Modeling in Mining Geophysics: When, Where, and How?

Michael W. Asten

Summary. Codes for 3-D electromagnetic (EM) modeling are limited by hardware and time constraints to models that only crudely resemble the geological structures encountered in mineral exploration. They are many orders of magnitude away from execution speeds that would allow routine inversion for models of realistic size and complexity. Nevertheless, important advances in the next few years will come from approximate 3-D methods. Examples are inversion of airborne and reconnaissance ground data in the inductive limit (high-frequency approximation), inversion by an EM analog of seismic migration, and modeling and inversion by fast algorithms allowing multiple plates in a conductive earth in the resistive limit (low-frequency approximation). Full 3-D modeling will be necessary for the development and validation of these approximations.

1 Introduction

The past 15 years have seen many developments in how we use and interpret 3-D electromagnetic (EM) data by modeling or inversion. A rough chronology of general practice might include the following milestones:

- 1982: Profiles of raw data, plotted by hand or from analog instruments; interpreted using nomograms or rules of thumb.
- 1988: Profiles plotted from digital recording, filtered, and enhanced; interpreted with the aid of modeling software such as PLATE (Dyck and West, 1984).
- 1994: Profiles transformed to images of conductivity (or conductance) versus depth by assuming a 1-D geometry at each point along a profile (Macnae et al., 1991; Smith and Buselli, 1991; Liu and Asten, 1994); interpreted with 3-D multiple-conductor software such as MULTILOOP (Lamontagne et al., 1988).

This paper uses parasection to denote a 2-D representation of data—in distance versus depth coordinates—that is made by a transformation based on an approximate (usually, layered) model. Parasections, also called images, contain artifacts caused by the ap-

BHP Research, 245 Wellington Road, Mulgrave, Victoria 3170, Australia.

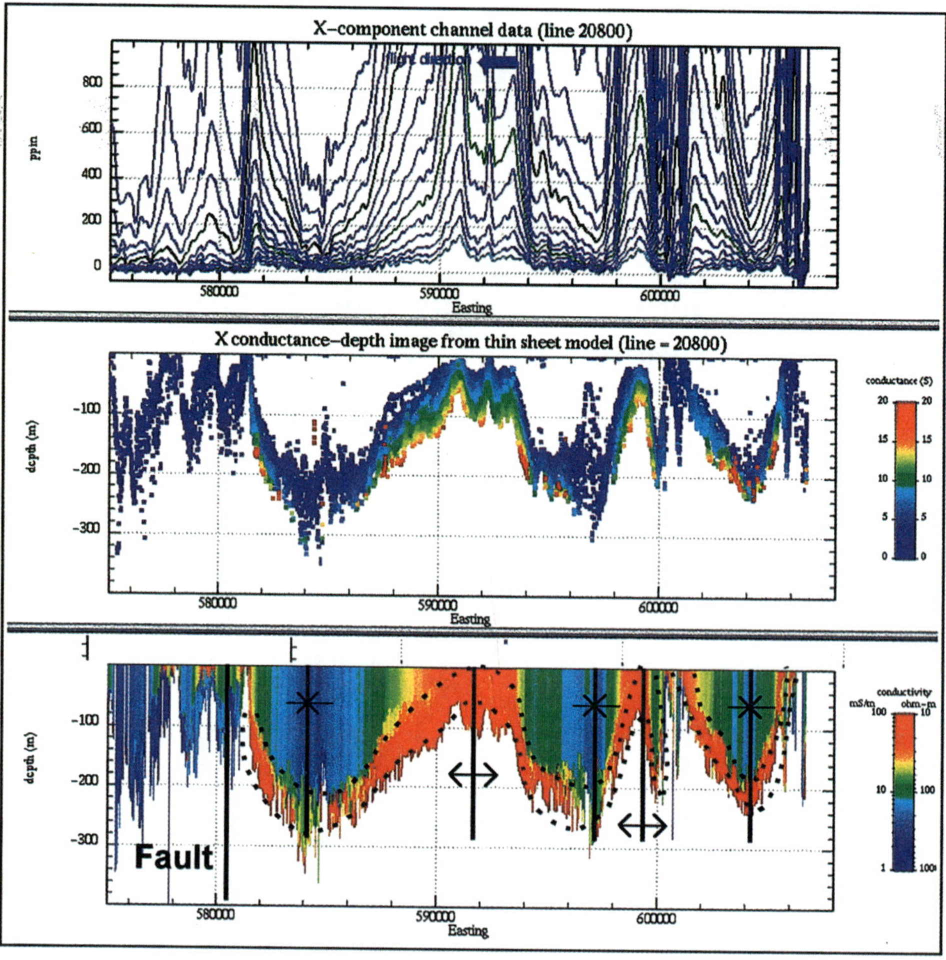

Figure 1. Airborne EM data and geological interpretation for a profile of length 30 km. Parasections are computed using 1-D conductive sheet solutions at each point of the profile (Liu and Asten, 1994). The vertical scale is stretched 25:1.

proximate model and should not be confused with true (geological) sections. Figure 1 illustrates current practice, in which a line of airborne EM data is transformed to a conductance-depth parasection, followed by a manual interpretation of folds and faults in the conductive horizon.

2 Challenge of realistic geometry

The challenge of fast 3-D inversion can be illustrated with SAMAYA, a program using a hybrid algorithm combining an integral equation with a finite-element method (Raiche and Sugeng, 1989). A small target of size 400 × 200 × 50 m (Fig. 2, left) can be modeled with 80 elements, and the execution time is 1.3 hours (Pentium 100-MHz processor) for each transmitter position. A real orebody, however, is more realistically of size 800 × 400 × 100 m (Fig. 2, right), and requires 8 times more elements and 512 times more execution time (to 27 days). Modeling a moving transmitter loop or iterative modeling for inversion increases the demands by one or two more orders of magnitude.

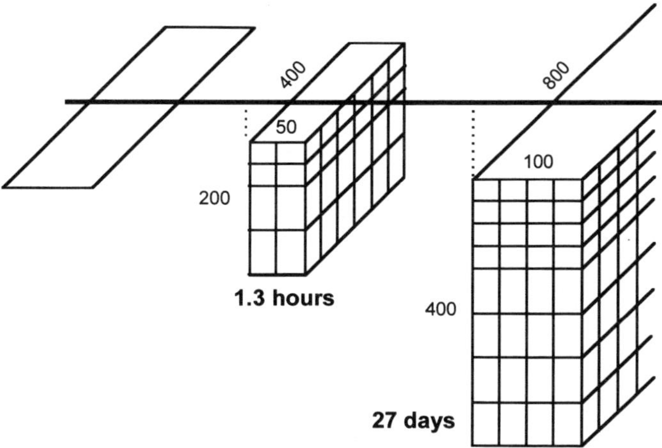

Figure 2. Two simple block models with common cell sizes for use in 3-D EM modeling. Dimensions are in meters. The fixed transmitter loop is shown at left.

The figures for other codes are in the same general range. For example, the program EM3D (Newman and Hohmann, 1988) is about as fast as SAMAYA; the program SYSEM (Xiong, 1992) is 5 times faster. We can expect further gains in speed from other algorithms—such as the multiple boundary element method of Liu and Lamontagne (1995; this volume)—but we are unlikely to achieve the necessary increases (three to five orders or magnitude) anytime soon. Even if we do so for simple models, the mineral industry will immediately demand the modeling of a halo around the principal conductor, introduction of multiple conductors, and inclusion of inhomogeneities in the overburden. The challenge for full 3-D modeling will always be there.

An attractive alternative to full 3-D modeling is to consider the EM interactions that are important in specific applications and to select model geometries and approximations to the physics that generate useful solutions. Some key features to include are

Geometry

- *Edge effects on induction and galvanic currents.* Edges are especially noticeable in near-surface conductors. Figure 3 shows a conductance-depth parasection created from an airborne EM survey over a flat-lying conductor. The image shows diffraction tails at the edges of the conductor (Liu and Asten, 1994).
- *Migration of currents in thick conductors.* Until recently, this phenomenon could be modeled only in spherical conductors (Dyck and West, 1984). Figure 4 shows an example of a conductor 100-m thick, excited by a large fixed transmitter loop. Responses at selected decay times show different profile shapes; inversion with the program FILAMENT (Duncan, 1987) yields equivalent current filaments at early and late decay times, as shown on the block model. The early-time current filament is oriented normal to the primary field; at late times the filament has migrated into the cross-section of largest area within the conductor. A field example of this behavior is given by Eadie (1987); Liu (1995) also provides an example of field and model data where migration of currents in a thick flat-lying conductor results in surface EM data sets that at first appear to be contradictory.

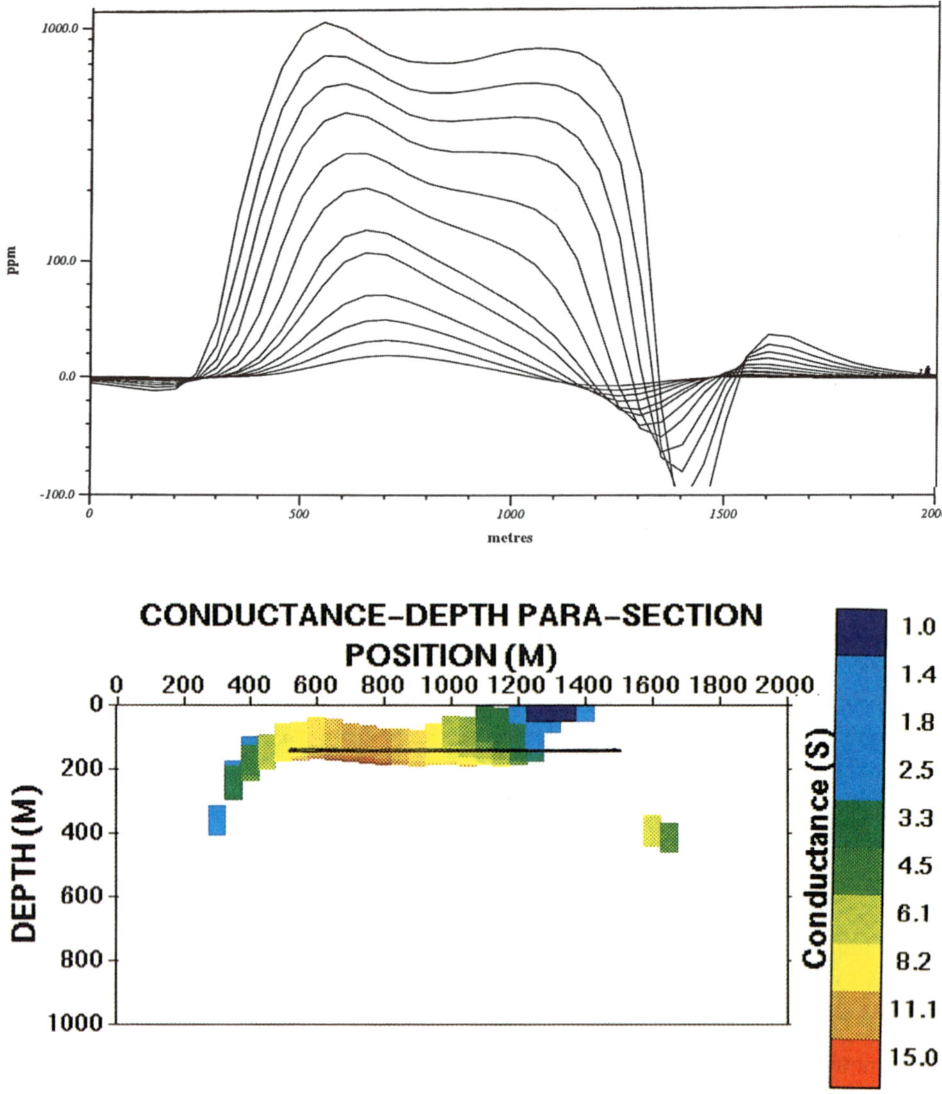

Figure 3. Model of a profile of airborne EM data over a flat-lying plate conductor and the conductance-depth parasection computed from the data. The modeled system uses a 75-Hz transmitter waveform with 1-ms half-sine pulse over a plate 1000 × 1000 m, edges at 500 and 1500 m (shown on the parasection), depth 140 m, of conductance 10 S. Diffraction tails on the parasection at each edge of the conductor are undesirable artifacts resulting from the use of a 1-D algorithm to process data from a 3-D model.

- *Enhanced response at early times.* Real massive sulfide conductors often produce anomalous early-time (or high-frequency) responses. I call this the halo effect. Figure 5 shows a field example of the response in a moving-loop survey of a massive sulfide conductor in a resistive earth. The data cannot be fitted to a simple plate response: The early-time response has amplitudes and decay characteristic of a 200-S conductor, whereas the late-time response is characteristic of a 1000-S conductor. The enhancement at early times exceeds that attributable to higher-order eigencurrents in a plate by a factor of three to five. Enhancement cannot

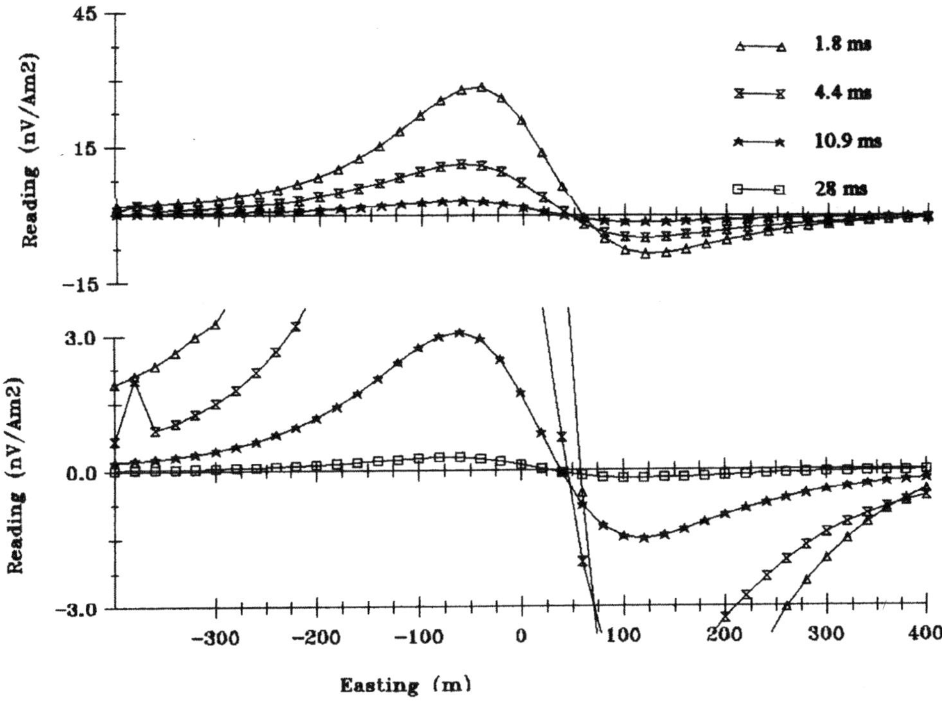

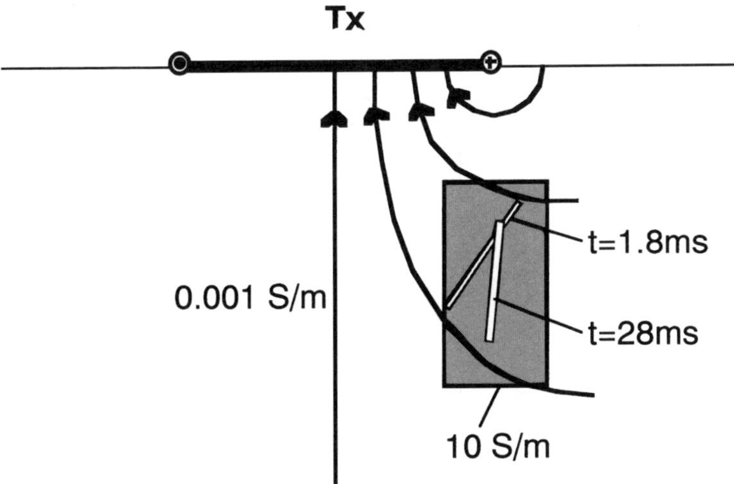

Figure 4. (Top) Modeled profiles for a conductor of strike length 200 m, depth extent 200 m, thickness 100 m, energized by a fixed (600 × 300 m) transmitter loop. (Bottom) Inversion of profiles to locate equivalent current filaments yields filament positions as marked on the model.

be attributed easily to galvanic excitation because the maximum response occurs when the transmitter is placed symmetrically over the target.

Cases in which massive sulfides responded as if less conductive at early times have been described by Parasnis (1971) and Betz (1976), who called it the "thick conductor effect." It appears to be common with sulfide conductors but less so with graphitic conductors (Gordon West, 1995, personal communication). Analog modeling with thick conductive slabs does not duplicate the effect (Yves

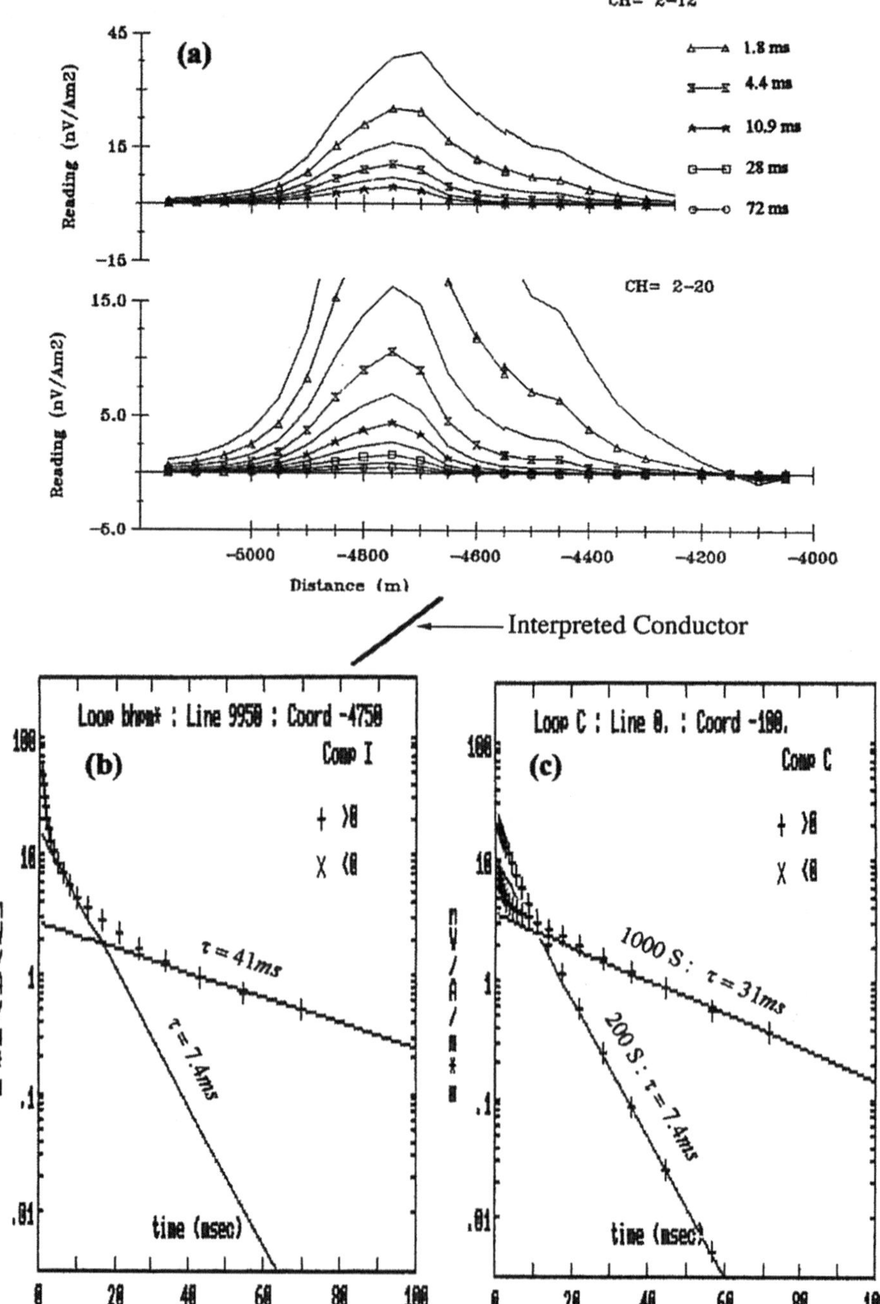

Figure 5. (a) Field-data profiles for a moving-loop survey over a massive sulfide conductor. The data can be modeled by a 200 × 200-m plate dipping 30° west, as shown. (b) The decay curve of field data at the anomaly maximum shows a time-constant increasing with time. At a delay time of 10 ms, the slope of the decay curve is equivalent to a time constant of 7.4 ms. At 72 ms, the time constant is 41 ms. (c) Decay curves of the plate model, for two different conductances (200 and 1000 S) are superimposed. Comparison with (b) suggests that the massive sulfide conductor responds as an equivalent 200-S conductor at early times, grading to an equivalent 1000-S conductor at late times.

Lamontagne, 1995, personal communication). My preferred explanation is that it is caused by a less-conductive halo about the core conductor, but whatever the physical explanation, the effect is sufficiently common to require that 3-D models be able to duplicate it.

Physics

- *EM screening.* Screening by conductive overburden and host is described by McNeill et al. (1984), West and Macnae (1991), Liu and Asten (1992), and McNeill (1995). Screening reduces the amplitudes of early-time responses and delays the peaks of responses in time. If conductive screening is not explicitly included in numerical modeling, it can be approximated by a convolution (Liu and Asten, 1992).
- *Inductive response with high contrasts between target and host.* Modeling inductive response accurately at both early and late times (from inductive through resistive limits) is difficult when there is a large contrast in conductivity between the host and the target. A good check on the accuracy is whether the solutions approach the free-space response when the host conductivity approaches zero. Today's 3-D software has only limited success with this condition. For example, the model in Figure 4, computed using SAMAYA, shows vortex currents in the body decaying with a time constant of 6.1 ms. By contrast, a plate of the same size and conductance (1000 S) modeled in free space with the same transmitter-receiver geometry gives a time constant of 19 ms [program OZPLATE, developed from Dyck and West, (1984)]. The degradation of inductive response in SAMAYA models at high conductivity ratios is characteristic of most (perhaps all) integral equation methods, and in fact, SAMAYA is better than most in achieving a useful result at the contrast of 10 000:1 (target:host).
- *Galvanic interactions (current channeling).* Galvanic excitation has been discussed by McNeill et al. (1984), Asten (1991), Hanneson (1992), and others references therein. It is a phenomenon most often observed with large fixed-loop transmitters and long offsets to receiver positions. Modern 3-D software includes this response, although there is evidence that results are not correct when a target touches or is partially within a conductive overburden.
- *Interaction of multiple conductors.* Few EM modeling programs can handle multiple conductors of realistic size requiring a large number of cells. A free-space approximation [MULTILOOP, Lamontagne et al. (1988)] is available. Modeling of multiple plates in a conductive space has become available only recently.

Because today's algorithms cannot model all of these phenomena routinely (i.e., without a tremendous amount of computation), it is necessary to consider the relevance of each for particular data. The next sections discuss this for typical air, ground, and borehole surveys.

3 Airborne time-domain EM modeling requirements

The chief characteristics of airborne data sets that place requirements on 3-D modeling are (1) the large amount of data (Fig. 1) and the moving-transmitter configuration, which require many modeling runs; (2) interactions at the edges of conductors, which produces diffraction-like artifacts on conductance-depth parasections (Fig. 3); and

(3) the limited time window of measurement (compared to ground EM systems), which makes it difficult to characterize the decay curve.

These characteristics suggest that the highest priority in processing and interpreting airborne EM data is an approximate 2-D or 3-D imaging method. Because airborne surveys have limited time windows and because relative depths are more important than absolute depths in interpretation of airborne data, I believe that a 3-D inductive-limit solution—i.e., a high-frequency or early-time approximation—may provide a fast inversion suited to airborne data. Such a solution would have limitations, particularly because airborne data generally have less high-frequency content than ground data. But inductive-limit models (like 1-D conductance-depth images) can map or transform EM data into 2-D or 3-D modes in a way that preserves relative relationships of conductor size and depth.

A useful pointer to a desirable tool of the future is the automatic depth picking available for airborne magnetic data, using the Naudy or Euler methods (Naudy, 1971; Reid et al., 1990). Figure 6 shows an example of the Naudy method; a similar presentation based on processing of airborne EM data would have obvious advantages.

There is a role for rigorous modeling of airborne EM surveys. In particular, the effect of overburden blanking in attenuating conductor response and the effect of enhancement of conductor response by conductor-host interactions are documented inadequately for the purposes of detailed "what if" survey design. Included in the category of conductor-host interactions are both the inductive effect of less-conductive halos and galvanic excitation of conductors, as discussed in Section 2.

4 Borehole time-domain EM modeling requirements

Some priorities in modeling borehole EM data are ability to handle detailed geometry, including boreholes that intersect conductors; ability to include multiple conductors (Fig. 7), with conductive overburden and host; and ability to produce accurate response in the resistive limit (at low frequencies or late times) to take advantage of the wide time window and low noise typical of borehole surveys.

The small amount of data in borehole EM profiles makes full 3-D EM modeling more feasible, but the need to model correctly the response of multiple intersecting conductors makes each run more demanding. Accurate modeling often requires a small mesh size near the borehole and consequent extended execution times.

Correct modeling of galvanic response at early times is also important because profiles near conductors supporting galvanic currents, determined by charges at boundaries, are different (sometimes of opposite sign) from profiles near conductors supporting inductive currents that close within the bodies (Irvine, 1987; Asten, 1991). My experience, however, is that this phenomenon is not common and, moreover, is limited by its rapid decay ($t^{-3.5}$ for an impulse-response EM system) to times shorter than 2 or 3 ms.

The priorities for borehole surveys require that the next generation of borehole EM software be able to model multiple plates, including overburden and host blanking effects, in geometries where boreholes intersect one or more plates. This goal is easily achievable in the resistive limit by simply summing the response of multiple plates. Moreover, the influence of a conducting overburden or conducting host can be computed to first order by convolving its response with the free-space response (Liu and Asten, 1992). Inclusion of galvanic excitation is a more challenging requirement that programs such as LEROI (Raiche, personal communication, 1995) and VHPLATE (Walker, 1995)

Figure 6. Depth to basement color-coded from red (shallow) to blue (deep) computed from airborne magnetic data over the margin of a sedimentary basin, using the Naudy algorithm. The computed depths are superimposed on a gray-scale aeromagnetic image covering an area approximately 120 by 150 km.

are trying to address. As an intermediate step, the full response can be approximated by simply adding the inductive and galvanic responses, computed separately, using ideas developed by West and Edwards (1985) and illustrated by Asten (1991).

One also can recognize an occasional role for using the inductive limit in borehole EM modeling. King (1993) showed examples of highly conductive pentlandite-pyrrhotite mineralizations that are not detected by an impulse-response transient electromagnetic induction (TEM) system operating at a (rather high) 25-Hz transmitter frequency but that are clearly detected by a step-response system. The problem is illustrated in Fig. 7, where the lower body is sufficiently conductive to behave as a perfect reflector of the incident field. The lines of magnetic induction therefore do not penetrate the conductor,

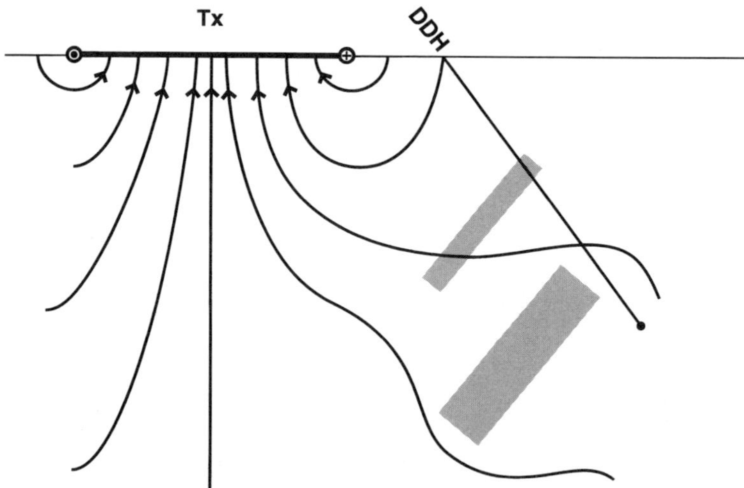

Figure 7. Effect on primary magnetic field produced by an upper body of moderate conductance and a lower body of very high conductance. The primary field penetrates the lesser body and generates induction currents detectable in time-domain EM by a decaying secondary field. The lower body of very high conductance acts as a reflector of primary-field energy and is detected best by the resulting perturbation of the primary magnetic field.

and there is no decaying secondary field detectable by a time-domain EM system. The physics to be approximated here is simple; it requires pure inductive-limit modeling, with the conductor behaving as a perfect reflector. The time decay of the EM response is irrelevant, and the target is best observed by the perturbation of the primary field (especially at the edges).

5 Ground time-domain EM modeling requirements

Ground EM data cover applications ranging from moving transmitter-loop reconnaissance surveys through detailed surveys using fixed transmitter loops designed for target selection and drill-hole siting. This range essentially is covered by the discussions of airborne EM for reconnaissance and borehole EM for detailed targeting. A particular requirement in ground surveys is modeling of thick and inhomogeneous conductors (with halo effect) to reconcile differences in responses seen on airborne and ground surveys. Such differences often can yield very different estimates of both geometry and conductivity of a target.

6 Conclusions

The next challenge in 3-D EM modeling for mining geophysics is to develop robust software that models the geometry of orebodies and the physics of EM interaction in ways that address the special needs of airborne, ground, or borehole surveys. Computational speed must be raised by several orders of magnitude.

Requirements for airborne EM data are very different from ground and borehole data. For airborne data, the greatest need is for fast approximate imaging methods (perhaps like seismic imaging or migration) that can remove edge effects from conductance-depth and conductivity-depth images. This might be achieved with inductive-limit modeling. For detailed ground surveys and borehole data, much remains to be learned from multiple-plate models in a conductive host, at the resistive limit. Full 3-D EM models will play an essential role in guiding and validating the development of these approximations.

The computational demands of full 3-D EM inversion for realistic models in mineral exploration are so large that such applications are probably a decade or more away. However, when I compare the primitive method available for modeling potential fields in my undergraduate studies 20 years ago with inversion software routinely used today, I am confident that practical full 3-D EM inversion will be a part of the mineral geophysicists' toolkit before I retire.

Acknowledgments

This paper is published with permission from BHP Minerals International. Figures 1, 3, 4, and 6 and many useful discussions were contributed by R. Brescianini, G. Liu, E. Jing, and A. Cooke. Professor G. West provided helpful comment on the subject of thick-conductor responses.

References

Asten, M. W., 1991, A study of galvanic response in the TEM response of conductive orebodies: Abstracts of 8th Conf., of Australian Soc. of Expl. Geophys., 24.

———1992, Interpretation of ground TEM data from conductive terrains: Expl. Geophys., **23**, 9–16.

Betz, J. E., 1976, Considerations behind the making of a well-rounded electromagnetic exploration system: Prospectors and Developers Convention, Toronto, *in* APEX MaxMin I-10 and I-8S EM systems operations manual: APEX Parametrics Ltd, Appendix.

Duncan, A. C., 1987, Interpretation of down-hole transient EM data using current filaments: Expl. Geophys., **18**, 36–38.

Dyck, A. V., and West, G. F., 1984, The role of simple computer models in interpretation of wide-band drill-hole electromagnetic surveys in mineral exploration: Geophysics, **49**, 957–980.

Eadie, T., 1987, The downhole EM response of the Hellyer ore deposit: Expl. Geophys., **18**, 255–264.

Hanneson, J. E., 1992, The transient EM step response of a dipping plate in a conductive half-space: Geophysics, **57**, 1116–1126.

Irvine, R. J., 1987, Drill-hole surveys at Thalanga, Queensland: Expl. Geophys., **18**, 285–293.

King, A., 1993, Deep DHEM at Sudbury, Canada: DHEM Workshop, Macquarie Univ., Sydney.

———1996, Deep drillhole electromagnetic surveys for nickel-copper sulfides at Sudbury, Canada: Expl. Geophys., **27**, 105–118.

Lamontagne, Y., Macnae, J., and Polzer, B., 1988, Multiple conductor modeling using program MULTILOOP: 58th Ann. Internat. Mtg., Soc. Expl. Geophys., Expanded Abstracts.

Liu, E. H., and Lamontagne, Y., 1995, EM modeling using surface integral equations, *in* Oristaglio, M., and Spies, B., Eds., Three dimensional electromagnetics: Schlumberger Doll Research, 121–130.

Liu, G., 1995, Three-dimensional modeling of transient electromagnetic data acquired in Queensland, Australia, *in* Oristaglio, M., and Spies, B., Eds., Three dimensional electromagnetics: Schlumberger Doll Research, 541–552.

Liu, G., and Asten, M. W., 1992, Fast approximate solutions of TEM response to a target buried beneath a conductive overburden: Geophysics, **58**, 810–817.

———1994, Conductance-depth imaging of airborne TEM data: Explor. Geophys., **24**, 655–662.

Macnae, J. C., Smith, R., Polzer, B. D., Lamontagne, Y., and Klinkert, P. S., 1991, Conductivity-depth imaging of airborne electromagnetic step-response data: Geophysics, **56**, 102–114.

McNeill, J. D., 1995, Optimal detection of TDEM anomalies under conductive overburden: Technical Note TN-28, Geonics Ltd.

McNeill, J. D., Edwards, R. N., and Levy, G. M., 1984, Approximate calculations of the transient electromagnetic response from buried conductors in a conductive half-space: Geophysics, **49**, 918.

Naudy, H., 1971, Automatic determination of depth on aeromagnetic profiles: Geophysics, **36**, 717–722.

Newman, G. A., and Hohmann, G. W., 1988, Transient electromagnetic responses of high-contrast prisms in a layered earth: Geophysics, **53**, 691–706.

Parasnis, D. S., 1971, Analysis of some multifrequency, multiseparation electromagnetic surveys: Geophys. Prosp., **19**, 163–179.

Raiche, A. P., and Sugeng, F., 1989, Predicting the transient EM response of complex structure using the compact finite element method: Expl. Geophys., **20**, 51–55.

Reid, A. B., Allsop, J. M., Granser, H., Millett, A. J., and Somerton, I. W., 1990, Magnetic interpretation in three dimensions using Euler deconvolution: Geophysics, **55**, 80–91.

Smith, R. S., and Buselli, G., 1991, Examples of data processed using a new technique for presentation of coincident- and in-loop impulse-response transient electromagnetic data: Expl. Geophys., **22**, 363–368.

Walker, P., 1995, VHPLATE: The development of a robust integral equation simulation algorithm, *in* Oristaglio, M., and Spies, B., Eds., Three dimensional electromagnetics: Schlumberger Doll Research, 103–110.

West, G. F., and Edwards, R. N., 1985, A simple parametric model for the electromagnetic response of an anomalous body in a host medium: Geophysics, **50**, 2542–2557.

West, G. F., and Macnae, J. C., 1991, Physics of the electromagnetic induction exploration method, *in* Nabighian, M. N., Ed., Electromagnetic methods in applied geophysics, Tulsa Soc. Expl. Geophys. **2A**, 5–46.

Xiong, Z., 1992, Electromagnetic modeling of 3-D structures by the method of system iteration using integral equations: Geophysics, **57**, 1556–1561.

3-D EM Inversion to the Limit

James Macnae
Andrew King[1]
Ned Stolz[2]
Philip Klinkert[3]

Summary. In a typical day of mineral exploration, an airborne electromagnetic (EM) system may cover 500 km of line, recording data at 10-m spacings. Wide-band data at the 50 000 stations then is interpreted to pick areas worthy of follow-up. In Australia, where conductive regolith is common, there may be 500 local anomalies in this data set. The first task, therefore, is to classify these anomalies by geometry and conductivity. Reduction of the original EM data at each station to only two parameters—the estimated free-space inductive and resistive limits—greatly reduces the complexity of the interpretation problem. The reduction involves stripping and correcting for background host effects. The inductive limit is a function of target geometry alone, whereas the resistive limit is linearly related to target conductivity for an isolated body, but also includes geometric effects. The calculation of forward and inverse 3-D conductivity models using these simple limits is possible in seconds on a fast personal computer. This allows for automated and interactive classification of the sources of anomalies. The results also can provide starting models for more careful inversion or modeling. The major limitation at present is that the method does not model current gathering, which takes place when conductors are in contact.

1 Introduction

Airborne electromagnetic (AEM) surveys in areas with extensive regolith, as in Australia, often show many anomalies. Regolith is a combination of weathered in-situ and transported material that varies laterally and tends to be conductive (Palacky, 1986). An AEM survey that covers 500 km of survey line (a typical day's work) can have 200 to 500 local anomalies. "Local" means that the widths of the anomalies are a few hundred meters or less. Regolith causes most of these anomalies; the real targets are conductors at depth.

The interpreter of such a daily set of data must classify the anomalies and select, say, 5 to 10 good anomalies and one or two candidates for best anomaly. "Good" and "best,"

[1]Cooperative Research Centre for Australian Mineral Exploration Technologies, Macquarie University, New South Wales 2109, Australia.
[2]Western Mining Ltd., P.O. Box 22, Leinster, Western Australia 6437, Australia.
[3]Geophysics Department, Anglo American Corporation, 45 Main St., Johannesburg, South Africa.

of course, refer to characteristics of the target of interest (e.g., an economic mineral body) and not just to physical properties such as highest conductivity, or mathematical properties such as largest time constant. For the EM method to be competitive with other exploration technologies—geochemistry, other geophysical methods, remote sensing, drilling—processing a daily survey ideally should be as fast as data acquisition.

The goal of processing is to produce a 3-D conductivity model that can be correlated with other physical and geological information. One usually settles, however, for 2-D sections stitched together from 1-D inversions of the data (Eaton and Hohmann, 1989; Stolz et al., 1995). Even this is time-consuming. For example, inversion to a five-layer model (from an unbiased starting point) takes about 20 s per station on a fast personal computer (PC) (Sattel, 1996), and thus would require about 280 hours to process the data from 50 000 stations collected in a day of AEM surverying. Lateral averaging can reduce the amount of data to be inverted but is not desirable for many reasons.

A fast alternative to 1-D inversion is the Maxwell receding-image solution, developed by Macnae and Lamontagne (1987) and called conductivity-depth imaging (CDI). CDI takes about 0.18 s per station, or about 1.5 hours to process 500 km of line on a fast PC. In areas of 3-D structure, CDI sections compare favorably to those generated by 1-D inversion (Eaton and Hohmann, 1989; Stolz et al., 1995). The CDI method described by Macnae and Lamontagne (1987) and Macnae et al. (1991) uses vertical smoothing constraints on the conductivity to stabilize the processing. One also could constrain CDI to fit the data with a specified number of layers, as in 1-D inversion, or by using maximum entropy regularization (Leggatt and Pendock, 1993).

Although the resulting maps or sections are useful starting points for geological interpretation (Wolfgram and Karlik, 1995), neither CDI nor 1-D inversion uses data fully when local conductive targets are present. What is needed is a fast, approximate method that identifies local anomalies and models their features with parameters that characterize local targets (size, conductance, dip, etc.). We have a developed a method that simplifies the computations by parameterizing both the model and the data.

2 Analyzing anomalies

The first task in interpretation is to identify all anomalies or potential anomalies. The simplest indentifier could be a bump selector on every channel in the data. Schemes that match profile shapes to those expected from sources of interest have proven to be more useful. For example, one-point anomalies are likely to be noise, and sources at depth will have smaller anomalies with greater half-width than shallow sources. Macnae and Osmakoff (1993) describe a scheme for picking anomalies within a range of widths or shapes, whether crossovers, peaks, or troughs.

Economic mines and mineral deposits are rare, and so, anomalies from economic sources must be rare, too. The job of a human interpreter or an automatic classification scheme is to select 5 to 10 good anomalies from the 200 to 500 in the daily data. If ranking targets is to take no more than 2 hours of the day on a computer, then about 10 to 20 s is available on average to assess each anomaly. A forward 3-D solution would have to run in less than 1 s to be useful (assuming that it takes 20 forward solutions to converge on a model fitting the data). No published 3-D solutions are close to this speed today. Parameterization therefore is needed to simplify and speed up the representation and calculation of anomalies from 3-D structures.

We believe, however, that model parameterization alone is insufficient. To meet time constraints, it is necessary to parameterize both the model *and* the data. The inductive and resistive limits [defined originally in frequency domain in older texts such as Grant and West (1965)] are a minimal set of parameters that usefully describe the data. If massive sulfides are the exploration target, the parameterized model can be a dipping tabular body under overburden. We found that these choices greatly simplify the computational problem. To assess the errors in the parameterization of data, the misfit between the actual data and the data predicted by the parameterization is easy to compute. It is harder to define a misfit of the parameterized model to the actual geological models, but one can calculate how well the data predicted by the parameterized model matches the (parameterized) data.

Classification of anomalies for further study can be based on quantitative criteria such as target depth (to confirm that the fitted source lies beneath regolith), target size, target conductance, and target dip in geologically favorable direction. Again, time contraints have dictated that data from only a single flight line can be involved in this automated process at any one time. In programming this process, it is useful to accumulate the anomalies into a grand list from which a number can be selected; regular human quality control also is needed to check the automatic selection. Plotting all selections on a map can allow interactive selection of zones of interest on the basis of geological criteria; this is also commonly done on contour maps of raw data.

3 Fast inversions to free-space models

Each anomaly is normally a combination of the response of the target of interest and the response of its host. The responses are not additive; there is nonlinear (galvanic and inductive) coupling which is frequency or time dependent. To approximate a pure target response, we have chosen to strip the effects of host conductivity in a process that also deconvolves the secondary response. First, the primary response is computed, as the magnetic-field step response of a half-space delayed by the overburden, using a receding-image solution (Macnae et al., 1991). This response then is convolved with the exponential response of a confined conductor to obtain overburden-blanked decays for a range of overburden conductances and target decays (time constants). Stripped field data are converted to the step response and matched to one of the calculated blanked decays. This comparison yields a pseudo-free-space response for the target at two limits: the inductive and resistive, which can be inverted easily for target geometry. We describe this process in more detail below (see also Macnae et al., 1996).

3.1 *Inductive limit in free space*

In the inductive limit for a body in free space, or in a very resistive host, currents induced on the surface of the conductive body cancel the primary field inside the body (Grant and West, 1965). Because the fields do not penetrate the body, the response is independent of the internal structure. Only the (external) shape of the body and its location relative to transmitter and receiver affect the response. The response in the inductive limit does not depend on time or frequency. Computing this response amounts to solving a potential-field problem with a boundary-value constraint. The problem is linear in the induced currents and relatively easy to solve.

The inductive limit is the limit of the response as frequency tends to infinity. For example, the frequency response of a loop (West and Macnae, 1991),

$$R(\omega) = A\frac{\omega^2\tau^2 + i\omega\tau}{1+\omega^2\tau^2},$$

is characterized by an amplitude A, which depends on the coupling between transmitter, loop, and receiver, and a time constant

$$\tau = L/R,$$

where L and R are the self-inductance and resistance of the loop, respectively, and where $i = \sqrt{-1}$. The inductive limit response is given by

$$\begin{aligned} \mathrm{IL} &= \lim_{\omega\to\infty} R(\omega) \\ &= \lim_{\omega\to\infty} A\,\frac{\omega^2\tau^2 + i\omega\tau}{1+\omega^2\tau^2} \\ &= A, \end{aligned}$$

which is independent of τ.

The step response in the time domain of a loop is a single exponential,

$$R(t) = Ae^{-t/\tau}.$$

The limit at zero time of this response is identical to the frequency-domain inductive limit,

$$\lim_{t\to 0} R(t) = A.$$

If a general response can be decomposed into a set of noninteracting eigencurrents, each characterized by a time constant,

$$\tau_j = L_j/R_j,$$

where L_j and R_j are the self-inductances and resistances of the jth current path, respectively, then the inductive limit is

$$\begin{aligned} \mathrm{IL} &= \lim_{\omega\to\infty} R(\omega) \\ &= \lim_{\omega\to\infty}\left[\sum_j A_j\frac{\omega^2\tau_j^2 + i\omega\tau_j}{1+\omega^2\tau_j^2}\right] \\ &= \sum_j A_j, \end{aligned} \tag{1}$$

where A_j depends on the coupling between transmitter, receiver, and body.

The time-domain response of an arbitrary conductor to a step change in the primary field (in free space) is a sum of exponential decays (Kaufman, 1978),

$$A(t) = \sum_j A_j e^{-t/\tau_j},$$

where each decay is associated with a noninteracting eigencurrent. In the inductive limit (the limit at zero time), the response is just the sum of the amplitude coefficients, A_j, as in the frequency domain.

Table 1. Frequency-domain response and time-domain response of an arbitrary body.

Frequency response	Step response
$R(\omega) = \sum_j A_j \frac{\omega^2\tau_j^2 + i\omega\tau_j}{1+\omega^2\tau_j^2}$	$R(t) = \sum_j A_j e^{-t/\tau_j}$
Inductive limit $= \lim_{\omega\to\infty} R(\omega)$ $= \lim_{\omega\to\infty}\left[\sum_j A_j \frac{\omega^2\tau_j^2 + i\omega\tau_j}{1+\omega^2\tau_j^2}\right]$ $= \sum_j A_j$	Inductive limit $= \lim_{t\to 0} R(t)$ $= \lim_{t\to 0}\sum_j A_j e^{-t/\tau_j}$ $= \sum_j A_j$
Resistive limit $= \frac{1}{i}\lim_{\omega\to 0}\frac{\partial}{\partial\omega} R(\omega)$ $= \frac{1}{i}\lim_{\omega\to 0}\frac{\partial}{\partial\omega}\left[\sum_j A_j \frac{\omega^2\tau_j^2 + i\omega\tau_j}{1+\omega^2\tau_j^2}\right]$ $= \sum_j A_j\tau_j$	Resistive limit $= \int_0^\infty R(t)\,dt$ $= \int_0^\infty \sum_j A_j e^{-t/\tau_j}\,dt$ $= \sum_j A_j\tau_j$

3.2 Resistive limit

The resistive limit in the frequency domain is the slope of the imaginary part of the response near zero frequency (Grant and West, 1965),

$$\begin{aligned}\mathrm{RL} &= \frac{1}{i}\lim_{\omega\to 0}\frac{\partial}{\partial\omega} R(\omega)\\ &= \frac{1}{i}\lim_{\omega\to 0}\frac{\partial}{\partial\omega}\left[A\frac{\omega^2\tau^2 + i\omega\tau}{1+\omega^2\tau^2}\right]\\ &= A\tau.\end{aligned}$$

In the time domain, the resistive limit can be shown to be equal to the area under the decay curve:

$$\mathrm{RL} = \int_0^\infty R(t)\,dt \int_0^\infty Ae^{-t/\tau}\,dt = A\tau.$$

In the resistive limit, the response varies linearly with conductivity (Lamontagne, 1975) and therefore is an additive for bodies that are not in contact; it reduces to a dc resistivity (or magnetometric resistivity) problem (West and Edwards, 1985). For good conductors, the resistive-limit response is very stable to estimate from data, because it depends mainly on long time constants of the decay. These slow decays allow an accurate deconvolution of the actual waveform (to recover the step response); also, their free-space response can be accurately estimated from the raw data, as described later. Table 1 and Fig. 1 summarize some characteristics of the inductive and resistive limits. No normalization to primary or other field strength is present.

4 Estimating the free-space response

Responses of a target in the inductive and resistive (free-space) limits can be estimated from raw data. It is necessary, however, to account for effects of the host environment

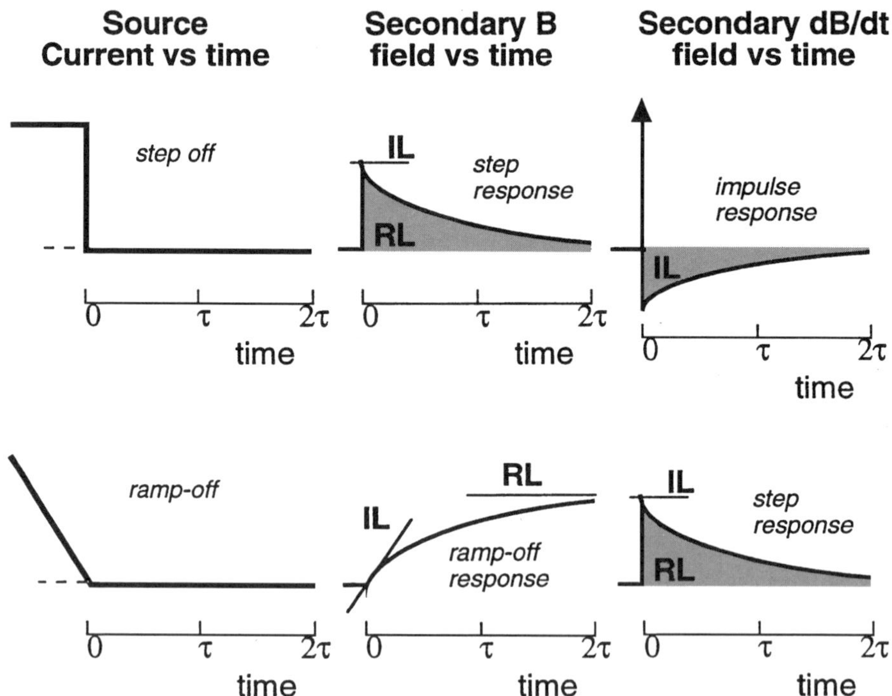

Figure 1. Equivalents for simple time-domain waveforms of the frequency-domain inductive IL and resistive RL limits. Shading refers to the area under the decay curve. Short horizontal lines represent the amplitude at either the zero or the infinite time limits; the short tilted line indicates the slope as the zero time limit is approached.

on the target's response. For example, the effect of a layered earth on a local target can be approximated by convolving the target response (a sum of exponentially decaying currents) with the layered earth's impulse response. The modified target responses can be computed in advance for a range of exponential time constants and host conductivities. Examples are shown in Fig. 2, for the step response of a loop conductor underneath overburden. The panels show decays for different loop time constants, increasing to the right, and for different overburden conductances, increasing downward. In each panel, the exponential response of the loop has been convolved with the overburden's response. The decays are plotted on log–log axes. The time range is 0.05 to 5 ms; all decays are normalized to better illustrate the shape. Overburden blanking is naturally more significant as conductance increases.

Fitting actual target decays to a family of modified target responses (convolved exponentials) implicitly yields inductive- and resistive-limit responses. When processing field data in this fashion, it is important to account for the actual system waveform and to allow for variations in the system geometry during the flight.

4.1 *Arbitrary waveforms*

EM systems in use today have different waveforms and sampling schemes, which may be time and temperature dependent. Physical intuition suggests, however, that data from a modern transient electromagnetic induction instrument can be summarized by a small number of fundamental parameters that are independent of the response

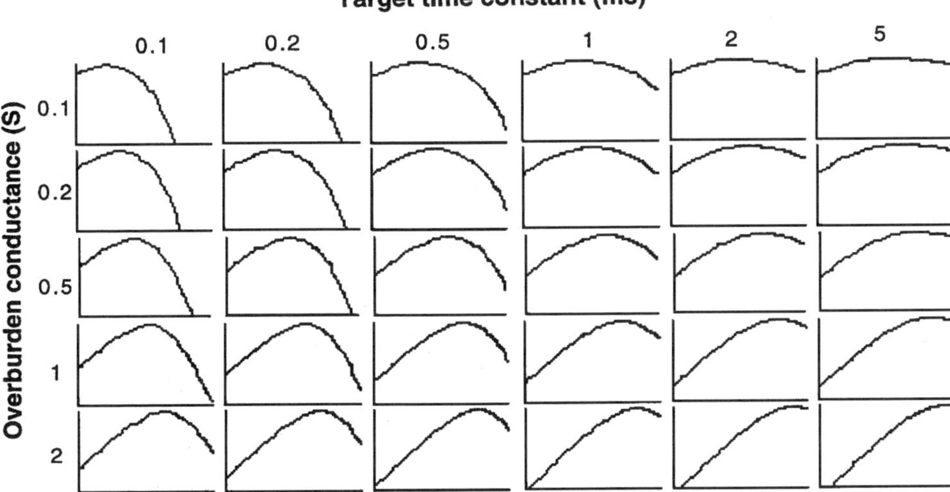

Figure 2. Plots of stripped amplitude versus time for the step response of a simple loop conductor under overburden. Overburden conductance increases downward whereas loop time constant increases to the right. Both axes in each plot are logarithmic and the peak amplitude is scaled to fit the plotting window.

of a particular system. This may be accomplished, for example, by deconvolving the transmitter waveform out of the data and converting data to either the magnetic step response or the wide-band frequency-domain response.

Deconvolution requires an accurate representative transmitter waveform recorded at the receiver. Airborne systems usually measure this directly by taking readings at high altitude. It is desirable to repeat these recording during the survey so that variations with time and temperature can be monitored. For example, the start of the exact "off" time of an AEM system waveform often is not well defined, because electronic time in the transmitter and receiver is synchronized to zero at the start of each transmitter pulse, and the length of the transmitter pulse may drift by as much as 10 or 20 μs as temperature changes. For ground EM systems designed to measure only in the transmitter off time, it may not be possible to record the waveform directly, and it must be estimated from the measured peak currents and ramp widths.

We have found that the most stable and efficient method of deconvolving the response is by linear inversion to a set of basis functions that are the convolution of the ideal response with the actual system response. If the step response of the conductor is expanded as a sum of exponentials,

$$A(t) = \sum_j A_j e^{-t/\tau_j},$$

then the impulse response will be

$$A(t) = \sum_j \frac{A_j}{\tau_j} e^{-t/\tau_j}$$

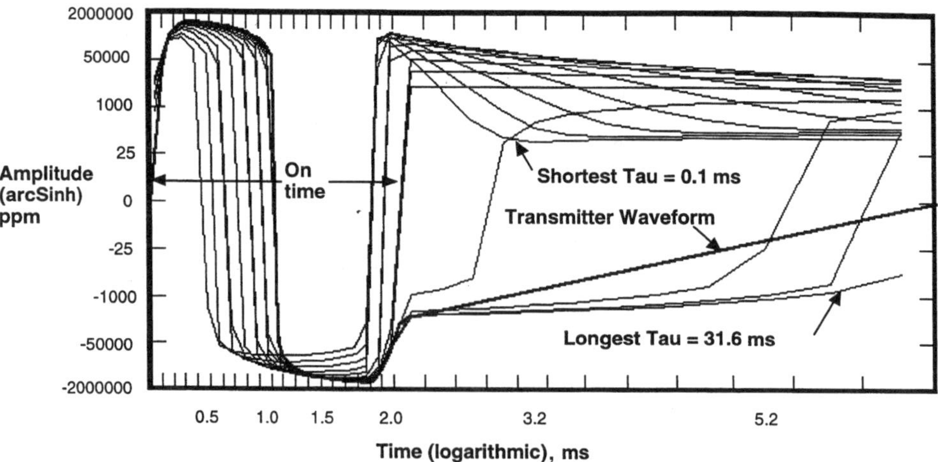

Figure 3. Results of convolution of exponential decays with a waveform similar to that of the old input EM system (Telford et al., 1976), with the presence of a small ramp in the nominal "off-time." The basis functions derived from this are plotted on a vertical arcsinh scale which is similar to a lin-log axis. The time scale shown extends from zero at the start of the transmitter on-time to 6.67 ms.

and the response of the body to an arbitrary exciting waveform $w(t)$ will be

$$R(t) = \sum_j \frac{A_j}{\tau_j} e^{-t/\tau_j} * w(t)$$

$$= \sum_j A_j \, B(t, \tau_j), \tag{2}$$

where the A_j are the original step-response coefficients and the $B(t, \tau_j)$ are convolutions of the exponential loop impulse responses with the exciting waveform (Fig. 3).

Fitting the data with the sum of basis functions in Eq. (2) implicitly deconvolves the system response. The coefficients of the basis functions can be obtained by minimizing the error of fit to the data, subject to some stabilizing constraint. When the coefficients are determined, the EM data have been transformed from the time domain into an amplitude–time-constant domain (tau domain), where representation of responses is much simpler. For example, the conversion to the step response follows trivially from Table 1. Also, modifying the exponential decays to include overburden effects, as in Fig. 2, approximately removes the effects of overburden. The inductive and resistive limits then can be extracted easily from the response, and automatic (approximate) interpretation methods applied.

The number and the range of exponential time constants to be used in the process can be deduced from a singular value decomposition analysis of the basis functions and a knowledge of the system noise levels (Stolz and Macnae, 1998). We have found that the required range of time constants extends slightly beyond the range of time delays sampled by the EM system, with about half as many time constants required as there are instrument time windows. This type of analysis also can be applied to survey design by showing what range of time constants (and hence earth conductivities) can be defined by a system with a waveform of specific period that measures decays over a certain time interval with a particular number of windows.

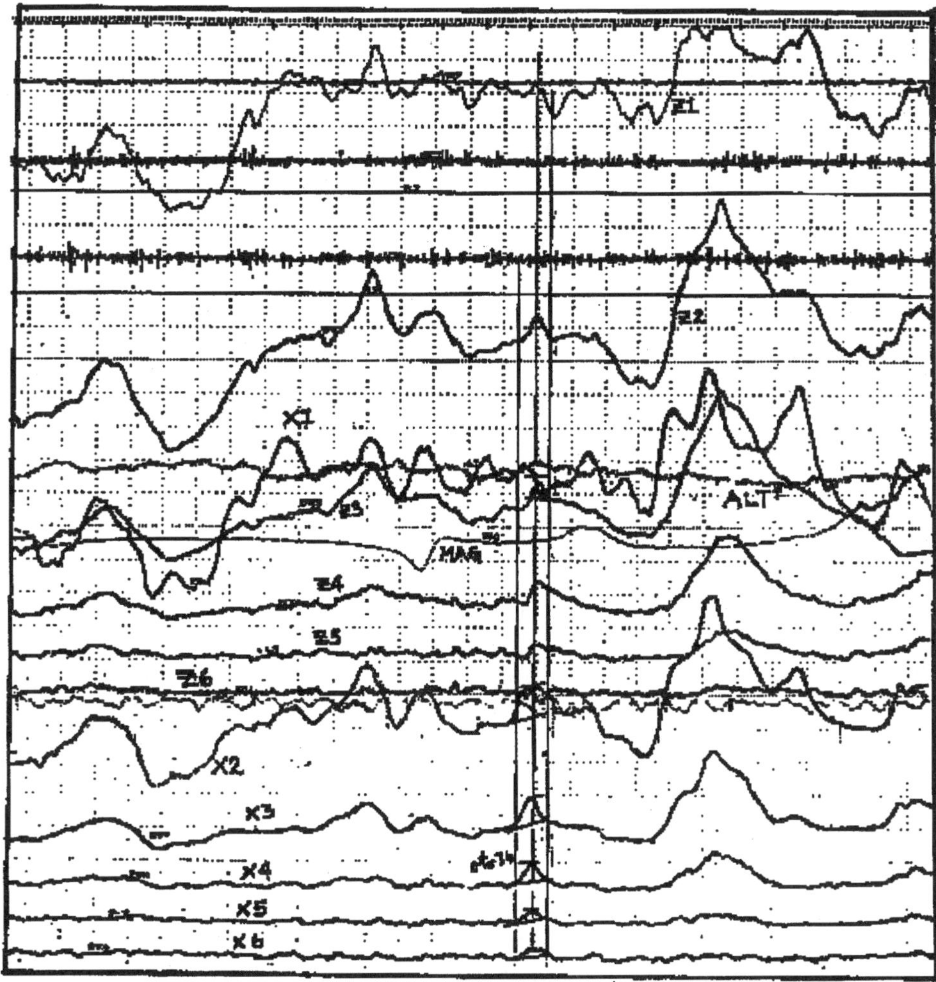

Figure 4. Example of raw data taken from analogue strip chart. Lines represent different time windows in the x and z EM components, altitude, magnetic field, and a 50-Hz powerline monitor. Several local anomalies are evident. At the hand-picked local anomaly indicated by the vertical lines, the absence of response on the earliest time channels shows the presence of overburden blanking.

Transformation of the data to a standard reduces the volume of data considerably, and simplifies the processing and interpretation software, because algorithms need only operate on one type of data. It also allows EM surveys collected with different waveforms or different instruments to be readily compared or merged. The advantage of the step-response representation for modeling is that simpler scaling may be achieved than is possible for other waveforms (West and Macnae, 1991).

4.2 Geometric variation

Processing must be able to handle variations in system geometry. For one fixed-wing AEM system in common use, the geometric variables include transmitter orientation $\pm 6^\circ$ from horizontal, receiver (bird) motion (x, y, z) variation $\pm(5, 10, 15$ m, respectively), and receiver (bird) rotations of as much as 10° (Sattel, 1996). Some of these

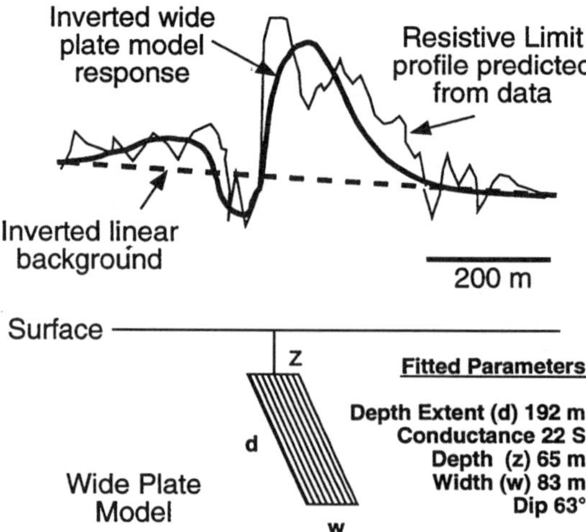

Figure 5. Example of inverted fit to predicted free-space resistive limit for anomaly indicated by vertical lines in Fig. 4; the resistive limit appears fairly noisy on the plot.

variations can be estimated using late-time primary field if on-time data are recorded, but ideally they should be measured directly in future EM systems. A critical consideration for automatic fitting of towed-bird AEM data is that a few degrees of receiver tilt may result in an order-of-magnitude change in the amplitude of the x-component at late time.

4.3 Allowance for other physical processes

The modeling and inversion also must allow for other systematic effects. For example, a physical process omitted in the approximations made above is current gathering. This process is essentially galvanic and is important in conductive environments or when inhomogeneous bodies are present. Research on methods to model such effects have been under way for a number of years with sponsorship through the Australian Mineral Industries Research Association project P407. (The results of this research are constrained to be confidential for two years following project completion.)

Variations in magnetic permeability are critical in the accurate fitting of in-phase data from a helicopter EM system (Fraser, 1981; Sattel, 1996) but are much less important in the fitting of quadrature or off-time AEM responses (Nabighian and Macnae, 1991). Where weathering has led to the development of shallow, fine-grained iron oxides, induced superparamagnetism bedevils the interpretation of coincident and closely separated loop ground AEM data (Buselli, 1982; Lee, 1984a,b). IP effects cause responses that cannot be fitted by any nondispersive conductivity structure (Flis et al., 1989; Smith and West, 1989) and are most evident in ground data with receivers inside the loop. However, IP effects also are reported to clearly affect data in resistive areas for some AEM systems (Smith and Klein, 1996).

5 Processing summary and example

We have applied the processing sequence outlined above to an AEM survey from Australia. At each station, arbitrary waveform data were transformed to the time-constant

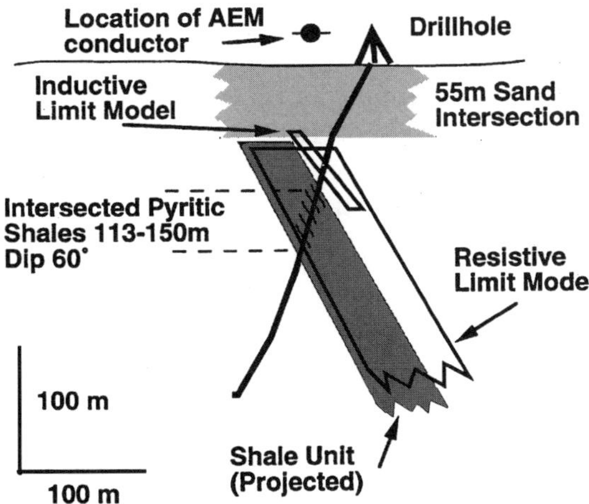

Figure 6. Comparison of inverted inductive- and resistive-limit parameters with drilling.

domain by determining amplitude coefficients for each exponential decay. The basis functions were selected with decay-time constants that have an effect on the data above noise level. (These generally cover a slightly wider time range than the range of time delays measured by a time-domain waveform or, in frequency domain, cover a time range greater than the inverse angular frequency range of the instrument $1/\omega$.) These tau-domain data were used to generate a CDI section, and local anomalies were selected. For a profile across each anomaly, the tau-domain data are stripped of background response and further reduced to an IL and a RL profile. This was followed by both automatic and interactive inversion of the limit responses to get simplified geometry and conductance parameters. Figures 4–6 (from Macnae et al., 1996) show an example of raw Spectrem AEM data (Macnae et al., 1991, 1996), the fit of inversion, and a section to show the correspondence between the inversions and drilling.

6 Conclusions

The interpretation of AEM surveys requires that tens of thousands of decays and hundreds of anomalies be examined every day. This means that less than 1 minute on average is available for the classification of each anomaly if interpretation is to keep up with data acquisition. The important parameters required to grade each anomaly for further examination can be obtained very quickly with a few simplifying assumptions. Extraction of the inductive- and resistive-limit free-space responses of a target, and simple, potential-field-type inversion of these for geometry and target conductance now is being done routinely. The limiting responses can, in principle, be obtained as a byproduct of the deconvolution to a step response or other system independent standard form. This is desirable because it allows standardized processing for different data, reduces data volume, and allows direct comparisons of data from different instruments.

References

Buselli, G., 1982, The effect of near-surface superparamagnetic material on electromagnetic measurements: Geophysics, **47**, 1315–1324.

Eaton, P. A., and Hohmann, G. W., 1989, A rapid inversion technique for transient electromagnetic soundings: Phys. Earth Plan. Int., **53**, 384–404.

Flis, M. F., Newman, G. A., and Hohmann, G. W., 1989, Induced polarization effects in time-domain electromagnetic measurements: Geophysics, **54**, 514–523.

Fraser, D. C., 1981, Magnetite mapping with a multicoil airborne electromagnetic system: Geophysics, **46**, 1579–1593.

Grant, F. S., and West, G. F., 1965, Interpretation theory in applied geophysics: McGraw-Hill Book Co.

Kaufman, A. A., 1978, Frequency and transient responses of methods of electromagnetic fields created by currents in confined conductors: Geophysics, **43**, 1002–1010.

Lamontagne, Y., 1975, Application of wideband, time-domain EM measurements in mineral exploration. Ph.D. thesis, Univ. of Toronto.

Lee, T., 1984a, The effect of a superparamagnetic layer on the transient response of a ground: Geophys. Prosp., **32**, 480–496.

——— 1984b, The transient electromagnetic response of a magnetic or superparamagnetic ground: Geophysics, **49**, 854–860.

Leggatt, P. B., and Pendock, N., 1993, Conductivity-Depth-Imaging of Airborne Electromagnetic Step-Response Data using maximum entropy, *in* Mohammal-Djafari, A., and Dermoment, G. (eds.) Maximum entropy and Bayesian methods. Kluwer.

Macnae, J. C., and Lamontagne, Y., 1987, Imaging quasi-layered conductive structures by simple processing of transient electromagnetic data: Geophysics, **52**, 545–554.

Macnae, J. C., and Osmakoff, A., 1993, Automatic EM anomaly selection: SEG Conference abstracts, Moscow, Soc. Expl. Geophys.

Macnae, J. C., Smith, R., Polzer, B. D., Lamontagne, Y., and Klinkert, P. S., 1991, Conductivity-depth imaging of airborne electromagnetic step-response data: Geophysics, **56**, 102–114.

Macnae, J. C., King, A., and Klinkert, P. S., 1996, A fast method for approximate modeling of localised airborne electromagnetic anomalies: Geophysics, submitted.

Nabighian, M. N., and Macnae, J. C., 1991, Time domain electromagnetic prospecting methods, *in* Nabighian, M. N., Ed., Electromagnetic methods in applied geophysics, Soc. Expl. Geophys. **2**, 427–521.

Palacky, G. J., 1986, Geological background to resistivity mapping: Geol. Surv. Canada.

Sattel, D., 1996, Resolving Shallow Conductivity Structures—The Lawlers regolith and the feasibility of an EM gradiometer, Ph.D. thesis, Macquarie Univ.

Smith, R. S., and Klein, J., 1996, A special circumstance of airborne induced-polarization methods: Geophysics, **61**, 66–73.

Smith, R. S., and West, G. F., 1989, Field examples of negative coincident-loop transient electromagnetic responses modeled with polarizable half-planes: Geophysics, **54**, 1491–1498.

Stolz, E. M., and Macnae, J. C., 1998, Evaluating EM waveforms by singular value decomposition of exponential basis functions: Geophysics, **63**, 54–64.

Stolz, E. M., Raiche, A., Sugeng, F., and Macnae, J. C., 1995, Is full 3-D inversion necessary for interpreting EM data?: Expl. Geophys., **26**, 167–171.

Telford, W. M., Geldart, L. P., Sheriff, R. E., and Keys, D. A., 1976, Applied geophysics: Cambridge Univ. Press.

West, G. F., and Edwards, R. N., 1985, A simple parametric model for the electromagnetic response of an anomalous body in a host medium: Geophysics, **50**, 242–257.

West, G. F., and Macnae, J. C., 1991, Physics of the electromagnetic induction exploration method, *in* Nabighian, M. N., Ed., Electromagnetic methods in applied geophysics: Soc. Expl. Geophys. **2**, 5–45.

Wolfgram, P., and Karlik, G., 1995, Conductivity depth transform of Geotem data: Expl. Geophys., **26**, 179–185.

Three-Dimensional Modeling of Transient Electromagnetic Data from Queensland, Australia

Guimin Liu

Summary. An airborne transient electromagnetic (EM) survey for mineral deposits in Queensland, Australia, showed an anomaly, which later was resurveyed with moving in-loop and large fixed-loop transient EM systems. The moving in-loop profiles show a broad positive anomaly at early times and a corresponding negative anomaly at late times, possibly caused by a shallow polarizable source. The fixed-loop profiles show a strong crossover anomaly with the possibility of a deep conductor. To determine whether the anomaly was caused by a paleochannel or a deep thick vertical conductor with potential mineralization, 3-D numerical modeling was done with the program SYSEM.

A comparison of modeling results and field data led to the conclusion that the source of the EM anomaly was likely to be a polarizable paleochannel. This was proved by subsequent drilling. The strong anomaly observed in the fixed-loop data and the shift of its position for two transmitter positions were caused mainly by channeling currents concentrating near the far edge of the paleochannel away from the transmitter. Numerical results show that a long strike length of a paleochannel is required to cause strong channeling currents and the associated fixed-loop anomaly. The low on the profiles due to the IP effect is, however, not much affected by the strike length. The moving in-loop profiles for the paleochannel model show a positive anomaly at early times and a negative anomaly at late times consistent with the field data. The moving in-loop profiles for the thick vertical conductor show a sharp peak at early times, caused by horizontal current flow induced at the top of the conductor, and an M-shaped anomaly at late times caused by induced currents circulating in the vertical plane of the conductor at late times. These patterns do not appear in the field data.

1 Introduction

A reconnaissance airborne transient electromagnetic (EM) survey for mineral deposits in central Queensland, Australia, showed an anomaly that was studied further with ground transient EM systems. The ground surveys were done in both moving in-loop

BHP Research—Melbourne Laboratories, 245 Wellington Road, Mulgrave, Victoria, Melbourne 3170, Australia.

and large fixed-loop configurations. The ground data confirmed the anomaly observed in the airborne data; but it was not clear whether the anomaly was caused by a shallow paleochannel or a deep thick vertical body with potential mineralization. This paper describes results from 3-D numerical modeling that helped resolve the ambiguity.

2 Ground TEM surveys and geological models

Figure 1 shows a map of the survey lines. The moving in-loop data were acquired on line 7850E with a loop size of 100 × 100 m. The receiver was located at the center of the transmitter loop. The fixed-loop data were acquired on a few lines parallel to 7850E with two transmitter loop positions. Line 7800E was the line nearest to 7850E and was used in this study. Data on other lines were similar. The transmitter loop had a size of 600 × 300 m in the fixed-loop surveys. All survey data shown here were acquired with the vertical component sensor of the SIROTEM Mark3 system.

Figure 2a shows the moving in-loop data. Only data of channels 9–26, which were used in the modeling, are shown. The delay times of these channels after shutoff of the transmitter current are listed in Table 1. Earlier channels were saturated and are disregarded. This is also true for the fixed-loop data. In Fig. 2a, the early-time profiles show a broad positive anomaly peaking at position 912 300 m. The late time profiles show a negative anomaly corresponding to the positive peak on the early-time profiles. Figure 3a shows the fixed-loop data for the transmitter located at the loop-1 position. The late-time profiles have a strong anomaly with zero crossovers clustered close to position 912 400 m. There is a dropout of a few points on the profile of channel 9, which probably was caused by the saturation of the SIROTEM instrument. Figure 4a shows the fixed-loop data for the transmitter located at the loop-2 position. The late-time profiles

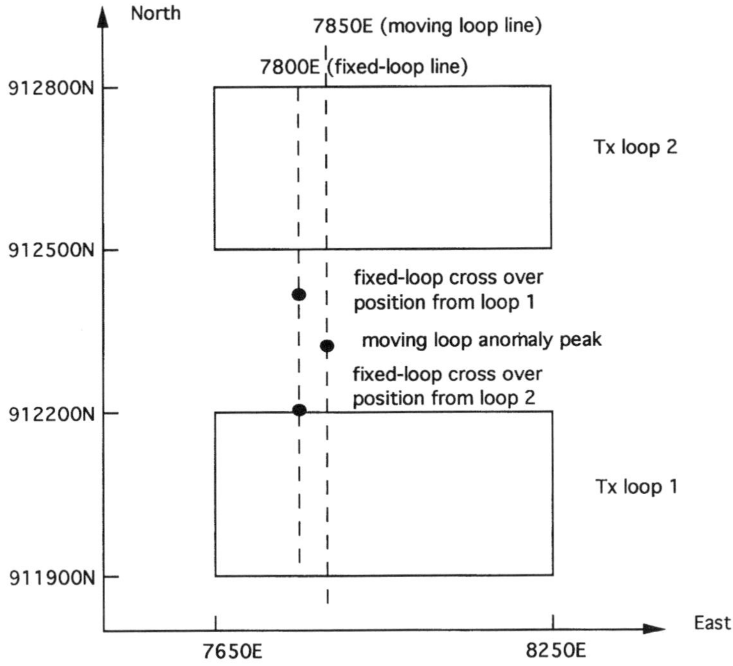

Figure 1. Plan map of survey lines at a site in Queensland, Australia.

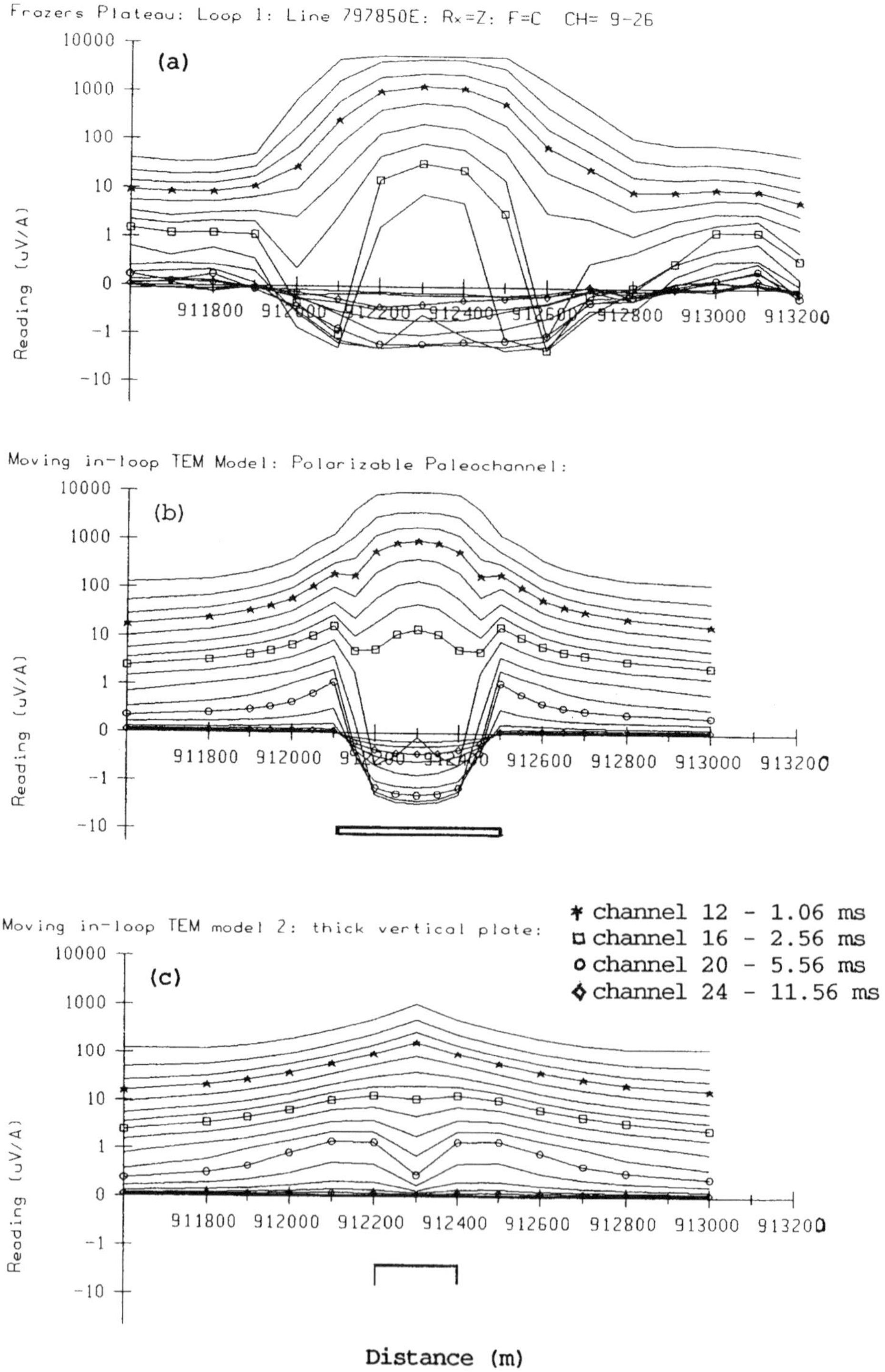

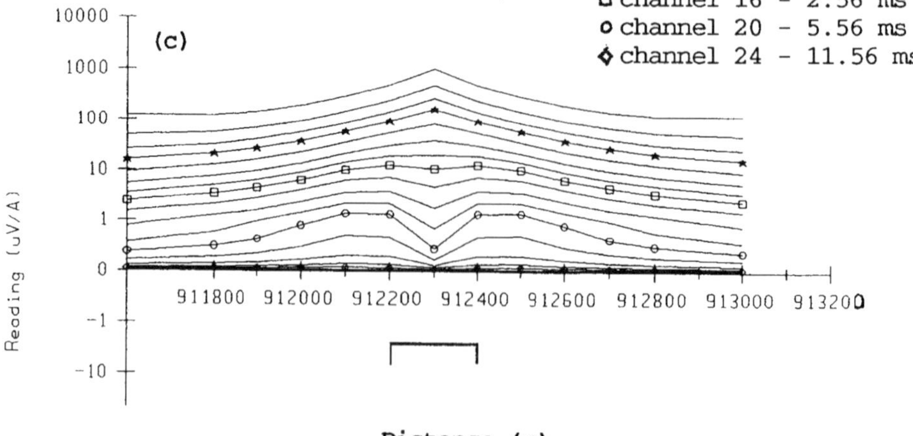

Figure 2. (a) Moving in-loop Sirotem data, (b) moving in-loop profiles of the paleochannel model shown in Fig. 5a, and (c) moving in-loop profiles of the thick vertical conductor model shown in Fig. 5b.

Table 1. Timing of channels 9-26 after termination of ramp turnoff of transmitter current used in field-data acquisition and modeling

Channel	Time (ms)	Channel	Time (ms)
9	0.456	18	3.956
10	0.656	19	4.756
11	0.856	20	5.556
12	1.056	21	6.756
13	1.356	22	8.356
14	1.756	23	9.956
15	2.156	24	11.556
16	2.556	25	13.956
17	3.156	26	17.156

also have a strong anomaly with zero crossovers clustered close to position 912 200 m, which is 200 m south of the crossover in the loop-1 data.

Two geological models likely to cause the observed EM anomaly are shown in Fig. 5. One is a polarizable paleochannel described electrically by a 40-m-thick, 400-m-wide slab of 5-ohm-m resistivity located at the Earth's surface. The strike of the paleochannel is parallel to the east-west direction; it is assumed to have a length of 2000 m. The background is a half-space of 200-ohm-m resistivity. The paleochannel is assumed to be polarizable because the moving in-loop data show a negative anomaly at late times. Late-time negative anomalies in moving in-loop data often have been attributed to near-surface polarizable conductors (Smith and West, 1988; Hohmann and Newman, 1990). I used various polarizable two-layered Earth models to simulate the negative response at the center of the anomaly at late times. Only some models with a polarizable upper layer could generate a negative response comparable to that shown in Fig. 2a. The IP effect of the paleochannel model is described by a Cole-Cole model with chargeability of 0.25, time constant of 0.0001 s, and frequency constant of 0.25.

The second model is a 200-m-wide vertical body (nonpolarizable) of 5-ohm-m resistivity (a possible mineralized body) buried at 70-m depth. The body has a depth extent of 500 m and a strike length of 1000 m in the east-west direction. This model was considered a possible cause of the anomaly in the fixed-loop data, although it is unlikely to generate the late-time negative anomaly in the moving in-loop data. Both models are centered at 7850 m in the east-west direction, and 912 300 m in the north-south direction.

3 Comparison of model results and field data

The modeling was done with the SYSEM package. SYSEM does a fast block iterative solution of the EM volume integral equation (Xiong, 1992). Only the 3-D body in the layered earth needs to be discretized in the computation. The numerical solution first is computed in the frequency domain at a selected spectrum of frequencies, and then is transformed to the time domain by a sine transform.

Figure 2b shows the profiles of the polarizable paleochannel model for the simulated moving in-loop survey. Qualitatively, the model profiles agree with the early-time positive anomaly and the late-time negative anomaly of the field data in Fig. 2a. The profile shape of the model response at intermediate channel 16 also compares well with that of the field data that show a peak anomaly flanked by troughs. However, the profile

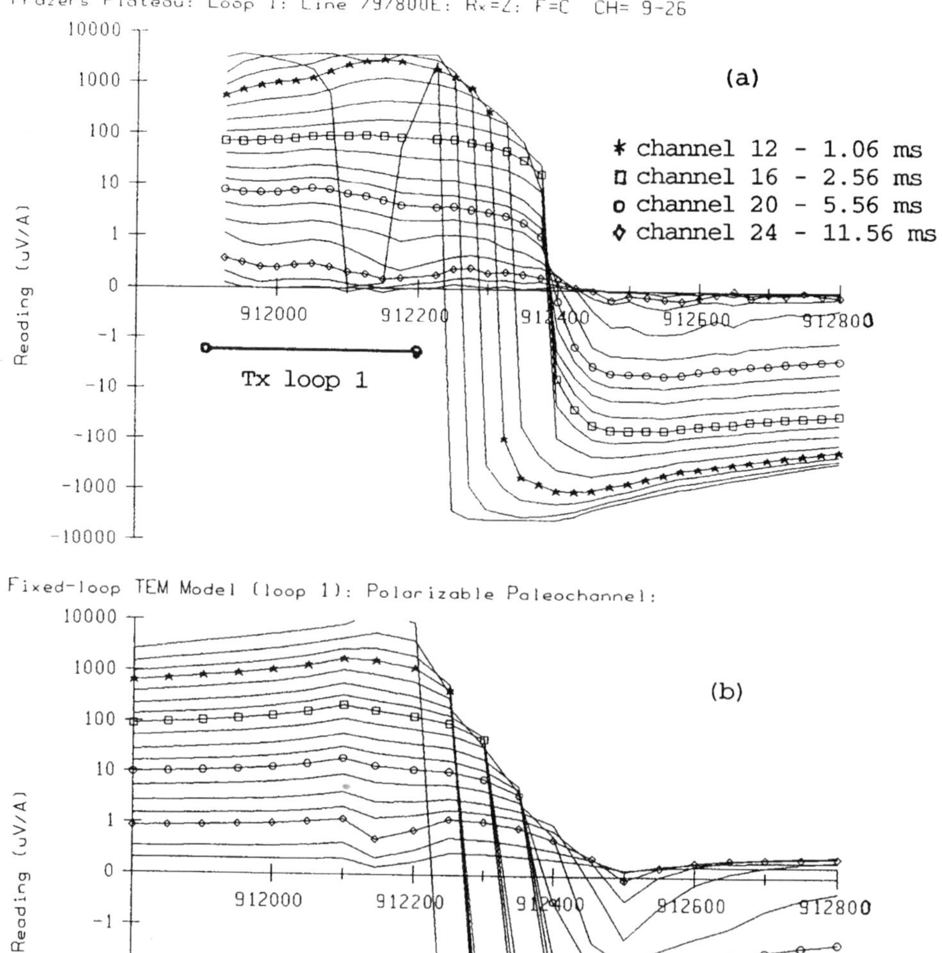

Figure 3. (a) Fixed-loop Sirotem data with the transmitter at the loop-1 position, (b) fixed-loop profiles of the paleochannel model shown in Fig. 5a, and (c) fixed-loop profiles of the thick vertical conductor model shown in Fig. 5b. *(Continues)*

of channel 20 of the model results show two peaks at the sides of a negative anomaly. These two peaks correspond to the sharp edges of the paleochannel model and are not observed in the field data. It is possible that a paleochannel that thickens gradually from the edge to the center would not generate such peaks. The anomaly in the model results is narrower than that of the field data, possibly because the paleochannel model used here is too narrow and too sharp at the edges. Although the model profiles mimic theshape of the field data, the amplitudes of the model response are larger than those of the field data at early times. This is caused mainly by the step current turnoff used in the modeling, which generates a larger amplitude at early times than does the ramp turnoff used in practical instruments.

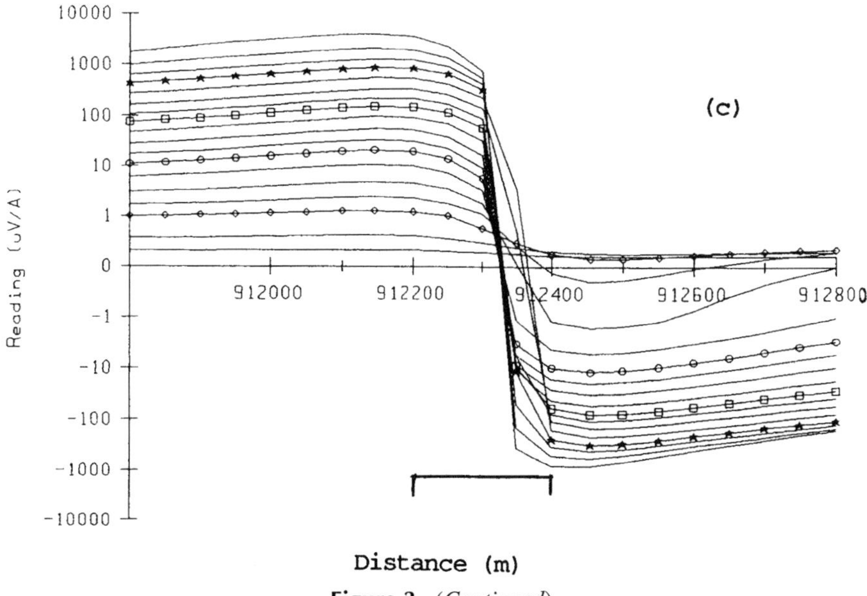

Figure 3. *(Continued)*

Figure 2c shows the profiles of the thick vertical conductor model for the simulated moving in-loop survey. Early channel profiles show a sharp peak over the conductor, which is probably due to the horizontal current loops flowing at the top of the vertical body at early times. Late-time profiles show an M-shaped anomaly typical of vertical conductors, which suggests that the induced currents circulate in a vertical plane in the conductor at late times. A comparison of Figs. 2c and 2a shows that these model profiles do not resemble the observed field data. Therefore, the vertical conductor model is inconsistent with the observed moving in-loop data.

Figure 3b shows the fixed-loop profiles of the paleochannel model for the transmitter located at the loop-1 position. The zero crossovers on the profiles migrate outward with the delay time and become somewhat stationary at late times at a position close to the edge of and inside the paleochannel. Actually, it is the point of greatest slope on the profile that becomes stationary in position at late times because the background half-space is conductive and has a nonzero response. The outward migration of the point of greatest slope corresponds to the outward movement of the induced currents in the paleochannel, which become fixed in a position near the edge at late times. This behavior also was observed in physical scale modeling by Spies and Parker (1984) and was referred to as channeling of induced currents.

A comparison of Figs. 3b and 3a shows that the model results closely resemble the field data in terms of the migration of the crossovers and the profile shape. However, the sharp negative peak at the edge at the position of 912 500 m on the model profiles is not shown clearly in the field data. This is probably caused by a smooth pinch-out of the actual paleochannel at the edge rather than a sharp step edge as assumed in the model.

There is a response low at the position of 912 150 m on the late-time model profiles in Fig. 3b. This response low on the profile is over the area where the transmitter loop overlaps with the polarizable paleochannel and is attributed to the IP effects of the paleochannel. This compares well with the response low in the field data at the same position in Fig. 3a. For an extensive polarizable conductive overburden, Smith and West (1988) demonstrated that there is a response low over the entire area covered by the transmitter.

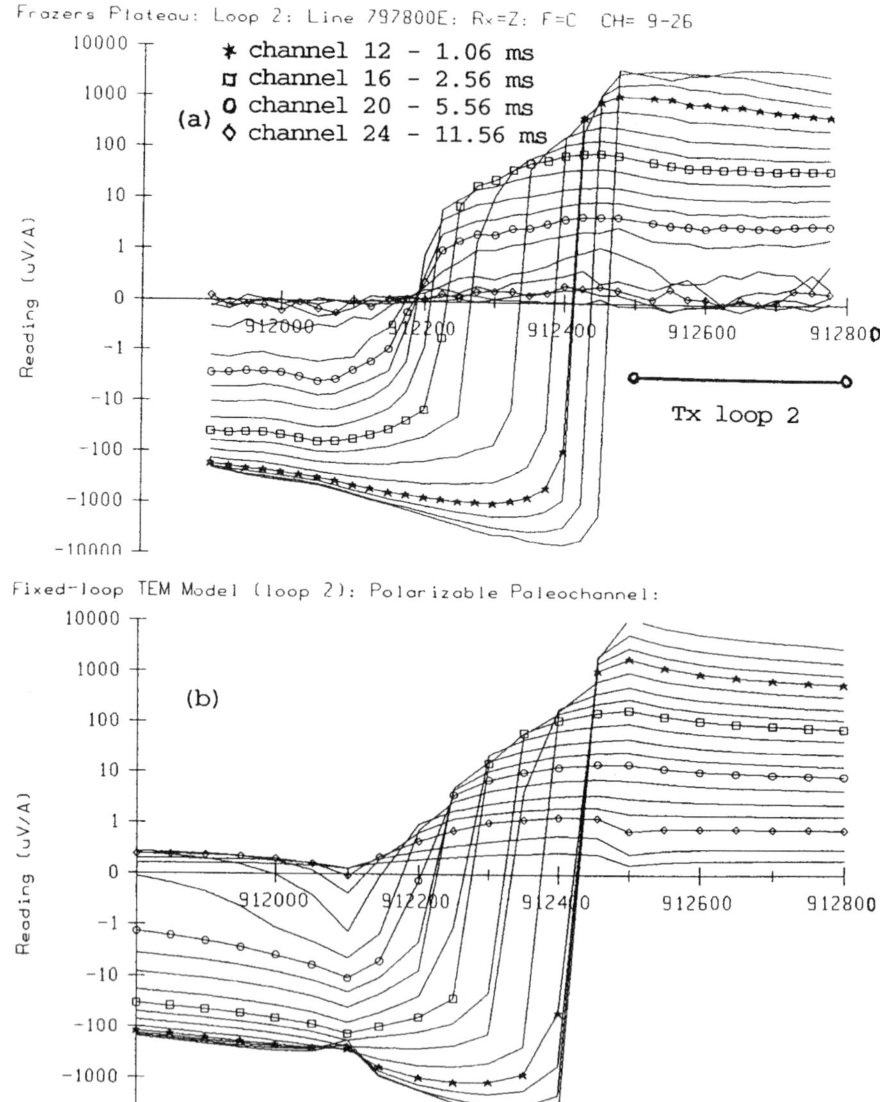

Figure 4. (a) Fixed-loop Sirotem data with the transmitter at the loop 2 position, (b) fixed-loop profiles of the paleo-channel model shown in Fig. 5a, and (c) fixed-loop profiles of the thick vertical conductor model shown in Fig. 5b. (*Continues*)

These features of the comparison of field data and model results in Figs. 3a and 3b are even clearer in Figs. 6a and 6b, which show the data and the model results on a linear scale. The model results in Fig. 6b have a striking resemblance to the field data in Fig. 6a from early to late channels. If the paleochannel is not polarizable, numerical model results show a small positive peak at the southern edge of the slab at the position of 912 100 m because of channeling currents flowing in one direction in the slab as a current sheet.

Figure 3c shows the model profiles over the thick vertical conductor when the transmitter is located at the loop-1 position. There is an indication of the outward migration of the crossovers in two early-time channels caused by the quick outward and downward movement of the induced currents in the somewhat conductive background half-space

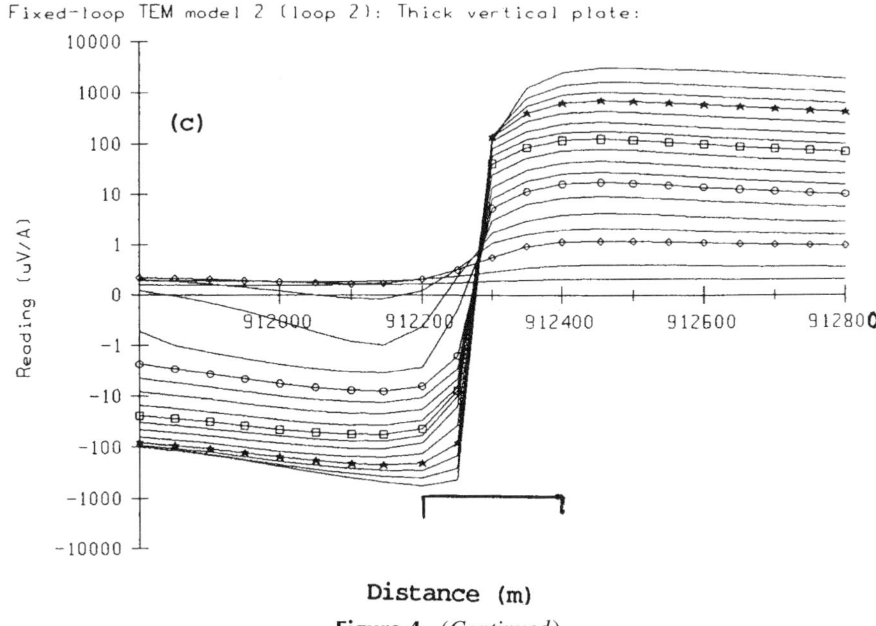

Figure 4. *(Continued)*

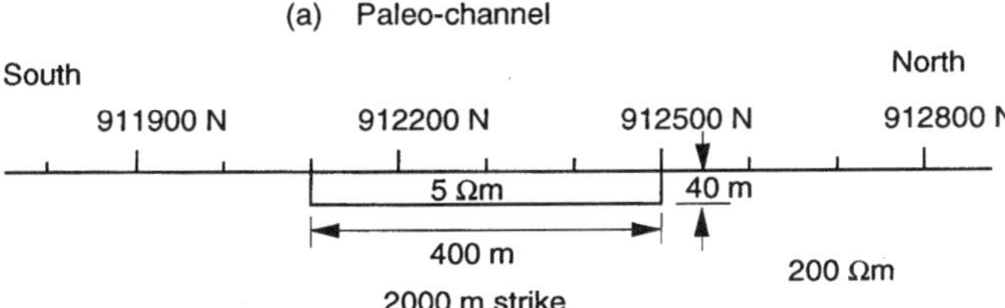

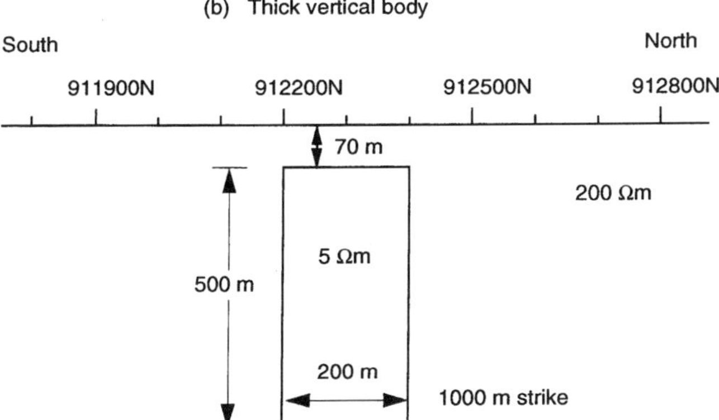

Figure 5. (a) Cross-section of the paleochannel model, (b) cross-section of the thick vertical conductor model on line 7850E.

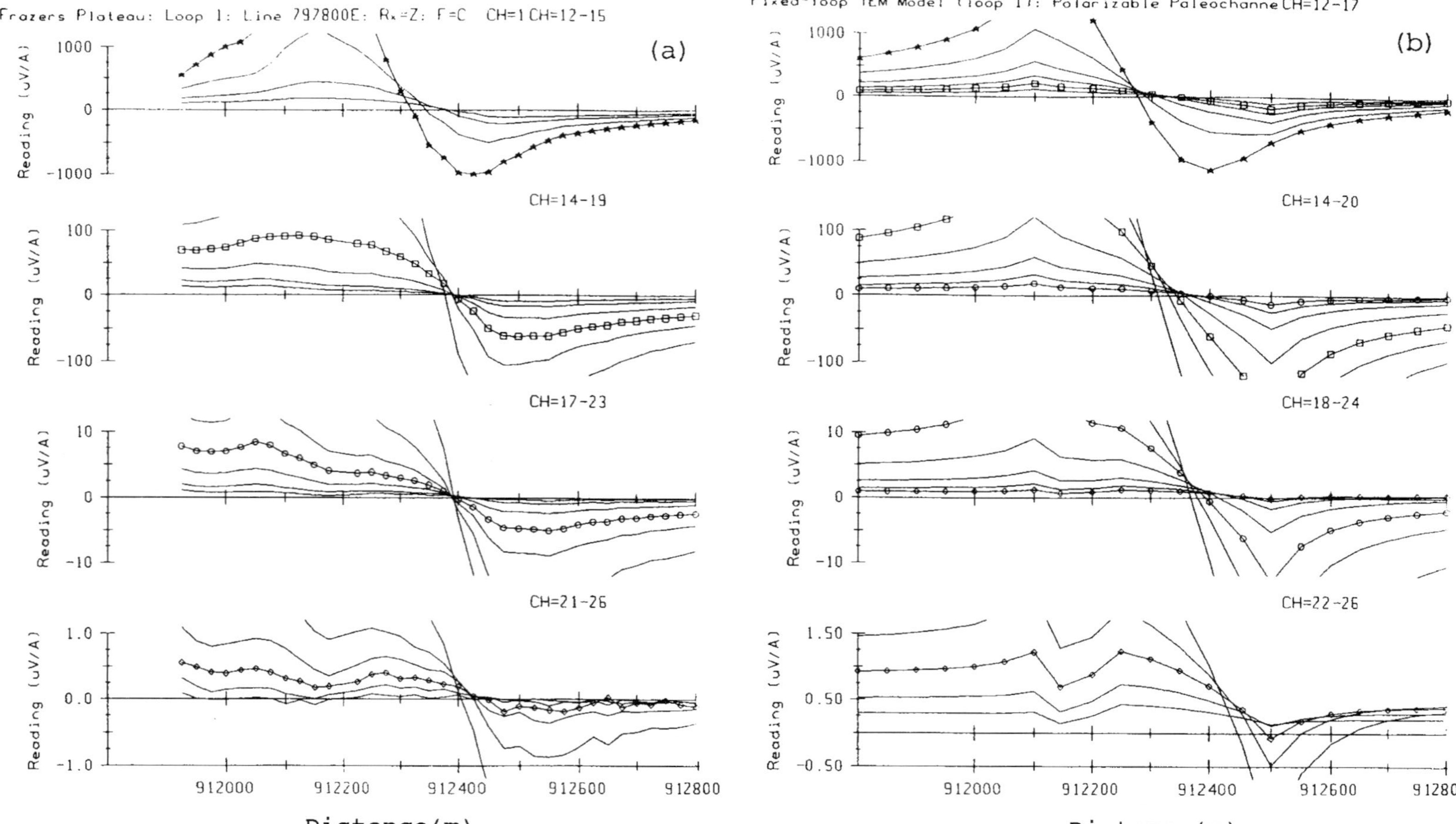

Figure 6. (a). Fixed-loop Sirotem data of channels 12-26 on linear scale with the transmitter at the loop-1 position, (b) fixed-loop profiles of the paleochannel model shown in Fig. 5a.

(Nabighian, 1979). For later channels the points of greatest slope are fixed at a position between the center of the conductor and its outer edge. This is due to the induced currents that circulate in the vertical plane of the conductive body. A comparison of Figs. 3c and 3a shows that the migration of the crossovers in the field data cannot be explained by the thick vertical conductor model.

Figure 4b shows the profiles of the paleochannel model for the transmitter located at the loop-2 position. The outward migration of the crossovers is similar to that in Fig. 3b for the transmitter located at the loop-1 position. However, these crossovers cover a larger distance because all of the paleochannel lies outside the transmitter loop 2. It is clear that the model profiles in Fig. 4b agree well with the field data in Fig. 4a. The points of greatest slope on the model profiles are clustered at the position of 912 200 m at late times and agree with that of field data. This position is shifted 200 m from the position of 912 400 m in the data collected for the loop-1 position. The shift of the crossover position is caused by the shift of the concentration of the channeling currents from one edge of the paleochannel to the other edge because of the different loop positions.

There is a small positive peak at the position of 912 450 m in the field and model data (Figs. 4a and 4b), which is likely due to the channeling-current sheet in the paleochannel. This peak is still present in the model results when the paleochannel is assumed to be nonpolarizable. The field data and the model results in Figs. 4a and 4b are replotted on a linear scale in Figs. 7a and 7b. It is clearer from the linear-scale plots that the model results resemble the field data closely from the early-to-late channels.

Figure 4c shows the model profiles over the thick vertical conductor when the transmitter is located at the loop-2 position. Similar to the case of the loop-1 position, the points of greatest slope are fixed at a position between the center of the conductor and its outer edge. No significant migration of the crossovers occurs in this model. The difference of the crossover position on the model profiles for the two transmitter positions is about 50 m.

The fixed-loop profiles for a thick vertical plate buried in a weakly conductive host are similar to those for a thin vertical plate. Misinterpretation of a thick conductor as a thin conductor will result in errors in the estimates of the depth and the dip of the conductor. To resolve the ambiguity, moving in-loop surveys should be carried out. The peak in the early-time channel profiles (Fig. 2c) can be used to distinguish a thick conductor from a thin one.

This comparison led to the conclusion that the source of the observed anomaly is likely to be a polarizable paleochannel. After the study, a hole was drilled at the center of the anomaly which confirmed that the source of the anomaly was indeed a paleochannel.

4 Effect of strike length of paleochannel

The strike length of a paleochannel can have a significant effect on the observed EM response. Figure 8 shows the computed fixed-loop profiles for a paleochannel of 600 m strike length, which is much shorter than that of the model shown in Fig. 5a. Other model parameters are unchanged. A comparison of Fig. 8 and Fig. 3b shows that reducing the strike length makes the crossover anomaly disappear almost completely on the late-time channels. Hence it is concluded that a large strike length of a paleochannel is required to cause strong channeling currents and the associated fixed-loop anomaly. This is consistent with the conclusion that Spies and Parker (1984) drew for the case of

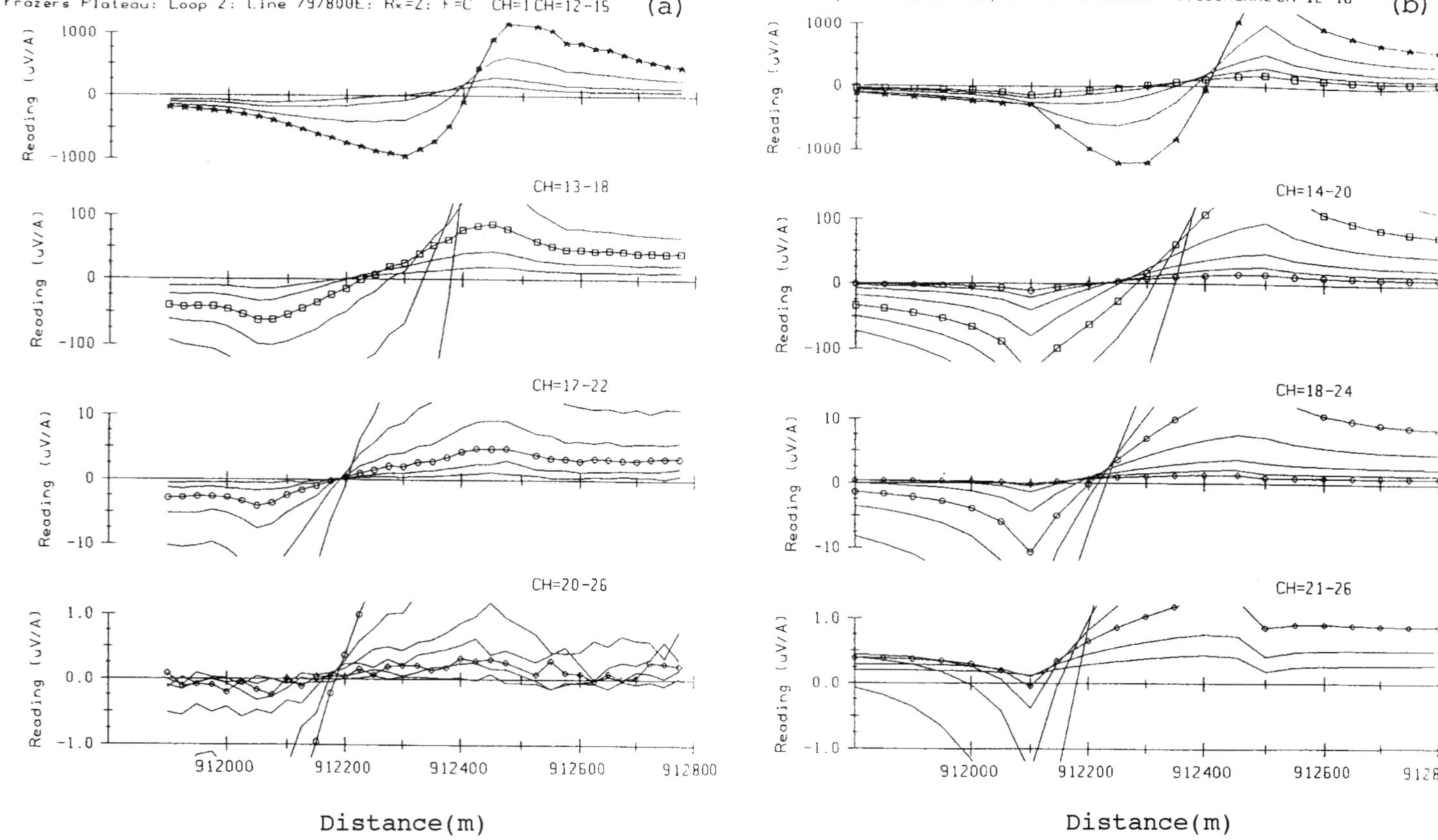

Figure 7. (a). Fixed-loop Sirotem data of channels 12-26 on linear scale with the transmitter at the loop-2 position, (b) fixed-loop profiles of the paleochannel model shown in Fig. 5a.

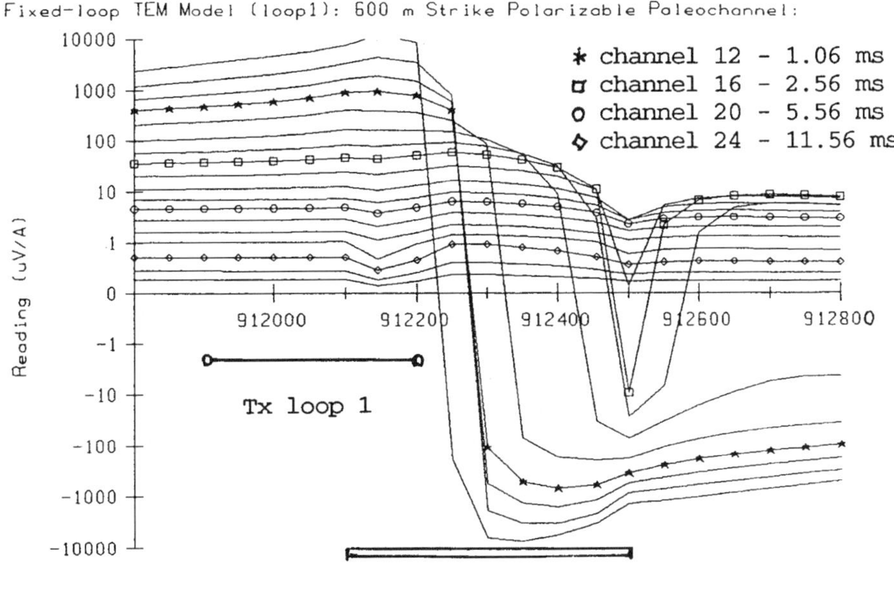

Figure 8. Fixed-loop profiles of a paleochannel model of 600 m strike length. Other model parameters are identical to those in Fig. 5a.

a vertical conductor buried in contact with a conductive overburden. The low on the profile, resulting from the IP effect at the position of 912 150 m, is still present on the late-time profiles in Fig. 8.

5 Conclusions

From the numerical modeling study, it was concluded that the source of the observed anomaly in the field survey was likely to be a polarizable paleochannel. This conclusion was proved to be correct by subsequent drilling at the center of the anomaly. The strong crossover anomaly observed in the fixed-loop data and the shift of its position for different transmitter positions were mainly caused by the channeling currents concentrating near the far edge of the paleochannel away from the transmitter. Numerical modeling shows that a long strike length of a paleochannel is required to cause strong channeling currents and the associated fixed-loop anomaly. The response low on the profiles resulting from the IP effect is, however, not affected much by the strike length.

The moving in-loop profiles for the paleochannel model show a positive anomaly at early times and a negative anomaly at late times consistent with the field data. The moving in-loop profiles for the thick vertical conductor show a sharp peak at early times, caused by horizontal current flow induced at the top of the conductor, and an M-shaped anomaly at late times resulting from induced currents circulating in the vertical plane of the conductor at late times. This is inconsistent with the observed data.

Additionally, the model results show that it can be difficult to distinguish between a thick and a thin vertical conductor from the fixed-loop profiles. These profiles have crossovers clustered close to the center of the thick plate and are similar to those of a thin vertical plate. To resolve the ambiguity, moving in-loop surveys should be carried out.

Acknowledgments

I wish to thank Zonghou Xiong of CSIRO, Australia, developer of the modeling package SYSEM for his assistance on its usage. I also wish to thank Mike Asten, Robyn Scott, and Richard Irvine for helpful discussions on possible geological models. I am also grateful to BHP Minerals Exploration for the permission to publish this paper.

References

Hohmann, G. W., and Newman, G. A., 1990, Transient electromagnetic responses of surficial, polarizable patches: Geophysics, **55**, 1098–1100.

Nabighian, M. N., 1979, Quasi-static transient response of a conducting half-space—an approximate representation: Geophysics, **44**, 1700–1705.

Smith, R. S., and West, G. F., 1988, An explanation of abnormal TEM responses: coincident-loop negatives, and the loop effect: Expl. Geophys, **19**, 435–446.

Spies, B. R., and Parker, P. D., 1984, Limitations of large-loop transient electromagnetic surveys in conductive terrains: Geophysics, **49**, 902–912.

Xiong, Z., 1992, Electromagnetic modelling of 3-D structures by the method of system iteration using integral equations: Geophysics, **57**, 1556–1561.

Three-Dimensional Transient Electromagnetic Modeling and Its Application to Geothermal Exploration

Toru Mogi[1]
Tatsuya Kajiwara[2]
Elena Y. Fomenko[3]

Summary. Our 3-D finite-element modeling scheme for transient electromagnetic (TEM) data consists of two parts: (1) calculation of the primary electric field of an electric bipole source on a layered earth, and (2) calculation of the secondary electric field caused by induced scattering currents in 3-D anomalous regions. The magnetic field is computed from derivatives of the electric field. The scheme works with harmonic fields; transient (vertical magnetic) fields are obtained by Fourier transformation over a wide frequency range. An aysmptotic boundary condition on the secondary field terminates the finite-element grid.

We constructed a 3-D model of the Mori geothermal field in northern Japan from TEM surveys with long-offset electric bipole sources at 86 sites and from a shallow controlled-source magnetotelluric (CSMT) survey. Joint inversion of the TEM and CSMT data gave a preliminary layered-earth model. Modeling of 3-D variations about the layered structure suggests that the reservoir should be developed around the vent of the caldera with hot fluid being supplied from the southwest of the vent.

1 Introduction

Geothermal fields have complex resistivity structures requiring 3-D modeling (Mogi et al., 1995). Hohmann (1975) pioneered 3-D electromagnetic (EM) modeling for geophysical applications with an integral-equation method. Pridmore et al. (1981) developed the finite-element method (FEM), which they formulated for the electric-field components. Because electric-field components are discontinuous at resistivity boundaries and standard FEM basis functions are continuous, a large number of small elements were needed to achieve reasonable accuracy. Lee et al. (1981) tried to overcome this problem with a hybrid method that used finite elements for a localized region containing the anomalous structure and an integral equation for the fields at the boundaries. The hybrid method obtained the boundary values by iterating between solution of

[1]Department of Mining, Kyushu University, Fukuoka 812-81, Japan.
[2]JMC Geothermal Engineering Co. Ltd., Takizawa, Iwate 020-01, Japan.
[3]Institute of Geoelectromagnetic Research, Troitsk, Moscow Region 142092, Russia.

the finite-element equations and the integral equation. Gupta et al. (1987) proposed a modification to solve for the boundary values directly.

We have developed a 3-D FEM for the secondary electric field which uses an asymptotic boundary condition originally proposed by Weaver and Brewitt-Taylor (1978) for 2-D finite-difference calculations and refined by Zhdanov et al. (1982). The boundary condition reduces considerably the size of the grid and, therefore, the computer time and memory needed for the finite-element solution. Our formulation is for harmonic fields; transient results are obtained by an inverse Fourier transformation.

2 Finite-element equations

Maxwell's equations with harmonic $\exp(i\omega t)$ time dependence, and without displacement currents, are

$$\nabla \times \mathbf{E} + i\omega\mu\mathbf{H} = -\mathbf{M}^i, \tag{1}$$

$$\nabla \times \mathbf{H} - \sigma\mathbf{E} = \mathbf{J}^i, \tag{2}$$

where ω is angular frequency, μ is magnetic permeability, and $\mathbf{M}^i$, $\mathbf{J}^i$ are impressed magnetic and electric sources. When there are no magnetic sources, the electric field $\mathbf{E}$ satisfies the equation

$$\nabla \times \nabla \times \mathbf{E} + i\omega\mu\sigma\mathbf{E} = -i\omega\mu\mathbf{J}^i. \tag{3}$$

Equation (3) allows a decomposition of $\mathbf{E}$ into two components,

$$\mathbf{E} = \mathbf{E}^p + \mathbf{E}^s, \tag{4}$$

where $\mathbf{E}^p$ is the primary field excited by the external source in the host medium (e.g., a layered earth), and $\mathbf{E}^s$ is the secondary field caused by scattering currents that arise at inhomogeneities in the host. The primary field satisfies

$$\nabla \times \nabla \times \mathbf{E}^p + i\omega\mu\sigma_h\mathbf{E}^p = -i\omega\mu\mathbf{J}^i, \tag{5}$$

whereas the secondary field satisfies

$$\nabla \times \nabla \times \mathbf{E}^s + i\omega\mu\sigma\mathbf{E}^s = -i\omega\mu\Delta\sigma\mathbf{E}^p, \tag{6}$$

where $\Delta\sigma$ is the difference in conductivity between the inhomogeneity and the host medium (σ_h).

The primary field, $\mathbf{E}^p$, in a layered earth has an integral representation in terms of the vector potential $\mathbf{A}$ (Wait, 1982), where

$$\mathbf{E} = \left(\nabla\nabla \cdot \mathbf{A} + k_i^2\mathbf{A}\right)/y_i; \quad \mathbf{H} = \nabla \times \mathbf{A}; \tag{7}$$

$k_i = (i\omega\mu\sigma_i)^{1/2}$; $y_i = \sigma_i - i\omega\epsilon$; and σ_i is the conductivity of the ith layer. If the vector potential for the primary field $\mathbf{A}^p$ is itself split into two parts—the vector potential $\mathbf{A}^U$ for an electrical dipole source in a whole space and the vector potential $\mathbf{A}^L$ for the field scattered by a layered earth—then,

$$\mathbf{A}^U = (M_x, M_y, M_z)\frac{e^{ikR}}{4\pi R} = (M_x, M_y, M_z)\frac{1}{4\pi}\int_0^\infty \frac{\lambda}{u_0} e^{-u_0(z-z')} J_0(\lambda\rho)\, d\lambda, \tag{8}$$

where

$$\rho^2 = (x - x')^2 + (y - y')^2,$$
$$R^2 = \rho^2 + (z - z')^2, \text{ and}$$
$$u_i = \left(\lambda^2 - k_i^2\right)^{1/2}.$$

If the dipole source is oriented in the x-direction, then the (x, z) components of the vector potential A_x^L and A_z^L in the ith layer are given by

$$A_{xi}^L = \frac{M_x}{4\pi} \int_0^\infty \frac{\lambda}{u_i} Q_i^+(z, \lambda) J_0(\lambda\rho)\, d\lambda, \tag{9}$$

$$A_{zi}^L = \frac{M_x}{4\pi} \frac{\partial}{\partial x} \int_0^\infty \frac{1}{\lambda} G_i^+(z, \lambda) J_0(\lambda\rho)\, d\lambda, \quad i = 1, \ldots, N. \tag{10}$$

The functions Q_i^+, G_i^+ contain upward- and downward-propagating plane-wave amplitudes in each layer.

After the primary electric field has been computed, the secondary field, $\mathbf{E}^s$, can be obtained by solving Eq. (6) with the FEM. We use brick-like finite elements with a node at each corner (eight nodes). The field in each element is approximated by

$$\mathbf{E}^s \approx \mathbf{E}^e = \sum_{j=1}^{8} N_j^e(x, y, z)\, \mathbf{E}_j^e, \tag{11}$$

where N_j^e are trilinear shape functions and E_j^e are the unknown vector secondary field values at each node. The finite-element representation for the full field is obtained by summing the local representation (Eq. 11) over all elements. Test vector fields in each element, $\mathbf{U}^e$, are defined as the product of a shape function N_j^e and a unit vector at each node

$$\mathbf{U}_{j,\alpha}^e = N_j^e(x, y, z)\, \hat{\mathbf{i}}^{(\alpha)}, \tag{12}$$

where $\hat{\mathbf{i}}^{(\alpha)}$ is a unit vector in the x-, y-, or z-directions. A full test vector field, $\mathbf{U}$, is obtained by summing Eq. (12) over all elements that share a (global) node.

The finite-element equations are obtained in Galerkin's method by making the error or residual $\mathbf{R}$ in the approximation of Eq. (6)—i.e., the difference between the left-hand and right-hand sides of this equation when $\mathbf{E}^s$ is approximated by Eq. (11)—orthogonal to the space spanned by the test vectors, $\mathbf{U}$:

$$\int \mathbf{U} \cdot \mathbf{R}\, dv = 0, \quad \text{for all } \mathbf{U}. \tag{13}$$

Integrating this equation by parts (and assuming that boundary terms vanish) gives the finite-element equations (Reddy et al., 1977; Gupta et al., 1989):

$$\int (\nabla \times \mathbf{U}) \cdot (\nabla \times \mathbf{E}^e)\, dv + i\omega\mu \int \sigma \mathbf{U} \cdot \mathbf{E}^e\, dv = i\omega\mu \int \Delta\sigma \mathbf{U} \cdot \mathbf{E}^p\, dv. \tag{14}$$

The integral is taken over the entire domain but can be assembled piecewise from integrals over individual elements.

It is necessary to specify boundary conditions for the electric field at the edges of the finite-element grid. We use boundary conditions that model the asymptotic behavior of the secondary field far away from the inhomogeneities. This idea was developed for 2-D models by Weaver and Brewitt-Taylor (1978) and extended to 3-D models by

Zhdanov et al. (1982) and Zhdanov and Spichak (1992). A simple boundary condition of this type is

$$\left(1 - ikr + \frac{\partial}{\partial r}\right)\mathbf{E}^s = 0 \tag{15}$$

We apply this boundary condition in elements that terminate the grid. The code has been verified for a conductive-prism model with plane-wave excitation (Mogi, 1996).

To model TEM surveys, we first compute the vertical magnetic field (by differentiating the electric field) in the frequency domain over a wide frequency range. The transient response to the shutoff of a steady current (in the bipole source) then is obtained from the cosine transform (Newman et al., 1986),

$$h(t) = -\frac{2}{\pi}\int_0^\infty \frac{\mathrm{Im}[H(\omega)]}{\omega}\cos(\omega t)\,d\omega, \tag{16}$$

by numerical evaluation of the integral with Anderson's (1974) digital filter.

3 Response for simple model

We calculated a 3-D conductive-prism model shown in Fig. 1. In the FEM calculation, the air and earth areas were divided into 4096 elements. There were 16 blocks in each direction; in the vertical direction, 4 planes of elements were in the air and 12 blocks were in the earth. A horizontal electrical source parallel to the x-direction was situated 5 km from the edge of the prism. The calculation was done for 46 frequencies in the range of 10^{-4} to 10^5 Hz (five frequencies at each decade). The distance between the prism and the grid boundaries was varied with frequency. The calculation was performed of a Dell OmniPlex590 personal computer (PC) with 32 MB of RAM. The computing time was around 22 hours.

Figure 2 shows transient curves computed at several points on the model. This figure also shows the transient curve for a layered structure with the same conductivities. Comparing the 3-D and 1-D curves shows the 3-D effect as a shift of the curve to a later time for the conductive-prism model.

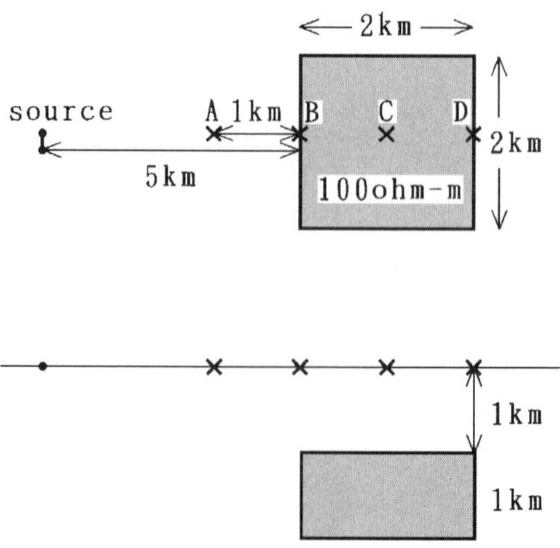

Figure 1. Conductive-prism model.

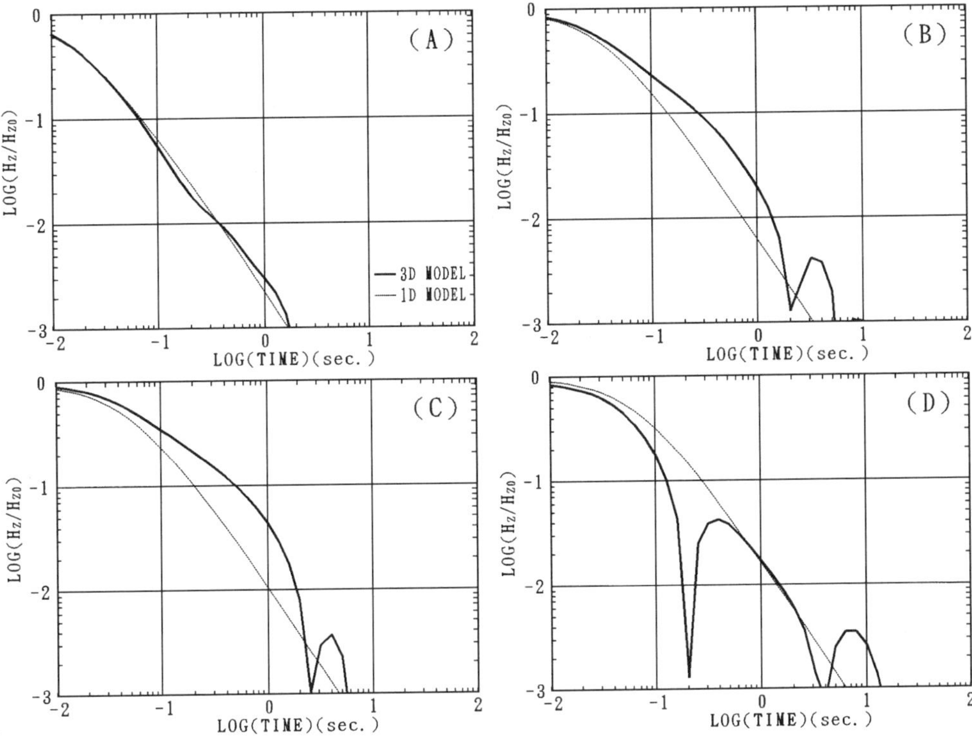

Figure 2. TEM responses for the conductive-prism model.

4 3-D resistivity model of Mori geothermal field

The Mori geothermal field is located in southwest Hokkaido, northern Japan. The geothermal manifestation is distributed around the Nigorikawa caldera, formed around 12 000 years ago (Sato, 1969). The Mori geothermal power plant with 50-MW capacity has been operating since 1982 at the north of the caldera. A transient electromagnetic (TEM) survey was carried out at 86 sites around the geothermal field in 1993. Acquisition, processing, and 1-D inversion of these data were reported by Mogi et al. (1994) and Kajiwara et al. (1995).

We modeled the Mori field with the 3-D FEM scheme described above. The area modeled is shown in Fig. 3. The grid consisted of 7200 elements ($18 \times 18 \times 20$ in $x \times y \times z$). Three blocks of elements at each side of the model were buffers to diminish the effect of grid boundaries. Four blocks in the z-direction represented the air layer; 16 blocks were in the earth. The area modeled is a flat basin, and so, we were not concerned about topography.

Figure 4 shows the resistivity model as slices at every 200 m level in shallower depth, at every 250 or 500 in deeper part. The initial model was constructed from the local 1-D inversions and from drilling data including electrical logs. Trial-and-error matching between the measured vertical magnetic transient curve and the computed curve was done until a reasonable fit was obtained. Figure 5 shows a comparison of modeled and measured curves. The final model (Fig. 4) shows a low-resistivity layer (<10 ohm-m) that appears shallower than 500 m below sea level across the region. The resistivity is generally higher in the deeper part. Resistivities are between 50 and 100 ohm-m north of the caldera and are more than 100 ohm-m south of the caldera at depth below 1750 m

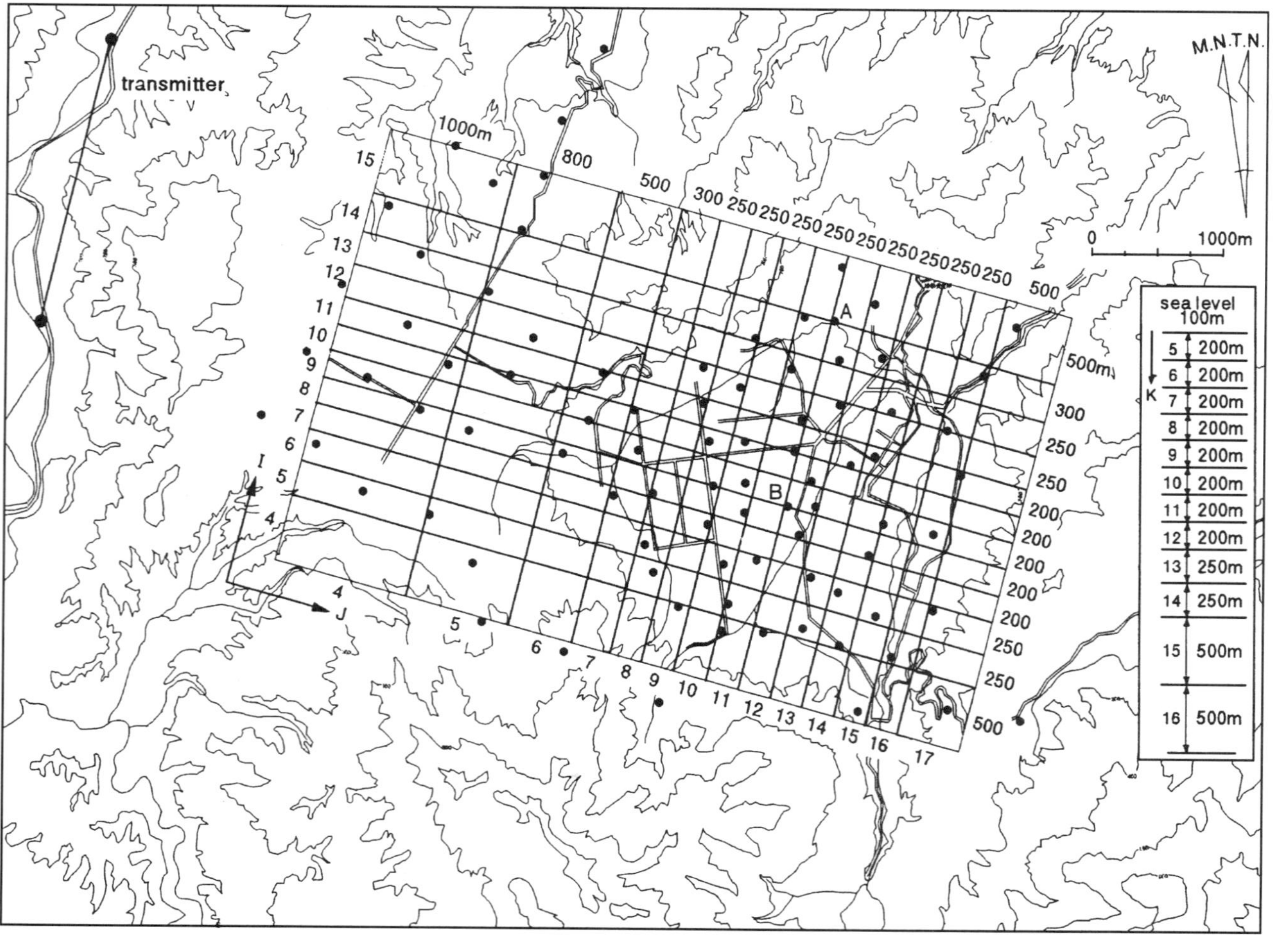

Figure 3. Modeling area in Mori geothermal field.

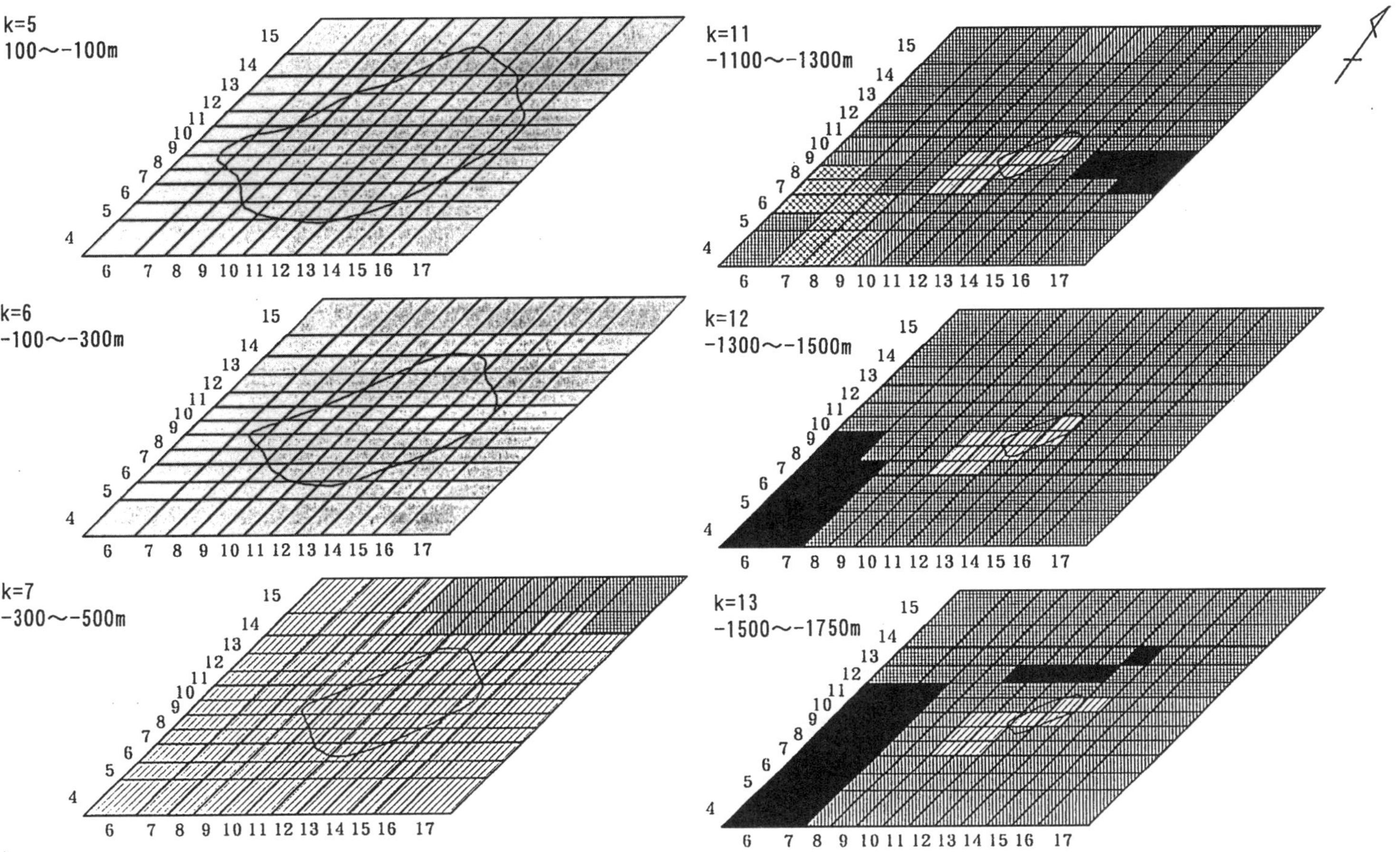

Figure 3. 3-D resistivity model of Mori geothermal field.

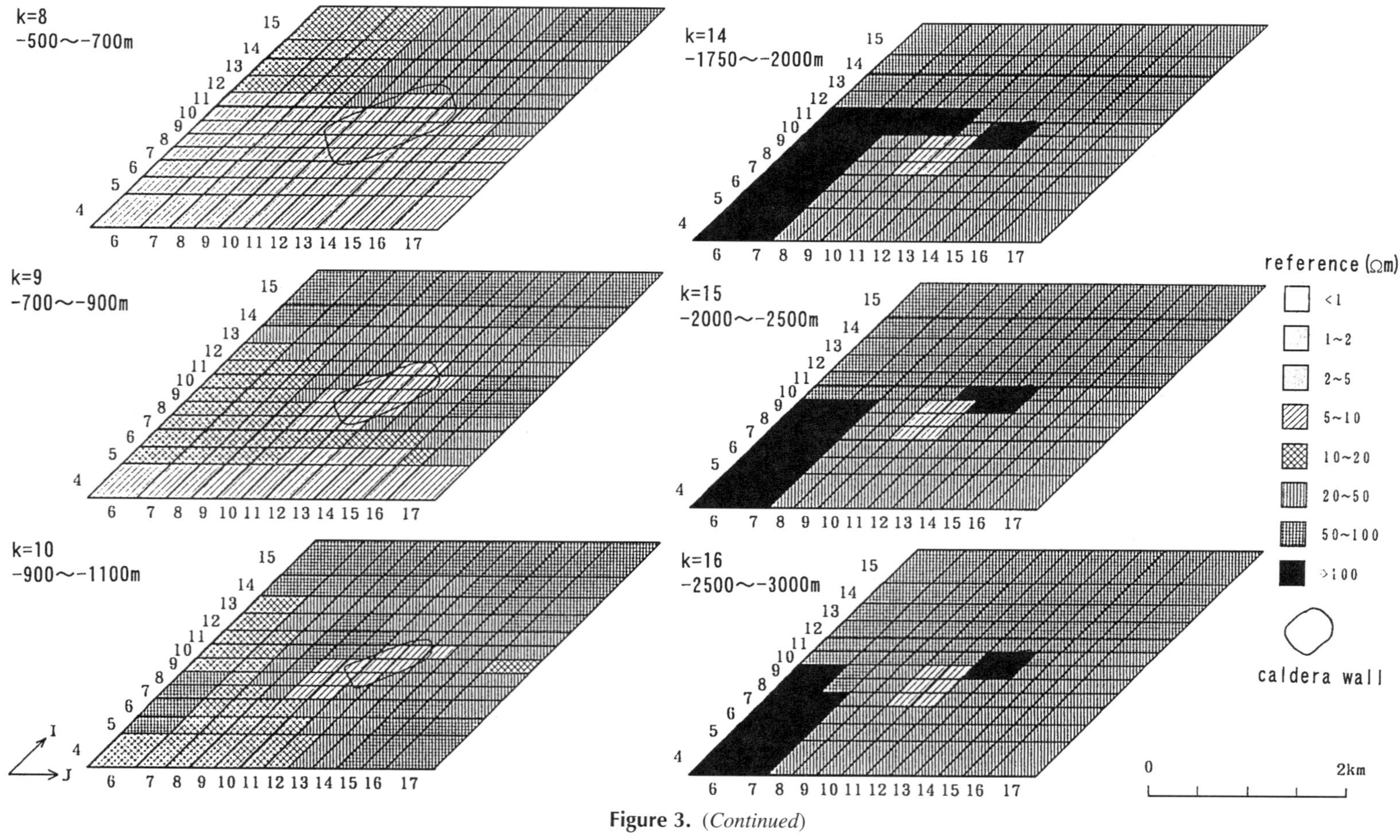

Figure 3. *(Continued)*

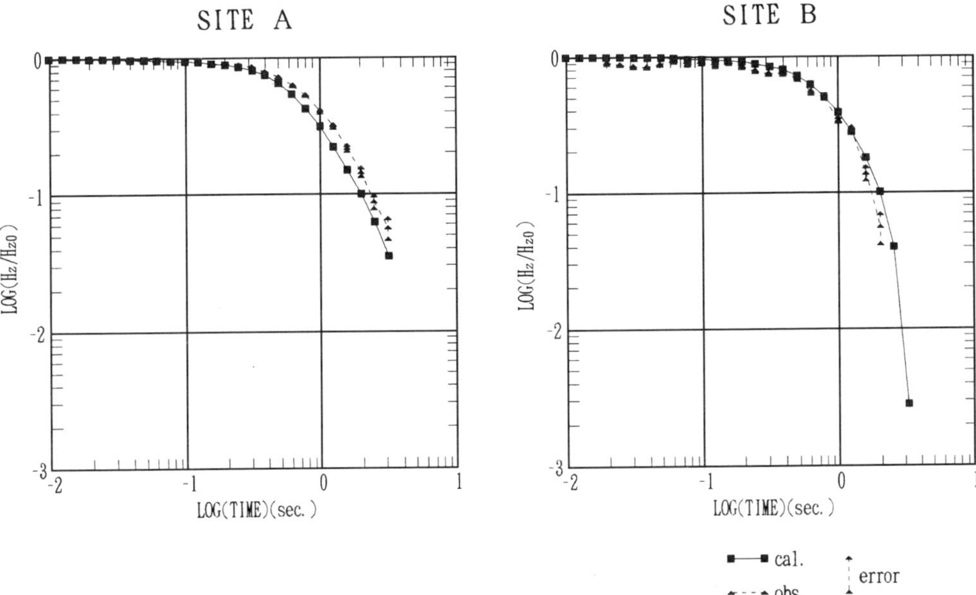

Figure 5. Comparison of measured transient curve with computed curve for 3-D model at sites A and B shown in Fig. 3.

below sea level. The center part of the caldera corresponding to the vent shows less than 10 ohm-m up to 1750 m below sea level, but shows more than 100 ohm-m at deeper levels. A low-resistivity zone (less than 5 ohm-m) is apparent southwest of the vent.

Drilling data indicate that most of the low-resistivity zones in the shallower part and south of the caldera correspond to altered zones. The low-resistivity zone in the southwest of the vent seems to correspond to the zone of upflow of geothermal fluid. The 3-D resistivity modeling helps to delineates the extent of these altered zones and the area of upflow in the geothermal field.

5 Summary

We have developed a 3-D FEM for modeling TEM data. The formulation requires a small memory size and can run in reasonable time on a PC. The small finite-element grid is terminated by an asymptotic boundary condition. Modeling of the Mori geothermal field gave results useful for understanding the extent of a geothermal development around the caldera.

Acknowledgments

The field data used in this study were obtained in the joint research work among Laboratory of Geothermal Engineering, Department of Mining, Kyushu University; Laboratory of Physical Geology, Department of Geology and Mineralogy, Kyoto University; Central Research Institute of Electric Power Industry; and JMC Geothermal Engineering Co., Ltd., cooperating with Japan Metals & Chemicals Co. Ltd. and Dohnan Geothermal Energy Co. Ltd. We would like to thank the members of the research team and the above organizations.

References

Anderson, W. L., 1974, Electromagnetic fields about a finite electric wire source: US Geological Survey open file report GD-74-04, PB-232 199.

Gupta, P. K., Bennett, L. A., and Raiche, A. P., 1987, Hybrid calculation of the three-dimensional electromagnetic response of buried conductors: Geophysics, **52**, 301–306.

Gupta, P. K., Raiche, A. P., and Sugeng, F., 1989, Three-dimensional time-domain electromagnetic modeling using a compact finite-element frequency-stepping method: Geophys. J., **96**, 457–468.

Hohmann, G. W., 1975, Three-dimensional EM modeling: Geophysics, **40**, 309–324.

Kajiwara, T., Mogi, T., Takahashi, M., Nishimura, S., Katsura, I., Suzuki, K., Kusunoki, K., and Nishida, J., 1995, Geoelectrical structure by CSMT and TDEM method at the Mori geothermal field, Hokkaido Proc. 93th SEGJ Conf., 181–185 (in Japanese).

Lee, K. H., Pridmore, D. F., and Morrison, H. F., 1981, A hybrid three-dimensional electromagnetic modeling scheme: Geophysics, **68**, 796–803.

Mogi, T., 1996, Three dimensional modelling of magnetotelluric data using finite element method: J. Appl. Geophys., **35**, 185–189.

Mogi, T., Kajiwara, T., Takahashi, M., Ikeuchi, K., Suzuki, I., Nishimura, S., Katsura, I., Nishida, J., Arsadi, E., Suzuki, K., Kusunoki, K., Matatoshi, S., Jomori, A., and Shirai, N., 1994, CSMT and TDEM survey in the Mori geothermal field, Hokkaido, north Japan: Proc. 90th SEGJ Conf., 374–378 (in Japanese).

Mogi, T., Nishimura, S., Katsura, I., Suzuki, K., and Widarto, D. S., 1995, Geothermal heat source assessment using electromagnetic prospecting: Proc. World Geotherm. Congr., **95**, 1131–1136.

Newman, G. A., Hohmann, G. W., and Anderson, W. L., 1986, Transient electromagnetic response of a three-dimensional body in a layered earth: Geophysics, **51**, 1608–1627.

Pridmore, D. F., Hohmann, G. W., Ward, S. H., and Sill, W. R., 1981, An investigation of finite-element modeling of electric and electromagnetic data in three dimensions. Geophysics, **46**, 1009–1024.

Reddy, I. K., Rankin, D., and Phillips, R. J., 1977, Three-dimensional modelling in magnetotelluric and magnetic variational sounding: Geophys. J. Roy. Astr. Soc., **51**, 313–325.

Sato, H., 1969, Newly obtained ^{14}C data concerning volcanic activity in Hokkaido: Chisitsu News, Geol. Surv. Japan, **178**, 30–25 (in Japanese).

Wait, J. R., 1982, Geo-electromagnetism: Academic Press Inc.

Weaver, J. T., and Brewitt-Taylor, C. R., 1978, Improved boundary conditions for the numerical solution of E-polarization problems in geomagnetic induction: Geophys. J. Roy. Astr. Soc., **54**, 309–317.

Zhdanov, M. S., and Spichak, V. V., 1992, Mathematical modeling of electromagnetic fields in three-dimensional inhomogeneous media: Nauka, Moscow (in Russian).

Zhdanov, M. S., Golubev, N. G., Spichak, V. V., and Varentsov, I. M., 1982, The construction of effective methods for electromagnetic modelling: Geophys. J. Roy. Astr. Soc., **68**, 589–607.

Electromagnetic Imaging of Fissured Crystalline Bedrock in Hydrogeology

Sophie Hautot[1]
Pascal Tarits[1]
Corinne Tarits[2]

Summary. We interpret regional electromagnetic (EM) data and chemical analyses of water, sampled in springs and seepages, from an area in St. Anne, near Brest (Finistére, France). The area is covered by a thin homogeneous arenaceous soil. We surveyed the region with both a controlled-source EM system (Geonics EM31) and a very-low-frequency (VLF) system. The controlled-source survey was sensitive mainly to the soil layer; the VLF system, to structures at depth. Inversion of the VLF data shows evidence of two different structural units below the soil. The first, which is resistive, is interpreted as a massive crystalline gneissic structure. In this zone, the water drainage is restricted to the topsoil. In the southwest, there is a heterogeneous conductive domain that we interpret as a fracture network able to drain water along the structural strike. Differences in water quality in the St. Anne area are associated with the different electrical structures at depth.

1 Introduction

The quality of underground water usually is determined by chemical analysis of water sampled in springs and streams or in collecting wells. Extrapolation of these results to the hydrogeological system, however, is difficult when there are complex subsurface structures. The problem is of particular importance for the aquifers of fissured crystalline environments. We are studying the usefulness of 3-D electromagnetic (EM) imaging in such complex hydrogeological environments.

We collected and analyzed hydrogeochemical and EM data at a test area in Brest (Finistére, France). We propose a drainage system for the area based on a joint analysis of these data. The area is located to the northwest of the Massif Armoricain in France. The gneiss de Brest is the only geological formation. The gneiss was derived from an Ordovician quartzite and presents a foliation direction N050. The crystalline structures are stepped by faults (directions N130, N10, N70) which belong to the fault complex of the Pays du Léon. The bedrock is covered unevenly by arenaceous soils

[1]Laboratoire de Géophysique Marine, UMR6538, Universite de Bretagne Occidentale, 6 Avenue Le Gorgeu, Brest cedex F-29285, France.
[2]Department of Earth Sciences & ISAMOR, Universite de Bretagne Occidentale, 6 Avenue Le Gorgeu, Brest cedex F-29285, France.

(Hallegouet, 1971). The morphology is dominated by a main valley, roughly east-west, the St. Anne valley (Fig. 1) and six additional smaller valleys. Natural springs are found either at emergences along the valley borders or in the valley along the slopes (Fig. 1). Most of the springs above the valley have collecting wells. Interpretation of EM data [very-low-frequency (VLF) and controlled-source soundings] determined the subsurface distribution of electrical conductivity in the region. This parameter is very sensitive to the presence of free water in geological structures.

2 Controlled-source EM survey

We surveyed the north of the main valley (Fig. 1b) with a controlled-source system (Geonics EM31) and with a VLF instrument. The EM31 system uses a coil transmitter that generates a primary magnetic field at 9.8 kHz. The distance between transmitter and receiver coils is 3 m. The ratio of the secondary EM field to the primary magnetic field is derived from the signal measured at the receiver coil. The ratio is proportional to the apparent electrical conductivity σ_a of the earth beneath the coils. The coils may be used in the vertical RV mode (with coil axes vertical) and in the horizontal RH mode (with axes horizontal). The RV and RH modes provide estimates of the electrical conductivity to depths of about 3 to 6 m. We generally used the RV mode.

Apparent conductivity σ_a was measured at 202 sites, shown as black dots on Fig. 2. This figure also shows the apparent conductivity contoured in milli-Siemens per meter (mS/m). The apparent conductivity varies from about 3 to about 12 mS/m across the area (resistivities of about 330 to 85 ohm-m). Average apparent conductivity decreases regularly from northwest to southeast. At the few sites where we used the RH mode as well, we found apparent conductivities equal to 6–7 mS/m (resistivities of about 150 ohm-m).

3 VLF survey

The VLF EM method makes use of powerful radio-frequency transmissions between about 15 and 30 kHz. We mostly used the frequencies 19.6, 18.3, and 20.3 kHz, generated by stations in France, the United Kingdom, and Italy, respectively, at directions N005, N108, and N112 to the survey area. The penetration depth of the EM field at these frequencies is about 40–60 m in this area.

The VLF instrument measured the apparent resistivity ρ_a, the phase lag ϕ between the horizontal electric and magnetic fields, and the tilt angle (calculated from the ratio of the vertical component to the horizontal component of the magnetic field). Figure 3 shows ρ_a, ϕ, and the tilt angle for the transmitters north and east of the area (N- and E-transmitters). The apparent resistivity ranges from about 50 ohm-m to more than 1000 ohm-m. The highest values are found in the north-northeast, on the plateau. The tilt angle exhibits complicated patterns, different from one mode to the other. It suggests the presence of conductive paths across the gneiss massif either toward or along the valley.

4 Three-dimensional interpretation

As expected in such a region, the data indicate the presence of complex 3-D features at depth. There is a striking difference between the weak variation across the area of the apparent conductivity from the EM31 data (Fig. 2), which senses the shallow structure, and the strong variation in the VLF data (Fig. 3), which senses deeper. A 3-D analysis of the VLF data is therefore necessary to image the deeper structure.

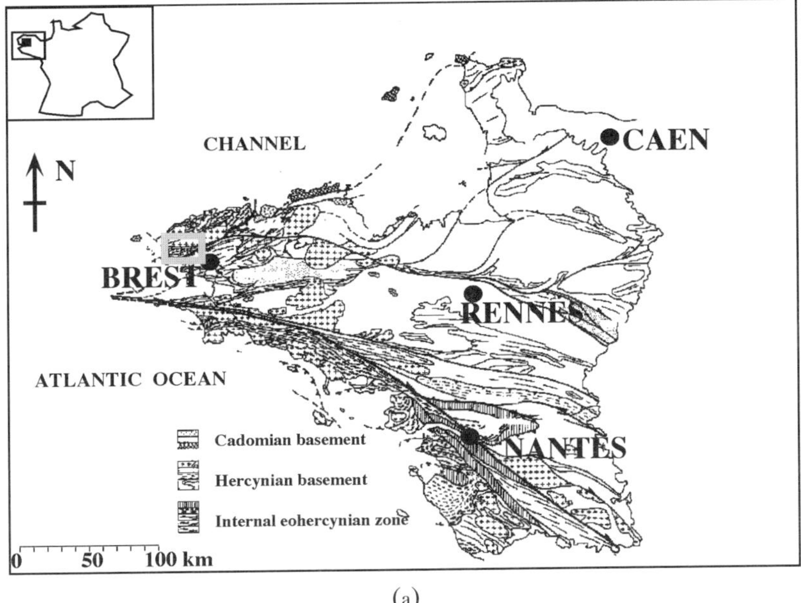

(a)

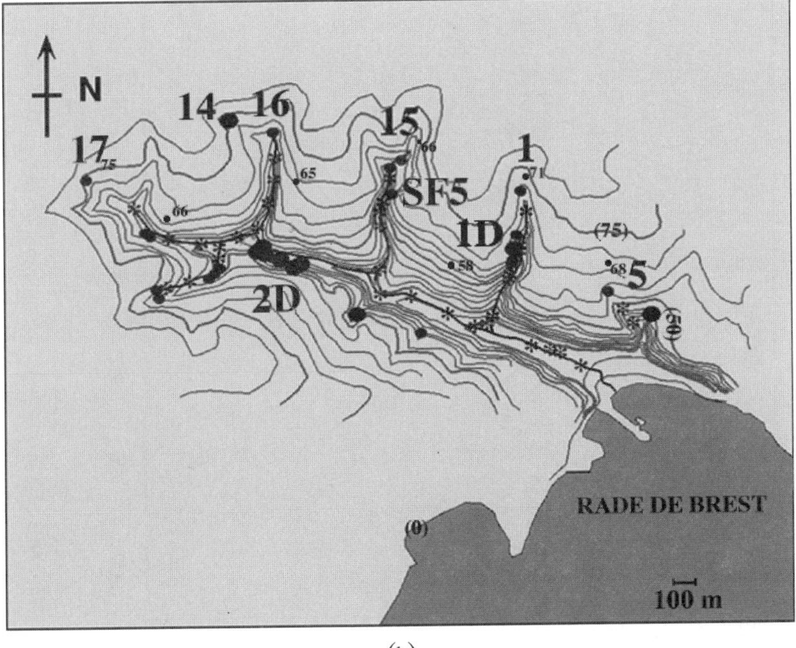

(b)

Figure 1. (a) Simplified geological map of Brittany, France. (b) Topography of St. Anne area. Contour lines are labeled in meters. Numbers stand for main natural springs and collecting wells. The small (big) solid circles are the water samples from the February experiment with mean electrical conductivities less (more) than 550 S/cm. Stars are locations of water samples taken from natural streams.

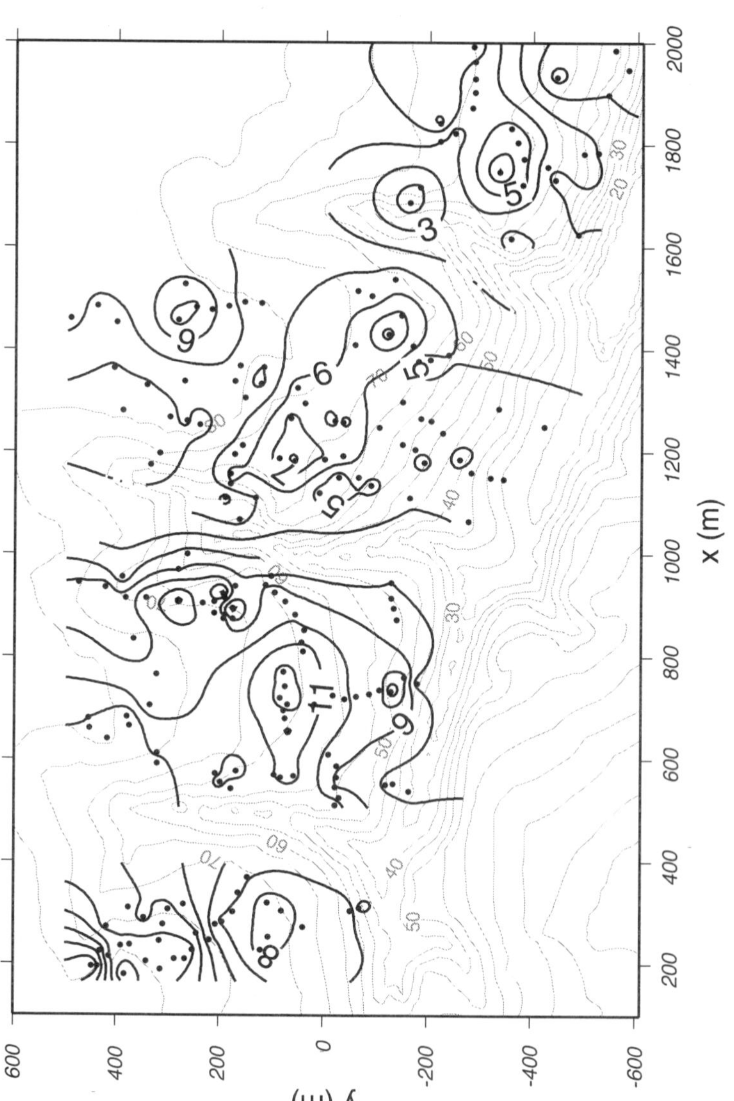

Figure 2. EM31 sites and the apparent conductivity contoured in milli-Siemens per meter.

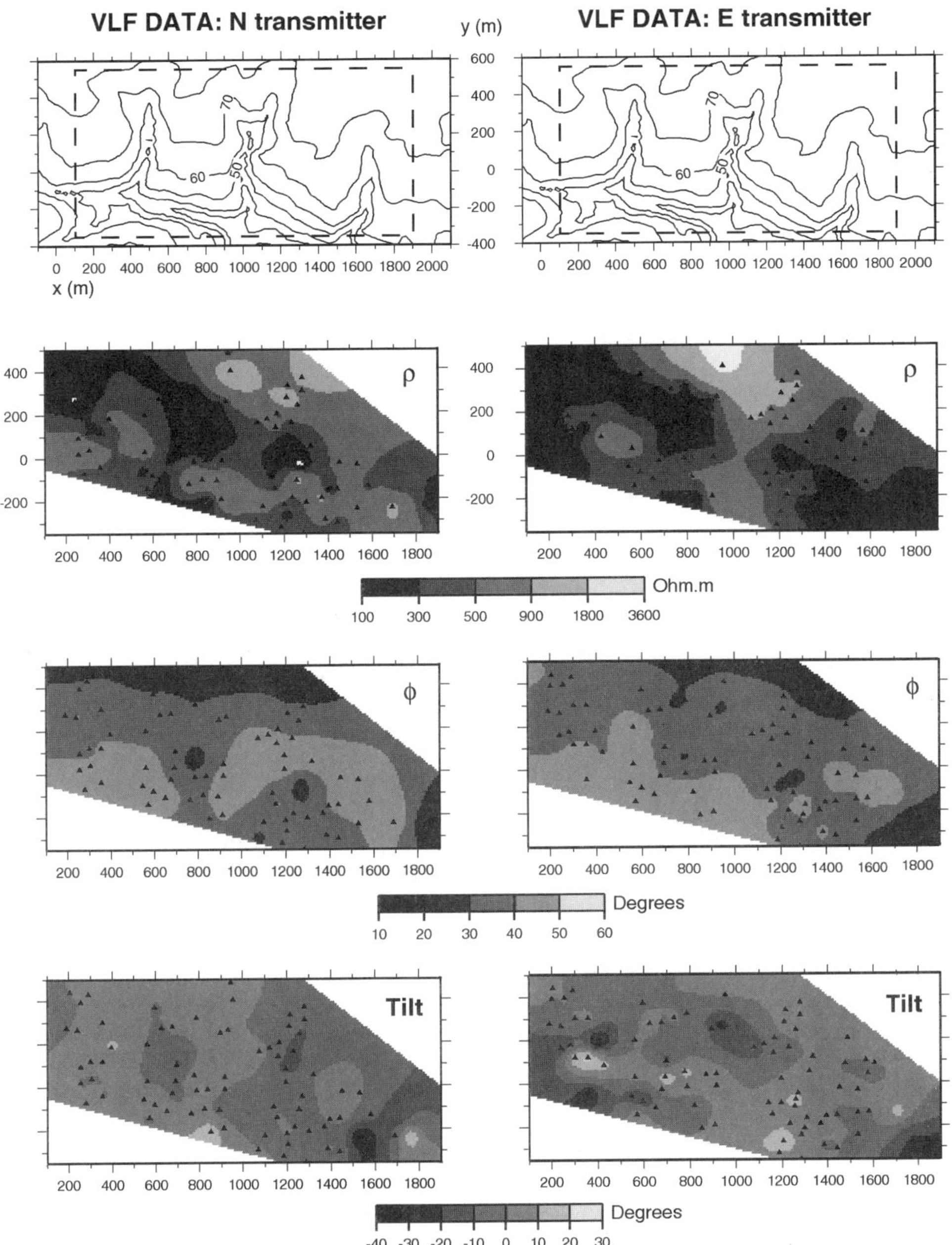

Figure 3. Areas surveyed and (from top to bottom) apparent resistivity ρ_a, phase lag ϕ, and tilt angle from VLF measurements with transmitter north (N, left) and east (E, right) of the area.

We successfully interpreted the VLF data with a 3-D inversion that minimized the misfit between data and a parameterized model. Forward modeling was performed with an algorithm developed by Mackie et al. (1993) to compute 3-D magnetotelluric (MT) responses. We modified the 3-D forward code to test the influence of the topography on the results of the inversion.

The starting models without topography consist of homogeneous layers. The thickness and the resistivity of the first layer was set to 7 m and 150 ohm-m, based on the

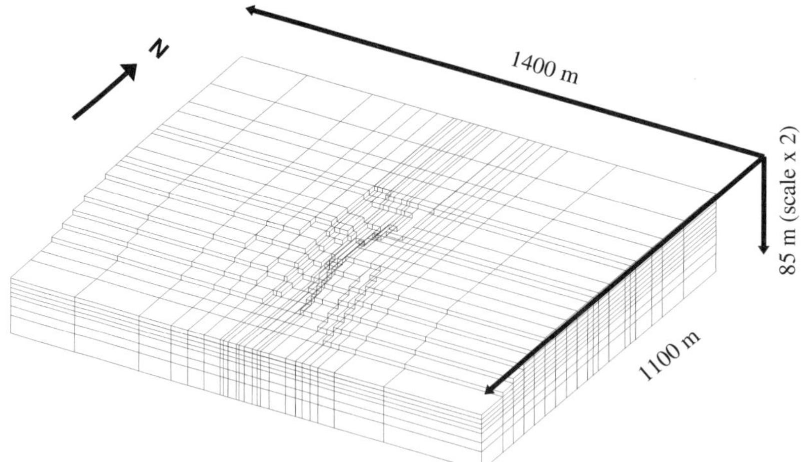

Figure 4. Central part of the model (with topography) for inversion. Thickness of layers 1 to 5 are 7, 7, 14, 14, and 7 m, respectively.

EM31 data. The model is divided into rectangular cells of variable dimensions (not less than $20 \times 100 \times 7$ m). The total size of the model is $2970 \times 3600 \times 140$ m (in x-, y-, z-directions) (Fig. 4). Only its central part ($1170 \times 1800 \times 79$ m) is used in the optimization procedure with a starting value equal to 150 ohm-m. The rest is considered as homogeneous and resistive (3000 ohm-m) crystalline basement (Fig. 4).

The misfit is the sum of the weighted squared differences between the data—the impedance Z recalculated from ρ_a, ϕ, and tilt angle for both transmitters E and N—and the values from the modeling. The solutions were regularized by the simultaneous minimization of the misfit function and the conductivity contrast between two successive cells in a given layer. The error bars on the data were derived from the scatter of the measurements. This uncertainty was unrealistically small at some sites. Hence we set the relative error to 0.05 for the impedance and the absolute error to 5° for the tilt angles when these uncertainties were too small.

4.1 Results

We carried out different tests, with and without topography (the main and the secondary valleys). The topographic effect is visible on the EM31 data (Fig. 2), though it is weak. It is unclear on the VLF data (Fig. 3). Figure 5 shows the inverted model when the topography is not included. The medium was split into four heterogeneous layers topped with a 7-m-thick homogeneous layer (Fig. 4) the characteristics of which were deduced from the EM31 data. This layer is not shown on Fig. 5.

The combination of VLF data from two almost perpendicular directions and the variability of the data over the area did not allow a model with a single heterogeneous layer. It was necessary to have a model with several layers even though the VLF data, available in a narrow frequency band, do not have much depth resolution. The true depth of the structures probably is not estimated properly. Further investigations with additional controled-source equipment and electric soundings will be carried out to improve the resolution in depth.

Figure 6a shows the calculated ρ_a, ϕ, and tilt angle for the model in Fig. 5. Figure 6b shows the misfit between the model and the data. In general, the fit is quite good. The

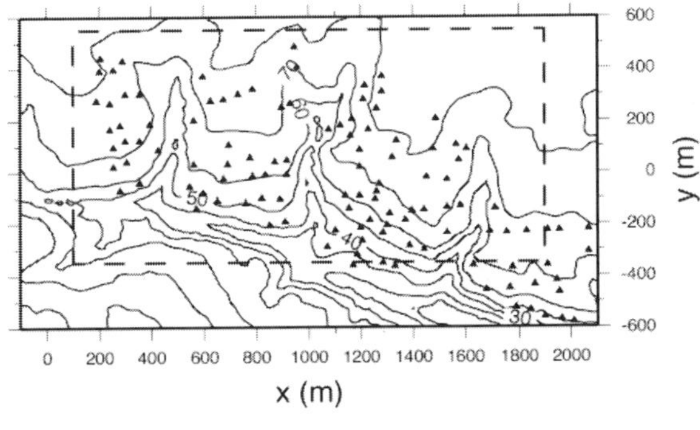

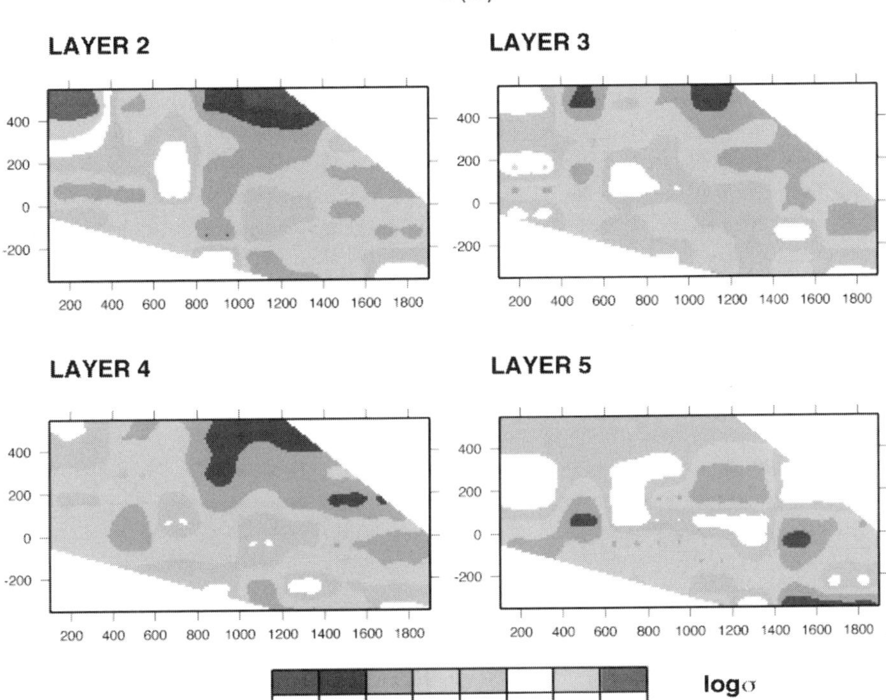

Figure 5. (top) Delineation of model (identical to area shown on Fig. 3); (bottom) regularized optimum conductivity model, layers 2, 3, 4, and 5 of thicknesses 7, 14, 14, and 7 m, respectively.

model, however, could not be adjusted to the data on or near the sides. These data have been weighted down so that the rest of the calculation will not be biased. Some misfits are still large in a few sites (Fig. 6b), most likely because the model cannot account for features that are too small.

We modified the model (Fig. 5) to include the main valley and one secondary valley to test the influence of the topography (Fig. 4). We used a different gridding with meshes from 10 × 10 m to 300 × 100 m to describe the topographic features. The comparison between the response functions for the model with or without topography is shown in Fig. 7. We only display the area surrounding the secondary valley. Figure 7a shows results for the N-transmitter and Fig. 7b shows the E-transmitter. The influence of the topography varies according to the type of data (i.e., apparent resistivity, phase, or tilt) and according to the transmitter direction. It generally is localized near the edges of the valleys. It seems that the topography effect may explain some of the misfit features in Fig. 6b.

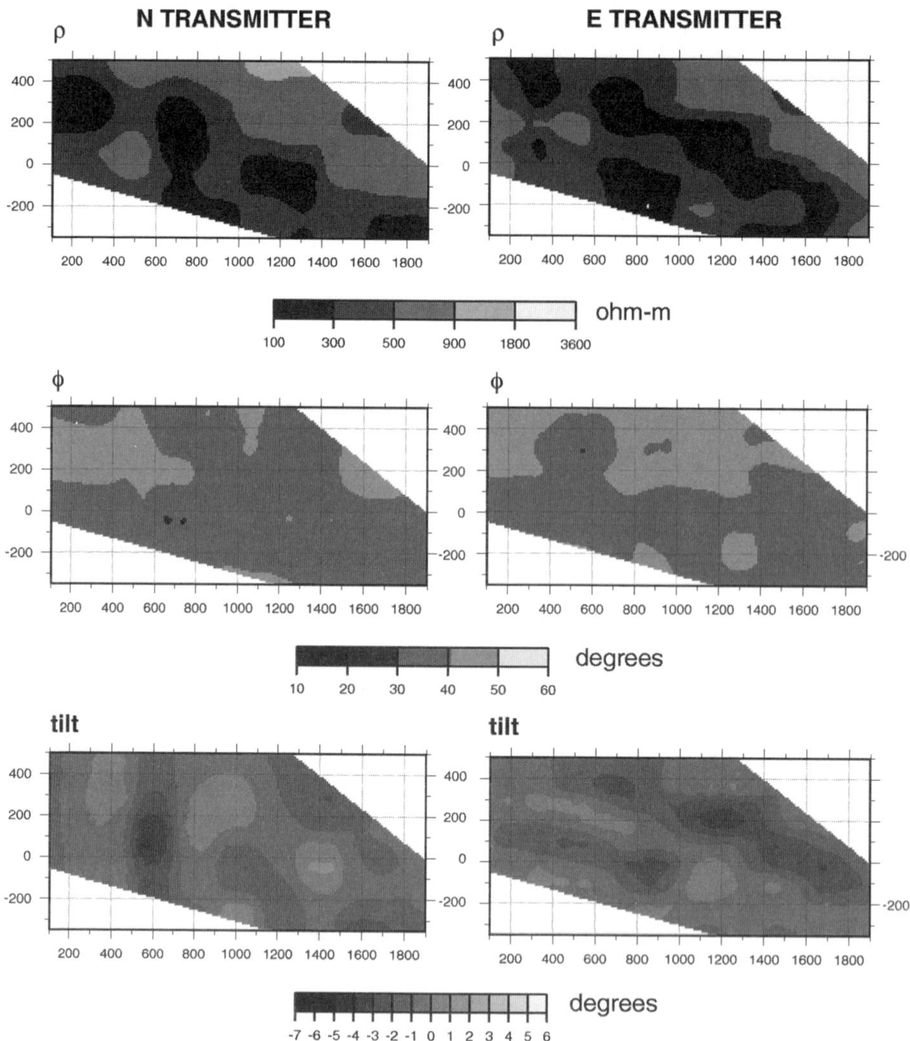

Figure 6a. Maps of calculated apparent resistivity and tilt angle for the N-transmitter and the E-transmitter.

The model in Fig. 5 shows two main domains: a resistive zone to the northeast and a more conductive zone to the west. The resistive zone seems to persist in layers 2 to 4. The strike of the transition from the resistive body to the more conductive medium is roughly southeast-northwest. The northeastern part of the resistive block is not resolved because there are no data there (Fig. 5a). The conductive zone contains complicated successions of conductive and more resistive features (in layers 2 and 3). A striking feature is the very conductive area to the northwest. This structure is tapered in layers 3 and deeper. Layer 5 has mostly small structures, a likely result of instabilities due to a poor resolution. On average, this layer is more conductive than the ones above. The details given by the small structures in all layers presumably are unreliable. This is visible in Fig. 8 which shows a model obtained without regularization. Some of the features observed in Fig. 5 have changed. The general trends, however, are preserved.

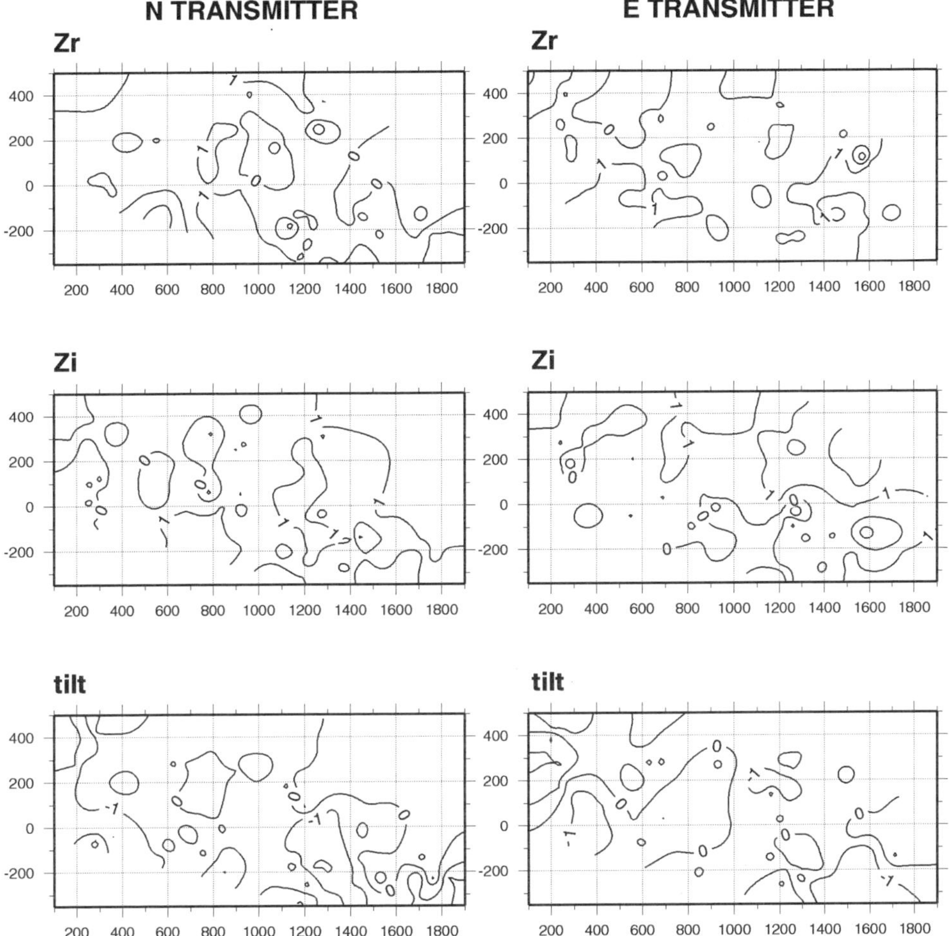

Figure 6b. Log of misfit m between model impedance (Z_r, Z_i) and tilt angles and the data for both transmitters. For a given data d_i at a site i and the corresponding model value c_i, the misfit m_i is $m_i = (d_i - c_i)^2 * w_i$. The weight w is $1/\text{variance}(d_i)$.

5 Hydrogeochemistry

Five field experiments were carried out between 8 February 1995 and 24 November 1995 (Tarits et al., 1995a,b) and on 6 January 1996 in the St. Anne Valley. The purpose of the study was to characterize the origin of the water and its quality in this area of fractured crystalline bedrock. A total of 82 water samples were collected during these experiments. The samples were taken in natural springs, in seepages, in collecting wells, and in streams. Rainwater and seawater samples were collected as well.

Temperature, pH data, and electrical conductivity were measured in the field. In February, the temperatures ranged from 7.1°C to 12.9°C. The pH varied from 6.02 to 7.03. The electrical conductivities were centered around 400–450 S/cm. A significant number of values ranged from 550 to 650 S/cm (Fig. 1b).

The water samples were analyzed for the major ionic species Ca, Na, Mg, K, SO_4, alkalinity, NO_3, and Cl. The alkalinity ($HCO_3 + CO_3$), noted “alk,” was determined by titration. The cations Ca, Na, Mg, and K were measured using an inductively coupled

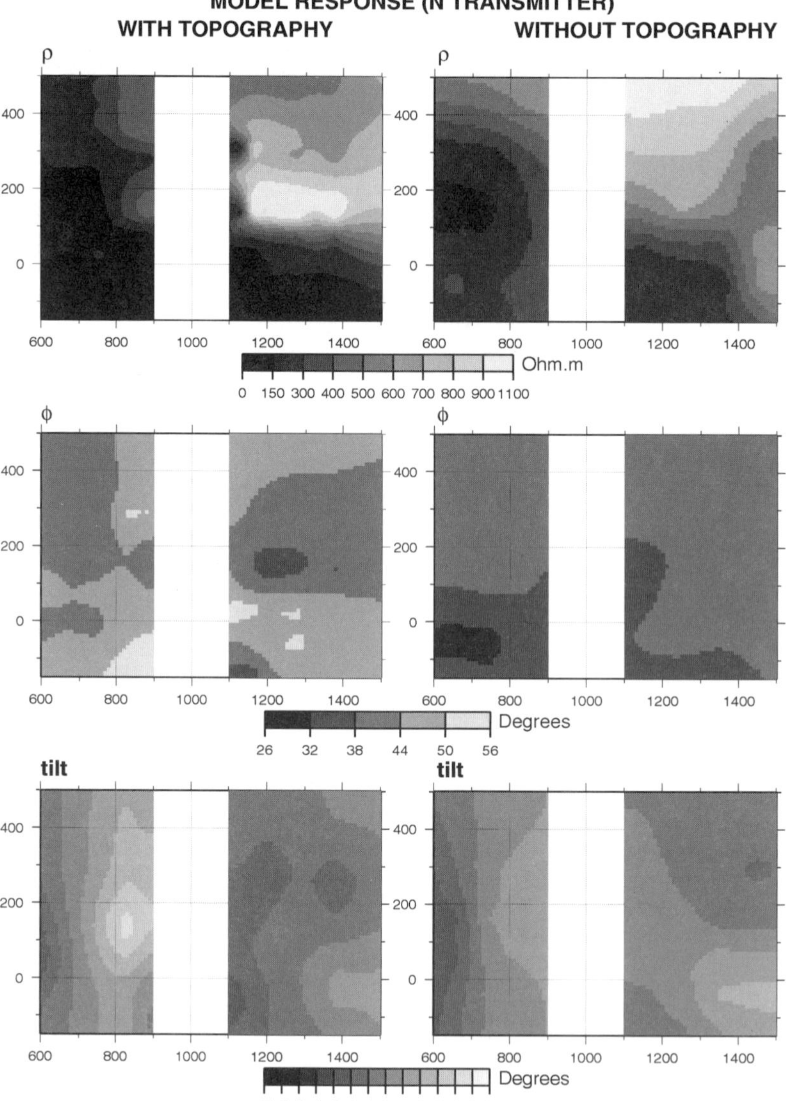

Figure 7a. Influence of topography on EM data [(top) apparent resistivity, (center) phase, (bottom) tilt], using N-transmitter. The response functions for the model presented in Fig. 5 modified to include the topography are on the left. The response function for the model without topography are on the right. We removed the part corresponding to the secondary valley (the white band) from the plots because there are no data there.

plasma atomic emission spectrometer (ICP-AES) and the anions SO_4, NO_3, and Cl by ionic chromatography. All samples have an NaCl-type natural structure (Fig. 9). Na is the predominant cationic species; Cl and NO_3 are the predominant anionic species.

The samples are the result of the mixing of waters of poles from different origins. Analysis of the variance-covariance matrix of the chemical elements for all samples is used to characterize the poles. These poles are the eigenvectors of the system and may be determined by principal-component analysis (PCA) (Albarède, 1995). Figure 10a shows the result of this analysis for Na, Ca, Mg, K, alk, and Cl with the whole set of raw data (from the five experiments). The first three components explain 91% of the

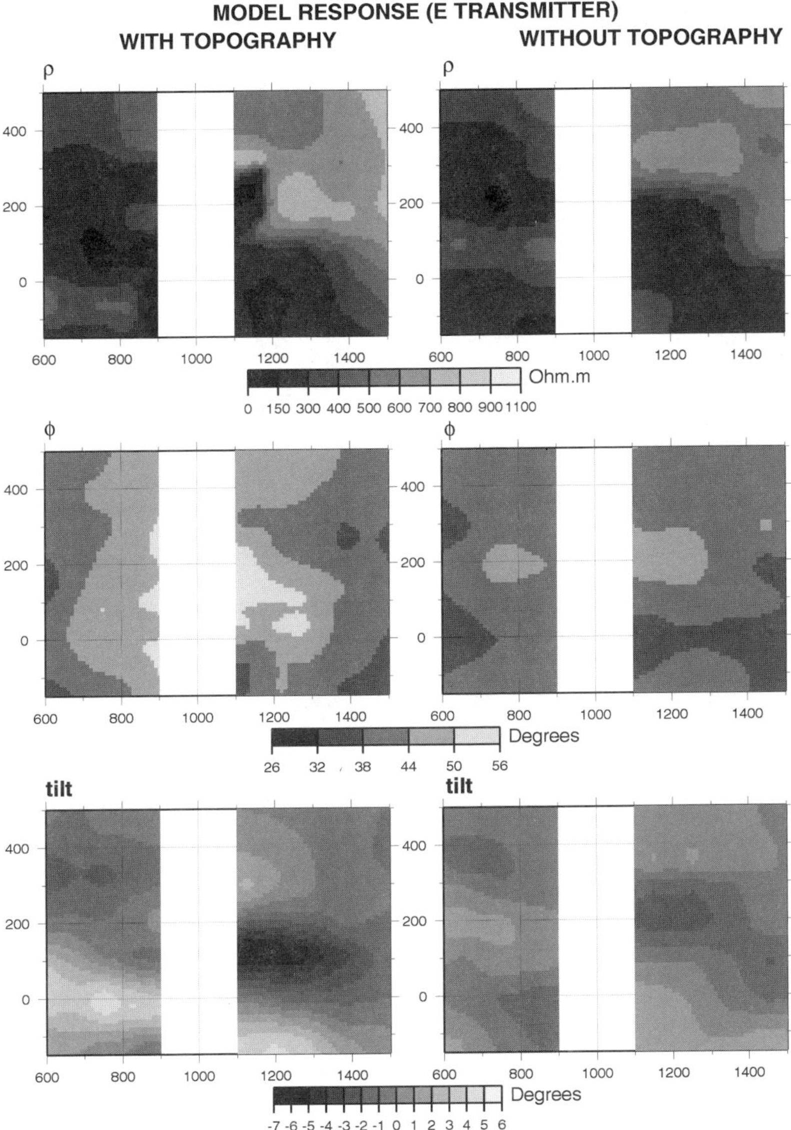

Figure 7b. Influence of topography on EM data, using E-transmitter. See Fig. 7a for details.

variance. The components are plotted pairwise (Fig. 10a). Component 1 is dominated by Na, Ca, Mg, and Cl and component 2 is dominated by K and alk (see upper-left plot). On the lower right, K and alk are close to zero. This component is orthogonal to the plane defined by components 1 and 3 and is well individualized. This is in agreement with the hypothesis that K may be considered as a runoff tracer (Pichon et al., 1995).

The correlation between Na, Ca, Mg, and Ca is complex. Na and Cl are the major constituents in the chemical composition of the seawater. In Fig. 10a, (Na, Cl) should represent the sea pole. Cl also may be found in industrial-pollution products. In this region, however, there is no clear evidence of such pollution. K is only associated with alk and never with the other species. In addition, incident concentration of chlorides has very little influence when the soil is not covered by forests (Pichon et al., 1995). The forest surface is small and covers only the bottom of the valleys. Then, Cl is

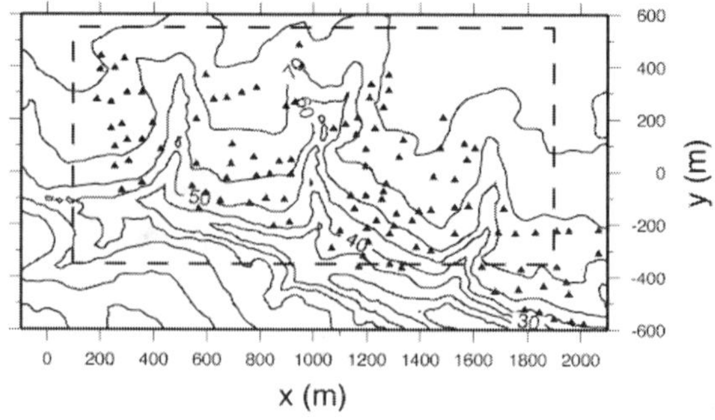

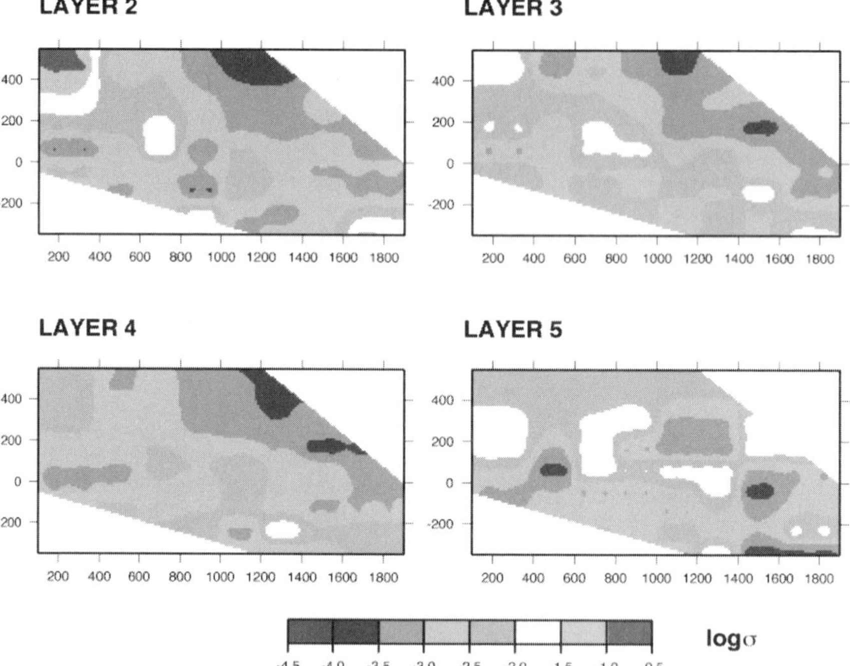

Figure 8. Same as Fig. 5 for a nonregularized model (see text for details).

likely to come mainly from the sea (through the rain). In this area, Cl accumulation can be considered as a qualitative tracer of the residence time of the sample in the soil.

Na as well as Mg and Ca can be found in manures and fertilizers, common in this agricultural area. We actually observe in some of the samples an excess of Na with respect to the very stable ratio Na/Cl from seawater and rainwater. Na, Mg, and Ca should be associated with the subsurface transit of the water in the soil along with Cl or in the fractured bedrock. The PCA is very stable over time. The result of the analysis done for each experiment individually is similar to the result obtained when all of the samples are taken into account. Figure 10b shows the PCA done in November. We only present the plane defined by components 1 and 3. Once we know the principal components, it is possible to characterize individual samples with respect to the significant poles. Figure 10b shows the relative composition in the PCA basis, of the November water samples normalized to one standard deviation.

The rainwater samples are the light blue dashes, and are all away from the principal components. The other water samples are divided into three groups. The green triangles

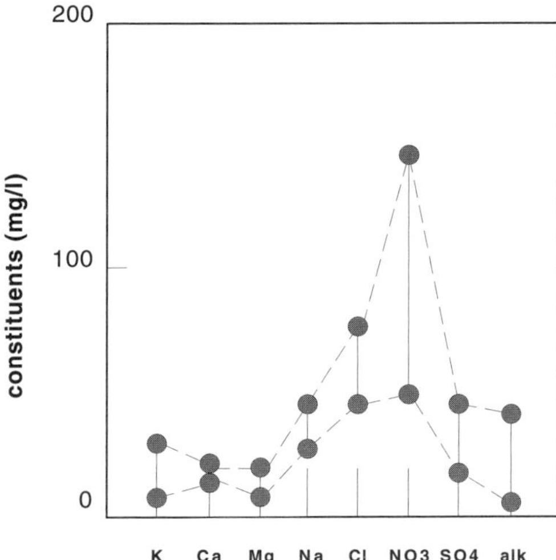

Figure 9. Chemical diagram for water samples of a selected experiment (February 1995). Alk = $HCO_3 + CO_3$. Each element is represented by two dots that correspond to the extreme values measured during the February experiment.

are remotely associated with principal component 1 (Na-Cl, Mg-Ca). Samples 17 or 15 (Fig. 1b) are representative of this group. The group seems to result from runoff or/and soil drainage. The pink diamonds are located in the upper-right quadrant (Fig. 10b). All samples are related to the sea pole (Na, Cl). This suggests that these waters, such as 16, 1, or SF5 (Fig. 1b), may be the result of either a subsurface transit longer than for the previous group or a soil drainage associated with a circulation in the fissured bedrock. Samples 1-D and 5 (Fig. 1b) are the blue circles in the upper-right quadrant. They are significantly influenced by the sea pole (Na, Cl). These samples have a specific behavior, however, because the catchments are the closest to the sea among all samples and are exposed directly to the sea spray. This could explain the large concentration of (Na, Cl), not necessarily connected to a long residence time. The last group in Fig. 10b are the water samples 2-D (Fig. 1b). The samples (blue circles, lower-right quadrant) are clearly associated with the (Mg, Ca) pole. These waters may have transited in the soil as well as in fractures as suggested by their geological settings. These waters are the only ones associated with gneiss outcrops in the St. Anne valley.

6 Discussion

The EM31 EM data mapped a thin (a few meters or so) homogeneous conductive layer that we interpreted to be the arenaceous soil. This layer tops a medium with more contrasting features revealed by the VLF EM data. The resistive structure to the northeast (Fig. 5) is likely to be a massive crystalline block. The conductive zone in the southwest half of the figure could be interpreted as an intensely fractured and fissured zone.

We have drawn the main structural directions proposed by Hallegouet (1971) on the electrical structures obtained from the 3-D EM interpretation (Fig. 11). For sake of clarity, only layer 2 from Fig. 5 is presented. The main northwest-southeast fault separates the northeast resistive block from the more conductive medium to the southwest. The strike in the electrical structures is present in layers 3 and 4 as well (Fig. 5). The conjugated fault (northeast-southwest) crosses the resistive zone along a strike that

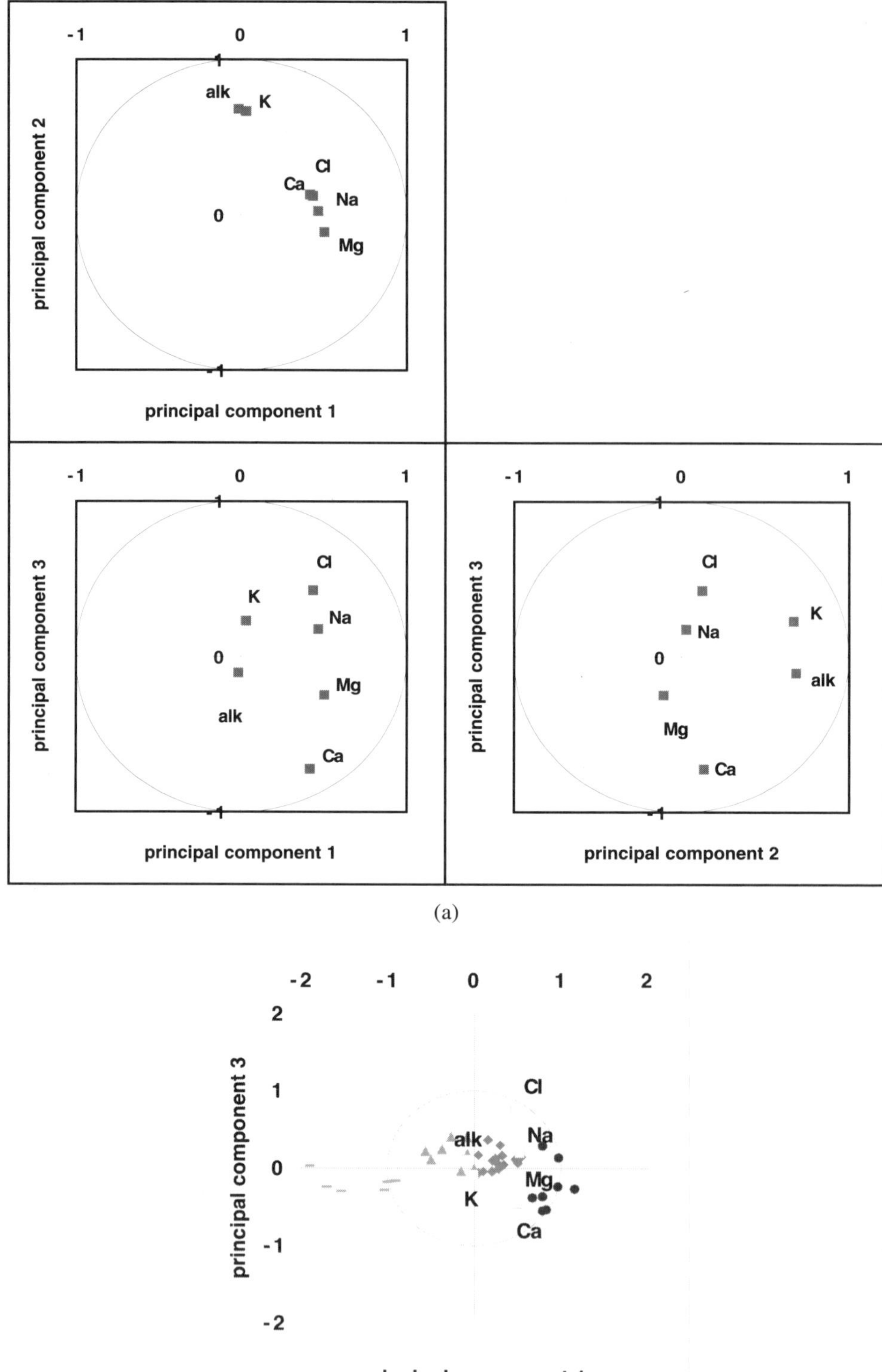

Figure 10a. (a) Results from PCA on raw data for the five experiments in the St. Anne area. (b) PCA performed on November data. Only components 1 and 3 are plotted. Composition of the samples is shown in the principal components basis. Symbols are described in text.

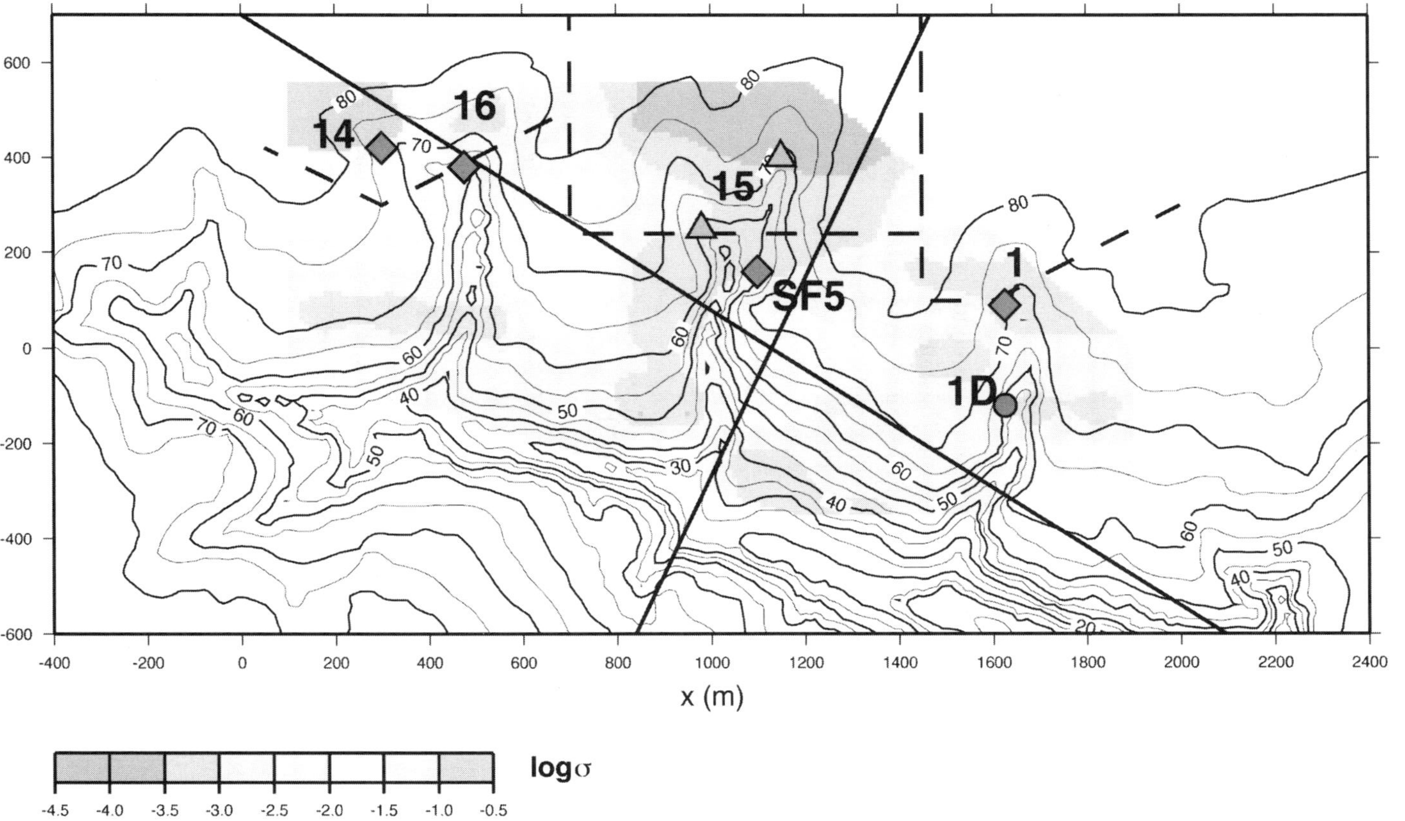

Figure 11. Synthesis of EM and hydrogeochemical results. Layer 2 from Figure 5 is in the background. Zones of the model without data have been removed. Color and symbols for the springs are the same as used in Fig. 9b. We reported the main faults (solid lines) and a schematic representation of the topographic catchment areas for the uphill springs (dashed lines).

is present in the model but shifted eastward. Some features seems to correlate with this fault in layers 3 and 4 (Fig. 5), but they are poorly resolved.

The results from the hydrogeochemistry (PCA shown in Fig. 10b) are in Fig. 11 with the colors and symbols used in Fig. 10b. We also show the approximate location of the topographic catchment areas for the uphill springs 1, 15, and 14–16.

The catchment areas are located on the main electrical structures, namely a fairly homogeneous resistive zone (above 15), a conductive zone to the northwest (above 14–16) and a zone (above 1) not well defined because of the lack of data. It seems to be a resistive structure with a more conductive zone to the west.

Samples 15 belong to the green triangles group (Fig. 10b) corresponding to soil draining or to runoff. They are associated with the resistive zone. Samples 14–16 belong to the pink diamonds group (Fig. 10b), influenced by both a transit in the fractured bedrock and in the soil as for the previous group. They are related to the northwest conductive zone. The conductive zone vanishes in depth (Fig. 5) which suggests that the water flow is rather shallow. Sample 1 is associated with the zone bordered by the northeast–southwest fault. It belongs to the pink diamonds group.

SF5 and 1-D correspond to springs down in the valleys. In the field, we noted a very narrow (about 10–20 m wide) linear morphological feature above each spring. These features might be associated with the beginning of the conductive zone southwest of the main fault. We cannot draw conclusions from the EM data because the scale is much too small.

7 Conclusions

A 3-D survey with EM31 and VLF data successfully mapped the general structures of the St. Anne valley area below a thin and homogeneous arenaceous topsoil. The structures revealed by the 3-D imaging are well correlated to the main structural strikes. In addition, correlation to the hydrogeochemistry data in the area proved useful to characterize the origin of the waters observed at emergences. It is striking that the different water qualities are associated with different electrical structures in depth.

The results of this preliminary study with limited data are encouraging. It suggests that denser surveys and additional chemical data will thoroughly describe the flow pattern, the chemical interaction, and the mixing of the meteoric water in a fissured crystalline bedrock.

Acknowledgments

J. Cotten did the IPC-AES analyses. J. Y. Cabon did the ionic chromatography analyses. We thank M. L. Guillermic and M. Wanner for their assistance in the field work. We thank the Association Syndicale d'Adduction d'Eau de Plouzané as well as the Farmers from the St. Anne valley area for allowing us to work on their property.

References

Albarède, F., 1995, Introduction to geochemical modelling: Cambridge Univ. Press.

Fischer, G., and LeQuang, B. V., 1981, Topography and minimization of the standard deviation in one-dimensional magnetotelluric modelling, Geophys. J. Roy. Astr. Soc., 67, 279-292.

Hallegouet, B., 1971, Le Bas-Léon (Finistère, France), etude géomorphologique, Ph.D. thesis, Université de Bretagne Occidentale, Brest, France.

Mackie, R. L., Madden, T. R., and Wannamaker, P. E., 1993, Three-dimensional magnetotelluric modeling using difference equations—Theory and comparisons to integral equation solutions, Geophysics, **58**, 215–226.

Pichon, A., Marc, V., and Travi, Y., 1995, Influence de l'interception des pluies sur le traçage chimique de l'eau en zone forestière méditérranéenne (BVRE du Réal Collobrier, Var, France), C.R. Acad. Sci., **321-IIa(2)**, 137–144.

Tarits, C., Guillermic, M. L., Hautot, S., Thonon, P., and Tarits, P., 1995a, Monitoring of nuclear and toxic waste sites with chemistry and electromagnetic soundings, XXI General Assembly IUGG, Abstracts, A416.

———1995b, Correlation between hydrogeochemistry and electromagnetic imaging: Application to hydrology of fissured crystalline bedrock, EOS, trans. AGU, 76(46), Fall meet. suppl., 253.

PART VII

BOREHOLE GEOPHYSICS AND LOGGING

Electrical drill hole methods . . . fall into two groups: single-hole methods or conventional well-logging, and hole to surface and hole to hole borehole geophysics. Either of these methods may employ electrodes or loops and measurement techniques commonly used in surface geophysics.

Glenn and Hohmann (1981)

Measurement of Surface and Borehole Electromagnetic Fields in 2-D and 3-D Geology

Michael Wilt[1], Clifford Schenkel[1],
Brian Spies[2], Carlos Torres-Verdin[2]
David Alumbaugh[3]

Summary. A 2-D oil-field survey and a 3-D environmental study illustrate the problems of collecting and interpretating electromagnetic data in crosshole and surface-to-borehole geometries. The first example employs both methods to map the development of a steam flood in a shallow heavy oil reservoir at the Lost Hills oil field in central California. Measurements were made from observation wells that straddle the steam injector in a plane orthogonal to the regional strike, and from a surface-to-borehole line that extends between the observation boreholes and into an undeveloped portion of the field. In a second example, a crosswell experiment was designed to image 250 000 liters of saltwater injected into a shallow aquifer near the University of California at Berkeley campus. A five-spot well pattern was drilled for the test, and saltwater was injected in the central well at a depth of 30 m. Crosswell data were collected, using all five boreholes, before and after injection.

Bias and uncertainty of calibration are at least as important as random noise in these data. The bias can be caused by ground loops, cultural noise, systematic drift, and errors in sensor placement. Repeatability and reciprocity tests can often quantify this measurement error.

Imaging schemes also introduce errors through geometrical assumptions. The Lost Hills data were imaged in a 2.5-D model (point sources in 2-D geometry), which appears to represent the background geology well, but begins to break down within the steam injection zone. Although a 3-D code was used to image the data collected at Richmond, artifacts were still present in the image. Subsequent analysis has shown that these are likely due to sensor placement errors as a result of borehole deviations.

1 Introduction

Recent field tests of low-frequency crosshole electromagnetic (EM) tomography have yielded impressive results. In one experiment, crosshole and surface-to-borehole

[1]Lawrence Livermore National Laboratory, P.O. Box 808 L-156, Livermore, CA 94550, USA.
[2]CRC AMET, Macquarie University, Sydney NSW 2109, Australia.
Schlumberger-Doll Research, Old Quarry Road, Ridgefield, CT 06877, USA.
[3]Sandia National Laboratories, P.O. Box 5800, Albuquerque, NM 87185, USA.

measurements were used to characterize the background resistivity structure in an oil field and to locate zones of high temperature associated with enhanced oil recovery operations (Wilt et al., 1995a). In a second example, crosshole EM was used to track a small plume of injected saltwater (Alumbaugh, 1993; Wilt et al., 1995b).

There are several obstacles to the practical implementation of this promising method. Successful field surveys so far have had generous budgetary and time constraints. Field sites were conveniently located and data were collected largely by graduate students. In commercial surveys, by contrast, field data must be collected and processed in a timely manner and applications will likely be in noisier environments with access and time constraints.

We review the general characteristics of crosswell EM systems and discuss sources of noise and error in field measurements. We then focus our attention on field data collected at the two experiments described above with a prototype system developed by Lawrence Livermore and Lawrence Berkeley National Laboratories (LLNL and LBNL). We describe some common problems of data collection and some methods to assess measurement error. We also examine images reconstructed from data collected at these two sites. These images demonstrate how well the geology can be resolved by 2-D and 2.5-D imaging schemes.

2 Systems and hardware limitations

2.1 High-frequency cross-borehole EM systems

Early cross-borehole EM systems evolved from radio-wave studies in boreholes and mines, carried out intermittently since the 1920s (Rust, 1940). Lytle and others in the late 1970s laid the theoretical foundation of geotomographic imaging, which raised the interest of the geophysical community (Lytle et al., 1979). The earlier systems operated at frequencies above 10 MHz, and were used to probe very-high-resistivity rocks, such as salt and granite (e.g., Holser, et al., 1972; Cook, 1977). The earliest oil-field crosswell experiments used frequencies in the 1- to 30-MHz range (e.g., Witterholt and Kretzschmar, 1982), but these surveys covered relatively short distances. Recent applications include detection and imaging of geological inhomogeneities in coal seams using the radio imaging method (RIM). This technique is based on the attenuation of 50-kHz to 20-MHz EM signals in the coal-seam waveguide (Stolarczyk, 1990; Thomson et al., 1990; and Vozoff et al., 1993). A fundamental limitation to high-frequency measurements is the small penetration depth in conductive rocks. Existing systems typically can penetrate no more than a few tens of meters in most oil fields.

In most oil fields, however, the radio- and radar-frequency methods cannot probe beyond a few tens of meters, at most, because of the strong attenuation. Only low frequencies (in the kilohertz range) can propagate the hundreds of meters that is required for oil-field investigations. This complicates matters because the simple wave concepts that are quite effective at interpreting high-frequency data do not apply at these lower frequencies. The physics of low-frequency EM is best described by diffusion, which makes interpretation more complex.

2.2 Cross-borehole EM at inductive frequencies

Cross-borehole field systems designed for frequencies below 50 kHz are being developed by LLNL/LBNL and Oyo Corporation, among others. The Oyo system features slim borehole tools (4 cm) and a three-component receiver (Osato and Takasugi, 1992; Gasnier et al., 1994). It operates between 4 Hz and 18 kHz and is designed for geotechnical and mining applications. The LLNL/LBNL system was designed for reservoir characterization and process monitoring in oil fields. This system features 10-cm borehole tools and operates at frequencies from 40 Hz to 100 kHz.

Borehole transmitters at inductive frequencies are solenoids with magnetically permeable cores. The highest moment (product of the effective coil area, the number of turns, and the current) to be achieved thus far is approximately 5000 A-m^2 at 10 Hz, reducing to 1500 A-m^2 at 1 kHz and 100 A-m^2 at 100 kHz. The source moment declines with frequency because of eddy current and hysteresis losses in the core. Fortunately, noise levels of high-quality magnetic-field sensors decrease with increasing frequency (Nekut, 1994). Practical noise levels for current-feedback induction coils are of the order of 50 μV/(Hz)$^{1/2}$ at 10 Hz, 10 μV/(Hz)$^{1/2}$ at 1 kHz and 1 μV/(Hz)$^{1/2}$ at 10 kHz.

Figure 1 shows cross-borehole magnetic-field amplitudes as a function of borehole separation and frequency for a 10 ohm-m uniform earth using a vertical magnetic dipole (VMD) source with a moment of 1000 A-m^2. The dashed lines are the borehole separations expressed as skin depths; the shaded region at the lower right corner is noise limit for existing borehole coils. The figure shows the resolution and range limits of crosshole EM for effective imaging. Assuming a lower limit of 100 Hz the frequency range extends from the upper curve to the shaded region at the base of the figure. For example, at a borehole separation of 100 m the figure shows that good-quality data can be collected at frequencies from 100 Hz to almost 100 kHz. We note that effective imaging is possible at two skin depths or greater and that high-resolution imaging requires 10 skin depths or more (Nekut, 1994). At larger separations the range is confined to the lower frequencies and is much reduced; this is caused mainly by the

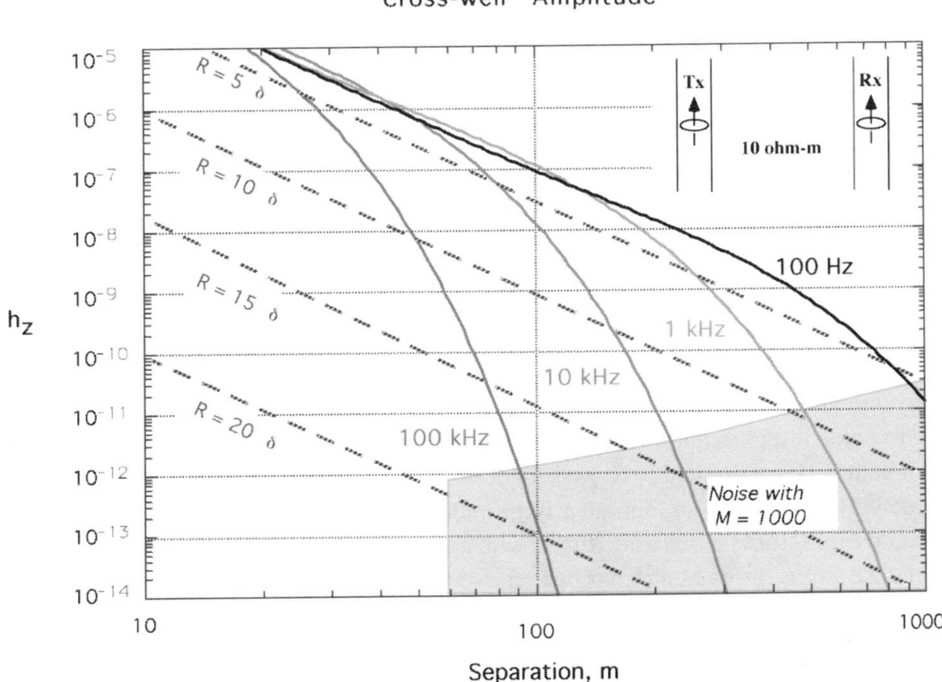

Figure 1. Crosswell magnetic-field signal amplitudes in a 10 ohm-m Earth. The dashed lines represent the distance between the wells expressed as skin depth.

geometric decay of the primary field. At larger separations, the signal is already small because of geometric spreading, any significant additional attenuation by the medium drops the signal level below the noise threshold.

Because the effective operating frequency range for crosshole EM depends on a combination of the formation resistivity and borehole separation, this type of figure is very useful for determining operating limits of the system (Spies, 1992). It shows that with current technology it is possible to make crosswell EM measurements between boreholes several hundreds of meters apart.

Sensitivity studies of the response of interwell heterogeneities suggest that the highest possible frequency should be used to image the conductivity distribution (Spies and Habashy, 1995). At higher frequencies the interwell medium is more uniformly sampled and the energy is more focused between sources and receivers. This results in better image definition. The high rate of attenuation at the highest frequencies, however, often makes it impossible to obtain adequate coverage for imaging. It is therefore desirable to collect data at several frequencies to obtain coverage and sensitivity.

2.3 Surface-to-borehole EM

Surface-to-borehole EM techniques were initially developed to search for ore deposits not intersected by drilling. A surface transmitter can have a large source moment and therefore a much greater range than a borehole source. In fact, these measurements have been made at depths as great as 3000 m (Spies, 1995). Although early surface-to-borehole EM systems were harmonic, broadband transient surface-to-borehole systems have become more popular because of their relative ease of interpretation (Noakes, 1951; and Ward and Harvey, 1954; Nabighian, 1984). Recent applications include characterization and cleanup monitoring at environmental sites (Tseng et. al., 1995).

In the surface-to-borehole geometry, the source can be moved on the Earth's surface to attain 3-D coverage. Also, only one borehole is required. However, the method has lower sensitivity than crosshole techniques and tomographic inversion is more difficult because it is necessary to incorporate near-surface structure and the air–earth interface into the inversion.

2.4 LLNL/LBNL prototype EM system

The LLNL/LBNL system is a single-component monochromatic frequency-domain system designed for crosshole and surface-to-borehole EM (Fig. 2). Data are collected by positioning transmitter and receiver tools at a number of levels that encompass the depths of interest between the boreholes. A typical data set consists of several thousand measurements.

The transmitter station generates high-power signals at the surface and sends them down standard logging cable to be broadcast using a vertical-axis coil. The borehole source coil consists of a magnetically permeable core (mu-metal or ferrite) wrapped with 100 to 300 turns of wire and tuned with a capacitor to resonate at a fixed frequency. In recent designs the oscillator has been moved from the surface into the transmitter tool with dc power supplied down the cable. In addition, we now switch the tuning capacitors remotely, making it possible to change the transmitter frequency from the surface.

A surface-based loop transmitter is used for the surface-to-borehole configuration. This transmitter is operated in the same manner as the borehole source (i.e., tuned with capacitors) but, because of the large surface area, the moment is up to 1000 times larger

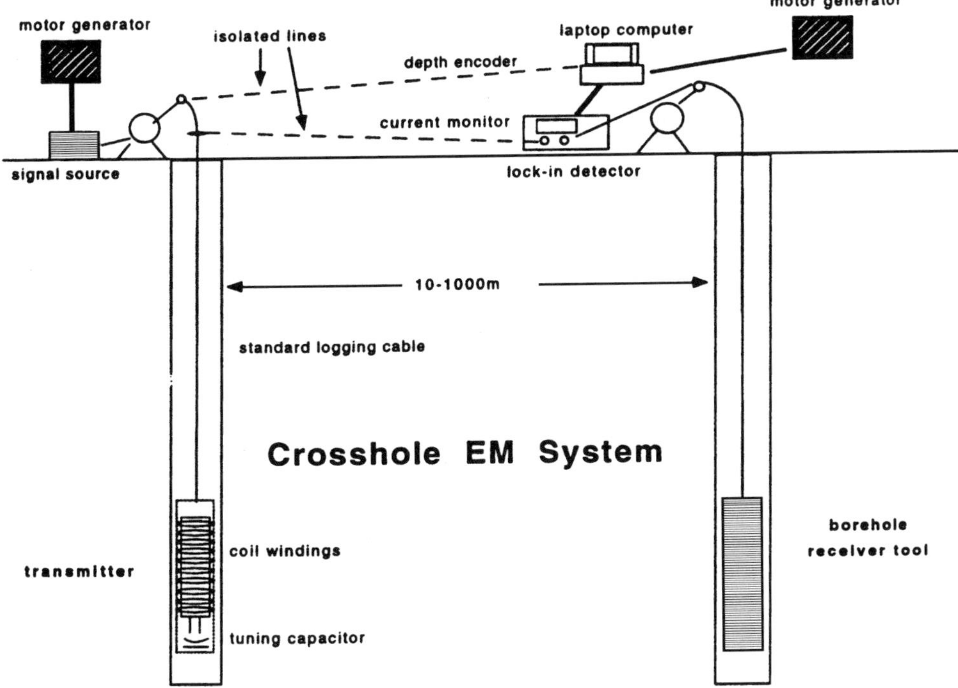

Figure 2. Schematic diagram of LLNL/LBL field system.

than the borehole source, allowing for much greater range. Vertical magnetic fields are detected in boreholes with a commercial current-feedback induction coil, and the signal is transmitted up the logging cable for measurement with a lock-in detector. The phase reference signal for the lock-in amplifier is a measurement of the transmitter current, or its magnetic flux. Wheel encoders are used to keep track of tool depths and a portable computer is used to control the data acquisition.

With this simple analog system, we have collected high-quality data at a number of field sites at borehole separations from 10 to 300 m, using frequencies from 40 Hz to 100 kHz (Wilt et al., 1995b). We attained our data accuracy by careful attention to isolation and local grounding of the transmitter and receiver sections. Each unit has a separate generator for power supply, and a local ground. The transmitter and receiver modules are connected for phase reference and depth control using optically isolated cables.

3 Sources of data error

There are three major sources of error in EM field surveys: external and system noise, sensor placement errors, and calibration errors. Each of these can be significant, and often appear as biases rather than random errors. Although the relative size of these errors may vary, there is always some component of each type in field data. In addition, because the errors often are not random, inversion schemes that assume a Gaussian noise distribution may not be effective.

3.1 Examples of data collection errors

Figure 3 shows two crosshole magnetic-field amplitude profiles collected three months apart at the same location within the Lost Hills oil field (Wilt et al., 1995a). The profiles

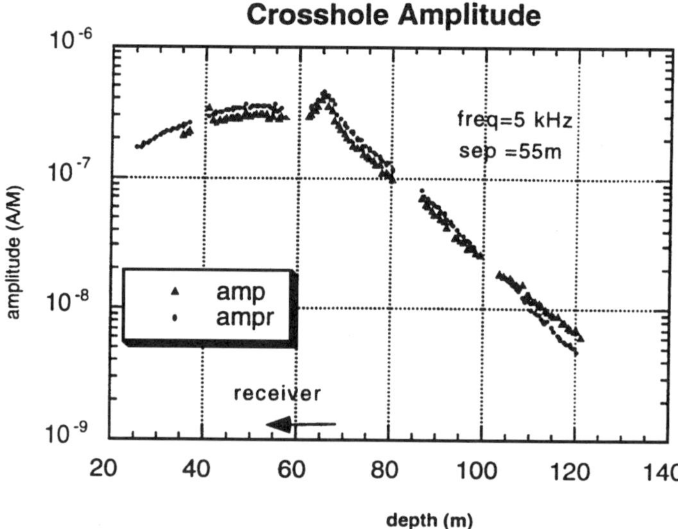

Figure 3. Repeated amplitude profiles showing an example of systematic drift.

are made by fixing the receiver in one borehole and making continuous measurements as the transmitter moves from the bottom to the top of the selected interval in the second well. The second amplitude profile is initially larger and parallel to the original but the slope changes, causing the profiles to cross at a depth of about 100 m. These profiles show both static offset and drift errors, resulting in an average difference of more than 20%. Sometimes these errors are the result of inadequate instrument warmup, but often they are due to a systematic ground loop, which is very difficult to correct. In this case the noise was due to a faulty power-supply ground.

Ground-loop noise constitutes transmitter signal that, by some means other than mutual coupling between sensors, appears in the detection circuit. Typical routes are capacitive coupling of cables and improper grounding and isolation of system components. This noise is particularly troublesome because it constitutes a leakage of the signal that is often dependent on the particular configuration of equipment; it is therefore occasionally repeatable and quite difficult to remove.

Note that the profiles in Fig. 3 have blank regions and an anomalous bump at a depth of 65 m; these are warts typical of many field data sets. The blank regions are no-signal zones where the transmitter has traversed steel centralizers used in emplacing the fiberglass casing. The bump is likely a near-wellbore effect caused by conductive pyritized clay; it is also evident in borehole induction logs. Although the blank zones should not have a large effect on crosswell inversion, the bump poses a problem for imaging because of its small size, high conductivity, and its close proximity to the borehole.

In the crosshole amplitude and phase profiles of Fig. 4, at 20 kHz, the signal becomes weaker at the end of the profile because of greater source-receiver separation, and the noise in the data is noticeably larger. This illustrates the issue of finite dynamic range. At higher frequencies, the signal attenuaton with distance is far greater than the simple geometric falloff characteristic of lower-frequency data. This makes recovery of these data difficult because of the large dynamic range requirement of the recording system. Because wide-aperture data are critical for determination of horizontal boundaries, it is better to use a lower frequency than to collect limited aperture data at a higher frequency.

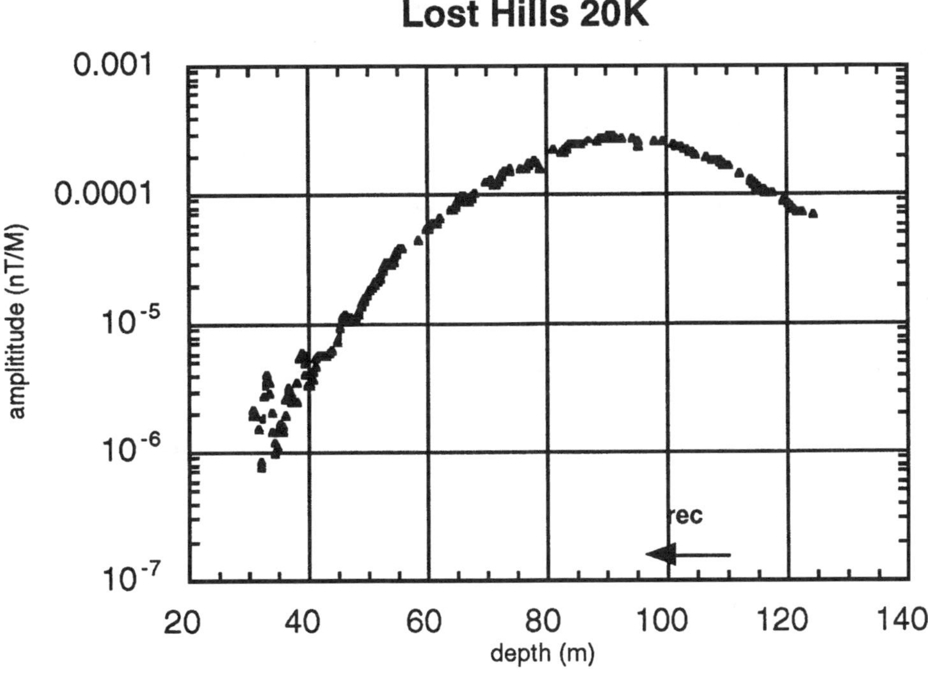

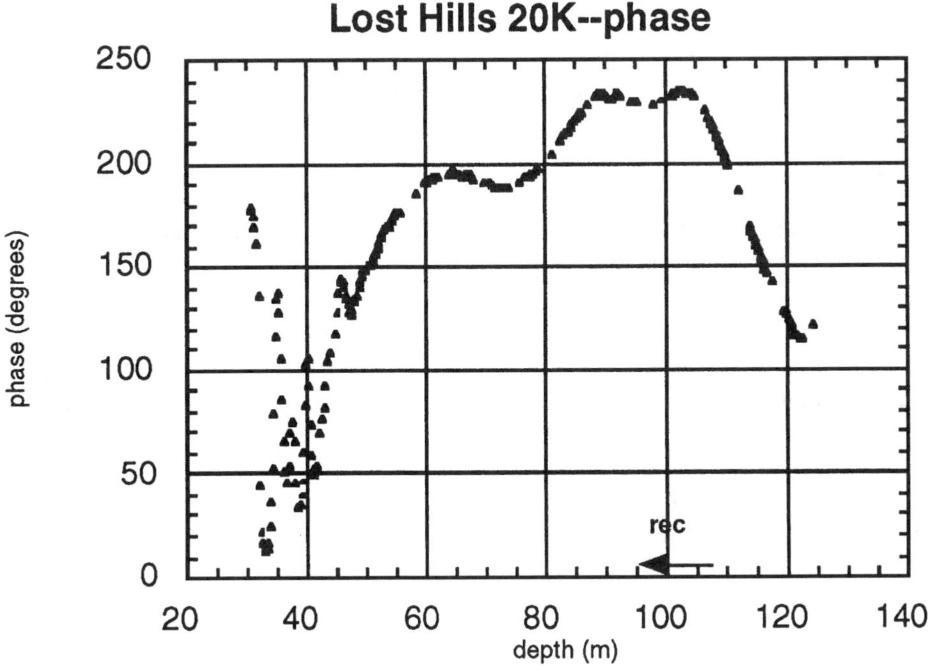

Figure 4. Amplitude and phase profiles from Lost Hills illustrating limited dynamic range and variable S/N.

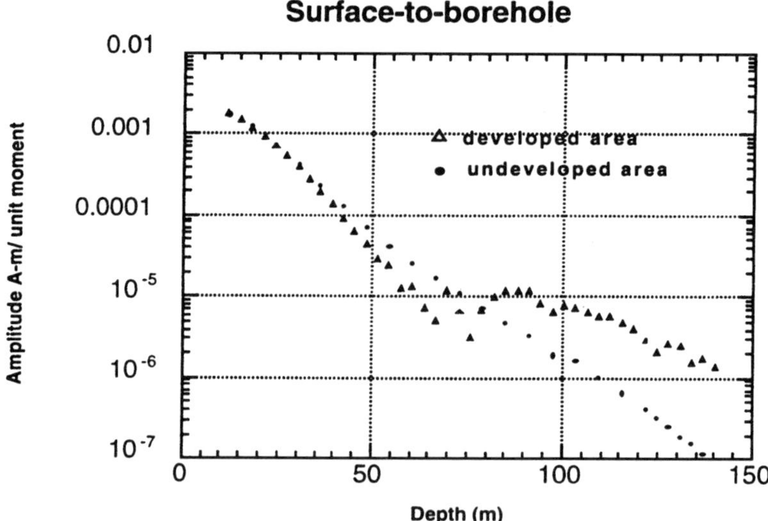

Figure 5. Surface-to-borehole amplitude measurements at Lost Hills. The upper profile was collected over the developed (noisy) portion of the field, the lower profile in an undeveloped field area.

Figure 5 shows two surface-to-borehole profiles collected at Lost Hills that demonstrate cultural noise effects on field data. In this case the same borehole receiver positions are employed with different but equidistant surface transmitters. In the upper profile the transmitter is deployed in a developed part of the oil field, near steam pipes and well heads; the lower profile is for a transmitter in an undeveloped section of the field with much less cultural noise. The amplitude profiles are similar at shallower depths, but below 80 m, the upper profile flattens whereas the lower one continues to decrease in size. Because, in our experience, surface-to-borehole amplitudes invariably decline with increasing separation, we suspect that the transmitter in the developed part of the field (profile A) is somehow coupling into the wellheads and surface piping to produce a secondary source. We note that this behavior is evident only in some data within the developed part of the field and it is quite difficult to predict.

Another common source of error is sensor positioning and orientation. We typically assume that boreholes are straight and vertical and this is often invalid, especially in hard-rock geology. Although errors may be corrected with a borehole deviation survey, this is rarely done because the deviations are assumed to be minor. An example of the effect of minor downhole positioning errors is given in the analysis of the Richmond field data below.

3.2 *Quantifying data error*

Error estimates are crucial to provide confidence in the interpretation and for stability of numerical inversions. Several tests, routinely conducted in the field, assist in determining the error; these are data repeatability, reciprocity, and system calibration. Experience indicates that the weakest of these tests is repeatability. Measurements that repeat to a high degree indicate only that the signal level is sufficiently strong and that any bias or systematic errors are consistent. Short-term repeatability tests do not uncover biases or long-term instabilities and often impart a false sense of security. The tests seem to be more rigorous if data are repeated at longer time intervals, because short-term biases may emerge. For example, Alumbaugh (1993) reported that back-to-back measurements in his Richmond Field Station experiment repeated to better than 0.5%, but the one-week repeatability errors were non-Gaussian and distributed about a mean of 2%; this is suggestive of some type of systematic drift.

The reciprocity test is far more rigorous. Switching transmitter and receiver tools often helps to uncover ground loops and tool placement errors. This test is not always made because of the difficulty and time required in moving the system. It should be done periodically, with any system, to ensure data integrity. In our experience, when the system is operating properly, reciprocity errors are about twice as large as short-term repeatibility errors.

Another source of error emerges from the calibration process. To calibrate the system, we deploy the coils on the Earth's surface parallel to each other, and vary the source-receiver separation from 5 m to about 40 m. The resulting amplitude and phase profiles are then fit, in a least-squares sense, to a homogeneous half-space model, where the resistivity is known or estimated from surface geophysics. The calibration constants then are set equal to the amplitude and phase correction factors needed to produce an optimal fit. The calibration errors include positioning and alignment uncertainties, measurement errors, and the affect of near-surface geologic complexities. We find that our system calibrations typically agree to within 1 or 2%.

To estimate data errors in our crosswell and surface-to-borehole field data, we use the reciprocity error (when available) or the long-term repeatability tests (overnight at least) when reciprocals are not measured. We therefore neglect calibration errors.

3.3 *Quantifying imaging errors*

Another source of error arises from assumptions made in the interpretation process. These include an inappropriate model geometry, imprecise knowledge of borehole spacing and trajectory, or invalid approximations in the modeling or inversion schemes. Figure 6 illustrates the effect of limited strike length on crosswell response. Here the 3-D sensitivity function (Spies and Habashy, 1995) has been plotted in plan view (x, y) for the vertical magnetic fields produced by a 1-kHz VMD source at the same level in boreholes 100 m apart in a 10-ohm-m whole space. The sensitivity function forms a cigar-shaped anomaly with maximum values nearest the borehole but significant sensitivity throughout the region connecting the source and receiver wells. The width of this anomaly affects the validity of using a 2-D interpretation to image 3-D data. Sensitivity profiles in the strike (y) direction at various distances from the source or receiver are plotted in the lower half of the figure. These profiles indicate that geological features near the source or receiver require only a short strike length to appear infinite or 2-D, whereas midway between the source and the receiver the strike length should be approximately equal to the borehole separation for a 2-D assumption to be valid. Spies and Habashy (1995) also show that the sensitivity anomalies thin and the profile widths decrease with increasing frequency. This suggests that using a 2.5-D interpretation (on 3-D data) may be more valid at higher frequency.

Alumbaugh and Morrison (1995) discuss a method of uncovering error caused by assumptions employed by the imaging scheme. They show that when the appropriate model geometry is used, the amplitude difference between the predicted and the observed data is random. However, if the model geometry is inappropriate, then the error

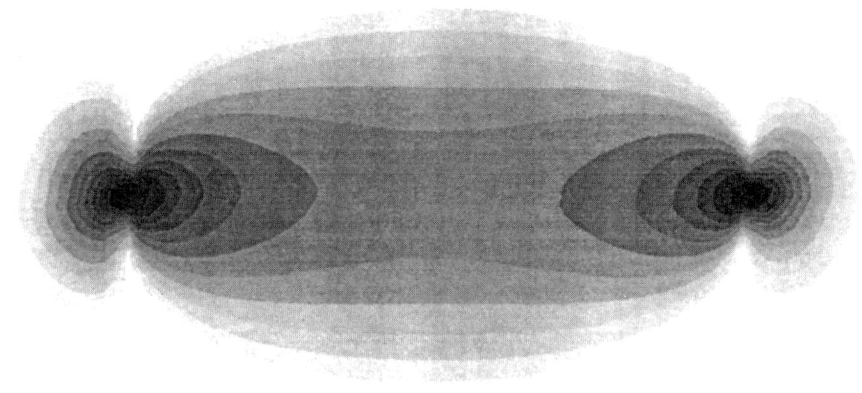

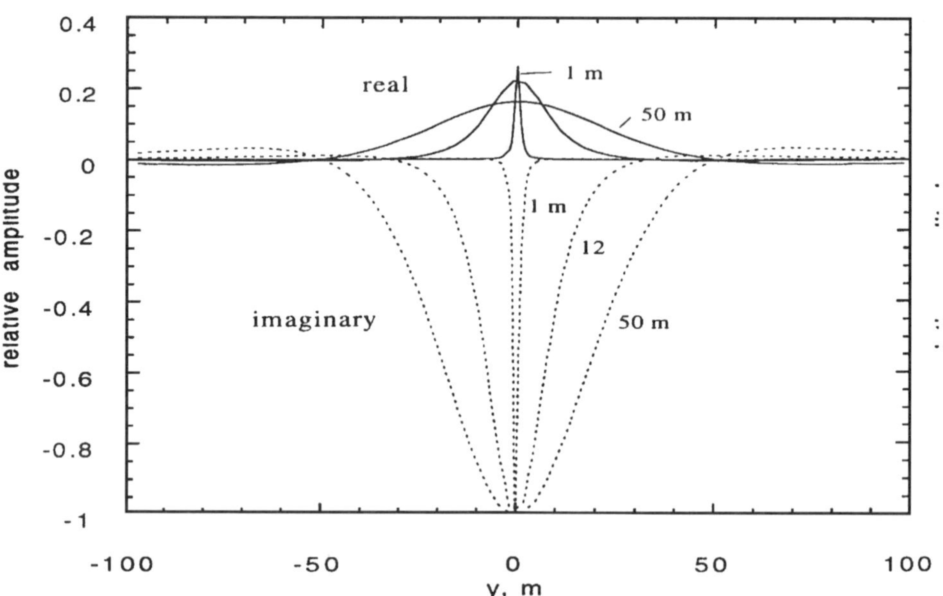

Figure 6. (Top) Plan view of crosswell sensitivity pattern between wells 100 m apart, 1 kHz, log scale. The sensitivity is highest immediately adjacent to the source and receiver. (Bottom) Sensitivity profiles in the strike direction at various distances from the source or receiver (after Spies and Habashy, 1995).

is nonrandom in regions where the model is invalid. We illustrate this below, using images constructed from various data sets.

4 Field example 1: Lost Hills (2-D data collection)

Mobil Exploration and Production U.S. operates several steam-flood projects in central California. EM technology has been applied in a test at their Lost Hills #3 pilot in an effort to track the injected steam (Wilt et al., 1995a). Two fiberglass-cased monitoring wells, drilled for the crosswell surveys, were positioned orthogonal to the regional strike; the crosshole EM data therefore roughly follow the assumption of a Cartesian 2-D geometry.

Field data were interpreted by applying a 2.5-D extended-Born inversion code developed by Torres-Verdín and Habashy (1994). The resistivity image for the preinjection crosshole data collected in 1993 is given in Fig. 7a. The blue sections in the image represents higher-resistivity zones associated with heavy-oil sands; the red areas are lower-resistivity silt and shale beds with an average value of 3 ohm-m. The initial image clearly shows the configuration of the upper oil sand and less clearly the middle and lower sands. This is consistent with borehole logs that indicate that the lower sands are less continuous. We note that the resistivity of the upper oil sand (20 ohm-m) is significantly lower than that measured in borehole induction logs (100 ohm-m). This may be the result of the low-contrast assumption in the 2.5-D code used in the inversion. The residual amplitudes plotted in Fig. 7b show a generally random distribution which indicates that the geology is consistent with a 2.5-D model. The notable exception is that, when both the source and the receiver are within the upper resistive oil sand (depths of 60–75 m), the residuals tend to be larger. This suggests that either the inversion code is having some difficulty imaging this resistive formation or there is some 3-D complexity within this region.

After steaming, the resistivity image in Fig. 8a is visibly different (from Fig. 7a) only at depths between 80 and 110 m. In this zone the resistivity has decreased significantly because of the steam injection. In all other parts of the image, the resistivities are similar, and in areas above the oil reservoir, the before-and-after data agree to within a few percent. However, notice in Fig. 8b that the residuals show a less random nature than those in Fig. 7b. This is particularly evident when the source and the receiver are between 80 and 110 m, where the largest changes in resistivity are observed, and when the source and the receiver are at maximum vertical separation. This implies that the geometry of the steam plume is too complex to be modeled accurately by the 2.5-D program or that the resistivity contrast is too large for the low-contrast assumption used with the extended-Born inversion. Clearly, some care must be exercised in geological interpretation of the image.

5 Field example 2: Richmond (3-D data collection)

Our second field example is a saltwater injection monitoring experiment at the University of California (UC) Richmond Field station, located 7 km north of the UC Berkeley campus. Five 60-m-deep wells were employed in the crosswell experiment; four of these are located at the corners of a 40 × 50 m polygon with the injection well positioned at its center. Measurements were made with the VMD source in the central injector and receivers in the four outer wells (Wilt et al., 1995a,b).

The field data were interpreted with a 3-D inversion scheme described by Newman and Alumbaugh (this volume). The forward modeling incorporated within the inversion employs a full 3-D vector wave equation formulation (Alumbaugh and Newman, 1994; Alumbaugh et al., 1995). This eliminates problems resulting from an assumption of a 2-D model geometry or a low-resistivity contrast.

In Fig. 9a, we display a 3-D image of the Richmond well field for the period after saltwater was injected. Just to the north of the central injection well, the image depicts a zone of high conductivity at a depth of 30 m. This anomaly corresponds to the injected saltwater plume, and its northern offset agrees with cone penetrometer and dc resistivity surveys; the survey therefore was successful in its primary objective. The image also shows unexpected regions of high conductivity between the receiver wells. In addition, the residuals for the injector-northwest well pair shown in Fig. 9b indicate a nonrandom

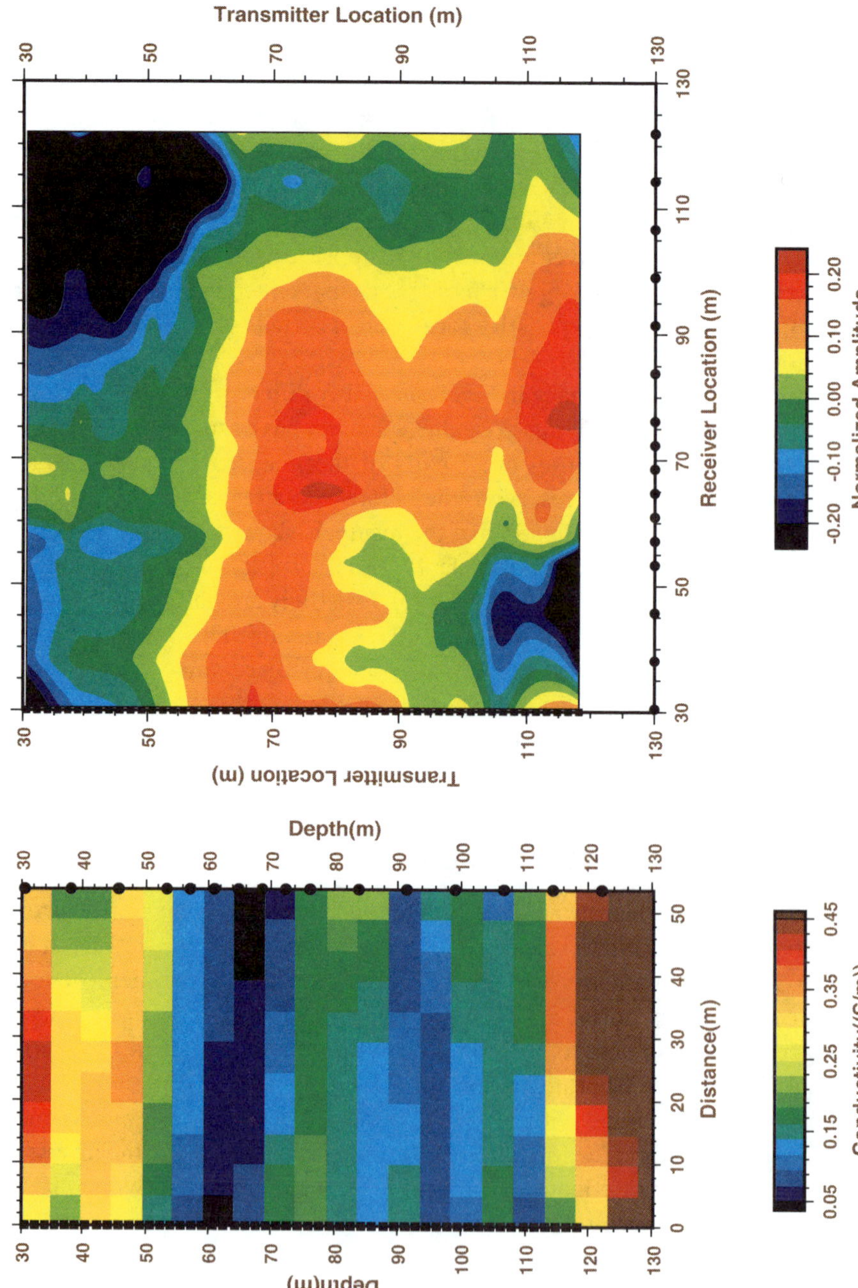

Figure 7. (a) Inversion of Lost Hills data, before steam injection; (b) residuals between the measured and the predicted data normalized by the measured data.

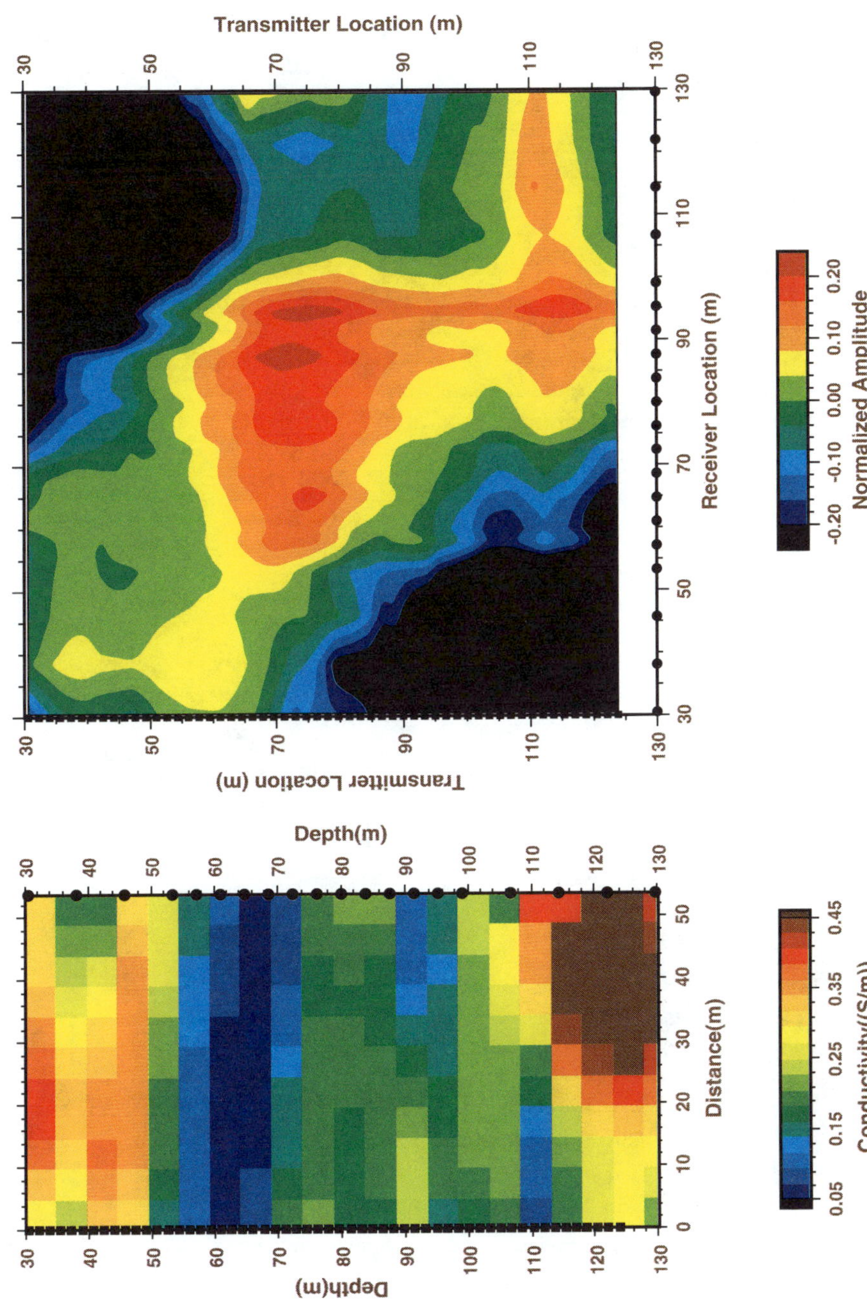

Figure 8. (a) Inversion of Lost Hills data, after steam injection; (b) residuals between the measured and the predicted data normalized by the measured data.

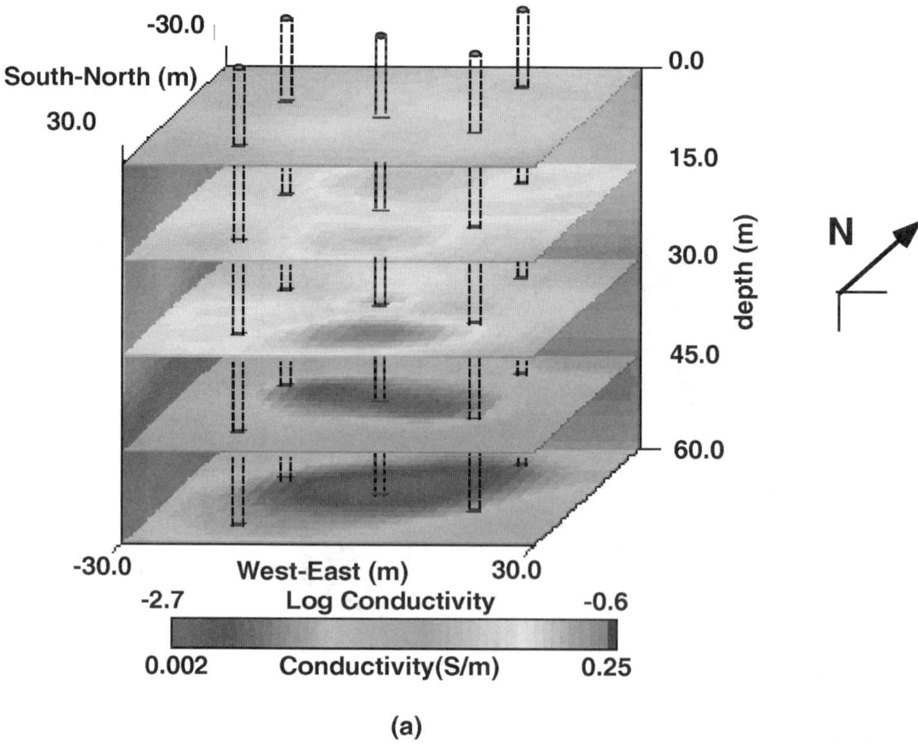

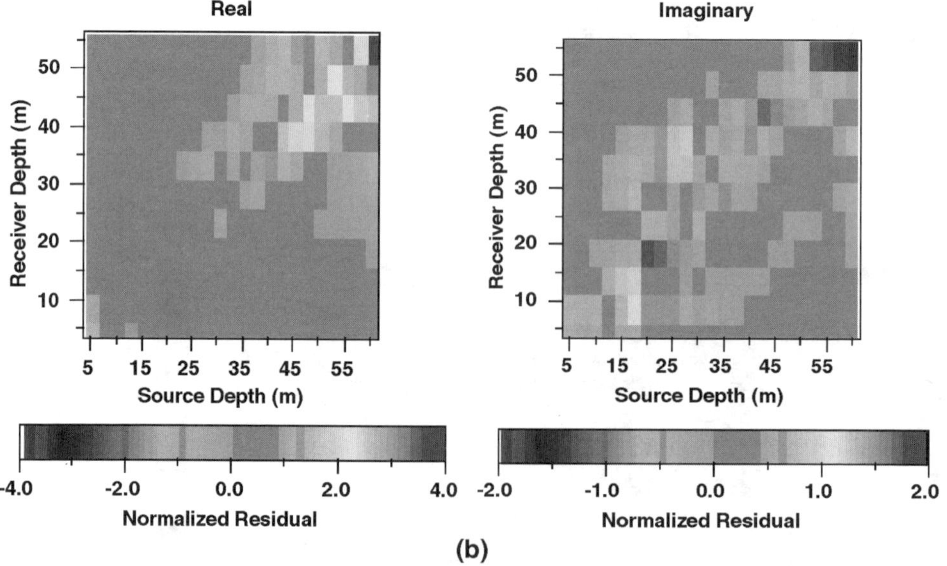

Figure 9. (a) Image of postinjection data collected at the Richmond Field Station. The image was produced using the 3-D inversion scheme described by Newman and Alumbaugh (this Volume). (b) Residual error between model-predicted data for image and measured data for the center-northwest well pair.

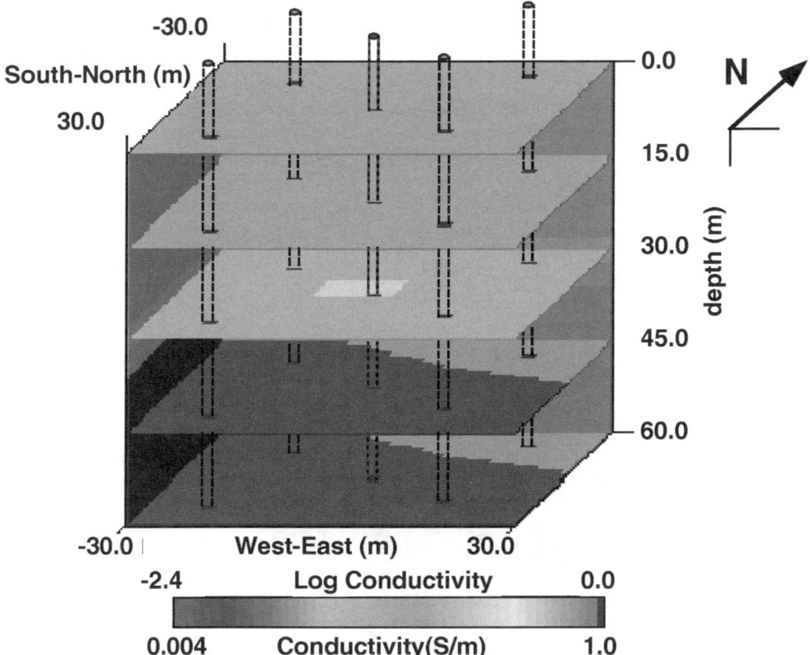

Figure 10. Model used to simulate the Richmond Field Station experiment.

distribution. This, coupled with the fully 3-D nature of the forward modeling, suggests that there is some type of error or bias in the data. In addition, the residuals are larger with increasing source and receiver depth; this is suggestive of a sensor positioning error, caused by borehole deviation, for example.

To examine if positioning errors, caused by deviated boreholes, could be the cause of imaging errors, a synthetic data set has been calculated for the model shown in Fig. 10 using the same code employed within the inversion scheme (see Alumbaugh and Newman, 1996). Two data sets were calculated; the first uses vertical boreholes and in the second the boreholes are deviated linearly with depth, toward the injection well at deviation angles between 0.2° and 2°. Both data sets use the same source and receiver sampling interval as in the Richmond experiment; both inversions assume vertical boreholes. Inverting the data with the correct sensor location results in the image shown in Fig. 11. Notice that the residuals are distributed randomly and there are no artifacts present in the images. The image corresponding to the deviated boreholes is shown in Fig. 12. In this case the residuals exhibit a nonrandom pattern, similar to those observed in the data (Fig. 9), and the image now is plagued by artifacts.

Thus sensor positioning errors, caused by borehole deviations, seems to be a valid explanation for the artifacts observed in Fig. 9. We have not yet done the borehole deviation surveys for the Richmond data and do not know the extent of the problem and therefore are unable to make necessary corrections.

6 Discussion and conclusions

Noise and uncertainty are always associated with field data. We can manage these problems by applying good field practice as well as understanding the measurement system. Experienced operators often can recognize noise and take steps to minimize

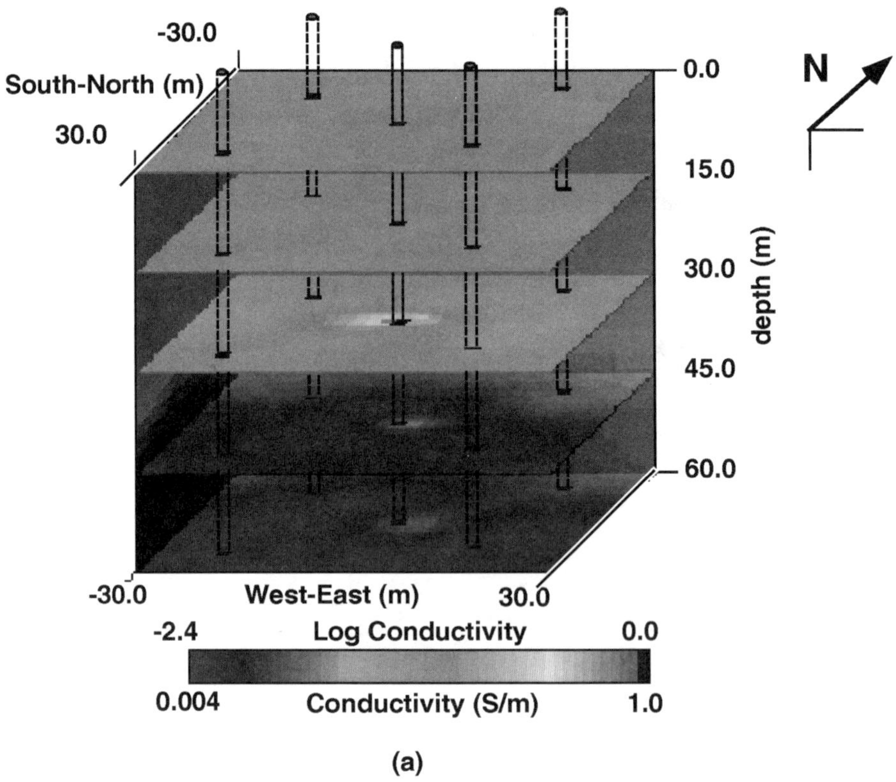

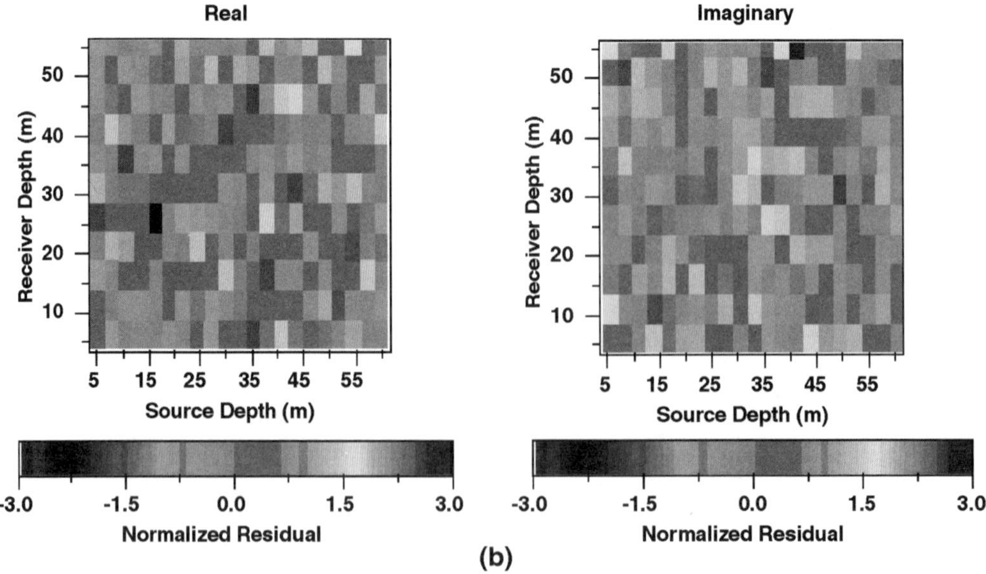

Figure 11. (a) Image of synthetic data with the correct sensor locations; (b) residual error between model-predicted data for image and synthetic data for center-northwest well pair.

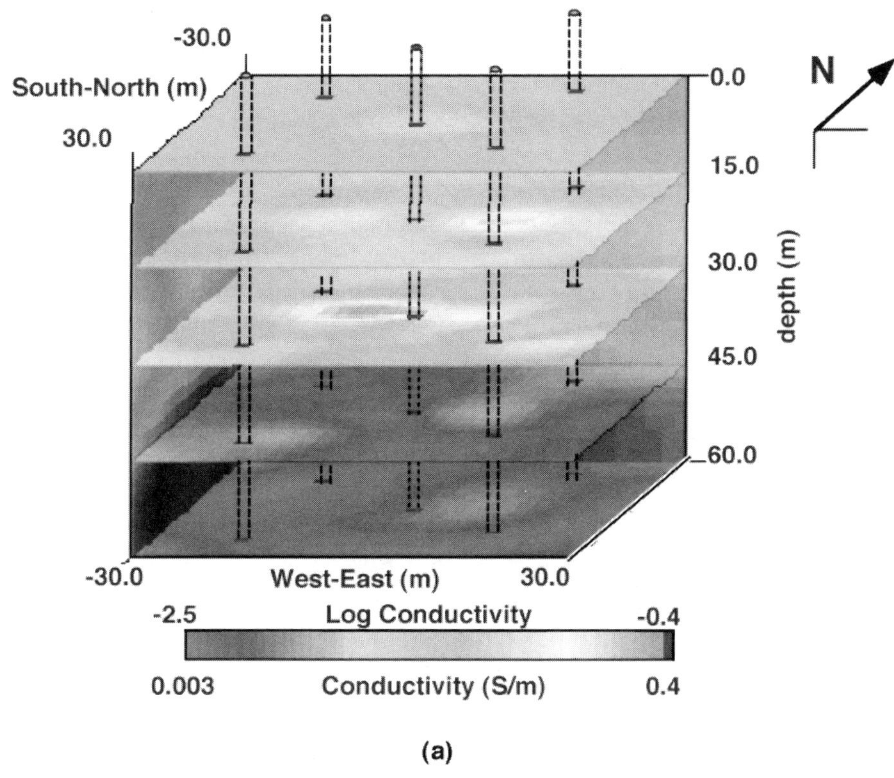

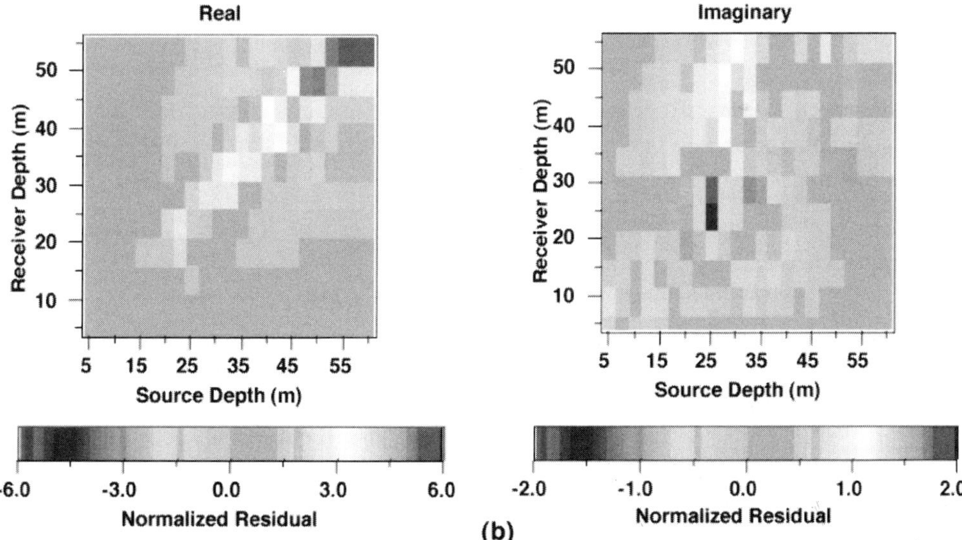

Figure 12. (a) Image of the synthetic data with incorrect sensor locations. The bottom of northwest and northeast wells were deviated toward the center well by 1 m, the southeast well by 2 m, and the southwest well by 0.2 m. (b) Residual error between model-predicted data for image and the synthetic data for the center-northwest well pair.

it in field data. It also can be very helpful to model the field experiment before data are collected. Based on resistivity logs, geological knowledge, and borehole locations, 3-D computer simulations can assist in sensor placement and recording aperture and operating frequencies. In addition, such modeling makes it much easier to evaluate field data on site, because the expected response has been determined beforehand. Finally, the synthetic data can be imaged prior to the survey to determine the amount of image distortion that will occur because of incorrect assumptions within the imaging schemes.

The geometry and spatial aperture in cross-borehole acquisition is limited severely by the placement of the boreholes. Fully 3-D data acquisition usually is not possible, except in some shallow studies. Fortunately, we know that the practical length for interwell geological structure to appear infinite is of the order of the borehole separation. We therefore can apply 2.5-D interpretation algorithms with some confidence. In addition, by employing multicomponent receivers, it is possible to infer the strike direction and detect 3-D features that are out of the plane of the two wells.

Acknowledgments

We acknowledge Mobil Exploration and Production U.S. for access to the Lost Hills field and for generously providing borehole logs and geologic and hydrologic information to the EM project. We also acknowledge members of the LBNL/LLNL EM consortium for access to the Richmond Field Station data set. These members include Schlumberger-Doll Research, Western Atlas, Oyo, Texaco, Noranda, EPA, and Nitetsu Mining. Partial support for this research was provided by DOE through the heavy-oil program under contract ENG-7408.

We also gratefully acknowledge the contribution made by the reviewers.

References

Alumbaugh, D. L., 1993, Iterative electromagnetic Born inversion applied to Earth conductivity imaging: Ph.D. thesis, Univ. of California at Berkeley.

Alumbaugh, D. L., and Morrison, H. F., 1995, Theoretical and practical considerations for crosswell electromagnetic tomography assuming a cylindrical geometry: Geophysics, **60**, 846–870.

Alumbaugh, D. L., and Newman, G. A., 1994, Fast frequency-domain electromagnetic modeling of a 3-D Earth using finite differences: 1994 Ann. Mtg., Soc. Expl. Geophys., Expanded Abstracts, 369–373.

———1996, 3-D massively parallel electromagnetic inversion, Part b: Analysis of a cross well EM experiment: Radio Sci.

Alumbaugh, D. L., Newman, G. A., Prevost, L., and Shadid, J. N., 1995, Three-dimensional wide band electromagnetic modeling on massively parallel computers: Radio Sci., in press.

Cook, J. C., 1977, Borehole-radar exploration in a coal seam: Geophysics, **42**, 1254–1257.

Gasnier, S., Shima, H., Sakashita, S., and Abdelhadi, A., 1994, Development of multi-component multi-frequency data acquisition system for borehole EM tomography: 56th Mtg. Eur. Assoc. Expl. Geophys., Expanded Abstracts, Session I030.

Holser, W. T., Brown, R. J. S., Roberts, F. A., Fredriksson, O. A., and Unterberger, R. R., 1972, Radar logging of a salt dome: Geophysics, **37**, 889–906.

Lytle, R. J., Laine, E. F., Lager, E. F., and Davis, D. T., 1979, Cross-borehole electromagnetic probing to locate high-contrast anomalies: Geophysics, **44**, 1667–1676.

Nabighian, M. N., 1984, Foreword and introduction—Time-domain electromagnetic methods of exploration: Geophysics, **49**, 849–853.

Newman, G. A., and Alumbaugh, D. L., 1996, 3-D electromagnetic modeling and inversion on massively parallel computers: this volume.

Noakes, J., 1951, An electromagnetic method of geophysical prospecting for application to drill holes: Ph.D. thesis, Univ. of Toronto.

Nekut, A. G., 1994, Electromagnetic ray-trace tomography: Geophysics, **59**, 371–377.

Osato, K., and Takasugi, S., 1992, The possibility of subsurface borehole electromagnetic surveys in geothermal reservoirs, Part 2: Development of borehole tool and field test, *in* Asakura, N., Ed., Geotomography: Soc. Expl. Geophys. Japan, 343–360.

Rust, W. M. J., 1940, Typical electrical prospecting methods: Geophysics, **5**, 243–249.

Spies, B. R., 1992, Survey design considerations for cross-well electromagnetics: 62nd Ann. Internat. Mtg., Soc. Expl. Geophys., Expanded Abstracts, 498–501.

Spies, B. R., 1995, Electrical and electromagnetic borehole measurements: A review: Surv. Geophys., **17**, 517–556.

Spies, B. R., and Habashy, T. M., 1995, Sensitivity analysis of crosswell electromagnetics: Geophysics, **60**, 834–845.

Stolarczyk, L. G., 1990, Radio imaging in seam waveguides, *in* Ward, S. H., Ed., Geotechnical and environmental geophysics: Soc. Expl. Geophys., 187–209.

Thomson, S., Hatherly, P., and Liu, G., 1990, The RIM in-mine method—Theoretical and applied studies of its mine exploration capabilities: Coal J., **29**, 33–40.

Torres-Verdín, C., and Habashy, T. M., 1994, Rapid 2.5-dimensional forward modeling and inversion via a new nonlinear scattering approximation: Radio Sci., **29**, 1051–1079.

Tseng, H. W., Becker, A., Wilt, M. J., and Deszc-Pan, M., 1995, A Surface-to-borehole electromagnetic survey. 65th Ann. Internat. Mtg., Soc. Expl. Geophys., Expanded Abstracts, 246–249.

Vozoff, K., Smith, G. H., Hatherly, P. J., and Thomson, S., 1993, An overview of the radio imaging method in Australian coal mining: First Break, **11**, 13–21.

Ward, S. H., and Harvey, H. A., 1954, Electromagnetic surveys of diamond drill holes: Can. Min. Manual, 19–30. Published by National Business Publications, Gardenvale, Quebec.

Wilt, M., Morrison, H. F., Becker, A., Tseng, H.-W., Lee, K. H., Torres-Verdín, C. and Alumbaugh, D. L., 1995a, Crosshole electromagnetic tomography—A new technology for oil field characterization: The Leading Edge, **14**, 173–178.

Wilt, M. J., Alumbaugh, D. L., Morrison, H. F., Becker, A., Lee, K. H., and Deszcz-Pan, M., 1995b, Crosswell electromagnetic tomography: System design considerations and field results: Geophysics, **60**, 871–885.

Witterholt, E. J., and Kretzschmar, J. L., 1982, The application of crosshole electromagnetic wave measurements to mapping of a steam flood: 33rd Ann. Mtg., Petr. Soc., Carad. Instit. Mining Paper 82-33-81.

Out-of-Plane Effects in Crosshole Radio-Frequency Tomography

P. K. Fullagar[1]
G. A. Pears[2]

Summary. Radio-frequency tomography (RFT) normally is assumed to sense the geological structure in a plane between the transmitters and receivers, loosely termed the *image plane*. In practice, out-of-plane objects also affect the tomograms. We illustrate these effects on synthetic crosshole tomograms generated for a conductive sphere in a more resistive host. The sources and receivers are vertical magnetic dipoles.

The tomographic image of a highly conductive sphere located wholly or partially between the sources and the receivers is a conductive feature the shape of which is governed by the conductivity of the host. As the sphere moves around one of the boreholes at a fixed distance, its expression on the tomogram is transformed from a conductive feature to a resistive feature. This reversal also occurs as the conductive sphere moves radially away from the transmitter. The period of these reversals can be related to the change in path length from source to sphere to receiver. Thus, if only amplitudes are recorded, an out-of-plane conductor could be misinterpreted as a resistive object in the plane; the two cases could probably be differentiated if phase data were also recorded.

The influence of the sphere on the tomogram is also negligible at certain azimuths and radial separations determined by the host conductivity and source frequency. A good conductor close to a borehole could be missed by an RFT survey if it were located in such a null. The risk of this occurrence is greatly reduced if multifrequency data are collected. Also, the shape of the tomographic expression of the sphere is insensitive to azimuth and radial separation.

1 Introduction

Radio-frequency tomography (RFT) between boreholes or openings in coal mines has been used to detect disruptions in coal seams and identify potential hazards (Rogers et al., 1987; Vozoff et al., 1993). Use of RFT to delineate ore zones in metalliferous mines is currently being studied (Thomson and Hinde, 1993; Wedepohl, 1993;

[1]CSIRO Exploration & Mining, P.O. Box 883, Kenmore, Queensland 4069, Australia.
[2]Department of Physics, University of Queensland, St. Lucia, Queensland 4072, Australia.

Fullagar et al., 1996). When processing tomographic data it is normally assumed that the anomalous structures lie in the image plane, which is the locus of straight lines drawn between all transmitter-receiver pairs. Here, by simulating tomographic data between vertical boreholes with a conductive sphere placed at different azimuths around the transmitter hole and at different radial distances, we show that out-of-plane effects can be important.

2 Sphere-in-host model

A conductive sphere in a resistive, isotropic medium is a simple analogue of a compact sulfide body. The electromagnetic fields scattered by a sphere excited by a dipole source can be computed analytically with modal expansions (March, 1953). For this study, we used software provided by Schlumberger Technology Corp. [see Habashy et al. (1995)]. Only the vertical component of the magnetic field—which is easiest to measure in practice—was used for the analysis. Absorption tomograms were generated from the synthetic data using the program MIGRATOM (Jackson and Tweeton, 1994).

The model image plane was bounded by two 150-m-long vertical holes separated by $L = 100$ m. Transmitters and receivers were distributed at 5-m intervals (Fig. 1). A sphere of radius 7.5 m and conductivity 10 S/m was initially centered in the image plane (0° azimuth) midway down the transmitter array. The host medium's conductivity was either 0.005 S/m or 0.1 S/m; its relative permittivity was 3. Thus the model could represent a small pod of massive sulfide ore within more resistive host rocks. Tomographic data sets were computed at 5° increments as the sphere was moved around the transmitter hole at a fixed radius (Fig. 2), and as the sphere was moved radially away from the transmitter hole.

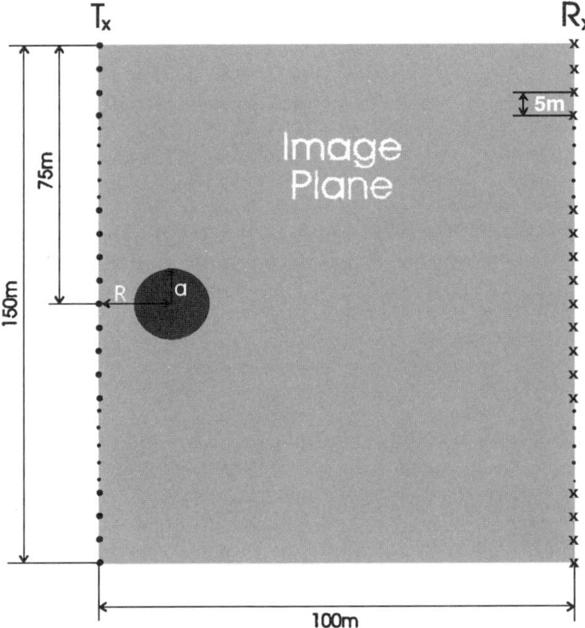

Figure 1. Section showing transmitter and receiver locations and a sphere, radius a, centered in the image plane, distance R from the transmitter axis.

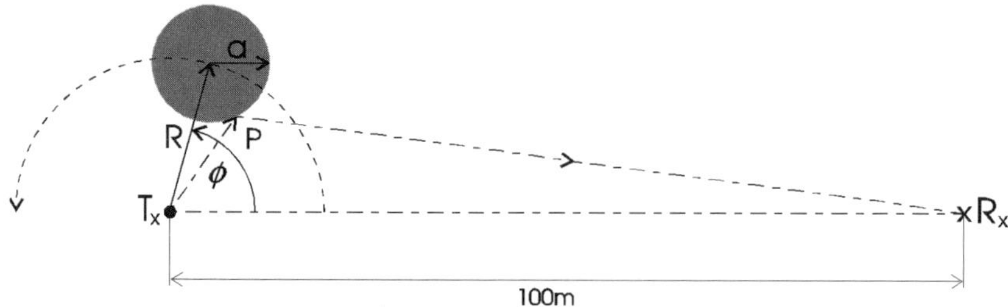

Figure 2. Plan view of an out-of-plane sphere. ϕ denotes the azimuth of the sphere center with respect to the image plane; P denotes the hypothetical point of reflection.

The source frequency was set at 300 kHz, which is within the range of existing systems. The field at 300 kHz in a medium of 0.005–0.1 S/m is entirely diffusive ($\sigma/\omega\varepsilon \gg 1$). Therefore, the absorption coefficient, α, is given by

$$\alpha = \sqrt{\frac{\omega\mu_0\sigma}{2}} = 1/\delta, \tag{1}$$

where δ is the skin depth. The skin depths at 300 kHz were 2.9 m and 13.0 m in the conductive (0.1 S/m) and resistive (0.005 S/m) hosts, respectively; the corresponding host wavelengths ($\lambda = 2\pi/\alpha$) were 18.3 m and 81.7 m. The transmitter-receiver separations are extreme in the conductive host ($L/\delta = 35$), but the results presented here are nonetheless indicative of measurable out-of-plane effects.

Reduction of the synthetic data prior to tomographic reconstruction uses the expression for a vertical magnetic dipole in a uniform medium in the far field (e.g., Ward and Hohmann, 1988), viz.,

$$|H_z| = A_0 \frac{\sin^2\theta}{r} e^{-\alpha r}, \tag{2}$$

where θ is the polar angle of the raypath with respect to the transmitter axis, and r is the separation of transmitter and receiver. A_0 denotes frequency-dependent source strength, determined empirically in this study from the primary field. Taking logarithms of equation (2) and inverting gives the net absorption τ in dB,

$$\tau = \alpha_a r = 20\log_{10}\left(\frac{A_0 \sin^2\theta}{H_z r}\right), \tag{3}$$

where α_a is the apparent absorption coefficient in decibels per meter. Images generated by MIGRATOM (with absorption as the analogue of slowness) assumed straight rays and far-field conditions as is common practice.

3 Results

Responses were calculated for successive cases as the sphere was rotated around the center of the transmitter array at intervals of 5°. The procedure was repeated for different separations, R, between 8 and 20 m (Figs. 1 and 2). Thus the sphere was never in the far field of the magnetic dipole: $R/\lambda \leq 1$.

There is no radio propagation through the sphere ($\delta = 0.3$ m). Thus, when the sphere intersects the image plane, the received signals arise from scattering, with extinction (radio shadowing) a special case. At larger azimuths, which are the focus of this paper, the total vertical magnetic field, H_z, can be visualized as the superposition of a primary signal, H_z^P, and its reflection off a near-perfect conductor.

3.1 In-plane targets

The attenuation tomogram for the sphere centred between the boreholes provides a reference against which to gauge changes as the sphere moves out of the image plane. The in-plane images for a sphere centered 12 m from the transmitter array in both hosts are presented in Fig. 3. Although the presence of the conductive sphere is obvious in these tomograms, its location is not accurate. The sphere has been effective in shielding the receivers in the more conductive host ($2a/\lambda = 0.82$), and the tomographic anomaly is compact (Fig. 3a), in keeping with the more focused, beamlike sensitivity between source and receiver at higher $\omega\sigma$ (Alumbaugh, 1993; Spies and Habashy, 1995). The anomaly for the 0.005-S/m host (Fig. 3b) is smeared over 40 m and influenced by interference of signals that have diffracted around the surface of the sphere ($2a/\lambda = 0.18$). The field behavior is qualitatively similar to radar diffraction around tunnels (Lytle et al., 1979).

3.2 Angular dependence

The cylindrical symmetry of the dipole source dictates that surface currents induced in the sphere are independent of azimuth. Consequently, the variations in received signal must be attributed to the changing face that the sphere presents to the receivers and to the changing range (Fig. 2). As the sphere is rotated about the transmitter axis, the polarity of its tomographic anomaly gradually changes from conductive to resistive. Figures 4 and 5 present a subset of the tomograms produced at different azimuths when $R = 12$ m in the 0.005-S/m host. The tomograms in Fig. 4 record the transition of the sphere's image from a conductive feature to a predominantly resistive feature. Figure 5 depicts the tomograms at 90°, 115° (maximum resistive anomaly), and 180°, the azimuth range over which the tomographic expression of the sphere is emphatically resistive.

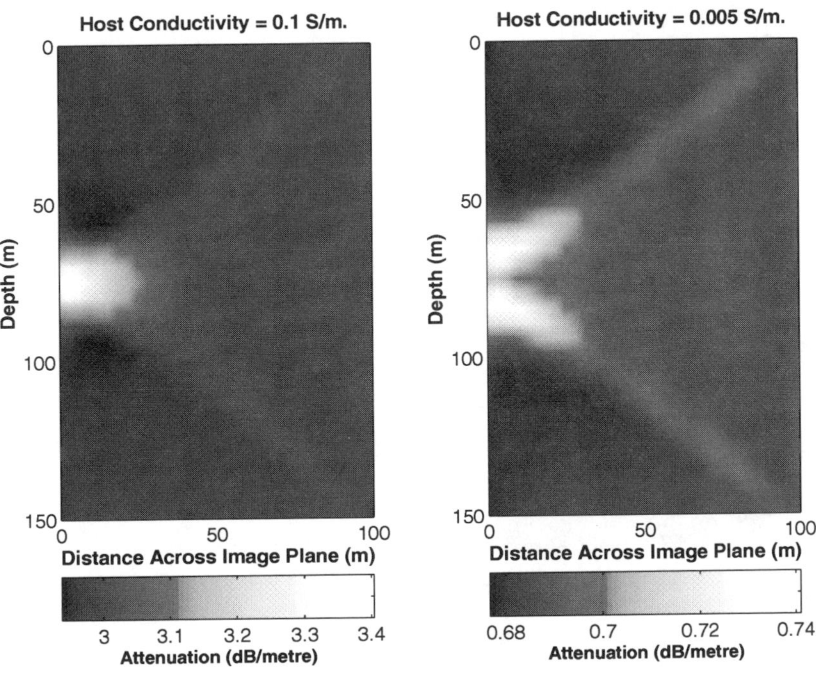

Figure 3. Tomograms for an in-plane sphere centered 12 m from the transmitter axis in (a) 0.1-S/m host and (b) 0.005-S/m host.

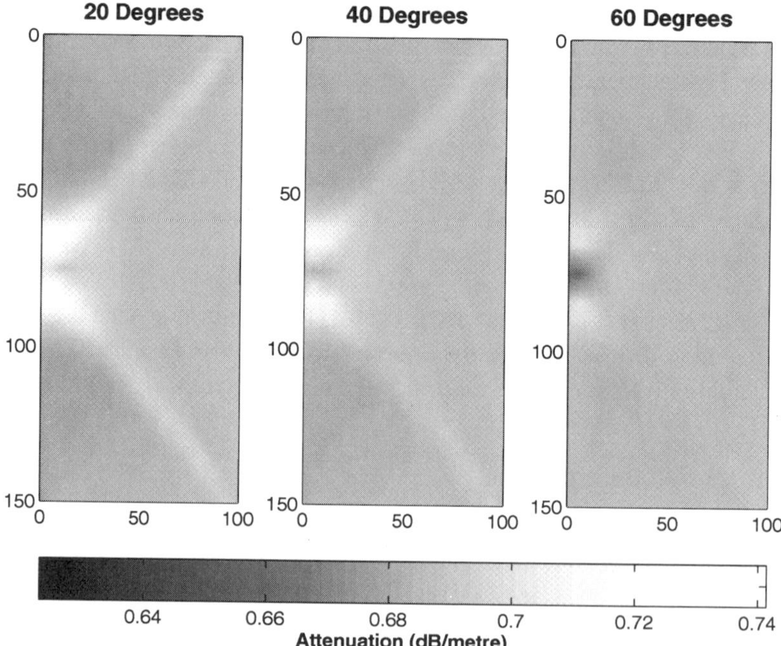

Figure 4. Absorption tomograms at 300 kHz for a 10-S/m sphere centered 12 m from transmitter hole in a 0.005-S/m host at azimuths 20°, 40°, and 60°. The anomaly is conductive at 20°, but predominantly resistive at 60°.

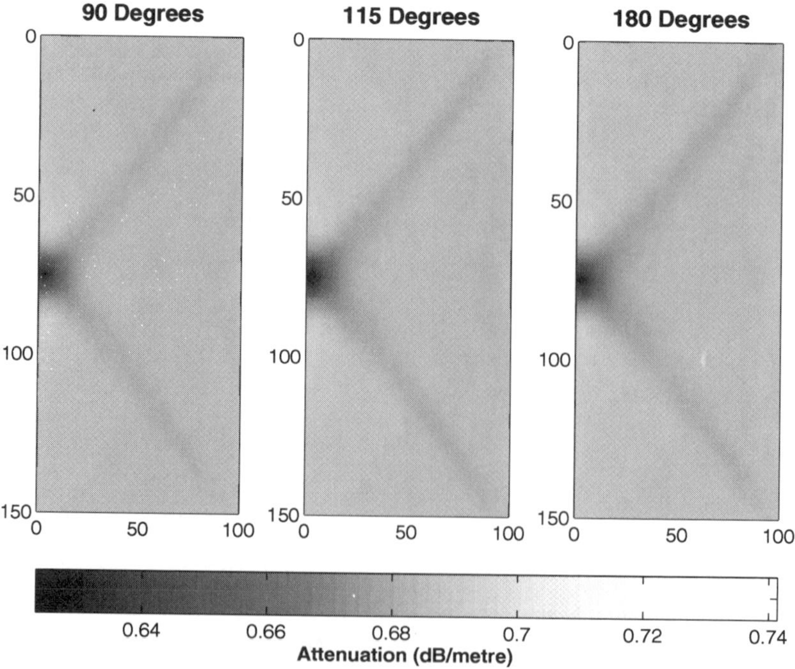

Figure 5. Absorption tomograms at 300 kHz for a 10-S/m sphere centered 12 m from transmitter hole in a 0.005-S/m host at azimuths 90°, 115°, and 180°. The resistive anomaly is most intense at 115°.

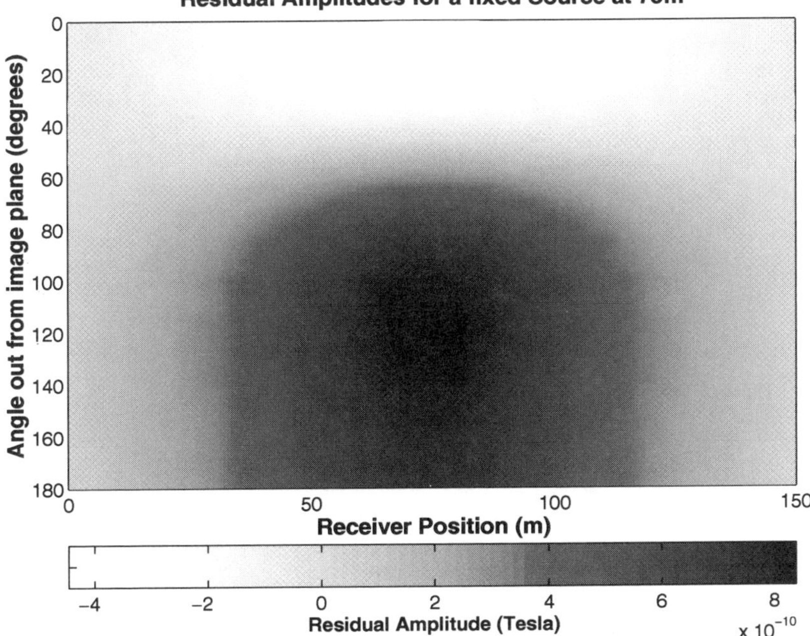

Figure 6. $\Delta|H_z|$ at 300 kHz in 0.005-S/m host as a sphere at range $R = 12$ m is rotated about a dipole source at a depth of 75 m.

To more fully define the azimuthal dependence of the responses, the amplitude difference, $\Delta|H_z|$, between total and primary vertical magnetic-field amplitudes, $|H_z|$ and $|H_z^P|$, can be plotted as a function of azimuth and receiver depth, where

$$\Delta|H_z| = |H_z| - \left|H_z^P\right|. \tag{4}$$

In Fig. 6, $\Delta|H_z|$ is plotted for a single source midway down the transmitter array (at the same level as the sphere center) in the 0.005-S/m host for azimuths 0–180°.

In Figs. 4 and 6 the transition from a conductive to a resistive image at 75-m depth is seen to occur at an azimuth of about 40°. This corresponds approximately to $\sin^{-1}(a/R)$, the azimuth at which the sphere is tangential to the image plane. The resistive anomaly is most intense at azimuth 115°, and gradually decays as the sphere rotates farther behind the transmitter hole. The intensity of the peak resistive anomaly at $\phi = 115°$ is almost identical to the intensity of the conductive anomalies at develop at depths of 65 m and 85 m when the sphere is in-plane (Fig. 3b). Given that a resistive anomaly signifies reinforcement of fields, the detectability of the sphere (as gauged by the S/N ratio) improves when it is located out-of-plane in this case.

The $\Delta|H_z|$ map for the sphere at separation 12 m rotated around a source at 75-m depth in the 0.1-S/m host conductivity is shown in Fig. 7. The conductive anomaly produced by the in-plane sphere is more intense than the maximum resistive anomaly, which in this case is associated with a sphere at an azimuth of about 75°. Thus in-plane targets are more readily detectable in the more conductive host, consistent with the tightly focused sensitivity at higher $\omega\sigma$. At azimuths greater than 120°, the anomaly becomes conductive again, albeit much reduced in intensity. The transitions between conductive and resistive anomalies in the tomograms occurred at azimuths of approximately 55° and 110°.

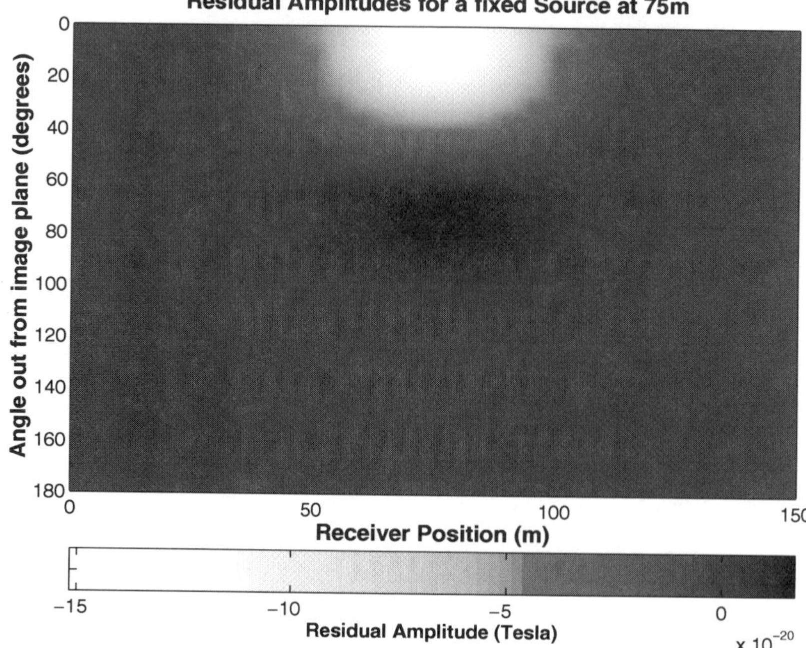

Figure 7. $\Delta|H_z|$ at 300 kHz in 0.1-S/m host as a sphere at range $R = 12$ m is rotated about a dipole source at a depth of 75 m.

The faster oscillation of the polarity of the anomaly as the azimuth increases in the more conductive host is caused by the shorter wavelength in that medium, and hence larger phase shift for given difference in path length. When the sphere is entirely out of plane, ($\phi > 60°$), the alternations in anomaly polarity can be related to differences in path length between direct and reflected rays (Fig. 8). This simple straight-ray analysis explains the data reasonably well, despite the diffusive behavior of the field (Nekut, 1994). The absolute phase difference between primary and secondary fields is governed by the near-field interaction between the source dipole and the sphere as well as by path lengths. Phase changes from reflection *per se* will be minor: when the source is at the same depth as the sphere, the vertical magnetic field will be unchanged in phase after reflection from the surface of the sphere.

The effect of path length on amplitude also is more pronounced in the more conductive host because of the higher absorption. Thus the sphere's influence on the total field is minor when it is positioned behind the transmitter ($\phi > 100°$) in the conductive host (Fig. 7). By contrast, the decrease in resistive-anomaly amplitude is much more gradual between azimuths 115° and 180° in the 0.005-S/m host (Fig. 6).

3.3 Radial dependence

The sign of the tomographic expression of a conductive sphere also alternates as the sphere moves radially away from the transmitter hole. The variations in the character of the tomographic image were more obvious in the more conductive host, again because of the shorter wavelength. For instance, at an azimuth of 90°, the anomaly on the tomogram altered from resistive to conductive at a separation, R, of about 14 m in the conductive host (Fig. 9). The conductive anomaly at the larger separations was strongly attenuated relative to the resistive anomaly at small R.

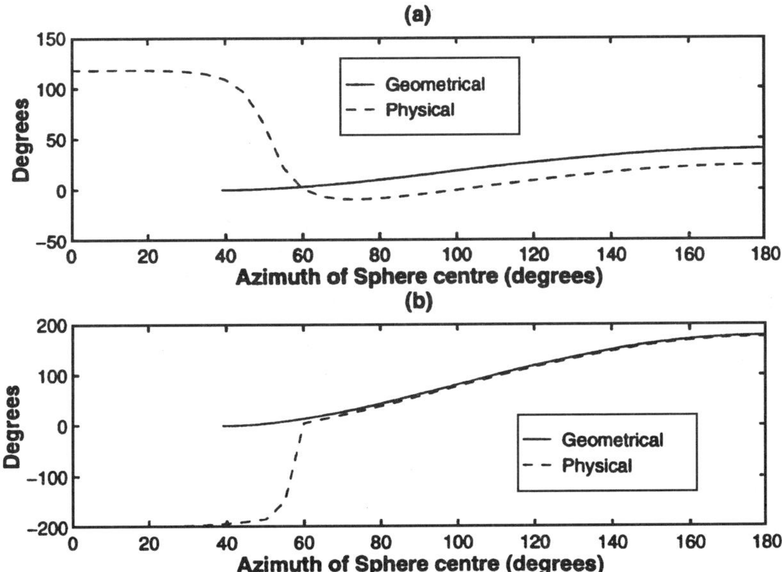

Figure 8. Comparison between actual phase difference between primary and secondary H_z fields (dashed) and the path-length difference between direct and reflected rays (converted to equivalent phase) as a 10-S/m sphere is rotated around the transmitter hole at range $R = 12$ m in (a) 0.005-S/m host and (b) 0.1-S/m host.

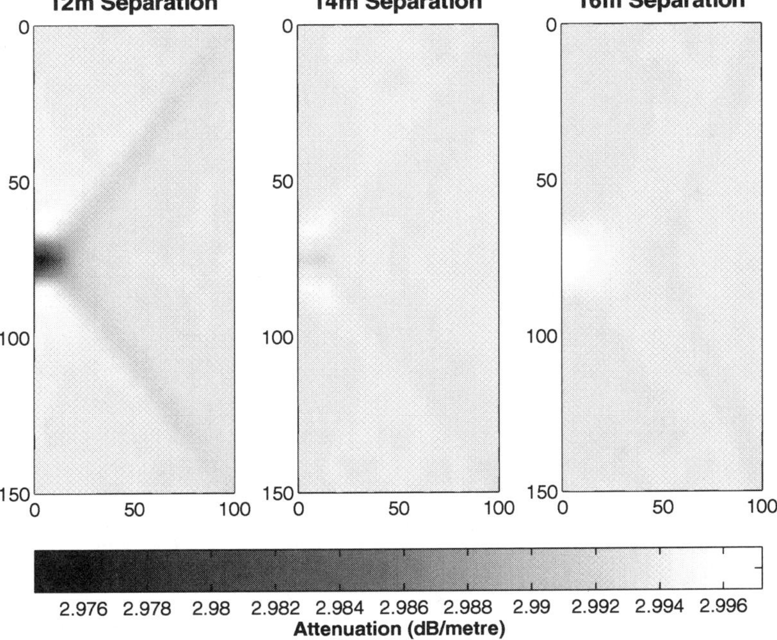

Figure 9. Absorption tomograms at 300 kHz for a 10-S/m sphere at an azimuth of 90° at different radial separations, R, in a 0.1-S/m host. The sphere passes through a null in response at about 14 m.

3.4 Discrimination between in-plane and out-of-plane targets

A resistive tomographic anomaly located close to a borehole has at least two possible interpretations (Fig. 10): it could be caused by a resistive body between the transmitters and receivers ($\phi \approx 0°$), or to a conductive body positioned behind the borehole. A way of discriminating between these two cases would be very useful. Phase measurements may allow bodies in the image plane to be differentiated from bodies out of the plane. Parallel scan phase profiles (with transmitter and receiver at the same depth) proved to be most diagnostic: the point-to-point phase differences are related to conductivity variations because the transmitter-receiver geometry is invariant.

Signals that have propagated through a resistive in-plane sphere will exhibit a phase lag with respect to the primary field (Fig. 11a), whereas an in-plane conductive sphere will produce a phase advance. When a conductive sphere is located behind the transmitters, on the other hand, the total phase may either slightly lag or lead the primary field, depending on the sphere's position. Figure 11b depicts the total phase for a conductive sphere directly behind the transmitters; in this case, a slight phase lead is observed. In general, a strong total phase anomaly will indicate an in-plane target, and could therefore help to resolve the ambiguity in amplitude tomograms.

4 Conclusions

RFT can help to delineate orebodies, especially massive sulfides, in metalliferous mines. Interpretation of 2-D tomograms must, however, take into account the possibility of

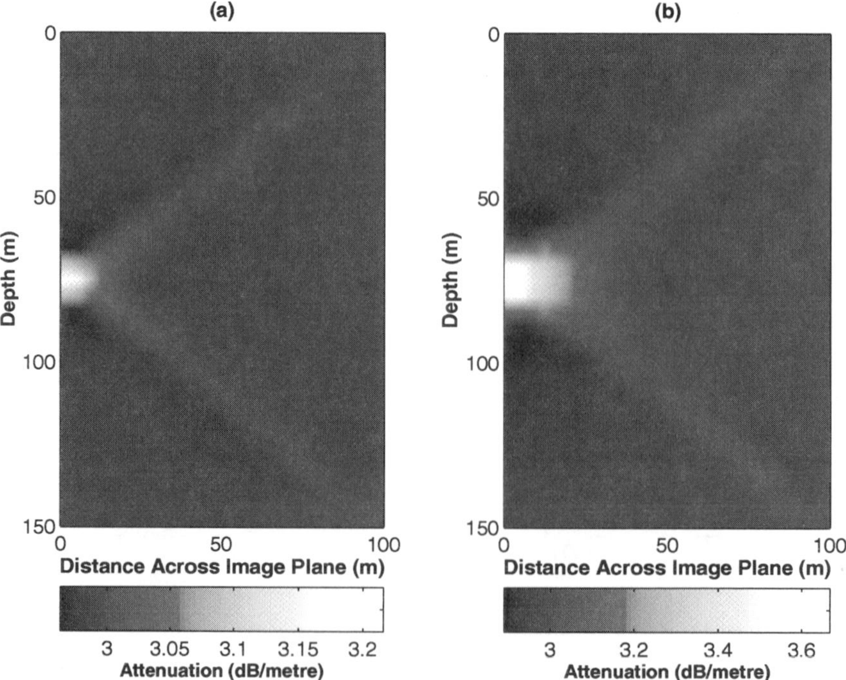

Figure 10. In-plane versus out-of-plane tomographic ambiguity: (a) absorption tomogram for a resistive sphere (0.001 S/m, radius 7.5 m) centered 8 m from the transmitter hole in the image plane ($\phi = 0°$); (b) absorption tomogram for a conductive sphere (10 S/m, radius 7.5 m) centered 8 m behind the transmitter hole ($\phi = 180°$).

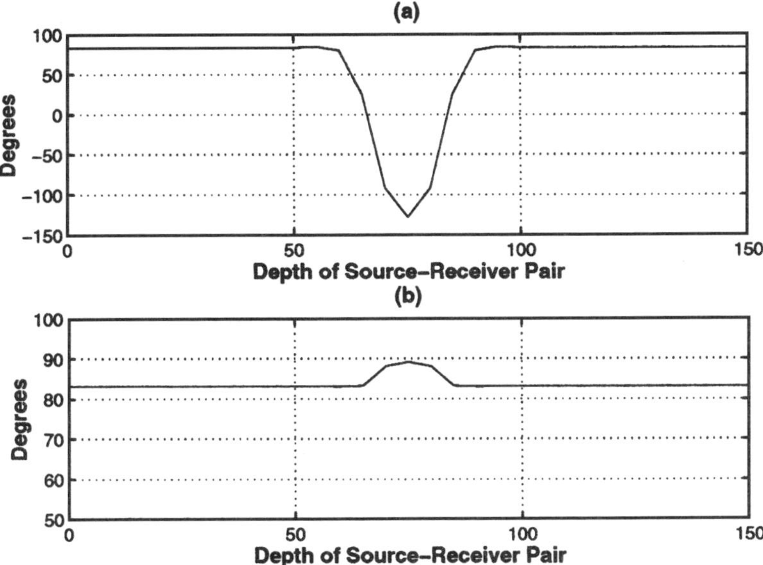

Figure 11. H_z phase profiles for transmitter-receiver pairs at the same depth for (a) in-plane resistive sphere ($\phi = 0°$, $R = 8$ m), imaged in Fig. 10a, and (b) conductive sphere behind the transmitter hole ($\phi = 180°$, $R = 8$ m), imaged in Fig. 10b. Host conductivity = 0.1 S/m.

3-D effects. In particular, highly conductive bodies close to either the transmitters or the receivers cause anomalies even if they are not in the image plane.

When a highly conductive sphere was located in the image plane, its tomographic image was a conductive feature the shape of which was governed by the conductivity of the host. The shape was compact in a conductive host, but smeared and modulated by interference effects in a more resistive host. As the sphere was rotated around one borehole, the shape of its tomographic expression proved to be insensitive to azimuth, but its polarity (sign of the contrast) with respect to the host alternated from conductive to resistive. A similar reversal occured as the sphere moved radially away from the transmitter axis at fixed azimuth. When the sphere was out of plane, the changes in polarity and intensity of the tomographic anomalies correlated with the change in the path length between transmitter, sphere, and receiver.

As the host conductivity increased, the maximum out-of-plane resistive anomaly decreased in intensity and the sphere position to which it corresponded migrated toward the image plane, consistent with the localization of sensitivity to a narrow beam between transmitter and receiver at high $\omega\sigma$. The shorter wavelengths in the more conductive host also can result in more than one polarity change of tomographic anomaly with azimuth.

Because of these effects, an out-of-plane conductor could be misinterpreted as a resistive object in the plane if only amplitudes are interpreted. However, the in-plane and out-of-plane cases probably could be differentiated on the basis of their phases. It is also possible that a good conductor close to a borehole could be missed by an RFT survey, if it happens to sit in a null in which the tomographic anomaly changes polarity. Because the position of the null depends on frequency, the use of multifrequency data sets would reduce the chances of an undetected target.

Acknowledgments

We gratefully acknowledge the support of Tarek Habashy and Brian Spies of Schlumberger-Doll Research, Ridgefield, Connecticut, in providing us with a copy of their program, SPHERE. One of us, G.A.P., received technical advice from Dr. Lynn Hastie, University of Queensland, and financial support from the University of Queensland, CSIRO Exploration and Mining (Brisbane), and Noranda (Montreal).

References

Alumbaugh, D. L., 1993, Iterative electromagnetic Born inversion applied to earth conductivity imaging: Ph.D. thesis, Univ. of California at Berkeley.

Fullagar, P. K., Zhang, P., Wu, Y., and Bertrand, M.-J., 1996, Trial of radio frequency tomography for exploration and delineation of massive sulphide deposits in the Sudbury Basin: Soc. Expl. Geophys., Expanded Abstracts, 2065–2068.

Habashy, T. M., Spies, B. R., and Groom, R. W., 1995, Computation of the electromagnetic response of a conductive sphere in a uniform conducting earth at arbitrary frequency (in prep).

Jackson, M. J., and Tweeton, D. R., 1994, MIGRATOM—geophysical tomography using wavefront migration and fuzzy constraints: U.S. Bur. Mines, Dept. of Interior Report of Investigations RI 9497.

Lytle, R. J., Laine, E. F., Lager, D. L., and Davis, D. T., 1979, Cross-borehole probing to locate high contrast anomalies: Geophysics, **44**, 1667–1676.

March, H. W., 1953, The field of a magnetic dipole in the presence of a conducting sphere: Geophysics, **16**, 671–684.

Nekut, A. G., 1994, Electromagnetic ray trace tomography: Geophysics, **59**, 371–377.

Rogers, P. G., Edwards, S. A., Young, J. A., and Downey, M., 1987, Geotomography for the delineation of coal seam structure: Geoexpl., **24**, 301–328.

Spies, B. R., and Habashy, T. M., 1995, Sensitivity analysis of crosswell electromagnetics: Geophysics, **60**, 834–845.

Thomson, S., and Hinde, S., 1993, Bringing geophysics into the mine—radio attenuation imaging and mine geology: Expl. Geophys., **24**, 805–810.

Vozoff, K., Smith, G. H., Hatherly, P. J., and Thomson, S., 1993, An overview of the radio imaging method in Australian coal mining: First Break, **10**, 13–21.

Ward, S. H., and Hohmann, G. W., 1988, Electromagnetic theory for geophysical applications, *in* Nabighian, M. N., Ed., Electromagnetic Methods in Applied Geophysics, Vol. 1, Theory: Soc. Expl. Geophys.

Wedepohl, E., 1993, Radio wave tomography—imaging ore bodies using radio waves: 3rd Technical Mtg., S. African Geophys. Assn., Expanded Abstracts, 85–88.

Occam's Inversion of 3-D Electrical Resistivity Tomography

Douglas J. LaBrecque, Gianfranco Morelli,[1]
William Daily, Abelardo Ramirez,[2]
Paul Lundegard[3]

Summary. Electrical resistivity tomography (ERT) images the electrical properties of the subsurface from dc resistivity measurements between surface and borehole electrodes. We experiment with 3-D inversion of ERT using finite-element forward solution and a conjugate-gradient inverse routine. The algorithm finds the smoothest model (Occam's inversion) that fits the data to a given prior error level. The algorithm takes 10 to 20 iterations to converge but requires only a single forward solution per iteration and does not require direct solution of a large system of equations.

Inversion of data from two sites is shown. The first site tests the ability of ERT to monitor leaks around large metal tanks at the Hanford Reservation in Washington State. Data were collected and inverted from 16 wells placed around a circular tank. The tank is of heavy-gauge steel covered with concrete, is 15 m in diameter, and extends 2 m below the ground surface. The 3-D algorithm was modified to allow the smoothness operator to be decreased at the tank boundary. The 3-D inversion was necessary to produce an accurate picture of the leak.

At a second site, ERT was used to monitor the injection of air from a vertical well at a shallow petroleum remediation site. Using a cone penetrometer, three electrode strings were placed in the ground on the corners of a right triangle. The background of the site was assumed to be layered. Results of 3-D and 2-D inversion agreed well when the regions of interest were approximately 2-D. Air injection caused large changes in resistivity. At early times, these were confined to an area near the injection point. Later, the changes were along a dipping, tabular region. At the latest times, there is evidence of mixing of brackish water at the depth of the injection point with freshwater in a shallower aquifer on the site. This mixing would have decreased the resistivity and thus the apparent size and magnitude of the zone of influence of sparging.

1 Introduction

Electrical resistivity tomography (ERT) has been demonstrated to be a tool that can image in-situ remediation processes (Ramirez et al., 1993) and conduct long-term

[1]SteamTech Environmental Services Bakers Field, CA 93309, USA.
[2]Lawrence Livermore National Laboratory, L-156, 7000 East Avenue, Livermore, CA 94550, USA.
[3]UNOCAL CERT, 376 S. Valencia Avenue, Brea, CA 92621, USA.

monitoring. Both of these are time-varying processes and require the collection and efficient interpretation of multiple 3-D data sets. ERT surveys differ from surface surveys in that ERT surveys require rapid collection of large data sets. A surface survey may collect a few tens or hundreds of measurements per day. A typical plane ERT data set has about 400 data points collected in about 20 minutes. Because we are monitoring time-varying processes, we often measure 20 to 40 planes per day. This requires an automated acquisition system and a sophisticated multidimensional interpretation scheme.

Our previous work (Schima et al., 1994; LaBrecque et al. 1996b) successfully used 2-D Occam's inversion on a number of sites. Increasingly, we have begun working on sites that require full 3-D inversion. At the first site discussed below, the Hanford Reservation, the site itself is strongly 3-D. The region to be imaged is covered by a large metal tank.

At the second site, the purpose of the survey was to monitor the movement of air injected into a shallow aquifer. This site was used to compare the relative strengths and weaknesses of 2-D and 3-D inversion. Results calculated using the 3-D inverse routine described here are compared with the inverse results from the 2-D code described by LaBrecque et al. (1996a).

2 Forward modeling

Our inversion algorithm requires a forward solution of the potential equation

$$\frac{\partial}{\partial x}\left(\sigma \frac{\partial}{\partial x} V\right) + \frac{\partial}{\partial y}\left(\sigma \frac{\partial}{\partial y} V\right) + \frac{\partial}{\partial z}\left(\sigma \frac{\partial}{\partial z} V\right) = I(x, y, z), \tag{1}$$

where V is the scalar electrical potential and $I(x, y, z)$ is the distribution of electrical-current sources and sinks. We use the finite-element method (FEM) with rectangular, hexahedral elements (Pridmore et al., 1981) to convert the differential Eq. (1) into a system of linear equations. This system of equations then is solved iteratively using the diagonally weighted preconditioned-conjugate-gradient method (Pini and Gambolati, 1990). The forward modeling algorithm is a modified version of one written by Qin (1995).

3 Inversion

A 3-D inversion often is strongly underdetermined. We implemented a regularized solution (Tikhonov and Arsenin, 1977) that jointly minimizes the misfit of the forward model to the field data and a stabilizing functional of the parameters. This regularized, nonlinear inversion process must be carried out iteratively. Furthermore, each of the nonlinear iterations involves the approximate solutions of a very large system of linear equations; solving these linear equations is itself an iterative process.

Therefore, we refer to two types of iteration: *nonlinear* and *conjugate-gradient*. With each nonlinear iteration, the algorithm steps from one set of parameters to the next. This requires at least one forward solution. The amount and direction of change are dependent on a sensitivity matrix (defined below), the elements of which change from iteration to iteration.

Conjugate-gradient iterations are used to find an approximate solution to a linear system of equations using the conjugate-gradient method. In our algorithm, this does not require additional forward solutions and the elements of the sensitivity matrix will remain the same. Several conjugate-gradient iterations must be performed for each nonlinear iteration.

3.1 Nonlinear iterations

Constable et al. (1987) coined the term Occam's inversion to describe a method of finding the smoothest possible model that fit the data to an a priori Chi-squared statistic. To find the optimal value of the parameter vector, **P**, our algorithm seeks the maximum value of α, the stabilization parameter, for which minimizing

$$\Psi(\mathbf{P}) = \chi^2(\mathbf{P}) + \alpha \mathbf{P}^T \underset{\sim}{\mathbf{R}} \mathbf{P}, \tag{2}$$

gives

$$\chi^2(\mathbf{P}) = \chi^2_{\text{prior}}. \tag{3}$$

In Eq. (2), we have chosen to use $\underset{\sim}{\mathbf{R}}$, the solution roughness, as the stabilizing functional. This is approximated by

$$\underset{\sim}{\mathbf{R}} = \underset{\sim}{\mathbf{x}}^T \underset{\sim}{\mathbf{x}} + \underset{\sim}{\mathbf{y}}^T \underset{\sim}{\mathbf{y}} + \underset{\sim}{\mathbf{z}}^T \underset{\sim}{\mathbf{z}}, \tag{4}$$

where $\underset{\sim}{\mathbf{x}}$, $\underset{\sim}{\mathbf{y}}$, and $\underset{\sim}{\mathbf{z}}$ are matrices first-order difference operators in the x-, y-, and z-directions. In Eqs. (2) and (3), χ^2 is given by

$$\chi^2(\mathbf{P}) = [\mathbf{D} - \mathbf{F}(\mathbf{P})]^T \underset{\sim}{\mathbf{W}} [\mathbf{D} - \mathbf{F}(\mathbf{P})], \tag{5}$$

where **D** is the vector of known data values, **F**(**P**) is the forward solution, and $\underset{\sim}{\mathbf{W}}$ is a data-weight matrix. The diagonal elements of $\underset{\sim}{\mathbf{W}}$ are the reciprocals of the data variances, and the nondiagonal elements are zero. This assumes noncorrelated data errors. We take χ^2_{prior} to be equal to the number of data points. Thus, on average, the fit of each data point to the forward model is one standard deviation. Model studies of 2-D ERT inversion (LaBrecque et al., 1996a) showed empirically that this is an optimal value for χ^2.

The parameters, **P**, are the natural logarithms of the conductivity of the FEM elements. In the foreground—the part of the FEM mesh between the boreholes—each parameter corresponds to a single finite element. In the background—the region away from the boreholes—we lump several finite elements together into a single parameter.

Each nonlinear iteration can be expressed as

$$\mathbf{P}_{k+1} = \mathbf{P}_k + \Delta \mathbf{P}_k, \tag{6}$$

where $\mathbf{P}_k$ is the vector of parameters from the previous iteration and $\Delta \mathbf{P}_k$ is the parameter-change vector which is found by solving the linear system

$$\Delta \mathbf{P}_k = \left(\underset{\sim}{\mathbf{A}}_k^T \underset{\sim}{\mathbf{W}} \underset{\sim}{\mathbf{A}}_k + \alpha \underset{\sim}{\mathbf{R}}\right)^{-1} \left(\underset{\sim}{\mathbf{A}}_k^T \underset{\sim}{\mathbf{W}} \Delta \mathbf{D}_k - \alpha \underset{\sim}{\mathbf{R}} \mathbf{P}_k\right), \tag{7}$$

where $\underset{\sim}{\mathbf{A}}_k$ is the sensitivity matrix at the kth iteration and

$$\Delta \mathbf{D}_k = \mathbf{D}_k - \mathbf{F}(\mathbf{P}). \tag{8}$$

The elements of the sensitivity matrix, $a_{i,j}$, are

$$a_{i,j} = \frac{\partial F_i(\mathbf{P}_k)}{\partial p_j}, \tag{9}$$

where p_j is the jth element of $\mathbf{P}_k$ and $F_i(\mathbf{P}_k)$ is the forward solution for the ith data point.

3.2 *Conjugate-gradient iterations*

Solving Eq. (7) exactly is not practical because the system is very large (often 40 000 $\times$ 40 000), full, and ill-conditioned. Instead, we use the conjugate-gradient method described by Mackie and Madden (1993) to give a stable, approximate solution.

Before beginning conjugate-gradient iterations, the parameter-change vector is initialized to

$$\Delta\mathbf{P}_k = 0. \tag{10}$$

In the following equations, the subscript, ℓ, indicates the conjugate-gradient iteration number for variables internal to the conjugate-gradient algorithm. The algorithm begins by initializing the vectors

$$\mathbf{r}_0 = \underset{\sim}{\mathbf{A}}_k^T \underset{\sim}{\mathbf{W}} \Delta\mathbf{D}_k - \alpha \underset{\sim}{\mathbf{R}} \mathbf{P}_k, \tag{11}$$

and

$$\mathbf{p}_0 = \mathbf{r}_0, \tag{12}$$

and the scalar

$$\beta_0 = 0. \tag{13}$$

The conjugate-gradient residual vector, $\mathbf{r}_\ell$, is the misfit of the forward solution to the data mapped into solution minus the solution roughness and $\mathbf{p}_\ell$ is a trial solution to the linear system of equations given by Eq. (7). The variable β_ℓ ensures that the solution vectors $\mathbf{p}_\ell$ are mutually conjugate (Press et al., 1992).

For $\ell = 1, 2, 3 \ldots$, we have

$$\mathbf{p}_\ell = \beta_{\ell-1}\mathbf{p}_{\ell-1} + \mathbf{r}_{\ell-1}, \tag{14}$$

and the estimate of the parameter-change vector is

$$\Delta\mathbf{P}_k = \Delta\mathbf{P}_k + \lambda\mathbf{p}_\ell, \tag{15}$$

where

$$\lambda = \frac{\mathbf{r}_{\ell-1}^T \mathbf{r}_{\ell-1}}{\mathbf{p}_\ell^T (\underset{\sim}{\mathbf{A}}_k^T \underset{\sim}{\mathbf{W}} \underset{\sim}{\mathbf{A}}_k + \alpha\underset{\sim}{\mathbf{R}})\mathbf{p}_\ell}. \tag{16}$$

For the next iteration, the values of $\mathbf{r}_\ell$ and β_ℓ are

$$\mathbf{r}_\ell = \mathbf{r}_{\ell-1} - \lambda\left(\underset{\sim}{\mathbf{A}}_k^T \underset{\sim}{\mathbf{W}} \underset{\sim}{\mathbf{A}}_k + \alpha\underset{\sim}{\mathbf{R}}\right)\mathbf{p}_\ell, \tag{17}$$

and

$$\beta_\ell = \frac{\mathbf{r}_\ell^T \mathbf{r}_\ell}{\mathbf{r}_{\ell-1}^T \mathbf{r}_{\ell-1}}. \tag{18}$$

The matrix $\underset{\sim}{\mathbf{R}}$ is very sparse, and so, the product $\mathbf{R}\mathbf{p}_\ell$ in Eqs. (16) and (17) can be calculated very inexpensively. However, the matrix $\underset{\sim}{\mathbf{A}}_k^T \underset{\sim}{\mathbf{W}} \underset{\sim}{\mathbf{A}}_k$ is full and too large to be calculated and stored directly. One of the advantages of the conjugate-gradient solver is that only the product of the system with a vector is needed. Therefore, we begin by calculating $\underset{\sim}{\mathbf{A}}_k \mathbf{p}_\ell$ which is a vector of length equal to the number of data points. This vector then is multiplied times $\underset{\sim}{\mathbf{W}}$ and then finally times $\underset{\sim}{\mathbf{A}}_k^T$. Our algorithm differs from that of Mackie and Madden in the method of making these multiplications. The method used here is similar to that described by Rodi (1976) and is discussed below.

A second way that our algorithm differs from that of Mackie and Madden (1993) is that we use more conjugate-gradient iterations than they did. For the MT inverse problem, Mackie and Madden found that the nonlinear inversion routine converged well with three conjugate-gradient iterations. We usually require between 10 and 60 conjugate-gradient iterations for the final data misfit to converge to the value of χ^2_{prior}. Third, we use smoothness instead of comparison with an a priori model to stabilize the inverse solution.

3.3 *Calculating sensitivities*

We can write the FEM approximation to the potential Eq. (1) as

$$\underset{\sim}{\mathbf{G}}\mathbf{V} = \delta\mathbf{I}, \tag{19}$$

where $\underset{\sim}{\mathbf{G}}$ is the FEM stiffness matrix, $\mathbf{V}$ is the vector of potentials at the nodes of the FEM mesh and $\delta\mathbf{I}$ is a point source of current at one of the nodes. For each data point i, we can define a pair of vectors $\mathbf{M}_i$ and $\mathbf{S}_i$ with lengths equal to the number of FEM nodes such that the elements of $\mathbf{M}_i$ are zero except for the locations of the measurement electrodes and $\mathbf{S}_i$ is nonzero only at the electric-current source and sink. So, we can write

$$\mathbf{V}_k = \underset{\sim}{\mathbf{G}}^{-1}\mathbf{S}_i, \tag{20}$$

and the FEM forward solution corresponding to data point i is

$$F_i(\mathbf{P}_k) = \mathbf{M}_i^T \underset{\sim}{\mathbf{G}}^{-1}\mathbf{S}_i = \mathbf{M}_i^T \mathbf{V}. \tag{21}$$

For a point source of current, derivative of the potential with respect to parameter j can be found by taking the partial of Eq. (19) with respect to the jth parameter

$$\underset{\sim}{\mathbf{G}}\frac{\partial \mathbf{V}}{\partial p_j} + \left(\frac{\partial \underset{\sim}{\mathbf{G}}}{\partial p_j}\right)\mathbf{V} = 0; \tag{22}$$

therefore,

$$\frac{\partial \mathbf{V}}{\partial p_j} = -\underset{\sim}{\mathbf{G}}^{-1}\left(\frac{\partial \underset{\sim}{\mathbf{G}}}{\partial p_j}\right)\mathbf{V}. \tag{23}$$

From Eqs. (9), (20), (21), and (23), the sensitivity coefficient for data point i and parameter j is

$$a_{i,j} = -\mathbf{M}_i^T \underset{\sim}{\mathbf{G}}^{-1}\left(\frac{\partial \underset{\sim}{\mathbf{G}}}{\partial p_j}\right)\underset{\sim}{\mathbf{G}}^{-1}\mathbf{S}_i. \tag{24}$$

Note that, because $\underset{\sim}{\mathbf{G}}^{-1}$ is symmetric,

$$\mathbf{M}_i^T \underset{\sim}{\mathbf{G}}^{-1} = (\underset{\sim}{\mathbf{G}}^{-1}\mathbf{M}_i)^T. \tag{25}$$

Now note that $(\underset{\sim}{\mathbf{G}}^{-1}\mathbf{S}_i)$ is the vector of potentials at each node for the source distribution $\mathbf{S}_i$ and that this entire vector is calculated automatically for every FEM forward solution.

Furthermore, we can calculate the vector $(\underset{\sim}{\mathbf{G}}^{-1}\mathbf{M}_i)$ in the same way by simply placing a distribution of sources at the same positions as the receiver points. We normally use every electrode as both a receiver point and a transmitter point, and so, the vectors $(\underset{\sim}{\mathbf{G}}^{-1}\mathbf{S}_i)$ and $(\underset{\sim}{\mathbf{G}}^{-1}\mathbf{M}_i)$ are calculated and stored as part of the forward solution.

If the parameters are the logarithm of the conductivity of a single FEM element, or a few adjacent elements, then the matrix $\partial\underset{\sim}{\mathbf{G}}/\partial p_j$ is extremely sparse. Thus, calculating the sensitivity coefficients using Eq. (24) requires only a fraction of the effort required to calculate a forward solution. Because the conjugate-gradient method requires only the product of vector times the sensitivity matrix or its transpose, these can be constructed in a piecewise fashion without ever creating or storing the entire sensitivity matrix.

3.4 *Choice of smoothness factor*

Occam's inversion relies on devising an algorithm to choose α. Choosing the correct value of α is critical both for achieving rapid convergence of the nonlinear inversion and for finding a good final parameter estimate. In 3-D inversion, one is limited to methods that do not require a large number of forward solutions or to solving a system of equations of size equal to the number of parameters. The method described here is approximate but, in our tests, has converged to the optimal value of α. With our method, we make an initial guess for α for the first iteration, and a new value of α is estimated at the end of each nonlinear iteration. The estimate uses the assumption that the relation between α and χ^2 can be approximated by the rational function

$$\chi^2 \cong \frac{b\alpha}{\alpha + a}, \tag{26}$$

where a and b are constants. The constant, b, is the value of χ^2 for a homogeneous half-space. The constant, a, is estimated from the values of α and χ^2 of the previous iteration. If the misfit is χ_k^2 and the desired misfit is $\chi^2(P) = \chi^2_{\text{target}}$, then the new estimate of α, α_{k+1}, is

$$\alpha_{k+1} = \frac{b/\chi_k^2 - 1}{b/\chi^2_{\text{target}} - 1}\alpha_k. \tag{27}$$

The value of χ^2_{target} is chosen as the larger of χ^2_{prior} and $\chi_k^2/2$.

This approach is simplistic, but the solution usually converges to the correct value of α in 10 to 20 iterations. However, care must be taken in choosing the initial value of α. During the early iterations when χ^2 is larger than χ^2_{prior}, the magnitude of α calculated from Eq. (27) can only decrease with each iteration. The magnitude of α cannot increase until an iteration is reached where $\chi^2 < \chi^2_{\text{prior}}$. Therefore, it is better to begin with too large of an estimate of α than too small an estimate.

3.5 *Robust inversion*

The least-squares method used in previous ERT modeling (LaBrecque et al., 1996a) is efficient if the errors are normally distributed and the correct data variances are known (Hampel et al., 1986). However, if the data variances are not known or there are outliers in the data, the results are poor. We implemented a variation of the least-absolute-deviations method described by Mostellar and Tukey (1977). After each inversion iteration, the weights are decreased for those data that are poorly fitted by the inverted model.

The weights of data with good fits are kept at their original values. The algorithm, described more fully by LaBrecque and Ward (1990), can be expressed mathematically as

$$ {}_{\text{new}}W_{i,i} = \begin{cases} {}_{\text{old}}W_{i,i}, & {}_{\text{trial}}W > {}_{\text{old}}W_{i,i}; \\ {}_{\text{trial}}W, & {}_{\text{trial}}W < {}_{\text{old}}W_{i,i}, \end{cases} \tag{28} $$

where ${}_{\text{old}}W_{i,i}$ is the diagonal term of the weighting matrix from the previous iteration, ${}_{\text{new}}W_{i,i}$ is the updated weight, and ${}_{\text{trial}}W_{i,i}$ is given by

$$ {}_{\text{trial}}W_{i,i} = \frac{{}_{\text{old}}W_{i,i}^{\frac{1}{2}}}{e_i} \left(\frac{\sum_j \left({}_{\text{old}}W_{j,j}^{\frac{1}{2}} e_j \right)}{\sum_j \left({}_{\text{old}}W_{j,j}^{\frac{1}{4}} e_j^{\frac{1}{2}} \right)} \right), \tag{29} $$

where e_i is the misfit of the ith data point which is given by

$$ e_i = \| D_i - F_i(\mathbf{P}) \|. \tag{30} $$

The use of this method for 2-D and 3-D inversions is described by Morelli and LaBrecque (1996). One of the concerns is that this reweighting scheme may bias the solution by eliminating data points that are not noisy but simply are different from the initial guess. Keep in mind that the results of any nonlinear inversion are influenced by the starting model. It is not clear whether the reweighting scheme makes this problem significantly worse. However, it is always a good idea to try a number of different starting models.

4 Leak detection at Hanford Reservation

The field experiment described here was carried out to evaluate ERT for detecting leaks and delineating the resulting plumes emanating from steel storage tanks. Our strategy was to produce ERT images of resistivity under the storage tank and look for changes that could be attributed to spillage of the tank contents. To test this strategy, we produced tomographs of the soil under a tank before and during the controlled release of sodium chloride brine below one edge of the tank. Figure 1 shows the layout for the leak-detection experiment. ERT data were collected using 16 boreholes in an octagonal pattern around the tank. Each borehole contained eight electrodes at even intervals from the surface to 10.7 m in depth.

This case was a difficult one to model numerically. First, the tank could not be approximated by a 2-D earth. Second, the tank was extremely conductive. This created difficulties for both the forward and the inverse solvers. The inverse solver was modified to allow the roughness operator to be discontinued at the tank boundaries. If this was not done, the roughness at the tank–soil boundary dominated the inverse solution. Because the roughness at the tank boundary is so great, the chosen value of α would be very small and most of the parameters in the mesh would no longer be stabilized.

Figure 2 shows a composite of four ERT images. The first image (Fig. 2a) shows the background resistivity. The tank is approximated by a zone of 10^{-3} Ω·m. The remaining background is fairly homogeneous. The remaining three images show the shapes of the regions of decreased resistivity between the background and later times. In Fig. 2b, we see a small region of decreased resistivity below the leak point. This region formed after about 640 liters had been spilled. Three days after the leak started, about 2100 liters

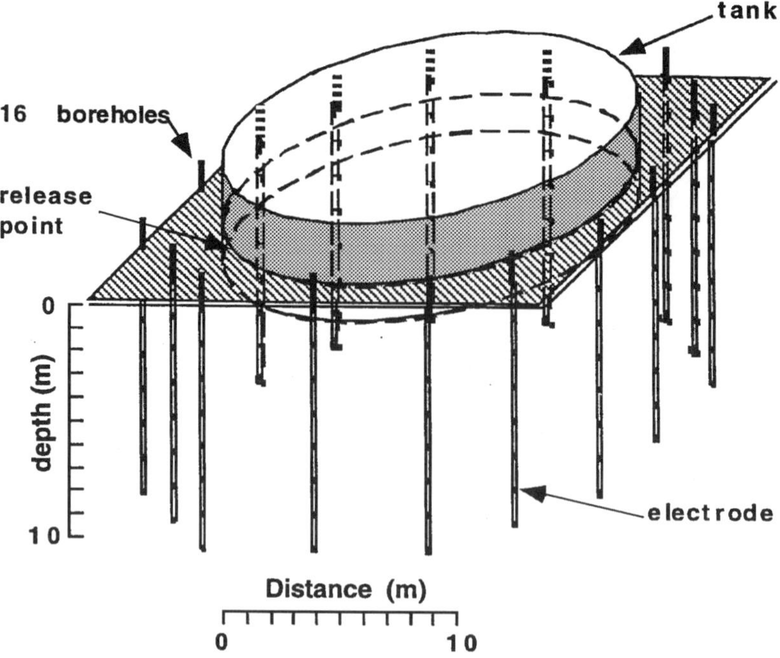

Figure 1. Layout of tank and electrodes for leak test at Hanford Reservation.

had been spilled and a distinct plume had formed and was moving under the tank. By the final day, 4000 liters of brine were released and the plume had expanded beyond the survey area.

5 Monitoring air sparging

The field site is a vacant lot in central California and lies near the Pacific Ocean. Soil on the site was contaminated by petroleum hydrocarbons leaking from a pipeline just west of the site. A layout of the sparge and the sampling wells is shown in Fig. 3. The ERT electrodes were placed by pushing a hollow 1-inch (2.54-cm) OD steel rod into the ground using a trailer-mounted cone penetrometer. The electrodes consisted of 10-cm lengths of 0.95-cm-diameter copper tubing along a polyvinyl chloride–covered multiconductor cable. The electrode string was placed in the center of the steel rod and the rod was removed, leaving only the electrodes and a disposable metal tip in the ground. Each electrode string contained 14 electrodes at 0.51-m intervals. The deepest electrodes were placed 7 m below the ground surface. Three additional electrodes were placed along the surface at the points midway between the ERT strings (Fig. 3).

The geology of the site consists mainly of poorly sorted, medium-grained sands. Prior to the ERT surveys, it was thought that the site was relatively homogeneous. Logs of drilling cuttings did not show any structures that would affect the air and fluid flow on the site. However, a continuous core taken on site showed two thin clay layers between 3 and 4 m in depth.

Figure 4a shows a 3-D ERT image of resistivity before sparging (air injection) started. The site is strongly layered geoelectrically. The lowermost zone is highly conductive, between 1 and 2 Ω·m. Water samples from the top of this zone gave resistivities of 0.56 Ω·m. The next layer is more resistive, between 10 and 50 Ω·m. The boundary between these two layers is roughly at the position of a clay layer in the continuous core.

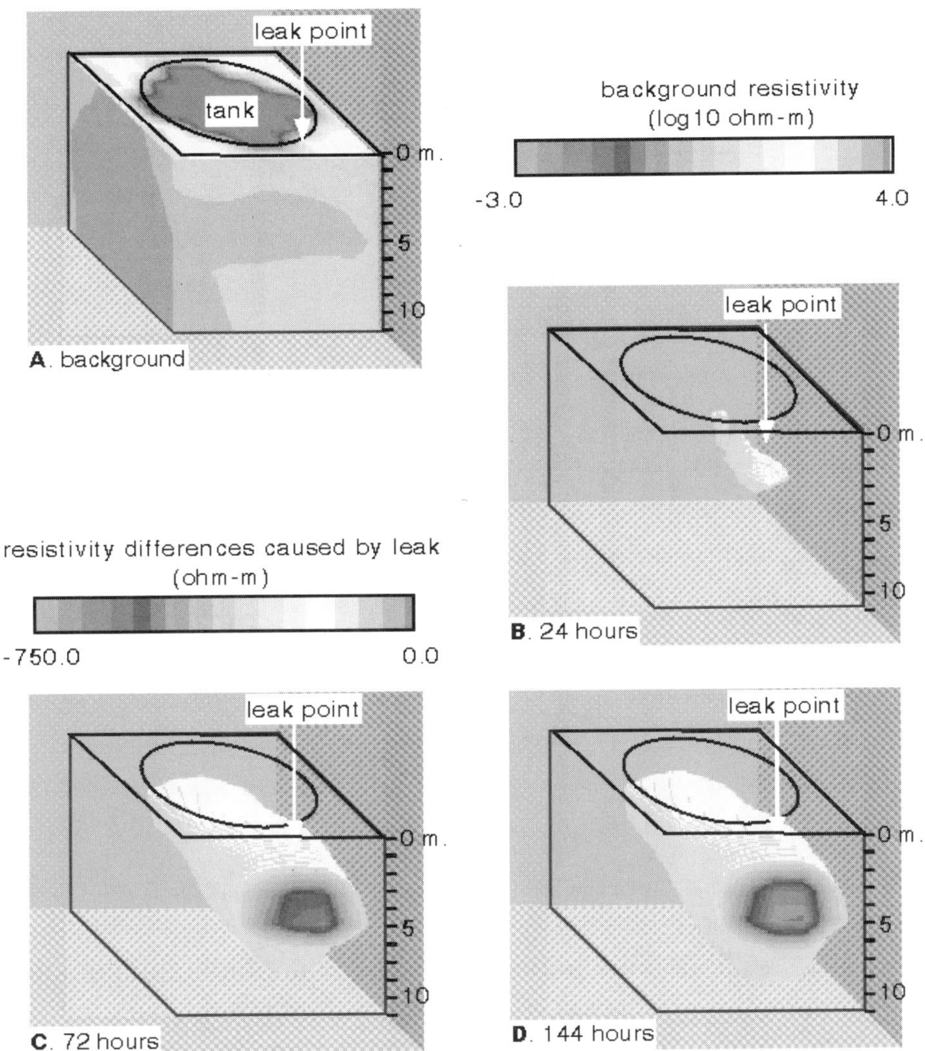

Figure 2. ERT images from Hanford showing (a) background resistivity image, and (b) change in resistivity from the background 24 hours after the leak started (640 liters spilled); (c) 72 hours after the leak started (2100 liters spilled); and (d) 144 hours after the leak started (4000 liters spilled). Images (b), (c), and (d) are transparent where resistivity differences are smaller than 150 Ω·m.

The uppermost layer has been interpreted as the unsaturated zone. Resistivities in this zone are as high as 2000 Ω·m. These may reflect zones of hydrocarbon contamination in the unsaturated zone.

In Fig. 4b, the 2-D ERT image of the plane ERT-1–ERT-2 is shown for comparison. The algorithm used to invert the 2-D data is described by LaBrecque et al. (1996a) and Morelli and LaBrecque (1996). Because the background resistivity structure is nearly 1-D, the two results agree very well. Note that the boundary between the brackish-water and freshwater layers is more gradual near the center of the 3-D image. We can be reasonably certain that the true boundary between these layers is sharply defined. Note that the boundary is abrupt at all three electrode strings. Disturbing the soil column by placing the electrode strings should have made the boundary more gradual, not sharper. Furthermore, it is very unlikely that all three electrode strings would be placed at points where the boundary is sharp if boundary is gradual in most places.

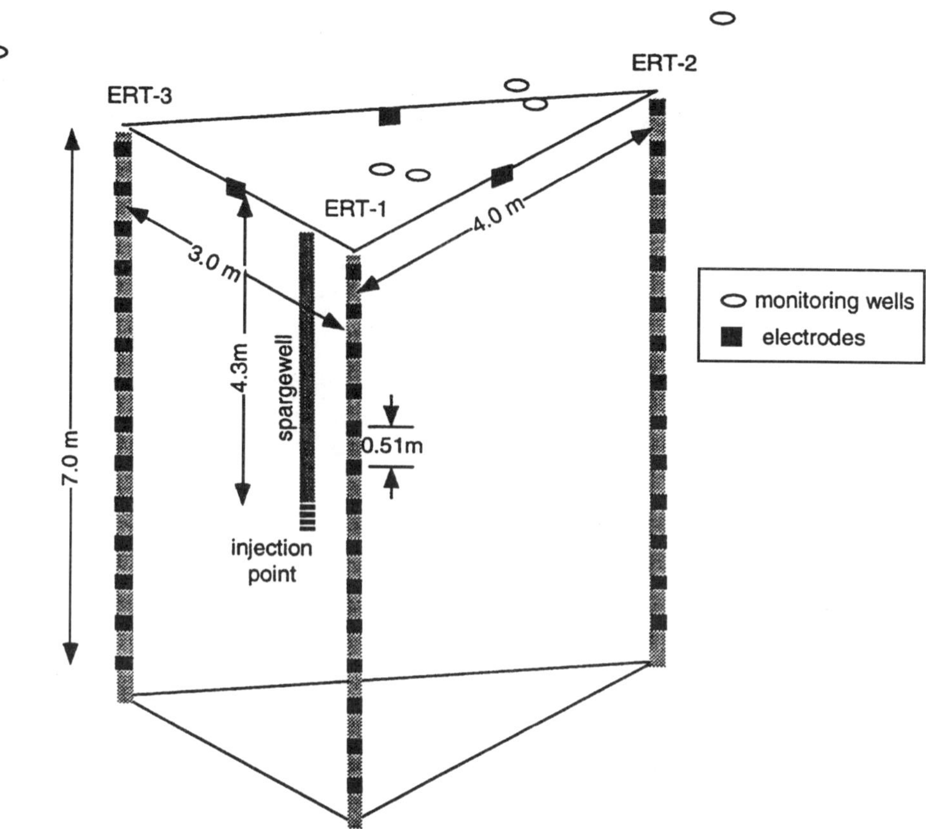

Figure 3. Configuration of ERT holes at air-sparging field site.

Decreasing resolution away from the electrodes is a common problem in both 2-D (LaBrecque et al., 1996a) and 3-D ERT images. Morelli and LaBrecque (1996) discuss a method that has improved the resolution in the 2-D algorithm but has not been implemented in the 3-D code. We have tried a number of regularizing functionals in both our 2-D and 3-D inversion codes. We have yet to find any substantial advantage of one over another. Any regularization functional that we apply will create artifacts in the final solution. We have found that artifacts are surprisingly similar for different regularizing functionals.

Two aspects of this site made data interpretation difficult: First, there is a large contrast in resistivities; second, during air injection, there were large, fairly rapid changes in resistivity near the electrode strings during the sparging process. Because a full 3-D data set requires about 1 hour to collect, some data values often changed by factors of two. Much of the field site also saw factor-of-two, or more, changes in actual resistivity during the survey. Ideally, we would implement a 4-D inversion routine that would incorporate time as well as spatial dimensions. Our alternative was to implement the robust-inversion 3-D scheme described in the theory section. This scheme was able to treat the rapidly changing data points as outliers and reduce the magnitudes of their weights. The method was so successful that it has become a routine part of our data interpretation (Morelli and LaBrecque, 1996).

The percent changes in resistivities after about 4 hours of air injection are shown in Fig. 5. In both the 3-D images (Fig. 5a) and the 2-D images (Fig. 5b), the changes

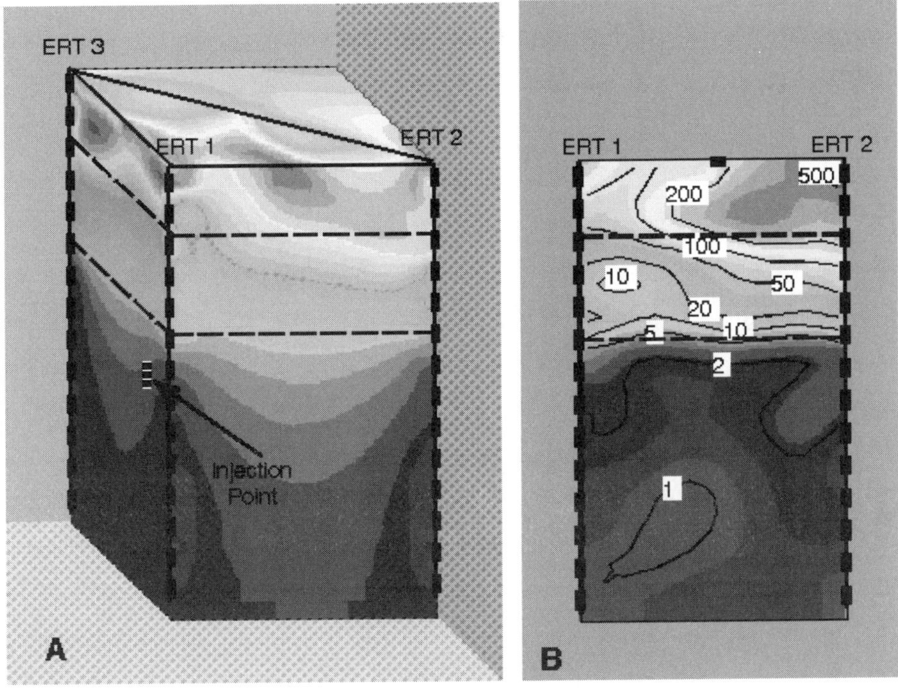

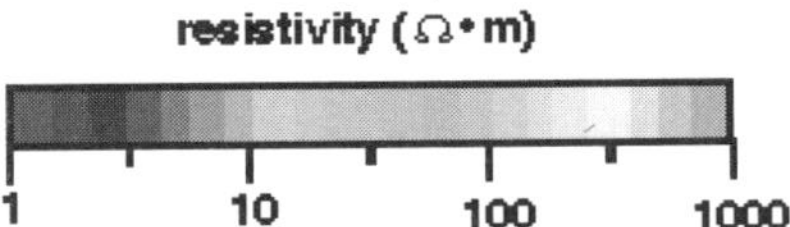

Figure 4. Background ERT images from a field site in California. Shown are (a) 3-D inversion code results and (b) 2-D inversion results from plane ERT-1–ERT-2; Note that the estimated locations of the water table and brackish-water–freshwater interface are shown as dashed lines.

in resistivity are small, mostly less than 50%, and are largely confined to an area near the injection point. The agreement between the two methods is fairly good despite the highly 3-D nature of the flow region.

After 20 hours of air injection, the 3-D image (Fig. 6a) shows increased resistivity along a dipping, tabular region that extends upward from just above the injection zone to the water table. In this region, the resistivities have increased dramatically (100–1500%). In Fig. 6a, the resistivity changes below 100% have been made transparent. This cutoff was chosen to make the anomalous region easier to see in gray-scale images. When the cutoff is reduced to 25%, the resistive region is slightly larger but essentially the same shape. Note that the anomalous region corresponds to a zone of roughly 50% saturation.

There are also some small regions in the vadose zone that show significant changes over time. Changes at the surface probably reflect the drying of water placed on the surface electrodes at the start of the survey. There are also changes near the electrode strings ERT-1 and ERT-3.

We would expect to see the largest changes near the injection point. In the 3-D results, we see no anomalous change near the injection point. This region is one of rather low resolution; in the background image (Fig. 4), the layer boundary is sharply defined as seen in the 2-D image (Fig. 4b) but appears gradual in the 3-D image (Fig. 4). Because

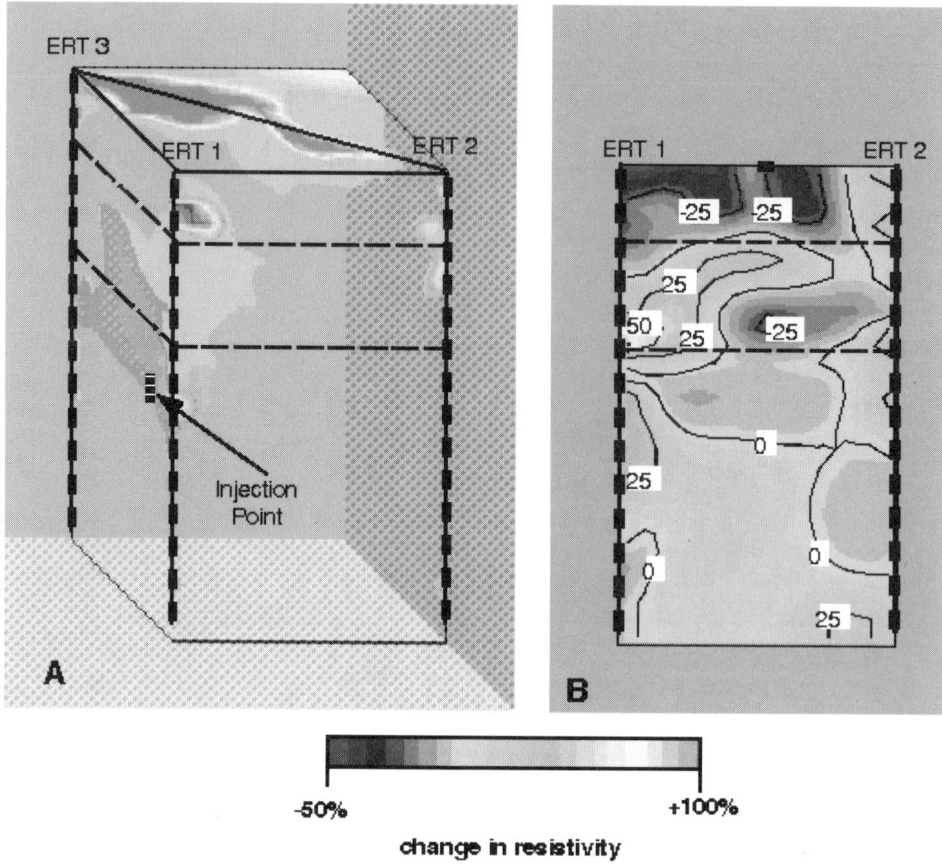

Figure 5. Percent changes in resistivity from background ERT images (Fig. 4) after about 4 hours of air injection: (a) 3-D inversion results; in the solid region shown, the resistivity increased by 25% or more; regions of less than 25% change are transparent. (b) The 2-D inversion results from plane ERT-1–ERT-2; estimated locations of water table and brackish-water–freshwater interface are shown as dashed lines.

of this smoothing, the background resistivities in this region are overestimated by as much as a factor of three. A change in resolution will change how sharply the layers are defined and create apparent (though false) changes in resistivity. The resolution changes with the data errors or the accuracy of the estimates of data errors (LaBrecque et al., 1996a). Despite these problems, a large region of increased resistivity should have been visible. We believe that the flow is occurring along relatively small channels. In particular, the holes for the ERT survey appear to form conduits for flow from the brackish aquifer to the upper aquifer. Increases in resistivity along ERT-1 may result from this airflow or from changes in the resolution from one data set to the next or a combination of both.

The shape of the anomalous regions in the 2-D images (Fig. 6b) agree well with the anomalies in the 3-D images. However, the changes are much smaller in the 2-D images. This is not surprising because 2-D inversion tends to underestimate the resistivity contrast of 3-D bodies (LaBrecque and Ward, 1990). There is also indication of flow near the injection zone that was not resolved in the 3-D images.

Figure 7a shows the 3-D images after 44 hours of air injection. The zone of increased resistivity is confined to a tabular region just below the water table and the area

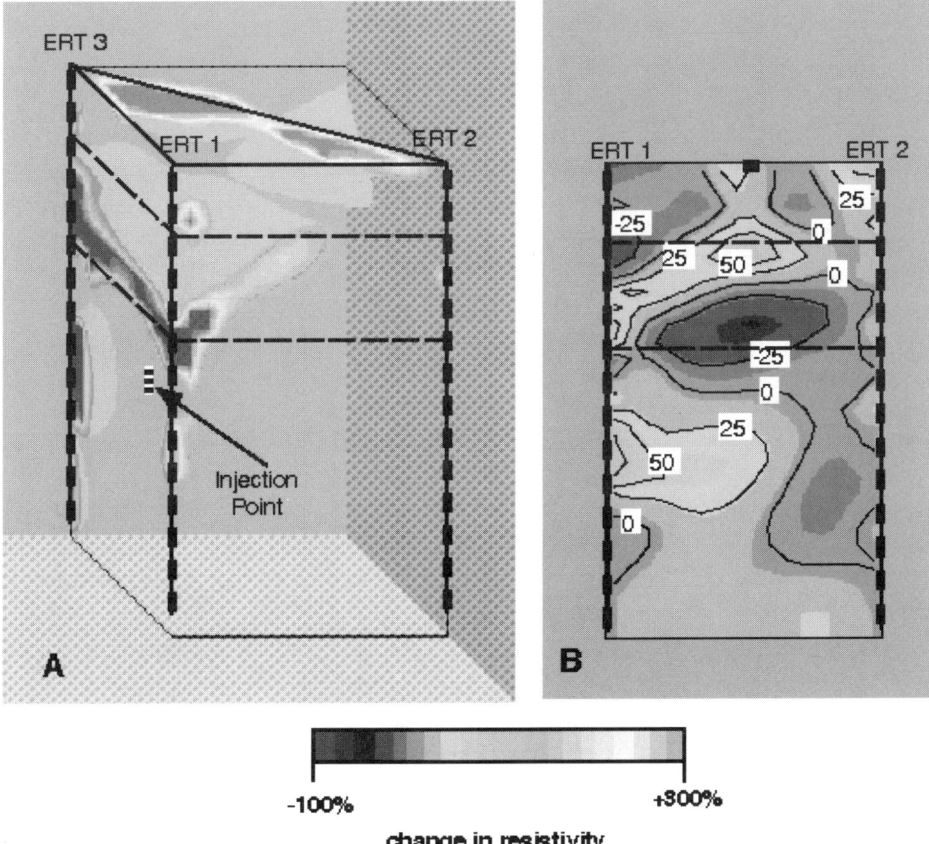

Figure 6. Percent changes in resistivity from background ERT images (Fig. 4) after about 20 hours of air injection: (a) 3-D inversion results; in the solid region shown, the resistivity increased by 100% or more; regions of less than 100% change are transparent. (b) The 2-D inversion results from plane ERT-1–ERT-2; estimated locations of water table and brackish-water–freshwater interface are shown as dashed lines.

immediately around ERT-1. In the upper part of the section, the 2-D images (Fig. 5a) agree very well with the 3-D images. However, there is a broad zone of increased resistivity at the base of the 2-D image. This zone is most likely an inversion artifact, possibly due to the 3-D nature of the anomalous regions or the lack of resolution near the base of the 2-D image. We also see a region of decreased resistivity at the base of the freshwater layer. This effect may represent the mixing of brackish waters into the freshwater aquifer induced by sparging. Because the level of total dissolved solids in the brackish layer is more than an order of magnitude larger than that in the upper aquifer, mixing of only a few percent of brackish water with the freshwater would produce these effects. If this mixing occurs, then both the extent and the magnitude of the air saturation are underestimated in these images.

6 Conclusions

We successfully developed a robust 3-D ERT inversion code. The method was used successfully to monitor time change in resistivity at the two sites shown. At the first site, the presence of a large metal tank inside the image area made the use of 2-D inversions

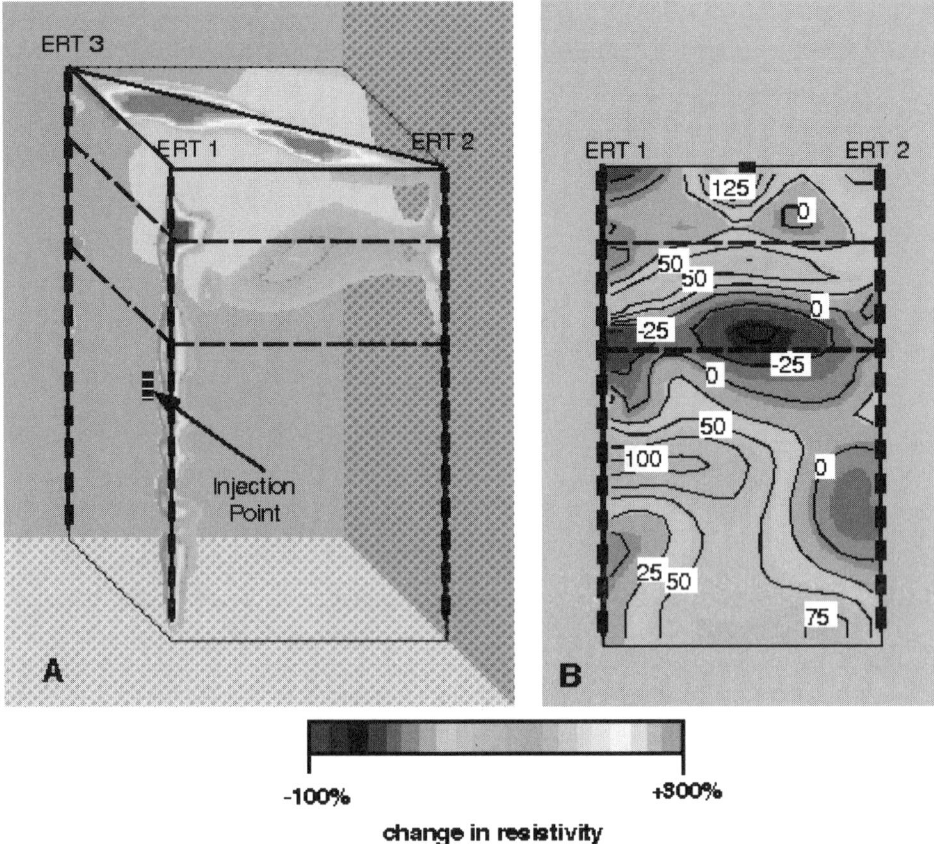

Figure 7. Percent changes in resistivity from background ERT images (Fig. 4) after about 44 hours of air injection: (a) 3-D inversion results; in the solid region, the resistivity increased by 25% or more; regions of less than 25% change are transparent. (b) The 2-D inversion results from plane ERT-1–ERT-2; estimated locations of water table and brackish-water–freshwater interface are shown as dashed lines.

impossible. By correctly modeling the tank in the 3-D forward model, leaks of brine could be seen beneath the tank.

At the second site, we had the opportunity to compare 2-D and 3-D inversions. The comparisons show both the need for 3-D inversion and the need for improvements in the present 3-D code. The background at the site was nearly 1-D. Thus, the 2-D and 3-D inversions gave similar results. However, the 3-D inversions did a poor job of representing the transition from a middle layer that ranged from 10 to 50 Ω·m to a lower layer with resistivity between 1 and 2 Ω·m. This created a zone in the upper part of the conductive layer that could not be imaged reliably.

The 3-D images during air sparging showed that the flow in the upper aquifer was confined to a tabular region that extended outward with time. The 2-D inversions showed very similarly shaped regions of flow, but the changes in resistivity were much lower. We feel that this is an artifact of the 3-D nature of the flow and that, in this case, the 3-D routine probably gave more accurate results.

One of our ERT cables was a conduit for airflow during the experiment. Much of the flow near the injection point was probably confined to small channels or pathways.

Finally, there is some evidence that the air sparging may have induced mixing of water between the upper freshwater aquifer and the deeper brackish aquifer.

Acknowledgments

We wish to recognize the contributions of the following colleagues at Lawrence Livermore National Laboratory (LLNL): John Carbino (LLNL) provided technical assistance before and during field operations; Jane Beatty provided programming support. Debra Iwatate of Westinghouse Hanford Company coordinated site activities and helped plan the field tests at Hanford. We wish to recognize Dave Yasamura of UNOCAL for providing field and technical support for work at the air-sparging site. Much of this work was performed under the auspices of the US Department of Energy by the LLNL under Contract W-7405-Eng-48. Additional funding was provided by UNOCAL-CERT.

References

Constable, S. C., Parker, R. L., and Constable, C. G., 1987, Occam's inversion: A practical algorithm for generating smooth models from electromagnetic sounding data: Geophysics, **52**, 289–300.

Hampel, F. R., Ronchetti, E. M., Rousseeuw, P. J., and Stahel, W. A., 1986, Robust statistics: John Wiley & Sons, Inc.

LaBrecque, D. J., and Ward, S. H., 1990, Two-dimensional cross-borehole resistivity model fitting, *in* S. H. Ward, Ed., Geotechnical and environmental geophysics, III, 51–74.

LaBrecque, D. J., Milletto, M., Daily, W., Ramirez, A., and Owen, E, 1996a, The effects of noise on Occam's inversion of resistivity tomography data: Geophysics, **61**, 538–548.

LaBrecque, D. J., Morelli, G., and Lundegard, P., 1996b, Monitoring air sparging in complex aquifers, in Proc. Symp. Appl. Geophys. Eng. Environ. Probl., 733–742.

Mackie, R. L., and Madden, T. R., 1993, Three-dimensional magnetotelluric inversion using conjugate gradients: Geophys. J. Internat., **115**, 215–229.

Morelli, G., and LaBrecque, D. J., 1996, Advances in ERT inverse modeling: European J. of Environ. and Eng. Geophys., **1**, 171–186.

Mostellar, F., and Tukey, J. W., 1977, Data analysis and regression: Addison-Wesley Publ. Co., 365–369.

Pini, G., and Gambolati, G., 1990, Is a simple diagonal scaling the best preconditioner for conjugate gradients on supercomputers?: Adv. Water Resour., **13**, 147–153.

Press, W. H., Teukolsky, S. A., Vetterling, W. T., and Flannery, B. P., 1992, Numerical recipes, 2nd ed.: Cambridge Univ. Press.

Pridmore, D. F., Hohmann, G. W., Ward, S. H., and Sill, W. R., 1981, An investigation of finite-element modeling for electrical and electromagnetic data in three dimensions: Geophysics, **46**, 1009–1024.

Qin, J., 1995, Three-dimensional dc resistivity forward and inversion by finite element method: Ph.D. dissertation, Univ. of Arizona, Tucson.

Ramirez, A., Daily, W., LaBrecque, D., Owen, E., and Chesnut, D., 1993, Monitoring an underground steam injection process using electrical resistance tomography: Water Resour. Res. **29**, 73–87.

Rodi, W. L., 1976, A technique for improving the accuracy of finite element solutions for magnetotelluric data: Geophys. J. Roy. Astr. Soc., **44**, 483–506.

Schima, S. A., LaBrecque, D. J., and Lundegard, P., 1994, Using resistivity tomography to track air sparging, in Proc. 1994 Symp. Appl. Geophys. Eng. Environ. Probl., 757–774.

Tikhonov, A. N., and Arsenin, V. Y., 1977, Solutions of ill-posed problems, Fritz, J., Ed.: John Wiley & Sons, Inc.

A Cubic-Hole Finite Element for 3-D Resistivity Modeling

Jianhua Li
Ganquan Xie

Summary. We have developed a new 3-D finite element for a cubic block with a vertical hole. The cubic-hole element, which has 24 nodes, is tailored to 3-D finite-element modeling of dc flow in formations penetrated by a borehole. It improves the condition number of the finite-element matrix by representing the borehole directly and avoiding elements with a large ratio of length to width. The new element assembles easily with conventional cubic elements and is suitable for domain decomposition and parallelization. We have incorporated this element into an iterative resistivity imaging algorithm.

1 Introduction

Finite-element modeling in geophysics often has to represent structures penetrated by a long, thin cylindrical hole. Standard 3-D cubic elements have difficulty matching the cylindrical hole structure and can lead to ill-conditioned finite-element matrices (Mezua et al., 1995; Xie et al., 1995). By a combined analytical and numerical approach (see Li et al., 1987; Xie and Li, 1988, 1989; Li, 1992; Xie et al., 1995; Li and Srivastav, 1997), we have constructed a new element for a cubic block with a vertical hole in it. This cubic-hole element, which has 24 nodes, is tailored to modeling the potential equation for dc flow, but also can be adapted to the wave and diffusion equations.

We first give a geometric description of the new element and derive its stiffness matrix. We then show how the element can be incorporated into a nonlinear resistivity inversion built around the finite-element method.

2 Cubic-hole element

Figure 1 shows the geometry of the new cubic-hole element; Figs. 2 and 3 show plan views of meshes with the new element (and conventional cubic elements) and with only conventional cubic elements. The new element is clearly more convenient

Earth Sciences Division, Building 90, Lawrence Berkeley National Laboratory, 1 Cyclotron Road, Berkeley, CA 94720, USA.

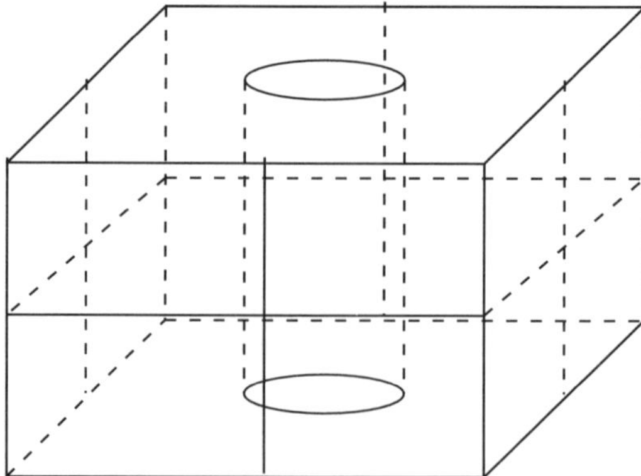

Figure 1. New cubic-hole element.

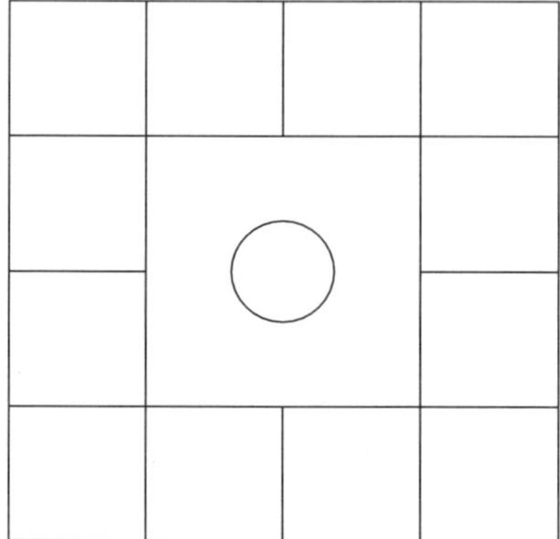

Figure 2. Plan view of mesh assembling the new cubic-hole element and the conventional cubic elements for structure with a hole.

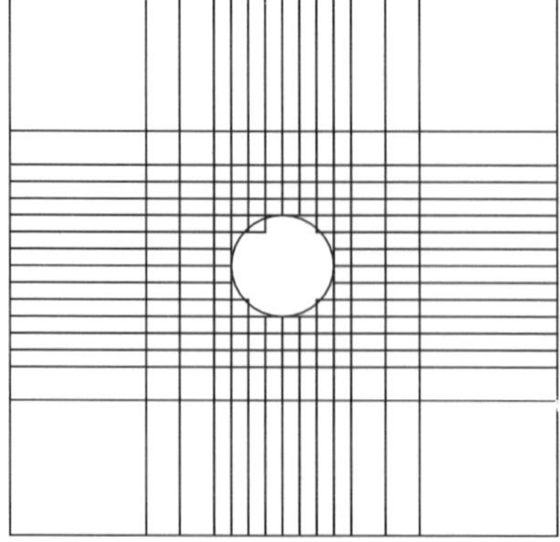

Figure 3. Plan view of the mesh using only the standard cubic elements for structure with a hole.

than standard elements for representing structures with long cylindrical holes. Fewer elements are required in the mesh surrounding the hole; moreover, the elements have better aspect ratios, which improves the condition number of the finite-element matrix.

To derive the stiffness matrix of the new element, consider the problem of dc flow, governed by the equation for the electrical potential ϕ,

$$-\nabla \cdot (\sigma \nabla \phi) = J, \tag{1}$$

where J is a current source and σ is the electrical conductivity. We need Green's function, G, for this equation, which is defined by

$$-\nabla \cdot (\sigma \nabla G) = \delta. \tag{2}$$

Applying Green's identity to domain Ω with outward normal n on the boundary $\partial\Omega$,

$$\int_{\Omega} [\nabla \cdot (\sigma \nabla \phi) - \nabla \cdot (\sigma \nabla G)] = \int_{\partial\Omega} \sigma \left(\frac{\partial \phi}{\partial n} G - \frac{\partial G}{\partial n} \phi \right) ds, \tag{3}$$

gives the integral equation equivalent to equation (1):

$$a\phi(r) - \int_{\partial\Omega} \sigma \left(\frac{\partial \phi}{\partial n} G - \frac{\partial G}{\partial n} \phi \right) ds = \int_{\Omega} GJ dv, \tag{4}$$

where a is a constant depending on the location of r on $\partial\Omega$: $a = 1$ if $r \notin \partial\Omega$, $a = 0.5$ if r lies on a face or a cylindrical surface of $\partial\Omega$, $a = 0.25$ if r lies on an edge of $\partial\Omega$, and $a = 0.125$ if r coincides with a vertex point of $\partial\Omega$.

To derive the element stiffness matrix of the cubic-hole element, we use finite elements to discretize the integral equation and then eliminate the internal nodes. For the cubic-hole element, $\partial\Omega$ consists of a cylindrical surface, four lateral faces, and the top and bottom faces (with holes). We divide the cylindrical surface into four quarter-circular elements. There are 9 nodes in each quarter-circular element, (θ_i, z_i), $i = 1, 2, \ldots, 9$. The shape function of the quarter-circular element is

$$\phi(\theta, z) = \sum_{i=1}^{9} p_i(\theta, z) \phi_i, \tag{5}$$

where p_i is the second-order finite-element basis function (Xie and Li, 1981). The shape function for the lateral faces is

$$\phi(x_j, x_k) = \sum_{i=1}^{9} q_i(x_j, x_k) \phi_i, \tag{6}$$

where q_i, the basis function of the 9-node element of the lateral face, is a second-order piecewise polynomial (Xie and Li, 1975). For the left and right lateral faces, $(x_j, x_k) = (y, z)$; for the front and back lateral faces, $(x_j, x_k) = (z, x)$; and for the top and bottom faces, $(x_j, x_k) = (x, y)$. The integrals over the top or bottom faces can be split:

$$\int_{\text{Face with hole}} \sigma \left(\frac{\partial \phi}{\partial n} G - \frac{\partial G}{\partial n} \phi \right) ds$$

$$= \int_{\text{Face with no hole}} \sigma \left(\frac{\partial \phi}{\partial n} G - \frac{\partial G}{\partial n} \phi \right) ds - \int_{\text{hole}} \sigma \left(\frac{\partial \phi}{\partial n} G - \frac{\partial G}{\partial n} \phi \right) ds. \tag{7}$$

Let $\phi_e = \{\phi_1, \phi_2, \ldots, \phi_{24}\}$ be the vector of the electrical potentials at the nodes on the cubic-hole element, and $\phi_c = \{\phi_1, \phi_2, \ldots, \phi_{24}\}$ be the vector of electrical potentials at nodes on the cylindrical surface. Substituting the shape functions into the surface integral Eq. (4), gives the matrix equation

$$\begin{bmatrix} \underset{\sim}{\mathbf{K}}_{cc} + a\underset{\sim}{\mathbf{I}}_c & \underset{\sim}{\mathbf{K}}_{ce} \\ \underset{\sim}{\mathbf{K}}_{ec} & \underset{\sim}{\mathbf{K}}_{ee} + a\underset{\sim}{\mathbf{I}}_e \end{bmatrix} \begin{bmatrix} \phi_c \\ \phi_e \end{bmatrix} = \begin{bmatrix} \mathbf{GJ}_c \\ \mathbf{GJ}_e \end{bmatrix}. \tag{8}$$

Here, $\underset{\sim}{\mathbf{I}}$ is the unit matrix, $\underset{\sim}{\mathbf{K}}$ is the stiffness matrix, the subscript c indicates 24 nodes of cylindrical surface, and the subscript e indicates 24 element nodes; $\mathbf{GJ}_c$ and $\mathbf{GJ}_e$ stand for vectors containing values of the right-hand side of Eq. (4) evaluated at the designated nodes.

Eliminating the auxiliary-node vector, we get

$$\underset{\sim}{\mathbf{U}}_{ee}\phi_e = \mathbf{Q}_e, \tag{9}$$

where

$$\underset{\sim}{\mathbf{U}}_{ee} = \underset{\sim}{\mathbf{K}}_{ee} + a\underset{\sim}{\mathbf{I}}_e - \underset{\sim}{\mathbf{K}}_{ec}(\underset{\sim}{\mathbf{K}}_{cc} + a\underset{\sim}{\mathbf{I}}_c)^{-1}\underset{\sim}{\mathbf{K}}_{ce}, \tag{10}$$

$$\mathbf{Q}_e = \mathbf{GJ}_e - \underset{\sim}{\mathbf{K}}_{ec}(\underset{\sim}{\mathbf{K}}_{cc} + a\underset{\sim}{\mathbf{I}}_c)^{-1}\mathbf{GJ}_c. \tag{11}$$

$\underset{\sim}{\mathbf{U}}_{ee}$ is the stiffness matrix of the new cubic element.

2.1 3-D dc resistivity imaging

One of our goals in developing the new element was to improve algorithms for 3-D resistivity imaging with borehole measurements (see Dey and Morrison, 1979; Petrick et al., 1981; Rijo, 1984; Park and Van, 1991; Shima, 1992; Lee and Xie, 1993; Sasaki, 1994). To derive the basic equations, consider a variation of the conductivity $\sigma \rightarrow \sigma + \delta\sigma$ about a background medium (e.g., a layered medium) and the corresponding variation in the potential $\phi \rightarrow \phi + \delta\phi$. Substituting in Eq. (1) and dropping second-order terms $(\delta\sigma\,\delta\phi)$ gives

$$\nabla \cdot (\sigma \nabla \delta\phi) = -\nabla \cdot (\delta\sigma \nabla\phi). \tag{12}$$

Let G^H be Green's function of the background medium with a vertical hole, which satisfies the following potential equation about a point source at the receiver location x_r:

$$\nabla \cdot (\sigma \nabla G^H) = \delta(x - x_r), \tag{13}$$

with boundary condition

$$\frac{\partial G^H}{\partial n} = 0$$

on the hole surface. Applying Green's identity to Eqs. (12) and (13) gives the integral equation

$$\int_\Omega [\nabla \cdot (\sigma \nabla \delta\phi) G^H - \nabla \cdot (\sigma \nabla G^H)]\, dv = \int_{\partial\Omega} \sigma \left(\frac{\partial \delta\phi}{\partial n} G^H - \frac{\partial G^H}{\partial n} \delta\phi \right) ds, \tag{14}$$

which, after the terms are rearranged, becomes an integral equation for the perturbation

$$\delta\phi(x_r) = -\int_\Omega \delta\sigma \nabla G^H \cdot \nabla\phi\, dv. \tag{15}$$

Equation (15) is the fundamental equation for 3-D resistivity inversion; it relates perturbations in the potential (about some known solution) to perturbations in the conductivity

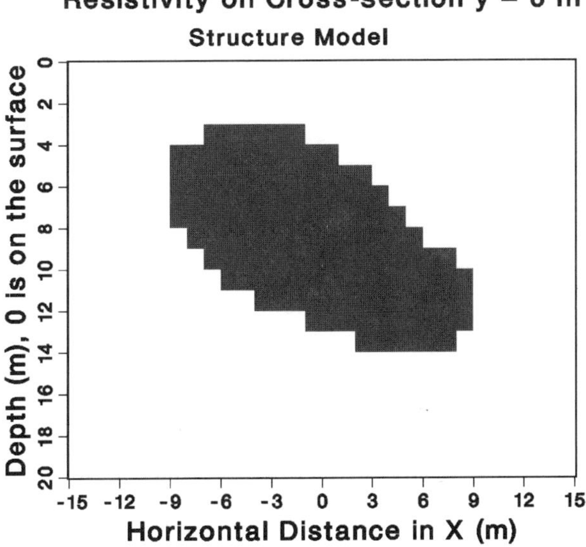

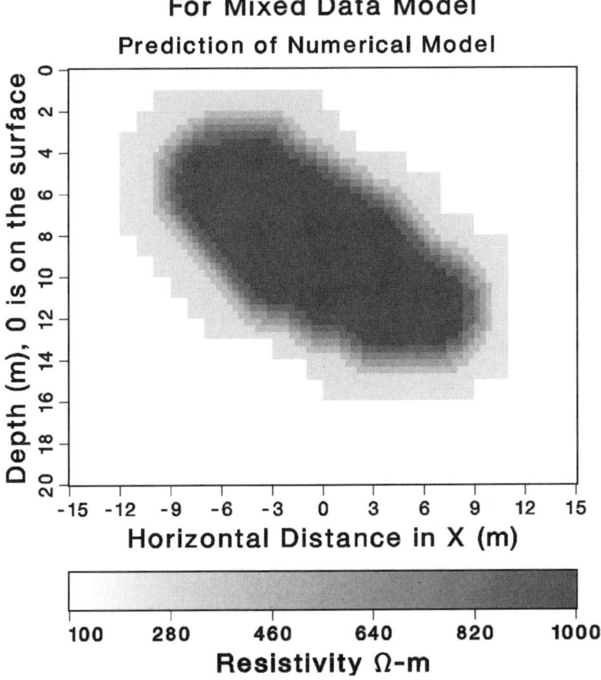

Figure 4. Resistivity imaging by 3-D nonlinear dc inversion.

(about some known model). For any problem involving a structure with one or more holes, the new cubic-hole element allows an efficient numerical solution for both the known potential ϕ and Green's function G^H.

3 Numerical simulations

We have developed an inversion algorithm to solve Eq. (15) in different configurations. Sources and receiver can be located on the surface or in one or more boreholes. The first simulation was designed to check the capability of the algorithm to model a complicated underground geometric structure. The synthetic experiment shows that a data set collected using surface-to-borehole and crosshole configurations can give a high-resolution image (see Li et al., 1995). Around the borehole, we use the new cubic-hole elements (Fig. 4). The bottom panel on Fig. 4 shows the numerical results; the top panel shows the resistivity structure.

The second simulation shows the flexibility of the new element. It involves the problem of detecting a leak in a grout block. The size of the grout block is 20 m wide, 10 m long, and 2.5 m thick; the top of the block is 5 m below the surface. The resistivity of the background is 50 ohm-m and of the grout is 10 000 ohm-m. The hole in the block has 0.3-m radius and 2.5-m thickness. The center of the hole is at location (0, 3.75, 6.25) m. The cubic-hole element is used only to model the hole in the block. There is no material filling the hole.

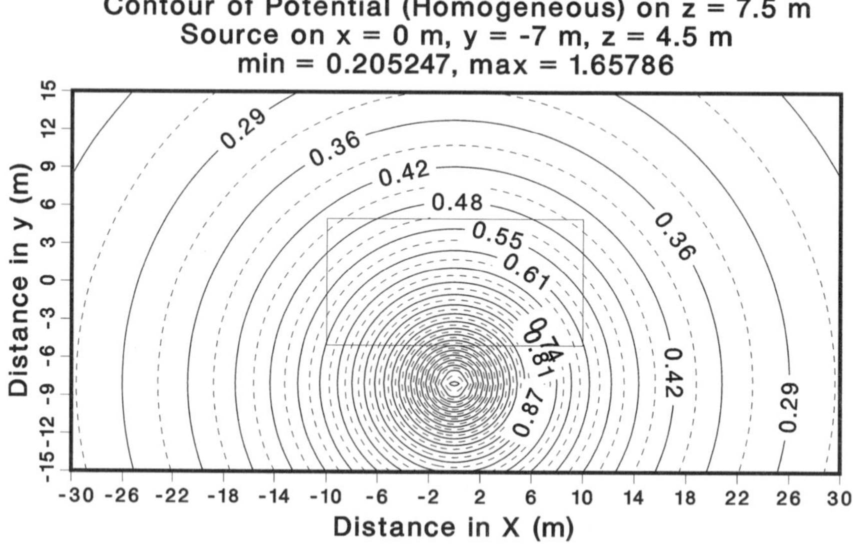

Figure 5. Potential distribution for homogeneous media.

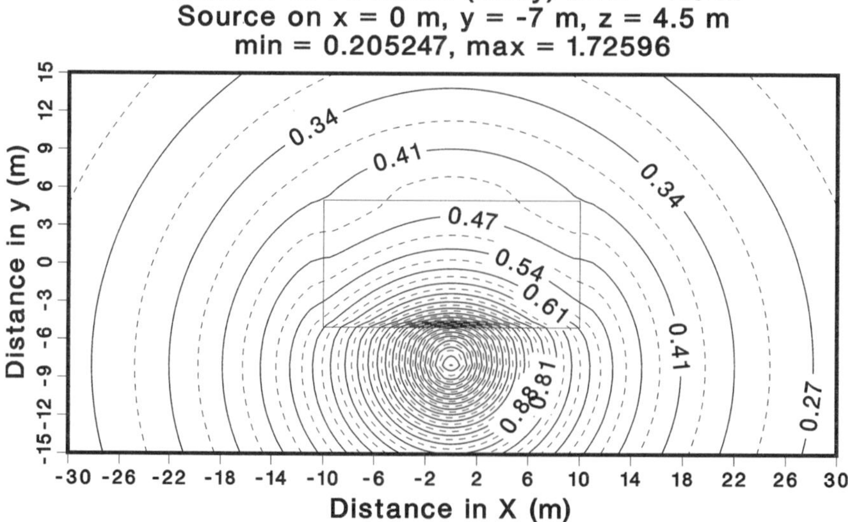

Figure 6. Potential distribution for the model with the resistivity block alone.

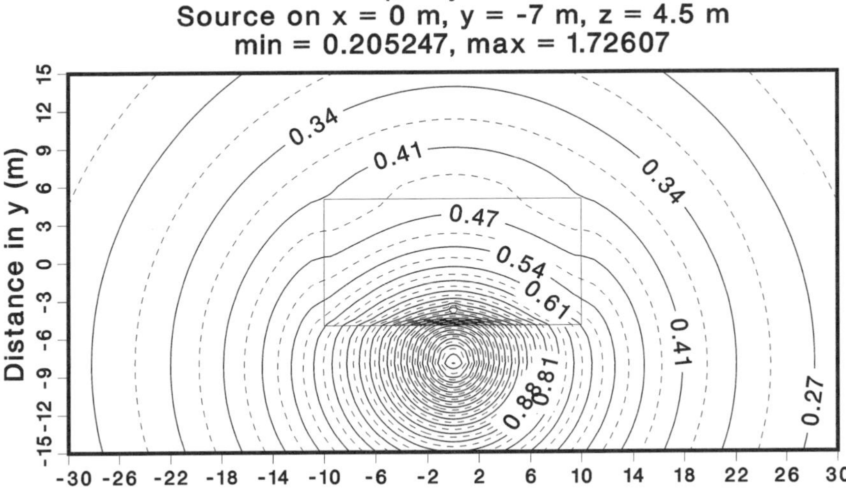

Figure 7. Potential distribution for model with resistivity block with the hole.

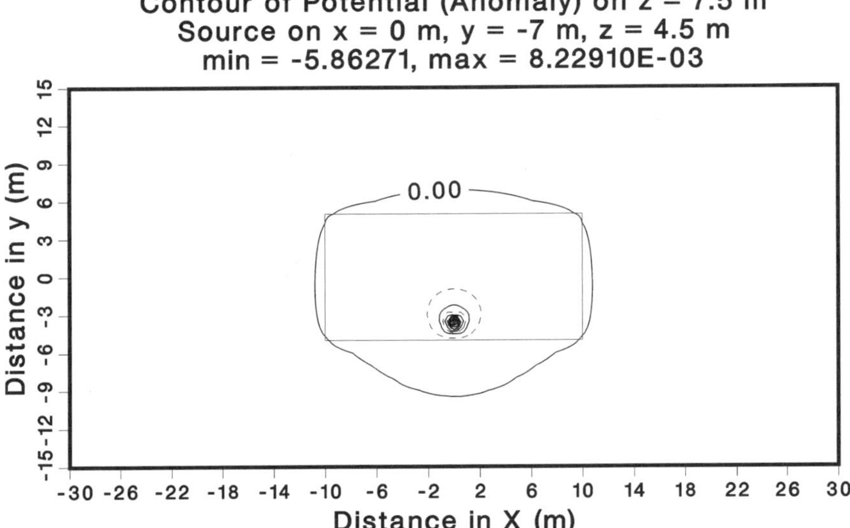

Figure 8. Normalized percentage potential difference between both models with or without the hole.

Figure 5 shows the contour of the potential distribution of homogeneous media. Figures 6 and 7 show potential distributions on the resistive block with or without the hole. Figure 8 shows the percentage difference of electrical potential between the two cases. The potential computed for the mode without the hole minus the potential with the hole, and then the difference is normalized by the potential without the hole. The source is located at $x = 0$ m, $y = -7$ m, $z = 4.5$ m, and the potentials shown are on the $z = 7.5$-m plane, at the bottom of the resistive block. The potential difference and the gradient of the potential difference are maximum at the hole, indicating the sensitivity of the solution to the presence of the hole. The size of the difference is about 5% (Fig. 8), which should be detectable with appropriate measurements, especially if sources at different locations are used.

4 Conclusions

We have developed a new cubic-hole element that is convenient for representing cylindrical holes (e.g., boreholes) in geological structures. The new element can be integrated with conventional cubic elements and gives stiffness matrices with good condition numbers. For example, in one of our tests, the condition number of the matrix that used a cubic-hole element to represent the hole was 10^3 compared to a condition number of about 10^9–10^{12} for a matrix that used only conventional elements. The new element can improve the accuracy of both forward modeling and inversion for many different geophysical problems.

Acknowledgments

This work was supported by the Office of Basic Energy Sciences, Engineering and Geosciences Division, Office of Technology Development, and the Office of Oil, Gas, and Shale Technologies, Fossil Energy Division, of the U.S. Department of Energy under Contract No. DE-AC03-76SF00098 and DOE Massively Parallel computer allocation. The authors would like to thank Dr. Bruce Curtis and consultants of the National Energy Research Supercomputer Center for their help.

References

Dey, A., and Morrison, H. F., 1979, Resistivity modeling for arbitrarily shaped three-dimensional structure: Geophysics, **44**, 753–780.

Lee, K. H., and Xie, G. Q., 1993, A new approach to imaging with low-frequency electromagnetic fields: Geophysics, **58**, 780–796.

Li, J., 1992, Integral equation methods for mixed boundary problem of fracture mechanics: Ph.D. dissertation, State Univ. of New York at Stony Brook.

Li, Y., and Oldenburg, D. W., 1992, Approximate inverse mappings in DC resistivity problems: Geophys. J. Internat., **109**, 343–362.

Li, J., and Srivastav, R. P., 1997, Computing the singular behavior of solutions of Cauchy singular integral equations with variable coefficients: Appl. Math. Lett., **10**, 3, 57–62.

Li, J., Longji, T., and Yafei, O., 1987, A numerical algorithm for solving inverse problems of the two-dimensional elastic wave equation: Num. Comput. Computer Appl. (Chinese), **8**, 35–43.

Li, J., Lee, K. H., Javandel, I., and Xie, G., 1995, Nonlinear three-dimensional resistivity inverse imaging for direct current data: 65th Ann. Internat. Mtg., Soc. Expl. Geophys., Expanded Abstracts, 250–253.

Mezua, E., Hormaza, M. V., Hernandez, A., and Ajuria, M., 1995, A method for the improvement of 3D solid finite-element meshes: Advances Eng. Software, **22**, 45–53.

Park, S. K., and Van, G. P., 1991, Inversion of pole-pole data for 3-D resistivity structure beneath arrays of electrodes: Geophysics, **56**, 951–960.

Petrick., W. R., Jr., Sill, W. R., and Ward, S. H., 1981, Three-dimensional resistivity inversion using alpha centers: Geophysics, **46**, 1148–1163.

Rijo, L., 1984, Inversion of three-dimensional resistivity and induced-polarization data: 54th Ann. Internat. Mtg., Soc. Expl. Geophys., Expanded Abstracts, 113–117.

Sasaki, Y., 1994, 3-D resistivity inversion using the finite-element method: Geophysics, **59**, 1839–1848.

Shima, H., 1992, 2-D and 3-D resistivity image reconstruction using crosshole data: Geophysics, **57**, 1270–1281.

Xie, G., and Li, J., 1975, The 3-D finite element method in the elastic structure: J. Math. Practice and Recognition, **3**.

———1981, A fast convergent finite element method for computation of nonlinear magnetic induction and corresponding generalizations: Advances Computer Method for PDE, **4**; IMACS, 48–52.

———1988, New iterative method for solving inverse scattering problem of 3-D wave equation: Sci. Sin. A, **31**, 1195–1202.

———1989, Nonlinear integral equation of coefficient inversion of acoustic wave equation and TCR iteration: Sci. China, Ser. A, Math. Phys. Astron. Technol. Sci., **32**, 513–523.

Xie, G., Chen, Y. M., and Li, J., 1987, Gauss-Newton-regularizing method for solving coefficient inverse problem of partial differential equation and its convergence: J. Comput. Math., 5, 38–49.

Xie, G., Lee, K. H., and Li, J., 1995, A new parallel 3-D numerical modeling of the electromagnetic field: 65th Ann. Internat. Mtg., Soc. Expl. Geophys., Expanded Abstracts, 821–824.

3-D Modeling of Resistivity Devices

T. Tamarchenko[1]
M. Frenkel[2]
A. Mezzatesta[2]

Summary. Resistivity logging plays a fundamental role in distinguishing between oil-saturated and water-saturated rocks. Accurate modeling of resistivity devices with complex electrode configurations in complicated borehole environments must take into account 3-D geometries. We present an efficient finite-difference solution to the dc modeling problem that can simulate resistivity devices in 3-D media. The algorithm uses a conjugate gradient solver with preconditioning. Examples indicate that the code is very accurate in the presence of very large contrasts in conductivity (up to 10^{11}). Convergence of our iterative solver is excellent, but it is a function of conductivity contrast in the model. We present some applications of practical importance, including the response of a microresistivity tool to a tilted layer.

1 Introduction

Resistivity logging is an important formation measurement that allows for distinguishing between oil- and water-saturated rocks. Accurate modeling of resistivity instrument responses in deviated and horizontal boreholes requires full 3-D capabilities from the modeling software. Even for straight boreholes, 2-D modeling might not be sufficient. This is true for microresistivity instruments that have electrodes of different, rather complicated shapes, usually positioned on a pad. Simulation of through-casing resistivity response to the inhomogeneities in the casing involves 3-D modeling as well.

In mathematical terms, the resistivity modeling problem is a boundary-value problem for a partial differential operator. To solve this problem, one can use integral equations (Hohmann, 1975; Weidelt, 1975; Wannamaker et al., 1984), finite differences (FD) (Dey and Morrison, 1979; Weidelt, 1994), or finite elements (Livelybrooks, 1993). The recent advances in the area of 3-D electromagnetic (EM) modeling are directed toward increasing the efficiency of the numerical techniques (Habashy et al., 1993; Bakhmutsky, 1994; Singer, 1995). After the work of Druskin and Knizhnerman (1992), who proved the stability of the Lanczos solution in computer arithmetic, and the estimation of the

[1]ENRON, 1400 Smith St., Houston, TX 70002, USA.
[2]Western Atlas Logging Services, Houston, Texas, USA.

speed of convergence for operator function calculations, conjugate gradient (CG) type methods have been used (Spitzer, 1993; Wurmstich and Spitzer, 1994.

In the papers referenced above, the EM field is calculated for point or line sources. To accurately simulate the response of laterolog-type devices or microresistivity tools, the geometry of volume electrodes needs to be taken into account. Barminsky et al. (1991) presented the solution for volume electrodes in a cylindrically layered formation, based on the analytical expression for Green's function. Kosenkov and Tamarchenko (1992) derived the 2.5-D solution for volume electrodes considering vertically homogeneous formations. A 3-D FD solution using a multigrid approach, described by Brandt (1977), was implemented by F. Bostick (personal communication, 1993). Our paper presents an efficient FD solution based on the CG technique and allowing for the simulation of resistivity instruments in 3-D media.

2 Simulation of microresistivity tools in 3-D geometry

2.1 Boundary-value problem

Consider a resistivity logging instrument modeled as an insulating cylinder with metal electrodes on its surface, each of which is represented by a number of rectangles. We consider a cylindrical coordinate system (ρ, φ, z) in which the z-axis coincides with the axis of the mandrel. Outside the mandrel, the potential $u(\rho, \varphi, z)$ satisfies the following partial differential equation:

$$\frac{\partial}{\partial \rho}\rho\sigma\frac{\partial u}{\partial \rho} + \frac{\partial}{\partial \varphi}\frac{\sigma}{\rho}\frac{\partial u}{\partial \varphi} + \frac{\partial}{\partial z}\rho\sigma\frac{\partial u}{\partial z} = 0, \tag{1}$$

where σ represents the formation conductivity. The electronics of the tool are described by the so-called system equations, relating the electrode potentials to the normal currents flowing from its surface. Generally, the system equations can be written in the form of the matrix equation,

$$\underset{\sim}{\mathbf{A}}\mathbf{V} + \underset{\sim}{\mathbf{B}}\mathbf{I} = \mathbf{F}, \tag{2}$$

where $\mathbf{V} = (u_1, u_2, \ldots, u_M)^T$ and $\mathbf{I} = (i_1, i_2, \ldots, i_M)^T$ are the vectors of potentials on the electrodes and of total currents flowing through the surface of each electrode, respectively; M is the number of electrodes in the tool, and $\underset{\sim}{\mathbf{A}}$ and $\underset{\sim}{\mathbf{B}}$ are given matrices defining the tool's electronics together with vector $\mathbf{F}$. On the insulating part of the mandrel, the current normal to the surface of the mandrel is equal to zero, i.e.,

$$\left.\frac{\partial u(x)}{\partial \rho}\right|_{x \in S_{\text{insulator}}} = 0. \tag{3}$$

The solution vanishes at infinity:

$$u(x) \to 0, \quad \text{when } |x| \to \infty. \tag{4}$$

Equations (1) through (4) define the solution uniquely, but Eq. (2) needs to be reformulated to obtain a boundary-value problem. There is a well-known way to use the admittance matrix to obtain the currents and potentials for the given system Eq. (2). Let us consider the particular case when $\underset{\sim}{\mathbf{B}}$ is a null matrix, $\underset{\sim}{\mathbf{A}}$ is a unitary matrix, and $\mathbf{F}$ is a basis vector $\mathbf{F}_i = (0, 0, \ldots, 0, 1, 0, \ldots, 0)^T$ that has all zero components except for the component number i. Let us denote $\mathbf{V}_i$ and $\mathbf{I}_i$ as the solutions of Eqs. (1)–(4). If we solve M problems like this for $i = 1, \ldots, M$, we obtain the matrix $\underset{\sim}{\mathbf{A}} = (\mathbf{I}_1, \mathbf{I}_2, \ldots, \mathbf{I}_M)$

of the currents that correspond to the unitary matrix $\underset{\sim}{\mathbf{V}}$ of the potentials. This matrix $\underset{\sim}{\mathbf{A}}$ is called the admittance matrix. It follows from the linearity of the problem that for any vector of potentials and the corresponding vector of currents, the following equation is satisfied: $\underset{\sim}{\mathbf{I}} = \underset{\sim}{\mathbf{A}}\underset{\sim}{\mathbf{V}}$.

Our objective is to calculate matrix $\underset{\sim}{\mathbf{A}}$, which means solving the boundary-value problem defined by Eqs. (1), (3), (4), and Dirichlet boundary conditions on the surface of the electrodes.

2.2 FD approximation

We bound the modeling region with a cylindrical surface on which we apply zero Dirichlet conditions. Grids are defined in the radial, azimuthal, and vertical directions. At each grid node, the following linear equation is written, relating a maximum of seven unknown potentials:

$$\begin{aligned}
&-a_{i,j,k}(u_{i+1,j,k} - u_{i,j,k}) + a_{i-1,j,k}(u_{i,j,k} - u_{i-1,j,k}) \\
&-b_{i,j,k}(u_{i,j+1,k} - u_{i,j,k}) + b_{i,j-1,k}(u_{i,j,k} - u_{i,j-1,k}) \\
&-c_{i,j,k}(u_{i,j,k+1} - u_{i,j,k}) + c_{i,j,k-1}(u_{i,j,k} - u_{i,j,k-1}) = 0,
\end{aligned} \tag{5}$$

where $u_{i,j,k}$ is the value of the potential at the point (ρ_i, φ_j, z_k). These equations represent the Kirchoff's law—balance of currents. The values of the conductances $a_{i,j,k}$, $b_{i,j,k}$, $c_{i,j,k}$ in the radial, azimuthal, and vertical directions are calculated as volume integrals of the 3-D conductivity over the corresponding cell:

$$a_{i,j,k} = \int_{\delta z_k} \int_{\delta \varphi_j} \frac{d\varphi\, dz}{\int \frac{d\rho}{\rho\sigma}}, \tag{6}$$

$$b_{i,j,k} = \int_{\delta z_k} \int_{\delta \rho_i} \frac{dz\, d\rho}{\rho \int_{\Delta\varphi_j} \frac{d\varphi}{\sigma}}, \tag{7}$$

$$c_{i,j,k} = \int_{\delta \rho_i} \int_{\delta \varphi_j} \frac{\rho\, d\rho\, d\varphi}{\int_{\Delta z_k} \frac{dz}{\sigma}}, \tag{8}$$

where

$$\Delta\rho_i = [\rho_{i-1}; \rho_i], \quad \Delta\varphi_j = [\varphi_{j-1}; \varphi_j], \quad \Delta z_k = [z_{k-1}; z_k], \tag{9}$$

$$\delta\rho_i = \left[\frac{\rho_{i-1} + \rho_i}{2}; \frac{\rho_i + \rho_{i+1}}{2}\right]; \quad \delta\varphi_j = \left[\frac{\varphi_{j-1} + \varphi_j}{2}; \frac{\varphi_j + \varphi_{j+1}}{2}\right];$$

$$\delta z_k = \left[\frac{z_{k-1} + z_k}{2}; \frac{z_k + z_{k+1}}{2}\right]; \tag{10}$$

and ρ_i, φ_j, z_k are the nodes of radial, azimuthal, and vertical grids, respectively. These integrals are calculated analytically whenever possible; otherwise, a numerical technique is used.

FD equations look different for the grid nodes on the surface of the insulating part of the mandrel where there is no current flowing inside the mandrel. For each of these nodes, there are four neighboring nodes on the surface of the mandrel, each of which

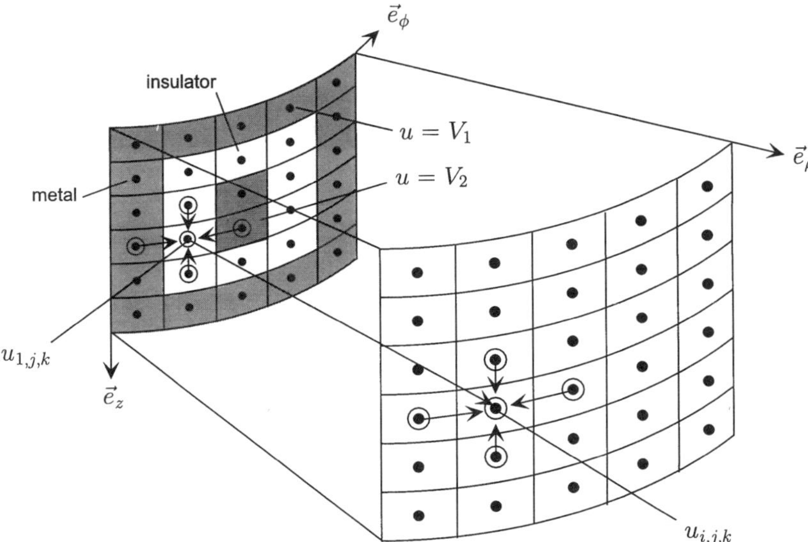

Figure 1. FD grid in cylindrical coordinate system (ρ, φ, z).

belongs either to the metal part of the mandrel (where the potential is known) or the insulating part (where the potential is unknown). The equations for the nodes that do not belong to the mandrel but have a neighbor on the metal part look different as well. For example, at the node $(1, j, k)$ shown in Fig. 1, the balance of currents is as follows:

$$\begin{aligned} -a_{1,j,k}(u_{2,j,k} - u_{1,j,k}) + b_{1,j,k}u_{1,j,k} - b_{1,j-1,k}u_{1,j,k} \\ -c_{1,j,k}(u_{1,j,k+1} - u_{1,j,k}) + c_{1,j,k-1}(u_{1,j,k} - u_{1,j,k-1}) = b_{1,j,k}V_1 - b_{1,j-1,k}V_2. \end{aligned} \tag{11}$$

2.3 *Solution to the linear system*

After the FD discretization is applied to the problem, a large sparse system of linear equations is obtained. The matrix of this system (matrix $\underset{\sim}{\mathbf{G}}$) is symmetric and positive definite. A sparse storage method is used to keep the nonzero elements of the matrix in the computer memory. To achieve sufficient accuracy in the FD approximation, a dense grid is required in the vicinity of the electrodes. At the same time, the modeling region may need to be relatively large. Additional complexity of the problem can be generated by high conductivity contrasts. As a result, the number of unknowns can reach a few hundred thousand. The condition number of the linear system matrix can reach hundreds of million. Here we compare Jacobi and incomplete Cholesky preconditionings. Jacobi preconditioning is simply scaling with the matrix $\underset{\sim}{\mathbf{G}}_d = \mathrm{diag}\{\underset{\sim}{\mathbf{G}}\}$, which makes the diagonal elements of the new matrix (matrix $\underset{\sim}{\mathbf{G}}^*$) equal to 1, i.e.,

$$\underset{\sim}{\mathbf{G}}^* = \underset{\sim}{\mathbf{G}}_d^{-1/2} \cdot \underset{\sim}{\mathbf{G}} \cdot \underset{\sim}{\mathbf{G}}_d^{-1/2}. \tag{12}$$

This preconditioning compresses the spectrum of the matrix without increasing the computational expenditures. Incomplete Cholesky decomposition is discussed in Meijerink and van der Vorst (1981). The idea is to use matrix $\underset{\sim}{\mathbf{P}} = (\underset{\sim}{\mathbf{L}} + \underset{\sim}{\mathbf{D}})\underset{\sim}{\mathbf{D}}^{-1}(\underset{\sim}{\mathbf{L}}^T + \underset{\sim}{\mathbf{D}})$ as a preconditioner, where $\underset{\sim}{\mathbf{L}}$ is a lower triangular matrix containing three diagonals of matrix $\underset{\sim}{\mathbf{G}}$ and $\underset{\sim}{\mathbf{D}}$ is a diagonal matrix chosen so that $\underset{\sim}{\mathbf{D}} = \mathrm{diag}(\underset{\sim}{\mathbf{G}}_d - \underset{\sim}{\mathbf{L}}\underset{\sim}{\mathbf{D}}^{-1}\underset{\sim}{\mathbf{L}}^T)$. Every CG iteration now involves calculation of $\mathbf{P}^{-1}$ multiplied by a vector. The implementation of this preconditioning requires $24 \times N$ multiplications and additions per CG iteration if N is the dimension of matrix $\underset{\sim}{\mathbf{G}}$. The number of operations per CG iteration without preconditioning is equal to $21 \times N$. With this slight increase of computational cost, the convergence rate increases significantly, which is illustrated by Fig. 2. The curves on Fig. 2 were calculated for a model including a borehole and a resistive layer with contrast of 20. Figure 3 shows the condition number of the preconditioned matrix as a function of resistivity contrast in the model. The IC preconditioning reduces the condition numer of the matrix 1000–2500 times, which means that IC preconditioning

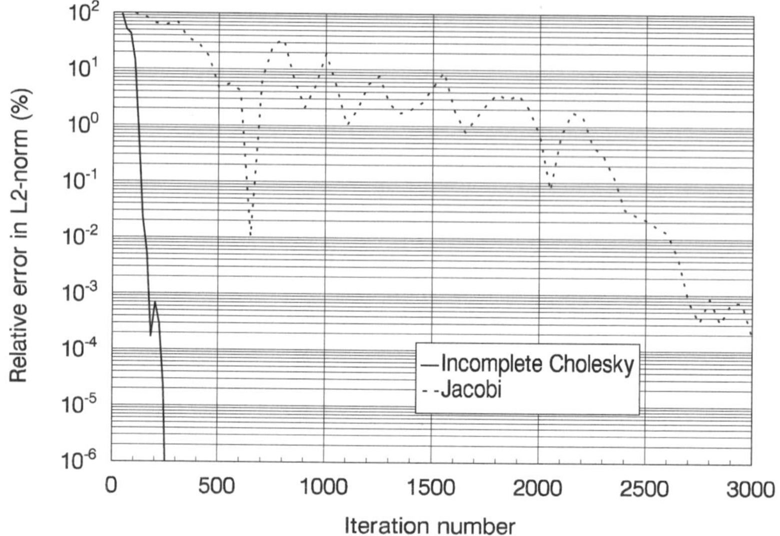

Figure 2. Convergence of iterative solutions obtained with Jacobi scaling and incomplete Cholesky preconditioning.

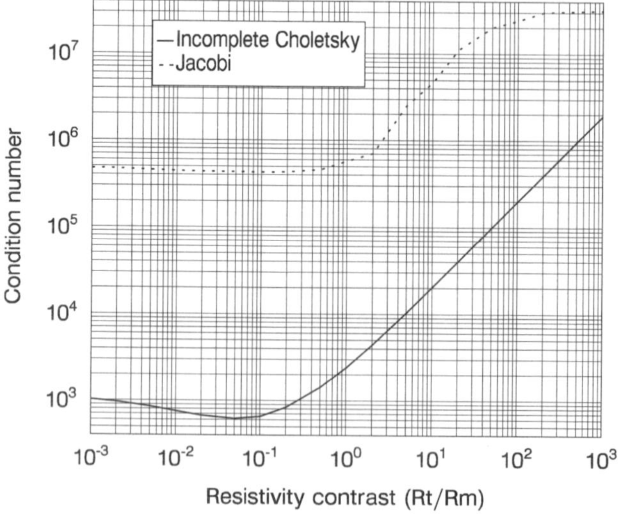

Figure 3. Convergence of iterative solutions vs resistivity contrast.

converges 10–50 times faster than Jacobi scaling. Typically, IC takes 100–600 iterations to converge with the accuracy of 10^{-10}.

2.4 *Code validation*

Several examples have been produced to validate the code.

2.4.1 Comparison against a 2-D code. The code was compared against a 2-D solution for the cases in which the model and the electrode configuration are azimuthally symmetric. Figure 4 shows the laterolog (LL7) response in a formation, including a borehole, a layer, and an invasion zone. The borehole radius is 0.1 m, mud resistivity is 1 ohm-m, shoulder resistivity is 0.5 ohm-m, invasion radius is 0.5 m, invasion resistivity is 5 ohm-m, and resistivity of the layer is 100 ohm-m. The tool configuration consisting of seven electrodes is shown in Fig. 1. The solution for horizontal bed (solid line) was compared against a 2-D modeling code based on a semianalytical hybrid method. The curves overlap completely.

2.4.2 Validation in the presence of high-resistivity contrasts. The next example demonstrates the ability of the code to produce accurate results for the models with high contrasts. Again the 2-D code mentioned in the preceding paragraph was used as a reference. The formation now consists of a 3-m horizontal layer of 10 000 ohm-m surrounded by 1 ohm-m formation (Fig. 5). The borehole of 0.09-m radius is cased. The conductance of the casing is 2×10^5 s, which corresponds to the resistivity of 0.5×10^{-7} ohm-m and a thickness of 1 cm. There is a cylindrical cement layer around the casing with resistivity of 0.1 ohm-m and thickness 0.1 m. A tool consists of one current electrode and three measuring electrodes at 0.75 m, 1 m, and 1.25 m from the current electrode. The apparent resistivity, defined as the ratio between the potential at the middle electrode and the second difference of the measured potentials normalized

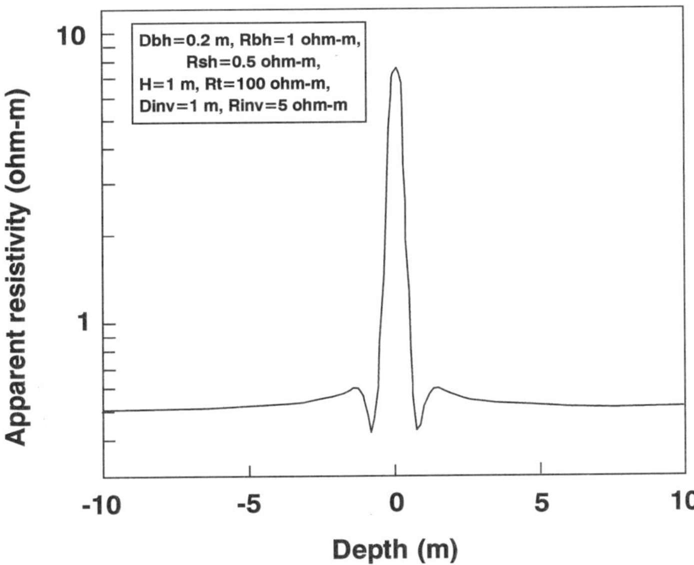

Figure 4. Comparison of 3-D and 2-D codes: LL7 response to a formation, including a borehole and a bed with invasion. Borehole diameter = 0.2 m, borehole resistivity = 1 ohm-m, invasion diameter = 1 m.

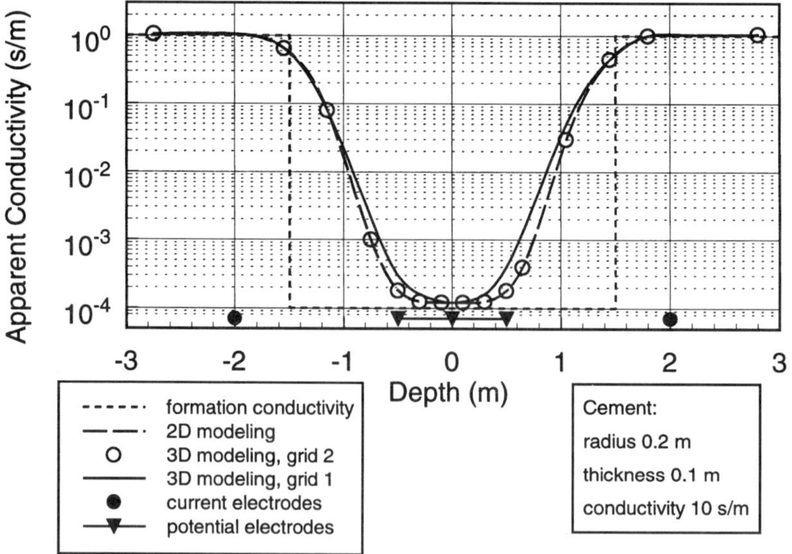

Figure 5. Comparison of 3-D and 2-D codes: Validation for high contrasts.

by the k-factor, was simulated. Grid 1 had 43 nodes in the radial direction extending up to 15 km. The vertical grid consisted of 300 nodes and extended to 15 km above and below the tool. The 4-m interval between two current electrodes was subdivided into 200 segments. Apparently, the results obtained with this grid differ significantly from the 2-D modeling results. Grid 2 uses 400 subdivisions of the 4-m interval between the current electrodes and produces accurate results. The number of iterations in this case was 350, the computational time 1 min per logging point on an IBM RISC/6000 processor. Note that that second differences calculated in this case are six orders of magnitude smaller than the potentials themselves. Nevertheless, the results presented in Fig. 5 do not exhibit any numerical artifacts.

2.4.3 Validation of ability of code to model complicated shapes of the electrodes. The following test was designed to validate the absence of programming errors in the code for complex electrode shapes. In this example, the behavior of the solution is expected to be caused by the symmetries in the electrode configurations. This type of behavior needs to be verified in the numerical solution. Figure 6 shows the unfolded surface of the mandrel and the corresponding grid. The tool consists of one large metal electrode (0.1×0.628 m) and 12 small sensor electrodes (buttons) of the same size (26.17×6.67 mm). The configuration has a variety of symmetries which means that, in the admittance matrix, certain elements are equal to each other in the homogeneous formation. Figure 7 presents 12×12 matrix $\underset{\sim}{\mathbf{G}} = \{g_{i,j}\} = \mathbf{A}^{-1}$ calculated for this configuration. As expected, there are equal elements in this matrix. The equal elements are denoted by the same letter. Such equalities are satisfied for any grid when the convergence of the CG iterative process is achieved.

2.5 Numerical examples

The new code can be used for various applications of practical importance. Some examples a presented in this section.

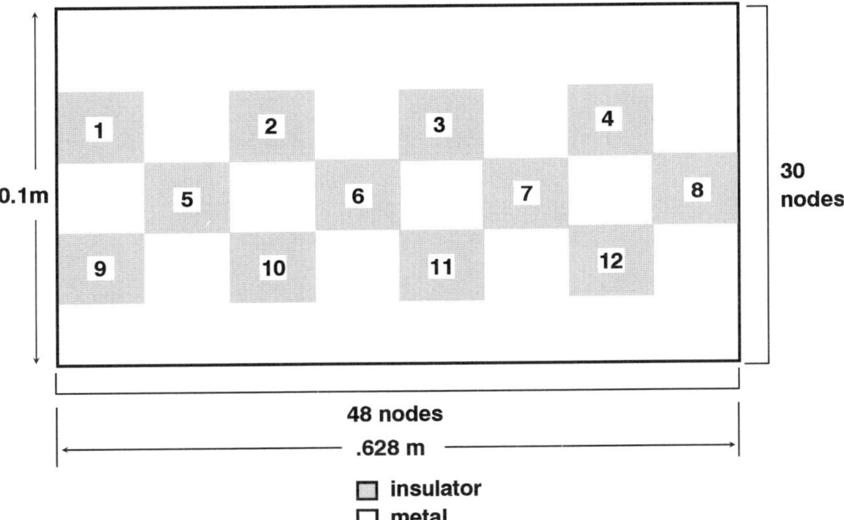

Figure 6. Tool configuration with azimuthally distributed electrodes.

a	b	c	b	d	e	e	d	f	g	c	g
b	a	b	c	d	d	e	e	g	f	g	c
c	b	a	b	e	d	d	e	c	g	f	g
b	c	b	a	e	e	d	d	g	c	g	f
d	d	e	e	a	h	i	h	d	d	e	e
e	d	d	e	h	a	h	i	e	d	d	e
e	e	d	b	i	h	a	h	e	e	d	d
d	e	e	d	h	i	h	a	d	e	e	d
f	g	c	g	d	e	e	d	a	b	c	b
g	f	g	c	d	d	e	e	b	a	b	c
c	g	f	g	e	d	d	e	c	b	a	b
g	c	g	f	e	e	d	d	b	c	b	a

Figure 7. Simulation of the resistivity tool with azimuthally distributed electrodes in a 1-ohm-m formation. a = 28.654 V, b = 0.34989 V, c = 0.34958 V, d = 0.37530 V, e = 0.35684 V, f = 0.60214 V, g = 0.34979 V, h = 0.36449 V, i = 0.36424 V.

2.5.1 Effect of offset on array of azimuthally distributed electrodes. The 0.1-m-radius tool with azimuthally distributed electrodes described in Section 2.4 and shown in Fig. 6 is positioned in an 0.2-m-radius borehole. The mud resistivity is 1 ohm-m, and the resistivity of the formation is 100 ohm-m. The apparent resistivity of each sensor electrode is calculated as the ratio of the current flowing from its surface in the 1 ohm-m homogeneous medium to the current in the inhomogeneous formation. The curves presented in Fig. 8 correspond to eight sensor electrodes; the distance between the center of the tool and the center of the borehole is offset. The discrepancies between the curves corresponding to electrodes 1 and 8, 2 and 7, 3 and 6, and 4 and 5 are caused by coupling between the electrodes; the curves for electrodes 9, 10, 11, and 12 overlap those for 1, 2, 3, and 4. This modeling shows the different sensitivity of the electrodes to the resistivity of the formation.

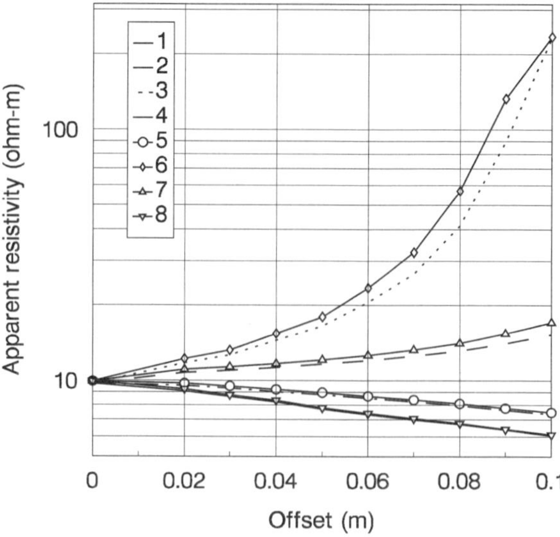

Figure 8. Response of tool with azimuthally distributed electrodes in a formation with the borehole: tool radius-0.1 m, borehole radius-0.2 m, mud resistivity-1 ohm-m, formation resistivity-100 ohm-m.

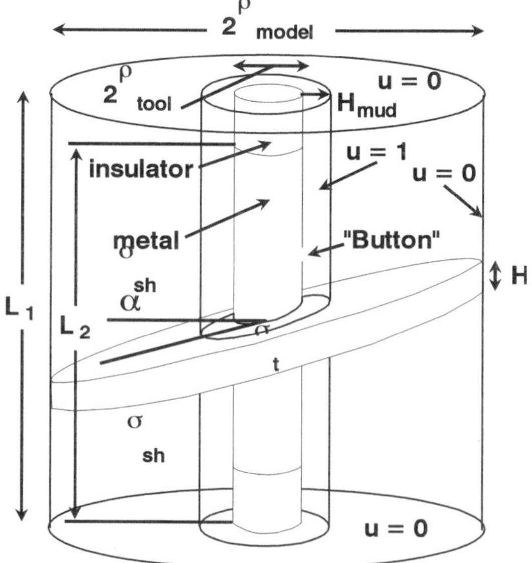

Figure 9. Tilted-bed geometry.

2.5.2 Microresistivity tool response to a thin tilted layer. Responses for a microresistivity device in a formation including borehole and a tilted layer, were calculated using the numerical approach described above. The tool (Fig. 9) consists of a 2-m-long metal electrode on an insulating mandrel. A small square section of this electrode, called a button, is used to measure the current. The button (4 × 4 mm) is positioned in the middle of the electrode. The radius of the mandrel is 0.1 m, the borehole radius is 0.101 m, and the mud resistivity is 1 ohm-m. A 10-mm-thick bed intersects the borehole at different angles. The conductivity of the bed is equal to 0.01 S/m, and the shoulder conductivity is 1 S/m. Figure 10 shows the apparent conductivity profile

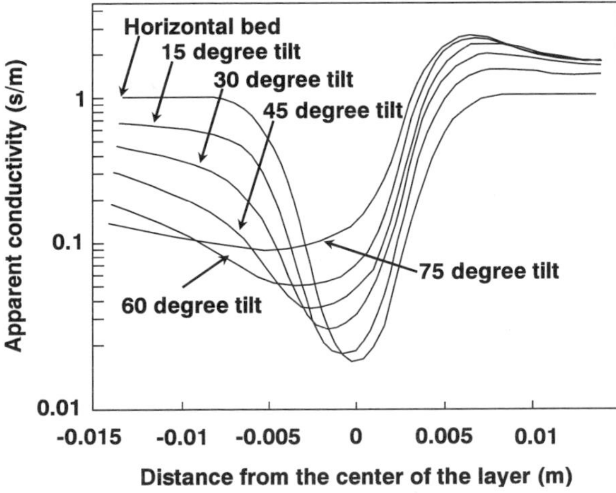

Figure 10. Response of small sensor electrode to a tilted resistive layer.

as the tool moves along the borehole axis. The apparent conductivity is calculated by taking the ratio of the current from the button in the formation to the current of the button in the homogeneous formation of 1 S/m. It took 11 min to calculate each profile on an IBM RS/6000 processor, model 3BT. The number of unknowns were 125 000.

3 Conclusions

An efficient and accurate numerical solution for resistivity well-logging tools has been developed and implemented in a computer program. The solution allows for simulating the behavior of real instruments with complex electrode geometries in a 3-D Earth.

Acknowledgments

The authors are grateful to Western Atlas Logging Services for permission to publish the work. We would like to thank our colleagues, Dr. B. Singer and Dr. Q. Zhou, for reviewing the paper, as well as the editors.

References

Bakhmutsky, M., 1994, Advanced method of numerical modeling in 3D dc earth sounding problem: 12th Workshop on Electromagnetic Induction in the Earth, poster P. 7a-13 sponsored by International Union of Geodesy and Geophysics.

Barminsky, A. G., Kucherov, R. A., Tanzurenko, N. I., and Tolmachev, Y. Y., 1991, Mathematical and experimental modeling of microlaterolog devices: The Log Analyst, **8**, 639–644.

Brandt, A., 1977, Multi-level adaptive solutions to boundary-value problems: Math. Comput., **31**, 333–390.

Dey, A., and Morrison, H. F., 1979, Resistivity modeling for arbitrary shaped three-dimensional structures: Geophysics, **44**, 753–780.

Druskin, V., and Knizhnerman, L., 1992, Using operator series of orthogonal polynomials calculating functions of self-adjoint operator: CAM, **42**, 221–231.

Habashy, T., Groom, R., and Spies, B., 1993, Beyond the Born and Rytov approximation: A nonlinear approach to electromagnetic scattering: J. Geophys. Res., **98**, 1759–1993.

Hohmann, G. W., 1975, Three-dimensional induced polarization and electromagnetic modeling: Geophysics, **40**, 309–324.

Kosenkov, O. M., and Tamarchenko, T. V., 1992, Mathematical modeling of electromagnetic logging sondes with volume electrodes in two- and three-dimensional geometry: Geol. Geophys., **3**, 128–136.

Livelybrooks, D., 1993, Program 3Dfeem: A multi-dimensional electromagnetic finite element model: Geophys. J. Internat., **114**, 443–458.

Meijerink, J. A., and van der Vorst, H. A., 1981, Guidlines for the usage of incomplete decomposition in solving sets of linear equations as they occur in practical problems: J. Comput. Phys., **44**, 134–155.

Singer, B. S., 1995, Methods for solution of Maxwell's equations in nonuniform media: Geophys. J. Internat., **120**, 590–598.

Spitzer, K., 1993, Dreidimensionale geoelektrisce Modellrechnugen nach der Methode der Finiten Differenzen: Bericht 111551, Niedersachsisches Landesamt fur Bodenforchung-Geowissenschaftliche Gemeinchaftsaufgaben, Ph. Thesis.

Wannamaker, P. E., Hohmann, G. W., and San Filipo, W. A., 1984, Electromagnetic modeling of three-dimensional bodies in layered earths using integral equations: Geophysics, **49**, 60–74.

Weidelt, P., 1975, Electromagnetic induction in three-dimensional structures: J. Geophys., **41**, 85–109.

———1994, Finite-difference modeling of 3-D structures with arbitrary anisotropic conductivity: 12th Workshop on Electromagnetic Induction in the Earth, poster P. 7A-19, sponsored by International Union of Geodesy and Geophysics.

Wurmstich, B., and Spitzer, K., 1994, Comparison of two finite-difference approaches and five solution algorithms for 3-D resistivity modeling: 64th Ann. Internat. Mtg., SEG Expanded abstracts, 381–383.

Modeling Induction Logs in 3-D Geometries

M. van der Horst[1]
V. Druskin[2]
L. Knizhnerman[3]

Summary. The spectral Lanczos decomposition method gives a very efficient finite-difference solution of Maxwell's equations in the diffusive limit. We have adapted this method to simulate borehole induction logging with an array tool used in oil and gas exploration. The tool has a magnetic point dipole transmitter operating at several frequencies and receiver dipoles located at different distances from the transmitter. Numerical examples show that effects of the borehole, invasion, and eccentricity of the tool in the hole are important to the proper interpretation of real logs.

1 Introduction

Induction logging (Doll, 1949; Moran and Kunz, 1962) is one of the standard methods for discriminating between hydrocarbon-bearing and water-bearing (or shale) zones in wells drilled for oil and gas. An induction tool uses a time-varying magnetic-dipole source (a small coil) to induce eddy currents in the surrounding rocks and an array of magnetic-dipole (coil) receivers to record the total magnetic field (of the source and eddy currents). The measured response is converted into an estimate of the formation conductivity. Frequently, a qualitative interpretation of the log can indicate the location of resistive hydrocarbon-bearing zones and conductive water-bearing or shale zones. A quantitative interpretation can provide a more accurate determination of the hydrocarbon saturation and other reservoir characteristics.

One of the complications in quantitative logging is that, during and after drilling, the borehole fluid (drilling mud) can displace the original formation fluids. This process, called invasion, creates an altered zone near the borehole, whose response along with that of the borehole itself can mask the response of the virgin formation. The total response depends very much on the conductivities, volumes, and shapes of the different zones. Modeling of complicated 3-D formation configurations is one way to study the sensitivity of induction tools to these environmental effects. Ultimately, we aim at incorporating 3-D effects in the inversion of field data.

[1]RTS, Shell Research, Volmerlaan 6, Rijswijk 2280 AB, The Netherlands.
[2]Schlumberger-Doll Research, Ridgefield, CT 06877-4108, USA.
[3]Central Geophysical Expedition, Narodnogo Opolcheniya Street, House 40, Moscow 123298, Russia.

2 Modeling induction logs

Modern induction logging tools carry arrays of transducers consisting of a single transmitter operating at several frequencies ranging from approximately 10 to 100 kHz and multiple receivers at distances from approximately 10 cm to 3 m from the transmitter. As a first approximation, it is common to assume that the formation is layered horizontally and penetrated by the borehole at a particular angle (today's oil and gas wells are drilled at all angles from vertical to horizontal). Well-known and fast numerical schemes are available to model this configuration when the effects of the borehole and invaded zones are neglected (Hardman and Shen, 1986; Kennedy et al., 1986). Fast and accurate numerical schemes are also available to compute the response in a formation including the borehole and invasion zones when the logging direction is perpendicular to the formation layering (Chew et al., 1984; Tamarchenko and Druskin, 1993). A full 3-D electromagnetic (EM) solution is needed to compute the response in a formation including layering, borehole, and invasion when the induction tool is oriented in an arbitrary direction.

The spectral Lanczos decomposition method (SLDM) (Druskin and Knizhnerman, 1994) allows a very efficient 3-D finite-difference solution of Maxwell's equations in the diffusive limit, and has been tested extensively for modeling surface EM surveys (Hördt et al., 1992). We have adapted this method to modeling induction logs. For the source we use a magnetic point dipole $\mathbf{M}$, which is a good approximation to the small transmitter coil used in induction tools. Maxwell's equations in the diffusive regime (neglecting the displacement current) then are

$$\nabla \times \mathbf{H} = \sigma \mathbf{E} \tag{1}$$

$$\nabla \times \mathbf{E} = \mu \frac{\partial}{\partial t} \mathbf{H} + \mu \mathbf{M}. \tag{2}$$

SLDM approximates the spatial derivatives in Maxwell's equations with finite differences on a staggered (Yee-Lebedev) grid (Druskin and Knizhnerman, 1994). The result is a matrix ordinary differential equation in time. The solution of this matrix system is written as the product of a function of the system matrix times the vector of initial conditions. At fixed frequency ω, this matrix functional is the standard matrix resolvent giving the solution for the harmonic amplitude of the magnetic-field strength. The matrix equation in the frequency domain is

$$\underset{\sim}{\mathbf{A}}\mathbf{H} - i\omega \mathbf{H} = \mathbf{M}, \tag{3}$$

where

$$\underset{\sim}{\mathbf{A}}\mathbf{H} = \frac{1}{\mu} \hat{\nabla} \times \left(\frac{1}{\sigma} \dot{\nabla} \times \mathbf{H} \right), \tag{4}$$

$\hat{\nabla}$ and $\dot{\nabla}$ refer to the finite-difference operators on the different (staggered) grids in the Yee-Lebedev scheme.

In the SLDM, the formal solution

$$\mathbf{H} = (\underset{\sim}{\mathbf{A}} - i\omega \underset{\sim}{\mathbf{I}})^{-1} \mathbf{M} \tag{5}$$

is approximated numerically by applying the Lanczos method to generate the eigenvectors of $\underset{\sim}{\mathbf{A}}$. The basis vectors $\{\mathbf{q}_1, \mathbf{q}_2, \ldots, \mathbf{q}_m\}$ of the Krylov subspace K^m after m

iterations,

$$K^m = \text{span}\{\mathbf{M}, \underset{\sim}{\mathbf{A}}\mathbf{M}, \ldots, \underset{\sim}{\mathbf{A}}^{m-1}\mathbf{M}\} \tag{6}$$

are computed by Gram-Schmidt orthogonalization of the (real and frequency-independent) vectors $\{\mathbf{M}, \underset{\sim}{\mathbf{A}}\mathbf{M}, \ldots, \underset{\sim}{\mathbf{A}}^{m-1}\mathbf{M}\}$. The orthogonalization is carried out by a three-term recursion whose coefficients form the tridiagonal matrix $\underset{\sim}{\mathbf{T}}$. The orthogonal basis vectors $\mathbf{q}_i$ form the matrix $\mathbf{Q}$. The SLDM approximation of the solution (5) thus is

$$\mathbf{H} = \|\mathbf{M}\|\mathbf{Q}(\underset{\sim}{\mathbf{T}} - i\omega\underset{\sim}{\mathbf{I}})^{-1}\mathbf{e}_1 \tag{7}$$

with $\mathbf{e}_1 = (1, 0, \ldots, 0)^T$.

Most of the computational effort resides in computing the matrices $\mathbf{Q}$ and $\underset{\sim}{\mathbf{T}}$; only the computation of $(\underset{\sim}{\mathbf{T}} - i\omega\underset{\sim}{\mathbf{I}})^{-1}$ requires an actual value for the frequency. But this expression can be solved directly, so that solutions for additional frequencies are obtained without much additional cost.

3 Numerical examples

Figure 1 shows the formation model that we use. The formation itself is horizontally layered; there is a cylindrical borehole with a single, piston-like invaded zone around the borehole in some layers (we assume that horizontal interfaces prevent fluids from flowing from one layer to another). In principle, the SLDM allows far more complicated

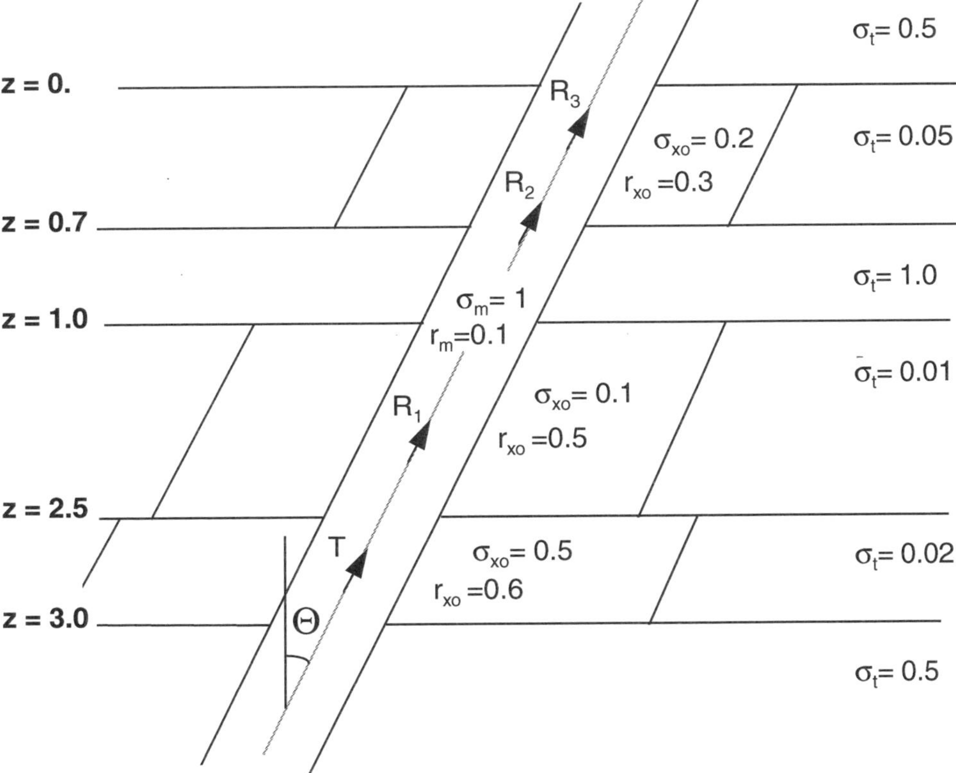

Figure 1. Formation model showing borehole conductivity σ_m (S/m), borehole radius r_m (m), invasion conductivity σ_{xo} (S/m), invasion radius r_{xo} (m), true formation conductivity σ_t (S/m), location of bed boundaries z (m), and dip angle θ (degrees). All radii are with respect to the borehole axis.

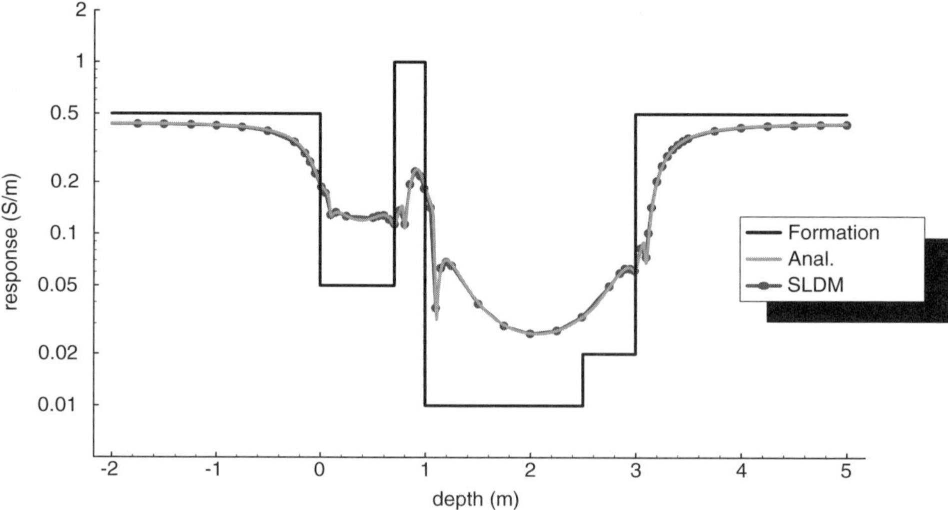

Figure 2. Response (apparent conductivity) of the AIT subarray A3H (spacing, ca. 0.3 m; frequency, ca. 100 kHz) in a highly deviated well (70°) in the formation of Fig. 1 but without borehole and invasion. The response has been calculated with a 1-D analytical method (Anal.) and the SLDM. The depth indicated is the true vertical position of the transmitter.

geometries (including gravity segregration, layers with a finite lateral extension, water fingering) but here we keep it simple so as not to obscure the more relevant ingredients of the method.

We use a model of the Array Induction Tool [AIT*; see Barber and Rosthal (1991)], which has one transmitter and several receiver subarrays. The transmitter operates sequentially at several different frequencies for every logging position. A single receiver subarray consists of two receivers, whose responses are combined in order to cancel the direct or vacuum term. The magnetic dipole moments of the receivers and transmitter are always pointed in the logging direction which is specified by the dip angle θ denoting the relative orientation of the borehole with respect to the horizontal layers. We concentrate in the examples on the imaginary component of the measured magnetic field which is multiplied by a constant to give the apparent conductivity (see Doll, 1949; Moran and Kunz, 1962).

To compute the response with the SLDM, the formation is discretized (using an interface program with the formation parameters of Fig. 1 as input) into small rectangular blocks with sides of a few centimeters. We first tested the SLDM on the 1-D formation model obtained by omitting the borehole and invasion zones in Fig. 1. The dip angle, θ, is 70°, corresponding to a highly deviated well. We consider the AIT subarray A3H with receiver/transmitter spacings of approximately 0.3 m and the transmitter operating at a frequency of approximately 100 kHz. Figure 2 compares the logs derived from the SLDM with those from an accurate, analytical solution (Hardman and Shen, 1986). The logs show the apparent conductivity as obtained from the unprocessed or unfocused response of the subarray. The depth displayed is the true vertical depth of the transmitter with respect to the formation layering. For the computation with the SLDM, we used a spatial grid (x, y, z) of size $67 \times 34 \times 85$ (in the y-direction, mirror symmetry was used). Overall, the agreement between the two solutions is very good. Similar results

*Trademark of Schlumberger. For the modeling results presented in this chapter, we have used the actual tool specifications of the AIT. However, the dimensions (spacings, numbers, frequencies, etc.) mentioned in the text are only approximate.

are found for the other subarrays and frequencies, which are not shown. Computation of each logging point required about 30–90 CPU minutes on an IBM RS/6000 (590). A smaller grid can be used if a less accurate result suffices, reducing the computer time significantly.

The convergence of the response at a specific logging depth is illustrated in Fig. 3 as function of the iteration order m, viz. Eq. (6). In general, the convergence is best for the receivers close to the transmitter and for the highest operating frequency of the transmitter.

A second test of the SLDM is made by comparing the results in a vertical well ($\theta = 0$) with the results of a 2-D semianalytical/numerical (hybrid) computer algorithm. The results for the subarray A7H with a spacing of approximately 1 m and operating frequency of approximately 50 kHz are shown in Fig. 4. Again, a good agreement is

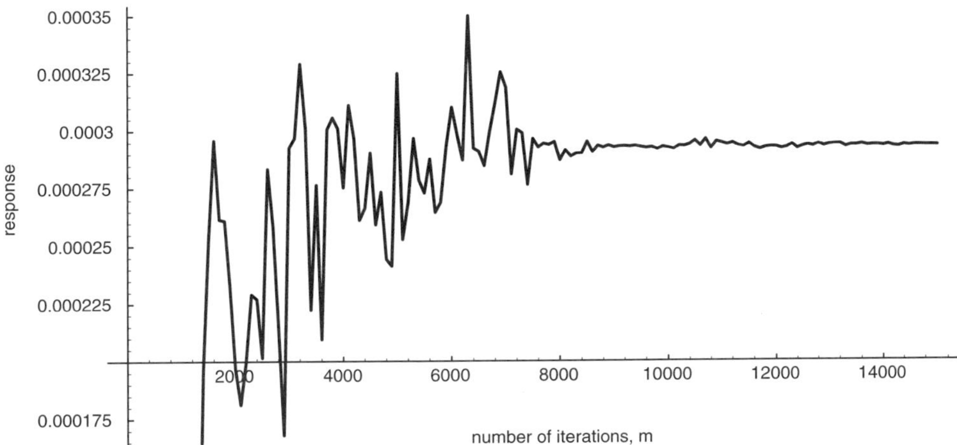

Figure 3. Example of convergence rate of computed response (arbitrary units) by the SLDM as a function of the number of iterations m. The result shown is the response of the AIT subarray A8L (spacing, ca. 2 m; frequency, ca. 25 kHz) modeled in the formation as used in Fig. 2.

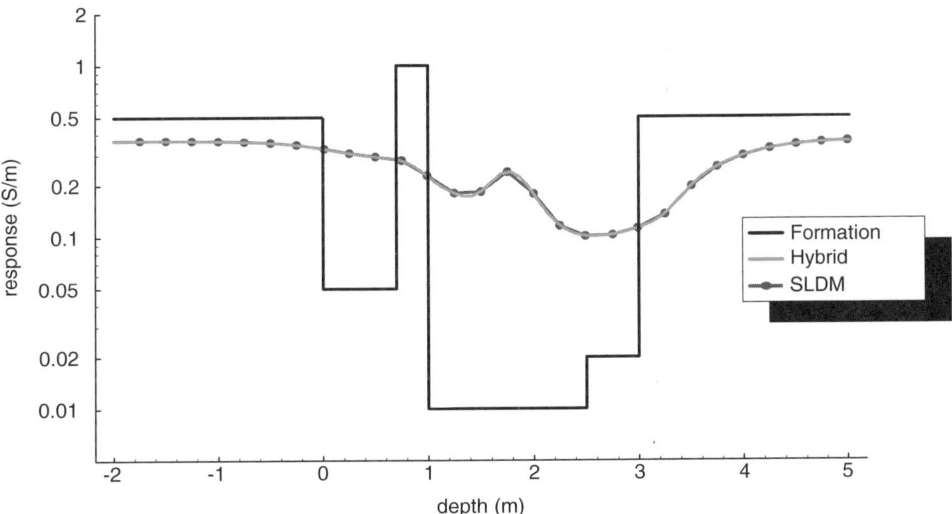

Figure 4. Response (apparent conductivity) of the AIT subarray A7H (spacing, ca. 1 m; frequency, ca. 50 kHz) in a vertical well (dip, 0°) in the formation from Fig. 1 (including borehole and invasion). The response has been calculated with a 2-D hybrid method and SLDM. The depth indicated is the true vertical position of the transmitter.

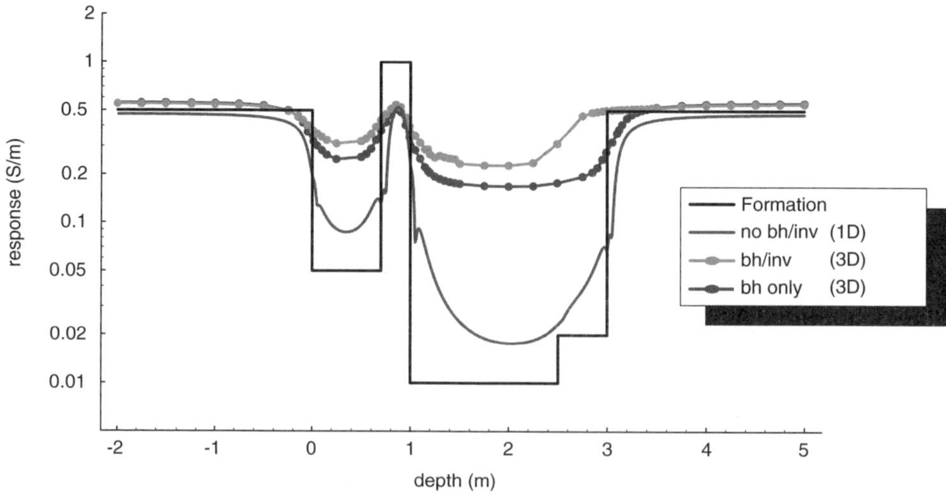

Figure 5. Response (apparent conductivity) of AIT subarray A1 (spacing, ca. 0.1 m; frequency, ca. 100 kHz) in a highly deviated well (70°) in the formation from Fig. 1 without borehole (bh) and invasion zone (inv), with both borehole and invasion zone, and with borehole only.

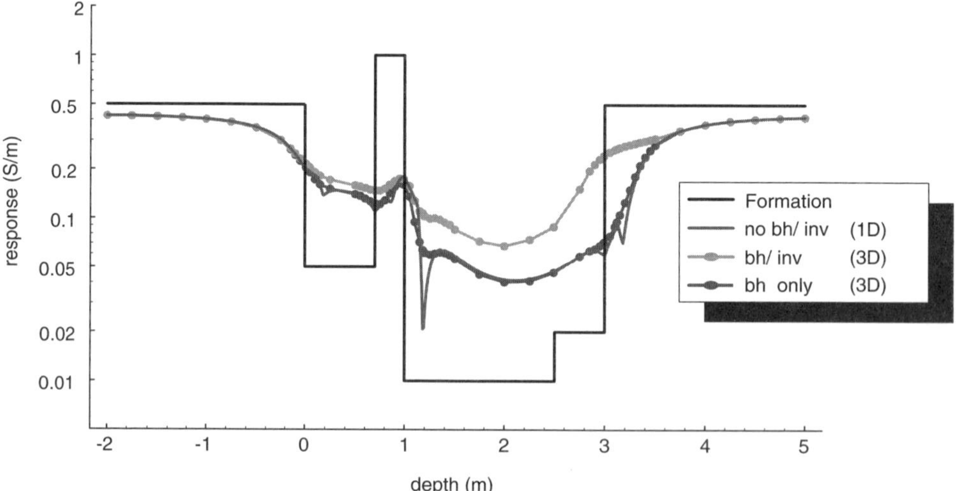

Figure 6. Response (apparent conductivity) of AIT subarray A5H (spacing, ca. 0.5 m; frequency, ca. 50 kHz) in a highly deviated well (70°) in the formation from Fig. 1.

found for both the subarray shown and all of the other subarrays and frequencies not shown.

These tests give us sufficient confidence that the SLDM can be used for computing the response of induction tools. Therefore, we next apply the method to solve some genuine 3-D problems. First, we repeat the modeling exercise of Fig. 2 for a highly deviated well with a dip angle of 70° but now including the borehole and invasion zones. The response as function of depth is shown in Figs. 5–7 for three AIT subarrays with different spacings and operating frequencies. To illustrate the effect of the borehole and invasion zones, the 1-D results (no borehole and invasion) and the effect of the borehole only (borehole included but no invasion) also are shown. As can be seen, the effect of borehole and invasion zones is significant for the relatively short subarray A1 (spacing, ca. 0.1 m; frequency, ca. 100 kHz) whereas the borehole/invasion effect is

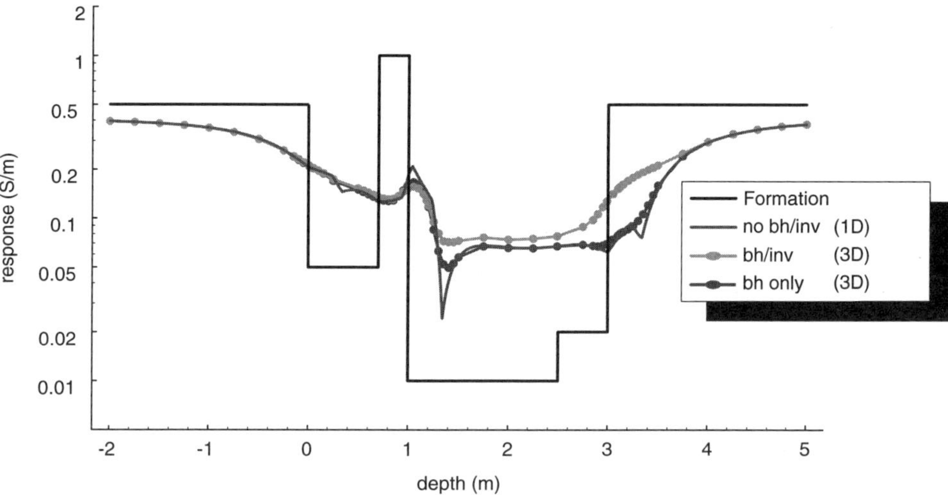

Figure 7. Response (apparent conductivity) of AIT subarray A7L (spacing, ca. 1 m; frequency, ca. 25 kHz) in a highly deviated well (70°) in the formation from Fig. 1.

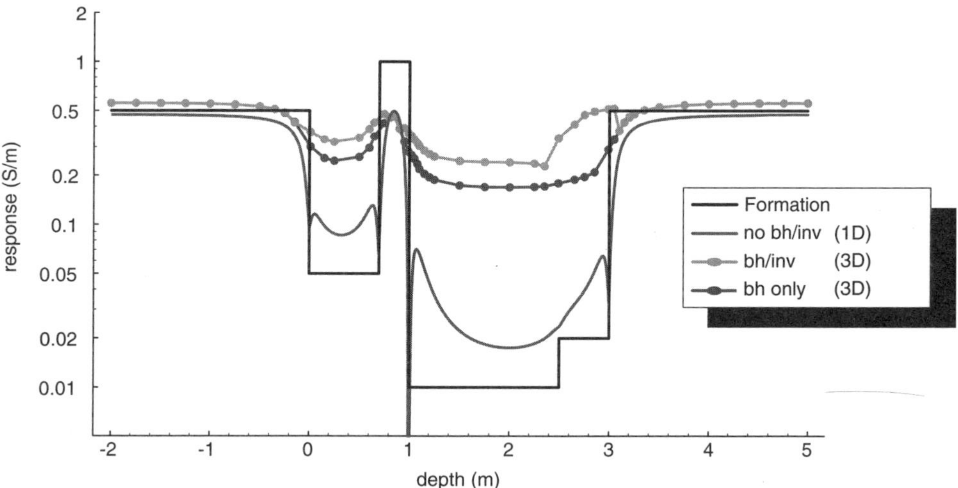

Figure 8. Response (apparent conductivity) of AIT subarray A1 (spacing, ca. 0.2 m; frequency, ca. 100 kHz) in a horizontal well (dip, 90°) in the formation from Fig. 1 without borehole (bh) and invasion zone (*inv*), with both borehole and invasion zone, and with borehole only.

smaller for the longer subarrays A5H and A7L. The longer subarrays show a reduced influence of the nearby regions in the formation. Furthermore, for the A1, the largest environmental effect is caused by the borehole and to a lesser extent by the invasion, whereas for the subarrays A5H and A7L the largest effect is caused by the invasion. For the longer subarrays the borehole only reduces the sharp polarization horns present in the 1-D response, which are caused by a charge buildup on the horizontal interfaces. In fact, the borehole shields the transmitter and receivers from a direct contact with the horizontal interfaces when crossing the interfaces but has only minimal effect farther away from the interfaces.

Today, many wells are drilled with large, almost horizontal sections to optimize hydrocarbon production in a relatively thin reservoir. In Figs. 8–10, the results are shown for the formation of Fig. 1 with a dip angle of 90° and for the same subarrays as

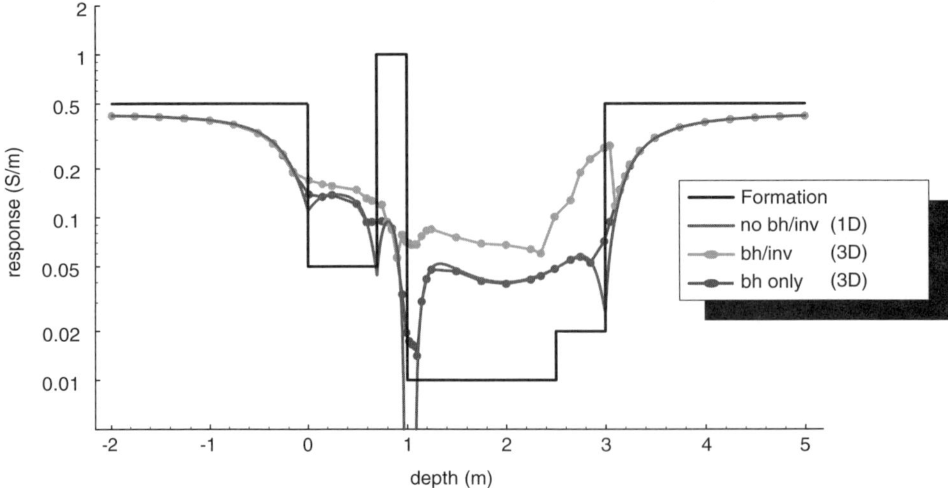

Figure 9. Response (apparent conductivity) of AIT subarray A5H (spacing, ca. 0.5 m; frequency, ca. 50 kHz) in a horizontal well (dip, 90°) in the formation from Fig. 1.

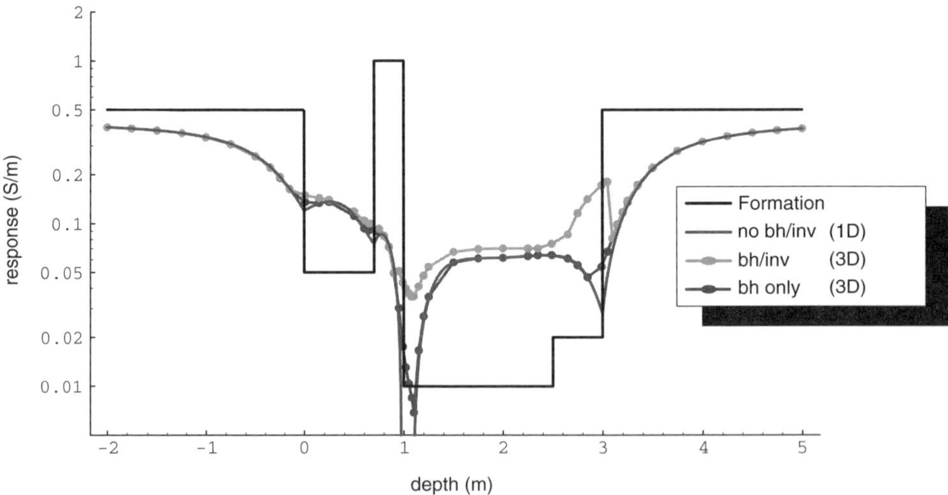

Figure 10. Response (apparent conductivity) of AIT subarray A7L (spacing, ca. 1 m; frequency, ca. 25 kHz) in a horizontal well (dip, 90°) in the formation from Fig. 1.

were used for Figs. 5–7. The logs shown are obtained as if the tool had been lowered horizontally in the formation surrounded by the borehole and invasion zone for the formation layer in which the tool is located. Comparing the result with the 1-D response, a large borehole and invasion effect is visible especially for the shorter subarrays.

In Figs. 11–13 the preceding results for a horizontal well are compared with the logs obtained for a formation that differs from the formation of Fig. 1 by a mud conductivity 10 times higher (10 S/m) and all invasion conductivities 5 times higher. Comparing the results with the results of Fig. 8–10, it is clear that in particular the short subarray A1 suffers from a larger borehole/invasion effect and is dominated by the borehole/invasion conductivities. For the longer subarrays, A5H and A7L, this effect is smaller. Comparing the results for the highly deviated well (Figs. 5–7) with the results for the horizontal well

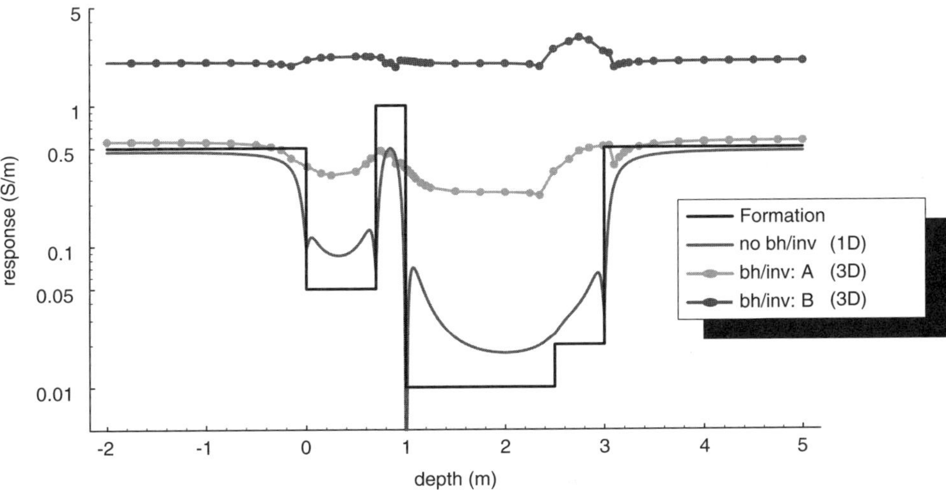

Figure 11. Response (apparent conductivity) of AIT subarray A1 (spacing, ca. 0.2 m; frequency, ca. 100 kHz) in a horizontal well (dip, 90°) in the formation from Fig. 1 with the borehole/invasion included (bh/inv: A) and without the borehole/invasion included and in the same formation but with a borehole conductivity 10 times higher and all invasion conductivities 5 times higher (bh/inv: B). The depth indicated is the true vertical position of the transmitter.

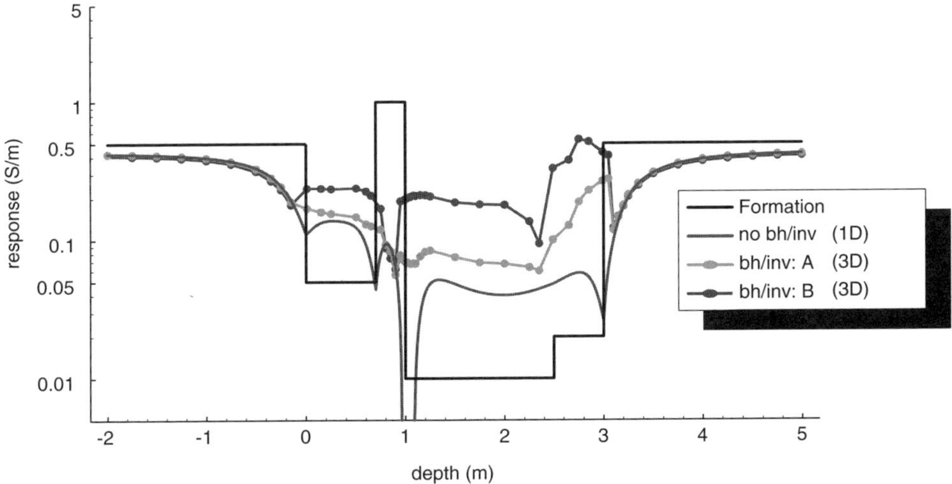

Figure 12. Response (apparent conductivity) of AIT subarray A5H (spacing, ca. 0.5 m; frequency, ca. 50 kHz) in a horizontal well (dip, 90°) in the formation from Fig. 1. See Fig. 11 for description of boreholes and invasion zones.

(Figs. 8–10), it can be seen that the latter generally shows more pronounced peaks. This is because of a larger charge buildup on the horizontal interfaces, which is proportional to the $\sin\theta$, when the coils cross a horizontal interface.

Our final results, in Figs. 14–16, show the effect of eccentricity of the tool in the borehole. In these runs, the tool is positioned 7 cm from the borehole axis, as if it were lying on the bottom of the borehole. Results are shown for the subarrays A1, A5H, and A7L. As expected, the longer subarrays are much less affected by the eccentricity than the short subarray A1.

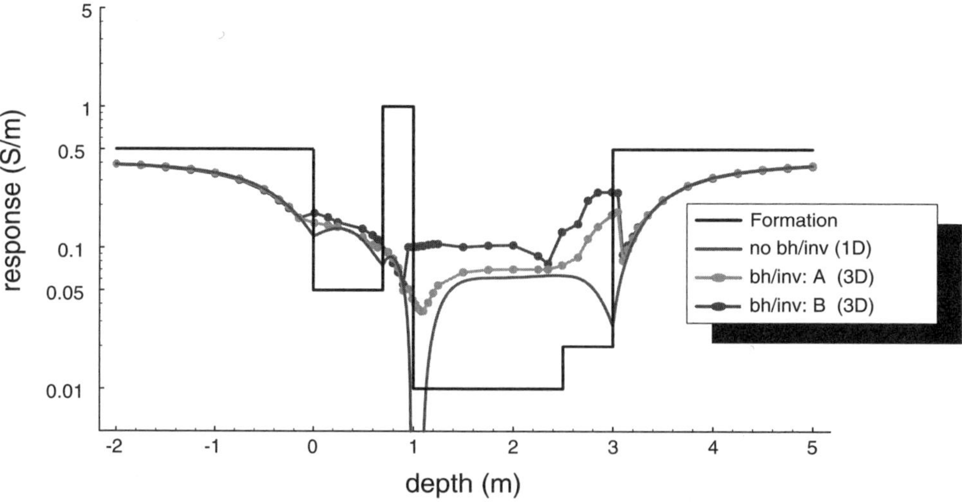

Figure 13. Response (apparent conductivity) of AIT subarray A7L (spacing, ca. 1 m; frequency, ca. 25 kHz) in a horizontal well (dip, 90°) in the formation from Fig. 1. See Fig. 11 for description of boreholes and invasion zones.

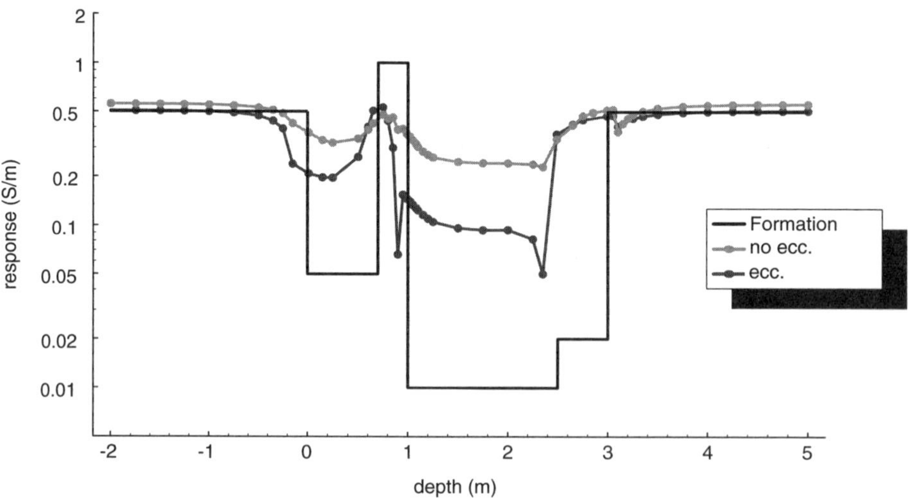

Figure 14. Response (apparent conductivity) of AIT subarray A1 (spacing, ca. 0.2 m; frequency, ca. 100 kHz) in a horizontal well (dip, 90°) in the formation from Fig. 1 without eccentricity [as in Fig. 11 with the tool located on the borehole axis, (*no ecc.*)] and with eccentricity [the tool having an offset of 7 cm down from the borehole axis (*ecc.*)]. The depth is the true vertical position of the transmitter.

4 Conclusions

We have developed 3-D finite-difference modeling of induction logging using the SLDM. The method was found to yield accurate results in tests on simpler (1-D and 2-D) formation geometries that can be modeled with programs that have been checked carefully. Although the method requires significant computer resources for fully 3-D models, it is very useful for studying the behavior of induction tools in complicated geometries. We have computed results including borehole, invasion, and eccentricity in highly deviated and horizontal wells. The high accuracy of the SLDM means that it can be used to benchmark faster approximate methods for 3-D modeling. These faster

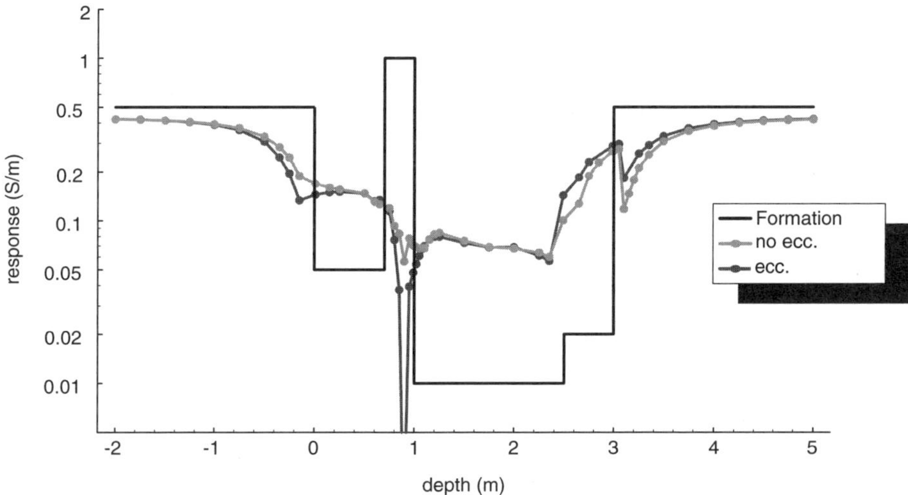

Figure 15. Response (apparent conductivity) of AIT subarray A5H (spacing, ca. 0.5 m; frequency, ca. 50 kHz) in a horizontal well (dip, 90°) in the formation from Fig. 1. See Fig. 14 for description of eccentricities.

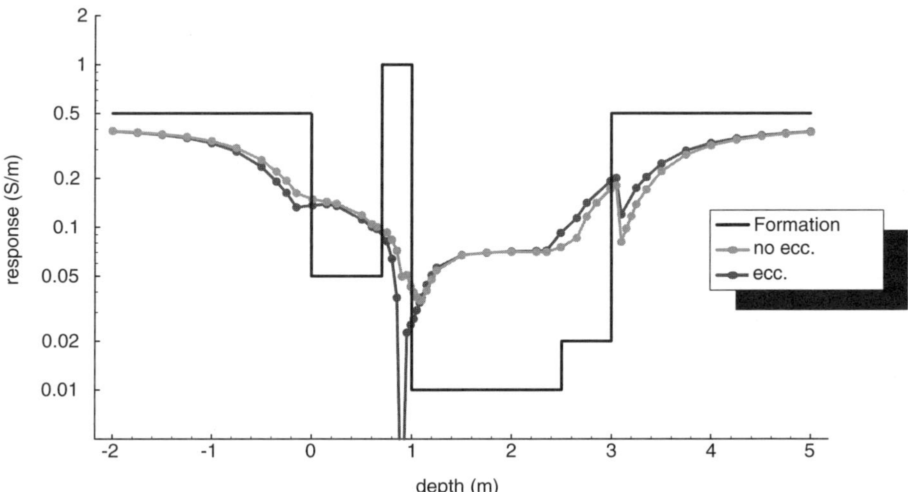

Figure 16. Response (apparent conductivity) of AIT subarray A7L (spacing, ca. 1 m; frequency, ca. 25 kHz) in a horizontal well (dip, 90°) in the formation from Fig. 1. See Fig. 14 for description of eccentricities.

methods can be used in an automated inversion scheme for field data. Also, a faster version of SLDM is being investigated (Druskin and Knizhnerman, 1995).

References

Barber, T. D., and Rosthal, R. A., 1991, Using a multiarray induction tool to achieve high-resolution logs with minimum environmental effects: 66th Ann. Technical Conference and Exhibition, Soc. Petr. Eng., 637–651.

Chew, W. C., Barone, S., Anderson, B., Hennessy, C., 1984, Diffraction of axisymmetric waves in a borehole by bed boundary discontinuities: Geophysics, **49**, 1586–1595.

Doll, H. G., 1949, Introduction to induction logging and application to logging of wells drilled with oil based mud: J. Petr. Tech., **1**, 148–162.

Druskin, V., and Knizhnerman, L., 1994, Spectral approach to solving three-dimensional Maxwell's diffusion equations in the time and frequency domain: Radio Sci., **29**, 937–953

———1995, Krylov subspace approximation of eigenpairs and matrix functions in exact and computer arithmetic: Numer. Linear Algebra Appl., **2**, 205–217.

Hardman, R. H., and Shen, L. C., 1986, Theory of induction sonde in dipping beds: Geophysics, **51**, 800–809.

Hördt, A., Druskin, V., Knizhnerman, L., and Strack, K.-M., 1992, Interpretation of 3-D effects in deep transient electromagnetic soundings in the Münsterland area (Germany): Geophysics, **57**, 1127–1137.

Kennedy, W. D., Curry, S. M., Gill, S. P., Morrison, H. F., 1986, Induction log response in deviated boreholes: Presented at 27th Ann. Logging Symposium, Soc. Prof. Well Log Analysts.

Moran, J. H., and Kunz, K. S., 1962, Basic theory of induction logging and application to study of two-coil sondes: Geophysics, **27**, 829–858.

Tamarchenko, T., and Druskin, V., 1993, Fast modeling of induction and resistivity logging in the model with mixed boundaries: Presented at 34th Ann. Logging Symposium, Soc. Prof. Well Log Analysts.

PART VIII

EQUIPMENT

During 1969 and 1970, a versatile set of EM equipment was developed in the electronics laboratory of Kennecott Exploration Services as part of its geophysical research program. The objective of this work was to construct a compact receiver which would precisely measure amplitude and phase over a broad frequency range, with a variety of transmitter-receiver coupling configurations.... Because the receiver is a vector voltmeter, the equipment is called the vector EM, or VEM, system.

Hohmann, Van Voorhis, and Nelson (1978)

ARLETT: A Prototype Three-Component Borehole Electromagnetic System

Bernard Bourgeois and Dominique Legendre[1]
Marc Lambert[2]
Grant Hendrickson[3]

Summary. ARLETT is a three-component surface-to-borehole harmonic electromagnetic (EM) system developed by BRGM mainly for deep mineral exploration. The probe contains three magnetometers and two inclinometers, which makes it possible to project the EM fields onto a fixed coordinate system and to independently calculate the borehole trajectory. The system has a low internal noise level, high spectral resolution, and effective noise-rejection and signal-processing capabilities.

Conductive targets can be detected directly on the recorded logs of the total EM field, particularly using the phases of the field components and the ellipticity. Locating these targets requires reduction of the data to remove the primary field (of the transmitting loop) and retain the (usually much smaller) secondary field of the target. The reduced data then are matched to the field of a simple 3-D current filament.

The ARLETT system has been tested in the Sudbury Basin, which has highly conductive targets in a resistive environment, and in the Iberian Pyrite Belt, which has moderately conductive targets in a (less) conductive host. The frequency range appears sufficiently wide to distinguish between low-, intermediate-, and high-frequency ranges of the induction response.

1 Introduction

Surface-to-borehole electromagnetic (EM) systems—both transient and harmonic—were developed in the early 1970s and were widely used for mineral exploration during the 1980s in North America and Australia (Dick, 1991). These were, however, single-component (axial) systems that did not allow a precise localization of a detected conductor; only an approximate location was possible with data from several transmitter positions. The early 1990s saw the development of three-component EM

[1]BRGM, Research Division, Department of Geophysics and Geological Imaging, BP 6009, 45060 Orléans cedex 2, France.
[2]BRGM, Research Division, Department of Geophysics and Geological Imaging, BP 6009, 45060 Orléans cedex 2, France; currently at Laboratoire des Signaux et Systèmes (CNRS/Supélec), Plateau de Moulon, 91192 Gif-sur-Yvette cedex, France.
[3]Delta Geoscience Ltd., 852 Tsawwaasn Beach Road, Delta, British Columbia V4M2J3, Canada.

Table 1. Review of the main three-component borehole EM systems

System	Description
Transient EM	
Crone	2-Component (transversal) probe with orientation module, plus axial probe
Geonics	3-Component, gimbal-mounted probe
Lamontagne	3-Component probe with orientation module (?)
Frequency EM	
EMI	3-Component multifrequency probe with full orientation module[a]
Boliden	3-Component (2 frequencies), gimbal-mounted probe
BRGM/Iris Instruments	3-Component multifrequency probe with full orienation module

[a]Impractical for mineral exploration because of the large diameter of its receiver probe (89 mm).

receivers aimed at providing a direct and low-cost method of target localization. We know of only six three-component EM systems (including ARLETT) that are commercially available or under development—three transient systems and three harmonic systems (Table 1). The success of these systems will depend on the availability of procedures for processing and interpretation of the numerous and complex data they collect.

The ARLETT system, developed mainly for deep mineral exploration, is aimed at detecting and locating conductive masses in the vicinity of barren exploration boreholes, and possibly at outlining conductive orebodies intersected by drilling. Other possible applications are in geotechnical engineering, hydrogeology, and geothermal exploration. Because routine 3-D modeling of data is still impractical, we have investigated simplified interpretation schemes; so far, satisfactory results have been obtained in interpreting the reduced EM fields by using simple current filaments in a way similar to that described by Barnett (1984), Boyd and Wiles (1984), and Fullagar (1987).

2 ARLETT three-component EM system

The ARLETT system (Fig. 1) developed by BRGM operates in the 50–2000-Hz frequency range, using the primary field created by a medium-size transmitting loop (ca. 300-m sides); the loop is fed by a lightweight 1-kW transmitter (TX 1000 of Iris Instruments) powered by a 3-kW portable generator. On an 8–10-ohm loop, the transmitter supplies 10–13 A at low frequency (<200 Hz) and only 2–3 A in the upper ARLETT frequency range (≈ 2000 Hz). Although 32 ARLETT frequencies are available, only six standard frequencies extending from 61 to 1953 Hz in geometric progression of ratio 2 are used routinely (Table 2).

The ARLETT probe—2.8-m length, 53.5-mm diameter, pressure tested to 100 bars (optionally 54.5 mm for up to 150 bars)—is connected to the surface unit by a standard four-conductor logging cable (1600 m max.) and from there to a portable microcomputer by an RS232 link. Unlike single-axis probes, which measure only the component of the field along the axis of the borehole (Bourgeois et al., 1991), the ARLETT probe measures all three components of the alternating magnetic field in the ground. This field (total field) is the sum of the primary field created by the source and the secondary

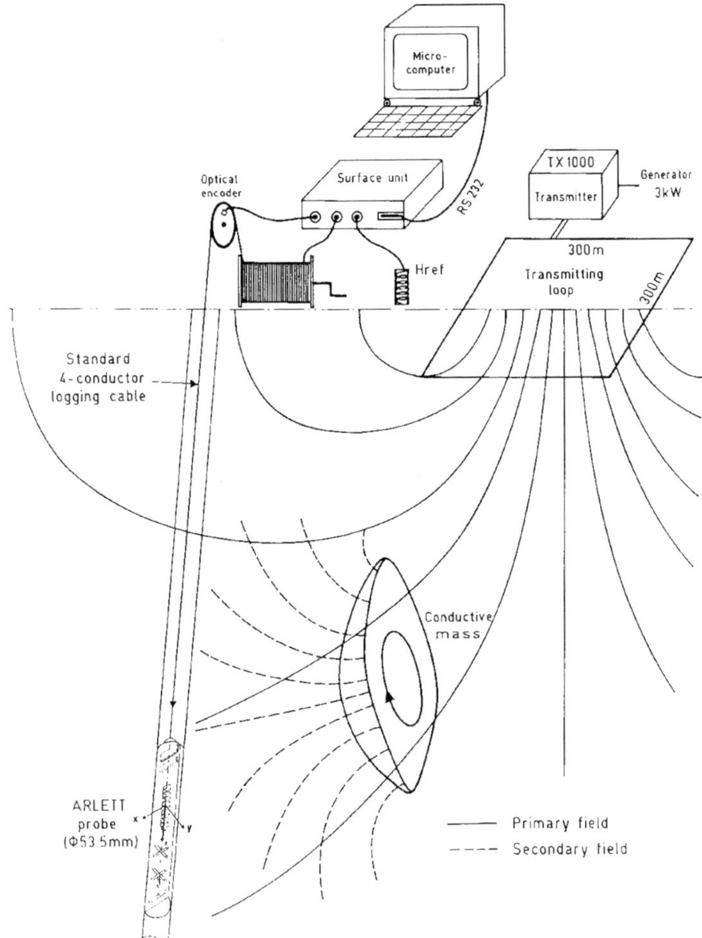

Figure 1. Configuration of ARLETT system.

fields created by induction currents inside conductive bodies. The probe also measures an alternating reference signal received in analog form via the logging cable; this reference can be either absolute, given by the current in the loop, or relative, given by a magnetic sensor near the borehole collar (useful when using a surface loop to survey a borehole drilled in an underground mine).[4]

In addition to the three EM sensors (Fig. 2), the probe includes a temperature sensor for correcting possible drifts and a full orientation module (three magnetometers and two transverse accelerometers[5]) for deriving the local magnetic north and the vertical (see below).

The probe CPU (1) performs analog-to-digital (A/D) conversion (with automatic gain control), numerical filtering (1- to 16-bit stacking) and fast Fourier transform

[4]The option of using absolute reference is recent, which explains why all of the field tests presented here only used a relative reference.

[5]Contrary to what may be interpreted from Fig. 2, an accelerometer to measure the axial component of gravity (g_z) must be transverse to the probe axis. The present probe does not include an axial accelerometer because no component small enough was available at reasonable price at the time the prototype was built. The axial component g_z thus is deduced from g_x and g_y, assuming that the gravity magnitude is constant at the surface of the Earth (9.8 m/s^2) and that the orientation of the borehole (descending or, less commonly, ascending) is known throughout.

Table 2. Standard ARLETT frequencies, and resolutions for an acquisition of 4096 samples

Frequency (Hz)			Spectral resolution	
Working	Sampling	K[a]	Absolute (Hz)	Relative (%)
61.0352	1041.67	17.0667	0.25	0.40
122.070	1041.67	8.53333	0.25	0.20
244.141	1041.67	4.26667	0.25	0.10
488.281	1302.08	2.66667	0.32	0.06
976.563	2604.17	2.66667	0.64	0.06
1953.13	5208.33	2.66667	1.27	0.06

[a]Ratio of sampling frequency to working frequency.

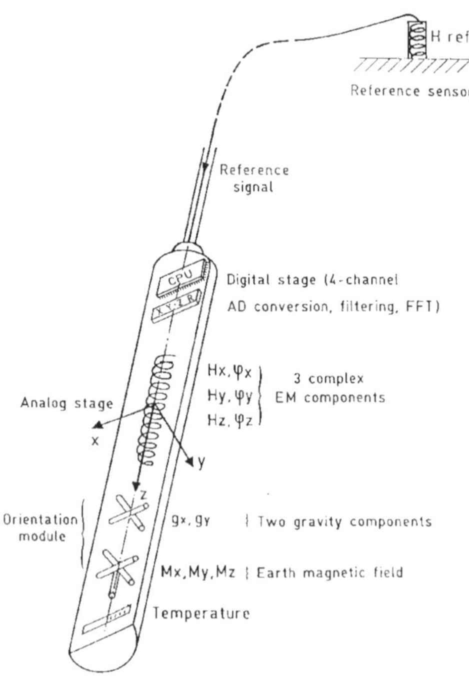

Figure 2. Details of ARLETT probe.

(FFT) for the four EM channels, (2) retrieves the temperature and orientation readings, and (3) transfers the digital information to the surface. For the EM acquisition, pre-FFT oversampling (sampling rate × 64) guarantees less than 0.01% aliasing under all conditions, whereas FFT on 4096 samples (the maximum number possible) provides excellent spectral resolution ($\Delta f/f < 0.5\%$ in all cases; Table 2). To speed up EM acquisition, sampling of a given time series (except the first one) is performed during the FFT of the previous time series (task overlapping); this reduces the duration of an EM cycle to the maximum duration of the two tasks. In practice the maximum cycle duration is about 4.5 s for 4096 samples at the lowest ARLETT frequency (54 Hz).

The surface microcomputer retrieves the digital data from the probe (i.e., four complex EM fields, five orientation parameters, temperature) and from an optical depth encoder; averages the EM cycles by cross-power spectra relative to the reference signal; controls noise, coherency, and dispersion of the EM measurements; applies EM

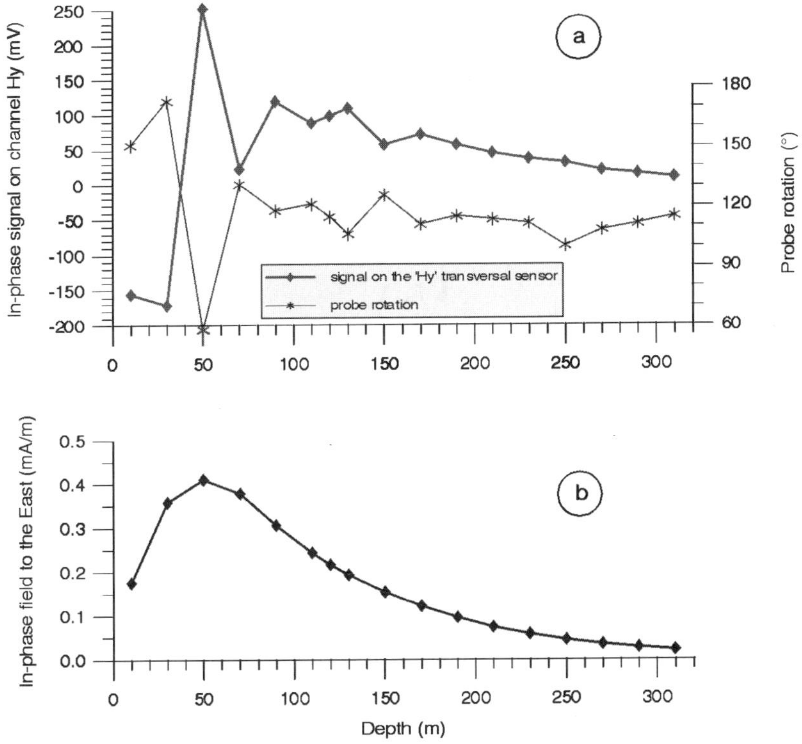

Figure 3. Example of EM field derotation: (a) nonderotated in-phase field (unprocessed signal from the transversal component Hy) mainly reflecting probe rotation; (b) derotated in-phase field (east component) free of probe rotation effect.

sensor calibrations; derotates the EM fields; and enables real-time graphical control of the EM logs being recorded.

2.1 Orientation module

The purpose of the orientation module is to project the measured EM fields onto a fixed coordinate system (derotation) and compute the borehole trajectory necessary for later reduction of EM data.

The measured three components of the EM field are related to the intrinsic coordinate system of the probe, which rotates significantly during depth changes (and also progressively tilts as the borehole deviates). Thus, they mainly reflect probe rotation and cannot be used directly (Fig. 3a). The gravity and magnetic vectors (**g** and **M**) given by the orientation module enable the recovery of a fixed coordinate system (Fig. 4) based on the local magnetic north (**Nm**), the magnetic east (**Em**), and the downward vertical (**Vb**). This right-hand coordinate system is fixed effectively in the absence of local deviations of magnetic north (see later). The measured EM components then are projected onto this system, or onto any other fixed system (geographical, local grid) derived from it. This projection, known as derotation, produces new field components that are free of the probe rotation effect (Fig. 3b). The fact that this derotation can be applied even in the case of completely vertical boreholes gives ARLETT a major advantage over gimbal-mounted systems.

The successive attitudes of the probe enable recovery of the borehole trajectory, i.e., the x-, y-, z-coordinates of each measuring station that are necessary for further EM

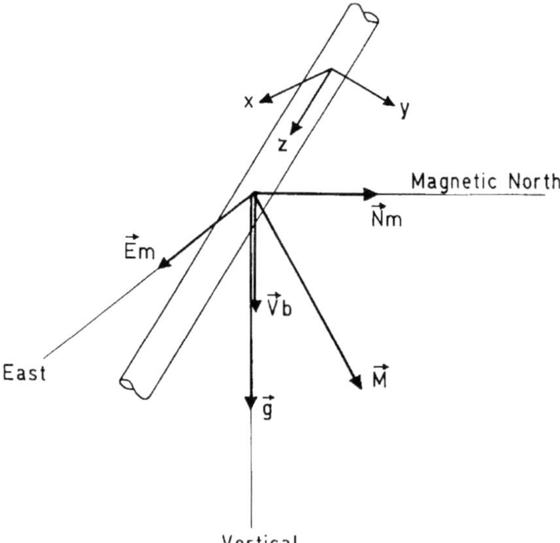

Figure 4. Principle of ARLETT orientation module. The rotating intrinsic coordinate system of the probe (x, y, z) is projected onto a fixed system based on magnetic north, magnetic east, and the downward vertical (**Nm**, **Em**, **Vb**).

reduction. This total independence from any other trajectory measurement (e.g., Tropari or gyro survey) gives greater processing rapidity and, in certain cases, greater reliability. ARLETT data can be processed immediately after recording and, if sufficiently clean (i.e., little or no editing required), the logs of the total and reduced fields can be produced within half an hour after completing acquisition.

It is clear that an orientation reference based on magnetic north can be a problem in the presence of highly magnetic masses that locally deviate the magnetic north. Nevertheless the problem generally can be resolved by a simple interpolation of the probe azimuths within the disturbed zone, as we see later.

2.2 Data acquisition

The acquisition process consists of taking measurements at several successive frequencies (generally the six standard frequencies) at discrete depths along the borehole. A station interval of 20 m generally is employed as being suitable for most economic targets; this can be reduced to a few meters for sampling short-wavelength anomalies produced by targets very close to the borehole. Each reading at a given frequency and depth begins with an orientation plus temperature measurement, followed by the EM acquisition that consists in stacking several EM cycles.

Because the high spectral resolution of ARLETT gives very low-noise EM signals, little stacking is necessary to obtain very good stability (four to eight cycles are generally sufficient). Acquisition times are thus relatively short: if six frequencies are used, a complete measurement at each depth takes an average of 7 minutes (including about 1 minute for probe displacement between two depths). ARLETT's real-time control of the logs is another facility that reduces overall logging time; for example, a large sampling interval can be used when lowering the probe (e.g., 40 m in a long borehole) to view the general shape of the logs, and a smaller interval (10 m or less) then can be applied when raising the probe to control and better define any anomalous variation.

The performance of the system is illustrated by one of the deepest ARLETT measurements ever made. This was recorded at a depth of 1300 m with a loop offset by 475 m (borehole 52820, see Fig. 19). At 1953 Hz—the least favorable frequency regarding the source power, with only 2.3 A in the loop—only 11 stacks (taking less than 1 minute) were required to obtain a reliable reading with about 1% noise (coherency above 99.98%).

2.3 Data interpretation

The interpretation of three-component borehole EM data is realized in two stages: first, the detection of characteristic anomalies on the recorded logs of the total EM field along the borehole; second, the localization of conductive targets responsible for these anomalies. The adopted approaches to the interpretation of ARLETT data have been developed both from 3-D numerical modeling [using the EM3D software of the University of Utah; Newman and Hohmann (1988)] and from various field tests (see later). The geoelectric model used for simulation (Fig. 5) is based on the context of the Iberian Pyrite Belt and features a moderately conductive target hosted by a less conductive country rock (contrast 1/100); the computation was performed at 2000 Hz for 1-A in the transmitting loop.

2.3.1 Target detection. Target detection is performed on the amplitudes and phases of the three derotated total (EM) magnetic components at several frequencies (and possibly on other parameters derived from them, such as the ellipticity). It is easy to understand that the detection capability of a three-component tool such as ARLETT is about three times greater than that of a single-axis tool, because the best anomaly often is given by a transversal component rather than the axial component (depending on target attitude and position relative to borehole).

Phases, which represent the ratio of quadrature to in-phase field components, are found to be generally more sensitive than amplitudes to the proximity of conductors. This is explained by the fact that, for the ARLETT frequency range and for common ground materials, the inductive responses are situated mainly in the low-to-intermediate induction domain. In this domain, the secondary field created by a conductor has a significant quadrature component in relation to the host-rock primary field and thus has a major effect on the total field phases. This relative sensitivity is evident on Fig. 6 at the upper limit of the ARLETT range (2000 Hz, intermediate induction in the target), and is also obvious in the lower part of the ARLETT range, i.e., in the low induction domain where the quadrature secondary response is dominant.

The horizontal ellipticity, which is uniformly null in any horizontally layered medium,[6] is also extremely sensitive to conductive targets (see Fig. 7). A particular advantage of this parameter is that it is invariant with rotations of the coordinate system.

2.3.2 Target localization. The major objective of a three-component tool such as ARLETT is the localization of a detected target using the directional information contained in the measured EM vectors. Modeling shows that this localization cannot

[6]This is only true for a circular loop; a square loop gives nonzero ellipticities very close to the loop vertices.

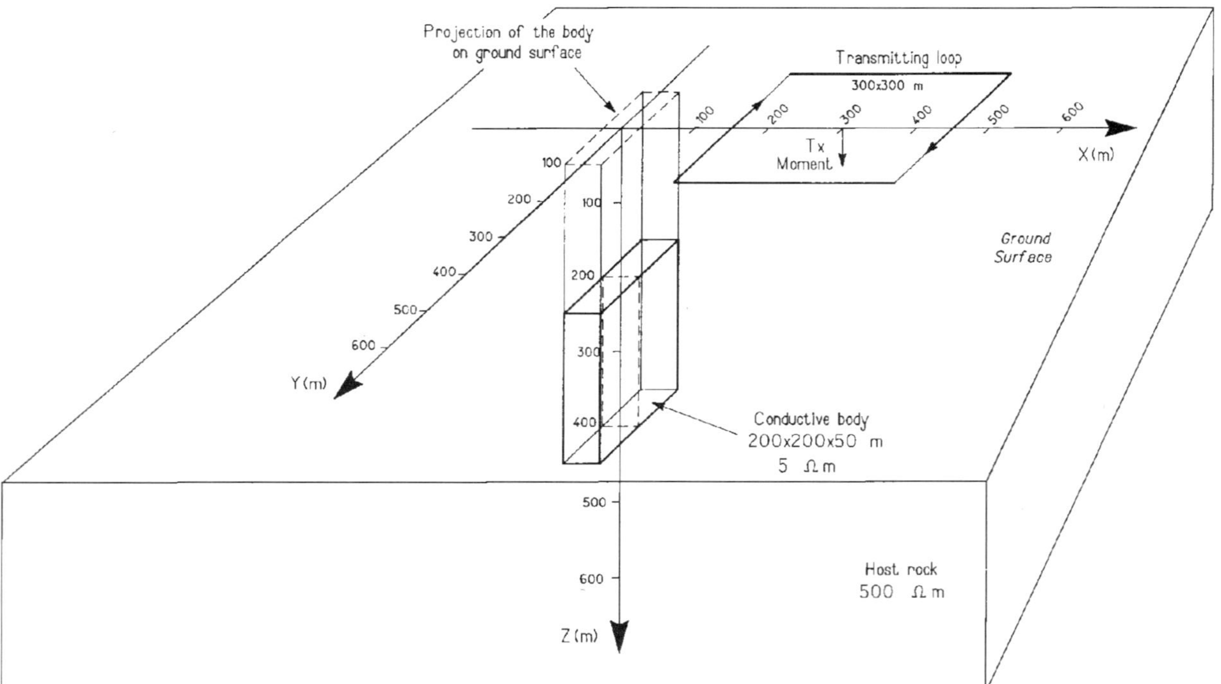

Figure 5. Model used for the 3-D computation presented in this paper, featuring a vertical plate-like conductor (50 × 200 × 200 m) of 5 ohm-m resistivity, located at 200-m depth in a 500-ohm-m half-space. The loop (300 × 300 m) is offset 300 m from the body center. Computation was performed at 2000 Hz, using 5 × 20 × 20 m rectangular cells and 5-m-side cubic subcells, for 1 A in the transmitting loop.

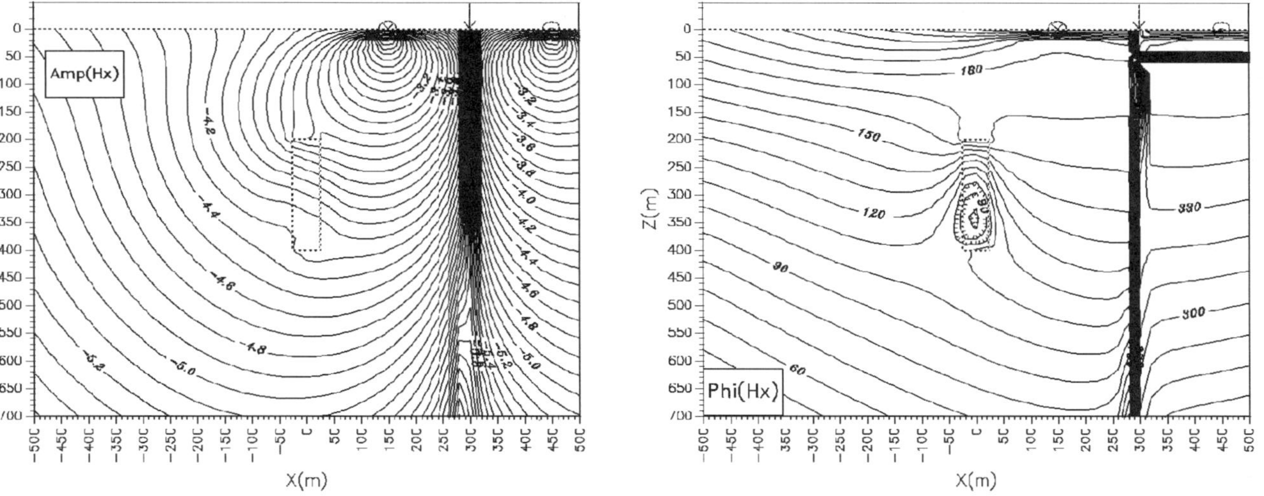

Figure 6. Simulated amplitudes (left) and phases (right) of total magnetic-field components Hx, Hy, Hz calculated with the geoelectric model of Figure 5 (at 2000 Hz and for 1 A in the transmitting loop), in the $Y = 60$ m vertical plane through the body. Amplitudes are contoured at logarithmic intervals of 0.1 $\log_{10}$ (A/m), and phases are contoured at 10° intervals. The response of the conductor appears to be limited to about 100 m on the shadow side of the conductor relative to the transmitting loop. The phases generally show stronger and more extended responses than the amplitudes. In this particular configuration, the component most sensitive to the conductor is Hy. (*Continues.*)

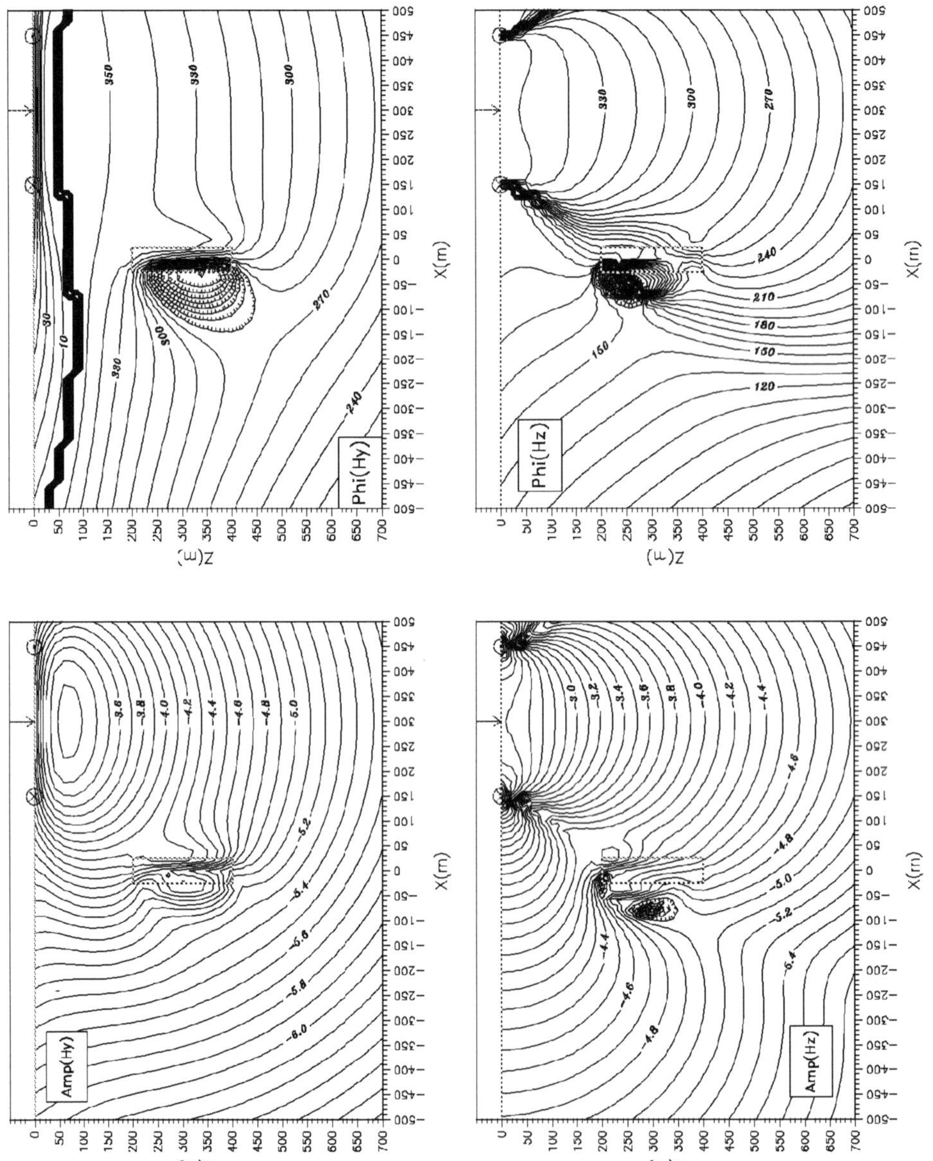

Figure 6. *(Continued.)*

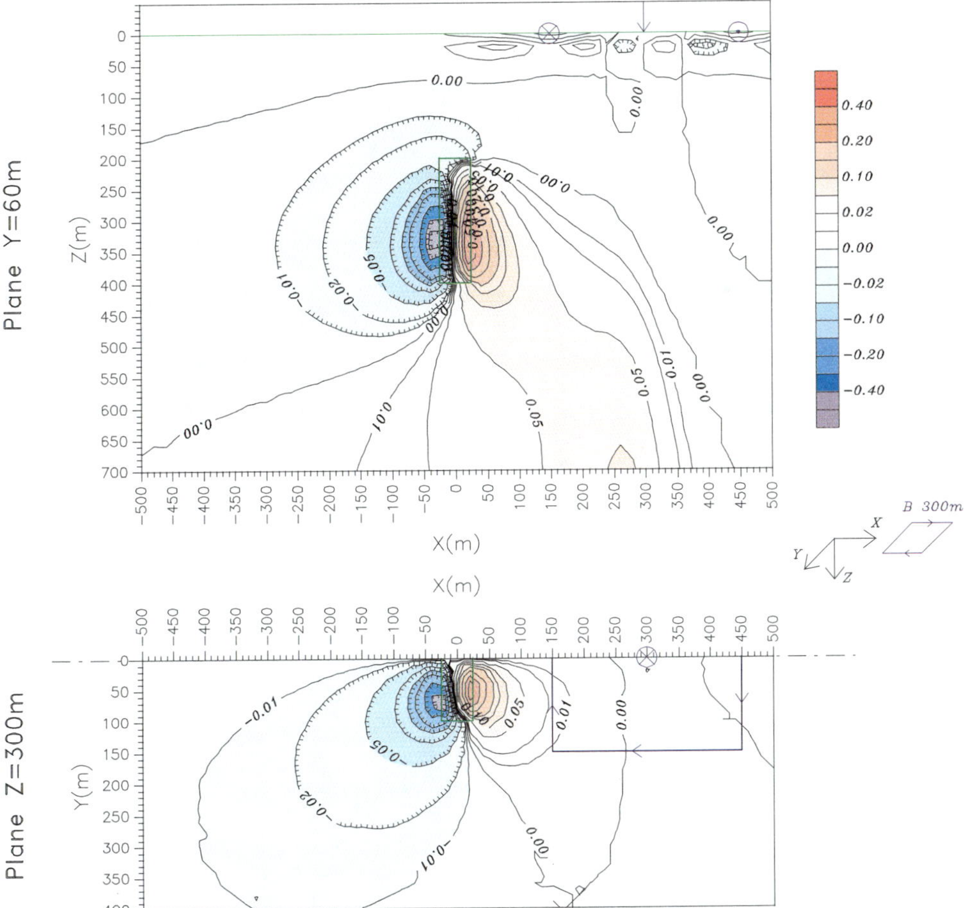

Figure 7. Horizontal ellipticity of total magnetic field calculated with the geoelectric model of Fig. 5 (at 2000 Hz) in two planes through the conductive body: (a) the vertical plane $Y = 60$ m; (b) the horizontal plane Z = 300 m. Positive values of the horizontal ellipticity (in red) indicate clockwise description of the ellipse (seen from above), whereas negative values (in blue) indicate counterclockwise description. Although the response of the body is not very strong (a few percent), it appears to be quite extended; this is explained by the fact that horizontal ellipticity in a homogeneous or horizontally layered half-space is almost uniformly null. Note the difference in sign to either side of the conductor, positive values being observed on the loop side.

be done by simple visualization of the measured total EM field, because this essentially is imprinted with the direction of the transmitting loop (Fig. 8, left). Reduction is thus necessary to remove as much of the primary field as possible, in order to approach the secondary field of the detected target (Fig. 8, right). For this operation the x-, y-, z-coordinates of the measuring stations and loop vertices, as well as the transmitted current, need to be known.

At present, two types of reduction are used routinely: (1) reduction by the calculated free-space primary field, and (2) reduction by the measured field at low frequency. Although very similar in many cases, the latter is much easier to use because no information concerning the transmitter (geometry, current intensity) is necessary. However, these two reductions are well suited only if the host rock is not too conductive and if the frequency to be processed is not too high (see later). In other cases, induction in the host is no longer negligible, and a more suitable reduction has to be applied, such

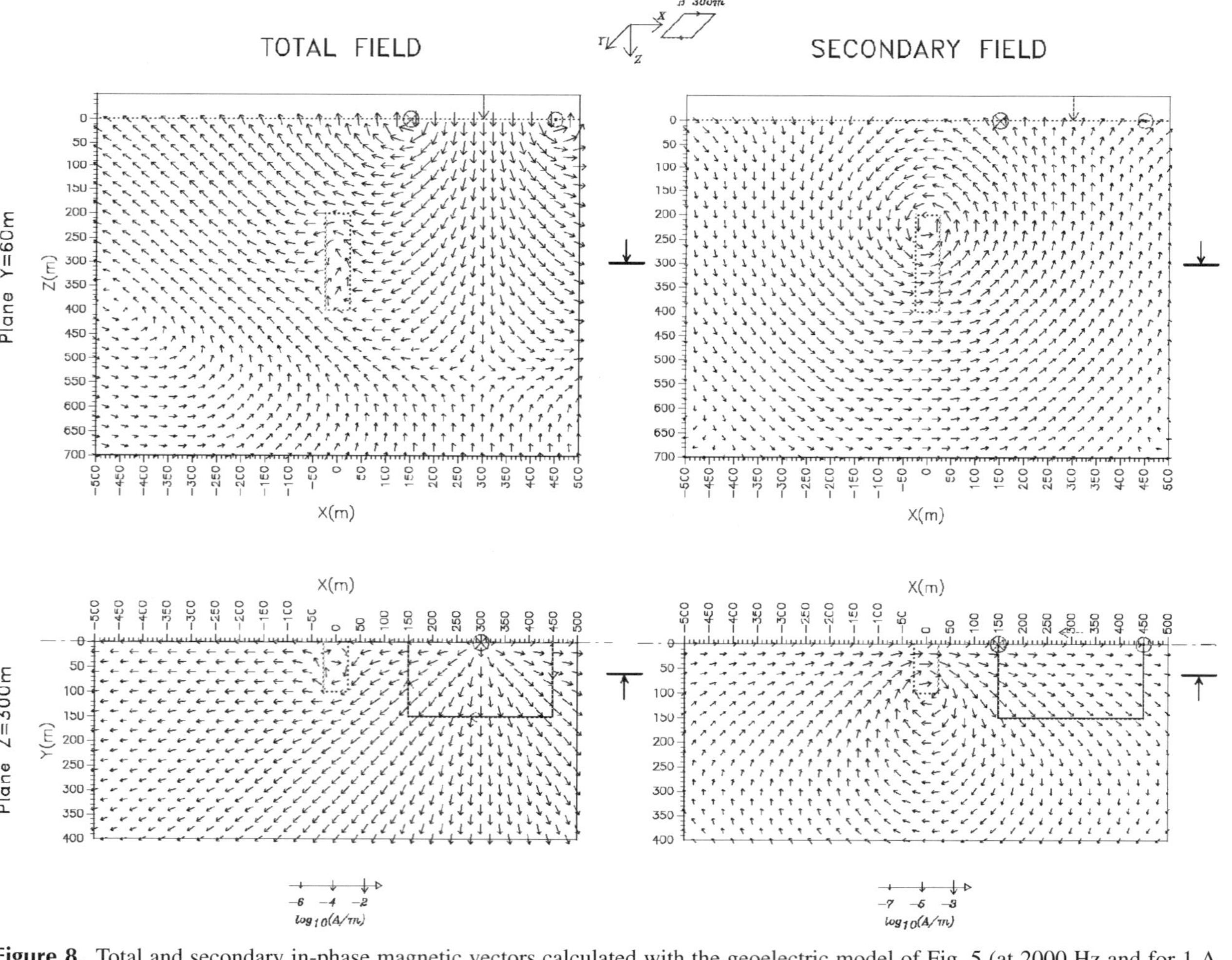

Figure 8. Total and secondary in-phase magnetic vectors calculated with the geoelectric model of Fig. 5 (at 2000 Hz and for 1 A in the transmitting loop), in two planes through the conductor ($Y = 60$, $Z = 300$ m). The total field (left) appears to be essentially influenced by the transmitting loop, being slightly distorted only in the immediate vicinity of the conductor. Consequently, the total field vectors cannot be used for locating a target more than a few meters away. The secondary field (right) is seen to be only influenced by the conductor, with the node of the field vortexes giving the approximate location of a current flowing at the surface of the target.

as the reduction by the primary field of the source computed on a conductive (possibly layered) half-space.

Once a satisfactory reduction has been applied, localization is performed on the reduced field equated to the secondary field of the target. For the moment, our interpretation process is very rudimentary and consists in looking for a simple 3-D electrical circuit of given intensity (in freespace) that accounts for the reduced vectors. The fit between the simulated and the reduced data is assessed either in vector form or in log form. This procedure gives the bearing, the structural attitude, and an estimate of the distance to the closest edge of the target (see below). Automatic fitting for single or multiple circuits currently is being developed, and more sophisticated procedures are being investigated to enhance the usefulness of the data and obtain more complete information concerning the target.

3 Field validation of the ARLETT system

The ARLETT system has been tested and validated in two very different environments: (1) the Sudbury Basin in the Canadian Shield, where the orebodies are extremely conductive (10^{-4} to 10^{-2} ohm-m) in a very resistive environment ($\approx 10^4$ ohm-m), thus giving tremendous contrasts ($>10^6$); and (2) the Iberian Pyrite Belt where the orebodies are only moderately conductive (a few ohm-m) in a conductive host rock (a few hundreds of ohm-meters), thus giving only modest contrasts (≈ 100).

The three examples presented below have been chosen to represent different borehole configurations with respect to a known or suspected orebody. Two are from tests made on Inco's MacConnell and Kirkwood exploration sites in the Sudbury Basin—one in a borehole crossing a very conductive (and also magnetic) deposit (borehole 80578), and the other in a barren borehole located at a significant distance (>100 m) from a suspected orebody (borehole 52820; far-miss configuration). The third example is from a test made in a barren borehole passing very close to a known pyrite orebody on Seiemsa's exploration site at Herrerias in the Iberian Pyrite Belt (borehole H12; near-miss configuration).

3.1 Borehole 80578: Borehole intersecting a conductive and magnetic deposit

The test in borehole 80578 (Sudbury Basin) reveals the pitfalls of magnetic north deviation, and how these can be circumvented when processing derotation and trajectory, and also demonstrates the good signal resolution of the ARLETT system.

3.1.1 Magnetic north deviations. Borehole 80578 intersects a highly magnetic and conductive pyrrhotite body which, between 240-m and 260-m depth, imparts a strong anomaly to the measured magnitude and dip of the Earth's magnetic field (Fig. 9a). A direct calculation of the trajectory (Fig. 9b, curves with stars) shows a quite unrealistic twist that is caused by a magnetic-north deviation inside and around the pyrrhotite body—obviously it is not the borehole that describes a 360° right turn, but the magnetic north that describes a 360° left turn. The problem was resolved by interpolating the probe azimuths between the upper and lower edges of the magnetic anomaly. The difference Δd between the direct reading and the interpolated value of the probe azimuth at each disturbed station then was used to (1) reorient the deviated magnetic north back to its normal direction (i.e., the regional

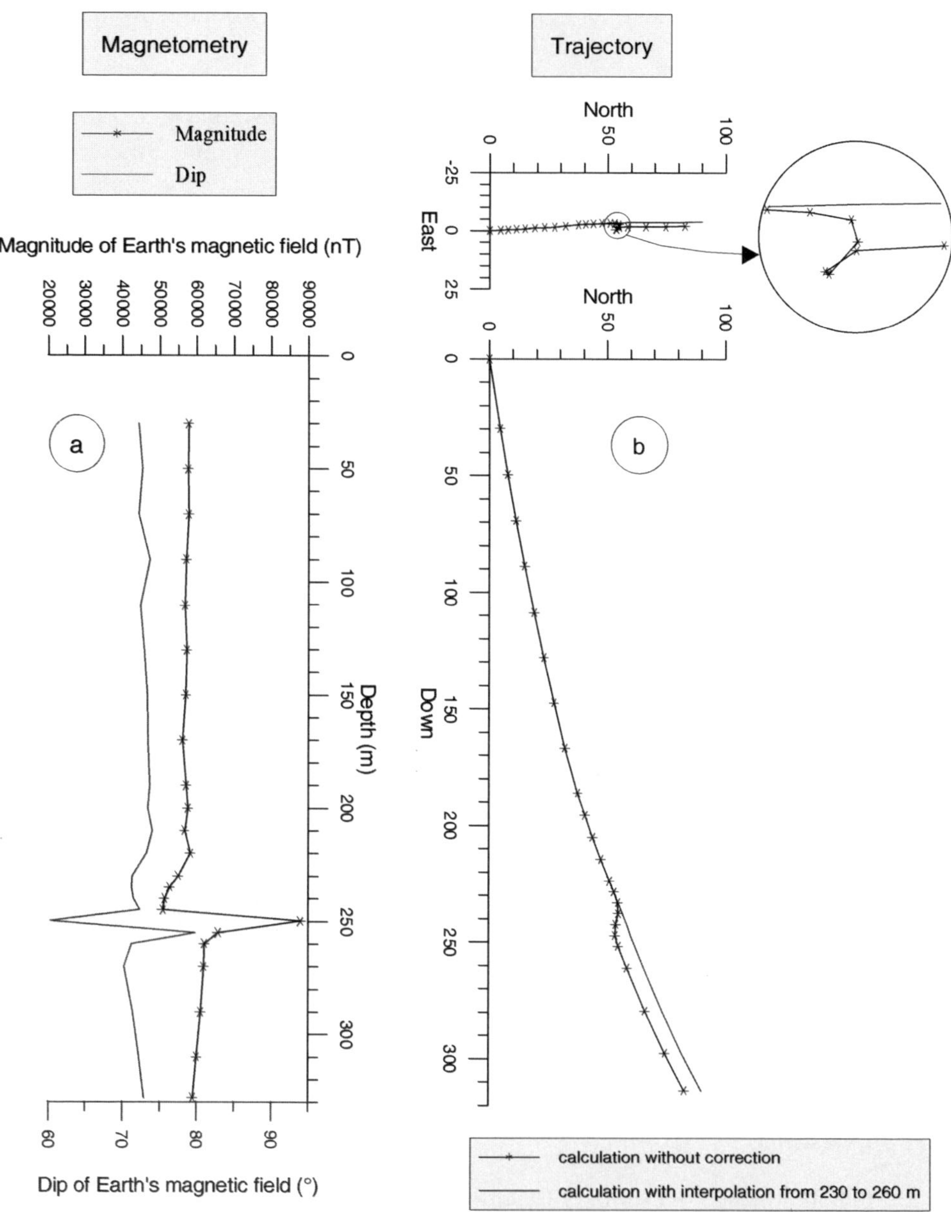

Figure 9. Trajectory calculation for a borehole crossing a magnetic deposit (borehole 80578): (a) recorded magnitude and dip of Earth's magnetic field; (b) calculated borehole trajectories (see text for explanation).

magnetic north),[7] and (2) rotate the coordinate system used for projecting the EM fields, so that the derotation can be effectively made toward fixed directions. The resultant trajectory (Fig. 9b, plain curves) is much more realistic than that given by the direct calculation.

[7]This also gives a local value of magnetic declination, which is different from the surface normal value. ARLETT thus provides a full 3-D magnetometry log (magnitude, dip, and declination of the Earth's magnetic field) that could be interpreted profitably, even though the declination derived by interpolation of borehole azimuth is not very accurate.

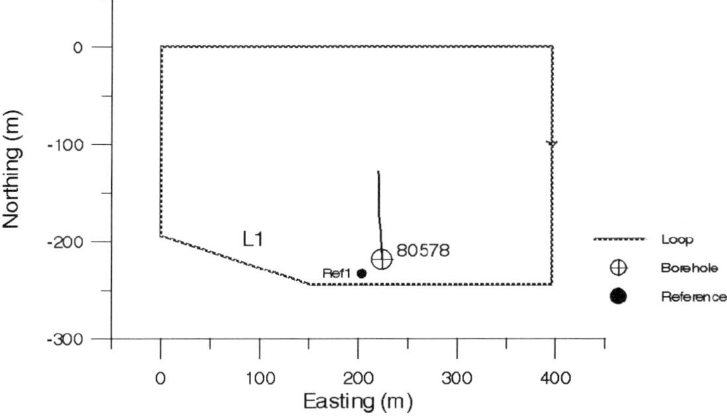

Figure 10. Loop and borehole layout at the MacConnell test site (Sudbury Basin) for the test in mineralized borehole 80578 with transmitting loop L1.

It is thus clear from this test that orientation data based on magnetic north can be suspect in the presence of a highly magnetic mass because of magnetic deviation. Nevertheless, this problem generally can be resolved (as here) simply by interpolating the probe azimuth through the magnetic anomaly. Such an interpolation is justified as long as the borehole horizontal projection does not present significant concavity changes within the disturbed zone, which is generally the case if the zone is relatively short.

3.1.2 ARLETT signal resolution. The source/borehole configuration used for surveying borehole 80578 is shown in Fig. 10. Two points should be noted with this configuration: (1) The reference used was relative, consisting of a magnetic sensor placed at about 10 m from the closest edge of the loop, and hence the phase reference can be considered as absolute; and (2) the loop surrounded the borehole (i.e., an in-loop configuration), which has the effect of simplifying the shape of the primary vertical field (no vertical field reversal). The phases and amplitudes of the derotated EM fields (after having applied the correction Δd to reorient the coordinate axes within the magnetically disturbed zone) are shown in Fig. 11 for a section of borehole 80578 across the deposit (230 to 290 m). The reduced sampling interval of 5 m in this section gives relatively good definition of the response within the orebody. Despite a strong amplitude attenuation, the data remain consistent within the conductive body; the fact that the curves at the different frequencies give a coherent pattern proves that the ARLETT system has resolved the signal and not merely picked up noise.

3.2 Borehole H12: Barren borehole in a near-miss configuration

The ability of ARLETT to detect and locate a nearby relatively shallow conductor of moderate contrast in a conductive environment is shown by this test in barren borehole H12. Massive pyrite already had been intersected between 100- and 150-m depth in borehole H3 (about 100 m east of borehole H12) and between 120- and 130-m depth in borehole H15 (about 200 m east of H12). The distance of the target from the borehole was not known, although it was considered to be relatively small because H12 intersected disseminated pyrite (up to 10%) at several places between 85- and 150-m depth.

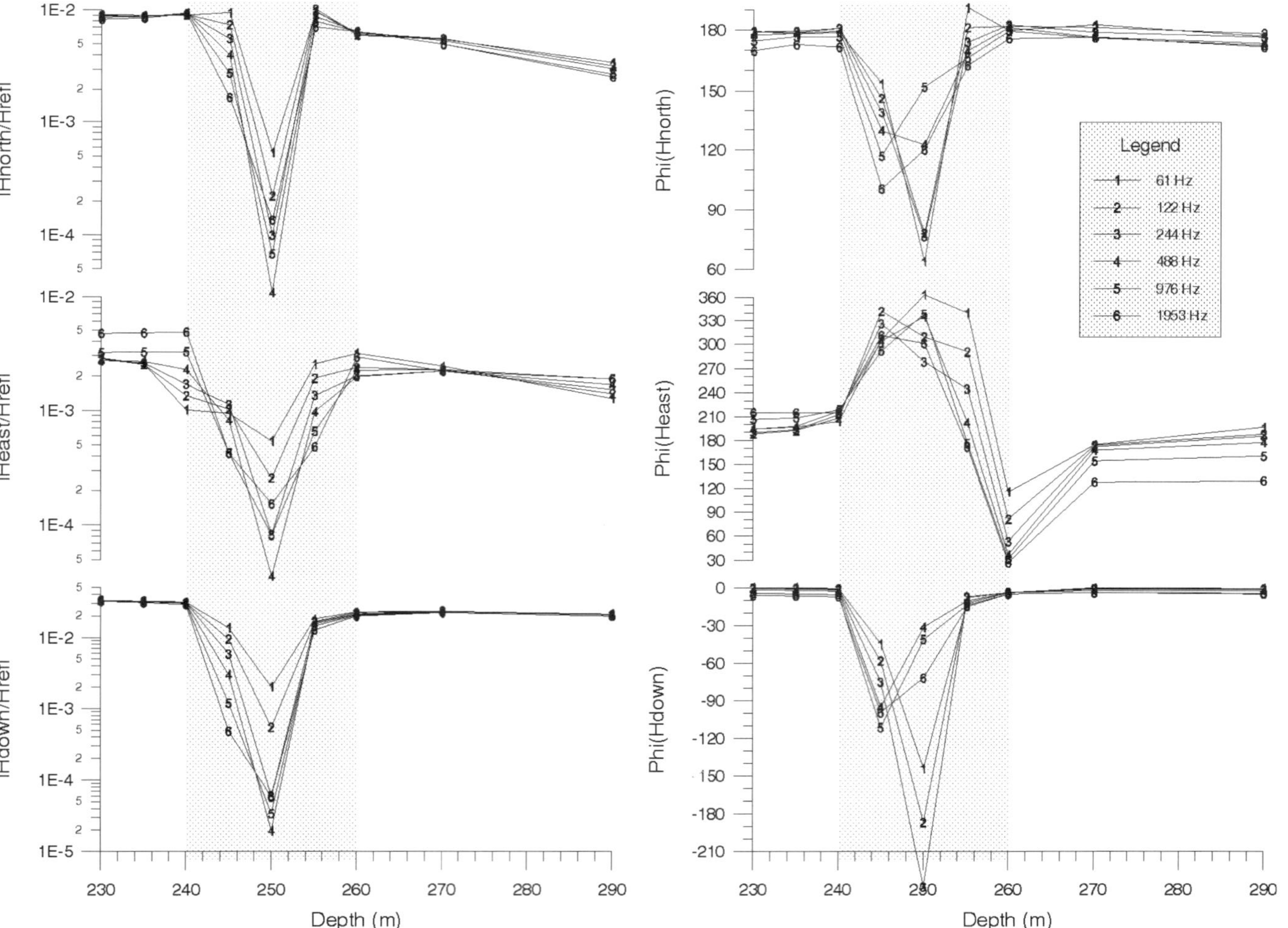

Figure 11. Normalized amplitudes (left) and phases (right) of the measured (EM) magnetic field across the very conductive deposit intersected by borehole 80578 (after derotation and correction for magnetic-north deviation). The orebody intersection is indicated in gray.

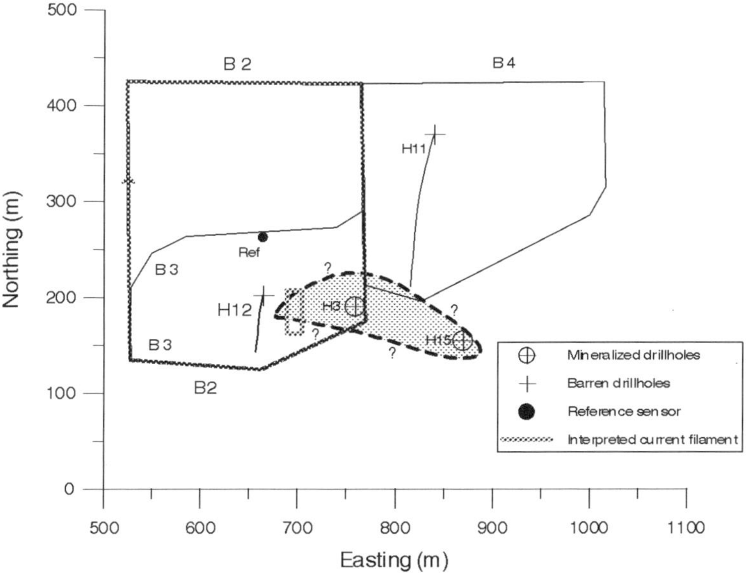

Figure 12. Loop and borehole layout at the Herrerias test site (Iberian Pyrite Belt) for the test in barren borehole H12 with transmitting loop B2. The locations of two nearby boreholes intersecting massive pyrite are shown, as is the surface projection of the inferred orebody.

The source/borehole layout for this test is shown in Fig. 12, featuring an in-loop configuration and a relative reference given by a magnetic sensor at about 100 m from a loop edge (which can result in a global phase shift of several degrees at 1953 Hz in this conductive environment). The amplitudes and phases of the three total-field components measured with this array at 20-m-depth intervals are presented in Fig. 13.

The east component in this case is the most sensitive to the pyrite body. Both the phases and the amplitudes of this component react strongly to the proximity of the conductor, showing significant anomalies at around 100- to 150-m depth: (1) on the amplitude, at all frequencies except the lowest (69 Hz), the response is a strong crossover centered at 150-m depth; (2) on the phase, the response is a strong minimum (>30°) at the lowest frequency (69 Hz)[8] which changes progressively to a strong maximum at high frequency (1953 Hz); (3) the frequency giving the maximum amplitude-crossover (976 Hz) shows only an intermediate response on the phase. The north and down components of the field also give a slight reaction to the orebody: the phases, but not the amplitudes, show anomalies around 170-m depth, especially at the intermediate frequencies 488 and 976 Hz.

The fact that the maximum target response on Fig. 13 (and also on Figs. 15 and 16) is observed on curves 3 and 4 indicates that, for the detected target, the ARLETT frequency range corresponds to the transition zone between the low and high induction domains. Thus the resonance frequency of the target, i.e., the frequency at which the quadrature response is maximum and the in-phase increase as a function of frequency

[8]This demonstrates, on experimental data, that phases are more sensitive than amplitudes, especially at low induction numbers.

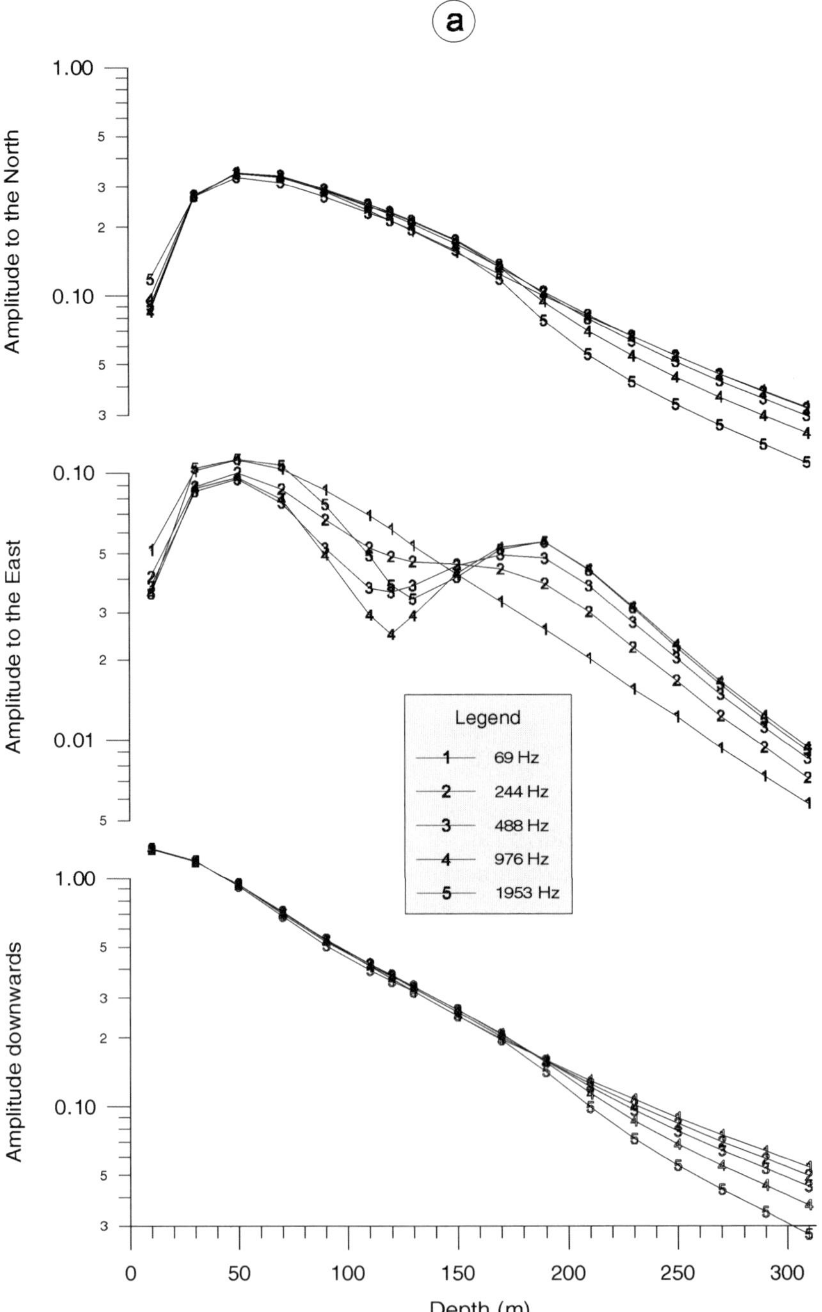

Figure 13. Borehole H12/loop B2: (a) Amplitudes (normalized to H_{ref}) in logarithmic scale and (b) phases of measured magnetic field (note that normalization gives dimensionless amplitudes). (*Continues*)

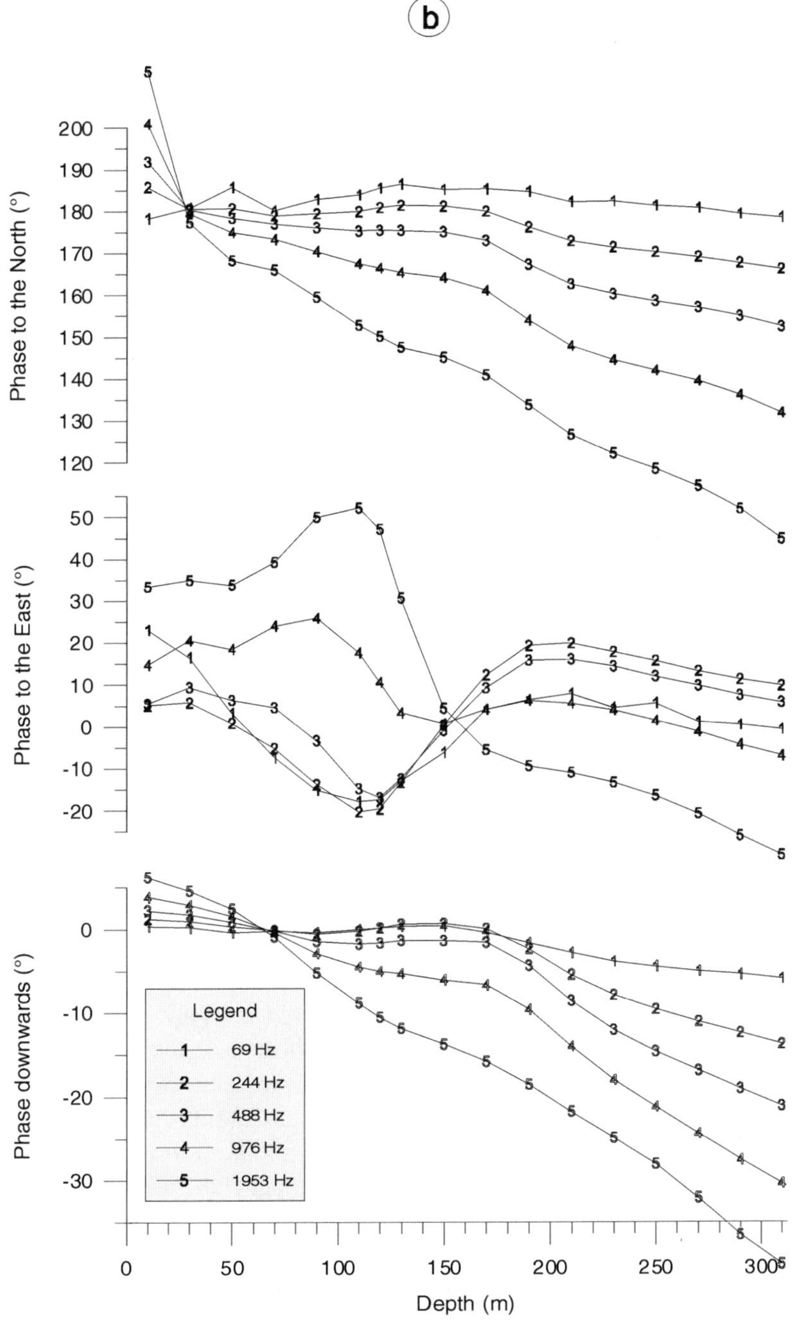

Figure 13. *(Continued)*

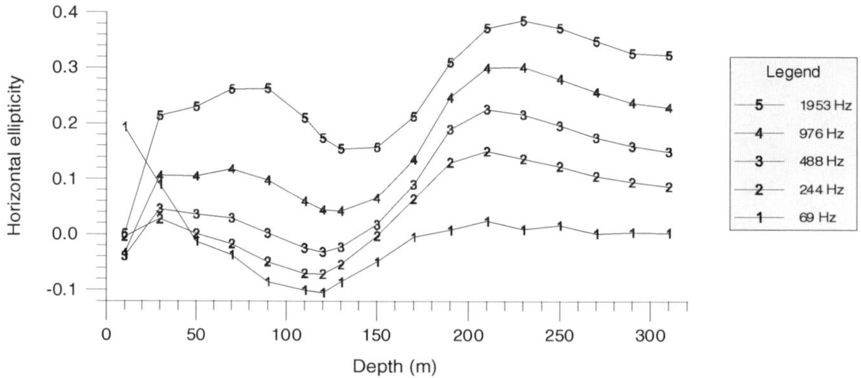

Figure 14. Borehole H12/loop B2: Horizontal ellipticity of measured (total) magnetic field.

begins to slow, is around 500–1000 Hz. The induction number y at the resonance is known to lie between 2 and 5, which can give an idea concerning either the size a or the resistivity ρ of the orebody from the formula $a \approx 356\, y\sqrt{\rho/f}$.[9]

The regular and rather rapid phase decrease (average $20^\circ/100$ m at 1953 Hz) for all of the components outside the area of target influence (Fig. 13b) denotes a rather conductive host (300–500 ohm-m), in agreement with the geoelectric environment.

The total field horizontal ellipticity (Fig. 14) reflects the proximity of the conductor at all frequencies by a well-defined minimum, the amplitude of which slowly increases with increasing frequency (up to about 15–20% ellipticity from 488 to 1953 Hz). This response also shows a slight frequency-dependent shift in depth (from 120 to 150 m) which possibly is related to target attitude (dip) with respect to the borehole.

Another presentation of the data set of Fig. 13 is given in Fig. 15, which shows the in-phase and quadrature field components (real and imaginary parts of the total measured field). The in-phase fields (Fig. 15a) are seen to be very similar to the amplitudes, which is normal because the phases of the measured field are globally closer to 0° than to 90°. Because of induction in the conductive host rock, the absolute value of the in-phase field to the north and down progressively attenuates as frequency increases. At the lowest frequency (curve 1, Fig. 15a), the in-phase field is close to the calculated free-space primary field (thick gray curve), showing that no in-phase induction from either the host or the target is measurable at this frequency; the measured field at 69 Hz thus can be used legitimately in this case instead of the calculated free-space field for the free-space reduction.

The quadrature components (Fig. 15b) apparently do not reflect the source primary field as strongly as the in-phase components because the free-space primary field is totally in phase and thus has zero quadrature. Nevertheless the primary field of a source on a conductive half-space (as here) does have a significant quadrature response, and so, in practice, the quadrature response cannot be considered as the secondary field of the target, particularly at high frequency. Like the in-phase response, the quadrature response

[9] Assuming that the value of y at the resonance is 3 and that the resonance frequency is about 700 Hz, the above formula can be written $a \approx 40\sqrt{\rho}$. For a target resistivity of 1 ohm-m, this gives a target size of 40 m.

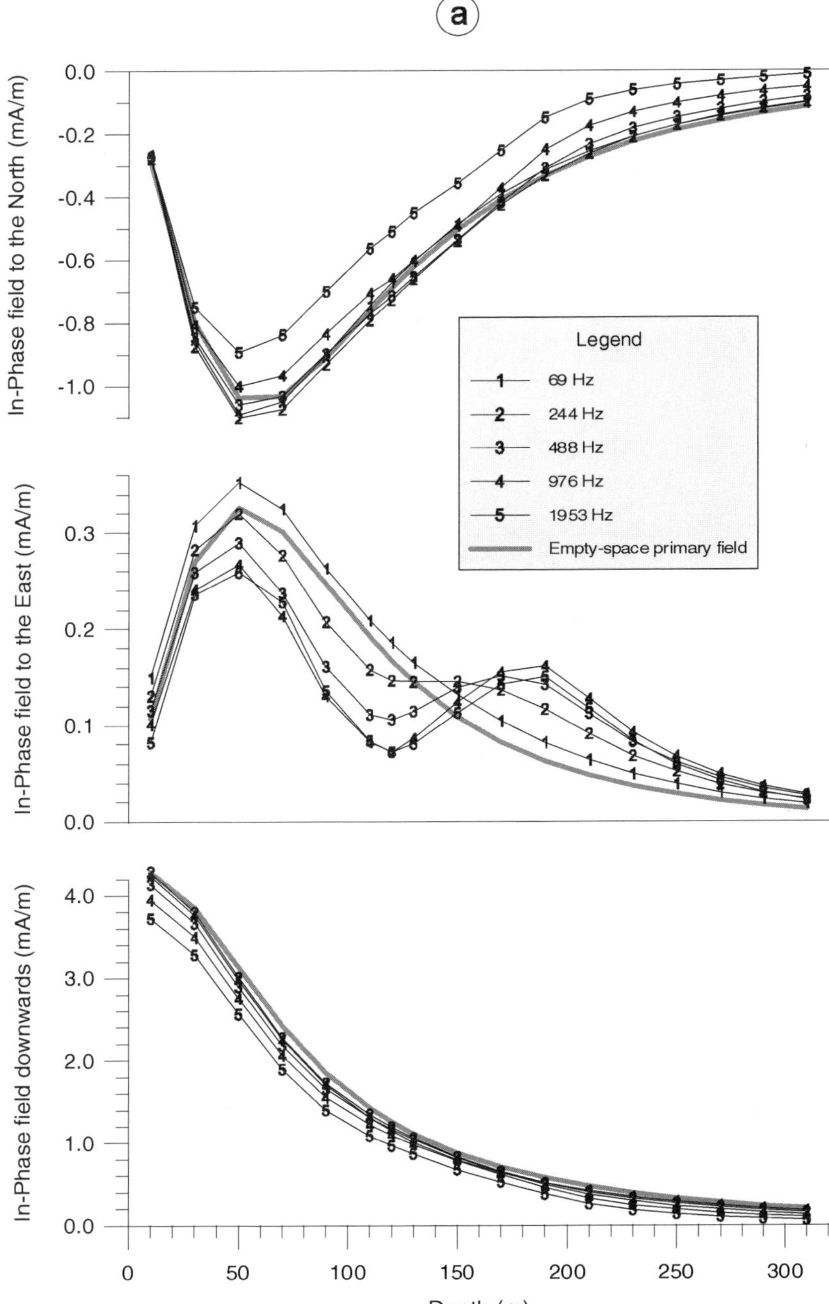

Figure 15. Borehole H12/loop B2: (a) In-phase and (b) quadrature parts of the total magnetic field normalized to 1 A in the transmitting loop. Also shown in (a) is the primary field of the transmitting loop calculated in free space (independent of frequency). (*Continues*)

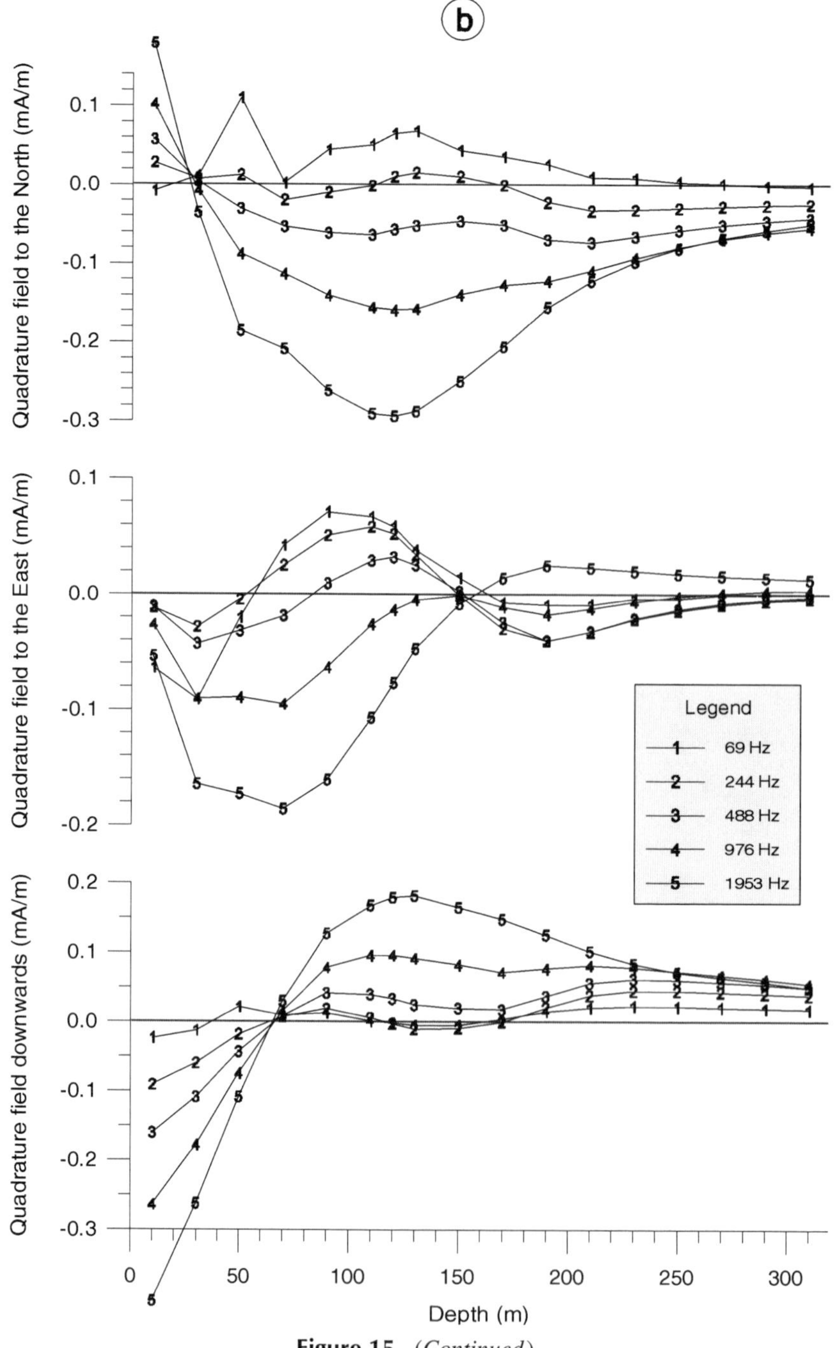

Figure 15. *(Continued)*

at high frequency needs reduction by a realistic conductive host to be used for interpretation of a target.

In the present case the quadrature field was not used for localizing the target because, at all frequencies except the lowest, it was evident that the target response was superposed on a background response reflecting induction in the host rock (e.g., in the north component at 488 Hz, the positive anomaly resulting from the target is clearly superposed on a large minimum resulting from the host). Although the quadrature field

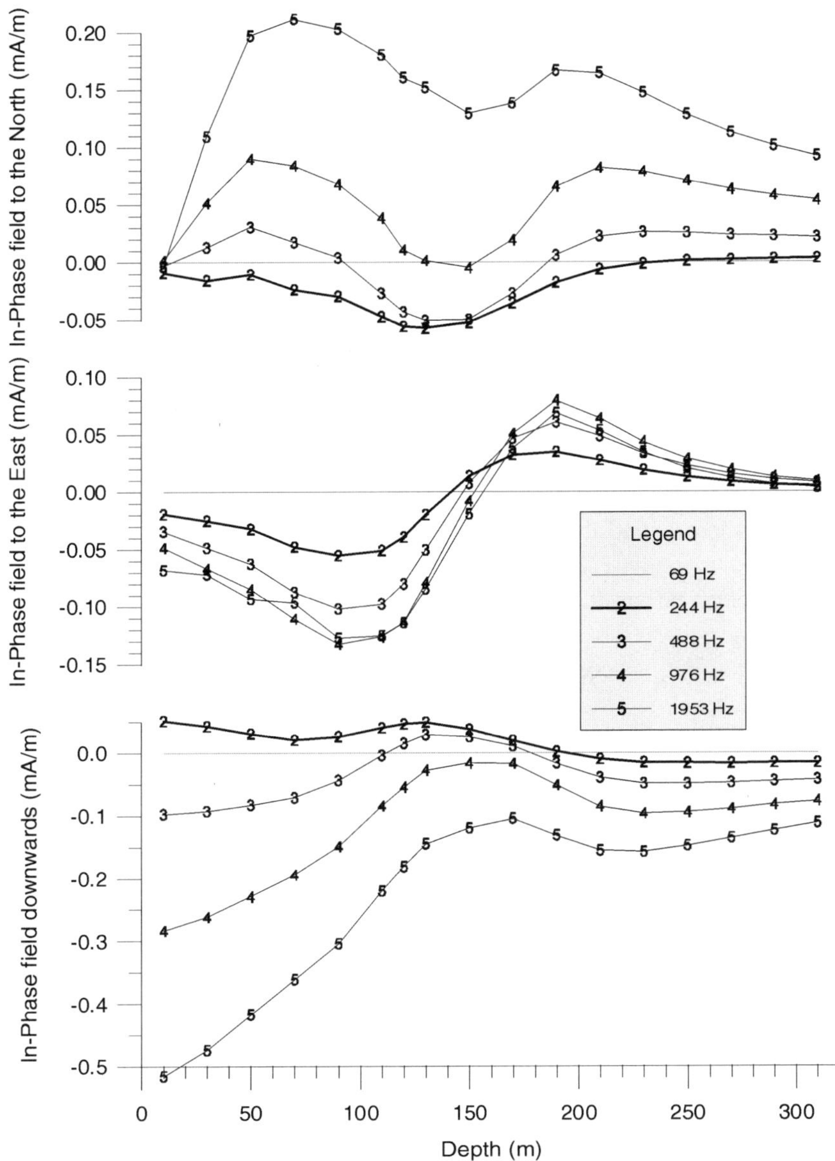

Figure 16. Borehole H12/loop B2: In-phase fields reduced by the lowest frequency (69 Hz).

at the lowest frequency (curve 1) could be equated to the secondary field of the target for localization, this was not used (at least at the beginning) because the in-phase field was less noisy.

The low-frequency reduction of the in-phase fields (Fig. 16) appears, in the present case, to give appropriate results for the frequency 244 Hz (curve 2) at which in-phase induction remains negligible in the host rock while being already significant in the conductive target. The reduced curves at this frequency mainly feature closed anomalies (either extremum or crossover) centered at about 150-m depth and decreasing asymptotically to zero at both extremities of the profile. The vertical channel, however, requires a small additional reduction because it shows a regular increase when approaching the surface.

At higher frequency, because of induction in the host rock, the reduced in-phase field still contains a strong residual influence of the transmitting loop on the north and down components. Curves 4 and 5 (Fig. 16) show reversed and attenuated images of the primary field, representing negative induced loops in the upper ground.

The reduced in-phase field at 244 Hz thus was assumed to be an approximation of the secondary field of the target and subsequently was used for its localization. To use higher frequencies (either in-phase or quadrature) for target localization would require a more suitable reduction taking into account a realistic conductive host (possibly layered). The primary field of such a host would, of course, be different at each frequency and have nonzero quadrature components. Because determining the precise host model can be laborious, it is more practical to apply free-space reduction (or its low-frequency substitute) to the in-phase field at any available frequency that is free of host-rock induction (such as the 244 Hz used here).

The reduced in-phase field at 244 Hz is represented in Fig. 17a as 3-D vectors along borehole H12. Current filament modeling (in freespace) was carried out to interpret these reduced vectors, and the model that was found (by manual adjustment) to give the best fit is shown in Fig. 17b. Although the fit is better controlled in log form (Fig. 18), the vector presentation has the advantage of being more intuitive during the initial stages of modeling. The achieved fit between the reduced in-phase field and the simulated field, though still imperfect, is acceptable. The horizontal components are fairly well reproduced, apart from a slight mismatch close to the surface on the north channel (obviously a near-surface effect) and a small amplitude shift on the east channel. The vertical component is not as well modeled: the vertical data, as we saw above, require additional leveling because curve 2 (Fig. 18, bottom) obviously still contains a small quantity of the downward free-space primary field (Fig. 15a). If such a regularization were to be done, the fit probably would be slightly improved by the use of automatic instead of manual fitting.

Figure 18 also shows the total quadrature field at 69 Hz, which has been reversed and edited to remove the major spike at 50 m on the north component. Even though this quadrature field is slightly noisier than the reduced in-phase field at 244 Hz, it has the same shape. Thus the interpretation obtained for the in-phase field is also suitable for the quadrature field—the fact that the sign is reversed is indifferent to the interpretation because it only means a reversed current in the filament.

Note that the eastward bearing of the interpreted circuit is absolutely consistent with the massive sulfides intersected by borehole H3. If the bearing of the circuit is modified significantly (Fig. 17c), the fit is no longer acceptable, which suggests that there is not much ambiguity concerning the determined direction of the target. Similarly, it can be shown that there is little ambiguity concerning the dip of the target (in this case 75° N, which is in good agreement with the geology)—a reverse dip cancels the fit. The modeling also gives a good determination of the distance to the closest edge of the target (here 25 m). What we cannot as yet determine with this simple modeling is the size of the target. We obtained a strike length of only 15 m, whereas massive sulfides were intersected as far as 100 m away in borehole H3. Possible equivalences between a larger circuit with less current can be expected to provide more realistic results in this respect.

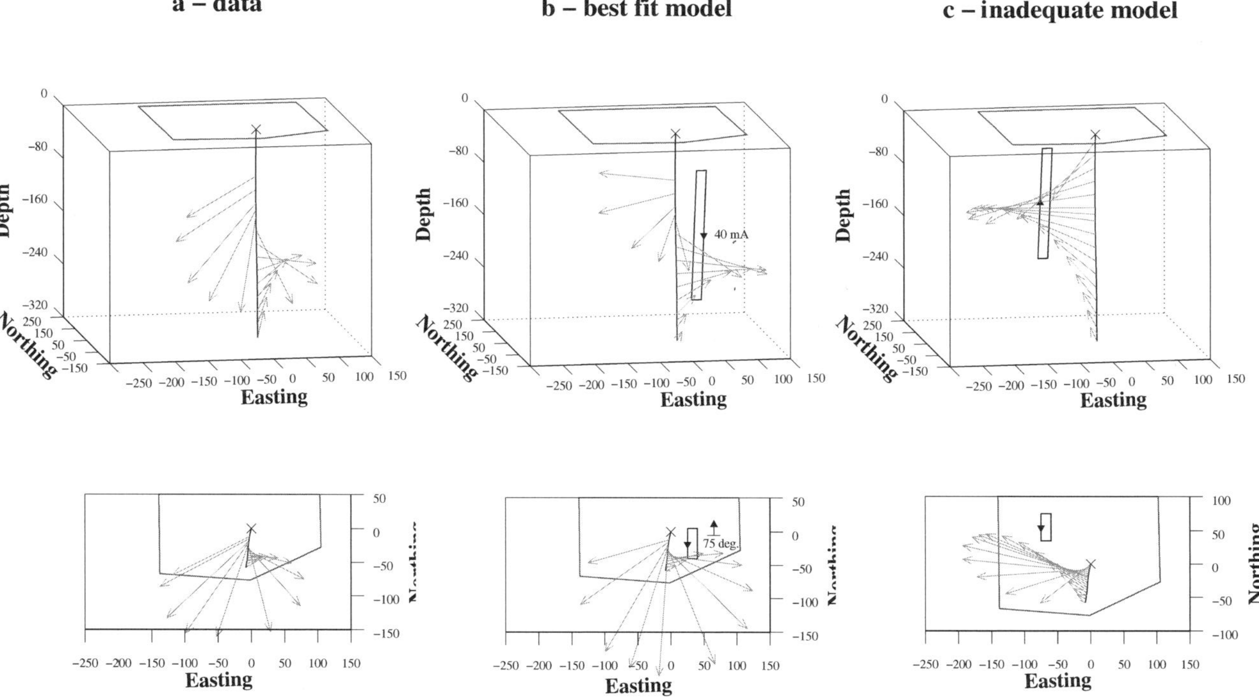

Figure 17. Borehole H12/loop B2: Three sets of vector plots, each showing a 3-D representation (above) and a horizontal projection (below): (a) in-phase field at 244 Hz reduced by 69 Hz; (b) field calculated in free space with the electrical circuit shown in black, giving an acceptable fit with the experimental vectors in (a); (c) another field calculated for a different position of the current filament giving absolutely no match with the experimental vectors in (a).

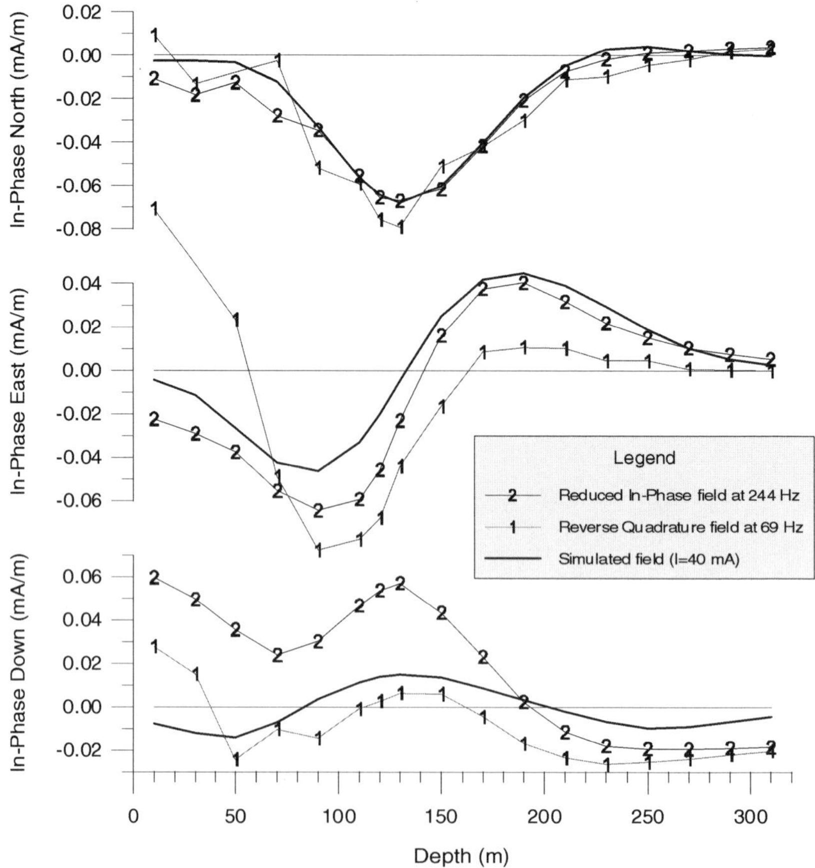

Figure 18. Borehole H12/loop B2: Comparison (in log form) of reduced in-phase field at 244 Hz with simulated field of Fig. 17b.

3.3 Borehole 52820: Barren borehole in a far-miss configuration

The test in borehole 52820 (Sudbury Basin) illustrates the detection and localization of a distant, deep, and very conductive target in a resistive environment, under difficult conditions (very close to high-voltage 60-Hz power lines). The borehole is barren, but a vein-type orebody was suspected in its vicinity, and small ore veins—possibly related to a nearby orebody—had been intersected in the neighboring borehole 85548 at about 550-m depth.

The layout for this test is given in Fig. 19. The reference was relative, given by a magnetic sensor placed at about 50 m from the closest edge of the loop (thus theoretically providing a near-absolute phase reference). The borehole was outside the loop, which has the effect of reversing the primary vertical field at depth, here at 300–350 m (see Fig. 20a). The borehole was logged to a depth of 1340 m (the present ARLETT record) at station intervals of 40 m, which were reduced to 20 m for detailed logging in 35% of cases. Because the depth scale is compressed, the logs have a noisy appearance with anomalies of less than 100 m appearing as spikes.

Figure 20 shows the total in-phase and quadrature fields measured with this array. Of particular interest here is the very good quality of the data recorded at the lowest frequency (61 Hz), despite the interference caused by the proximity of the high-voltage

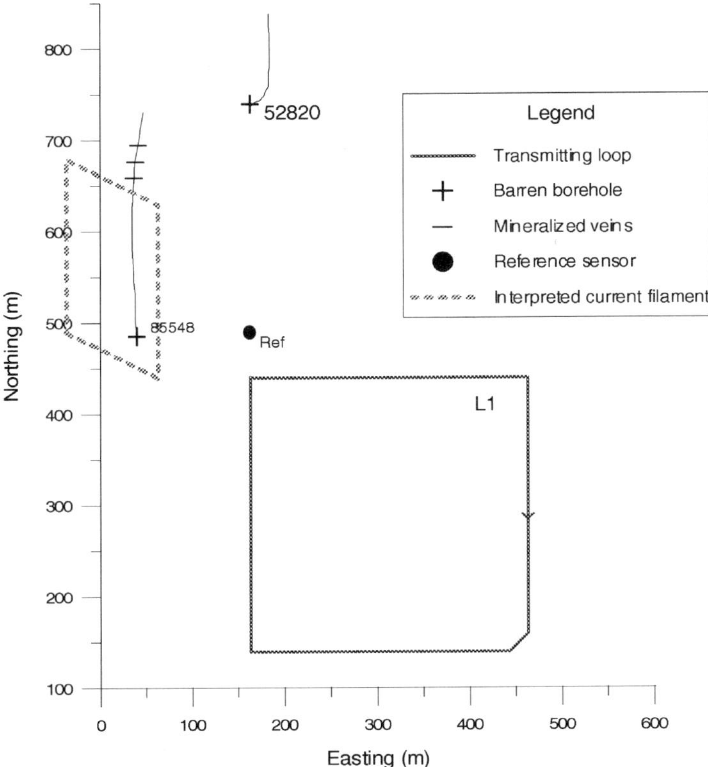

Figure 19. Loop and borehole layout at the Kirkwood test site (Sudbury Basin) for the test in barren borehole 52820 with transmitting loop L1. The location of a distant borehole (85548) that intersected small veins of ore is shown.

60-Hz power lines. The 61-Hz measurements required numerous repetitions (20 to 30 stacks instead of the normal 6 to 10), but the corresponding curves show no more evidence of noise than the other frequencies and are placed coherently with respect to these. The ARLETT system thus demonstrates excellent signal-processing and noise-rejection capabilities.

All three components of the in-phase field show strong and extended anomalies (Fig. 20a) centered at around 600- to 700-m depth and decreasing asymptotically to zero at both extremities of the profile; the east in-phase additionally shows a shallower response between 100- and 500-m depth. These anomalies show up very well against the calculated free-space field (thick gray curve). Considering their extensions and shapes, they are attributed to at least one (and probably two) large, distant conductive targets.

In contrast to the situation in borehole H12, the in-phase curves are all closely parallel (apart from curves 5 and 6 for the horizontal channels below 900 m). The in-phase target response is thus approximately constant at even the lowest ARLETT frequency. From the theoretical standpoint, this means that the target reacts in the high induction domain (inductive limit) where the in-phase response is known to be saturated (no increase of the response when frequency is increased) and where the quadrature response is known to be negligible (West and Macnae, 1991).[10] A practical

[10]Taking an induction number y equal to 50 and a target size a between 100 and 500 m, the formula $y = a/\delta$ [where $\delta \approx 356(\rho/f)^{1/2}$] gives a resistivity ρ between 2. 10^{-3} and 5. 10^{-2} ohm-m, which confirms the high conductivity of the target.

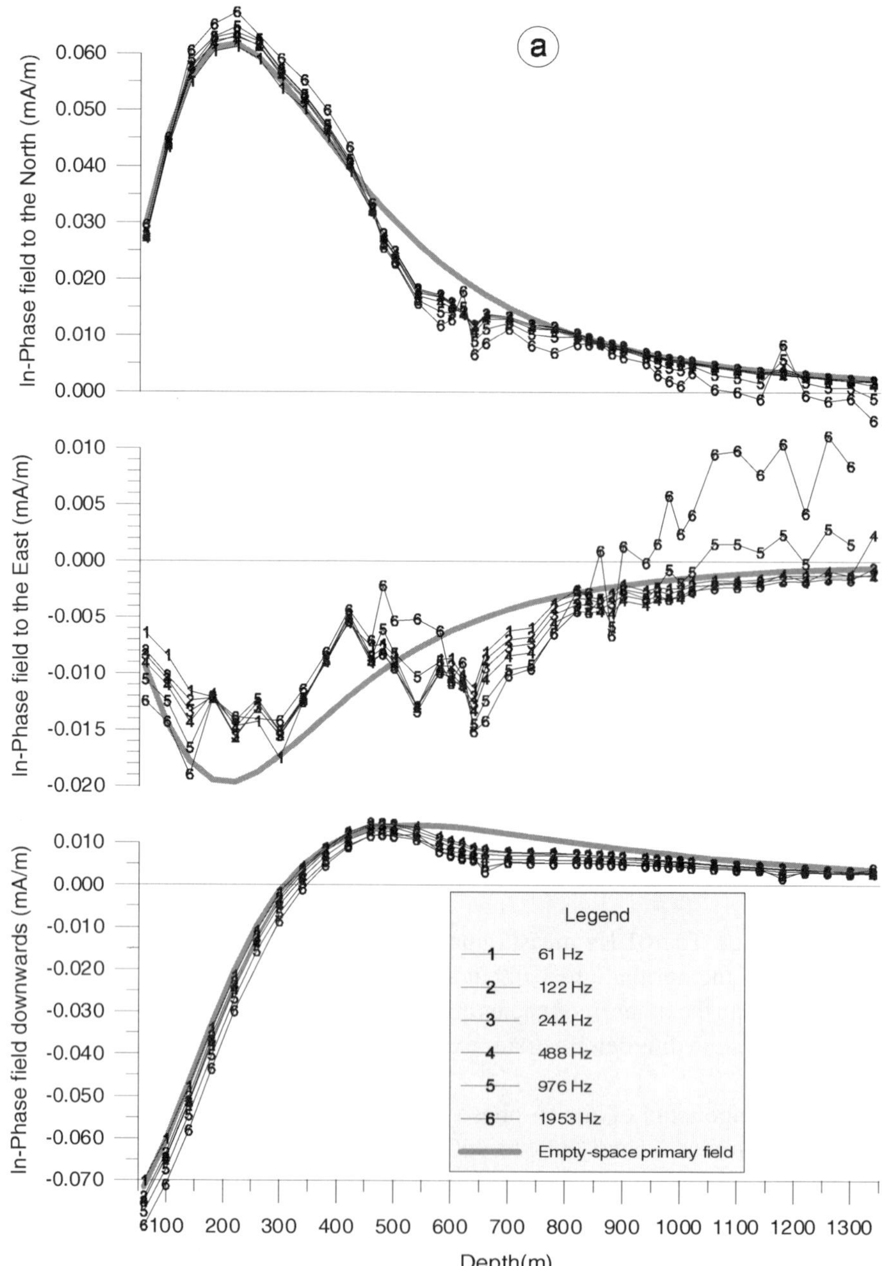

Figure 20. Borehole 52820/loop L1: (a) In-phase and (b) quadrature parts of the total magnetic field normalized to 1 A in the transmitting loop. Also shown in (a) is the primary field of the transmitting loop calculated in freespace (independent of frequency). (*Continues*)

consequence of this target saturation is that the lowest ARLETT frequency (61 Hz) cannot be used for reducing the in-phase field because the effect of the target would subtract from itself and give unusable residuals. The reduction thus has to be done by the calculated free-space primary field (Fig. 20a). The result of this reduction is shown in Fig. 21.

The pattern of the in-phase curves being closely parallel is also the result of the very resistive host, which gives only little quadrature response and almost no in-phase response. Consequently, very little attenuation is observed at high frequency. Once

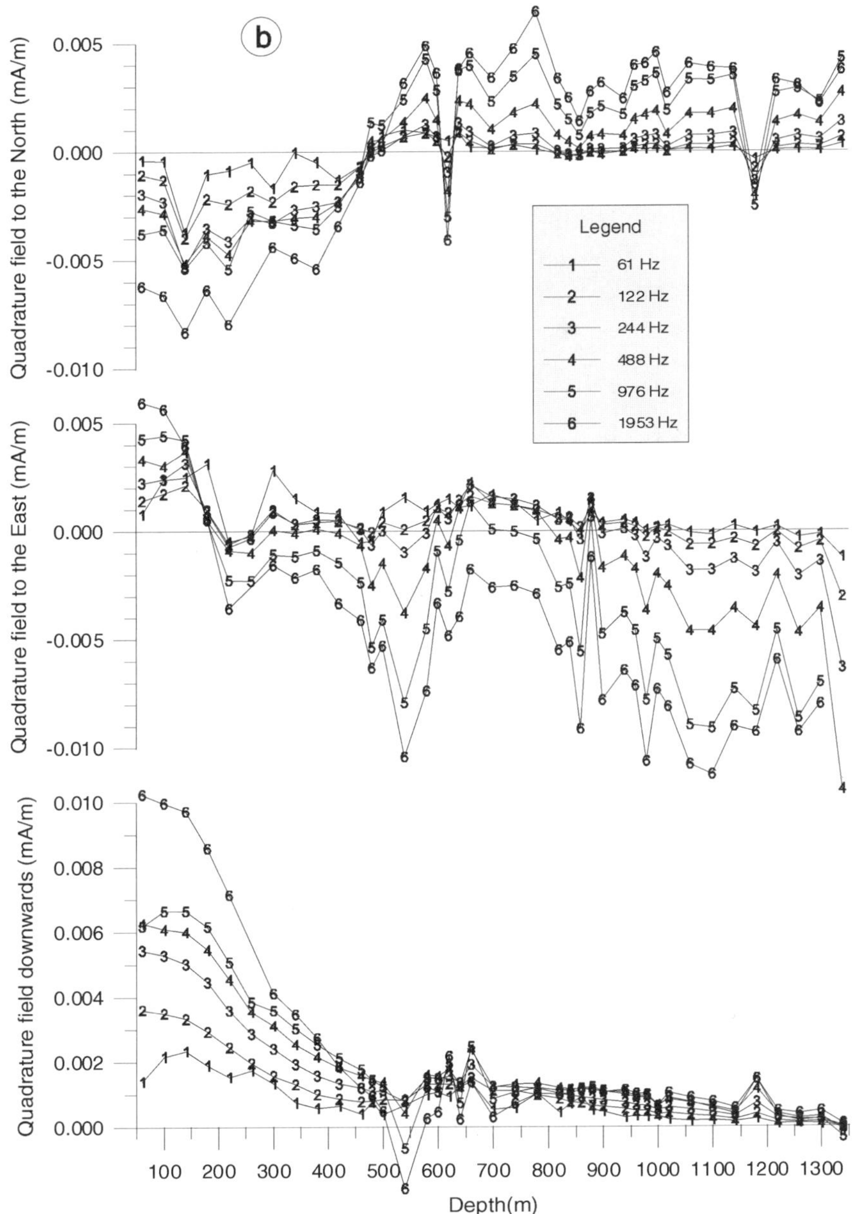

Figure 20. *(Continued)*

reduced by the free-space primary field, the in-phase field at any frequency thus can be considered as reflecting only concealed targets. The discrepancies observed at high frequency on the in-phase horizontal components (departures from the grouped curves) obviously reflect poor conductors very close to the borehole. Such conductors react at the low to intermediate induction domain and thus give an in-phase response that increases with increasing frequency.

As can be expected from the high conductivity of the target, the three components of the quadrature field (Fig. 20b) do not show a clear correlation (or anticorrelation) with the in-phase response of the target(s) (Fig. 21). The observed quadrature anomalies are likely to reflect the nearby poor conductors mentioned above, because these (reacting in the low to intermediate induction domain) give comparable quadrature and in-phase

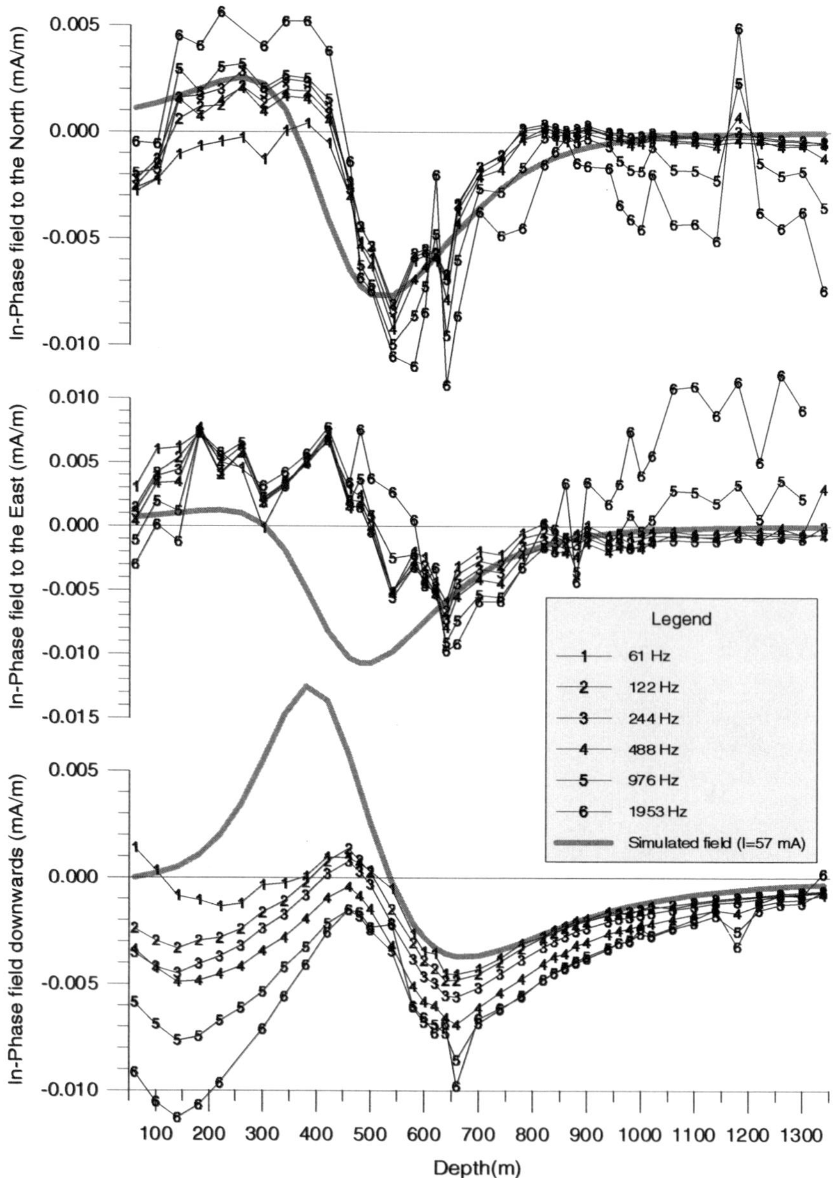

Figure 21. Borehole 52820/loop L1: In-phase field reduced by the calculated free-space primary field compared to the field simulated with the model of Fig. 22b.

responses. It is concluded that the quadrature field is not related to the highly conductive target(s) looked for here, and thus cannot be used for interpreting the target(s).

The field finally selected for target interpretation is the in-phase field at 61 Hz reduced by the calculated free-space field (curve 1, Fig. 21). The lowest available frequency was selected because it gives the same response as the other frequencies for the main target, but minimizes the response of the poor conductors. This frequency also gives the best return to zero close to the surface in the down component, and so circumvents the problem of the minimum observed in this component at 150-m depth which could be due to a secondary anomaly or to an incomplete reduction.

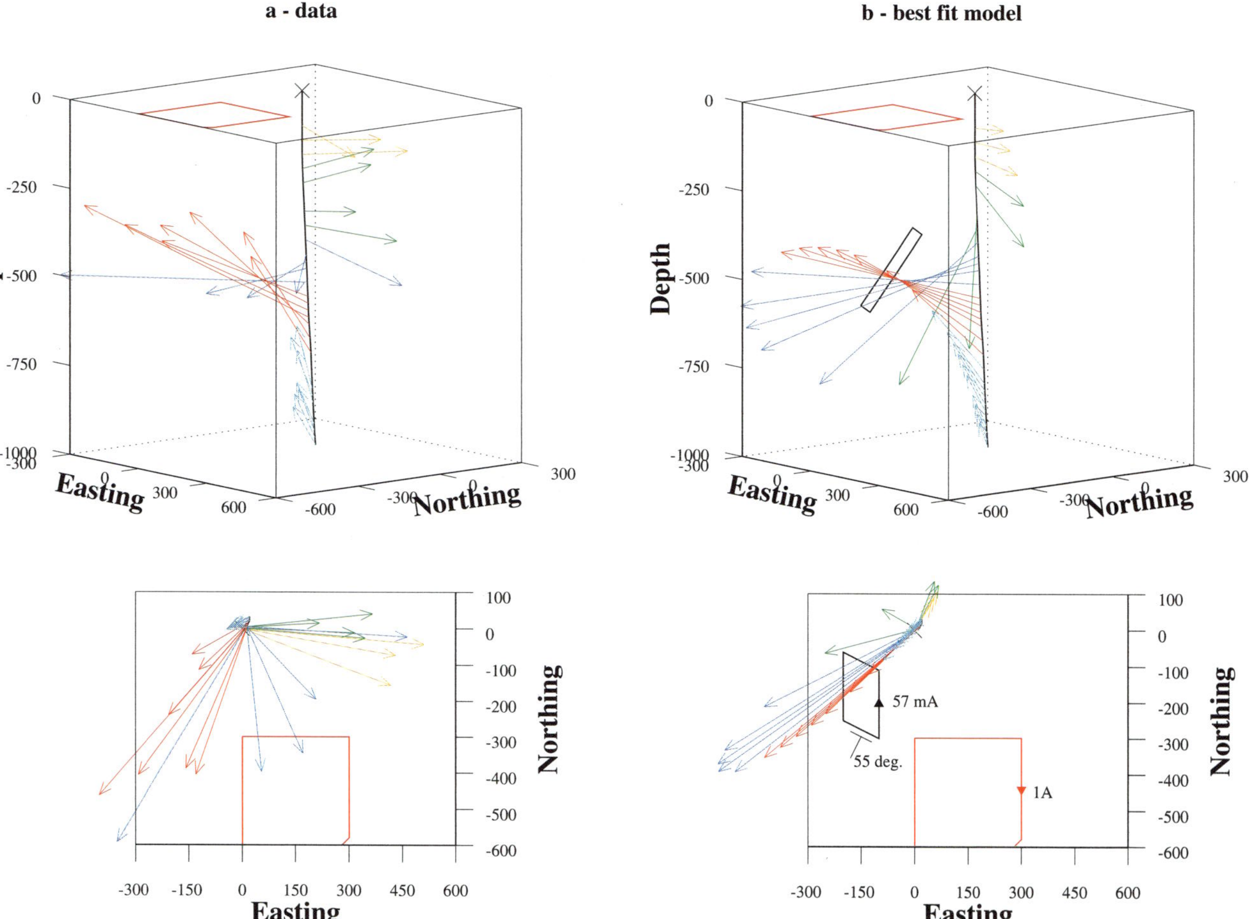

Figure 22. Borehole 52820/loop L1: Two sets of vector plots, each showing a 3-D representation (above) and a horizontal projection (below): (a) in-phase field at 61 Hz reduced by the free-space primary field; (b) field calculated in free space with the electrical circuit shown in black.

The reduced in-phase field at 61 Hz is represented in Fig. 22a as 3-D vectors along the borehole, and the result of current filament modeling (in freespace) is shown in Fig. 22b (as well as in Fig. 21, thick gray curve). The fit between the data and the simulation is mediocre but acceptable at and below the depth of the target (500-m depth and below—blue and red vectors), but is very poor in the uppermost 500 m (orange and green vectors). This clearly shows that a second shallower target also may be present and indicates that modeling should include several current filaments. Despite these uncertainties, the location (about 150 m to the southwest) and dip (55° south-southwest) of the interpreted target circuit were confirmed by the Canadian mining company Inco. From the strength of the recorded anomalies in this resistive setting, it is likely that ARLETT would have located such a highly conductive orebody at an even greater distance (perhaps up to 300 m).

4 Conclusions

ARLETT is a new three-component harmonic EM system for surface-to-borehole measurements; it has sensors that allow a full orientation of the probe and estimation of the borehole trajectory. The system has been tested successfully in the Sudbury Basin and the Iberian Pyrite Belt. Routine onsite data processing includes rotation of the measured fields into an absolute coordinate system, calculation of the borehole trajectory and polarization ellipse parameters, and computation of the secondary field by substraction of a free-space or low-frequency primary field. More sophisticated reductions, such as in a homogeneous or layered conductive half-space, are possible at the office. Interpretation of the reduced field (assumed to be the secondary field of a target) by a 3-D current filament is still rudimentary but already gives the bearing to the target, its structural attitude, and an estimate of its distance. The procedure is being improved by the development of automatic fitting for 3-D current filaments. The data are, of course, ideal for later full 3-D forward modeling.

Acknowledgments

Development of the ARLETT method has been partly supported by a European Community Brite-Euram II research program (contract BRE2-CT92-0299, 1992–1995). We would like to thank the Canadian mining company Inco for funding the Sudbury field surveys and allowing the data to be published. Similarly, we would like to thank the Spanish mining company Seiemsa for the facilities and permissions in relation to the field tests in the Iberian Pyrite Belt. This paper is BRGM contribution 97020.

References

Barnett, C. T., 1984, Simple inversion of time-domain electromagnetic data: Geophysics, **49**, 925–933.

Bourgeois, B., Straub, A., and Valla, P., 1991, Surface-to-borehole frequency EM method, a test study in the Iberian Pyrite Belt: Proc., 4th International MGLS/KEGS Symp. on Borehole Geophysics for Minerals, Geotechnical and Groundwater Applications, Soc. Petrol. Well Log Analysts, Aug. 18–22, 1991, Toronto, 97–105.

Boyd, G. W., and Wiles, C. J., 1984, The Newmont drill-hole EMP system—Examples from eastern Australia: Geophysics, **49**, 949–956.

Dyck, A. V., 1991, Drill-hole electromagnetic methods, *in* Nabighian, M. N., Ed., Electromagnetic methods in applied geophysics, **II** B: Soc. Expl. Geophys. Investigations in Geophysics, **3**, 881–930

Fullagar, P. K., 1987, Inversion of down-hole TEM data using circular current filaments: Expl. Geophys., **18**, 341–344.

Newman, G. A., and Hohmann, G. W., 1988, Transient electromagnetic responses of high contrast prisms in a layered earth: Geophysics, **53**, 691–706.

West, G. F., and Macnae, J. C., 1991, Physics of the electromagnetic induction exploration method, *in* Nabighian, M. N., Ed., Electromagnetic methods in applied geophysics: **II** A: Soc. Expl. Geophys., Investigations in Geophysics, **3**, 5–45.

Use of 3-D Modeling in Design of a New Type of Near-Surface Survey

V. Rath
T. Radic
Y. Krause[1]

Summary. We are developing a new type of near-surface electromagnetic (EM) survey based on measurements of the magnetic field and its derivatives in the megahertz frequency range. The vertical derivatives of the horizontal magnetic fields can replace the electric fields in plane-wave methods. These derivatives are difficult to determine at low frequencies, but are large enough to be measured in the frequency range considered.

We used 3-D modeling to answer several questions concerning this technique. Because the derivative is to be determined in a vertical hole, the measuring procedure itself may introduce a distortion of the fields. It was shown that this effect is significant and has to be taken into account in the design of the sensor or the measuring procedure. Other modeling projects included: (1) the resolution of the method as compared with more traditional techniques of resistivity determination, (2) the influence of small-scale inhomogeneities, and (3) the use of tensorial measurements in complicated areas.

Under realistic conditions the derivatives to be determined are never larger than a few percent of the total field. The modeling requires high accuracy that can be reached only with very dense discretizations and an optimized iteration strategy.

1 Introduction

Electromagnetic (EM) methods are being used to map man-made targets (such as waste dumps) in the near surface. Unlike many geological structures, these targets are often inherently 3-D. The scale of the target also makes it possible to acquire large and dense data sets, provided that the appropriate measuring devices are at hand. A new type of survey for such applications, called radio-magnetic sounding (RMS) [see Radic (1994) where it is called LFM], is being developed by the EM group at the Institut für Angewandte Geophysik, TU Berlin. It uses measurements of the vertical gradient of the magnetic field (i.e., field differences) in the frequency band between 10 kHz and 2 MHz. The penetration depth is a few meters or less. Measurements can be made at radio frequencies that produce magnetic fields large enough to yield accurate

[1]Institut für Angewandte Geophysik, Technische Universität Berlin, Sekr. ACK2, Ackerstr. 71-76 Berlin 13355, Germany.

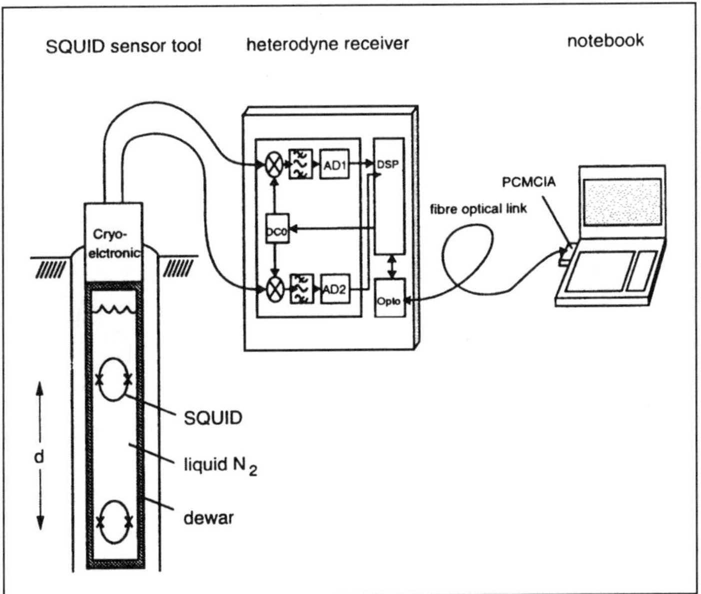

Figure 1. General layout of RMS equipment. Two sensors with a small distance d of 5–20 cm are installed in a vertical hole wide enough to take the dewar. The necessary depth for placement of the sensor as well as the optimal value of d are problems to be investigated.

values of the gradients. In Europe or other industrialized areas, transmitters reaching a magnetic field strength of >10 pT are numerous, so that the frequency band can be sampled quasi-continuously. Because the vertical derivative of the magnetic field is related to the electric field (see Section 2), results may be interpreted with standard plane-wave (magnetotelluric) interpretation techniques. In practice, high-T_C SQUID sensors are used in the equipment. The general layout is shown in Fig. 1. More detailed information can be found in Radic (1994).

Three-dimensional modeling has been a valuable tool in development of the RMS method and we describe some of its uses. Two codes were used, depending on the type of model: a modified version of the finite-difference code of Mackie et al. (1993a,b) and the integral equation program of Wannamaker (1991). In some cases it was informative to cross-check results from both codes.

2 RMS method

The idea of measuring derivatives of the magnetic field is not new (Meyer, 1965, 1966): There have been numerous attempts to utilize vertical gradients [see Spitzer (1991) and references there] at low frequencies or long periods. But at periods $T > 10$ s, the depth interval Δz needed to achieve reasonable accuracy in the derivative (computed by a finite difference) is greater than 1000 m, so that this method is impractical. At the high frequencies considered here, accuracy can be obtained with a much smaller Δz. Estimates given by Radic (1994) show that even for unfavorable conditions (half-space of $\rho = 100$ ohm-m at $f = 20$ kHz with a sensor distance of $d = 10$ cm, and a rather pessimistic primary field strength B_0 of 3 pT) the field differences of ≈ 1 fT are still above the noise level of high-T_C SQUID sensors.

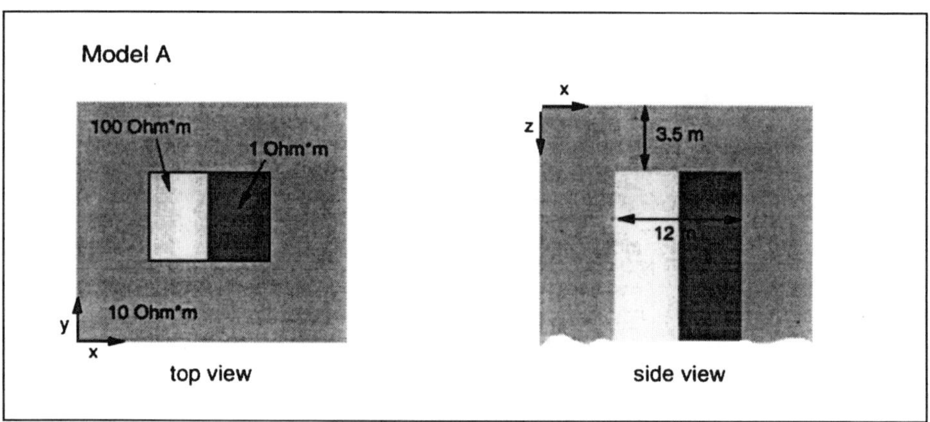

Figure 2. Simple model used to demonstrate general features of the RMS technique.

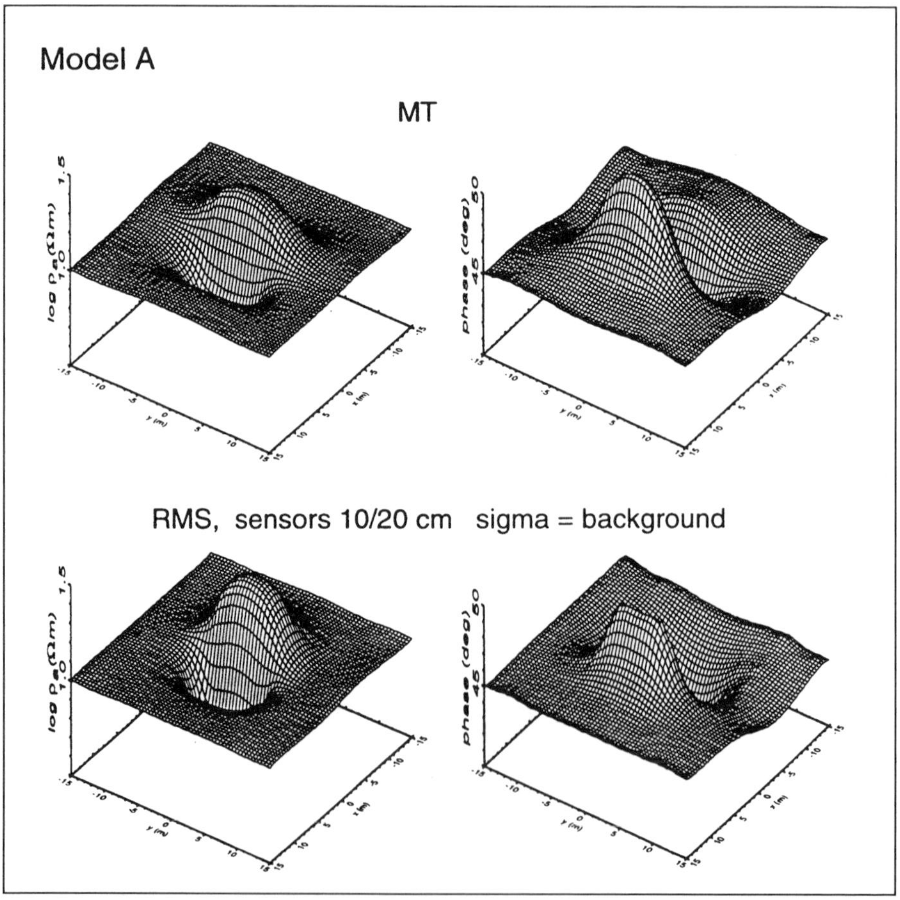

Model A

MT

RMS, sensors 10/20 cm sigma = background

Figure 3. Responses for the model shown in Fig. 2. *Top*: Traditional MT ρ_a and ϕ. *Bottom*: RMS response; $\sigma^* = \sigma_0$ and a sensor distance of 10 cm with $z_0 = 10$ cm was assumed, with the top sensor placed 10 cm below the surface.

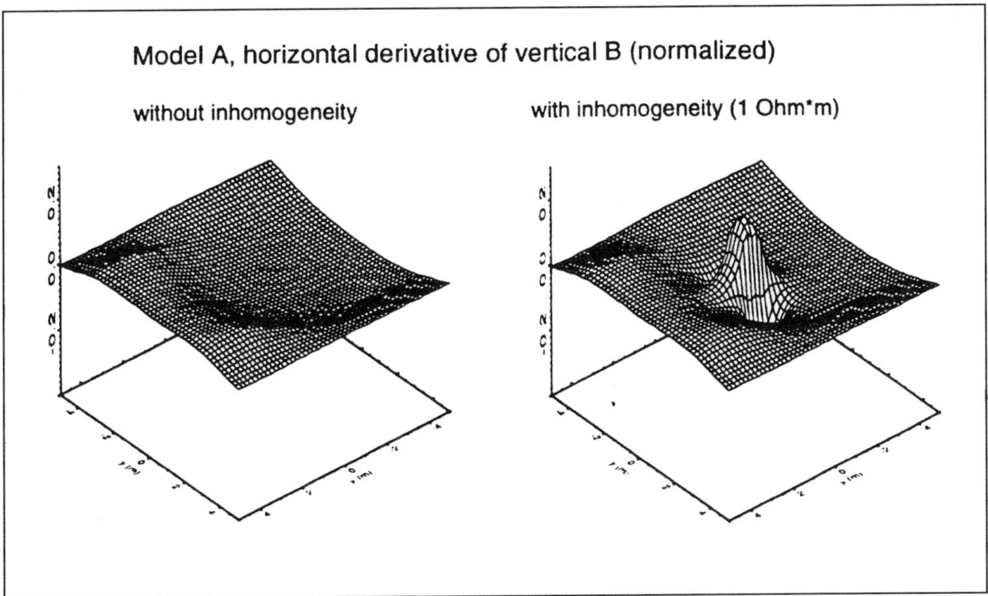

Figure 4. Response for model shown in Fig. 2 with and without small-scale inhomogeneity (1 ohm-m). Shown is $\partial \mathrm{Re}(B_z)/\partial y$ normalized by $\partial \mathrm{Re}(B_y)/\partial z$ for $B_y = 1$ as incident field.

Magnetotelluric (MT) data usually are reduced to impedances,

$$Z_{ij} = E_i/B_j,$$

which are ratios of the electric, E, and corresponding orthogonal magnetic field, B, measured at the surface of the Earth. The electric and magnetic fields have very different spatial behavior. Magnetic fields are usually very smooth, whereas electric fields often have large gradients, which correspond to conductivity discontinuities in the near surface or at depth. The presence of small-scale inhomogeneities may cause erratic behavior of the sampled values. Because of this, measurements of electric fields often must be smoothed spatially, or can end up undersampled. Measurements of magnetic fields seem to be less sensitive to distortions by near-surface inhomogeneities (Meyer, 1966).

According to Maxwell's equations,

$$\nabla \times \mathbf{B} = \mu(\sigma - \iota\omega\epsilon)\mathbf{E}, \tag{1}$$

(σ = conductivity, $\mu \approx \mu_0$ = permeability, and ϵ = permittivity). The electric fields can be replaced by the corresponding components of $\nabla \times \mathbf{B}$, e.g.,

$$E_x = \frac{1}{\mu\sigma^*}\left(\frac{\partial B_z}{\partial y} - \frac{\partial B_y}{\partial z}\right) \tag{2}$$

where σ^* is an appropriate value $\sigma - \iota\omega\epsilon$. We assume that $\sigma \gg \omega\epsilon$, as usual in the audio MT and MT frequency ranges. Consequently, the divisor σ^* contains only conductivity as a physical parameter. For the high frequencies considered here, displacement currents can be neglected only under favorable conditions (high σ, no water present).

One problem of the RMS technique is the choice of σ^*. For the examples in this paper, σ^* was set to the background conductivity σ_0 (usually 10 ohm-m in this study). This amounts to treating the measurement as a relative one that is only as good as the chosen background conductivity. Practical methods of determining a useful σ^* are still being investigated. Nevertheless, the method has several features attractive for near-surface investigations.

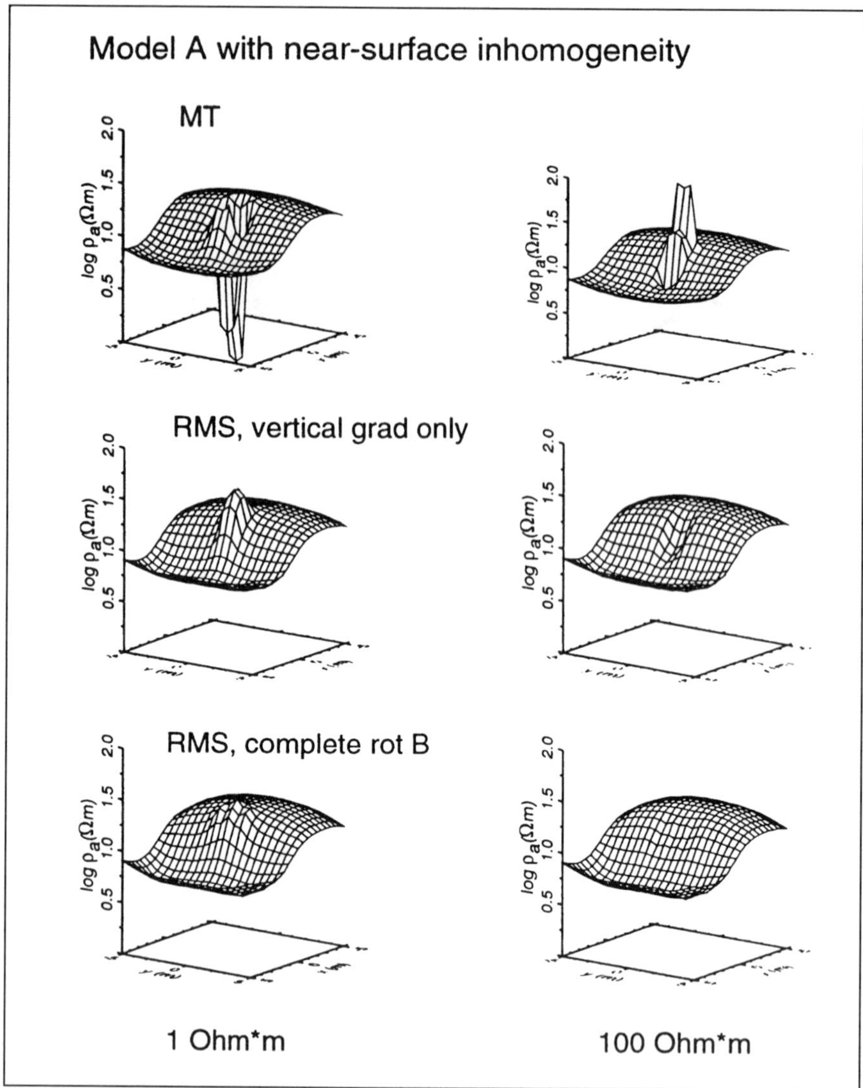

Figure 5. Calculated ρ_a at a frequency of 10 kHz for model A (Fig. 2) with a small-scale inhomogeneity of 1 ohm-m (*left*) and 100 ohm-m (*right*) added in the center of the structure. From top to bottom, values were calculated from classical MT, using Eq. (3), and Eq. (2). In all three cases, $\sigma^* = \sigma_0$ was assumed. The robustness against the effect of small-scale features is clearly visible.

The RMS response of a target (see Fig. 2) is shown in Fig. 3. Apparent resistivities and phases are given for MT and RMS techniques based on $\sigma^* = \sigma_0$. When there are no near-surface inhomogeneities, both methods lead to nearly identical results. As shown in the next section, in the presence of small-scale near-surface scatterers, this is no longer true.

3 Role of lateral gradients of B_z

The first applications of the RMS method assumed that some terms in the full $\nabla \times \mathbf{B}$ were much smaller than the others. Considering the x-component,

$$\frac{\partial B_z}{\partial y} - \frac{\partial B_y}{\partial z} \approx -\frac{\partial B_y}{\partial z}. \tag{3}$$

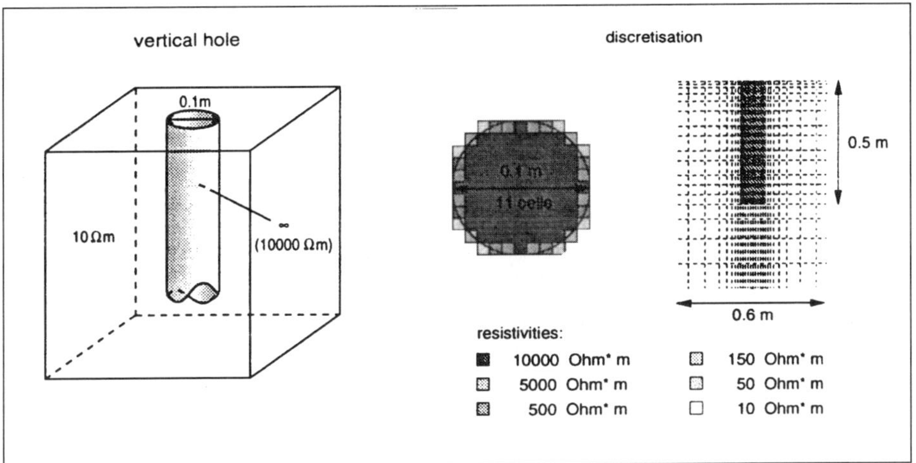

Figure 6. General geometry of the hole and its discretization used in the calculations. The specific resistivity of air was set to 10 000 ohm-m or more.

Other components were treated similarly; i.e., the conductivity is considered to be nearly 1-D. This restriction originally was motivated by practical considerations, because at long periods it was only possible to measure vertical gradients in a (vertical) well of adequate depth. Though from the viewpoint of interpretation, any type of measured data could be used for inversion or modeling, it seems that possibly important information is lost, especially if targets are near-surface and very inhomogeneous structures. To check this, several numerical calculations were made to find out under what conditions this is true.

The importance of the complete $\nabla \times \mathbf{B}$ becomes clear when small-scale inhomogeneities are present. To demonstrate this, small 1.2 m × 1.2 m inhomogeneities of 1 and 100 ohm-m with a vertical extent of 30 cm were placed in the center of the model shown in Fig. 2. In Fig. 4 the ratio of the two terms on the left side of Eq. (3) is given for the target structure in Fig. 2 with and without the surficial body. A moderate response is observed in the first case (<0.1), and a much larger one in the second case. Apparent resistivities (ρ_a) for MT and the two variants of RMS are shown in Fig. 5. Phases exhibit similar behavior in all cases. In the ρ_a results for MT, the typical discontinuous character is evident. The behavior is already much smoother in the one-term RMS, but the values obviously are biased. It is clear that the influence of small-scale structure is reduced with the second RMS technique using the full $\nabla \times \mathbf{B}$. However, this comparison is not really fair to MT, because the electric fields (not the magnetic derivatives) in this example were calculated as point receivers, whereas, in reality, finite electrode dipoles are used.

The importance of lateral gradients means that further development of the RMS technique will have to incorporate the lateral gradients. In principle, this could be done by designing an appropriate gradiometer or by estimating the gradients from dense measurement arrays. Both possibilities will be investigated further.

4 Influence of the measuring configuration

The RMS technique requires a measuring hole, which should be wide enough to take the sensor and not too deep for practical reasons. This air-filled hole is already a small-scale anomalous body affecting the accuracy of the measured gradients. For this reason, 3-D

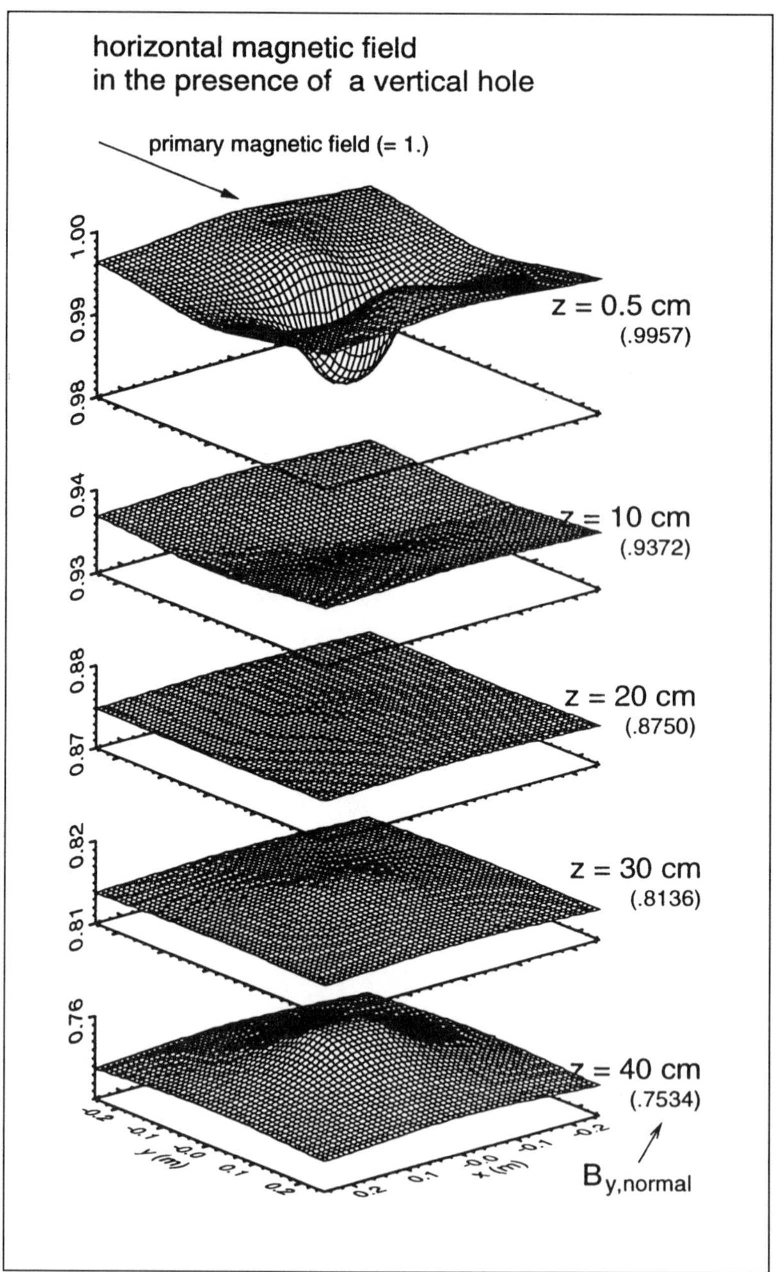

Figure 7. Magnetic fields B_y (real part) for the model shown in Fig. 6 at a frequency of 1 MHz and a hole depth of 0.5 m.

modeling was used to check whether the magnetic field is disturbed significantly by the presence of this hole. Remember that, for the situation modeled here, the vertical extent of the hole has the same order of magnitude as the induction depth ($\lambda \approx 1.5$ m), whereas its lateral extent is much smaller. The general geometry of the problem as well as its discretization near the anomalous structure are shown in Fig. 6. To reduce the resistivity contrasts and possibly the staircase effect resulting from the Cartesian grid, border cells were associated with values near the geometric mean of the resistivities according to the volumes occupied by air and earth.

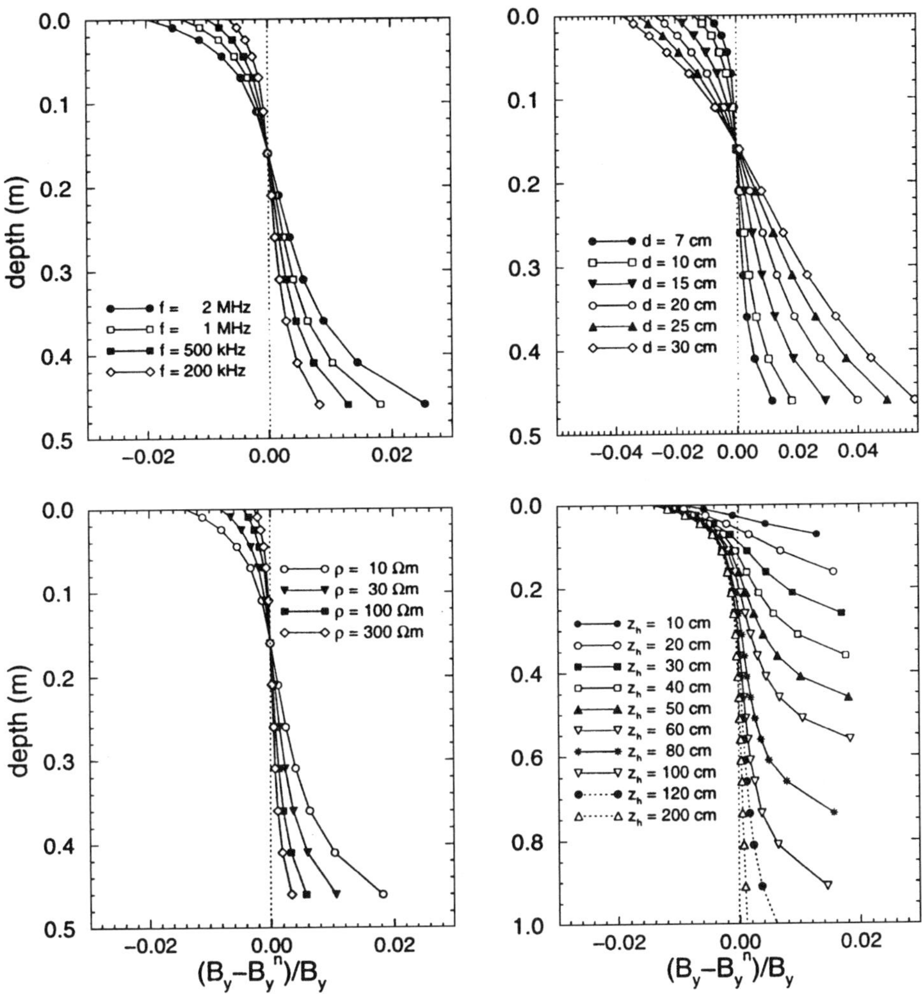

Figure 8. Influence of frequency, half-space conductivity, and geometry on B_y. In each subfigure the deviation from the corresponding half-space response is shown. If not indicated otherwise, the standard parameters are $f = 1$ MHz, diameter $d = 0.1$ m, depth $z_h = 0.5$ m, and a resistivity of 10 ohm-m; only the real part is shown.

It is clear from the results shown in Fig. 7 that most of the effect of the hole is concentrated near the surface, so that putting the device at a depth >10 cm could reduce the bias significantly. The distortion of B_y at the center of the hole seems to have a minimum at a depth somewhere between 20 and 30 m. In this configuration, this should be the optimum depth for the placement of the device. Some more systematic results are given in Fig. 8. It is clear that, to reduce its effect, the hole should be as narrow and as deep as possible. Preferentially, the sensor should be placed in the nearly linear area near $\approx z_h/3$. Frequency and conductivity dependence can be considerable.

5 Conclusions

Modeling has helped us to understand the behavior of a new type of survey that measures derivatives of the magnetic field. We were able to check and even correct some heuristic assumptions that we originally made about the method. The improved understanding of the physics generated improvements on the use and interpretation of this technique under realistic conditions. Our experience with 3-D modeling of the RMS technique indicates that an increase in accuracy of the available codes would be highly desirable. When vertical gradients have to be simulated, accuracy has to be higher than in standard MT modeling ($<10^{-2}$). For example, $\Delta B_y / B_y$ (real part) based on the depth interval used in practical measurements usually will not exceed 0.05. To get tolerable errors in the derivatives, an accuracy of four to five decimal places in the calculated fields is required. To reach this with the finite-difference code of Mackie et al. (1993) considerable care has to be taken in the discretization of the model, especially when the influence of small-scale inhomogeneities is included. Even simple structures require very large discrete models. To achieve high accuracy in the values of the magnetic field in the air, in the presence of high contrasts nearby, the iteration correcting for nonvanishing $\nabla \cdot \mathbf{B}$ had to be applied in short cycles restarting the implemented biconjugate gradient iteration approximately every 10th to 20th iteration. Even with these modifications it sometimes took more than 500 iterations to reach acceptable accuracy. Experiments with more effective iterative schemes [e.g., implementing minimum residual smoothing (Zhou and Walker, 1994) or another variant of the quasi-minimal residual method (Freund and Nachtigal, 1991)] and preconditioners are currently under way. We also are studying the direct computation of anomalous fields with the finite-difference algorithm.

Acknowledgments

This work was funded by the German BMBF under contracts 13N6446/8 and 13N6435/9. Special thanks to R. Mackie and T. Madden for providing us with the original version of their 3-D MT modeling code, and P.E. Wannamaker for his 3-D integral equation program.

References

Freund, R., and Nachtigal, N., 1991, QMR: A quasi-minimal residual method for non-Hermitian linear systems: Numer. Math., **60**, 315–339.

Mackie, R. L., and Madden, T. R., 1993, Conjugate direction relaxation solutions for 3-D magnetotelluric modeling: Geophysics, **58**, 1052–1057.

Mackie, R. L., Madden, T. R., and Wannamaker, P. E., 1993, Three-dimensional magnetotelluric modeling using difference equations—theory and comparison to integral equation solutions: Geophysics, **58**, 215–226.

Meyer, J., 1965, Übertragung des Cagniardschen Verfahrens auf den Fall der Beobachtung nur magnetischer Größen, *in* Kolloquium Elektromagnetische Tiefenforschung, Goslar: Inst. Geophys. Meteorol. TH Braunschweig, 133–146.

———1966, Die magneto-tellurische Tiefensondierung und ihr erdmagnetisches Analogon: Gerlands Beitr. Geophys., **75**, 289–300.

Radic, T., 1994, Konzept einer LFM-Apparatur zur Messung des oberflächennahen spezifischen Widerstandes unter Anwendung des magnetischen Gradientenverfahrens im LW- und MW-Frequenzbereich, *in* Bahr, K., and Junge, A., Eds., Kolloquium Elektromagnetische Tiefenforschung, Höchst/Taunus: Deuts. Geophys. Gesellschaft, 470–479.

Spitzer, K., 1991, Ein triaxiales Magnetometer zur fortlaufenden Registrierung erdmagnetischer Variationen in tiefen Bohrungen—erste Erprobung in der Vorbohrung des Kontinentalen Tiefbohrprogramms: Ph.D. thesis, Georg-August-Univ. zu Göttingen.

Wannamaker, P. E., 1991, Advances in three-dimensional magnetotelluric modeling using integral equations: Geophysics, **56**, 1716–1728.

Zhou, L., and Walker, H. F., 1994, Residual smoothing techniques for iterative methods: SIAM J. Sci. Stat. Comput., **15**, 297–312.

PART IX

GENERAL

Interaction of Electromagnetic Fields and a Model of the Human Head

Atef Z. Elsherbeni[1]
Joseph S. Colburn and Yahya Rahmat-Samii[2]
Clayborne D. Taylor, Jr.[3]

Summary. Antennas for personal communications devices transmit and receive electromagnetic (EM) waves (at microwave frequences) near the operator's head. We have used the finite-difference time-domain method to model this EM radiation as it interacts with a model of the human head, and have generated animated movies to visualize this interaction. We have studied two antennas in detail—a 900-MHz terrestrial wireless antenna and a 1.8-GHz handset-to-satellite link. In one of our studies, the model of the human head was derived from magnetic resonance images; in another study, the model came from pictures of anatomical cuts. Movies showing the evolution of the fields in these models are available at the Society of Exploration Geophysicists world wide web site.

1 Introduction

The growth of mobile communications has generated considerable interest in the interaction of the radiated electromagnetic (EM) fields with humans. Not only is the operator's influence on the antenna's gain, radiation pattern, and input impedance an issue for the performance of the hand-held devices, but there are also concerns about how the radiated EM energy affects human tissue.

Many authors have studied this interaction using different numerical techniques and different (approximate) models for the handset and the human head (Dimbylow, 1993; Toftgård et al., 1993; Dimbylow and Gandhi, 1993; Chen and Wang, 1994; Chuang, 1994; Martens, 1994; Colburn and Rahmat-Samii, 1995a,b; Gutschling and Weiland, 1995; Jensen and Rahmat-Samii, 1995). The finite-difference time-domain (FDTD) method (Yee, 1966; Taflove and Brodwin, 1975; Luebbers et al., 1992) is well suited for this problem because it can model EM fields in general geometries easily and efficiently. Also, because it computes the evolution of the fields at all points in space,

[1]Department of Electrical Engineering, University of Mississippi, University, MS 38677, USA; on leave at the Department of Electrical Engineering, University of California, Los Angeles.
[2]Department of Electrical Engineering, University of California, Los Angeles, Los Angeles, CA 90095-1594, USA.
[3]Department of Electrical Engineering, University of Mississippi, University, MS 38677, USA.

the FDTD method allows a full visualization of the EM interaction. We have used the FDTD method and modern EM visualization techniques to study the interaction of two different handset antennas with the human head.

In our first example, a simple dipole antenna models a handset radiator at 900 MHz. The head model is created from actual magnetic resonance imaging (MRI) scans transformed onto an FDTD grid. The computational head model includes eight different tissue types, for which the electrical properties are prescribed at 900 MHz. In our second example, a circular polarized helix antenna represents the radiating element at 1.8 GHz. These models are good examples for cellular and satellite telecommunications. For each model, we compute near-field distributions, specific absorption rates (SAR), and far-field patterns. We have cast the time-domain simulations (of the dipole antenna interacting with the human head) into movie files that can be played on personal computers (PCs) under either Windows/DOS or MacOS (Elsherbeni and Taylor, 1995). Interested readers can obtain these movie files on the world wide web at `http://www.seg.org/books/3dem/elsher/elsher.html`, or at `http://www.olemiss.edu/~atef/head/head.html`.

2 Numerical approach

2.1 FDTD method

The FDTD method has an extensive literature, so we will give only a brief description here. The FDTD method steps Maxwell's equations in time on a discrete grid; its simplicity lends itself for use in computing EM fields in complex geometries. On a rectangular grid, centered difference approximations for the spatial and temporal derivatives in Maxwell's equations result in explicit equations for values of all field components at the next time step in terms of their values at present and past steps. The centered spatial differences operate naturally on a discretization of space into unit cells, termed Yee cells, which are rectangular parallelepipeds of dimensions Δx, Δy, and Δz. The electric-field components lie along edges of the Yee cells, and the magnetic-field components are perpendicular to cell faces in a staggered fashion. Centered differences in time are constructed such that the electric and magnetic fields are computed half a time step apart, which allows leap-frogging between the electric and the magnetic fields.

The material characteristics of each unit cell can be assigned independently to allow for modeling of inhomogeneous media. To simulate infinite space in a finite computational domain, the FDTD grid is terminated by a special (absorbing) boundary conditions to simulate outgoing waves in free space. The computation starts by introducing a prescribed electric field into the grid and continues until the fields in the domain go to zero or a steady state is achieved, depending on the type of excitation used. Frequency-domain quantities can be determined by Fourier transformation of the temporal values. Subunit cell features, like thin wires, can be handled by modifying the difference equations locally.

2.2 Power absorption and SAR

Various quantities can serve to assess the absorption of energy in tissue. The real power delivered to the antenna (P_{del}) is the sum of the power radiated to the far-field (P_{rad})

and the power absorbed within the lossy tissue (P_{abs}). The quantities (P_{abs}) and (P_{rad}) can be determined by

$$P_{abs} = \frac{1}{2}\int_V \sigma |\mathbf{E}|^2 dV, \tag{1}$$

$$P_{rad} = \frac{1}{2}\,\mathrm{Re}\left\{\int_S \mathbf{E} \times \mathbf{H}^* \cdot \hat{n}\, dS\right\}, \tag{2}$$

where **E** and **H** are the frequency-domain electric and magnetic fields (phasors) and σ is the medium's conductivity. The symbols V, S, and $\hat{n}$ represent the volume containing the tissue, a surface completely surrounding the antenna, and the unit outward normal to the surface, respectively.

SAR is a measure of the power absorbed per unit mass of tissue. This quantity is defined as

$$\mathrm{SAR} = \frac{\sigma}{2\rho}|\mathbf{E}|^2 \tag{3}$$

where ρ is the material density. The values of σ and ρ for the different tissue types used in the computational head models are listed in Table 1 at 900 MHz and 1.8 GHz. The SAR averaged over any 1 g of tissue for 30 minutes or more should remain under 1.6 mW/g (American National Standards Institute, 1991; Fischetti, 1993).

3 Dipole antenna

We present first a study of a dipole antenna next to an adult head. An important part of the study is the generation of a realistic computational model for the human-head model. We describe a procedure that generates such a model automatically from MRI. The procedure allows one to adjust the level of discretization to the physical resolution desired or to the computer resources available.

3.1 Construction of a 3-D human-head model

The input for the head model was obtained from the World Wide Web in the form of MRI scans (Johnson and Becker, 1995). Figure 1 shows a sample scan. The circle at the top of this figure is at the intended location of the antenna, which is transmitting in the simulations.

The MRI scans were downloaded as graphical interchange files (GIF) in binary format. The set consists of 54 axial slices through a human head. For each axial slice, three separate MRI scans were acquired, each concentrating on a different tissue types. The three types of MRI scans are designated as PD-, T1-, and T2-weighted (see Johnson and Becker, 1995) for explanation of the scan types.

We used the *Xview* (see Bradley) program to convert and save each image in a postscript format. Figure 2 shows a section of one of the MRI postscript files, which are in hexadecimal format. Each pixel is represented by two hexadecimal characters which are converted to decimal and fall within the range of (0,255). Each MRI images occupies a 256 × 256 pixel grid; each pixel has a value in the range (0,255). The physical material type at each pixel location then is determined by the value of the pixel in the MRI image. Table 2 is a pixel-range-to-tissue-type conversion based on the T2-weighted image.

Table 1. Material parameters at 900 and 1800 MHz

		900 MHz		1800 MHz	
Tissue type	ρ, g/cm^3	σ, S/m	ϵ_r	σ, S/m	ϵ_r
Bone	1.85	0.105	8.0	0.15	8.0
Skin/fat	1.10	0.60	34.5	0.57	32.0
Muscle	1.04	1.21	58.5	1.76	56.0
Brain	1.03	1.23	55.0	1.58	53.0
Humour	1.01	1.97	73.0	2.27	74.0
Lens	1.05	0.8	44.5	1.19	42.0
Cornea	1.02	1.85	52.0	2.29	51.0

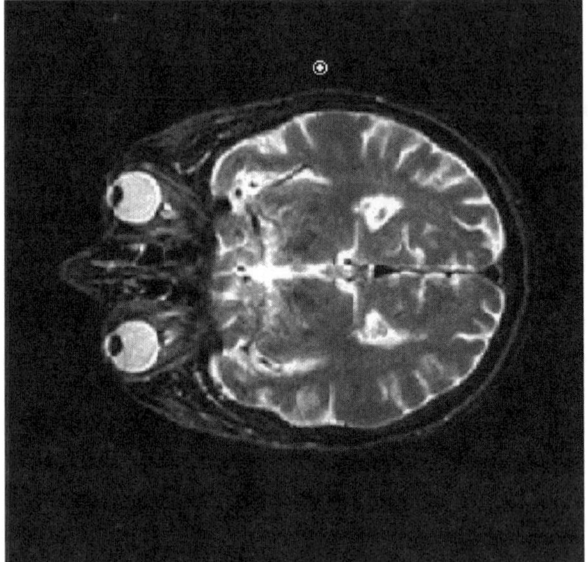

Figure 1. MRI slice of the head at 11.5 cm up from the chin.

```
256 256 8 % dimensions of data
[256 0 0 -256 0 256] % mapping
matrix
{currentfile pix readhexstring pop}
image

00000000000000000202000000000000000000000000000000000
00000000000000000000000000000000000000000000000000000
00000000000000000000000000000000000000000000000000000
000000000000000000000000000000001010323246463d3d2c2c2
35353838323222f2f32322f2f2a2a24241e1e21211e1e191913131
0000000000000000000002020202000002021c1c3a3a43434b4b7
272727272c2c383827271e1e40405c5c56563d3d1e1e020200001
101002020b0b27273d3d4646464654545c5c40402a2a272705050
```

Figure 2. Section of the postscript code that describes the MRI scan.

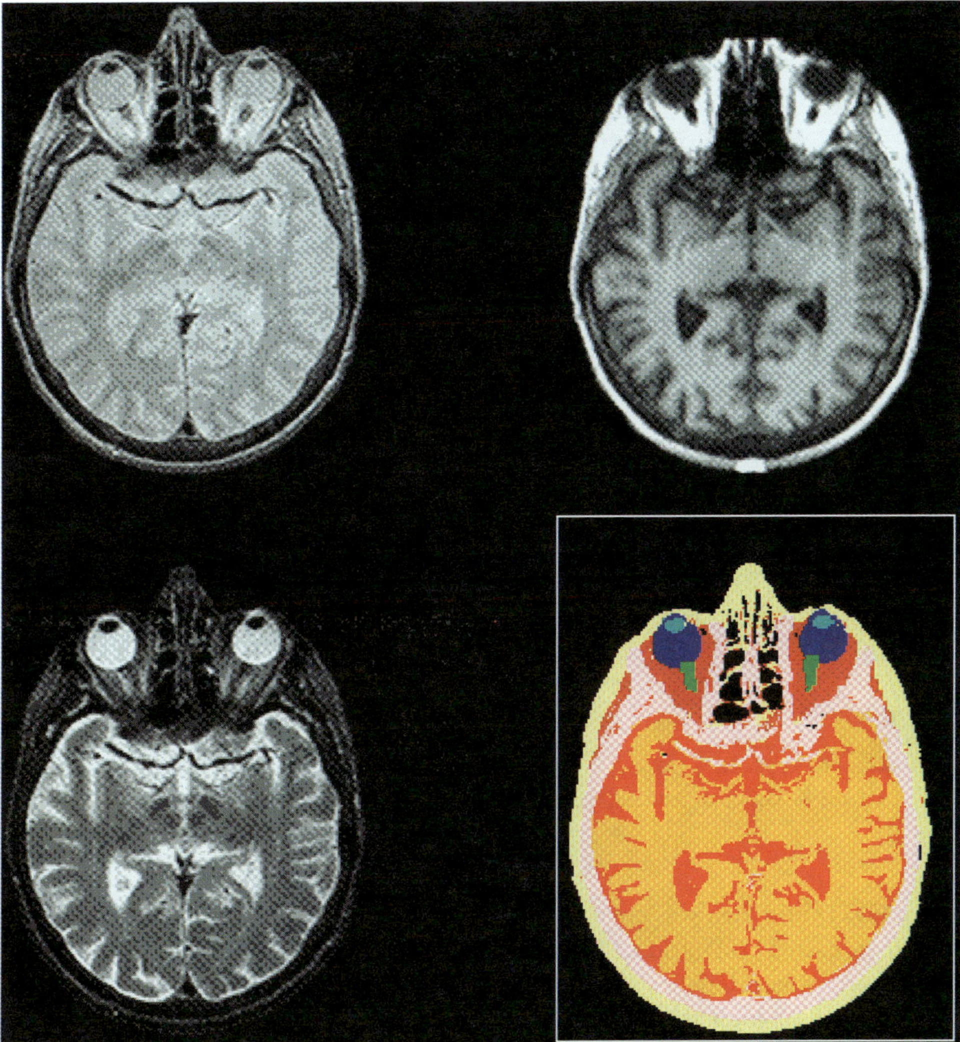

Figure 3. Three MRI scans for a given axial cut through the head: (upper left) PD-weighted image, (upper right) T1-weighted image, (lower left) T2-weighted image), and (lower right corner) corresponding discretized image.

Initially only the T2 image scans were considered. For every pixel in the head, a material type was assigned on the basis of the pixel value in the T2 image file. This was the first approximation of the human head, which consisted only of pixels assigned a physical location and material type. Next, an interactive program refined the first head model with data from the PD and T1 images. The program would first display the three different weighted MRI scans as shown in Fig. 3 along with the current head model. Next, the user was allowed to choose a material type and "paint" any desired corrections to that material-type placement with the mouse cursor to any of the 54 layer scans. This painting to correct the material-type designation could be done either directly on the head model or on any one of the MRI scans. These refinements were automatically entered in the head model. We found that the interactive program was a very efficient method for refining initial models.

Table 2. Material gray scale and respective color

Tissue type	Gray level	Color value
Air	0–60	Black
Skin	60–90	Yellow
Bone	90–110	White
Muscle	110–130	Dark red
Fat	130–160	Light gray
Brain	160–200	Dark gray
Eye	200–220	Blue
Blood	220–255	Red

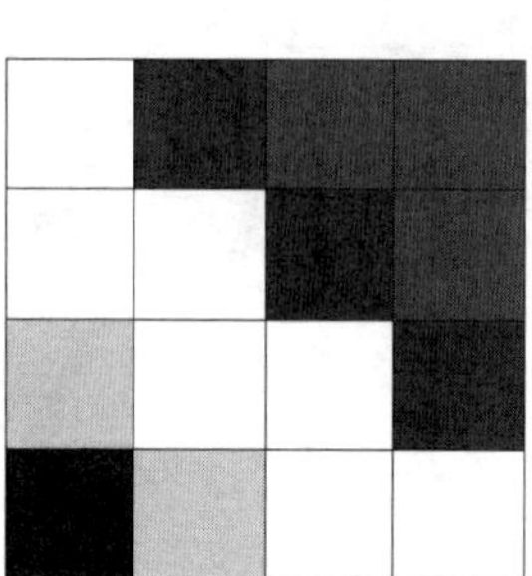

Figure 4. Group of material types in one cell.

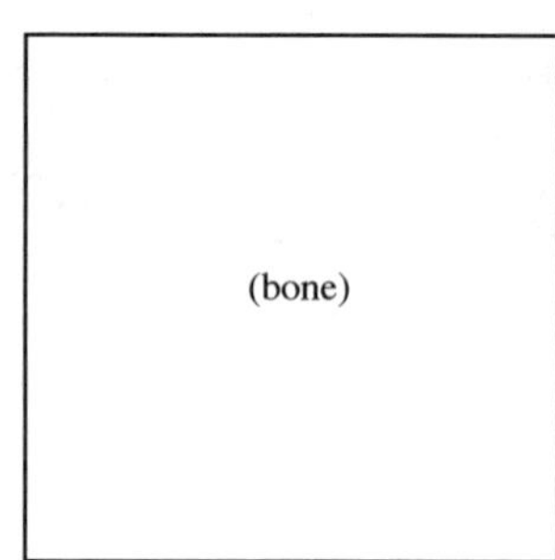

Figure 5. Resulting bone cell type.

An FDTD grid then was created for the head model. Because of limited computer resources, each FDTD grid cell incorporated multiple pixels from the head model. The material characteristics assigned to each FDTD cell were taken as the material type that had the highest occurrence in the volume of that unit cell. Figures 4 and 5 illustrate the quantization process in two dimensions. With this procedure, the same head model can generate FDTD grids of different resolution on the basis of available computer resources. This process of dynamically setting the simulation resolution is a key feature in the modeling technique.

The materials used to make up the FDTD head model have electrical properties that are frequency dependent. Table 1 lists the electrical properties of the material of the adult human head at 900 MHz and 1800 MHz (Dimbylow, 1993).

3.2 Visualization of computed data

To visualize the results of the simulations, we developed an animated surface plotting program using the GL graphical programming language. The program can create movie files showing the propagation of the electric or magnetic fields through and around the head. These movie files were generated on a Sillicon Graphics workstation and were converted to a format suitable for viewing on PCs. Samples of these movie files can be found on the network at `http://www.seg.org/books/3dem/elsher/elsher.html`, or at `http://www.olemiss.edu/~atef/head/head.html`.

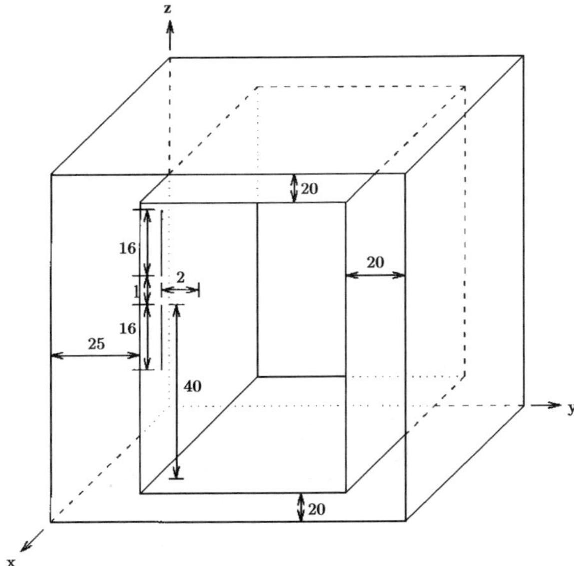

Figure 6. Positioning of antenna and artificial box surrounding the head model in the FDTD mesh domain. Dimensions are in terms of cell numbers.

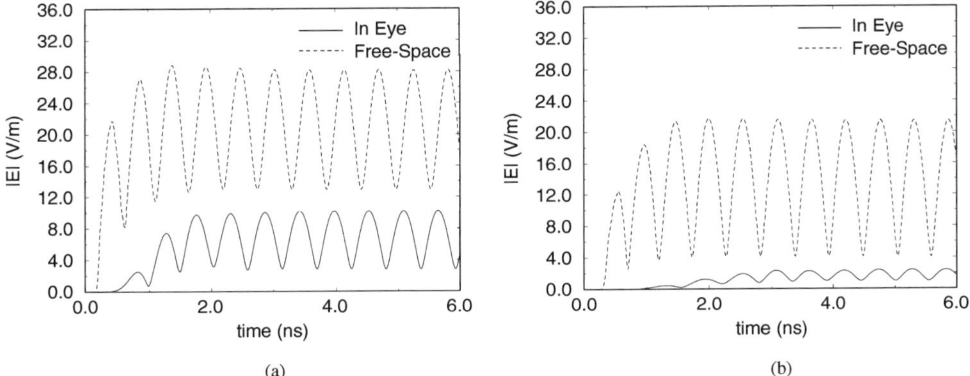

Figure 7. Computed electric field for dipole antenna at eye locations with and without the computational head model present: (a) right eye, (b) left eye.

3.3 Dipole antenna geometry and results

For these dipole simulations, a $55 \times 38 \times 54$ unit-cell head model was used. Considering the head size to be 17.5 cm $\times$ 11.5 cm $\times$ 20 cm, the physical dimensions of the unit cell were 3.125 mm $\times$ 3.125 mm $\times$ 3.636 mm. The boundary condition used in these dipole simulations was Liao's third order. A 20-cell boundary buffer on all sides of the computational domain was used, except the side closest to the antenna where a 25-cell buffer was used, as shown in Fig. 6.

The dipole antenna was 12 cm long with a 0.6-mm radius and was located approximately 6 mm away from the head model. The dipole antenna was excited by a 900-MHz sinusoidal voltage source, with the amplitude appropriately chosen to deliever 0.6 W.

For this configuration, the total electric fields induced in the right and the left eyes are shown in Figs. 7a and 7b. In these figures, the corresponding field values at the same positions also are shown for the dipole antenna radiating in free space (dashed

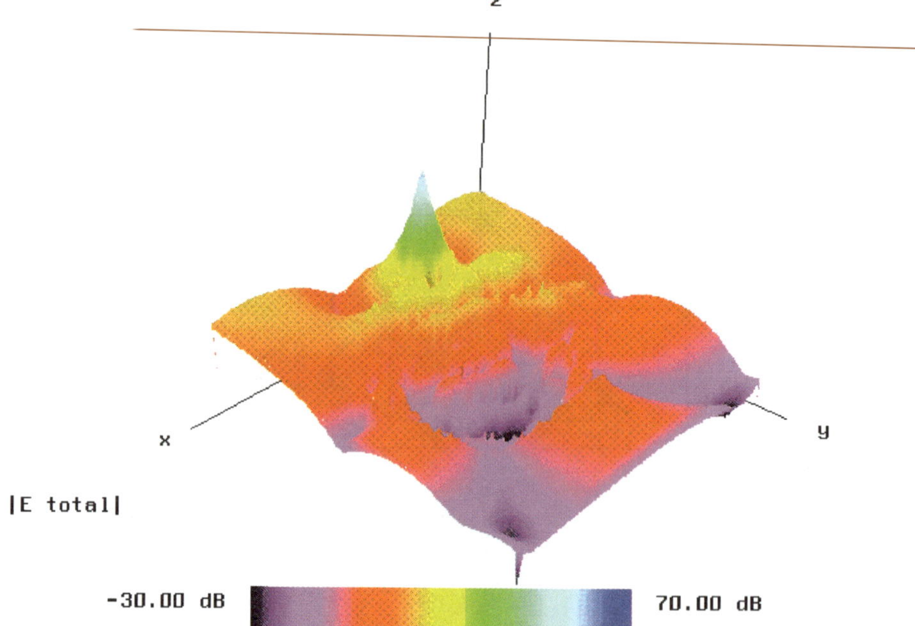

Figure 8. Electric field across a slice of the head model at 12 cm up from the chin with a dipole antenna close to the left ear; eyes are facing the y-axis.

line), illustrating the effects of the head model on the radiation from the antenna. One can also see from these figures that the steady-state response was achieved after a very short period of time. The electric- and magnetic-field distributions across an x-y plane cut 12 cm up from the chin are shown in Figs. 8 and 9, respectively. These two figures are single frames from the movie files, illustrating the time interaction between the radiation from the dipole antenna and the head model. The movie files show a complex standing-wave pattern in the brain area, and a very obvious surface wave that surrounds the head. The computed SAR distribution in the x-y plane cut containing the voltage source is shown in Fig. 10.

4 Helix antenna

We also studied a square, thin-wire helix antenna, as a model for a personal satellite transceiver handset. The purpose was to look for possible effects of the operator on the radiation characteristics of this circularly polarized antenna, and to see how the radiated fields from this antenna acted on the head model. This study used a slightly different head model.

4.1 Biological tissue modeling

In Jensen and Rahmat-Samii (1995), a computational head model was included in the FDTD simulations in order to ascertain the impact that the user's presence has on the radiation characteristics of the transceiver at 915 MHz. To construct that head model, a grid with a 6.56-mm spatial resolution was placed on cross-sectional images of the head

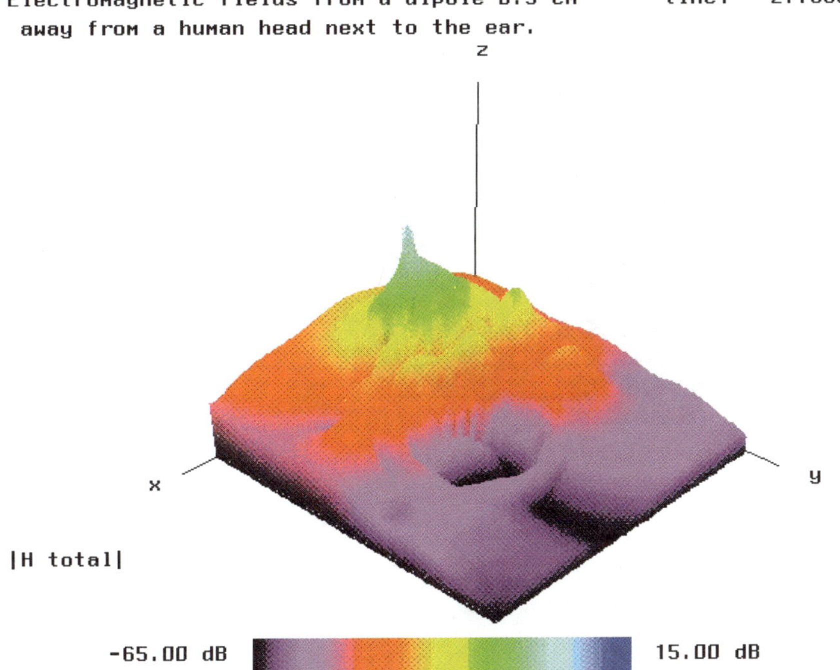

Figure 9. Magnetic field across a slice of the head model at 12 cm up from the chin with a dipole antenna close to the left ear; eyes are facing the y-axis.

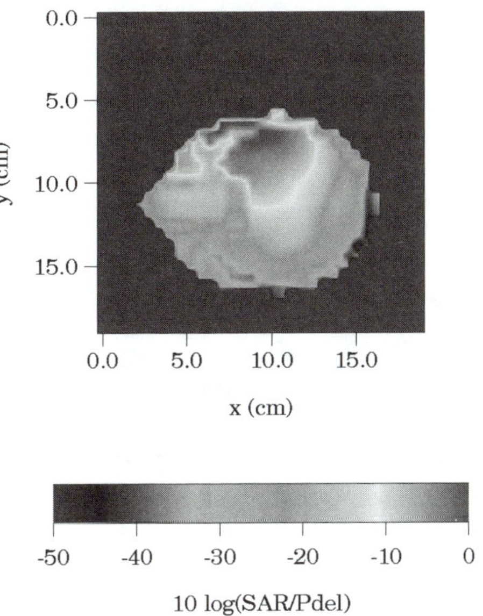

Figure 10. Computed SAR for an x-y plane cut of the head model that contains the voltage source that excites the antenna shown in Fig. 6 at 900 MHz.

Figure 11. Midsagital view of FDTD head model.

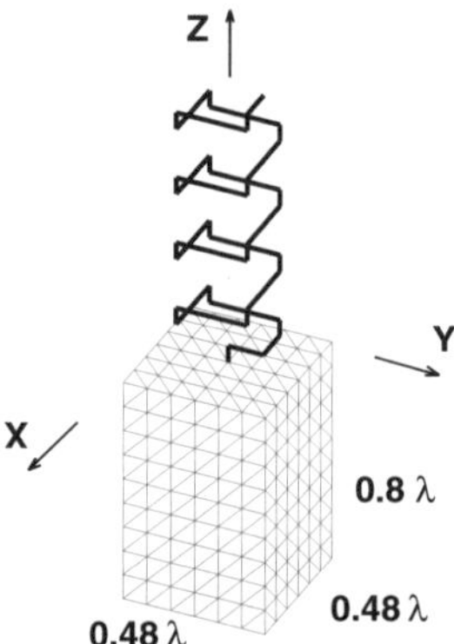

Figure 12. Schematics of a thin-wire square-helix antenna mounted on a $0.48\lambda \times 0.48\lambda \times 0.8\lambda$ conducting box.

obtained from an anatomy atlas (Eycleshymer and Schoemaker, 1911). Each cell in the grid then was assigned permittivity and conductivity classifications corresponding to the type of tissue that filled the *majority* of the cell. Figure 11 illustrates a midsagittal cross-section of the head model.

We used a similar computational phantom head in our simulation of a satellite communication transceiver near a user. Although the same structure and approximate head size were used, the electrical parameters of the head model were changed to those of head tissue at 1.8 GHz (Dimbylow, 1993) as shown in Table 1. The FDTD grid size used in the computations was 3.33 mm.

4.2 Antenna geometry and results for satellite-handset link

Figure 12 shows a modified-helix structure in which the thin-wire element is constructed only of straight wire segments. This square-helix geometry has been seen to exhibit pattern and circuit characteristics similar to those of a standard circular-helix structure (Colburn and Rahmat-Samii, 1995b). In the square-helix structure, all vertically oriented

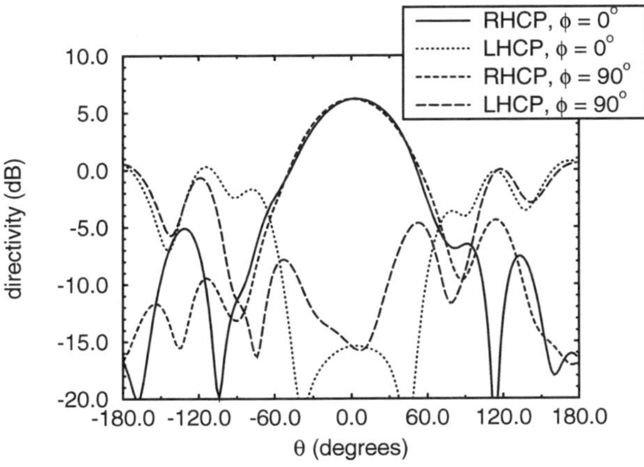

Figure 13. FDTD-computed far-field pattern for square-helix antenna illustrated in Fig. 12.

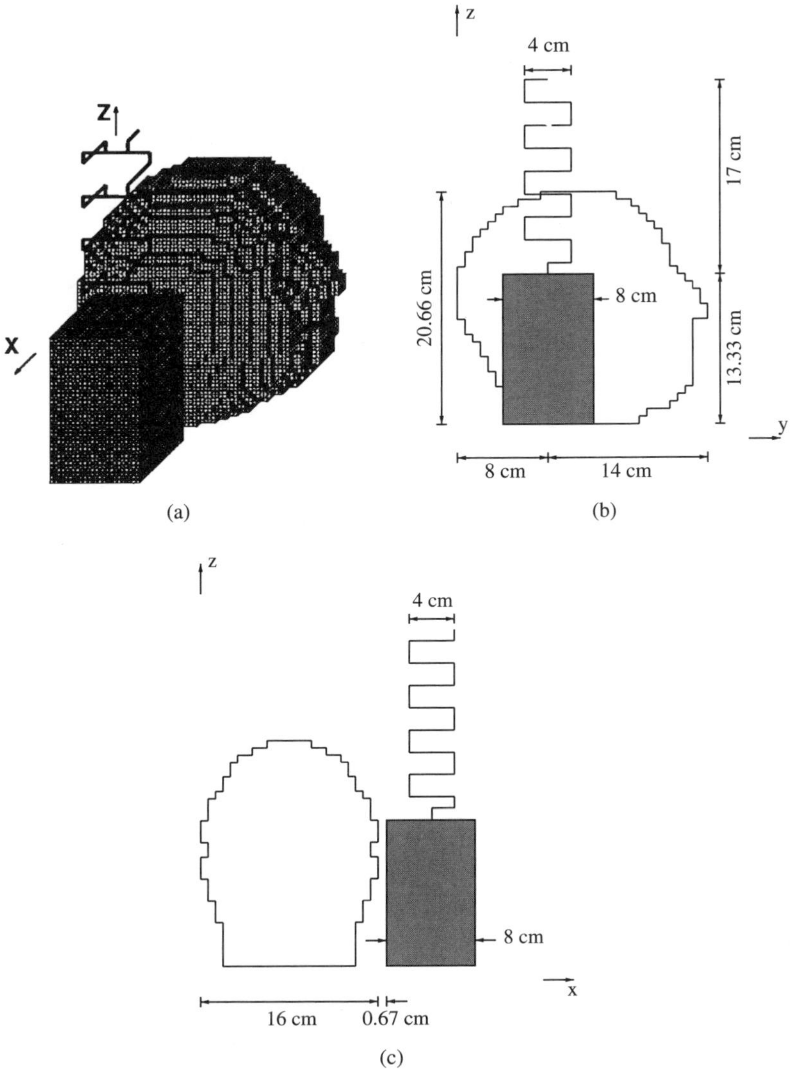

Z
X
(a)

z
4 cm
17 cm
20.66 cm
8 cm
13.33 cm
y
8 cm
14 cm
(b)

z
4 cm
8 cm
x
16 cm
0.67 cm
(c)

Figure 14. (a) Illustration of square-helix antenna on handset positioned near the head model; (b) y-z plane cut and (c) x-z plane cut of head model and handset.

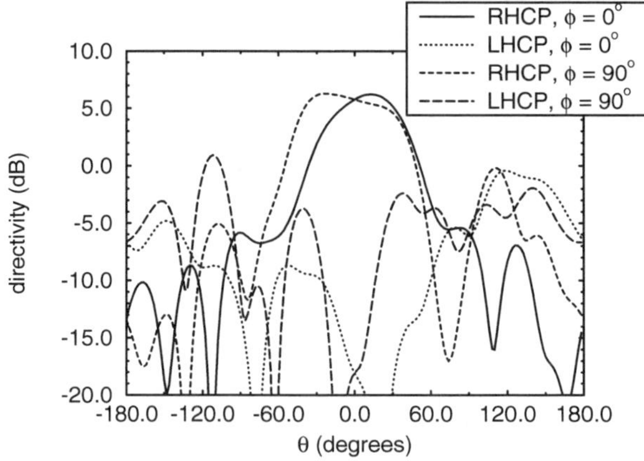

Figure 15. FDTD-computed far-field pattern for square-helix structure near the head model illustrated in Fig. 14a.

wire segments are 0.06λ in length and all horizontally oriented wire segments are 0.24λ in length. This square-helix structure has a geometry that is easier to integrate with the handset, and can be simulated on a rectangular FDTD grid. Figure 13 is a plot of the computed far-field pattern in two ϕ cuts in terms of right-hand circularly polarized (RHCP) and left-hand circularly polarized (LHCP) components for the square-helix structure shown in Fig. 12 using a Cartesian-grid–based FDTD code.

Figure 14a illustrates the computational head model positioned near the transceiver handset shown in Fig. 12. The frequency of interest was 1.8 GHz, which fixes the FDTD grid size of $0.02\,\lambda$ to 3.33 mm and the size of the transceiver box to $8 \times 8 \times 13.33$ cm. In the computations performed, the handset and head model were spaced two cells apart. Figures 14b and 14c are schematics in the y-z and x-z cuts of the exact geometry analyzed.

Figure 15 is a plot of the FDTD-computed far-field pattern for the square-helix structure positioned next to the phantom head seen in Fig. 14. The pattern plots in Fig. 15 can be compared directly to the pattern plots for the handset itself shown in Fig. 12. The computations indicate that approximately 13% of the total delivered power is lost in the head model, which is less than the losses reported by Jensen and Rahmat-Samii (1995) and is mainly attributed to the greater distance between the head model and the antenna element. Besides the radiated-power loss due to absorption in the head model, it can be seen in Fig. 15 that there are significant distortions to the computed far-field pattern because of the presence of the head model. This results in a very poor axial ratio performance. The computed electric-field distribution in the vicinity of the handset is shown with (Fig. 16b) and without (Fig. 16a) the head model present in the computational domain. From Figs. 16a and 16b, one can see the perturbations of the fields caused by the head's presence. Figure 17 is a plot of the computed SAR averaged over 1 g of tissue in the computational head model for this case.

5 Summary and conclusions

The FDTD method can simulate very efficiently the interaction of EM fields from portable communications devices with the human operator. We used models of a human head, generated semiautomically from MRI scans, in FDTD simulations of radiation from dipole and helix antennas. The evolution of the EM fields in the models has been captured in movies that can be played on PCs. The simulations demonstrate that the

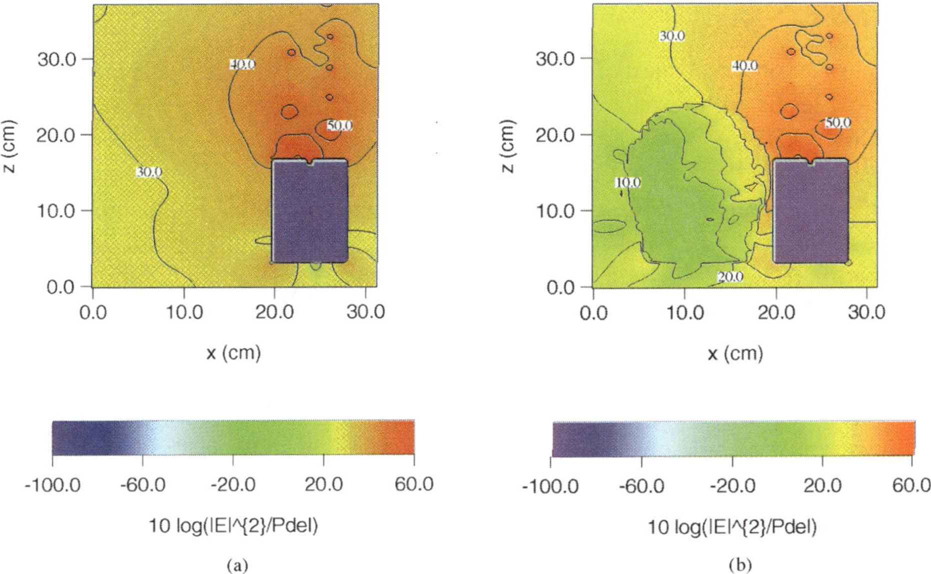

Figure 16. FDTD-computed fields in the vicinity of the handset: (a) without the head model in the computational domain; (b) with the head model in the computational domain.

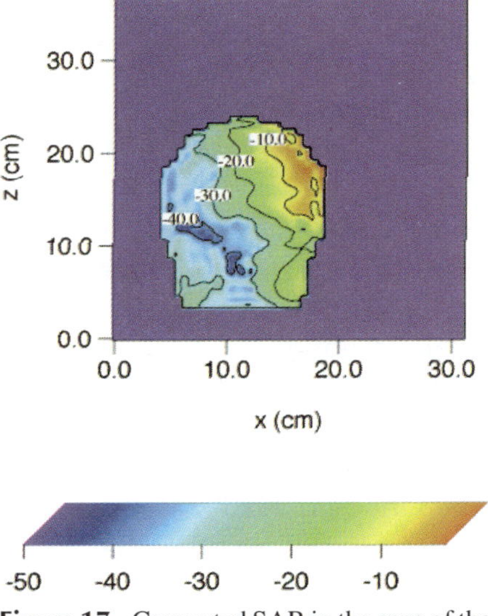

Figure 17. Computed SAR in the case of the square helix on the handset next to the head model.

tissue has a noticeable effect on the antenna radiation patterns, polarization state, and directivity. Most of the energy absorbed by the head is concentrated on the illuminated side of the skin and bone layers of the head model.

Acknowledgments

This work was supported in part by the Army Research Office under Grant No. DAAH04-94-G-0355 and Advanced Research Projects Agency Contract No. DAAB07-93-C-C501.

References

American National Standards Institute, 1991, IEEE C95.1–1991: IEEE standard for safety levels with respect to human exposure to radio frequency electromagnetic fields, 3 kHz to 300 GHz.

Bradley, J., X window system: `bradleycis.upenn.edu`.

Chen, H., and Wang, H., 1994, Current and SAR induced in a human head model by the electromagnetic fields irradiated from a cellular phone: IEEE Trans. Microwave Theory Tech., **42**, 2249–2254.

Chuang, H. R., 1994, Human operator coupling effects on radiation characteristics of a portable communication dipole antenna: IEEE Trans. Antennas Propag., **42**, 556–560.

Colburn, J. S., and Rahmat-Samii, Y., 1995a, Electromagnetic scattering and radiation involving dielectric objects: J. Elect. Waves Appl., **9**, 1249–1277.

———1995b, Analysis of a square helix applicable to the personal satellite communication handset, *in* 1995 IEEE AP-S Intl. Symp. Dig.: Conference Proceedings, **3**, 1418–1421.

Dimbylow, P. J., 1993, FDTD calculations of the SAR for a dipole closely coupled to the head at 900 MHz and 1.9 GHz: Phys. Med. Biol., **38**, 361–368.

Dimbylow, P. J., and Gandhi, O. P., 1993, Finite-difference time-domain calculations of SAR in a realistic heterogeneous model of the head for plane-wave exposure from 600 MHz to 3 GHz: Phys. Med. Biol., **38**, 361–368.

Elsherbeni, A. Z., and Taylor, C. D., Jr., 1995, Electromagnetic interaction between cellular phones and a human head: Presented at Internat. Symp. 3-D Electromagn. Sponsored by Schlumberger-Doll Research.

Eycleshymer, A. C., and Schoemaker, D. M., 1911, A cross-section anatomy: D. Appleton-Century Co.

Fischetti, M., 1993, The cellular phone scare: IEEE Spectrum, 43–47.

Gutschling, S., and Weiland, T., 1995, Detailed SAR distribution in high resolution human head models, *in* IEEE 1995 Internat. Symp. Electromagn. Compatibility: sponsor, 291–296.

Jensen, M. A., and Rahmat-Samii, Y., 1995, EM interaction of handset antennas and a human in person communications: Proc. IEEE, **83**, 7–17.

Johnson, K. A., and Becker, J. A., 1995, The whole brain atlas: `http://count51.med.harvard.edu/AANLIB/home.html`.

Luebbers, R., Chen, L., Uno, T., and Adachi, S., 1992, FDTD calculation of radiation patterns, impedance, and gain for a monopole antenna on a conducting box: IEEE Trans. Antennas Propag., **40**, 1577–1583.

Martens, L., 1994, Determine the EM fields induced by wireless telephones: Microwave & RF, Volume No. 33, 161–166,

Taflove, A., and Brodwin, M. E., 1975, Numerical solution of steady-state electromagnetic scattering problems using the time-dependent Maxwell's equations: IEEE Trans. Microwave Theory Tech., **23**, 623–630.

Toftgård, J., Hornsleth, S. N., and Andersen, J. B., 1993, Effects on portable antennas of the presence of a person: IEEE Trans. Antennas Propag., **41**, 739–746.

Yee, K. S., 1966, Numerical solution of initial boundary value problems involving Maxwell's equations in isotropic media: IEEE Trans. Antennas Propag., **14**, 302–307.

Nondestructive Evaluation of Corrosion Damage in Aging Aircraft

Fadil Santosa

Summary. We study an idealized problem of nondestructive evaluation relevant to the inspection of aircraft. The problem is to obtain quantitative information about corrosion damage in a plate from voltage readings on its accessible side. The corrosion damage is modeled as material loss occurring on the inaccessible side. The inverse problem is to reconstruct the material loss profile from the data. The computational model is based on the assumption that the plate's thickness is small. The voltage response is linearized in the loss profile, and an equation is derived relating the derivative of the profile to the voltage readings. Although this equation can be integrated directly, numerical examples indicate that a regularized least-squares solution gives more reasonable results.

1 Introduction

Corrosion is a problem commonly found in older aircraft. The process leading to the damage is still not well understood, although it is quite well documented (ASM, 1987). The estimated cost resulting from corrosion to the aircraft industry is tremendous, and, as the commercial and military fleets age, the problems posed by corrosion to the airworthiness of the fleet become more serious.

Table 1 gives an indication of the magnitude of the problem. In the commercial fleet, over 50% of some types of aircraft are over 20 years old. The problem is more serious on the military side. For example, the C-135 refueling and transport planes, which rolled off the factory in the late 1950s, must operate until the year 2040.

Several types of corrosion are found in older aircraft: pitting, general corrosion, and exfoliation [see ASM (1987) for a description]. Corrosion damage often occurs in inaccessible areas of the plane. Even when such areas are accessible, they are often hidden. The problem of corrosion in aircraft presents challenges and opportunities for analysis, numerical computation, and instrumentation.

Minnesota Center for Industrial Mathematics, School of Mathematics, University of Minnesota, Minneapolis, MN 55455, USA.

Table 1. Age of Jet airplane fleets

Type	Total	% >20 years
Boeing commercial airplanes[a]		
B-727	1182	50.2
B-737	1012	14.9
B-747	213	54.4
DC-9	1157	36.4
DC-10	315	29.5
Military airplanes[b]		
B-52	148	100
C-135	479	100

Note: [a]From *Boeing World Jet Airplane Inventory* for year ending 1993.
[b]From *Air Force Magazine*, 1993.

In addition to ultrasonics, two other methods are used commonly for nondestructive evaluation in the aircraft industry. The first uses eddy currents (Halmshaw, 1987), for which there are instruments of high sensitivity and accuracy. The probe typically consists of a coil, which is scanned over the work area being inspected. The method is used mainly for detection because there has not been a viable imaging method that takes eddy-current data and produces quantitative information about corrosion damage. We have recently developed a quantitative imaging technique for eddy-current inspection (Luong and Santosa, 1997). The second method uses thermography. A heat source heats up a work area rapidly; an infrared detector then measures the response of the surface. Thermal imaging has been shown to be promising for aircraft inspection (Valley et al., 1993). The main challenge is that it still is not possible to obtain quantitative information from thermal images.

This work, which is a synopsis of the work of Kaup et al. (1996), considers the simple problem of quantifying hidden corrosion in plates. Our goal is to demonstrate that a nondestructive evaluation technique based on electrical impedance tomography can be applied effectively to image corrosion damage in thin plates. Electrical impedance tomography uses static voltage and current measurements on the surface of a specimen to determine the conductivity distribution in the interior. We develop a method to determine material loss occurring on the inaccessible side of a thin plate by measuring voltages and currents on the opposite (accessible) side. The method leads to quantitative reconstruction of the loss profile. A drawback of this method is that the probe requires contact with the work area. This presents a practical challenge for aircraft applications.

The present method is similar to the classical potential-drop method (Halmshaw, 1987) which is devised to estimate the depth of a surface-breaking crack. However, in the potential drop method, the crack occurs on the same side as the voltage measurements. The method requires highly accurate voltage readings: four to five significant figures. With this level of accuracy, one is able to reconstruct material loss of around 5% plate thickness. Although such a level of measurement accuracy may appear difficult to achieve in general, we believe that because of the simplicity of the present procedure, data of this quality are possible.

2 Model

We model damage by corrosion as material loss, leading to an unknown profile of the inaccessible side of the plate. Consider an undamaged plate, Ω_0, given by

$$\Omega_0 = \{(x_1, x_2, x_3) : |x_1| \leq 1/2, |x_2| \leq 1/2, 0 \leq x_3 \leq a\}.$$

Corrosion has occurred on the upper, inaccessible surface $x_3 = a$. We assume that it has caused the plane $x_3 = a$ to become a surface $x_3 = a + \epsilon\theta(x_1, x_2)$. The damaged plate is therefore given by

$$\Omega_\epsilon = \{(x_1, x_2, x_3) : |x_1| \leq 1/2, |x_2| \leq 1/2, 0 \leq x_3 \leq a + \epsilon\theta(x_1, x_2)\}.$$

The goal is to determine the loss profile $\epsilon\theta(x_1, x_2)$.

To do this, we create a voltage potential by applying a current pattern to the bottom of the plate, $x_3 = 0$. The resulting voltage potential is also measured on the bottom. We seek to determine $\epsilon\theta(x_1, x_2)$ from these measurements of the voltage potential $u(x_1, x_2, 0) = g(x_1, x_2)$. Let $u(x_1, x_2, x_3)$ be the voltage potential in the damaged plate. Then, u satisfies Laplace's equation

$$\Delta u = 0 \quad \text{in} \quad \Omega_\epsilon. \tag{1a}$$

For simplicity (only), we assume that the sides of the plate are insulated[1]:

$$\frac{\partial u}{\partial x_i} = 0 \quad \text{on} \quad x_i = \pm 1/2, \quad i = 1, 2. \tag{1b}$$

The top of the plate, where corrosion has occurred, also is assumed to be insulated:

$$\frac{\partial u}{\partial \nu} = 0 \quad \text{on} \quad x_3 = a + \epsilon\theta(x_1, x_2). \tag{1c}$$

Here, ν is the outward normal to the surface $x_3 = a + \epsilon\theta(x_1, x_2)$. A more realistic boundary condition, which takes into account the chemical reduction and oxidation, is the focus of Vogelius and Xu (1995). Finally, on the bottom of the plate, we apply a current pattern to generate the voltage potential in the plate:

$$\frac{\partial u}{\partial x_3} = -\phi(x_1, x_2) \quad \text{on} \quad x_3 = 0. \tag{1d}$$

The solution to the boundary value problem (1) is unique up to a constant. To make the solution unique, we choose the normalization $\int_{-\frac{1}{2}}^{\frac{1}{2}} \int_{-\frac{1}{2}}^{\frac{1}{2}} u(x_1, x_2, 0)\, dx_1\, dx_2 = 0$.

This model is somewhat simplistic but serves as a good starting point for further research. A two-dimensional version of this problem was studied in detail by Kaup and Santosa (1995). We also note that the question of identifiability in two dimensions has been addressed in the work of Andrieux et al. (1993).

3 Linearization

Our reconstruction algorithm is based on a linearized relationship between the voltage data and the function describing the loss profile. Linearization is justified when material loss is small in relation to the plate thickness—for instance, of the order of 5–10%. Therefore, henceforth, we assume that $\epsilon \ll 1$.

[1] Indeed, this method can be modified for the case in which the sides of the plate are grounded.

The linearization procedure goes as follows: Let $\omega = \{(x_1, x_2) : |x_1| \le 1/2, |x_2| \le 1/2\}$.

1. **Green's identity.** Using Green's first identity, we can rewrite the boundary value problem (1) in variational form as

$$\int_{\Omega_\epsilon} (\nabla_x u \cdot \nabla_x v)\, dx = \int_{\partial\Omega_\epsilon} \left(\frac{\partial u}{\partial n} v\right) d\sigma_x$$
$$= \int_\omega \phi(x_1, x_2) v(x_1, x_2, 0)\, dx_1\, dx_2,$$

where $\omega = \{(x_1, x_2, x_3) : x_3 = 0, |x_1|, |x_2| \le 1/2\}$ and $v(x_1, x_2, x_3) \in H^1(\Omega_\epsilon)$ is an arbitrary test function.

2. **Change of coordinates.** The variational equation above requires integration over the unknown region. Therefore, we use a change of coordinates $z = \Phi_\epsilon(x)$ given by

$$z_1 = x_1, \quad z_2 = x_2, \quad z_3 = x_3 - \epsilon\theta(x_1, x_2)\gamma(x_3),$$

where $\gamma(t)$ is a smooth function that satisfies $\gamma(t) = 0$ for t near 0, and $\gamma(t) = 1$ for t near a. For small ϵ, this is a regular coordinate transformation that maps the domain Ω_ϵ to the reference (undamaged) domain Ω_0.

3. **Jacobian.** The Jacobian matrix for this transformation is

$$\left[\frac{d\Phi_\epsilon}{dx}\right] = \underset{\sim}{\mathbf{I}} - \epsilon \underset{\sim}{\mathbf{J}}; \qquad \underset{\sim}{\mathbf{J}} = \begin{bmatrix} 0 & 0 & 0 \\ 0 & 0 & 0 \\ \frac{\partial\theta}{\partial x_1}\gamma & \frac{\partial\theta}{\partial x_2}\gamma & \theta\dot{\gamma} \end{bmatrix}.$$

The corresponding determinant is

$$\det\left[\frac{d\Phi_\epsilon}{dx}\right] = 1 - \epsilon\theta\dot{\gamma}.$$

We also have

$$\det\left[\frac{d\Phi_\epsilon^{-1}}{dz}\right] = (1 + \epsilon\theta\dot{\gamma}) \circ \Phi_\epsilon^{-1} + \mathcal{O}(\epsilon^2).$$

4. **Expansion.** For small ϵ, we formally expand the voltage potential u as

$$u = u_0 + \epsilon u_1 + \mathcal{O}(\epsilon^2).$$

The term u_0 is the solution to the boundary-value problem for the undamaged reference body. Similarly, the term $\epsilon u_1 \approx u - u_0$ is the $\mathcal{O}(\epsilon)$ voltage change resulting from the surface perturbation $\epsilon\theta$.

5. **Harmonic test function.** We now choose $v(x_1, x_2, x_3)$ to be harmonic in Ω_0 and to satisfy the boundary conditions

$$\frac{\partial v}{\partial x_i} = 0 \quad \text{on} \quad x_i = \pm 1/2 \quad i = 1, 2$$
$$\frac{\partial v}{\partial x_3} = 0 \quad \text{on} \quad x_3 = a.$$

Putting these together, we obtain the identity

$$\int_\omega u_1(x_1, x_2, 0)\frac{\partial v}{\partial x_3}(x_1, x_2, 0)\,dx_1\,dx_2$$
$$= \int_\omega \theta(x_1, x_2)\left(\frac{\partial u_0}{\partial x_1}\frac{\partial v}{\partial x_1} + \frac{\partial u_0}{\partial x_2}\frac{\partial v}{\partial x_2}\right)(x_1, x_2, a)\,dx_1\,dx_2. \qquad (2)$$

We use equation (2) as the basis for our solution of the linearized inverse problem. We can tacitly assume from now on that θ is small, and that $u_1 = u - u_0$. In this equation, v is an arbitrary harmonic function satisfying the boundary conditions. The left side represents data, whereas the right side represents an operator acting on the unknown θ. The linearized inverse problem is the determination of θ in equation (2).

It is worth noting that the choice of both u_0 and v is at our disposal. The background field u_0 is determined by the applied current $\phi(x_1, x_2)$ and the undamaged domain Ω_0. We will choose ϕ and v with a view toward obtaining the simplest formula.

4 Thin-plate approximation

To determine for $\theta(x_1, x_2)$ from equation (2), we need to have at our disposal a set of harmonic test functions $v(x_1, x_2, x_3)$. To evaluate the right side of equation (2), each test function will need to be integrated against the partial derivatives of the background field $u_0(x_1, x_2, x_3)$. Each test function gives rise to a single equation. The unknown loss profile is found by solving this set of equations.

To arrive at a much simpler procedure, we consider the regime in which $a \ll 1$, that is, the situation in which the plate thickness is small. With this approximation, the right side of equation (2) becomes more explicit. Recall that the linearization is valid for $\epsilon \ll a$. Therefore, the thin-plate approximation is justified, provided $\epsilon/a \ll 1$ and at the same time $a \ll 1$.

We begin by scaling the variables:

$$\bar{x} = (x_1, x_2), \quad x_3' = x_3/a.$$

This effectively transforms the plate domain $\Omega_0 = \omega \times [0, a]$ to $\Omega_0' = \omega \times [0, 1]$. Let $v(\bar{x}, x_3')$ represent a harmonic function in Ω_0' with homogeneous Neumann boundary conditions on all sides except the bottom ($x_3' = 0$) where we set

$$\frac{\partial v}{\partial x_3} = a\zeta. \qquad (3)$$

The function $\zeta(\bar{x})$ is assumed to have compact support inside ω; its exact form is specified later.

We expand the functions $v(\bar{x}, x_3')$ in (even) powers of the small parameter a:

$$v(\bar{x}, x_3') = v_0(\bar{x}, x_3') + a^2 v_2(\bar{x}, x_3') + a^4 v_4(\bar{x}, x_3') + \cdots.$$

By equating powers of a in Laplace's equation and the boundary conditions for v, we arrive at the equation satisfied by the lowest-order term v_0. It can be shown that it is independent of the variable x_3' and solves the boundary-value problem

$$\left(\frac{\partial^2 v_0}{\partial x_1^2} + \frac{\partial^2 v_0}{\partial x_2^2}\right) = \zeta(\bar{x}) \quad \text{for} \quad \bar{x} \in \omega \qquad (4a)$$

with

$$\frac{\partial v_0}{\partial \nu} = 0 \quad \text{for} \quad \bar{x} \in \partial\omega. \tag{4b}$$

By incorporating the identity (3) into equation (2), and replacing v by v_0 on the right side, we obtain

$$a \int_\omega u_1(x_1, x_2, 0)\zeta(x_1, x_2)\, dx_1\, dx_2$$
$$\approx \int_\omega \theta(x_1, x_2)\left[\frac{\partial u_0}{\partial x_1}(x_1, x_2, a)\frac{\partial v_0}{\partial x_1}(x_1, x_2) + \frac{\partial u_0}{\partial x_2}(x_1, x_2, a)\frac{\partial v_0}{\partial x_2}(x_1, x_2)\right] dx_1\, dx_2. \tag{5}$$

The consequences of this approximate identity are studied next.

5 Method for reconstructing profile loss in three dimensions

We use the notation

$$\bar{\nabla} = \left(\frac{\partial}{\partial x_1}, \frac{\partial}{\partial x_2}\right) \quad \text{and} \quad \bar{\Delta} = \frac{\partial^2}{\partial x_1^2} + \frac{\partial^2}{\partial x_2^2}.$$

The identity

$$\bar{\nabla} \cdot [\theta(\bar{\nabla} u_0)v_0] = (\bar{\nabla}\theta \cdot \bar{\nabla} u_0)v_0 + \theta(\bar{\Delta} u_0)v_0 + \theta(\bar{\nabla} u_0 \cdot \bar{\nabla} v_0) \tag{6}$$

holds for fixed x_3.

Using the thin-plate approximation of Section 4, we find that

$$u_0(x_1, x_2, x_3) \approx w(x_1, x_2), \quad \text{where} \quad \bar{\Delta} w = -(1/a)\phi(x_1, x_2), \tag{7}$$

with the boundary condition $\partial w/\partial \nu = 0$ on $\partial\omega$. Let θ be a loss profile with compact support. We select input current $\phi(x_1, x_2)$ with the property

$$\{\operatorname{supp} \phi\} \bigcap \{\operatorname{supp} \theta\} = \emptyset.$$

In practice, this means that our inspection method is designed to image an area where it is suspected a priori that some damage has occurred, by setting up current sources and sinks away from the damaged area. Therefore, the term $\theta(\bar{\Delta} u_0)$ is always zero (approximately).

Using this fact and substituting the identity (6) into equation (5), and integrating by parts, we get the approximate identity

$$a \int_\omega u_1(x_1, x_2, 0)\zeta(x_1, x_2)\, dx_1\, dx_2 \approx -\int_\omega \left(\bar{\nabla}\theta \cdot \bar{\nabla} u_0\big|_{x_3=a}\right) v_0\, dx_1\, dx_2. \tag{8}$$

Next, we narrow our choice of ϕ further. Observe from equation (7) that because of the square-plate geometry, if we set ϕ to be independent of x_2, then we can make $\partial w/\partial x_2 = 0$. A natural choice is

$$\phi(x_1, x_2) = \delta\left(x_1 + \tfrac{1}{2} - t\right) - \delta\left(x_1 - \tfrac{1}{2} + t\right) \tag{9}$$

(see Fig. 1). This current pattern generates, approximately, a constant current sheet directed in the x_1 direction. The desired background potential u_0 is computed using

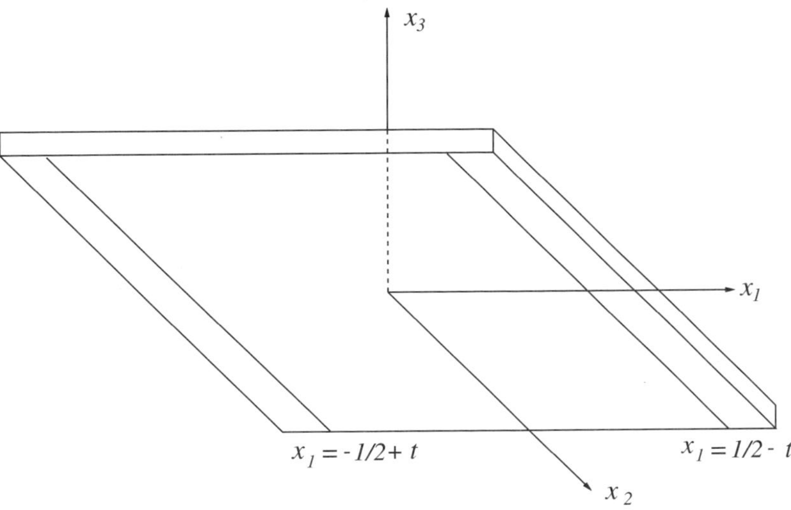

Figure 1. Current is applied along $|x_1| = \frac{1}{2} - t$. The damage to be imaged is on the top side of the plate (not shown).

equation (7), leading to

$$\frac{\partial u_0}{\partial x_1}(\bar{x}, x_3) \approx \frac{dw}{dx_1}(x_1) = -\frac{1}{a}, \qquad |x_1| < \tfrac{1}{2} - t.$$

With this choice of current pattern, equation (8) reduces to

$$a \int_\omega u_1(x_1, x_2, 0)\zeta(x_1, x_2)\, dx_1\, dx_2 \approx \frac{1}{a} \int_\omega \frac{\partial \theta}{\partial x_1} v_0\, dx_1\, dx_2. \tag{10}$$

Our method is to set, formally, $v_0 = \delta(x_1 - \xi_1)\delta(x_2 - \xi_2)$, corresponding to $\zeta = \bar{\Delta}[\delta(x_1 - \xi_1)\delta(x_2 - \xi_2)]$. We thus obtain

$$\bar{\Delta} u_1(\xi_1, \xi_2, 0) \approx \frac{1}{a^2} \frac{\partial \theta}{\partial x_1}(\xi_1, \xi_2). \tag{11}$$

This last formula may be viewed as a direct, approximate reconstruction scheme. Data collected are represented by $u_1(\xi_1, \xi_2, 0)$. We can view the equation as a first-order differential equation for θ.[2] Null values for θ on an appropriately chosen curve permit us to integrate the equation.

A direct marching scheme based on equation (11), despite its simplicity, generally does not perform well. As a remedy, we propose a regularization of the inverse problem that is quite effective in removing some of the undesirable characteristics of the direct marching algorithm.

The main reason why the direct marching algorithm does not perform well is that the data g may not be in the range of a profile θ that has compact support. If it is not exactly satisfied for the data, some quite disturbing shadow artifacts may occur. Ideally, we know that

$$\operatorname{supp} \theta \subset\subset \omega_0.$$

[2]It was pointed out by a referee that equation (11) can be derived by assuming an ansatz for the current distribution and an integrations-by-parts argument. The ansatz, appropriate for thin conductive sheets, can be found in Moon and Spencer (1961).

If the data g are inconsistent, we can project them so that they are the range of profiles θ that are supported in ω_0. Such a procedure, however, still retains the unwanted artifacts.

A second reason that the reconstruction from direct marching is often poor is that by the choice of input electrodes, and hence the background field u_0, the profile along a line $x_2 = \alpha$, $\theta(x_1, x_2 = \alpha)$, is independent from its values along a neighboring line $x_2 = \beta$. Therefore, depending on the quality of the data g, we can lose smoothness in the reconstructed θ along the x_2 axis.

To overcome these difficulties, we propose to regularize the reconstruction with a smoothing operator. If we are given $u_1(x_1, x_2, 0) = g(x_1, x_2)$ as data, we suggest solving

$$\begin{aligned} &\min \left\{ \left\| a^2 \bar{\Delta} g - \frac{\partial \theta}{\partial x_1} \right\|^2 + \lambda \| \bar{\nabla} \theta \|^2 \right\} \\ &\text{subject to } \theta = 0 \quad \text{on} \quad \partial \omega_0. \end{aligned} \tag{12}$$

The term involving $\bar{\nabla}\theta$ has the effect of smoothing the reconstructed θ. This smoothing is achieved at the cost of a loss of fit to the data.

The procedure above can be given a partial-differential-equation description. The minimizer of equation (12) solves the boundary-value problem

$$\begin{aligned} &(1+\lambda)\frac{\partial^2 \theta}{\partial x_1^2} + \lambda \frac{\partial^2 \theta}{\partial x_2^2} = a^2 \frac{\partial}{\partial x_1} \bar{\Delta} g, \\ &\theta = 0 \quad \text{on} \quad \partial \omega_0. \end{aligned} \tag{13}$$

We propose to use the L-curve method discussed by Hansen (1992) to find the parameter λ. Note that, by adding the regularization, we have made the reconstruction algorithm less direct. In effect, we need to solve an elliptic boundary-value problem, which, after discretization, translates to a sparse-matrix inversion.

6 Numerical experiments

We report some results from numerical calculations using the algorithm described. In all of the computations, the plate thickness a is 0.10 (aspect ratio of 1 to 10). Data for inversion are generated using the boundary-element method and represent the full 3-D, nonlinear relationship between the loss profile and the measured voltage. Instead of the Neumann problem (1), we alter the problem slightly and replace the boundary conditions at $x_1 = \pm 1/2$ with homogeneous Dirichlet conditions. This was done to allow the use of a conjugate residual algorithm [see Eisenstat et al. (1983)] to solve the resulting square linear system coming from the boundary-element method.[3] Instead of delta-function input currents, we use an approximation based on piecewise linear functions. We emphasize that the change in the boundary condition and the approximation in the input current patterns do not affect the inversion algorithm in an essential way.

Current is distributed along strips parallel to the x_2-axis on the bottom of the plate. The strips are of width 0.1 and centered at $x_1 = \pm 0.425$. The distribution is independent

[3]Recall that the original boundary-value problem was unique up to a constant. Using the normalization described in Section 2, we would get a nonsquare linear system, which would have to be appropriately transformed before application of a conjugate residual method.

of x_2 in the form of a symmetric triangle reaching the maximum height of 1. We absorb the parameter ϵ into θ and specify the loss profile $\theta(x_1, x_2)$ at the nodes of a uniform mesh of size $1/40$. Values of θ at all other points are obtained using bilinear interpolation.

The data needed for inversion are computed by first selecting a loss profile $\theta(x_1, x_2)$ and solving for the corresponding boundary values $u(x_1, x_2, 0)$. Then, we compute the corresponding boundary values for the undamaged plate. The difference in the boundary voltages between the damaged plate and the undamaged plate represents our data. Note that the data computed this way are truly 3-D and depend nonlinearly on the loss profile, whereas our reconstruction method is based on linearization and a two-dimensional approximation. Data are presumed known on a 27×27 square array of points over the region $[-0.325, 0.325] \times [-0.325, 0.325]$.

As an example, we choose a square damage that is off-center relative to the measurement array. The depth of the profile is 0.01 (10% thickness loss). Images of the true profile and its reconstruction with $\lambda = 1/16$ are displayed in Fig. 2. The value λ is arrived at by inspecting the L-curve, which is shown in Fig. 3. For this example, the L-curve does have a recognizable elbow, which appears near $\lambda = 1/16$.

The surface plot of the reconstruction is given in Fig. 4. Note that the true depth of the loss profile, which is 0.01 (10% material loss), is well estimated by the reconstruction. Observe the shadow artifacts and the overshoot in the recovered profile.

When we pick smaller values of λ, the shadow artifacts are more pronounced. The case $\lambda = 0$ corresponds to the unregularized case, i.e., the direct scheme discussed earlier. We found that setting $\lambda = 0$ produces poor reconstructions. An attempt to remove the striping artifacts by preprocessing the data was not successful. We found that an effective way to minimize this artifact and to control the smoothness of the reconstruction is to add the regularization mentioned earlier.

We ran several more examples with similar findings. Experiments in which small amounts of noise were added to the data also were conducted. The use of regularization allowed reconstruction of comparable quality. Finally, we note that other regularization can be implemented. Indeed, perhaps a more appropriate regularization is the total variation in the function θ. This particular regularization has been shown to be quite effective in electrical impedance tomography (Dobson and Santosa, 1994).

7 Discussion

We have developed a method for obtaining quantitative information about corrosion damage from measurements of voltages and currents on the boundary. The computational algorithm uses a thin-plate approximation and linearizes the voltage readings in the loss profile. The (regularized) reconstruction algorithm is quite effective. Although we did not implement a procedure that automatically determines the optimal regularization parameter, we believe that this can be done. The question of obtaining data of the quality required is more difficult to answer. Preliminary results from a laboratory experiment conducted by Ian Hall (personal communication, 1995) indicated that the high-accuracy data needed for our method posed certain difficulties. However, more recent indications are that it may be possible to obtain data of sufficiently high quality in the laboratory setting. On the other hand, the ability of the method to detect small material loss, of the order of 5 to 10%, suggests that research in instrumentation for this method may be worthwhile.

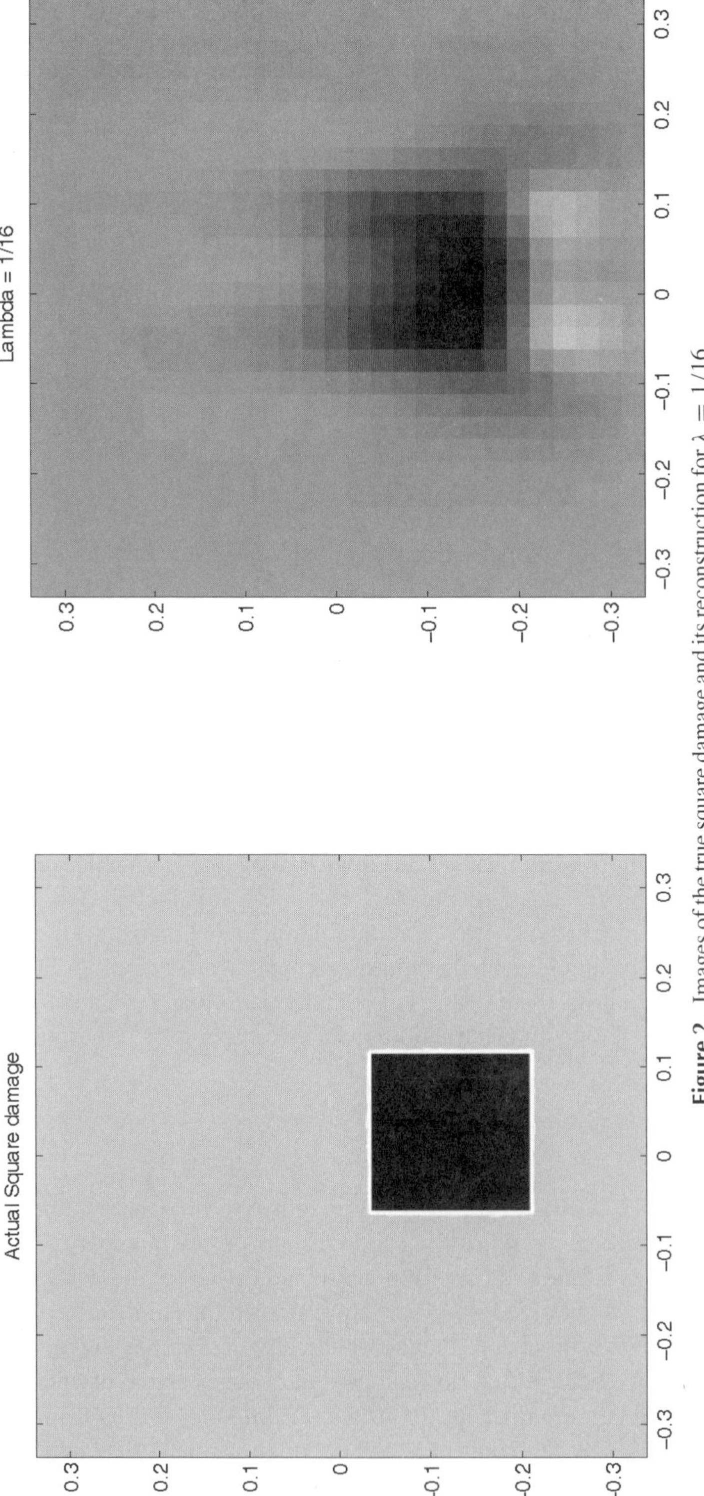

Figure 2. Images of the true square damage and its reconstruction for $\lambda = 1/16$.

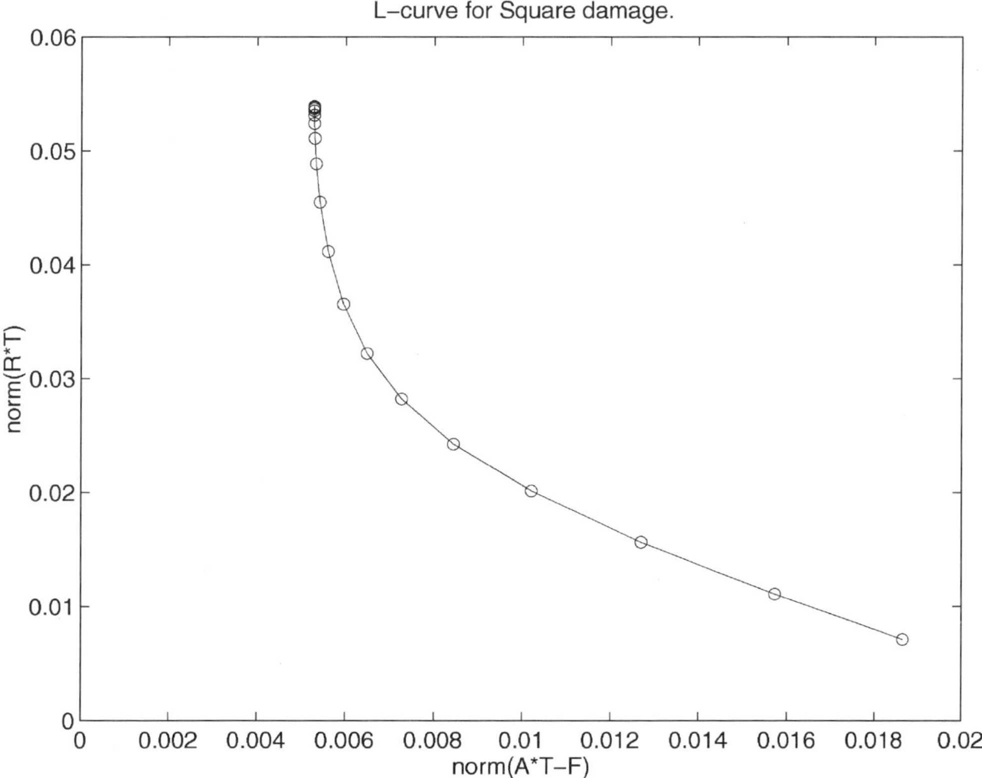

Figure 3. The L-curve associated with the square damage is a plot of the data misfit (horizontal) vs the smoothness of the solution (vertical) for various values of λ. The elbow of the curve is near $\lambda = 1/16$.

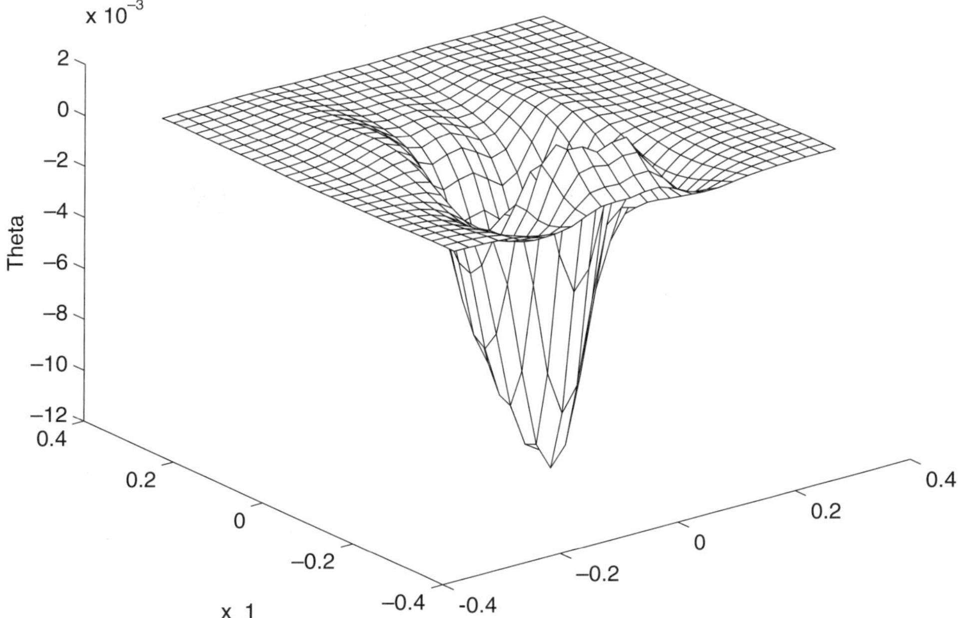

Figure 4. Plot of reconstructed square damage. Note that the thickness loss, which is 0.01, is estimated accurately.

References

ASM, 1987, ASM metals handbook, Vol. 13, Corrosion: ASM Internat.

Andrieux, S., Ben Abda, A., and Jaoua, M. 1993, Identifiabilité de frontière inaccessible par des mesures de surface: C.R. Acad. Sci. Paris, **316-1**, 429–434.

Dobson, D., and Santosa, F., 1994, An image enhancement technique for electrical impedance tomography: Inverse Problems, **10**, 317–334.

Eisenstat, S., Elman, H., and Schultz, M., 1983, Variational iterative methods for non-symmetric systems of linear equations: SIAM J. Numer. Anal., **20**, 345–357.

Halmshaw, R., 1987, Non-destructive testing: Wiley-Interscience.

Hansen, P., 1992, Analysis of discrete ill-posed problems by means of the L-curve: SIAM Rev. **34**, 561–580.

Kaup, P., and Santosa, F., 1995, Nondestructive evaluation of corrosion damage using electrostatic boundary measurements: J. Nondestruct. Eval. **14**, 127–136.

Kaup, P., Santosa, F., and Vogelius, M., 1996. A method for imaging corrosion damage in thin plates from electrostatic data: Inverse Problems, **12**, 279–293.

Luong, B., and Santosa, F., 1997, Quantitative imaging of corrosion using eddy current methods. To appear in SIAM J. Appl. Math.

Moon, P., and Spencer, D., 1961, Field theory for engineers: D. Van Nostrand Co.

Valley, M., Del Grande, N., and Kobayashi, A., 1993, Nondestructive inspection of aging aircraft, SPIE Proc. **2001**, SPIE, Bellingham, WA.

Vogelius, M., and Xu, J., A nonlinear elliptic boundary value problem related to corrosion modeling: Q. Appl. Math. Mech.

Appendix: Published Works of Gerald W. Hohmann

Articles in Journals

Wang, T., and Hohmann, G. W., 1993, A finite-difference, time-domain solution in three-dimensional electromagnetic modeling: Geophysics, **58**, 797–809.

Olsen, K. B., and Hohmann, G. W., 1992, Adaptive noise cancellation for time-domain EM data: Geophysics, **57**, 466–469.

Boschetto, N. B., and Hohmann, G. W., 1991, Controlled-source audiofrequency magnetotelluric responses of three-dimensional bodies: Geophysics, **56**, 255–264.

Wannamaker, P. E., and Hohmann, G. W., 1991, Electromagnetic induction studies: Reviews of Geophysics, **29**, 405–415.

Hohmann, G. W., and Newman, G. A., 1990, Transient electromagnetic responses of surficial, polarizable patches: Geophysics, **55**, 1098–1100.

Pellerin, L., and Hohmann, G. W., 1990, Transient electromagnetic inversion—a remedy for magnetotelluric static shifts: Geophysics, **55**, 1242–1250.

Adhidjaja, J. I., and Hohmann, G. W., 1989, A finite-difference algorithm for the transient electromagnetic response of a 3-dimensional body: Geophys. J. Int., **98**, 233–242.

Eaton, P. A., and Hohmann, G. W., 1989, A rapid inversion technique for transient electromagnetic soundings, Phys. Earth Planet. Int., **53**, 384–404.

Flis, M. F., Newman, G. A., and Hohmann, G. W., 1989, Induced-polarization effects in time-domain electromagnetic measurements: Geophysics, **54**, 514–523.

Newman, G. A., Anderson, W. A., and Hohmann, G. W., 1989, Effect of conductive host rock on borehole transient electromagnetic responses: Geophysics, **54**, 598–608.

Adhidjaja, J. I., and Hohmann, G. W., 1988, Step responses for two-dimensional electromagnetic models: Geoexpl., **25**, 13–35.

Newman, G. A., and Hohmann, G. W., 1988, Transient electromagnetic responses of high-contrast prisms in a layered earth: Geophysics, **53**, 691–706.

Eaton, P. A., and Hohmann, G. W., 1987, Evaluation of electromagnetic methods in the presence of geologic noise: Geophysics, **52**, 1106–1126.

Newman, G. A., Anderson, W. L., and Hohmann, G. W., 1987, Interpretation of transient electromagnetic soundings over 3-dimensional structures for the central-loop configuration: Geophys. J. Roy. Astron. Soc., **89**, 889–914.

Gunderson, B. M., Newman, G. A., and Hohmann, G. W., 1986, Three-dimensional transient electromagnetic responses for a grounded source: Geophysics, **51**, 2117–2130.

McCracken, K. G., Oristaglio, M. L., and Hohmann, G. W., 1986a, Comparison of electromagnetic exploration systems: Geophysics, **51**, 810–818.

McCracken, K. G., Oristaglio, M. L., and Hohmann, G. W., 1986b, Minimization of noise in electromagnetic exploration systems: Geophysics, **51**, 819–832.

Newman, G. A., Hohmann, G. W., and Anderson, W. L., 1986, Transient electromagnetic response of a three-dimensional body in a layered earth: Geophysics, **51**, 1608–1627.

Adhidjaja, J. L., Hohmann, G. W., and Oristaglio, M. L., 1985, Two-dimensional transient electromagnetic responses: Geophysics, **50**, 2849–2861.

Bartel, D. C., and Hohmann, G. W., 1985, Interpretation of Crone pulse electromagnetic data: Geophysics, **50**, 1488–1499.

Gallagher, P. R., Ward, S. H., and Hohmann, G. W., 1985, Model study of a thin plate in free space for the EM37 transient electromagnetic system: Geophysics, **50**, 1002–1019.

Newman, G. A., Wannamaker, P. E., and Hohmann, G. W., 1985, On the detectability of crustal magma chambers using the magnetotelluric method: Geophysics, **50**, 1136–1143.

SanFilipo, W. A., Eaton, P. A., and Hohmann, G. W., 1985, Effect of a conductive half-space on the transient electromagnetic response of a three-dimensional body: Geophysics, **50**, 1144–1162.

Eaton, P. A., and Hohmann, G. W., 1984, Influence of a conductive host on two-dimensional borehole transient electromagnetic responses: Geophysics, **49**, 861–869.

Oristaglio, M. L., and Hohmann, G. W., 1984, Diffusion of electromagnetic fields into a two-dimensional earth: A finite-difference approach: Geophysics, **49**, 870–894.

Silva, J. B. C., and Hohmann, G. W., 1984, Airborne magnetic susceptibility mapping: Exploration Geophysics, **15**, 1–13.

Tripp, A. C., and Hohmann, G. W., 1984, Block diagonalization of the electromagnetic impedance matrix of a ymmetric buried body using group theory: Inst. Elect. and Electron. Eng. Trans. Geoscience and Remote Sensing, **GE-22**, 62–69.

Tripp, A. C., Hohmann, G. W., and Swift, C. M., Jr., 1984, Integral equation solution for the transient electromagnetic response of a three-dimensional body in a conductive half-space: Geophysics, **49**, 1708–1717.

Silva, J. B. C., and Hohmann, G. W., 1983, Nonlinear magnetic inversion using a random search method: Geophysics, **48**, 1645–1658.

Hohmann, G. W., 1983, 3-dimensional EM modeling, Geophys. Surv., **6**, 27–53.

SanFilipo, W. A., and Hohmann, G. W., 1983, Computer simulation of low-frequency electromagnetic data acquisition: Geophysics, **48**, 1219–1232.

Sandberg, S. K., and Hohmann, G. W., 1982, Controlled-source audiomagnetotellurics in geothermal exploration: Geophysics, **47**, 100–116.

Pridmore, D. F., Hohmann, G. W., Ward, S. H., and Sill, W. R., 1981, Investigation of finite-element modeling for electrical and electromagnetic data in three dimensions: Geophysics, **46**, 1009–1024.

Silva, J. B. C., and Hohmann, G. W., 1981, Interpretation of three-component borehole magnetometer data: Geophysics, **46**, 1721–1731.

Ting, S. C., and Hohmann, G. W., 1981, Integral equation modeling of three-dimensional magnetotelluric response: Geophysics, **46**,182–197.

Stodt, J. A., Hohmann, G. W., and Ting, S. C., 1981, The telluric-magnetotelluric method in 2-dimensional and 3-dimensional environments: Geophysics, **46**, 1137–1147.

Fox, R. C., Hohmann, G. W., Killpack, T. J., and Rijo, L., 1980, Topographic effects in resistivity and induced-polarization surveys: Geophysics, **45**, 75–93.

Hohmann, G. W., Van Voorhis, G. D., and Nelson, P. H., 1978, A vector EM system and its field applications: Geophysics, **43**, 1418–1440.

Ward, S. H., Campbell, R. E., Corbett, J. D., Hohmann, G. W., Moss, C. K., and Wright, P. M., 1977, The frontiers of mining geophysics: Geophysics, **42**, 878–886.

Hohmann, G. W., 1975, Three-dimensional induced polarization and electromagnetic modeling: Geophysics, **40**, 309–324.

Ward, S. H., Ryu, J., Glenn, W. E., Hohmann, G. W., Dey, A., and Smith, B. D., 1974, Electromagnetic methods in conductive terranes: Geoexpl., **12**, 121–183.

Hohmann, G. W., 1973, Electromagnetic coupling between grounded wires at the surface of a two-layer earth: Geophysics, **38**, 854–863.

Hohmann, G. W., 1971, Electromagnetic scattering by conductors in the earth near a line source of current: Geophysics, **36**, 101–131.

Hohmann, G. W., Kintzinger, P. R., Van Voorhis, G. D., and Ward, S. H., 1970, Evaluation of the measurement of induced electrical polarization with an inductive system: Geophysics, **35**, 901–915.

Articles in Monographs

Hohmann, G. W., 1990, Three-dimensional IP models: *in* Fink, J. B., McAlister, E. O., Sternberg, B. K., Wieduwilt, W. G., and Ward, S. H., Ed., Induced polarization; applications and case histories, Investigations in Geophysics, **4**, 150–178, SEG.

Hohmann, G. W., and Raiche, A. P., 1988, Inversion of controlled-source electromagnetic data: *in* Nabighian, M. N., Ed., Electromagnetic methods in applied geophysics—Theory, Investigations in Geophysics, **3**, 469–503, SEG.

Hohmann, G. W., 1988, Numerical modeling for electromagnetic methods of geophysics: *in* Nabighian, M. N., Ed., Electromagnetic methods in applied geophysics—Theory, Investigations in Geophysics, **3**, 313–363, SEG.

Ward, S. H., and Hohmann, G. W., 1988, Electromagnetic theory for geophysical applications: *in* Nabighian, M. N., Ed., Electromagnetic methods in applied geophysics—Theory, Investigations in Geophysics, **3**, 131–311, SEG.

Glenn, W. E., and Hohmann, G. W., 1981, Well logging and borehole geophysics in mineral exploration: *in* Skinner, B. J., Ed., Economic Geology, Seventy-Fifth Anniversary Volume; 1905–1980, 850–862, Econ. Geol. Publ. Co.

Hohmann, G. W., and Ward, S. H., 1981, Electrical methods in mining geophysics: *in* Skinner, B. J., Ed., Economic Geology, Seventy-Fifth Anniversary Volume; 1905–1980, 806–828, Econ. Geol. Publ. Co.

Index